Springer

*Berlin
Heidelberg
New York
Barcelona
Hong Kong
London
Milan
Paris
Singapore
Tokyo*

Gunter Faure

Origin of Igneous Rocks

The Isotopic Evidence

With 420 Figures and 60 Tables

Springer

Author

Prof. Gunter Faure

Faculty of Geological Sciences
Ohio State University
329 Mendenhall Lab
125 S. Oval Mall
Columbus, Ohio 43210, U.S.A.

Library of Congress Cataloging-in-Publication Data

Faure, Gunter.
    Origin of igneous rocks : the isotopic evidence / Gunter Faure.
        p. cm.
    Includes bibliographical references and index.

    1. Rocks, Igneous. I. Title.

QE461 .F355 2001
522'.1--dc21

00-064091

ISBN 978-3-642-08728-8

Springer-Verlag Berlin Heidelberg New York
a member of the BertelsmannSpringer Science+Business Media GmbH
© Springer-Verlag Berlin Heidelberg 2001
Softcover reprint of the hardcover 1st edition 2001

Cover Design: Erich Kirchner, Heidelberg
Dataconversion: Büro Stasch · Uwe Zimmermann, Bayreuth

Dedicated to Patrick M. Hurley, Harold W. Fairbairn, and William H. Pinson Jr., who laid the foundation for the study of the isotope composition of strontium in terrestrial and extraterrestrial materials.

Preface

This book is intended for graduate students of the Earth Sciences who require a comprehensive examination of the origins of igneous rocks as recorded by the isotope compositions of the strontium, neodymium, lead, and oxygen they contain. Students who have not had a formal course in the systematics of radiogenic isotopes can acquire a basic understanding of this subject by a careful study of Chap. 1. Additional information is readily available in a textbook by Faure (1986). The primary purpose of this book is to demonstrate how the isotope composition of Sr, Nd, Pb, and O in igneous rocks has been used to shed light on the origin of igneous rocks and hence on the activity of the mantle and on its interactions with the continental and oceanic crust.

The presentations are based on the premise that igneous and metamorphic rocks form as a direct consequence of the dynamic processes of the mantle and of the resulting interactions between the mantle and the crust. Accordingly, Chap. 2 to 6 examine specific types of igneous rocks that form in particular tectonic settings. Each of these chapters starts with questions about the properties of the mantle and crust, and about the relation between the tectonic setting and the rock-forming processes that take place in that setting. This introduction is followed by interpretations of the isotope compositions of Sr, Nd, Pb, and O in the rocks at selected sites around the world. Each chapter ends with a summary of the insights that have been achieved including an assessment of the level of understanding of the rock-forming processes that operate in the tectonic settings examined in that chapter.

The most basic question considered in this book concerns the range of physical, chemical, and isotopic properties of the mantle of the Earth. The evidence overwhelmingly indicates that the mantle is heterogeneous with respect to all of its properties and that it is mechanically unstable. As a result, parts of the mantle are in convective motion and thereby cause the geologic activity that occurs on the surface of the Earth. The convection of the mantle exposes rocks to changes in pressure, temperature, and chemical composition and causes magmas to form by partial melting. Tectonic processes in the mantle also cause uplift, rifting, volcanic eruptions, sea-floor spreading, continental drift, subduction, and orogenesis on the surface of the Earth.

The objective of Chap. 7 to 8 is to examine the petrogenesis of rocks in the Precambrian and to document the growth of the continental crust by the formation of volcano-sedimentary complexes and related granite batholiths along continental margins. The evidence supports the conclusion that large stratified mafic intrusives occupy magma chambers from which basalt magma was erupted to the surface resulting in the formation of large plateaus of flood basalt. The study of greenstone belts of North America presented in Chap. 8 indicates that they are the remnants of lava flows and associated sedimentary rocks that formed in island arcs associated with subduction zones. These volcano-sedimentary complexes were deformed and metamorphosed during the closure of marine basins and the collisions of microcontinents. The extreme deformation and regional recrystallization of the rocks, caused the expulsion of fluids, the re-equilibration of the isotope compositions of Sr, and hence resulted in the resetting of their Rb-Sr dates.

The examination of the geologic history of the continental crust, as recorded by the isotopic data brings into focus the question concerning the origin of granite. The evidence confirms the view that the granitic rocks of the continental crust have formed by remelting and/or recrystallizaton of volcano-sedimentary rocks whose prior crustal residence ages increase with decreasing ages of the granitic rocks. The record is noisy and does not support sweeping conclusions or narrowly defined classification schemes. Every generalization concerning the origin of granite can be contradicted by citing exceptions. There are no simple answers in the Earth Sciences. We must accept complexity.

The text is augmented by more than 400 original diagrams. In addition, all chapters include summaries in which important conclusions are emphasized and unresolved problems are identified. All $^{87}Sr/^{86}Sr$ ratios have been adjusted to 0.7080 for the Eimer and Amend $SrCo_3$ and/or to 0.71025 for the NBS 987 isotope reference standard. The adjusted isotope ratios are presented graphically in diagrams in order to reveal their relationship to other chemical or physical parameters. In addition, the figures have descriptive captions that facilitate browsing and reinforce the text. Last but not least, the extensive list of references, author and geographic indexes, and a geologic timescale will enhance the encyclopedic function of this book by helping users to find the information they are looking for. The detailed Table of Contents takes the place of a subject index.

In conclusion, I express my admiration for my many colleagues whose work I have attempted to summarize in this book. Their ingenuity, skill, and hard work have generated the insights that have taught us how the Earth works. I also thank T. M. Mensing, L. M. Jones, S. Chaudhuri, C. I. Cholukwa, and P. C. Lightfoot for information and reprints of their papers.

I am especially grateful to Dr. Wolfgang Engel, Editor of Springer Verlag, for his steadfast support during the many years it took me to write this book and to Peter J. Wyllie for his encouragement along the way. I also owe a large debt of gratitude to my technical support team consisting of Betty Heath, word processor, and Karen Tyler, illustrator, who took a personal interest in this project and helped me to complete it. The final mauscript benefited greatly from the editorial work of Uwe Zimmermann to whom I express my thanks and appreciation.

Gunter Faure
Paradox Ranch
15619 Myers Rd.
Marysville, Ohio 43040

Contents

Chapter 1

Chemical Properties and Isotope Systematics

The origin of igneous rocks is now understood in considerable detail based on interpretations of mineralogical, chemical, and isotopic data. The insights that have been gained about the petrogenesis of different kinds of rocks have enhanced our understanding of how the Earth works on a global scale, and how it has evolved since it formed about 4.5×10^9 years ago. The evidence that has provided these insights has come from many sources, including the concentrations of trace elements and the isotope compositions of strontium (Sr), neodymium (Nd), lead (Pb), hafnium (Hf), and osmium (Os) all of which have radiogenic isotopes. In addition, a complete listing of all sources of isotopic data includes certain elements of low atomic number (e.g. H, C, O, N, and S) whose stable isotopes are fractionated by physical, chemical, and biological processes. The resulting theory explains igneous activity as a manifestation of the complex dynamic processes that occur in the mantle of the Earth. The objective of this book is to use the isotope compositions of Sr, Nd, and Pb in volcanic and plutonic rocks to demonstrate how isotopic data have been used to explain the origin of igneous rocks. This chapter contains brief summaries of relevant facts and theories that will facilitate the scientific journey we are about to undertake. Additional information is available in a textbook by Faure (1986).

1.1 Chemical Properties of Rb and Sr

Rubidium and strontium are related to each other because naturally-occurring ^{87}Rb decays to stable ^{87}Sr with a half-life of 48.8×10^9 years (Steiger and Jäger 1977). Since both elements are lithophile and occur in many common and accessory rock-forming minerals, this parent-daughter relationship provides a useful *geochronometer* for the measurement of geologic time as recorded in common terrestrial and extraterrestrial rocks and their constituent minerals. In addition, the decay of ^{87}Rb to ^{87}Sr has continually changed the isotope composition of Sr in geological reservoirs which therefore contain Sr having characteristic isotope compositions that can be used to study the *origin* of igneous rocks. The resulting insight into the sources of magma and

their subsequent evolution has also shed light on the growth of the continental crust throughout geologic time and on the complementary differentiation of the upper mantle.

The concentrations of Rb and Sr in rocks and minerals are routinely measured in geochemical and isotopic studies of igneous, sedimentary, and metamorphic rocks. The techniques used for this purpose include x-ray fluorescence (XRF), atomic absorption (AA), inductively-coupled plasma spectrometry (ICP), isotope dilution (ID), and, more rarely, flame photometry and neutron activation. Because of the relative ease with which these elements can be determined and because of their importance for dating and in petrogenetic studies, the abundances of Rb and Sr in a wide variety of rocks are now well known. Compilations of the chemical compositions of different rock types, including the concentrations of Rb and Sr, have been published by Turekian and Wedepohl (1961), Vinogradov (1962), Taylor (1964), Turekian (1964), Ronov and Yaroshevsky (1969, 1976), Shaw et al. (1967, 1986), Wedepohl (1969 to 1978), and by Taylor and McLennan (1985).

Rubidium is an alkali metal whose chemical properties listed in Table 1.1a are similar to those of its congeners: lithium, sodium, potassium, and cesium. All of the alkali metals form cations with a charge of 1+ because they each have a single electron in an s orbital outside of noble-gas cores. The alkali metals have low electronegativities and form highly ionic bonds in crystals with anions such as O^{2-}. The ionic radii of the alkali metals increase with increasing atomic number and are relatively large compared to those of other metallic elements in the periodic table.

The ionic radius of Rb^+ is 1.68 Å (1 Å = 10^{-8} cm) when it resides in a lattice site that has a coordination number of eight. In twelve-fold coordinated sites the radius of Rb^+ is 1.81 Å (Whittacker and Muntus 1970). The Rb ion is only 6 to 8% larger than K^+ and can replace it in all of its minerals. However, Rb^+ is 46 to 58% larger than Na^+ and therefore does not replace it extensively. As a result, Rb is a dispersed lithophile trace element that occurs in all K-bearing minerals, including the feldspars, micas, clay minerals, and K-bearing evaporite minerals. During the progressive crystallization of

Table 1.1.
Physical and chemical properties of the alkali metals and the alkaline earths

A. Alkali metals, elements	Li	Na	K	Rb	Cs
Atomic number	3	11	19	37	55
Naturally-occurring isotopes	2	1	3	2	1
Atomic weight (^{12}C)	6.941	22.9898	39.0983	85.4678	132.905
Electron formula	[He]2s[a]	[Ne]3s[a]	[Ar]4s[a]	[Kr]5s[a]	[Xe]6s[a]
Ionic radius (Å) and coordination number()[a]	0.68 (4) 0.82 (6)	1.10 (6) 1.24 (8)	1.59 (8) 1.68 (12)	1.68 (8) 1.81 (12)	1.96 (12)
Electronegativity[b]	1.0	0.9	0.8	0.8	0.7
Percent ionic character[c] of bond with O^{2-}	82	83	87	87	89
B. Alkaline earths, elements	**Be**	**Mg**	**Ca**	**Sr**	**Ba**
Atomic number	4	12	20	38	56
Naturally occurring isotopes	1	3	6	4	7
Atomic weight (^{12}C)	9.01218	24.305	40.08	87.62	137.33
Electron formula	[He]2s[b]	[Ne]2s[b]	[Ar]2s[b]	[Kr]2s[b]	[Xe]2s[b]
Ionic radius (Å) and coordination number()[a]	0.35 (4)	0.80 (6)	1.08 (6) 1.20 (8)	1.21 (6) 1.33 (8)	1.50 (8) 1.68 (12)
Electronegativity[b]	1.5	1.2	1.0	1.0	0.9
Percent ionic character[c] of bond with O^{2-}	63	71	79	82	84

[a] Pauling (1960).
[b] Whittacker and Muntus (1970).
[c] Electronegativity of oxygen is 3.5.

magma, Rb^+ is concentrated into the residual liquid relative to K^+. However, Rb is not sufficiently enriched in residual magmatic fluids or brines to form its own minerals.

Strontium is a member of the alkaline earths which consist of beryllium, magnesium, calcium, strontium, barium, and radium. Each of these elements has two electrons in an *s* orbital outside of noble-gas cores and forms cations with a charge of 2+. The alkaline earths (except Be) have low electronegativities and form highly ionic bonds with anions. The ionic radii and other relevant properties of the alkaline earths are listed in Table 1.1b. The radius of Sr^{2+} in six-fold coordinated sites is 1.21 Å and 1.33 Å when the coordination number is eight. The Sr^{2+} ion is only 10 to 12% larger than Ca^{2+} and can replace it in Ca-bearing rock-forming minerals including plagioclase, calcite, aragonite, apatite, sphene, and others. However, the substitution is restricted because Sr^{2+} prefers eight-fold coordinated sites whereas Ca^{2+} occupies sites having coordination numbers of either six or eight. For example, Sr^{2+} replaces Ca^{2+} much more extensively in aragonite (eight-fold coordination) than in calcite (six-fold coordination).

Strontium also occurs in microcline and orthoclase, which capture it in preference to K^+, because the charge-to-radius ratio of Sr^{2+} (coordination of eight) is significantly larger than that of K^+ (1.5 for Sr^{2+} and 0.63 for K^+). However, Sr^{2+} does *not* replace K^+ in micas and clay minerals because K^+ occupies sites in the tetrahedral layers of these minerals that have coordination numbers of twelve and seem to be too large for Sr^{2+}.

The data in Table 1.2 indicate that the rock-forming minerals can be subdivided into four categories based on their Rb and Sr concentrations and resulting Rb/Sr ratios:

a High Rb, low Sr, Rb/Sr > 1.0: biotite, muscovite, (phlogopite) orthoclase/microcline, glauconite, illite, some kaolinites, sylvite, and carnallite.
b Low Rb, high Sr, Rb/Sr < 0.010: plagioclase, calcite, apatite, sphene, and smectites (including montmorillonite and nontronite), gypsum, and anhydrite.
c Low Rb, low Sr, Rb/Sr < 1.0: hornblende, pyroxene (cpx and opx), olivine, garnet, quartz, and halite as well as magnetite, ilmenite, rutile.
d High Rb, high Sr, Rb/Sr < 1.0: sanidine, nepheline, leucite.

The concentrations of Rb and Sr in a particular mineral in igneous rocks vary widely, commonly by two or three orders of magnitude, depending on the environment in which the mineral formed. For example, feldspars and micas in late-stage magmatic differentiates tend to have higher Rb concentrations but lower Sr concentrations than feldspars or micas in main-stage differentiates (Wedepohl 1978).

The preferential enrichment of different minerals in Rb or Sr causes the Rb/Sr ratios of common igneous and

Table 1.2. Average Rb and Sr concentrations and Rb/Sr ratios of minerals. The number of samples included in the average is stated in parentheses and the range of values is indicated where the information is available

Mineral	Rb (ppm)		Range		Sr (ppm)		Range		Rb/Sr	Reference[a]
A. Silicate minerals in igneous and metamorphic rocks										
Biotite	550	(206)	122	– 2 525	31.1	(378)	0.867	– 676	17.7	1, 19
Muscovite	476	(23)	125	– 1 000	46.0	(139)	2.0	– 398	10.3	1, 10, 23
Phlogopite	–				70.7	(8)	3.9	– 180.7	–	1
K-feldspar	561	(144)	53	– 1 650	396	(652)	3	– 5 100	1.42	1, 19, 23
Sanidine	186	(4)	152	– 215	315	(4)	127	– 449	0.590	2
Plagioclase	14.1	(28)	0.14	– 56	566	(440)	0.65	– 5 000	0.0091	1, 19
Amphibole	76.8	(55)	0.2	– 167	106	(186)	3	– 1 060	0.725	1
Kaersutite	5.8	(3)			581	(3)			0.010	3
Pyroxene	1.28	(27)	0.033	– 3.6	89.1	(222)	3.51	– 853	0.014	1
Clinopyroxene	1.7	(2)	2.3	– 1.1	6.32	(45)	0.226	– 21	0.27	1, 4
Orthopyroxene	0.36	(1)			2.02	(1)			0.18	4
Olivine	0.13	(11)	0.004	– 0.55	5.51	(30)	0.162	– 21.0	0.024	1
Tourmaline	1.3	(23)	0.1	– 6	601	(23)	38	– 2 469	0.0021	25
Scheelite	0.5	(3)	0.4	– 06	604	(3)	387	– 832	0.00083	25
Actinolite	1.6	(2)	0.8	– 2.3	364	(2)	301	– 426	0.0044	25
Piedmontite (epidote)	0.5	(1)			1 150	(1)			0.00043	25
Zircon	21	(3)	0.27	– 5.8	50.4	(3)	0.87	– 77.0	0.42	21
Garnet	1.9	(11)	0.019	– 3.9	19.3	(63)	0.98	– 330	0.098	1, 21
Nepheline	85	(8)	50	– 120	209	(8)	80	– 430	0.407	1
Leucite	–				482	(5)	110	– 1 400	–	1
Sphene	2.7	(1)			1980	(13)	0	– 6 700	0.0014	1, 21
Epidote	31	(4)	4	– 59	8518	(4)	5 900	– 11 900	0.0036	19
Melilite (carbonatite)	–				9 200	(5)	8 400	– 10 900	–	1
Quartz	0.67	(14)	0.001	– 2.5	0.32	(10)	0.03	– 1.0	2.1	2, 6, 11
B. Oxide minerals										
Ilmenite	0.48	(4)	0.15	– 0.76	1.99	(4)	0.21	– 4.57	0.24	24
C. Phosphate minerals										
Apatite (igneous)	1.55	(3)	0.31	– 2.18	1 329	(23)	35	– 73 558	0.0011	1, 21
Apatite (metamorphic)	–				2 872	(25)	8	– 9 839	–	1
Apatite (pegmatite)	–				483	(18)	4.9	– 4 320	–	1
Apatite (carbonatite)	–				6 160	(11)	4 370	– 7 680	–	1, 5
D. Clay minerals										
Glauconite	198	(56)	14	– 283	32.5	(86)	3.8	– 475	6.1	1, 13–16
Kaolinite	47.5	(10)	3.2	– 128.9	44.8	(10)	0.91	– 174.4	1.06	10
Smectite	31.0	(7)	6.01	– 60.8	184.6	(13)	98.1	– 312	0.17	10, 12
Nontronite	56.9	(7)	48.1	– 61.0	472.8	(7)	422	– 499	0.12	20
E. Carbonate minerals										
Calcite (carbonatite)	–				2 170	(4)	1 220	– 3 400	–	1
Calcite (travertine)	–				923	(92)	20	– 3 900	–	7, 17
Calcite (veins, Archean greenstones)	0.04	(3)	0.02	– 0.06	253	(4)	136	– 600	0.00016	22
Calcite (disseminated, Archean greenst.	0.58	(2)	0.51	– 0.64	79	(2)	67.6	– 90.0	0.0073	22
F. Miscellaneous										
Fluorite	69.7	(1)			539.6	(1)			0.129	18
Cryolite	0.1	(2)	0.1	– 0.1	22.0	(2)	5.9	– 38.0	0.0045	18
Sylvite			47	– 252	1.0	(0.2)			47 – 1 260	1, 9
Carnallite			990	– 1 670	1.0	(4)	0.2	– 2.0	495 – 8 350	1, 9
Gypsum and anhydrite	–				1 753	(46)			–	7, 8

[a] References: 1. Wedepohl (1978), 2. Nash and Crecraft (1985), 3. Othman et al. (1990), 4. Stuckless and Irving (1976), 5. Nash (1972), 6. Rossman et al. (1987), 7. Barbieri et al. (1979), 8. Stueber et al. (1984), 9. Braitsch (1971), 10. Clauer (1979), 11. Worden and Compston (1973), 12. Schultz et al. (1989), 13. Grant et al. (1984), 14. Laskowski et al. (1980), 15. Morton and Long (1980), 16. Harris and Fullagar (1991), 17. Demovic et al. (1972), 18. Blaxland (1976), 19. Franklyn et al. (1991), 20. Clauer et al. (1982), 21. Barrie and Shirey (1991), 22. Brooks et al. (1969a), 23. Heller et al. (1985), 24. Burton and O'Nions (1990), 25. Kerrich et al. (1987).

metamorphic rocks to vary depending on the abundances of the minerals that carry these elements. This is an important phenomenon because the Rb/Sr ratio of a volume of rock determines the rate at which the isotope composition of Sr changes with time by the radioactive decay of ^{87}Rb to ^{87}Sr.

1.2 Isotope Systematics of Rb and Sr

The radioactive decay of ^{87}Rb to ^{87}Sr continually changes the isotope composition of Sr in Rb-bearing rocks and minerals. Consequently, Sr that resides in Rb-rich rocks of the continental crust acquires a characteristic isotope composition that differs from that of Sr in the Rb-poor rocks of the mantle. This fact is the key to the interpretation of the isotope composition of Sr in igneous rocks.

1.2.1 Isotope Abundances

Rubidium has two naturally-occurring isotopes $^{85}_{37}$Rb and $^{87}_{37}$Rb whose abundances in terrestrial and extraterrestrial rocks and minerals have the same numerical values throughout the solar system. The Subcommission on Geochronology of the International Union of Geological Sciences therefore recommended a value of 2.59265 for the ^{85}Rb/^{87}Rb ratio (Steiger and Jäger 1977.)

The corresponding isotopic abundances of ^{85}Rb and ^{87}Rb are obtained as shown in Table 1.3.

Strontium has four naturally-occurring isotopes ($^{84}_{38}$Sr, $^{86}_{38}$Sr, $^{87}_{38}$Sr, $^{88}_{38}$Sr) all of which are stable. However, the isotope composition of Sr in terrestrial and extraterrestrial rocks and minerals varies widely because ^{87}Sr is produced by beta decay of ^{87}Rb. Therefore, the abundance of radiogenic ^{87}Sr in all Rb-bearing systems *increases* with the passage of time. The abundances of the non-radiogenic Sr isotopes *decrease* proportionately even though the number of these isotopes in a unit weight of sample remains constant. The ratios of the abundances of the non-radiogenic Sr isotopes are: ^{86}Sr/^{88}Sr = 0.119400 and ^{84}Sr/^{86}Sr = 0.056584 (Steiger and Jäger 1977). Therefore, the abundances of the isotopes of Sr in seawater (^{87}Sr/^{86}Sr

= 0.70916) can be calculated from these ratios as illustrated above for Rb (see Table 1.4).

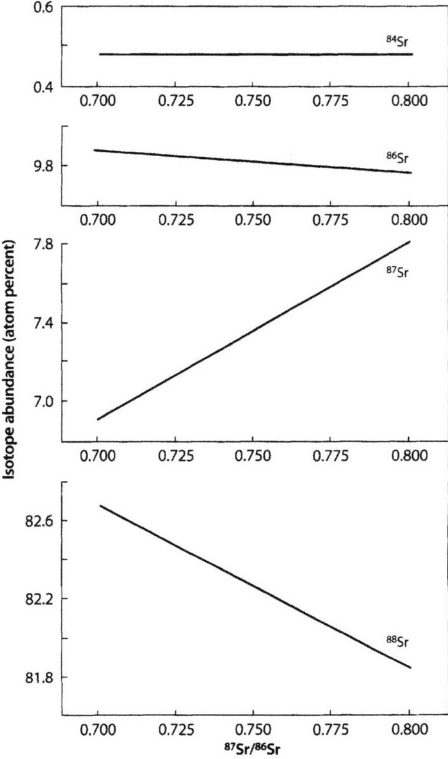

Fig. 1.1. Changes of the abundances (in percent by number of atoms) of the naturally-occurring isotopes of Sr with increasing abundance of radiogenic ^{87}Sr expressed as the atomic ^{87}Sr/^{86}Sr ratio. Even though the abundances of the non-radiogenic isotopes (^{84}Sr, ^{86}Sr, and ^{88}Sr) decrease with increasing ^{87}Sr/^{86}Sr, the number of these isotopes in a unit weight of sample remains constant

Table 1.4. Abundances of the isotopes of Sr in seawater (^{87}Sr/^{86}Sr = 0.70916)

Ratio	Value	Isotope	Abundance (% by number)
^{84}Sr/^{88}Sr	0.006756[a]	^{84}Sr	0.5579
^{86}Sr/^{88}Sr	0.119400	^{86}Sr	9.8610
^{87}Sr/^{88}Sr	0.084673[b]	^{87}Sr	6.9929
^{88}Sr/^{88}Sr	1.000000	^{88}Sr	82.5880
Sum	1.210829		99.9998

[a] ^{84}Sr/^{88}Sr = (^{84}Sr/^{86}Sr) × (^{86}Sr/^{88}Sr).
[b] ^{87}Sr/^{88}Sr = (^{86}Sr/^{88}Sr) × (^{87}Sr/^{86}Sr).

Table 1.3. Isotopic abundances of ^{85}Rb and ^{87}Rb for the value of 2.59265 which is recommended by the Subcomission on Geochronology of the International Union of Gological Sciences (Steiger and Jäger 1977)

Ratio	Value	Isotope	Abundance (% by number)
^{85}Rb/^{87}Rb	2.59265	^{85}Rb	72.1654
^{87}Rb/^{87}Rb	1.00000	^{87}Rb	27.8346
Sum	3.59265		100.0000

The change of the abundances of the isotopes of Sr with increasing abundance of ^{87}Sr (from ^{87}Sr/^{86}Sr = 0.700 to 0.800) is illustrated in Fig. 1.1 based on the calculations demonstrated above.

1.2.2 Decay of ^{87}Rb to ^{87}Sr

The decay of ^{87}Rb to ^{87}Sr is represented by the equation:

$$^{87}\text{Rb} \rightarrow {}^{87}\text{Sr} - \beta^- + v + Q \tag{1.1}$$

where β^- is a negatively charged beta particle, v is a neutrino, and Q is the total decay energy consisting of the sum of the kinetic energies of the β^- particle and of the neutrino released by each decay event. The decay of ^{87}Rb to ^{87}Sr is *not* followed by the emission of gamma rays and the total decay energy is $Q = 0.275$ million electron volts (MeV).

The rate of decay of a radioactive nuclide like ^{87}Rb is governed by the Law of Radioactivity derived by Rutherford and Soddy (1902) on the basis of experimental results with thorium salts. The law states that the rate of decay of a radioactive nuclide is proportional to the number of radioactive atoms remaining at any time t:

$$-\frac{dN}{dt} = \lambda N \tag{1.2}$$

where

- dN/dt = rate of decay
- N = number of parent atoms remaining
- λ = decay constant

By rearranging Eq. 1.2 we obtain:

$$\frac{dN}{N} = -\lambda dt \tag{1.3}$$

which can be integrated to yield:

$$\ln N = -\lambda t + C \tag{1.4}$$

where $\ln N$ is the natural logarithm of N, and C is the constant of integration. If $t = 0$, $N = N_0$ and therefore $C = \ln N_0$. Consequently, the integrated equation is:

$$\ln N - \ln N_0 = -\lambda t$$

$$\ln\left(\frac{N}{N_0}\right) = -\lambda t$$

$$\frac{N}{N_0} = e^{-\lambda t}$$

$$N = N_0 e^{-\lambda t} \tag{1.5}$$

The decay constant (λ) can be expressed in terms of the half-life ($T_{1/2}$) defined as the time required for one half of a given number of radioactive parent atoms to decay. Therefore, after one half-life, $N = N_0 / 2$. Substituting into Eq. 1.5 yields:

$$\frac{N_0}{2} = N_0 e^{-\lambda T_{1/2}}$$

$$-\ln 2 = -\lambda T_{1/2}$$

$$T_{1/2} = \frac{\ln 2}{\lambda} \tag{1.6}$$

Equation 1.6 can be used to calculate the half-life of a radionuclide from its decay constant or vice versa.

The decay of a radioactive parent nuclide (N) to a stable radiogenic daughter (D^*) in a closed system is constrained by the requirement that the number of radiogenic daughter atoms present at any time t is equal to the number of parent atoms that have decayed ($N_0 - N$). Therefore,

$$D^* = N_0 - N \tag{1.7}$$

Substitution of Eq. 1.5 for N yields:

$$D^* = N_0 - N_0 e^{-\lambda t}$$

$$D^* = N_0(1 - e^{-\lambda t}) \tag{1.8}$$

This equation indicates that $D^* = N_0$ when t becomes very large. In other words, all of the radioactive parent atoms ultimately decay to form stable radiogenic daughters as t goes to infinity.

Equation 1.8 cannot be used to describe the decay of radionuclides in rocks and minerals because, in most cases, the initial number of radioactive parent nuclides (N_0) is not known. However, N_0 can be replaced in Eq. 1.7 by $N_0 = N e^{\lambda t}$ obtained from Eq. 1.5.

Therefore,

$$D^* = N e^{\lambda t} - N$$

$$D^* = N(e^{\lambda t} - 1) \tag{1.9}$$

If the system in which the radioactive decay is occurring initially contains a certain number of daughter atoms (D_0), the total number of daughters at any time t is:

$$D = D_0 + D^*$$

$$D = D_0 + N(e^{\lambda t} - 1) \tag{1.10}$$

Equation 1.10 contains two measurable quantities (D and N) and two constants (D_0 and λ) and therefore can be solved for t which is the time elapsed since the radiogenic daughter began to accumulate.

In the Rb-Sr decay scheme, ^{87}Rb is the radioactive parent whereas ^{87}Sr is the stable radiogenic daughter. Therefore, from Eq. 1.10:

$$^{87}\text{Sr} = {}^{87}\text{Sr}_0 + {}^{87}\text{Rb}\,(e^{\lambda t} - 1) \tag{1.11}$$

Dividing each term by the constant number of ^{86}Sr atoms yields:

$$\frac{^{87}\text{Sr}}{^{86}\text{Sr}} = \left(\frac{^{87}\text{Sr}}{^{86}\text{Sr}}\right)_0 + \frac{^{87}\text{Rb}}{^{86}\text{Sr}}(e^{\lambda t} - 1) \tag{1.12}$$

where the decay constant of ^{87}Rb is 1.42×10^{-11} yr^{-1} and the corresponding half-life is 48.8×10^9 yr (Steiger and Jäger 1977). The atomic ^{87}Sr/^{86}Sr ratios of Sr in rock and mineral samples are measured by mass spectrometry, whereas the atomic ^{87}Rb/^{86}Sr ratio can be calculated from the measured Rb and Sr concentrations by the application of Avogadro's Law.

If (Rb) is the concentration of this element in units of μg g^{-1}, the number of ^{87}Rb atoms per gram of sample is:

$$^{87}\text{Rb} = \frac{(\text{Rb}) \times \text{Ab}^{87}\text{Rb} \times \text{A}}{\text{At.Wt.Rb}} \tag{1.13}$$

where
- At.Wt.Rb = atomic weight of Rb
- Ab^{87}Rb = abundance of ^{87}Rb expressed as a decimal fraction
- A   = Avogadro's number

The number of ^{86}Sr atoms per gram of sample is calculated similarly, except that the abundance of ^{86}Sr (Ab^{86}Sr) and the atomic weight of Sr (At.Wt.Sr) depend on the abundance of ^{87}Sr and hence on the ^{87}Sr/^{86}Sr ratio of each sample as indicated in Fig. 1.1. Therefore, the atomic ^{87}Rb/^{86}Sr ratio is calculated from the Rb/Sr concentration ratio by the relation

$$\left(\frac{^{87}\text{Rb}}{^{86}\text{Sr}}\right)_{\text{atom.}} = \left(\frac{\text{Rb}}{\text{Sr}}\right)_{\text{conc.}} \frac{\text{Ab}^{87}\text{Rb} \times \text{At.Wt.Sr}}{\text{At.Wt.Rb} \times \text{Ab}^{86}\text{Sr}} \tag{1.14}$$

and Avogadro's number cancels. If the ^{87}Sr/^{86}Sr ratio of the Sr is 0.70916 (seawater), the abundances of the naturally occurring isotopes and the corresponding atomic weight of Sr are shown in Table 1.5.

The atomic mass unit (amu) is defined as 1/12 of the mass of $^{12}_{6}$C whose mass is defined as 12.0000 amu. Since the abundance of ^{87}Rb and the atomic weight of Rb are

Table 1.5. Abundances of the naturally occurring isotopes and the corresponding atomic weight of Sr for the ^{87}Sr/^{86}Sr ratio of 0.70916 (seawater)

Isotope	Abundance	Atomic mass (amu)	Abundance × mass
^{84}Sr	0.005579	83.9134	0.468152
^{86}Sr	0.098610	85.9092	8.471506
^{87}Sr	0.069929	86.9088	6.077445
^{88}Sr	0.825880	87.9056	72.599476
Sum	0.999998		87.616579

constant, substitution of the appropriate values into Eq. 1.14 yields:

$$\left(\frac{^{87}\text{Rb}}{^{86}\text{Sr}}\right)_{\text{atom.}} = \left(\frac{\text{Rb}}{\text{Sr}}\right)_{\text{conc.}} \frac{0.278346 \times 87.616579}{85.46776 \times 0.098610}$$

$$\left(\frac{^{87}\text{Rb}}{^{86}\text{Sr}}\right)_{\text{atom.}} = \left(\frac{\text{Rb}}{\text{Sr}}\right)_{\text{conc.}} 2.8936 \tag{1.15}$$

The value of the conversion factor increases by only 0.2% to 2.8994 when the ^{87}Sr/^{86}Sr ratio increases from 0.70916 to 0.8000.

Equation 1.12 can be simplified by the approximation:

$$e^{\lambda t} \approx 1 + \lambda t$$

$$e^{\lambda t} - 1 \approx \lambda t$$

which, combined with Eq. 1.15, yields the linear equation in the slope-intercept form:

$$\frac{^{87}\text{Sr}}{^{86}\text{Sr}} = \left(\frac{^{87}\text{Sr}}{^{86}\text{Sr}}\right)_0 + 2.89\left(\frac{\text{Rb}}{\text{Sr}}\right)_{\text{conc.}} \lambda t \tag{1.16}$$

where
- y-coordinate = ^{87}Sr/^{86}Sr
- x-coordinate = t

- slope = $2.89\left(\dfrac{\text{Rb}}{\text{Sr}}\right)_{\text{conc.}} \lambda$

- intercept = $(^{87}\text{Sr}/^{86}\text{Sr})_0$

Equation 1.16 indicates that the ^{87}Sr/^{86}Sr ratio of Sr in a Rb-bearing system increases with time at a rate that depends on the Rb/Sr ratio of the system. If $t = 4 \times 10^9$ yr and Rb/Sr = 1.0, the ^{87}Sr/^{86}Sr ratio calculated from Eq. 1.16 is only about 0.5% lower than the value given by Eq. 1.12. Therefore, Eq. 1.16 was used in Fig. 1.2 to illus-

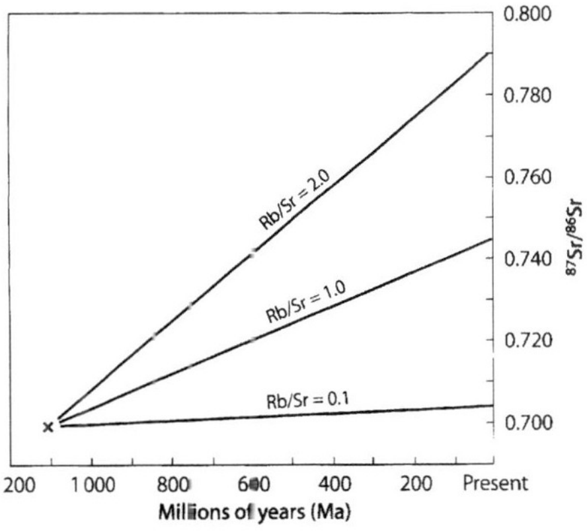

Fig. 1.2. Time-dependent evolution of the isotope composition of Sr in systems having different Rb/Sr ratios. The diagram illustrates the increase of the $^{87}Sr/^{86}Sr$ ratios of specimens of comagmatic rocks that crystallized at 1100 million years (Ma) from a homogeneous magma whose $^{87}Sr/^{86}Sr$ ratio was 0.700. The present $^{87}Sr/^{86}Sr$ ratios and Rb/Sr ratios of these rock specimens can be used to determine the coordinates of the point of convergence. In the simple case under consideration, the coordinates are the $^{87}Sr/^{86}Sr$ of the Sr in the magma (initial $^{87}/Sr^{86}Sr$) and the time elapsed since crystallization of the magma, i.e. the *age* of the rocks (based on Eq. 1.16)

trate the time-dependent increase of the $^{87}Sr/^{86}Sr$ ratio of different systems having Rb/Sr ratios of 2.0, 1.0, and 0.1. However, Eq. 1.12 should be used whenever accurate results are required.

1.2.3 Isotope Fractionation

In order to measure the isotope composition of Sr, a small amount of Sr salt extracted from the sample is deposited on a metal filament (Ta, Re, or W) in the source of a mass spectrometer. After the source housing has been evacuated, the filament is heated electrically in order to form Sr^+ ions representing all of the isotopes of Sr in the sample. However, the relative abundances of the Sr isotopes in the ion beam are not the same as those in the sample because of isotope fractionation during evaporation and thermal ionization. Consequently, the measured $^{87}Sr/^{86}Sr$ ratios differ from the true value and commonly vary with time during the measurement which may take several hours. Under these conditions, the average $^{87}Sr/^{86}Sr$ ratio recorded at the end of the "run" may have been changed by isotope fractionation relative to the true value in the sample.

Several different procedures and measurement strategies have been tried to reduce the fractionation during the isotope analysis of Sr. These have included the design of the ion source (single, double, or triple filaments), the choice of metal for the filament (Ta, Re, or W), the

composition of the Sr salt (chloride, nitrate, phosphate, oxalate), and the use of various additives (phosphoric acid, silica gel, Ta-oxide powder); but none of these measures has succeeded in completely eliminating isotope fractionation. Therefore, some analysts have standardized all of the parameters that affect fractionation and have adopted a rigorous schedule during the run, thereby hoping to improve the reproducibility of the results. However, the fractionation of Sr isotopes in the mass spectrometer may also be affected by the presence of impurities (Al, Fe, Mg, Na, and others) that are not removed with constant efficiency during the chemical separation of Sr from the same rock powder. Consequently, standardizing the timing of different steps in the operation of the mass spectrometer does not always produce the same result in the analysis of replicate analyses of the same sample.

In order to eliminate the effects of both natural and instrumental fractionation of Sr isotopes, Herzog et al. (1958) proposed that a correction could be applied to measured $^{87}Sr/^{86}Sr$ ratios derived from the value of the measured $^{86}Sr/^{88}Sr$ ratio. The correction is based on the assumption that all samples of terrestrial and extraterrestrial Sr have the same $^{86}Sr/^{88}Sr$ ratio and that isotope fractionation is proportional to the mass differences between the isotopes of an element. The value of the $^{86}Sr/^{88}Sr$ ratio originally reported by Nier (1938) was 0.1194 and therefore this value is used as the reference (Steiger and Jäger 1977). Since the mass difference between ^{87}Sr and ^{86}Sr is only about one half the mass difference between ^{86}Sr and ^{88}Sr, the correction factor (f) is calculated by dividing the measured $^{86}Sr/^{88}Sr$ ratio (M) by a number that lies half-way between 0.11940 and M:

$$f = \frac{M}{\dfrac{M + 0.1194}{2}} \tag{1.17}$$

Faure (1961) and Faure and Hurley (1963) adopted this method of correcting measured $^{87}Sr/^{86}Sr$ ratios for the effects of isotope fractionation and demonstrated a significant improvement in the reproducibility of replicate determinations of this ratio. The procedure was explained again by Faure and Powell (1972) and has been used by all analysts measuring $^{87}Sr/^{86}Sr$ ratios ever since.

This correction also eliminates the effects of fractionation of Sr isotopes in nature, although the actual occurrence of Sr-isotope fractionation has never been demonstrated unequivocally (Morse 1983). The fractionation correction described above is based on an assumed linear relation between fractionation of two isotopes and the difference in their masses. Hart and Zindler (1989a) considered an "exponential" fractionation law, but the "linear" fractionation correction in Eq. 1.17 is still widely used.

1.2.4 Interlaboratory Sr-isotope Standards

Although the fractionation correction improves the reproducibility (precision) of measured $^{87}Sr/^{86}Sr$ ratios, systematic differences in the isotope ratios obtained in different laboratories persist. In order to reduce these interlaboratory differences, most analysts report their results for certain standard samples that are analyzed repeatedly in the course of a particular investigation. The two interlaboratory standards used most frequently are a Sr carbonate reagent formerly marketed by the Eimer and Amend Co. (Lot Number 492327) and NBS 987 distributed by the U.S. National Bureau of Standards.

The Eimer and Amend Sr carbonate (E&A) was originally analyzed by Aldrich et al. (1953) and by Herzog et al. (1953) in the course of a study to measure the isotope composition of Sr from different sources. They chose this Sr compound partly because A. O. Nier, with whom L. T. Aldrich had worked at the University of Minnesota, had originally used metallic Sr obtained from Eimer and Amend to measure the isotope composition of Sr (Nier 1938). Subsequently, the same Sr reagent was used in the laboratory of P. M. Hurley at M.I.T. to prepare a standard Sr solution for calibration of Sr-spike solutions by isotope dilution (Herzog and Hurley 1955). This solution was analyzed periodically by Faure (1961) in order to determine the reproducibility of measured $^{87}Sr/^{86}Sr$ ratios, and a summary of the results was later published by Faure and Hurley (1963).

The E&A Sr carbonate was distributed informally in the 1960s to isotope laboratories and has been widely used as an interlaboratory standard by isotope geochemists studying the isotope composition of Sr in volcanic rocks, marine carbonates, water on the continents, and in many other projects that rely on measurements of the absolute values of $^{87}Sr/^{86}Sr$ ratios. Most well-adjusted mass spectrometers yield $^{87}Sr/^{86}Sr$ ratios close to 0.70800 and this value has been adopted as the reference value. Accordingly, measured $^{87}Sr/^{86}Sr$ ratios are adjusted to be compatible with an $^{87}Sr/^{86}Sr$ ratio of 0.70800 for E&A.

The E&A Sr carbonate continues to be used almost half a century after it was originally adopted. However, because the isotope composition of Sr in this compound has never been properly certified and because the supply is limited, the National Bureau of Standards of the USA undertook to prepare a more appropriate interlaboratory Sr-isotope standard. The compound was obtained by the National Bureau of Standards from Spex Industries, Inc. in Metuchen, New Jersey. According to the Certificate of Analysis issued on November 8, 1971, its $^{87}Sr/^{86}Sr$ ratio is 0.71014 ±0.00020 after correcting it for isotope fractionation to $^{86}Sr/^{88}Sr = 0.1194$. The certificate also reported a value of 0.70794 for the $^{87}Sr/^{86}Sr$ ratio of the E&A Sr carbonate. However, more recent

measurements indicate that the $^{87}Sr/^{86}Sr$ ratios of the two standards are:

- E&A: 0.70800
- NBS 987: 0.71025

when both are corrected for isotope fractionation to $^{86}Sr/^{88}Sr = 0.11940$.

1.3 The Rb-Sr Method of Dating

The decay of ^{87}Rb and growth of ^{87}Sr expressed by Eqs. 1.12 and 1.16 is used to date certain selected igneous and high-grade metamorphic rocks and some of their constituent minerals. The method is based on measurements of the isotope composition of Sr and of the concentrations of Rb and Sr from which the $^{87}Rb/^{86}Sr$ ratio can be calculated (Dodson 1970; Birck 1986; Pin et al. 1994; Ludwig 1997). The value of the initial $^{87}Sr/^{86}Sr$ ratio is determined by analyzing two or more specimens of comagmatic igneous rocks or minerals that can be assumed to have the same age. Alternatively, Rb-rich minerals with large $^{87}Sr/^{86}Sr$ ratios, such as muscovite and biotite, can be dated by assuming a value of the initial $^{87}Sr/^{86}Sr$ ratio provided that the numerical value of the date calculated from Eqs. 1.12 and 1.16 is not sensitive to the choice of initial $^{87}Sr/^{86}Sr$ ratios.

The validity of age determinations of rocks and minerals by the Rb-Sr method depends on the following assumptions:

1. The rocks or minerals to be dated did not gain or lose Rb or Sr after their original crystallization.
2. When a suite of rocks or minerals is dated as a set, each specimen must have had the same initial $^{87}Sr/^{86}Sr$ ratio.
3. Similarly, each rock or mineral being dated in a set must have crystallized at the same time.
4. The value of the decay constant (or half-life) of ^{87}Rb is known accurately.
5. The measured values of the $^{87}Sr/^{86}Sr$ ratios and concentrations of Rb and Sr are free of systematic analytical errors.

The first three assumptions specify a simple model for the origin and subsequent geologic history of the rocks to be dated. This model is applicable to suites of differentiated igneous rocks (plutonic and volcanic) and to high-grade metamorphic rocks (schists and gneisses) (e.g. Burton and O'Nions 1990). However, conformance to this simple model cannot be taken for granted, but must be verified in each case by appropriate mineralogical and petrographic examination of the specimens to be dated and by using other complementary isotope geochronometers. Consequently, dating of igneous and metamorphic rocks is a formidable task that should not be undertaken lightly. Sedimentary rocks (clastic or chemically precipitated) are datable only under exceptional circumstances.

The increase of the $^{87}Sr/^{86}Sr$ ratios of a set of rocks or minerals in accordance with Eq. 1.16 is illustrated in Fig. 1.2. This diagram can be viewed as a model of the time-dependent evolution of the isotope composition of Sr in specimens of comagmatic igneous rocks which crystallized from an isotopically-homogeneous magma during an interval of time that was short compared to the age of the rocks. Under these conditions, all of the rock specimens in the set can be assumed to have had the same initial $^{87}Sr/^{86}Sr$ ratio at the time of their formation at 1 100 million years ago (Ma). After crystallization, the $^{87}Sr/^{86}Sr$ ratios of the rock specimens increased at different rates depending on their Rb/Sr ratios, provided that each rock remained closed to Rb and Sr. Therefore, each rock specimen has a different $^{87}Sr/^{86}Sr$ ratio at the present time. After the present Rb/Sr ratios of the rocks have been determined, the slopes of the isotope evolution lines can be extrapolated to obtain the coordinates of the point of convergence. These coordinates are the initial $^{87}Sr/^{86}Sr$ ratio and the time elapsed since the rock specimens formed by crystallization of their parental magma. In the simple case considered here, this time is the age of the rocks. Alternatively, the Rb-Sr systems in Fig. 1.2 may also be the constituent minerals of a coarse-grained igneous rock.

In current practice, age determinations of igneous rocks by the Rb-Sr method are actually based on Eq. 1.12, which is interpreted as a straight line in coordinates of $^{87}Sr/^{86}Sr$ and $^{87}Rb/^{86}Sr$, such that the slope is $e^{\lambda t} - 1$ and the intercept on the vertical axis is $(^{87}Sr/^{86}Sr)_0$. All Rb-Sr systems that have the same initial $^{87}Sr/^{86}Sr$ ratio and the same age (t) satisfy Eq. 1.12 and are represented by points that lie along a straight line provided they have remained closed to Rb and Sr since their formation. The straight line formed by such Rb-Sr systems is called the *isochron* because all Rb-Sr systems on the line have existed for the same length of time. The goodness of fit of the data points to a straight line is a test of the assumptions that all of the rock samples actually have the same initial $^{87}Sr/^{86}Sr$ ratio and the same age and have remained closed to Rb and Sr. The slope (m) of the best-fit line is:

$$m = e^{\lambda t} - 1$$

which is solved for t:

$$t = \frac{1}{\lambda}\ln(m+1) \qquad (1.18)$$

The value of t obtained from Eq. 1.18 is a *date* in the past. The geological significance of this date depends on the extent to which the model on which Eq. 1.12 is based actually represents the geologic history of the rocks that form the isochron. Therefore, a Rb-Sr isochron *date* is the *age* of the rocks *only* when all specimens actually had the same initial $^{87}Sr/^{86}Sr$ ratio at the time of their formation and remained closed to Rb and Sr since that time.

Linear data arrays in coordinates of $^{87}Sr/^{86}Sr$ and $^{87}Rb/^{86}Sr$ ratios can also be generated by mixing of two components having different $^{87}Sr/^{86}Sr$ and Rb/Sr ratios (Haack 1990). In addition, isochrons formed by the constituent *minerals* of igneous or metamorphic rocks may have lower slopes than the whole-rock isochron because the $^{87}Sr/^{86}Sr$ ratios of the minerals may have been homogenized during thermal alteration at some time after their formation (Faure 1986).

The $^{87}Sr/^{86}Sr$ and $^{87}Rb/^{86}Sr$ ratios of the rock samples, used in Fig. 1.2 to illustrate the time-dependent evolution of the isotope composition of Sr, were used again in Fig. 1.3 to form a whole-rock Rb-Sr isochron having a slope $m = 0.01575$ that yields a value of t by substitution into Eq. 1.18:

$$t = \frac{\ln(0.01575 + 1)}{1.42 \times 10^{-11}} = 1\,100 \times 10^6 \text{ years}$$

The intercept on the vertical axis, the initial $^{87}Sr/^{86}Sr$ ratio, is equal to 0.700 in this case. The isochron method is preferred for the interpretation of $^{37}Sr/^{86}Sr$ and Rb/Sr ratios of comagmatic igneous rocks because straight

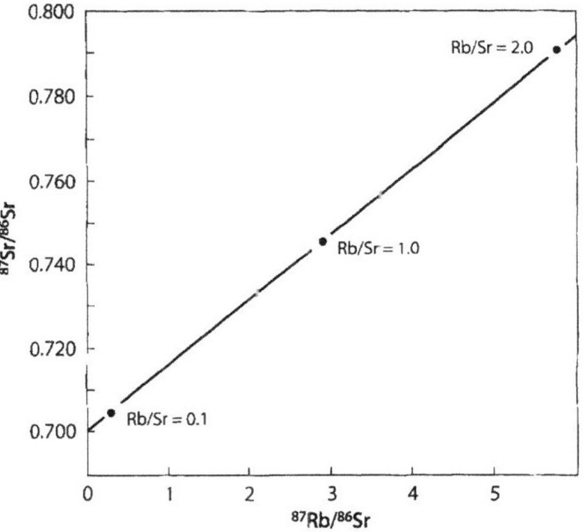

Fig. 1.3. Rb-Sr isochron formed by three whole-rock specimens of igneous rocks that crystallized from the same magma at the time. Rock samples that satisfy these conditions lie along a straight-line isochron in accordance with Eq. 1.12. The slope of the isochron yields the age of the rocks by Eq. 1.18 and the intercept on the vertical axis is the $^{87}Sr/^{86}Sr$ ratio of the magma at the time the rock samples crystallized from it. The three data points used to draw the isochron in this diagram are identical to the rock samples used in Fig. 1.2 to illustrate the evolution of the isotope composition of Sr

lines can be fitted to data points by statistical methods, whereas the convergence of isotope-evolution lines is difficult to treat statistically.

The Rb-Sr method can also be used to date metamorphic and certain kinds of sedimentary rocks (Clauer 1979; Harris and Fullagar 1991). These applications are based on the same principles and assumptions outlined above, but require careful selection of samples to assure that they satisfy the necessary assumptions for dating because metamorphic and sedimentary rocks may have complex histories that include mixing, complete or partial re-equilibration of Sr isotopes, and changes in the concentrations of Rb and Sr. Therefore, the interpretation of Rb-Sr data derived from such rocks is often difficult and may require additional information from complementary isotopic dating methods.

In summary, dating of rocks is a difficult task because even igneous rocks may not have the simple histories on which isotopic age determinations are based. These complexities will be discussed in subsequent chapters as appropriate. Additional information about isotopic dating methods is available in the textbook by Faure (1986).

The dates determined by isotopic methods are expressed in units of 10^3 years, 10^6 years, or 10^9 years. The North American Commission on Stratigraphic Nomenclature has defined these amounts of time as follows:

- 10^3 years = ka (kilo annum) or thousands of years ago
- 10^6 years = Ma (mega annum) or millions of years ago
- 10^9 years = Ga (giga annum) or billions of years ago

where "annum" is the Latin word for year. Because of the way in which these units are defined, they should not be used with the word "ago" or be combined with the phrase "before the present." In cases where an *interval* of time is referred to, the length of time should be expressed in terms of years (y or yr), thousands of years (ky, or k.y., or k.yr.), or millions of years (my, or m.y., or m.yr.). For example, a geologic event that occurred at 1 100 Ma, may have lasted 10 m.y. (AAPG Bulletin 67:841–875, 1983).

1.4   Fitting of Isochrons

Rock or mineral samples that have the same age (t), the same initial $^{87}Sr/^{86}Sr$ ratio, and have remained closed to Rb and Sr since the time of their formation are represented by points that form a straight line in coordinates of the $^{87}Sr/^{86}Sr$ and $^{87}Rb/^{86}Sr$ ratios. Both of these coordinates are determined by measurements and therefore have random (or "normal") as well as systematic analytical errors, where the word "error" means the deviation of a measured value from its true value (Brooks et al. 1972). If the measured values of the coordinates have significant *systematic* errors, the slope and intercept of the

isochron derived from the data are inaccurate. Therefore, conscientious efforts must be made to calibrate the equipment and procedures used to measure the concentrations of Rb and Sr and the isotope composition of Sr so that systematic analytical errors are eliminated.

The random analytical errors of the coordinates are used to calculate weighting factors in the regression to determine the best slope and intercept of the isochron. In addition, a statistical evaluation of the goodness of fit is used to determine whether the scatter of the data points above and below the best straight line is consistent with the magnitudes of the random analytical errors. Consequently, the magnitudes of random errors need to be determined by replication of analyses. In cases where the scatter of data points exceeds the random analytical errors, the validity of the slope and intercept of the straight line derived from the data are in doubt and the line is referred to as an "errorchron" (Brooks et al. 1972) or "scatterchron" (Wendt 1993).

The use of random analytical errors in the fitting of Rb-Sr isochrons inevitably leads to the paradoxical situation that improvement of the precision of the measurements (i.e. reduction in the random analytical errors) causes isochrons to be rejected because the random analytical errors do not account for all of the scatter of the data points. In such cases, it becomes apparent that real rocks and minerals do not actually satisfy the assumptions required for dating by the Rb-Sr method. The distinction between *isochrons* and *errorchrons* can be manipulated by "enhancing" the analytical errors. However, this practice defeats the purpose of the statistical evaluation of Rb-Sr data sets and is not recommended.

Samples that deviate from the best-fit line by a larger amount than is consistent with random analytical errors are said to have "geological error" (Brooks et al. 1972), i.e. they may have different ages and different initial $^{87}Sr/^{86}Sr$ ratios than the other samples in the suite and/or they may not have remained closed to Rb and/or Sr. Therefore, samples of igneous rocks to be dated by the Rb-Sr method should be carefully selected in the field and should be examined in thinsection to exclude specimens that may not be cogenetic or may have been altered by regional or contact metamorphism, tectonic deformation, hydrothermal alteration, or chemical weathering. Even more stringent restrictions apply to Rb-Sr dating of metamorphic and sedimentary rocks.

Statistical procedures for fitting isochrons and for evaluating the goodness of fit of the data points have been presented by many authors including McIntyre et al. (1966), York (1966, 1967, 1969), Purdy and York (1966), Brooks et al. (1968, 1972), Williamson (1968), Wendt (1969), Giletti (1974), Chayes (1977), Roddick (1978, 1987), Dodson (1973, 1976, 1979, 1982), Titterington and Halliday (1979), Field and Råheim (1979), Cameron et al. (1981) (with a critical comment by Austrheim 1983),

Vugrinovich (1981), Butler (1982), Rock and Duffy (1986), Ganguly and Ruiz (1986), Rock et al. (1987), Cavazzini (1988), Castorina and Riccardo (1989), Kostitsyn (1989), Kalsbeek and Hansen (1989), Haack (1990), Kent et al. (1990), Provost (1990), and Wendt (1993). Harmer and Eglington (1990) published a useful review of the statistical principles used to fit straight lines to data points. Ludwig (1992) developed a program for fitting isochrons on IBM-PCs based on the equations of York (1966, 1969). Kullerud (1991) reconsidered the use of analytical errors to obtain weighting factors and emphasized the importance of determining these errors accurately in sets of samples having a wide range of Rb and Sr concentrations. Additional shortcomings of the Rb-Sr isochron method of dating have been pointed out by Schleicher et al. (1983) and Lutz and Srogi (1986).

Brooks et al. (1972) examined the different statistical procedures proposed by McIntyre et al. (1966), York (1966, 1969), Brooks et al. (1968), and by Wendt (1969). Using simulated test data they demonstrated that all four models of McIntyre et al. (1966) yield identical results in those cases where the observed scatter of the data points can be accounted for solely on the basis of random analytical errors. In such cases, the line defined by the data is a true *isochron* because the data satisfy the assumptions for dating. In cases where the scatter *exceeds* analytical error, the slope and intercept of the best fit line (errorchron or scatterchron) vary depending on the statistical model that is used, and therefore may not be reliable. The dates and initial ^{87}Sr/^{86}Sr ratios derived from errorchrons may, nevertheless, convey geologically useful information, especially in cases where additional facts or judgments support them, such as the results of dating by other isotopic methods.

All of the weighted regression procedures reviewed by Brooks et al. (1972) make use of a statistical index to test the goodness of fit of the data points to the best straight line. The index used by McIntyre et al. (1966) is the mean square of weighted deviates (*MSWD*). The numerical value of *MSWD* for a set of analytical data can be used to decide whether the scatter of data points is consistent with the magnitudes of the analytical errors (Wendt and Carl 1991). If the regression is based on a very large number of data sets and if the analytical errors have been calculated from a very large number of duplicates, then a value of unity or less for *MSWD* signifies that the line is an *isochron*. However, when the number of data sets and duplicates is less than in the ideal case, the limiting value of *MSWD* increases. The magnitudes of the limiting values of *MSWD* for different numbers of data sets and duplicates are given in an F-distribution table in Appendix 3 of Brooks et al. (1972). For example, in case a regression is based on five data sets and five duplicates, the limiting value of *MSWD* is 5.41 at the 95% confidence level. Limiting values of *MSWD* for different numbers of duplicates and data sets

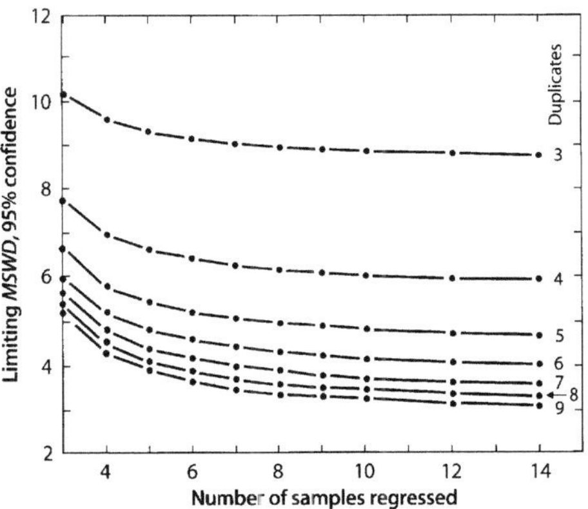

Fig. 1.4. Limiting values of the mean square of weighted deviates (*MSWD*, McIntyre et al. 1966) for increasing number of data sets and duplicates used in the "least-square cubic" regression of McIntyre et al. (1966) and York (1966). If the *MSWD* calculated for a data set is equal to or less than the theoretical value, the scatter of data points is consistent with the analytical errors and the regression line is an *isochron*. If the *MSWD* is greater than the predicted value, the regression line is an *errorchron* whose slope and intercept may be inaccurate and require confirmation by other evidence unrelated to the regression (*Source:* Brooks et al. 1972, Appendix 3)

are presented in Fig. 1.4 based on Appendix 3 of Brooks et al. (1972).

The data in Fig. 1.4 indicate that the distinction between *isochrons* and *errorchrons* for a given number of data sets depends on the number of duplicates used to evaluate analytical errors. For example, *MSWD* = 6.0 for ten samples and four duplicates indicates that the regression line is an *isochron*. However, if the number of duplicates is greater than 4, *MSWD* = 6.0 indicates that the line is an *errorchron* and that *geological error* is present. In other words, as analytical errors *decrease*, the tolerance for scatter *decreases* and regression lines become *errorchrons* or *scatterchrons*.

Geological errors may be associated with all of the samples in a set or with only one or two of the samples. The former case may arise in dating volcanic rocks which may have different initial ^{87}Sr/^{86}Sr ratios because the magma from which they formed may have assimilated crustal rocks containing varying amounts of radiogenic ^{87}Sr. If only one or two specimens in a suite of samples deviate significantly from the best-fit line, they can be excluded from the regression. The remaining samples may then yield an *isochron* from which a reliable date can be derived. The geological error recognized in the deviant samples is a clue that they may have been affected by geological events not recorded in the other samples. Therefore, samples having geological error deserve additional study to determine the cause for their deviation from the isochron.

1.5   The Sm-Nd Method of Dating

The sensitivity of Rb-Sr dates to alteration mandates that such rocks should also be dated by the Sm-Nd method because it is less vulnerable to alteration. Samarium (Sm) and neodymium (Nd) are both rare-earth elements and occur in rock-forming minerals in low but easily measurable concentrations (Faure 1986).

Samarium ($Z = 62$) has seven stable and naturally occurring isotopes whose abundances are: $^{144}\mathrm{Sm} = 3.1\%$; $^{147}\mathrm{Sm} = 15.0\%$; $^{148}\mathrm{Sm} = 11.2\%$; $^{149}\mathrm{Sm} = 13.8\%$; $^{150}\mathrm{Sm} = 7.4\%$; $^{152}\mathrm{Sm} = 26.7\%$; $^{154}\mathrm{Sm} = 22.8\%$. Samarium-147 is radioactive and decays by alpha emission to stable $^{143}_{60}\mathrm{Nd}$ with a half-life of 1.06×10^{11} years ($\lambda = 6.54 \times 10^{-12}\,\mathrm{yr}^{-1}$).

Neodymium ($Z = 60$) also has seven stable isotopes whose abundances are: $^{142}\mathrm{Nd} = 27.1\%$; $^{143}\mathrm{Nd} = 12.2\%$; $^{144}\mathrm{Nd} = 23.9\%$; $^{145}\mathrm{Nd} = 8.3\%$; $^{146}\mathrm{Nd} = 17.2\%$; $^{148}\mathrm{Nd} = 5.7\%$; $^{150}\mathrm{Nd} = 5.6\%$. The abundances of the Nd isotopes in rocks and minerals change with time because of the formation of radiogenic $^{143}\mathrm{Nd}$ by alpha decay of $^{147}\mathrm{Sm}$.

The $^{143}\mathrm{Nd}/^{144}\mathrm{Nd}$ ratios of rocks and minerals are related to the $^{147}\mathrm{Sm}/^{144}\mathrm{Nd}$ ratios and to time by the equation:

$$\frac{^{143}\mathrm{Nd}}{^{144}\mathrm{Nd}} = \left(\frac{^{143}\mathrm{Nd}}{^{144}\mathrm{Nd}}\right)_0 + \frac{^{147}\mathrm{Sm}}{^{144}\mathrm{Nd}}(e^{\lambda t} - 1) \qquad (1.19)$$

This equation is analogous to Eq. 1.12 for the Rb-Sr system and is the basis for the Sm-Nd isochron method of dating. Sets of whole-rock samples having the same age (t) and the same initial $^{143}\mathrm{Nd}/^{144}\mathrm{Nd}$ ratio define the Sm-Nd isochron in coordinates of $^{143}\mathrm{Nd}/^{144}\mathrm{Nd}$ and $^{147}\mathrm{Sm}/^{144}\mathrm{Nd}$. The slope ($m$) of the isochron is related to the age of the rocks that define the isochron by:

$$m = e^{\lambda t} - 1 \qquad (1.20)$$

$$t = \frac{1}{\lambda}\ln(m + 1)$$

In some cases, Sm-Nd whole-rock isochron dates are older than the Rb-Sr isochron dates of the same rock specimens because Sm and Nd tend to be less mobile during thermal or hydrothermal alteration. For this reason, the Sm-Nd method is preferred for dating metavolcanic rocks of Archean and Proterozoic age.

1.6   The Epsilon Notation

The epsilon notation was introduced by DePaolo and Wasserburg (1976a,b) in order to facilitate the interpretation of measured $^{143}\mathrm{Nd}/^{144}\mathrm{Nd}$ ratios of basaltic volcanic rocks derived from magma sources in the mantle.

The epsilon parameter $\varepsilon^0(\mathrm{Nd})$ compares the measured $^{143}\mathrm{Nd}/^{144}\mathrm{Nd}$ ratio of a rock or mineral sample to the present $^{143}\mathrm{Nd}/^{144}\mathrm{Nd}$ ratio of a hypothetical Chondritic Uniform Reservoir called "CHUR" and is defined by the equation:

$$\varepsilon^0(\mathrm{Nd}) = \left[\frac{(^{143}\mathrm{Nd}/^{144}\mathrm{Nd})_{\mathrm{meas.}}}{(^{143}\mathrm{Nd}/^{144}\mathrm{Nd})^0_{\mathrm{CHUR}}} - 1\right]10^4 \qquad (1.21)$$

where:

$\left(\dfrac{^{143}\mathrm{Nd}}{^{144}\mathrm{Nd}}\right)^0_{\mathrm{CHUR}}$ = 0.511847 or 0.512638 (depending on the way the isotope fractionation correction is applied) is the present value of this ratio in CHUR (DePaolo 1988).

The epsilon value can be recalculated so that it refers to a time in the past. The appropriate value of the $^{143}\mathrm{Nd}/^{144}\mathrm{Nd}$ ratio of CHUR is obtained from Eq. 1.12:

$$\left(\frac{^{143}\mathrm{Nd}}{^{144}\mathrm{Nd}}\right)^0_{\mathrm{CHUR}} = \left(\frac{^{143}\mathrm{Nd}}{^{144}\mathrm{Nd}}\right)^t_{\mathrm{CHUR}}$$
$$+ \left(\frac{^{147}\mathrm{Sm}}{^{144}\mathrm{Nd}}\right)^0_{\mathrm{CHUR}}(e^{\lambda t} - 1) \qquad (1.22)$$

where

$\left(\dfrac{^{147}\mathrm{Sm}}{^{144}\mathrm{Nd}}\right)^0_{\mathrm{CHUR}}$ = value of this ratio at the present time = 0.1967 (DePaolo 1988; Wasserburg et al. 1981)

$\left(\dfrac{^{143}\mathrm{Nd}}{^{144}\mathrm{Nd}}\right)^t_{\mathrm{CHUR}}$ = $^{143}\mathrm{Nd}/^{144}\mathrm{Nd}$ of CHUR at some time t in the past

$\left(\dfrac{^{143}\mathrm{Nd}}{^{144}\mathrm{Nd}}\right)^0_{\mathrm{CHUR}}$ = 0.511847 or 0.512638 (value of this ratio at the present time)

Epsilon values referred to a time in the past are designated by $\varepsilon^t(\mathrm{Nd})$ where t is the specified age.

The $^{143}\mathrm{Nd}/^{144}\mathrm{Nd}$ ratio of the chondritic uniform reservoir is affected by the way in which the measured isotope ratios of Nd are corrected for isotope fractionation. Most isotope geochemists base the isotope fractionation correction of measured $^{143}\mathrm{Nd}/^{144}\mathrm{Nd}$ ratios on a value of 0.7219 for the $^{146}\mathrm{Nd}/^{144}\mathrm{Nd}$ ratio, whereas others use $^{146}\mathrm{Nd}/^{142}\mathrm{Nd} = 0.636151$ as the reference ratio. Wasserburg et al. (1981) demonstrated the relation among the isotope ratios shown in Table 1.6. Both schemes yield identical $\varepsilon^0(\mathrm{Nd})$ values provided that the $^{143}\mathrm{Nd}/^{144}\mathrm{Nd}$ ratios of the rock samples were corrected

Table 1.6. Relation among the isotope ratios $^{143}Nd/^{144}Nd$, $^{146}Nd/^{144}Nd$, and $^{146}Nd/^{142}Nd$ (Wasserburg et al. 1981)

	Reference ratio	Present $^{143}Nd/^{144}Nd$ of CHUR
$^{146}Nd/^{144}Nd$	0.7219	0.512638
$^{146}Nd/^{142}Nd$	0.636151	0.511847

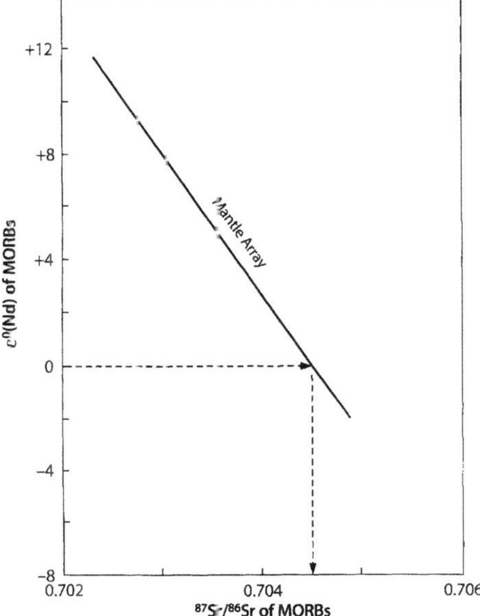

Fig. 1.5 Schematic representation of the "Mantle Array" defined by the isotope compositions of Nd and Sr in mid-ocean ridge basalts (MORB) from certain locations. The measured $^{143}Nd/^{144}Nd$ ratios of MORBs have been expressed in terms of epsilon values relative to the Chondritic Uniform Reservoir (CHUR) based on Eq. 1.20. A value of zero for ε^0(Nd) implies that MORB and CHUR have the same $^{143}Nd/^{144}Nd$ ratio and that no differentiation has occurred. Therefore, the Sr in MORBs having ε^0(Nd) = 0 presumably also originated from an Undifferentiated Reservoir called UR. The Mantle Array indicates that UR has a present $^{87}Sr/^{86}Sr$ ratio of 0.7045 ±0.0005. Therefore this value is used to define the epsilon parameter for Sr in Eq. 1.23 (*Source:* DePaolo 1988)

to the same reference value as the $^{143}Nd/^{144}Nd$ of CHUR used to calculate the epsilon parameter.

The ε^0(Nd) values of some mid-ocean ridge basalts (MORB) vary *inversely* with their $^{87}Sr/^{86}Sr$ ratios in the so-called "Mantle Array" shown schematically in Fig. 1.5. This correlation between the Sm-Nd and the Rb-Sr systems is caused by differences in the geochemical properties of Sm and Rb during magma formation by partial melting of rocks in the mantle. Whereas Rb is concentrated in the silicate liquid, Sm remains in the solid phases. Therefore, partial melting in the mantle causes the Rb/Sr ratio of the residual rocks to *decrease*, whereas their Sm/Nd ratios *increase*. Consequently, the so-called depleted regions of the mantle have comparatively low $^{87}Sr/^{86}Sr$ ratios but elevated $^{143}Nd/^{144}Nd$ ratios. The Mantle Array may be the result of mixing of magmas derived from depleted and undepleted source rocks in the mantle.

MORBs whose ε^0(Nd) value is equal to zero contain Nd whose $^{143}Nd/^{144}Nd$ ratio is identical to that of CHUR. Therefore, the magma sources of such MORBs are *undifferentiated* with respect to the chondritic reservoir. Presumably, these MORBs also contain Sr derived from an *undifferentiated reservoir*. The Mantle Array in Fig. 1.5 indicates that the Sr in the Undifferentiated Reservoir called UR is characterized by:

$$\left(\frac{^{87}Sr}{^{86}Sr}\right)^0_{UR} = 0.7045$$

(DePaolo 1988). The $^{87}Rb/^{86}Sr$ ratio of UR is calculated on the basis of the assumption that the $^{87}Sr/^{86}Sr$ ratio of UR at the time of formation of the Earth at 4.55 Ga was identical to that of basaltic achondrite meteorites:

$$\left(\frac{^{87}Sr}{^{86}Sr}\right)_{BABI} = 0.69899$$

where BABI stands for Basaltic Achondrite Best Initial. Therefore, the $^{87}Rb/^{86}Sr$ ratio of UR can be calculated from Eq. 1.12 for $t = 4.55 \times 10^9$ yr:

$$0.7045 = 0.69899 + \left(\frac{^{87}Rb}{^{86}Sr}\right)^0_{UR}(e^{\lambda t} - 1)$$

$$\left(\frac{^{87}Rb}{^{86}Sr}\right)^0_{UR} = 0.0827$$

$$\left(\frac{Rb}{Sr}\right)^0_{UR} = \frac{0.0827}{2.89} = 0.0286$$

The Undifferentiated Reservoir is also referred to as the Bulk Earth, but the validity of this concept has become somewhat tarnished by the fact that the correlation between $^{143}Nd/^{144}Nd$ and $^{87}Sr/^{86}Sr$ ratios of MORBs is not as well constrained as it originally appeared to be (Faure 1986; DePaolo 1988).

By analogy with the Sm-Nd system, the ε^0(Sr) parameter is defined as:

$$\varepsilon^0(Sr) = \left[\frac{(^{87}Sr/^{86}Sr)_{meas.} - (^{87}Sr/^{86}Sr)^0_{UR}}{(^{87}Sr/^{86}Sr)^0_{UR}}\right]10^4 \quad (1.23)$$

where

$\varepsilon^0(\mathrm{Sr})$ = epsilon value at the present time

$\left(\dfrac{^{87}\mathrm{Sr}}{^{86}\mathrm{Sr}}\right)_{\mathrm{meas.}}$ = measured value of this ratio at the present time after correcting it for isotope fractionation to $^{86}\mathrm{Sr}/^{88}\mathrm{Sr} = 0.11940$

$\left(\dfrac{^{87}\mathrm{Sr}}{^{86}\mathrm{Sr}}\right)^0_{\mathrm{UR}}$ = value of this ratio of UR at the present time and assumed to be equal to 0.7045, corrected to $^{86}\mathrm{Sr}/^{88}\mathrm{Sr} = 0.11940$

Epsilon values for Sr can also be calculated for any time in the past, such as at the time of formation of an igneous rock. In this case, the measured $^{87}\mathrm{Sr}/^{86}\mathrm{Sr}$ ratio is replaced by the initial ratio (determined from an isochron) and the $^{87}\mathrm{Sr}/^{86}\mathrm{Sr}$ ratio of UR is recalculated to the appropriate time based on Eq. 1.12:

$$\left(\frac{^{87}\mathrm{Sr}}{^{86}\mathrm{Sr}}\right)^0_{\mathrm{UR}} = \left(\frac{^{87}\mathrm{Sr}}{^{86}\mathrm{Sr}}\right)^t_{\mathrm{UR}} + 0.0827(e^{\lambda t} - 1)$$

Epsilon values that represent the $^{87}\mathrm{Sr}/^{86}\mathrm{Sr}$ ratio of a Rb-Sr system in the past are designated by $\varepsilon^t(\mathrm{Sr})$.

A positive value of $\varepsilon^0(\mathrm{Sr})$ implies that the Rb-Sr system has been enriched in radiogenic $^{87}\mathrm{Sr}$ compared to the hypothetical Undifferentiated Reservoir, whereas a negative value of $\varepsilon^0(\mathrm{Sr})$ indicates depletion in $^{87}\mathrm{Sr}$. Enrichment or depletion of igneous rocks in radiogenic $^{87}\mathrm{Sr}$ relative to UR may be caused either by appropriate changes in the Rb/Sr ratio of their magma sources prior to partial melting, or to processes that affected the magma after its formation, such as assimilation of old crustal rocks, or mixing of magmas derived from different sources.

The $^{87}\mathrm{Rb}/^{86}\mathrm{Sr}$ ratios of terrestrial samples can also be referred to the $^{87}\mathrm{Rb}/^{86}\mathrm{Sr}$ ratio of the Undifferentiated Reservoir by means of the relation:

$$f(\mathrm{Rb/Sr}) = \frac{(^{87}\mathrm{Rb}/^{86}\mathrm{Sr})_{\mathrm{meas.}} - (^{87}\mathrm{Rb}/^{86}\mathrm{Sr})_{\mathrm{UR}}}{(^{87}\mathrm{Rb}/^{86}\mathrm{Sr})_{\mathrm{UR}}} \quad (1.24)$$

Therefore, Rb-Sr systems that are enriched in Rb or depleted in Sr relative to UR have positive f values whereas systems that are depleted in Rb or enriched in Sr have negative values of f.

The epsilon notation for expressing isotope compositions of Sr, as well as $f(\mathrm{Rb/Sr})$ values, depend on the $^{87}\mathrm{Sr}/^{86}\mathrm{Sr}$ and $^{87}\mathrm{Rb}/^{86}\mathrm{Sr}$ ratios of the hypothetical Undifferentiated Reservoir. These values should therefore be specified whenever data are presented in terms of epsilon values or in terms of enrichment factors.

1.7 Mixtures of Two Components

Geological processes occurring in a variety of environments in the crust of the Earth and on its surface cause mixing of materials with different concentrations and different isotope compositions of elements having radiogenic isotopes (e.g. Sr, Nd, Pb, Hf, and Os). Examples of such processes include:

1. Assimilation of granitic rocks of the continental crust by basaltic magma originating in the mantle;
2. Mixing of two magmas derived from different sources;
3. Co-deposition of detrital sediment derived from young volcanic and old crustal rocks;
4. Mixing of water masses that have interacted with different kinds of rocks in the subsurface or on the surface of the Earth (e.g. Franklyn et al. 1991).

1.7.1 Two-component Mixtures

The concentrations of Rb, Sr, and of other *conservative* elements of two-component mixtures depend on their concentrations in the components and on the proportions by weight or volume of these components. If [X] is the concentration of a conservative element in two components A and B, then:

$$[\mathrm{X}]_\mathrm{M} = [\mathrm{X}]_\mathrm{A}\, f_\mathrm{A} + [\mathrm{X}]_\mathrm{B}(1 - f_\mathrm{A}) \quad (1.25)$$

where:

- $[\mathrm{X}]_\mathrm{M}$ = concentration of element X in a mixture M of components A and B,
- $[\mathrm{X}]_\mathrm{A,B}$ = concentration of element X in the components A and B, respectively, and
- f_A = weight or volume fraction of component A in the mixture M.

The mixing parameter f_A is defined by the equation:

$$f_\mathrm{A} = \frac{W_\mathrm{A}}{W_\mathrm{A} + W_\mathrm{B}} \quad (1.26)$$

where W_A and W_B are the weights of A and B in a given mixture. For example, if one gram of a mixture contains 0.3 g of A and 0.7 g of B, the corresponding value of $f_\mathrm{A} = 0.3 / (0.3 + 0.7) = 0.3$. Similarly, the mass fraction of B in the same mixture is $f_\mathrm{B} = 0.7 / (0.3 + 0.7) = 0.7$, or $f_\mathrm{B} = 1 - f_\mathrm{A}$. The mixing parameter is expressed in terms of weights when the concentrations of element X are measured in weight units, such as $\mu g\ g^{-1}$. If the concentrations are expressed in volume units, such as $\mu g\ mL^{-1}$, the mixing parameter is the corresponding volume fraction.

The concentration of a second conservative element Y in a mixture M of two components A and B is expressed by:

$$[Y]_M = [Y]_A f_A + [Y]_B(1 - f_A) \qquad (1.27)$$

Solving Eqs. 1.26 and 1.27 for f_A and equating the results yields:

$$\frac{[Y]_M - [Y]_B}{[Y]_A - [Y]_B} = \frac{[X]_M - X]_B}{[X]_A - X]_B}$$

$$[Y]_M = \frac{([X]_M - X]_B)([Y]_A - [Y]_B)}{[X]_A - [X]_B} + [Y]_B$$

which reduces to the equation of a straight line in the slope-intercept form:

$$[Y]_M = \frac{[X]_M([Y]_A - [Y]_B)}{[X]_A - [X]_B}$$
$$+ \frac{[X]_A[Y]_B - [X]_B[Y]_A}{[X]_A - [X]_B} \qquad (1.28)$$

where the slope is:

$$m = \frac{[Y]_A - [Y]_B}{[X]_A - [X]_B}$$

and the intercept on the y-axis is:

$$b = \frac{[X]_A[Y]_B - [X]_B[Y]_A}{[X]_A - [X]_B}$$

Consequently, the concentrations of Rb, Sr, and of other conservative elements in mixtures of two components lie along straight lines that connect the points whose coordinates are the concentrations of elements X and Y in components A and B.

The co-linearity of the concentrations of two conservative elements in two-component mixtures expressed by Eq. 1.28 can be used to test real data to determine whether a suite of samples may have formed by mixing of two components. If the concentrations of two conservative elements form a straight line, the samples *may* be two-component mixtures. However, the goodness of fit of concentration data derived from two-component mixtures may be less than perfect because:

1. The composition of the components is variable;
2. The chemical elements under consideration are not conservative; and
3. Additional components may be present.

In spite of these limitations, mixing is an important consequence of geological activity and is reflected by systematic relationships among the concentration of conservative elements.

1.7.2 Isotopic Mixtures of Sr and Nd

When two components A and B have different concentrations and isotope compositions of an element such as Sr, the resulting mixtures exhibit a range of both chemical and isotopic compositions. The Sr concentrations of two-component mixtures are expressed in accordance with Eq. 1.25:

$$[Sr]_M = [Sr]_A f_A + [Sr]_B(1 - f_A) \qquad (1.29)$$

The $^{87}Sr/^{86}Sr$ ratios of such mixtures are related to the isotope ratios and concentrations of Sr of the components by an equation derived by Faure (1986):

$$\left(\frac{^{87}Sr}{^{86}Sr}\right)_M = \left(\frac{^{87}Sr}{^{86}Sr}\right)_A f_A \frac{[Sr]_A}{[Sr]_M}$$
$$+ \left(\frac{^{87}Sr}{^{86}Sr}\right)_B (1 - f_A)\frac{[Sr]_B}{[Sr]_M} \qquad (1.30)$$

The equation contains weighting factors for the Sr contributed by each component ($[Sr]_A/[Sr]_M$, $[Sr]_B/[Sr]_M$) as well as the mixing parameter f_A that expresses the abundances of the components A and B in a given mixture M.

Equations 1.29 and 1.30 can be used to calculate the $^{87}Sr/^{86}Sr$ ratios and Sr concentrations of two-component mixtures for selected values of f_A (or f_B) from <1 to >0. Fig. 1.6a shows the results of such a calculation for the case where the coordinates of components A and B are:

- $[Sr]_A = 100$ ppm, $(^{87}Sr/^{86}Sr)_A = 0.8000$
- $[Sr]_B = 500$ ppm, $(^{87}Sr/^{86}Sr)_B = 0.7000$

For example, if $f_A = 0.4$, Eq. 1.29 yields:

- $[Sr]_M = 100 \times 0.4 + 500 \times 0.6 = 340$ ppm

By substituting $[Sr]_M = 340$ ppm into Eq. 1.30 and setting $f_A = 0.4$ as before, we obtain:

$$\left(\frac{^{87}Sr}{^{86}Sr}\right)_M = 0.800 \times 0.4 \times \frac{100}{340}$$
$$+ 0.700 \times 0.6 \times \frac{500}{340} = 0.7117$$

The resulting mixing curve is the positive branch of a hyperbola in contrast to the straight lines (Eq. 1.28) that represent two-component mixtures in terms of the concentrations of two conservative elements.

The curvature of the mixing hyperbola depends on the ratio of the Sr concentrations of the two components. In the case depicted in Fig. 1.6a, the hyperbola is

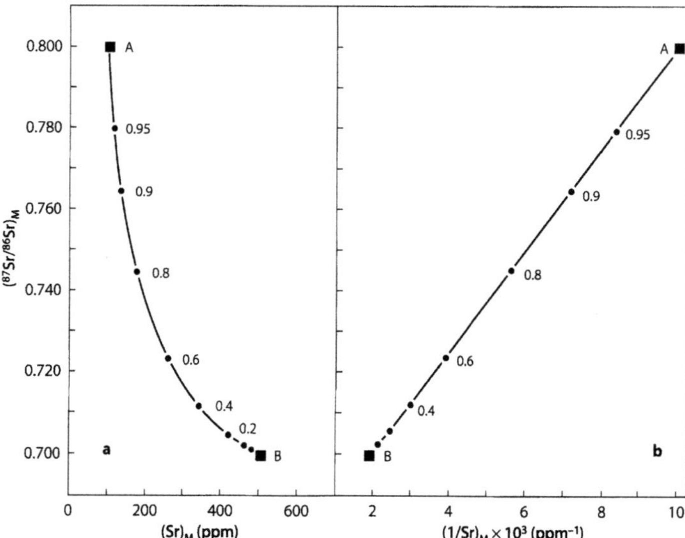

Fig. 1.6.
a Mixtures of Components A and B having different $^{87}Sr/^{86}Sr$ ratios and Sr concentrations calculated by means of Eqs. 1.29 and 1.30. The values of the mixing parameter f_A (Eq. 1.25) are indicated along the curve. The resulting mixing curve is the positive branch of a hyperbola; **b** The mixing hyperbola in Part a has been converted into a straight line by plotting the reciprocals of $(Sr)_M$ in accordance with Eq. 1.31

convex *downward* because $Sr_A/Sr_B < 1.0$. If $Sr_A/Sr_B > 1.0$, the mixing hyperbola is convex *upward*. If $Sr_A/Sr_B = 1.0$, the mixing hyperbola turns into a straight line. An example of such an occurrence is presented in Sect. 3.2 for a suite of samples of hydrothermal waters discharged by hotsprings on the seafloor of the Mariana Trough in the Pacific Ocean based on data by Kusakabe et al. (1990).

Equations 1.29 and 1.30 can be combined by eliminating f_A from both equations. The resulting equation, derived by Faure (1986) is:

$$\left(\frac{^{87}Sr}{^{86}Sr}\right)_M = \frac{a}{[Sr]_M} + b \qquad (1.31)$$

where:

$$a = \frac{[Sr]_A [Sr]_B \left[\left(\dfrac{^{87}Sr}{^{86}Sr}\right)_B - \left(\dfrac{^{87}Sr}{^{86}Sr}\right)_A\right]}{[Sr]_A - [Sr]_B}$$

$$b = \frac{[Sr]_A \left(\dfrac{^{87}Sr}{^{86}Sr}\right)_A - [Sr]_B \left(\dfrac{^{87}Sr}{^{86}Sr}\right)_B}{[Sr]_A - [Sr]_B}$$

Equation 1.31 is a hyperbola in coordinates of the $^{87}Sr/^{86}Sr$ ratio and the Sr concentration as shown in Fig. 1.6a. However, the equation can be transformed into a straight line by inverting the Sr concentrations. In other words, the x-coordinate is set equal to $1/(Sr)_M$. In that case, a is the slope of the straight line and b is the inter-

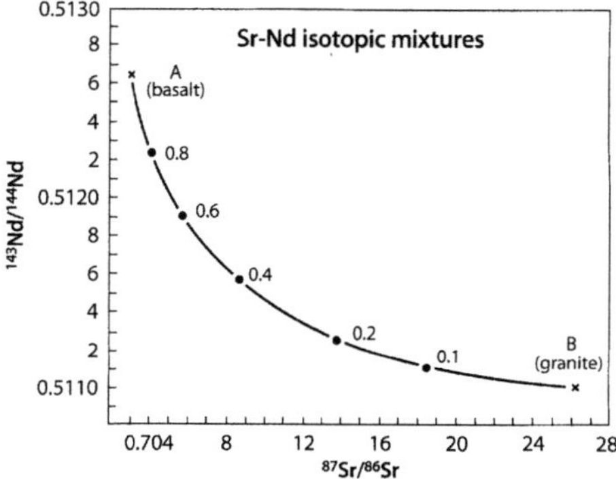

Fig. 1.7. Mixing hyperbola of two components A (basalt) and B (granite) defined in the text. The $^{87}Sr/^{86}Sr$ and $^{143}Nd/^{144}Nd$ ratios of all mixtures of these components form points that lie on the hyperbola

cept on the y-axis. The conversion of the mixing hyperbola into a straight line is illustrated in Fig. 1.6b.

Igneous rocks may form by mixing of magmas, as a result of assimilation of crustal rocks by mantle-derived magmas, or by melting of heterogeneous source rocks containing Sr, Nd, and Pb having different isotope compositions. The resulting isotopic mixtures of Sr-Nd, Sr-Pb, Nd-Pb, etc. can be modeled using Eqs. 1.29 and 1.30. The hyperbola in Fig. 1.7 describes such mixtures in coordinates of the $^{87}Sr/^{86}Sr$ and $^{143}Nd/^{144}Nd$ ratios formed by combining two components in varying proportions (Component A = basalt; Component B = granite):

$[Sr]_A = 450$ ppm; $\quad \left(\dfrac{^{87}Sr}{^{86}Sr}\right)_A = 0.70300$

$[Nd]_A = 35$ ppm $\quad \left(\dfrac{^{144}Nd}{^{143}Nd}\right)_A = 0.51264$

$[Sr]_B = 100$ ppm; $\quad \left(\dfrac{^{87}Sr}{^{86}Sr}\right)_B = 0.7260$

$[Nd]_B = 45$ ppm; $\quad \left(\dfrac{^{144}Nd}{^{143}Nd}\right)_B = 0.51100$

If $f = 0.8$:

$[Sr]_M = 450 \times 0.8 + 100 \times 0.2 = 380$ ppm

$$\left(\dfrac{^{87}Sr}{^{86}Sr}\right)_M = 0.7030 \times 0.8 \times \dfrac{450}{380}$$

$$+ 0.7260 \times 0.2 \times \dfrac{100}{380} = 0.70421$$

$[Nd]_M = 35 \times 0.8 + 45 \times 0.2 = 37$ ppm

$$\left(\dfrac{^{144}Nd}{^{143}Nd}\right)_M = 0.51264 \times 0.8 \times \dfrac{35}{37}$$

$$+ 0.51100 \times 0.2 \times \dfrac{45}{37} = 0.51224$$

Eqations 1.29, 1.30 and $f = 0.6$ yield: $[Sr]_M = 310$ ppm, $(^{87}Sr/^{86}Sr)_M = 0.70596$, $[Nd]_M = 39$ ppm, $(^{143}Nd/^{144}Nd)_M = 0.51188$.

Such hyperbolas represent igneous or metamorphic rocks that contain components derived both from the mantle and from the continental crust. The mantle component may be represented by magmas that formed by partial melting of ultramafic rocks in the asthenospheric or lithospheric mantle. In addition, magmas derived by remelting of mantle-derived volcanic rocks of basaltic composition may also contain Sr, Nd, and Pb (as well as Hf and Os) whose isotope compositions are indistinguishable from those that characterize magma sources in the mantle.

The systematics of mixing two components containing *two* elements having *different* isotope compositions have been presented by Vollmer (1976), Langmuir et al. (1978), Faure (1986), and are discussed in Sect. 3.2.

1.8 The U-Pb Methods of Dating

Uranium has two long-lived naturally-occurring radioactive isotopes ($^{238}_{92}U$ and $^{235}_{92}U$) both of which decay to stable isotopes of Pb. (Table 1.7). Each of the two U isotopes supports a series of unstable daughters including ^{234}U, which is a daughter of ^{238}U. A third series results from the decay of $^{232}_{90}Th$, the only long-lived naturally-occurring radioactive isotope of Th. The isotopes of U and Th reach a state of secular equilibrium with their short-lived daughters such that all members of the chain decay at the same rate as their respective parents. In cases where secular equilibrium was established and has been maintained to the present time, the rate of production of the stable Pb isotopes is equal to the rate of decay of the parent nuclide at the head of the decay series. Therefore, the U,Th-Pb decay systematics can be treated as though the parent isotope had decayed directly to the stable daughter:

- $^{238}U \longrightarrow {}^{206}Pb$
- $^{235}U \longrightarrow {}^{207}Pb$
- $^{232}Th \longrightarrow {}^{208}Pb$

Although the decay of the ^{232}Th to ^{208}Pb ($\lambda = 4.9475 \times 10^{-11}$ yr^{-1}) can be used to date Th-bearing minerals, such dates do not always agree with U-Pb dates of the same sample. For this and other reasons, the Th-Pb method of dating is no longer widely used.

The growth of the radiogenic isotopes of Pb by decay of the isotopes of U is expressed by equations:

$$\frac{^{206}Pb}{^{204}Pb} = \left(\frac{^{206}Pb}{^{204}Pb}\right)_0 + \frac{^{238}U}{^{204}Pb}(e^{\lambda_1 t} - 1) \tag{1.32}$$

(The $^{238}U/^{204}Pb$ ratio is also symbolized by μ).

Table 1.7.
Physical properties of the naturally occurring isotopes of U and Pb

Isotope	Stable or unstable	Mass (amu)	Abundance (%)	Decay constant (yr^{-1})
$^{238}_{92}U$	unstable	238.050785	99.2743	1.55125×10^{-10}
$^{235}_{92}U$	unstable	235.043924	0.7200	9.84850×10^{-10}
$^{206}_{82}Pb$	stable[a]	205.974440	24.1	–
$^{207}_{82}Pb$	stable[a]	206.975871	22.1	–
$^{208}_{82}Pb$	stable[a]	207.976627	52.4	–
$^{204}_{82}Pb$	stable[a]	203.973020	1.4	–

[a] Radiogenic.

$$\frac{^{207}Pb}{^{204}Pb} = \left(\frac{^{207}Pb}{^{204}Pb}\right)_0 + \frac{^{235}U}{^{204}Pb}(e^{\lambda_2 t} - 1) \qquad (1.33)$$

These equations are used to calculate dates of certain U-bearing minerals that:

1. Admit U but exclude Pb;
2. Resist chemical alteration;
3. Occur in a wide variety of rocks.

The minerals that meet these criteria include zircon, baddeleyite, monazite, sphene, apatite, and rutile. Whole-rock samples are generally not suitable for dating by the U-Pb method because both elements may be lost by hydrothermal alteration and chemical weathering. However, igneous and metamorphic rocks have been dated successfully by the Pb-Pb method which relies on the relation between the $^{207}Pb/^{206}Pb$ ratios of U-bearing rocks and their ages (Sect. 1.8.2).

1.8.1 The U-Pb Concordia

In most cases, dates based on the decay of ^{238}U to ^{206}Pb and of ^{235}U to ^{207}Pb are discordant (i.e. they do not agree). The reason for the discordance is that the minerals may have lost radiogenic Pb during episodes of thermal metamorphism or by continuous diffusion at elevated temperature. If the loss of Pb occurred in the recent past, the $^{207}Pb/^{206}Pb$ ratio of the remaining Pb may not have been changed significantly. Therefore, dates calculated from the $^{207}Pb/^{206}Pb$ ratio are older and hence more reliable than dates calculated from the individual decay schemes (Faure 1986). The problem of the discordance of U-Pb dates was solved by Wetherill (1956, 1963) by means of the concordia diagram derived from Eq. 1.32 and 1.33.

The date based on decay of ^{238}U to ^{206}Pb is calculated from Eq. 1.32 which can be rearranged to yield the ratio of radiogenic ^{206}Pb to ^{238}U:

$$e^{\lambda_1 t} - 1 = \frac{^{206}Pb/^{204}Pb - (^{206}Pb/^{204}Pb)_0}{^{238}U/^{204}Pb} = \frac{^{206}Pb^*}{^{238}U} \qquad (1.34)$$

Similarly for the decay of ^{235}U to ^{207}Pb (Eq. 1.33)

$$e^{\lambda_2 t} - 1 = \frac{^{207}Pb/^{204}Pb - (^{207}Pb/^{204}Pb)_0}{^{235}U/^{204}Pb}$$

$$e^{\lambda_2 t} - 1 = \frac{^{207}Pb^*}{^{235}U} \qquad (1.35)$$

where the asterisk identifies the radiogenic Pb isotopes. Equations 1.34 and 1.35 are used to calculate the $^{206}Pb^*/^{238}U$ and $^{207}Pb^*/^{235}U$ ratios by substituting the same value of t

Table 1.8. Coordinates for construction of the U-Pb concordia curve for 500-million-year increments of time ($\lambda_1 = 1.55125 \times 10^{-10}$ yr^{-1}; $\lambda_2 = 9.8485 \times 10^{-10}$ yr^{-1})

$t\,(10^6\,\mathrm{yr})$	$^{206}Pb^*/^{238}U\,(y)$	$^{207}Pb^*/^{235}U\,(x)$
0	0	0
500	0.080649	0.636279
1 000	0.167803	1.677410
1 500	0.261987	3.380991
2 000	0.363766	6.168525
2 500	0.473753	10.729710
3 000	0.592611	18.193083
3 500	0.721055	30.405247
4 000	0.859857	50.387759

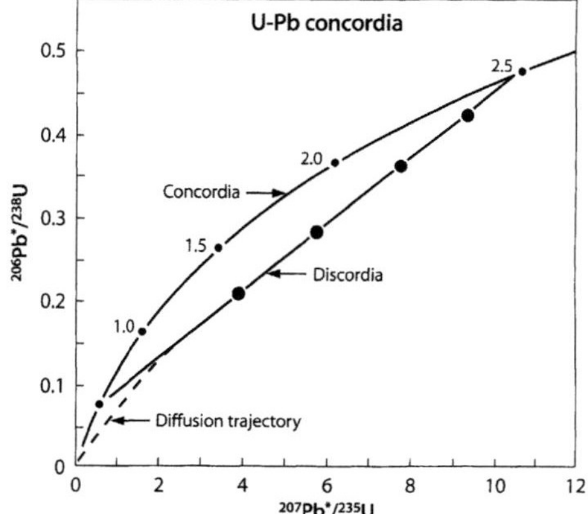

Fig. 1.8. U-Pb concordia diagram based on data in Table 1.8. The discordia line is defined by U-bearing mineral fractions that yield discordant U-Pb dates because of varying losses of radiogenic Pb. The original crystallization age of the minerals is calculated from the coordinates of the upper point of intersection. The lower point of intersection indicates the time elapsed since Pb-loss (or U-gain) only in cases where the loss (or gain) was episodic rather than continuous. The *dashed line* represents the diffusion trajectory based on the work of Tilton (1960)

into both equations. The results in Table 1.8 generate the concordia curve which represents U-Pb systems that yield concordant U-Pb dates (Fig. 1.8).

All U-Pb systems that have remained closed to U and Pb since the time of their formation have $^{206}Pb^*/^{238}U$ and $^{207}Pb^*/^{235}U$ ratios that define a point on concordia and therefore yield concordant dates. When a U-bearing mineral loses radiogenic Pb during an episode of thermal metamorphism, the point representing it leaves concordia and moves along a straight line towards the origin. The resulting chord in Fig. 1.8 is called *discordia* because the U-Pb systems that define it yield discordant dates.

Data points representing mineral fractions that yield discordant U-Pb dates are used to construct the discordia chord, which is then extrapolated until it intersects concordia in two points. The time elapsed since crystallization of the mineral fractions is calculated from the coordinates of the *upper* point of intersection on concordia. If the U-bearing minerals of a volume of rocks crystallized at the same time as the other minerals, then the U-Pb date derived from the upper point of intersection is the age of the rocks.

In many cases, the U-Pb concordia age of refractory U-bearing minerals (e.g. zircon) *exceeds* the date derived by the whole-rock Rb-Sr isochron method not only because zircon crystals res st alteration, but also because variable losses of Pb (as well as gain of U) are restored on the concordia diagram (e.g. Beakhouse et al. 1988). For this reason, U-Pb concordia dates of zircon and other U-bearing minerals have contributed greatly to reconstructions of the geological history of the Earth during the Archean and Proterozoic Eons.

However, in some cases U-Pb concordia dates of zircon and other refractory U-bearing minerals are older than the age of rocks in which they occur. This situation arises when zircons in igneous or metamorphic rocks contain cores representing detrital grains that were present in the sedimentary protoliths. For this reason, zircon crystals are examined microscopically to determine whether cores are present before they are selected for analysis. When cores are identified, the grains are crushed in order to separate the overgrowths from the cores and both are analyzed separately for U-Pb dating on concordia diagrams.

The interpretation of the *lower* intercept of the discordia chord with concordia is ambiguous. If Pb loss (or U gain) was episodic, the lower intercept yields the time elapsed since the alteration of the U-Pb systematics. However, when U-bearing minerals exist at elevated temperatures for long periods of geologic time, they lose radiogenic Pb by diffusion. Tilton (1960) demonstrated that the diffusion trajectory on the concordia diagram is a straight line for most of its length, but turns toward the origin as the Pb loss approaches 80%. Therefore, a linear extrapolation of the diffusion trajectory may yield a fictitious date derived from the lower intercept on concordia illustrated in Fig. 1.8.

The U-Pb dating method was originally based on analyses of several hundred milligrams of zircon grains or other U-bearing minerals that were separated from large blocks of rock in order to constrain the discordia chord. Krogh (1973) made a significant improvement in the existing technique by digesting single zircon crystals in small Teflon-lined bombs. This procedure reduced the amount of contamination of the samples by decreasing the amount of reagents needed. Subsequently, Krogh (1982a,b) made additional improvements by sorting zircon grains magnetically and by removing their altered surface layers by abrading them with pyrite particles in a stream of air. These treatments significantly improved the concordance of the U-Pb dates and reduced the errors arising from long extrapolations of the discordia to the points of intersection with concordia. As a result, T. E. Krogh and his colleagues of the Jack Satterly Geochronology Laboratory at the Royal Ontario Museum in Toronto have been able to date single zircon grains or fragments of grains weighing only a few micrograms with a precision of better than ±0.1% (e.g. Krogh 1993). As a result of the refinements of the analytical techniques by Krogh and others, U-Pb concordia dates have provided definitive ages of rock-forming events during the Archean and Proterozoic Eons.

The U-Pb method was also used by Krogh et al. (1993a,b) to date shocked zircon grains in a sample of impact breccia of the Chicxulub impact crater on the Yucatan Peninsula of Mexico and from the K-T boundary layers on Haiti and in Colorado, USA. The results indicate that 18 of 36 zircon grains from these three sites define a discordia chord with an upper intercept age of 544.5 ±4.7 Ma. This date is the age of granitic basement rocks that underlie the sedimentary rocks of the Yucatan Peninsula. The zircons in the K-T boundary layer in the Raton Basin of Colorado also yielded a Pb-loss age (lower concordia intercept) cf 65.5 ±3 Ma (Krogh et al. 1993b). These results strongly support the conclusion that ejecta of the Chicxulub Crater having a primary age of 545 Ma exist in the Raton Basin of Colorado about 2 000 km distant and that these zircons lost radiogenic Pb at the time of the formation of the impact crater dated independently by Swisher et al. (1992) and Sharpton et al. (1992) by means of the ^{40}Ar/^{39}Ar method. The studies of the impact at Chicxulub and of its aftermath have revolutionized geology by demonstrating that the Earth has been profoundly affected by extraterrestrial processes (Powell 1998).

1.8.2 The Pb-Pb Method

Most igneous and metamorphic rocks contain Pb that is incorporated at the time of their formation. These rocks also contain U which decays to form radiogenic ^{206}Pb and ^{207}Pb. As a result, the ^{206}Pb/^{204}Pb and ^{207}Pb/^{204}Pb ratios of such rocks increase with time at rates that depend on their U/Pb ratios. The isotope ratios of Pb of a suite of cogenetic rock samples define a straight line in coordinates of ^{207}Pb/^{204}Pb and ^{206}Pb/^{204}Pb provided that all samples:

1. Formed at the same time;
2. Initially contained Pb having the same isotope composition; and
3. Remained closed to U and Pb throughout their history.

Table 1.9. Numerical values of the slopes of Pb-Pb isochrons for specified values of the age ($\lambda_1 = 1.55125 \times 10^{-10}$ yr^{-1}; $\lambda_2 = 9.8485 \times 10^{-10}$ yr^{-1}

t (10^6 yr)	Slope (Pb-Pb isochron)	t (10^6 yr)	Slope (Pb-Pb isochron)
0	0	2 500	0.16426
500	0.057219	3 000	0.222656
1 000	0.072500	3 500	0.305829
1 500	0.093597	4 000	0.421737
2 000	0.122986	4 500	0.596707

The equation for the resulting Pb-Pb isochron is derivable from Eqs. 1.32 and 1.33:

$$\frac{^{207}Pb/^{204}Pb - (^{207}Pb/^{204}Pb)_0}{^{206}Pb/^{204}Pb - (^{206}Pb/^{204}Pb)_0} = \left(\frac{^{235}U}{^{238}U}\right)\frac{(e^{\lambda_2 t} - 1)}{(e^{\lambda_1 t} - 1)} \quad (1.36)$$

where the present $^{235}U/^{238}U$ ratio is a constant equal to 1/137.88 and the slope (m) is:

$$m = \frac{1}{137.88}\frac{(e^{\lambda_2 t} - 1)}{(e^{\lambda_1 t} - 1)} \quad (1.37)$$

Therefore, the age of the rocks that define a Pb-Pb isochron can be determined from its slope by solving Eq. 1.37 for t. This method of dating is insensitive to losses of U and Pb in the recent past and only requires measurements of the isotope ratios of Pb. In addition, the method does not require nor does it provide information about the initial isotope ratios of the Pb. Equation 1.37 is most easily solved for t by interpolating in Table 1.9 containing numerical values of the slope of Pb-Pb isochrons for known values of t. Table 1.9 can also be used to determine dates from the radiogenic $^{207}Pb/^{206}Pb$ ratios of U-bearing minerals discussed in Sect. 1.8.1.

The Pb-Pb method was used by Patterson (1956) to determine an age of 4.55 ±0.05 × 10^9 years for the Earth based on isotope ratios of Pb in a small suite of meteorites. The Pb-Pb method is still widely used to date igneous and metamorphic rocks of Precambrian age. For example, Moorbath et al. (1973) obtained a well defined Pb-Pb isochron for banded ironstones at Isua in southern West Greenland corresponding to an age of 3 760 ±70 Ma. This value is in good agreement with a whole-rock Rb-Sr isochron date of 3 611 ±140 Ma ($\lambda = 1.42 \times 10^{-11}$ yr^{-1}) derived from granitic gneisses at Isua by Moorbath et al. (1972). The rocks of this part of Greenland are among the oldest on the Earth.

1.9 Oxygen Isotope Composition

Oxygen is the most abundant chemical element in the crust of the Earth because it occurs in all silicate, oxide, carbonate, phosphate, and sulfate minerals. The isotope composition of oxygen in these minerals varies widely depending on many factors including the temperature of crystallization and the isotope composition of oxygen in the reservoir with which the oxygen was in isotopic equilibrium. In general, silicate minerals in rocks of the crust of the Earth are enriched in ^{18}O whereas meteoric water and seawater on the surface of the Earth are depleted in this isotope. Consequently, the isotope composition of oxygen in igneous and metamorphic rocks may be altered by isotope-exchange reactions with water at elevated temperature. For these reasons, the isotope composition of oxygen is useful in the study of the origin of igneous and metamorphic rocks (Valley et al. 1986; Hoefs 1996).

Oxygen ($Z = 8$) has three stable isotopes whose abundances are: $^{18}_8O = 0.200\%$; $^{17}_8O = 0.038\%$; $^{16}_8O = 99.762\%$. The isotope composition of O in nature varies as a result of isotope fractionation during isotope exchange reactions, phase changes, and as a result of kinetic effects. The isotope composition of O is expressed in terms of the $\delta^{18}O$ parameter defined as:

$$\delta^{18}O = \left[\frac{R_{spl} - R_{std}}{R_{std}}\right] \times 10^3\text{‰} \quad (1.38)$$

where $R = {}^{18}O/^{16}O$ (atomic), and the subscripts "spl" = sample and "std" = standard. The standard used for this purpose is "Standard Mean Ocean Water" (SMOW) prepared and maintained by the International Atomic Energy Agency in Vienna, Austria. Therefore, the $\delta^{18}O$ parameter is the permil difference between the $^{18}O/^{16}O$ ratio of a sample and SMOW. Rocks composed of silicate, carbonate, or oxide minerals have a range of positive $\delta^{18}O$ indicating that they are enriched in ^{18}O relative to SMOW. The $\delta^{18}O$ values of meteoric water are negative because water on the surface of the Earth is depleted in ^{18}O during evaporation of water vapor from the surface of the oceans and during condensation events as air masses move from the equatorial region in a northeasterly direction in the northern hemisphere and in a southwesterly direction in the southern hemisphere.

Taylor (1980) observed that the $^{87}Sr/^{86}Sr$ ratios of igneous rocks in some cases correlate positively with the isotope composition of O expressed as the $\delta^{18}O$ parameter. Such correlations may arise by several different processes during the formation of magmas and their subsequent evolution:

1. Mixing of magmas having different isotope compositions of Sr and O;
2. Assimilation of crustal rocks by mantle-derived magmas;
3. Partial melting of source rocks that are physical mixtures of crustal and mantle components;
4. Selective alteration of the isotope composition of Sr (and of other elements) by interactions between magma and the country rock without detectable assimilation;
5. Any combination of these processes.

The chemical compositions of igneous rocks that form by these processes are generally *not* explainable as simple mixtures of two components because of the removal of cumulate minerals that form during fractional crystallization of hybrid magmas. The isotope compositions of Sr, Nd, Pb, etc. of hybrid magmas are not appreciably changed by crystallization of minerals because the differences in their isotopic masses are small, because isotope fractionation effects decrease with increasing temperatures, and because all measured isotope ratios are routinely corrected for isotope fractionation regardless of its origin (Sect. 1.2.3). Even the isotope composition of O of hybrid magmas is primarily determined by the differences in the $\delta^{18}O$ values of the components and by the proportions of mixing rather than by subsequent fractional crystallization (Taylor 1980). The isotope composition of O in the highly differentiated gabbro intrusion at Skaergaard in East Greenland (Sect. 5.4.6) was investigated and interpreted by Taylor and Epstein (1963), Taylor and Forester (1979), and Norton and Taylor (1979).

1.10 Fractional Crystallization of Magma

When minerals crystallize from a cooling magma, the concentrations of Rb and Sr of the remaining residual liquid change depending on whether the minerals accept or exclude these elements. The transfer of Rb and Sr from a silicate melt into the minerals that are crystallizing from it can be described by means of distribution coefficients combined with the Rayleigh equation.

1.10.1 Distribution Coefficients

Trace elements may enter a mineral crystallizing from a silicate melt by substituting for the ion of a major element. The extent of substitution depends on many factors pertaining to the ions, the environmental conditions, and the properties of the site in the crystal lattice. In general, these factors combine to give rise to three possible scenarios:

1 The ions of a trace element may be *captured* by growing crystals in preference to the ions of the major element;
2 Crystals may *camouflage* a trace element by allowing it to enter without discrimination;
3 A trace element may be excluded and is *admitted* only to a limited extent.

Goldschmidt (1937) originally attempted to predict the replacement of ions in crystals solely on the basis of their radii and charges. However, the resulting "rules" were only qualitative at best and were violated in many

cases (Shaw 1953; Ringwood 1955; Burns and Fyfe 1967; Philpotts 1978). Consequently, the partitioning of trace elements between crystals and silicate liquids is expressed more effectively by means of the distribution coefficient (D) defined by the relation:

$$D_l^s = \frac{C^s}{C^l} \qquad (1.39)$$

where s and l stand for *solid* and *liquid*, respectively, C is the concentration of any trace element and the replacement reaction is assumed to be at equilibrium. Distribution coefficients are determined either by analyzing phenocrysts and matrix in porphyritic volcanic rocks or by analyzing crystals grown from melts in the laboratory.

Numerical values of distribution coefficients of Rb^+ and Sr^{2+} and many other trace elements in the common rock-forming minerals have been compiled by Irving (1978), Allègre and Hart (1978), Cox et al. (1979), Henderson (1982), Wilson (1989), Foley and Van der Laan (1994), and other recently published textbooks in igneous petrology.

Trace elements having $D > 1.0$ are said to be *compatible* whereas those having $D < 1.0$ are *incompatible*. Trace elements which are compatible in the crystals forming from a magma are enriched in the solids and depleted in the residual liquid, whereas incompatible trace elements tend to be concentrated into the remaining liquid.

Rubidium is an *incompatible* trace element in all major rock-forming minerals. Consequently, it is progressively concentrated into the residual liquid during fractional crystallization of magma. Late-forming crystals of certain minerals (e.g. mica and feldspar) therefore have higher Rb concentrations than early-formed crystals of the same minerals.

Strontium is *compatible* in K-feldspar, plagioclase, apatite, and sphene but *incompatible* in micas, most amphiboles, pyroxenes, and olivine. Therefore, the Sr concentration of residual magmas *increases* when olivine or pyroxene are the only crystallizing minerals. When plagioclase begins to crystallize, it takes up Sr from the magma and causes its concentration in the residual liquid to *decrease*.

The measured values of the distribution coefficients of Rb and Sr in some minerals listed in Table 1.10 vary by a factor of ten or more because their magnitudes depend not only on the properties of the ions and on the character of the lattice sites in the crystals, but also on the chemical composition of the melt, as well as on the temperature and pressure. The chemical composition of the melt becomes the dominant factor in silica-rich magmas because of the increasing polymerization of Si-Al-O units. Consequently, such liquids reject trace elements, causing their crystal/liquid distribution coefficients to in-

Table 1.10. Average distribution coefficients for Rb and Sr in minerals crystallizing from magmas of different composition

Mineral	D(Rb)	Range		D(Sr)	Range		Reference[a]
A. Basalt and andesite							
Olivine	0.0079	0.000179	– 0.011	0.0171	0.000191	– 0.02	1, 2, 3, 28
Orthopyroxene	0.033	0.022	– 0.061	0.012	0.0026	– 0.02	1, 2, 3, 5
Diopside	0.015			0.11	0.078	– 0.12	1, 7, 8
Clinopyroxene	0.029	0.0010	– 0.1	0.13	0.054	– 0.608	2, 3, 5, 9, 10, 11, 12, 28
Augite	0.031			0.12			1
Amphiboles	0.225	0.027	– 0.41	0.605	0.19	– 1.02	2, 5
Hornblende	0.31	0.29	– 0.33	0.47	0.46	– 0.48	1, 11
Plagioclase	0.099	0.0262	– 0.50	1.89	1.18	– 3.06	1, 2, 10, 11, 13, 14, 28
Mica	3.2			0.21			5
Phlogopite	3.08			0.08			1, 2
Garnet	0.042			0.012			1
Apatite	–			5.05	3.7	– 6.4	5
Whitlockite	–			1.00 (1)			15
Perovskite	–			0.734			27
B. Dacite and rhyolite							
Orthopyroxene	0.09	0.0005	– 0.29	0.030	0.009	– 0.05	2
Hypersthene	0.0027			0.085			1
Clinopyroxene	0.061	0.032	– 0.09	0.516			1, 2
Hornblende	0.014			0.022			1
Plagioclase	0.070	0.024	– 0.46	9.1	1.5	– 40	1, 2, 16, 17, 18, 19
Alkali feldspar	0.58	0.11	– 1.04	5.45	2.7	– 26	1, 2, 17, 20
Sanidine	0.70	0.28	– 2.4	4.7	2	– 7.3	16, 19, 21, 22
Quartz	0.014	0.012	– 0.016	–			16
Garnet	0.0087	0.0085	– 0.009	0.017	0.015	– 0.02	1, 2
Biotite	3.21	2.24	– 5.3	0.21	0.08	– 0.53	1, 2, 16, 17, 21
Muscovite	1.54			0.104			17
C. Alkalic rocks							
Olivine	0.045	0.02	– 0.08	0.012	0.003	– 0.02	4, 6, 23
Orthopyroxene	0.022			0.017			6
Clinopyroxene	0.049	0.004	– 0.04	0.14	0.08	– 0.16	4, 6, 23, 24
Magnetite	0.35	0.14	– 0.47	0.69	0.68	– 0.70	6, 23, 24
Amphibole	0.11	0.09	– 0.14	–			6
Hornblende	1.9			0.3			23
Kaersutite	0.2			0.61			4
Plagioclase	0.16	0.03	– 0.39	3.1	1	– 5	6, 24, 25
Alkali feldspar	0.3			10			23
Sanidine	0.21	0.16	– 0.25	1.5	0.9	– 2.8	6, 21, 25
Anorthoclase	0.15	0.11	– 0.18	4.9	2.82	– 30	24, 25, 26
Melilite (Ak 12 to 90)	–			0.84	0.62	– 1.12	8, 27
Biotite	2.6	1.9	– 3.2	0.46	0.21	– 0.7	4, 23
Apatite	–			2.4	1	– 5	4, 24
D. Undifferentiated							
Plagioclase	0.069	0.0294	– 0.188	1.78	1.27	– 2.84	29
K-feldspar	0.659			3.87			29
Clinopyroxene	0.0497	0.0129	– 0.284	0.166	0.00187	– 0.516	29
Orthopyroxene	0.0217	0.0148	– 0.0287	0.0172	0.0104	– 0.0241	29
Micas	2.418	0.936	– 3.26	0.291	0.0812	– 0.672	29
Hornblende	0.294	0.0448	– 0.427	0.459	0.188	– 0.641	29
Garnet	0.00851			0.0154			29
Olivine	0.00984	0.00839	– 0.0113	0.0139	0.00937	– 0.0185	29

[a] References: 1. Arth and Hanson (1975), 2. Henderson (1982), 3. Shimizu et al. (1982), 4. Irving and Price (1981), 5. Irving and Frey (1984), 6. LeMarchand et al. (1987), 7. Grutzeck et al. (1974), 8. Kuehner et al. (1989), 9. Shimizu (1974), 10. Sun et al. (1974), 11. Dostal et al. (1983a), 12. Ray et al. (1983), 13. Warren (1983), 14. Drake and Weill (1975), 15. Irving (1978), 16. Nash and Crecraft (1985), 17. Mittlefehldt and Miller (1983), 18. Dupuy (1972), 19. Stix and Gorton (1990), 20. Long (1978), 21. Mahood and Hildreth (1983), 22. Leeman and Phelps (1981), 23. Villemant et al. (1981), 24. McDonough and Nelson (1984), 25. Berlin and Henderson (1969), 26. Mahood and Stimac (1990), 27. Nagasawa et al. (1980), 28. Hart and Brooks (1974), 29. Philpotts and Schnetzler (1970).

crease and to vary significantly in response to slight changes in the composition of the liquid even though pressure and temperature remain constant (Mahood and Hildreth 1983). For this reason, the data in Table 1.10 are grouped by composition of the rocks into: (*A*) basalt and andesite; (*B*) dacite and rhyolite; (*C*) alkalic rocks.

Other potential sources of error in the measurement of distribution coefficients based on analyses of phenocrysts and matrix in volcanic rocks include: (*1*) disequilibrium between the crystals and the liquid (Dasch 1969; Berlin and Henderson 1969; Mazzone and Grant 1988; Geist et al. 1988a); (*2*) kinetic disequilibrium caused by local depletion of the melt in trace elements in the vicinity of rapidly growing crystals (Albarède and Bottinga 1972; Lindstrom 1983; Lesher 1986); (*3*) zoning of the crystals; (*4*) presence of inclusions; (*5*) incomplete separation of crystals from the matrix prior to wet chemical analysis; and (*6*) low precision of electron and ion microprobes for trace elements (Mahood and Stimac 1990). In addition, the distribution coefficients of some minerals (e.g. alkali feldspar, clinopyroxene, melilite, and others) appear to vary systematically with the composition of the crystals (Sun et al. 1974). However, the compositional variation may be masked by temperature effects. For example, the temperature dependence of $D(Sr)$ in plagioclase may be an artifact caused by the temperature dependence of the chemical compositions of this mineral (Blundy and Wood 1991).

Although Sr^{2+} is able to replace Ca^{2+} in plagioclase, the Sr concentration of plagioclase does *not* increase

with the Ca content as expected (Heier 1962). The data in Fig. 1.9 indicate that the Sr concentration of plagioclase of *intermediate* compositions reaches a maximum and that its concentration in Ca-rich plagioclases actually *decreases* with increasing anorthite content (Ewart and Taylor 1969). This apparent anomaly was explained by Blundy and Wood (1991) by demonstrating that albite is more *elastic* than anorthite (based on comparisons of the bulk modulus, sheer modulus, and Young's modulus) and that Sr^{2+} and Ba^{2+}, both of which are larger than Ca^{2+} and Na^+, are more easily accommodated in Na-plagioclase than in Ca-plagioclase because of the greater elasticity of the albite lattice.

1.10.2 Rubidium and Strontium Concentrations of Plutonic Igneous Rocks

The Rb and Sr concentrations of the rock-forming minerals (Table 1.2) and the values of their distribution coefficients (Table 1.10) help to explain the concentrations of Rb and Sr in plutonic igneous rocks ranging in composition from ultramafic to granitic (Table 1.11). Ultramafic rocks (except anorthosites) have low concentrations of Rb and Sr compared to mafic (gabbro, diorite) and felsic rocks (monzonite, granite). Figure 1.10a indicates that the average Rb concentrations of plutonic igneous rocks increase with increasing silica concentration (degree of differentiation), whereas the Sr concentrations reach a maximum in diorites and granodiorites (SiO_2 between about 52% and 67%) and then decline with increasing silica concentration. Consequently, the Rb/Sr ratios of the mafic rocks in Fig. 1.10b rise only slightly with increasing degree of differentiation because initially the concentrations of Rb and Sr *both* rise. However, when the Sr concentrations begin to decline after the main stage of plagioclase crystallization, the Rb/Sr ratios increase rapidly with increasing degree of differentiation beyond silica concentrations of about 65 to 70%.

The concentrations of Rb and Sr in each of the rock types listed in Table 1.11 in many cases vary by factors of ten or more. Consequently, the *average* concentrations are subject to sampling bias and are not necessarily representative of all occurrences of a particular rock type. The wide range of Rb and Sr concentrations results from the formation of magmas by varying degrees of partial melting of different kinds of source rocks in the lower crust or upper mantle. Once formed, the magmas may evolve by fractional crystallization under a variety of conditions and by assimilation of different kinds of crustal rocks. As a result, igneous rocks may acquire regional or local characteristics in their trace-element concentrations.

The Rb and Sr concentrations of the Kiglapait layered intrusion in Labrador (Morse 1969) have been plot-

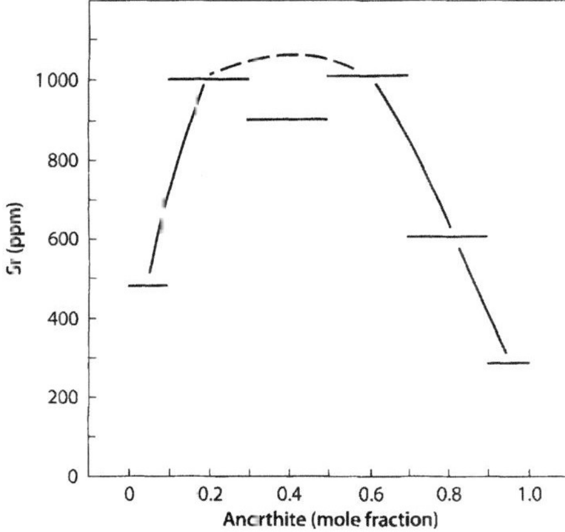

Fig. 1.9. Variation of the average Sr concentrations of plagioclase based on data compiled from the literature. The Sr concentration of plagioclase increases with increasing anorthite content up to a mole fraction of about 0.2 and has a maximum at mole fractions of about 0.2 and 0.6. At anorthite mole fractions greater than about 0.6 the Sr content of plagioclase actually decreases with increasing Ca concentrations (*Source:* Data from Wedepohl 1978)

Table 1.11.
Average concentrations of Rb and Sr in plutonic igneous rocks. The number of samples included in the averages is given in parentheses

Rock type	Rb (ppm)	Sr (ppm)	Rb/Sr	SiO$_2$ (%)	Ref.[a]
A. Ultramafic rocks					
Dunite	0.39 (9) 0.072–2.42	4.6 (19) 0.12–14.7	0.0085	40.2	1
Pyroxenite	1.65 (3) 0.38–3.47	64 (16) 0.23–199	0.016	50.5	1
Anorthosite	4.5 (18) 1.6–14.7	667 (78) 156–1441	0.0067	54.5	1, 7, 8
Peridotite	1.27 (11) 0.093–4.48	19 (22) 0.4–50.4	0.068	43.5	1
B. Mafic to felsic rocks					
Gabbro	32 (331)	293 (101) 41–860	0.11	48.4	1
Diorite	88 (21)	472 (79) 173–870	0.19	51.9	1
Granodiorite	122 (9)	457 (149) 40–1100	0.27	66.9	1
Monzonite and quartz monzonite	136	167 (271) 29–876	0.81	69.2	1
Granite	230 (504)	147 (512) 2.16–917	1.56	72.1	1
I type granite	132	253	0.52		9
S type granite	180	139	1.3		9
C. Alkalic rocks					
Alkalic ultramafics (Kola Penin., Russia)	80	1 300	0.062	–	2
Alkali gabbro	–	1 367 (49) 446–2195	–	46.5	1
Nepheline syenite	364 85–950	1098 47–3500	0.33	55.4	1, 2
Syenite	136 (14)	553 (84) 5.2–2924	0.25	59.4	1
Lamprophyre	115	1 010	0.11	46.3	5
Kimberlite	68 (39) 63–162	879 (34) 48–1883	0.077	35.0	1, 3, 6
Lamproite	272 (22) 50–614	1 633 (22) 549–3150	0.18	53.3	5, 6
Carbonatite	–	2 350 (28) 300–3910	–	–	1
Calciocarbonatite	14 (6) 4–35	7 270 (66) 0–27 820	0.0019	2.72	4
Magnesiocarbonatite	31 (4) 2–80	5 830 (29) 507–12 680	0.0053	3.63	4

[a] References: 1. Wedepohl (1978), 2. Gerasimovsky (1974), 3. Heinrichs et al. (1980), 4. Woolley and Kempe (1989), 5. Bergman (1987), 6. Fraser et al. (1986), 7. Geist et al. (1990), 8. Kolker et al. (1990), 9. Taylor and McLennan (1985).

ted in Fig. 1.11 vs. log PCS (percent solidified by volume) because these rocks provide an excellent example of the effects of fractional crystallization on the concentrations of these elements (Morse 1981, 1982). The Kiglapait Intrusion (Sect. 7.1.4) ranges in composition from troctolite (olivine plus Ca-plagioclase) to ferrosyenite and formed by fractional crystallization during slow cooling of a low-K olivine tholeiite magma without significant interaction with the country rock. Figure 1.11 indicates that the Rb concentrations of these rocks remained virtually constant until more than 90% of the magma had crystallized and then rose steeply, especially in the

Fig. 1.10.
a Average Rb and Sr concentrations of mafic to felsic plutonic igneous rocks with increasing degree of magmatic differentiation represented by increasing SiO_2 concentrations; b The Rb/Sr ratio initially increased slowly during magmatic differentiation until Sr has been removed from the magma by crystallization of feldspars (*Source:* data from Table 1.10)

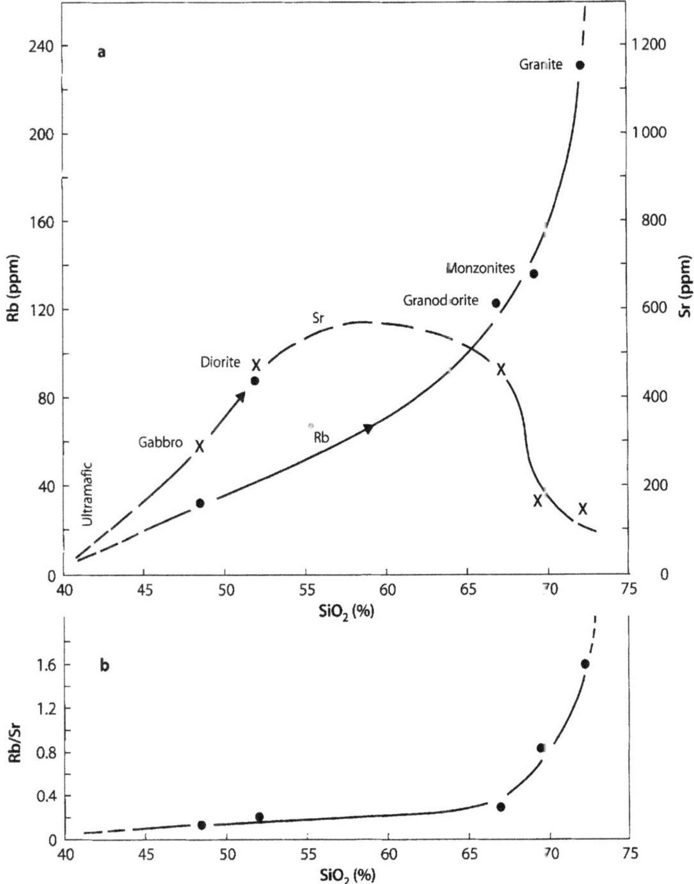

rocks which formed from the last 1% of the magma. The Sr concentrations of the rocks increased initially from about 310 to 430 ppm until about 35% of the magma had crystallized. Subsequently, the Sr concentrations of the rocks declined irregularly to 164 ppm until only 5% of the liquid remained. The rocks which formed from this liquid became progressively enriched in Sr, reaching a maximum value of 463 ppm, when only 0.5% of the original magma remained. The Sr concentrations of the very last differentiates decreased to about 170 ppm as the remaining liquid crystallized.

The distribution of Rb in the rocks of the Kiglapait Intrusion illustrates the progressive enrichment of an *incompatible* lithophile trace element into the residual liquid in the course of progressive fractional crystallization of magma in a closed magma chamber. This process can be modeled by the Rayleigh equation for $D(\text{Rb}) < 1.0$ (Gast 1968; Greenland 1970; Barca et al. 1988).

The behavior of Sr during fractional crystallization of magma is much more complicated than that of Rb because of the opposing effects of the crystallization of olivine or pyroxenes and plagioclase or apatite. The crystallization of olivine and pyroxenes forms rocks having low Sr concentrations but it enriches the residual magma in Sr, whereas the crystallization of plagioclase and apatite enriches the rocks in Sr but has the opposite effect on the remaining liquid. Consequently, the Sr concentration of differentiated igneous rocks is a sensitive indicator of sequential crystallization of different minerals from a cooling magma.

The alkalic rocks in Table 1.11 are strongly enriched in Sr compared to the mafic to felsic suites and also have somewhat higher Rb concentrations (Chap. 6). They characteristically contain more K_2O than Na_2O and occur in continental settings, some oceanic islands, and, more rarely, in subduction zones. According to Wilson

Fig. 1.11.
Concentrations of Rb and Sr in the rocks of the differentiated Kiglapait Intrusion in Labrador plotted vs. the logarithm of the percent solidified by volume (PCS). The rocks range in composition from troctolite (olivine + plagioclase) to ferrosyenite. The Rb concentrations remain nearly constant, but rise in the rocks formed from the final 10% of the residual magma. The concentrations of Sr in the rocks vary irregularly depending on their mineral composition. Crystallization of olivine and clinopyroxene produces rocks having low Sr concentrations but enriches the residual liquid in Sr, whereas crystallization of plagioclase and apatite have the opposite effect (*Source:* Morse 1981, 1982a)

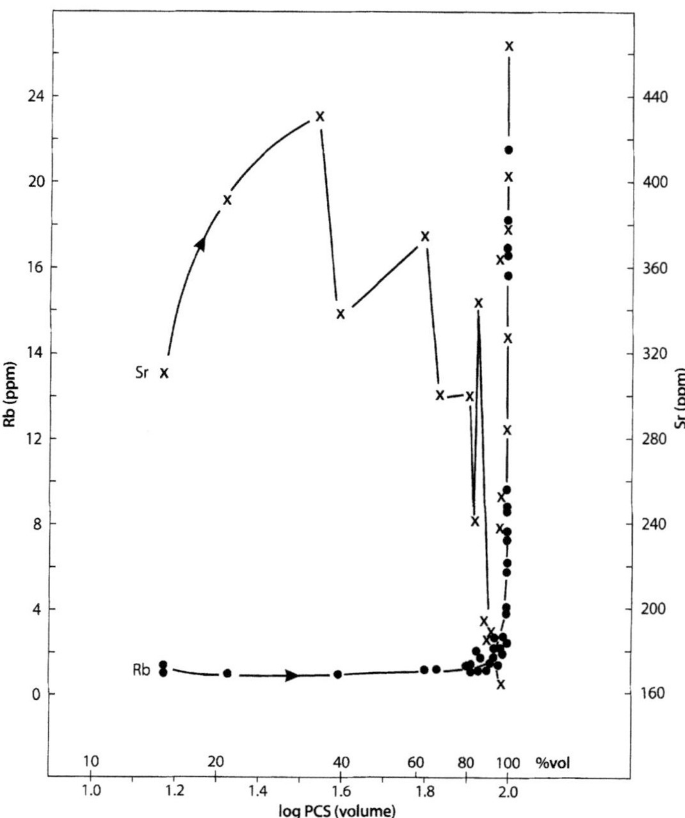

(1989, p 377), the enrichment of alkalic magmas in K and in incompatible trace elements indicates that they are the products of small degrees of partial melting of anomalous source rocks in the mantle. Although the alkalic rocks constitute only a small fraction of the volume of igneous rocks on the continents, their unusual mineral composition has attracted the attention of petrologists who have proposed a complex classification containing many exotic names (Sørensen 1974; Streckeisen 1976; Fitton and Upton 1987; Foley et al. 1987; Rock 1991).

Carbonatites are usually grouped with the alkalic rocks because they are commonly associated with them in the field (Heinrich 1966). However, most carbonatites are alkali-poor and are enriched instead in either Ca, Mg, or Fe (Woolley and Kempe 1989). Alkali carbonatite lavas occur on Oldoinyo Lengai, a volcano in Tanzania that erupted as recently as 1993; (Bell 1989; Dawson et al. 1994).

The carbonatites are strongly enriched in Sr (Van Gross 1975) with average concentrations of 7 270 and 5 830 ppm in calciocarbonatites and magnesiocarbonatites, respectively (Table 1.11). Their Rb concentrations are low, consistent with their depletion in alkali metals.

1.10.3 Trace-Element Modeling

Magmas may crystallize in such a way that the crystals remain in equilibrium with the residual liquid (equilibrium crystallization). Alternatively, the crystals may be separated from the liquid as soon as they form (fractional crystallization). During *equilibrium crystallization* a material balance is maintained such that the concentration of an element in a unit weight of the original magma C_0^l is related to its concentrations in the residual liquid (C^l) and in the crystals (C^s) by the mass balance equation:

$$C_0^l = C^l F + C^s (1 - F) \tag{1.40}$$

where F is the weight fraction of liquid remaining. According to Eq. 1.39, $C^s = DC^l$ and therefore:

$$C_0^l = C^l F + D C^l (1 - F)$$

$$\frac{C^l}{C_0^l} = [F + D(1 - F)]^{-1} \tag{1.41}$$

If the element of interest resides in the crystals of *several* minerals, its distribution between the solid phases and the remaining liquid is described by the *bulk distribution coefficient*:

$$\overline{D} = \sum_i w_i D_i \tag{1.42}$$

where w_i is the weight fraction of each of the i minerals and D_i is the distribution coefficient of the element in each of those minerals. Therefore, in the case of an igneous rock composed of crystals of several minerals:

$$\frac{C^l}{C_0^l} = [F + \overline{D}(1 - F)]^{-1} \tag{1.43}$$

Alternatively, C^l in Eq. 1.40 can be replaced by:

$$C^l = C^s / D$$

$$C_0^l = \frac{C^s F}{D} + C^s (1 - F)$$

$$C_0^l = \frac{C^s F + C^s D(1 - F)}{D}$$

$$\frac{C^s}{C_0^l} = \frac{D}{F + D(1 - F)} = \frac{D}{F(1 - D) + D} \tag{1.44}$$

If the element of interest is present in the crystals of several minerals, then D must be replaced by the bulk distribution coefficient $\overline{D}$ as defined by Eq. 1.42 (Wood and Fraser 1976; Henderson 1982).

In case the crystals are completely removed from the liquid as soon as they form (fractional crystallization), the concentration of a chemical element in the residual liquid changes in accordance with the Rayleigh equation:

$$\frac{C^l}{C_0^l} = F^{(D-1)} \tag{1.45}$$

where D must be replaced by the bulk distribution coefficient when the element of interest resides in the crystals of more than one mineral. Since $C^l = C^s / D$, the Rayleigh equation can be restated in terms of the concentration of the trace element in the solid phases:

$$\frac{C^s}{C_0^l} = DF^{(D-1)} \tag{1.46}$$

Both crystallization models assume that the distribution coefficients remain constant throughout the process. This assumption is invalid because distribution coefficients vary in response to changes in the chemical compositions of both the liquid and the solid phases and are also sensitive to temperature and pressure. In addition, bulk distribution coefficients depend on the changing proportions among the crystallizing phases. Nevertheless, the Rayleigh model has been widely used to explain the chemical evolution of cooling magmas and of the rocks that are produced (Gast 1968; Shaw 1970; Arth 1976; Allègre et al. 1977; Drake and Holloway 1978; Minster and Allègre 1978; Hanson 1978).

An example of the effects of fractional crystallization of magma is presented in Fig. 1.12 based on Eq. 1.46. The crystallizing phases are plagioclase and clinopyroxene and the bulk distribution coefficient for Rb is assumed to remain constant at $\overline{D}(Rb) = 0.062$. Consequently, the concentration of Rb in the solids increases as documented in natural systems by Figs. 1.10 and 1.11.

The bulk distribution coefficient of Sr is assumed to vary from 0.73 ($F = 1.0$ to 0.6) to 1.5 ($F = 0.6$ to 0.3), and finally to 2.9 ($F = 0.3$ to 0) because of changes in the values of $D(Sr)$ of plagioclase and clinopyroxene in accordance with Blundy and Wood (1991) and Sun et al. (1974). Consequently, the Sr concentration of the solids increases slowly at first ($F = 1.0$ to 0.6) and then rises rapidly at $F = 0.6$ as $\overline{D}(Sr)$ changes from 0.73 to 1.5. Continued crystallization depletes the liquid in Sr and causes the Sr content of the crystals to decline. At $F = 0.3$, $\overline{D}(Sr)$ rises to 2.9 and the Sr concentrations of both the residual liquid and the late-stage crystals approach zero as the final liquid crystallizes.

Although Fig. 1.12 is greatly oversimplified, it does reproduce the typical features of the distributions of Rb and Sr in differentiated igneous rocks. The model does not include the crystallization of other minerals (olivine, apatite, amphiboles and micas); it assumes that crystals are completely removed from the liquid and that the composition of the magma is not changed by assimilation of crustal rocks or by mixing with other magmas.

1.11 Assimilation and Fractional Crystallization

The chemical differentiation of magmas may take place during their formation, during subsequent migration toward the surface, and in magma chambers in the continental crust where they may form a variety of igneous rocks by fractional crystallization. The isotope compositions of Sr and of some other elements (e.g. Nd, Pb, Os, O, and S) of igneous rocks can be used to determine whether magmas assimilated minerals of the continental crust as they differentiated. The energy required to assimilate minerals of crustal rocks is provided partly by the heat transported by the magmas from source regions in the mantle into the crust and partly by the la-

Fig. 1.12.
Effect of fractional crystallization of magma on the Rb and Sr concentrations of the solids based on Rayleigh distillation (Eq. 1.49). The initial concentrations of the liquid C_0^l are 10 ppm Rb and 100 ppm Sr. The solids are plagioclase (45%) and clinopyroxene (55%). The distribution coefficients for Rb (0.10, plagioclase; and 0.04, cpx) were assumed to remain constant. The distribution coefficients for Sr were varied as follows: $F = 1.0$ to 0.6, D(Sr, plagioclase) = 1.5, D(Sr, cpx) = 0.10; $F = 0.6$ to 0.3, D(Sr, plagioclase) = 3.0, D(Sr, cpx) = 0.20; $F = 0.3$ to 0, D(Sr, plagioclase) = 6.0, D(Sr, cpx) = 0.30. The resulting bulk distribution coefficients ($\bar{D}$(Sr) and $\bar{D}$(Rb)) are shown on the diagram

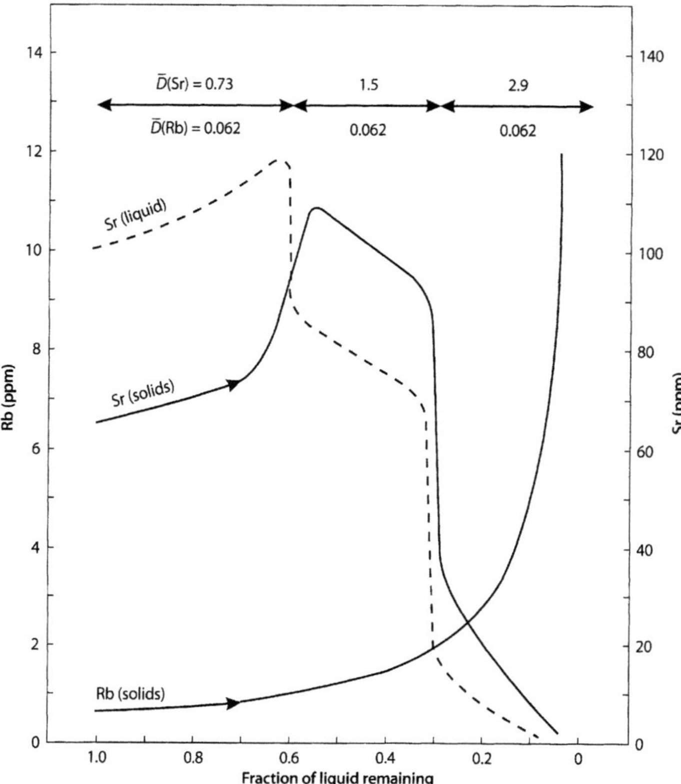

tent heat of crystallization of minerals that form in the cooling magma.

The evidence for melting and recrystallization of crustal xenoliths in basalt magmas has been documented by many investigators including Holmes (1936), Wyllie (1961), Sigurdsson (1968), Maury and Bizouard (1974), Al-Rawi and Carmichael (1976), McBirney (1979), Kays et al. (1981), and others. Since the liquidus temperature of basalt magma (about 1 200 °C) is higher than the melting temperature of feldspar and other minerals in crustal rocks, melting of crustal xenoliths and felsic wallrocks in contact with basalt magma is likely (Watson 1982). The process of simultaneous assimilation and fractional crystallization of magma (AFC) was modeled by Taylor (1980) and DePaolo (1981).

The change in the isotope composition of O of a basalt magma as a result of simultaneous assimilation and fractional crystallization depends on:

1. The isotope composition of the magma and of the country rock being assimilated (e.g. $\delta^{18}O = +5.7‰$ in the magma and +19.0‰ in the country rock);

2. The weight ratio of cumulate minerals crystallized by the magma and of the wallrock that is assimilated (e.g. 5 g of crystals per gram of assimilated wallrocks).

The weight ratio of cumulate to wallrock must be large enough to provide sufficient heat to melt one gram of wallrock. The magnitude of this ratio depends significantly on the temperature of the country rock because less heat is required to assimilate hot rocks than is required to assimilate cold rocks. Taylor (1980) also assumed that isotope fractionation of O between the cumulate crystals and the melt is negligible and that magma, the country rocks, and the cumulate crystals have virtually the same concentration of O. Model calculations by Taylor (1980) in Fig. 1.13 demonstrate that the $\delta^{18}O$ value of the hybrid magma increases stepwise and reaches +7.0‰ when 40% of the magma has crystallized, and rises to +11.0‰ when 89% has crystallized.

The effect of assimilation of crustal rocks by basalt magma on its $^{87}Sr/^{86}Sr$ ratio and Sr concentration can

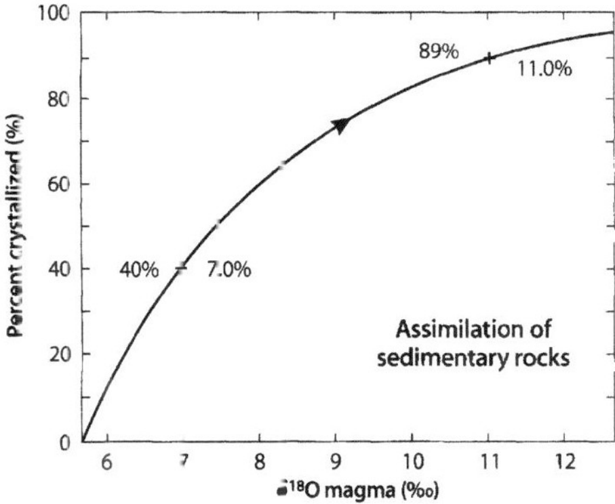

Fig. 1.13. Assimilation of sedimentary rocks $\delta^{18}O = +19.0‰$) by a basaltic magma $\delta^{18}O = +5.7‰$) that is crystallizing 5 g of cumulate minerals to generate enough heat to melt 1 g of sedimentary rocks. All components are assumed to have virtually the same oxygen concentration and the cumulate crystals have nearly the same $\delta^{18}O$ value as the magma. This process causes the $\delta^{18}O$ of the hybrid magma to increase from the initial value +5.7‰ to 7.0‰ when 40% of the magma has crystallized to +11.0‰ when 89% has crystallized (*Source:* adapted from Taylor 1980)

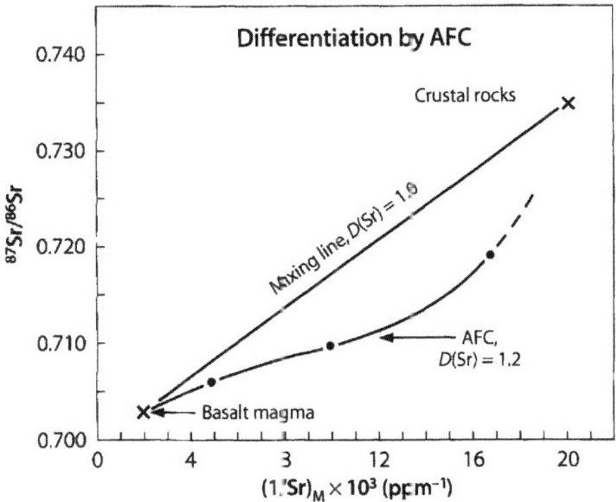

Fig. 1.14. Differentiation of basalt magma ($Sr = 500$ ppm, $^{87}Sr/^{86}Sr = 0.7030$) by simultaneous assimilation and fractional crystallization (AFC) of sedimentary rocks ($Sr = 50$ ppm, $^{87}Sr/^{86}Sr = 0.7350$). The ratio of assimilated wallrock to cumulate minerals crystallized is 1:5, and the distribution coefficient of Sr in the cumulate minerals is assumed to be 1.2. The resulting AFC trajectory is curved, whereas a two-component mixing line of the basalt magma and the sedimentary rocks is a straight line (*Source:* adapted from Fig. 4 of Taylor 1980)

also be modeled by the procedure developed by Taylor (1980) except that the concentrations of Sr in the magma, the country rock, and cumulate crystals must be considered. In most cases, the magma has a higher Sr concentration than the wallrock such that the concentration ratio (M/C) is:

$$\frac{M}{C} > 1.0 \tag{1.47}$$

where M and C represent the Sr concentrations of the magma and the country rock, respectively. In addition, the distribution coefficient $D(Sr)$ between the cumulate crystals and remaining liquid must be specified. The data in Table 1.10 indicate that olivine, pyroxene, amphiboles, and micas crystallizing from basalt discriminate against Sr (i.e. $D(Sr) < 1.0$), and that only plagioclase and apatite have Sr distribution coefficients greater than one. An example of the effect of AFC on the $^{87}Sr/^{86}Sr$ ratio of a basalt magma is provided in Fig. 1.14 based on calculations by Taylor (1980).

1.12 Meteorites and the Isotope Evolution of Terrestrial Strontium

According to the current theory of the origin of the solar system, the Earth formed in orbit around the Sun by accretion of planetesimals more than 4.5×10^9 years ago. The heat released by the impacts of planetesimals and by radioactive decay of unstable isotopes of K, U, Th,

Rb, and of other elements caused the Earth to melt. The resulting immiscible liquids (silicate, sulfide, and metallic iron) separated under the influence of gravity to form the metallic core, the silicate mantle, and blobs of sulfide that may be dispersed in the lower mantle. In addition, the noble gases, nitrogen, water vapor, and carbon dioxide escaped from the hot mantle and formed a dense atmosphere.

As the frequency of impacts diminished, the Earth began to cool allowing a thin silicate crust to form which reduced the amount of heat that was transferred to the atmosphere. As the atmosphere cooled, water vapor precipitated as rain when the relative humidity of the atmosphere reached 100% Eventually the surface of the Earth was covered by a layer of water a few kilometers deep containing dissolved carbon dioxide. The removal of water vapor and carbon dioxide from the atmosphere reduced the amount of greenhouse warming and allowed the surface temperature of the Earth to decline. Frequent eruptions of magma from the mantle formed volcanic islands which were eroded by chemical and mechanical weathering to form sediment and ions dissolved in the water. Convection in the form of rising columns of hot rocks in the mantle initiated plate tectonics very early in the history of the Earth and ultimately contributed to the growth of microcontinents composed of granitic rocks that could float on the denser rocks of the mantle. The early stage of evolution of the Earth probably occurred more than 4.0×10^9 years ago when impacts of large objects were still disrupting its surface during the Hadean Eon.

The theory of the origin of the Earth outlined above is consistent with the chemical and mineralogical composition of stony and iron meteorites. Most meteorites originated from the asteroids which themselves appear to be fragments of larger bodies. These so-called parent bodies of the meteorites were, in some cases, large enough to melt and to differentiate in their own gravitational fields to form cores of metallic iron surrounded by thick layers of silicate rocks. Subsequent collisions among these objects, caused by gravitational effects of Jupiter and Mars, scattered fragments of the parent bodies into the solar system where they impacted on the Earth and the Moon, as well as on Mars, Venus, Mercury and the other planets and their satellites.

The study of meteorites continues to provide information about the origin of planetary objects and their subsequent evolution. In fact, the carbonaceous chondrites contain grains that formed as a result of the explosion of the ancestral stars before the solar system had begun to form (e.g. Clayton 1993; Nicolussi et al. 1997; Hoppe et al. 1996; Amari et al. 1994).

Meteorites have also yielded determinations of the age of the solar system and of the Earth by various isotopic methods based on the Rb-Sr, Sm-Nd, U-Pb, and Re-Os decay schemes. For example, Sm-Nd mineral-isochron dates of six achondrite meteorites referenced by Faure (1986) range from 4.55 ±0.04 to 4.60 ±0.03 Ga with an average of 4.57 ±0.01 Ga ($2\bar{\sigma}$, $N = 6$).

The achondrite meteorites have low Rb/Sr ratios and are therefore difficult to date precisely by the Rb-Sr method. However, the achondrites are well suited for determinations of the $^{87}Sr/^{86}Sr$ ratio at the time silicate melts crystallized in the parent bodies of the meteorites. The definitive measurement was made by Papanastassiou and Wasserburg (1969) who reported an initial $^{87}Sr/^{86}Sr$ ratio of 0.698990 ±0.000047 for seven achondrite meteorites. This value is known as BABI which stands for: Basaltic Achondrite Best Initial.

The crystallization age of stony meteorites and their initial $^{87}Sr/^{86}Sr$ ratio are assumed to provide the starting point of the isotopic evolution of Sr in the Earth.

Accordingly, we assume that 4.5×10^9 years ago all parts of the Earth had the same $^{87}Sr/^{86}Sr$ ratio equal to 0.6990. Measurements to be discussed in Chap. 2 demonstrate that the $^{87}Sr/^{86}Sr$ ratios of the mantle of the Earth at the present time range from about 0.7020 to 0.7060. This observation is our first clue that the mantle does not have a monotonous chemical composition but consists of domains having different Rb/Sr ratios. The task we face is to document the range of $^{87}Sr/^{86}Sr$ ratios in different parts of the mantle and hence to develop a theory that explains how the mantle works.

1.13 Summary and Preview

The isotope compositions of Sr, Nd, Pb, and O of young volcanic rocks in the ocean basins, on oceanic islands, and on the continents convey information about the mantle of the Earth where magmas originate. The message contained in the isotope ratios of these elements is not easy to decode because, in many cases, the isotopic data can be interpreted in several different ways. Therefore, the decoding of isotope ratios must be constrained by the concentration patterns of trace elements and by an understanding of partial melting of magma sources and of fractional crystallization of magmas.

The isotope ratios of Sr, Nd, and Pb are also used to date rock-forming events and hence to reconstruct the evolution of igneous activity on a regional scale. Such reconstructions of the geologic history invariably demonstrate that igneous activity is a direct consequence of the interactions between the mantle and the overlying oceanic or continental crust. In a very real sense, the origin of igneous rocks is related to the tectonic activity of the mantle which causes rifting of the oceanic and continental crust and results in decompression melting and volcanic activity at the surface.

In the chapters that follow, we examine the isotopic compositions of volcanic and plutonic igneous rocks in different tectonic settings in order to determine the origin of the igneous rocks that form in these settings. In this way, we will discover how the Earth works.

Chapter 2

The Origin of Volcanic Rocks in the Oceans

The isotope ratios of Sr, Nd, and Pb of volcanic rocks in the oceans shed light on the sources of the magmas from which they formed and on their subsequent chemical evolution. The early work by Gast (1960), Faure (1961), and Faure and Hurley (1963) indicated that volcanic rocks in the oceans have low $^{87}Sr/^{86}Sr$ ratios compared to old granitic rocks of the continental crust and therefore originated from magma sources in the mantle having low Rb/Sr ratios. Subsequent studies have disclosed significant differences in the isotope compositions of Sr, Nd, and Pb between volcanic rocks of the mid-ocean ridges, oceanic islands, and island arcs. Consequently, volcanic rocks in the ocean basins must have originated from magma sources in the mantle in which these elements have different isotope compositions. Such differences are caused by variations of their respective parent/daughter elemental ratios existing for varying lengths of time in the past. In addition, magmas that formed by partial melting of different kinds of source rocks may mix prior to eruption to form "hybrid" magmas. Alternatively, different kinds of source rocks may themselves become mixed as a consequence of tectonic processes in the mantle. In the final analysis, volcanism in the ocean basins, as well as on the continents, is a manifestation of the tectonic activity in the mantle of the Earth.

The diversity of chemical and isotopic compositions of volcanic rocks in the oceans and on the continents has given rise to a large number of proposals concerning their origin. These will come up for discussion in the context of the different tectonic settings in which volcanic and plutonic rocks form. The presentation of the subject matter in this chapter proceeds from the Atlantic to the Pacific and to the Indian oceans. Islands selected for presentation are intended to exemplify different aspects of the petrogenetic process and to make the point that each island or group of islands differs in significant detail from others, regardless of proximity. The general conclusion is that the volcanic activity along mid-ocean ridges and on oceanic islands is a consequence of convection in the asthenospheric mantle (Saunders and Norry 1989).

2.1 Magma Formation in the Mantle

The structure of the mantle of the Earth has been defined on the basis of the reflection and refraction of seismic waves generated by earthquakes (Anderson 1992). The major subdivisions of the mantle and their mineralogical compositions are listed in Table 2.1. The geophysical data indicate that the *upper mantle* extends from the Mohorovičić discontinuity (at a depth of about 40 km under continents) to a depth of 400 km and is composed of peridotite (olivine + orthopyroxene), or eclogite (garnet + clinopyroxene), or of a mixture of peridotite and eclogite called fertile peridotite. The upper mantle is underlain by the *transition region* (400 to 650 km) composed of clinopyroxene, β-spinel, γ-spinel, garnet, and majorite (orthopyroxene with garnet structure). The *lower mantle* (650 to 2 740 km) makes up about 49% of the mass of the Earth and about 73% of the combined mantle and crust. It is composed of Fe-rich perovskite, magnesiowüstite, corundum, and stishovite. The lower mantle is underlain by a *thermal boundary layer* (200 to 300 km thick) characterized by a high thermal gradient caused by heat conduction from the core. Williams and Garnero (1996) proposed that this layer is partially molten in order to explain a 10% reduction of the velocity of P-waves in the so-called ultra-low velocity layer (ULVL). The occurrence of the ULVL in discontinuous patches appears to correlate with centers of volcanic activity at the surface. Therefore, this layer may be a source of mantle plumes; but plumes may also form in the asthenospheric mantle where subducted oceanic crust can cause instabilities by excessive heat production resulting from the radioactive decay of K, U, and Th (Hofmann and White 1982).

The occurrence of volcanic activity on the surface of the Earth is evidence that magma can form in the upper mantle and in the continental crust and does rise to the surface of the Earth. Melting starts when a volume of solid rock reaches its melting curve (solidus) in coordinates of pressure and temperature. In addition, melting is promoted by the presence of water, carbon dioxide, and

Table 2.1.
Major subdivisions of the mantle of the Earth and their mineralogical composition based on geophysical data (Anderson 1992)

Boundary or layer	Depth or thickness (km)	Mineral composition
Mohoroviciv discontinuity	0 – 40	Chemical boundary
Upper mantle	40 – 400	Olivine + orthopyroxene (peridotite) or garnet + clinopyroxene (eclogite), or eclogite-peridotite mixture
400 km discontinuity	400	Phase changes: Mg olivine to β-spinel, opx to majorite, plus garnet and clinopyroxene
Transition region	400 – 650	Clinopyroxene, β-spinel, γ-spinel garnet, majorite
650 km discontinuity	650 varies by 50–100 km	Phase change or chemical boundary within ~4 km
Lower mantle	650 – 2740	Perovskite: (Mg, Fe) SiO_3 including CaO and Al_2O_3 magnesiowüstite: (Mg,Fe)O corundum: Al_2O_3 stishovite: SiO_2
Thermal boundary layer of high density	200 – 300	Stishovite: SiO_2 perovskite: (Mg,Fe) SiO_3 magnesiowüstite: (Mg,Fe)O possibly also CaO, Al_2O_3, TiO_2
Core-mantle boundary (CMB)	2 900	Chemical boundary

other volatile elements or compounds whose presence decreases the melting temperature of rocks. The principal sites where partial melting occurs are (Sparks 1992):

1. Mid-ocean ridges
2. Plumes or hotspots
3. Subduction zones
4. Continental rift zones
5. Orogenic belts in the continental crust

The basic cause for the formation of magma at all of the sites listed above is *convection* in the asthenospheric mantle because convection transports heat from the interior towards the surface, moves volumes of rocks to regions of lower pressure or higher temperature, and causes rifting in the overlying brittle lithospheric plates. In many cases, melting occurs as a result of *decompression* rather than because of an increase in temperature. However, mantle rocks above subduction zones may also melt due to the release of aqueous fluids by the subducted oceanic crust, and rocks in the continental crust may melt locally because of the transfer of heat by basalt magma originating in the upper mantle.

The chemical composition of magmas that form in any of the tectonic settings listed above depends on the mineral composition of the source rocks, on the presence of volatiles, and on the extent of partial melting. The chemical composition of magmas is subsequently modified by fractional crystallization, and/or by assimilation of rocks with which they come in contact, and/or

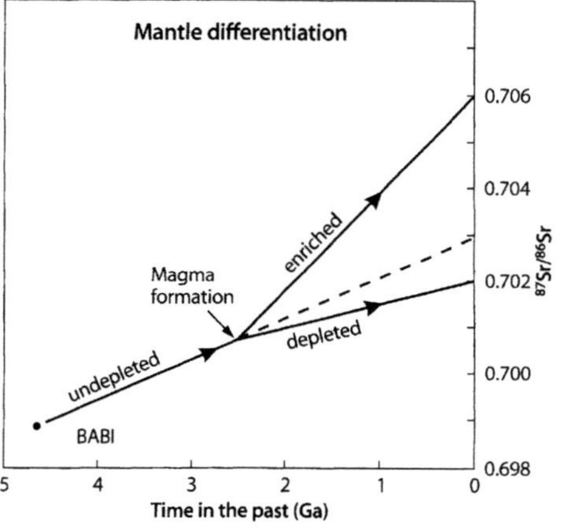

Fig. 2.1. Schematic diagram illustrating the isotope evolution of Sr in a small volume of rock in the upper mantle before and after formation of magma by partial melting. The $^{87}Sr/^{86}Sr$ ratio of a parcel of undepleted rocks in the upper mantle increases initially from 0.699 (Basaltic Achondrite Best Initial) at 4.55 Ga to 0.7008 at 2.5 Ga when magma forms by partial melting. Since the Rb/Sr ratio of the residual solids is less than that of the silicate liquid, the $^{87}Sr/^{86}Sr$ ratio of the residual solids (*depleted* mantle) subsequently increases more slowly than it did before melting occurred. The present value of the $^{87}Sr/^{86}Sr$ ratio in the *depleted* region of the upper mantle in the illustration is 0.7020. If the *enriched* silicate liquid in this illustration crystallized without assimilating Sr from the continental crust, its $^{87}Sr/^{86}Sr$ ratio at the present time is 0.7060. If the undepleted mantle rocks had remained undisturbed, their present $^{87}Sr/^{86}Sr$ ratio is 0.7030

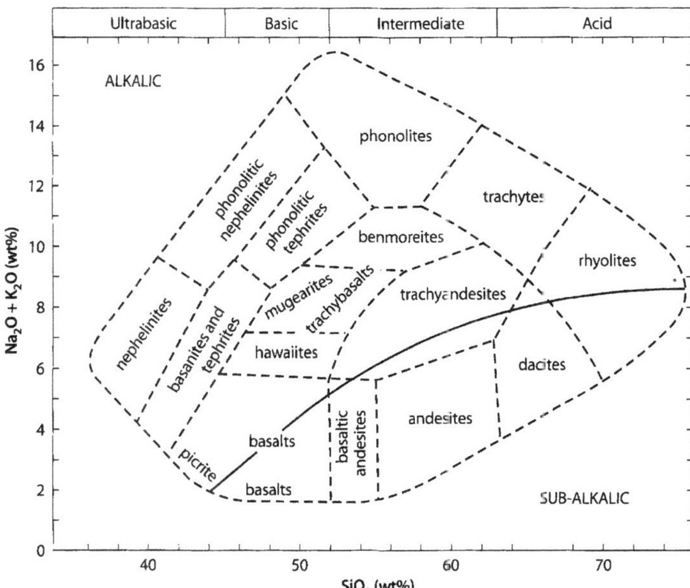

Fig. 2.2.
Classification of volcanic rocks based on their concentration of alkali metals and silica (*Sources:* Wilson 1989; Cox et al. 1979; Miyashiro 1978)

by mixing of magmas derived from different source rocks. In addition, the resulting igneous rocks may be altered by water-rock interaction involving seawater or groundwater of varying salinity. Consequently, the chemical compositions of igneous rocks and the isotope compositions of certain elements they contain may *not be directly related* in all cases to the chemical and isotope compositions of their source rocks in the mantle or in the continental crust.

When pyroxene and garnet melt in the mantle to form a silicate liquid of basaltic composition, Rb and Sr are strongly partitioned into the liquid phase. Although the details of this process depend on many factors, the Rb/Sr ratio of the resulting melt is commonly *greater* than the Rb/Sr of the rocks before melting (Faure and Hurley 1963). The subsequent movement of the magmas toward the surface of the Earth has caused the continental and oceanic crust to become enriched in Rb, Sr, and other incompatible elements that preferentially enter the liquid phase during magma formation. The residual mantle rocks are thereby depleted in Rb, Sr, and other incompatible elements. Figure 2.1 illustrates the effect of a decrease of the Rb/Sr ratio on the rate of change of the $^{87}Sr/^{86}Sr$ ratio in depleted regions of the upper mantle. The diagram demonstrates that the present $^{87}Sr/^{86}Sr$ ratios of *depleted* regions of the upper mantle are *lower* than those of undepleted regions.

These considerations indicate that the rocks of the mantle probably have a range of Rb/Sr and $^{87}Sr/^{86}Sr$ ratios depending on the occurrence and age of previous episodes of magma formation by partial melting. In addition, these parameters may also be affected locally by the presence of crustal rocks injected into the upper mantle in subduction zones (Hofmann and White 1982), by the presence of rising plumes of hot rocks originating from the undepleted lower mantle, or by the formation of Rb-rich veins by fluids arising from depth in the mantle. Therefore, the concentrations of Rb and Sr and the $^{87}Sr/^{86}Sr$ ratios of volcanic rocks in the ocean basins depend on many factors, including:

1. Prior history and composition of the magma sources;
2. Extent of partial melting during magma formation;
3. Mixing of magma derived from different sources;
4. Assimilation of rocks from the oceanic crust or the lithospheric mantle;
5. Fractional crystallization of the magma;
6. Alteration of the resulting igneous rocks by water-rock interaction.

The response of the Sm-Nd system to these kinds of processes differs from that outlined above for Rb and Sr (Chap. 1). Therefore, a combination of isotope ratios of Sr and Nd in igneous rocks is especially informative.

Volcanic rocks are classified both on the basis of compositional criteria (Streckeisen 1967; Irvine and Baragar 1971; Middlemost 1975, 1980) and in terms of the tectonic setting in which they form (Wilson 1989). The volcanic rocks in the ocean basins range widely in composition from sub-alkalic theoleiite basalts to alkali-rich rocks such as basanites and nephelinites, as well as hawaiites, mugearites, benmoreites, trachytes, and phonolites. The definition of these and other compositional varieties of volcanic rocks is outlined in Fig. 2.2 based on their total alkali and silica concentrations. The alkalic rocks on oceanic islands form primarily by very low degrees of partial melting (less than one percent) of depleted or undepleted source rocks rather than by fractional crystallization of basalt magma.

2.2   Mid-Ocean Ridge Basalt, Atlantic Ocean

Magma of basaltic composition forms under the mid-ocean ridges by decompression melting in the depleted lithospheric mantle (Wilkinson 1982; Wilson 1989; Hofmann 1997). The volume of magma so generated amounts to about 90% of the magma formed in the Earth (Sparks 1992). Melting starts at a depth of about 80 to 100 km when peridotites reach their solidus temperature at about 1 350 °C because of the decrease in pressure. The magma that is extracted from the mantle under mid-ocean ridges crystallizes to form the volcanic rocks of the oceanic crust whose average thickness is about 6 km. The extent of partial melting affects the concentrations of incompatible trace elements (including Rb and Sr) in the resulting magmas (Hart 1971). High degrees of partial melting (about 20% for normal MORBs) lead to low concentrations of incompatible

trace elements in the magma compared to magmas that represent small melt fractions which have higher concentrations of incompatible elements.

Mid-ocean ridge basalts typically have low $^{87}Sr/^{86}Sr$ ratios (0.7020 to 0.7030) and low concentrations of Rb and Sr (Allègre et al. 1983a,b; Ito et al. 1987; Hofmann 1988). The low $^{87}Sr/^{86}Sr$ ratios of MORBs from certain segments of the mid-ocean ridge system indicate that these rocks were derived from depleted magma sources in the upper mantle (Fig. 2.1). Their low Rb and Sr concentrations imply high degrees of partial melting (20 to 30%) without subsequent assimilation and fractional crystallization of the magma. However, the $^{87}Sr/^{86}Sr$ ratios as well as the Rb and Sr concentrations of MORBs vary significantly, especially in the vicinity of oceanic islands. Therefore, Sr in the mantle underlying the oceans is not isotopically homogeneous, but has different $^{87}Sr/^{86}Sr$ ratios depending on the Rb/Sr ratios and ages of the rocks from which MORB magmas originate.

2.2.1   North Atlantic Ocean

An excellent demonstration of the systematic regional variations of $^{87}Sr/^{86}Sr$ ratios of MORBs was provided by White and Schilling (1978). Their data in Fig. 2.3 show that the $^{87}Sr/^{86}Sr$ ratios of MORBs along the Mid-Atlantic Ridge between latitudes 29° N and 59° N vary significantly from 0.70233 (50°46' N; 29°42' W) to 0.70359

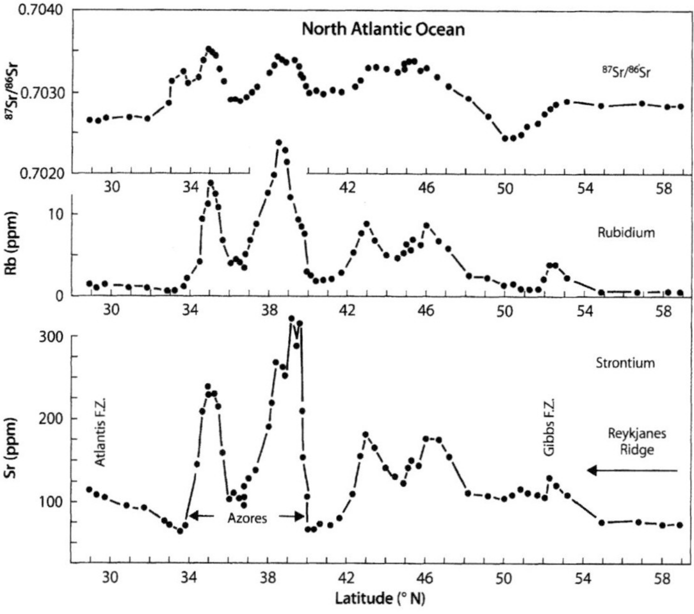

Fig. 2.3.
Longitudinal profiles of $^{87}Sr/^{86}Sr$ ratios and concentrations of Rb and Sr of mid-ocean ridge basalt (MORB) along the Mid-Atlantic Ridge between latitudes 29° N and 59° N. The data have been smoothed by presenting them as three-point moving averages. The $^{87}Sr/^{86}Sr$ ratios range from 0.70233 to 0.70359 and correlate closely with the concentrations of Rb and Sr. Elevated values of all three parameters are associated with the Azores Platform and occur also between 43° N and 47° N. The low $^{87}Sr/^{86}Sr$ ratios of normal MORBs indicate that these rocks originated from depleted source regions in the mantle that have had low Rb/Sr ratios for long periods of geologic time. MORBs having elevated $^{87}Sr/^{86}Sr$ ratios were either contaminated with radiogenic ^{87}Sr or originated from undepleted or Rb-enriched source regions in the mantle. The $^{87}Sr/^{86}Sr$ ratios were adjusted to a value of 0.7080 for the $^{87}Sr/^{86}Sr$ ratio of the Eimer and Amend Sr carbonate (*Source:* White and Schilling 1978)

(34°94' N, 36°61' W). In general, elevated $^{87}Sr/^{86}Sr$ ratios occur in three segments of the Mid-Atlantic Ridge included in the study of White and Schilling (1978). Two of these are associated with the Azores Platform (45° N and 41° N), whereas the third occurs between 43° and 47° N latitude. The concentrations of Rb and Sr of the MORBs in Fig. 2.3 vary in parallel with the $^{87}Sr/^{86}Sr$ ratios. This relationship requires the presence of two different kinds of source rocks having different $^{87}Sr/^{86}Sr$ ratios because neither partial melting in the mantle nor fractional crystallization of magma can alter the isotope composition of Sr. In any case, the effects of isotope fractionation in nature and in the mass spectrometer are reduced by the fractionation correction that is routinely applied to all measured $^{87}Sr/^{86}Sr$ ratios based on $^{86}Sr/^{88}Sr = 0.11940$ (Sect. 1.2.3). White and Schilling (1978) also excluded contamination by seawater because all of the rock specimens they analyzed contained less than 1% H_2O and because the $^{87}Sr/^{86}Sr$ ratios and Sr concentrations are not explainable in terms of interactions of the rocks with seawater.

The preferred explanation for the variation of $^{87}Sr/^{86}Sr$ ratios of MORBs along the Mid-Atlantic Ridge is based on the existence of plumes from the lower mantle rising into the lithosphere under the Azores and between latitudes 43° N and 47° N along the Mid-Atlantic Ridge. Similar variations in the $^{87}Sr/^{86}Sr$ ratios and in the concentrations of Rb and Sr of MORBs have been detected along the Reykjanes Ridge (Fig. 2.7) in the North Atlantic Ocean (Hart et al. 1973; O'Nions and Pankhurst 1974; Brooks et al. 1974; Machado et al. 1982), in the vicinity of Bouvet Island in the South Atlantic (Dickey et al. 1977; LeRoex et al. 1987), and along the Galapagos Ridge in the Pacific Ocean (Verma and Schilling 1982). In addition, the volcanic rocks of most oceanic islands have higher $^{87}Sr/^{86}Sr$ ratios and higher concentrations of Rb and Sr than normal MORBs (Halliday et al. 1995).

The $^{143}Nd/^{144}Nd$ ratios of MORBs along the Mid-Atlantic Ridge from 50° to 30° N are inversely correlated with the $^{87}Sr/^{86}Sr$ ratios (Dupré and Allègre 1980) and define the so-called Mantle Array illustrated in Fig. 2.4 based on data by Yu et al. (1997). The existence of this linear array of data points emphasizes the conclusion that magmas are derived from two kinds of source rocks having different Rb/Sr and $^{87}Sr/^{86}Sr$ ratios. The Rb-depleted magma source (low $^{87}Sr/^{86}Sr$ ratio) has an elevated $^{143}Nd/^{144}Nd$ ratio and hence a high Sm/Nd ratio, whereas the Rb-enriched plume rocks (high $^{87}Sr/^{86}Sr$ ratio) have a low $^{143}Nd/^{144}Nd$ ratio and are therefore depleted in Sm. The difference in the geochemical properties of the Sm-Nd and the Rb-Sr couples arises from their behavior during partial melting. Whereas partial silicate melts are enriched in Rb relative to the remaining solid phases, they are depleted in Sm (DePaolo 1979; Hart and Zindler 1986).

The mantle array in Fig. 2.4 is the result of mixing of magmas derived from the depleted lithospheric mantle

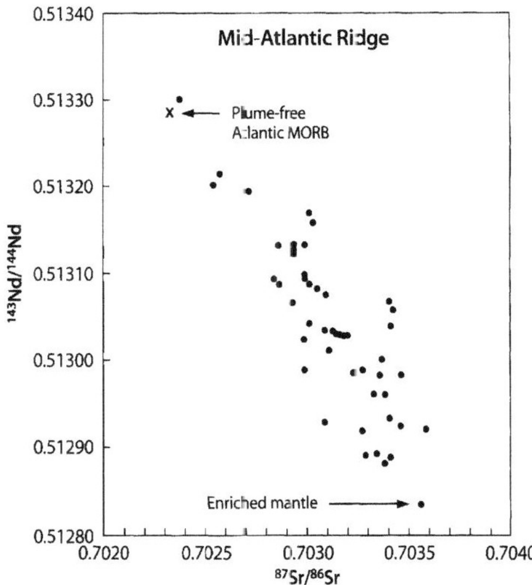

Fig. 2.4. Isotope ratios of Sr and Nd of basalts on the Mid-Atlantic Ridge between latitudes of 50° to 30° N. This segment of the ridge includes the Azores Plateau (29 to 32° N) along which the $^{87}Sr/^{86}Sr$ ratios rise above 0.7030. The $^{87}Sr/^{86}Sr$ ratios of MORBs along this segment of the MAR are inversely correlated with the $^{143}Nd/^{144}Nd$ ratios leading to the interpretation that they originated by mixing of magmas derived from Rb-depleted and Rb-enriched sources in the mantle. The $^{87}Sr/^{86}Sr$ ratios are relative to 0.70800 for E&A and the $^{143}Nd/^{144}Nd$ ratios were corrected for isotope fractionation to $^{146}Nd/^{144}Nd = 0.721903$. The plume-free Atlantic MORB (depleted mantle) is from Taylor et al. (1997) (*Source:* Data from Yu et al. 1997)

with magmas that formed by partial melting in plumes of asthenospheric mantle that penetrated into the overlying lithosphere along the Mid-Atlantic Ridge. Alternatively, these magmas may have formed by partial melting of mechanical mixtures of plume rocks and entrained blocks of lithospheric mantle.

2.2.2 Central Atlantic Ocean

South of the Azores, the Mid-Atlantic Ridge is off-set by numerous transform faults that manifest themselves as the Oceanographer, Atlantis, Kane, Vema, Romanche, and Ascension fracture zones, to name some of the most prominent of these topographic features. The petrogenesis of basalts at 36°47' N located a short distance north of the Oceanographer fracture zone was investigated by the French-American Undersea Study (FAMOUS) described by Langmuir et al. (1977). A small suite of basalts dredged from the vicinity of the Oceanographer fracture zone at 35° N originated from magma sources that contain an unusual component characterized by a high $^{87}Sr/^{86}Sr$ ratio of 0.70629 that Shirey et al. (1987) attributed to the presence of a block of subcontinental mantle that may have been entrained in the upwelling mantle at this site.

The presence of blocks of subcontinental mantle along the central segment of the Mid-Atlantic Ridge is supported by the rocks exposed on St. Peter-Paul islets located at about 1° N latitude north of the prominent Romanche fracture zones. Darwin landed here in 1831 and noticed that these islands are not of volcanic origin. Subsequent studies by Roden et al. (1984a), Bonatti (1990), and others have revealed that these islands are composed primarily of peridotites and represent uplifted blocks of subcontinental mantle that were left behind during the opening of the Atlantic Ocean.

MORBs collected between the Kane and Vema fracture zones from 24° to 10° N have primarily low $^{87}Sr/^{86}Sr$ ratios whose one-degree averages in Fig. 2.5 are less than 0.7030. Dosso et al. (1993) reported a value of only 0.702167 for a tholeiite recovered from the ridge axis at

17°27.60' N and 46°26.0' W at a water depth of 3 800 m. The concentrations of Rb and Sr of this specimen are 0.670 ppm and 112.8 ppm, respectively.

The segment of the Mid-Atlantic Ridge centered at latitude 14° N and extending from 12 to 16° N includes only one significant anomaly (Bougault et al. 1988; Dosso et al. 1991). In this region, the $^{87}Sr/^{86}Sr$ ratios of basalts rise to an average of 0.70294 ±0.00013 ($2\bar{\sigma}, N = 12$) at about 14° N latitude. One specimen collected at 14°16.20' N and 45°43.20' W has an anomalously high $^{87}Sr/^{86}Sr$ ratio of 0.706406 (omitted from the average) with Rb = 14.68 ppm and Sr = 1 260 ppm. The chemical composition of this specimen (SiO_2 = 46.60%, Na_2O = 3.40%, K_2O = 1.37%) identifies it as an alkali basalt (Dosso et al. 1993).

A second anomaly on the Mid-Atlantic Ridge at 1.7° N also has $^{87}Sr/^{86}Sr$ ratios that approach 0.7030 relative to 0.71025 for NBS 987. Schilling et al. (1994) attributed this anomaly to the presence of a plume that appears to have been the magma source that built up the Sierra Leone Rise east of the ridge and the complementary Ceará Rise in the western Atlantic. The $^{87}Sr/^{86}Sr$ ratios of this segment of the Mid-Atlantic Ridge between 4.91° N and 2.54° S are included in the histogram of Sr isotope ratios in Fig. 2.7.

2.2.3 South Atlantic Ocean

The $^{87}Sr/^{86}Sr$ ratios of MORBs in the South Atlantic rise gradually from 0.70219 ±0.00005 (3.4 to 6.3° S) to values approaching 0.70350 over the Tristan da Cunha Plume between 36° and 40° South (Hanan et al. 1986; Hanan and Graham 1996; Fontignie and Schilling 1996). The data displayed in Fig. 2.6 reveal a gradual increase of the $^{87}Sr/^{86}Sr$ ratios from north to south mirrored by a complementary decline of the $^{143}Nd/^{144}Nd$ ratios. The profile of $^{87}Sr/^{86}Sr$ ratios has positive spikes at 8.5° S (Ascension Island), 15.5° S (St. Helena), and at 24.8° S. The $^{143}Nd/^{144}Nd$ profile has sharp negative spikes close to the same locations.

These observations confirm the inverse correlation between the isotope ratios of Sr and Nd caused by the differences in the geochemical properties of the Rb-Sr and Sm-Nd couples and reflect the presence of plumes along the spreading ridge in the South Atlantic. Wilson (1963b) originally proposed that the plume underlying the Mid-Atlantic Ridge adjacent to Tristan da Cunha and Gough Island is the source of the chain of seamounts located along the Walvis Ridge which extends in a northeasterly direction toward the Etendeka basalts (Early Cretaceous) in Namibia (Sect. 5.7.2). In addition, the volcanic rocks on the Rio Grande Rise in the western South Atlantic link the Early Cretaceous Paraná basalts of Brazil to the Tristan Plume (Sect. 5.7.1). These connections imply that the mantle plume under Tristan da Cunha has been active for about 140 million years.

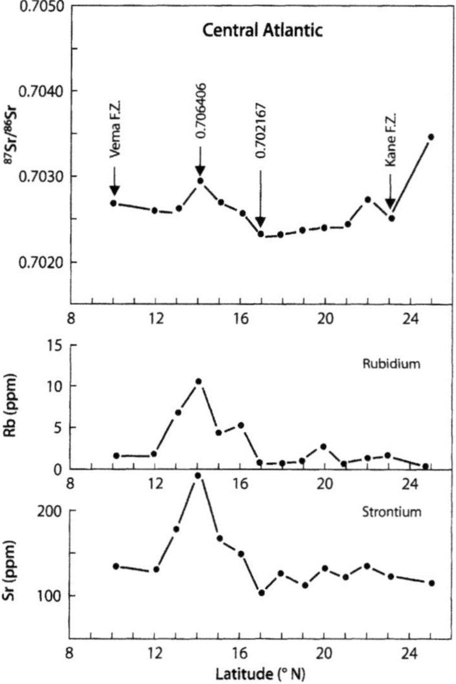

Fig. 2.5. Longitudinal profiles of $^{87}Sr/^{86}Sr$ ratios and concentrations of Rb and Sr of MORBs along the Mid-Atlantic Ridge from 25° N (about 2° north of the Kane fracture zone) to the Vema fracture zone at 10° N. The data were averaged in one-degree increments and were plotted on the same scale as the data in Fig. 2.3. This segment of the Mid-Atlantic Ridge has generally low $^{87}Sr/^{86}Sr$ ratios with only one anomalous region centered at 14° N latitude where the concentration of Rb and Sr also rise (*Source:* data from Dosso et al. 1993)

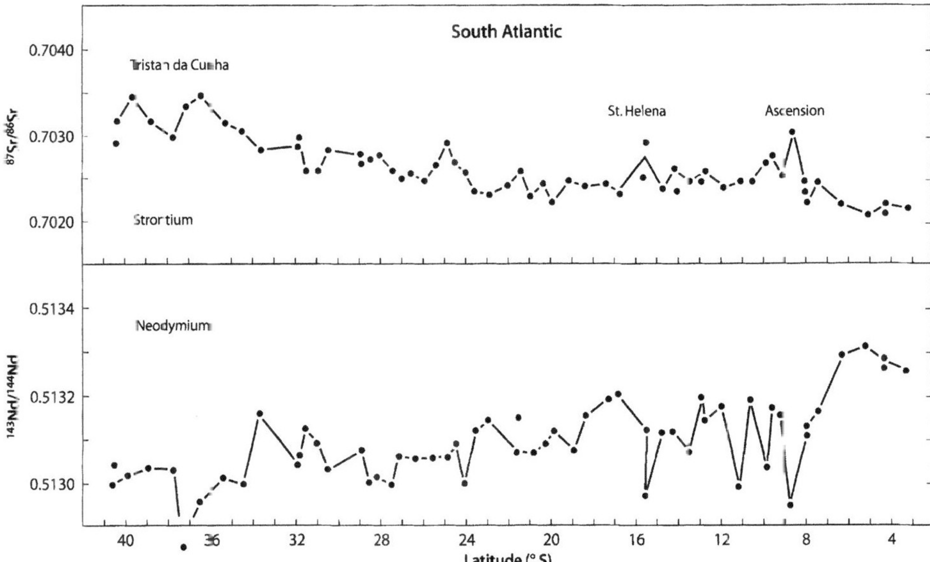

Fig. 2.6. Profiles of $^{87}Sr/^{86}Sr$ and $^{143}Nd/^{144}Nd$ ratios of MORBs along the spreading ridge in the South Atlantic Ocean. The $^{87}Sr/^{86}Sr$ ratios increase gradually from north to south and approach high values close to 0.70350 over the Tristan da Cunha-Bouvet Island Plumes. Positive spikes of $^{87}Sr/^{86}Sr$ ratios occur in MORBs adjacent to Ascension Island and St. Helena. The MORBs at these sites have anomalously low $^{143}Nd/^{144}Nd$ ratios consistent with the geochemical differences of the Rb-Sr and Sm-Nd couples (*Source:* data from Fontignie and Schilling 1996)

The spectrum of $^{87}Sr/^{86}Sr$ ratios of MORBs on the Mid-Atlantic Ridge in Fig. 2.7 extends from 0.7021 to 0.7037 with a few higher values greater than 0.7040. The elevated $^{87}Sr/^{86}Sr$ ratios of some plume MORBs in Fig. 2.7 may be attributed to interaction with seawater. However, the analysts made an effort to exclude altered samples and leached the powdered rock samples with dilute acid prior to analysis in order to remove secondary minerals (Cohen et al. 1980; Cohen and O'Nions 1982a). The longitudinal profiles (Figs. 2.4, 2.5, and 2.6) indicate that the $^{87}Sr/^{86}Sr$ ratios rise in certain segments of the ridge (e.g. the Azores and Tristan da Cunha) where magmas are forming in plumes of undepleted or Rb-enriched rocks that have risen from the asthenosphere.

The spectrum of $^{87}Sr/^{86}Sr$ ratios of basalt along the Mid-Atlantic Ridge in Fig. 2.7 has three modes representing normal MORBs (0.7025 ±0.0010) and plume MORBs (07035 =0.0010) derived from enriched sources (e.g. Azores and Tristan da Cunha). The third mode between 0.7028 to 0.7030 characterizes transitional MORBs (e.g. the anomaly at 14° N) derived from somewhat less enriched or undepleted source rocks. The existence of magma sources in the Atlantic Ocean having a range of

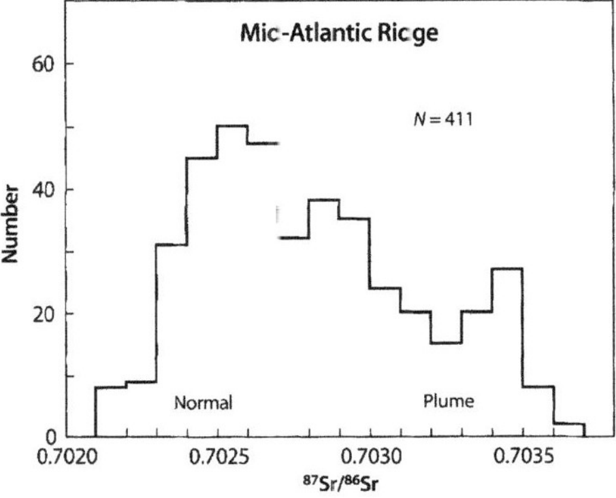

Fig. 2.7. Histogram of $^{87}Sr/^{86}Sr$ ratios of MORBs along the Mid-Atlantic Ridge from 60° N to 44° S. Volcanic rocks on oceanic islands and seamounts were excluded. Plume MORBs having $^{87}Sr/^{86}Sr$ ratios close to or greater than 0 7030 occur primarily on the Azores Platform, close to Ascension and St. Helena, and in the vicinity of Tristan da Cunha-Bouvet Island. All $^{87}Sr/^{86}Sr$ ratios are relative to 0.7080 for E&A or 0.71025±0.00001 for NBS 987 (*Sources:* data from White and Schilling 1978; White and Hofmann 1982; LeRoex et al. 1987; Castillo and Batiza 1989; Dosso et al. 1993; Schilling et al. 1994; Fontignie and Schilling 1996; Yu et al. 1997)

elevated $^{87}Sr/^{86}Sr$ ratios indicates that not all mantle plumes are alike, presumably because of differences in the prior histories of the rocks of which they are composed (Hofmann and White 1982). The lack of resolution of these modes implies a high degree of mixing of magmas derived from different sources in the underlying mantle. The distinctive character of the $^{87}Sr/^{86}Sr$ ratios of MORBs along the Atlantic Ocean becomes more apparent when they are compared to the $^{87}Sr/^{86}Sr$ ratios of volcanic rocks on the oceanic islands that formed along the Mid-Atlantic Ridge (e.g. Fig. 2.14). The classification of MORBs on the basis of geochemical and isotopic criteria has been discussed by Bryan et al. (1976), Sun et al. (1979), and Schilling et al. (1983).

2.2.4   The Walvis Ridge in the South Atlantic

Volcanoes that originally formed on the Mid-Atlantic Ridge or over a midplate hotspot may be carried away from their magma source by seafloor spreading and thus form chains of submerged seamounts and oceanic islands (Epp and Smoot 1989). Significant examples of this phenomenon include the New England Seamounts (Sect. 5.10.3) in the North Atlantic and the Walvis Ridge in the South Atlantic. Other examples, to be considered later, include the seamounts associated with the islands of Fernando de Noronha and Trinidade in the equatorial and southern Atlantic, respectively. In some cases, the volcanic activity recorded by the seamounts extends onto the adjacent continent (e.g. the Cameroon Line, Sect. 5.8.2).

The Walvis Ridge is a prominent topographic feature on the bottom of the South Atlantic Ocean that extends from the vicinity of Tristan da Cunha towards the Early Cretaceous basalt plateau formed by the Etendeka Group in Namibia (Sect. 5.7.2). O'Connor and Duncan (1990) demonstrated that the ages of volcanic rocks recovered from Walvis Ridge increase with distance from Tristan da Cunha in a northeasterly direction and imply a rate of seafloor spreading of 3.1 cm yr^{-1}.

The initial $^{87}Sr/^{86}Sr$ ratios of selected samples of basalt glass and plagioclase obtained from DSDP cores 525A (29°04.2' S, 2°59.1' E), 527, and 528 (28°31.5' S, 2°19.4' E) on Walvis Ridge range from 0.70391 to 0.70507 relative to 0.7080 for E&A (Richardson et al. 1982). The high $^{87}Sr/^{86}Sr$ ratios of these submarine basalts are compatible with the $^{87}Sr/^{86}Sr$ ratios of volcanic rocks on Tristan da Cunha (0.70505 to 0.70513) reported by O'Nions and Pankhurst (1974) and White and Hofmann (1982). Therefore, the $^{87}Sr/^{86}Sr$ ratios of the basalts from the Walvis Ridge confirm the proposal by Wilson (1963b) that it records the movement of the African Plate relative to the stationary Tristan Plume.

2.3   Iceland

The rocks that form the Mid-Atlantic Ridge from the Charlie Gibbs Fracture zone north through the Reykjanes Ridge, Iceland, and Jan Mayen in Fig. 2.8 have been studied intensively for decades (Fitch et al. 1965; Lussiaa-Berdou-Polve and Vidal 1973; Pedersen et al.

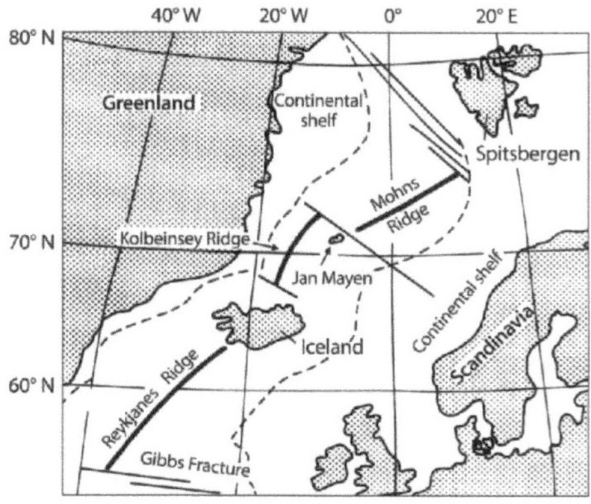

Fig. 2.8. Map of the North Atlantic Ocean showing the Reykjanes Ridge, Iceland, Jan Mayen, and Spitsbergen. Many concepts concerning the petrogenesis of volcanic rocks have been developed as a consequence of the intensive study of the rocks in this region

1976; Haase et al. 1996). As a result, much of our present understanding of the physical and chemical properties of the upper mantle under the oceans was developed here. Therefore, a review of the early work on the petrogenesis of the Icelandic lavas is an appropriate starting point for a discussion of the origin of oceanic island basalt (OIB).

2.3.1   The Tholeiites of Iceland

The volcanic rocks of Iceland consist primarily of Late Tertiary to Recent tholeiite basalts with lesser amounts of alkali basalts, rhyolites, and obsidians. The history of volcanic activity has been reconstructed primarily by K-Ar dating of whole-rock specimens (Gale et al. 1966; Moorbath et al. 1968; Saemundsson and Noll 1974; McDougall et al. 1976a,b, 1977; Mussett et al. 1980; Saemundsson et al. 1980). The isotope compositions of S of the volcanic rocks and associated gases were measured by Torssander (1988, 1989), whereas Galimov and Gerasimovsky (1978) reported $\delta^{13}C$ values of igneous rocks on Iceland and Kurz et al. (1985) presented data on the isotope composition of He. The first measurements of isotope ratios of Pb were made by Welke et al. 1968) and, most recently, by Hanan and Schilling (1997).

The $^{87}Sr/^{86}Sr$ ratios of the volcanic rocks of Iceland and of the Reykjanes Ridge in Fig. 2.9 are significantly higher than those of normal MORBs (Hart et al. 1973; O'Nions and Pankhurst 1974). The very existence of Iceland astride the Mid-Atlantic Ridge implies abnormally high production rates of magma and suggests that the underlying magma sources differ locally from the mantle elsewhere along the ridge.

Fig. 2.9.
Variation of $^{87}Sr/^{86}Sr$ ratios of volcanic rocks along the Mid-Atlantic Ridge north of latitude 52° including Iceland and Jan Mayen. The increase of this ratio in tholeiites of the Reykjanes Ridge indicates that the magmas that feed the volcanoes of Iceland and Jan Mayen originated from asthenospheric plumes or that the magmas were contaminated with Sr derived from old differentiated rocks that may underlie these islands. However, other explanations for the elevated $^{87}Sr/^{86}Sr$ ratios are possible and are discussed in the text. The $^{87}Sr-^{86}Sr$ ratios were standardized to c.70800 E&A. In addition, the data of Hart et al. (1973) were increased by 0.000176 in order to match them to the $^{87}Sr/^{86}Sr$ ratios of O'Nions and Pankhurst (1974) (*Sources*: Hart et al. 1973, *open circles*; O'Nions and Pankhurst 1974, *solid circles*; Pedersen et al. 1976, Jan Mayen)

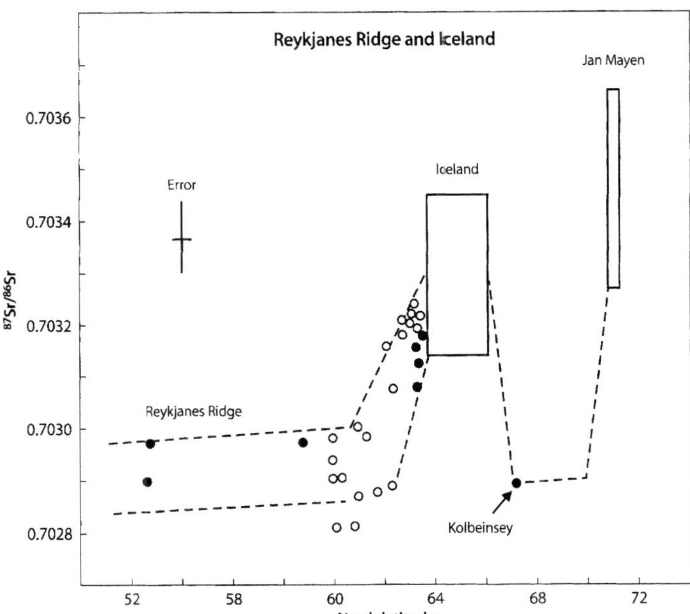

The elevated $^{87}Sr/^{86}Sr$ ratios of the volcanic rocks in Iceland are attributable to one or several of the following causes.

1. The magma sources in the mantle under Iceland have higher Rb/Sr ratios than the magma sources of normal MORBs because they are either undepleted, less depleted, or enriched in Rb and other incompatible elements.
2. Magma was derived by partial melting of basalt or felsic differentiates in the oceanic crust under Iceland.
3. Normal MORB magma was contaminated by assimilation of old crustal rocks of sialic composition under Iceland.
4. The isotope composition of Sr in the volcanic rocks of Iceland was altered by assimilation of sediment by basalt magma and/or by interaction of the volcanic rocks with seawater or mixtures of seawater and meteoric water.
5. Decay of ^{87}Rb to ^{87}Sr after crystallization of differentiated volcanic rocks having high Rb/Sr ratios.

Large-scale contamination of magma by marine sediment or seawater can be excluded because this process is unlikely to produce the narrow range of $^{87}Sr/^{86}Sr$ ratios that characterizes the volcanic rocks of Iceland displayed in Fig. 2.10. In situ decay of ^{87}Rb to ^{87}Sr can also be ruled out because the age of the oldest rocks on Ice-

land is only 16.0 ±0.3 Ma (Moorbath et al. 1968) and because the Rb/Sr ratios of the dominant basalts are so low that any increase of the $^{87}Sr/^{86}Sr$ ratio is insignificant in most cases.

However, the relatively rare rhyolite flows and granophyres in Iceland have higher Rb/Sr ratios than the basalts and, in some cases, do have elevated present $^{87}Sr/^{86}Sr$ ratios. For example, Åberg et al. (1987) were able to date the granophyric and gabbroic rocks of the Austurhorn Intrusion on the southeast coast of Iceland by the Rb-Sr isochron method because the Rb/Sr ratios of these rocks range from about 0.10 to 1.27 and their $^{87}Sr/^{86}Sr$ ratios have increased by about 0.1% since crystallization at 12.8 ±5.3 Ma. In addition, O'Nions and Pankhurst (1973) reported $^{87}Sr/^{86}Sr$ ratios of the Austurhorn and related granophyric rocks, all of which contained some in situ produced radiogenic ^{87}Sr. However, the relatively old granophyric differentiates in Iceland are a minor exception to the more important phenomenon that rhyolites and alkalic volcanics on Iceland have similar $^{87}Sr/^{86}Sr$ ratios as the basaltic rocks (Moorbath and Walker 1965).

The most likely explanation for the elevated $^{87}Sr/^{86}Sr$ ratios of the volcanic rocks on Iceland compared to those of normal MORBs is that the magma is generated within a *plume* rising from the asthenospheric mantle. The existence of such plumes was originally postulated by Wilson (1963a) and was later used by Morgan (1971) to explain the origin of linear chains of islands in the Pacific Ocean, including the Hawaiian-Emperor, Tuamotu-

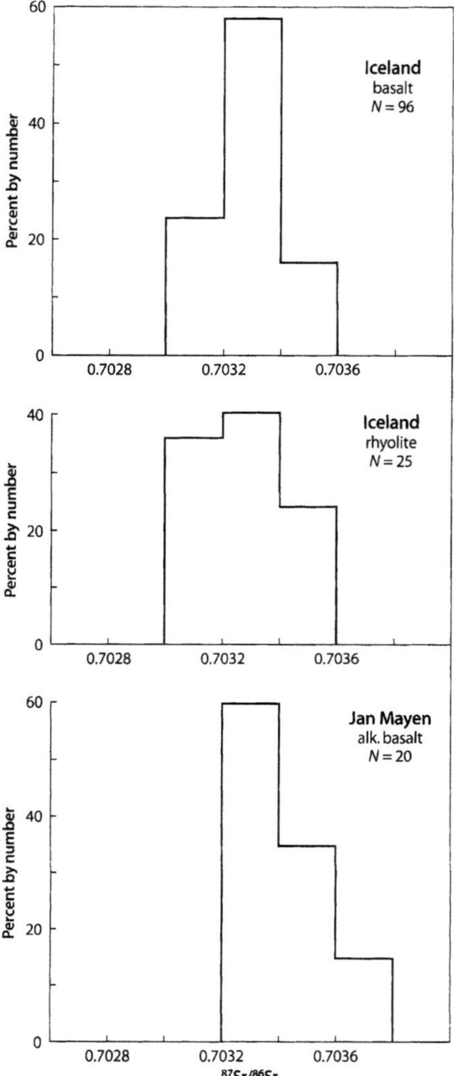

Fig. 2.10. $^{87}Sr/^{86}Sr$ ratios of volcanic rocks of Iceland and Jan Mayen. The volcanic rocks of Iceland, including tholeiite basalts as well as rhyolites, granophyres, and alkalic rocks, have similar $^{87}Sr/^{86}Sr$ ratios in spite of differences in their chemical compositions. In addition, the $^{87}Sr/^{86}Sr$ ratios of the alkali-rich volcanic rocks of Jan Mayen do not differ significantly from the $^{87}Sr/^{86}Sr$ ratios of Icelandic tholeiites and rhyolites. All $^{87}Sr/^{86}Sr$ ratios in this diagram are relative to 0.7080 for E&A. The data of Hart et al. (1973) were increased by 0.000176 to match them to the $^{87}Sr/^{86}Sr$ ratios of O'Nions and Pankhurst (1974) (*Sources:* Hart et al. 1971; Hart et al. 1973; O'Nions and Grönvold 1973; O'Nions and Pankhurst 1973, 1974; O'Nions et al. 1973; Lussiaa-Berdou-Polve and Vidal 1973; Sun and Jahn 1975b; Pedersen et al. 1976; Zindler et al. 1979; Wood et al. 1979; Cohen and O'Nions 1982a; Åberg et al. 1987)

Line, and Austral-Gilbert-Marshall Chains. The plumes that rise like "thunderheads" are composed of rocks whose $^{87}Sr/^{86}Sr$ ratios are higher than those of rocks in the lithospheric mantle. As the plumes ascend to regions of lower pressure, partial melting occurs within them and the resulting magmas give rise to volcanic activity on the surface.

Schilling (1973a) proposed that the increase in the concentrations of La, P_2O_5, K_2O, and TiO_2 of the basalt along the Reykjanes Ridge from latitude 60° N to 64° N is caused by mixing of magmas originating from such a plume under Iceland and from the depleted mantle of the lithosphere. The magma-mixing hypothesis received support from O'Nions and Pankhurst (1973) who demonstrated that the initial $^{87}Sr/^{86}Sr$ ratios of igneous rocks in Iceland have decreased from 0.7036 at 15 Ma to 0.7033 at 2 Ma. Such temporal variations in the $^{87}Sr/^{86}Sr$ ratio may be caused by changing proportions of magmas originating from the plume (high $^{87}Sr/^{86}Sr$) and from the depleted mantle (low $^{87}Sr/^{86}Sr$) under the ridge. Subsequently, Sun et al. (1975) confirmed that the isotope ratios and concentrations of Pb in MORBs along the Reykjanes Ridge can be accounted for by two-component mixing.

In addition, Langmuir et al. (1978) tested the mixing hypothesis for the petrogenesis of Icelandic volcanics by means of both chemical and isotopic data. Their interpretation indicated that mixing of rocks or magmas may be taking place, but the details are more complex than expected. In particular, the chemical composition of the volcanic rocks also depends on variations in the extent of melting of the source rocks and on fractional crystallization of the resulting magmas.

The occurrence of secular changes in the chemical and isotopic compositions of igneous rocks in Iceland suggests that mantle plumes do not necessarily rise as continuous "thunderheads" but perhaps more in the form of "blobs" (Schilling et al. 1982; Allègre et al. 1984). Therefore, the intensity of volcanic activity and the composition of lavas above a plume may vary with time as blobs rise toward the surface. The blobs may interact with the depleted rocks of the upper mantle in three possible ways (O'Nions and Pankhurst 1973):

1. Mixing of solids prior to melting.
2. Blending of magmas derived from the two kinds of sources.
3. Assimilation of rocks in the lithospheric mantle or in the oceanic crust by magmas originating from blobs or plumes.

However, the plume model is not necessarily the only possible explanation for the high $^{87}Sr/^{86}Sr$ ratios and compositional differences between normal MORBs and the volcanic rocks on Iceland (O'Hara 1973; Balashov 1979; Sleep 1984; Gerlach 1990; Mertz et al. 1991). For example, the step-like increase of the $^{87}Sr/^{86}Sr$ ratio along the Reykjanes Ridge (Fig. 2.7) could be caused by the presence of phlogopite in the mantle beneath Iceland and the northern part of the Reykjanes Ridge (Flower et al. 1975). The presence of Rb-rich phlogopite mica or of the amphibole kaersutite in the mantle facilitates partial melting because of the presence of water. The resulting magmas have elevated $^{87}Sr/^{86}Sr$ ratios because of diffusion of radiogenic ^{87}Sr from phlogopite and kaersutite crystals to other Rb-poor minerals in the magma sources (Nelson and Dasch 1976; Basu and Murthy 1977a; Hofmann and Hart 1978; Sneeringer et al. 1984).

Zindler et al. (1979) reported that the $^{143}Nd/^{144}Nd$ ratios of tholeiites on the Reykjanes Peninsula of South Iceland vary widely, whereas their $^{87}Sr/^{86}Sr$ ratios are restricted to a narrow range from 0.7031 to 0.7032. The isotopic heterogeneity of Nd in these basalts is not compatible with magma generation in a homogeneous mantle under Iceland. Consequently, Zindler et al. (1979) evaluated models of a vertically stratified heterogeneous mantle based on different scales of heterogeneity, different ages of the heterogeneity, and on the state of mixing. For example, in the LOM model (Large-scale heterogeneity, Old age of heterogeneity, Melt mixing) the mantle is assumed to be stratified with spinel lherzolite (olivine + orthopyroxene + clinopyroxene) overlying garnet lherzolite. The two layers have similar $^{87}Sr/^{86}Sr$ ratios but different $^{143}Nd/^{144}Nd$ ratios such that magmas formed by mixing of melts derived by different degrees of partial melting from the two types of source rocks have similar $^{87}Sr/^{86}Sr$ ratios but different $^{143}Nd/^{144}Nd$ ratios, depending on the proportions of mixing. However, the step-change in the $^{87}Sr/^{86}Sr$ ratios and concentrations of incompatible elements under the Reykjanes Ridge also implies lateral heterogeneity in the mantle that is the basis for the plume model advocated by Schilling (1973a).

A third alternative was proposed by Wood et al. (1979) and Wood (1981) who favored magma generation from regionally homogeneous but locally heterogeneous veined mantle rocks. This model is also capable of explaining the observed trace-element concentrations and isotope ratios of Sr and Nd of the basaltic rocks of Iceland.

The multiplicity of petrogenetic models that have been proposed to explain the origin of basalts in Iceland demonstrates how difficult it is to uniquely constrain the structure, age, and composition of magma sources in the mantle from the chemical and isotope composition of the rocks that are produced (Sleep 1992). Recent studies of the Icelandic volcanic field in Veidivötn by Mørk (1984) and at Krafla by Nicholson et al. (1991) emphasize the importance of magma mixing, assimilation of hydrothermally altered rocks, and fractional crystallization of magma extruded at these localities. The

evidence derived from the study of volcanic rocks in Iceland indicates that magmas are generated by varying degrees of partial melting within the plume and/or in the adjacent lithospheric mantle, with the possibility of subsequent differentiation by mixing, assimilation, and fractional crystallization. This scenario appears to be the most promising approach to explaining the compositional diversity of volcanic rocks on Iceland and elsewhere within the ocean basins (Hardarson and Fitton 1991; Elliott et al. 1991; Condomines et al. 1981; Allègre and Condomines 1982).

2.3.2 Rhyolites and Alkali-rich Rocks of Iceland

The occurrence of rhyolites on Iceland in close association with tholeiite basalt was first described by Bunsen (1851) more than a century ago and has posed one of the most challenging problems in petrology (Yoder 1973). Bunsen proposed that basalt and rhyolite in Iceland crystallized from different magmas and that intermediate rock types are the result of mixing between them. The bimodal distribution of silica concentrations of volcanic rocks in the oceans was further emphasized by Chayes (1963) and has been cited as evidence against the formation of different compositional varieties of igneous rocks by fractional crystallization of basalt magma.

The occurrence of alkali-rich rocks in Iceland, Jan Mayen, and on many other oceanic islands might be taken as evidence that the magmas had formed by partial melting of granitic basement rocks similar to those that occur in the continental crust. This hypothesis was tested by Moorbath and Walker (1965) based on the conclusion of Faure and Hurley (1963) that igneous rocks which contain a component of crustal Sr have higher $^{87}Sr/^{86}Sr$ ratios than igneous rocks which were derived from the mantle without crustal contamination. The results of Moorbath and Walker (1965) indicated that the $^{87}Sr/^{86}Sr$ ratios of felsic rocks on Iceland are *indistinguishable* from those of the basalts (Fig. 2.10.). Therefore, they concluded that the granophyres, rhyolites, and obsidians of Iceland originated by partial melting of basalt of the oceanic crust and *not* by melting of old sialic rocks under Iceland. This conclusion was confirmed by Hart et al. (1971) who demonstrated that the $^{87}Sr/^{86}Sr$ ratio of a composite sample of high-alkali rocks from Iceland is indistinguishable from the $^{87}Sr/^{86}Sr$ ratio of a composite of low-alkali rocks. Subsequently, Sigurdsson (1977) attributed the origin of the rhyolites to melting of plagiogranites formed previously by fractional crystallization of basalt magma.

The Snaefellsnes area of western Iceland and the Vestmann Archipelago along the southern coast in Fig. 2.11 contain alkali-rich basalts ranging in composition from alkali-olivine basalt to hawaiite and mugearite (Fig. 2.2). The Vestmann Archipelago includes the is-

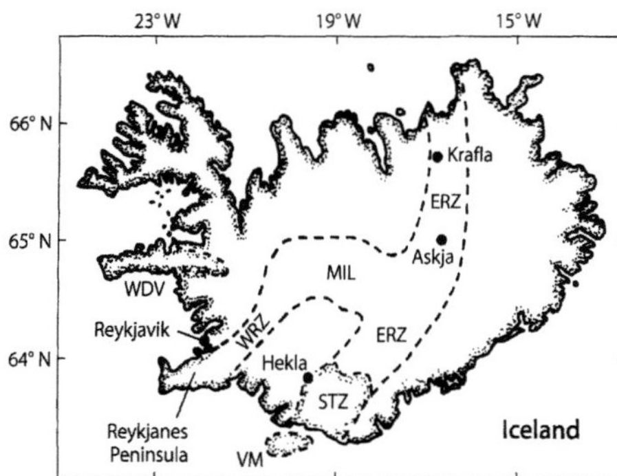

Fig. 2.11. Neovolcanic zones of Iceland identified by the abbreviations: *ERZ* = East Rift Zone; *MIL* = Mid-Iceland Belt; *WRZ* = West Rift Zone; *WDV* = West Diverging Volcanism; *STZ* = South Transitional Zone; and *VM* = Vestmann Islands. The volcanoes Krafla, Askja, and Hekla are also identified as well as the city of Reykjavik and the Reykjanes Peninsula. The stippling of the *STZ*, *VM*, and *WDV* indicates enrichment in alkali metals of the volcanic rocks. The volcanic rocks in these zones have different isotope ratios of Sr, Nd, and O indicating that they originated from different types of source rocks (*Source:* data from Hémond et al. 1988, 1993)

lands of Surtsey, which formed as a result of volcanic eruptions between November 1963 and June 1967, and Heimaey, where an eruption started on January 23, 1973.

O'Nions et al. (1973) reported that the alkali-rich volcanic rocks on the islands of Surtsey and Heimaey have constant $^{87}Sr/^{86}Sr$ ratios with a mean of 0.70313 ±0.00004 (corrected to 0.70800 for E&A). The authors attributed the alkali-enrichment of the lavas on Surtsey and Heimaey to a small degree of partial melting (4 to 6%), whereas the tholeiites formed by about 20 to 30% melt fractions.

The wide range of chemical compositions of the volcanic rocks on Iceland is well represented by the rocks of the volcano Krafla located in the northeastern axial rift zone of the island (Fig. 2.11). Nicholson et al. (1991) reported that the lavas erupted at this volcanic center during the past 70 000 years range from Mg-rich olivine tholeiites to rhyolites whose MgO concentrations approach zero. However, the data in Fig. 2.12 indicate that the $^{87}Sr/^{86}Sr$ ratios of the entire suite of volcanic rocks at Krafla range only from 0.70309 to 0.70326 and are independent of the MgO concentrations which vary from 9.73 to 0.04%. Nicholson et al. (1991) concluded that the compositional diversity of the volcanic rocks at Krafla can be attributed to fractional crystallization of basalt magma. However, the enrichment of the most highly differentiated rocks in certain incompatible trace elements including Rb, as well as the isotope compositions of O and Th, indicate that the magma under Krafla is also assimilating hydrothermally altered rocks exposed in the walls of magma chambers and fissures. These wallrocks are so young that their $^{87}Sr/^{86}Sr$ ratios have increased very little since the rocks were formed.

The study of Nicholson et al. (1991) therefore highlights the fact that the chemical and isotope compositions of the volcanic rocks on Iceland depend not only on the magma sources in the mantle and on the degree of partial melting, but are also modified to some extent by assimilation of hydrothermally altered rocks underlying the volcanic centers and by fractional crystallization of the resulting magmas.

The conclusion of Nicholson et al. (1991) that the magmas under Krafla assimilated hydrothermally altered rocks is consistent with the range of $\delta^{18}O$ values of Icelandic basalts and rhyolites in Fig. 2.13 based on data from Condomines et al. (1983), Nicholson et al. (1991), and Sigmarsson et al. (1992). The $\delta^{18}O$ values (relative to SMOW) of the Icelandic basalts range from about +5.5‰ down to about +3‰, whereas the felsic rocks have even lower $\delta^{18}O$ values between +4 and about +1.4‰.

The lowering of the $\delta^{18}O$ values of volcanic rocks on Iceland originally reported by Muehlenbachs et al. (1974) is consistent with the way oxygen isotopes are frac-

Fig. 2.12.
Relation between the $^{87}Sr/^{86}Sr$ ratio and MgO concentration of the volcanic rocks extruded by the volcano Krafla in northeastern Iceland. Since the $^{87}Sr/^{86}Sr$ ratios are constant and do not vary with the MgO concentrations, these rocks apparently formed by fractional crystallization of a magma that was isotopically homogeneous with respect to Sr. The elevated average $^{87}Sr/^{86}Sr$ ratio of the volcanic rocks of Krafla indicates that the magma originated from Rb-enriched source rocks, such as a plume or blob from the lower mantle (*Source:* Nicholson et al. 1991)

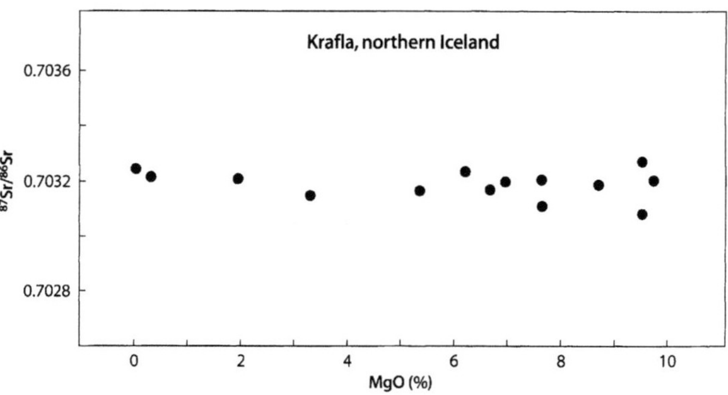

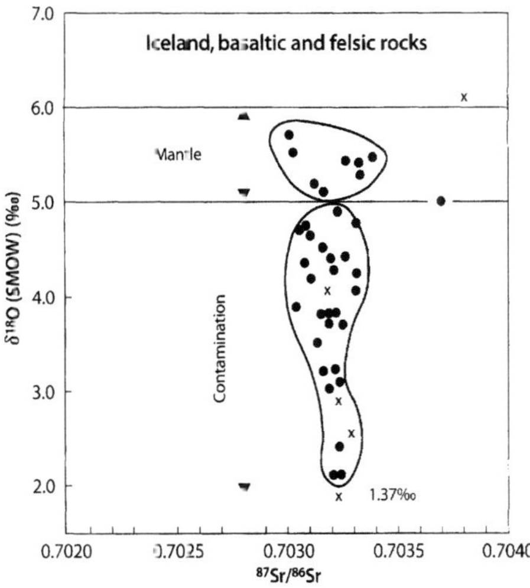

Fig. 2.13. Isotope compositions of Sr and O of basaltic rocks (*solid circles*) and felsic rocks (*crosses*) on Iceland. The $^{87}Sr/^{86}Sr$ ratios vary between narrow limits from about 0.7030 to 0.7034. However, the $\delta^{18}O$ values range widely from typical mantle values (+5 to +6‰) to low values of about +2.5‰. The variation of the $\delta^{18}O$ values is attributable to contamination of magmas by assimilation of hydrothermally altered volcanic rocks whose $^{87}Sr/^{86}Sr$ ratios did not differ s gnificantly from those of mantle-derived magmas (*Sources:* data from Condomines et al. 1983; Nicholson et al. 1991; Sigmarsson et al. 1992)

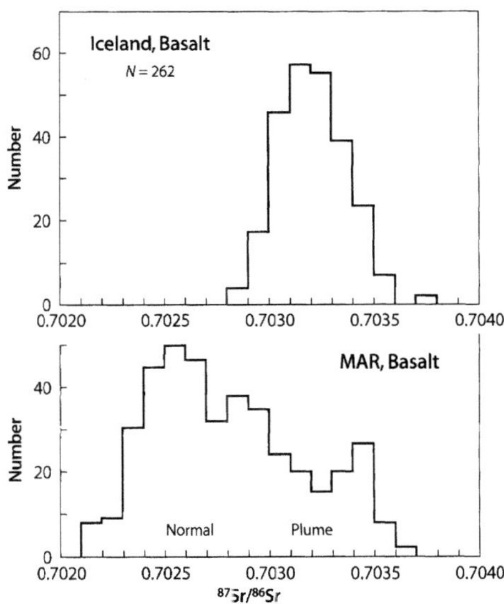

Fig. 2.14. Isotope ratios of Sr in the basalts of Iceland, (excluding the Reykjanes and Kolbeinsey Ridges) compared to basalts from the Mid-Atlantic Ridge (excluding oceanic islands and seamounts). The $^{87}Sr/^{86}Sr$ ratios of the Iceland basalts range from greater than 0.70280 to less than 0.70380, relative to 0.7080 for E&A and 0.71025 for NBS 987. Therefore, these data reveal the internal heterogeneity of the rocks in the plume that underlies Iceland (*Sources:* O'Nions et al. 1973; Hart et al. 1973; O'Nions and Pankhurst 1973; O'Nions and Grönvold 1973; Sun and Jahn 1975; Wood et al. 1979; Zindler et al. 1979; Condomines et al. 1983; Nicholson et al. 1991; Elliott et al. 1991; Sigmarsson et al. 1992; Hémond et al. 1993; Taylor et al. 1997)

tionated in nature. The upper range of $\delta^{18}O$ values (+5 to +6‰) is typical of igneous rocks derived from the mantle, whereas the $\delta^{18}O$ values of rocks in the continental crust range up to about +20‰ or more, depending on their mineral composition (Faure 1986). Compared to igneous rocks, meteoric water is depleted in ^{18}O because of isotope fractionation during evaporation of water and condensation of vapor and therefore has negative $\delta^{18}O$ values. When volcanic rocks, such as those on Iceland, interact with groundwater at elevated temperatures in systems having large water/rock ratios, the $\delta^{18}O$ values of the rocks are lowered and may approach those of the water. Therefore, hydrothermally altered rocks are depleted in ^{18}O and have lower $\delta^{18}O$ values than unaltered volcanic rocks derived from the mantle. When mantle-derived basalt magmas assimilate such hydrothermally altered volcanic rocks, the $\delta^{18}O$ values of the rocks that form from such magmas are lowered as demonstrated in Fig. 2.13.

The availability of a large number of measurements of isotopic ratios of Sr in the volcanic rocks of Iceland permits a meaningful comparison with the $^{87}Sr/^{86}Sr$ ratios of MORBs collected along the Mid-Atlantic Ridge. The $^{87}Sr/^{86}Sr$ ratios of Icelandic basalts in Fig. 2.14 range from greater than 0.7028 to less than 0.70380 and therefore overlap the $^{87}Sr/^{86}Sr$ of plume MORBs on the Mid-

Atlantic Ridge. However, the wide range of $^{87}Sr/^{86}Sr$ ratios of the Icelandic basalts suggests either an appreciable degree of heterogeneity of the magma sources or variable contamination of magmas by assimilation of rocks from the oceanic crust, or both. Hémond et al. (1988, 1993) investigated the apparent isotopic heterogeneity of basalts from different neovolcanic zones in Iceland identified in Fig. 2.11 and found significant variations of the average isotope ratios of Sr, Nd, and O (see also Furman et al. 1995).

In general, the alkali-rich rocks have high $^{87}Sr/^{86}Sr$ and low $^{143}Nd/^{144}Nd$ ratios indicating that they formed from more Rb-enriched source rocks in the Iceland Plume than Mg-rich picrites which originated from more Rb-depleted sources (Hémond et al. 1993). Consequently, the Iceland Plume does not have a uniform composition, but contains at least two kinds of sources characterized by differences in their Rb/Sr and Sm/Nd ratios (Hards et al. 1995). However, the data in Fig. 2.14 make clear that both magma sources in the Iceland Plume have higher $^{87}Sr/^{86}Sr$ ratios than normal MORBs on the Mid-Atlantic Ridge. In addition, the low $\delta^{18}O$ values of the lavas on the volcanoes Krafla and Askja and elsewhere in Iceland indicate that, in some cases, the elevated $^{87}Sr/^{86}Sr$ ratios are attributable to assimilation of hydrothermally altered volcanic rocks by magmas derived from the plume.

2.3.3 The Iceland Plume

The present diameter of Iceland is about 800 km, but the influence of the Iceland Plume extends to the west coast of Greenland, to the Outer Hebrides of Scotland and northern Ireland, as well as to the coast of Norway (Prestvik et al. 1999). The reason is that the expansion of the mushroom-shaped head of the Iceland Plume contributed to the stretching of the overlying lithosphere and provided the heat for magma formation in a large area that now borders the North Atlantic Ocean (Taylor et al. 1997).

A thick sequence of basalt lava flows along the Blosseville Coast and at other sites in East Greenland north of Scoresbysund are products of volcanic activity associated with the Iceland Plume at about 50 Ma. Thirlwall et al. (1994) used these rocks to study the interaction between magmas of the Iceland Plume and the continental lithosphere of East Greenland. The initial $^{87}Sr/^{86}Sr$ ratios of some of these rocks increase with rising concentrations of SiO_2 indicating that the magmas assimilated varying amounts of rocks from the continental crust of Greenland. The details of this study are presented in Sect. 5.4.8.

The studies of the volcanic rocks of Iceland have refined our understanding of the interaction of plumes and the overlying lithospheric mantle and oceanic crust. Solid plumes rising from depth in the asthenosphere have diameters measured in terms of tens to hundreds of kilometers. When the plumes reach the overlying lithosphere, they spread out like the caps of mushrooms. The lateral flow in the plume heads thins the overlying lithosphere by stretching and eroding it. In addition, plumes transport heat into the overlying lithosphere and cause melting by decompression of plume rocks and of entrained blocks of asthenospheric rocks. The resulting magmas may have a range of isotopic compositions of Sr, Nd, and Pb because they formed from heterogeneous mixtures of rocks in the head of the plume and/or because the magmas subsequently assimilated rocks of the lithospheric mantle, the oceanic crust, or even of continental crust depending on the circumstances.

The head of the Iceland Plume was able to penetrate the overlying lithosphere and thereby actively participated in the opening of the North Atlantic Ocean as its head expanded to about 2 400 km stretching from the west coast of Greenland to Norway (Thirlwall et al. 1994). The relation of Jan Mayen to the Iceland Plume was considered by Maaløe et al. (1986), Skogseid and Eldholm (1987), and Haase et al. (1996).

The history of the Iceland Plume and its effect on the opening of the Atlantic Ocean were the subject of the second Arthur Holmes European Research Meeting in Reykjavik in early July 1994. A collection of short reports presented at this meeting was subsequently published in 1995 in volume 152 of the "Journal of the Geological Society of London" (White and Morton 1995).

Most of the islands in the Atlantic Ocean (including the Canary Islands, the Azores, and the Cape Verde Islands) are not located on the Mid-Atlantic Ridge. In fact, Iceland appears to be the only volcanic island in the Atlantic Ocean that is actually positioned on the ridge. In addition, the volcanic rocks on most of the other islands in the Atlantic Ocean are predominantly alkali-rich and silica undersaturated, whereas those of Iceland are generally silica saturated and not exceptionally alkali rich. Therefore, Iceland is not typical of the other islands in the Atlantic Ocean.

2.4   Canary Islands

In contrast to Iceland which is located on the Mid-Atlantic Ridge, the Canary Islands in Fig. 2.15, as well as Madeira (Hughes and Brown 1972), formed on the passive margin of the African Plate. The island of Gran Canaria lies only about 150 km west of the continental margin of Africa which consists of metamorphic rocks ranging in age from 3.4 to 1.8 Ga. In addition, the oceanic crust in this area is overlain by marine and terrigenous sediment (Late Jurassic to Tertiary) up to 10 km thick. Therefore, the conditions on the Canary Islands are right for mantle-derived magmas to assimilate not only hydrothermally altered basalt (e.g. Iceland) but also terrigenous sediment derived from Africa.

2.4.1   Volcanic Rocks of Gran Canaria and Tenerife

The island of Gran Canaria is a large shield volcano composed of tholeiite basalt of Miocene age (McDougall and

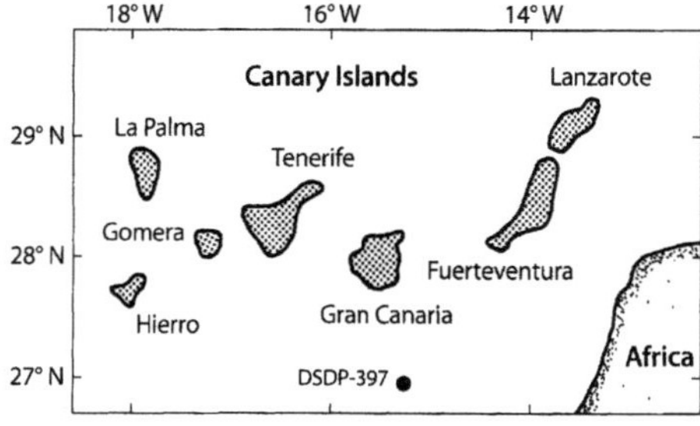

Fig. 2.15. Map of the Canary Islands off the west coast of North Africa (*Source:* adapted from Hoernle et al. 1991)

Schmincke 1976). The subaerial Miocene basalts (Guigui and Hogarzales Formations) are overlain by a sequence of late Miocene ignimbrites composed of rhyolite/trachyte (Mogan Formation) and of phonolite/trachytes (Fataga and Tejeda Formations) (Schmincke 1982, 1987). After a period of quiescence from about 9 to 6 million years, volcanic activity resumed during the Pliocene Epoch with the eruption of alkali-rich to tholeiite lavas of the El Tablero, Roque Nublo, and Los Llanos Formations. The most recent Quaternary lavas consist of basanites, tephrites, and nephelinites. (Thirlwall et al. 1997; Hoernle et al. 1991; Cousens et al. 1990).

The average initial $^{87}Sr/^{86}Sr$ ratios of the Miocene volcanic rocks of Gran Canaria decreased with time from 0.70334 ±0.00002 ($2\bar{\sigma}, N = 96$) in the Guigui, Hogarzales, and Mogan Formations to 0.70308 ± 0.00002 ($2\bar{\sigma}, N = 12$) in the Fataga and Tejeda Formations. The decrease of the $^{87}Sr/^{86}Sr$ ratio started at about 13 Ma in the uppermost lava flows of the Mogan Formation and signaled a change in magma sources from Rb-enriched rocks of a plume or blob under Gran Canaria to less enriched rocks.

The $\delta^{18}O$ values of clinopyroxene in the basalts of the Guigui and Hogarzales Formations in Fig. 2.16 are clustered between +5 to +5‰ but do increase to +6.81‰ in a few specimens (Thirlwall et al. 1997). The positive correlation between the isotope compositions of Sr and O is evidence that some Miocene magmas assimilated marine sediment enriched in ^{18}O. Hoernle et al. (1991) reported that Jurassic to Tertiary sediment recovered at DSDP 397 (located 100 km south of Gran Canaria) has $^{87}Sr/^{86}Sr$ ratios ranging from 0.709288 to 0.723619 relative to 0.710250 for NBS 987 with a weighted mean of 0.71374 and an average Sr concentration of 820 ppm. Thirlwall et al. (1997) assumed a $\delta^{18}O$ value of +22‰ for the terrigenous sediment in DSDP core 397 and determined that the magma of sample T 10 (Fig. 2.16) from Tazatico on Gran Canaria could have assimilated ~8% of such terrigenous sediment. Thirlwall et al. (1997) also noted that the evidence for contamination of the Miocene magmas by terrigenous sediment obscures information derivable from these rocks about their sources in the mantle. The isotope composition of Pb is especially sensitive to contamination because the addition of about 8% of sediment to basalt magma provides about 50% of the Pb it contains.

The Pliocene volcanic rocks of Gran Canaria range in composition from alkali-rich to alkali-poor and have a much smaller total volume than the preceding Miocene basalts and ignimbrites. The data of Hoernle et al. (1991) in Fig. 2.17 indicate that the silica concentrations of the

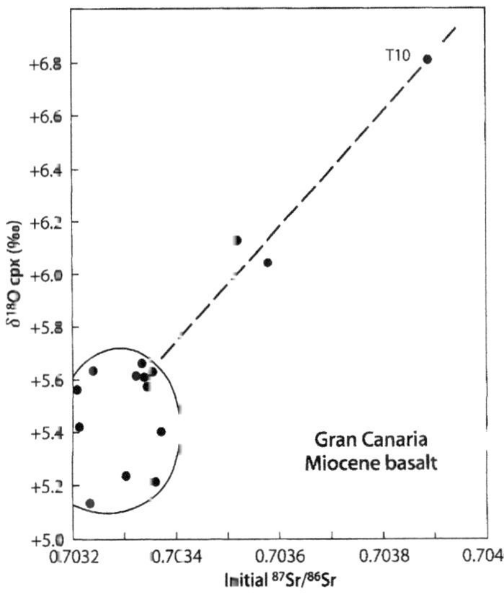

Fig. 2.16. Positive correlation between the isotope compositions of Sr and O in clinopyroxene Miocene basalts on Gran Canaria, Canary Islands. This evidence indicates that the magma of specimen T 10 assimilated about 8% of terrigenous sediment having elevated $^{87}Sr/^{86}Sr$ and $\delta^{18}O$ values (*Source:* data and interpretation from Thirlwall et al. 1997)

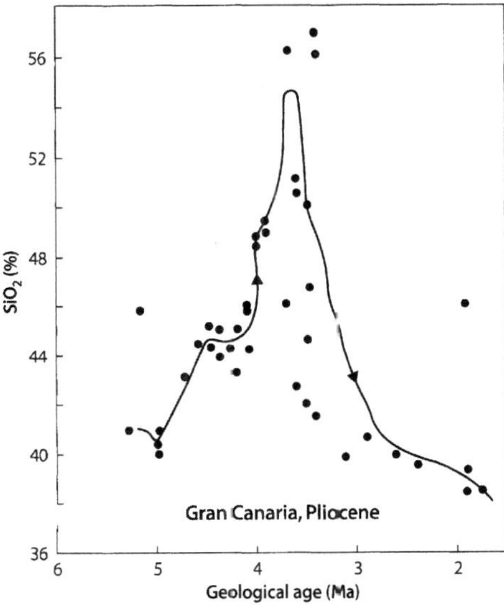

Fig. 2.17. Age-dependent variation of the concentration of SiO_2 in the Pliocene lavas on Gran Canaria. The chemical composition of the Pliocene lavas changed from undersaturated alkali-rich nephelinites and basanites to high-silica tholeiites and then back to low-silica melilitites and nephelinites. The systematic variation of chemical compositions of the Pliocene lavas on Gran Canaria may have resulted from varying degrees of partial melting in a rising blob of plume rocks (*Sources:* Hoernle et al. 1991; Hoernle and Schmincke 1993a)

Pliocene lavas increased with time from about 40% at about 5 Ma and approached 60% in a phonolite flow extruded at 3.5 Ma. Subsequently, the silica concentrations declined to about 38% at 1.7 Ma. Hoernle and Schmincke (1993a) explained this pattern of variation by means of decompression melting in a blob of plume rocks. Initially, silica-undersaturated lavas (nephelinite and basanite) formed by small degrees of partial melting in the upper margin of a rising blob. The melt zone then passed into the hot interior of the blob where tholeiite and alkali basalt magmas formed by larger degrees of partial melting than had occurred in the cooler margin of the blob. Eventually, the melt zone shifted to the lower margin of the rising blob to form a second suite of silica-undersaturated melilitites and nephelinites by small degrees of partial melting.

The petrogenetic model proposed by Hoernle et al. (1991) predicts that the silica-saturated lavas of Gran Canaria should have higher $^{87}Sr/^{86}Sr$ ratios than the alkali-rich rocks because they originated from Rb-enriched plume rocks in the center of the blob, whereas the alkalic rocks originated from mixtures of plume rocks and lithospheric mantle. This prediction is supported by the data in Fig. 2.18 which clearly shows that the $^{87}Sr/^{86}Sr$ ratios of lavas containing more than about 46% SiO_2 rise with increasing silica concentration. However, the positive correlation between SiO_2 concentrations and $^{87}Sr/^{86}Sr$ ratios is also explainable by assimilation of silica-rich rocks having elevated $^{87}Sr/^{86}Sr$ ratios (e.g. terrigenous sediment).

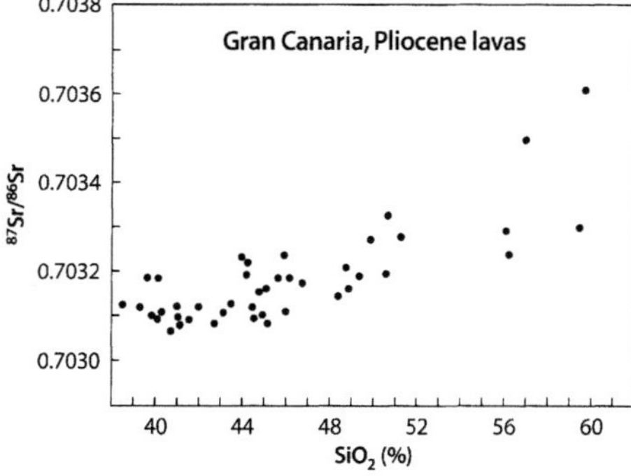

Fig. 2.18. Relation between the $^{87}Sr/^{86}Sr$ ratios and silica concentrations of lavas of Pliocene age on Gran Canaria. The distribution of data points indicates that rocks having $SiO_2 > 46\%$ have increasing $^{87}Sr/^{86}Sr$ ratios. This relation can be taken as evidence of assimilation of silica-rich rocks having elevated $^{87}Sr/^{86}Sr$ ratios (e.g. terrigenous sediment of Late Jurassic to Cretaceous age) by a mantle-derived magma. Alternatively, the relation between the silica concentration and the $^{87}Sr/^{86}Sr$ ratio may have been caused by different degrees of partial melting in the border zone and in the interior of a blob of plume rocks as proposed by Hoernle and Schmincke (1993a) (*Source:* data from Hoernle et al. 1991)

The Miocene basalts of Gran Canaria may also have originated by partial melting in a blob of plume rocks. In other words, the decrease of the $^{87}Sr/^{86}Sr$ ratios of alkali-rich lavas of the upper Mogan Formation and of the Fataga and Tejeda Formations (Cousens et al. 1990 and Hoernle et al. 1991) is also attributable to a change in magma sources from Rb-enriched rocks of a plume or blob to more depleted rocks of the lithospheric mantle. However, the positive correlation of $^{87}Sr/^{86}Sr$ ratios and $\delta^{18}O$ values in (Fig. 2.16) in some of the basalt flows of the Hogarzales Formation observed by Thirlwall et al. (1997) is evidence for assimilation of terrigenous sediment by some of the mantle-derived magmas.

The Quaternary cycle of volcanic activity on Gran Canaria has not yet progressed far enough to recognize the shift to higher $^{87}Sr/^{86}Sr$ ratios that signals the change from small degrees of partial melting in the border zone to high degrees of partial melting in the core of the plume or blob.

In conclusion, the data indicate that the magmas on Gran Canaria originated by varying degrees of partial melting of source rocks that had different $^{87}Sr/^{86}Sr$ ratios. Some of these magmas also assimilated varying amounts of terrigenous sediment that overlies the oceanic crust at this site. In addition, fractional crystallization of magma occurred in the upper mantle prior to eruption (Hoernle and Schmincke 1993b).

The island of Tenerife located northwest of Gran Canaria in Fig. 2.15 contains the Las Cañadas Volcanoes. The petrology of the lavas was described by Ridley (1970), whereas Oversby et al. (1971) reported isotope ratios of Pb in the volcanic rocks of Tenerife. More recently, Araña et al. (1994) found evidence for magma mixing in the trace element composition of the lavas on Tenerife, whereas Palacz and Wolff (1989) discussed the reasons for isotopic disequilibrium in the Granadilla Pumice.

2.4.2 Fuerteventura

The island of Fuerteventura (Fig. 2.15) began to form at about 80 Ma making it the oldest known oceanic island (LeBas et al. 1986; Hoernle and Tilton 1991). Its basal complex of alkali-rich volcanic rocks (70 to 25 Ma) is underlain by and interbedded with Late Cretaceous marine chalk. It was intruded by carbonatite-ijolite-syenite complexes at 60 and 30 Ma as well as by dikes and small mafic plutons. The subaerial basalts range in age from 20.6 Ma (Miocene) to Recent. Isotope compositions of O and Pb in the volcanic rocks of Fuerteventura were reported by Javoy et al. (1986) and Sun (1980), respectively.

The initial $^{87}Sr/^{86}Sr$ ratios of the rocks in the basal complex (including two carbonatites) are virtually constant and have an average value of 0.70327 ±0.00002

$(2\bar{\sigma}, N = 24)$ relative to 0.71025 for NBS 987 (Hoernle and Tilton 1991). Consequently, the isotope composition of Sr of these rocks provides no evidence for contamination of the basal complex by marine chalk and terrigenous sediment. The calcites in two carbonatites that formed at 60 and 30 Ma have extremely high Sr concentrations (14 280 to 36 000 ppm) and nearly identical $^{87}Sr/^{86}Sr$ ratios of 0.70321 ±0.00002 and 0.70328 ±0.00001, respectively.

The initial $^{87}Sr/^{86}Sr$ ratios of the subaerial basalts (16 to 0 Ma) are lower than those of the rocks in the basal complex and are also more variable. The basalt specimens analyzed by Hoernle and Tilton (1991) have an average initial $^{87}Sr/^{86}Sr$ ratio of 0.70306 ±0.0004 ($2\bar{\sigma}$, $N = 10$) relative to 0.71025 for NBS 987. The silica concentrations of the subaerial basalts range from 42.0 to 49.8% but do not correlate significantly with the $^{87}Sr/^{86}Sr$ ratios.

Hoernle et al. (1991) attributed the origin of volcanic and plutonic rocks of the basal complex of Fuerteventura (70 to 25 Ma) to partial melting of plume rocks that may now be located under the island of La Palma in Fig. 2.15 at the western end of the Canary Island chain (Schmincke 1982). The younger lavas on Fuerteventura (16 to 0 Ma) presumably originated from a mixture of plume rocks and depleted lithospheric mantle. The occurrence of the most recent volcanic activity on Fuerteventura implies that the Rb-enriched rocks of the plume head have flowed eastward with the movement of the African Plate and have mixed with depleted rocks of the lithospheric mantle.

2.4.3 Ultramafic Inclusions, Hierro

Ultramafic inclusions in lava flows are a potential source of information about the mantle from which the lavas originated. In principle, such inclusions may represent different kinds of rocks in the mantle and in the overlying crust:

1. Restites of mantle rocks from which a partial melt was previously removed;
2. Unaltered mantle rocks located above the melt zone in the mantle;
3. Cumulates that formed by fractional crystallization of mantle-derived magma;
4. Rocks derived from different levels in the oceanic or continental crust.

The ultramafic inclusions are typically composed of olivine, clinopyroxene, orthopyroxene, spinel, etc. most of which have low concentrations of Rb and Sr. Therefore, ultramafic inclusions are sensitive to contamination by the magma that transported them to the surface as well as by meteoric precipitation and/or groundwater.

The ultramafic inclusions on the island of Hierro (Fig. 2.15) analyzed by Whitehouse and Neumann (1995) were collected from an alkali basalt flow of Quaternary age about 300 m above sealevel in the northwestern part of the island. Prior to analysis, all of the rock powders were leached sequentially with cold 0.1 and 0.5 N HCl followed by washing with ultrapure water in order to remove contaminant Sr. The results indicate that leaching caused the $^{87}Sr/^{86}Sr$ ratios of the rocks to decrease as a result of removal of a contaminant. The analytical data of Whitehouse and Neumann (1995) indicate that seven leached harzburgites and one lherzolite have an average $^{87}Sr/^{86}Sr$ ratio of 0.70314 ±0.00007 ($2\bar{\sigma}$, $N = 8$) relative to 0.71025 for NBS 987. The average $^{87}Sr/^{86}Sr$ ratios (after leaching) of a second group of dunites, pyroxenites, and wehrlites (same lava flow) is 0.70300 ±0.00002 ($2\bar{\sigma}$, $N = 6$) and is indistinguishable from the $^{87}Sr/^{86}Sr$ ratio (0.70298 ±0.00001) of the host basalt. Neumann (1991), Hansteen et al. (1991), and Wulff-Pedersen et al. (1996) concluded that the inclusions in some cases, had been contaminated by infiltration of silicate melt. The origin of this highly siliceous melt was discussed by Neumann and Wulff-Pedersen (1997).

2.5 Islands of the Atlantic Ocean

The Atlantic Ocean contains a large number of island groups whose origin is attributable to magma formation in plumes that are not necessarily associated with the Mid-Atlantic Ridge. The islands to be considered here include the Azores, the Cape Verde Islands, Fernando de Noronha, Ascension, St. Helena, Tristan da Cunha, Gough, and Bouvet.

2.5.1 Azores Archipelago

The Azores Archipelago in Fig. 2.19 consists of nine islands that rise from a submarine plateau which straddles the Mid-Atlantic Ridge. The volcanic rocks on the Azore Islands range widely in chemical composition from alkali basalt to comendite and pantellerite (alkali-rich rhyolites). The petrogenesis of the basalt-rhyolite suites on the island of Terceira was recently discussed by Mungall and Martin (1994). The volcano Capelinhos on the island of Faial erupted for a period of 13 months from 1957 to 1958 (Machado et al. 1962). The history of volcanic activity on the eastern group of islands of the Azores (Santa Maria, São Miguel and the Formigas Islands) was discussed by Abdel-Monem et al. (1975) based on K-Ar dates, whereas Oversby (1971) reported isotope ratios of Pb on Faial. The first determinations of the $^{87}Sr/^{86}Sr$ ratios of volcanic rocks on the Azores by Faure and Hurley (1963), O'Nions and Pankhurst (1974), White

Fig. 2.19.
Map of the Azores in the North Atlantic Ocean. The Azores Plateau is defined by the depth contour at 2 000 m. The islands are identified by number: *1.* Corvo; *2.* Flores; *3.* Faial; *4.* Pico; *5.* Jorge; *6.* Graciosa; *7.* Terceira; *8.* São Miguel; *9.* Santa Maria (*Source:* adapted from Turner et al. 1997)

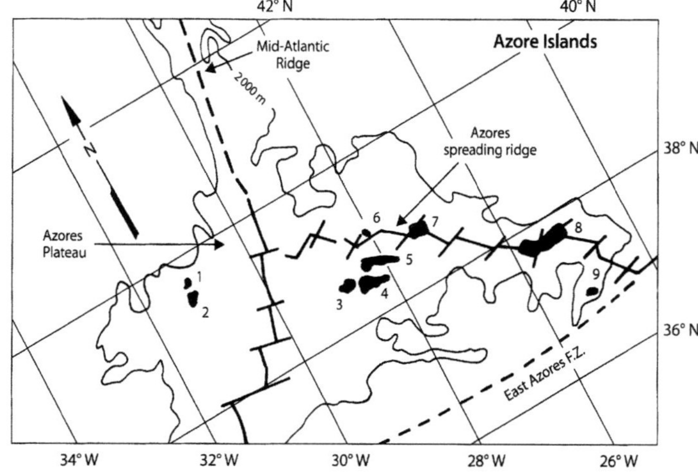

Fig. 2.20. Range of $^{87}Sr/^{86}Sr$ ratios of volcanic rocks (Pleistocene ▶ to Recent) on the Azore Islands in the Atlantic Ocean. The lavas of São Jorge, Graciosa, Terceira, Santa Maria, Flores, and Corvo originated from the Azore Plume whose $^{87}Sr/^{86}Sr$ ratio varies between 0.7033 and 0.7036 relative to 0.71025 for NBS 987. The volcanic rocks of Faial and Pico are enriched in radiogenic ^{87}Sr with isotope ratios between 0.7036 and 0.7040. The lavas of São Miguel have a multi-modal distribution of $^{87}Sr/^{86}Sr$ ratios between 0.7032 and 0.7054 because the lavas of the eastern end of the island originated by melting in a block of subcontinental lithospheric mantle, whereas those at the western end originated primarily from the Azores Plume (*Sources:* Hawkesworth et al. 1979a; White et al. 1979; White and Hofmann 1982; Widom et al. 1997; Turner et al. 1997)

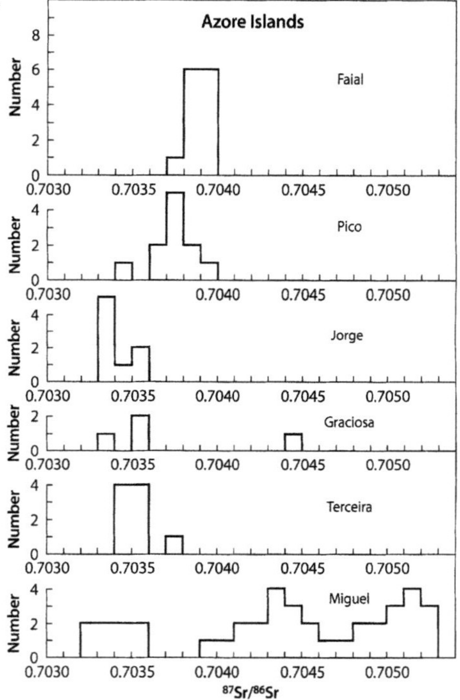

et al. (1976, 1979), Hawkesworth et al. (1979a), White and Hofmann (1982), and Dupré et al. (1982) yielded comparatively high values ranging from 0.70320 to 0.70525 consistent with the elevated $^{87}Sr/^{86}Sr$ ratios of basalt of the Azores Platform on the Mid-Atlantic Ridge in Fig. 2.3 (White and Schilling 1978). Hawkesworth et al. (1979a) demonstrated that the $^{87}Sr/^{86}Sr$ ratios of volcanic rocks on São Miguel, the largest of the Azore Islands, increase from west to east across the island. They also demonstrated that the high $^{87}Sr/^{86}Sr$ ratios on this island cannot be accounted for by assimilation of Sr from seawater or oceanic sediment and concluded that the regional trend in the $^{87}Sr/^{86}Sr$ ratios on São Miguel is caused by mixing of Sr derived from different magma sources in the mantle.

The $^{87}Sr/^{86}Sr$ ratios of the lavas on the Azore Islands range widely from 0.7032 to 0.7054. The histograms in Fig. 2.20 indicate that comparatively low $^{87}Sr/^{86}Sr$ ratios (0.7033 to 0.7036) characterize the lavas on São Jorge, Graciosa, Terceira, Santa Maria, Flores, and Corvo. These values are similar to those of MORBs on the Azores Plateau (Fig. 2.3) consistent with their derivation from the Azores Plume without significant contamination by Sr derived from other sources. The rocks on Faial and Pico have somewhat higher $^{87}Sr/^{86}Sr$ ratios (0.7036 to 0.7044) than the islands listed above.

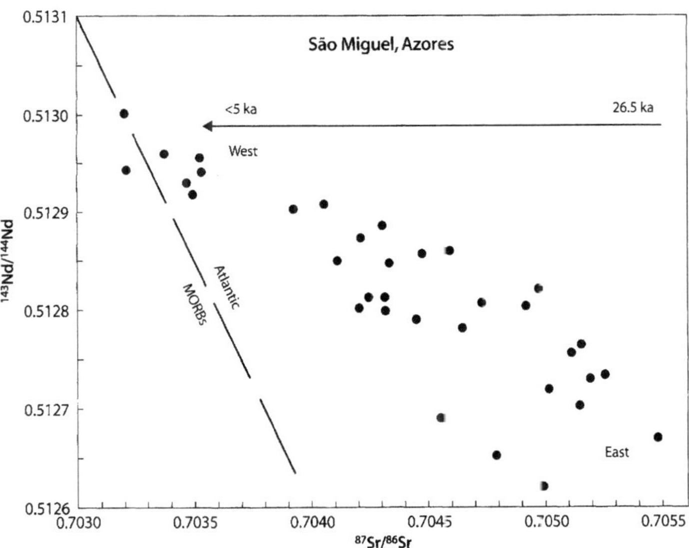

Fig. 2.21.
Isotope ratios of Sr and Nd of basalts on São Miguel, Azores. The isotopic compositions of Sr and Nd (as well as of Pb) vary systematically with distance across the island from west to east and with age from <5 to 26.5 ka. The range of $^{87}Sr/^{86}Sr$ ratios from 0.703202 to 0.705474 implied a change in the magma sources as the island was transported in a southeasterly direction away from the Mid-Atlantic Ridge (*Sources:* data from Widom et al. 1997; Turner et al. 1997; Hawkesworth et al. 1979a; White and Hofmann 1982)

São Miguel is noteworthy among the islands of the Azores Archipelago because the isotope ratios of Sr, Nd, and Pb of rocks on this island all increase from west to east (Turner et al. 1997; Widom et al. 1997). In addition, the lavas on São Miguel are more enriched in K and in the light rare earths than the rocks of the other islands (Schmincke 1973; Flower et al. 1976). The island is also cut by the Azores spreading ridge (Fig. 2.19) which is the boundary between the Eurasian and African Plates.

The $^{87}Sr/^{86}Sr$ and $^{143}Nd/^{144}Nd$ ratios of volcanic rocks on São Miguel form an array in Fig. 2.21 that has a distinctly shallower slope than the array of Atlantic MORBs in Fig. 2.4. Therefore, the oldest rocks (26.5 ka) at the eastern end of São Miguel formed from magma sources that had an anomalously high $^{87}Sr/^{86}Sr$ ratio of about 0.7055 and a low $^{143}Nd/^{144}Nd$ ratio of 0.51260 relative to $^{146}Nd/^{144}Nd = 0.72190$. As São Miguel was progressively displaced to the southeast by seafloor spreading, the $^{87}Sr/^{86}Sr$ ratios of the younger lavas decreased, whereas their $^{143}Nd/^{144}Nd$ ratios increased. The lavas that were erupted at the western end of the island less than 5000 years ago have $^{87}Sr/^{86}Sr$ and $^{143}Nd/^{144}Nd$ ratios that coincide with those of plume MORBs along the Mid-Atlantic Ridge (Figs. 2.4 and 2.20).

The $^{87}Sr/^{86}Sr$ ratios of the lavas on São Miguel do not correlate with the concentrations of SiO_2 or with the reciprocals of their Sr concentrations as expected in case of magmatic differentiation by AFC. However, Storey et al. (1989a) concluded that basalt magma in the Agua de Pau Volcano was contaminated by assimilating previously formed trachyte flows. More recently, Turner et al. (1997) postulated that the unusually high $^{87}Sr/^{86}Sr$ ratios (>0.7050) of the lavas on the east end of São Miguel require the presence of 5 to 10% of recycled upper crustal sediment in the magma sources. Similarly, Widom et al. (1997) attributed the elevated $^{87}Sr/^{86}Sr$ ratios of eastern São Miguel to the incorporation of about 5% of subducted terrigenous sediment into the magma from which these lavas originated.

The isotopic and trace-element data of the lavas on São Miguel are also attributable to melting of hydrothermally altered rocks in a block of *subcontinental* lithospheric mantle. Widom et al. (1997) proposed that a block of this material was originally delaminated from the African or Eurasian Plate during opening of the Atlantic Ocean and now resides as a localized contaminant within the lithospheric mantle under São Miguel. The heat of the Azores Plume caused melting of these rocks and the resulting magma then mixed with melts derived from the plume to produce the observed range of isotope ratios of Sr, Nd, and Pb.

Based on a review of isotope ratios of Sr-Nd-Pb and of trace-element concentrations, Davies et al. (1989) postulated that the magmas erupted on the Azore Islands represent mixtures of four components: (1) depleted lithospheric mantle (MORB source); (2) recycled oceanic crust; and (3) two different kinds of subcontinental lithospheric mantle. In general, the Azores are a good example of the phenomenon that the isotopic composition and trace element concentrations of volcanic rocks on oceanic islands represent mixtures of four or even five different components. This subject is considered in more detail in Sect. 2.6.1.

2.5.2   Cape Verde Archipelago

The Cape Verde Islands in Fig. 2.22 are located only about 500 km west of Senegal in West Africa and more than 2 000 km east of the Mid-Atlantic Ridge. The ten islands of this archipelago formed during the Tertiary Period by volcanic activity on a platform composed of oceanic crust, plutonic rocks, and sediment of Jurassic age (Mitchell et al. 1983). According to Duncan (1984), the plate on which the Cape Verde Islands are located drifted only about 200 km to the northeast in the past 50 to 60 million years. Consequently, the ages of the islands decrease from northeast to southwest making Fogo and Brava in Fig. 2.22 the only islands that have been volcanically active in historical time (Gerlach et al. 1988a).

The volcanic rocks of the Cape Verde Islands are highly alkali rich, range in composition from tholeiites to phonolite (Gunn and Watkins 1976), and include carbonatites that were studied by Allègre et al. (1971), Javoy et al. (1985), and Kogarko (1993). The measurements of $^{87}Sr/^{86}Sr$ ratios by Klerkx et al. (1974a) yielded comparatively low values between 0.7029 and 0.7032 for alkali-rich volcanic rocks relative to E&A = 0.7080. The $^{87}Sr/^{86}Sr$ ratios reported by Gerlach et al. (1988a) for volcanic rocks from five of the ten Cape Verde Islands range more widely from 0.70292 to 0.70393 (relative to 0.7080 for E&A) and have a distinctly bimodal distribution. The combined data of Klerkx et al. (1974a) and Gerlach et al. (1988a) in Fig. 2.23 indicate that the rocks of S. Vicente,

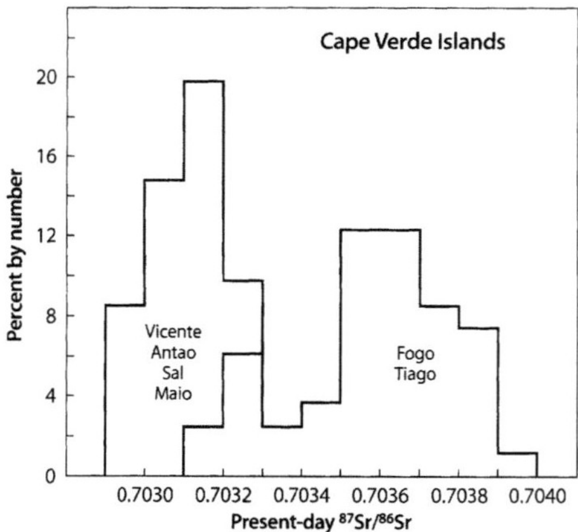

Fig. 2.23. Variation of present-day $^{87}Sr/^{86}Sr$ ratios of volcanic and plutonic rocks on the islands of the Cape Verde Archipelago. The rocks on the islands of S. Vicente, S. Antao, Sal, and Maio have relatively low $^{87}Sr/^{86}Sr$ ratios between 0.7029 and 0.7033. Most of the rock specimens from the southern islands Fogo and Tiago have distinctly higher $^{87}Sr/^{86}Sr$ ratios, but some values range down to 0.7031. All $^{87}Sr/^{86}Sr$ ratios have been adjusted to 0.7080 for the E&A standard (*Sources:* Klerkx et al. 1974a; Gerlach et al. 1988a)

S. Antao, Sal, and Maio have low $^{87}Sr/^{86}Sr$ ratios between 0.7029 and 0.7033, whereas those of the younger islands (Fogo and Tiago) have high $^{87}Sr/^{86}Sr$ ratios ranging from 0.7031 up to 0.7039.

The volcanic rocks on the older islands (S. Vicente, S. Antao, Sal, and Maio) with $^{87}Sr/^{86}Sr$ ratios between 0.7029 and 0.7033 crystallized from plume-derived magmas which, in some cases, mixed with partial melts derived from the depleted lithospheric mantle. The elevated $^{87}Sr/^{86}Sr$ ratios (and low $^{143}Nd/^{144}Nd$ ratios) of the lavas on S. Tiago and Fogo (Fig. 2.23) require the presence of a third Rb-enriched and Sm-depleted component among the magma sources in the mantle. Gerlach et al. (1988a) identified this component as *subcontinental* lithospheric mantle.

The Central Intrusive Complex on São Vicente includes malignites (aegirine-augite + nepheline + orthoclase), microsövite, and carbonatite (Gerlach et al. 1988a). The $^{87}Sr/^{86}Sr$ ratios of these rocks range from 0.70308 (malignite) to 0.70360 (carbonatite) and therefore these rocks were not derived from the same magma. Their Sr concentrations range from 939 ppm (carbonatite) to 4 397 ppm (microsövite). One specimen of carbonatite from the island of Fogo analyzed by Gerlach et al. (1988a) has $^{87}Sr/^{86}Sr$ = 0.70320 and Sr = 7986 ppm. The occurrence of carbonatites on the Cape Verde Islands as well as on Fuerteventura of the Canary Islands (Sect. 2.4.2) is noteworthy because carbonatites are more typically associated with alkali-rich rocks on the continents (e.g. the East African Rift valleys, Sect. 6.1.4).

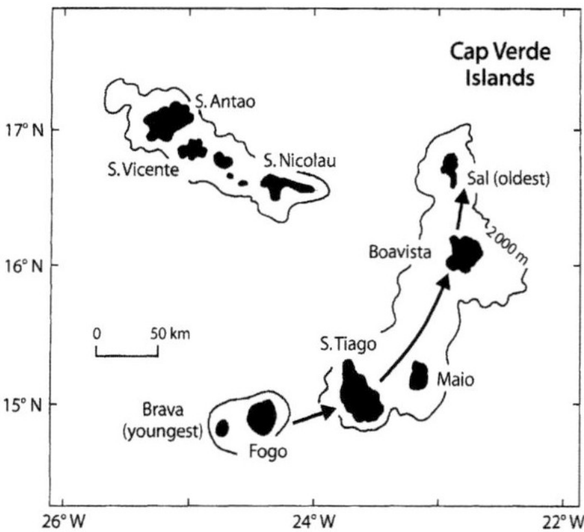

Fig. 2.22 Map of the Cape Verde Islands which rise from submerged lava plateaus on the floor of the Atlantic Ocean. The ages of the exposed volcanic rocks on the islands connected by arrows range from early Tertiary (50–60 Ma) to Recent. The arrows indicate the direction of movement of the African Plate over magma sources that presently underlie the islands of Fogo and Brava (*Source:* adapted from Gerlach et al. 1988a)

2.5.3   Fernando de Noronha

The South Atlantic Ocean contains a large number of small volcanic islands and seamounts identified in Fig. 2.24. Some of these were mentioned previously in Sect. 2.2.3 and 2.2.4 in connection with the study of MORBs on the Mid-Atlantic Ridge.

The island of Fernando de Noronha is located at 3°51' S and 32°25' W about 345 km off the northeast coast of Brazil in the western basin of the Atlantic Ocean. It is composed of highly differentiated volcanic rocks that range in composition from alkali basalt to phonolite and formed between 12.32 ±0.37 and 1.81 ±0.13 Ma (Gunn and Watkins 1976; Cordani 1967). The lava flows have been assigned to the Remedios, São Jose, and Quixaba Formations in order of decreasing age. The phonolites that cap the Remedios Formation yield a whole-rock Rb-Sr isochron date of 9.1 ±0.2 Ma with an initial $^{87}Sr/^{86}Sr$ ratio of 0.70384 (Gerlach et al. 1987). The isotope compositions of Sr, Nd, and Pb of these rocks were measured by Gerlach et al. (1987) who reported that the oldest alkali basalts and trachytes of the Remedios Formation have higher initial $^{87}Sr/^{86}Sr$ ratios (0.70457 to 0.70485) than younger phonolites and melilite basalts (0.70365 to 0.70418). The bimodal distribution of initial $^{87}Sr/^{86}Sr$ ratios requires that magmas were derived from at least two sources in the mantle, such as a plume and delaminated *subcontinental* lithospheric mantle.

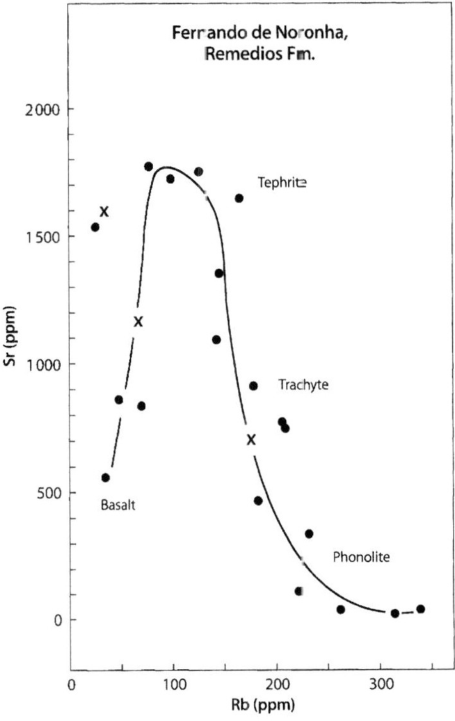

Fig. 2.25. Concentrations of Rb and Sr of the strongly differentiated alkali-rich lavas of the Remedios Formation (12.3 to 8.0 Ma) on Fernando de Noronha. The pattern of variation displayed by these rocks can result from progressive fractional crystallization of magma in a closed chamber or from varying degrees of partial melting of source rocks in the mantle. The *crosses* represent cognate inclusions in the lavas (*Source:* data from Gerlach et al. 1987)

The Rb and Sr concentrations of the volcanic rocks on Fernando de Noronha range widely in Fig. 2.25 in a pattern that can form both by progressive fractional crystallization of magma and by a time-dependent decrease in the degree of partial melting of magma sources in the mantle. The Rb/Sr ratios of the phonolites of the Remedios Formation rise to a value of 92.3 causing the $^{87}Sr/^{86}Sr$ ratio of this rock type to increase by 0.00037 per million years. Therefore, the $^{87}Sr/^{86}Sr$ ratios of magmas with Rb/Sr ratios of about 100 may increase measurably prior to eruption.

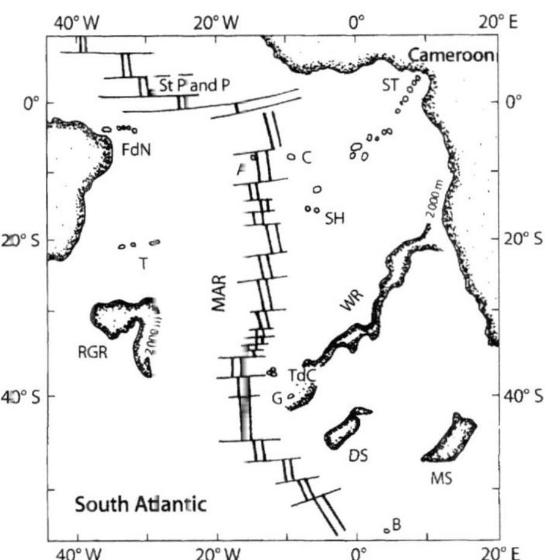

Fig. 2.24. Islands and seamounts of the South Atlantic Ocean: *MAR* = Mid-Atlantic Ridge; *St P and P* = St. Peter and St. Paul Rocks; *ST* = São Tomé; *FdN* = Fernando de Noronha; *A* = Ascension Island; *C* = Circe Seamount; *SH* = St. Helena; *T* = Trinidade; *RGR* = Rio Grande Rise; *TdC* = Tristan da Cunha; *WR* = Walvis Ridge; *DS* = Discovery Seamount (or Tablemount); *MS* = Meteor Seamount; *B* = Bouvet (*Source:* adapted from Fontignie and Schilling 1996)

2.5.4   Ascension Island

Ascension (7°57' S, 14°22' W) is a small volcanic island located about 120 km west of the Mid-Atlantic Ridge in the South Atlantic Ocean in Fig. 2.24. The island consists of a composite volcanic cone that rises to a height

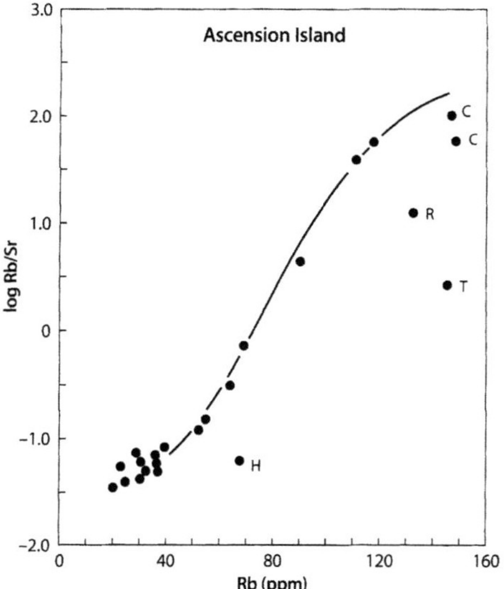

Fig. 2.26. Range of Rb/Sr ratios (expressed as $\log_{10}$) of volcanic rocks on Ascension Island as a function of the Rb concentrations. The Rb/Sr ratios vary from 0.032 (basalt) to 110 (comendite). Some of the rock specimens deviate from the pattern: H = hawaiite, T = trachyte, R = rhyolite, and C = comendite (alkali-rich rhyolite) (*Sources:* O'Nions and Pankhurst 1974; Harris et al. 1982, 19832; Weis et al. 1987a)

of 3 500 m above the seafloor and reaches an elevation of about 800 m above sealevel. The exposed lava flows range in composition from basalt (SiO_2 = 49.67%) to comendite (SiO_2 = 74.05%) and include alkali-rich rocks. Some of the lava flows contain coarse grained blocks of gabbro and granite. The K-Ar dates of the lavas are all less than 1.5 ±0.2 Ma (Harris et al. 1982, 1983; Weis et al. 1987a; Weaver et al. 1987).

The $^{87}Sr/^{86}Sr$ ratios of the volcanic rocks on Ascension Island measured by Faure and Hurley (1963) and Gast et al. (1964) range from 0.7025 (olivine-poor basalt) to 0.7073 (aegerine-riebeckite trachyte), whereas O'Nions and Pankhurst (1974) reported virtually constant $^{87}Sr/^{86}Sr$ ratios between 0.70276 and 0.70286 compared to 0.7080 for the E&A standard.

One of the noteworthy features of the lavas and plutonic inclusions of Ascension Island is the wide range of their Rb/Sr ratios in Fig. 2.26 from 0.032 (basalt and gabbro) up to 110 (comendite and granite) based on the data of Harris et al. (1982, 1983) and Weis et al. (1987a). In addition, the most Rb-rich rocks are silica saturated in contrast to the rocks on Fernando de Noronha and the Cape Verde Islands which are composed primarily of alkali-rich and silica-undersaturated rocks. The initial $^{87}Sr/^{86}Sr$ ratios of trachytes, rhyolites, comendites, and granites in Fig. 2.27 range up to 0.71212 at 1.5 Ma

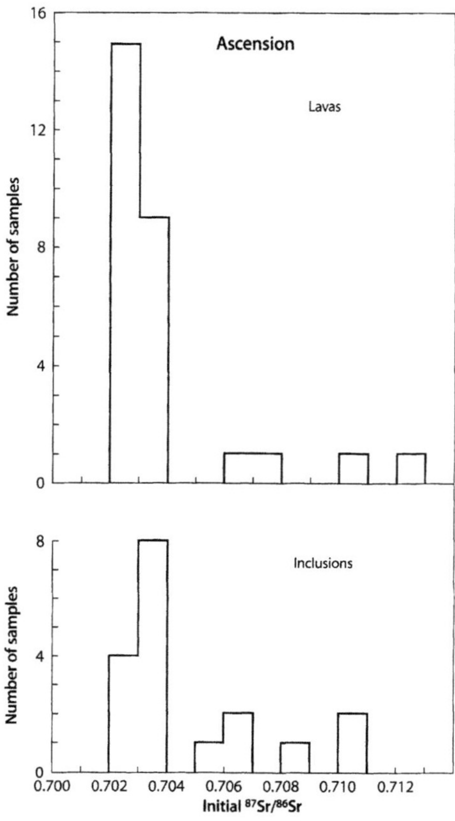

Fig. 2.27. Histogram of the initial $^{87}Sr/^{86}Sr$ ratios at 1.5 Ma of lavas and cognate inclusions of plutonic rocks on Ascension Island, South Atlantic Ocean. The rock specimens having elevated $^{87}Sr/^{86}Sr$ ratios have low Sr concentrations compared to basalt and were contaminated by assimilation of oceanic crust (Rb-enriched basalt and pelagic sediment) during late stages of differentiation (*Sources:* data from O'Nions and Pankhurst 1974; Harris et al. 1982, 1983; Weis et al. 1987a)

(Harris et al. 1982, 1983). Weis et al. (1987a) demonstrated that the ^{87}Sr-enrichment cannot be attributed only to in situ decay of ^{87}Rb because ages ranging from 1.7 to 84.5 Ma are required to increase the $^{87}Sr/^{86}Sr$ ratios of the different rock types from 0.7030 to the present values. Such old dates are unacceptable because the oldest known lavas on Ascension are only about 1.5 million years old.

The rocks having elevated initial $^{87}Sr/^{86}Sr$ ratios also have low Sr concentration between 2.0 and 54 ppm. Therefore, Harris et al. (1982, 1983) and Weis et al. (1987a) concluded that the low-Sr magmas on Ascension Island were contaminated by assimilating small amounts of hydrothermally altered rocks from the oceanic crust which increased their $^{87}Sr/^{86}Sr$ ratios. Subsequently, the $^{87}Sr/^{86}Sr$ increased even further by decay of ^{87}Rb in the Rb-rich magmas and in the rocks that formed from them. In other words, decay of ^{87}Rb can cause the $^{87}Sr/^{86}Sr$ ratios of magmas to increase significantly even before they are erupted (e.g. Cavazzini 1994).

The mafic high-Sr lavas of Ascension Island in Fig. 2.27 have an average $^{87}Sr/^{86}Sr$ ratio of 0.70297 ±0.00007 ($2\bar{\sigma}$, $N = 19$). This value characterizes the isotope composition of Sr in the plume from which these lavas originated and is consistent with the $^{87}Sr/^{86}Sr$ ratios of MORBs in the Mid-Atlantic Ridge near Ascension Island (Fig. 2.6). The $^{87}Sr/^{86}Sr$ ratios of the gabbro inclusions in Fig. 2.27 are similar to those of the mafic lavas which means that they could have formed by crystallization of magma at depth or that they are cumulates of minerals that were segregated from the parent magma.

The $\delta^{18}O$ values of lavas on Ascension Island (Weis et al. 1987a) vary only within narrow limits from about +5.23 to +6.77‰. In general, the oxygen isotope data exclude marine sediment and seawater as possible contaminants. Weis et al. (1987a) concluded that the elevated $^{87}Sr/^{86}Sr$ ratios of the Sr-poor rocks can be accounted for by the assimilation of less than 1% of altered oceanic crust (Sr = 100 ppm, $^{87}Sr/^{86}Sr$ = 0.709). In that case, the resulting change of the $\delta^{18}O$ values of the magmas would not be detectable. Similarly, the $^{87}Sr/^{86}Sr$ ratios of the high-Sr lavas would also not be affected by assimilation of oceanic crust.

Since the elevated $^{87}Sr/^{86}Sr$ ratios of the low-Sr rocks can be explained by contamination of their magmas, we may conclude that these rocks formed either by fractional crystallization of basalt magmas or that they originated by partial melting of basaltic rocks at the base of the volcanic pile as suggested by Moorbath and Walker (1965) for the rhyolites on Iceland. In addition, the isotope compositions of Sr and O exclude the possibility that the rhyolites and granites on Ascension Island originated from remnants of continental crust that might have been preserved since the opening of the Atlantic Ocean.

2 5.5 St. Helena

The island of St. Helena at 16°00′ S and 5°40′ W in the South Atlantic in Fig. 2.24 is known to the public because Napoleon Bonaparte was imprisoned there from October 15, 1815, to his death on May 5, 1821. The island came to the attention of the geochemical community when Gast (1969) reported that Pb in the volcanic rocks on St. Helena is anomalously enriched in the radiogenic isotopes.

St. Helena is part of a chain of seamounts and islands that extends in a northeasterly direction from the Mid-Atlantic Ridge to Cameroon in west-central Africa. The volcanic rocks of the so-called Cameroon Line are discussed in Sect. 5.8.2. The chain of islands and sea-

mounts connecting the volcanic province in Cameroon to St. Helena establishes a link between the volcanic activity on the mainland of Africa and the St. Helena Plume on the Mid-Atlantic Ridge A similar connection between volcanic provinces on continental margins of South Africa and South America and a mantle plume in the vicinity of Tristan da Cunha is discussed in Sect. 2.2.4, 5.7.1, and 5.7.2.

The volcanic rocks of St. Helena range in composition from basalt to phonolite and crystallized between 6.8 ±0.4 and 14.6+1.0 Ma (Baker et al. 1967; Baker 1969; Abdel-Monem and Gast 1967). The $^{87}Sr/^{86}Sr$ ratios of the volcanic rocks on St. Helena were originally measured by Hedge (1966), White and Hofmann (1982), and by Cohen and O'Nions (1982a). In addition, Grant et al. (1976) analyzed 50 specimens of volcanic rocks from St. Helena encompassing a wide range of compositions.

The data in Fig. 2.28a indicate that the average Rb/Sr ratios of the volcanic rocks on St. Helena increase with

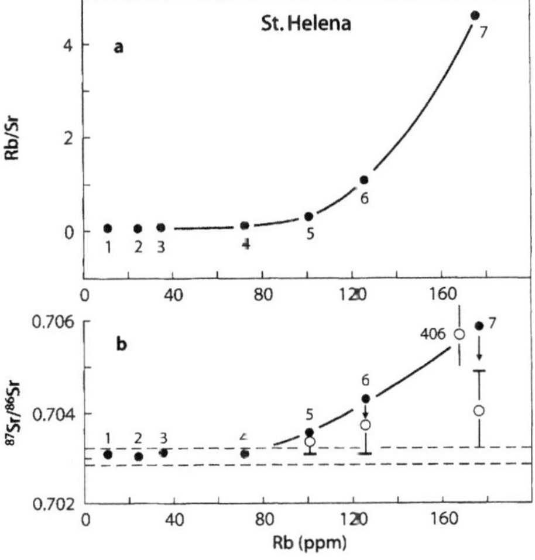

Fig. 2.28. a The average Rb/Sr ratios of the volcanic rocks of St. Helena increase non-linearly with increasing Rb concentration which serves as an index of differentiation. The numbers adjacent to the data points identify the rock type. *1*. ankaramite; *2*. basalt; *3*. trachybasalt; *4*. trachyandesite; *5*. trachyte; *6*. phonolitic trachyte; *7*. phonolite. The average Rb/Sr ratios of the ankaramites (*1*), basalts (*2*) and trachybasalts (*3*) range only from 0.025 to 0.046, whereas those of the more highly differentiated rocks rise to about 4.6 in the phonolites (*7*) and reach 14.0 in Sample *406* analyzed by Grant et al. (1976); b Average present $^{87}Sr/^{86}Sr$ ratios (*solid circles*) and age-corrected $^{87}Sr/^{86}Sr$ ratios (*open circles*) of the different volcanic rock types on St. Helena. The average present $^{87}Sr/^{86}Sr$ ratios of the trachytes, phonolitic trachytes, and phonolites are *not* significantly higher than the initial $^{87}Sr/^{86}Sr$ ratios of the ankaramites, basalts, and trachybasalts indicated by the dashed lines. However, the initial $^{87}Sr/^{86}Sr$ ratio of specimen 406 (phonolite, 0.70569 ±0.00062) does exceed the range of initial $^{87}Sr/^{86}Sr$ ratios of the less differentiated rocks on St. Helena (*Source:* Grant et al. 1976)

increasing Rb concentration (and hence with the degree of differentiation) from 0.025 (ankaramite) to 4.6 (phonolite). The measured average $^{87}Sr/^{86}Sr$ ratios of the ankaramite (1), basalt (2), trachybasalt (3), and trachyandesite (4) in Fig. 2.28b are virtually constant at 0.70308 ±0.00006 ($2\bar{\sigma}$, N = 41) relative to 0.7080 for E&A (Grant et al. 1976). However, the $^{87}Sr/^{86}Sr$ ratios of the more highly evolved rock types trachyte (5), phonolitic trachyte (6), and phonolite (7) in Fig. 2.28b rise with increasing average Rb/Sr ratios and reach an average value of 0.70610 ±0.00165 ($2\bar{\sigma}$, N = 6). A phonolite specimen (no. 406 in Fig. 2.28b) has the highest measured $^{87}Sr/^{86}Sr$ ratio of 0.70916.

Grant et al. (1976) demonstrated that the elevated $^{87}Sr/^{86}Sr$ ratios of the more highly differentiated rocks on St. Helena were caused by decay of ^{87}Rb in the rocks after their crystallization. Only sample no. 406 and a few other phonolites and phonolitic trachytes are anomalously enriched in ^{87}Sr after correction for in situ decay of ^{87}Rb. Grant et al. (1976) attributed the elevated age-corrected $^{87}Sr/^{86}Sr$ ratios of the phonolites and phonolitic trachytes on St. Helena to assimilation of hydrothermally altered rocks by residual magmas having low Sr concentrations. Harris et al. (1982, 1983) and Weis et al. (1987a) reached a similar conclusion for the low-Sr but silica-saturated lavas on Ascension Island (Sect. 2.5.4).

One possible implication of this interpretation is that the low-Sr magmas of St. Helena formed by fractional crystallization of basalt magmas and then assimilated hydrothermally altered rocks from the walls of magma chambers within the volcanic edifice or in the underlying oceanic crust. The parent basalt magmas and their high-Sr derivatives may also have assimilated wallrocks, but their $^{87}Sr/^{86}Sr$ ratios were not altered detectably because of buffering by their high Sr concentrations.

2.5.6 Tristan da Cunha and Inaccessible Island

The islands of Tristan da Cunha, Inaccessible, and Nightingale in Fig. 2.24 are located at 37°05′ S and 12°17′ W about 400 km east of the Mid-Atlantic Ridge in the South Atlantic Ocean. Each island is a separate volcano that rises from the seafloor at depths of more than 2 000 m. Tristan da Cunha (or Tristan) is a nearly circular volcanic cone having a diameter of about 12 km at sealevel and rising to a height of 2 600 m above the surface of the ocean. The volcanic rocks that form this island yielded K-Ar dates of less than 0.21 ±0.01 Ma consistent with the absence of significant erosion (McDougall and Ollier 1982). Age determinations by Miller (1964) indicated that the other two islands in this group are older than Tristan. The most recent volcanic eruption on Tristan occurred in 1961 and required the temporary evacuation of the small population to England (Baker et al. 1964).

The significance of Tristan da Cunha arises from the fact that it is located at the junction between the Walvis Ridge and the Mid-Atlantic Ridge (Sect. 2.2.4). The Walvis Ridge establishes a physical connection between the Tristan Plume on the Mid-Atlantic Ridge (Fig. 2.6) and the Etendeka basalt of Namibia (Sect. 5.7.2). The Tristan Plume also produced the continental basalt plateau in Paraná, Brazil, before the Atlantic Rift widened thereby causing Africa and South America to separate from each other (O'Connor and LeRoex 1992; Milner et al. 1995b; Milner and LeRoex 1996; Ewart et al. 1998 a,b).

The lavas exposed on Tristan da Cunha have high $^{87}Sr/^{86}Sr$ ratios ranging from 0.70504 to 0.70517 relative to 0.7080 for E&A (O'Nions and Pankhurst 1974). The concentrations of Rb and Sr of these lavas in Fig. 2.29 vary systematically from ankaramitic basanite to phonolite in a pattern that is typical of the products of fractional crystallization of a parent magma of basaltic composition. The $^{87}Sr/^{86}Sr$ ratios of these rocks are elevated with a mean of 0.70507 ±0.00002 ($2\bar{\sigma}$, N = 28) relative to 0.7080 for E&A (LeRoex et al. 1990; O'Nions and Pankhurst 1974). The chemical diversity of the lavas on Tristan da Cunha (basalt to phonolite), the pattern of variation of the Rb and Sr concentrations (Fig. 2.29), and the elevated but nearly constant $^{87}Sr/^{86}Sr$ ratios all favor the interpretation that the magmas originated from a homogeneous Rb-enriched source (i.e. the Tristan Plume) and subsequently differentiated by fractional crystallization without detectable assimilation of hydrothermally altered rocks.

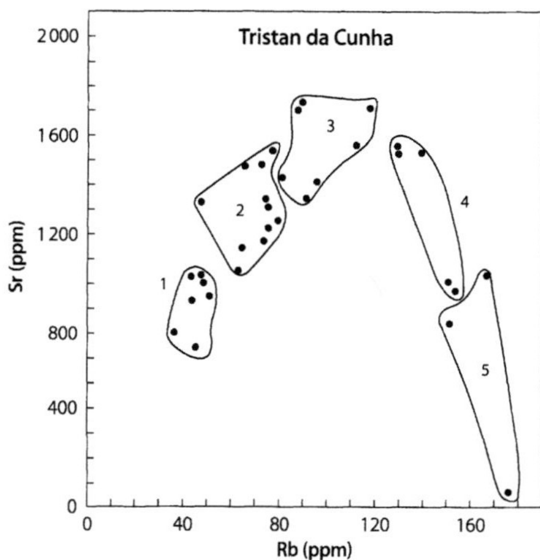

Fig. 2.29. Variation of Rb and Sr concentrations of volcanic rocks on Tristan da Cunha: *1.* ankaramitic basanites; *2.* basanite; *3.* phonotephrite; *4.* tephriphonolite; *5.* phonolite. The pattern of variation is consistent with fractional crystallization of basalt magma (*Source:* LeRoex et al. 1990)

The $^{87}Sr/^{86}Sr$ ratios of the lavas on Tristan differ somewhat from those of the seamounts on the Walvis Ridge (Sect. 2.2.4). Richardson et al. (1982) reported that the initial $^{87}Sr/^{86}Sr$ ratios of basalts erupted on the Walvis Ridge at about 70 Ma range from 0.70391 to 0.70512 relative to 0.7080 for E&A. The range of Sr isotope ratios is attributable either to the internal isotopic heterogeneity of the Tristan Plume or to mixing of magmas derived from the plume and from depleted source rocks of the lithospheric mantle. The possible time-dependent change of $^{87}Sr/^{86}Sr$ ratios of the Tristan Plume was also investigated by Milner and LeRoex (1996) in a study of the Okenyenya igneous complex in Namibia (Sect. 5.7.3). They showed that the initial $^{87}Sr/^{86}Sr$ ratios of a suite of undersaturated alkali-rich rocks of this complex at 128 Ma are similar to those of the lavas on Tristan da Cunha, whereas rocks of a tholeiite suite show signs of significant contamination by assimilation of crustal rocks.

Inaccessible Island, located about 27 km southwest of Tristan da Cunha, exposes a suite of alkali-rich volcanic rocks ranging in composition from alkali basalt to phonolitic trachyte similar to, but less alkaline than, the lavas of Tristan (Baker et al. 1964). The $^{87}Sr/^{86}Sr$ ratios of the lavas on Inaccessible Island range from 0.70412 to 0.70503 relative to 0.71025 ±0.00002 for NBS 987 (Cliff et al. 1992). The silica concentrations vary between 45.42% (alkali basalt) and 63.39% (phonolitic trachyte). The lavas have comparatively high Sr concentrations between 271 and 1 019 ppm making their magmas insensitive to the effects of assimilation of older rocks at the

base of the volcanic pile and in the underlying oceanic crust. For these and other reasons, Cliff et al. (1992) concluded that the magmas originated primarily from an enriched source in the mantle (i.e. the Tristan Plume). Lavas having low $^{87}Sr/^{86}Sr$ ratios (e.g. 0.7041) and elevated $^{143}Nd/^{144}Nd$ ratios (e.g. 0.51267) are hybrids derived from the depleted lithospheric mantle and the Tristan Plume.

2.5.7 Gough Island

Gough is a small volcanic island located at 40°19' S and 9°56' W in Fig. 2.24 about 400 km southeast of Tristan da Cunha. Its dimensions are 13 × 6 km and it rises to a height of 910 m above sealevel. The exposed volcanic rocks range in composition from picrite basalt to sodalite-bearing trachytes described by LeMaitre (1962) who found strong evidence for fractional crystallization. This conclusion was subsequently confirmed by Zielinski and Frey (1970).

The stratigraphy of the lava flows includes five units composed of basalt and trachyte which alternate sequentially starting with the Lower Basalts followed by the Lower Trachyte, Middle Basalts, Middle Trachytes, and Upper Basalts. The ages of the lava flows on Gough range from about 1.0 Ma (Lower Basalts) to about 0.13 Ma (Upper Basalts) (LeRoex 1985).

The $^{87}Sr/^{86}Sr$ ratios published by Gast et al. (1964) vary from 0.7042 to 0.7094 and suggest that the trachytes have

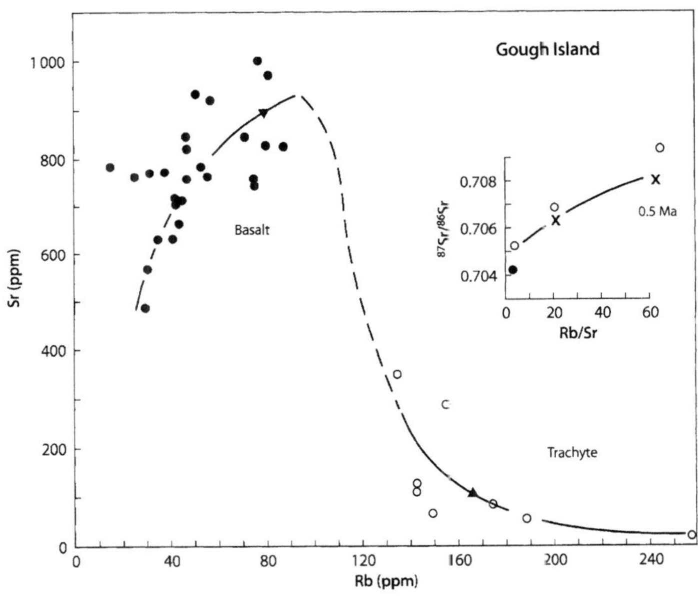

Fig. 2.30.
Rubidium and strontium concentrations of alkali basalts and trachytes on Gough Island in the South Atlantic Ocean. Although both rock types display evidence of fractional crystallization, they form separate clusters, and intermediate rock types are missing. The measured $^{87}Sr/^{86}Sr$ ratios of trachytes (open circles, inset) rise with increasing Rb/Sr ratios and hence with decreasing Sr concentration. A correction for in situ decay (crosses, 0.5 Ma) lowers the $^{87}Sr/^{86}Sr$ ratio only slightly. This evidence is attributable to contamination of the trachytes by marine Sr having $^{87}Sr/^{86}Sr = 0.70916$ (Source: Rb and Sr concentrations from LeRoex 1985; $^{87}Sr/^{86}Sr$ and Rb/Sr ratios of trachytes by Gast et al. 1964)

higher $^{87}Sr/^{86}Sr$ ratios than the basalts. The Rb/Sr ratios also increase from 0.06 (trachybasalt) to 65.5 (trachyte) consistent with the evidence for fractional crystallization cited by LeMaitre (1962), Zielinski and Frey (1970), and LeRoex (1985).

The Rb and Sr concentrations reported by LeRoex (1985) of the basalts and trachytes in Fig. 2.30 form two separate clusters. The Sr concentrations of the basalts rise up to 1 000 ppm with increasing Rb concentrations, whereas the Sr concentrations of the trachytes decrease to low values around 20 ppm as the Rb concentrations increase. Trachyte magmas having such low Sr concentrations are vulnerable to contamination by assimilation of rocks within the volcanic cone or from the oceanic crust. In addition, low-Sr trachytes may be contaminated by exposure to seawater or even rain. The elevated $^{87}Sr/^{86}Sr$ ratios of trachytes (Rb/Sr: 2 to 64.5) reported by Gast et al. (1964) are not decreased significantly by a correction for in situ decay assuming an age of 0.5 Ma (Fig. 2.30). Consequently, the elevated $^{87}Sr/^{86}Sr$ ratios of the trachytes are the result of contamination and are not attributable to the isotope composition of Sr in the magma sources in the mantle. The best available estimate of the $^{87}Sr/^{86}Sr$ ratio of the Gough Plume is provided by the basalts (Rb/Sr = 0.07) whose average $^{87}Sr/^{86}Sr$ ratio reported by Gast et al. (1964) is 0.7044 ±0.0005 (2$\bar{\sigma}$, N = 13). This value is not distinguishable from the $^{87}Sr/^{86}Sr$ ratio of the Tristan Plume. White and Hofmann (1982) reported an $^{87}Sr/^{86}Sr$ ratio of 0.70532 ±0.00006 relative to 0.7080 for E&A for one rock specimen from Gough.

2.5.8   Bouvet Island

The Mid-Atlantic Ridge joins the Atlantic-Indian Ridge and the American-Antarctic Ridge in a triple junction close to the island of Bouvet located at 54°26' S and 3°25' E in the South Atlantic Ocean in Fig. 2.24. The configuration of this important plate junction was described by Mitchell and Livermore (1998).

Bouvet is a small island (7 × 9.5 km) consisting of an ice-covered stratovolcano that formed between 6 and 1 Ma (LeRoex and Erlank 1982; Imsland et al. 1977). The volcano on Bouvet Island is still active and erupted recently (Baker and Tomblin 1964). The lavas exposed on Bouvet consist of hawaiite, mugearite, benmoreite, and rhyolite whose Sr concentrations in Fig. 2.31 decrease with increasing Rb concentrations. LeRoex and Erlank (1982) demonstrated that all of the rock types could have formed by fractional crystallization of a hawaiite parent magma derived from a mantle plume. The principal crystallizing phases were plagioclase, olivine, clinopyroxene, Fe + Ti oxides, and apatite.

The $^{87}Sr/^{86}Sr$ ratios of Sr-rich basalt and trachyte on Bouvet Island range narrowly from 0.70365 to 0.70375 relative to 0.7080 for E&A (O'Nions and Pankhurst 1974).

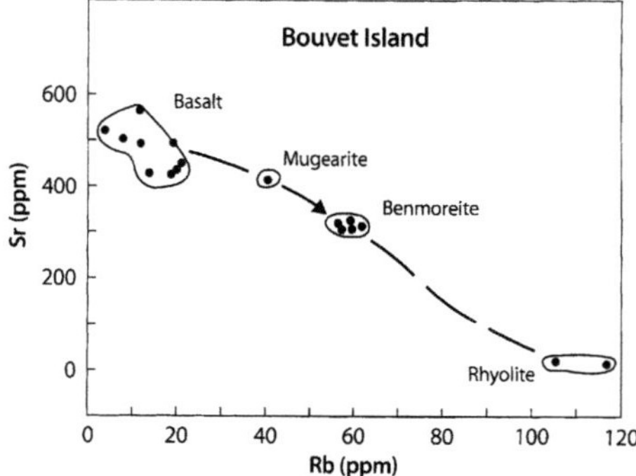

Fig. 2.31. Variation of the Rb and Sr concentrations of the volcanic rocks on Bouvet Island. The progressive Sr depletion of the rocks with increasing Rb concentration is due to fractional crystallization of basalt (hawaiite) magma (*Source:* O'Nions and Pankhurst 1974; LeRoex and Erlank 1982)

These values appear to be unaffected by contamination of the basalt magma (Sr = 482 ppm) and therefore represent the isotope composition of the local mantle plume.

Seven specimens of basalt and glass dredged from the seafloor at four locations in the vicinity of Bouvet Island have a range of $^{87}Sr/^{86}Sr$ ratios between 0.70281 to 0.70691 relative to 0.7080 for E&A (Dickey et al. 1977). The concentrations of water in these samples was 1.02% or less and the glasses appeared to be fresh and unaltered by seawater. Excluding the anomalous value of 0.70691, the remaining samples (0.70281 to 0.70372) could have originated from a mixture of magma sources including the Bouvet Plume (0.70370) and depleted source rocks of normal MORBs (0.7026).

2.6   Mantle Components and Plumes

The data presented in the preceding sections of this chapter demonstrate the existence of correlations between the isotope ratios of Sr and Nd. Some of the evidence is contained in Figs. 2.4, 2.6, and 2.21 which reveal the inverse correlation between $^{87}Sr/^{86}Sr$ and $^{143}Nd/^{144}Nd$ ratios of MORBs and OIBs in the Atlantic Ocean. Such correlations occur on all mid-ocean ridges and oceanic islands and extend also to correlations between the isotope ratios of Sr and Pb as well as Nd and Pb. Many authors have documented and commented upon these correlations (e.g. DePaolo 1979; Zindler et al. 1982; Allègre and Turcotte 1985; Zindler and Hart 1986; Hart et al. (1986); Hart and Zindler 1986, 1989b; Allègre et al. 1986; Hart 1988; Weaver 1991). Salters (1996) considered the origin of MORB magmas based on isotope ratios of Hf and Nd. The correlations of isotope ratios of Sr, Nd, Pb, and Hf in MORBs and OIBs is attributed to mixing of magmas derived from different components in the mantle or to the derivation of magmas from sources consisting of mechanical mixtures of these components.

The driving force for the formation of magma in the mantle is provided by plumes of solid rocks whose existence was first suggested by Wilson (1963a,b) and whose presence was subsequently documented by Morgan (1971, 1972a,b), Schilling (1973a,b), and Baksi (1999) based on topographic evidence in the ocean basins and by the correlated isotope compositions of Sr, Nd, and Pb of MORBs and OIBs. The existence of plumes along the Mid-Atlantic Ridge and associated with oceanic islands in the Atlantic Ocean has been confirmed by many authors some of whom are cited in Sect. 2.3 (Iceland), 2.4 (Canary Islands), and 2.5 (Islands of the Atlantic Ocean). However, Anderson (1996) and Sleep (1984) have questioned the existence of plumes.

2.6.1 Initiation of Plumes

The interpretation of the isotopic compositions and trace element concentrations of MORBs and OIBs supports the existence of plumes composed of rocks made buoyant by an increase in temperature caused by the heat released by the decay of U, Th, K, and Rb. These plumes rise from the asthenospheric part of the mantle until they encounter the rigid lithosphere. Therefore, the asthenosphere and the lithosphere affect the activity of plumes even though they are defined on the basis of mechanical rather than chemical properties. The $^{87}Sr/^{86}Sr$ ratios of normal MORBs on the Mid-Atlantic Ridge indicate that they originated from depleted magma sources believed to be the lithospheric mantle where large-scale melting occurs because of decompression under the ridge crest and because of heat provided by intruding plumes (Zindler et al. 1982).

The existence of plumes in the mantle has been confirmed both by numerical modeling as well as by laboratory experiments (van Keken 1997). The results of experimental studies summarized by Hill (1991) indicate that plumes are initiated by the formation of a large head that grows by entrainment of material through which it rises. Hart et al. (1992) actually deduced evidence for entrainment from the isotope ratios of Sr, Nd, and Pb of MORBs and OIBs. As the head rises, a conduit develops behind it (called the tail) through which hot material streams upward into the head. The conduit persists even after the head has flattened itself against the overlying lithospheric mantle.

The laboratory simulations of mantle plumes indicate that significant uplift and extension occurs when rising plumes interact with the lithospheric mantle. In addition, magmas can form both by decompression melting and by the transfer of heat into the lithospheric

mantle. Consequently, plumes provide a plausible explanation for the volcanic activity along mid-ocean ridges and on oceanic islands. Plumes may also cause large-scale rifting of continental crust that may evolve into ocean basins. The origin of plumes and the manifestations on the surface of the Earth caused by plume heads rising from depth have been discussed by Griffiths (1986), Griffiths and Campbell (1990, 1991), and Herzberg (1995).

The radioactive heating required to start a plume was attributed by Hofmann and White (1982) to the presence of subducted oceanic crust in the mantle. This topic has been the subject of a lively debate in the literature concerning the detection of recycled oceanic crust in oceanic basalt (Cohen and O'Nions 1982b; Weaver et al. 1986; Loubet et al. 1988; Hart and Staudigel 1989; Woodhead et al. 1993, and others cited by them). Another question that has been raised concerns the chemical alteration of the oceanic crust by dehydration during subduction (e.g. Davidson 1983; Chauvel et al. 1995). Nevertheless, in a recent review of the evidence concerning the geochemistry of the mantle, Hofmann (1997) concluded that: "Recycling of oceanic crust and lithosphere plays an important role in generating mantle heterogeneities."

2.6.2 Mantle Components

The evidence that has been used to identify the different source components in the mantle consists primarily of the isotope ratios of Sr, Nd, and Pb of MORBs and OIBs and of the concentrations of large-ion lithophile (LIL) and high fieldstrength (HFS) elements, including the rare-earth elements (REE) (Weaver 1991). The different kinds of magma sources in the mantle are best defined by means of isotope ratios because they are not altered during partial melting of source rocks in the mantle and by subsequent fractional crystallization of the resulting magma. Consequently, the isotope compositions of Sr, Nd, and Pb of magmas that have not assimilated rocks of the oceanic crust and were not contaminated by seawater are identical to those of the source rocks in the mantle from which they formed.

Table 2.2 shows the isotopic compositions of Sr, Nd, and Pb of volcanic rocks in the ocean basins, which can be accommodated by four components identified by Zindler and Hart (1986) and Hart (1988).

Volcanic rocks on mid-ocean ridges and oceanic islands originate by mixing of magmas derived from two or more of these components identified in Fig. 2.32a and b.

The isotope ratios of Sr, Nd, and Pb of *normal MORBs* define the DMM component which consists of the depleted suboceanic lithospheric mantle from which melts were previously extracted at 1 to 2 Ga. The isotope compositions of Sr, Nd, and Pb of *plumes* along the Mid-Atlantic Ridge are defined by the plume MORBs such as the basalts of the Azores Platform and the lavas of Tristan

Table 2.2.
Isotopic compositions of Sr, Nd, and Pb of volcanic rocks in the ocean basins, which can be accommodated by four components identified by Zindler and Hart (1986) and Hart (1988)

Component	Name	$^{87}Sr/^{86}Sr$	$^{143}Nd/^{144}Nd$	$^{206}Pb/^{204}Pb$
DMM	Depleted-MORB mantle	0.7026	0.51315	18.4
EM1	Enriched mantle 1	0.7055	0.51235	17.6
EM2	Enriched mantle 2	0.7075	0.51265	19.2
HIMU	Magma source having a high $(^{238}U/^{204}Pb)$ ratio (μ)	0.7028	0.51285	20.5 – 21.5

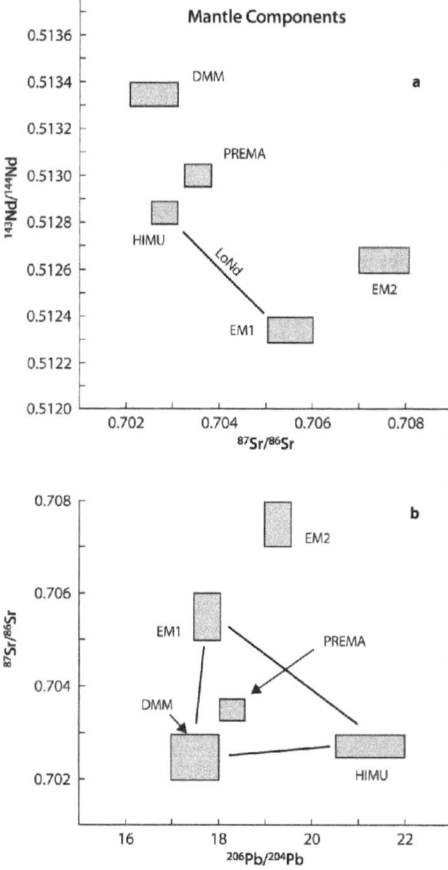

Fig. 2.32 a,b. Isotope ratios of Sr, Nd, and Pb of mantle compositions identified from the systematic variations of these ratios among MORBs and OIBs. *DMM* = Depleted MORB mantle; *HIMU* = magma source having a high $^{238}U/^{204}Pb$ (μ) ratio; *EM1* and *EM2* = enriched mantle 1 and 2; *PREMA* = prevalent mantle formed by mixing of EM1-HIMU-DMM. The straight line formed by mixtures of HIMU and EM1 is called the LoNd Array. The numerical values of the isotope ratios of the mantle components are approximate (*Sources:* Zindler and Hart 1986; Hart 1988; Hart and Zindler 1989b)

da Cunha. The data clearly show that plume MORBs are enriched in ^{87}Sr, ^{206}Pb, and ^{207}Pb and are depleted in ^{143}Nd compared to normal MORBs and that plumes must therefore consist of rocks having elevated Rb/Sr and U/Pb ratios, but lower Sm/Nd ratios, compared to the DMM component. These chemical characteristics fit the basalt and pelagic marine sediment of subducted oceanic crust as originally proposed by Hofmann and White (1982). The so-called Mantle Array defined by MORBs along mid-ocean ridges results from the mixing of magma derived from the DMM and the EM1 component which resides in the heads of plumes or, in some cases, takes the form of blocks of enriched mantle rocks at the base of the depleted lithospheric mantle.

Since the isotope compositions of many MORBs appear to be the result of mixing of DMM with EM1 and HIMU, Zindler and Hart (1986) called this magma source the "prevalent" mantle or PREMA and considered that it may be a fifth component in the mantle. Figure 2.32b demonstrates the relation of PREMA to DMM, EM1, and HIMU. The interpretation presented above suggests that the PREMA component is the source of mixed magmas that form by partial melting of plume heads and adjacent depleted lithospheric mantle. The curvature of the mixing lines joining these components depends on the concentrations of the elements (e.g. Sr, Nd, and Pb). The mixing lines in Fig. 2.32a and b have been drawn as straight lines for the sake of simplicity, implying that the Sr/Nd and Sr/Pb concentration ratios of the components are equal to one (Langmuir et al. 1978).

Hart et al. (1986) demonstrated that rock samples having the lowest $^{143}Nd/^{144}Nd$ ratios on certain islands exhibit an inverse linear correlation with their $^{87}Sr/^{86}Sr$ ratios and thereby define the low-Nd or LoNd Array identified in Fig. 2.32a. The specimens that define the LoNd Array are also collinear in coordinates of $^{208}Pb/^{204}Pb$ vs. $^{206}Pb/^{204}Pb$, $^{207}Pb/^{204}Pb$ vs. $^{206}Pb/^{204}Pb$, $^{87}Sr/^{86}Sr$ vs. $^{208}Pb/^{204}Pb$, and $^{143}Nd/^{144}Nd$ vs. $^{208}Pb/^{204}Pb$. The islands and seamounts that define the LoNd Array include St. Helena, the New England Seamounts, and Walvis Ridge in the Atlantic Ocean, Tubuai and San Felix in the Pacific Ocean, as well as the Comoro Islands and D5 (Hamelin and Allègre 1985) in the Indian Ocean.

An important question is how these components came into existence and where they reside in the present mantle. Hart (1988) associated the DMM component with the Rb-depleted oceanic lithosphere. The origin of the HIMU component is not as well constrained as that of DMM. Hart (1988) cited proposals from the literature attributing the origin of HIMU to: (1) transfer of Pb into the core; (2) recycling of old oceanic crust by subduction; and (3) U-enrichment of mantle domains by metasomatism. Although each of these hypotheses can explain an increase of the U/Pb ratio of certain mantle domains, recycling of oceanic crust and U-metasomatism are favored. However, Hart (1988) pointed out that recycled oceanic crust is enriched in both U and Rb which causes the isotope ratios of both Pb and Sr to rise with time. Since the HIMU component actually has a *low* $^{87}Sr/^{86}Sr$ ratio (Fig. 2.32a,b), oceanic crust must lose Rb (but not U) by dehydration and recrystallization during subduction into the mantle. Hanyu and Kaneoka (1997) reported that HIMU basalts have low $^3He/^4He$ ratios which confirms the derivation of magma from recycled material that lost most of its original ^3He content. The HIMU component (whatever its origin) is prevalent in the lavas of St. Helena, Ascension, Guadelupe, the Canaries, and Azores in the Atlantic Ocean and on the islands of Tubuai, Mangaia, and Rurutu in the Pacific Ocean (Hart and Zindler 1989b; Chauvel et al. 1992; Vidal 1992).

The EM1 component consists of subducted oceanic crust and pelagic marine sediment and, in some cases, resides in the heads of plumes such as those underlying Iceland, Tristan da Cunha, Hawaii, and many other oceanic island groups. The isotope composition of EM2 in Fig. 2.32a,b is characterized by high $^{87}Sr/^{86}Sr$ (>0.707), low $^{143}Nd/^{144}Nd$ (<0.5127), and moderate $^{206}Pb/^{204}Pb$ (~17 to 18) ratios. This combination of isotope ratios is consistent with the composition of continental rocks that were subducted into the mantle. Examples of EM2-rich lavas occur on São Miguel of the Azores of the Atlantic Ocean and on the islands of the Samoan and Marquesas Archipelagos in the Pacific Ocean (Hart 1988). The origin of the mantle components is discussed again in Sect. 2.10.5 based on evidence from the lavas of Polynesia.

2.6.3 Interactions of Plumes and Ridges

In some cases, the volcanic activity along mid-ocean ridges is maintained by magma that originates from plumes located several hundred kilometers from the ridge (Schilling 1991; Kincaid et al. 1996). This situation arises as a result of migration of the ridge axis so that it is no longer centered over the plume. In such cases, the flow of magma from the plume to eruption sites at the ridge axis may be maintained by plume channels which can be understood as lateral elongations of the conduit illustrated in Fig. 2.33. The diagram demonstrates that the flow of the plume is diverted by the overlying litho-

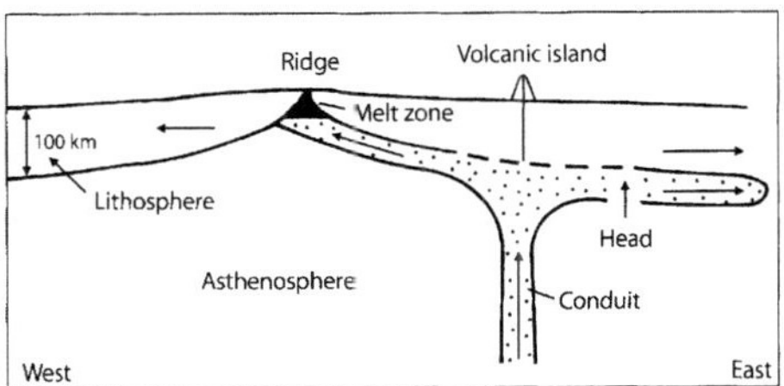

Fig. 2.33. Model of the interaction of an off-axis plume with a mid-ocean ridge and the oceanic lithosphere. The head of the plume continues to flow to the ridge crest by means of a channel within the rheological boundary layer at the base of the lithosphere. In addition, part of the plume head flows away from the ridge in the direction of motion of the overlying lithospheric plate. Magma generated in the plume head and in the adjacent lithospheric mantle can penetrate the oceanic lithosphere to form a sea-floor volcano which may extend above the surface of the water as an oceanic island (*Source:* adapted from Kincaid et al. 1996)

spheric plate. Under suitable rheological conditions, a plume channel may develop that permits part of the buoyant plume to flow towards the ridge axis where melting occurs in the melt-triangle region. Experimental results referenced by Kincaid et al. (1996) indicate that the establishment and persistence of the plume channel depends on the presence of an upward-sloping rheological boundary layer at the base of the lithosphere and on the buoyancy of the plume rocks. Examples of magmatic interactions between plumes located east of the Mid-Atlantic Ridge and the ridge axis occur near the seamount Circe and adjacent to the islands of St. Helena and Tristan da Cunha identified in Fig. 2.24 (Schilling et al. 1985; Humphris et al. 1985; Hanan et al. 1986). At these locations, MORBs along the Mid-Atlantic Ridge have anomalously high La/Sm and $^{87}Sr/^{86}Sr$ as well as low $^{143}Nd/^{144}Nd$ ratios (Fig. 2.6).

The phenomenon described above also permits the opposite process, namely that plume heads and magma derived from them under spreading ridges flow along the base of the lithospheric plates away from the ridge axis or even along it (Kenyon and Turcotte 1987). Eruption of this material may cause volcanic activity on the seafloor at some distance from the ridge and may result in the formation of seamounts (Castillo and Batiza 1989). Alexander and Macdonald (1996) observed that volcanic cones less than 70 m high are concentrated within 10 km of the East Pacific Rise and that only larger cones occur at greater distances. The authors proposed that the small seamounts form by the eruption of small volumes of plume-derived magmas that "leak" through the relatively thin lithosphere close to the ridge. At distances greater than about 10 km from the ridge axis, the increasing thickness of the lithosphere does not permit small volumes of magma to penetrate. Instead, the flow of plume-derived magma under the lithosphere is channelized such that large volcanic edifices can form on the seafloor hundreds of kilometers from the ridge.

The occurrence of seamounts and oceanic islands that are displaced from mid-ocean ridges can therefore be explained in three different ways based on the evidence from the Atlantic Ocean and the ideas discussed above:

1. Oceanic islands and chains of seamounts form by volcanic activity over a plume located beneath a mid-ocean ridge and are subsequently displaced from their site of origin by seafloor spreading. In this case, only the island situated on the ridge axis remains volcanically active.
2. Volcanic activity may occur on the seafloor adjacent to a spreading ridge at a site above an off-axis plume that may also supply magma to the central rift of the ridge.
3. Volcanoes located hundreds of kilometers from a spreading ridge may erupt plume-derived magma that flowed in channels from a plume located at or close to a mid-ocean ridge.

The petrogenetic scenarios identified above are all related to the presence of a spreading ridge because the evidence considered so far was taken from examples in the Atlantic Ocean which is dominated by the Mid-Atlantic Ridge. We next turn to the volcanic activity in the basin of the Pacific Ocean where oceanic islands and seamounts are not as closely related to mid-ocean ridges as those of the Atlantic Ocean (Pringle et al. 1993).

2.7 MORBs of the Pacific Ocean

The East Pacific Rise in Fig. 2.34 extends north from about 50° S and 120° W in the southern Pacific Ocean, enters the Gulf of California, and emerges from beneath the American Plate in the form of the Gorda and Juan de Fuca Rises off the coast of Oregon and Washington (USA) and the province of British Columbia (Canada). The chemical compositions of tholeiite basalts of the East Pacific Rise, as well as those of the Mid-Atlantic and Indian Ocean Ridges, were described by Engel and Engel (1964) and Engel et al. (1965), whereas Bender et al. (1984) and Goldstein et al. (1991) investigated the petrogenesis of basalts in the Tamayo region of the East Pacific Rise and the Juan de Fuca-Gorda Rise, respectively. The geophysics and geochemistry of the Juan de Fuca Ridge were the subject of a set of reports edited by Brett (1987) that included a study by Hegner and Tatsumoto (1987) on the isotope compositions of Sr and Pb in MORBs and sulfide minerals along the southern part of the ridge.

The $^{87}Sr/^{86}Sr$ ratios of tholeiites and glass from the East Pacific Rise reported by Subbarao (1972) and from the Gorda and Juan de Fuca Rises by Hedge and Peterman (1970) are generally less than 0.7030 relative to

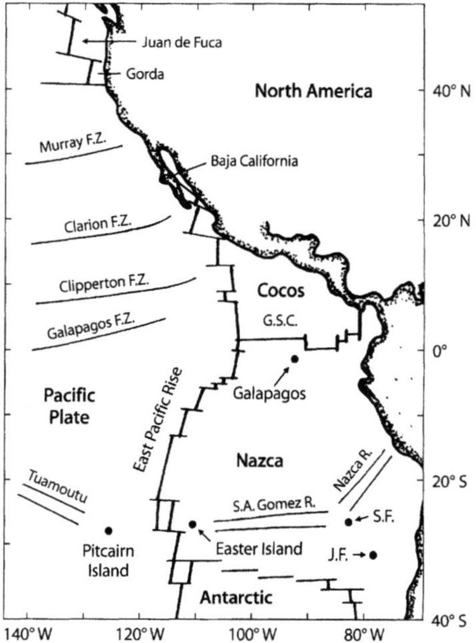

Fig. 2.34. East Pacific Ridge and related oceanic islands including the Galapagos Islands, Easter Island, Pitcairn Island, and San Felix Island (*Sources:* adapted from White et al. 1987; Macdougall and Lugmair 1986; National Geographic Society 1989)

0.7080 for E&A. Subsequent measurements by Eaby et al. (1984) and Hegner and Tatsumoto (1987) displayed in Fig. 2.35 confirm the remarkable homogeneity of the $^{87}Sr/^{86}Sr$ ratios of MORBs on the East Pacific Rise (0.70255 ±0.00003) and its extension into the Gorda and Juan de Fuca Ridges (0.70251 ±0.00002) compared to MORBs on the Mid-Atlantic Ridge. These results imply that the magma sources underlying the East Pacific Ridge are homogeneous with respect to the isotope composition of Sr (Macdougall and Lugmair 1985). In addition, White et al. (1987) reported that the $^{143}Nd/^{144}Nd$ ratios of East Pacific MORBs likewise vary only between narrow limits from 0.51300 to 0.51325. Whereas the $^{206}Pb/^{204}Pb$ ratios range from about 18.0 to 19.3. The isotope ratios of Pb reported by White et al. (1987) are in good agreement with those published earlier by Hegner and Tatsumoto (1987), Hamelin et al. (1984), Dupré et al. (1981), Cohen et al. (1980), and Church and Tatsumoto (1975).

The Juan de Fuca Ridge is noted for the occurrence of massive hydrothermal sulfide deposits occurring in the median valley (Kim and McMurtry 1991). The $^{206}Pb/^{204}Pb$ ratios of the sulfides at Bent Hill in the Middle Valley range from 18.43 to 18.86 and are similar to the $^{206}Pb/^{204}Pb$ ratios of MORBs on the Juan de Fuca

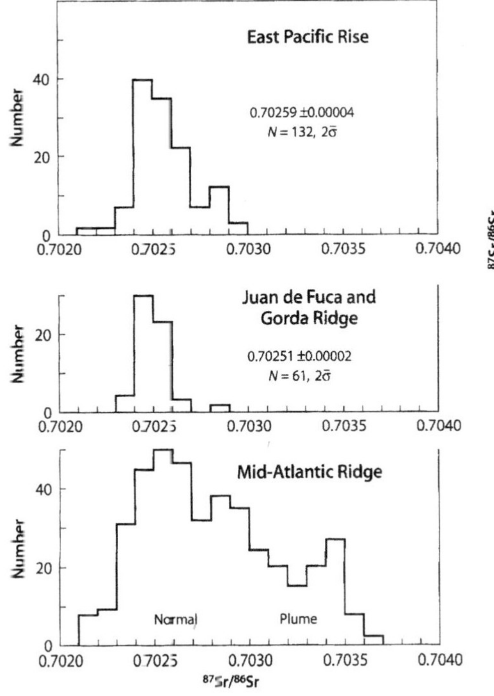

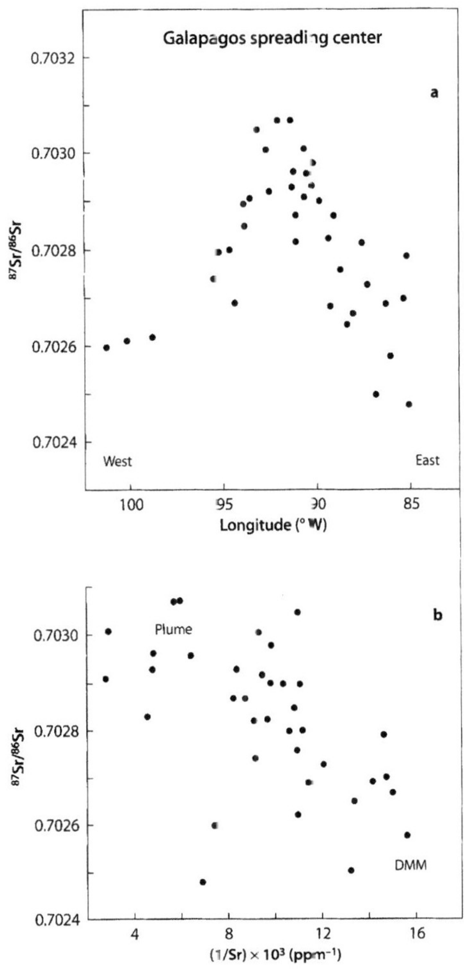

Fig. 2.35. Sr in isotope ratios of MORBs on the East Pacific Rise and the Gorda/Juan de Fuca Ridge (Fig. 2.34) compared to the $^{87}Sr/^{86}Sr$ ratios of MORBs on the Mid-Atlantic Ridge (Fig. 2.7). Samples from seamounts and oceanic islands were excluded and all of the $^{87}Sr/^{86}Sr$ ratios were corrected to 0.7080 for E&A or 0.71025 for NBS 987. The sources of the data for MORBs of the Mid-Atlantic Ridge are listed in the caption of Fig. 2.7 (*Sources:* East Pacific Rise: Dupré et al. 1981; White and Hofmann 1982; Hamelin et al. 1984; Macdougall and Lugmair 1985, 1986; White et al. 1987; Fontignie and Schilling 1991; Gorda/Juan de Fuca Ridge: Eaby et al. 1984; White et al. 1987; Hegner and Tatsumoto 1987)

Fig. 2.36. a Variation of the $^{87}Sr/^{86}Sr$ ratios of MORBs along the Galapagos Spreading Center from about 101° W longitude eastward to 85° W longitude. The pattern of variation is attributable to the changing proportions of magma derived from the Galapagos Plume and from depleted sources in the lithospheric mantle; **b** Strontium-isotope mixing diagram (Eq. 1.3., Sect. 1.7.2) of MORBs of the Galapagos Spreading Center. Although the data points scatter widely, a negative correlation of the $^{87}Sr/^{86}Sr$ and 1/Sr ratios is suggested. Consequently, the plume component has an elevated $^{87}Sr/^{86}Sr$ ratio (>0.7031) and Sr concentration (>300 ppm) compared to the depleted-mantle component (DMM) (*Source:* data from Verma and Schilling 1982)

Ridge (Stuart et al. 1999). Therefore, these authors concluded that the Pb was leached from the oceanic crust by hydrothermal fluids.

In a similar study of sulfides at 21° N on the East Pacific Rise, Vidal and Clauer (1981) also reported uniform $^{206}Pb/^{204}Pb$ ratios (about 18.47) similar to the $^{206}Pb/^{204}Pb$ ratio of the local MORBs (18.35 to 18.58), whereas the $^{87}Sr/^{86}Sr$ ratios of the sulfides range widely from 0.70565 to 0.70806 and are intermediate between the $^{87}Sr/^{86}Sr$ ratios of the local MORBs (0.70235 to 0.70262) and seawater (0.70916) relative to 0.71025 for NBS 987. The isotope composition of S of the sulfide deposit at 21° N on the East Pacific Rise was measured and discussed by Kerridge et al. (1983), whereas Lalou et al. (1984) dated sulfide and silica deposits of the East Pacific Rise.

The interaction of low-temperature fluids (<150 °C) with MORBs along the eastern flank of the Juan de Fuca

Ridge caused the formation of phyllosilicates (saponite, celadonite, chlorite, and talc), tektosilicates (quartz, zeolites), as well as of hematite, iddingsite, and carbonate minerals (Hunter et al. 1999). The same authors also reported that the alteration of MORBs is accompanied by an increase of the $^{87}Sr/^{86}Sr$ ratios of unleached samples from 0.70247 to 0.70647 relative to 0.71025 for NBS 987. The $\delta^{18}O$ values of the altered rocks increase from +6.1‰ to +19.3‰ for the clay matrix of a highly altered hyaloclastic breccia which also yielded the highest $^{87}Sr/^{86}Sr$ ratio (0.70647). The $\delta^{18}O$ values of unaltered basalts along the East Pacific Rise were measured by Barrett and Friedrichsen (1987).

The Galapagos Spreading Center (GSC in Fig. 2.34) extends east from a triple junction on the East Pacific Rise and separates the Nazca Plate in the south from the Cocos Plate in the north. The MORBs along this ridge are exceptional because their $^{87}Sr/^{86}Sr$ ratios vary systematically. Measurements by Verma and Schilling (1982) in Fig. 2.36a show that the $^{87}Sr/^{86}Sr$ ratios of MORBs on the Galapagos Spreading Center increase from about 0.7025 at the triple junction to about 0.7031 near Darwin Island at about 92° W longitude and then decline farther east to 0.7026. The systematic variation of the $^{87}Sr/^{86}Sr$ ratios along the Galapagos Spreading Center was tentatively attributed by Verma and Schilling (1982) to mixing of magmas derived from the Galapagos Plume and from depleted mantle. In that case, the $^{87}Sr/^{86}Sr$ ratios and reciprocal Sr concentrations of the basalts should define a straight line, provided the Sr concentrations of the mixed magmas were not altered by subsequent fractional crystallization. The distribution of data points in Fig. 2.36b reveals that the $^{87}Sr/^{86}Sr$ and 1/Sr ratios are scattered, but are correlated as expected. Accordingly, the lavas derived from the plume component have $^{87}Sr/^{86}Sr$ > 0.7031 and Sr > 300 ppm, whereas MORBs that originated from the depleted-mantle component (DMM, Fig. 2.32) have $^{87}Sr/^{86}Sr$ < 0.7026 and low Sr concentrations from 60 to 140 ppm. The scattering of the data points in Fig. 2.36b is attributable to the heteroge-

neity of the magma sources and, to a lesser extent, to the effects of fractional crystallization of the mixed magmas (Muehlenbachs and Byerly 1982).

2.8 Islands of the East Pacific Ocean

The Galapagos Islands located a short distance south of the Galapagos Spreading Center in Fig. 2.34 are the principal archipelago in the East Pacific Ocean. In addition, Easter Island and Sala y Gomez are located along the Sala y Gomez Ridge in Fig. 2.34, whereas San Felix and San Ambrosio are intraplate volcanic islands. In addition, the Juan Fernández Islands are located along the Challenger Fracture Zone about 1 000 km south of San Felix and about 750 km off the coast of Chile.

2.8.1 Galapagos Islands

The Galapagos Islands in Fig. 2.37 are located on a submarine plateau close to the equator and include a large number of islands, the largest of which is Isabela. The lavas were erupted in Late Tertiary time and range in composition from tholeiite to alkali olivine basalt. Geist et al. (1986) reported that San Cristóbal Island (2.3 to 0.6 Ma) contains three distinct magma series: (1) Alkali basalts; (2) MORB-like tholeiites; and (3) OIB tholeiites enriched in incompatible elements. Each of these series ranges in composition from primitive basalts to evolved varieties including low-Mg basalts, hawaiites, and icelandites.

Age determinations by White et al. (1993) of volcanic rocks collected above sealevel indicate that subaerial volcanic activity occurred simultaneously on several is-

Fig. 2.37.
Map of the principal islands of the Galapagos Archipelago. The contour lines delineate the regional variation of $^{87}Sr/^{86}Sr$ ratios of the lavas on the Galapagos Islands. The geographic pattern of variation of $^{87}Sr/^{86}Sr$ ratios has been attributed to magma formation in a concentrically-zoned mantle plume ($^{87}Sr/^{86}Sr$ > 0.7030) that contained depleted asthenospheric mantle rocks in its core ($^{87}Sr/^{86}Sr$ < 0.7030). The penetration of asthenospheric rocks into the core of a rising diapir of hot and undepleted rocks was predicted by Griffiths (1986) on theoretical grounds (*Source:* adapted from White and Hofmann 1978 and White et al. 1993)

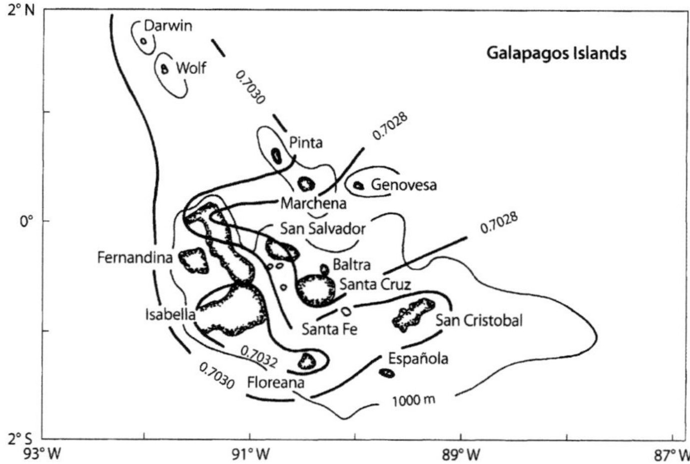

lands starting at about 2.8 Ma and continued to the present. The island of Fernandina west of Isabela (Fig. 2.37) is the youngest volcano among the Galapagos Islands with several volcanic eruptions in historical time. Isabela consists of six large shield volcanoes several of which have also been active recently.

The ^{87}Sr/^{86}Sr ratios of volcanic rocks on the Galapagos Islands reported by Stueber and Murthy (1966) and by Hedge (1978) vary from 0.7024 to 0.7038 relative to E&A = 0.7080 and thus encompass the range from normal MORBs to typical plume-derived volcanic rocks. Although the existence of a mantle plume, originally proposed by Morgan (1971), is a plausible explanation for the elevated ^{87}Sr/^{86}Sr ratios of the lavas on the Galapagos Islands, Geist et al. (1988b) pointed out several unusual features:

1. The islands are clustered on a submarine plateau and do not form a linear chain as expected if the Nazca Plate has moved across the head of a plume.
2. The ages of the lavas on the different islands do not vary systematically in accordance with the direction of movement of the Nazca Plate on which they formed.
3. The lavas on the islands located in the center of the archipelago (e.g. Marchena, Genovesa, San Salvador, and Santa Cruz) have low ^{87}Sr/^{86}Sr ratios, whereas the islands on the periphery have high ^{87}Sr/^{86}Sr ratios (e.g. Isabela, Fernandina, and Floreana) based on the data of White and Hofmann (1978) and White et al. (1993).

The observed bimodal distribution of ^{87}Sr/^{86}Sr ratios in the lavas of the Galapagos Islands in Fig. 2.38 differs from the pattern expected for volcanic islands that erupted magma generated in a plume. In that case, the rocks on the central islands should have elevated ^{87}Sr/^{86}Sr ratios, whereas the peripheral islands should have low ^{87}Sr/^{86}Sr ratios because magmas that originated from the borders of a plume formed by partial melting of depleted rocks in the oceanic lithospheric mantle. Geist et al. (1988b) explained this anomaly by postulating that the rising Galapagos Plume head entrained blocks of Rb-depleted asthenospheric mantle which became concentrated in the center of the plume head, thus forming a torus. This phenomenon of plume dynamics is consistent with modeling by Griffiths (1986) and explains the horseshoe-shaped distribution of ^{87}Sr/^{86}Sr ratios of lavas on the Galapagos Islands (Fig. 2.37). Accordingly, the first lavas to be erupted on the Galapagos Plateau presumably originated from the enriched outer part of the plume followed later by magmas that formed in its depleted core.

The isotope ratios of Sr and Nd in Fig. 2.38 are consistent with the explanation proposed by Geist et al. (1988b) because the rocks of the outer islands have high

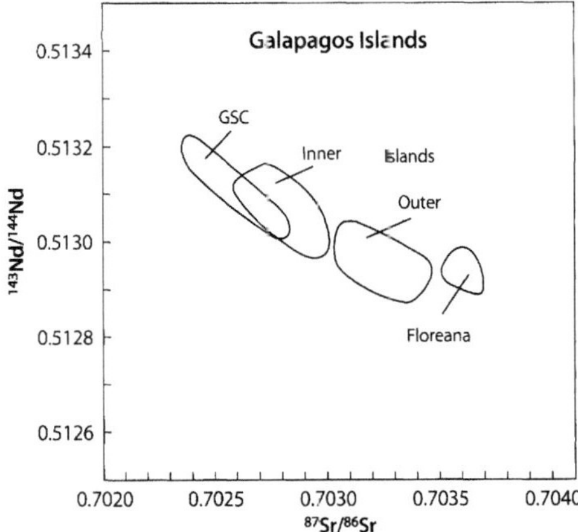

Fig. 2.38. Sr-Nd isotope ratio correlation diagram for the lavas of the Galapagos Islands and the Galapagos Spreading Center (*GSC*). The rocks of the inner islands originated from depleted magma sources in the core of the Galapagos Plume, whereas magmas erupted on the outer islands on the periphery of the archipelago formed from the enriched rocks of the Galapagos Plume. The isotope ratios of Sr and Nd of MORBs on the Galapagos Spreading Center reveal the influence of a magma component derived from the Galapagos Plume. The lavas of the "outer" islands have ^{87}Sr/^{86}Sr > 0.7030, whereas those of the 'inner" islands have ^{87}Sr/^{86}Sr < 0.7030 (*Sources:* White et al. 1987, 1993)

^{87}Sr/^{86}Sr and low ^{143}Nd/^{144}Nd ratios compared to the inner islands. In addition, the isotope ratios of MORBs on the Galapagos spreading ridge suggest that some specimens contain a magma component derived from the Galapagos Plume. The evidence for interaction between the Galapagos Plume and the ridge implies that the head of the plume (or magmas derived from it) extends north from the Galapagos Islands for a distance of more than 250 km.

2.8.2 Easter Island and Sala y Gomez

The East Pacific Rise appears to split into two parallel branches between latitudes 22°31.8' S and 26°42' S. The two branches define the Easter Microplate (Herron 1972; Naar and Hey 1989). The isotope compositions of Sr in MORBs from the borders of the Easter Microplate reported by Fontignie and Schilling (1991) are included in Fig. 2.35 with other measurements of ^{87}Sr/^{86}Sr ratios of MORBs on the East Pacific Rise.

Easter Island (called Rapa Nui in Polynesian) is located at 27°07' S and 109°27' W about 350 km east of the East Pacific Rise on the Sala y Gomez Ridge (Haase and Devey 1996; Bonatti et al. 1977; Baker et al. 1974 and Fig. 2.34). The island is about 24 km long and 9 km wide and contains several volcanoes all of which are now extinct. Poike, on the eastern end of the island, was active from 2.54 ±0.28 to 0.75 ±0.15 Ma, according to whole-rock K-Ar dates reported by Clark and Dymond (1977). These dates overlap those of alkali basalts on the island of Sala y Gomez (26°28' S, 106°28' W) for which Clark and Dymond (1977) obtained dates ranging from 1.94 ±0.07 to 1.31 ±0.06 Ma. Therefore, these age determinations indicate that volcanic eruptions occurred simultaneously on both islands.

The simultaneous occurrence of volcanic activity on Easter Island and on Sala y Gomez is not consistent with the proposal that the seamounts and islands on the Sala y Gomez Ridge formed over a plume head on or near the East Pacific Rise and were subsequently displaced from it by the eastward movement of the Nazca Plate (Wilson 1963a; Morgan 1971, 1972a,b). If the Sala y Gomez Ridge is a plume trace that recorded the motion of the Nazca Plate over a plume head, then the volcanic rocks on the island of Sala y Gomez should be *older* than those on Easter Island because Sala y Gomez is located 475 km east of Easter Island. Therefore, Clark and Dymond (1977) supported the proposal of Herron (1972) that the Sala y Gomez Ridge is a fracture along which magma leaked upward and formed both islands and the related seamounts that now define the ridge.

The volcanic rocks on Easter Island are highly differentiated ranging in composition from tholeiites to hawaiites, mugearites, benmoreites, trachytes, and rhyolites (comendites). The silica concentrations of these rocks increase from 42.80% (tholeiite) to 72.7% (comendite glass). Consequently, most of the rocks on Easter Island described by Baker et al. (1974) and Clark and Dymond (1977) are quartz normative and probably formed by differentiation of tholeiite magmas.

Sala y Gomez is a very small island with a surface area of only about 0.3 km^2. It is the summit of a large volcanic seamount that extends for 30 to 50 km southwest and southeast below the surface of the ocean. The rocks exposed on Sala y Gomez consist of mugearite flows separated by a layer of calcareous marine sediment (Fisher and Norris 1960). Chemical analyses of lavas published by Clark and Dymond (1977) and Engel and Engel (1964) identify them as alkali basalt and mugearites based on the classification in Fig. 2.2.

The $^{87}Sr/^{86}Sr$ ratios of rocks on Easter Island and on Sala y Gomez are similar and range from 0.70293 to 0.70328 (Easter Island) and from 0.703025 to 0.70327 (Sala y Gomez). Consequently, the rocks on these islands are significantly enriched in radiogenic ^{87}Sr compared to MORBs on the East Pacific Rise. The only rocks along the eastern segment of the Easter Microplate that have high $^{87}Sr/^{86}Sr$ ratios comparable to those on Easter Island and Sala y Gomez are two picrites recovered at 27°31.8' S and 112°45.6' W whose average $^{87}Sr/^{86}Sr$ ratio is 0.70321 ±0.000007 relative to 0.70800 for E&A (Fontignie and Schilling 1991). In addition, Macdougall and Lugmair (1986) recorded elevated $^{87}Sr/^{86}Sr$ ratios of

0.702840 ±0.000026 and 0.702884 ±0.000026 for two MORBs at a site on the eastern border of the Easter Microplate close to the picrites analyzed by Fontignie and Schilling (1991).

In summary, the available information suggests that the volcanic activity along the Sala y Gomez Ridge was caused by a leaky fracture and that the magmas that were erupted along it originated from a plume located along the East Pacific Rise in the vicinity of Easter Island. Consequently, the Sala y Gomez Ridge must be distinguished from the Walvis Ridge in the South Atlantic Ocean (Sect. 2.2.4) and also from the chain of the Hawaiian Island both of which formed by the progressive displacement of volcanoes which were originally located over a plume head or hotspot.

2.8.3 San Felix and San Ambrosio Islands

The islands of San Felix and San Ambrosio, located at 26°25' S and 79°59' W, were visited in 1923 by Bailey Willis (Willis and Washington 1924). They are small (<4 km^2) and consist of interbedded lava flows and pyroclastics. The volcanic activity appears to be extinct, although Willis and Washington (1924) reported seeing active fumaroles and fresh-looking pahoehoe flows on San Felix.

Isotope ratios of Sr, Nd, and Pb of basanite and tephrite measured by Gerlach et al. (1986) included $^{87}Sr/^{86}Sr$ ratios between 0.703983 and 0.704122 relative to 0.70800 for E&A. The $^{143}Nd/^{144}Nd$ ratios of the basanites on San Felix (0.512552 to 0.512585) are unusually low. For this reason, Hart et al. (1986) used them in defining the LoNd array (Fig. 2.32, Sect. 2.6.1).

The elevated $^{87}Sr/^{86}Sr$ and $^{206}Pb/^{204}Pb$ (18.913 to 19.312) ratios of the volcanic rocks on San Felix and San Ambrosio Islands indicate that the magmas erupted at this site originated from a mixtures of the DMM, EM1, and HIMU components (Gerlach et al. 1986). The alkali-enrichment and elevated Sr concentrations of the lavas (K_2O = 2.88%, Rb = 63.2 ppm, Sr = 1 341 ppm) are attributable to a small degree of partial melting of the source rocks.

Although the islands are small, they are in fact the summits of volcanic mountains each rising about 4 000 m above the bottom of the ocean. The large volume of magma that was erupted at this site implies that it originated from a large plume and that the magmas penetrated the lithospheric plate to reach the bottom of the ocean.

2.8.4 Juan Fernández Islands

The Juan Fernández Islands are located at 33°37' S and 78°50' W on an aseismic ridge north of the Challenger Fracture Zone (Fig. 2.34). This group of islands includes

Robinson Crusoe (12×5 km), Selkirk (10×6 km), and Santa Clara (3×1 km) islands. Robinson Crusoe Island contains an extensive section of interbedded basalt flows and pyroclastics divided into three units (Gerlach et al. 1986):

- Bahia del Padre (youngest)
  ~250 m, pyroclastics and some olivine basalt flows
- Puerto Ingles, 3.1–3.5 Ma
  ~1 200 m, olivine basalt, picrite, and pyroclastics
- Punta Larga (oldest)
  >800 m, basalt flows with minor interbedded pyroclastics

The $^{87}Sr/^{86}Sr$ ratios of the volcanic rocks of the Puerto Ingles unit range from 0.703512 to 0.703762 relative to 0.70800 for E&A with a mean of 0.70366 ± 0.00007 ($2\bar{\sigma}$, $N = 7$) that is indistinguishable from the average $^{87}Sr/^{86}Sr$ ratio of the older Punta Larga unit of 0.70363 ± 0.00001 (Gerlach et al. 1986). The high $^{87}Sr/^{86}Sr$, low $^{143}Nd/^{144}Nd$ (0.512844), and high $^{206}Pb/^{204}Pb$ ratios (19.11) are indications that these lavas originated from enriched sources in the mantle consisting of a mixture of DMM, EM1, and HIMU components. The lavas on Robinson Crusoe Island have significantly lower concentrations of K_2O ($0.89 \pm 0.19\%$), Rb (5.6 ± 5 ppm), and Sr (501 ± 52) ppm) and higher concentrations of SiO_2 ($47.3 \pm 6\%$) compared to those on San Felix and San Ambrosio and therefore formed by a higher degree of partial melting.

2.9 Seamounts of the Eastern Pacific Ocean

The seamounts of the eastern Pacific Ocean occur in close proximity to the East Pacific Rise (Zindler et al. 1984; Graham et al. 1988) and as linear chains such as the Pratt-Welker Chain in the northeastern Pacific (Subbarao et al. 1973; Subbarao and Hekinian 1978; Cousens et al. 1984, 1985; Cousens 1988; Allan et al. 1989). The volcanic rocks on the seamounts of the eastern Pacific Ocean vary widely in chemical composition (Engel and Engel 1964) and in the isotope ratios of Sr, Nd, and Pb. Consequently, seamounts are a manifestation of the heterogeneity of the magma sources in the suboceanic mantle and of the complexity of magma evolution by mixing, assimilation, and fractional crystallization.

2.9.1 Seamounts of the East Pacific Rise

The seamounts adjacent to the East Pacific Rise between latitudes 9° to 14° N formed on young oceanic crust (<6.8 Ma) and are composed of basalt ranging from tholeiitic to alkali-rich varieties. The samples studied by Zindler et al. (1984) display a range of alteration manifested by the presence of smectite, calcite, and ferro-

manganese material. As a result of the alteration, the $^{87}Sr/^{86}Sr$ ratios increased significantly in most cases because of the addition of Sr from seawater. However, the $^{143}Nd/^{144}Nd$ ratios of the rocks were not affected because of the low concentration of Nd in seawater. After leaching the powdered rocks with 6.2 N HCl at 80 °C overnight, Zindler et al. (1984) reported $^{87}Sr/^{86}Sr$ ratios ranging from 0.70251 (Seamount 8) to 0.70314 (Seamount 6) relative to 0.7080 for E&A.

The results for Seamount 6 are especially significant because the $^{87}Sr/^{86}Sr$ ratios of the basalts on this volcano range widely from 0.70275 to 0.70314 and because the Rb concentrations also vary from 4.37 to 34.43 ppm. The difference in the $^{87}Sr/^{86}Sr$ ratios indicates that the lavas originated from heterogeneous sources in the mantle which contained at least two components. Since the concentrations of Rb (and K) in Fig. 2.39 of the lavas on Seamount 6 are positively correlated with the $^{87}Sr/^{86}Sr$ ratios, the range of chemical compositions of the lavas is not attributable to fractional crystallization, but was more likely caused by differences in the chemical composition of the different magma sources involved in the petrogenesis of these rocks.

Zindler et al. (1984) observed that the lavas on these seamounts are also positively correlated on Rb-Sr and Sm-Nd isochron diagrams (Sect. 1.3 to 1.5) and considered whether the linear data arrays should be interpreted as isochrons or mixing lines. Regardless of which interpretation is correct in this case, the existence of a mantle component having elevated $^{87}Sr/^{86}Sr$ ratios (>0.7032)

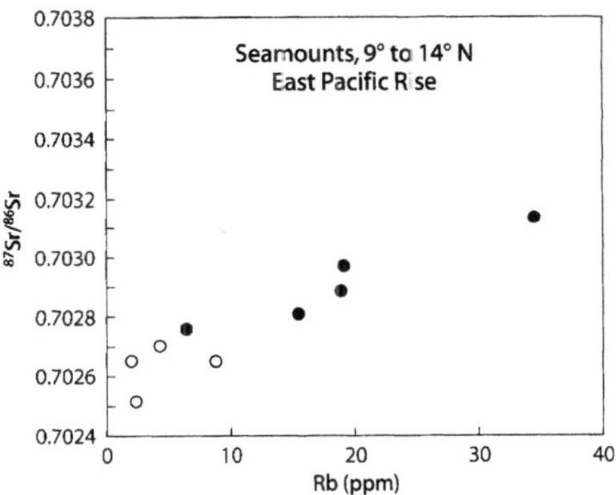

Fig. 2.39. Correlation of Rb concentrations with $^{87}Sr/^{86}Sr$ ratios of basalt samples dredged from seamounts associated with the East Pacific Rise. All samples were leached with hot HCl and the $^{87}Sr/^{86}Sr$ ratios are relative to c.7080 for E&A. The *solid circles* represent Seamount 6 at 12°45' N and 102°30' W, whereas the *open circles* represent three other seamounts in this area. The correlation of chemical and isotopic parameters indicates that the lavas on Seamount 6 originated from heterogeneous magma sources in the mantle rather than by fractional crystallization of magma derived from an isotopically homogeneous source (*Source:* Zindler et al. 1984)

indicates that its Rb/Sr ratio was increased at some time in the past, followed by a period of incubation when the $^{87}Sr/^{86}Sr$ ratio increased to its present value to become the EM1 component whose presence is reflected by the elevated $^{87}Sr/^{86}Sr$ ratios of Seamount 6.

According to Zindler et al. (1984), the EM1 component could have originated as:

1. Basaltic melts that were trapped in the mantle;
2. Remnants of oceanic crust that had been altered previously by dehydration and recrystallization during subduction;
3. Ultramafic mantle rocks that had been enriched in Rb by prior metasomatism like the peridotites of St. Peter and St. Paul Islands (Sect. 2.2.2) in the equatorial Atlantic Ocean (Roden et al. 1984a).

The proximity of these seamounts to the East Pacific Rise eliminates the need to postulate a separate plume for each of the large number of small seamounts that are scattered along it. Instead, the seamounts formed from small batches of magma that leaked through fractures in the lithospheric mantle within a short distance of the Rise (Alexander and Macdonald 1996; Sect. 2.6.3).

2.9.2 Seamounts of the Juan de Fuca Ridge

Basalt specimens dredged from the Explorer Ridge and from other sites on the western flank of the Juan de Fuca Ridge are similar to normal MORBs. The $^{87}Sr/^{86}Sr$ ratios of these rocks range from 0.7023 to 0.7026 and the concentrations of Rb and Sr are uniformly low. Therefore, Cousens et al. (1984) concluded that the lavas

erupted in this area originated by varying degrees of partial melting of a depleted source in the mantle (DMM component). However, a MORB specimen from one of the Dellwood Knolls (50°46'38" N, 130°25'12" W) has an elevated $^{87}Sr/^{86}Sr$ ratio of 0.70279 after leaching with HCl, suggesting that the EM1 component contributed to the formation of this seamount. However, farther south on Cobb Seamount (47°05' N, 130°45' W) transitional basalts have low $^{87}Sr/^{86}Sr$ ratios consistent with those of normal MORBs (Subbarao et al. 1973).

2.9.3 Pratt-Welker Seamount Chain

The Pratt-Welker Seamount chain in Fig. 2.40 extends from the Queen Charlotte Islands off the coast of British Columbia (Canada) in a northwesterly direction for about 1100 km. This chain includes the Kodiak, Giacomini, and Hodgkins Seamounts (Subbarao et al. 1973), as well as the Bowie and Tuzo Wilson Seamounts (Cousens et al. 1985; Cousens 1988; Allan et al. 1993). A possible explanation for the origin of these seamounts is that they were formed above the Pratt-Welker Plume located a short distance south of the Queen Charlotte Islands and are being transported towards the Aleutian Trench along the southwest coast of Alaska (Cousens et al. 1985). The unusual aspect of the volcanic rocks on some of the Pratt-Welker Seamounts is illustrated in Fig. 2.41a which demonstrates that they are alkali-rich like typical OIBs, but have low $^{87}Sr/^{86}Sr$ ratios like normal MORBs.

Another noteworthy feature of the Bowie Seamount is the correlation of the $^{87}Sr/^{86}Sr$ and $^{206}Pb/^{204}Pb$ ratios of the rocks in Fig. 2.41b. Therefore, the magmas that formed this seamount were mixtures derived from the

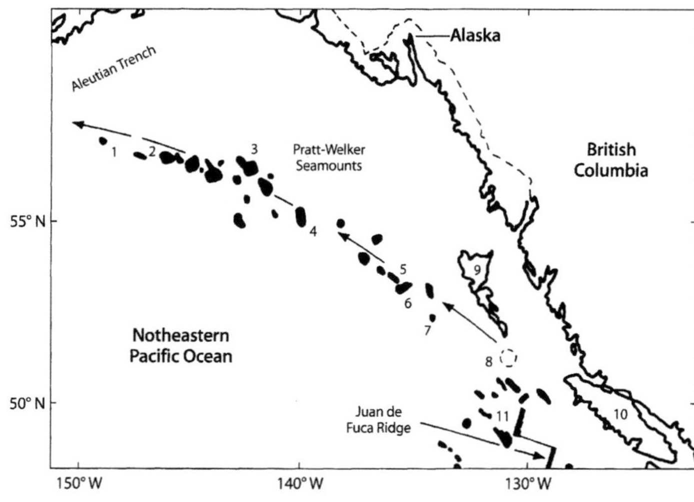

Fig. 2.40.
The Pratt-Welker chain of seamounts in the northeastern Pacific Ocean. The seamounts (*black*) and islands (*white*) mentioned in the text are identified by number: *1*. Kodiak; *2*. Giacomini; *3*. Pratt; *4*. Welker; *5*. Hodgkins; *6*. Bowie; *7*. Oshawa; *8*. Tuzo Wilson Seamounts; *9*. Queen Charlotte Islands; *10*. Vancouver Islands; *11*. Explorer Seamount. Several coastal islands were omitted for the sake of clarity. The arrows indicate the direction of motion of the Pacific Plate which is transporting these seamounts into the Aleutian Trench (*Source:* adapted from Cousens et al. 1985)

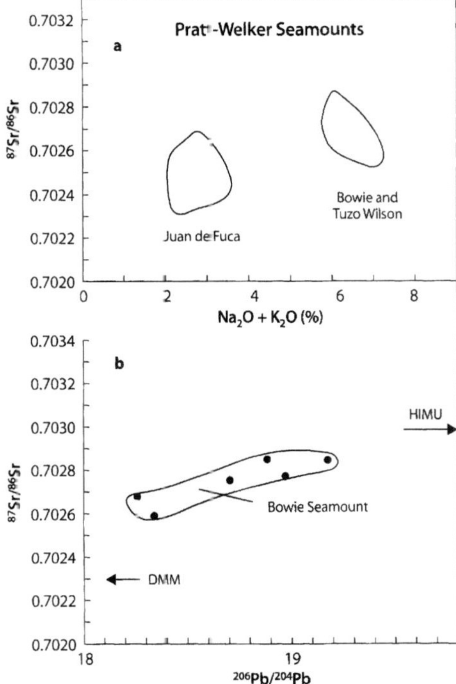

DMM and the HIMU components. Calculations by Cousens (1988) indicated that the abundance of the DMM component ranged from 60 to 80%.

Some aspects of the origin of the Pratt-Welker Seamounts remain uncertain. The samples dredged from the Tuzo Wilson Seamounts (Cousens et al. 1985) are fresh, glassy pillow basalts of Recent age, as expected if these seamounts are located above or close to a mantle plume south of the Queen Charlotte Islands (Fig. 2.39). However, the origin of the Bowie Seamount (no. 6 in Fig. 2.40) is less certain because, even though the magnetic anomaly patterns indicate an age of 16 Ma (Late Miocene) for the oceanic crust on which it formed, the summit shows evidence of Recent volcanic activity (Herzer 1971). If the Bowie Seamount formed above a plume at the site presently occupied by the Tuzo Wilson Seamounts, it has been displaced from its source by about 625 km in 16×10^6 yr at a rate of about 4 cm yr^{-1}. The antiquity of the Bowie Seamount is consistent with its flat top (Herzer 1971) but the source of the Recent volcanic activity is not clear. An alternative explanation for the origin of some or all of the Pratt-Welker Seamounts is that they formed on the crest of the Juan de Fuca Ridge or its extension known as the Explorer Ridge and have been displaced from it by seafloor spreading.

Fig. 2.41. a Alkali concentrations and $^{87}Sr/^{86}Sr$ ratios of alkali-rich volcanic rocks (hawaiites, tephrites, mugearites) on the Tuzo Wilson and Bowie Seamounts (Pratt-Welker Chain) and of MORBs on the Juan de Fuca Ridge between latitudes 44°37′ and 49°03′ N. The similarity of the $^{87}Sr/^{86}Sr$ ratios indicates that both suites of volcanic rocks originated primarily from depleted source rocks in the mantle (DMM component) in spite of the evident alkali enrichment of the seamount lavas; **b** The positive correlation of $^{87}Sr/^{86}Sr$ and $^{206}Pb/^{204}Pb$ ratios of the alkali-rich lavas on the Bowie Seamount is the result of mixing of magmas derived from the DMM and the HIMU components (*Sources:* data from Cousens et al. 1985; Cousens 1988)

2.10 The Island Chains of Polynesia

The linear chains of volcanic islands of Polynesia in Fig. 2.42 trend southeast to northwest and include, from north to south, the Marquesas Islands, the Tuamotu Islands, the Duke of Gloucester Islands, the Society Islands, and the Austral-Cook Islands all of which are located on the Pacific Plate and are not associated with a spreading ridge. In general, the volcanic rocks of Polynesia have high $^{87}Sr/^{86}Sr$ ratios compared to normal MORBs be-

Fig. 2.42.
Location of the linear chains of islands of Polynesia including Pitcairn Island at the southeastern end of the Duke of Gloucester Chain (*Source:* adapted from Duncan and McDougall 1976)

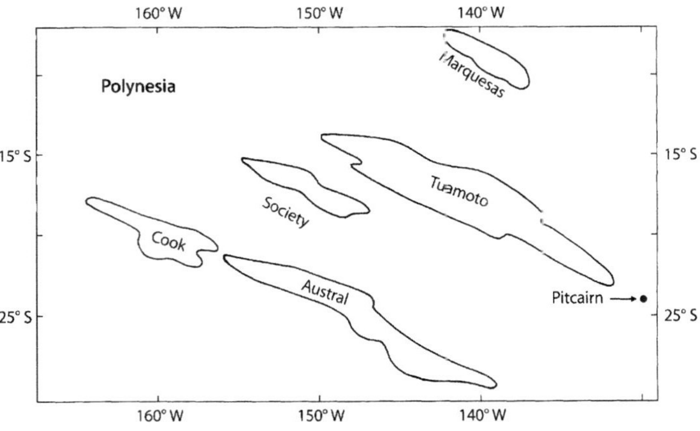

cause they originated from plumes having higher Rb/Sr ratios than the magma sources of MORBs. Age determinations by Duncan and McDougall (1976) indicate that the ages of the islands increase from southeast to northwest implying that the islands formed above stationary plumes and were subsequently displaced in a northwesterly direction by the movement of the Pacific Plate (Mayes et al. 1990).

2.10.1 Pitcairn and the Gambier Islands

Pitcairn Island in Fig. 2.43 is notorious as the refuge of Fletcher Christian and his followers who may have chosen this small island (~5 km^2) because of its remote location and steep unapproachable coast (Nordhoff and Hall 1936). In any case, Pitcairn at 24°04' S and 130°06' W and Mangareva in the Gambier Islands are the only volcanic islands among the coral atolls of the Duke of Gloucester chain of islands (Fig. 2.43) that stretch for 1750 km in a northwesterly direction parallel to and south of the Tuamotu Ridge (Rocaboy et al. 1987; Dupuy et al. 1993).

Pitcairn Island, at first glance, appears to be related to the East Pacific Rise by its association with the Tuamotu Ridge which may be viewed as being complementary to the Sala y Gomez Ridge (Fig. 2.34). Both ridges arise in the vicinity of the Easter Microplate but extend in opposite directions. These hypothetical relations of Pitcairn Island to the East Pacific Rise were disproved by Duncan et al. (1974) who reported that lavas from surface exposures on Pitcairn Island crystallized between 0.93 and 0.45 Ma.

The young age of the lavas is not consistent with the hypothesis that Pitcairn Island formed on the crest of the East Pacific Rise, because the island is much too far (about 1700 km) west of the Rise. Therefore, Pitcairn Island presumably formed over a nearby mantle plume and was subsequently displaced from it in a northwest-

erly direction consistent with the direction of motion of the Pacific Plate.

Duncan et al. (1974) also cited K-Ar dates of basalts recovered from the Gambier Islands of the Duke of Gloucester Chain (Fig. 2.43). The oldest date of 8 Ma was obtained for basalt in a drill core recovered at Mururoa Atoll located 850 km northwest of Pitcairn Island measured along the Duke of Gloucester Chain. Accordingly, the spreading rate implied by these results is 11 cm yr^{-1}. Duncan et al. (1974) pointed out that other investigators have reported similar spreading rates for other island chains in the Pacific Ocean ranging from 9 cm yr^{-1} (Austral Islands) to 15 cm yr^{-1} (Hawaiian Islands).

The volcanic rocks on Pitcairn Island are alkali-rich and range in composition from olivine basalt to hawaiite, mugearite, and some trachyte. The rocks were mapped by Carter (1967) and are arranged here in order of increasing age on the basis of K-Ar dates published by Duncan et al. (1974):

- Christians Cave Formation (youngest)
- Adamstown volcanics
- Pulawana volcanics
- Tedside volcanics (oldest)

The ^{87}Sr/^{86}Sr and ^{143}Nd/^{144}Nd ratios of these units form a linear array in Fig. 2.44 indicating that the magmas originated from mixtures of the DMM and the EM1 components located in the head of the Pitcairn Plume and the overlying suboceanic lithospheric mantle (Woodhead and McCulloch 1989). The clustering of isotope ratios in Fig. 2.44 indicates that the proportions of the mantle components changed with time such that the EM1 component initially dominated in the Tedside magmas, whereas the DMM component later dominated in the Adamstown and Christians Cave magmas. Two samples of the Pulawana unit have intermediate ^{87}Sr/^{86}Sr and ^{143}Nd/^{144}Nd ratios consistent with their intermediate age. Woodhead and McCulloch (1989) concluded that EM1 was derived from pelagic sediment after it had been modified by passing through a subduction zone. Accordingly, these authors proposed that the Tedside volcanics formed by partial melting of EM1 in the form of a sediment residuum formed between 1 and 2 Ga by subduction of pelagic sediment. The Adamstown and Christians Cave units formed subsequently by partial melting of altered oceanic crust that formed the Pitcairn Plume after the sediment residuum had become exhausted during the eruption of the Pulawana lavas. Therefore, the Pitcairn Plume was originally composed primarily of recycled oceanic crust with minor amounts of sediment residuum.

A cluster of six seamounts was discovered in 1989 at about 25°15' S and 129°20' W southeast of Pitcairn Island (Stoffers et al. 1990). Dredge samples recovered from Volcanoes 1, 2, and 5 contained fresh basalt glass suggest-

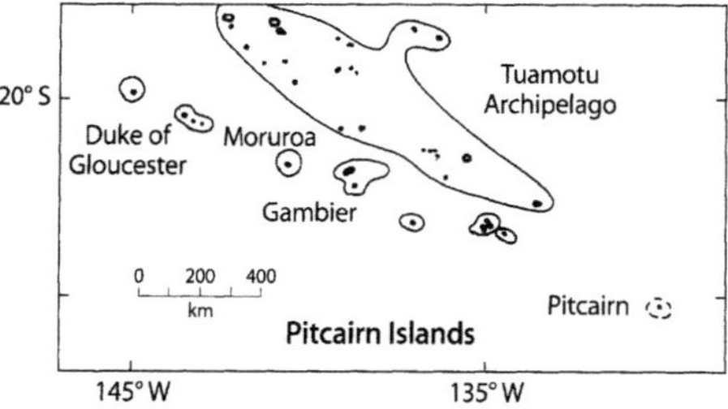

Fig. 2.43. Pitcairn Island in the Duke of Gloucester Chain of islands in Polynesia. The 2000-m depth contour is indicated for the Pitcairn, Gambier, and Duke of Gloucester Islands. The Tuamotu Archipelago is only identified diagramatically (*Source:* adapted from Duncan et al. 1974)

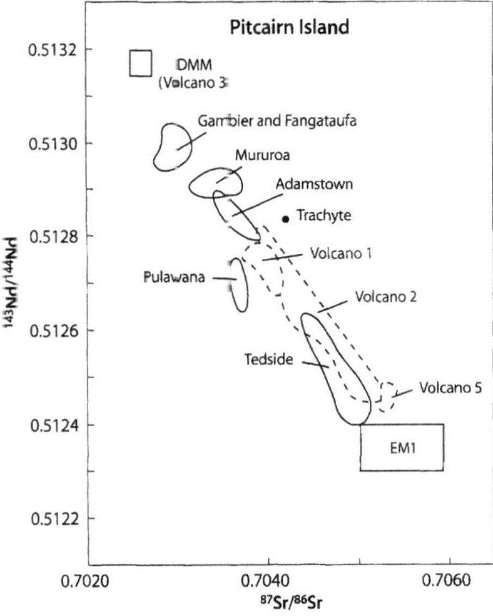

Fig. 2.44. Range of isotope ratios of Sr and Nd in the volcanic rocks of Pitcairn Island and of the Pitcairn Seamounts. The isotope ratios of the lavas on Pitcairn Island evolved with time from the Tedside lavas (0.95 to 0.76 Ma) to the Pulawana lavas (0.67 to 0.62 Ma) and the Adamstown and Christians Cave volcanics (0.63 to 0.45 Ma). The isotope ratios of the youngest seamounts (*Volcanoes 1, 2, and 5*) encompass the same range as the lavas on Pitcairn Island. A tholeiite on *Volcano 3* has Sr and Nd isotope ratios of normal MORBs and apparently formed at 20 to 30 Ma when this volcanic mountain was located on the East Pacific Rise (*Source:* data from Woodhead and McCulloch 1989; Woodhead and Devey 1993)

ing that these volcanoes may have recently erupted magma from the Pitcairn Plume. The other seamounts (3, 4, and 6) have flat tops covered by coral sand. Therefore, they may actually have formed on the East Pacific Rise more than 20 to 30 millions of years ago (Woodhead and Devey 1993).

The isotope ratios of Sr and Nd of the lavas on Volcano 1 cluster close to the Pulawana lavas of Pitcairn Island in Fig. 2.44, whereas those of Volcano 2 span the range from the Adamstown to the Tedside lavas. One specimen of tholeiite (MORB) on Volcano 3 originated from depleted magma sources in the mantle unlike all other samples from this area. The lavas of Volcano 5 analyzed by Woodhead and Devey (1993) have elevated ^{87}Sr/^{86}Sr ratios and low ^{143}Nd/^{144}Nd ratios similar to those of the EM1 components. In this way, the isotope ratios of Sr and Nd of volcanic rocks on the Pitcairn Seamounts 1, 2, and 5 are complementary to those of the lavas on Pitcairn Island. The presence of MORBs on Volcano 3 confirms that this seamounts is not genetically related to the Pitcairn Plume because it originated on the East Pacific Rise and was transported westwards by

the movement of the Pacific Plate. Woodhead and Devey (1993) did not report the isotope ratios of Sr and Nd in rocks of Volcano 4 and 6.

The isotope ratios of Sr, Nd, and Pb of basalt on Mururoa, Fangataufa, and Gambier Islands reported by Dupuy et al. (1993) differ from those on Pitcairn Island: ^{87}Sr/^{86}Sr = 0.70298 to 0.70368; ^{143}Nd/^{144}Nd = 0.51303 to 0.51288; ^{206}Pb/^{204}Pb = 18.99 to 19.62 relative to 0.71025 for NBS 987 (Sr) and corrected to ^{146}Nd/^{144}Nd = 0.7219 (Nd). Nevertheless, the isotope ratios of Sr and Nd of the basalt on these islands are collinear with those of the lavas on Pitcairn Island in Fig. 2.44 but approach the isotope composition of the DMM component more closely. Dupuy et al. (1993) concluded that the lavas on the islands of Mururoa, Fangataufa, and Gambier (all of which are older than the lavas of Pitcairn Island) formed by varying degrees of partial melting in the suboceanic lithosphere as it passed over the plume located southeast of the present Pitcairn Island.

2.10.2 Marquesas Islands

The Marquesas Archipelago of Polynesia in Fig. 2.45 is located at about 9°24' S and 140°00' W and consists of about 20 islands and several seamounts whose ages increase from 1.4 Ma (Fatu Hiva) in a northwesterly direction to 6.3 Ma (Eiao) at the end of the chain (Duncan and McDougall 1974; Duncan et al. 1986). The chemical compositions of the volcanic rocks of these islands range from tholeiite to trachytes and phonolites and their ^{87}Sr/^{86}Sr ratios extend from 0.70288 to 0.70656 (Duncan and Compston 1976; Vidal et al. 1984; Liotard et al. 1986; Duncan et al. 1986; Dupuy et al. 1987; Woodhead 1992).

The volcanic activity of Ua Pou (Fig. 2.45) lasted from 5.6 to 1.8 Ma and encompassed the entire range of rock types from early-formed tholeiites to the later phonolite. Duncan et al. (1986) pointed out that the ^{87}Sr/^{86}Sr ratios of the rocks on Ua Pou *increased* with time such that the late-forming alkali-rich rocks have higher ^{87}Sr/^{86}Sr

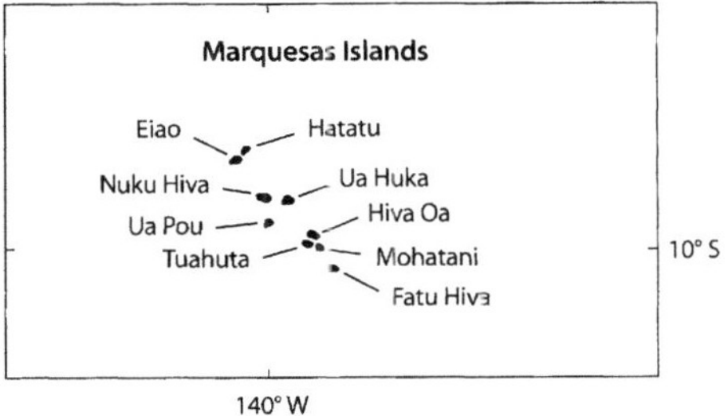

Fig. 2.45. The Marquesas Islands of Polynesia in the South Pacific Ocean. Only the islands mentioned in the text are identified by name (*Source:* adapted from Dupuy et al. 1987)

ratios than the alkali-poor tholeiites which are the dominant rock type in the submarine portion of the island. This trend applies not only to Ua Pou, but also to Hiva Oa and Nuku Hiva for which sufficient data are available (Woodhead 1992). The differences in the ^{87}Sr/^{86}Sr ratios indicate that the island-building tholeiites and the late-stage alkali-rich volcanics of the Marquesas Islands are *not* the products of fractional crystallization of the same parent magma, but actually originated from different sources. In this case, the tholeiite magma formed by partial melting in the suboceanic lithospheric mantle which had lower Rb/Sr ratios than the magma sources of the alkali-rich rocks. This relationship is *opposite* to that in the Hawaiian Islands where the early-formed shield-building tholeiites have higher ^{87}Sr/^{86}Sr ratios than the late-stage alkali-rich rocks. (Woodhead 1992).

The isotope ratios of Sr and Nd of the lavas on the Marquesas Islands in Fig. 2.46 can be explained by mixing of three components including DMM, HIMU, and a Rb-enriched magma source (EC) characterized by ^{87}Sr/^{86}Sr > 0.710 and ^{143}Nd/^{144}Nd ≈ 0.5122. Dupuy et al. (1987) proposed that this enriched component is either a small mass of subcontinental lithosphere or subducted oceanic crust and marine sediment. Trace-element ratios favor the second alternative, but allowance must be made for alteration of the marine sediment by dehydration and recrystallization during subduction. Tholeiites

on the island of Ua Pou have low ^{87}Sr/^{86}Sr (<0.7032), but high ^{206}Pb/^{204}Pb ratios (19.5 to 20.0) and plot close to the HIMU component in Fig. 2.46.

The progressive increase of the ages of the volcanic rocks on this chain of islands (Duncan and McDougall 1974) indicates that the Marquesas Islands have been displaced by the northwesterly movement of the Pacific Plate from a magma source that is located in the mantle southeast of the island of Fatu Hiva (Fig. 2.45). This magma source may be a plume that caused partial melting in the lithospheric mantle to produce voluminous shield-building tholeiite basalts followed by partial melting of plume rocks (subducted oceanic crust and sediment residuum) to form the late-stage alkali-rich lavas. Woodhead (1992) and Dupuy et al. (1987) concluded that the chemical and isotope compositions of the late stage alkali basalts are the result of mixing of magmas derived from the different source components and did not form by varying degrees of partial melting of the same magma source. This explanation of the origin of the volcanic rocks on the Marquesas Islands may apply also to the lavas on the other island chains of Polynesia.

The prevalence of shield-building tholeiites having low ^{87}Sr/^{86}Sr and high ^{143}Nd/^{144}Nd ratios means that the Marquesas Plume was not strong enough to penetrate the Rb-depleted rocks of the lithosphere. Instead, the

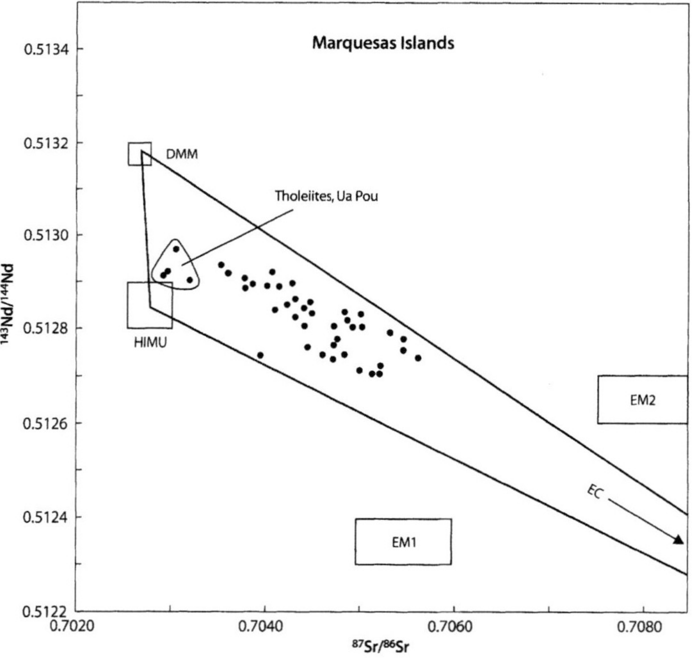

Fig. 2.46.
Isotope ratios of Sr and Nd in volcanic rocks on the islands of the Marquesas Archipelago of Polynesia. The range of isotope compositions is attributable to mixing of magmas derived from three components in the mantle: DMM, HIMU, and an enriched component (EC). Dupuy et al. (1987) tentatively identified EC as subducted oceanic crust and oceanic sediment having ^{87}Sr/^{86}Sr > 0.710 and ^{143}Nd/^{146}Nd = 0.5122 (*Sources:* data from Vidal et al. 1984; Dupuy et al. 1987; Woodhead 1992)

head of the plume expanded along the underside of the overlying lithosphere allowing heat to be transferred into it. Consequently, the early-formed tholeiites originated primarily from the lithosphere, and therefore their isotope ratios of Sr and Nd approach those of MORBs (e.g. Ua Pou, Fig. 2.46). The alkali-rich lavas that were erupted after a period of quiescence formed as a result of small degrees of partial melting of rocks within the plume itself. Woodhead (1992) therefore used the Marquesas Islands to characterize the properties of "weak" plumes in contrast to "strong" plumes that penetrate into the lithosphere causing the early-formed magmas to have high $^{87}Sr/^{86}Sr$ and low $^{143}Nd/^{144}Nd$ ratios because they contain an appreciable fraction of plume-derived melt.

2.10.3 Society Islands

The Society Islands in Fig. 2.47 are located south of the Tuamotu Archipelago and include Mehetia, Taiarapu, Tahiti, Moorea, Huahine, Raiatea, Tahaa, Bora Bora, and Maupiti (from southeast to northwest). The ages of volcanic rocks on these islands increase in a northwesterly direction from 0.43 ±0.04 Ma (Taiarapu) to 4.26 ±0.12 Ma (Maupiti) and yield an average velocity of 11.1 cm yr^{-1} for the Pacific Plate at this location. (Duncan and McDougall 1976). The lavas that are exposed on these islands are generally alkali basalts and related differentiates including phonolites and benmoreites (Dostal et al. 1982; Devey et al. 1990).

The $^{87}Sr/^{86}Sr$ ratios range widely from 0.70360 (Tahiti; Duncan and Compston 1976) to 0.70694 (Tahaa; White and Hofmann 1982). More recent studies by Hémond et al. (1994a) and Devey et al. (1990) have established that the late-stage alkali-rich lavas of the Society Islands have higher $^{87}Sr/^{86}Sr$ ratios than the shield-building tholeiites.

Although the volcanoes on the Society Islands are extinct, a group of seamounts located southeast of Tahiti are still active. They include Teahitia, Rocard, Cyana, and Moua Pihaa. Rock samples recovered from these seamounts and from the small island of Mehetia consist primarily of alkali basalt, but include a tholeiite having an $^{87}Sr/^{86}Sr$ ratio of 0.702806 relative to 0.708000 for E&A (Devey et al. 1990). Lavas with Rb concentrations greater than 100 ppm on Rocard Seamount have a high average $^{87}Sr/^{86}Sr$ ratio of 0.70594 ±0.00005 ($2\bar{\sigma}$, $N = 8$), whereas all alkali basalts on Tahiti, Moorea, Mehetia, Teahitia, and Cyana analyzed by Devey et al. (1990) and Hémond et al. (1994a) yield 0.70459 ±0.000098 ($2\bar{\sigma}$, $N = 35$) relative to 0.70800 for E&A and 0.71025 for NBS987. Alkali basalts on Moua Pihaa Seamount analyzed by these authors have a comparatively low average $^{87}Sr/^{86}Sr$ ratio of 0.70361 ±0.00006 ($2\bar{\sigma}$, $N = 10$).

The isotope ratios of Sr and Nd of lavas on the seamounts of the Society Islands in Fig. 2.48 form a linear array consistent with their derivation from two mantle components consisting of Rb-depleted and Rb-enriched rocks in the mantle. The Rb-enriched component (EC) may be a plume composed in part of subducted oceanic crust (Devey et al. 1990). The depleted source may be the lithospheric mantle where melting occurred because of heat provided by the plume. The $^{206}Pb/^{204}Pb$ ratios reported by Devey et al. (1990) range from 18.993 (Cyana) to 19.221 (Rocard) indicating that the contribution of the HIMU component to the lavas on the seamounts of the Society Islands is less than in the tholeiites on Ua Pou in the Marquesas Islands.

2.10.4 Austral and Cook Islands

The Austral Islands in Fig. 2.47 form a chain that includes Marotiri at the southeastern end followed in a northwesterly direction by the islands of Rapa, Raivavae, Tubuai, Rurutu, and Iles Maria. The seamount Macdonald, located southeast of Marotiri, is an active volcano which has erupted several times in the 20th century (Dupuy et al. 1988a; Hémond et al. 1994a).

The Austral Islands continue in a northwesterly direction and merge with the chain of Cook Islands which include Mangaia, Rarotonga, Mauke, Atiu, Mitiaro, Takutea, Aitutaki, and the Suvorov Islands, all of which are identified in Fig. 2.47. However, K-Ar dates compiled by Chauvel et al. (1997), including the work of Turner and Jarrard (1982), Matsuda et al. (1984), Duncan and McDougall (1976), and Dalrymple et al. (1975a), indicate that these islands form three separate arrays in coordinates of age vs. distance. The slopes of these arrays yield an average spreading rate of 11 cm yr^{-1} in agreement with the spreading rates indicated by the Pitcairn-Gambier, Marquesas, and Society Islands.

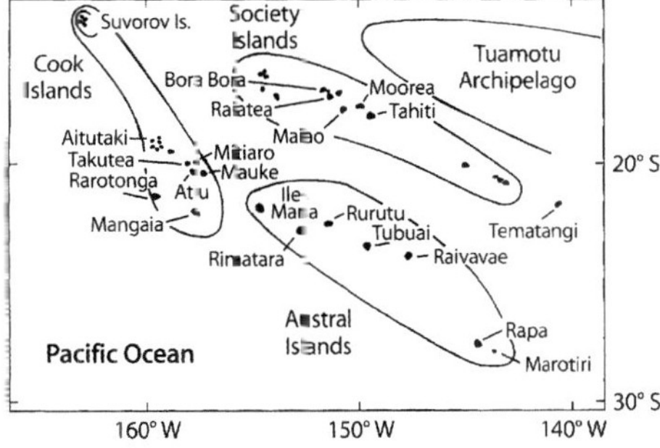

Fig. 2.47. Island groups in the South Pacific, including the Tuamotu Archipelago as well as the Society, Austral, and Cook Island chains (*Source:* adapted from National Geographic Society 1990)

Fig. 2.48.
Isotope ratios of Sr and Nd of lavas on the seamounts located southeast of Tahiti in the Society Islands (Fig. 2.47). The linear distribution of data points is consistent with mixing of magma derived from two components. The enriched component (EC) may be subducted oceanic crust in the plume, whereas the depleted source may be the lithospheric mantle (*Source:* White and Hofmann 1982; Devey et al. 1990; Hémond et al. 1994)

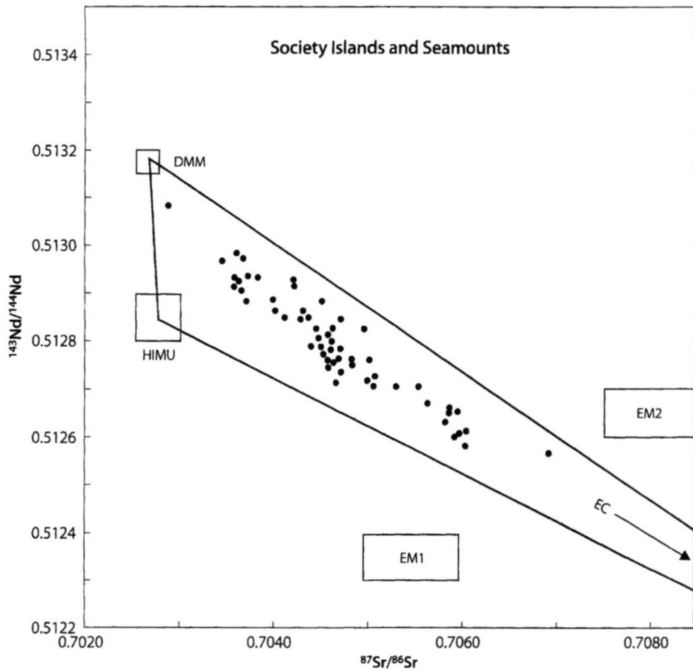

Fig. 2.49.
Isotope ratios of Sr and Nd of volcanic rocks on the Austral and Cook Islands, Polynesia. Certain localities are identified by number: 1. Tubuai, Rurutu (older suite) and Mangaia; 2. Rurutu (younger suite). The distribution of data points indicates that the lavas on Tubuai, Mangaia, and Rurutu (older suite) originated from the HIMU component in the mantle. The Cook Islands are consistently enriched in the Rb-enriched component (EC) compared to the Austral Islands. (*Sources:* Austral Islands, Palacz and Saunders 1986; Chauvel et al. 1992, 1997; Hémond et al. 1994a. Cook Islands, Palacz and Saunders 1986; Nakamura and Tatsumoto 1988; Woodhead 1996)

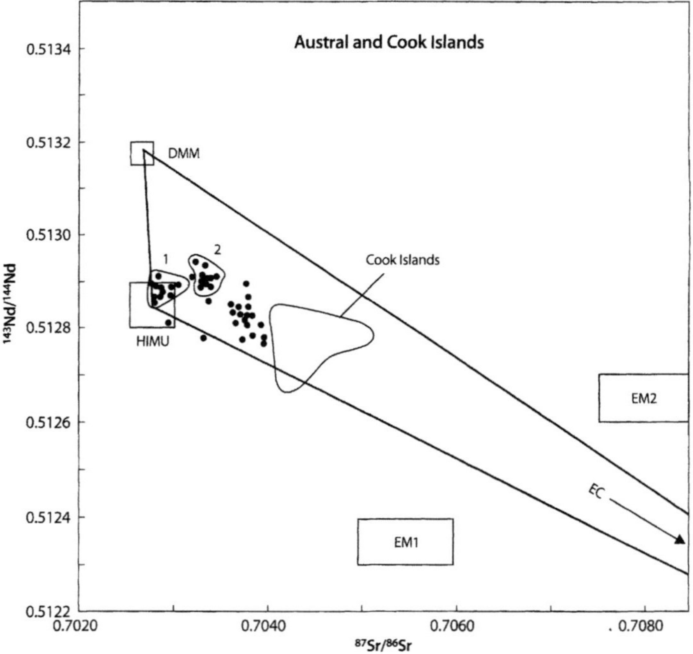

The $^{87}Sr/^{86}Sr$ and $^{143}Nd/^{144}Nd$ ratios of the Austral Islands and Macdonald Seamount in Fig. 2.49 are clustered close to the HIMU component. The lavas on Tubuai (Vidal et al. 1984; Chauvel et al. 1992) and the older volcanics on Rurutu (13 to 10.8 Ma, Chauvel et al. 1997) actually overlap the isotope ratios of Sr and Nd of HIMU (Hart 1988). The range of $^{87}Sr/^{86}Sr$ and $^{143}Nd/^{144}Nd$ ratios confirms that the lavas on the other Austral Islands were derived from mixtures of HIMU and a Rb-enriched component (EC).

The $^{87}Sr/^{86}Sr$ and $^{143}Nd/^{144}Nd$ ratios of the lavas on the Cook Islands form a tight cluster in Fig. 2.49 indicating that they too originated from magmas containing contributions from HIMU and EC in the mantle (Palacz and Saunders 1986; Nakamura and Tatsumoto 1988). Some of the lavas of Rarotonga actually transcend the HIMU-EC mixing line, perhaps because of a small contribution from EM1.

The hypothetical mixing lines from HIMU and DMM to EC in Figs. 2.46 (Marquesas), 2.48 (Society), and 2.49 (Austral and Cook Islands) were purposely drawn to have the same slopes in order to emphasize that the same enriched component may be present in the magma sources of these Polynesian islands. The isotope ratios of Sr and Nd of this enriched component differ from those assigned by Hart (1988) to EM1 and EM2.

The identity of the EC component of Polynesia is indicated in Fig. 2.50 by the $^{206}Pb/^{204}Pb$ ratios of the lavas which are strongly clustered about a value of about 19.0. The only exceptions are the lavas of Tubuai (Vidal et al. 1984; Chauvel et al. 1992), the older volcanics on Rurutu (13 to 10.8 Ma, Chauvel et al. 1997) the lavas of Mangaia (Palacz and Saunders 1986; Nakamura and Tatsumoto 1988; Woodhead 1996), and the shield-building tholeiites on Ua Pou (Marquesas Islands; Dupuy et al. 1987). The $^{206}Pb/^{204}Pb$ ratios of these lavas rise to values of 21.0 or higher indicating that they contain significant contributions from the HIMU component (Sect. 2.6.1). Nevertheless, the isotope ratios of Pb in the Polynesian lavas in Fig. 2.50 indicate clearly that the Rb-enriched component is EM2. Pitcairn Islands and the associated seamounts in Fig. 2.44 and the lavas of Rarotonga (Cook Islands) are an exception to this generalization because the enriched component in these lavas is EM1.

2.10.5 Lessons about the Mantle from the Lavas of Polynesia

The islands of Polynesia are the summits of large shield volcanoes whose elevations above the local seafloor range from 3 to 4 km and whose circumference at the base is much larger than that of the islands. In most cases, the lavas that formed the large subaqueous parts of these volcanoes differ in chemical and isotopic composition from the most recently erupted lavas that are exposed on the islands. The two stages of volcanic activity are separated by an interval of time of sufficient duration to permit extensive erosion of the early-formed lavas. In addition, the interruption of volcanic activity lasted long enough in some cases for the original volcanic edifice to collapse to form a caldera (e.g. Pitcairn Island, Sect. 2.10.1). For this reason, the volcanic rocks of the younger suite are sometimes described as the "post-caldera lavas."

In most cases, the youngest lava flows on the islands of Polynesia are more alkali-rich and have different isotope ratios of Sr, Nd, and Pb than the older suite of rocks. In addition, the rocks of both suites may have a range of chemical compositions either as a result of differentiation by fractional crystallization in a magma chamber, or as a result of changes in the degree of partial melting, or because of assimilation of rocks from the oceanic crust.

The island of Rurutu (Austral Islands) provides a good example of the bimodal character of the volcanic

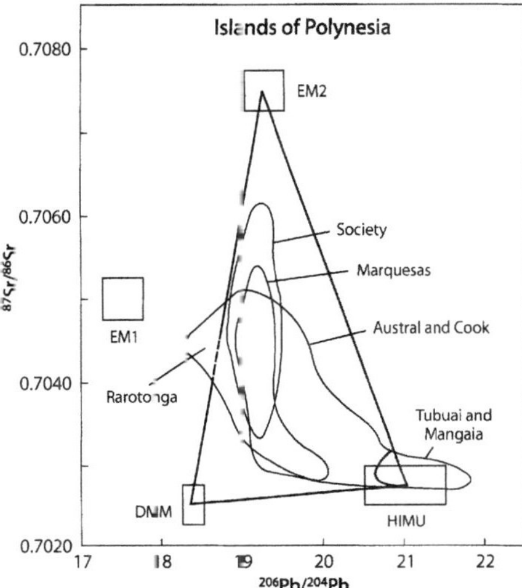

Fig. 2.50. Isotope ratios of Sr and Pb of volcanic rocks on the islands and seamounts of Polynesia. The volcanic rocks of Tubuai and Rurutu (older suite) in the Austral Islands and of Mangaia (Cook Islands) originated form the HIMU magma source in the mantle. The lavas on the other Austral and Cook Islands as well as those on the Marquesas and Society Islands are primarily mixtures of magmas derived from HIMU and EM2. The lavas of Rarotonga (Cook Islands) have low $^{206}Pb/^{204}Pb$ and high $^{87}Sr/^{86}Sr$ ratios consistent with the presence of the EM1 component in the magma sources. This diagram reveals that the enriched source (EC) in the lavas of the Marquesas and Society Islands (Figs. 2.46 and 2.48) is EM2 rather than EM1 (*Sources:* Marquesas Islands, Dupuy et al. 1987; Duncan et al. 1986; Vidal et al. 1984. Society Islands, Hémond et al. 1994a; Devey et al. 1990. Austral Islands, Chauvel et al. 1992, 1997. Cook Islands, Palacz and Saunders 1986; Nakamura and Tatsumoto 1988; Woodhead 1996)

rocks of the Polynesian Islands. In addition, the lavas on this island raise questions about the sources of the magmas that were erupted during the two stages of volcanic activity. The old volcanic suite on Rurutu consists primarily of basalt with some alkali basalt and hawaiite all of which were extruded between 13 and 10.8 Ma. The young volcanic rocks, which formed about 9 million years later between 1.8 and 1.1 Ma, consist of alkali-rich basanite, hawaiite, and nephelinite. They are separated from the old suite by a layer of coral limestone that is about 100 m thick (Chauvel et al. 1997).

Age determinations compiled by Chauvel et al. (1997) indicate that the old volcanic suite on Rurutu lies on the Tubuai trend that includes Mangaia, Tubuai, Raevavae, and Rapa all of which formed above the plume that is presently generating the volcanic activity of the Macdonald Seamount (Fig. 2.47). The young lavas on Rurutu originated from a different plume that produced the Atiu trend of the Cook Islands including Mauke, Atiu, and the older suite of lavas on Aitutaki. Therefore, the two suites of lavas on Rurutu appear to have originated from two different plumes that gave rise to the Tubuai trend of the Austral Islands and to the Atiu trend of the Cook Islands even though the island of Rurutu is clearly a member of the Austral Islands.

This confusing situation is made even more complicated by the fact that the isotope ratios of Sr, Nd, and Pb of the lavas of the southern Austral Islands (Macdonald, Rapa, and Marotiri) differ systematically from those of the lavas on the northern Austral Islands (old suite on Rurutu, Tubuai, Rimatara, and Mangaia). In addition, the isotope ratios of the young suite of lavas on Rurutu do *not* actually agree with those of the southern Cook Islands (Mauke, Atiu, and Aitutaki). Therefore, Chauvel et al. (1997) considered the possibility that the volcanic activity on the Austral and Cook Islands was caused by one large and isotopically heterogeneous superplume that penetrated the lithosphere in different places and caused simultaneous volcanic activity on different islands. In this case, the isotope ratios of Sr, Nd, and Pb of lavas erupted on different islands at the same time would depend on the particular magma sources in the plume and or lithospheric mantle, whereas the chemical composition of the lavas would depend on:

1. The chemical composition of the respective magma sources;
2. The degree of partial melting;
3. The contamination of the resulting magma by assimilation of rocks from the lithospheric mantle and/or the oceanic crust; and
4. The differentiation of magma by fractional crystallization in a magma chamber.

The existence of superplumes in the Pacific Ocean was supported by Kogiso et al. (1997a) and Tatsumi et al. (1998) following earlier proposals by Larson (1991a,b) based on the presence of seafloor plateaus such as the Ontong Java and Manihiki Plateaus in the equatorial Pacific Ocean west of Polynesia (Mahoney and Spencer 1991; Mahoney et al. 1993). Tatsumi et al. (1998) postulated that the presence of basalt lavas having elevated $^{206}Pb/^{204}Pb$ ratios in Polynesia suggests that the Late Tertiary volcanic activity of this region is the surface expression of a superplume that originated during the Cretaceous Period from oceanic crust that had been subducted into the lower mantle. According to this interpretation of both geophysical and geochemical data, the plumes associated with Marquesas, Tuamotu, Pitcairn-Gambier, Society, and Austral/Cook Chains all arose from the same superplume whose presence had caused regional uplift and the formation of basalt plateaus during the Cretaceous Period (Sect. 2.11).

On the basis of the trace element concentrations and isotope ratios of Pb of the young lavas on Rurutu, Chauvel et al. (1997) concluded, in agreement with Green and Wallace (1988), that the magma had been contaminated by assimilating carbonatites that had previously intruded the lithospheric mantle during the first phase of volcanic activity. The actual presence of carbonatite veins in the lithospheric mantle under the Austral Islands was inferred by Hauri et al. (1993), Hémond et al. (1994a), and Schiano et al. (1992). Hauri et al. (1993) based their conclusions concerning the metasomatic alteration of the lithospheric mantle by carbonatite liquids on the study of a xenolith collected on Tubuai. The occurrence of carbonatites on oceanic islands has also been recorded on Fuerteventura (Canary Islands, Sect. 2.4.2) and São Vicente (Cape Verde Islands, Sect. 2.5.2). Chauvel et al. (1997) estimated that the young lavas on Rurutu contained less than about 10% of the carbonatic liquid in order to explain the observed Ce/Pb, Ti/Eu, and Zr/Hf ratios of these lavas using the distribution coefficients of Green (1994). However, the isotope ratios of Sr, Nd, and Pb that identify the HIMU, EM1, and EM2 components reflect the ancestry of the *plume* rocks rather than of the *carbonatites* because the carbonatites in the lithospheric mantle are not old enough to have acquired the isotope ratios of the mantle components. Therefore, the assimilation of carbonatite by basalt magma originating from plumes may affect the trace element composition of the rocks but not their isotope ratios of Sr, Nd, and Pb.

The petrogenesis of the lavas on Tubuai differs considerably from that of the lavas on Rurutu even though both islands are part of the Austral Chain and are separated by only about 250 km. Tubuai contains two deeply eroded volcanoes that were active between about 12.4 and 9.9 Ma when the older suite of lavas was being erupted on Rurutu. The lavas on Herani Volcano on Tubuai are alkali-rich and range in composition from alkali basalt to basanite leading up to analcitite and nephelinites at the top of the section (Caroff et al. 1997). The lavas contain inclusions of ultramafic rocks studied by Hauri et al. (1993). There is no evidence of a more recent episode of volcanic activity on Tubuai similar to that on Rurutu.

Caroff et al. (1997) used the chemical compositions of the lavas on Tubuai to demonstrate that the alkali-rich lavas that form the summit of the Tubuai shield volcano formed by decreasing degrees of partial melting of spinel lherzolites at the base of the lithospheric mantle above the plume. The degree of melting decreased with time resulting in the sequential formation of basanites, analcitites, and nephelinites. Some of the ascending alkali-rich magmas may have released carbonatic fluids that formed carbonatite veins in the lithosphere overlying the plume as postulated by Hauri et al. (1993). The important point of the petrogenetic model deduced by Caroff et al. (1997) is that the alkali-rich lavas that were erupted near the end of the volcanic activity on Tubuai formed by decreasing degrees of partial melting rather than by fractional crystallization of an alkali-basalt magma.

2.10.6 Mantle Components of Polynesia

The evident efficiency of mixing of magmas, presumably derived from the different components in the mantle of Polynesia, caused Chauvel et al. (1992) to propose that these components are the products of a single evolutionary process and occur together rather than as discrete and separate entities in the mantle.

Accordingly, Chauvel et al. (1992) suggested that the HIMU component evolved because of an increase of the $^{238}U/^{204}Pb$ ratio of oceanic crust caused by loss of Pb as a result of hydrothermal alteration of MORBs shortly after they were extruded along spreading ridges. Since the concentration of Pb in pelagic and terrigenous sediment is much higher than that of the altered oceanic crust, the isotope composition of Pb in mixtures of subducted oceanic crust and associated sediment is dominated by the sediment components. Chauvel et al. (1992) demonstrated that subducted mixtures of altered oceanic crust and *terrigenous* sediment acquire the Pb isotope compositions of EM2 in about two billion years, whereas mixtures of oceanic crust and *pelagic* sediment form EM1. Subducted oceanic crust *not* associated with pelagic or terrigenous sediment evolves into the HIMU component in the mantle. The calculations by Chauvel et al. (1992) also indicated that the mixtures of oceanic crust plus 5% of *pelagic* sediment acquire the $^{87}Sr/^{86}Sr$ and $^{143}Nd/^{144}Nd$ ratios of the EM1 component after an incubation time of about two billion years whereas mixtures of oceanic crust plus 5% of *terrigenous* sediment evolve into EM2 in the same period of time.

The model of Chauvel et al. (1992) postulates that the heads of the plumes that underlie the active volcanic centers of Polynesia are composed of heterogeneous mixtures of oceanic crust and altered sediment of various kinds that were subducted together a long time ago (e.g. at 2 Ga). The isotope compositions of Sr, Nd, and Pb of the lavas that originate from these plumes depend on the particular source rocks that melted to form the magmas. In other words, the mixing of magmas derived from different components takes place at the *source* rather than during their migration to the surface. These insights apply not only to the lavas on the islands of Polynesia, but to all oceanic islands and seamounts. In addition, subduction of oceanic crust and overlying sediment is continuing at the present time and will affect the chemical and isotopic compositions of volcanic rocks in the future.

The identification of recycled marine and continental sediment in the volcanic rocks of oceanic islands continues to be a topic of intense debate exemplified by the papers of Hofmann and White (1982), Cohen and O'Nions (1982b), Weaver et al. (1986), Loubet et al. (1988), Hart and Staudigel (1989) Dupuy et al. (1989), Woodhead et al. (1993), Chauvel et al. (1995), Woodhead (1996), and Thirlwall et al.(1997).

2.10.7 Mantle Isochrons

The $^{87}Sr/^{86}Sr$ ratios of the volcanic rocks on some of the islands of Polynesia correlate positively with their $^{87}Rb/^{86}Sr$ ratios (Duncan and Compston 1976; Nakamura and Tatsumoto 1988). Such correlations were also observed by Leeman (1974), Sun and Hanson (1975a), and by Brooks et al. (1976a,b) for basalts from continental and oceanic regions. These kinds of correlations arise by mixing of magmas derived from sources in the mantle having different Rb/Sr ratios and hence different $^{87}Sr/^{86}Sr$ ratios. The slopes of such linear data arrays in coordinates of $^{87}Sr/^{86}Sr$ and $^{87}Rb/^{86}Sr$ may indicate the incubation time of the magma sources in the mantle, provided the following assumptions are satisfied:

1. The Rb/Sr ratio of the melt fraction was identical to that of the solids before melting. (True only when the melt fraction is large.)
2. The $^{87}Sr/^{86}Sr$ ratio of the melt was identical to that of the solids before melting. (Requires "equilibrium" melting.)
3. The Rb/Sr ratio of the magma was not changed by fractional crystallization.
4. The $^{87}Sr/^{86}Sr$ and $^{87}Rb/^{86}Sr$ ratios of the magma were not changed by assimilation of oceanic crust, or by interaction with seawater.
5. The $^{87}Sr/^{86}Sr$ ratios of the volcanic rocks were corrected for decay of ^{87}Rb after eruption and crystallization.

Most of these assumptions are not satisfied by volcanic rocks, meaning that the slopes of straight lines defined by OIBs on Rb-Sr isochron diagrams cannot be used to measure incubation times of the Rb-enriched component in their magma sources. For example, Duncan and Compston (1976) cited a report by Shimizu (1974) that the Rb/Sr ratio of a 20% partial melt formed at 25 kbar is 18% higher than that of the parent material. In cases where this is true, the dates indicated by OIBs in the Rb-Sr isochron diagram are *less* than the incubation age of their magma sources in the mantle.

A test of the validity of the so-called mantle isochrons is provided by data in Fig. 2.51 for volcanic rocks in Polynesia reported by Duncan and Compston (1976), Vidal et al. (1984), and by Palacz and Saunders (1986) which indicate that 77% of the specimens lie within a band defined by two arbitrarily drawn parallel lines. Two samples may have been contaminated by seawater, whereas twenty three specimens (ten are off-scale) appear to have been enriched in Rb either during partial melting or by subsequent fractional crystallization of the magma. However, even the specimens within the band scatter significantly and do not form an isochron in the strict definition of that term (Sect. 1.3). Moreover, the data

points representing the Austral Islands (except those from the island of Rapa, but including Mangaia) form a tight cluster, whereas those from the Society, Cook, and Marquesas Islands scatter widely.

Duncan and Compston (1976) derived a date of 880 ±220 Ma for the magma sources of the Marquesas and Society Islands with an initial $^{87}Sr/^{86}Sr$ ratio of 0.70204. The data array in Fig. 2.51 yields an approximate date of 810 Ma. Both dates underestimate the incubation age of the magma sources because of the probable enrichment of the melt in Rb. If the slope is increased by 18%, the date rises to about 960 Ma, or about 1.0 Ga in round numbers. Taken at face value, this result implies that the Rb-enriched components in the mantle underlying Polynesia have remained closed to Rb and Sr for about one billion years prior to partial melting.

2.10.8 The Dupal Anomaly

The heterogeneity of isotope compositions of Sr and Pb of the magma sources under Polynesia has been observed elsewhere in the southern hemisphere. Dupré and

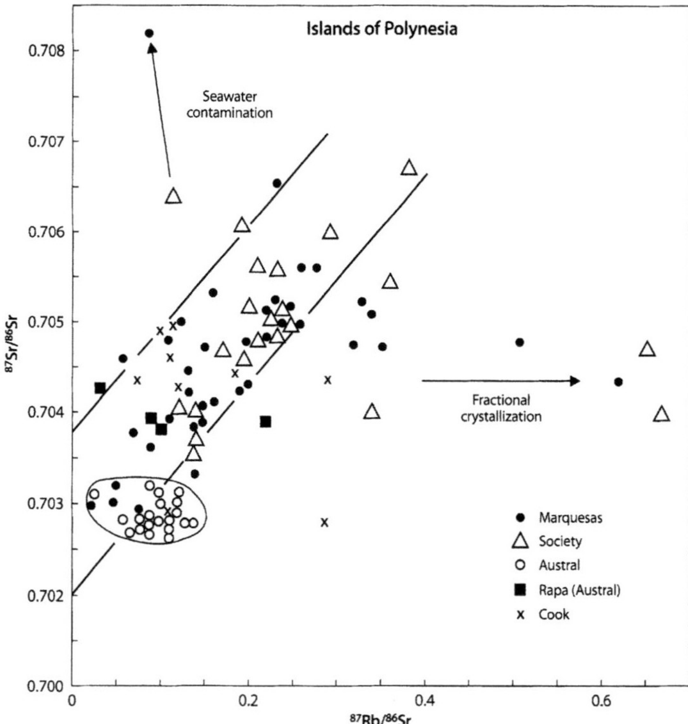

Fig. 2.51.
Correlation between $^{87}Sr/^{86}Sr$ and $^{87}Rb/^{86}Sr$ ratios of volcanic rocks from the islands of Polynesia in the South Pacific. The data points representing rocks from the Marquesas and Society Islands have a wide range of $^{87}Sr/^{86}Sr$ and $^{87}Rb/^{86}Sr$ ratios, whereas those of the Austral Islands are tightly clustered. The arrows indicate the effects of differences in the degree of partial melting and/or a fractional crystallization of magma on the Rb/Sr ratios of the rocks, and the effects of seawater contamination on their $^{87}Sr/^{86}Sr$ ratios. Such linear arrays can be interpreted as "mantle isochrons" discussed in the text, but are probably the result of mixing of magmas derived from Rb-enriched and Rb-depleted sources in the mantle. Nevertheless, the Rb-enriched magma source in this case must have remained closed to Rb and Sr for up to one billion years in order to acquire elevated $^{87}Sr/^{86}Sr$ ratios (*Sources:* Duncan and Compston 1976; Vidal et al. 1984; Duncan et al. 1986; Palacz and Saunders 1986)

Allègre (1983) reported that volcanic rocks in the Indian Ocean are variably enriched in radiogenic isotopes of Sr and Pb and attributed this phenomenon to mixing of magmas derived from different sources in the mantle. In addition, they proposed that the volcanic rocks of certain oceanic islands have characteristic isotope compositions of Sr and Pb that define global patterns.

The existence of global patterns in the isotope compositions of Sr and Pb, proposed by Dupré and Allègre (1983), was supported and elaborated by Hart (1984) who used the characteristic isotope compositions of Sr and Pb of volcanic rocks in the southern hemisphere to define an anomaly he called *Dupal* after B. Dupré and C. Allègre. Hart (1984) defined the isotope compositions of Pb in the Dupal anomaly by the deviation of the $^{207}Pb/^{204}Pb$ and $^{208}Pb/^{204}Pb$ ratios from the Northern Hemisphere Pb Reference Lines (NHRL) and expressed the $^{87}Sr/^{86}Sr$ ratios in terms of ΔSr defined as:

$$\Delta Sr = [(^{87}Sr/^{86}Sr)_{DS} - 0.7]10^4 \qquad (2.1)$$

The subscript DS stands for "data set" because ΔSr is to be calculated from the average $^{87}Sr/^{86}Sr$ ratio of the rocks on a particular island, regardless of internal isotope variations. For example, the Marquesas Islands have an elevated average $^{87}Sr/^{86}Sr$ ratio of 0.70462 for 46 specimens. Therefore the Dupal parameter for the Marquesas Islands is:

$$\Delta Sr = (0.70462 - 0.7)10^4 = 46.2$$

However, Duncan et al. (1986) pointed out that only the late-stage alkali-rich rocks have high $^{87}Sr/^{86}Sr$ ratios and that averages of isotope ratios are subject to sampling bias. For example, the average $^{87}Sr/^{86}Sr$ ratio of 18 specimens from Ua Pou in the Marquesas Islands is 0.70468 ($\Delta Sr = 46.8$), but the tholeiites taken alone yield only $\Delta Sr = 30.0$, whereas the alkali-rich rocks have $\Delta Sr = 50.1$. Evidently, only the alkali-rich rocks contain Sr that is anomalously enriched in radiogenic ^{87}Sr and therefore originated from the Dupal source.

The isotope ratios of Pb that define the Northern Hemisphere Pb Reference Line are (Hart 1984):

$$\frac{^{207}Pb}{^{204}Pb} = 0.1084\left(\frac{^{206}Pb}{^{204}Pb}\right) + 13.491 \qquad (2.2)$$

$$\frac{^{208}Pb}{^{204}Pb} = 1.209\left(\frac{^{206}Pb}{^{204}Pb}\right) - 15.627 \qquad (2.3)$$

The magnitude of the Dupal component in a given data set is expressed as the difference between average $^{207}Pb/^{204}Pb$ and $^{208}Pb/^{204}Pb$ ratios of a data set and the Northern Hemisphere Pb Reference Line (Hart 1984):

$$\Delta\frac{^{208}Pb}{^{204}Pb} = \Delta 8/4 = [(^{208}Pb/^{204}Pb)_{DS} - (^{208}Pb/^{204}Pb)_{NHRL}] \times 100 \qquad (2.4)$$

$$\Delta\frac{^{207}Pb}{^{204}Pb} = \Delta 7/4 = [(^{207}Pb/^{204}Pb)_{DS} - (^{207}Pb/^{204}Pb)_{NHRL}] \times 100 \qquad (2.5)$$

where DS = data set and NHRL = northern hemisphere Pb reference line.

For example, Vidal et al. (1984) reported the following average isotope ratios of Pb in the lavas on Tubuai (Austral Islands):

$$\frac{^{206}Pb}{^{204}Pb} = 21.104; \qquad \frac{^{207}Pb}{^{204}Pb} = 15.770; \qquad \frac{^{208}Pb}{^{204}Pb} = 40.354$$

The equation for the NHRL indicates that the $^{207}Pb/^{204}Pb$ and $^{208}Pb/^{204}Pb$ ratios that correspond to $^{206}Pb/^{204}Pb = 21.104$ are:

$$\left(\frac{^{207}Pb}{^{204}Pb}\right)_{NHRL} = 0.1084 \times 21.104 + 13.491 = 15.778$$

$$\left(\frac{^{208}Pb}{^{204}Pb}\right)_{NHRL} = 1.209 \times 21.104 + 15.627 = 41.141$$

The Dupal parameters for Pb in the lavas of Tubuai are:

$$\Delta 7/4 = (15.770 - 15.778) \times 100 = -0.8$$

$$\Delta 8/4 = (40.354 - 41.141) \times 100 = -78.7$$

Hart (1984) reported strongly positive values of $\Delta 7/4$ and $\Delta 8/4$ for islands in the Indian Ocean (Amsterdam, Christmas, Crozet, Kerguelen, Marion-Prince Edward, and Réunion), in the South Atlantic Ocean (Fernando de Noronha, Tristan da Cunha, Gough, Bouvet, Rio Grande Rise, and Walvis Ridge), and in the South Pacific (Easter, San Felix, Fiji, Marquesas, Rapa, and Rarotonga). In general, oceanic basalts on the islands of the southern hemisphere have positive values of $\Delta 7/4$ and $\Delta 8/4$.

In this regard, the lavas of Tubuai are exceptional, presumably because their elevated average $^{206}Pb/^{204}Pb$ ratio (HIMU component) causes their $\Delta 7/4$ and $\Delta 8/4$ parameters to have negative values. Similarly, their low average $^{87}Sr/^{86}Sr$ ratio (0.70277, Vidal et al. 1984) yields $\Delta Sr = (0.70277 - 0.7) \times 10^4 = 27.7$ which is relatively low compared to that of the Marquesas Islands ($\Delta Sr = 46.2$). The significance of the Dupal phenomenon may be that it reflects the upwelling of the lower asthenospheric mantle activated by heat emanating from the core as discussed briefly in Sect. 2.1 (Castillo 1988; Staudigel et al. 1991).

2.11 Ontong Java and Manihiki Plateaus

The western Pacific Ocean contains the Ontong Java and Manihiki Plateaus as well as the Magellan, Shatsky, and Hess rises identified in Fig. 2.52. The Ontong Java Plateau has an area of 1.5×10^6 km^2 compared to only 0.55×10^6 km^2 for the Manihiki Plateau. Both plateaus rise between 2 and 3 km above the surrounding ocean floor and are underlain by oceanic crust having a thickness from 20 to 40 km (Mahoney and Spencer 1991; Hussong et al. 1979). Petrographic studies of drill cores indicate that the plateaus are composed of MORB-like tholeiite basalts of Cretaceous age whose average Rb and Sr concentrations are 1.55 and 81.0 ppm, respectively, based on analyses by Mahoney and Spencer (1991). In addition, Beiersdorf et al. (1995) reported the eruption of large volumes of hyaloclastite on the Manihiki Plateau.

The isotope ratios of Sr and Nd of basalts on the Ontong Java and Manihiki Plateaus are similar to those of OIBs rather than those of MORBs. However, these plateaus do not appear to be associated with hot-spot tracks in the form of island or seamount chains that record the movement of the Pacific Plate over hotspots in the mantle. In addition, the Ontong Java and Manihiki Plateaus differ from the Iceland, Azores, and Galapagos Plateaus because they are not associated with spreading ridges. Instead, the Ontong Java and Maniki Plateaus

are the oceanic equivalents of continental flood basalts (Chap. 5).

The basalts of the southern Ontong Java Plateau are exposed on the island of Malaita of the Solomon Islands (Tejada et al. 1996). The rocks consist of Early Cretaceous tholeiite basalt overlain by marine carbonate rocks and alkali basalts of Late Cretaceous to Eocene age (Hughes and Turner 1977). These rocks were subsequently intruded by alkali-rich alnoite (Sect. 3.9.2) at 34 Ma (Davis 1978). A specimen of the tholeiite basalt of Malaita has an elevated initial ^{87}Sr/^{86}Sr ratio (0.70439 at 120 Ma) consistent with the lavas of oceanic islands (Bielski-Zyskind et al. 1984).

The isotope ratios of Sr and Pb in Fig. 2.53 indicate that the tholeiite basalts of the Ontong Java and Manihiki Plateaus formed by melting of mixtures of subducted pelagic sediment (EM1) and oceanic basalt crust (HIMU) with varying amounts of entrained rocks of the asthenospheric mantle (Richards et al. 1989). These rocks presumably occurred in a large plume head that caused uplift and fracturing of the overlying lithosphere prior to large-scale melting and outpouring of tholeiite basalt during a relatively short interval of time (e.g. <5 million years). Large lava plateaus composed of tholeiite also occur on continents (e.g. Paraná, Brazil; Etendeka Group, Namibia; Ferrar dolerites, Transantarctic Mountains). These and other continental basalt plateaus formed during the Mesozoic and Cenozoic Eras as a consequence of rifting of continents (Chap. 5). The enormous volume of lava contained within the Ontong Java Plateau (50×10^6 km^3; Schubert and Sandwell 1989) requires a source volume of at least 250×10^6 km^3 assuming 20% partial melting. Therefore, Tatsumi et al. (1998) included the Ontong Java Plateau among the examples of superplumes in the Pacific Ocean.

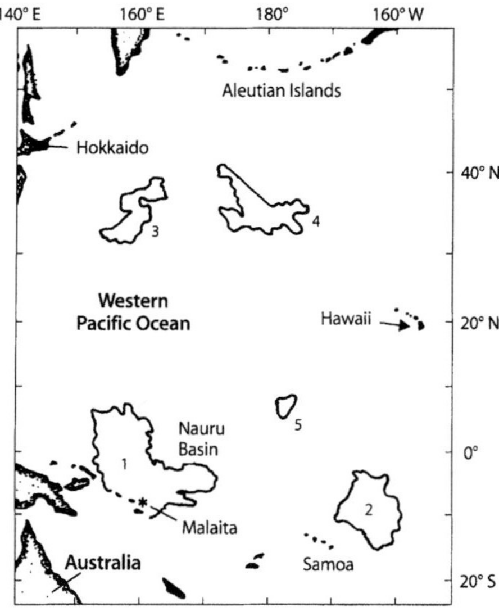

Fig. 2.52. Location of basalt plateaus in the western Pacific Ocean: 1. Ontong Java; 2. Manihiki; 3. Shatsky; 4. Hess; 5. Magellan (*Source:* adapted from Mahoney and Spencer 1991)

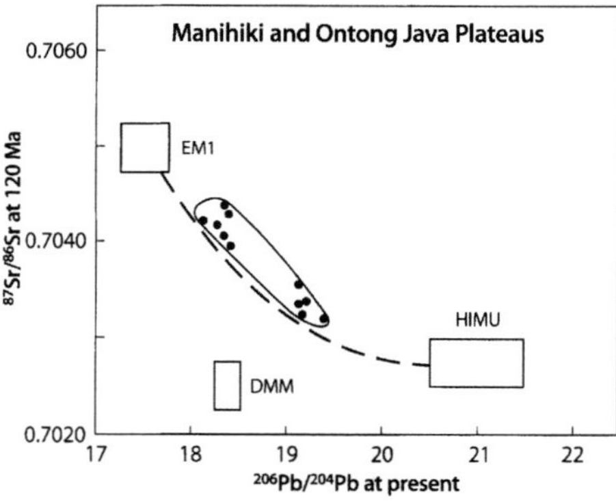

Fig. 2.53. Isotope ratios of Sr (at 120 Ma) and Pb (present) of Early Cretaceous tholeiites of the Ontong Java and Manihiki Plateaus in the Pacific Ocean. The data points define a mixing hyperbola indicating that the magmas originated by partial melting of subducted pelagic sediment and oceanic crust in the heads of large plumes (*Source:* data from Mahoney and Spencer 1991)

2.12 The Hawaiian Islands and Emperor Seamounts

The Hawaiian Islands and the related Emperor Seamounts are prime examples of the interaction of a large stationary plume and the overlying Pacific Plate which is moving over it. However, when viewed in detail, the process at work here differs significantly from that seen in the islands chains of Polynesia.

2.12.1 The Hawaiian Islands

The volcanoes of the Hawaiian Islands in Fig. 2.54 form two curved but parallel lines identified as the Kea trend and the Loa trend (Jackson et al. 1972). The Kea trend contains the volcanoes Kilauea, Mauna Kea, Kohala, Haleakala, West Maui, and East Molokai. The Loa trend is defined by Loihi, Mauna Loa, Hualalai, Kahoolawe, Lanai, West Molokai, Oahu, and Kauai. The Loa trend is continued by the seamounts and islands of the Hawaiian Ridge and the Emperor Chain which together extend for over 6 000 km and include 107 volcanoes (Bargar and Jackson 1974). The ages of the Hawaiian volcanoes and seamounts increase linearly with distance from 0 Ma for Kilauea to 64.7 Ma for Suiko Seamount and imply a rate of motion of the Pacific Plate of 8.2 cm yr^{-1} (Greene et al. 1978; Dalrymple et al. 1980a,b, 1981).

The volcanic rocks extruded by the Hawaiian volcanoes have been subdivided into three suites on the basis of their chemical composition and stratigraphic position (MacDonald and Katsura 1964; Clague and Frey 1982):

1. Nephelinitic suite, posterosional stage (youngest)
2. Alkalic suite, caldera filling stage
3. Tholeiite suite, shield building stage (oldest)

The isotope compositions of Sr, Pb, Nd, Os, and Hf as well as trace-element concentrations have been used to constrain the magma sources and to assess the effects of partial melting, fractional crystallization, and magma mixing on the chemical compositions of the Hawaiian volcanic rocks (Leeman et al. 1980; Clague and Frey 1982; Wright 1984; Feigerson 1984; Clague 1987; Wyllie 1988; Frey and Rhodes 1993). Recent studies of U-series nuclides in lava flows that were erupted in historical time on the island of Hawaii have shed light on the mechanics and rates of melt formation and extraction of magma. (Hémond et al. 1994b; Sims et al. 1995; Cohen and O'Nions 1994; Reinitz et al. 1991; Newman et al. 1984; Condomines et al. 1976). The effect of partial melting of ultramafic rocks in the mantle on the U-series radionuclides has been discussed by Somayajulu et al. (1966), Allègre et al. (1982), Krishnaswami et al. (1984), Condomines et al. (1988), Williams and Gill (1989), Gill and Condomines (1992), Spiegelman and Elliott (1993), and Beattie (1993a,b).

The cause for the volcanic activity that formed the Hawaiian Islands in the middle of the Pacific Ocean was originally attributed by Wilson (1963a,b) to melting over a rising current in the mantle followed by the displacement of the resulting volcanoes to form a chain of islands. Subsequently, Morgan (1972a,b) postulated that the melting occurred in a plume of hot rock rising from the lower mantle. The first measurements of ^{87}Sr/^{86}Sr ratios of volcanic rocks from the Hawaiian Islands by Gast (1960), Faure (1961), and by Faure and

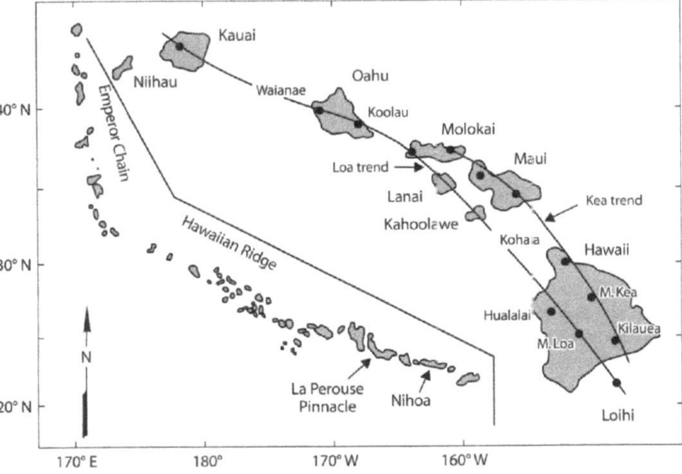

Fig. 2.54.
Map of the Hawaiian Islands and of the seamounts of the Hawaiian Ridge and Emperor Chain. The volcanoes of the Hawaiian Islands form two linear belts identified as the Loa and Kea trends. The ages of the rocks increase from 0 Ma (Loihi, Mauna Loa, Kilauea) to 65 Ma (Suiko Seamount). According to Dalrymple and Clague (1976), the bend in the Hawaii-Emperor Chain occurred at about 43 Ma (adapted from Stille et al. 1986)

Hurley (1963) indicated that the rocks had originated from magma sources in the mantle having low Rb/Sr ratios. In addition, Faure and Hurley (1963) found a small difference in the $^{87}Sr/^{86}Sr$ ratios of basalt on Hawaii and Maui, whereas Lessing and Catanzaro (1964) demonstrated that the $^{87}Sr/^{86}Sr$ ratios of Hawaiian volcanics decrease with increasing K/Rb ratios. In the following year, Powell et al. (1965a) reported that alkali-rich basalts of Oahu, containing less than about 45% SiO_2, have *lower* $^{87}Sr/^{86}Sr$ ratios than more silica-rich alkali basalts and tholeiites on Oahu and Hawaii. Hedge (1966) likewise concluded that the tholeiites on Hawaii have higher $^{87}Sr/^{86}Sr$ ratios than melilite basalts on Oahu, although the difference barely exceeded the analytical error.

The indication of real differences in the $^{87}Sr/^{86}Sr$ ratios among the volcanic rocks of the Hawaiian Islands has been confirmed by more recent studies based on data displayed in Fig. 2.55. The early measurements of Gast (1960), Faure and Hurley (1963), Lessing and Catanzaro (1964), Powell et al. (1965a), and Hedge (1966) were excluded. Lanphere et al. (1980a) compiled a list of $^{87}Sr/^{86}Sr$ ratios of volcanic rocks from the Hawaiian Islands published before 1977, whereas Stille et al. (1986) tabulated isotope ratios of Sr, Nd, Hf, and Pb for volcanic rocks from the principal Hawaiian Islands. In addition, Tatsumoto (1966) and Chen (1987) reported isotope ratios of Pb in Hawaiian basalts. The correlations of the $^3He/^4He$ ratio with isotope ratios of Pb and with other isotopic and chemical parameters were discussed by Eiler et al. (1998) and by Kurz and Kammer (1991). These and many other studies on the petrology, chemical composition, trace-element concentrations, isotope ratios of Sr, Nd, Pb, O, He, Os, and Hf have repeatedly confirmed the theory that the volcanic rocks of the Hawaiian Islands originated from a large asthenospheric plume and from the surrounding suboceanic lithospheric mantle (Hauri 1996; Bennett et al. 1996).

According to this theory, the shield-building tholeiites on Oahu and on the other Hawaiian islands formed by decompression melting in the head of the plume composed of the EM1 component and entrained asthenospheric mantle rocks (Wyllie 1988; Watson and MacKenzie 1991; Iwamori 1993). The resulting volcanic activity on the seafloor caused the formation of large shield volcanoes composed of tholeiite basalt having elevated $^{87}Sr/^{86}Sr$ ratios, but low $^{143}Nd/^{144}Nd$ and low $^{206}Pb/^{204}Pb$ ratios. After an interval of volcanic inactivity, erosion, and caldera formation, magma formation resumed by a lesser degree of partial melting of source rocks that included a larger proportion of DMM in the lithospheric mantle. The final stage of volcanic activity resulted in the extrusion of comparatively small volumes of peralkaline nephelinites and melilitites that formed by very small degrees of partial

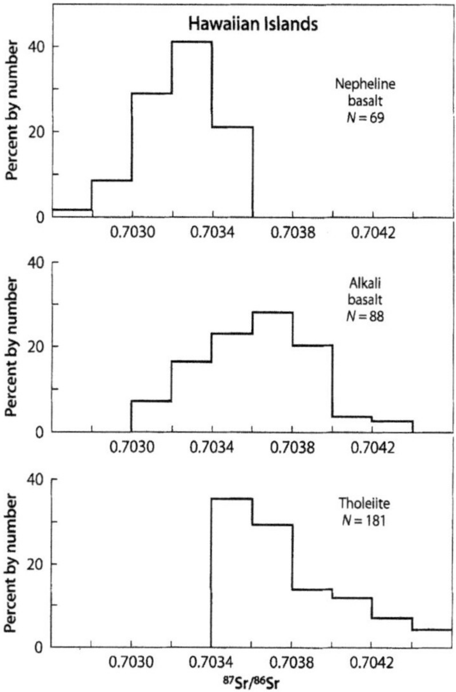

Fig. 2.55. Distribution of $^{87}Sr/^{86}Sr$ ratios of the principal volcanic series on the Hawaiian Islands relative to 0.7080 for E&A and 0.71025 for NBS 987. Most of the shield-building tholeiites have $^{87}Sr/^{86}Sr$ ratios between 0.70340 and 0.70380, but some have higher ratios up to about 0.70450. The $^{87}Sr/^{86}Sr$ ratios of the caldera-filling alkali basalts (hawaiite, mugearites, trachyandesites, etc.) range from 0.70320 to 0.70380 with a few values greater than 0.70400. The posterosional nepheline-melilite basalts have distinctly lower $^{87}Sr/^{86}Sr$ ratios than the tholeiites with most values between 0.7030 and 0.70360. The differences in the isotope composition of Sr indicates that the tholeiites and nepheline basalts originated from distinctly different magma sources. The $^{87}Sr/^{86}Sr$ ratios of the alkali basalt are intermediate between those of the tholeiites and of the nepheline basalts (*Sources:* Staudigel et al. 1984; White and Hofmann 1982; O'Nions et al. 1977; Stille et al. 1983, 1986; Lanphere 1983; Powell and DeLong 1966; Lanphere and Dalrymple 1980; Roden et al. 1984b; O'Neil et al. 1970; Lanphere et al. 1980a; Feigenson et al. 1983; Chen and Frey 1983, 1985; Hegner et al. 1986; Clague et al. 1982; Maaløe et al. 1992; West et al. 1987; West and Leeman 1987; Feigenson 1984; Hofmann et al. 1984)

melting of source rocks consisting primarily of the DMM component. The changing chemical and isotopic compositions of the lavas extruded on Oahu and the other Hawaiian islands were controlled by the movement of the Pacific Plate over the stationary head of the Hawaiian Plume.

The shield-building tholeiites in Fig. 2.55 have relatively low average concentrations of Rb and Sr but a high average $^{87}Sr/^{86}Sr$ ratio compared to other varieties of volcanic rocks in the Hawaiian islands:

- Rb $= 5.2$ (89) ppm
  $(0-15.1)$

- Sr $= 378$ (89) ppm
  $(159-540)$

- $^{87}Sr/^{86}Sr = 0.70379$ (181)
  $0.70310-0.70461$

Alkali basalts (hawaiites, mugearites, trachyandesites, etc.) have significantly higher average Rb and Sr concentrations, but have a *lower* average $^{87}Sr/^{86}Sr$ ratio than the tholeiites.

- Rb $= 22.4$ (76) ppm
  $2.7-84.4$

- Sr $= 765$ (74) ppm
  $51-1681$

- $^{87}Sr/^{86}Sr = 0.70342$ (88)
  $0.70305-0.70440$

The posterosional nepheline- or melilite-bearing basalts on Maui, Oahu, and Kauai are distinguished from the alkali basalts by having higher average Sr concentrations but a lower average $^{87}Sr/^{86}Sr$ ratio than the tholeiites.

- Rb $= 23.3$ (55) ppm
  $3-72$

- Sr $= 844$ (55) ppm
  $51.-2781$

- $^{87}Sr/^{86}Sr = 0.70324$ (65)
  $0.70270-0.70357$

The three lava series on Oahu were analyzed by Stille et al. (1983, 1986), Lanphere and Dalrymple (1980), Roden et al. (1984b, 1994), and Frey et al. (1994). Those on Maui were studied by Chen and Frey (1983, 1985), Hegner et al. (1986), and by West and Leeman (1987), whereas Maaløe et al. (1992), Clague and Dalrymple (1988), and Feigenson (1984) worked on the nepheline basalts and tholeiites of Kauai. The evidence for mixing of magmas derived from different sources was especially well presented in a study by Feigenson (1984) who showed that the $^{87}Sr/^{86}Sr$ ratios of the tholeiites on Kauai are positively correlated with the reciprocals of the Sr concentrations, as expected for two-component mixtures.

The volcanic activity that gave rise to the Hawaiian Islands is still occurring on the "big island" of Hawaii where the volcanoes Mauna Loa and Kilauea (Hofmann et al. 1984) continue to erupt (Wright et al. 1975; Wright 1984; Tilling et al. 1987; Rhodes et al. 1989; Kennedy et al. 1991; Hémond et al. 1994b). In addition, a submarine volcano named Loihi is forming off the south coast of Hawaii (Fig. 2.54) and appears to be destined to become

the next Hawaiian island (Lanphere 1984; Staudigel et al. 1984), whereas the extinct volcano Kohala on the island of Hawaii was investigated by Feigenson et al. (1983); Lanphere and Frey (1987) and by Hofmann et al. (1987). Moreover, Clague et al. (1990) reported the presence of alkali flood basalts of Pliocene-Pleistocene age on the seafloor north of the Hawaiian Islands.

The differences in the $^{87}Sr/^{86}Sr$ ratios among the three volcanic series on Maui, Oahu, and Kauai also imply that the isotope composition of Sr of the magmas changed as a function of *time*. Chen and Frey (1983, 1985) developed a physical model to explain this trend for the volcanic rocks on the volcano Haleakala on East Maui. According to this model, the tholeiites (Honomanu Series) formed first by partial melting (about 14%) within a plume having relatively high $^{87}Sr/^{86}Sr$ ratios. The intrusion of the plume caused partial melting (0.1 to 1%) of the depleted rocks in the oceanic lithosphere to form melts with low $^{87}Sr/^{86}Sr$ ratios but elevated Rb and Sr concentrations. The two types of melt mixed to form the alkali basalts (Kula Series). The proportion of lithosphere-derived magma gradually increased as the degree of melting decreased causing the $^{87}Sr/^{86}Sr$ ratios of the alkali basalts to *decrease* with time. Finally, the posterosional alkali basalts of the Hana Series formed primarily from melts derived from the depleted lithospheric mantle, thus explaining their low $^{87}Sr/^{86}Sr$ ratios. The concentrations of incompatible elements, including Rb and Sr, increased during this process because the degree of partial melting decreased. As the volume of magma declined, its composition was progressively dominated by melting in the oceanic lithosphere because the volcano was no longer centered over the plume.

In a subsequent study of the lava flows of Haleakala on East Maui, West and Leeman (1987) confirmed that the $^{87}Sr/^{86}Sr$ ratios of the alkali basalts in the Kula Series decrease up-section implying a time-dependent, progressive change in the proportions of magma derived from the plume and from the adjacent lithospheric mantle. Similarity, the $^{87}Sr/^{86}Sr$ ratios of the volcanic rocks on West Maui, consisting of tholeiites (Wailuku Series), alkali basalt (Honolua Series), and posterosional alkalirich basalts (Lahaina Series) decrease in the same sense as those on Haleakala and Oahu (Hegner et al. 1986; West et al. 1987).

The derivation of the lavas on Oahu from magmas that originated from different source rocks in the mantle is clearly illustrated by the correlation of isotope ratios of Sr, Nd, and Pb. The three lava series on this island include the Koolau tholeiites, followed by the younger Waianae alkali basalts, and by the Honolulu nephelinites. The data reported by Stille et al. (1983) in Figs. 2.56 and 2.57 demonstrate that the isotope ratios of these lava series form clusters in Sr-Nd and Sr-Pb isotope mixing diagrams. The data clusters are linearly aligned between the DMM and EM1 components defined by Hart (1988),

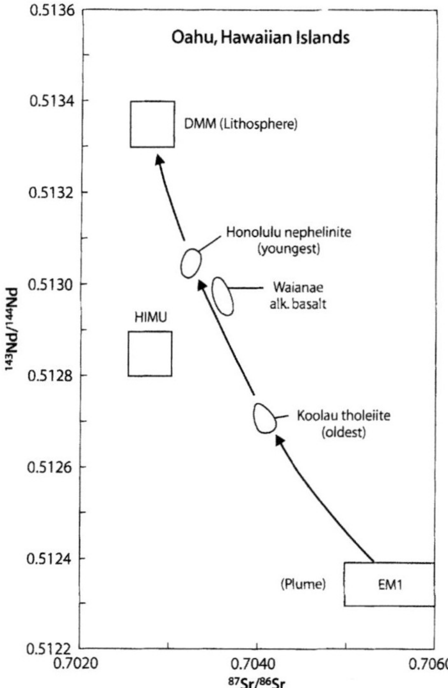

Fig. 2.56. Isotope ratios of Sr and Nd of the lavas on Oahu. The three lava series have progressively changing $^{87}Sr/^{86}Sr$ and $^{143}Nd/^{144}Nd$ ratios presumable because magma originated by partial melting of mixtures of EM1 and DMM in varying proportions. Accordingly, the tholeiites of the Koolau Series originated by large-scale melting within the plume (EM1) which intruded the lithospheric mantle (DMM). As the island of Oahu was transported away from the plume, the alkali-rich lavas originated by decreasing degrees of melting of source rocks that contained increasing proportions of the DMM component (*Sources:* Stille et al. 1983; Hart 1988)

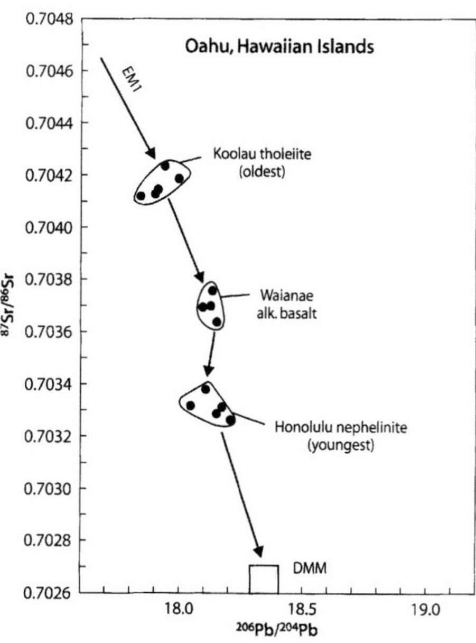

Fig. 2.57. Isotope ratios of Sr and Pb of the lavas on Oahu. The distribution of data points confirms the evidence provided by the $^{87}Sr/^{86}Sr$ and the $^{143}Nd/^{144}Nd$ ratios in Fig. 2.56 that the magmas formed by decreasing degrees of partial melting of source rocks composed of the components EM1 and DMM (*Sources:* Stille et al. 1983; Hart 1988)

thereby indicating that the magmas formed by varying degrees of partial melting of mixtures of these components whose proportion changed with time. The sequence of events is exactly as outlined by Chen and Frey (1983, 1985) for the petrogenesis of lavas on the volcano Haleakala on East Maui.

The sequence of events leading to the formation of the other Hawaiian islands was similar in principle to that outlined above in Figs. 2.56 and 2.57. However, the differences in the isotope ratios of the lava series on the other islands are not as great as on Oahu. For example, the isotope ratios of Sr, Nd, and Pb of the lava series on the island of Kahoolawe overlap extensively (West et al. 1987). Similarly, the isotope ratios of the tholeiites and alkali basalts on West Maui overlap, but differ from those of the late-stage basanites (Hegner et al. 1986). On the volcano Haleakala (East Maui) the isotope ratios of the tholeiites of the Honomanu Series differ from those of the younger alkali-rich lavas (Kumuiliahi, Kula, and Hana Series) studied by West and Leeman (1987) and by Hegner et al. (1986). These differences in the isotope ratios of Sr, Nd, Pb in the lava series of the Hawaiian Islands are presumably caused by the displacement of the individual volcanic islands from the hotspot above which they all formed.

The origin of the Hawaiian Islands differs from those of the island chains of Polynesia because the shield-building tholeiites of the Polynesian Islands have *lower* $^{87}Sr/^{86}Sr$ ratios than the later alkali basalts. For example, on Pitcairn Island (Fig. 2.44, Sect. 2.10.1) the Adamstown tholeiites have lower $^{87}Sr/^{86}Sr$ and higher $^{143}Nd/^{144}Nd$ ratios than the overlying Pulawana and Tedside alkali-rich lavas. The reason for this difference is that the plume that gave rise to the Hawaiian Islands was strong enough to penetrate into the lithosphere, thereby exposing the rocks in its head to large-scale decompression melting. The plumes of Polynesia were comparatively weaker and caused decompression melting in the lithospheric mantle by heating it from below and by causing fractures to form as a result of uplift.

In a further elaboration of the two-component plume model, Maaløe et al. (1992) proposed that the observed range of isotopic compositions of Sr, Pb, and Nd of the

Hawaiian volcanics can be explained by magma formation in a radially zoned plume derived from the lower mantle. If the lower mantle is compositionally layered, the lower layers of the mantle form the central core of a plume and the upper layers form the outer margin of a plume arising from such a source. Maaløe et al. (1992) suggested that the $^{87}Sr/^{86}Sr$ ratio of the core of the Hawaiian Plume is between 0.7030 and 0.7033, whereas the $^{87}Sr/^{86}Sr$ ratio of the outer part may range up to about 0.7047. According to Maaløe et al. (1992), tholeiite magmas formed by partial melting in all parts of the plume to yield $^{87}Sr/^{86}Sr$ ratios between 0.7035 and 0.7040. During the final posterosional phase, melting occurred only in the core of the plume to form nepheline basalt with low $^{87}Sr/^{86}Sr$ ratios. However, the concentric zonation of the head of the Hawaiian Plume may also have been caused by the entrainment of blocks of asthenospheric mantle into the rising head composed of subducted oceanic crust, and there is no compelling geochemical evidence that the Hawaiian Plume originated from stratified layers in the lower mantle.

2.12.2 Hawaiian Xenoliths

The magma sources in the mantle presently underlying the island of Hawaii can also be characterized by a study of the petrology, geochemistry, and the isotope composition of xenolithic inclusions contained in the basalt flows of the Hawaiian Islands (Shimizu 1975; Frey 1980; Frey and Roden 1987). The inclusions in the Honolulu Series (nephelinites) on the island of Oahu are composed of dunite, lherzolite, and garnet pyroxenite. These inclusions as well as their basalt host contain well crystallized calcite which was analyzed by O'Neil et al. (1970) to determine the isotope compositions of O and C. The results confirmed that the calcite was deposited by local groundwater as originally proposed by Hay and Iijima (1968).

O'Neil et al. (1970) also reported that the calcite in three garnet websterite (garnet-bearing pyroxenite composed of interlocking crystals of orthopyroxene and clinopyroxene) have high Sr concentrations (951 to 4 867 ppm) compared to the websterite inclusions (61 to 275 ppm on a calcite-free basis). The $^{87}Sr/^{86}Sr$ ratios of the calcite (0.7031 to 0.7032) are similar to those of the host basalt (0.7033 to 0.7036) and of the websterite inclusions (0.7029 to 0.7035) relative to 0.7080 for E&A (O'Neil et al. 1970). Therefore, the isotope ratios of Sr support the secondary origin of the calcite (indicated by the isotope ratios of O and C) and permit the conclusion that the nephelinite lavas formed by partial melting of garnet pyroxenite in the mantle.

There is, however, an interesting discrepancy in the Rb-Sr data of these ultramafic inclusions because the average Rb/Sr ratio (calcite-free) of the garnet websterite

inclusions (0.0048) is too low to account for their present average $^{87}Sr/^{86}Sr$ ratio (0.7032) even if decay started at 4.5 Ga with a primordial $^{87}Sr/^{86}Sr$ ratio (0.699). The present average $^{87}Sr/^{86}Sr$ ratio of these inclusions (0.7032) requires them to have an average Rb/Sr ratio of 0.023. Since the measured Rb/Sr ratio is only 0.0048, these rocks appear to be restites from which a melt fraction was previously extracted causing their Rb/Sr ratio to decrease by a factor of 0.21 from about 0.023 to 0.0048. O'Neil et al. (1970) suggested that the lowering of the Rb/Sr ratios of the ultramafic inclusions may have been caused by the selective melting of Rb-rich minerals such as phlogopite. Similarly, Morioka and Kigoshi (1975) concluded from an interpretation of Pb isotope ratios that another suite of lherzolite inclusions of Oahu also consists of refractory residue of a previous melt event. Alternatively, the websterite inclusions studied by O'Neil et al. (1970) may have formed as cumulates of the basalt magma that subsequently transported them to the surface. In that case, their low Rb/Sr ratios are a consequence of mineralogy and are not related to the $^{87}Sr/^{86}Sr$ ratios of the inclusions.

A suite of ultramafic inclusions from the Salt Lake Crater on the volcano Koolau on Oahu analyzed by Vance et al. (1989) included three specimens that contain veins of mica- and amphibole-bearing pyroxenite. The $^{87}Sr/^{86}Sr$ ratios of the inclusions range from 0.703188 (Garnet pyroxenite) to 0.704162 (lherzolite), whereas their $^{143}Nd/^{144}Nd$ isotope ratios vary in the opposite sense from 0.513092 (garnet pyroxenite) to 0.512955 (lherzolite). These ratios are indistinguishable from those of the nephelinites of the Honolulu Series in Fig. 2.56. The isotope ratios of Sr and Nd of the pyroxenite veins in each of three specimens are identical to those of their host inclusions which is remarkable because the host inclusions and the pyroxenite veins have quite different Rb/Sr ratios.

The Rb-Sr data in Table 2.3 of one of the veined lherzolite inclusions (69SAL110AB) analyzed by Vance et al. (1989) illustrate the situation.

This data indicate that the Rb/Sr ratio of the lherzolite host is much too low to be compatible with its $^{87}Sr/^{86}Sr$ ratio of 0.703325. Therefore, this inclusion is not a piece of the primordial mantle, but may be either the residue of a previous episode of magma formation, or it may be a cognate inclusion (i.e. a cumulate) of the magma that brought it to the surface. The presence of pyroxe-

Table 2.3. Rb-Sr data of one of the veined lherzolite inclusions (69SAL110AB) analyzed by Vance et al. (1989)

Material	$^{87}Sr/^{86}Sr$	Rb/Sr
Lherzolite	0.703325	0.00027
Pyroxenite vein	0.703263	0.00065

nite veins in this and other inclusions at the Salt Lake Crater favors the interpretation that the inclusion is a restite (i.e. melt residue) and that the depleted mantle was subsequently refertilized by the intrusion of a hydrous and alkali-rich silicate melt. A study by Sen et al. (1993) of the REE distribution in clinopyroxenes in ultramafic inclusions from the Honolulu Series on Oahu confirmed these conclusions.

To summarize, the study of ultramafic inclusions in the lavas of Oahu indicates that they are melt residues from the depleted lithospheric mantle which was subsequently altered by the formation of pyroxenite veins containing mica and amphiboles deposited by a hydrous and alkali-rich silicate melt. After eruption to the surface, the inclusions as well as the host lavas were exposed to groundwater which, in some cases, caused the deposition of secondary calcite. Nevertheless, the isotope ratios of Sr and Nd of the inclusions are similar to those of the volcanic rocks which therefore permits the inclusions to represent the magma sources in the lithospheric mantle under the Hawaiian Islands.

2.12.3 Emperor Seamounts

The $^{87}Sr/^{86}Sr$ ratios of tholeiites and alkali basalt of seamounts and islands on the Hawaiian Ridge and the Emperor Chain (Fig. 2.54) are shown graphically in Fig. 2.58 based primarily on the work of Lanphere et al. (1980a). The $^{87}Sr/^{86}Sr$ ratios of tholeiites of the Hawai-

ian Ridge are similar to those of the Hawaiian Islands and have an average value of 0.70384 ±17. However, the $^{87}Sr/^{86}Sr$ ratios of alkali basalts (not shown in Fig. 2.58) are more variable and range from 0.70341 (Gardner Pinnacles) to 0.70401 (Daikakuji Seamount). Following the bend in the Emperor Chain at 43 Ma (Dalrymple and Clague 1976), the $^{87}Sr/^{86}Sr$ ratios of the tholeiites decrease with increasing latitude from 0.70372 (Yuryaku Seamount) to 0.70336 ±7 (Suiko Seamount). The $^{87}Sr/^{86}Sr$ ratios of the alkali basalts also decline, but range more widely from 0.70326 (Kinmei Seamount) to 0.70428 for a sample of nepheline phonolite from Koko Seamount (not shown). The comparatively low $^{87}Sr/^{86}Sr$ ratios of tholeiites on the oldest Emperor Seamounts imply that the first magmas to form contained melt derived from the lithospheric mantle being heated by the approaching asthenospheric plume.

The Emperor Chain continues northward from Suiko Seamount to Meiji Seamount (DSDP 192A at about 53° N and 164° E) composed of altered basalt overlain by 1044 m of diatomaceous clay, chalk, and claystone ranging in age from Upper Cretaceous to Pleistocene. The age of the basalt is more than 61.9 ±5.0 Ma according to K-Ar and $^{40}Ar/^{39}Ar$ dating by Dalrymple et al. (1980a,b) consistent with the early Maastrichtian age of the oldest sediment.

The systematic variation of $^{87}Sr/^{86}Sr$ ratios of the tholeiites in the Emperor Chain indicates a change in the proportions of the source components of magma with time. The relatively low $^{87}Sr/^{86}Sr$ ratio of the tholeiites erupted before 38 Ma indicates that their magma sources contained a larger proportion of the DMM component than the tholeiites that were erupted from 32 Ma to the present.

2.13 Samoan Islands, Western Pacific

The islands of Samoa in Fig. 2.59 are located at about 14° S and 172° W in the western Pacific Ocean and are approximately collinear with the Austral/Cook Island chain (Sect. 2.10.4). The western part of the Pacific Ocean is the site of deep-sea trenches where the oceanic crust of the Pacific Plate is being subducted and where the resulting volcanic activity gives rise to island arcs. The islands of Samoa are located a short distance north of where the Kermadec-Tonga Trench turns sharply west. In this tectonic setting, the petrogenesis of the lavas on the islands of Samoa may be affected by the ongoing subduction of oceanic crust into the Kermadec-Tonga Trench which extends southwest from Samoa for a distance of more than 2500 km to the vicinity of the North Island of New Zealand.

The Samoan Archipelago consists of six islands containing eight volcanoes several of which have been active in historic time. The principal islands from southeast to northwest are the Manua Islands, (Ta'u, Ofu, and

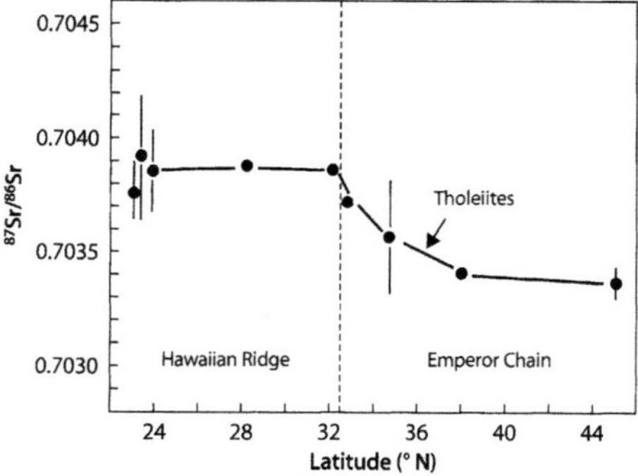

Fig. 2.58. Variation of $^{87}Sr/^{86}Sr$ ratios of tholeiites of islands and seamounts in the Hawaiian Ridge and Emperor Chain. The $^{87}Sr/^{86}Sr$ ratios of tholeiites in the Hawaiian Ridge are restricted to a narrow range with a mean of 0.70384 ±17 (1σ) for 10 specimens. The $^{87}Sr/^{86}Sr$ ratios of tholeiites in the Emperor Chain *decrease* with increasing latitude to a value of 0.70336 ±7 (1σ) for five specimens from Suiko Seamount at 44°46.6' N and 170°01.3' E. The change in the $^{87}Sr/^{86}Sr$ ratios of the tholeiites between 38 and 32 Ma indicates a shift in the composition of the magma sources during the formation of the seamounts. All $^{87}Sr/^{86}Sr$ ratios are compatible with 0.71025 for NBS 987 (*Sources*: Lanphere et al. 1980a; White and Hofmann 1982)

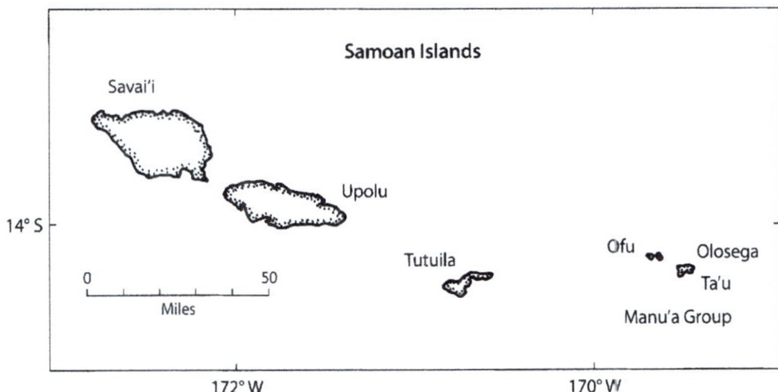

Fig. 2.59.
The Samoan Islands in the South Pacific Ocean (*Source:* adapted from Stice and McCoy, Jr. 1968)

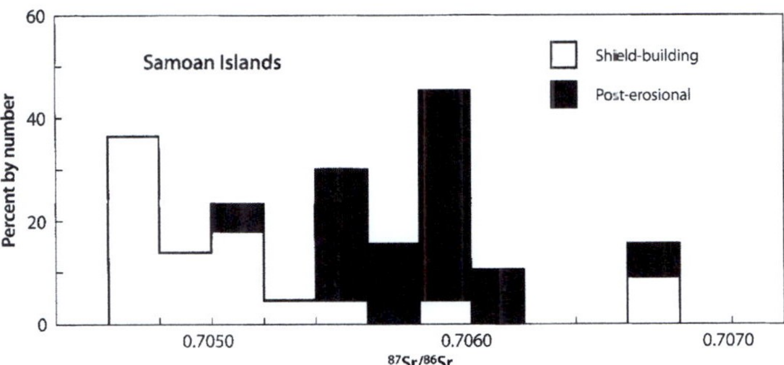

Fig. 2.60.
Range of $^{87}Sr/^{86}Sr$ ratios of the shield-building basalts of the Samoan Islands and of posterosional lavas. The distribution of data indicates that 73% of the specimens of shield-building basalts have $^{87}Sr/^{86}Sr < 0.7054$, whereas 95% of the specimens of posterosional lavas have $^{87}Sr/^{86}Sr > 0.7054$. All data are relative to 0.7080 for E&A or 0.71025 for NBS 987 (*Sources:* Palacz and Saunders 1986; Wright and White 1987)

Olosega), Tutuila, Upolu, and Savai'i. The rocks on these islands range in composition from tholeiitic basalt to highly undersaturated alkali-rich rocks including basanite, trachyte, and olivine nephelinite. The volcanic rocks of the Samoan Islands were erupted in three stages (Wright and White 1987):

1. Posterosional alkali olivine basalt, basanite, and olivine nephelinite (youngest);
2. Caldera-filling hawaiite, alkali olivine basalt, and trachyte;
3. Shield-building tholeiitic and alkalic basalt (oldest).

The ages of the shield-building basalts increase in a northwesterly direction from 1.0 to 2.8 Ma (Natland and Turner 1985; Duncan 1985). However, the progressive age trend of the Samoan Islands is obscured by the voluminous posterosional alkali-rich volcanics of late Pleistocene to historic age (Kear and Wood 1959). Although the age progression of the shield-building basalts is consistent with plume-related volcanic activity, the nearly simultaneous eruption of the posterosional volcanics indicates that these lavas originated along a rift zone that runs parallel to the island chain for at least 250 km (Kear and Wood 1959; Wright and White 1987).

The $^{87}Sr/^{86}Sr$ ratios of volcanic rocks on the Samoan Islands range widely from 0.70461 to 0.70742 relative to 0.7080 for E&A and 0.70125 for NBS 987 (Farley et al. 1992; Wright and White 1987; Palacz and Saunders 1986; White and Hofmann 1982; Hedge et al. 1972). However, the $^{87}Sr/^{86}Sr$ ratios of most shield-building tholeiites are lower than those of the posterosional lavas. The data in Fig. 2.60 demonstrate that about 73% of the shield-building basalts analyzed by Palacz and Saunders (1986) and Wright and White (1987) have $^{87}Sr/^{86}Sr < 0.7054$, whereas 95% of the specimens of the posterosional volcanics have $^{87}Sr/^{86}Sr > 0.7054$. The difference in the isotope compositions of Sr indicates that the shield-building basalts and posterosional lavas originated from different magma sources

The isotope ratios of Sr and Pb of volcanic rocks on the Samoan Islands in Fig. 2.61 form two overlapping clusters representing the shield-building and posterosional lavas. The shield-building tholeiites have lower $^{87}Sr/^{86}Sr$ but higher $^{206}Pb/^{204}Pb$ ratios than the more recently erupted posterosion alkali basalts. The distribution of data points representing the posterosional lavas on the islands of Samoa indicates that their magma sources contained a mixture of subducted terrigenous (EM2) and pelagic (EM1) sediment components with-

Fig. 2.61.
Isotope ratios of Sr and Pb of the lavas on Samoa in relation to the four mantle components. The shield-building tholeiite basalts of Samoa appear to have formed from mixtures of EM2 (sub-ducted terrigenous sediment) and DMM (depleted lithospheric mantle) with varying but minor proportions of HIMU (subducted oceanic crust). The magma sources of the most recently erupted posterosional alkali-rich lavas are mixtures primarily of EM1 (subducted pelagic sediment) and EM2 without significant involvement of HIMU or DMM (*Source:* Wright and White 1987)

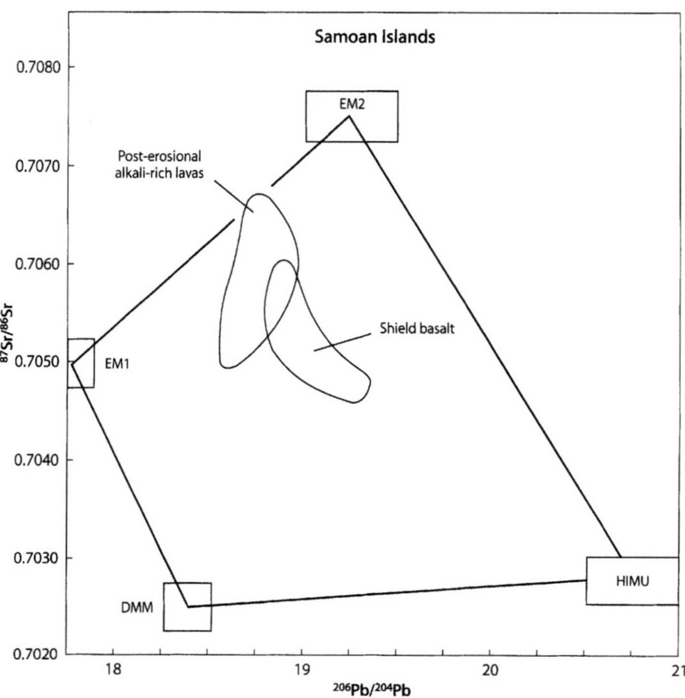

out significant participation of lithospheric mantle (DMM) and subducted oceanic crust (HIMU). The magma sources of the shield-building lavas likewise contained mixtures of EM1 and EM2 as well as varying but minor amounts of subducted oceanic crust (HIMU). Wright and White (1987) concluded that the posterosional volcanic activity was a result of rifting caused by the flexure of the Pacific Plate at the termination of Tonga-Kermadec Trench in agreement with Hawkins and Natland (1975). However, the shield-building tholeiites are attributable to magma generation in a plume presently located southeast of the Manua group of islands.

The lava flows of Pago Volcano on the north coast of Tutuila at Masefau Bay have unusually high ^{3}He/^{4}He ratios up to 23.9 R_A where R_A is the isotope ratio of atmospheric He which has a value of 1.4×10^{-6}. Consequently, Farley et al. (1992) concluded that the mantle under Samoa includes a magma-source component they called PHEM (primitive helium mantle). The existence of this component implies the presence of primitive undegassed mantle rocks under Samoa. Farley et al. (1992) tentatively identified PHEM as the undepleted asthenospheric mantle.

The western Pacific Ocean also contains chains of volcanic seamounts such as the Ratak Chain of the Marshall Islands located at 13°54' N and 167°39' E to the northwest of the Samoan chain of islands (Davis et al. 1989). In addition, McDougall and Duncan (1988) and Eggins et al. (1991) described the ages and petrogenesis of volcanic rocks dredged from the Tasmantid and Lord Howe Seamounts in the Tasman Sea off the coast of eastern Australia.

2.14 MORBs of the Indian Ocean

The Indian Ocean contains the same tectonic features that characterize the Atlantic and Pacific Oceans including spreading ridges, non-spreading ridges, submarine plateaus, volcanic islands, seamounts, deep-sea trenches, and islands composed of continental crust. In spite of the tectonic complexity of the Indian Ocean, the petrogenesis of the volcanic rocks that have formed there can be accounted for in most cases by the plume model of the mantle and by the different components of the magma sources defined previously for rocks of the Atlantic and Pacific Oceans. The origin of these rocks has also been discussed by Frey et al. (1979), Klein et al. (1991) and others on the basis of their chemical and mineralogical compositions.

Measurements of ^{87}Sr/^{86}Sr ratios of MORBs from the Mid-Indian Ocean Ridge by Subbarao and Hedge (1973) and Subbarao and Reddy (1981) yielded higher values

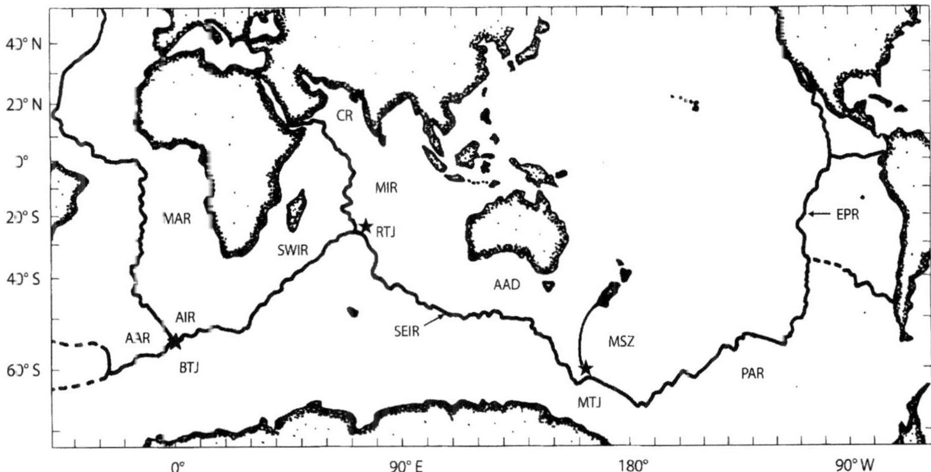

Fig. 2.62. Map of the mid-ocean ridges of the Indian Ocean in relation to those of the Atlantic and Pacific Oceans. *RTJ* = Rodriguez Triple Junction, *SWIR* = Southwest Indian Ridge; *MIR* = Mid-Indian Ridge; *CR* = Carlsberg Ridge, *SEIR* = Southeast Indian Ridge; *AAD* = Australian-Antarctic Discordance; *MTJ* = Macquarie Triple Junction; *MSZ* = Macquarie Shear Zone; *PAR* = Pacific-Antarctic Ridge; *EPR* = East Pacific Rise; *BTJ* = Bouvet Triple Junction; *AAR* = American-Antarctic Ridge; *AIR* = Atlantic-Indian Ridge; *MAR* = Mid-Atlantic Ridge (*Sources:* Klein et al. 1988; Ferguson and Klein 1993)

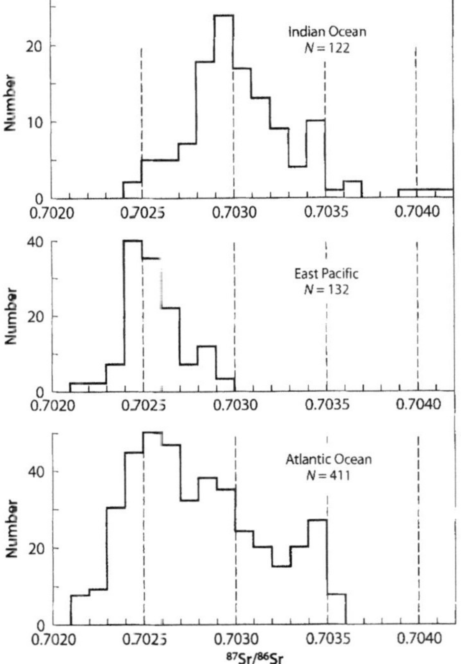

Fig. 2.63. Isotope ratios of Sr in MORBs of the Indian Ocean compared to those of the East Pacific Rise (Fig. 2.35) and the Mid-Atlantic Ridge (Fig. 2.7). Samples from non-spreading ridges, seamounts, and oceanic islands were excluded. All ⁸⁷Sr/⁸⁶Sr ratios are consistent with 0.70800 for E&A and 0.71025 for NBS 987. The comparison clearly shows that most MORBs in the Indian Ocean have higher ⁸⁷Sr/⁸⁶Sr ratios than MORBs in the other two oceans (*Sources:* data from Pyle et al. 1995 (Indian MORBs only); Mahoney et al. 1992 (excluding samples from 17° to 26° E on the Southwest Indian Ridge); Schilling et al. 1992 (East Sheba Ridge only); Mahoney et al.1989; Dosso et al. 1988; Klein et al. 1988 (Indian MORBs only); Hamelin et al. 1986; Michard et al. 1986; Price et al. 1986; Hamelin and Allègre 1935; Hedge et al. 1979)

(0.7032 to 0.7035) than had been reported for MORBs from the Mid-Atlantic Ridge and from the East Pacific Rise. Subsequent studies have confirmed that MORBs of the Indian Ocean have elevated ⁸⁷Sr/⁸⁶Sr ratios and the phenomenon is now recognized as the Dupal anomaly in the southern hemisphere of the Earth discussed in Sect. 2.10.8 (Dupré and Allègre 1983; Hart 1984). The boundary between the mantle domains of the Pacific and Indian Oceans occurs at about 125° E longitude along the Southeast Indian Ridge (SEIR), which separates Australia from Antarctica in this region (Klein et al.1988; Ferguson and Klein 1993; Lanyon et al. 1995; Mahoney et al. 1998).

The Southeast Indian Ridge in Fig. 2.62 extends in a southeasterly direction from the Rodriguez triple junction and terminates at the Macquarie triple junction located at about 60° S and 160° E. The northern branch of that triple junction is the Macquarie shear zone which merges with the Alpine Fault that runs along the west coast of the South Island of New Zealand. The third branch is the Pacific-Antarctic Ridge which trends to the northeast and joins the East Pacific Rise at about 36° S and 110° E (Fig. 2.34, Sect. 2.7). Work by Ferguson and Klein (1993) demonstrated that MORBs along a short segment of the Pacific-Antarctic Ridge at about

Fig. 2.64.
Isotope ratios of Sr and Nd in MORBs of the Indian Ocean. All $^{87}Sr/^{86}Sr$ ratios are relative to 0.7080 for E&A or 0.71025 for NBS 987 and have been corrected for decay of ^{87}Rb as necessary. The $^{143}Nd/^{144}Nd$ ratios are compatible with 0.51265 for BCR-1, 0.511959 for Johnson and Mathey 321, or 0.511850 for the LaJolla Nd standard. The distribution of data points indicates that the magmas originated from the DMM component with varying contribution from EM1 and EM2 as defined by Hart (19988) (*Sources:* data from Pyle et al. 1995 (excluding Pacific samples); Mahoney et al. 1992 (excluding MORBs from 17° to 26° E on the SW Indian Ridge) Schilling et al. 1992 (East Sheba only); Mahoney et al. 1989; Dosso et al. 1988; Klein et al. 1988 (west of Pacific-Indian Boundary) Hamelin et al. 1986; Michard et al. 1986; Price et al. 1986; Hamelin and Allègre 1985)

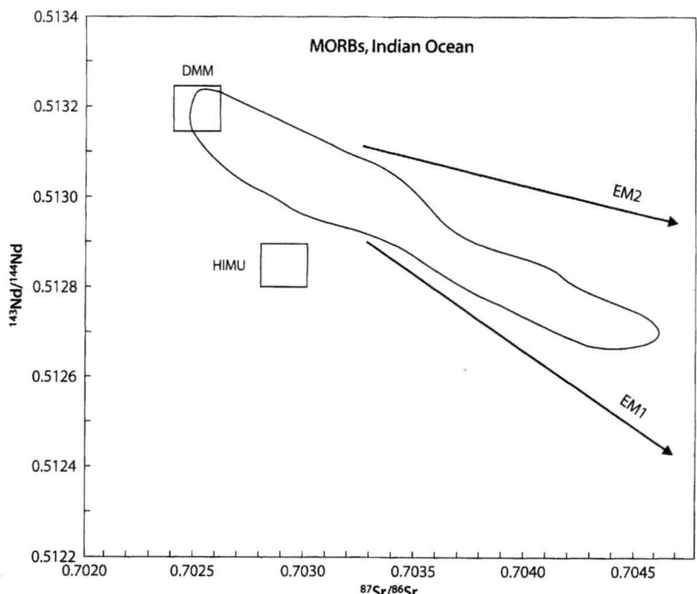

Fig. 2.65.
Isotope ratios of Sr and Pb of MORBs on the Southeast, Central, Southwest, and Carlsberg (East Sheba) spreading ridges in the Indian Ocean. Volcanic rocks on non-spreading ridges, seamounts, and oceanic islands were excluded. MORBs on the Southwest Indian Ridge (*SWIR*) between longitudes of 39° and 41° E have anomalously high $^{87}Sr/^{86}Sr$ ratios and low $^{206}Pb/^{204}Pb$ ratios extending to 16.867 (Mahoney et al. 1992). The mantle components were taken from Hart (1988) (*Sources:* Pyle et al. 1995 (excluding Pacific samples); Schilling et al. 1992 (East Sheba Ridge only); Mahoney et al. 1992 (Excluding MORBs on SWIR between 39° and 41° E longitude); Mahoney et al. 1989; Dosso et al. 1988; Klein et al. 1988 (west of Indian-Pacific boundary on the Southeast Indian Ridge); Hamelin et al. 1986; Michard et al. 1986; Price et al. 1986; Hamelin and Allègre 1985)

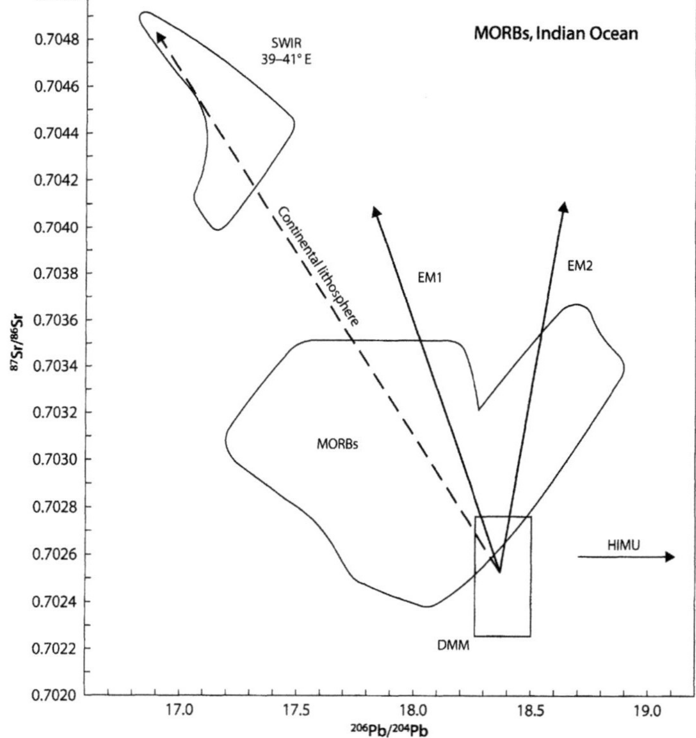

54° S and 171° W have low average $^{87}Sr/^{86}Sr$ ratios of 0.70241 ±0.00001 ($2\bar{\sigma}$, $N = 4$) relative to 0.710250 for NBS 987. These rocks also have $^{143}Nd/^{144}Nd = 0.513122$ and $^{206}Pb/^{204}Pb = 18.349$ consistent with the isotope compositions of normal MORBs derived from the DMM component in the mantle. Therefore, Ferguson and Klein (1993) noted that this segment of the Pacific-Antarctic Ridge is within the Pacific domain.

The Southeast Indian Ridge (SEIR) between Antarctica and Australia (Fig. 2.62) includes a 500-km segment along which the depth of water above the ridge exceeds 4300 m. This stretch of the ridge is known as the Australian-Antarctic Discordance (Sempéré et al. 1991). Klein et al. (1988) demonstrated that the isotope ratios of Sr, Nd, and Pb of MORBs at the Discordance change from values typical of Pacific MORBs to values that characterize MORBs of the Indian Ocean.

The *western* boundary of the mantle domain underlying the Indian Ocean occurs near the Du Toit Fracture Zone at about 26° E longitude on the Southwest Indian Ridge (SWIR; Fig. 2 62). The isotope ratios of Sr, Nd, and Pb of MORBs on the ridge that extends west of that longitude to the Bouvet triple junction (Sect. 2.5.8) approach the characteristic values of Atlantic MORBs (Hamelin and Allègre 1985; Mahoney et al. 1992).

The compilation of data in Fig. 2.63 demonstrates that the $^{87}Sr/^{86}Sr$ ratios of MORBs on the spreading ridges of the Indian Ocean are strongly clustered between 0.7028 and 0.7032. Some specimens of MORBs recovered from the Southeast Indian Ridge at about 78° E longitude (Dosso et al. 1988; Hamelin et al. 1986; Michard et al. 1986) and from the Southwest Indian Ridge between 39° and 41° E longitude (Mahoney et al. 1992; Hamelin and Allègre 1985) have high $^{87}Sr/^{86}Sr$ ratios that exceed 0.7040. The negative correlation of Sr and Nd isotope ratios in Fig. 2.64 shows that the magmas extruded along the spreading ridges in the Indian Ocean originated from mixed sources consisting of depleted MORB mantle (DMM) and Rb-enriched components represented by EM1 and EM2. In this regard, the MORB sources of the Indian Ocean resemble those of the OIBs on the Galapagos Islands (Fig. 2.38), on Pitcairn Islands (Fig. 2.44), on the Marquesas Islands (Fig. 2.46), on the Society Islands (Fig. 2.48), and on the Austral/Cook Islands (Fig. 2.49). However, the apparent absence of the HIMU component distinguishes the MORBs of the Indian Ocean from OIBs in the Pacific Ocean.

The $^{206}Pb/^{204}Pb$ ratios in Fig. 2.65 generally corroborate the conclusion that the MORBs of the Indian Ocean originated from mixed magma sources in the mantle. However, MORBs of the Southwest Indian Ridge between 39° and 41° E longitude have remarkably low $^{206}Pb/^{204}Pb$ ratios between 16.867 and 17.449 that are virtually unique among MORBs (Mahoney et al. 1992; Hamelin and Allègre 1985). Since the $^{87}Sr/^{86}Sr$ ratios of these rocks are anomalously high (0.70399 to 0.70488), the data

points form a separate cluster in Fig. 2.65. Low $^{206}Pb/^{204}Pb$ ratios of less than 17.5 occur also in MORBs in the area of the Rodriguez triple junction (Fig. 2.62, Michard et al. 1986; Price et al. 1986), along the Carlsberg Ridge (Mahoney et al. 1992), and along the Southeast Indian Ridge (Michard et al. 1986). However, the $^{87}Sr/^{86}Sr$ ratios of these MORBs are between 0.7028 and 0.7032. Mahoney et al. (1992) favored the explanation that the anomalous isotope ratios of MORBs on the Southwest Indian Ridge (39°–41° E) in Fig. 2.65 were caused by the presence of continental lithosphere eroded from India or Madagascar during the break-up of Gondwana in Early Cretaceous time. The dashed line in Fig. 2.65 is a hypothetical mixing line between the depleted MORB mantle and a component consisting of continental lithosphere based on isotope ratios assigned to that component by Michard et al. (1986): $^{87}Sr/^{86}Sr = 0.7055$ and $^{206}Pb/^{204}Pb = 16.5$.

2.15 Kerguelen and Heard Islands

The anomalous isotope ratios of Sr, Nd, and Pb of MORBs along the spreading ridges of the Indian Ocean raise a question about the composition of the underly-

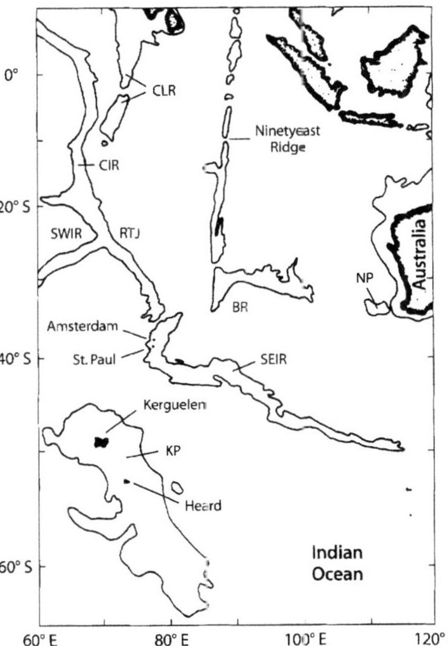

Fig. 2.66. Map of the eastern Indian Ocean showing the locations of the Kerguelen Plateau (with Kerguelen and Heard Islands), as well as the Ninetyeast Ridge, Broken Ridge, and the Naturaliste Plateau (*Source:* Mahoney et al. 1995)

ing mantle. A possible answer was proposed by Storey et al. (1988, 1989b) who suggested that the asthenospheric mantle under the Indian Ocean was contaminated by the large plume that formed the Kerguelen Plateau and the associated Kerguelen and Heard Islands. The apparent magnitude and duration of the Kerguelen Plume is enhanced by evidence that it also caused the formation of the Ninetyeast Ridge, Broken Ridge, and the Naturaliste Plateau (Fig. 2.66) all of which are now separated from the Kerguelen Plateau by the Southeast Indian Ridge (Class et al. 1993). Evidence reviewed by Davies et al. (1989) indicates that the Kerguelen Plateau formed between 118 and 95 Ma during the opening of the Indian Ocean when the subcontinent of India separated from the supercontinent Gondwana.

Other proposals concerning the origin of the Kerguelen Plateau referenced by Weis et al. (1989) include suggestions that it is:

1. A fragment of the continental crust of Gondwana.
2. A failed spreading ridge.
3. The product of magmatic activity related to a fault.
4. Composed of thick oceanic crust, uplifted either by isostatic forces or because of thermal expansion.
5. The result of intraplate volcanic activity related to the plume that produced the Ninetyeast Ridge (Fig. 2.66).

The importance of the Kerguelen Plume in the tectonic evolution of the Indian Ocean invites comparison with other plumes that have caused the formation of other archipelagoes in the Indian Ocean, including the Crozet Islands, the Prince Edward Islands, the Mascarene Islands, Rodriguez Island, and the Amsterdam-St. Paul Islands. The petrogenesis of volcanic rocks of the islands in the Indian Ocean was summarized and discussed by Frey et al. (1977, 1979), Alibert (1991), Duncan and Storey (1992), and Weis et al. (1992) among others.

Fig. 2.67.
Range of $^{87}Sr/^{86}Sr$ ratios of volcanic rocks associated with the topographic features in the Indian Ocean identified by name in Figure 2.66. The $^{87}Sr/^{86}Sr$ ratios of the igneous rocks on Kerguelen, Heard, and McDonald Islands as well as those on the submarine Kerguelen Plateau range primarily from low values of 0.7035 (Salters et al. 1992) up to 0.7060 (Barling and Goldstein 1990) consistent with values that characterize oceanic island basalts (*Sources:* data from Mahoney et al. 1995; Barling et al. 1994; Weis et al. 1993; Salters et al. 1992; Weis and Frey 1991; Barling and Goldstein 1990; Gautier et al. 1990; Davies et al. 1989; Weis et al. 1989; Storey et al. 1988; Mahoney et al. 1983; Dupré and Allègre 1983; Clarke et al. 1983; Reddy et al. 1978; Whitford and Duncan 1978; Hedge et al. 1973)

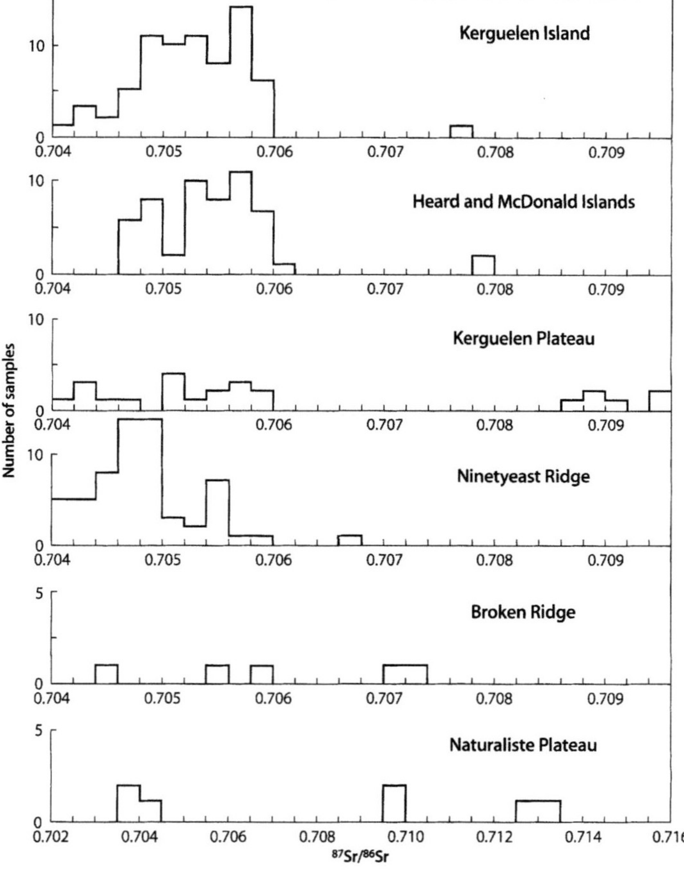

2.15.1 Kerguelen Islands

The Kerguelen Islands in Fig. 2.66 are located at about 49° S and 70° E at the northern end of the Kerguelen Plateau. The main island has an east-west diameter of about 130 km and contains four shield volcanoes (Loranchet, Cook, Jeanne d'Arc, and Courbet) composed of tholeiites overlain by alkali basalt, rhyolites, trachytes, and phonolites. K-Ar age determinations by Watkins et al. (1974), Dosso et al. (1979), and Giret and Lameyre (1983) indicated that the ages of the igneous rocks on Kerguelen Island extend from 40 Ma or older to 1.15 ±0.05 Ma or younger.

The Rallier du Baty Peninsula in the southwestern part of Kerguelen Island contains five plutonic ring complexes composed of syenites, alkali gabbros, and granite. The ages of these complexes range from 15.4 ±0.5 to 4.9 ±0.2 Ma and decrease sequentially from south to north. The age-corrected ^{87}Sr/^{86}Sr ratios of the syenites and gabbros on Rallier du Baty (0.70553 to 0.70767) are similar to those of lava flows elsewhere on Kerguelen Island (Dosso et al. 1979; Dosso and Murthy 1980).

The ^{87}Sr/^{86}Sr ratios of volcanic rocks on Kerguelen Islands in Fig. 2.67 vary widely but cluster between 0.7041 and 0.7061. The initial ^{87}Sr/^{86}Sr ratios of basalts that formed the submarine Kerguelen Plateau at about 114 Ma vary even more widely than those of the islands from about 0.70411 to 0.70950 (Davies et al. 1989; Weis et al. 1989; Mahoney et al. 1995).

2.15.2 Heard Island

The Heard and McDonald Islands are located at 53°06' S and 73°30' E at the southern end of the Kerguelen Plateau (Fig. 2.66). The main part of the island has an area of about 65 km² and consists primarily of the Newer lavas erupted by the volcano Big Ben (2 745 m) in late Pleistocene to Holocene time. In addition, Mt. Dixon (a smaller volcanic cone, 705 m) forms the Laurens Peninsula which is connected to the main part of the island by a narrow strip of land only a few hundred meters wide (Clarke et al. 1983).

The Newer lavas are underlain by the Drygalski Formation (late Miocene to early Pliocene) composed of clastic rocks and basalt flows. This formation was deposited unconformably on the Laurens Peninsula Limestone (middle Eocene to middle Oligocene, Quilty et al. 1983). The Newer lavas on Big Ben consist of two groups of flows characterized by basanites and by a succession of alkali basalt and trachybasalt. All of the volcanic rocks of the Newer lavas on Big Ben and on Mt. Dixon are alkali-rich and have a wide range of silica concentrations from 42.64% (basanite, Big Ben) to 61.06% (trachyte, Mt. Dixon). The ^{87}Sr/^{86}Sr ratios of the Newer lavas range widely from 0.704728 to 0.707974 relative to 0.71023 for NBS 987. (Barling and Goldstein 1990; Barling et al. 1994).

The isotope ratios of Sr, Nd, and Pb in the Newer lavas of Heard Island are well correlated and form two-component mixing hyperbolas in coordinates of ^{143}Nd/^{144}Nd vs. ^{206}Pb/^{204}Pb and ^{87}Sr/^{86}Sr vs. ^{206}Pb/^{204}Pb. The isotope ratios of Sr and Nd of the Heard lavas in Fig. 2.68 lie within the field of the volcanic and plutonic rocks on Kerguelen Island and form an extension of the field of MORBs in the Indian Ocean depicted in Fig. 2.64. In addition, the ^{87}Sr/^{86}Sr and ^{143}Nd/^{144}Nd ratios of the lavas on Heard define a mixing hyperbola whose end-member components are the "Depleted Heard" and "Enriched Heard" magma sources defined by Barling and Goldstein (1990). This hyperbola passes between EM1 and EM2 indicating that the "Enriched Heard" component may be characteristic of the Kerguelen Plume and is not simply a combination of EM1 and EM2 as defined by Zindler and Hart (1986) and Hart (1988).

The placement of the Enriched Heard component in Fig. 2.68 and of the data for basalt of the Kerguelen Pla-

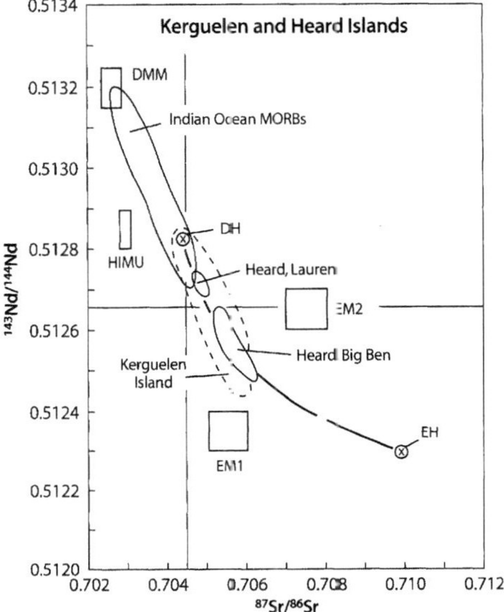

Fig. 2.68. Systematic relations of ^{87}Sr/^{86}Sr and ^{143}Nd/^{144}Nd ratios of alkali-rich lavas on Heard Island (Laurer and Big Ben) in relation to the volcanic rocks on Kerguelen Island and Indian Ocean MORBs (Figure 2.64). The lavas on Heard Island define a mixing hyperbola between the Depleted Heard (*DH*) and the Enriched Heard (*EH*) components. The continuity of isotope ratios of Indian Ocean MORBs with the lavas of Kerguelen and Heard Islands implies that the MORB sources in the mantle also contain small amounts of the enriched component (*Sources:* data from Dosso and Murthy 1980; Storey et al. 1988; Davies et al. 1989; Weis et al. 1989; Barling and Goldstein 1990; Gautier et al. 1990; Weis et al. 1993; Barling et al. 1994)

teau (ODP 738C) in Fig. 2.69 strongly suggests that the Rb-enriched component associated with the formation of the Kerguelen Plateau and the related islands resembles *continental* crust. Remnants of continental lithosphere may have originated from the separation of the Indian subcontinent from Antarctica at about 115 Ma that was caused partly by the interaction of the Kerguelen Plume with the continental lithosphere of Gondwana. However, there is no direct evidence for the existence of blocks of continental crust within the Kerguelen Plateau (Dosso and Murthy 1980). In addition, Barling et al. (1994) favored the view that the crustal component consists of marine sediment of crustal origin that was subducted into the mantle more than 600 million years ago. This component may have occurred in the large head of the Kerguelen Plume and caused the asthenospheric mantle under the Indian Ocean to be contaminated as proposed by Storey et al. (1989b). The collinearity of isotope ratios of Sr and Nd in MORBs of the Indian Ocean with the Kerguelen-Heard array in Figs. 2.68 and 2.69 suggests that even the magma sources of MORBs in the Indian Ocean contained the enriched component of the Kerguelen Plume.

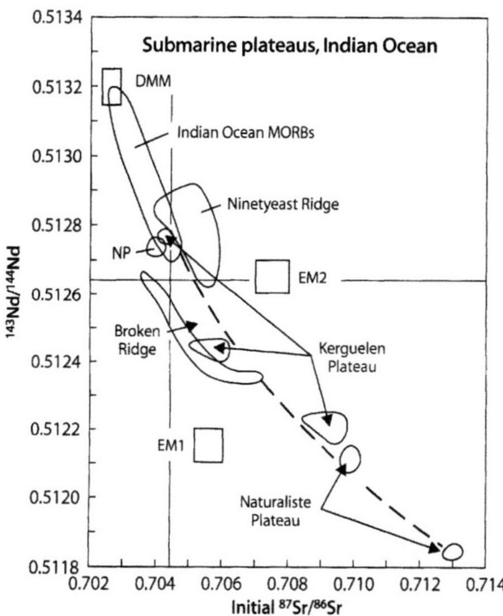

Fig. 2.69. Isotope ratios of Sr and Nd of volcanic rocks on the Kerguelen Plateau, Broken Ridge, Naturaliste Plateau, and Ninetyeast Ridge in the Indian Ocean. The data clusters are approximately aligned along a hypothetical mixing hyperbola that joins Indian Ocean MORBs to rocks on the Naturaliste Plateau that have high $^{87}Sr/^{86}Sr$ and low $^{143}Nd/^{144}Nd$ ratios. Therefore, all of the magma sources represented by these rocks contain a highly Rb-enriched component that may have originated from the continental crust of Gondwana (*Sources:* Mahoney et al. 1983, 1995; Salters et al. 1992; Weis and Frey 1991; Saunders et al. 1991)

2.15.3 Ninetyeast Ridge, Broken Ridge, and the Naturaliste Plateau

The Ninetyeast Ridge on Fig. 2.66 lies along the 90° E meridian between latitudes 9° N and about 30° S (Frey et al. 1991). It rises about 2 km above the surrounding seafloor and ranges in width from 100 to 200 km (Whitford and Duncan 1978). The age of basalt and overlying basal sediment increases from south to north, which led Duncan (1978) to conclude that the ridge had formed during the northward migration (at 9.4 ± 0.3 cm yr^{-1}) of the Indian-Ocean Plate over a hotspot in the mantle identified as the Kerguelen Plume. Additional $^{40}Ar/^{39}Ar$ dates by Duncan (1991) later confirmed these results.

Broken Ridge is a submarine plateau (1.5×10^5 km^2) located at about 30° S latitude between about 87° and 100° E longitude. This submarine plateau is composed of a basement of Early Cretaceous basalt overlain by Late Cretaceous to Eocene sediment known from DSDP and ODP cores. The underlying basalts were recovered by dredging along the southern slope of the plateau. (Mahoney et al. 1995).

The Naturaliste Plateau (Fig. 2.66) is located close to the southwestern corner of Australia, but is separated from it by the Naturaliste Trough. The plateau has an area of about 1.0×10^5 km^2 and rises about 2.5 km above the adjacent seafloor. The Naturaliste Plateau consists of a basalt basement overlain by several hundred meters of Cretaceous and Tertiary sediment recovered in DSDP drill cores at sites 258 and 264. (Storey et al. 1992).

Information reviewed by Mahoney et al. (1995) and by Gautier et al. (1990) indicates that all three features described above formed by volcanic activity that started at about 125 ±5 Ma during the initial rifting of Gondwana and the subsequent opening of the Indian Ocean. The earliest evidence for volcanic activity along this rift are the Bunbury volcanics (now located near the coast in southwestern Australia) and the Rajmahal basalts (now located in northeastern India and discussed in Sect. 5.6). In addition, alnöites (ultramafic lamprophyres) were intruded in the present Prince Charles Mountains of East Antarctica. All of these deposits record the magmatic activity caused by the Kerguelen Plume that is now located in the vicinity of Heard Island under the Kerguelen Plateau. Clearly, this is one of the superplumes which has affected the Indian Ocean for the last 125 million years (Storey et al. 1992).

The initial $^{87}Sr/^{86}Sr$ ratios of the volcanic rocks that form the submarine plateaus of the Indian Ocean vary widely and reach high values in excess of 0.7130 (Eltanin 55–12, Naturaliste Plateau; Mahoney et al. 1995). The isotope ratios of Sr and Nd in volcanic rocks on the submarine plateaus and ridges of the Indian Ocean in Fig. 2.69 cluster close to a hypothetical mixing hyperbola that relates Indian Ocean MORBs to a component in the mantle having elevated ^{87}Sr/^{86}Sr ratios (>0.7130) and low ^{143}Nd/^{144}Nd (<0.51180). This component has the isotopic signature of rocks in the continental crust. However, it is not clear whether this component exists in the form of blocks of continental crust within the lithospheric mantle (e.g. within or beneath the Kerguelen Plateau) or whether the enriched component consists of subducted sediment derived from rocks of the continental crust. In addition, it is not clear whether the enriched component was assimilated by magmas that originated by melting in the head of the Kerguelen Plume or whether this component was part of the plume head itself. In either case the wide range of isotope ratios of Sr and Nd in the volcanic rocks of the submarine plateaus and ridges of the Indian Ocean results from the presence of varying amounts of the enriched component in the magma sources.

2.16 Oceanic Islands of the Indian Ocean

The Indian Ocean contains a large number of volcanic islands, including Amsterdam and St. Paul, Rodriguez, the Mascarene and Comoro Archipelagos, the Prince Edward Islands, and the Crozet Islands. The petrogenesis of the volcanic rocks on some of these islands is still not well understood because the islands are remote and hence complete collections of rocks samples are difficult to obtain.

2.16.1 Amsterdam and St. Paul

Amsterdam and St. Paul Islands (Gunn et al. 1971; Girod et al. 1971) are located close to 38°30' S and 77°50' E and are the only exposed part of the Southeast Indian Ridge. Both islands are composed primarily of tholeiite basalt flows and pyroclastic deposits. The lavas on St. Paul Island range in age from 0.57 ±0.07 to 0.04 ±0.03 Ma based on K-Ar dates by Watkins et al. (1975).

The ^{87}Sr/^{86}Sr ratios of eight samples of the volcanic rocks on Amsterdam and St. Paul range from 0.7035 to 0.7040 relative to 0.7080 for E&A (Hedge et al. 1973). Subsequent analyses by Hamelin et al. (1986), Dupré and Allègre (1983), Dosso et al. (1988), and Michard et al. (1986) have confirmed these results for both islands.

The ^{87}Sr/^{86}Sr ratios of the volcanic rocks on Amsterdam and St. Paul Islands are significantly higher than those of the majority of the Indian Ocean MORBs included in Fig. 2.70. Consequently, the magmas that were extruded to form these islands probably originated from a plume located beneath the Southeast Indian Ridge. For this reason, Michard et al. (1986) concluded that the MORBs along the Southeast Indian Ridge within about 400 km of the islands are mixtures of a MORB component (^{87}Sr/^{86}Sr = 0.7028; ^{143}Nd/^{144}Nd = 0.51304, ^{206}Pb/^{204}Pb

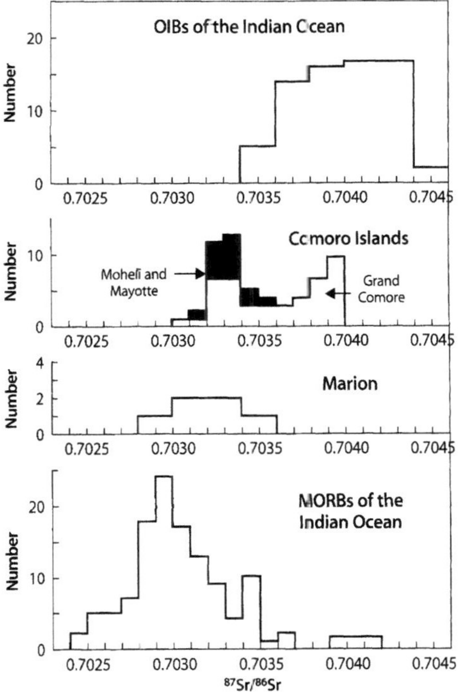

Fig. 2.70. The ^{86}Sr/^{87}Sr ratios of oceanic island basalts (OIBs) of the Indian Ocean compared to those of MORBs on the spreading ridges in the same ocean. The OIBs included here are from Amsterdam, St. Paul, Rodriguez, Réunion, Mauritius, and Crozet (*First panel*). The lavas on Grande Comore and Marion-Prince Edward Island are shown separately. Volcanic rocks on Kerguelen, Heard, Ninetyeast Ridge, Broken Ridge, and Naturaliste Plateau were excluded (*Sources:* Hamilton 1965; McDougall and Compston; Hedge et al. 1973; Ludden 1978; Dupré and Allègre 1983; Baxter et al. 1985; Hamelin et al. 1986; Michard et al. 1986; Dosso et al. 1988; Mahoney et al. 1989, 1992; Späth et al. 1996; Deniel 1998)

= 17.8) and a plume component (^{87}Sr/^{86}Sr = 0.7036, ^{143}Nd/^{144}Nd = 0.5129, ^{206}Pb/^{204}Pb = 18.7). A MORB sample recovered from the Southeast Indian Ridge at 38.982° S and 78.136° E about 60 km southeast of St. Paul has an anomalously high ^{87}Sr/^{86}Sr ratio of 0.704747 relative to 0.707982° S for E&A similar to basalt of the Kerguelen Plateau (Michard et al. 1986).

2.16.2 Rodriguez Island

The island of Rodriguez (Upton et al 1967) is located at 19°42' S and 63°25' E about 400 km west of the Central Indian Ridge and north of the Rodriguez triple junction (Fig. 2.62). The island has an area of close to 120 km^2 and rises to an elevation of 396 m above sealevel at Mt. Limon which is the summit of a volcano that rises from the seafloor. The present island is situated on a large submarine platform that may be the eroded top of an older volcanic edifice formed during an earlier stage of volcanic activity. The island is composed of alkali basalt yielding K-Ar dates of 1.32 ±0.03 and 1.54 ±0.05 Ma which indicate that volcanic activity on Rodriguez Island ceased in Plio-Pleistocene time. The island also contains deposits of cross-bedded calcarenite of Pleistocene age that overlie weathered basalt (McDougall et al. 1965; McBirney 1989a).

The average $^{87}Sr/^{86}Sr$ ratio of the two specimens of alkali basalt dated by McDougall et al. (1965) is 0.70375 ±0.00005 relative to 0.7080 for E&A (McDougall and Compston 1965). These values exceed those of most MORBs in the Indian Ocean and indicate the presence of a plume similar to that which underlies Amsterdam and St. Paul Islands (Morgan 1978). The petrology and geochemistry of Rodriguez Island were later discussed by Baxter et al. (1985) who concluded that the volcanic rocks consist of mildly alkaline differentiated basalt flows containing abundant inclusions of gabbro and a suite of megacrysts composed of plagioclase, kaersutite (amphibole), olivine, clinopyroxene, spinels, and apatite. The additional $^{87}Sr/^{86}Sr$ ratios they reported range from 0.703580 to 0.704057 which is characteristic of the OIBs on Amsterdam, St. Paul, and other volcanic islands of the Indian Ocean.

2.16.3 Mascarene Islands (Réunion, Mauritius, and Seychelles)

The Mascarene Plateau in the western Indian Ocean extends from about 2.5° S to 22.5° S between longitudes of 55° and 60° E. The plateau contains the Seychelles Islands at its northern end as well as Mauritius and Réunion at its southern end. According to Duncan (1990), the Mascarene Ridge formed as a result of volcanic activity on the Indian Plate as it moved northward over the stationary Réunion Plume (Sect. 5.5).

The island of Réunion contains two volcanoes: Piton des Neiges (3 069 m) which is extinct and Piton de la Fournaise (2 621 m) which is still active. The geology of the island and the origin of the volcanic rocks were described by Upton and Wadsworth (1965, 1966, 1972).

The subaerial shield-building basalts of Piton des Neiges were extruded episodically at 2 Ma, 1.2 to 1.0 Ma, and at 0.55 to 0.43 Ma (McDougall 1971; later confirmed by Gillot and Nativel 1982). After a short period of quiescence, alkali-rich andesites and basalts were erupted between 0.35 and 0.07 Ma (Zielinski 1975). The basalt lavas of Piton de la Fournaise range in age from older than 0.36 Ma to Recent. McDougall (1971) also reported that the center of volcanic activity on Réunion has been shifting to the southeast at rates between 2 and 7 cm yr^{-1} which implies movement of the island to the

northwest relative to a stationary magma source in the mantle.

The $^{87}Sr/^{86}Sr$ ratios of volcanic rocks on Réunion were first measured by Hamilton (1965) and McDougall and Compston (1965) who reported values ranging from 0.7034 to 0.7045 relative to 0.7080 for E&A, excluding a quartz syenite analyzed by Hamilton (1965) who reported a value of 0.7055. These data are in good agreement with more recent results published by Hedge et al. (1973), Ludden (1978), and Dupré and Allègre (1983) and with the $^{87}Sr/^{86}Sr$ ratios of the lavas on Amsterdam, St. Paul, and Rodriguez. The limited range of the $^{87}Sr/^{86}Sr$ implies that the Réunion Plume and the adjacent depleted lithospheric mantle contributed magma in fairly constant proportion. The $^{206}Pb/^{204}Pb$ ratios of the lavas on Réunion range from 18.654 to 19.000 (Oversby 1972; Dupré and Allègre 1983). The Réunion Plume has supplied magma not only to the volcanoes on the islands of Réunion and Mauritius but has left a long trail of volcanic deposits in the western basin of the Indian Ocean including the Mascarene Plateau, the Chagos-Laccadive Ridge, and the Deccan basalt of western India, to be discussed in Sect. 5.5 (Duncan 1990; White et al. 1990).

The island of Mauritius is located at 20°20' S and 57°30' E in the southern part of the submarine Mascarene Plateau. It is elliptical in shape and has a surface area of about 2 700 km^2. The island is entirely composed of volcanic rocks which have been subdivided into two series of distinctly different ages (McDougall and Chamalaun 1969):

- Younger Series (3.5 to 0.2 Ma)
  - Late lavas
  - Early lavas
- Older Series (7.8 to 6.8 Ma)

The petrology of the lavas on Mauritius was investigated by Baxter (1975, 1976, 1978). The $^{87}Sr/^{86}Sr$ ratios of these rocks range from 0.70368 to 0.70432 relative to 0.71025 for NBS 987 and 0.70804 for E&A (Mahoney et al. 1989; Hamelin and Allègre 1986). Mahoney et al. (1989) also reported $^{143}Nd/^{144}Nd$ ratios for these rocks between 0.512850 and 0.512902 relative to 0.512632 for BCR-1.

The Seychelles at the northern end of the Mascarene Plateau consist of about 40 mountainous islands scattered between 4° S and 11° S and from 46° to 56° E in an area of about 400 000 km^2. The largest of these islands is Mahé with a surface area of less than 150 km^2.

The Seychelles differ from Réunion and Mauritius because they are composed of *granitic rocks* and are not of volcanic origin (Wegener 1924). Miller and Mudie (1961) originally reported K-Ar dates of about 650 Ma for several varieties of granite on the island of Mahé. This date was later confirmed by Wasserburg et al. (1963) and by Weis and Deutsch (1984). The latter cited a study by Demaiffe and his colleagues indicating a date of 705 ±8 Ma and an initial ^{87}Sr/^{86}Sr ratio of 0.70408 ±0.0018 based on a Rb-Sr isochron defined by 23 whole-rock samples of granite from the Seychelles (Demaiffe et al. 1985). Information summarized by Dickin et al. (1986) supports the interpretation that the Seychelles microcontinent was originally part of the Indian Plate and became detached from it during the northward drift of that plate.

Some of the Seychelles (Silhouette and Ile du Nord) are composed of Late Cretaceous to Early Tertiary anarogenic alkali-rich syenites and granites. In addition, the island of Praslin contains dolerite dikes of Tertiary age (Baker and Miller 1963). Age determinations by the Rb-Sr method of the younger granites and syenites on Silhouette and Ile du Nord yielded a whole-rock Rb-Sr isochron date of 63.2 ±1.0 Ma with an initial ^{87}Sr/^{86}Sr ratio of 0.7056 ±0 0009 (Dickin et al. 1986). The authors interpreted this date as the time of rifting between the Seychelles microcontinent and the Indian Plate. The dolerite dikes are not datable by the Rb-Sr method because their Rb/Sr ratios do not have sufficient range. Nevertheless, Dickin et al. (1986) reported an initial ^{87}Sr/^{86}Sr ratio of 0.7152 ±0.0018 (2σ, $N = 3$) for three Late Cretaceous dolerites on the island of Praslin.

The initial ^{87}Sr/^{86}Sr and ^{143}Nd/^{144}Nd ratios of the syenites and granites reported by Dickin et al. (1986) place them into the field of volcanic rocks of Kerguelen Island. Therefore, these rocks presumably originated from magma sources in the mantle similar to those of many OIBs in the Indian Ocean. The elevated initial ^{87}Sr/^{86}Sr (0.7152) and low ^{143}Nd/^{144}Nd ratios of the dolerite dikes on Praslin were caused by extensive contamination of the dolerite magma by assimilation of rocks in the continental crust of the Seychelles microcontinent.

2.16.4 Comoros

The Comoros are located at the northern end of the Mozambique Channel that separates Madagascar from the mainland of Africa. The principal islands in this archipelago are Mayotte, Anjouan, Moheli, and Grande Comore all of which are of volcanic origin. Age determinations by Hajash and Armstrong (1972) yielded K-Ar dates ranging from 3.65 ±0.10 to 0.10 ±0.10 Ma and indicated that the centers of volcanic activity migrated in a northwesterly direction from Mayotte to Anjouan and to Grande Comore which is the youngest as well as the largest of the islands. The names of the islands were changed when they achieved independence: Njazidja (Grande Comore), Mwali (Moheli), Nzwani (Anjouan). However, Mayotte is still administered by France and retained its French name.

Grande Comore Island is composed of two large shield volcanoes known as Karthala (2 361 m) and LaGrille (1 087 m). Both have been active in historical time. The lavas of Karthala in the center of the island

encompass the main phase of shield formation (Old Karthala), a postcaldera suite of lavas (Recent Karthala), and the youngest flows erupted during historical time (Present Karthala). The lavas on Karthala are only mildly alkaline and range from picrites to trachybasalt. The LaGrille Volcano in the northern part of the island is older than Karthala and has erupted a differentiated suite of alkali-rich lavas ranging from picrite basalts to basanites and nephelinites (Deniel 1998).

The geology of all of the islands in the Comoros and the origin of the lavas were described by Esson et al. (1970), Nougier et al. (1986), and Späth et al. (1996). In addition, the petrology of the lavas on Grande Comore and Moheli was investigated by Strong (1972a,b). The lavas of the Comoros contain xenoliths composed of sandstone (Strong and Flower 1969) and, more rarely, of granitic rocks composed of quartz, K-feldspar, plagioclase, and zircon (Deniel 1998).

The ^{87}Sr/^{86}Sr ratios of lavas on Grande Comore range from 0.70315 (LaGrille) to 0.70396 (Recent Karthala) relative to 0.71024 ±0.00006 for NBS 987 (Deniel 1998). These values overlap with, but are somewhat lower than, those of other volcanic islands in the Indian Ocean included in Fig. 2.70 (e.g. Amsterdam, St. Paul, Rodriguez, Réunion, and Mauritius).

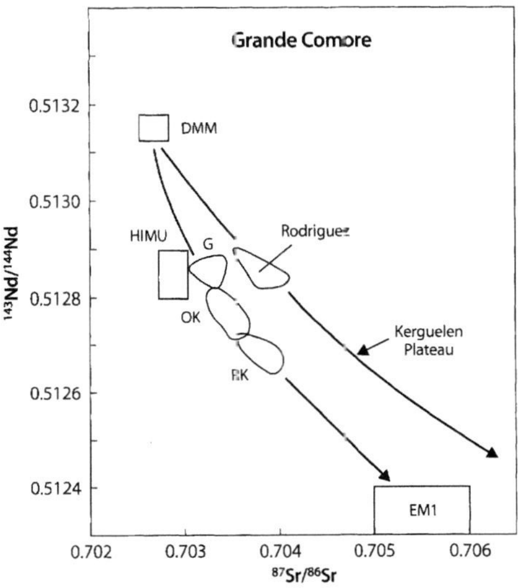

Fig. 2.71. Isotope ratios of Sr and Nd of lavas on Grande Comore compared to those of Rodriguez Island. G = LaGrille Volcano; OK = Old Karthala suite; RK = Recent as well as Present Karthala suite. The magmas erupted on Grande Comore originated from sources containing both the DMM and EM1 components. The proportion of magma derived from the EM1 component increased with time. The lavas of Rodriguez contain the Kerguelen component that may have contaminated the mantle underlying the Indian Ocean (*Sources:* Deniel 1998; Baxter et al. 1985)

The lavas of Grande Comore in Fig. 2.71 define a mixing hyperbola that extends from the DMM component to EM1 in coordinates of $^{87}Sr/^{86}Sr$ and $^{143}Nd/^{144}Nd$ ratios. The abundance of the EM1 component increases with time from the lavas of LaGrille to those of the Old Karthala and Recent Karthala which means that the LaGrille lavas are more closely related to the Rb-depleted magma source of MORBs (Späth et al. 1996).

The observed systematic variation of the isotope ratios of Sr and Nd of the lavas on Grande Comore also indicates that the most recently erupted lavas of Karthala contain more magma that formed in the head of the Comoros Plume. In addition, the systematic age progression of the Comoros (Hajash and Armstrong 1972; Emerick and Duncan 1982) implies that the plume is located close to Grande Comore and that the plate that overlies it is moving to the southeast. The $^{206}Pb/^{204}Pb$ ratios (18.981–19.863) reported by Deniel (1998) and Späth et al. (1996) are consistent with the magma sources indicated by the isotope ratios of Sr and Nd in Fig. 2.71, but reveal that these lavas also contain Pb derived partly from the HIMU component.

The lavas on Rodriguez Island analyzed by Baxter et al. (1985) deviate from the Sr-Nd mixing hyperbola of the rocks on Grande Comore in Fig. 2.71. Instead, they lie on the mixing hyperbola in Fig. 2.68 defined by the basalts of the Kerguelen Plateau (Mahoney et al. 1995). These results suggest that the volcanic rocks on Grande Comore originated from magma sources that differ from those that characterize the mantle under the Indian Ocean.

The lavas on Moheli and Mayotte in Fig. 2.72 have wide-ranging concentrations of Rb (11.2 to 164 ppm) and Sr (446 to 1 552 ppm). Späth et al. (1996) attributed the

Table 2.4. Summary of the isotope ratios of Sr, Nd, and Pb of LaGrille and Karthala lavas on Grande Comore and of the rocks on Moheli and Mayotte

Island/Volcano	$^{87}Sr/^{86}Sr$	$^{143}Nd/^{144}Nd$	$^{206}Pb/^{204}Pb$
Grande Comore			
LaGrille	0.70327 (7) ±0.00005	0.51285 (7) ±0.00001	19.192 (7) ±0.0098
Karthala (old)	0.70344 (5) ±0.00010	0.51276 (5) ±0.00004	19.606 (5) ±0.140
Karthala (Recent)	0.70385 (12) ±0.00005	0.51267 (12) ±0.00001	19.403 (11) ±0.050
Moheli	0.70333 (6) ±0.00009	0.51283 (5) ±0.00002	19.712 (2) ±0.414
Mayotte	0.70333 (9) ±0.00006	0.51282 (9) ±0.00002	19.425 (4) ±0.298

chemical composition of these lavas to the effects of fractional crystallization of LaGrille-type magma involving olivine, clinopyroxene, titanomagnetite, and amphibole followed by feldspar and apatite. The evidence in favor of fractional crystallization is consistent with the homogeneity of isotope ratios of Sr, Nd, and Pb reported by Späth et al. (1996) and Deniel (1998).

The summary of the available data in Table 2.4 confirms that the average isotope ratios of Sr, Nd, and Pb of LaGrille and Old Karthala lavas on Grande Comore and of the rocks on Moheli and Mayotte are indistinguishable from each other. However, the Recent lavas on Karthala are significantly enriched in radiogenic ^{87}Sr and depleted in radiogenic ^{143}Nd as indicated also in Fig. 2.71.

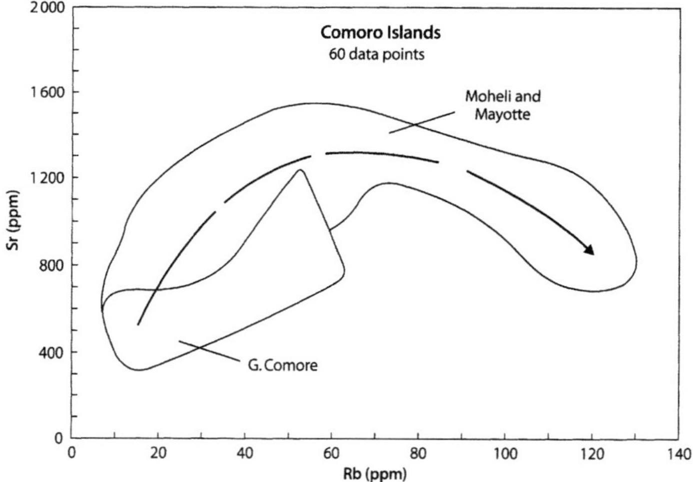

Fig. 2.72.
Rb and Sr concentrations of alkali-rich lavas on the Comoro Islands including Grande Comore, Moheli, and Mayotte. The observed pattern of Rb and Sr concentrations is attributable to the effects of fractional crystallization of alkali basalt magma (*Sources:* Späth et al. 1996; Deniel 1998)

2.16.5 Prince Edwards Islands (Marion and Prince Edward)

The Prince Edward Islands and the Funk Seamount are located about 180 km south of the spreading axis of the Southwest Indian Ridge at about 46° S and 38° E in the southern Indian Ocean. The islands are composed of basalt whose isotope ratios of Sr, Nd, and Pb vary only within narrow limits. Mahoney et al. (1992) reported ^{87}Sr/^{86}Sr ratios between 0.70295 and 0.70340 relative to 0.71025 for NBS 987 and 0.7080 for E&A. The volcanic rocks of the Funk Seamount contain kaersutite-bearing xenoliths analyzed by Reid and LeRoex (1988).

The isotope ratios of Sr and Nd of the Prince Edward Islands in Fig. 2.73 define a short segment of a mixing line involving DMM and an enriched mantle component. The low ^{87}Sr/^{86}Sr ratios of the lavas on these islands indicate that they originated primarily from depleted MORB sources with only small contributions from Rb-enriched rocks of the Marion Plume (Duncan 1990). The ^{205}Pb/^{204}Pb ratios of these rocks range from 18.567 to 18.633 (Mahoney et al. 1992).

2.16.6 Crozet Islands

The Crozet Islands (including Ile de l'Est, Ile de la Possession, Ile aux Cochons, Ile de Pingouins, and the Iles de Apotres) form a small archipelago located 1 600 km east of the Prince Edward islands at about 46° S and 51° E in the southern Indian Ocean. Ile de l'Est is a deeply dissected Tertiary shield volcano overlain unconformably by alkali-rich volcanic rocks (oceanites and ankaramites) ranging in age from 0.69 to 0 Ma (Gunn et al. 1970). Therefore the Crozet Archipelago marks the site of a mantle plume which contributed to the break-up of Gondwana and the subsequent evolution of the Indian Ocean together with the Kerguelen and Marion Plumes all of which are located in the southern Indian Ocean (Curray and Munasinghe 1991).

The ^{87}Sr/^{86}Sr ratios of ankaramite lavas on Ile de l'Est range from 0.7039 to 0.7041 relative to 0.7080 for E&A (Hedge et al. 1973; Dupré and Allègre 1983).

2.17 Gulf of Aden and the Red Sea

The Gulf of Aden and the Red Sea are extensions of the Carlsberg Ridge of the Indian Ocean (Sect. 2.14) and are the sites of rifting that has resulted in the break-up of the Afro-Arabian continent. The driving force for the break-up and the associated volcanic activity in the area of the triple junction has been the Afar Plume (Sect. 5.15.2). Therefore, this area provides opportunities to investigate the contrast between the Afar Plume and the sources

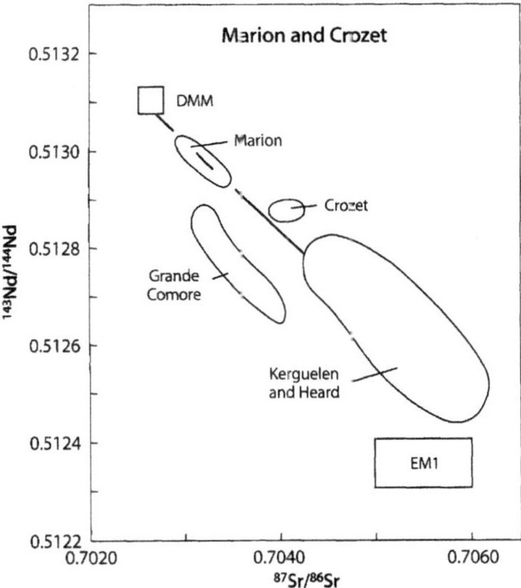

Fig. 2.73. Isotope ratios of Sr and Nd in volcanic rocks erupted on Marion and on the Crozet Islands in the southern Indian Ocean. The lavas on Marion are more similar to MCRBs derived from the DMM component than those of the Crozet Islands. However, both are aligned with the data array derived from the lavas of Kerguelen Island (*Sources:* Marion and Crozet: Mahoney et al. 1992; Grande Comore: Fig. 2.71; Kerguelen and Heard: Fig. 2.68)

of mid-ocean ridge basalt that underlie the Red Sea and the Gulf of Aden. Insights concerning this issue have been gained from interpretations of the isotope compositions of Sr, Nd, and Pb and trace-element concentrations of volcanic rocks of this area and from the xenoliths they contain.

2.17.1 Gulf of Aden

The Sheba Ridge of the Gulf of Aden extends westward from the Owen fracture zone in the Indian Ocean into the Gulf of Tadjoura and the Asal Rift of Djibouti and Ethiopia (Sect. 5.15.2). Schilling et al. (1992) reported isotope ratios of Sr, Nd, and Pb as well as concentrations of the rare-earth elements of basalt dredged from this ridge for a distance of about 800 km along its length. The ^{87}Sr/^{86}Sr ratios of the fresh interiors of basalt pillows along this traverse range from 0.702796 (49.39° E long., Sheba Ridge) to 0.704144 (Gulf of Tadjoura) relative to E&A = 0.7080. The profile of ^{87}Sr/^{86}Sr ratios in Fig. 2.74 is a three-point moving average along which Schilling et al. (1992) identified the boundary of the head of the Afar Plume. The ^{87}Sr/^{86}Sr ratios of normal MORBs extruded along the eastern part of the Sheba Ridge range only from 0.702843 to 0.703013 and have a mean of 0.70292 ±0.00004 (2$\bar{\sigma}$, N = 10). However, along the pro-

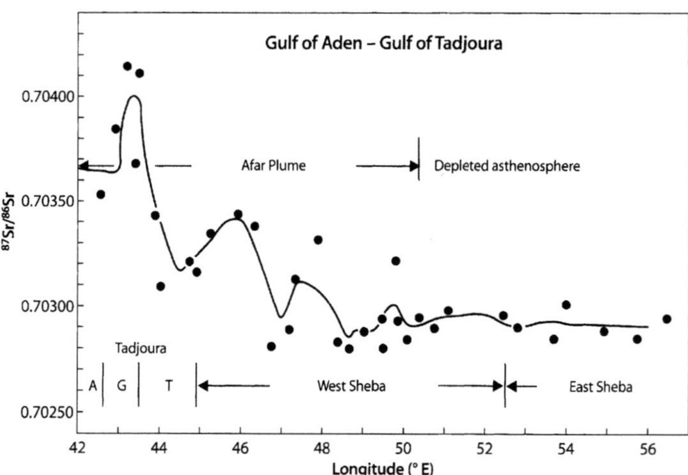

Fig. 2.74.
Profile of $^{87}Sr/^{86}Sr$ ratios of the interiors of basalt pillows along the spreading ridge in the Gulf of Aden (Sheba Ridge), the Gulf of Tadjoura, and the Asal Rift. T = Tadjoura Trough, G = Gulf of Tadjoura, A = Asal Rift. The line is a three-point moving average. The basalts extruded along the Sheba Ridge are MORBs derived from depleted lithospheric mantle with an average $^{87}Sr/^{86}Sr$ ratio of 0.70292 ±0.00004 ($2\bar{\sigma}$, N = 10) relative to E&A = 0.7080. The $^{87}Sr/^{86}Sr$ ratios within the head of the Afar Plume rises and falls suggesting that the plume head contains blocks of entrained asthenospheric mantle (*Source:* Schilling et al. 1992)

file of the head of the Afar Plume, the $^{87}Sr/^{86}Sr$ ratios rise and fall from 0.70280 to 0.70414, suggesting that the plume head contains entrained blocks of subducted oceanic crust and depleted mantle. In addition, the $^{87}Sr/^{86}Sr$ ratios of the basalts in the Gulf of Tadjoura and the Asal Rift were increased by assimilation of Sr from the lithospheric mantle (Sect. 5.15.2). The interpretation of Schilling et al. (1992) of isotope ratios and REE concentrations indicates that the center of the flattened head of the Afar Plume outlined in Fig. 5.93 is located at Lake Abhe (11°15' N, 41°25' E) in the Afar region of Ethiopia and that its radius is about 1 000 km.

2.17.2 Red Sea

The axial valley of the Red Sea is the site of volcanic activity investigated by Altherr et al. (1988), Eissen et al. (1989), and Volker et al. (1993). Eissen et al. (1989) reported that four samples of hand-picked fresh volcanic glass from the central rift at 18° N yielded an average $^{87}Sr/^{86}Sr$ ratio of 0.70290 ±0.00002 ($2\bar{\sigma}$, N = 4; E&A = 0.7080; NBS 987 = 0.71025). This value is similar to $^{87}Sr/^{86}Sr$ ratios recorded along the Mid-Atlantic Ridge (Sect. 2.2, Fig. 2.7) and the East Pacific Rise (Sect. 2.7, Fig. 2.35) and agrees with the average $^{87}Sr/^{86}Sr$ ratio of MORBs along the Sheba Ridge of the Gulf of Aden (Schilling et al. 1992).

The work of Altherr et al. (1988, 1990) and Volker et al. (1993) indicates that the $^{87}Sr/^{86}Sr$ ratios of MORBs of Recent age in the central valley of the Red Sea range from 0.70272 (Atlantis II Deep) to 0.70320 (Ramad Seamount) relative to 0.7080 for E&A and 0.71025 for NBS 987. The authors observed regional trends in the chemical composition and in the profile of $^{87}Sr/^{86}Sr$ ratios of MORBs extending from the Shaban Deep at the northern end of

the Red Sea to the Ramad Seamount and the Hanish-Zukur and Zubair Islands at the southern end (Fig. 2.75). The data of Altherr et al. (1988, 1990) yield an average $^{87}Sr/^{86}Sr$ ratio of 0.70319 ±0.00007 ($2\bar{\sigma}$, N = 8) for enriched (E) MORBs in the Shaban and Al Wajh Deeps (26°15' to 25°28' N). Farther south, normal (N) MORBs in the Nereus, Hadarba, Atlantis II, and Shagara Deeps (23°09' to 21°08' N) have a low average $^{87}Sr/^{86}Sr$ ratio of 0.70267 ±0.00017 ($2\bar{\sigma}$, N = 6). The $^{87}Sr/^{86}Sr$ ratios then rise to an average value of 0.70295 ±0.00006 ($2\bar{\sigma}$, N = 7) for MORBs in the Erba, Port Sudan, Volcano, and Suakin Deeps from 20°44' N to 18°09' N.

The basalts on the Ramad Seamount and on the Zubair and Hanish-Zukur Islands in the southern Red Sea are distinctly alkalic in composition and have a comparatively high average $^{87}Sr/^{86}Sr$ ratio of 0.70330 ±0.00005 ($2\bar{\sigma}$, N = 5). The regional variations of the chemical composition of MORBs in the Red Sea may be caused by prior metasomatic alteration or by differences in the degree of partial melting of the underlying mantle. The regional variation of the $^{87}Sr/^{86}Sr$ ratios of MORBs in the Red Sea reflects variations of the Rb/Sr ratios of the magma sources.

2.17.3 Islands in the Red Sea

Questions about the composition of magma sources in the mantle underlying the Red Sea can be answered by the study of xenoliths in basalt lava flows on some of the islands in the Red Sea. Zabargad Island (23°36' N, 36°12' E) is especially well suited for this purpose because it contains exposures of large blocks of periodotite up to about 2.0 km long and 0.75 km wide. The peridotite is in contact with granulites of the lower continental

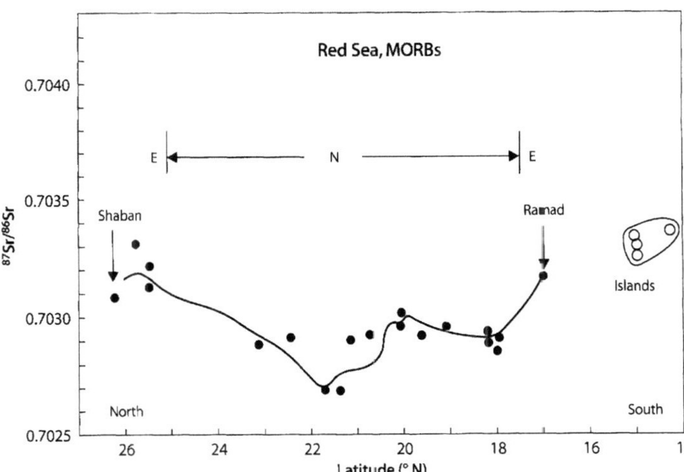

Fig. 2.75.
Profile of $^{87}Sr/^{86}Sr$ ratios of MORBs (*solid circles*) in the Red Sea from the Shaban Deep in the north to the Ramad Seamount in the south. The line is a three-point moving average based only on isotope analyses of volcanic glass. The regional variation of the $^{87}Sr/^{86}Sr$ ratio reflects differences in the Rb/Sr ratios of the magma sources in the mantle: E = enriched, N = normal MORBs. The Zubair and Hanish-Zukur Islands (*open circles* in the southern Red Sea consist of alkali basalt and hawaiites with elevated $^{87}Sr/^{86}Sr$ ratios compared to MORBs (*Sources: data from Altherr et al. 1988; Volker et al. 1993; Eissen et al. 1989. The $^{87}Sr/^{86}Sr$ ratios of Volker et al. (1993) were substituted for those of Altherr et al. (1988) in cases where both teams analyzed the same specimens)

crust, and the entire assemblage has been interpreted as a sliver of the lithospheric mantle and overlying continental crust that was uplifted into its present position during the opening of the Red Sea (Bonatti et al. 1986).

The $^{87}Sr/^{86}Sr$ ratios of spinel lherzolites, amphibole peridotites, and troctolitic gabbros on Zabargad Island reported by Brueckner et al. (1988) appear to have been altered because whole-rock samples of the ultramafic rocks have a range of $^{87}Sr/^{86}Sr$ ratios between 0.70235 to 0 70407 relative to E&A = 0.7080 even though all samples were cleaned with 2.5 N hydrochloric acid. In contrast, the $^{87}Sr/^{86}Sr$ ratios of clinopyroxenes and hornblendes of the ultramafic rocks range only from 0.70227 to 0.70302 and have an average value of 0.70273 ±0.00030 ($2\bar{\sigma}$, N = 5), whereas clinopyroxenes in the troctolitic gabbro have elevated $^{87}Sr/^{86}Sr$ ratios between 0.70350 and 0.70365.

The ultramafic rocks of Zabargad Island as well as diabase dikes (Petrini et al. 1988) and some MORB glasses from the axial valley of the Red Sea define a straight line on the Sm-Nd isochron diagram (not shown). The slope of this line corresponds to a date of 675 ±30 Ma (Brueckner et al. 1988). However, the troctolitic gabbros on Zabargad deviate significantly from the array. Nevertheless, Brueckner et al. (1988) concluded that the peridotites of Zabargad carry an imprint of the Pan-African event which affected the continental crust of northeast Africa and Arabia (Stern and Hedge 1985; Stoeser and Camp 1985; Kröner et al. 1979, 1987, 1991; Pallister et al. 1987).

The lesson to be learned from the ultramafic rocks on Zabargad Island is that these rocks are isotopically heterogeneous on a scale of meters and that they have been affected by a complex series of tectonic and metamorphic/metasomatic events starting with the Pan-

African event. Brueckner et al. (1988) concluded their paper with the sobering thought that the mantle rocks on Zabargad remind us "how complicated the mantle might be and how difficult it would be to unravel these complications from nodules or melts derived from the mantle."

The peridotites on Zabargad Island were intruded at about 5.0 Ma by basaltic dikes ranging in composition from picrite to alkali basalt. The dikes cut not only the peridotite and granulite of the island but also a sequence of limestone, sandstone, and black shale of Late Cretaceous or Paleocene age known as the Zabargad Formation. The $^{87}Sr/^{86}Sr$ ratios of 24 samples of the basaltic dikes range from 0.70326 to 0.70643 relative to E&A = 0.7080 (Petrini et al. 1988). The data points form a fan-shaped array in coordinates of $^{87}Sr/^{86}Sr$ and 1/Sr (not shown) which suggests that the magmas originally had an $^{87}Sr/^{86}Sr$ ratio of 0.70326 or less and subsequently assimilated Sr having an $^{87}Sr/^{86}Sr$ ratio of 0.70634 or higher. Petrini et al. (1988) concluded that the Zabargad Dikes could have formed by partial melting of peridotitic source rocks and that the rocks were subsequently altered by seawater at high temperature. The marine Sr was incorporated into scapolite and cannot be removed by leaching with hot 6 N HCl.

2.18 Summary: The Importance of Plumes

The volcanic activity along mid-ocean ridges and on oceanic islands and on seamounts occurs as a result of magma formation by decompression melting in plumes and in the adjacent lithospheric mantle. The low $^{87}Sr/^{86}Sr$ ratios of normal mid-ocean ridge basalts (MORBs) reveal that their magmas originated from Rb-depleted magma source of the lithospheric mantle. The variation of $^{87}Sr/^{86}Sr$ ratios of MORBs along mid-ocean ridges arises because of mixing of melts produced both in the lithospheric mantle and in the heads of plumes which contain heterogeneous assemblages of rocks having elevated $^{87}Sr/^{86}Sr$ ratios.

The evidence for mixing of magmas derived from different kinds of sources in the mantle is strengthened by the observed correlations of $^{87}Sr/^{86}Sr$ ratios of oceanic volcanic rocks with the isotope ratios of Nd and Pb (as well as with the isotope ratios of Hf and Os). The study of oceanic volcanic rocks in Sr-Nd-Pb isotope space indicates that their magma sources in the mantle can be described in terms of only four components:

- DMM = depleted MORB mantle
- EM1 = Rb-enriched mantle 1
- EM2 = Rb-enriched mantle 2
- HIMU = mantle having elevated isotope ratios of Pb as a result of high $^{238}U/^{204}Pb$ ratios (μ)

These components have originated as a consequence of the formation and subduction of oceanic crust throughout geologic time:

- DMM: Residual rocks in the lithospheric mantle from which magma was extracted during a previous episode of partial melting along a spreading ridge.
- EM1: Subducted oceanic crust and pelagic sediment whose isotope compositions of Sr, Nd, and Pb changed by decay of their respective parent elements during incubation times of 1 to 2 billion years.
- EM2: Subducted oceanic crust and terrigenous sediment whose isotope composition was similarly changed by incubation in the mantle.
- HIMU: Subducted oceanic crust composed of MORBs and related rocks. The U/Pb ratios of these rocks increased by loss of Pb either because of hydrothermal activity along spreading ridges or later by dehydration during subduction.

In this way, the chemical and isotope compositions of the components in the mantle are all related to plate tectonics. In addition, the characteristic isotope ratios of Sr, Nd, and Pb in EM1, EM2, and HIMU are attributable to chemical processing of oceanic crust during subduction followed by long-term decay of their respective parent elements during incubation in the mantle.

Mathematical modeling and laboratory experiments indicate that subducted oceanic crust in the asthenospheric mantle may form plume heads which rise toward the surface because of buoyancy caused by radioactive heating. Rising plume heads entrain blocks of asthenospheric mantle and are trailed by tails of asthenospheric mantle. When the head of a plume encounters the rigid lithosphere, it may spread out by streaming laterally for hundreds of kilometers. As a result, heat is transferred to the basal lithosphere thereby causing uplift, rifting, and large-scale decompression melting of lithospheric rocks. The resulting tholeiite basalt magmas are erupted on the seafloor to form submarine shield volcanoes whose summits may ultimately reach the surface of the ocean to form islands. Some volcanic islands are related genetically to mid-ocean ridges whereas others are caused by the presence of underlying plumes.

Oceanic islands associated with spreading ridges form on the ridge crest and are subsequently displaced by seafloor spreading. The volcanic activity of such islands dies out causing them to become seamounts with eroded summits. Other oceanic islands form by volcanic activity at some distance from spreading ridges where magma in the plume head is able to penetrate the overlying lithosphere. Such off-axis islands are located along "leaky fractures" and can remain volcanically active for millions of years until the supply of magma is finally exhausted. In some cases, the crest of a spreading ridge may shift (e.g. in the southern Atlantic Ocean), thus causing islands to be displaced from it. In such cases, the off-axis plumes may continue to provide magma to the ridge crest by maintaining channels along the base of the lithosphere.

The oceanic islands of Polynesia and the Hawaiian Islands in the Pacific Ocean exemplify intraplate volcanic activity caused by magma formation in stationary plumes. Under these conditions, chains of islands are formed as the Pacific Plate moves over the "hotspot" in a northwesterly direction. Consequently, the ages of the islands increase from southeast to northwest implying spreading rates of about 11 cm yr^{-1} in many cases. The plumes that have given rise to island chains may continue to cause volcanic activity on the seafloor southeast of the youngest island in the chain. Such active submarine volcanoes are known from the area southeast of Pitcairn Island and off the southeast coast of the island of Hawaii.

The chemical composition of lavas extruded on most oceanic islands has changed during their formation from silica-saturated tholeiites to silica-undersaturated alkali-rich rocks. The change in chemical composition of the lavas generally occurs after a lengthy period of volcanic inactivity, erosion, and caldera formation by partial collapse of the original volcanic edifice. The late-forming alkali basalts dominate the subaerially exposed summits of oceanic islands and obscure the far more voluminous older tholeiite basalts. In addition, the isotope ratios of Sr, Nd, and Pb of the alkali-rich lavas on oceanic islands differ in most cases from those of the tholeiites.

The differences in the isotope ratios of Sr, Nd, and Pb between the shield-forming tholeiites and the postcaldera alkali-rich lavas requires a change of their magma sources in the mantle. In general, the shield-forming tholeiites on the island chains of Polynesia have low $^{87}Sr/^{86}Sr$ ratios indicating that they originated by partial melting of Rb-depleted rocks in the lithospheric mantle as a result of heating by the rising head of a plume. The alkali-rich rocks on the islands of Polynesia have higher $^{87}Sr/^{86}Sr$ ratios than the shield-forming tholeiites because they originated by partial melting in the plume head. However, on the Hawaiian Islands the shield-building tholeiites originated from the plume head, whereas the late-stage alkali-rich lavas formed by small degrees of partial melting in the lithospheric mantle. The reason for the difference between the islands of Polynesia and Hawaii is that the Hawaiian Plume was strong enough to invade the overlying lithospheric mantle thereby causing decompression melting in its own head.

The opening of the Indian Ocean evolved from the break-up of Gondwana which was caused by powerful plumes that are still active under the Kerguelen Plateau, under Réunion, and under the Crozet Islands. In addition, the Afar Plume is presently splitting the Afro-Arabian Plate along the Gulf of Aden and the Red Sea.

The evidence presented in this chapter justifies the statement that the convection of the asthenospheric mantle in the form of rising plumes is responsible not only for causing the volcanic activity that is occurring along mid-ocean ridges and on many oceanic islands, but also for the movement of the lithospheric plates from spreading ridges to subduction zones.

Chapter 3

Subduction Zones in the Oceans

The formation of new oceanic crust at mid-ocean ridges requires a complementary mechanism by means of which lithospheric plates are consumed. The required destruction of lithospheric plates takes place in subduction zones located in the oceans and along the margins of some continents. Subduction zones are places where oceanic lithosphere sinks or is dragged back into the mantle and where magma is produced by partial melting within the mantle wedge above the descending lithospheric plate and/or in the plate itself. The resulting volcanic activity in the ocean basins is manifested by the formation of island arcs, whereas continental margins along which oceanic lithosphere is subducted under continental crust are uplifted and contain volcanic mountain ranges. All active subduction zones are sources of earthquakes that originate primarily within the Benioff zone associated with the down-going slab.

3.1 Petrogenetic Models

The two kinds of subduction zones mentioned above are illustrated in Fig. 3.1a and b. Partial melting in the mantle wedge is facilitated by the addition of water and other volatiles released during dehydration of the oceanic sediment (Clauer et al. 1982) and of the hydrothermally altered basaltic rocks of the subducted oceanic crust (Schiano et al. 1995; Hawkesworth 1982; Arculus and Johnson 1981; Baker 1973). The silica-bearing aqueous fluids released by the rocks of the down-going slab cause the formation of amphiboles, phlogopite, and other hydrous phases in the rocks of the mantle wedge, thereby lowering their melting temperature, reducing their density, and permitting diapirs to form. As the diapirs rise toward the surface, partial melting occurs within them as a result of decompression (Sparks 1992). Alternatively, metasomatized mantle rocks may be dragged downward by the motion of the descending slab. The amphiboles become unstable at a depth of about 100 km and their decomposition triggers partial melting (Tatsumi 1989). Another possibility is that partial melting may occur in the subducted oceanic crust (sediment and hydrothermally altered basalt), but the chemical compositions and abundances of radiogenic isotopes of volcanic rocks in island arcs indicate that only small amounts of bulk sediment contribute to the formation of magma.

The dehydration of serpentine at 350 °C and 12 kbar was investigated experimentally by Tatsumi et al. (1986). Their results in Fig. 3.2 indicate that the "mobility" of the elements included in the study increases with their ionic radii in six-fold coordinated sites (Whittacker and Muntus 1970). Accordingly, Cs is the most mobile element releasing nearly 60% of the original amount in the serpentine to the aqueous phase produced during the reaction:

$$\text{serpentine} \longrightarrow \text{forsterite} + \text{enstatite} + \text{water.}$$

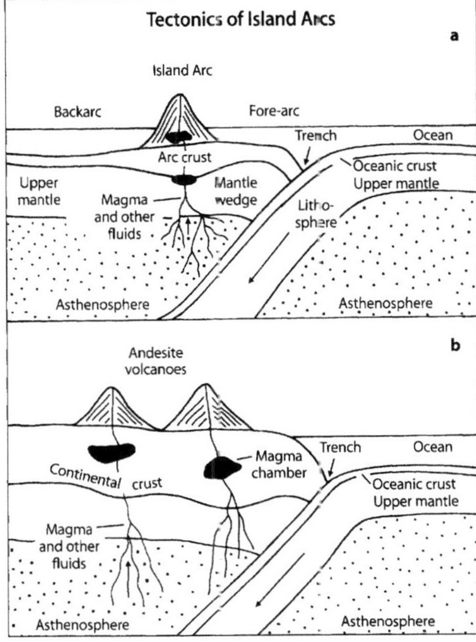

Fig. 3.1. a Schematic diagram of the principal features of an island arc associated with subduction of oceanic lithosphere under oceanic lithosphere; **b** Schematic diagram of the principal features of an active continental margin featuring subduction of oceanic lithosphere under continental lithosphere (*Sources:* Wilson 1989; Sparks 1992)

Fig. 3.2.
"Mobility" of trace elements in serpentine at 850°C and 12 kbar. The mobility is expressed as the amount of an element lost as a result of heating, expressed as the percent of its original concentration. The data show that the mobility of trace elements increases with their ionic radii (coordination number 6, Whittacker and Muntus 1970). Although both Rb and Sr are more mobile than the REEs (Sm-Yb), 2.6 times more Rb than Sr is lost from serpentine under these conditions (*Source:* Tatsumi et al. 1986)

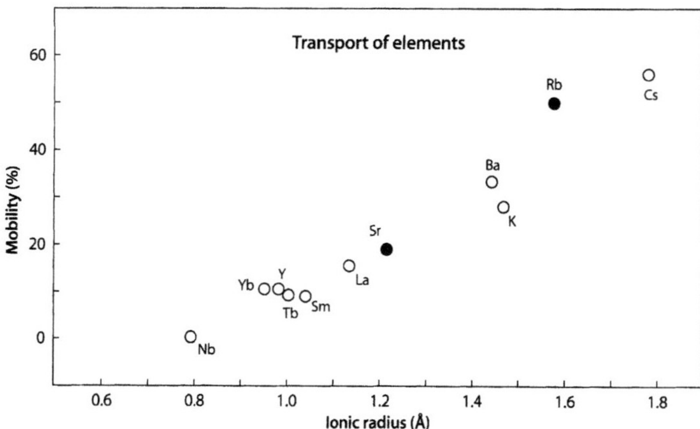

Fig. 3.3.
Cross-section of an intra-oceanic subduction zone showing dehydration of subducted oceanic crust under the fore-arc, metasomatic alteration of the overlying mantle wedge followed by downward transport of the metasomatized mantle rocks into the zone of partial melting under the volcanic island arc. Diapirs of magma with suspended crystals then rise from the melting zone toward the surface and are erupted through volcanic vents. Dehydration of the oceanic crust releases alkali metals, alkaline earths, and REEs which are transferred to the mantle wedge and ultimately enter the magmas that form in the melt zone (*Source:* Modified from Tatsumi et al. 1986)

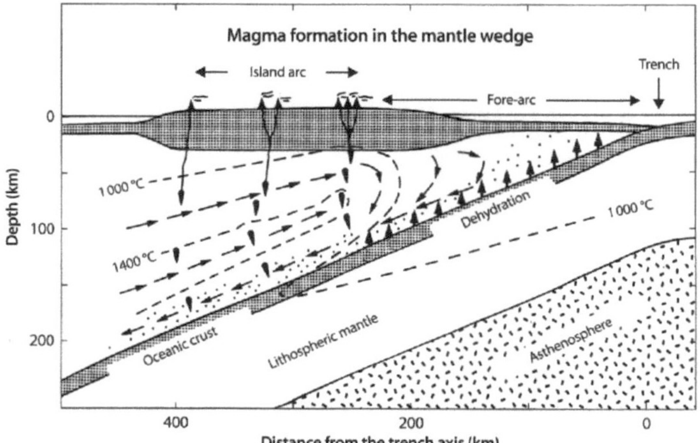

Similarly, 50% of the Rb and 19% of the Sr were lost from the serpentine in this experiment. The results therefore confirm that serpentine of the oceanic crust being subducted into the mantle releases an aqueous fluid containing alkali metals (Cs, Rb, K), alkaline earths (Ba, Sr) and REE (La-Lu). The alkali metals are more mobile under these conditions than the alkaline earths and are transferred with greater efficiency from the subducted oceanic crust to the rocks of the overlying mantle wedge. The REEs as a group have the lowest mobility, but the light REEs (La-Sm) are slightly more mobile than the heavy REEs (Tb-Lu).

Tatsumi et al. (1986) and Tatsumi (1989) pointed out that virtually all of the water from hydrous minerals is expelled before the subducted oceanic crust reaches a depth of 120 km. Therefore, the water is expelled under the fore-arc rather than under the island arc and rises into the mantle wedge where it causes metasomatic alteration of the rocks and enriches them in LIL elements before partial melting actually takes place. The altered rocks of the mantle wedge then must be transported downward by convection induced by the motion of the subducted oceanic lithosphere as shown in Fig. 3.3. Partial melting of the metasomatized mantle rocks ultimately occurs under the island arc at temperatures above 1 000 °C and at depths between 100 and 200 km. Diapirs of magma rise from the melting zone toward the surface where they feed the volcanoes of island arcs.

Kogiso et al. (1997b) subsequently reported that dehydration of amphibolite under the pressure and temperature conditions of the upper mantle causes the release of an aqueous fluid that is enriched in Pb, Nd, and

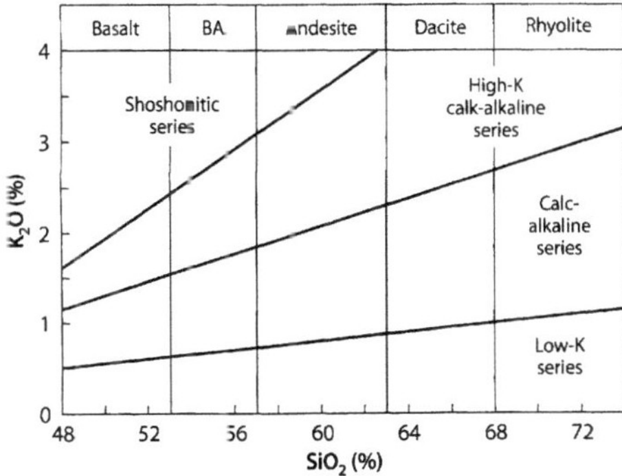

Fig. 3.4. Classification of volcanic rocks formed in subduction zones in the oceans and along active continental margins (*Sources:* adapted from Gill 1981; Wilson 1989)

Rb. The loss of Pb results in an increase of the U/Pb ratio of the dehydrated amphibolite and thus confirms the work of Chauvel et al. (1995) who concluded that the HIMU component in the mantle consists of subducted oceanic crust that has been incubated in the mantle long enough to become enriched in radiogenic ^{206}Pb and ^{207}Pb. In addition, the preferential loss of Rb during dehydration of amphibolite reported by Kogiso et al. (1997b) decreases its Rb/Sr ratio and causes the HIMU component to have a low average ^{87}Sr/^{86}Sr ratio between about 0.7028 and 0.7050 (Hart 1988).

Additional confirmation for this petrogenetic model was provided by You et al (1996) who studied the dehydration of sediment under hydrothermal conditions. Their results demonstrated that As, B, Cs, Li, Pb, and Rb are mobilized at about 300 °C and are therefore available for transport into the mantle wedge. The chemical composition of the aqueous fluid generated by dehydration of sediment (You et al. 1996) is similar to that released by altered basalt (Kogiso et al. 1997b). Consequently, both studies strongly support the view that aqueous fluids released by the down-going slab contribute to the formation of magmas in the mantle wedge above subduction zones.

The fluids are released not only because of the increase in temperature, but also because of the transformation of amphibolite to eclogite at 2.7 Giga Pascals (GPa), because of the breakdown of amphiboles at 3.5 GPa, and because of the decomposition of phlogopite at 6 GPa (Tatsumi and Kogiso 1997; Ernst 1999). The phase changes in subducted oceanic crust at still higher pressures and temperatures at 28 GPa and 1 500 °C were investigated by Irifune and Ringwood (1993).

The volcanic rocks that form on island arcs and along active continental margins have been subdivided in Fig. 3.4 into four series based on their concentrations of K₂O and SiO₂ (Gill 1981):

1. Low-K or island-arc tholeiite series
2. Calc-alkaline series
3. High-K calc-alkaline series
4. Shoshonitic or alkaline series

Each series contains a wide variety of rock types from basalt to rhyolite, although basalt dominates both the low-K and the shoshonitic series, whereas andesite dominates in the calc-alkaline series (Baker 1982; Wilson 1989).

The occurrence of large volumes of rhyolite in association with basalt in certain island arcs is not easy to explain because the rhyolites are much too voluminous to be the differentiation products of basalt magmas. Notable occurrences of rhyolite exist on the North Island of New Zealand, in Japan, and on some of the islands of Indonesia. Rhyolites also occur along the west coast of the American continents (Chap. 4) and with continental flood basalts (e.g. Karoo volcanic province, South Africa, Sect. 5.11.1). These and other occurrences of rhyolite deserve special attention because the isotope ratios of Sr, Nd, and Pb may shed light on the origin of these rocks.

The petrogenesis of volcanic rocks in subduction zones has been discussed by many authors including Cox et al. (1979), Wyllie (1981, 1982), Basaltic Volcanism Project (1981), Gill (1981), Mysen (1982), Thorpe (1982), Barker (1983), McBirney (1984), Maaløe (1985), Pitcher et al. (1985), Ellam and Hawkesworth (1988), Wilson (1989), Crawford (1989), Hess (1989), and McCulloch and Gamble (1991), among others. The occurrence and petrogenesis of alkali-rich shoshonites in subduction zones was discussed in a series of papers introduced by Box and Flower (1989). The examples presented in the subsequent sections of this chapter are intended to illustrate the relation between the tectonic setting in which magmas form and evolve in subduction zones, and the chemical compositions and isotope ratios of Sr, Nd, and Pb of the volcanic rocks that are erupted at the surface.

3.2 The Mariana Island Arc in the Pacific Ocean

The Mariana Islands in Fig. 3.5a,b are a part of the long trench-arc system in the western Pacific Ocean extending south from the Japan Trench and including the Izu, Bonin, Volcano, Mariana, Yap, and Palau deep-sea trenches. The Mariana Islands, including Guam and Saipan in Fig. 3.5b, are located west of the Mariana Trench and form an intra-oceanic island arc in which oceanic crust is being subducted under oceanic lithosphere. The Mariana Trough west of the islands is a back-arc basin separated from the Parece Vela Basin of the Philippine Sea by the South Honshu Ridge which may be a remnant arc (Karig 1972).

Fig. 3.5.
a Regional map of the system of deep-sea trenches extending south from the coast of Japan to the Mariana Trench and beyond; b Map of Mariana Islands and the associated Mariana Trough (back-arc basin) (*Source:* adapted from National Geographic Society 1990; Meijer 1976)

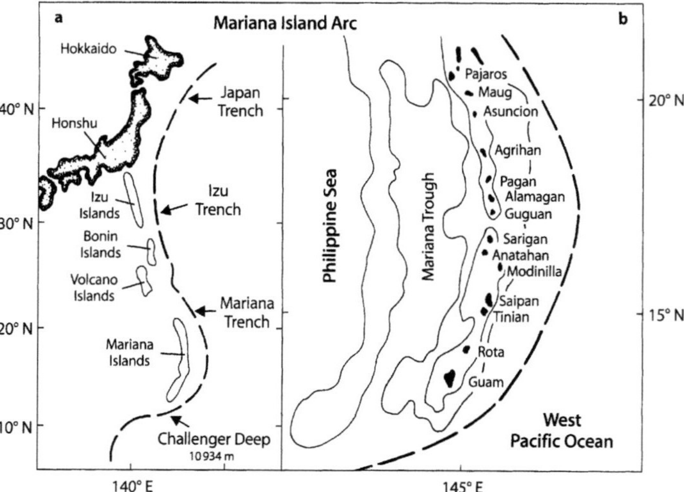

The Mariana Island arc consists of an old frontal arc (Eocene to Miocene) and of the active volcanic islands located west of it. The frontal arc includes Guam, Rota, Saipan, and Tinian, whereas the active arc consists of the islands extending from Anatahan north to Farallon de Pajaros (Uracas) identified in Fig. 3.5b. The volcanic rocks of both sets of islands consist primarily of basalt, but also include andesites, dacites, and shoshonites (Meijer 1976; Dixon and Batiza 1979; Stern 1978, 1979; Wood et al. 1981; Ito and Stern 1981; Stern and Bibee 1984; Stern et al. 1988). The history of volcanic activity extends over a period of about 40 million years with maxima between 35 and 24 Ma, 18 and 11 Ma, and from 6 Ma to the present (Lee et al. 1995).

3.2.1 Role of Subducted Sediment

Because of the absence of continental crust anywhere near the Mariana Island arc, subducted marine sediment and basalt altered by seawater are the only potential magma sources having elevated $^{87}Sr/^{86}Sr$ and low $^{143}Nd/^{144}Nd$ ratios. Ito and Stern (1986) reported that sediments (carbonates, cherts, clays, and tuffs) from DSDP cores (Legs 6 and 20) taken east of the Mariana Island arc have $^{87}Sr/^{86}Sr$ ratios between 0.70526 (tuff) and 0.71499 (clay). A composite of different kinds of sediment representing a thickness of 500 m in the western Pacific was found to have $^{87}Sr/^{86}Sr = 0.7083$ and Sr = 360 ppm. Similarly, Woodhead and Fraser (1985) and Woodhead (1989) reported 0.70822 for a composite of sediment from DSDP site 452. Lin (1992) arrived at similar estimates for bulk sediment from the western Pacific Ocean, namely: $^{87}Sr/^{86}Sr = 0.7083$, Rb = 32.5 ppm, Sr = 362 ppm. How-

ever, ooze and volcaniclastic sediment in the Mariana Trough west of the island arc has lower $^{87}Sr/^{86}Sr$ ratios between 0.70415 and 0.70630 (Woodhead and Fraser 1985; Woodhead 1989). The $^{87}Sr/^{86}Sr$ ratios of pelagic clay in the central Pacific Ocean range from 0.70859 to 0.7130 with a mean of 0.71045 (Ben Othman et al. 1989). Therefore, if subducted marine sediment contributed Sr and other elements to the magmas forming under the Mariana Island arc, the isotope compositions of Sr and of other elements (Nd, Pb, O, S, etc.) in the volcanic rocks should reflect that fact (Woodhead et al. 1987).

The $^{87}Sr/^{86}Sr$ ratios of volcanic rocks from the Mariana Islands in Fig. 3.6 are higher than 0.7030 but less than 0.7040, such that 77% of the analyzed specimens are clustered in the interval 0.7034 to 0.7036. The $^{87}Sr/^{86}Sr$ ratios of volcanic rocks from the seamounts of the southern Mariana Arc (Dixon and Stern 1983; Stern and Bibee 1984) are indistinguishable from those of the islands, and both are significantly higher than the $^{87}Sr/^{86}Sr$ ratios of MORB-like basalt in the back-arc basin located west of the Mariana Islands (Hart et al. 1972; Volpe et al. 1987; Stern et al. 1990; Gribble et al. 1996, 1998).

Several authors have demonstrated by means of mixing calculations that the proportion of subducted sediment in the magmas formed under the Mariana Island arc is quite small. For example, Meijer (1976) used the average concentrations and isotope ratios of Pb to show that sediment contributed less than 1% to the Mariana Arc lavas. Tatsumoto (1969), Church (1973), and Ben Othman et al. (1989) came to similar conclusions for volcanic rocks in other subduction zones. Later work in the Mariana Island arc by Stern (1979), Dixon and Batiza (1979), and by Dixon and Stern (1983) repeatedly reinforced the conclusion that the magmas under island arcs

Fig. 3.6.
Range of $^{87}Sr/^{86}Sr$ ratios of basalt (and hydrothermal barite) in the Mariana Island arc and the associated back-arc basin. The $^{87}Sr/^{86}Sr$ ratios of the basalts in the island arc and the seamounts are strongly clustered between 0.7032 and 0.7036, whereas those of the back-arc are much more MORB-like between 0.7026 and 0.7032. Barites at a hydrothermal vent in the Mariana Trough (back-arc basin) are precipitated from a mixture of hot vent water and seawater containing varying amounts of Sr derived from the rocks of the oceanic crust (*Sources:* Pushkar 1968; Hart et al. 1972 Meijer 1976; DePaolo and Wasserburg 1977; Stern 1979; Stern 1982; Dixon and Stern 1983; Stern and Bibee 1984; Woodhead and Fraser 1985; Ito and Stern 1986; Volpe et al 1987; Woodhead 1989; Kusakabe et al. 1990; McDermott and Hawkesworth 1991)

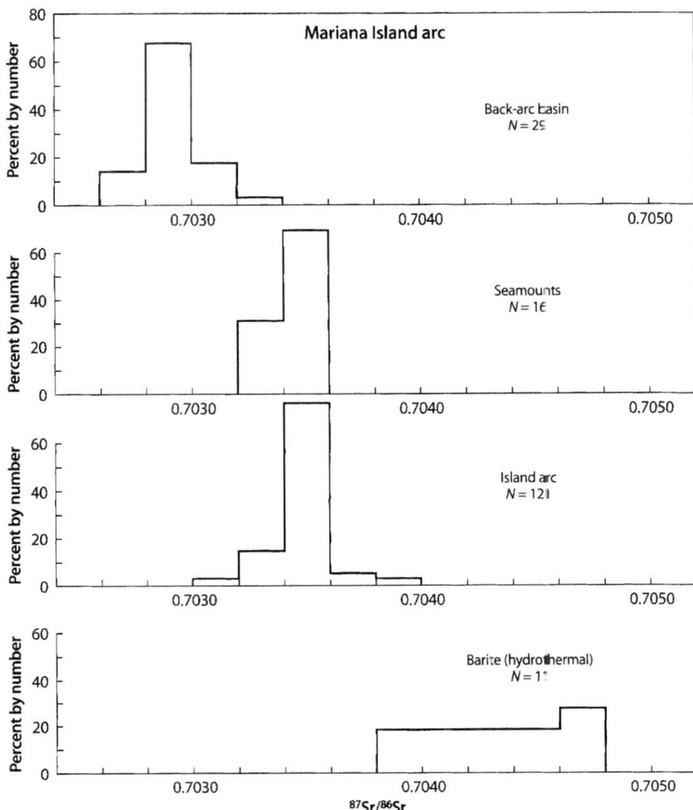

originate primarily from the mantle wedge and not from the subducted sediment (Stern 1982). The relevance of such mixing calculations is further diminished by the growing evidence that Sr and other elements are transferred selectively to the mantle wedge by aqueous fluids released by subducted marine sediment rather than by assimilation or melting of subducted sediment (Schiano et al. 1995).

The transfer of marine S and cosmogenic ^{10}Be from the sediment to the mantle wedge and hence into the magmas that form there (Woodhead and Fraser 1985; Woodhead et al. 1987; Alt et al. 1993) supports the theory that Sr is also transferred from the subducted sediment and basalt to the mantle wedge. However, it is surprising that the $^{87}Sr/^{86}Sr$ ratios of volcanic rocks on the Mariana Islands (600 km) are so strongly clustered into a narrow range from 0.7034 to 0.7036 (Fig. 3.6) given the wide range of $^{87}Sr/^{86}Sr$ ratios of the sediment in the Pacific Ocean.

The isotope ratios of Sr, Nd, and Pb of volcanic rocks on the Mariana Islands and on some of the associated seamounts are shown in Fig. 3.7a and b in relation to the mantle components of magmas that form oceanic islands (Hart 1988). In this case, the mantle wedge is the site of the DMM component, whereas the EM components are the subducted oceanic crust and associated sediment. In island arcs, these components are the sources of aqueous fluids that cause metasomatic alteration of the mantle wedge before melting occurs. The distribution of the isotope ratios of Sr, Nd, and Pb in Fig. 3.7a and b confirms that the lavas originated from depleted rocks of the lithospheric mantle with varying additions of Sr, Nd, and Pb derived from subducted marine sediment by transfer in aqueous fluids.

3.2.2 The K₂O-depth Relation

Dickinson and Hatherton (1967) postulated that the concentration of K_2O of lavas extruded in island arcs increases with increasing depth to the Benioff zone. This hypothesis has potentially important consequences for

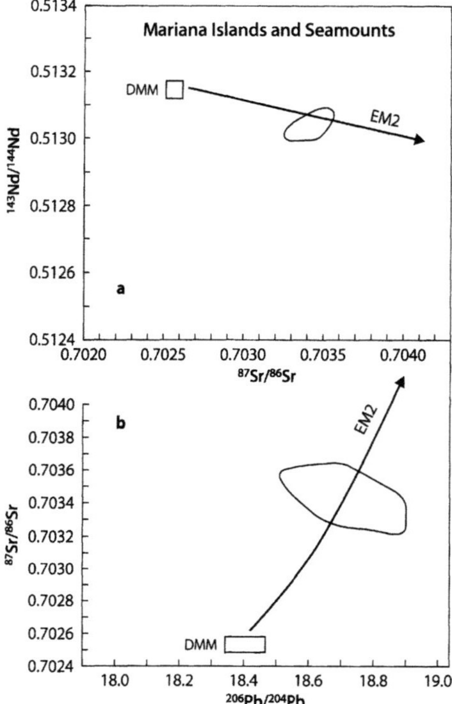

Fig. 3.7. a Variation of $^{87}Sr/^{86}Sr$ and $^{143}Nd/^{144}Nd$ ratios of volcanic rocks on the Mariana Islands and associated seamounts. The distribution of data points, compared to the magma-source components of oceanic island basalt, confirms that these lavas could have originated from the mantle wedge to which varying amounts of Sr and Nd from the downgoing slab had been added. The source of the Sr and Nd resembles the EM2 rather than EM1 or HIMU components; **b** Isotope ratios of Sr and Pb in lavas of the Mariana Islands. These data are consistent with the interpretation of the $^{87}Sr/^{86}Sr$ and $^{143}Nd/^{144}Nd$ ratios (*Sources:* Dixon and Stern 1983; Stern and Bibee 1984; Woodhead and Fraser 1985)

the origin of such rocks. The K-enrichment of the lavas is presumably caused by a progressive decrease in the degree of partial melting with increasing depth in response to smaller amounts of water released by the downgoing slab. If this hypothesis is correct, then the isotope ratios of Sr, Nd, and Pb in the K-rich lavas of rear-rank volcanoes should approach those of the uncontaminated mantle wedge and thus provide information about the magma sources of island-arc lavas. Data to be presented in subsequent sections of this chapter demonstrate that the functional relationship between K_2O and depth to the Benioff zone varies widely in different island arcs and that local factors tend to dominate (Nielson and Stoiber 1973). Consequently, the isotope ratios of Sr and Pb are not closely related to the concentrations of K_2O of arc lavas in most cases. The

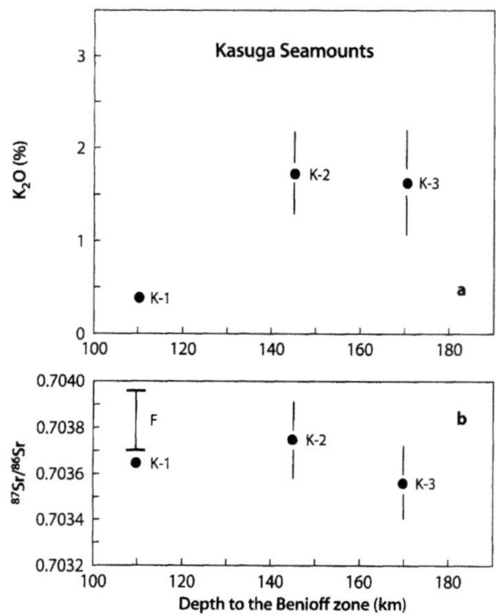

Fig. 3.8. a Variation of the average K_2O concentration of lavas of the Kasuga Seamounts with increasing depth to the Benioff zone of the Mariana Island arc. The error bars represent $2\bar{\sigma}$ values for eight lavas on *K-2* and nine on *K-3*; **b** Average $^{87}Sr/^{86}Sr$ ratios of lavas on the Kasuga Seamounts. The error bars represent $2\bar{\sigma}$ as above. The bar labeled *F* represents the range of $^{87}Sr/^{86}Sr$ ratios of lavas on Fukujin Seamount located adjacent to *Kasuga 1* (*Sources:* data from Stern et al. 1993)

systematic variations of the chemical compositions of lavas along cross-sections of subduction zones have been described by Kuno (1966), Sugimura (1967), and Matsuhisa (1977).

A set of three volcanic seamounts in the northern sector of the Mariana Islands has provided chemical and isotopic data relevant to the hypothesis of Dickinson and Hatherton (1967). The Kasuga Seamounts are aligned approximately at right angles to the trend of the Mariana Islands at about 21°45' N latitude. Kasuga 1, closest to the Mariana Trench, is extinct. Kasuga 2 and 3 are located 20 and 43 km southwest of Kasuga 1, respectively, and are either active or dormant. Kasuga 1 is located close to the active submarine volcano Fukujin. Stern et al. (1993) estimated that the depths to the Benioff zones at Kasuga 1 and at Fukujin is 110 km, 145 km at Kasuga 2, and 170 km at Kasuga 3. The petrology of the lavas on Fukujin was discussed by Jackson (1993) and others identified by Stern et al. (1993). The volcanic rocks on Kasuga 1 consist of low-K basaltic andesite, whereas K-rich basalts and absarokites occur on Kasuga 2 and 3.

The average concentrations of K_2O of the lavas on Kasuga 2 and 3 in Fig. 3.8a reported by Stern et al. (1993) are higher than those on Kasuga 1 (and on Fukujin) in general agreement with the hypothesis of Dickinson and Hatherton (1967). The average $^{87}Sr/^{86}Sr$ ratios in Fig. 3.8b decrease as expected with increasing depth to the Benioff zone, especially when the $^{87}Sr/^{86}Sr$ ratios of the lavas on Fukujin are combined with the single flow on Kasuga 1 analyzed by Stern et al. (1993).

3.2.3 Mariana Trough (Back-arc Basin)

The back-arc basin associated with the Mariana Island arc (Fig. 3.5) extends from about 13° N to 22° N for about 1 000 km in a north-south direction. The volcanic rocks that were erupted on the floor of the back-arc basin originated partly by decompression melting in the lithospheric mantle in response to rifting and partly because of fluxing by water derived from the subducted oceanic crust. The petrology of these rocks has been discussed by Hawkins and Melchior (1985) and by Hawkins et al. (1990).

The petrogenesis of volcanic rocks in the back-arc basin of the Mariana Island arc, as well as in other back-arc basins in the western Pacific Ocean, has received much attention from petrologists because of the interesting tectonic evolution of these basins:

1. Back-arc basins are regions of extension rather than compression.
2. Rifting in back-arc basins may lead to the development of spreading ridges.
3. Magmas initially form by decompression melting of depleted mantle rocks that experienced significant metasomatic alteration by fluids emanating from the subduction zone.
4. The character of the volcanic rocks in a back-arc basin may change both with time and in a geographic sense as the rifts widen such that early-formed lavas have island-arc character whereas lavas formed later along spreading ridges are increasingly MORB-like in their chemical and isotope compositions (e.g. Gribble et al. 1996, 1998).

Basalt from the Mariana Trough analyzed by Hart et al. (1972), Meijer (1976), Stern (1982), Ito and Stern (1986), and Volpe et al. (1987) in Fig. 3.6 yields an average $^{87}Sr/^{86}Sr$ ratio of 0.70292 ±0.0005(2σ̄) for 29 specimens, and average concentrations of Rb = 4.3 ppm and Sr = 176 ppm. Although the $^{87}Sr/^{86}Sr$ ratios of the lavas in the Mariana Trough in Fig. 3.6 are lower than those of the islands, they are not as low as those of MORBs. Therefore, most of these lavas did not form solely by decompression melting of depleted rocks of the mantle wedge. Instead, the elevated $^{87}Sr/^{86}Sr$ ratios of the back-arc lavas compared to those of MORBs (Fig. 2.7) indicate that the underlying mantle was variably contaminated prior to melting and eruption of the magmas or

that the lavas formed by mixing of magmas derived from MORB sources with magmas that originated from metasomatically-altered rocks of the mantle wedge adjacent to the subduction zone (Volpe et al. 1987).

The isotope ratios of Sr and Nd in Fig. 3.9a of volcanic glasses recovered from the Mariana Trough reported by Stern et al. (1990) and by Gribble et al. (1996, 1998) increase from south to north between latitudes 12° to 24° N, whereas those of Nd decrease. This pattern of variation implies a progressive change in the magma sources from MORB-like in the south to arc-like in the north. In other words, the isotope ratios of the basalts in the northern Mariana Trough are attributable to the addition of increasing amounts of Sr and Nd derived from subducted sediment to the rocks of the mantle

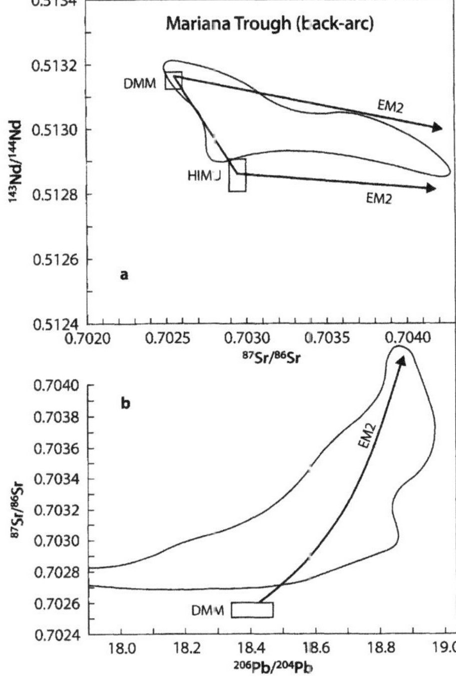

Fig. 3.9. a Isotope ratios of Sr and Nd of volcanic rocks in the northern Mariana Island arc. The distribution of data points indicates that these rocks originated from depleted lithospheric mantle (DMM component) to which varying amounts of Sr and Nd derived from subducted terrigenous sediment had been added. The Sr and Nd were transported into the mantle wedge by aqueous fluids released by dehydration of terrigenous sediment; b Isotope ratios of Sr and Pb of the same volcanic rocks shown in Part a of this diagram. The distribution of data points is consistent with the interpretation suggested in Part a, but some of the rocks have low $^{206}Pb/^{294}Pb$ ratios that are not compatible with Pb in the DMM and EM 2 components (*Source:* data from Stern et al. 1990; Gribble et al. 1996, 1998; Hart 1988)

wedge. The isotope ratios of Pb in Fig. 3.9b are consistent with that interpretation.

The regional increase of $^{87}Sr/^{86}Sr$ and $^{206}Pb/^{204}Pb$ ratios from south to north (and a concurrent decrease of the $^{143}Nd/^{144}Nd$ ratios) implies that larger amounts of aqueous fluid have interacted with the magma sources underlying the northern Mariana Trough compared to the southern part of the trough. Consequently, the degree of partial melting of the northern magma sources should also have increased. Gribble et al. (1996, 1998) actually determined that the degree of partial melting increased from about 13% in the south to about 28% in the northernmost segment of the Mariana Trough.

These authors also reported that the lavas extruded along the northern rift include a suite of differentiated calc-alkaline rocks (Fig. 3.4) ranging in composition from basalt to rhyolites that contain up to 74.65% SiO_2 and 1.56% K_2O. Similar results were reported by Lonsdale and Hawkins (1985). The average $^{87}Sr/^{86}Sr$ ratio of the rhyolites (0.70363) is similar to that of the calc-alkaline basalt (0.70379). However, a suite of 30 low-K basalts from the northern part of the basin has a lower average $^{87}Sr/^{86}Sr$ ratio (0.70301; range: 0.70273 to 0.70396) than the calc-alkaline lavas.

The Mariana Trough described above is one of the largest and most thoroughly studied back-arc basins in the Pacific Ocean. It is tectonically related to the Sumizu Rift located west of the Izu Islands and the Izu Trench which is a northward extension of the Mariana Trench (Fig. 3.5a). The isotope ratios of Sr, Nd, and Pb of MORB-like basalts in the Sumizu Back-arc were measured and interpreted by Hochstaedter et al. (1990).

Other back-arc basins in the western Pacific Ocean include the Lau Basin located west of the Tonga-Kermadec Trench (Gill 1976a; Volpe et al. 1988; Wright and White 1987), the Woodlark Basin west of the Solomon Islands (Trull et al. 1990; Muenow et al. 1991) and the Nauru Basin west of the Gilbert Island (Castillo et al. 1991). All three basins mentioned above are related to intra-oceanic subduction zones. Nevertheless, studies of the petrogenesis of the lava flows extruded in these basins indicate significant deviations from the norm as represented by the volcanic rocks of the Mariana Trough.

3.2.4 Hydrothermal Vents in the Mariana Trough

Before leaving the Mariana Island arc, we consider a report by Kusakabe et al. (1990) that hydrothermal activity occurs at two sites in the Mariana Trough. U-series disequilibrium dating by Moore and Stakes (1990) indicated ages between 0.5 to 2.5 years for barite-sulfide chimneys. The $^{87}Sr/^{86}Sr$ ratios of the barite chimneys range from 0.70385 to 0.70470 and are significantly higher than those of the basalts. The hydrothermal waters that deposit these chimneys have even higher

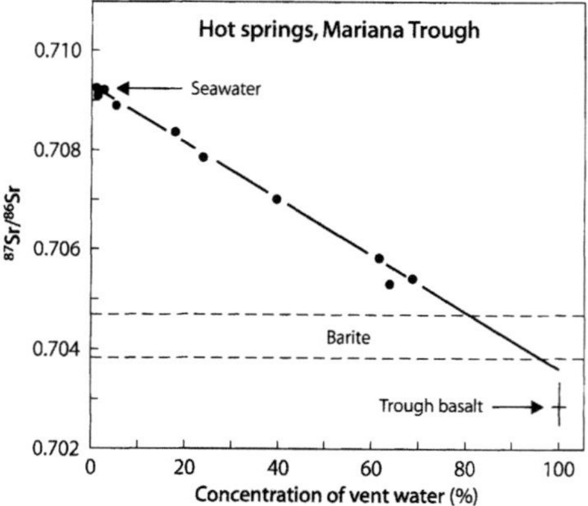

Fig. 3.10. Relation between the measured $^{87}Sr/^{86}Sr$ ratios of hydrothermal solutions discharged at barite chimneys in the Mariana Trough and the abundance of "vent water." The linear dependence between the $^{87}Sr/^{86}Sr$ ratios of the hydrothermal water and the concentration of vent water (calculated from their Mg^{2+} concentrations) implies that the Sr concentrations of the two components are similar (Eqs. 3.1 and 3.2). The diagram suggests that the barite precipitated when the concentration of vent water in the mixtures was between 95 and 80%, or when 5 to 20% of seawater had been entrained in the ascending vent water (*Source:* Kusakabe et al. 1990)

$^{87}Sr/^{86}Sr$ ratios between 0.70530 and 0.70928 and are considered to be mixtures of seawater and varying proportions of "vent water" that rises toward the seafloor from below. The concentrations of vent water in the hydrothermal solutions, analyzed by Kusakabe et al. (1990), were calculated from their Mg^{2+} concentrations based on the assumption that pure vent water contains no Mg^{2+}. According to this interpretation, the $^{87}Sr/^{86}Sr$ ratios of the hydrothermal solutions discharged in the Mariana Trough depend on the proportions of vent water and seawater.

The data in Fig. 3.10 reveal that the $^{87}Sr/^{86}Sr$ ratios of the hydrothermal solutions decrease linearly with increasing concentrations of vent water. The range of $^{87}Sr/^{86}Sr$ ratios of the barite suggests that this mineral began to precipitate when between 5 to 20% seawater had mixed with the pure vent water. The diagram also indicates that the $^{87}Sr/^{86}Sr$ ratio of the pure vent water is about 0.7036 and lies slightly above the $^{87}Sr/^{86}Sr$ ratios of the trough basalts whose $^{87}Sr/^{86}Sr$ ratios range from 0.70255 (Hart et al. 1972) to 0.70340 (Volpe et al. 1987) with a mean of 0.70290.

The linear dependence of the $^{87}Sr/^{86}Sr$ ratios of the hydrothermal waters in Fig. 3.10 with the concentrations of vent water implies that the two components of water had similar Sr concentrations. In that case, Eq. 1.30 reduces to:

$$\left(\frac{^{87}Sr}{^{86}Sr}\right)_m = \left(\frac{^{87}Sr}{^{86}Sr}\right)_v f_v + \left(\frac{^{87}Sr}{^{86}Sr}\right)_{sw} (1 - f_v) \qquad (3.1)$$

where m signifies a mixture of vent water (v) and seawater (sw). Equation 3.1 can be rearranged to yield:

$$\left(\frac{^{87}\mathrm{Sr}}{^{86}\mathrm{Sr}}\right)_m = \left[\left(\frac{^{87}\mathrm{Sr}}{^{86}\mathrm{Sr}}\right)_v - \left(\frac{^{87}\mathrm{Sr}}{^{86}\mathrm{Sr}}\right)_{sw}\right] f_v + \left(\frac{^{87}\mathrm{Sr}}{^{86}\mathrm{Sr}}\right)_{sw} \quad (3.2)$$

which is a straight line in coordinates of $(^{87}\mathrm{Sr}/^{86}\mathrm{Sr})_m$ and f_v (the concentration of vent water in the mixture).

3 2.5 Shoshonites of the Northern Seamount Province

Although the volcanic rocks of intra-oceanic island arcs typically have low concentrations of alkali elements, K-rich shoshonitic rocks (Fig. 3.4) do occur in the seamounts north of the Mariana Islands between latitudes 23 and 26° N (Box and Flower 1989). The occurrence of shoshonitic rocks in this segment of the Mariana-Volcano island arc is documented in Fig. 3.11a,b based on Rb and Sr concentrations reported by Bloomer et al. (1989) and Lin et al. (1989), and discussed by Stern et al. (1988). The enrichment of these rocks in Rb, Sr and other LIL elements (Ba, La, Ce) cannot be attributed to fractional crystallization because these rocks have low average SiO_2 concentrations ranging from 47.4% to 53.89% (Bloomer et al. 1989). In addition, the LIL element enrichment of the shoshonitic rocks is not caused by variations in the composition of the subducted lithosphere or by low degrees of melting of depleted rocks in the mantle wedge. Therefore Bloomer et al. (1989) and Lin et al. (1989) attributed the occurrence of the shoshonitic rocks to magma formation in LIL-enriched OIB-like source rocks in the mantle wedge. These kinds of source rocks may have been exposed to partial melting by the northward propagation of back-arc spreading in the Mariana Trough. Therefore, eruption of shoshonitic lavas in the Northern Seamount Province is associated with an early stage of back-arc development (Stern et al. 1988).

3.2.6 Boninites of the Bonin and Izu Islands

The Volcano Islands (including Iwo Jima), the Bonin Islands (Ami Jima and Haha Jima) and the Izu Islands (Oshima, Nishino Shima, etc.) in Fig. 3.5a extend northward from the Mariana Islands towards Honshu of the Japanese Islands. All of these islands are located west of a system of deep-sea trenches along which oceanic lithosphere of the North Pacific Ocean is being subducted. The rocks on these islands range in composition from basalt through andesite to rhyolites and their $^{87}\mathrm{Sr}/^{86}\mathrm{Sr}$ ratios range from about 0.7033 to 0.7039 based on measurements by Pushkar (1968) and Stern

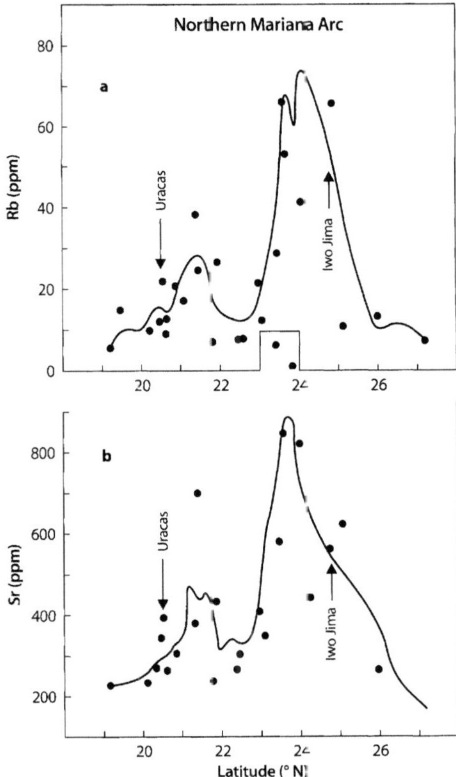

Fig. 3.11. Variations of average concentrations of Rb and Sr of basaltic volcanic rocks in the Northern Seamount Province located between Uracas of the Mariana Islands and Iwo Jima of the Volcano Islands (latitude 23 to 26° north). a Rb; b Sr. The concentrations of K, Ba, La and Ce of the shoshonites are also elevated. The *line* is a three-point moving average (*Sources:* Bloomer et al. 1989; Lin et al. 1989; Stern et al. 1988)

(1982). Pushkar (1968) also reported unusually high values of about 0.7053 for the Bonin Islands and 0.7050 for andesite from Tori Shima of the Izu Islands. These high values were confirmed more recently by Cameron et al. (1983) who reported an initial $^{87}\mathrm{Sr}/^{86}\mathrm{Sr}$ ratio of 0.70503 for a boninite (>26 Ma) from the island of Chichi Jima.

The volcanic rocks in the Bonin Islands (Ogasawara) include varieties that are simultaneously enriched in both SiO_2 and MgO and therefore have been variously classified as basaltic andesites, high-Mg andesites, komatiitic basalt, siliceous high-Mg basalt, etc. These rocks constitute a distinct compositional variety known as boninite described in a book edited by Crawford (1989). Boninites are defined as volcanic rocks containing SiO_2>53% and having magnesium numbers >0.60, where the Mg number is expressed as $Mg / Mg + Fe^{2+}$ and Fe^{2+} represents total Fe. Boninites occur not only on Ogasawara Island in the Bonin Islands, but also in the Mariana Trench, in the Mariana fore-arc, and on Guam, as well as at many other sites around the world. The potential economic importance of boninitic rocks arises from their occurrence in the Stillwater Complex of Montana (Sect. 7.1.4), the Bushveld Complex in South Africa (Sect. 7.2), and the Troodos Massif on Cypress all of which are enriched in platinum group elements (PGE).

Boninites are derived from rocks in the mantle that have been depleted by one or several previous extractions of basaltic magma. These unusually refractory source rocks can yield another melt fraction because of the addition of water derived from the subducted oceanic crust. The LIL element concentrations and $^{87}Sr/^{86}Sr$ ratios of boninites depend on the extent to which these elements and radiogenic ^{87}Sr are transferred into the magma source by the aqueous fluid emanating from the rocks of the subducted plate.

In the course of a study of Eocene boninites from the Mariana fore-arc, Stern et al. (1991) determined that in this case between 70 and 92% of the Sr in the rocks had been added by an aqueous fluid to depleted mantle rocks (harzburgite) of the mantle wedge. This subduction component also provided 56 to 95% of the Pb, but much less of the Nd in the boninites. The $^{87}Sr/^{86}Sr$ ratios of the boninites in the Bonin Islands (corrected for in situ decay of ^{87}Rb to 40 Ma) range only from 0.70324 to 0.70338 and are very similar to the $^{87}Sr/^{86}Sr$ ratios of the volcanic rocks extruded on the Mariana Islands in Recent times (Fig. 3.6).

3.3 The Aleutian Islands

The Aleutian Islands in Fig. 3.12 extend from the Alaska Peninsula across the northern Pacific Ocean almost to the Kamchatka Peninsula of Asia. This chain of volcanic islands differs from the Mariana Islands because the western sector of the Aleutian Islands is an intra-oceanic arc, whereas the eastern islands and the Alaska Peninsula are located on the Bering Shelf. Consequently, the volcanic rocks of the Aleutian Islands provide an opportunity to investigate:

1. The possible contamination of magma by assimilation of rocks from the continental crust of the Bering Shelf.
2. The role of subducted sediment and altered basalt in the formation of arc magmas.
3. The effect of distance from the trench (or depth of the Benioff zone) on the chemical composition of magmas.
4. Time-dependent changes in the composition of lava flows extruded at a particular site.

The petrogenesis of volcanic rocks on the Aleutian Islands has been extensively investigated on the basis of isotopic and geochemical data by many scientists, including Arculus et al. (1977), Kay et al. (1978), Perfit et al. (1980a), McCulloch and Perfit (1981), Morris and Hart (1983), Myers et al. (1985, 1986), Myers and Marsh (1987), Nye and Reid (1986), Von Drach et al. (1986), and Kay and Kay (1988a). In addition, Perfit et al. (1980b) and White and Patchett (1984) included the Aleutian Island

Fig. 3.12.
Map of the Aleutian Islands in the North Pacific Ocean including the rear-rank islands of Amak and Bogoslof as well as the Pribilof and St. George Islands located in the back-arc basin. The eastern Aleutian Islands are located on the continental shelf of the Bering Sea, whereas the volcanoes of the western Aleutian Islands rise from the abyssal plain of the Bering Sea (*Source:* adapted from von Drach et al. 1986)

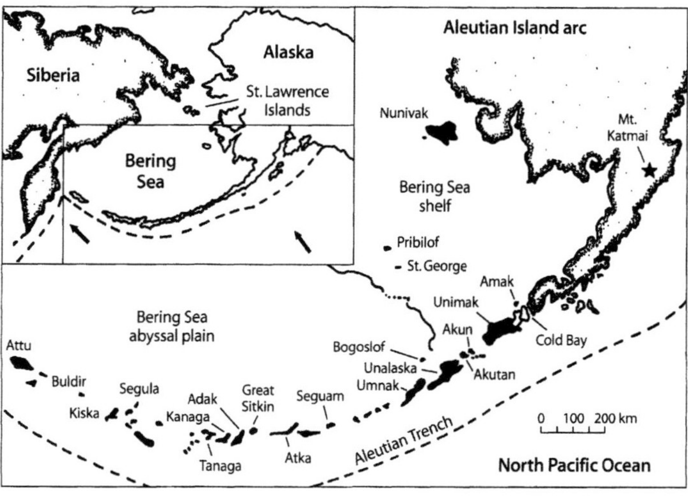

arc in their more general discussions of the petrogenesis of island-arc basalt, whereas Kay and Kay (1988b) provided an overview of the studies concerning the origin of Aleutian magmas.

3.3.1 Lateral Homogeneity of $^{87}Sr/^{86}Sr$ Ratios

The $^{87}Sr/^{86}Sr$ ratios of volcanic rocks in the Aleutian Islands in Fig. 3.13 lie primarily between 0.7030 and 0.7034 and do not vary along the length of the arc as one might expect. About 44% of the $^{87}Sr/^{86}Sr$ ratios measured in rocks of the western intra-oceanic islands (Kiska, Segula, Adak, Sitkin, and Umnak) and almost 79% of the rocks from the eastern islands (Unalaska, Akutan, Unimak, Amak, and the Alaska Peninsula) lie between 0.7030 and 0.7034. Therefore, magmas generated under the Bering Shelf are not significantly contaminated with radiogenic ^{87}Sr derived from the rocks that form the shelf.

Figure 3.13 also demonstrates that about 56% of the volcanic rocks extruded on the islands of the Bering Sea (back-arc basin) have $^{87}Sr/^{86}Sr$ ratios between 0.7028 and 0.7030. The comparatively low isotope ratios of Sr in the rocks of these islands (Pribilof and Nunivak) indicate that the magmas extruded at these sites were not directly affected by the subducted oceanic lithosphere and originated by decompression melting of lithospheric mantle under the Bering Sea (Stuart et al. 1974). The $^{87}Sr/^{86}Sr$ ratios of volcanic rocks in the back-arc basin of the Aleutian Island arc have the same distribution as those in the Mariana Trough (Fig. 3.6). The only known exception is the volcanic rocks from St. Lawrence Island (Fig. 3.12) whose $^{87}Sr/^{86}Sr$ ratios are anomalously high (0.70349 and 0.70385, Von Drach et al. 1986; relative to 0.71025 for NBS 987). This island (like the Pribilof Islands and Nunivak) is located on the Bering Shelf about 350 km from its edge.

3.3.2 Amak and Bogoslof Islands.

The volcanic rocks on Bogoslof (Arculus et al. 1977) and Amak Islands (Morris and Hart 1983) in Fig. 3.12 are noteworthy because both are located behind the volcanic front. The volcanic rocks on these two islands have $^{87}Sr/^{86}Sr$ ratios that are, in part, lower than those of the principal Aleutian Islands. In addition, the volcanic rocks of Amak and Bogoslof Islands are enriched in Rb, Sr and other LIL elements compared to the rocks of the islands along the volcanic front. Therefore, the lavas on these islands provide another opportunity to evaluate the effect of depth to the Benioff zone on the composition of magmas.

Amak Island is located about 50 km behind the volcanic front at Cold Bay and about 160 km above the Benioff zone, whereas Cold Bay is only about 90 km above the Benioff zone. The data of Morris and Hart (1983) in Fig. 3.14a indicate that the average $^{87}Sr/^{86}Sr$ ratio of the rocks on Amak Island is 0.70313 ±0.00004 ($2\bar{\sigma}$) and that the $^{87}Sr/^{86}Sr$ ratios are independent of SiO_2. The $^{87}Sr/^{86}Sr$ ratios of andesites at Cold Bay increase with increasing concentrations of SiO_2 and range from 0.70321 (SiO_2 = 56.5%) to 0.70352 (SiO_2 = 59.2%). The correlation between $^{87}Sr/^{86}Sr$ ratios and SiO_2 concentrations implies that time-dependent changes occurred in the magma sources under Cold Bay.

Figure 3.14b reveals that the andesites of Amak Island have higher concentrations of Rb and Sr than the andesites of Cold Bay (Morris and Hart 1983). Additional data by Kay et al. (1978) and Von Drach et al. (1986) confirm this observation. Similarly, the volcanic rocks on Bogoslof Island (located about 40 km north of Umnak Island) not only have a low average $^{87}Sr/^{86}Sr$ ratio of 0.70295 ±0.00004 (Arculus et al. 1977; Kay et al. 1978; Von

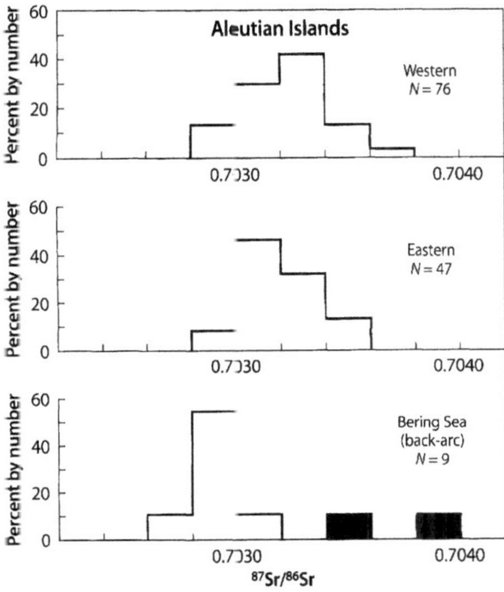

Fig. 3.13. Range and distribution of $^{87}Sr/^{86}Sr$ ratios of volcanic rocks on the Aleutian Islands and in the related back-arc basin. The western islands (Kiska, Segula, Adak, Atka, Umnak, Bogoslof, and Sitkin) form an intra-oceanic island arc. The eastern islands (Unimak, Unalaska, Akutan, Amak, and the Alaska Peninsula) are underlain by the continental crust of Alaska. However, the $^{87}Sr/^{86}Sr$ ratios of volcanic rocks from the western and eastern parts of the arc are indistinguishable. The $^{87}Sr/^{86}Sr$ ratios of rocks on the islands in the Bering Sea (back-arc; Pribilof Islands, Nunivak, and DSDP 191) are distinctly lower than those of island-arc volcanics, except for those on St. Lawrence (shaded). All $^{87}Sr/^{86}Sr$ ratios are compatible with 0.7080 for E&A and 0.71025 for NBS 987 (Sources: Kay et al. 1978; McCulloch and Perfit 1981; Morris and Hart 1983; White and Patchett 1984; von Drach et al. 1986; Myers et al. 1985; Arculus et al. 1976; Nye and Reid 1986)

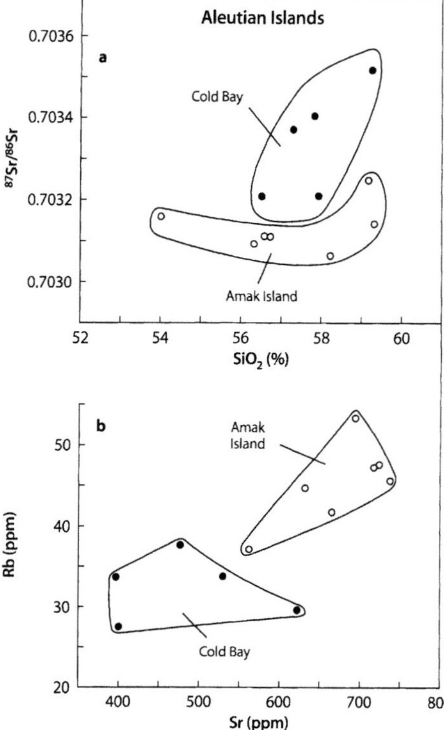

Fig. 3.14. a Variation of $^{87}Sr/^{86}Sr$ ratios and SiO_2 concentrations of andesites at Cold Bay and nearby Amak Island in the Aleutian Island arc. At Cold Bay, the $^{87}Sr/^{86}Sr$ ratios increase with SiO_2 concentrations and with the sequence of extrusion of the lava flows, suggesting that these time-dependent variations were caused by changes of the magma sources. On Amak Island, the $^{87}Sr/^{86}Sr$ ratios remained constant at 0.70313 ±0.00006 (2$\bar{\sigma}$) presumably because the magma was derived from a "plum" of enriched or undepleted mantle without additions of radiogenic ^{87}Sr from the wet sediment subducted into the Aleutian Trench; **b** Rubidium and Sr concentrations of andesites at Cold Bay and Amak Island. The concentrations of these elements in the andesites at Cold Bay are lower than those of Amak Island and do not vary appreciably with the concentration of SiO_2. The elevated Rb and Sr concentrations of the rocks on Amak Island may be caused by a lower degree of partial melting related to the position of Amak Island about 50 km behind the volcanic front. The positive correlation of the Rb and Sr concentrations of the rocks on Amak Island may be caused by fractional crystallization of magma or by differences in the extent of melting of the source rocks (*Source:* Morris and Hart 1983)

Drach et al. 1986), but also have elevated average concentrations of K_2O (1.60 to 2.47%), Rb (70 ppm) and Sr (735 ppm) compared to the volcanic rocks of Umnak Island which contain only about 22 ppm Rb and about 382 ppm Sr on the average.

The silica concentrations of lavas extruded on Bogoslof Island actually decreased from 61.00% in 1796, to 51.54% in 1883, and finally to 46.00% in 1927 (Arculus

et al. 1977). However, the change in chemical composition cannot be attributed to fractional crystallization alone because the lavas that were erupted in 1796 have an elevated $^{87}Sr/^{86}Sr$ ratio of 0.70343 compared to 0.70295 ±0.00004 reported for the later flows. The unusually high concentration of K_2O of these rocks is caused by or reflects the presence of K-rich amphibole phenocrysts. The low $^{87}Sr/^{86}Sr$ ratios of the most recent lavas indicate that magma was derived from the lithospheric mantle without significant additions of radiogenic ^{87}Sr from the subducted sediment and altered basalt.

These observations focus attention on the occurrence of alkali-rich volcanic rocks on some of the islands of the Aleutian Arc and other island arcs elsewhere (Marsh and Carmichael 1974; DeLong et al. 1975; and Sect. 3.2.2). The apparent increase of the K_2O content of andesites in island arcs with increasing depth to the Benioff zone is consistent with the hypothesis of Dickinson and Hatherton (1967) which was confirmed by Hart et al. (1970a,b). However, Nielson and Stoiber (1973) demonstrated that the relationship between the concentrations of K_2O of andesite in island arcs and the depth to the Benioff zones is not constant, but varies widely at different locations on the Earth. Therefore, the occurrence of alkali-rich rocks in island arcs is dependent not only on the tectonic setting or on the effects of fractional crystallization of magma, but may also require the presence of unusual source rocks that form alkali-rich magmas.

3.3.3 The Plum-pudding Model and Sediment Recycling Revisited

The isotope compositions of Sr in island-arc basalts in the Mariana Islands, the Aleutian Islands, and in other island arcs are similar to those of oceanic island basalts (e.g. White and Patchett 1984). Therefore, Morris and Hart (1983) proposed that island-arc basalts at Cold Bay and Amak Island in the Aleutians are derived from magmas that form by partial melting of "plums" of enriched (e.g. veined) rocks embedded within the depleted rocks of the lithospheric mantle wedge. Partial melting occurs preferentially in the plums because they are less refractory than the depleted rocks of the lithospheric mantle. This idea was later used by Stern et al. (1993) to explain the heterogeneity of lavas extruded on the Kasuga Seamounts in the Mariana Island arc.

Morris and Hart (1983) demonstrated that island-arc basalts are enriched only in Cs when they are compared to OIBs instead of to MORBs. The enrichment of island-arc basalts in Cs is caused by the efficient scavenging of this element from the subducted sediments by aqueous fluids, as originally demonstrated by Tatsumi et al. (1986) in Fig. 3.2.

The importance of sediments in the generation of island-arc basalt in the Aleutian Islands was debated by

Perfit and Kay (1986) and Morris and Hart (1986). Without going into the details of this discussion, at issue here is the fundamental quest on of the extent to which detrital sediment derived from the continental crust is recycled in subduction zones. Some authors (e.g. Armstrong 1971), postulated a significant contribution of material from the subducted sediment to island-arc basalts, whereas others (e.g. Kay et al. 1978) have suggested that only a small fraction of the sediment is recycled (Sect. 3.2.). If the sediment serves only as a source of water (and perhaps Cs), as implied by the plum-pudding model of Morris and Hart (1983), then very little subducted sediment is returned to the crust and most of it is transported into the mantle. However, if plumes under oceanic islands and plums under subduction zones are themselves composed in part of previously subducted and recrystallized sediment, as postulated by Hofmann and White (1982), then basalts on oceanic islands and in island arcs are recycling sediment on a large scale.

The extent to which young sediment at the top of the down-going slab contributes to magmas in subduction zones can be judged by the presence of cosmogenic ^{10}Be in freshly extruded volcanic rocks in island arcs. Both Tera et al. (1986) and Monaghan et al. (1987) detected cosmogenic ^{10}Be in historic lava flows in the Aleutian Arc and thereby confirmed that this radionuclide, whose half-life is 1.5×10^6 yr, is transferred from the uppermost sediment layers of the subducted plate into magmas that are subsequently erupted as lava flows on the surface.

Since subducted sediment does contribute ^{10}Be to the volcanic rocks in the Aleutian Islands and in other volcanic island arcs, Kay and Kay (1988a) considered how much sediment must be consumed in the formation of magmas in order to explain the excess amounts of certain elements that characterize island-arc basalts compared to MORBs. They concluded that less than one quarter of the subducted sediment is needed to explain the elevated K content of the lava flows on the Aleutian Islands (assuming a rate of subduction of 5 cm yr^{-1} and a production rate of 400 km^3 of crust/km of arc). However, the sediment is not transferred into the magmas in bulk, but certain elements are transferred selectively (Tatsumi et al. 1986).

Kay and Kay (1988a) also considered the possibility that magmas, formed in the mantle wedge, are subsequently contaminated by assimilation of rocks in the crust under the Aleutian Islands. Evidence in favor of this process comes from the chemical compositions and ^{87}Sr/^{86}Sr ratios of cumula e xenoliths in the basaltic lavas of Adak Island (Kay and Kay 1986). However, xenoliths from the volcano Mt. Moffet on Adak Island record differentiation of basalt magma in a periodically refilled magma chamber at the base of the crust (Conrad and Kay 1984). The absence of xenoliths of sedimentary rocks suggests that the lower crust and upper mantle are composed of mafic igneous rocks and that sedimentary rocks are absent.

Sediment from the Gulf of Alaska, the Atka Basin, and from the Aleutian Abyssal Plain (ranging in age from Eocene to Recent) contains about 60 ppm Rb and about 230 ppm Sr (Von Drach et al. 1986). The ^{87}Sr/^{86}Sr ratios of the sediment range from 0.70525 to 0.70719, depending primarily on the abundance of volcaniclastic compared to terrigenous detritus. The average ^{87}Sr/^{86}Sr ratio of the sediment, weighted by the Sr concentration, is 0.70636. Kay and Kay (1988a) published additional trace-element concentrations (not including Rb and Sr) in sediment from the North Pacific Ocean.

The distribution of data points in Fig. 3.15a and b clearly demonstrates that the magmas erupted on the Aleutian Islands did not originate by partial melting of depleted mantle alone. As in the case of the Mariana Islands (Fig. 3.7), the volcanic rocks of the Aleutians con-

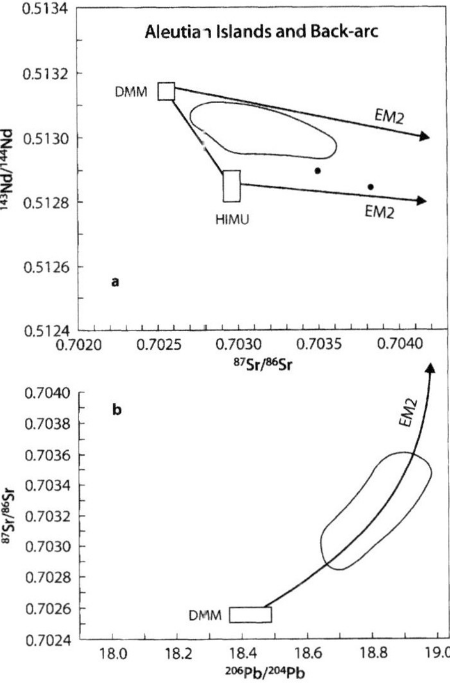

Fig. 3.15. a Isotope ratios of Sr and Nd of volcanic rocks on the Aleutian Islands and in the back-arc basin associated with them. The distribution of data points is consistent with the theory that these rocks originated from the depleted magma sources of the mantle wedge to which varying amounts of Sr and Nd derived from terrigenous sediment had been added; b The isotope ratios of Sr and Pb are consistent with the petrogenesis of the volcanic rocks stated above (*Sources:* Kay et al. 1978; McCulloch and Perfit 1981; Morris and Hart 1983; von Drach et al. 1986; Nye and Reid 1986)

tain varying amounts of Sr, Nd, and Pb that could have been derived from subducted sediment having elevated $^{87}Sr/^{86}Sr$ and $^{206}Pb/^{204}Pb$ ratios but lower $^{143}Nd/^{144}Nd$ ratios than the magma sources of MORBs. These data do not necessarily contradict the proposal of Morris and Hart (1983) that magma in the Aleutians formed by preferential melting of "plums" in the mantle wedge that include subducted sediment. However, in that case, special circumstances are required (i.e. the existence of "plums" in the mantle wedge) to explain the origin of the volcanic rocks in the Aleutian Islands, whereas the transfer of selected elements from the down-going slab into the mantle wedge is a plausible consequence of subduction and therefore explains the origin of volcanic rocks on all island arcs.

The Aleutian Island arc continues westward beyond the island of Attu (Fig. 3.12) to the Commander Islands off the coast of Kamchatka. The islands of Medny and Bering consist of differentiated calc-alkaline volcanic rocks ranging from basalt to rhyolite and including granodiorite plutons and aplite veins. Borsuk et al. (1983) reported that the initial $^{87}Sr/^{86}Sr$ ratios of basalts and rhyolites on Medny Island range from 0.7037 to 0.7055

with a mean of 0.7044 ±0.0004 ($2\bar{\sigma}, N = 9$) compared to 0.7031 for the granodiorite and aplite. The basalts on Bering Island have an average initial $^{87}Sr/^{86}Sr$ ratio of 0.7034 ±0.0004 ($2\bar{\sigma}, N = 3$). Borsuk et al. (1983) explained the $^{87}Sr/^{86}Sr$ ratios of the basalt-rhyolite suite on Medny as a result propylitization of the rocks by seawater having an $^{87}Sr/^{86}Sr$ ratio of 0.7091. Consequently, the principal conclusion is that the isotope ratios of the lavas on the Commander Islands are similar to those of the Aleutian Islands and therefore originated by the same petrologic processes.

3.4   Kamchatka and the Kuril Islands

The Kamchatka Peninsula in Fig. 3.16 extends south from the mainland of Siberia into the North Pacific Ocean. It contains active volcanoes which have erupted basalt and andesite lavas that formed as a consequence of subduction of the Pacific Plate into the Kuril Trench located off the east coast of southern Kamchatka. The Kuril Trench continues south from the southern tip of Kamchatka for a distance of about 1 200 km to the island of Hokkaido of Japan. The volcanoes that are associated with the Kuril Trench form the Kuril Island arc including from north to south: Paramushir, Onekotan, Simushir, Urup, Iturup, and Kunashir.

3.4.1   Kamchatka Peninsula

The southern part of the Kamchatka Peninsula is the site of westward subduction of the Pacific Plate into the Kuril Trench (Fig. 3.16). The geology of this peninsula includes pre-Cretaceous metamorphic rocks of crustal origin as well as Oligocene to Miocene plutons and calc-alkaline lavas. Volcanic activity continues at the present time at the volcanoes Bezymyanna, Khangar, and Ksudach (Vinogradov et al. 1988). Therefore, the tectonic setting of the Kamchatka Peninsula is similar to that of the Alaska Peninsula where oceanic crust is being subducted under continental crust.

The initial $^{87}Sr/^{86}Sr$ ratios of Miocene granitoid plutons on the Kamchatka Peninsula range from 0.70346 (plagioclase, Akhomten Pluton) to 0.70463 (amphibole, Lunton Pluton) based on analyses by Vinogradov et al. (1988). Additional measurements of $^{87}Sr/^{86}Sr$ ratios were reported by Hedge and Gorshkov (1977), whereas Bibikova et al. (1983) measured isotope ratios of Pb in andesites of Bezymyanna Volcano.

The Quaternary lavas of Khangar Volcano in the Central Kamchatka Range are primarily dacitic in composition, but range from andesite to rhyolite. The last eruption of this volcano occurred about 6 500 years ago (Vinogradov et al. 1988). The flows contain abundant xenoliths of diorite, granodiorite, and gabbro. The age

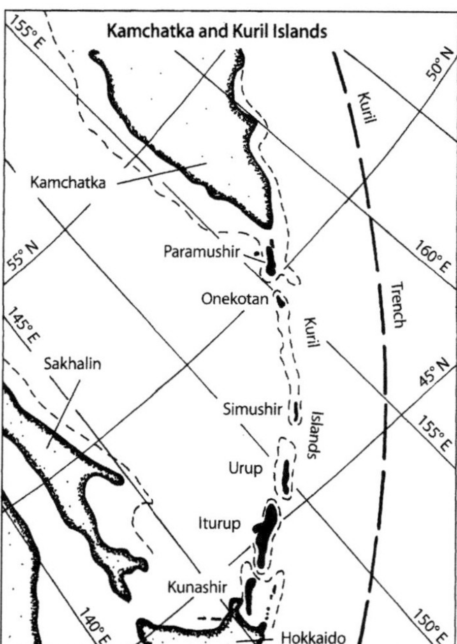

Fig. 3.16. Map of the Kamchatka Peninsula and the Kuril Islands. The active volcanoes on both erupt basalt, andesite, and related calc-alkaline lavas that originate from the mantle wedge above the subduction zone associated with the Kuril Trench (*Source:* adapted from National Geographic Society 1990)

of a granodiorite xenolith (about 14 Ma) is indistinguishable from that of the Miocene Akhomten Pluton. Therefore, Vinogradov et al. (1988) concluded that the xenoliths in the lava flows of Khangar Volcano originated from Miocene plutons rather than from Precambrian basement rocks.

An important exception to the dearth of ultramafic inclusions in the lava flows of island arcs are the spinel peridotite and pyroxenite inclusions described by Kepezhinskas et al. (1996) from the Valovayam volcanic field west of the Komandorsky Basin which is located north of the junction between the Aleutian and Kamchatka Arcs. Some of the inclusions contain felsic veins that formed by interaction of the ultramafic rocks with a siliceous melt whose composition was quite different from the host basalt. The products of this interaction include pargasitic amphibole, albite-rich plagioclase, Al-rich augite, and garnet (Carroll and Wyllie 1989). Kepezhinskas et al. (1996) suggested that the siliceous melts originated by melting of basalt in the subducted oceanic crust (i.e. they are adakites). These results therefore draw attention to the possibility that the metasomatic alteration of the mantle wedge above subduction zones is caused not only by aqueous fluids but also by silica-rich melts derived from the subducted oceanic crust. Similarly, the presence of siliceous glass in mantle xenoliths on the Canary Islands was reported by Neumann (1991) and Neumann and Wulff-Pedersen (1997).

3.4.2 The Kuril Islands

The Kuril Islands are located west of the Kamchatka-Kuril Trench into which the oceanic crust of the northern Pacific Plate is being subducted. The Kuril Islands contain several large active volcanoes including Vernadski on Paramushir and Mendeleev on Kunashir (Fig. 3.16).

The Quaternary lavas of the Kuril Islands range in composition from basalt to rhyolite and belong to different suites based on their K_2O concentrations (Fig. 3.17). The lavas on the northern Kuril Islands (e.g. Paramushir) are more K-rich than those on the southern islands (e.g. Kunashir). However, the 61 Ma lavas on Tanfil' eva in the Lesser Kurils east of the island of Kunashir are actually shoshonitic in composition with K_2O concentrations up to 4.7% (Bailey et al. 1987).

The $^{87}Sr/^{86}Sr$ ratios of the lavas on the Kuril Islands measured by Bailey et al. (1987) and Zhuravlev et al. (1985, 1987) range from 0.70284 (basalt, Beliankin submarine volcano) to 0.70407 (rhyolite, Mendeleev Volcano, Kunashir Island) and 74% of the analyzed lavas from the Kuril Islands in Fig. 3.17a have $^{87}Sr/^{86}Sr$ ratios between 0.7030 and 0.7036 relative to 0.7080 for E&A. The $^{87}Sr/^{86}Sr$ ratios of lavas on the Aleutian Islands and on the Alaska Peninsula in Fig. 3.17b have very similar

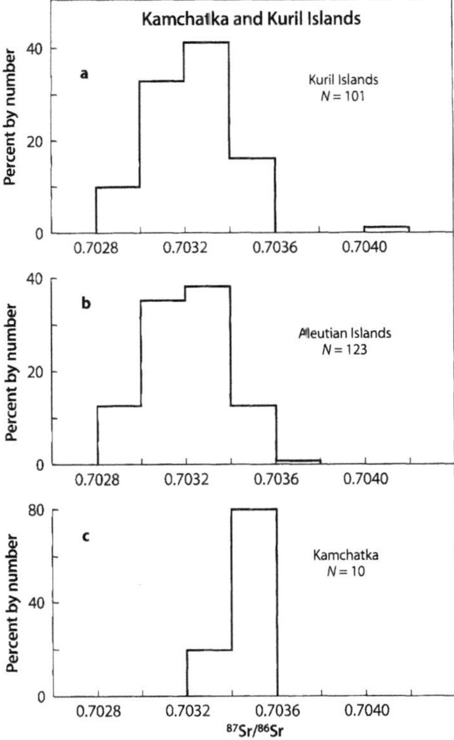

Fig. 3.17. a Range of $^{87}Sr/^{86}Sr$ ratios of the Quaternary lavas on the Kuril Islands relative to 0.7080 for E&A (*Sources:* Zhuravlev et al. 1987; Bailey et al. 1987); b Range of $^{87}Sr/^{86}Sr$ ratios of the Quaternary lavas on the Aleutian Islands and the Alaska Peninsula relative to 0.7080 for E&A and 0.71025 for NBS 987 (*Sources:* Same as those for Fig. 3.13); c Range of $^{87}Sr/^{86}Sr$ ratios of Quaternary lavas on the Kamchatka Peninsula relative to 0.7080 for E&A (*Source:* Bailey et al. 1987)

distributions. However, 80% of the Quaternary lavas on Kamchatka in Fig. 3.17c have slightly higher $^{87}Sr/^{86}Sr$ ratios between 0.70340 and 0.70360 than the lavas of the Kuril Islands.

The isotope compositions of Sr in the volcanic rocks of the Kuril Islands vary both along the length of the island arc and across it. Zhuravlev et al. (1985, 1987) subdivided the Kuril Islands into northern, central, and southern segments and classified the volcanoes in each segment into "frontal" and "back-arc." The resulting pattern suggests that lavas of frontal volcanoes in the northern and southern segments of the island arc have higher average $^{87}Sr/^{86}Sr$ ratios than those in the back row (Table 3.1).

However, in the central segment, volcanoes in the front row and in the back row on Simushir and Ushishir Islands have identical low $^{87}Sr/^{86}Sr$ ratios between

Table 3.1. The isotope compositions of Sr in the volcanic rocks of the Kuril Islands. The Kuril Islands were subdivided into northern, central, and southern segments and the volcanoes in each segment were classified into "frontal" and "back-arc" (Zhuravlev et al. 1985, 1987)

Segment	Front row	Back row
Northern	0.70322 (2) ±0.00008 (2$\bar\sigma$)	0.70303 (7) ±0.00008
Central	0.70302 (3) ±0.00012	0.70308 (8) ±0.00009
Southern	0.70352 (4) ±0.00037	0.70311 (3) ±0.00018

0.70290 and 0.70311 relative to 0.7080 for E&A. In addition, the average $^{87}Sr/^{86}Sr$ ratios of lava flows on the Kuril Islands increase from north to south. The data of Bailey et al. (1987) and Zhuravlev et al. (1987) for Quaternary lavas yield averages of 0.70318 ±0.00005(2$\bar\sigma$, N = 21) for Paramushir in the north and 0.70346 ±0.00010 (2$\bar\sigma$, N = 15) for Kunashir in the south relative to 0.7080 for E&A. The average $^{87}Sr/^{86}Sr$ ratios of the lavas on Kunashir have decreased with time from 0.70377 ±0.00018 (2$\bar\sigma$, N = 5) at 20 Ma, to 0.70365 ±0.00016 (2$\bar\sigma$, N = 5) at 11 Ma, and to 0.70346 ±0.00010 at 0 Ma (Bailey et al. 1987).

The longitudinal variation of the $^{87}Sr/^{86}Sr$ ratios of Quaternary lavas on the Kuril Islands is inconsistent with the thickness of the continental crust underlying the islands because the crustal thickness decreases from north to south, whereas the $^{87}Sr/^{86}Sr$ ratios increase (Bailey et al. 1987). The island of Paramushir in the north is underlain by about 30 km of crust composed of basalt overlain by granitic/metamorphic rocks and sediment, whereas at Kunashir in the south the crust is probably less than 20 km thick and consists primarily of basalt with lesser amounts of granitic/metamorphic rocks and sediment (Zhuravlev et al. 1987). The lack of correlation of $^{87}Sr/^{86}Sr$ ratios with crustal thickness indicates that assimilation of crustal rocks played only a minor role in the petrogenesis of these rocks and that magmas originated primarily by partial melting in the mantle wedge above the subduction zone. In addition, the similarity of $^{87}Sr/^{86}Sr$ ratios among different rock types, demonstrated in Fig. 3.18 for lavas an Paramushir and Atlasova Islands, indicates that they could have formed by fractional crystallization of basaltic magma under closed conditions rather than by assimilation of crustal rocks. Therefore, Bailey et al. (1987) concluded that the magmas originated by partial melting of rocks in the mantle wedge that had been contaminated by ^{87}Sr-bearing fluids derived from subducted marine sediment. However, $\delta^{18}O$ values of basalts, andesites, and rhyolites on the Kuril Islands range widely from +5.4 to +10.8‰ and correlate with the concentration of H_2O+ which varies up to 1.78%. Pokrovskiy and Zhuravlev (1991) concluded

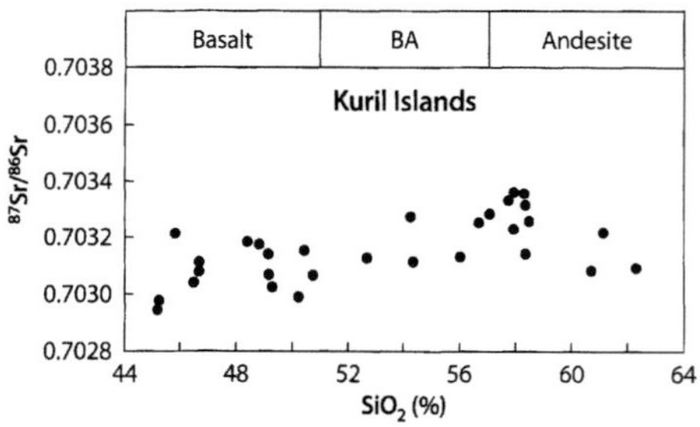

Fig. 3.18. $^{87}Sr/^{86}Sr$ ratios and SiO_2 concentrations of lavas (<0.01 Ma) on Paramushir and Atlasova Islands of the northern Kuril Islands. The absence of a systematic variation between these parameters indicates that the basaltic andesites and andesites may have formed by fractional crystallization of basaltic magma without significant assimilation of rocks from the crust underlying these islands (*Source:* Bailey et al. 1987)

that the high water content is evidence that the rocks were hydrothermally altered.

3.5   The Japanese Islands

The principal Japanese Islands (Hokkaido, Honshu, Shikoku, and Kyushu) form a mature island arc that has been active since Cretaceous time. On the main island of Honshu in Fig. 3.19, volcanic activity and igneous intrusions occurred episodically from the Cretaceous Period to the present (Seki 1978, 1981) and encompassed a wide range of rock types including rhyolites and granites. The volcanic activity on the Japanese Islands is caused by magma formation associated with subduction of oceanic lithosphere under continental crust (Tatsumi 1989). Therefore, these islands provide an opportunity to consider the history of Mesozoic igneous activity based on Rb-Sr dating of the rhyolitic volcanic rocks and associated granites and the origin of the large volume of granitic and associated gabbroic rocks based on their initial $^{87}Sr/^{86}Sr$ ratios and Sr concentrations.

3.5.1   Granite Plutons on the Japanese Islands

The occurrence of rhyolite lava flows and granitic plutons on the Japanese Islands raises the fundamental question of the origin of granitic rocks. Therefore, isotopic and geochemical data were used as soon as they became available to study the origin of granite in Japan and elsewhere (Gorai 1950, 1960; Gorai et al. 1972). These investigations extended the classical work of Eskola (1932), Read (1949), and Perrin (1954), among many others, on the origin of granite and the formation of the continental crust. This problem has also been studied intensively by means of the physical chemistry of the

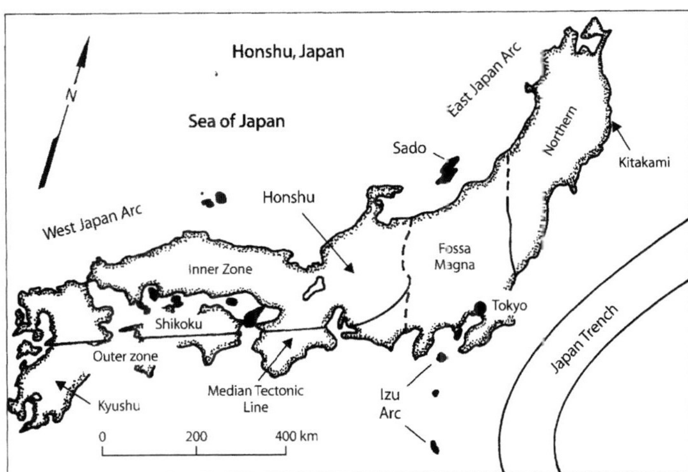

Fig. 3.19.
The principal islands of Japan including Honshu, Kyushu, and Shikoku. (Hokkaido, north of Honshu, is not shown). Honshu, the largest island, is divided into three regions referred to in the text as the northern, central, and southwestern regions. The island is composed of a basement complex of igneous, sedimentary, and metamorphic rocks of Paleozoic age intruded by granitic plutons some of which are related to and overlain by volcanic rocks of Late Cretaceous to Recent age ranging in composition from basalt to rhyolite (*Source:* adapted from Nohda and Wasserburg 1981)

relevant silicate systems (e.g. Bowen 1928; Yoder 1973; Wyllie 1984) and was the subject of the Third Hutton Symposium edited by Brown and Piccoli (1995).

The petrogenesis of granitic igneous rocks can be investigated by means of their initial ^{87}Sr/^{86}Sr ratios obtained from whole-rock isochrons (Faure and Hurley 1963). Compilations of such data by Faure and Powell (1972) and by Gorai et al. (1972) indicated that the initial ^{87}Sr/^{86}Sr ratios of granitic intrusives world-wide range widely, but tend to decrease with age, and partly overlap with the initial ^{87}Sr/^{86}Sr ratios of mafic volcanic and plutonic rocks. Gorai et al. (1972) deduced from these results that granitic magmas had either formed by partial melting of continental tholeiite basalts or that both granites and tholeiites originated from enriched source rocks in the lower crust or upper mantle. Faure and Powell (1972) reported that the initial ^{87}Sr/^{86}Sr ratios of about 50% of the 131 granitic intrusives included in their survey are similar to those of basaltic rocks of comparable ages. Therefore, they concluded that these rocks could have originated by partial melting in the upper mantle or by differentiation of basaltic magmas that had not assimilated detectable amounts of crustal rocks. However, the data do not exclude the possibility that granitic magmas having low initial ^{87}Sr/^{86}Sr ratios formed by partial melting of relatively young basaltic or andesitic volcanic rocks as proposed by Gorai et al. (1972) for the granitic rocks of Japan and by Moorbath and Walker (1965) for the rhyolites of Iceland (Sect. 2.3.2).

A significant factor in the petrogenesis of Cretaceous and Neogene granites on Honshu is the presence of continental crust represented by cobbles of granitic rocks in the Kamiaso conglomerate (Permian) and by exposures of Precambrian granitic gneisses on the Hida Peninsula on the north coast of the island (Hayase and Nohda 1969). Shibata and Adachi (1974) reported whole-rock Rb-Sr dates of 2055 ± 25 Ma and 1884 ± 40 Ma ($\lambda = 1.42 \times 10^{-11}$ yr^{-1}) for cobbles of granitic rocks in the Kamiaso conglomerate. The measured ^{87}Sr/^{86}Sr ratios of these cobbles range widely from 0.7163 to 1.0166 relative to 0.7080 for E&A. The Hida gneisses are likewise Precambrian in age (Shibata and Nozawa 1986) and have initial ^{87}Sr/^{86}Sr ratios between 0.70589 and 0.72007 relative to 0.71025 for NBS 987 (Ishizaka and Yamaguchi 1969; Shibata et al. 1970, 1988; Arakawa 1984).

The relation of the Cretaceous granites of Honshu and neighboring Kyushu to the Precambrian basement rocks was indicated by Shibata and Ishihara (1979) who reported that the initial ^{87}Sr/^{86}Sr ratios of these granites correlate with the thickness of the continental crust and rise to values in excess of 0.710 relative to 0.7080 for E&A. Similarly, Yanagi (1975) concluded from a consideration of Rb/Sr ratios that Cretaceous granites on Honshu and Kyushu originated from magma sources in the crust. Whole-rock Rb-Sr dates of granites on the Japanese Islands were reported by Nohda (1973), Shibata (1974), Ishizaka and Yanagi et al. (1977), Kagami et al. (1988), Arakawa and Shimura (1995), and others listed in the captions to Figs. 3.20 and 3.21.

The initial ^{87}Sr/^{86}Sr ratios of the Cretaceous granites of Honshu and of associated gabbro plutons scatter widely in Fig. 3.20. An important aspect of these data reported by Shibata and Ishihara (1979) and by Kagami et al. (1985) is that the initial ^{87}Sr/^{86}Sr ratios of the gabbros range just as widely as the initial ^{87}Sr/^{86}Sr ratios of the Cretaceous granites. Therefore, Kagami et al. (1985) concluded that both rock types originated from the continental crust. The heterogeneity of the initial Sr isotope ratios of the Cretaceous granites and gabbros is not consistent with the expected consequences of crustal as-

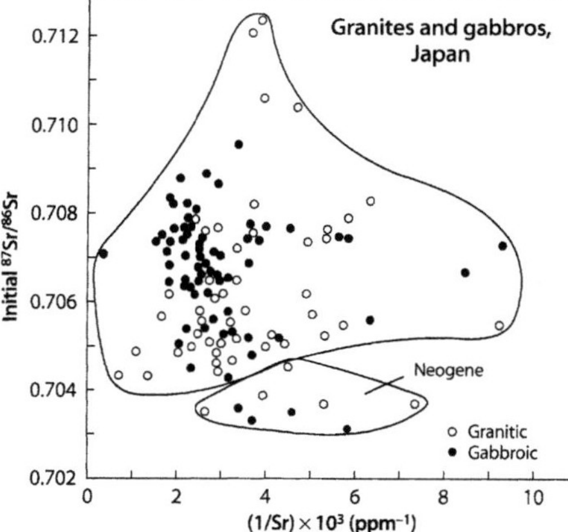

Fig. 3.20. Sr-isotope mixing diagram for granitic and gabbroic intrusives on the Japanese Islands. The granitic rocks include granodiorites, tonalites, diorites, monzonites, and quartz diorites, whereas the gabbros encompass quartz gabbros, metadiabase, hornblende gabbro, and norite. The granitic and gabbroic rocks analyzed by Shibata and Ishihara (1979) range in age from Cretaceous to Neogene (120 to 12 Ma), whereas the age of the gabbroic rocks studied by Kagami et al. (1985) was assumed to be 400 Ma, equivalent to the oldest rocks in southwestern Honshu according to Nozawa (1983). The scatter of data points representing both granitic and gabbroic rocks does not support the crustal-assimilation hypothesis. Instead, these rocks formed primarily by remelting of rocks at the base of the crust (*Sources:* Shibata and Ishihara 1979; Kagami et al. 1985)

similation by mantle-derived basalt magma which should have generated linear mixing patterns in coordinates of $^{87}Sr/^{86}Sr$ and 1/Sr.

The initial $^{87}Sr/^{86}Sr$ of Neogene granites (40 to 10 Ma) and related gabbro plutons of Honshu vary regionally and reflect the tectonic structure of the island. In southwestern Honshu, the initial $^{87}Sr/^{86}Sr$ ratios of the Neogene granites decrease from south to north across the Median Tectonic Line (Fig. 3.19). According to Shibata and Ishihara (1979), the initial $^{87}Sr/^{86}Sr$ ratios of Neogene granites south of the Median Tectonic Line range from 0.7062 to 0.7077, whereas in the Green Tuff Belt north of the Line the Neogene granites have, in many cases, lower initial $^{87}Sr/^{86}Sr$ ratios from 0.7037 to 0.7070. Accordingly, the Neogene granites in Fig. 3.20 define a small data field that is distinct from that of the Cretaceous granites.

3.5.2 Cretaceous and Miocene Rhyolites of Honshu

Rhyolitic welded tuffs and lava flows of Cretaceous age occur in several districts of southwestern Honshu, including Yamaguchi, Himeji, Koto, Nohi, and Okayama (Seki 1978, 1981). Whole-rock samples of the volcanic rocks and of the associated granites in each district define isochrons (broadly speaking) that yield dates and initial $^{87}Sr/^{86}Sr$ ratios for these rocks by the methods

Fig. 3.21.
Isotope ratios of Sr and Nd of welded tuffs of Cretaceous age in the Inner Zone of southwestern Honshu and of Miocene granitic rocks and gabbros of the Outer Zone of Kyushu, Shikoku, and Honshu (Fig. 3.19). Both groups of rocks could have formed by mixing of magmas derived from the mantle wedge and the continental crust or by assimilation of crustal rocks by mantle-derived magmas. However, the apparent absence of rocks having low initial $^{87}Sr/^{86}Sr$ ratios can also be taken as evidence that the rhyolitic magmas are predominantly of crustal origin and that basalt magmas derived from the mantle wedge under Honshu were not directly involved in their formation (*Sources:* Terakado and Nakamura 1984; Terakado et al. 1988)

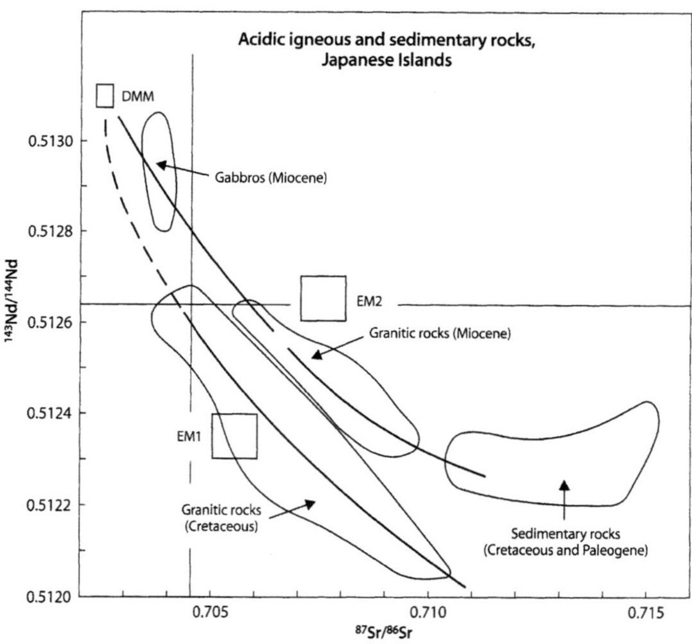

described in Sec. 1.3 and 1.4. The initial ^{87}Sr/^{86}Sr ratios of the Cretaceous and Tertiary felsic volcanics of southwestern and northern Honshu reported by Seki (1978), Terakado and Nakamura (1984), and Terakado et al. (1988) range widely from 0.70348 to 0.71045 relative to 0.7080 for E&A and 0.71025 for NBS 987.

The felsic volcanics and associated granite and gabbro plutons of Honshu analyzed by Terakado and Nakamura (1984) and by Terakado et al. (1988) define discrete data fields in Fig. 3.21 depending on their lithology and age. The elevated initial ^{87}Sr/^{86}Sr and low ^{143}Nd/^{144}Nd ratios of the rhyolites and granites demonstrate a strong crustal imprint for both the Cretaceous and Neogene (Miocene) suites. The distribution and shape of the data fields in Fig. 3.21 can be explained by progressive assimilation of crustal rocks by magmas arising from within the mantle wedge beneath Honshu or by partial melting of heterogeneous mixtures of crustal rocks having a range of chemical compositions and ages. The large gap between the isotope ratios of Sr and Nd in the mantle wedge (DMM component) and the rhyolite tuffs favors the crustal origin of these rocks, although magmas from the mantle wedge probably provided the heat that caused rhyolite magmas to form in the crust (Terakado and Nakamura 1984; Seki 1978).

The intensive study of the isotope composition of felsic volcanic rocks on Honshu has revealed that tuffs located east of 136° E longitude have higher initial ^{87}Sr/^{86}Sr on average than those west of this line. Seki (1978) interpreted this step-like change as evidence for the existence of a lithologic or tectonic boundary in the crustal rocks of Honshu. Such regional variations of initial ^{87}Sr/^{86}Sr ratios of continental volcanic rocks are an important phenomenon that has also been observed on other continents, (e.g. in western North America; Kistler and Peterman 1973, 1978)

3.5.3 Cenozoic Basalts and Andesites of Honshu

The Late Tertiary and Quaternary volcanic rocks of northern Honshu (Shuto 1974) range in age from early Miocene to Pleistocene (20 to 1.0 Ma) and include not only basalt and andesite, but also dacite and rhyolite which were deposited on a basement complex of granitic and metamorphic rocks and sedimentary formations of Paleozoic and Mesozoic age (Kaneoka et al. 1978). The ^{87}Sr/^{86}Sr ratios of these rocks in Fig. 3.22a and b range from 0.70382 to 0.70561 relative to 0.7080 for E&A (Togashi et al. 1985; Hedge and Knight 1969). The central and northern regions of Honshu contain a large number of volcanoes of Quaternary age whose presence is related to the triple junction in central Honshu where the Philippine Plate, the Pacific Plate, and the Eurasian Plate join. The volcanic rocks extruded by these volcanoes range in composition from basalt to andesite and

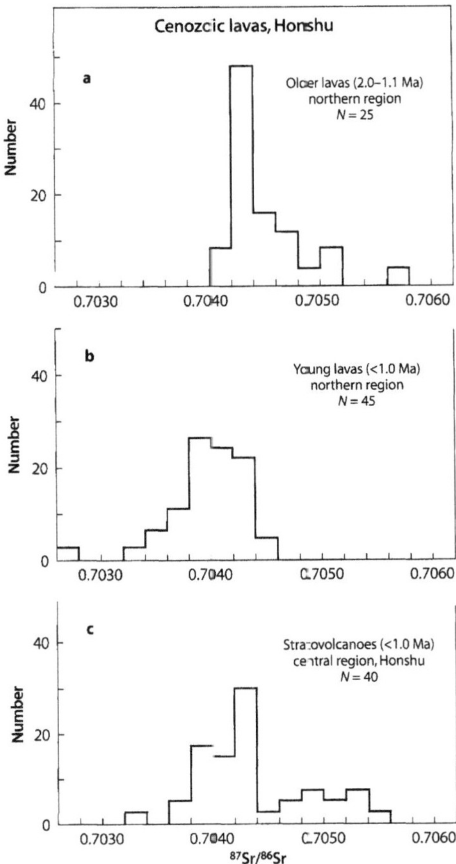

Fig. 3.22 a,b,c. Range of variation of ^{87}Sr/^{86}Sr ratios of Pleistocene to Recent volcanic rocks (basalt to rhyolite) in the northern and central regions of Honshu, Japan. The wide range and elevated values of the ^{87}Sr/^{86}Sr ratios are partly attributable to assimilation of crustal rocks by magmas ascending from the mantle wedge above the subduction zone associated with the Japan Trench east of Honshu (*Sources:* Shuto 1974; Ishizaka et al. 1977; Nohda and Wasserburg 1981; Togashi et al. 1985)

dacite. Basalts are restricted to the volcanic front, whereas andesites and dacites occur throughout the volcanic arc.

The ^{87}Sr/^{86}Sr and ^{143}Nd/^{144}Nd ratios of basalts and andesites in Fig. 3.23 extruded by the stratovolcanoes in the Fossa Magna identified in Fig. 3.19 (Hakone, Asama, Eboshi, Myoko) and in northern Honshu (Kampu, Hachimantai, Moriyoshi) are attributable to magma formation in the mantle wedge with varying additions of Sr and Nd from rocks of the continental crust (Nohda and Wasserburg 1981). Two samples of granitic basement rocks of Japan have high ^{87}Sr/^{86}Sr and low ^{143}Nd/^{144}Nd ratios (Table 3.2)

Fig. 3.23.
Isotope ratios of Sr and Nd of Quaternary andesites on Honshu, Japan. The distribution of data points demonstrates that the magmas could have formed by partial melting of depleted rocks in the mantle wedge (represented by the DMM component) to which Sr and Nd derived from subducted sediment had been added. In addition, the resulting magmas may have assimilated rocks from the continental crust of Honshu. The $^{143}Nd/^{144}Nd$ ratios were recalculated to be compatible with 0.512638 for the $^{143}Nd/^{144}Nd$ ratio of the chondritic reservoir (Wasserburg et al. 1981) (*Source:* Nohda and Wasserburg 1981)

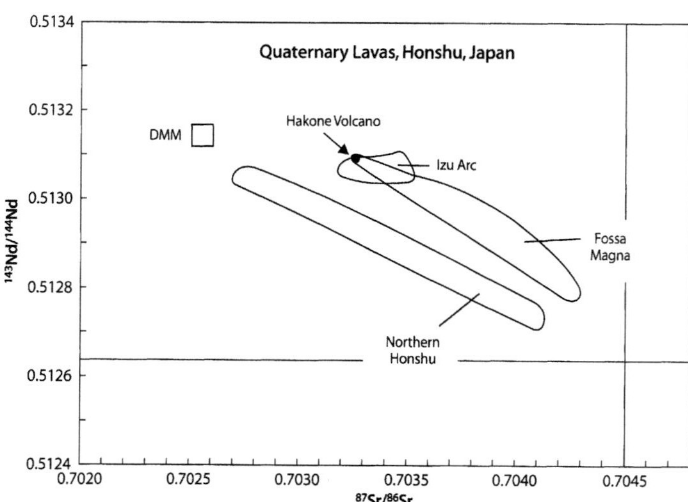

Table 3.2. Granitic basement rocks of Japan with high $^{87}Sr/^{86}Sr$ and low $^{143}Nd/^{144}Nd$ ratios

Sari granodinorite, Fossa Magna, Honshu	0.71079	0.51241[a]
Yatsushiro gneiss, Kyushu	0.71155	0.51222

[a] Recalculated to $^{143}Nd/^{144}Nd = 0.512638$ for CHUR (Wasserburg et al. 1981).

These results support the conclusion of Nohda and Wasserburg (1981) that the Quaternary lavas of the Fossa Magna region of Honshu were contaminated with Sr and Nd derived from the granitic basement rocks. The $^{87}Sr/^{86}Sr$ ratios in Quaternary lavas of northern Honshu reported by Nohda and Wasserburg (1981) decline from 0.70407 (Hachimantai) to 0.70355 (Moriyoshi) and 0.70270 (Kampu) on the west coast of northern Honshu. Low $^{87}Sr/^{86}Sr$ ratios (0.7030) have also been reported for volcanic rocks on the island of Oshima in the Sea of Japan off the west coast of Hokkaido (Hedge and Knight 1969; Notsu 1983).

The Rb-Sr systematics of Quaternary andesites and basalts of the volcano Iizuna in the northern part of the Fossa Magna are remarkable because of what they reveal about the petrogenesis of these lavas. The $^{87}Sr/^{86}Sr$ and Rb/Sr ratios of andesites and basalts on this volcano form two parallel linear arrays in Fig. 3.24. Most of the andesites analyzed by Ishizaka et al. (1977) have higher $^{87}Sr/^{86}Sr$ ratios (relative to 0.7080 for E&A) than the basalts. Therefore, the andesites originated from different magma sources or formed from basalt magmas that assimilated crustal rocks but did not form by fractional crystallization of basalt magmas.

The negative correlation of $^{87}Sr/^{86}Sr$ and Rb/Sr ratios, as well as the systematic difference of $^{87}Sr/^{86}Sr$ ra-

tios of basalts and andesites on Iizuna Volcano, violate one of the cardinal assumptions of dating by the Rb-Sr method; namely, that samples taken from a comagmatic suite of igneous rocks all have the same initial $^{87}Sr/^{86}Sr$ ratio (Sect. 1.3). In cases like Iizuna Volcano, the date determined from the slope of an isochron based on andesites alone underestimates the age of these rocks and may even be negative, i.e. refer to the future. In addition, when andesites and basalts are combined in the same isochron diagram, the data points scatter and generate an errorchron (or scatterchron) rather than an isochron (Sect. 1.4). The date derived from such an errorchron deviates from the age of the rocks by amounts of time that depend on the particular choice of rock samples. Therefore, the volcanic (and plutonic) rocks of Japan demonstrate the possible failure of one of the key assumptions of the Rb-Sr method of dating.

An additional point of interest is that the concentrations of Rb and Sr and the $^{87}Sr/^{86}Sr$ ratios of the basalts and andesites on Iizuna Volcano vary systematically up-section. The stratigraphic variation of the $^{87}Sr/^{86}Sr$ ratios in Fig. 3.25 means that the magmas assimilated crustal rocks having elevated $^{87}Sr/^{86}Sr$ ratios and that the abundance of this component decreased with depth below the top of the magma chamber. The magmas probably also differentiated by fractional crystallization which occurred primarily in the upper (cooler) part of the magma chamber. During stage II in Fig. 3.25, the $^{87}Sr/^{86}Sr$ ratios and Sr concentrations of sequentially erupted flows decreased up-section, whereas their Rb concentrations increased. The pattern of variation in the lavas erupted during stage I appears to differ from that in stage II, but the available data are insufficient to be certain.

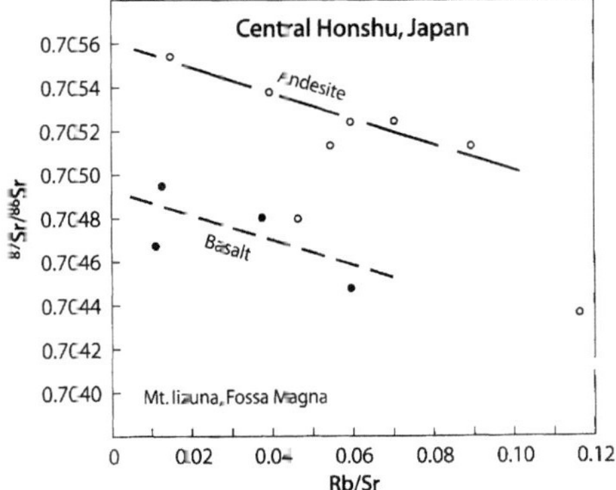

Fig. 3.24. Negative correlation of $^{87}Sr/^{86}Sr$ and Rb/Sr ratios of basalt and andesite on Iizuna Volcano, central Honshu, Japan. The basalts have consistently lower $^{87}Sr/^{86}Sr$ ratios than most of the andesites, indicating that the two suites originated from different magmas having different $^{87}Sr/^{86}Sr$ and Rb/Sr ratios (*Source:* Ishizaka et al. 1977)

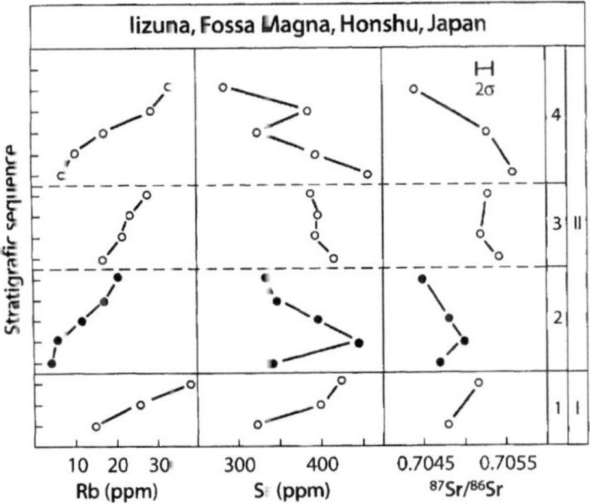

Fig. 3.25. Stratigraphic variation of the concentrations of Rb and Sr and $^{87}Sr/^{86}Sr$ ratios of Pleistocene to Holocene basalts (*solid circles*) and andesites (*open circles*) on Iizuna Volcano on the north coast of central Honshu, Japan. The cyclical changes in these parameters define a series of "steps" (1, 2, 3 and 4) in the two "stages" (I and II) of volcanic activity. The time-dependent variation of the $^{87}Sr/^{86}Sr$ ratios indicates that the magmas were evolving during each step by assimilating rocks of the continental crust (*Sources:* Ishizaka et al. 1977; Yanagi et al. 1978)

The evidence in Fig. 3.25 presented by Yanagi and Ishizaka (1978) and Ishizaka et al. (1977) for the lavas of Iizuna is an example of volcanic activity arising from a magma chamber in which magma evolved by fractional crystallization and wallrock assimilation and which was periodically refilled after being emptied by eruption (O'Hara 1977). The andesite and basalt lavas erupted by Iizuna presumably originated from the same crustal

magma chamber suggesting that the andesite magmas assimilated more crustal rocks than the basalts or that the andesites and basalts originated from different kinds of source rocks in the underlying mantle wedge.

The southern part of the Fossa Magna in Fig. 3.19 is an area of great complexity because it is the site where the Izu Island arc enters Honshu Island. The lavas on the islands of the Izu Arc (Oshima, Miyake-Jima, Hachijo Jima) consist of tholeiite basalt whose $^{87}Sr/^{86}Sr$ and $^{143}Nd/^{144}Nd$ ratios in Fig. 3.23 are more similar to those of MORBs than are those of volcanic rocks on the island of Honshu. Similar results were reported by Stern (1982) and Pushkar (1968). The difference in the $^{87}Sr/^{86}Sr$ ratios of these lavas compared to those on Honshu is related to the fact that along the Izu Arc oceanic crust is being subducted under oceanic crust instead of under continental crust as in the case of the principal Japanese Islands. The low $^{87}Sr/^{86}Sr$ ratios of the lavas on the islands of the Izu Arc contrast markedly with the high average $^{87}Sr/^{86}Sr$ ratio (0.70767) of andesites on Mt. Akagi located along the southern coast of central Honshu (Notsu 1983).

3.5.4 The Volcanic Arc of Southwestern Honshu and the Ryukyu Islands

The subduction of the Philippine Sea Plate into the Nankai-Ryukyu Trench in Fig. 3.26 causes the volcanic activity on southwestern Honshu, Kyushyu, and the Ryukyu Islands which extend south to Okinawa and Taiwan (Notsu et al. 1990). The Philippine Sea Plate being subducted into the Nankai Trench reaches a depth of only about 80 km and does not actually extend under the volcanic front located along the northern region of southwestern Honshu. However, farther south in the Ryukyu Island arc, the Philippine Sea Plate does extend under the islands to a depth of about 150 km (Notsu et al. 1987a; Nakada and Kamata 1991).

The $^{87}Sr/^{86}Sr$ ratios of Quaternary lavas extruded by volcanoes in southwestern Honshu and Kyushu range widely from 0.70362 (Abu Volcano) to 0.70657 (Gembudo Volcano). The variation of this ratio occurs both on a regional as well as on a local scale (Notsu et al. 1990). For example, the $^{87}Sr/^{86}Sr$ ratios of lavas on Abu Volcano on the north coast of southwestern Honshu range from 0.7036 to 0.70531 (Kurasawa 1984a). These results indicate that the different rock types extruded by this volcano did not form by fractional crystallization of isotopically homogeneous basalt magmas. The $^{87}Sr/^{86}Sr$ ratios of volcanic rocks of Kimpo Volcano on Kyushu reported by Notsu et al. (1990) also vary widely from 0.70377 to 0.70537. The wide range of isotope compositions of Sr of the lavas on Abu Volcano (Honshu) as well as on Kimpo and other volcanoes on Kyushu is attributable to assimilation of crustal rocks by mag-

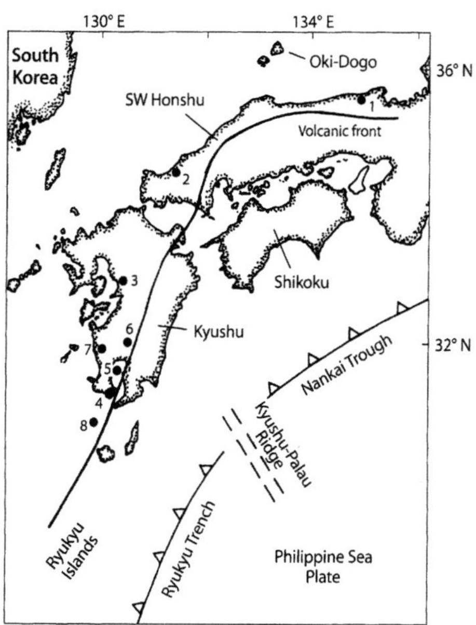

Fig. 3.26. Volcanic activity in southwestern Honshu, Kyushu, and the Ryukyu Islands. The numbered sites identify volcanoes mentioned in the text: 1. Gembudo; 2. Abu; 3. Kimpo; 4. Ata; 5. Aira; 6. Kirishima; 7. Sendai; 8. Kikai (adapted from Notsu et al. 1990)

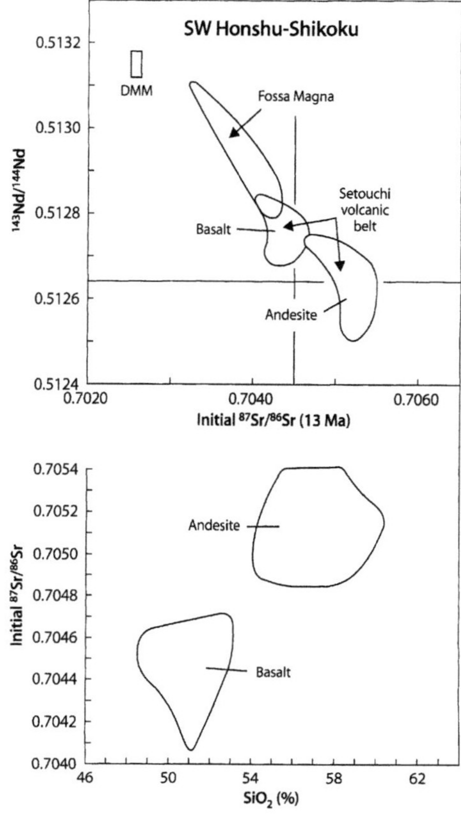

Fig. 3.27. a Isotope ratios of Sr and Nd of basalts and andesites in the Setouchi volcanic belt between southwestern Honshu and the island of Shikoku, Japan. The isotope ratios of these rocks extend the trend of Quaternary andesites in the Fossa Magna area of central Honshu to higher $^{87}Sr/^{86}Sr$ and lower $^{143}Nd/^{144}Nd$ ratios; b Initial $^{87}Sr/^{86}Sr$ ratios and silica concentrations of the basalts and andesites in the Setouchi volcanic belt of Japan. The difference in $^{87}Sr/^{86}Sr$ ratios proves that the andesites are not differentiates of basalt magma, but either originated from a different source or formed from magmas that assimilated larger amounts of crustal rocks than the basalt magmas (*Sources:* Ishizaka and Carlson 1983; Nohda and Wasserburg 1981; Hart 1988)

mas derived from the mantle wedge above the subduction zone (Yanagi et al. 1988; Kurasawa et al. 1986; Notsu et al. 1990).

The isotope ratios of Sr and Nd of basalts and andesites on Shodoshima and Takamijima Islands in Fig. 3.27a in the channel between southwestern Honshu and Shikoku (Setouchi volcanic belt) extend into the crustal quadrant of the Sr-Nd isotope mixing diagram (Ishizaka and Carlson 1983). In addition, the isotope ratios of these lavas continue the field of the Quaternary volcanic rocks of the Fossa Magna region of central Honshu (Nohda and Wasserburg 1981). The andesites in the Setouchi volcanic belt are not differentiates of basalt magma because they have consistently higher initial $^{87}Sr/^{86}Sr$ ratios than the basalts (Fig. 3.27b). Ishizaka and Carlson (1983) considered that andesitic magmas formed by partial melting of hydrous rocks at the top of the mantle wedge above the subduction zone. However, the andesites could also have originated from basalt magmas that rose from the mantle wedge and subsequently assimilated rocks of the continental crust of Japan. Calculations by Ishizaka and Carlson (1983) indicated that the isotope composition of Sr of the andesites in the Setouchi Belt can be accounted for by the addition of 10 to 20% by weight of the Sari granite (Sr = 186 ppm, 0.71079) in the Fossa Magna of Honshu and the Yatsushiro granite

(Sr = 347 ppm, 0.71155) on Kyushu analyzed by Nohda and Wasserburg (1981).

The involvement of granitic basement rocks in the petrogenesis of Cenozoic volcanic rocks on the Japanese Islands is confirmed by the positive correlation of $^{87}Sr/^{86}Sr$ ratios and $\delta^{18}O$ values reported by Matsuhisa and Kurasawa (1983). Their data in Fig. 3.28 for suites of basalt, andesite, and dacites on volcanoes of Honshu and Kyushu demonstrate that these lavas have been simultaneously enriched in ^{18}O and ^{87}Sr both of which presumably originated from granitic rocks of the continen-

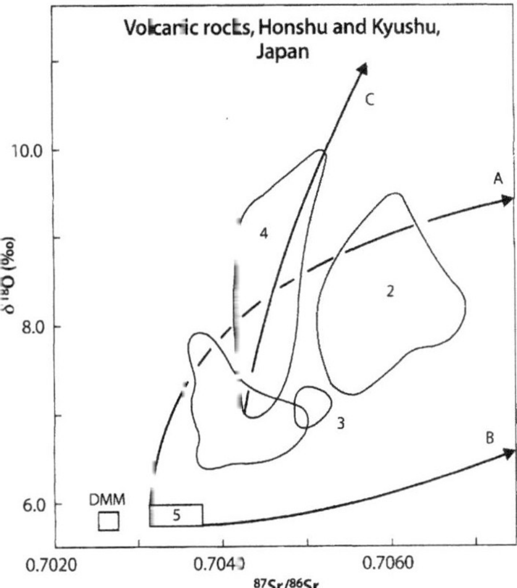

Fig. 3.28. Isotope compositions of Sr and O in Cenozoic volcanic rocks of Honshu and Kyushu Japan. The volcanoes included in the diagram are: *1.* Yatsugataké, central Honshu; *2.* Kiso-Ontaké, central Honshu; *3.* Daisen, southwestern Honshu; *4.* Unzen, Kyushu; *5.* Primitive island-arc basalts. The mixing lines labeled *A* and *B* represent mixtures of primitive mantle-derived magmas with melts of the continental crust ($^{87}Sr/^{86}Sr = 0.712$, $\delta^{18}O = +10.0‰$) having Sr concentrations of 200 ppm (*A*) and 500 ppm (*B*). A third mixing trajectory (*C*) represents mixtures of andesite magma ($\delta^{18}O = +7.0‰$; $^{87}Sr/^{86}Sr = 0.70425$; Sr = 600 ppm) with metasedimentary rocks ($\delta^{18}O = +15‰$ $^{87}Sr/^{86}Sr = 0.722$; Sr = 50 ppm). The data for Aonoyama and Aso Volcanoes are not shown (*Source:* Matsuhisa and Kurasawa 1983)

tal crust. The scatter of the data arises because the isotope compositions of Sr and O of mixtures depend on many factors, including:

1. The isotope compositions of Sr and O in the end-members.
2. The range of isotope compositions of the crustal component (assuming that the mantle-derived magmas are significantly more homogeneous in their isotope compositions).
3. The proportions of mixing.
4. The Sr concentrations of the primitive basalt magma and of the crustal component.
5. The range of Sr concentrations of the crustal component (Sect. 1.7.1 and 1.7.2).

The trend of the data for the volcanic rocks on Honshu and Ryukyu is similar to that of the Miocene basalt on Gran Canaria (Fig. 2.16; Thirlwall et al. 1997), but opposite to that of basalt on Iceland (Fig. 2.13; Condomines et al. 1983; Nicholson et al. 1991; Sigmarsson et al. 1992).

In spite of the scatter of the data, the mixing trajectories in Fig. 3.28 fit the data fields well enough to strengthen the conclusion that magmas derived from the mantle wedge in the Japanese Islands incorporated crustal rocks either by assimilation, or by magma mixing, or both. Therefore, the volcanic rocks of the Japanese Islands differ in origin from the lavas of the Mariana and Aleutian Islands where magma contamination by crustal rocks either did not take place or was limited in scope, respectively.

3.5.5 Sea of Japan

The Sea of Japan in Fig. 3.19 is a complex back-arc basin that includes the Japan Basin in the north and the Yamato Basin adjacent to northern Honshu. These two basins are separated by the Yamato Bank and related seamounts. Age determinations by the $^{40}Ar/^{39}Ar$ method of Kaneoka et al. (1990) on volcanic rocks from the chain of Yamato Seamounts and a review of previously published dates indicates that the Yamato Basin formed between 17 and 25 Ma. The Japan Basin in the northern area of the Sea of Japan may be slightly older than the Yamato Basin. The $^{87}Sr/^{86}Sr$ ratios of volcanic rocks dredged from the Yamato Bank and the seamounts range widely from 0.7033 to 0.7063 relative to 0.7080 for E&A and corrected for decay (Ueno et al. 1974). Other measurements by Kaneoka (1990) and Kaneoka et al. (1990) yielded values between 0.70352 and 0.70363 relative to 0.71025 for NBS 987, but without correction for in situ decay of ^{87}Rb.

The volcanic rocks dredged from the Sea of Japan have higher $^{87}Sr/^{86}Sr$ ratios than volcanic rocks from other back-arc basins such as the Mariana Trough and the Sumizu Rift described in Sect. 3.2.3. In addition, specimens of plutonic igneous and metamorphic rocks having K-Ar dates greater than 100 Ma have been recovered from this area (Kaneoka et al. 1990). Therefore, the Yamato Bank is probably underlain by continental crust separated from Honshu Island as a result of the opening of the Yamato Basin.

The opening of the Sea of Japan was recorded in the isotope compositions of Sr and Nd of volcanic rocks that formed along the west coast of northern Honshu between about 30 Ma and the present. Nohda and Wasserburg (1986) and Nohda et al. (1988) reported that the initial $^{87}Sr/^{86}Sr$ ratios of Tertiary andesites and basalts from an area near Akita City on the west coast of northern Honshu decreased from about 0.70547 at 27.3 Ma to 0.70336 at 6 Ma. Their data in Fig. 3.29 indicate that, before opening of the Sea of Japan at about 15 Ma, the volcanic rocks had high $^{87}Sr/^{86}Sr$ ratios (~0.7055) implying significant crustal involvement in their petrogenesis. After the opening of the Sea of Japan by back-arc spreading, the $^{87}Sr/^{86}Sr$ ratios of the volcanic rocks decreased to about 0.7033, presumably because of thin-

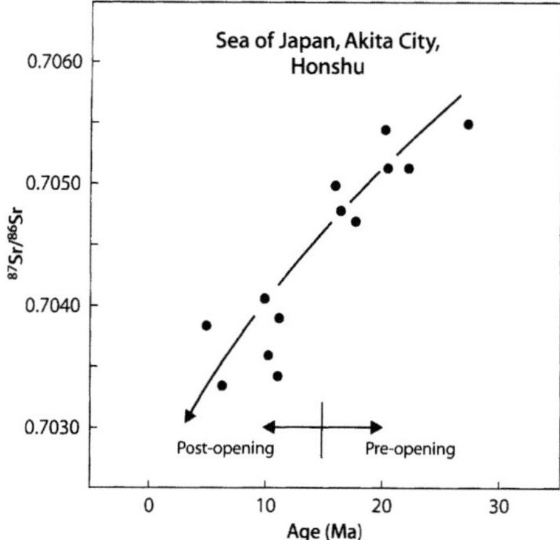

Fig. 3.29. Decrease of the initial $^{87}Sr/^{86}Sr$ ratios of andesites near Akita City on the west coast of northern Honshu between 30 Ma and the present. The time-dependent decrease of the isotope composition of Sr and Nd in these volcanic rocks was caused by crustal thinning associated with the opening of the Sea of Japan by back-arc spreading at about 15 Ma (*Sources:* Nohda and Wasserburg 1986; Nohda et al. 1988)

ning of the continental crust allowing magmas derived from depleted mantle to become dominant over the crustal component.

The volcanic rocks on the islands in the Sea of Japan include alkali-rich varieties on the Oki-Dogo (Fig. 3.26) and Jejudo Islands (Kurasawa 1967, 1984b; Nakamura et al. 1989) as well as calc-alkaline lavas on Sado and Oshima-Oshima Islands (Shuto 1974; Hedge and Knight 1969). Sado Island, located off the west coast of northern Honshu (Fig. 3.19), contains large amounts of rhyolite of early and late Miocene age. The initial $^{87}Sr/^{86}Sr$ of these rocks are unusually high and range from 0.7068 (rhyolite) to 0.7096 (dacite) relative to 0.7080 for E&A. Shuto (1974) attributed the origin of these lavas to magma formation by partial melting of eclogite in the uppermost mantle under Sado Island. The elevated $^{87}Sr/^{86}Sr$ ratios of the volcanic rocks of Sado Island further emphasize the importance of continental crust in the petrogenesis of Cenozoic volcanic rocks in northeastern Honshu (Nohda and Wasserburg 1981).

3.5.6 Petrogenesis in the Japanese Island-Arc System

The petrogenesis of the late Mesozoic-to-Recent volcanic rocks of the Japanese island-arc system is attributable to magmatic processes in subduction zones where oceanic lithosphere is being subducted under continental crust. These processes include:

1. Transfer of mobile elements during dehydration of subducted sediment and hydrothermally altered basalt of the subducted oceanic crust to the rocks of the overlying mantle wedge.
2. Alteration of the rocks in the mantle wedge by metasomatism resulting in partial melting either because of the depression of the melting temperature, or as a result of decompression of rising diapirs of metasomatized mantle rocks, or because of convection of altered mantle rocks downward into regions of higher temperature and pressure.
3. The resulting magmas are enriched in radiogenic ^{87}Sr (and depleted in ^{143}Nd) compared to normal MORBs and are subsequently changed chemically and isotopically by interactions with the unaltered rocks of the mantle wedge through which they rise toward the surface.
4. Magmas arising from the mantle wedge may pool at the base of the continental crust or at higher levels within it and evolve by fractional crystallization, by assimilation of a wide range of crustal rocks, and/or by mixing with crustal melts. Therefore, the lava flows formed by the sequential eruption of batches of magma have a range of $^{87}Sr/^{86}Sr$ ratios and chemical compositions depending on the chemical evolution of the pooled magma and the extent of mixing with new batches of magma arriving from depth.
5. The movement of magma from the mantle may transport sufficient heat into the continental crust to cause partial melting of crustal rocks. The $^{87}Sr/^{86}Sr$ ratios of the resulting granitic magmas depend on the ages and Rb/Sr ratios of their source rocks and on the extent of mixing with magmas derived from sources in the mantle (Huppert and Sparks 1988).

Although all island arcs and compressive continental margins contain the necessary conditions for magma formation by a combination of these processes, the $^{87}Sr/^{86}Sr$ ratios of the resulting volcanic rocks do not by themselves uniquely identify the processes that contributed to their petrogenesis. The origin of volcanic rocks in island arcs can be constrained only when the isotope compositions of several elements (e.g. Sr, Nd, Pb, Hf, Os, and O) and the concentrations of trace elements (e.g. LIL, REE, and HFS) are combined with other geological and geophysical information (e.g. flow stratigraphy, composition of sediment prior to subduction, crustal thickness, depth to the Benioff zone, etc.)

3.6 Tonga-Kermadec-New Zealand

The Tonga-Kermadec Trench in Fig. 3.30 was originally continuous with the Vanuatu Trench from which it is now separated by the Hunter Fracture Zone. The Tonga-Kermadec Trench extends south from the Samoan Is-

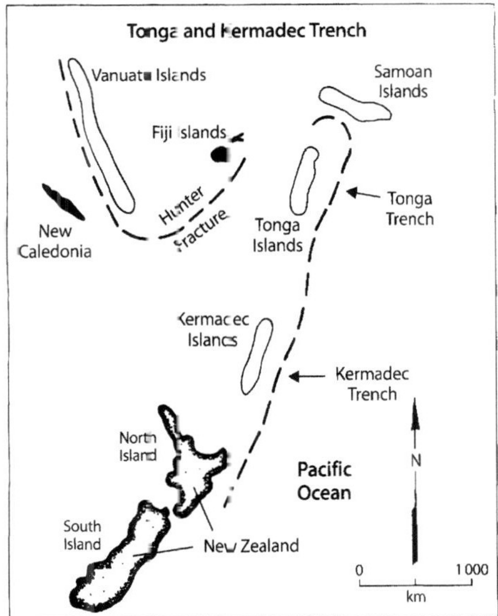

Fig. 3.30. Map of the Tonga-Kermadec-New Zealand island-arc system including also the Samoan (Sect. 2.13), Fiji, Tonga, and the Vanuatu Islands (*Source:* adapted from National Geographic Society 1990)

lands (Sect. 2.13) for a distance of about 3 000 km to the North Island of New Zealand. The volcanic activity of the Tonga and Kermadec Islands is caused by the subduction of the Pacific Plate under oceanic crust like that occurring along the Mariana Trench (Coleman 1973). However, the tectonic setting of the Taupo Volcanic Zone on the North Island of New Zealand involves subduction of the Pacific Plate under the continental crust of New Zealand and thus resembles the tectonic setting of the Japanese Island arc. Therefore, the isotope ratios of Sr, Nd, and Pb and the trace-element concentrations of the volcanic rocks in the Taupo Volcanic Zone are expected to differ from those of the Kermadec-Tonga Islands depending on the extent to which the continental crust of New Zealand has affected the composition of magmas produced in the underlying mantle.

3.6.1 The Fiji Islands

The principal islands of the Fiji Archipelago are Viti Levu, Vanua Levu, and Kandavu, but the total number of islands in this group is very large. The geologic history of Viti Levu has been divided into three periods characterized by island-arc tholeiites of the first period, calc-alkaline volcanics in the second period, and alkali-rich shoshonites in the third period (Rodda 1967; Dickin-

son 1967; Dickinson et al. 1968; Gill 1970, 1976b, 1984, 1987; Gill and Whelan 1989a,b; Rogers and Setterfield 1994). The ages of the rocks on the Fiji Islands range from late Eocene to late Pliocene based both on K-Ar dating (McDougall 1963a; Gill and McDougall 1973) and on paleontological evidence.

The secular variation of the chemical composition of the volcanic rocks in the Fiji Islands indicates that the magma sources changed with time presumably reflecting the tectonic evolution of the Fiji Arc. The concentrations of Rb and Sr both increased from the tholeiites to the andesites and to the alkali-rich shoshonites. Initial results by Gill (1970) seemed to indicate that these rocks all have about the same $^{87}Sr/^{86}Sr$ ratio of 0.7040 relative to 0.7080 for E&A. However, later work by Gill (1984) revealed that the $^{87}Sr/^{86}Sr$ ratios of the volcanic rocks on Viti Levu Island increase from 0.70347 (tholeiites) to 0.70368 (andesites) and 0.70384 (shoshonites), relative to 0.71025 for NBS 987. Similarly, on Kandavu Island the $^{87}Sr/^{86}Sr$ ratios of the andesite suite range from 0.70308 to 0.70357, whereas the shoshonites have an average $^{87}Sr/^{86}Sr$ ratio of 0.70433 (Gill and Compston 1973).

The isotope ratios of Sr and Nd in the lavas on Viti Levu form a cluster in Fig. 3.31 that lies close to the isotope ratios of the depleted MORB mantle (DMM). Accordingly, these lavas originated by partial melting of depleted rocks in the mantle wedge with varying additions of Sr and Nd derived from sediment of the subducted Pacific Plate (Gill 1984; Rogers and Setterfield 1994).

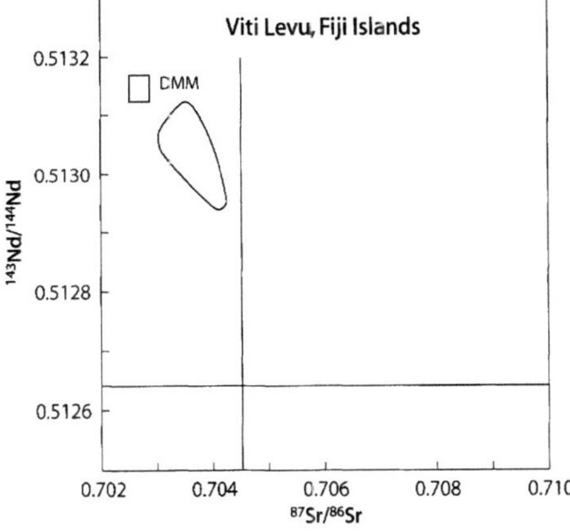

Fig. 3.31. Isotope ratios of Sr and Nd in volcanic rocks on the island of Viti Levu in the Fiji Archipelago. The proximity of the data cluster to DMM is an indication that the lavas on this island originated from the lithospheric mantle in the mantle wedge above a subduction zone (*Sources:* data from Gill 1984; Rogers and Setterfield 1994)

3.6.2 Tonga-Kermadec Island Arc

The Tonga and Kermadec Islands in Fig. 3.30 are located west of the deep-sea trench that bears their name and consist of volcanic islands some of which have subsided to become coral atolls. The Tonga Islands extend from about 15° to 23° S latitude at about 175° W longitude. The principal islands are, from north to south: Tafahi and Niuatoputapu Islands, the Vava'u Group (18°40' S), the Ha'apai Group (20°00' S), and the Tongatapu Group (21°00' S) which includes island of Eua.

The Kermadec Islands occur both north and south of 30° S latitude at about 178° W longitude. The major island groups from north to south are the Raoul Islands, the Macauley Islands, Curtis Island, and L'Esperance Rock. The geology of the Tonga and Kermadec Islands and the petrogenesis of the volcanic rocks was discussed by Ewart and Bryan (1972), Oversby and Ewart (1973), Ewart et al. (1977), Falloon and Crawford (1991) and McDermott and Hawkesworth (1991) based primarily on isotope compositions of Sr, Nd, and Pb as well as on trace element concentrations.

The $^{87}Sr/^{86}Sr$ ratios of the low-K basalts and andesites on the Tonga and Kermadec Islands in Fig. 3.32 are lower than those of the volcanic rocks in the Taupo Volcanic Zone of New Zealand and are more tightly clustered. The data in Fig. 3.32 demonstrate that more than 83% of the samples on the Tonga Islands have $^{87}Sr/^{86}Sr$ ratios between 0.7034 and 0.7040. The volcanic rocks on the Kermadec Islands have slightly lower $^{87}Sr/^{86}Sr$ ratios such that more than 83% lie in the interval 0.7032 to 0.7038. Therefore, the Sr isotope ratios of the volcanic rocks on the Tonga and Kermadec Islands have about the same range and distribution as those on the Mariana Islands (Fig. 3.6) and on the western Aleutian Islands (Fig. 3.13), both of which are typical intra-oceanic island arcs. The volcanic rocks of the Taupo Volcanic Zone (basalts and rhyolites) betray their crustal heritage by having elevated $^{87}Sr/^{86}Sr$ ratios which, in addition, vary more widely than those of volcanic rocks from the oceanic island arcs (Ewart and Stipp 1968; McDermott and Hawkesworth 1991; Ewart and Hawkesworth 1987; Stern 1982).

The volcanic rocks of the Kermadec-Tonga Islands differ from those of the Mariana and Aleutian Islands by having lower concentrations of Rb, Sr, and other LIL elements (Table 3.3).

Evidently, intra-oceanic island arcs are not alike, but differ in significant ways in the chemical and isotopic properties of the rocks that are produced.

Ewart and Hawkesworth (1987) noted that the Tonga Islands form a double chain in which the volcanically active islands occur along the western side of the Tonga Ridge, whereas the Kermadec Islands occur as a single chain of active volcanoes. In addition, the dip of the Benioff zone

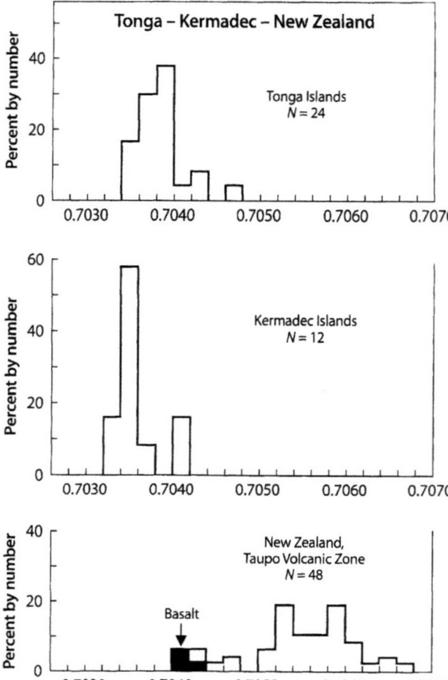

Fig. 3.32. Range of $^{87}Sr/^{86}Sr$ ratios of volcanic rocks in the Tonga Islands, the Kermadec Islands, and the Taupo Volcanic Zone, North Island, New Zealand. The Kermadec and Tonga Islands are typical intra-oceanic island arcs associated with the Tonga-Kermadec Trench which continues south into the Hikurangi Trough located east of the North Island of New Zealand. The $^{87}Sr/^{86}Sr$ ratios of andesites, rhyolites, and ignimbrites in the Taupo Volcanic Zone are not only higher than those of basalts and andesites from Kermadec and Tonga, but also have a wider range. Both features may be caused by extensive assimilation of crustal rocks (greywackes and argillites) by the parental magmas that formed in the mantle wedge under the North Island of New Zealand or by remelting of older volcanic rocks of the oceanic crust (*Sources:* Ewart and Stipp 1968; McDermott and Hawkesworth 1991; Ewart and Hawkesworth 1987; Stern 1982)

Table 3.3. Concentrations of Rb and Sr of the volcanic rocks of the Kermadec-Tonga Islands, the Mariana Islands, and the Aleutian Islands

	Rb (ppm)	Sr (ppm)
Aleutian Islands[a] (basalt and basalt andesite)	19.1 (36) 8.1–51	515 (36) 320–850
Mariana Islands (basaltic rocks)	14.1 (51) 5.7–43.5	318 (52) 135–452
Tonga-Kermadec (basalt and basalt andesite)	4.1 (44) 0.6–11.0	180 (45) 100–235

[a] Omitting Bogoslof and Amak Islands, and the Alaska Peninsula.

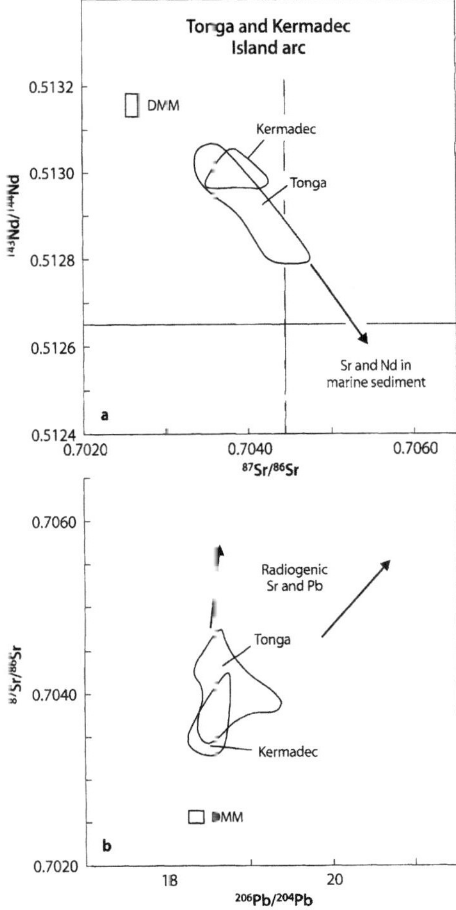

Fig. 3.33. a Isotope ratios of Sr and Nd of lavas on the Tonga and Kermadec Islands in relation to the DMM component (depleted MORB mantle). The distribution of data points is consistent with the theory that the magmas originated from depleted mantle to which Sr and Nd derived from sediment on the down-going slab had been added; b Isotope ratios of Sr and Pb of lavas on the Tonga and Kermadec Islands. The isotope composition of Pb is compatible with the petrogenesis of volcanic rocks on island arcs (*Source:* data from Ewart and Hawkesworth 1987; Hart 1988)

steepens from 43–45° under the Tonga Islands to 55–60° under the Kermadec Islands, but the Benioff zone is not continuous between them. The subduction rates also differ and have been estimated at 9.1 cm yr^{-1} for the Tonga Islands, but only 4.7 cm yr^{-1} for the Kermadec Islands. Ewart and Hawkesworth (1987) concluded from these and other differences that the Tonga segment of this intra-oceanic island-arc system is more advanced in its evolution than the Kermadec segment.

A suite of boninites (Sect. 3.2.6) from the Tonga Trench and fore-arc have variable $^{87}Sr/^{86}Sr$ ratios between 0.70294 and 0.70454 and comparatively low average concentrations of Rb (6.2 ppm) and Sr (139 ppm) reported by Falloon et al. (1989) and Falloon and Crawford (1991). The $^{87}Sr/^{86}Sr$ ratios do not correlate with the concentrations of these elements or with their Rb/Sr ratios. Nevertheless, the magma that produced these rocks originated from a depleted mantle source in which melting had been induced by the addition of an aqueous fluid containing LIL elements. This explanation of the petrogenesis of the boninites in the Tonga Trench and fore-arc is similar to that of boninites elsewhere.

The isotope ratios of Sr, Nd, and Pb in the volcanic rocks of the Tonga and Kermadec Islands in Fig. 3.33a,b are consistent with the theory that these elements were added to the mantle wedge by aqueous fluids emanating from the sediment on the subducted Pacific Plate. The lavas on the Tonga Islands contain more Sr, Nd, and Pb derived from the subducted sediment than the lavas on the Kermadec Islands. However, the isotope ratios of the rocks on both groups of islands are aligned with the DMM component (Depleted MORB Mantle) which here represents the rocks of the mantle wedge. The Sr and Pb that were derived from the subducted sediment were enriched in radiogenic ^{87}Sr and ^{206}Pb, whereas the Nd was depleted in ^{143}Nd relative to the isotope compositions of these elements in DMM.

3.6.3 Taupo Volcanic Zone, New Zealand

New Zealand consists primarily of two large islands. The North Island is composed predominantly of volcanic rocks (Tertiary to Recent) which are underlain by a complex of older argillites and greywackes. The South Island is composed of igneous and metasedimentary rocks that range in age from Late Proterozoic to Cretaceous (Adams 1975; Pickett and Wasserburg 1989; Weaver and Pankhurst 1991). In addition, volcanic activity occurred during the Cretaceous Period on Mt. Somers and at Dunedin (Barley et al. 1938; Price and Compston 1973; McDougall and Coombs 1973) and in Tertiary time on the Banks Peninsula (Stipp and McDougall 1968; Barley et al. 1988). The South Island is cut by the Alpine Fault (Sect. 2.14) which runs parallel to the New Zealand Alps along the west coast of the South Island (Hurley et al. 1962a).

The volcanic rocks of the Taupo Volcanic Zone in the North Island of New Zealand (Fig. 3.30) range in composition from basalt to andesite, dacite, and rhyolite (including ignimbrites and pumice) all of which are less than two million years in age (Ewart et al. 1975). According to Healy (1962), the rhyolites, ignimbrites, and pumice (17 640 km^3) are much more voluminous than the andesites (882 km^3) which, in turn, are more voluminous

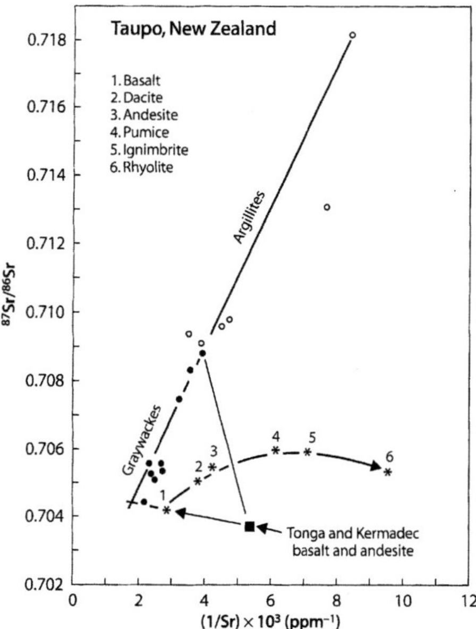

Fig. 3.34. Petrogenesis of the volcanic rocks in the Taupo Volcanic Zone, North Island, New Zealand. The Mesozoic greywackes (*solid circles*) and argillites (*open circles*) form a linear mixing array indicating that they are primarily two-component mixtures of volcanogenic and terrigenous detritus. The different compositional varieties of volcanic rocks in the Taupo Volcanic Zone (*stars*) are identified by number. Basalt and andesite of the Tonga-Kermadec Island arc (*solid square*) represent volcanic rocks that formed in an oceanic rather than in a continental setting. The basalt, dacite, and andesite of the Taupo Volcanic Zone formed from oceanic basalt and andesite by assimilation of varying amounts of greywacke. The pumice, ignimbrite, and rhyolite could have formed by partial melting of greywacke (Ewart and Stipp 1968) but probably originated by partial fusion of volcanic rocks of the oceanic crust (Blattner and Reid 1982; Graham et al. 1992) (*Source:* Ewart and Stipp 1968)

than the basalts (44.1 km^3). The data in Fig. 3.34 indicate that the rocks of the andesite-dacite-rhyolite (ignimbrite and pumice) suite have elevated $^{87}Sr/^{86}Sr$ ratios between 0.7044 to 0.7067 with an average of 0.7056 ±0.00016 ($2\bar{\sigma}, N = 44$) adjusted to 0.7080 for E&A. The basalts have a distinctly lower average $^{87}Sr/^{86}Sr$ ratio of 0.7042 ±0.0001 ($2\bar{\sigma}, N = 4$) which, nevertheless, is higher than the $^{87}Sr/^{86}Sr$ ratios of basalts in intra-oceanic island arcs (e.g. the Tonga, Kermadec, Mariana, and Aleutian Islands). Ewart and Stipp (1968) concluded from these data that the andesite-dacite-rhyolite suite of rocks could not have formed by differentiation of basalt magmas represented by the small amount of basalt in the Taupo Volcanic Zone.

The greywackes and argillites of Mesozoic age that underlie the volcanic rocks of the North Island form a straight line in Fig. 3.34 indicating that they are mix-

tures of two components consisting of volcanogenic sediment ($^{87}Sr/^{86}Sr < 0.7045$, Sr > 560 ppm) and terrigenous detritus of sialic composition ($^{87}Sr/^{86}Sr > 0.718$, Sr < 120 ppm), where the isotope ratios are present rather than initial values (Ewart and Stipp 1968). Since the average $^{87}Sr/^{86}Sr$ ratio of the greywackes (0.7062 ±0.0010, $2\bar{\sigma}, N = 9$) is similar to the average $^{87}Sr/^{86}Sr$ ratio of the andesite-dacite-rhyolite suite (0.7056 ±0.00016), Ewart and Stipp (1968) proposed that the voluminous rhyolite-ignimbrite suite formed by partial melting of a mixture of greywacke and argillite and that the andesites formed from basalt magma by assimilation of sedimentary rocks. Peterman et al. (1967) likewise tested the hypothesis that the greywackes of Mesozoic and Cenozoic age in western Oregon and California were the source of magmas of granitic composition. The average initial $^{87}Sr/^{86}Sr$ ratio (0.7059 ±0.0002) of Jurassic and Cretaceous greywackes in Oregon and California is quite similar to the average $^{87}Sr/^{86}Sr$ ratio of greywackes in New Zealand (0.7062 ±0.0010) and both match the Sr isotope ratio of the associated igneous rocks well enough to permit the conclusion that greywackes may have contributed to the formation of granitic magmas at these locations.

The different kinds of volcanic rocks in the Taupo Volcanic Zone in Fig. 3.34 (identified by number) are represented by average values of the coordinates to avoid crowding of the diagram. In addition, the diagram contains a data point (solid square) representing the basalts and andesites of the Tonga-Kermadec Islands which are similar to the parental magmas that formed above the subduction zone under the North Island of New Zealand before they encountered the Mesozoic greywacke-argillite complex and/or the still older basement rocks of the continental crust.

Two mixing lines have been drawn from the Tonga-Kermadec basalts to encompass the data points of the greywackes. The points representing the basalt-dacite-andesite suite (1, 2, 3) of the Taupo Volcanic Zone lie inside the resulting triangle in Fig. 3.34. Therefore, these rocks could have formed from island-arc magmas that assimilated varying amounts of sediment (or older rocks).

The interpretation of the Sr data in Fig. 3.34 does not explain the origin of the rhyolite-ignimbrite suite (4, 5, 6), because the data points representing these rocks lie outside of the mixing triangle formed by the Tonga-Kermadec basalt and the greywackes of the North Island of New Zealand. However, the Sr data permit the rhyolite-ignimbrite suite to have formed either by partial melting of the greywacke-argillite complex, as originally proposed by Ewart and Stipp (1968), or by differentiation of andesite magmas. However, the latter scenario is unrealistic because the volume of rhyolite and ignimbrites is approximately 20 times larger than that of andesites. In addition, no complementary mafic differentiates have been found in the Taupo area.

A large number of oxygen isotope analyses and concentrations of SiO_2 and K_2O reported by Blattner and Reid (1982) later demonstrated that the rhyolite-ignimbrite suite could not have formed by partial fusion of the greywacke-argillite complex. Therefore, the most likely remaining alternative is that the rhyolite-ignimbrite suite formed by partial melting of volcanic rocks of the oceanic crust under the Taupo Volcanic Zone followed by minor assimilation of crustal rocks (Graham et al. 1992).

The presence of acid volcanic rocks in the Taupo Volcanic Zone distinguishes it from intra-oceanic island arcs and can be attributed to special conditions associated with subduction of oceanic lithosphere under continental crust. The best explanation for the origin of the acid volcanic rocks in the Taupo Volcanic Zone (partial melting of volcanic rocks of the oceanic crust) is similar to the origin of rhyolites in Iceland (Sect. 2.3.2) but differs from the petrogenesis of granites and rhyolites in Japan which originated by partial melting of heterogeneous mixtures of crustal rocks.

3 6.4 Mount Taranaki

Mount Taranaki (Mt. Egmont) is a large stratovolcano located on the west coast of the North Island of New Zealand. It is composed of basalt and andesite flows all of which are less than 10 000 years old and some of which formed during the last eruption in A.D. 1755. Mount Taranaki is located about 140 km west of the Taupo Volcanic Zone and about 180 km above the Benioff zone associated with the same subduction zone that has caused volcanic activity at Taupo. Therefore, the lavas of Mount Taranaki provide another opportunity to examine the relation between the depth to the Benioff zone and the chemical composition and $^{87}Sr/^{86}Sr$ ratios of the volcanic rocks.

The data of Price et al. (1992) indicate that the basalt-andesite suite of Mount Taranaki has higher concentrations of Rb and Sr than the equivalent rocks in the Taupo Volcanic Zone. The $^{87}Sr/^{86}Sr$ ratios vary only within narrow limits (0.70441 to 0.70476) and have a mean value of 0.70458 ±0.00006 ($2\bar{\sigma}$, $N = 12$) relative to 0.71025 for NBS 987. In addition, the lavas of Mount Taranaki are consistently enriched in K_2O compared to equivalent rocks at Mount Ruapehu at the southern end of the Taupo Volcanic Zone.

Therefore, the lavas of Mount Taranaki display the same enrichment in LIL elements (K, Rb, Cs, Sr, Ba etc.) noted previously on Amak and Bogoslof Islands in the Aleutian Chain (Sect. 3.3.2) and on the Kasuga Seamounts (Sect. 3.2.2). Price et al. (1992) suggested that the enrichment in LIL elements of the Taranaki lavas was caused by a lower degree of partial melting at greater depth in the mantle wedge than at Taupo. Subsequently, the magma at Mount Taranaki evolved both chemically and

isotopically by a combination of assimilation of crustal rocks and fractional crystallization (AFC).

The petrogenesis of the igneous rocks at Dunedin and on the Banks Peninsula of the South Island was discussed by Price and Compston (1973) and by Barley et al. (1988), respectively.

3.7 The Sunda Islands of Indonesia

The Sunda Arc of Indonesia in Fig. 3.35a includes primarily the islands of Sumatra, Java, Bali, Lombok, Sumbawa, Sumba, Flores, and Timor (van Bemmelen 1949). These islands consist in part of continental crust that thins from west (Sumatra) to east (Java). The Sunda Arc continues as the Banda Island arc which turns north toward Irian Jaya. The tectonic setting of the Sunda Island arc was discussed by Katili (1975) and Hamilton (1979).

3.7.1 Sunda Arc

The volcanic activity along the Sunda Arc is caused by the subduction of oceanic lithosphere of the Indian Ocean into the Java Trench (Fig. 3.35a). The sediment that has accumulated on this plate originated in part from the Precambrian shield of Western Australia. The Indian Ocean Plate is moving north at a rate of about 6 cm yr^{-1} and has been traced to a depth of about 600 km in the Java-Bali sector of the Sunda Arc (Fitch and Molnar 1970). The continental crust thins eastward from about 30 km in Sumatra to only about 15 km in the Flores Sea north of the islands of Sumbawa and Flores. Volcanic activity in the Sunda Arc started in mid-Tertiary time and perhaps even earlier in the Late Mesozoic Era (Katili 1975; Hamilton 1979).

The island of Sumatra contains volcanoes of Quaternary age several of which have been active in historic time (Leo et al. 1980). The lavas consist primarily of andesite, but include rhyolitic and andesitic tuffs at the Toba Caldera, in the Padang area of central Sumatra, in the Pasumah Highlands, and at Lampung Bay at the south end of the island (Fig. 3.35a). The volcanic activity appears to be concentrated along the Semangko Fault zone that extends along the length of the island from northwest to southeast.

The $^{87}Sr/^{86}Sr$ ratios of the Quaternary lavas of Sumatra reported by Whitford (1975) range from 0.70447 (basaltic andesite, Merapi) to 0.71384 (rhyolitic ignimbrite, Toba Caldera) relative to 0.71025 for NBS 987. The ^{87}Sr enrichment of the rhyolitic ignimbrite at the Toba Caldera suggests a crustal source for this rock type. However, the $^{87}Sr/^{86}Sr$ ratios of two rhyolite samples in the Padang area (0.7066) do not differ appreciably from those of seven andesitic rocks (0.7060 ±0.0004, $2\bar{\sigma}$) compared to 0.7080 for E&A. Therefore, Leo et al. (1980) suggested

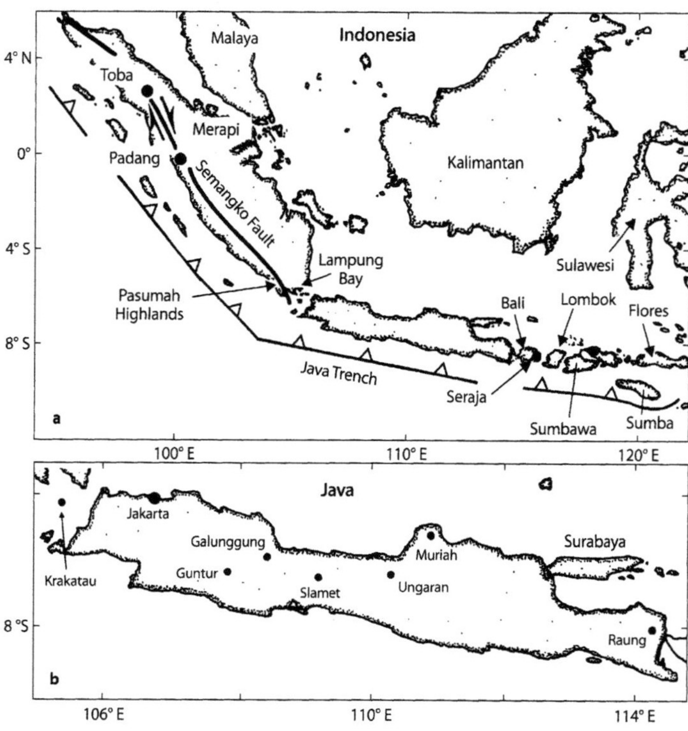

Fig. 3.35.
a The islands of Indonesia; **b** The island of Java. The volcanoes identified by name in *Parts a* and *b* are mentioned in the text. Many other volcanoes of this region are not identified (*Sources:* adapted from Whitford 1975; Leo et al. 1980)

that the rhyolites in the Padang area either formed by fractional crystallization of andesite magma or originated by fractional melting of andesite or volcanogenic sediment at $P > 7$ kbar in the presence of water (Yoder 1973).

The $^{87}Sr/^{86}Sr$ ratios of the lavas of Sumatra, taken as a group, are higher than those of intra-oceanic island arcs (e.g. Mariana Islands, Aleutians, and Tonga-Kermadec Islands), but are similar to those of the Taupo Volcanic Zone of New Zealand and of the Japanese Islands where oceanic lithosphere is also being subducted under continental crust.

The chemical composition of the volcanic rocks on the island of Java were discussed by Whitford (1975), Whitford et al. (1979), Nicholls and Whitford (1978), and Nicholls et al. (1980). The data indicate that the average Rb and Sr concentrations have characteristic values that vary with the silica concentrations (Whitford et al. 1979). The tholeiite suite occurs only along the southern coast of Java at a depth to the Benioff zone of about 150 km, whereas the calc-alkaline suite has a wide distribution across Java. However, high-K calc-alkaline lavas occur only where the depth to the Benioff zone is greater than about 250 km. The most K-rich lavas on the volcano Muriah (Fig. 3.35b) on the north coast of Java (Whitford 1975) have the highest concentrations of Rb and Sr and are

associated with the greatest depth to the Benioff zone of 360 km. These results strongly support the thesis of Dickinson and Hatherton (1967) and Hatherton and Dickinson (1969) that the K_2O concentration of lavas in island arcs increases with increasing depth to the Benioff zone.

The causes for this relationship were reviewed by Whitford et al. (1979) based on previous suggestions in the literature. They concluded that the primary magmas were olivine tholeiites that formed by partial melting in the mantle wedge in such a way that their silica content and the degree of melting both decreased with depth. For example, the olivine tholeiites of Galunggung and Guntur (Fig. 3.35b) formed by 20 to 25% partial melting. Alkali olivine basalt magma of Slamet and Ungaran (Fig. 3.35b) originated by 5 to 15% melting. Similarly, the high-K lavas of Muriah represent 5% melts generated in the mantle wedge. Melting was triggered by the transfer of an aqueous fluid containing alkali metals and alkaline earths from the subducted sediment to the overlying mantle wedge (Tatsumi 1989; Vukadinovic and Nicholls 1989).

The $^{87}Sr/^{86}Sr$ ratios of lavas on the island of Java range from 0.70402 (Raung, Fig. 3.35b; Whitford 1975) to 0.70626 (Slamet; Vukadinovic and Nicholls 1989) relative to 0.71025 for NBS 987. Whitford (1975) demonstrated that the $^{87}Sr/^{86}Sr$ ratios of the lavas increase somewhat with

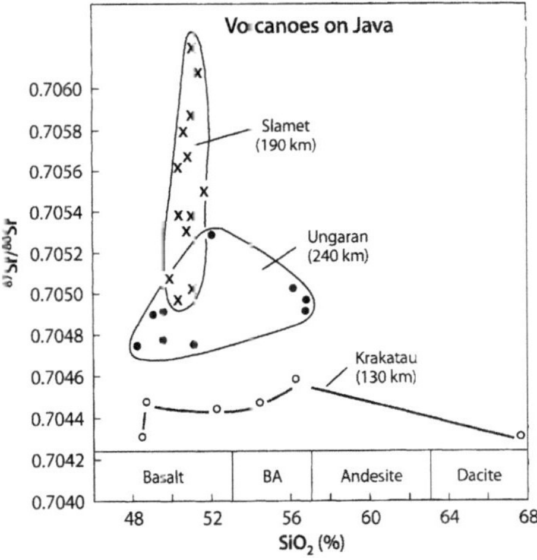

Fig. 3.36. Variation of $^{87}Sr/^{86}Sr$ ratios of lavas having different SiO_2 concentrations on three volcanoes of Java, Indonesia. The data indicate a spectrum of relationships between $^{87}Sr/^{86}Sr$ ratios and silica concentrations exemplified by the three chosen volcanoes. The lavas on Gunung Slamet (Gunung = mountain) have a range of $^{87}Sr/^{86}Sr$ ratios, but nearly constant silica concentrations, whereas G. Krakatau exhibits the opposite trend of nearly constant $^{87}Sr^{86}Sr$ ratios and a wide range of SiO_2 concentrations. The rocks on Gunung Ungaran are intermediate between these extremes by suggesting a weak positive correlation between the two parameters. The depth to the Benioff zone (indicated in parentheses) does not exert a detectable influence on the $^{87}Sr/^{86}Sr$ ratios or on the silica concentrations of the lavas at these locations (*Sources:* Whitford 1975; Vukadinovic and Nicholls 1989)

increasing depth to the Benioff zone, but the relationship is complicated by variations of this ratio along the island arc and by differences among lavas extruded at some of the volcanoes of Java. For example, the data in Fig. 3.36 indicate that the lavas of Slamet Volcano (Fig. 3.35b) have a wide range of $^{87}Sr/^{86}Sr$ ratios (0.70488 to 0.70626, Vukadinovic and Nicholls 1989) but a narrow range of silica concentrations. The reverse applies to the lavas of Krakatau (Fig. 3.35b) whose $^{87}Sr/^{86}Sr$ ratios are nearly constant (0.70430 to 0.70458, Whitford 1975), whereas their silica concentrations vary widely from 48.4 to 67.8%. The lavas on Ungaran (Fig. 3.35b) are intermediate and suggest a weak positive correlation between their $^{87}Sr/^{86}Sr$ ratios and silica concentrations.

The $^{87}Sr/^{86}Sr$ ratios of volcanic rocks on Java are lower on average than those of Sumatra and decrease along the island arc from western Java to the island of Sumbawa located about 1 500 km east of Krakatau (Whitford 1975; Foden and Varne 1980). The lowest $^{87}Sr/^{86}Sr$ ratios occur in the lavas of Seraja Volcano (Fig. 3.36) on Bali for which Whitford (1975) reported values of 0.70380 and 0.70394 relative to 0.71025 for NBS 987.

The isotope ratios of Sr and Nd in Quaternary lavas of Java in Fig. 3.37 are consistent with petrogenesis by

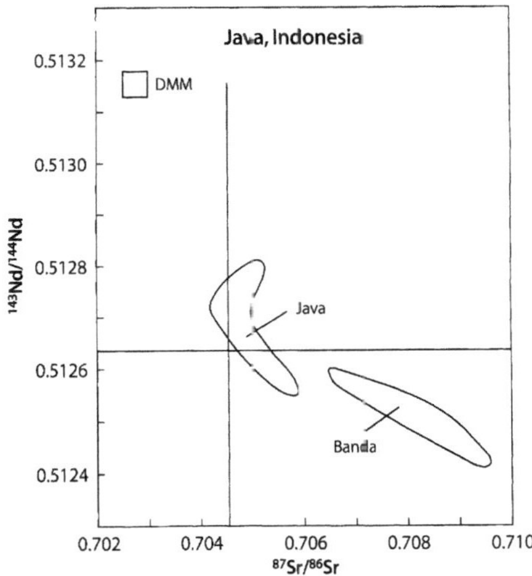

Fig. 3.37. Isotope ratios of Sr and Nd of Quaternary lavas on Java (Sunda Arc) and the islands of Damar, Teun, and Serua (Banda Arc). The lavas of the Banda Arc have a crustal imprint resulting from the transfer of Sr and Nd from subducted terrigenous sediment to the overlying mantle wedge (*Source:* Whitford et al. 1981)

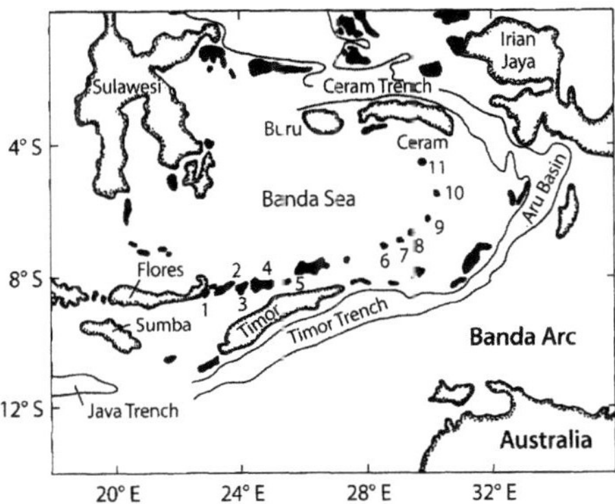

Fig. 3.38. The Banda Island arc of Indonesia. The islands identified by number are: *1.* Solor; *2.* Pantar; *3.* Pura Beser; *4.* Alor; *5.* Atauro; *6.* Damar; *7.* Teun; *8.* Nila; *9.* Serua; *10.* Manuk; *11.* Banda (*Source:* adapted from Whitford et al. 1977)

partial melting in the mantle wedge after additions of varying amounts of Sr and Nd from subducted oceanic sediment (Whitford et al. 1981). The lavas on some of the islands of the southwestern Banda Arc, to be discussed below, have elevated $^{87}Sr/^{86}Sr$ and low $^{143}Nd/^{144}Nd$ ratios compared to the lavas on Java because the sediment subducted into the Timor Trench in Fig. 3.38 originated by erosion of Precambrian rocks of Australia.

3.7.2 Banda Arc

The islands of the Banda Arc in Fig. 3.38 reside on oceanic crust north and west of the Timor Trench which is subducting sediment derived from the continental crust of Australia. Therefore, the magmas of the Banda Arc have a strong crustal isotopic signature (Whitford and Jezek 1979) including the oxygen isotope compositions of andesites reported by Magaritz et al. (1978) which are consistent with mixing of mantle-derived magmas with a sialic component. The $^{87}Sr/^{86}Sr$ ratios of lavas on the islands of the Banda Arc range widely from 0.70451 (tholeiite basalt, Banda Islands) to 0.70947 (andesite, Serua Island) relative to 0.7080 for E&A (Whitford and Jezek 1979). Even higher $^{87}Sr/^{86}Sr$ ratios of 0.71574 and 0.71754 occur on Ambon Island south of Ceram at the northern end of the Banda Arc (Fig. 3.39).

The average Sr concentrations of andesitic lavas (SiO_2 = 53 to 63%) in Fig. 3.39 increase from about 353 ppm on Solor (no. 1) to about 555 ppm on Damar (no. 6), and then decline with increasing distance from Solor to about 170 ppm in the Banda Islands. The only known exceptions to this pattern are the lavas of Pura Beser and Atauro both of which are identified in Fig. 3.38. The average Rb concentrations vary similarly along the length of the Banda Arc.

The $^{87}Sr/^{86}Sr$ ratios reported by Whitford et al. (1977) and Whitford and Jezek (1979) in Fig. 3.40 reveal that the lavas of Damar, Manuk, and the Banda Islands form distinct collinear clusters that define a mixing line which extrapolates to MORB-like values of $^{87}Sr/^{86}Sr \approx 0.7029$

and Sr ≈ 100 ppm. The lavas of Pantar and Solor also lie close to that line, whereas the lavas of Teun, Nila, and Serua contain excess radiogenic ^{87}Sr and cluster above the Damak-Manuk-Banda array. The lavas from Atauro and Pura Beser, whose Sr concentrations deviate from the pattern of Sr concentrations in Fig. 3.38, also deviate from the linear data array in Fig. 3.40.

One basalt specimen from the Banda Islands and two from the island of Ambon located south of Ceram at the far end of the Banda Arc form a separate array in Fig. 3.40 that also extrapolates to MORB-like values. The origin of these basalts and of the lavas that form the Damar-Manuk-Banda array can be explained by the addition of varying amounts of Sr with a high $^{87}Sr/^{86}Sr$ ratio to magmas that originated from sources in the mantle wedge above the Benioff zone located between 100 and 200 km below the Banda Arc. The lavas of Serua and Atauro are remarkable because of their wide range of $^{87}Sr/^{86}Sr$ ratios. The volcanic rocks of Teun and Nila, located between Damar and Serua (Fig. 3.38)

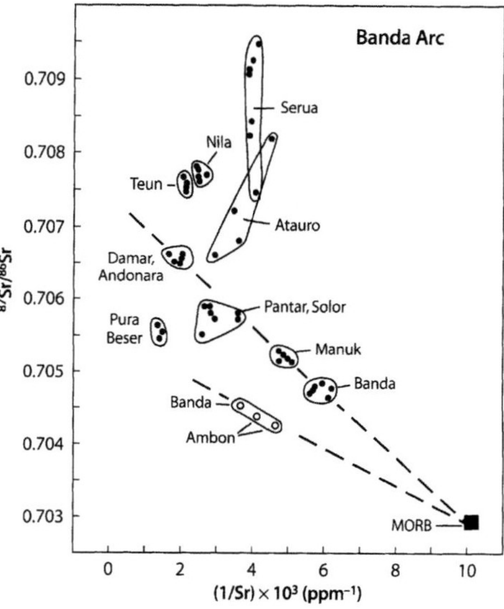

Fig. 3.40. Sr-isotope mixing diagram of the calc-alkaline lavas of the eastern Sunda and Banda Arcs. The volcanic rocks of Damar, Manuk, and the Banda Islands form three distinct collinear clusters that define a mixing line which extrapolates to MORB-like values of $^{87}Sr/^{86}Sr \approx 0.7029$ and Sr ≈ 100 ppm. Basalts on the Banda Islands and on Ambon define a separate line that also extends to MORB. The lavas of Serua, Nila, and Teun contain excess ^{87}Sr relative to the Damar-Manuk-Banda array. The islands of Atauro and Pura Beser identified in Fig. 3.38 also deviate from this array. The linear patterns seen on this diagram were caused by additions of varying amounts of Sr having elevated $^{87}Sr/^{86}Sr$ ratios to magmas generated by partial melting in the mantle wedge (*Sources:* Whitford et al. 1977; Whitford and Jezek 1979)

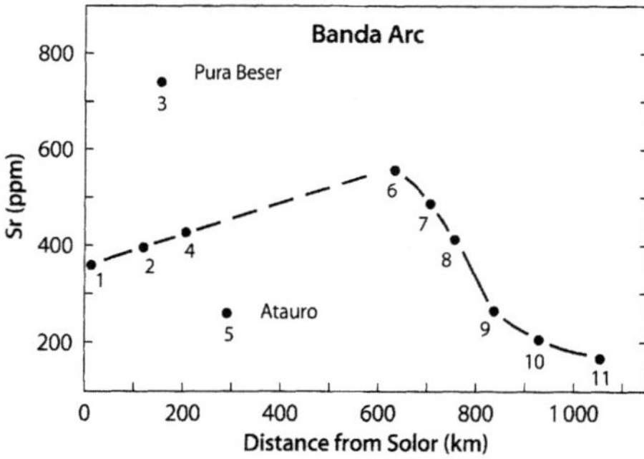

Fig. 3.39. Variation of average Sr concentrations of andesitic lavas (SiO_2 = 53 to 63%) on the islands of the Banda Arc with increasing distance east and northeast of the island of Solor located just east of Flores (Fig. 3.38). The average concentrations of Rb vary similarly along the length of this island arc. The lavas of Pura Beser and Atauro deviate markedly from the regional pattern. The islands are identified by number: *1.* Solor; *2.* Pantar; *3.* Pura Beser; *4.* Alor; *5.* Atauro; *6.* Damar; *7.* Teun; *8.* Nila; *9.* Serua; *10.* Manuk; *11.* Banda (*Sources:* Whitford et al. 1977; Whitford and Jezek 1979)

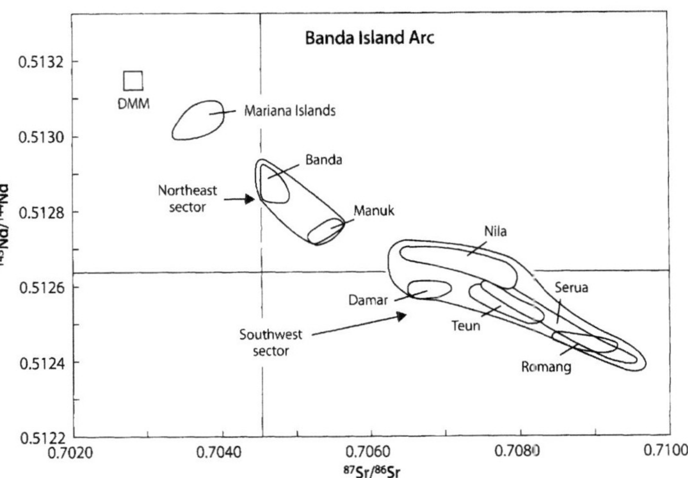

Fig. 3.41.
Isotope ratios of Sr and Nd in lavas on the islands of the Banda Arc. The magmas of the Banda Islands and of the island of Manuk in the north east sector of the arc are consistent with melting of metasomatically altered rocks of the mantle wedge. The lavas on the islands of the southwest sector contain excessive amounts of crustal Sr and Nd derived from subducted sediment that originated from Australia (*Sources:* Vroon et al. 1993; Hart 1988)

also contain excess ^{87}Sr relative to the Damar-Manuk-Banda array.

The presence of Sr and Nd derived from crustal sources in the lavas of the Banda Arc is clearly indicated by the range of $^{87}Sr/^{86}Sr$ and $^{143}Nd/^{144}Nd$ ratios in Fig. 3.41 (Vroon et al. 1993). The lavas on the Banda Islands and on Manuk in the northeast sector of the arc are typical of other island arcs (e.g. the Mariana Arc) with $^{87}Sr/^{86}Sr$ ratios from 0.7045 to 0.7055 and $^{143}Nd/^{144}Nd$ ratios between 0.51273 and 0.51291 relative to 0.710250 ±0.000026 for NBS 987 and 0.511853 ±0.000012 for the LaJolla Nd standard, respectively. The volcanoes on the islands of the southwestern segment of the Banda Arc have anomalously high $^{87}Sr/^{86}Sr$ ratios (0.7065 to 0.7095) and low $^{143}Nd/^{144}Nd$ ratios (0.51252 to 0.51240) which characterize rocks of the continental crust. Nevertheless, all of the lavas of the Banda Arc form a single array that could be generated by the addition of varying amounts of crustal Sr and Nd to magmas derived from the lithospheric mantle wedge represented by DMM component.

The magmas extruded by volcanoes in the northeast segment of the Banda Arc conform to the standard model of petrogenesis in island arcs. However, the volcanic rocks of the southwest group of islands require special circumstances to explain the extreme enrichment in crustal Sr and Nd. Vroon et al. (1993) emphasized that the oceanic crust being subducted into the Timor Trench contains a significant amount of continental sediment derived from Australia. They calculated that from 0.5 to 5% of this sediment appears to have been added to the magmas at depth, presumably by mixing of arc magmas with melts derived from the subducted oceanic crust.

3.7.3 Sulawesi and Kalimantan

The islands of Sulawesi (Celebes) and Kalimantan (Borneo) in Fig. 3.42 are located in a tectonically complex region dominated by subduction, volcanic activity, and the development of back-arc basins (Williams and Harahap 1987; Spadea et al. 1996; Hall and Blundell 1996).

The volcanic rocks of central Sulawesi range in composition from basalt to andesite (Oligocene to Pliocene) which are overlain by Miocene to Pliocene molasse sediment and by Quaternary tuff. In addition, the older volcanic rocks and the molasse sediment were intruded by plutons of Miocene to Pliocene age composed of gabbro, clinopyroxenite, and lamprophyre as well as by syenite and granodiorite. K-Ar age determinations of minerals by Elburg and Foden (1999) yielded dates ranging from 19.24 ±0.22 Ma (andesite) to 1.08 ±0.07 Ma (tuff), whereas Bergman et al. (1996) reported that the Tertiary volcanic rocks of central Sulawesi have elevated $^{87}Sr/^{86}Sr$ (>0.710) and low $^{143}Nd/^{144}Nd$ ratios similar to the lavas of the southwest sector of the Banda Arc in Fig. 3.41.

The initial $^{87}Sr/^{86}Sr$ and $^{143}Nd/^{144}Nd$ ratios of Tertiary volcanic rocks (andesites, trachyandesites, and rhyodacites) as well as gabbro, granodiorite, and syenite plutons of central Sulawesi in Fig. 3.43 range widely from 0.703337 to 0.72212 and from 0.512971 to 0.512000, respectively. In addition, the $^{206}Pb/^{204}Pb$ ratios (not shown) range from 18.284 to 19.090.

The isotope compositions of Sr, Nd, and Pb of the Tertiary igneous rocks of central Sulawesi are characteristic of magmas derived from heterogeneous crustal rocks. Only one sample of a medium-K andesite analyzed by Elburg and Foden (1999) has isotope compositions

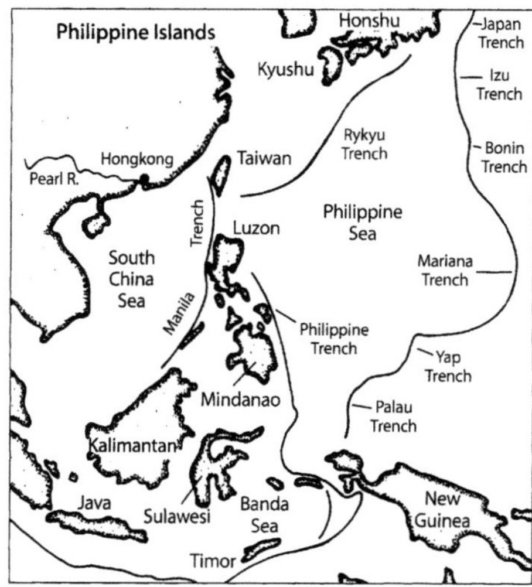

Fig. 3.42. Tectonic setting of Sulawesi and Kalimantan (Indonesia) and of the Philippine Islands. The latter are bracketed by the Philippine Trench on the east and by the Manila Trench on the west. These trenches link the Ryukyu Trench and the Japanese Islands in the north to the Sunda and Banda Arcs in the south. In addition, the locations of the Izu-Bonin-Mariana-Yap-Palau Trench system is indicated (*Source:* adapted from the National Geographic Society 1990)

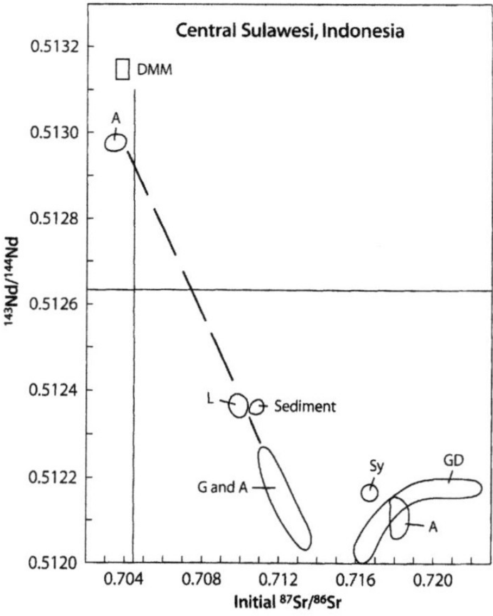

Fig. 3.43. Isotope ratios of Sr and Nd of a diverse group of Tertiary volcanic and plutonic rocks in central Sulawesi, Indonesia: A = andesite, L = lamprophyre, G = gabbro, Sy = syenite, GD = granodiorite. The heterogeneous isotopic compositions as well as the elevated initial ^{87}Sr/^{86}Sr and low ^{143}Nd/^{144}Nd ratios are typical of igneous rocks that contain a crustal component or formed by remelting of crustal rocks (*Source:* Elburg and Foden 1999)

of Sr and Nd (0.70337 and 0.512971, respectively) consistent with its derivation from the lithospheric mantle. Bergman et al. (1996) as well as Elburg and Foden (1999) observed that Cretaceous and Paleogene sedimentary rocks have lower ^{87}Sr/^{86}Sr and higher ^{143}Nd/^{144}Nd ratios than most of the volcanic and plutonic rocks of central Sulawesi in Fig. 3.43. Consequently, the young sedimentary cover rocks are not the source of the crustal component of the Tertiary magmas in central Sulawesi.

Therefore, the presence of crustal Sr and Nd in the volcanic and plutonic rocks of central Sulawesi must be attributed to melting of older granitic basement rocks or to extensive assimilation of such rocks by andesite magmas originating from the underlying lithospheric mantle. The crustal rocks may have been subducted under Sulawesi at about 20 Ma as postulated by Hall (1996) based in part on the isotopic data of Bergman et al. (1996) and on a reconstruction of the tectonic history of Southeast Asia.

The characteristic crustal isotope compositions of the Tertiary igneous rocks on Sulawesi reinforce the evidence derived from the southwestern sector of the Banda Arc (Fig. 3.41) that the petrogenesis of the Tertiary volcanic rocks of this region was affected by the northward motion of Australia. The introduction of the component of continental crust into the magma sources in the lithospheric mantle occurred either by subduction of terrigenous sediment or by subduction of a sliver of the continental crust of Australia. In either case, the crustal component may have melted in this case to form the magmas represented by the Tertiary rocks of Sulawesi.

3.8 Philippine Islands and the Luzon Arc

The tectonic setting of the Philippine Islands in Fig. 3.42 is characterized by two trench systems in which oceanic crust is subducted in opposite directions. In the Philippine Trench, located east of the Philippine Islands, oceanic lithosphere of the Philippine Sea is subducted westward (Hickey-Vargas 1998). On the west side of the Philippine Islands, oceanic lithosphere of the South China Sea is subducted eastward into the Manila Trench. Consequently, the frequent volcanic eruptions on the Philippine Islands are a direct consequence of their unstable tectonic setting.

The Philippine Islands and associated deep-sea trenches also link the island arcs of Japan and the Ryukyu Islands in the north (Honma et al. 1991; Shinjo 1999) to the Sunda and Banda Arcs in the south. In addition, Fig. 3.42 includes the Japan-Izu-Bonin-Mariana-Yap-Palau trench system (Sect. 3.2) and thereby relates the volcanic rocks on the Philippine Islands to the lavas on the other island arcs in the western Pacific Ocean.

3.3.1 Taiwan and the Luzon Arc

The volcanic activity on Taiwan and on the chain of small islands extending south from Taiwan towards the island of Luzon in the Philippines is the result of subduction of oceanic lithosphere of the South China Sea eastward into the Manila Trench (Yen 1977; Juang and Chen 1989). The islands include Lu-Tac, Lan-Yu, Hsiao-Lan Yu, Batan, Babuyan, and Calayan. The volcanic rocks in these islands are composed primarily of andesite whose ages

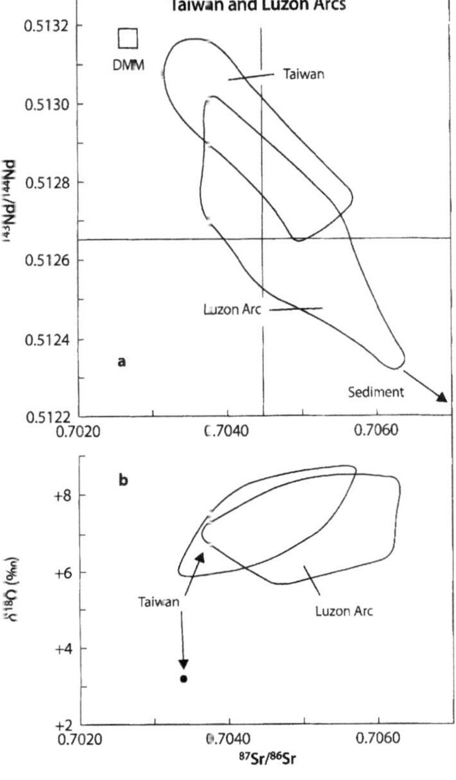

Fig. 3.44. a Isotope ratios of Sr and Nd of volcanic rocks (17 Ma to Recent) of the Coastal Range of Taiwan and on the small islands extending south from Taiwan toward the island of Luzon (Fig. 3.42). The negative correlation of ^{87}Sr/^{86}Sr and ^{143}Nd/^{144}Nd ratios is consistent with contamination of the magma sources by transfer of Sr and Nd from subducted sediment to the rocks of the mantle wedge; b Contrary to expectation, the δ^{18}O values of the lavas on Taiwan and on the islands of the Luzon Arc range primarily from +6.0 to +8.5‰ and therefore are not attributable to assimilation of marine sediment by the andesite magmas. The isotope compositions of Sr, Nd, and O of these lavas indicate that the magmas originated from source rocks that contained Sr and Nd (but not O) derived from subducted sediment (*Source:* Chen et al. 1990)

range from 17 Ma (Coastal Range, eastern Taiwan) to Recent in the Batan-Babuyan Ridge north of Luzon (Chen et al. 1990).

The ^{87}Sr/^{86}Sr ratios of the volcanic rocks in the Coastal Range of Taiwan in Fig. 3.44a vary from 0.70337 to 0.70559 relative to 0.71025 for NBS 987 and correlate inversely with the ^{143}Nd/^{144}Nd ratios as expected. The isotope ratios of Sr and Nd of andesites on the small islands of the Luzon Arc extend the field of variation of the rocks on Taiwan (Chen et al. 1990). However, the measured δ^{18}O values of these lavas in Fig. 3.44b range only from +6.0 to +8.5‰ and are not well correlated with the ^{87}Sr/^{86}Sr ratios as expected if the andesite magmas had assimilated crustal rocks or oceanic sediment on the subducted South China Plate. A correction by Chen et al. (1990) based on the volatile content of the andesites reduced the δ^{18}O values of these rocks to values between +5.2 and +6.4‰ which is characteristic of the upper mantle. Consequently, the oxygen isotope compositions provide evidence that the heterogeneity of isotope ratios of Sr and Nd of the lavas of the Luzon Arc is attributable to the transfer of Sr and Nd (but not O) from the subducted sediment to the mantle wedge.

Sediment from the South China Sea west of the Manila Trench has ^{87}Sr/^{86}Sr ratios between 0.71177 to 0.71526 relative to 0.71025 for NBS 987 (Chen et al. 1990). The elevated ^{87}Sr/^{86}Sr ratios are caused by the presence of a component of terrigenous sediment derived from southern China by the Zhu (Pearl) River which enters the South China Sea at Hong Kong (Fig. 3.42). The sediment samples analyzed by Chen et al. (1990) form a linear array in coordinates of ^{87}Sr/^{86}Sr and 1/Sr ratios (not shown), which suggests that they are mixtures in varying proportions of marine carbonate (^{87}Sr/^{86}Sr = 0.70916; Sr ≈ 360 ppm) and terrigenous sediment having a high ^{87}Sr/^{86}Sr ratio and a low Sr concentration (^{87}Sr/^{86}Sr > 0.715, Sr < 145 ppm).

3.8.2 Bataan Arc, Luzon

The islands of the Philippines in Fig. 3.42 include Luzon, Mindanao, Mindoro, Marinduque, Masbate, Samar, and many others. The geology of these islands is complicated by the presence of accreted terranes (Encarnación et al. 1993) and by intermittent volcanic activity associated with subduction along both the west and east coasts of Luzon. The petrogenesis of the volcanic rocks of Oligocene to Recent age on Luzon and the neighboring islands to the south has been studied by many investigators including DeBoer et al. (1980), Mukasa et al. (1987), Defant et al. (1988), and Miklius et al (1991).

The island of Luzon in Fig. 3.45 contains the Bataan-Mindoro Arc along the west coast in the central part of the island and the Bicol Arc along the east coast in southern Luzon (Knittel and Defant 1988). The Bataan Arc has

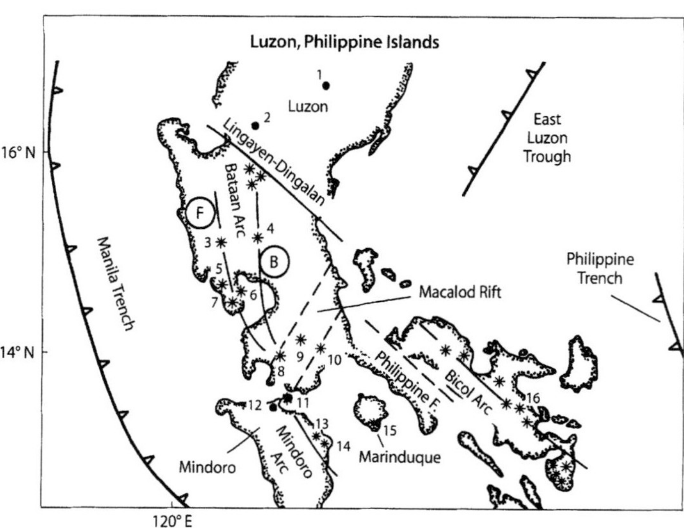

Fig. 3.45.
Tectonic setting of the island of Luzon in the Philippines. The island is situated between the Manila Trench on the west and the Philippine Trench on the east. Volcanic activity caused by subduction occurs in the Bataan-Mindoro Arc, the Bicol Arc, and in the Macalod Corridor which is probably a rift. The Bataan Arc has a front row (*F*) and a back row (*B*) of volcanoes. Luzon is divided into three segments by the Lingaen Dingalan and Philippine Faults. The volcanoes and sites of plutonic rocks mentioned in the text are identified by number. *1.* Cordon syenite; *2.* Agno Batholith; *3.* Pinatubo; *4.* Arayat; *5.* Natib; *6.* Limay; *7.* Marivales; *8.* Taal; *9.* Mapinggon; *10.* San Cristobal; *11.* Verde Island; *12.* Puerto Galera; *13.* Macapili; *14.* Dumali; *15.* Marlanga; *16.* Tiwi (*Source:* adapted from Knittel and Defant 1988)

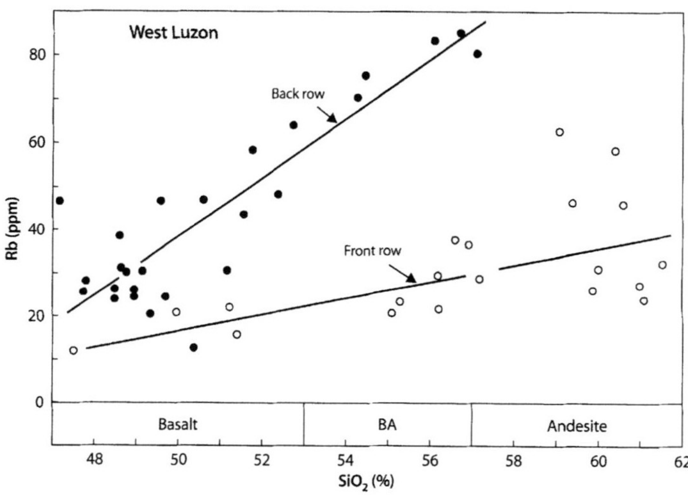

Fig. 3.46.
Variation of Rb and silica concentrations of volcanic rocks in the Bataan Arc of western Luzon in the Philippines. The data demonstrate that the lavas of the rear-rank volcano Arayat have higher concentrations of Rb (and other LIL elements) than the front-rank volcanoes Natib, Marivales, Orion, Limay, Pinatubo, and Taal. The *lines* were fitted by eye (*Sources:* Knittel and Defant 1988; Bau and Knittel 1993)

developed as a result of subduction of oceanic crust of the South China Sea eastward into the Manila Trench located west of Luzon. This arc is interrupted at its southern end by the Macalod Corridor (rift) and then continues south as the Mindoro Arc. The Bataan Arc includes the volcanoes Pinatubo, which erupted in 1992, and Taal, which has erupted explosively 26 times since 1572 (Mukasa et al. 1994). The most recent eruption of Taal in September 1965 caused an area of about 60 square kilometers to be covered with a layer of volcanic ash about 25 cm thick and resulted in more than 50 deaths (Moore et al. 1966).

The Bataan volcanic arc consists of two subparallel segments located at different distances above the Benioff zone. The front row of volcanoes along the west coast of central Luzon includes the volcanoes Natib, Marivales, Orion, and Limay (Fig. 3.45). Behind this segment is a second array of volcanoes including Amorong and Arayat. The volcanoes in the back row are younger (1.7 to 0.1 Ma) than those in the front row (7 to 0 Ma). In addition, the lavas of the volcanoes in the back row in Fig. 3.46 have higher concentrations of Rb and Sr than those in the front row (Knittel and Defant 1988; Bau and Knittel 1993). The enrichment of the lavas on the volcano Arayat (back row) in Rb, Sr, and other LIL elements is presumably caused by a smaller degree of partial melting in the mantle wedge and by the greater thickness of the wedge compared to the volcanoes in the front row. Elevated LIL element concentrations have also been noted in the lavas of volcanoes behind the volcanic front in Java (Whitford et al. 1979).

The volcanic rocks of the Bataan Arc (Fig. 3.45) range in composition from basalt to dacite, and their age is less than 2.8 Ma. However, the volcano Arayat in the back row of the Bataan Arc, San Cristobal and Mapinggon in the Macalod Rift, San Andrés on Verde Island, as well as Dumali, Marlanga, and Macapili in the Mindoro Arc have erupted high-K calc-alkaline lavas. Although the Sr concentrations of the lavas of the Bataan-Mindoro Arc range widely from 183 ppm (basalt, Limary Volcano) to 1 310 ppm (high-K basaltic andesite, Macapili Volcano), their ^{87}Sr/^{86}Sr ratios are strongly clustered between 0.704 and 0.705 and do not correlate convincingly with reciprocal Sr concentrations as expected for two-component mixtures.

In spite of the tectonic complexity of Luzon and the other Philippine islands, the isotope compositions of Sr of plutonic and volcanic rocks are remarkably uniform. The data compiled by Knittel and Defant (1988) indicate that the initial ^{87}Sr/^{86}Sr ratios of most of the plutonic rocks range only from about 0.7032 to 0.7045 and appear to be independent of their ages and chemical compositions. For example, the rocks of the Cordon syenite complex (25.1 Ma, SiO$_2$ = 44.8 to 56.0%) in northern Luzon (Fig. 3.45) have a mean initial ^{87}Sr/^{86}Sr ratio of 0.70362 ±0.00007 (2$\bar{\sigma}$). The quartz diorites of the nearby Agno Batholith (25.0 Ma, SiO$_2$ = 58.7 to 71.8%) yield a very similar average initial ^{87}Sr/^{86}Sr ratio of 0.70364 ±0.00015 (2$\bar{\sigma}$).

The only known examples of plutonic igneous rocks with high ^{87}Sr/^{86}Sr ratios in the northern Philippines are granite plutons in the Puerto Galera area of Mindoro Island for which Knittel and Defant (1988) reported an initial ^{86}/Sr^{86}Sr ratio of 0.71507. The high-K calc-alkaline lavas of the Mindoro Arc also have elevated ^{87}Sr/^{86}Sr ratios between 0.7051 and 0.7054. However, these rocks have high Sr concentrations (510–1 310 ppm), whereas the Puerto Galera granite contains only 54 ppm Sr on the average. Therefore, assimilation of this granite by a typical andesitic magma of the Bataan Arc (^{87}Sr/^{86}Sr = 0.7045 ±0.00015; Sr = 339 ±48) would produce mixtures having Sr concentrations between 240 and 260 ppm for ^{86}Sr/^{86}Sr ratios between 0.7051 and 0.7054. The predicted Sr concentrations are much lower than the measured Sr concentrations of the volcanic rocks in the Mindoro Arc. Accordingly, Knittel and Defant (1988) concluded that the ^{87}Sr/^{86}Sr ratios of the young volcanic rocks (<5 Ma) of Luzon are controlled by the transfer of Sr from the subducted plates into the overlying mantle wedge rather than by remelting or assimilation of old sialic rocks of the continental crust.

The isotope ratios of Sr, Nd, and Pb of the lavas on the Taal, Laguna de Bay, and Arayat Volcanoes in the back row of the Bataan Arc in Fig. 3.47a and b are remarkably homogeneous (Mukasa et al. 1994). In addition, the data fields in Sr-Nd and Pb-Sr isotope coordinates are located between the DMM component representing the wedge of lithospheric mantle and EM2 which consists of subducted oceanic crust and associated terrigenous sediment. Therefore, the lavas erupted on these volcanoes originated from the rocks of the mantle wedge which had been altered by additions of Sr, Nd, and Pb derived from the subducted oceanic crust. Mukasa et al. (1994) considered that the subducted oceanic crust originated from the South China Sea (Fig. 3.42) with terrigenous sediment derived from the mainland of Asia.

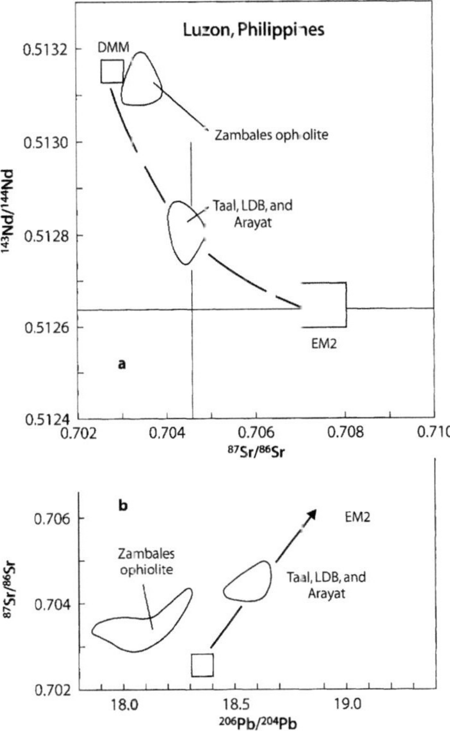

Fig. 3.47 a,b. Isotope ratios of Sr, Nd, and Pb of volcanic rocks on the volcanoes Taal, Laguna de Bay, and Arayat in the back row of the Bataan Arc on the island of Luzon in the Philippines. The data fields are located between DMM (mantle wedge) and EM2 (subducted oceanic crust + continental sediment). In addition, *Parts a* and *b* contain data fields representing igneous rocks in the Zambales ophiolite (44.2 ±0.5 Ma) that underlies the present volcanoes of the Bataan Arc (*Sources:* Encarnación et al. 1993; Mukasa et al. 1994; Encarnación et al. 1999)

The Bataan Arc along western Luzon is underlain by the Zambales ophiolite. Encarnación et al. (1993) determined concordant U-Pb dates of 44.2 ±0.9 Ma (Eocene) for two zircon fractions from tonalite in the Acoje block of this ophiolite (tonalites are granodiorites containing Ca-rich oligoclase and andesine but are virtually lacking in K-feldspar). The results in Fig. 3.47a,b demonstrate that igneous rocks of the Zambales ophiolite (including tonalites, diorites, gabbro, and peridotites) define small data fields close to the DMM component in the Sr-Nd and Pb-Sr isotope-mixing diagrams. However, the Zambales ophiolite is not considered to be the source of the lavas erupted by the present volcanoes along the Bataan Arc. Instead, the igneous rocks of the Zambales ophiolite formed during an earlier episode of igneous activity along a spreading ridge (Encarnación et al. 1999).

3.9 Other Oceanic Island Arcs in the Pacific and Atlantic Oceans

In the foregoing sections of this chapter most of the tectonic and compositional variants of petrogenesis in oceanic island arcs have been illustrated with data from selected localities in the region of the Pacific and Indian Oceans. In this section we consider data from other sites that have not yet come up for discussion. These include the island arcs of New Britain, the Solomon Islands, the Vanuatu Islands, the South Sandwich, and the South Shetland Islands.

3.9.1 The New Britain Arc, Papua New Guinea

The island of New Britain is located north of eastern Papua New Guinea together with New Ireland and other islands of the Bismarck Archipelago identified in Fig. 3.48. The volcanic activity along the northern coast of New Britain and on some of the islands in the Bismarck Sea is caused by subduction of the Solomon Sea Plate (at 6.2 to 12.4 cm yr^{-1}) into a trench located south of New Britain (Heming 1974; Johnson 1976). The resulting volcanic rocks range in composition from basalt to rhyolite of the calc-alkaline to high-K calc-alkaline suites (Fig. 3.4). The volcanic activity of the New Britain Arc extends westward to the small islands off the north coast of Papua New Guinea in Fig. 3.48.

Volcanic centers also occur in the Fly Highlands in the interior of the island of Papua New Guinea (Mackenzie and Chapell 1972; Hamilton et al. 1983a). The location of these volcanoes appears to be controlled by the presence of continental crust (20 to 30 km) representing the northern edge of the Australian Craton. One of the volcanic centers in the Fly Highlands contains a significant deposit of gold in the Porgera Intrusive Complex at 5°28' S and 143°05' E which was intruded at 6.0 ±0.3 Ma into sedimentary rocks of Cretaceous age. The rocks of the Porgera Complex are highly differentiated and range from mafic cumulates to feldspar porphyry dikes. The gold and related mineralization are of hydrothermal origin including high-grade gold-bearing veins containing tellurides of Au and Ag. Richards et al.

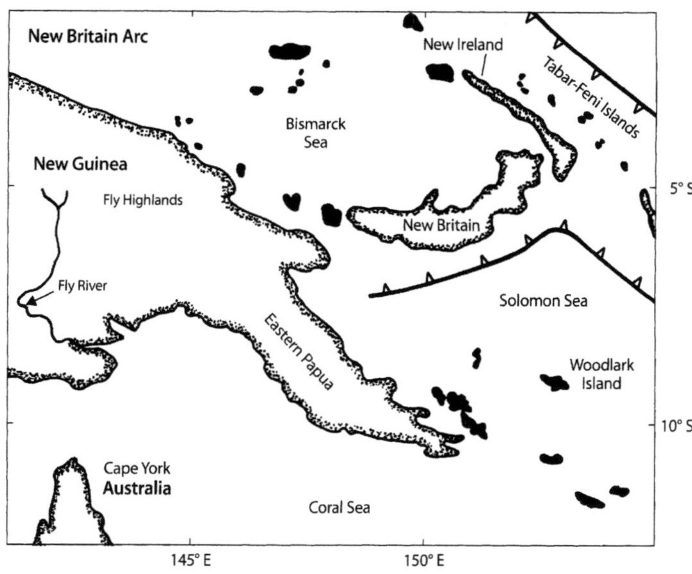

Fig. 3.48. Papua New Guinea and the New Britain Arc. The Late Cenozoic to Recent volcanic activity along the north coast of New Britain and on the islands in the Bismarck Sea was caused by subduction of the Solomon Sea Plate northward under New Britain. The volcanic activity in the Fly Highlands and in eastern Papua New Guinea is caused by presence of the northern edge of the Australian continental crust under these regions. The Tabar-Feni Group of islands results from subduction from the Pacific Plate westward into the Kilinailau Trench. However, subduction is not occurring at the present time (*Source:* adapted from National Geographic Society 1990)

1991) reported initial isotope ratios of Sr (0.703508 to 0.704130), Nd (0.512924 to 0.5123985), and Pb (^{206}Pb/^{204}Pb: 18.642 to 18.684) for unaltered intrusive rocks of the complex. These isotopic data are consistent with the derivation of the Porgera magma from the lithospheric mantle after prior metasomatic alteration by aqueous fluids emanating from subducted oceanic crust. The gold and associated metals were either derived from the Porgera magma or from the carbonaceous mudstones and siltstones (Cretaceous) which were intruded by the Porgera magma.

Volcanic rocks ranging in age from Eocene (45 Ma) to Recent occur as well in eastern Papua New Guinea and on the islands of the Solomon Sea (Fig. 3.48). This volcanic province is divided by the Owen Stanley Mountains into northern and southern sub-provinces. The northern sub-province includes Cape Nelson as well as Goodenough, Ferguson, and Normanby Islands, whereas the southern sub-province is confined to the southern coast of eastern Papua New Guinea. The rocks of this sub-province overlie submarine tholeiite basalts and related rocks of Eocene age. The volcanic activity in this area is of Miocene to Recent age, but is not associated with subduction occurring at the present time (Smith and Compston 1982).

New Ireland in Fig. 3.48 is separated from New Britain by a transform fault and is not part of the New Britain-New Guinea volcanic arc. The volcanoes of New Ireland became extinct at 15 Ma. Volcanic activity of Quaternary age subsequently developed on the islands of Tabar, Lihir, Tanga, and Feni northeast of New Ireland. Although this chain of islands is located west of the Kilinailau Trench, no subduction is occurring at the present time (Page and Johnson 1974; Kennedy et al. 1990).

The volcanoes of New Britain and of the offshore islands are located above a steeply-dipping Benioff zone whose depth approaches 500 km under the Mundua Islands located only about 200 km north of New Britain in the Bismarck Sea. The ^{87}Sr/^{86}Sr ratios of the lavas of New Britain and of the related islands in the Bismarck Sea are strongly clustered in Fig. 3.49a between 0.7034 and 0.7038 and are not significantly affected by the depth to the Benioff zone (Page and Johnson 1974). Their average Sr concentrations rise from 439 ppm in basalt to 495 ppm in basaltic andesite and then decline with increasing SiO$_2$ to 232 ppm in rhyolites. The relatively narrow range of ^{87}Sr/^{86}Sr ratios and the regular variation of Sr concentrations suggest that fractional crystallization of magmas derived from the mantle wedge played an important role in the petrogenesis of these rocks (Peterman et al. 1970a, Peterman and Heming 1974). The full range of possible petrogenetic models was considered by DePaolo and Johnson (1979).

The ^{87}Sr/^{86}Sr ratios of the lavas at the volcanic centers in the Fly Highlands and in the northern sub-province of eastern Papua New Guinea in Fig. 3.49b and c are dis-

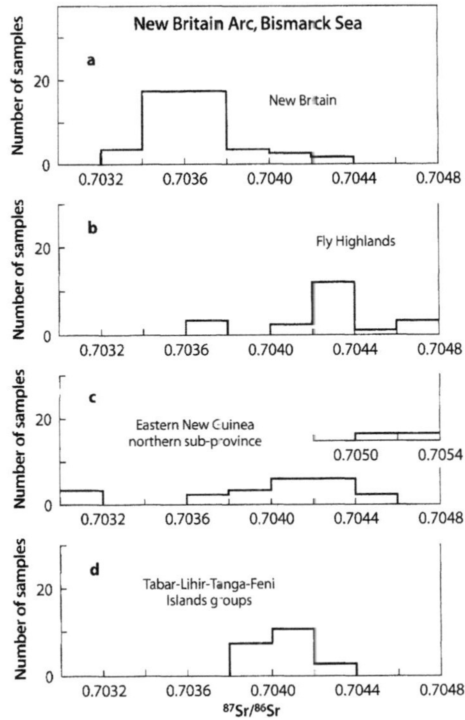

Fig. 3.49. Range of ^{87}Sr/^{86}Sr ratios of Late Tertiary to Recent lava flows in New Britain and related territories of Papua New Guinea; a North coast of New Britain and islands in the Bismarck Sea; b Fly Highlands in the central region of Papua New Guinea; c Northern sub-province of eastern Papua New Guinea and islands in the Solomon Sea; d Island groups located northeast of New Ireland (*Sources:* Peterman et al. 1970a; Peterman and Heming 1974; Page and Johnson 1974; DePaolo and Johnson 1979; Smith and Compston 1982; Kennedy et al. 1990)

tinctly higher than those of the New Britain Arc. Fifty percent or more of the specimens analyzed from each of these areas have ^{87}Sr/^{86}Sr ratios between 0.7040 and 0.7044. The enrichment in radiogenic ^{87}Sr may indicate that mantle-derived magmas either assimilated rocks of the continental crust or that magmas were generated by partial melting of enriched source rocks (Page and Johnson 1974; Hamilton et al. 1983a). Although, the volcanoes in the Fly Highlands and in eastern Papua New Guinea are not presently underlain by a Benioff zone, Hamilton et al. (1983a) suggested that magma may have formed as a result of uplift in Pliocene time following active subduction during the Cretaceous Period.

The Quaternary volcanic activity on the Tabar-Feni Island group is not directly related to that of the New Britain Arc. The lavas are Na-rich undersaturated rocks of the shoshonite suite. About 90% of the rocks analyzed by Page and Johnson (1974) and by Kennedy et al. (1990) have $^{87}Sr/^{86}Sr$ ratios between 0.7038 and 0.7042 relative to 0.7080 for E&A. The Sr concentrations of these rocks range from 948 to 1780 ppm which is typical for undersaturated alkali-rich rocks. Kennedy et al. (1990) considered the possibility that the magmas on the island of Lihir had assimilated about 4% marine carbonate having 2000 ppm Sr and $^{87}Sr/^{86}Sr = 0.70905$. However, the limited range of isotope compositions of Sr, Nd, and Pb favors more plausible explanations of magma formation by small-scale decompression melting of source rocks in the mantle that had been enriched in Rb and other alkali elements during preceding subduction (Kennedy et al. 1990).

3.9.2   The Alnöite of Malaita, Solomon Islands

The Solomon Islands, like the Tabar-Feni Islands, are located along the boundary between the Pacific Plate and the Indo-Australian Plate (Fig. 3.50). The Pacific Plate adjacent to the Solomon Islands contains the Ontong Java Plateau (Sect. 2.11) where the oceanic crust locally reaches a thickness of up to 42 km (Neal and Davidson 1989). The principal Solomon Islands (Bougainville, Choiseul, New Georgia, Santa Isabel, Guadalcanal,

Malaita, and San Cristóbal) form the eastern border of the Solomon-Woodlark Sea which is a back-arc basin.

The island of Malaita in Fig. 3.50 is especially noteworthy because it contains alnöite intrusives and because it exposes a cross-section of the oceanic crust of the Ontong Java Plateau (Sect. 2.11) which was obducted over the oceanic crust of the Solomon Arc, starting at about 10 Ma (Bielski-Zyskind et al. 1984). Consequently, Malaita is not a volcanic island like the other Solomon Islands and alnöite is an alkali-rich rock unrelated to the calc-alkaline rocks in the island arcs. The alnöite of Malaita is mentioned here, nevertheless, because it is a noteworthy type of rock and because Malaita does occur in close proximity to the island arcs of the western Pacific Ocean. Therefore, what follows is intended to complement the account of the Ontong Java Plateau in Sect. 2.11 and the description in Sect. 6.7 of the alkalirich rocks and carbonatites on Alnö Island off the coast of Sweden where alnöites were first recognized.

Alnöites were defined by Rock (1986) as ultramafic lamprophyres containing melilite but lacking feldspar. The alnöites of Malaita occur in pipe-like bodies that were explosively emplaced into a folded sequence of limestones and mudstones of Late Cretaceous to Early Tertiary age. The sedimentary rocks are underlain by tholeiite basalt (Alite volcanics) of the oceanic crust of the Ontong Java Plateau. The alnöite intrusion was emplaced at 34 Ma according to a U-Pb date determined by Davis (1978) from a zircon xenocryst.

The alnöite of Malaita is characterized by low concentrations of SiO_2 (36.1%), Al_2O_3 (6.90%), CaO (8.18%), Na_2O (0.84%), and high concentrations of TiO_2 (2.87%), MgO (23.0%), P_2O_5 (1.35%), LOI (4.58%), whereas K_2O (1.97%), total Fe as Fe_2O_3 (12.1%), and MnO (0.20%) are comparable to high-K basalt like those on nearby Lihir Island (Kennedy et al. 1990).

The alnöite contains very large phenocrysts of garnet (up to 8.2 kg) and subcalcic diopside (up to 2.5 kg) as well as clinopyroxene-ilmenite intergrowths, bronzite, ilmenite, phlogopite, and zircon (Nixon et al. 1980; Neal and Davidson 1989). In addition, the alnöite intrusion contains ultramafic xenoliths (lherzolite) derived from the lithospheric mantle at depths between 60 and 110 km (Nixon and Boyd 1979) and analyzed by Bielski-Zyskind et al. (1984).

The different kinds of phenocrysts in the alnöite have different $^{87}Sr/^{86}Sr$ ratios indicating that they formed sequentially from a magma whose $^{87}Sr/^{86}Sr$ ratio was changing. The augite crystals analyzed by Neal and Davidson (1989) have the lowest average $^{87}Sr/^{86}Sr$ ratio of 0.70325 ±0.00015 ($2\bar{\sigma}$, $N = 5$) relative to 0.71025 for NBS 987. The diopside and phlogopite crystals have higher average $^{87}Sr/^{86}Sr$ ratios of 0.70381 ±0.00012 ($2\bar{\sigma}$, $N = 6$) and 0.70383 ±0.00015 ($2\bar{\sigma}$, $N = 6$), respectively. The average $^{87}Sr/^{86}Sr$ ratio of the alnöite matrix itself is 0.70442 ±0.00020 ($2\bar{\sigma}$, $N = 4$) and is higher than the

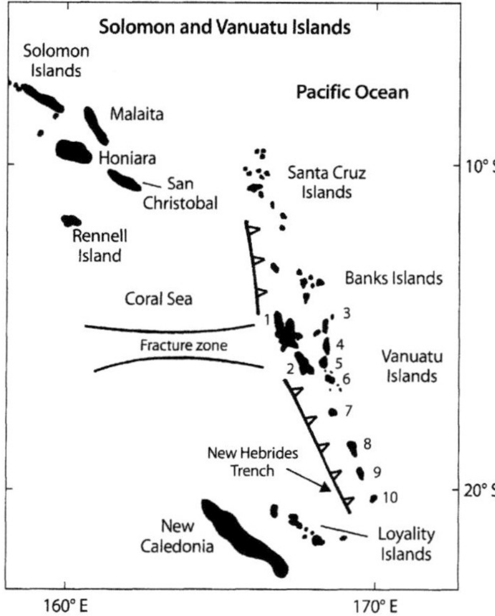

Fig. 3.50. The Vanuatu Island arc in the western Pacific Ocean. The principal islands are identified by number: *1.* Espiritu Santo; *2.* Malekula; *3.* Maewo; *4.* Pentecost; *5.* Ambrim; *6.* Epi; *7.* Efate; *8.* Eromanga; *9.* Tanna; *10.* Aneityum (*Source:* adapted in part from Briqueu and Lancelot 1983)

^{37}Sr/^{86}Sr ratios of any of the phenocrysts analyzed by Neal and Davidson (1989).

The crystallization of the different kinds of phenocrysts enriched the residual magma in Sr because their crystal-liquid partition coefficients (Sect. 1.10.1) are all less than one. This explains why the alnöite matrix analyzed by Neal and Davidson (1989) has a high average Sr content of 1293 ±186 ppm ($2\bar{\sigma}$, $N = 4$), whereas the Sr concentrations of the phenocrysts are all less than about 90 ppm. The average Rb concentration of the alnöite matrix is comparatively low at 57.3 ±12.2 ppm ($2\bar{\sigma}$, $N = 4$) because Rb was taken up by phlogopite (D_1^x(Rb) = 3.06; Schnetzler and Philpotts 1970) which has an average Rb concentration of 404 ±15 ppm ($2\bar{\sigma}$, $N = 6$).

Neal and Davidson (1989) proposed that the original magma formed by decompression melting in a rising diapir of the asthenospheric mantle. The ascent of the diapir was halted when it reached the lithosphere under the Ontong Java Plateau and its chemical composition then changed by formation of the large phenocrysts. At the same time, the ^{87}Sr/^{86}Sr ratio of the magma increased because of assimilation of seawater-altered basalt having ^{87}Sr/^{86}Sr = 0.710 and Sr = 160 ppm. The augite crystals formed first, followed by diopside and phlogopite. Garnet, which comprises 27% of the phenocryst minerals, crystallized with augite and diopside, but its ^{87}Sr/^{86}Sr ratio was not reported. Ultimately, fracturing of the lithosphere under the Ontong Java Plateau allowed the residual alnöite magma and associated phenocrysts to be injected into the oceanic crust.

The ^{87}Sr/^{86}Sr ratios of lherzolite and garnet-pyroxenite xenoliths reported by Bielski-Zyskind et al. (1984) are nominally higher than those of the alnöite samples they analyzed. Therefore these xenoliths do not represent the source rocks from which the proto-alnöite magma originated, but are accidental inclusions entrained in the magma on its way to the surface.

Andesite and dacite of Quaternary age from Bougainville, analyzed by Page and Johnson (1974), have ^{87}Sr/^{86}Sr ratios of 0.7037 and 0.7039, respectively, relative to 0.7080 for E&A. The ^{87}Sr/^{86}Sr ratios and chemical compositions of these lavas reported by Taylor et al. (1969) are consistent with subduction of oceanic crust into the Solomon Trench west of Bougainville.

3.9.3 The Vanuatu Island Arc

The Vanuatu (New Hebrides) Islands (Espiritu Santo, Malekula, Maewo, Obe, Pentecost, Ambrim, Epi, Tongoa, Efate, Eromanga, Tanna, and Aneityum) in Fig. 3.50 are located southeast of the Solomon Islands between latitudes 14 to 20° S. The volcanic activity of the Vanuatu Islands is related to subduction of oceanic crust of the Coral Sea eastward into the New Hebrides Trench located west of the islands. The volcanic rocks of the

Vanuatu Islands have been described by Colley and Warden (1974) and Gorton (1977) among others. Although volcanic activity may have started in Eocene time, the oldest rocks on the northern islands (Espiritu Santo, Malekula, Maewo, and Pentecost) are submarine lavas and volcaniclastic sediment of Oligocene to Miocene age (Gorton 1977). The islands in the central and southern part of the chain (Obe to Aneityum) formed by volcanic activity during Pleistocene to Recent time.

The chemical compositions of Quaternary lavas on the central islands (Obe, Ambrim, Epi and Tongoa) range from calc-alkaline to high-K calc-alkaline. Data by Colley and Ash (1971) reported by Gorton (1977) also demonstrate the occurrence of high-K calc-alkaline lavas on Eromanga Island. Gorton (1977) determined that the magmas of the high-K suite formed at a depth of about 50 km by about 10% partial melting of moderately hy-

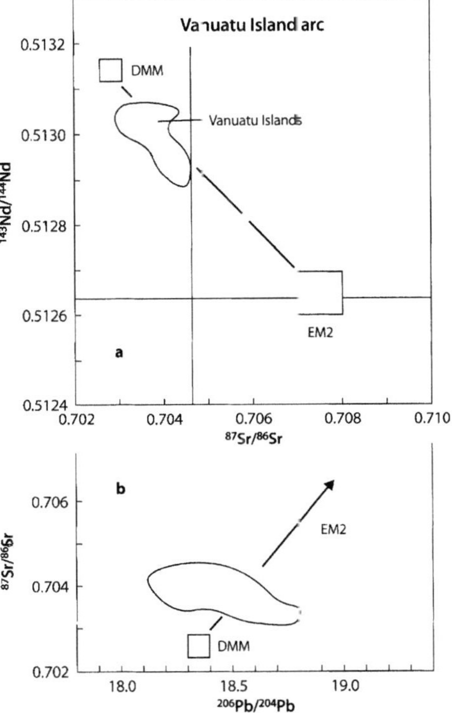

Fig. 3.51 a,b. Isotope ratios of Sr, Nd, and Pb of volcanic rocks on the Vanuatu Island arc. The ^{87}Sr/^{86}Sr and ^{143}Nd/^{144}Nd ratios in Part *a* form a small data field aligned between the DMM and EM2 components. The same lavas in Part *b* are likewise a mixture of DMM and EM2. Therefore, the isotope ratios of Sr, Nd, and Pb are consistent with the standard model of the petrogenesis of volcanic rocks in oceanic island arcs (*Sources:* Peate et al. 1997)

drous rocks in the mantle wedge. Therefore, the chemical compositions of these lavas can be accounted for by the standard model of petrogenesis in intra-oceanic subduction zones (Dupuy et al. 1982).

Briqueu et al. (1982) reported that the $^{87}Sr/^{86}Sr$ ratios of the lavas on the northern islands of Epi, Tongoa, and Ambrim range from 0.7037 to 0.7043, but decrease to about 0.7030 on the southern islands of Eromanga, Tanna, Aneityum, and Futana. They attributed the longitudinal variation of $^{87}Sr/^{86}Sr$ ratios among the Vanuatu Islands to the subduction of the d'Entrecasteaux fracture zone into the New Hebrides Trench adjacent to the northern islands (Briqueu and Lancelot 1983; Peate et al. 1997).

The oceanic crust of the Coral Sea being subducted into the New Hebrides Trench consists of gabbro and altered tholeiite basalt of Eocene age overlain by about 650 m of volcaniclastic sediment and nannofossil chalk ranging in age from late Eocene to Recent. The initial $^{87}Sr/^{86}Sr$ ratios of the basalt and gabbro (after leaching with 6 N HCl) range from 0.70250 to 0.70351 relative to 0.71025 for NBS 987 and are significantly lower than those of unleached whole-rock samples (Briqueu and Lancelot 1983). These results indicate that the basaltic rocks originally were normal MORBs derived from depleted source rocks in the mantle and that they were subsequently altered by interacting with seawater. The $^{87}Sr/^{86}Sr$ ratios of the sedimentary rocks (whole rocks, unleached) vary from 0.70355 for volcanic conglomerate to 0.70750 in clay containing Mn micronodules and volcanic glass shards. The sediment contains a component of marine Sr that is the principal source of radiogenic ^{87}Sr in the lavas of the Vanuatu Island arc.

The isotope ratios of Sr, Nd, and Pb of the volcanic rocks on the islands of the Vanuatu Arc in Fig. 3.51a,b define small data fields aligned between the DMM and EM2 components (Peate et al. 1997). In this regard, the volcanic rocks of the Vanuatu Island arc are similar to those of the Mariana Islands (Fig. 3.7), Aleutian Islands (Fig. 3.15), Fiji Islands (Fig. 3.32), Tonga and Kermadec Islands (Fig. 3.34), northeast Banda Arc (Fig. 3.41), and the Bataan Arc, Philippines (Fig. 3.41). At all of these sites, oceanic crust is being subducted in an oceanic setting without significant contamination of magmas by rocks of the continental crust. Instead, at these sites the rocks of the mantle wedge have been contaminated by the transfer of Sr, Nd, and Pb by aqueous fluids originating from the EM2 component which consists of subducted oceanic crust and overlying marine sediment derived in part from continental sources.

3.9.4 Scotia Arc, South Atlantic Ocean

The South Sandwich Islands in Fig. 3.52 form the Scotia Arc in the South Atlantic Ocean between South America and the Antarctic Peninsula. The names of the islands

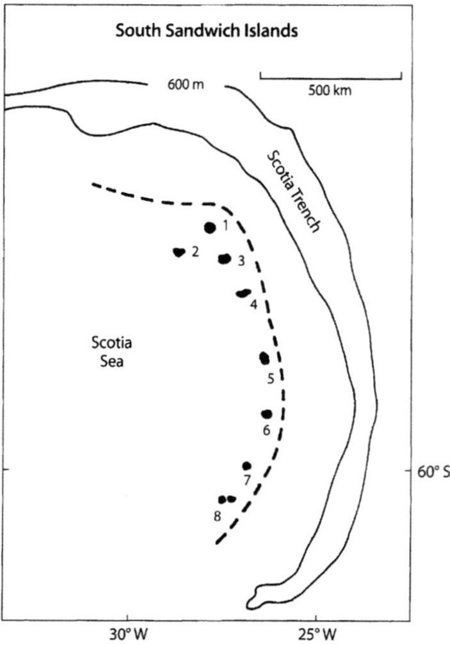

Fig. 3.52. The South Sandwich Islands in the Atlantic Ocean. The islands in this arc are identified by number: 1. Zavodovski; 2. Leskov; 3. Visokoi; 4. Candlemas; 5. Saunders; 6. Montagu; 7. Bristol; 8. Thule. The volcanic activity results from westward subduction of oceanic lithosphere into the Scotia Trench (*Source:* adapted from Hawkesworth et al. 1977)

from north to south are: Zavodovski, Leskov, Visikoi, Candlemas, Saunders, Montagu, Bristol, and Thule. These islands are located west of the Scotia Trench into which oceanic crust of the South American Plate is being subducted from the east at about 5.4 cm yr^{-1}. The Benioff zone dips west and is located at a depth of about 180 km under the islands. The lavas on the South Sandwich Islands range in age from about 4 Ma to Recent (Hawkesworth et al. 1977). The Scotia Sea west of the islands is a back-arc basin containing an active spreading ridge that strikes approximately north-south (not shown in Fig. 3.49). The spreading ridge, named the Scotia Sea Rise by Hawkesworth et al. (1977), is truncated at its northern and southern ends by transform faults that strike east-west.

The lavas on the South Sandwich Islands range in composition from basalt to rhyolite and belong to the low-K calc-alkaline series (Fig. 3.4). Their $^{87}Sr/^{86}Sr$ ratios in Fig. 3.53a are strongly clustered between 0.7038 and 0.7040 relative to 0.7080 for E&A. In some cases, samples were boiled in deionized water (Gledhill and Baker 1973) or treated with HCl (Hawkesworth et al.

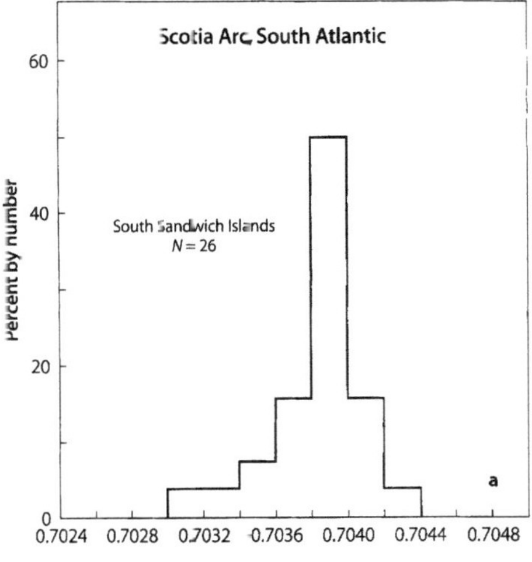

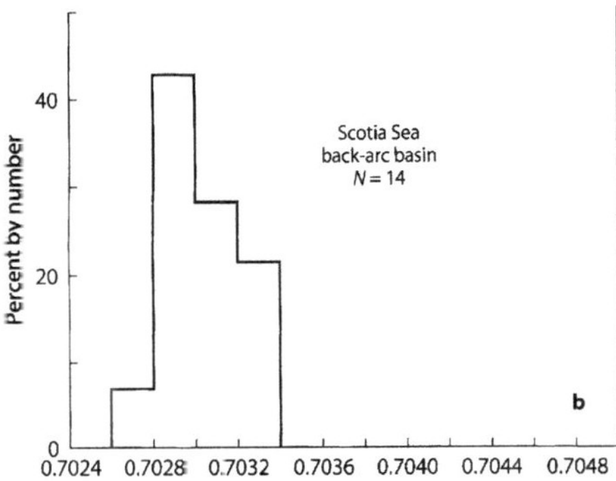

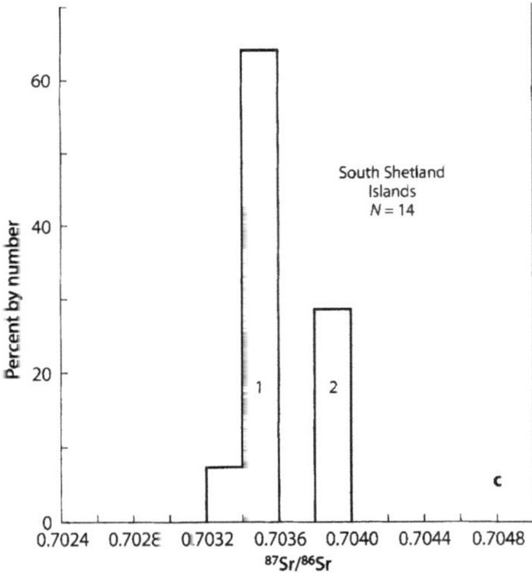

1977) in order to remove alteration products prior to isotope analysis. The elevated $^{87}Sr/^{86}Sr$ ratios of the lavas on the South Sandwich Islands (compared to normal MORBs) are attributable to the transfer of Sr from the subducted sediment into the rocks of the overlying mantle wedge. The isotope compositions of Pb in lavas from the Scotia Arc were reported by Barreiro (1983).

The tholeiite basalts in Fig. 3.53b from the Scotia Sea Rise in the back-arc basin have significantly lower $^{87}Sr/^{86}Sr$ ratios than the lavas on the South Sandwich Islands. Nearly 50% of the analyzed samples have $^{87}Sr/^{86}Sr$ ratios between 0.7028 and 0.7030. This result is typical of volcanic rocks in other back-arc basins (Sect. 3.2.3) and indicates that magmas formed by decompression melting of rocks in the mantle wedge without significant additions of Sr by aqueous fluids arising from the subducted sediment (Saunders and Tarney 1979).

3.9.5 South Shetland Islands, Antarctic Peninsula

The South Shetland Islands are located west of the northern tip of the Antarctic Peninsula and include King George and Livingston Islands, both of which are composed of continental crust (dated by Tanner et al. 1982 and Pankhurst et al. 1980) and are located between the South Shetland Trench and the Bransfield Strait (back-arc). K-Ar dates of low-K tholeiites and andesites reported by Pankhurst and Smellie (1983) indicate that the volcanic activity on the South Shetland Islands occurred from 130 to 30 Ma and migrated northeastward along the arc. After a hiatus of about 28 million years, volcanic activity resumed at about 2 Ma with the eruption of more alkali-rich basalts. The volcanically active islands (Bridgeman, Penguin, and Deception) lie along the northwestern side of Bransfield Strait and are associated with a narrow trough up to 2 km deep that extends from Deception Island towards Bridgeman Island. The volcanic activity on these islands has been attributed to seafloor spreading in Bransfield Strait even though the rate of subduction in the South Shetland Trench decreased significantly at about 4 Ma. Therefore, the lavas on the volcanic islands may be recording the early stage of magma formation in a back-arc basin (Weaver et al. 1979).

Fig. 3.53. a $^{87}Sr/^{86}Sr$ ratios of volcanic rocks on the South Sandwich Islands in the southern Atlantic Ocean; b $^{87}Sr/^{86}Sr$ ratios of volcanic rocks dredged from the spreading ridge in the Scotia Sea which is a back-arc basin west of the South Sandwich Islands. The difference in $^{87}Sr/^{86}Sr$ ratios of the rocks on the island arc and in the associated back-arc basin is unmistakable; c $^{87}Sr/^{86}Sr$ ratios of the South Shetland Islands off the west coast of the Antarctic Peninsula. The $^{87}Sr/^{86}Sr$ ratios of Deception and Bridgeman Islands (1) are lower than those of Penguin Island (2) (Sources: Gledhill and Baker 1973; Hawkesworth et al. 1977; Cohen and O'Nions 1982b; Saunders and Tarney 1979; Weaver et al. 1979)

The lavas of Deception and Bridgeman Islands are subalkaline and range in composition from basalt to rhyolite (Deception), whereas the lavas of Bridgeman and Penguin Islands consist of basaltic andesite. The $^{87}Sr/^{86}Sr$ ratios in Fig. 3.53c range from 0.70336 (Deception) to 0.70396 (Penguin) relative to 0.7080 for E&A (Weaver et al. 1979). The average $^{87}Sr/^{86}Sr$ ratio of the lavas on Penguin Island (0.70389 ±0.00005, no. 2 in Fig. 3.53c) is higher than those of Bridgeman (0.70351 ±0.00006) and Deception Islands (0.70343 ±0.00004, no. 1 in Fig. 3.53c). The $^{87}Sr/^{86}Sr$ ratios of the lavas on these islands are compatible with those of island-arc volcanics, but are significantly higher than those of normal MORBs. Therefore, the magma sources below the South Shetland Islands were altered during an earlier period of active subduction in the South Shetland Trench by transfer of radiogenic ^{87}Sr and LIL elements from subducted sediment.

3.10 The Lesser Antilles, Caribbean Sea

The Lesser Antilles extend north from Grenada, 150 km off the coast of Venezuela, and include a large number of islands, including St. Vincent, St. Lucia, Martinique, Dominica, Guadeloupe, St. Kitts, Antigua, and Barbuda identified in Fig. 3.54. The distance from Grenada in the south to the northernmost island of Anguilla is about 675 km. The northern end of this island arc splits into two branches. The islands of the eastern branch (Anguilla, St. Martin, St. Barthelemey, Barbuda, Antigua, east Guadeloupe, and Marie Galante) are composed of Cenozoic volcanics overlain by sedimentary rocks, whereas the islands of the western branch (St. Kitts, Nevis, Montserrat, and western Guadeloupe) as well as the southern islands (Dominica, Martinique, St. Lucia, St. Vincent, and Grenada) contain volcanic rocks of Recent age and, in many cases, have active volcanoes (Pushkar 1968).

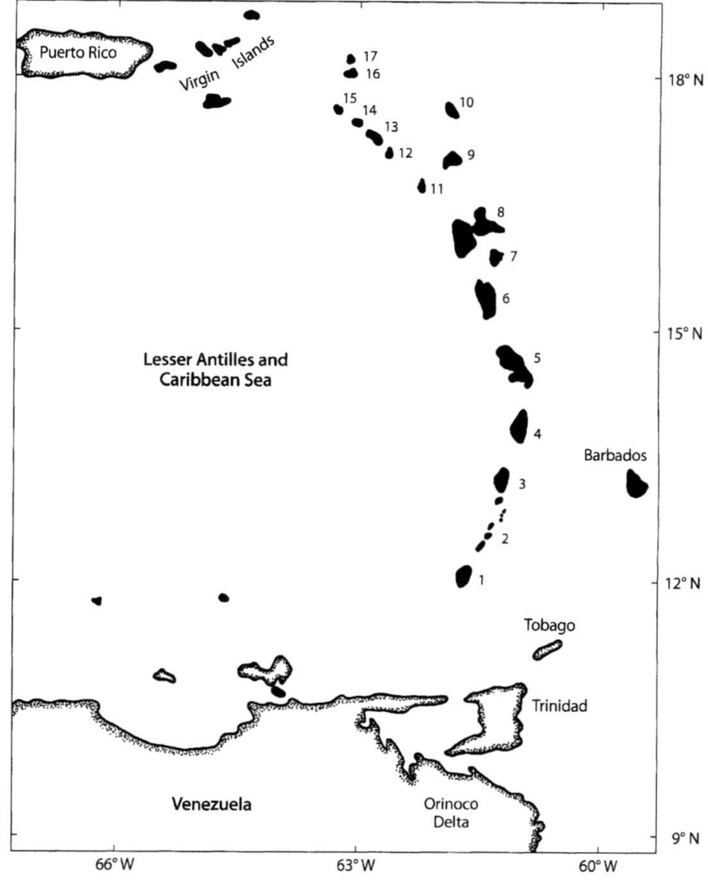

Fig. 3.54.
The Lesser Antilles in the Caribbean Sea. The islands are identified by number: *1*. Grenada; *2*. Grenadines; *3*. St. Vincent; *4*. St. Lucia; *5*. Martinique; *6*. Dominica; *7*. Marie Galante; *8*. Guadeloupe; *9*. Antigua; *10*. Barbuda; *11*. Montserrat; *12*. Nevis; *13*. St. Kitts; *14*. St. Eustacius; *15*. Saba *16*. St. Martin; *17*. Anguilla (*Source:* adapted from National Geographic Society 1990)

The volcanic activity of the Lesser Antilles results from subduction of oceanic lithosphere of the Atlantic Ocean into the Puerto Rico Trench which is presently recognizable only north of the arc. Fisher and Hess (1963) suggested that the Puerto Rico Trench formerly extended along the eastern side of the Lesser Antilles Arc to a location near the mouth of the Orinoco River of Venezuela. Sediment deposited by this river filled the southern end of the trench and was later compressed during subduction to form the Barbados Ridge including the island of Barbados.

In this tectonic setting, the lavas on the southern islands (e.g. Grenada, St. Vincent, St. Lucia, Martinique, and Dominica) contain Sr, Nd, and Pb derived from the sediment that originated from the Precambrian rocks of the Guyana Shield of Venezuela. In addition, the isotope ratios of the volcanic rocks of the southern islands appear to have changed with time perhaps depending on the composition of sediment being subducted. Therefore, the volcanic rocks of the southern islands have a crustal imprint because of the continental provenance of the subducted sediment, whereas the volcanic rocks of the northern islands are similar to volcanic rocks on intra-oceanic island arcs elsewhere (e.g. Tonga-Kermadec, Aleutian, and Mariana Islands) (Hawkesworth and Powell 1980).

Pushkar (1968) reported that two samples of calcareous ooze taken at the northern end of the Barbados Ridge east of the island of Dominica have high Sr concentrations of 812 and 3766 ppm (dry weight) and $^{87}Sr/^{86}Sr$ ratios of 0.7083 and 0.7087 relative to E&A = 0.7080. In addition, White et al. (1985) reported $^{87}Sr/^{86}Sr$ ratios ranging from 0.7090 to 0.7256 for sediment on the Barbados Ridge and the Demerara Plain at the mouth of the Orinoco River and east of the Lesser Antilles. The occurrence of sediment enriched in radiogenic ^{87}Sr east of the southern end of the Lesser Antilles is therefore well established.

The chemical composition and petrogenesis of the lavas on the islands of the Lesser Antilles, with particular emphasis on Grenada, have been discussed by Cawthorn et al. (1973), Arculus (1976, 1978), Brown et al. (1977a), Maury and Westercamp (1985), and others identified in this section.

3.10.1 Islands of the Lesser Antilles

The lavas extruded on the southern islands (Grenada, St. Vincent, St. Lucia, Martinique and Dominica) have elevated $^{87}Sr/^{86}Sr$ ratios in Fig. 3.55 compared to the northern islands because the southern segment of the Puerto Rico Trench east of the Lesser Antilles contains sediment deposited by the Orinoco River of Venezuela, whereas the subducted sediment in the northern segment consists of a mixture calcareous ooze and volcano-

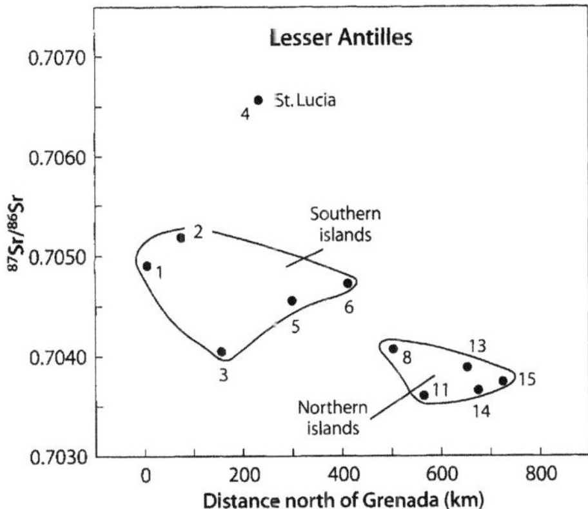

Fig. 3.55. Average $^{87}Sr/^{86}Sr$ ratios of volcanic rocks on the islands of the Lesser Antilles. In spite of real variations of these ratios within several of the islands and in spite of overlap of the $^{87}Sr/^{86}Sr$ ratios between islands, average values suggest that the southern islands (1. Grenada; 2. Grenadines; 3. St. Vincent; 4. St. Lucia; 5. Martinique; 6. Dominicia) collectively have higher $^{87}Sr/^{86}Sr$ ratios than the northern islands (8. Guadeloupe; 11. Montserrat; 13. St. Kitts; 14. St. Eustacius; 15. Saba). The islands are numbered as in Figure 3.54, and distances were measured from the south coast of Grenada (*Sources:* White and Patchett 1984; Hawkesworth et al. 1979b; Pushkar et al. 1973; Pushkar 1968; Hedge and Lewis 1971; White and Dupré 1986; Davidson 1986)

genic sediment with subsidiary amounts of terrigenous detritus. The average $^{87}Sr/^{86}Sr$ ratio of basalts and andesites in the Grenadine Islands (no. 2, Fig. 3.55) is 0.7052 ±0.00016 ($2\bar{\sigma}$, $N = 6$; Pushkar 1968; Hedge and Lewis 1971) which is marginally higher than the average $^{87}Sr/^{86}Sr$ ratio of the rocks on Grenada (0.70488 ±0.00011, ($2\bar{\sigma}$, $N = 55$, Hawkesworth et al. 1979b). However, basalts and andesites from Soufriere Volcano and from other sites on St. Vincent (no. 3) have comparatively low $^{87}Sr/^{86}Sr$ ratios with an average of 0.7040 for 27 specimens. The calc-alkaline rocks of St. Lucia (no. 4) north of St. Vincent have a wide range of $^{87}Sr/^{86}Sr$ ratios between 0.7035 (andesite, Vigie quarry, Pushkar 1968) to 0.7092 (andesite, Qualibou Volcano, Pushkar et al. 1973). The wide range of $^{87}Sr/^{86}Sr$ ratios of the lavas on St. Lucia was confirmed by Davidson (1987). Because of the isotopic heterogeneity of Sr in the volcanic rocks of St. Lucia, average values are likely to be biased by sample selection. Nevertheless, the data of Pushkar (1968), Pushkar et al. (1973), and White and Dupré (1986) yield a mean of 0.7066 for 13 specimens. The elevated $^{87}Sr/^{86}Sr$ ratios continue to Martinique (no. 5) and Dominica (no. 6) where seven rock and mineral specimens have an average $^{87}Sr/^{86}Sr$ ratio of 0.7047.

The $^{87}Sr/^{86}Sr$ ratios of the islands in the northern segment of the Lesser Antilles in Fig. 3.55 range from 0.70359 (basaltic andesite, Montserrat; White and Patchett 1984) to 0.7046 (dacite, St. Kitts; Pushkar 1968) and therefore

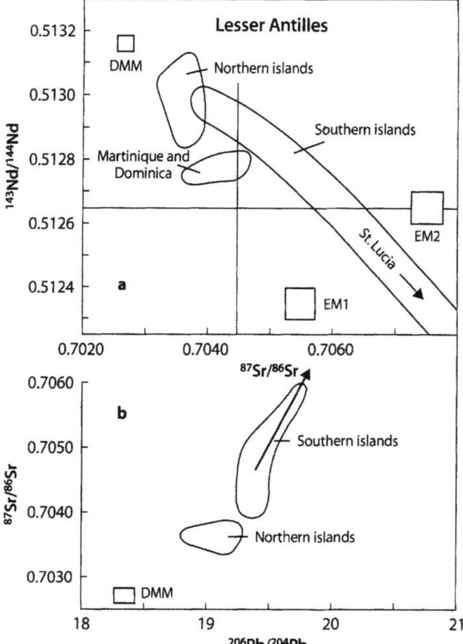

Fig. 3.56. a Isotope ratios of Sr and Nd of late Tertiary lavas on the islands of the Lesser Antilles in the Caribbean Sea. The southern islands (Grenada, Kick-em-Jenny, St. Vincent, St. Lucia, Martinique, and Dominica) have higher $^{87}Sr/^{86}Sr$ and lower $^{143}Nd/^{144}Nd$ ratios than the northern islands (Guadeloupe, Montserrat, Redonda, St. Kitts, St. Eustatius, Statia, and Saba). However, both data sets are aligned between the depleted-mantle component (DMM) and sediment derived from the continental crust. The rocks on Martinique and Dominica deviate from the pattern presumably because their $^{143}Nd/^{144}Nd$ ratios have low values; **b** Isotope ratios of Sr and Pb of volcanic rocks in the Lesser Antilles. The $^{206}Pb/^{204}Pb$ ratios are unusually high because the sediment from which Sr, Nd, and Pb were derived originated from the Precambrian shield of Venezuela (*Sources:* White and Dupré 1986; Davidson 1987; Hart 1988)

overlap the $^{87}Sr/^{86}Sr$ ratios in the southern islands. Nevertheless, the average $^{87}Sr/^{86}Sr$ ratio of 19 volcanic rocks on these islands (0.70381) is distinctly lower than the average $^{87}Sr/^{86}Sr$ ratios of the southern islands, except for St. Vincent.

The isotope ratios of Sr and Nd of the lavas on the northern and southern islands of the Lesser Antilles reported by White and Dupré (1986) and by Davidson (1987) define two fields in Fig. 3.56a. The alignment of the northern islands relative to the DMM mantle components is consistent with the theory of petrogenesis of island-arc volcanics based on the addition of varying amounts of Sr and Nd from subducted marine sediment (EM1) to the rocks of the mantle wedge. The magmas on the southern islands (especially St. Lucia) incorporated sediment derived from the continental crust (EM2)

Table 3.4. Average $^{206}Pb/^{204}Pb$ ratio of the southern islands (Grenada, St. Vincent, St. Lucia, Martinique, and Dominica) and the northern islands (Guadeloupe, Montserrat, St. Kitts, St. Eustatius, and Saba) of the Lesser Antilles (White and Dupré (1986)

	$^{206}Pb/^{204}Pb$
Northern islands	19.011 ±0.076 (2σ, N = 11)
Southern islands	19.378 ±0.073 (2σ, N = 19)

which increased the $^{87}Sr/^{86}Sr$ ratios but decreased their $^{143}Nd/^{144}Nd$ ratios of the volcanic rocks on these islands. The volcanic rocks on Martinique and Dominica define a third field in Fig. 3.56a, perhaps because the Sr and Nd that were added to their magma sources originated from subducted sediment resembling EM1 rather than EM2.

The lavas of the Lesser Antilles also have unusually high $^{206}Pb/^{204}Pb$ ratios ranging from 18.847 (Saba) to 19.745 (St. Lucia). In addition, the data of White and Dupré (1986) indicate that the average $^{206}Pb/^{204}Pb$ ratio of the southern islands (Grenada, St. Vincent, St. Lucia, Martinique, and Dominica) is higher than that of the northern islands (Guadeloupe, Montserrat, St. Kitts, St. Eustatius, and Saba; see Table 3.4).

Accordingly, the $^{87}Sr/^{86}Sr$ and $^{206}Pb/^{204}Pb$ ratios define two separate fields in Fig. 3.56b much like the Sr and Nd isotope ratios in Part a. The lavas of Martinique and Dominica do not deviate from the Sr-Pb array defined by the southern islands. Therefore, the deviation of the Sr-Nd isotope ratios of Martinique and Dominica in Fig. 3.53a is attributable to low $^{143}Nd/^{144}Nd$ ratios rather than to low $^{87}Sr/^{86}Sr$ ratios of the lavas on these islands.

The process by means of which crustal sediment was incorporated into the magmas that formed under the southern islands has been the subject of much speculation (e.g. Hawkesworth et al. 1979b; White and Dupré 1986; Davidson 1985, 1986, 1987; Chabaux et al. 1999; and others). The magmas could have been contaminated at the source by sediment that was injected into the mantle wedge, or magmas from the mantle wedge could have assimilated sediment in the overlying oceanic crust, or the magmas could have mixed with partial melts of sedimentary and volcanic rocks of the oceanic crust. These issues as well as the evidence for or against differentiation of the magmas by fractional crystallization are best considered by example of specific islands.

3.10.2 Grenada

The volcanic rocks of Grenada consist of an unusual assemblage of silica-undersaturated alkali-rich rocks (ankaramites, basanitoids, alkali-olivine basalts) and calc-alkaline andesites and dacites. These rocks range in age from about 25 Ma (Miocene) to Recent and were deposited on a basement of deformed volcaniclastic and sedimentary rocks of Eocene to Oligocene age. The $^{87}Sr/^{86}Sr$ ratios of the volcanic rocks of Grenada range from 0.70394 to 0.70579 relative to E&A = 0.7080 based on measurements by Hawkesworth et al. (1979b).

The alkali-olivine basalts define a linear array in coordinates of $^{87}Sr/^{86}Sr$ and 1/Sr in Fig. 3.57, whereas

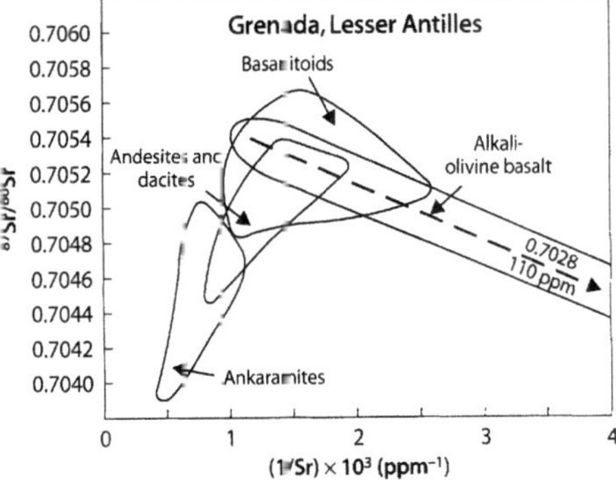

Fig. 3.57. Strontium-isotope mixing diagram of the alkali-rich undersaturated volcanic rocks of Grenada, Lesser Antilles. The distribution of data points suggests that the petrogenesis of each of the major rock types requires a unique set of circumstances. The alkali olivine basalts may have formed by contamination of MORB-like magmas derived from the mantle wedge by a component having $^{87}Sr/^{86}Sr$ ratios between 0.705 and 0.706. The basanitoids overlap the array of alkali basalts, whereas the andesites and ankaramites are characterized by high Sr concentrations and form separate arrays on the mixing diagram (*Source:* Hawkesworth et al. 1979b)

the ankaramites, andesites, and basanitoids form three overlapping clusters. The linear array of the alkali-olivine basalts extrapolates to MORB-like values of $^{87}Sr/^{86}Sr$ = 0.7028 at about 110 ppm Sr. Therefore, the alkali basalts may be the descendants of magmas that formed in the mantle wedge after it had been contaminated with Sr having an elevated $^{87}Sr/^{86}Sr$ ratio. The relation of the ankaramites, andesites, and basanitoids to the alkali basalts is not immediately apparent in Fig. 3.57. Hawkesworth et al. (1979b) demonstrated that the $^{87}Sr/^{86}Sr$ ratios of the ankaramites increase with rising silica concentrations from 47.72 to 51.55%. These observations indicate that the ankaramites, andesites, basanitoids, and alkali basalts on Grenada were not derived by fractional crystallization of a common magma and that special circumstances apply to the petrogenesis of each suite.

The wide range of $^{87}Sr/^{86}Sr$ ratios (0.70451 to 0.70579) of the andesites excludes the possibility that they formed by fractional crystallization of a common magma. However, they may be the products of magma mixing or contamination, in which case one of the components required by the data had a low $^{87}Sr/^{86}Sr$ ratio and a high Sr content. The ankaramites satisfy this requirement, whereas detrital sediment of continental origin has the high $^{87}Sr/^{86}Sr$ ratios and low Sr content required of the other component (White et al. 1985). Alternatively, the high-Sr andesites may be descendants of ankaramite magmas, whereas the low-Sr andesites and dacites may be derivatives of the alkali basalt magmas. The Sr data alone are not sufficient to distinguish among these alternatives.

The isotope ratios of Sr and Nd in Fig. 3.58 demonstrate that all of the volcanic rocks on Grenada could

Fig. 3.58. Isotope ratios of Sr and Nd in the volcanic rocks of Grenada, Lesser Antilles. The distribution of data points indicates that the ankaramites, alkali basalts, basanitoids, and andesites on this island originated from the mantle wedge to which varying amounts of Sr and Nd derived from subducted sediment (similar to the EM2 component) had been added. Alternatively, magmas from the mantle wedge could have assimilated subducted sediment or mixed with another magma formed by partial melting of sediment. The volcanic rocks of Grenada have higher $^{87}Sr/^{86}Sr$ and lower $^{143}Nd/^{144}Nd$ ratios than the lavas on the Mariana Islands which are the result of subduction of oceanic crust under oceanic lithosphere (*Sources:* Hawkesworth et al. 1979b; White and Dupré 1986; Hart 1988)

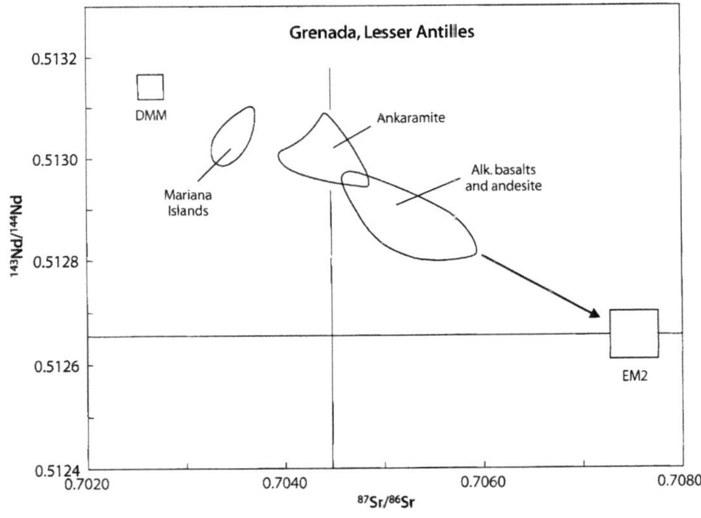

have originated from magma sources consisting of depleted mantle to which had been added varying amounts of Sr and Nd derived from subducted continental sediment and oceanic crust (EM2). In fact, Sigurdsson et al. (1973) postulated that the high alkali content of the undersaturated rocks of Grenada resulted from the preferential transfer of alkali metals to the rocks of the mantle wedge. Subsequently, Hofmann and Feigenson (1983) used REE data to constrain the degree of partial melting and the mineral composition of the mantle rocks under Grenada.

3.10.3 Martinique

The volcanic rocks of Martinique in the southern group of the Lesser Antilles Islands include basalt and andesite with minor rhyolite whose ages range from about 37 Ma (Oligocene) to Recent. The isotope ratios of Sr, Nd, and Pb of these rocks reported by Davidson (1986) vary widely and confirm that the lavas originated from depleted sources in the mantle wedge to which had been added varying amounts of Sr, Nd, and Pb derived from subducted terrigenous sediment.

The $\delta^{18}O$ values of volcanic rocks on Martinique range from +7.0 to +14.0‰ and thereby demonstrate the significant enrichment of these rocks in ^{18}O compared to MORBs whose $\delta^{18}O$ values are restricted to the range +5.0 to +6.0‰ (Davidson 1986). The positive correlation between $\delta^{18}O$ values and $^{87}Sr/^{86}Sr$ ratios of volcanic rocks on Martinique in Fig. 3.59 is not attributable to metasomatic alteration of the mantle wedge by aqueous fluids released by dehydration of subducted sediment, because the $\delta^{18}O$ value of water of the metasomatic fluid approached the $\delta^{18}O$ of the rocks in the mantle wedge as a result of high-temperature isotope re-equilibration in a system having a low water-to-rock ratio. However, silicate minerals (e.g. feldspar, quartz, etc.) have elevated $\delta^{18}O$ values ranging up to about +16‰ (Taylor and Sheppard 1986). Therefore, Davidson (1986) concluded that magmas derived from the metasomatized mantle wedge assimilated terrigenous sediment in the crust underlying Martinique.

In conclusion, the petrogenesis of the volcanic rocks on the northern islands of the Lesser Antilles is consistent with the theory of magma formation in island arcs. However, the isotope ratios of Sr, Nd, and Pb of lavas on the southern islands bear a discernible relationship to the isotope composition of the terrigenous sediment that is being deposited east of the Puerto Rico Trench (White et al. 1985). Therefore, White and Dupré (1986) and Davidson (1987) emphasized that the volcanic rocks on the southern islands of the Lesser Antilles contain a component derived from Precambrian rocks of the continental crust. The oxygen-isotope compositions of lavas

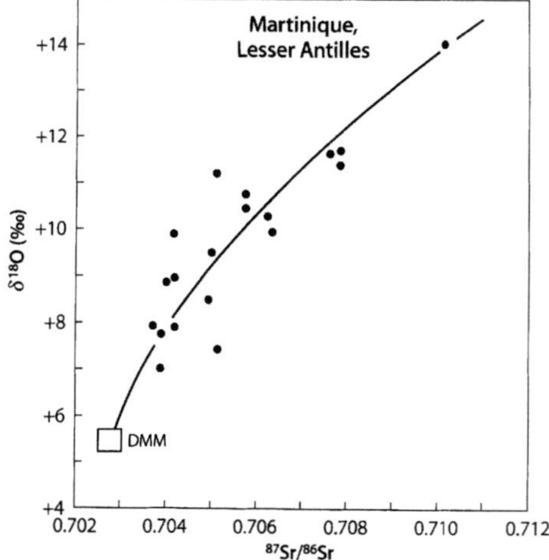

Fig. 3.59. Isotope ratios of O and Sr in lavas on Martinique, Lesser Antilles. The positive correlation indicates that magmas arising in the mantle wedge assimilated terrigenous sediment before being erupted at the surface (*Source:* Davidson 1986)

on Martinique reported by Davidson (1985, 1986) strongly indicate that the magmas also assimilated sediment before being erupted on the surface. This conclusion is consistent with the petrogenesis of lavas on Grenada proposed by Thirlwall and Graham (1984) and was confirmed by Smith et al. (1997) based on a study of the isotope composition of boron in the volcanic rocks of Martinique.

3.11 Mesozoic Basalt, Caribbean Sea

The Caribbean Sea in Fig. 3.60 has a complicated tectonic history (Burke 1988; Dengo and Case 1990; Bonini et al. 1984). It contains a large number of islands including the Greater Antilles (Cuba, Hispaniola, Puerto Rico, and Jamaica), the Lesser Antilles (Sect. 3.10), and a series of islands along the northern coast of Venezuela (including Curaçao and Tobago) which contain volcanic rocks of Mesozoic age.

The Greater Antilles Islands are composed of continental crust and have extensive geologic and tectonic histories. The $^{87}Sr/^{86}Sr$ ratios and concentrations of Rb and Sr of granitic plutons of Cretaceous to Early Tertiary age on Puerto Rico were measured by Jones and Kesler (1980). In addition, Wadge and Wooden (1982) reported the isotope compositions and concentrations of Sr in Cenozoic alkali-rich volcanic rocks from volcanic centers in Honduras, Nicaragua, Costa Rica, Jamaica, Haiti, and the Dominican Republic. These centers

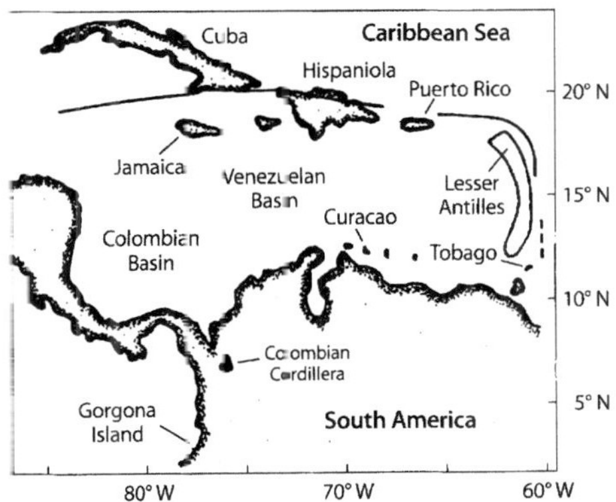

Fig. 3.60. Map of the Caribbean Sea including the Greater Antilles (Puerto Rico, Hispaniola, Cuba, and Jamaica), the Lesser Antilles, Curaçao, Tobago, and Gorgona (*Source:* Modified from Sinton et al. 1998)

are located along the boundaries of the Caribbean Plate where it is in contact with the North American Plate in the north and with the Cocos Plate in the west (Burke 1988). The rocks range in composition from alkali basalt to trachyandesites, trachytes, nephelinites, and limburgites. The $^{87}Sr/^{86}Sr$ ratios of these rocks vary widely from 0.70266 to 0.70632 relative to 0.71025 for NBS 987 (Wadge and Wooden 1982) and have high Sr concentrations ranging up to 3 536 ppm for a nephelinite at Thomazeau in Haiti. The petrogenesis of alkali-rich igneous rocks is the subject of Chap. 6 and therefore will not be pursued further at this time.

The southern part of the Caribbean Plate contains a submarine basalt plateau of late Mesozoic age (88 to 90 Ma) that formed in the eastern Pacific Ocean and was tectonically emplaced between North and South America (Sinton et al. 1998). The occurrence of Late Cretaceous basalt in the Caribbean Basin was originally reported by Donnelly et al. (1973) and was investigated more recently by Sinton et al. (1997), Alvarado et al. (1997), and Kerr et al. (1996a). The basalt flows of the plateau formed by partial melting in a plume rather than in the mantle wedge above a subduction zone. (see also Sect. 4.1.1).

Basalt lavas of Late Cretaceous age have also been reported from the islands of Aruba, Curaçao, Bonaire, Blanquilla, and Tobago located off the coast of Venezuela along the southern border of the Venezuelan Basin (Fig. 3.60) as well as on the island of Hispaniola (Lapierre et al. 1999). The volcanic rocks exposed on Curaçao consist of basalt lavas and hyaloclastites that were erupted at about 90 Ma (Sinton et al. 1998) and have an estimated thickness of about 5 km. The origin of these rocks was discussed by Kerr et al. 1996a).

3.11.1 Gorgona Island

The petrologic province of Cretaceous basalts extends from the Caribbean Basin to the Cordillera of Colombia (Kerr et al. 1997) and to the island of Gorgona located in the eastern Pacific Ocean close to the west coast of Colombia (Fig. 3.60). This tiny island has received much attention from petrologists because it is the only known occurrence of quench-textured, Mg-rich komatiite basalt of Phanerozoic age. This Mg-rich variety of basalt characterizes the so-called greenstone belts of Precambrian shields and is discussed in Sect. 8.1.4 of this book.

The presence of komatiites on Gorgona Island was reported by Gansser et al. (1979) and was subsequently investigated by Echeverría (1980). The Cretaceous age of the volcanic rocks on Gorgona was indicated by Walker et al. (1991) based on age determinations by the Re-Os method. This result was later confirmed by Sinton et al. (1998) who obtained a date of 87.9 ±2.1 Ma by the $^{40}Ar/^{39}Ar$ method. The origin of the komatiites on Gorgona Island was attributed by Arndt et al. (1997) to partial melting in the head of a mantle plume, based in part on isotopic and geochemical data published by Dupré and Echeverría (1984) and Kerr et al. (1996b).

3.11.2 Tobago

The presence of Cretaceous volcanic rocks on the islands along the northern coast of Venezuela and in the mountains of western Colombia records the former existence of a Mesozoic island arc that was dispersed following its collision with the northern margin of the South American Plate. One of the fragments of the Mesozoic island arc now forms the island of Tobago. Others are preserved in the chain of islands extending from Aruba to Curaçao, Bonaire, and the island of Blanquilla.

The Mesozoic volcanic rocks of Tobago consist of the North Coast Schist, ultramafic to mafic plutonic rocks, and the Tobago Volcanic Group. Information reviewed by Frost and Snoke (1989) indicates an Early Cretaceous or even Late Jurassic age for the protoliths of the North Coast Schist. The plutonic and volcanic rocks have yielded $^{40}Sr/^{39}Ar$ dates between 108 to 101 Ma, whereas at least one younger dike has an age of 91 ±2 Ma.

The initial isotope ratios of Sr and Nd of most of the Cretaceous rocks on Tobago in Fig. 3.61 vary only between narrow limits. Frost and Snoke (1989) concluded that the entire complex of rocks had formed in an island-arc setting. The source of Sr and Nd that were added to the former mantle wedge of this Mesozoic subduction zone originated from the EM1 component rather than from EM2 as in the case of the Tertiary lavas of

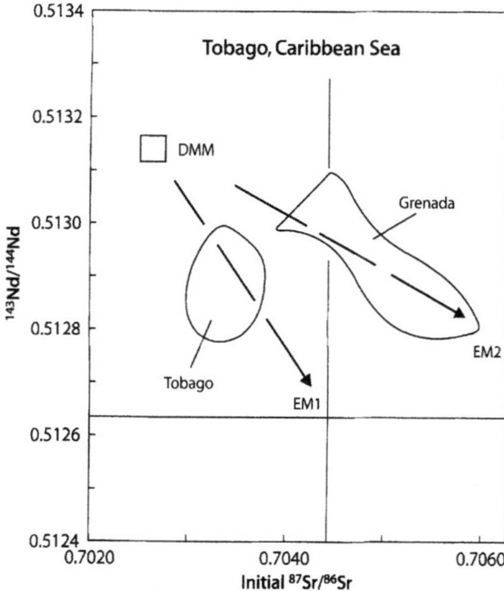

Fig. 3.61. Initial isotope ratios of Sr and Nd of volcanic and plutonic rocks on the island of Tobago which is a fragment of a Cretaceous subduction zone. The isotope ratios of Sr and Nd are consistent with the theory that the magmas originated from the mantle wedge with additions of water-soluble elements derived from subducted sediment that did not contain a terrigenous component like that which presently affects the isotope composition of Sr and Nd of Late Tertiary lavas on Grenada (*Sources:* data from Frost and Snoke 1989; Hawkesworth et al. 1979b; White and Dupré 1986; Hart 1988)

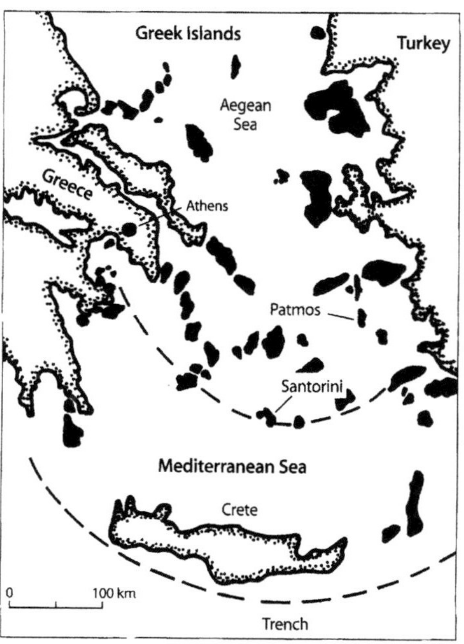

Fig. 3.62. The Hellenic Island arc in the Aegean Sea. The volcanic activity on Patmos, Santorini, and the other islands of the Aegean Sea is caused by the subduction of the African Plate into a trench located south of Crete (*Source:* adapted from Wyers and Barton 1987)

Grenada. In other words, the sediment that was subducted into the trench of the Mesozoic subduction zones did not originate from continental sources. Therefore, the ancestral island arc of Tobago existed in a different kind of tectonic and geographic setting than the present island arc of the Lesser Antilles.

3.12 Mediterranean Sea

Volcanic activity on the islands in the Mediterranean Sea is caused by subduction of the African Plate northward under the Eurasian Plate and by rifting associated with the complex interactions of the two plates. The petrogenesis of volcanic rocks in this area is affected not only by transfer of alkali metals and Sr from the subducted sediment into the overlying mantle wedge, but also by subsequent assimilation of crustal rocks by mantle-derived magmas, by mixing of magmas derived from sources in the mantle and the continental crust, and by fractional crystallization (Pe and Gledhill 1975; Mezger et al. 1985; Katerinopoulos et al. 1998). The island arcs of the Mediterranean Sea include the Greek islands of the Aegean Sea and the Lipari Islands in the Tyrrhenian Sea.

3.12.1 Hellenic Island Arc

The Greek islands in the Aegean Sea in Fig. 3.62 formed as a result of volcanic activity along the Hellenic Arc in Neogene to Recent time. The last eruption in this area occurred in 1950 on the island of Santorini (Barton et al. 1983: Wyers and Barton 1987).

The island of Patmos is located about 100 km north of the volcanic front and was active from about 6.0 to 4.5 Ma. The lavas on this island are alkali-rich and range in composition from nepheline trachybasalt to trachyte, phonolite, and rhyolite. Data from Wyers and Barton (1987) in Fig. 3.63a indicate that Sr concentrations of the alkali-rich lavas on Patmos decrease non-linearly from about 1500 ppm at $SiO_2 \approx 50\%$ to about 500 ppm at $SiO_2 \approx 66\%$, whereas their Rb concentrations rise from about 100 ppm ($SiO_2 \approx 50\%$) to about 250 ppm ($SiO_2 \approx 66\%$). The concentrations of both elements are exceptionally high compared to the rocks on other islands arcs.

The $^{87}Sr/^{86}Sr$ ratios of the lavas on Patmos in Fig. 3.63b range from 0.70489 to 0.70763 and increase with rising concentration of SiO_2. The correlation of the $^{87}Sr/^{86}Sr$ ratios and the silica concentrations of the lavas indicates that assimilation of crustal rocks or mixing of magmas

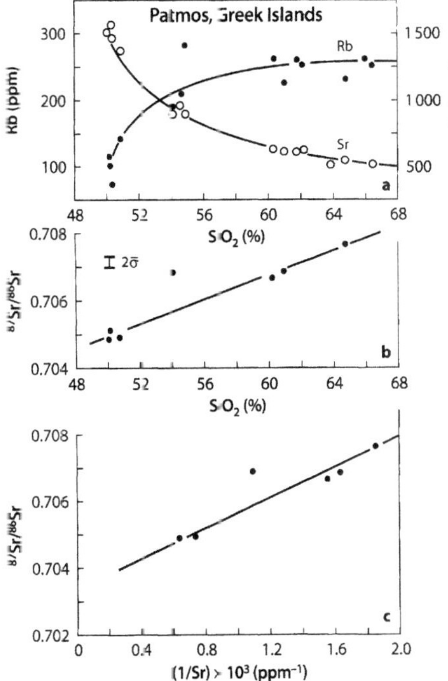

Fig. 3.63. a Non-linear variation of the Rb and Sr concentrations with increasing SiO$_2$ content on the lavas of Patmos, Hellenic Island arc; b Positive correlation between the ^{87}Sr/^{86}Sr ratios of the SiO$_2$ concentrations of the lavas on Patmos; c Sr-isotope mixing diagram for the lavas on Patmos. The collinearity of these data is consistent with magma mixing, whereas the non-linear variation of Rb and Sr concentrations (Part a), and the correlation of ^{87}Sr/^{86}Sr ratios and SiO$_2$ concentrations are evidence for assimilation of crustal rocks (Source: Wyers and Barton 1987)

derived from different sources has occurred at this site. On the Sr-isotope mixing diagram in Fig. 3.63c, the data points form a straight line as expected for two-component mixtures. The data suggest that the crustal component had ^{87}Sr/^{86}Sr > 0.7080 and Sr < 500 ppm, whereas the mantle-derived component had ^{87}Sr/^{86}Sr < 0.7040 and Sr > 1500 ppm. The high Sr content of the mantle-derived magma component implies a small degree of partial melting in the mantle wedge.

The pattern of variation of Rb and Sr concentrations with SiO$_2$ (Fig. 3.63a), the positive correlation of ^{87}Sr/^{86}Sr ratios and SiO$_2$ (Fig. 3.63b), and the linear increase of ^{87}Sr/^{86}Sr ratios with 1/Sr (Fig. 3.63c) are not consistent with fractional crystallization of magma. Wyers and Barton (1987) reached the same conclusion from a consideration of trace-element concentrations in the lavas of Patmos and deduced that magma mixing, assimilation, and fractional crystallization had occurred concurrently.

The ^{87}Sr/^{86}Sr ratios of calc-alkaline lavas on Santorini (basalt to rhyolite) and Milos (andesite to rhyolite) range from 0.70472 to 0.70573 and from 0.70540 to 0.70662, respectively, but do not correlate with the Rb/Sr ratios and with the SiO$_2$ concentrations of the rocks (Briqueu et al. 1986). The high ^{87}Sr/^{86}Sr ratios of some of the analyzed samples are due to alteration by seawater which increased the ratios by about 0.0008 relative to unaltered samples. Therefore, Barton et al. (1983) considered a value of 0.70494 ±0.00022 (2$\bar{\sigma}$) to be representative of the unaltered lavas of Santorini, whereas the unaltered rocks on Milos have a range of ^{87}Sr/^{86}Sr ratios from 0.70540 to 0.70620. Pe (1975) reported similar ^{87}Sr/^{86}Sr ratios for the lavas on Aegina (0.7041–0.7068) and Methana (0.7058–0.7067).

The constant ^{87}Sr/^{86}Sr ratios of lavas on Santorini suggest that they formed by fractional crystallization of magma and that assimilation of crustal rocks was not the cause for their chemical diversity. On the other hand, the ^{87}Sr/^{86}Sr ratios of the magma (0.70494) is unusually high compared to those of other oceanic island arcs and may not be attainable by transfer of ^{87}Sr from subducted oceanic crust to the mantle wedge alone. Therefore, Barton et al. (1983) considered the possibility that the magma sources under Santorini consist of "enriched" mantle having an unusually high ^{87}Sr/^{86}Sr ratio. However, oxygen isotope compositions of the lavas on Santorini (δ^{18}O = +7.2 to +9.1‰; Hoefs 1978) require the involvement of crustal rocks in the formation of this magma. A similar conclusion is justified for the magma on Milos, even though oxygen isotope compositions are lacking, because the ^{87}Sr/^{86}Sr ratios of unaltered lavas on Milos are even higher than those on Santorini.

Volcanic rocks (andesite-dacite-rhyolite) of middle Miocene age (14 Ma) at Oxylithos on the island of Evia (Fig. 3.62) have high but constant initial ^{87}Sr/^{86}Sr ratios averaging 0.70950 ±0.00006 (2$\bar{\sigma}$, N = 4) relative to 0.7080 for E&A (Pe-Piper and Piper 1994). The same authors also reported an average ^{87}Sr/^{86}Sr ratio of 0.70887 ±0.00001 (N = 2) for andesites on the island of Skyros located about 40 km northwest of Evia in the Aegean Sea.

3.12.2 Eolian Island Arc

The Lipari Islands north of Sicily in the southern part of the Tyrrhenian Sea in Fig. 3.64 form the Eolian Island arc which includes Lipari, Stromboli, and Vulcano. The volcanic activity that formed these islands resulted from the collision of the African Plate with the Eurasian Plate in middle Miocene time in a complex tectonic pattern revealed in Fig. 3.64 based on the work of Barberi et al. (1973). In addition, rifting on the island of Sicily caused the volcanic activity of Mt. Etna and of the Iblean Mountains. Rifting also occurred in the Sicily Channel between Sicily and North Africa to form the Pantelleria

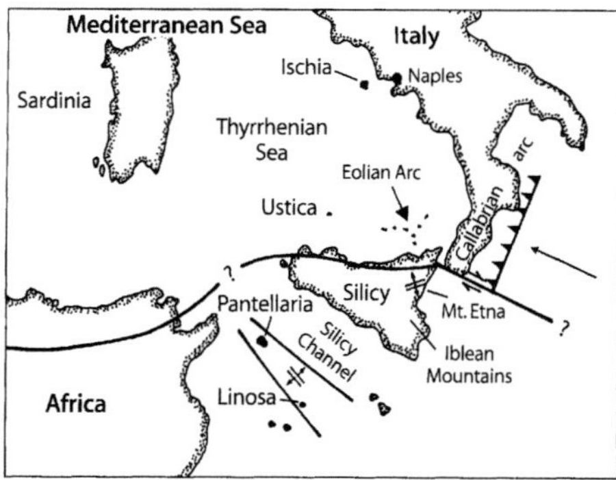

Fig. 3.64. The Eolian Island arc in the Tyrrhenian Sea. Volcanic activity in this tectonically complex area is caused by interactions between the African and the Eurasian Plates. The Eolian Island arc has formed as a result of subduction of the African Plate, whereas the volcanic activity on Sicily (Mt. Etna, Iblean Mountains) and in the Sicily Channel (Pantellaria, Linosa) is the result of crustal extension (*Source:* adapted from Barberi et al. 1973)

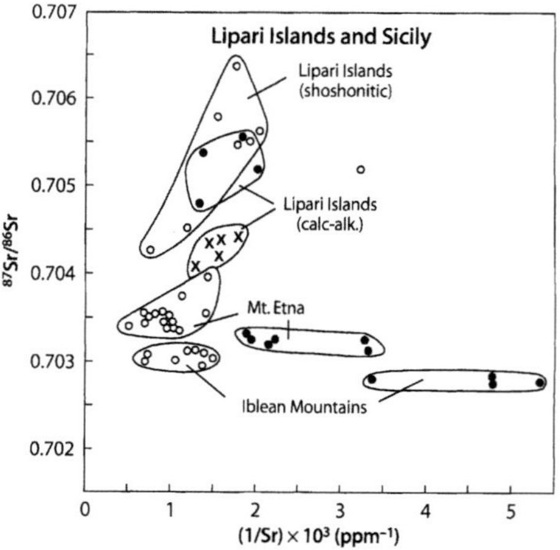

Fig. 3.65. The volcanic rocks of the Lipari Islands (Eolian Island arc) and of Mt. Etna and the Iblean Mountains of Sicily. Solid circles: basalt; open circles: alkali-rich rocks; crosses: andesites. The tholeiite basalts of the Iblean Mountains and Mt. Etna have low $^{87}Sr/^{86}Sr$ ratios indicating that they formed by decompression melting in the lithospheric mantle without contamination by crustal rocks. The alkali-rich rocks on Mt. Etna and the Iblean Mountains have higher $^{87}Sr/^{86}Sr$ ratios and Sr concentrations than the associated tholeiite basalts. The $^{87}Sr/^{86}Sr$ ratios of the shoshonitic lavas of the Lipari Islands overlap those of the calc-alkaline rocks and therefore could have originated from the same magma sources (*Sources:* Barberi et al. 1969, 1974; Klerkx et al. 1974b; Carter and Civetta 1977; Carter et al. 1978)

Rift which contains the volcanic islands of Pantelleria and Linosa. Northward subduction of the African Plate is also the cause for the volcanic activity on the islands of Sardinia, Pontine, and Ischia in the Tyrrhenian Sea. (Hurley et al. 1966; Barberi et al. 1969, 1973). The alkali-rich volcanic rocks that characterize the Mediterranean region in general and the Roman volcanic province of Italy in particular are discussed in Sect. 6.5.

The volcanic rocks on the Lipari Islands consist of an older calc-alkaline series and younger shoshonitic rocks that occur on Lipari, Vulcano, and Stromboli (Barberi et al. 1974; Klerkx et al. 1974b). The $^{87}Sr/^{86}Sr$ ratios in Fig. 3.65 of the alkali-rich shoshonitic rocks (Fig. 3.4) reported by Klerkx et al. (1974b) range widely from 0.7045 (leucite-tephrite, Vulcano) to 0.7064 (nepheline phonotephrite, Stromboli) and exceed the range of $^{87}Sr/^{86}Sr$ ratios of the calc-alkaline rocks. Therefore, Barberi et al. (1974) concluded that both rock suites had formed from magmas that originated in the underlying mantle wedge and were subsequently contaminated by assimilating rocks of the continental crust having elevated $^{87}Sr/^{86}Sr$ ratios. The shoshonitic magmas may have formed at greater depth by a smaller degree of partial melting than the magmas of the calc-alkaline suite.

3.12.3 Sardinia and Ischia

Volcanic activity related to the compressive boundary between the African and the Eurasian Plate also occurred in northwestern Sardinia from Oligocene to middle Miocene (13–24 Ma). Dupuy et al. (1974) reported that the calc-alkaline lavas range in composition from ba-

salt to rhyolite and have a wide range of initial $^{87}Sr/^{86}Sr$ ratios from 0.7040 to 0.7077 relative to 0.7080 for E&A. The elevated and variable $^{87}Sr/^{86}Sr$ ratios are consistent with magma formation in the mantle wedge under Sardinia followed by assimilation of continental crust or by magma mixing and fractional crystallization.

Elevated $^{87}Sr/^{86}Sr$ ratios between 0.7061 and 0.7079 relative to 0.7080 for E&A were also reported by Hurley et al. (1966) for trachytes on the island of Ischia (Fig. 3.64). These rocks belong to the Campanian province of K-rich lavas whose $^{87}Sr/^{86}Sr$ ratios increase regionally from south to north and reach a surprisingly high average value of 0.7108 ±0.0005 $(2\bar{\sigma}, N = 5)$ relative to 0.7080 for E&A at Vico Volcano near Rome (Sect. 6.5.1). The enrichment of these rocks in ^{87}Sr and K is a serious problem to be discussed in Chap. 6 in the context of other ultrapotassic alkali-rich lavas that occur in East Africa, Spain, and in western North America.

3.12.4 Sicily

Mt. Etna on the east coast of Sicily is an active volcano located on the African Plate (Barberi et al. 1973). The rocks on Mt. Etna consist of olivine tholeiite overlain by a suite of alkali-rich rocks ranging in composition from alkali basalt to hawaiite, mugearite, and benmoreite (Fig. 2.2), all of which are less than 100 000 years old (Carter and Civetta 1977; Carter et al. 1978a).

The tholeiite basalts on Mt. Etna have a low average $^{87}Sr/^{86}Sr$ ratio of 0.70325 ±0.00006 ($2\bar{\sigma}, N = 6$) and those of the Iblean Mountains, about 60 km south of the summit of Mt. Etna, have an even lower average $^{87}Sr/^{86}Sr$ ratio of 0.70278 ±0.00003 ($2\bar{\sigma}, N = 4$) relative to 0.7080 for E&A (Carter and Civetta 1977). Therefore, the tholeiite basalts of the Iblean Mountains originated from depleted source rocks in the mantle without significant contamination by crustal rocks. However, the data in Fig. 3.65 demonstrate that the alkali-rich lavas at both locations have slightly higher and more variable $^{87}Sr/^{86}Sr$ ratios than the associated tholeiites.

3.12.5 Pantelleria and Linosa

The islands of Pantelleria and Linosa in the Sicily Channel are the result of volcanic activity (<1 Ma) caused by rifting. The rocks on these islands are generally alkali-rich and have a wide range of $^{87}Sr/^{86}Sr$ ratios from 0.7032 (trachyte, Pantellaria) to 0.7068 (picrite basalt, Linosa) relative to 0.7091 for seawater. Rocks having low $^{87}Sr/^{86}Sr$ ratios originated from sources in the lithospheric mantle without contamination (Barberi et al. 1969). Lavas having elevated $^{87}Sr/^{86}Sr$ ratios formed from magma that was contaminated by assimilation of crustal rocks.

3.13 Summary: Petrogenesis in Subduction Zones

The isotope ratios of Sr, Nd, and Pb of basalt and andesite in oceanic island arcs superficially resemble those of oceanic island basalts because they appear to be mixtures of the DMM and the EM components. The apparent similarity is misleading because volcanic rocks on island arcs originate by a completely different process than oceanic island basalts. The DMM component under island arcs is the wedge of lithospheric mantle above the Benioff zone. The EM components are the downgoing slab of oceanic crust and associated pelagic or terrigenous sediment which together release an aqueous fluid into the overlying mantle wedge. The fluid contains alkali metals and other elements, as well as Sr, Nd, and Pb whose isotope compositions depend on the provenance of the subducted marine sediment and on the alteration of the basalt by seawater. The aqueous fluid emanating from the subducted oceanic crust alters the chemical, mineralogical, and isotope composition of the mantle wedge. As a result, partial melting occurs in the wedge, and the magmas, so formed, are erupted at the

surface to form volcanic islands that extend parallel to the associated ocean trench where the process starts.

According to this theory, the subducted oceanic crust does not melt in most cases. Instead, the addition of Sr, Nd, and Pb takes place by aqueous transfer from the subducted oceanic crust (EM component) to the mantle wedge (DMM component) before melting occurs. The resulting magmas in intra-oceanic island arcs, in most cases, do not assimilate crustal rocks although they may differentiate by fractional crystallization while in transit to the surface or while being stored temporarily in magma chambers beneath volcanoes. Consequently, the isotope ratios of Sr, Nd, and Pb in volcanic rocks of intra-oceanic islands occupy comparatively small data fields that are aligned between the DMM and the EM components in Sr-Nd and Pb-Sr mixing diagrams (e.g. Mariana Islands, Aleutian Islands, Fiji and the Tonga-Kermadec Islands, the northeastern Banda Arc, the volcanoes on the Bataan Arc of western Luzon in the Philippines, the island arcs of New Britain, Vanuatu, Scotia and South Shetland Islands, and the northern islands of the Lesser Antilles).

In cases where oceanic crust is subducted under continental crust, the isotope ratios of Sr, Nd, and Pb of the resulting volcanic rocks range more widely and exceed the isotope ratios of the EM components. In such cases, basaltic and andesitic magmas that formed in the metosomatically altered mantle wedge may assimilate wallrocks in crustal magma chambers or may mix with magmas that formed by melting of heterogeneous crustal rocks. The resulting volcanic rocks have a wide range of chemical compositions extending from uncontaminated basalt to more highly contaminated intermediate and felsic rocks including rhyolites. In addition, the isotope ratios of Sr, Nd, and Pb correlate with chemical parameters such as the concentrations of SiO_2 or MgO or with $\delta^{18}O$ values. Examples of volcanic rocks with crustal isotope signatures include the volcanoes of Kamchatka, Japan, Indonesia, and New Zealand. However, not all of the lava flows erupted at these sites are necessarily contaminated, because magma may move rapidly from its source in the mantle wedge to the surface without interacting extensively with the overlying rocks. On the other hand, some magmas in intra-oceanic island arcs have crustal isotope signatures because the subducted sediment originated by erosion of Precambrian granitoids on nearby continents (e.g. the southwestern Banda Arc and the southern islands of the Lesser Antilles).

A puzzling feature of some island arcs is the occurrence of alkali-rich volcanic rocks. In some cases, the presence of K-rich lavas correlates with increasing depth to the Benioff zone and hence is attributable to the effects of small degrees of partial melting. However, the occurrence of shoshonites in some island arcs (e.g. the northern Mariana Arc) may require special circumstances such as prior alkali metasomatism of the rocks in the mantle wedge, or the presence of plume rocks embedded within the mantle wedge, or the subduction of alkali-rich marine sediment.

Nevertheless, the standard model for petrogenesis associated with subduction zones in the oceans provides plausible explanations for the isotope compositions of volcanic rocks in a variety of tectonic settings, and clearly distinguishes them from the volcanic rocks that form along oceanic spreading ridges and on chains of islands related to asthenospheric plumes. The insights gained in this chapter will next be severely tested by the study of igneous rocks that have formed along the western margins of the American continents where oceanic crust is being subducted under thick continental crust.

Chapter 4

The Compressive Margin of the American Continents

Subduction of oceanic lithosphere under continental crust causes volcanic activity and uplift of compressive continental margins exemplified by the west coast of the American continents. The subsequent erosion of the uplifted continental margins has locally uncovered the plutonic rocks which form the roots of the volcanoes at the surface. Such tectonic processes have occurred throughout the history of the Earth and have produced igneous rocks whose ages range from Archean to Recent. Therefore, the continents have preserved a much more varied and extensive record of igneous activity than the present ocean basins whose ages do not exceed the Mesozoic Era.

The Tertiary volcanic rocks in the mountains along the compressive margin of the American continents include very large deposits of rhyolite tuffs (ignimbrites) whose origin has been the subject of a lengthy debate. The resolution of this problem arises from a combination of detailed fieldwork and the interpretation of chemical and isotopic data. Although felsic volcanics occur on oceanic islands (e.g. Iceland, Chap. 2) and on island arcs (e.g. the Japanese Islands, Sunda Arc, and the North Island of New Zealand, Chap. 3), the ignimbrites of South America, Central America, and North America are far more voluminous and are characteristic of subduction zones where oceanic lithosphere is subducted under the continental lithosphere.

4.1 Andes of South America

The volcanic activity of the Andes in Fig. 4.1 is caused by the eastward subduction of the Nazca Plate under the continental lithosphere of South America (Harmon and Barreiro 1984). The thickness of the continental crust along the west coast of South America varies regionally from north to south and is reflected by differences in the isotope compositions of Sr, Nd, Pb, and O in the calc-alkaline lavas that have been extruded. Therefore, the Andes are divided into four zones from north to south identified in Table 4.1. The Late Tertiary lavas of the Northern and Southern Volcanic Zones of the Andes in Fig. 4.2 have low and narrowly distributed $^{87}Sr/^{86}Sr$ ratios compared to the lavas of the Central Volcanic Zones whose $^{87}Sr/^{86}Sr$ ratios range up to 0.7149. The differences in the $^{87}Sr/^{86}Sr$ ratios of Neogene volcanic rocks in the three volcanic zones included in Fig. 4.2 correlate with their average crustal thicknesses listed in Table 4.1. This feature suggests that the magmas in the Central Volcanic Zone, where the continental crust reaches its greatest average thickness of about 70 km, were contaminated by assimilating granitic basement rocks. The isotope ratios of Sr of the Late Tertiary lavas in the Northern and Southern Volcanic Zones are consistent with the standard petrogenetic model for the formation of volcanic rocks in subduction zones by metasomatic alteration of the magma sources in the mantle wedge above the Benioff Zone without extensive assimilation of crustal rocks.

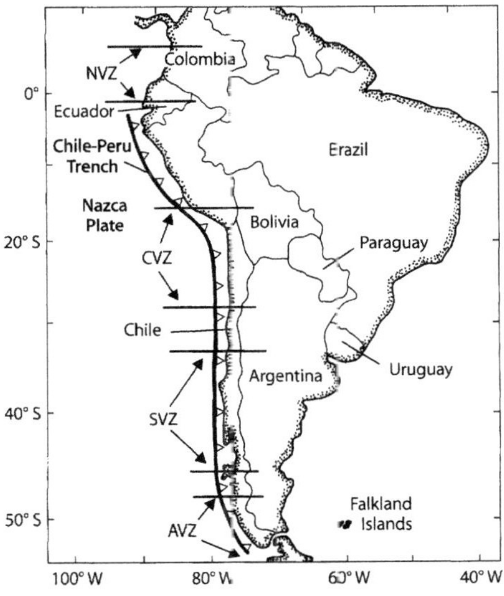

Fig. 4.1. Map of South America showing the Northern Volcanic Zone (*NVZ*), Central Volcanic Zone (*CVZ*), Southern Volcanic Zone (*SVZ*), and Austral Volcanic Zone (*AVZ*), of the Andes Mountains. The volcanic activity of the Andes north of latitude 46° S is caused by subduction of the Nazca Plate eastward into the Chile-Peru Trench. South of that latitude in the AVZ the Antarctic Plate is being subducted (*Source:* adapted from the National Geographic Society 1990)

Table 4.1.
Summary of active volcanic zones in the Andes of South America (*Sources:* Deruelle et al. 1983; Stern et al. 1984; Harmon et al. 1984; Notsu et al. 1987b)

Region	Latitude	Countries	Crustal thickness (km)	$^{87}Sr/^{86}Sr$
Northern	5°N to 2°S	Colombia Ecuador	~45	0.7036–0.7043
Central	16 to 28°S	S. Peru Bolivia N. Chile N. Argentina	~70	0.7054–0.7149
Southern	33 to 46°S	Chile Argentina	~50–30	0.7036–0.7048
Austral	49 to 55°S	Chile Argentina	<30	0.7027–0.7054

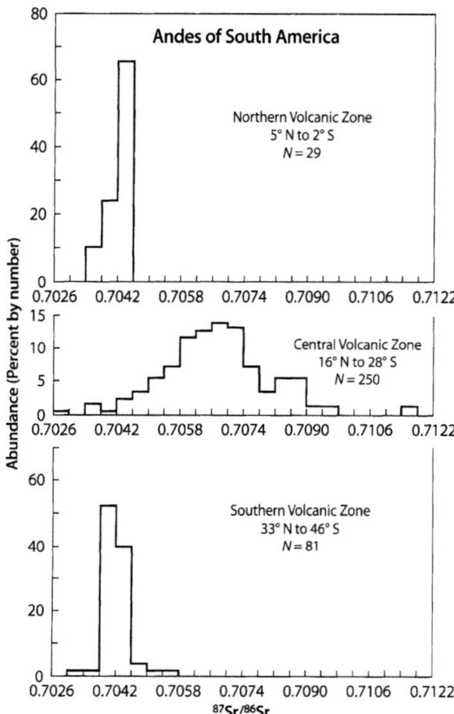

Fig. 4.2. Range of $^{87}Sr/^{86}Sr$ ratios of Late Tertiary calc-alkaline lavas in the Andes of South America. The $^{87}Sr/^{86}Sr$ ratios of volcanic rocks in the northern and southern zones range primarily from 0.7038 to 0.7048, whereas those of the central zone vary widely and have a distinctly higher mean value. The elevated and variable $^{87}Sr/^{86}Sr$ ratios of the lavas in the Central Volcanic Zone indicate that magmas in this region interacted with the granitic rocks of the continental crust. The volcanic rocks in the northern and southern zones could have formed from magmas that originated in the continental lithosphere after it had been enriched in Rb, Sr, and other LIL elements derived from the subducted oceanic crust of the Nazca Plate (*Sources:* Noble et al. 1975; McNutt et al. 1975; James et al. 1976; Klerkx et al. 1977; Francis et al. 1977, 1980, 1989; Hawkesworth et al. 1979c; Thorpe et al. 1979a,b, at 12 Ma; Briqueu and Lancelot 1979; Hawkesworth et al. 1982; James 1982; Notsu and Lajo 1984; Hickey et al. 1986; Notsu et al. 1987b; Rogers and Hawkesworth 1989)

4.1.1 Petrogenesis in the Northern Andes of Colombia

The Colombian Andes in Fig. 4.1 consist of the Western, Central, and Eastern Cordillera which trend north-south and are separated from each other by the Magdalena Valley (located between the Central and Eastern Cordillera) and by the Cauca-Patia Graben (between the Western and Central Cordillera). The Eastern Cordillera consists of sedimentary rocks of Paleozoic to Cenozoic age overlying Precambrian basement rocks. The Central Cordillera is composed of Paleozoic schists intruded by younger plutons and overlain by Mesozoic sedimentary and volcanic rocks which also occur in the Western Cordillera. The Mesozoic volcanic rocks include low-K tholeiites of Cretaceous age which are related to the Caribbean submarine plateau and to the Cretaceous island arc mentioned in Sect. 3.11 (McCourt et al. 1984).

The Cretaceous tholeiite basalts of the diabase group have a narrow compositional range (SiO_2 = 47.2 to 51.7%; Sr = 30 to 116 ppm). Their initial $^{87}Sr/^{86}Sr$ ratios vary from 0.70346 to 0.70410 at 136 Ma (Millward et al. 1984) and overlap the initial $^{87}Sr/^{86}Sr$ ratios of the Mesozoic tholeiites of Tobago (0.70303 to 0.70364; Frost and Snoke 1989) as well as the Sr isotope ratios of the Cretaceous Duarte igneous complex on Hispaniola (0.7028 to 0.70365; Lapierre et al. 1999). However, the $^{87}Sr/^{86}Sr$ ratios of basalts, picrites, and komatiites of Gorgona Island range more widely from 0.7025 to 0.7070 (Echeverría and Aitken 1986). The isotope compositions of Pb and Nd of the Cretaceous volcanic rocks of Gorgona Island were measured by Dupré and Echeverría (1984) and Arndt et al. (1997), respectively.

The plutonic rocks of the Central and Western Cordillera of Colombia range in age from Early Triassic to Miocene. McCourt et al. (1984) recognized five episodes of igneous activity based primarily on K-Ar dates:

- Late Tertiary: 21 ±1 to   6 ±1 Ma
- Early Tertiary: 57 ±2 to 30 ±2 Ma
- Cretaceous: 119 ±10 to 72 ±2 Ma
- Jurassic to Early Cretaceous: 174 ±10 to 142 ±6 Ma
- Triassic: 248 ±10 to 214 ±7 Ma

The Santa Barbara Batholith (Triassic) yielded a whole-rock Rb-Sr isochron date of 211 ±51 Ma and an initial $^{87}Sr/^{86}Sr$ ratio of 0.7047 ±0.0002 (16 samples, $MSWD = 1.4$). The El Tampor stock (Late Cretaceous) was dated at 94 ±16 Ma and has an initial $^{87}Sr/^{86}Sr$ ratio of 0.7038 ±0.0001 (10 samples, $MSWD = 0.3$; McCourt et al. 1984). The initial $^{87}Sr/^{86}Sr$ ratios of these plutonic rocks are similar to those of the volcanic rocks in the Central Cordillera in spite of compositional differences. For example, the Cretaceous El Tampor stock is composed of calc-alkaline granitoids, whereas the associated Cretaceous lava flows are low-K tholeiites (McCourt et al. 1984).

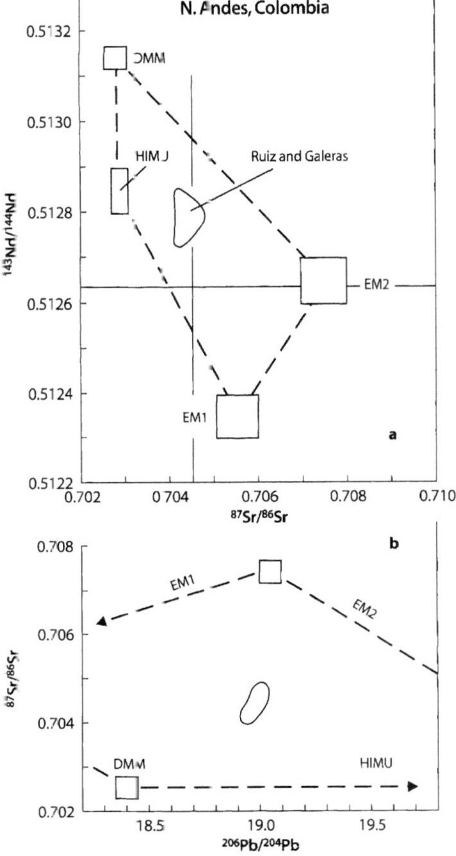

Fig. 4.3 a,b. Isotope ratios of Sr, Nd, and Pb of Late Tertiary andesites on the volcanoes Ruiz and Galeras of the Central Cordillera of Colombia. The DMM component is the mantle wedge, EM1 and EM2 are Sr and Nd derived from subducted oceanic crust. The $^{87}Sr/^{86}Sr$ ratios are relative to 0.7080 for E&A and the $^{143}Nd/^{144}Nd$ ratios were corrected for isotope fractionation to $^{146}Nd/^{144}Nd$ = 0.7219 (*Source:* James and Murcia 1984)

Andesites of Tertiary age have been erupted at several volcanic centers in the Central Cordillera, including the volcanoes Ruiz at about at 5° N latitude and Galeras at about 1.25° N. The isotope ratios of Sr, Nd, and Pb of andesites on these volcanoes in Fig. 4.3a,b occupy small data fields that lie within boundaries defined by the mantle components DMM, EM1, EM2, and HIMU (James and Murcia 1984). As before, DMM represents the depleted lithospheric rocks of the mantle wedge, the enriched components (EM1 and EM2) represent subducted sediment of pelagic and continental origin, respectively, which release an aqueous fluid containing Sr, Nd, and Pb, whereas HIMU is basalt of the subducted oceanic crust which contributes Pb having an elevated $^{206}Pb/^{204}Pb$ ratio. According to the theory of petrogenesis in subduction zones, magmas form by partial melting of the metasomatically altered rocks of the mantle wedge rather than by melting of the sediment and basalt of the subducted oceanic crust. The position of the data fields of Colombian andesites in Fig. 4.3a,b leads to the conclusion that EM1, EM2, and HIMU all contributed Sr, Nd, and Pb to the mantle wedge prior to magma formation. Therefore, the isotope compositions of these elements are consistent with the standard model of subduction-zone petrogenesis without requiring input from the continental crust.

However, the $\delta^{18}O$ values of the andesites on the volcanoes Ruiz and Galeras (not shown) are correlated with the $^{87}Sr/^{86}Sr$ ratios even though they range only from +6.52 to +7.84‰. Therefore, James and Murcia (1984) concluded that the isotopic data permit a limited amount of assimilation of a crustal component in addition to contamination of the magma source in the mantle wedge. However, the isotope ratios of Sr, Nd, and Pb of the Colombian andesites in Fig. 4.3a,b do not exceed the isotope ratios of EM1 and EM2, implying that the isotopic compositions of these elements were not significantly affected by assimilation of crustal rocks.

4.1.2 Crustal Contamination of Magmas in the Central Volcanic Zone

The Late Tertiary andesites of the Central Volcanic Zone in southern Peru, Bolivia, and the northern parts of Chile and Argentina (Table 4.1, Fig. 4.1) have a wide range of $^{87}Sr/^{86}Sr$ ratios (0.7054 to 0.7149) because magmas arising from the mantle wedge in this sector of the Andes were contaminated by assimilating rocks of the continental crust (Harmon et al. 1984; Thorpe et al. 1984; Deruelle et al. 1983).

The complexity of the igneous activity in the Central Volcanic Zone is clearly revealed by the results of a study by McNutt et al. (1975, 1979) of rocks between latitudes 26 to 29° S. The ages of rocks in Fig. 4.4a indicate that igneous activity started along the coast at about 190 Ma

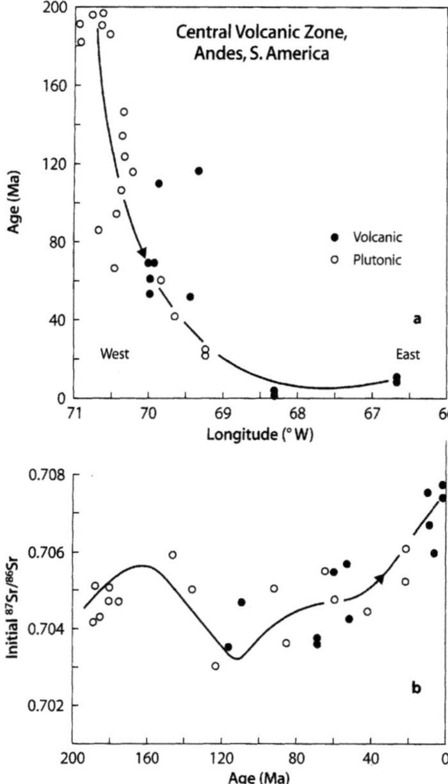

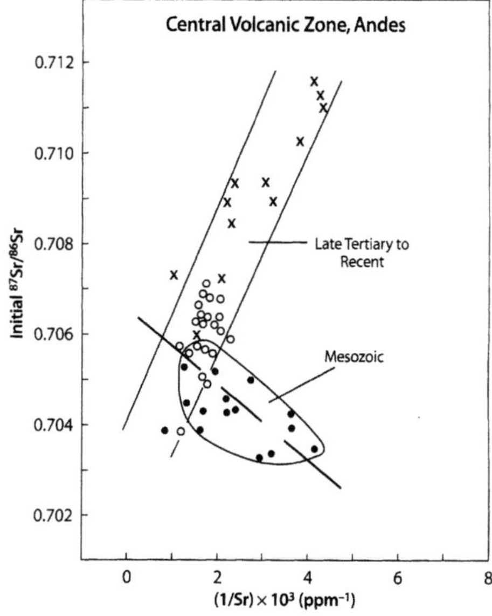

Fig. 4.5. Difference in petrogenesis of Mesozoic plutonic and volcanic rocks and the calc-alkaline lavas of Late Tertiary to Recent age in the Central Volcanic Zone of the Andes. *Closed circles:* average values of Mesozoic rocks; *open circles:* late Tertiary to Recent lavas, northern Chile; *crosses:* lavas on Cerro Galan, northwestern Argentina. During the Mesozoic Era, magmas formed primarily by partial melting of lithospheric mantle above the subduction zone, whereas the Recent lavas are the result of assimilation of crustal rocks and crystallization of plagioclase by the mantle-derived magmas (*Sources:* Rogers and Hawkesworth 1989; Francis et al. 1980, Cerro Galan)

Fig. 4.4. a Eastward progression of igneous activity in the Central Volcanic Zone of the Andes from the west coast starting at about 190 Ma. Late Tertiary volcanic activity (0 to 10 Ma) has occurred about 500 km east of the coastal batholith; **b** Variation of initial $^{87}Sr/^{86}Sr$ ratios of plutonic and volcanic rocks with time. Since Late Cretaceous time (100 to 120 Ma), the initial $^{87}Sr/^{86}Sr$ ratios have increased from about 0.7032 to 0.7077 thereby suggesting increasing involvement of rocks of the continental crust in the formation of magmas (*Source:* McNutt et al. 1975)

and then gradually shifted to the east such that the most recent volcanic activity (0 to 5 Ma) has occurred about 500 km east of the coast (Drake et al. 1982). The eastward shift of the centers of igneous activity implies that magmas were generated at increasing depth and that they encountered an increasing thickness of continental crust before they were erupted at the surface. The time-dependent change of the tectonic setting of igneous activity is reflected by systematic variations of the initial $^{87}Sr/^{86}Sr$ ratios of the igneous rocks in this transect of the Andes. The Mesozoic granitoids (180 to 190 Ma) in Fig. 4.4b have an average initial $^{87}Sr/^{86}Sr$ ratio of 0.7047 ±0.0003 (2$\bar{\sigma}$, N = 6) relative to 0.7080 for E&A. Subsequently, the initial $^{87}Sr/^{86}Sr$ ratios of quartz diorite

and granodiorites (110 to 120 Ma) decreased to the remarkably low value of about 0.7022, suggesting that magma formed by decompression melting of depleted rocks in the lithospheric mantle. Starting at about 110 Ma, the $^{87}Sr/^{86}Sr$ ratios of plutonic and volcanic rocks, forming at increasing distances from the coast, have been rising. The average $^{87}Sr/^{86}Sr$ ratio of volcanic rocks erupted during the last ten million years between 66°40' and 68°20' W longitude is 0.7071 ±0.0006 (2$\bar{\sigma}$, N = 5) relative to 0.7080 for E&A. The rise of the $^{87}Sr/^{86}Sr$ ratios from 0.7022 at about 110 Ma to 0.7071 in the most recent past indicates increasing involvement of crustal rocks in the formation of these magmas.

The geographic and time-dependent changes in the initial $^{87}Sr/^{86}Sr$ ratios of plutonic and volcanic rocks along a transect of the Andes between 22 and 23° S were subsequently confirmed by the work of Rogers and Hawkesworth (1989). Their results reveal the same eastward migration of igneous activity from the coast to the interior and the concurrent increase of initial $^{87}Sr/^{86}Sr$ ratios reported by McNutt et al. (1975) and illustrated in Fig. 4.4a and b.

Rogers and Hawkesworth (1989) made the point illustrated in Fig. 4.5 that two different petrogenetic processes have occurred in the central Andes. The first process started at about 200 Ma and resulted in progressively increasing Sr concentrations with only moderate increases of initial $^{87}Sr/^{86}Sr$ ratios. Subsequently, the Sr concentrations of Late Tertiary to Recent lavas decreased as their $^{87}Sr/^{86}Sr$ ratios increased up to about 0.7115 for the dacites at Cerro Galan in northwestern Argentina analyzed by Francis et al. (1980, 1989). Rogers and Hawkesworth (1989) attributed the rising $^{87}Sr/^{86}Sr$ ratios and decreasing Sr concentrations of the Quaternary lavas in the Central Volcanic Zone of the Andes to assimilation of crustal rocks and crystallization of plagioclase by the contaminated magmas. The relatively low $^{87}Sr/^{86}Sr$ ratios and rising Sr concentrations of the Mesozoic and Early Tertiary rocks resulted from magma formation in the lithospheric mantle above the Benioff zone.

The lavas near the town of Arequipa and in the Cordilliera el Barroso in southern Peru provide additional evidence of crustal contamination of the magmas from which they formed. The $^{87}Sr/^{86}Sr$ ratios of the Arequipa volcanics range from 0.70669 to 0.70821 while the $^{87}Sr/^{86}Sr$ ratios of the Barroso volcanics extend from 0.70544 to 0.70716 relative to 0.7080 for E&A (James 1982) and 0.71025 for NES 987 (Briqueu and Lancelot 1979). The initial $^{87}Sr/^{86}Sr$ ratios of these rocks and those of Cerro Galan in northern Argentina (Francis et al. 1980, 1989) in Fig. 4.6a and b correlate positively with Rb/Sr ratios and with reciprocal Sr concentrations (not shown). These relations result from assimilation of crustal rocks and/or by mixing of magmas having different $^{87}Sr/^{86}Sr$ ratios and Sr concentrations. The lava flows of the Arequipa and Barroso volcanics in Fig. 4.6a form distinctly different linear arrays even though the distance between these volcanic centers is only about 175 km and the rocks are similar in age (Arequipa volcanics: <1 Ma; Barroso volcanics: Pliocene to Pleistocene, James et al. 1976; James 1982). Therefore, the magmas from which these lavas formed appear to have interacted with different kinds of crustal rocks. In addition, the $^{87}Sr/^{86}Sr$ ratios of the lavas on Cerro Galan (10.4 to 0 Ma in northern Argentina) also increased with rising Rb/Sr ratios and silica concentrations from basalt (0.70569), to basaltic andesite (0.70637), to andesite (0.70824), and to dacite (0.71041) relative to 0.7080 for E&A (Fig. 4.6b, Francis et al. 1980, 1989). Therefore, these lavas likewise formed by contamination of mantle-derived magmas with crustal rocks characterized by elevated $^{87}Sr/^{86}Sr$ ratios (0.71 to 0.78) and low Sr concentrations (<300 ppm).

The correlation of $^{87}Sr/^{86}Sr$ ratios of volcanic rocks with their Rb/Sr ratios occurs in many lavas of the Central Volcanic Zone of the Andes. Such arrays are not isochrons but are products of two-component mixing. For example, the linear arrays formed by the Arequipa

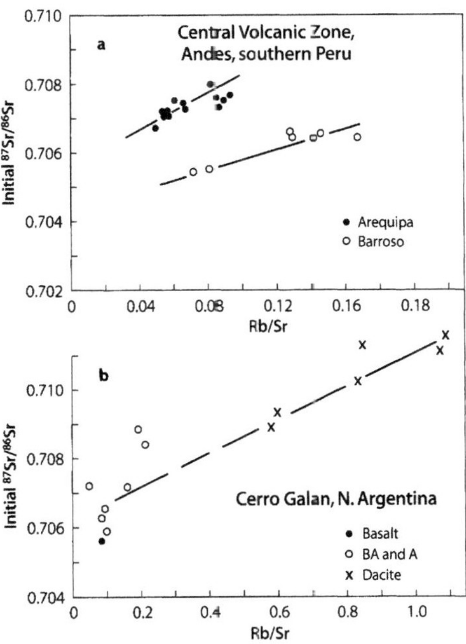

Fig. 4.6. **a** Correlation of $^{87}Sr/^{86}Sr$ and Rb/Sr ratios of the Arequipa and Barroso volcanics in southern Peru. These linear arrays are the result of assimilation of crustal rocks and fractional crystallization of magmas derived from the subcrustal lithosphere (*Sources:* James et al. 1976; James 1982); **b** Correlation of initial $^{87}Sr/^{86}Sr$ and Rb/Sr ratios of lavas on Cerro Galan in northern Argentina. The $^{87}Sr/^{86}Sr$ ratios of these rocks increase with rising silica concentration indicating that the chemical compositions of these rocks are controlled more by assimilation of crustal rocks than by fractional crystallization of magma (*Source:* Francis et al. 1980). Similar results were later published by Francis et al. (1989)

and Barroso volcanics of southern Peru and by the lavas of Cerro Galan in Northern Argentina in Fig. 4.6a and b are mixing lines whose slopes have no specific time significance (Francis et al. 1977, 1980). Although Brooks et al. (1976b) discussed the possibility that such arrays are "mantle isochrons" (Sect. 2.10.7), James (1982) considered them to be the result of assimilation of crustal rocks with elevated $^{87}Sr/^{86}Sr$ ratios and low Sr concentrations by magmas that originated in the mantle wedge. Pankhurst (1977) likewise refuted the validity of mantle isochrons.

The isotope ratios of Sr and Nd of the volcanic rocks on the Barroso and Arequipa Volcanoes in Fig. 4.7 exemplify the extensive crustal contamination of mantle-derived magmas in the Central Volcanic Zone of the Andes. The data fields of the lavas on these volcanoes are located far from the DMM component in the fourth quadrant of the Sr-Nd isotope mixing diagram and beyond the locations of EM1 and EM2. Therefore, the isotope ratios of Sr and Nd of the Barroso and Arequipa

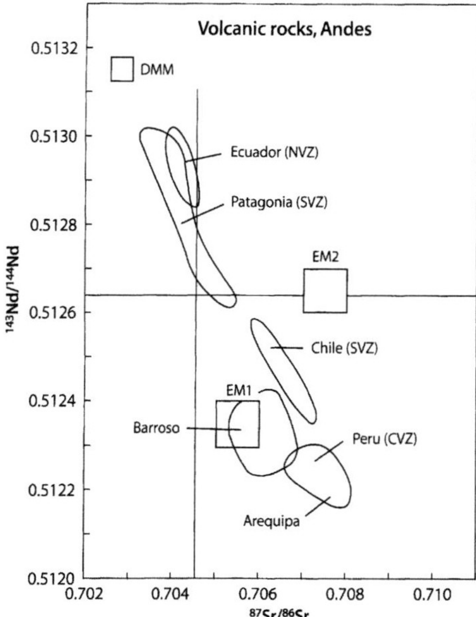

Fig. 4.7. Isotope ratios of Sr and Nd of Late Tertiary andesites and basalts of the Andes of South America including lavas of the Northern Volcanic Zone (*NVZ*) in Ecuador, the Central Volcanic Zone (*CVZ*) in Peru (Barroso and Arequipa Volcanoes), and from the Southern Volcanic Zone (*CVZ*) of Patagonia and southern Chile. The isotopic data reveal the full range of petrogenesis from magma formation in the mantle wedge without crustal assimilation in Ecuador to significant assimilation of crystalline basement rocks in southern Peru (*Sources:* James 1982; Hawkesworth et al. 1979b; Hart 1988)

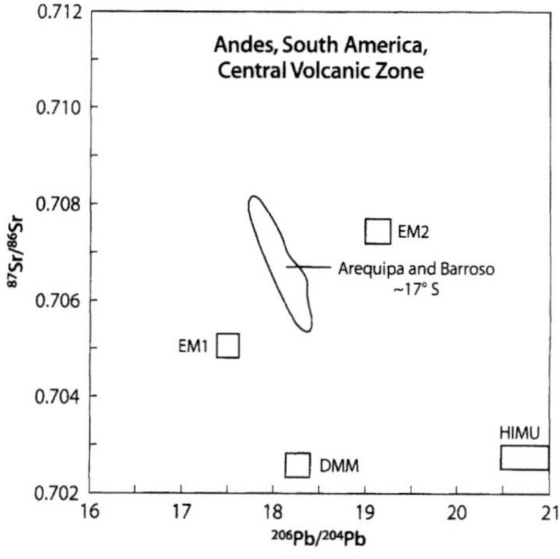

Fig. 4.8. Isotope ratios of Sr and Pb in the Quaternary basalts and andesites of the Arequipa and Barroso Volcanoes in southern Peru. The principal mantle components of oceanic island basalt are indicated for reference. The alignment of the data array indicates that the mantle-derived magmas assimilated crustal rocks having elevated $^{87}Sr/^{86}Sr$ but low $^{206}Pb/^{204}Pb$ ratios (*Sources:* Cobbing et al. 1977; Shackleton et al. 1979; Tilton and Barreiro 1980; James 1982; Hart 1988)

lavas are not attributable to metasomatic alteration of the mantle wedge alone, but also require inputs of Sr and Nd from the rocks of the continental crust.

The data in Fig. 4.7 also emphasize the difference between the lavas of the Central Volcanic Zone and those of the Northern and Southern Volcanic Zones. The isotope ratios of Sr and Nd of the lavas of Ecuador (Northern Volcanic Zone) in Fig. 4.7 differ markedly from those of the Peruvian lavas and are consistent with their derivation from the mantle wedge with minimal contamination by Sr and Nd of the continental crust. The lavas of Patagonia and southern Chile in the Southern Volcanic Zone of the Andes have a wide range of Sr and Nd isotope ratios implying that the extent of crustal contamination of the magmas ranged from insignificant in Patagonia to quite extensive in southern Chile (Hawkesworth et al. 1979b; James 1982).

The isotope ratios of Sr and Pb in the lavas on the Arequipa and Barroso Volcanoes in southern Peru define a mixing array in Fig. 4.8 which indicates that the crustal component had a lower $^{206}Pb/^{204}Pb$ ratio than the mantle component (James 1982). The crustal Pb may

have originated from the Charcani gneiss and/or the Mollendo Granulites which represent the crystalline basement complex of this area. Tilton and Barreiro (1980) actually reported average $^{206}Pb/^{204}Pb$ ratios of 16.9576 ±0.063 (2$\bar{\sigma}$, $N = 9$) for the Charcani gneiss and 16.176 ±0.217 (2$\bar{\sigma}$, $N = 4$) for the Mollendo Granulites both of which are Precambrian in age. The $^{206}Pb/^{204}Pb$ ratios of both rock units are lower than the $^{206}Pb/^{204}Pb$ ratio of MORBs to which Hart (1988) assigned an average $^{206}Pb/^{204}Pb$ ratio of ~18.3. Crustal contamination of the Neogene lavas of the Central Volcanic Zone of the Andes is also indicated by their elevated $\delta^{18}O$ values (Francis et al. 1989; Harmon and Hoefs 1984; Deruelle et al. 1983; James 1982, 1981; Harmon et al. 1981; Magaritz et al. 1978).

Therefore, all of the isotopic data considered here consistently demonstrate that the magmas in the Central Volcanic Zone of the Andes assimilated significant amounts of Precambrian granitoids of the continental crust prior to their eruption at the surface. The ore deposits of Mesozoic to Early Tertiary age in the coastal area of the central Andes as well as the Mesozoic sedimentary rocks of the high Andes also contain mixtures of mantle-derived and crustal Pb (Macfarlane et al. 1990). Similarly, Mukasa (1986a) reported that regional variations in the isotope composition of Pb in feldspars of the Peruvian Coastal Batholith (188 to 37 Ma) are attributable to assimilation of Precambrian granitoids by mantle-derived magmas. The isotope compositions of Pb in volcanic and plutonic rocks of the Andes were fur-

...er investigated by McNutt et al. (1979), Mukasa and Tilton (1984, 1985), Mukasa and Henry (1990), and by other authors referred to by them.

4.1.3 Ignimbrites of the Central Volcanic Zone

The volcanic rocks in the Central Volcanic Zone of the Andes include extensive deposits of ignimbrite (dacite to rhyolite) upon which the cones of andesitic lavas were subsequently built. Thorpe et al. (1979a,b), Klerkx et al. (1977), and James et al. (1976) reported that the ignimbrites of northern Chile, southwestern Bolivia, and southern Peru are strongly enriched in Rb and depleted in Sr compared to andesites and have Rb/Sr ratios between 0.072 and 6.56. However, the initial $^{87}Sr/^{86}Sr$ ratios of the ignimbrites analyzed by Thorpe et al. (1979a,b) and James et al. (1975) (0.70516 to 0.70883) are comparable to those of the andesites in Fig. 4.2. Therefore, the ignimbrites either formed by fractional crystallization of andesitic magma (Thorpe et al. 1979a,b) or by remelting of relatively young mantle-derived volcanic or plutonic rocks at depth in the crust (Klerkx et al. 1977) similar to the rhyolites in Iceland (Sect. 2.3.2).

A third alternative is that the ignimbrites of the Central Volcanic Zone of the Andes originated from magmas that assimilated substantial amounts of crustal rocks. Evidence for this alternative was presented by Hawkesworth et al. (1982) from the Purico-Chascon volcanic complex in northern Chile at 22°57' S and 67°45' W near the border between Bolivia and Chile (Fig. 4.1). The complex consists of a large ignimbrite shield which is about 25 km wide and extends even farther in the north-south direction. The volcanic zones of Purico and Chascon are located on top of the ignimbrite shield and are composed of andesites and dacites (<1.5 Ma). The silica concentrations of these lavas range from 56.4 to 70.4%, Sr from 194 to 539 ppm, and Rb from 46 to 245 ppm. The $^{87}Sr/^{86}Sr$ ratios of the andesites and dacites vary from 0.70703 to 0.70960 relative to 0.71025 for NBS 987 (Hawkesworth et al. 1982) and have overlapping ranges. However, the lavas on Purico have slightly higher $^{87}Sr/^{86}Sr$ ratios than those on Chascon.

The isotope ratios of Sr and Nd of the andesites and dacites of the Purico-Chascon Volcanoes in Fig. 4.9 overlap those of the underlying ignimbrites and exceed the isotope ratios of the EM components (Hawkesworth et al. 1982; Thorpe et al. 1976, 1979a,b). Therefore, the andesite-dacite as well as most of the ignimbrites at this location contain Sr and Nd derived from the underlying continental crust because transfer of Sr and Nd from subducted sediment to the magma sources in the mantle wedge cannot account for the isotope ratios observed in these volcanic rocks.

A second example of ignimbrite petrogenesis in the Central Volcanic Zone of the Andes was provided by

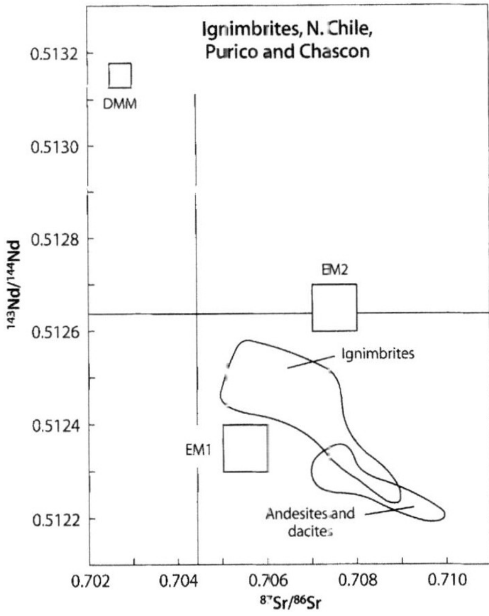

Fig. 4.9. Isotope ratios of Sr and Nd of Late Tertiary ignimbrites and overlying andesites and dacites of the Purico-Chascon volcanic complex in the Central Volcanic Zone of northern Chile (*Sources:* Hawkesworth et al. 1982; Thorpe et al. 1976, 1979a,b; Hart 1988)

Francis et al. (1989) from the Cerro Galan Caldera complex in northwestern Argentina. The volcanic rocks at this location are underlain by late Precambrian and early Paleozoic igneous and metamorphic basement rocks (600 to 365 Ma). Volcanic activity started at 14.5 Ma and continued episodically until 2.2 Ma when a major eruption discharged more than 1 000 km³ of dacitic ignimbrite which exceeded the volume of all previous ignimbrites erupted in this area between 6.0 and 4.2 Ma.

The ignimbrites erupted at Cerro Galan have high initial $^{87}Sr/^{86}Sr$ ratios (0.71071 to 0.71168), high $\delta^{18}O$ values (+11.2 to +12.7‰) and low $^{143}Nd/^{144}Nd$ (0.51215 to 0.51225) ratios leaving no room for doubt that the magmas either assimilated rocks from the continental crust or formed originally by partial melting of crustal rocks (Fig. 4.9). The ignimbrites have only a narrow compositional range implying that the dacitic magmas which produced them did not differentiate extensively by fractional crystallization in contrast to rhyolite ignimbrites in Mexico and California.

4.1.4 Crystalline Basement Rocks of Peru

The isotope ratios of Sr, Nd, Pb, and O provide evidence that the andesite magmas in the Central Volcanic Zone of the Andes assimilated igneous and metamorphic rocks of the crystalline basement complex which range in age and composition from Precambrian granulites and metasediments to early Paleozoic granitic intrusives (Pitcher and Bussell 1977; Pitcher 1978; Shackleton et al. 1979). Measurements of whole-rock $^{87}Sr/^{86}Sr$ ratios and age determinations by the Rb-Sr method of basement rocks in the Central Volcanic Zone were published by Stewart et al. (1974), McNutt et al. (1975), Klerkx et al. (1977), Cobbing et al. (1977), and Halpern (1978). The basement rocks of the Southern Volcanic Zone in central and southern Chile were dated by Halpern and Carlin (1971), Halpern (1973), Munizaga et al. (1973), Halpern and Fuenzalida (1978), Hervé et al. (1984), Miller (1984), and others.

The basement rocks exposed in the Central Volcanic Zone between the coast and the Andes of Peru constitute the Arequipa Massif (Shackleton et al. 1979). It is composed of the Mollendo granulite gneiss whose metamorphic age is about 1 900 Ma (Dalmayrac et al. 1977). In addition, the massif includes the Precambrian staurolite-andalusite schists of Ocoña (Cobbing et al. 1977). The Mollendo gneisses have comparatively low average Sr concentrations (163 ppm), but high $^{87}Sr/^{86}Sr$ ratios ranging from 0.716 to 0.796 with a mean of 0.7532 for 13 specimens weighted in accordance with the Sr concentrations.

The Precambrian gneisses of the Arequipa Massif were intruded by granitic rocks of the Atico complex of early Paleozoic age. These rocks have distinctly higher average Sr concentrations (369 ppm) than the older granulitic gneisses and lower $^{87}Sr/^{86}Sr$ ratios between 0.709 and 0.832, yielding a weighted average of 0.7170 for 9 specimens.

The igneous and metamorphic rocks of the Mollendo gneiss, Ocoña schist, and Atico granitoids are unconformably overlain by low-grade metasedimentary rocks of the Marcona Formation which is intruded by the San Nicolas Batholith. The rocks of the San Nicolas Batholith range in composition from adamellite to diorite and have uniformly high Sr concentrations with an average of 467 ppm. The $^{87}Sr/^{86}Sr$ ratios of the San Nicolas granitoids range from 0.7117 to 0.7133 with a weighted mean of 0.7118.

The different components of the Arequipa Massif form a linear array in coordinates of the average $^{87}Sr/^{86}Sr$ and 1/Sr ratios in Fig. 4.10. In addition, the late Cenozoic volcanic rocks of southern Peru at Arequipa and Barroso (James et al. 1976; James 1982; Briqueu and Lancelot 1979) and at Purico and Chascon (Hawkesworth et al. 1982) also plot on this line as do the volcanic rocks from the Northern Volcanic Zone in Ecuador (Francis et al. 1977). The collinearity of these data points demonstrates a long history of interaction between mantle-derived magmas, represented by the Ecuadoran lavas, with rocks of the continental crust of South America, exemplified by the Mollendo granulite gneisses. Accord-

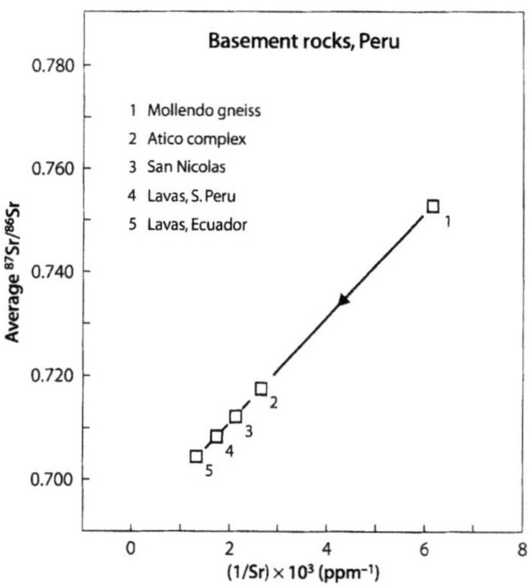

Fig. 4.10. Relationship between the crystalline basement rocks of the Arequipa Massif and the Cenozoic calc-alkaline volcanic rocks of southern Peru and Ecuador. *1.* Mollendo granulite gneiss, ~1900 Ma; *2.* Atico granitoids, ~540 Ma; *3.* San Nicolas Batholith, ~390 Ma; *4.* Andesites, southern Peru, <5 Ma; *5.* Andesites, Ecuador, <5 Ma. The average Sr concentrations of these rock units increase as the $^{87}Sr/^{86}Sr$ ratios decrease with decreasing geologic age. The systematic relationships revealed in this diagram suggest that the rocks of the continental crust represented by the Mollendo granulite gneisses (*1*) contributed Sr to the Atico granitoids (*2*) and that both may have contributed to the formation of the San Nicolas Batholith (*3*) Subsequently, all three units of the Arequipa Massif may have been assimilated by andesite magmas like those of Ecuador to form the andesites of southern Peru which have a stronger crustal imprint than those of Ecuador (*Sources:* Cobbing et al. 1977; Shackleton et al. 1979; Francis et al. 1977; James et al. 1976; Briqueu and Lancelot 1979; Hawkesworth et al. 1982)

ing to this scenario, the magmas that produced the Attico complex of granitoids and the San Nicolas Batholith originated in the lithospheric mantle and assimilated varying amounts of continental crust. Subsequently, all of the components of the Arequipa Massif contributed Sr to magmas arising from the mantle wedge to produce the calc-alkaline lavas of southern Peru whose average $^{87}Sr/^{86}Sr$ ratio (0.7075) is significantly higher than that of the Ecuadoran lavas (0.7042). Therefore, the data in Fig. 4.10 demonstrate that the chemical and isotope compositions of the Cenozoic and Quaternary lavas of the Central Volcanic Zone of the Andes depend strongly on assimilation of crustal rocks by magmas arising from the lithospheric mantle. However, other scenarios, e.g. remelting of rocks of the lower crust, are not necessarily excluded.

The coastal batholith of Peru is a major component of the pre-Andean basement rocks of this region. It extends for a distance of about 1 600 km along the coast of Peru from about 4 to 18° S latitude. The batholith has been subdivided into five major units identified from north to south as the Piura, Trujillo, Lima, Arequipa, and Toquepala segments. Age determinations reviewed by Mukasa (1986b) indicate that the emplacement of different parts of the coastal batholith may have started at about 190 Ma (Early Jurassic) and continued episodically until about 35 Ma (Oligocene). The U-Pb zircon dates obtained by Mukasa (1986b) for the Lima, Arequipa, and Toquepala segments generally confirm previously determined K-Ar and Rb-Sr dates of the coastal batholith of Peru. The Linga Group of monzonites and diorites in the Arequipa segment yielded a whole-rock Rb-Sr isochron date of 68 ± 3 Ma ($\lambda = 1.42 \times 10^{-11}$ yr^{-1}) and an initial ^{87}Sr/^{86}Sr ratio of 0.70526 $\pm$0.00008 relative to 0.71025 for NBS 987 (LeBel et al. 1985).

4 1.5 Southern Volcanic Zone

The ^{87}Sr/^{86}Sr ratios of the Quaternary volcanic rocks in the Southern Volcanic Zone of the Andes (33 to 46° S) in Fig. 4.1 range primarily from 0.7038 to 0.7046. The isotope composition of Sr in these rocks does not correlate with the thickness of the continental crust even though the latter decreases from about 50 km in the northern part of this zone to about 30 km in the south (Dalziel 1981; Notsu et al. 1987b). In addition, there is no evidence that the ^{87}Sr/^{86}Sr ratios of the lavas record the fact that the age of the Nazca Plate being subducted into the Peru-Chile Trench is decreasing from north to south, nor does the subduction of the Chile Rise between 46 to 48° S latitude seem to affect the chemical composition or ^{87}Sr/^{86}Sr ratios of the Quaternary lavas in southern Chile.

Nevertheless, the complementary studies of Hickey et al. (1986) and Notsu et al. (1987b) reveal the existence of systematic variations of the ^{87}Sr/^{86}Sr ratios of Quaternary lavas along strike of the southern Andes. The data in Fig. 4.11 indicate the occurrence of elevated ^{87}Sr/^{86}Sr ratios between latitudes 33 to 34° S and between 42 to 44° S. Since the systematic increases of the ^{87}Sr/^{86}Sr ratios from about 0.7040 to 0.7045 are not related to changes in crustal thickness or sediment composition on the Nazca Plate, Hickey et al. (1985) and Notsu et al. (1987b) attributed the variations of the ^{87}Sr/^{86}Sr ratio in the lavas of the Southern Volcanic Zone to local causes such as: (1) variation in the composition of the rocks in the mantle wedge; (2) differences in the efficiency with which LIL elements are transferred to the mantle wedge by aqueous fluids emanating from the subducted oceanic crust; (3) variations in the interaction of mantle-derived magmas with rocks of the overlying continental crust; and (4) regional variations in the Rb/Sr ratios and ages of the lower continental crust. Notsu et al. (1987b) concluded from a review of the relevant literature that neither the ^{143}Nd/^{144}Nd ratios nor the δ^{18}O values display the wave-like pattern of variation of the ^{87}Sr/^{86}Sr ratios.

The volcanic rocks that were extruded in Pleistocene to Recent time in the San Pedro-Pellado complex at about 36°00' S in southern Chile provide insights into the petrogenesis of lavas in the Southern Volcanic Zone (Davidson et al. 1987, 1988). The lavas erupted at this site range in composition from basaltic andesite to andesite, dacite, and rhyolite whose silica concentrations increase from 53.65% (basaltic andesite) to 71.28% (rhyolite and tuff). The isotope ratios of Sr, Nd, and Pb of the lavas are nearly constant with average values of 0.70395 $\pm$0.00007, 0.51276 $\pm$0.0003, and 18.615 $\pm$0.012 ($2\bar{\sigma}$), respectively (Davidson et al. 1987). These values are consistent with the standard model of petrogenesis of volcanic rocks derived from the mantle wedge in a subduction zone.

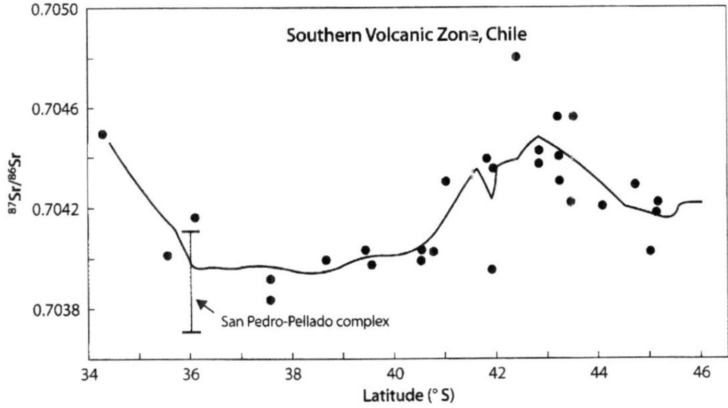

Fig. 4.11. Latitudinal variation of average ^{87}Sr/^{86}Sr ratios of Quaternary basalt, basaltic andesite, and andesite in the Southern Volcanic Zone of the Andes. In this segment of the Andes, the Nazca Plate is being subducted eastward into the Peru-Chile Trench. The ^{87}Sr/^{86}Sr ratios of the volcanic rocks have maxima at 33 to 34° S and at 42 to 44° S. The variations of the ^{87}Sr/^{86}Sr ratios do not correlate with the thickness of the continental crust or with the age of the Nazca Plate being subducted. The line is a three point moving average (Source: Hickey et al. 1986; Notsu et al. 1987b; Davidson et al. 1987)

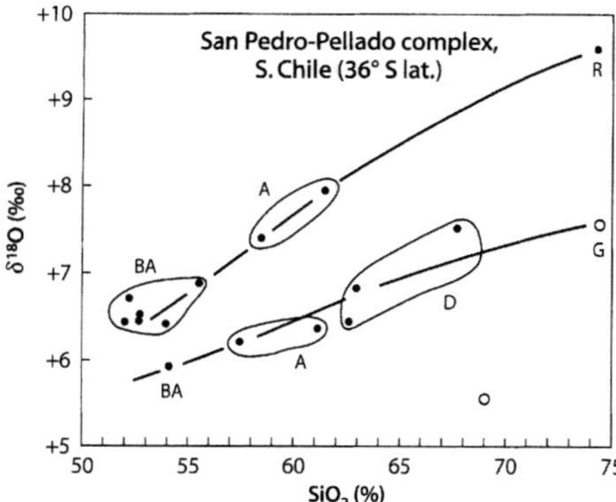

Fig. 4.12. Correlation of δ^{18}O values and silica concentrations of lavas erupted in the San Pedro-Pellado volcanic complex of southern Chile in Pleistocene to Recent time. BA = basaltic andesite; A = andesite; D = dacite; R = rhyolite; G = basement granite (*Source:* Davidson et al. 1988)

The δ^{18}O values of the lavas of the San Pedro-Pellado complex in Fig. 4.12 vary significantly from +5.94‰ (basaltic andesite) to +9.59‰ (rhyolite) and correlate positively with the concentrations of SiO$_2$. This evidence confirms the conclusion based on trace-element concentrations that the mantle-derived basaltic magmas erupted at San Pedro-Pellado did assimilate crustal rocks as they differentiated by fractional crystallization (Davidson et al. 1987, 1988). The isotope ratios of Sr, Nd, and Pb of the lavas do not reveal the contamination of the magmas in this case because the isotope composition of these elements in the granitic basement rocks underlying the San Pedro-Pellado volcanic complex are similar to those of the magmas (Davidson et al. 1987, 1988). The geochemistry and petrogenesis of other volcanic centers in the Southern Volcanic Zone of the Andes have been discussed by Hickey et al. (1984), Stern et al. (1984), Hildreth et al. (1984), Frey et al. (1984), Lopez-Escobar et al. (1985), Grunder (1987), Gerlach et al. (1988b), and Hildreth and Moorbath (1988). Most of these authors present evidence for the interaction of magmas derived from the mantle wedge with rocks of the overlying continental crust.

Unaltered Cretaceous flood basalts in central Chile at about 33° S latitude have low initial ^{87}Sr/^{86}Sr ratios at 117 Ma ranging from 0.70360 to 0.70390 with a mean of

0.70384 ±0.00016 (2$\bar{\sigma}$, N = 6; Åberg et al. 1984). In a related study, Levi et al. (1988) concluded that the chemical compositions of volcanic rocks of Mesozoic to Tertiary age along five traverses across the Andes of central Chile (25 to 35° S latitude) do not vary as expected with increasing distance from the Chile-Peru Trench (and hence with decreasing age) because of episodic changes in the tectonics of subduction.

4.1.6 The Austral Andes

The Austral Volcanic Zone of the Andes (Fig. 4.1) is a small volcanic province at the southern tip of South America that consists of six volcanoes between latitudes 49 and 55° S. The southernmost volcano on Cook Islands in the Strait of Magellan is actually located on a corner of the Scotia Plate (Sect. 3.9.4) rather than on the South American Plate (Stern et al. 1984). The names of the volcanoes on the South American mainland are (from north to south): Lautaro, Nunatak, Aguilera, Mano del Diablo, and Mt. Burney (Charrier et al. 1979). The first four occur within the Patagonian ice cap and only their summits are exposed.

The basement rocks in southern Chile include granitic gneisses of the Magellan Basin for which Halpern (1973) reported a whole-rock Rb-Sr isochron date of 317 ±156 Ma ($\lambda = 1.42 \times 10^{-11}$ yr^{-1}). The present average ^{87}Sr/^{86}Sr ratio of seven specimens of granodiorite gneiss in the Magellan Basin dated by Halpern (1973) is 0.7166 ±0.0016 (2$\bar{\sigma}$, N = 7). These results indicate that the granitic gneisses that underlie the Magellan Basin are significantly younger than those of the Arequipa Massif of Peru and have lower ^{87}Sr/^{86}Sr ratios. The gneisses are overlain by a complex suite of volcanic and sedimentary rocks of Jurassic to Cretaceous age (Suárez and Pettigrew 1976; Winn 1978). Both rock units were subsequently intruded by plutonic rocks of the Andean and Cordillera Darwin suites whose ages range from Cretaceous to Mid-Tertiary (Hervé et al. 1981, 1984; Stern and Stroup 1982).

The calc-alkaline lavas of the Austral Andes range in composition from adakitic andesites to dacites and thus differ from the olivine basalts that predominate in the Southern Volcanic Zone. The trace-element composition of adakitic andesites suggests that they formed by partial melting of subducted oceanic crust and lithospheric mantle which had recrystallized to garnet, amphibole,

Table 4.2.
Average chemical and isotope compositions of the lavas on the volcanoes of the Austral Andes (Stern et al. 1984)

Volcanoes	Rb (ppm)	Sr (ppm)	^{87}Sr/^{86}Sr	^{143}Nd/^{144}Nd	δ^{18}O (‰)
Lautaro, Nunatak, Aguilera	573	533	0.70523	0.512617	+7.6
Mt. Burney	14	602	0.70416	0.512754	+6.7
Cook Island	3	2003	0.70274	0.513160	+5.6

and pyroxene but lacked olivine and plagioclase (Defant and Drummond 1990, 1993; Stern and Kilian 1996).

Table 4.2 shows the average chemical and isotope compositions of the lavas on the volcanoes of the Austral Andes varying systematically from north to south (Stern et al. 1984).

The systematic variations of the isotope ratios of Sr, Nd, and O of the Quaternary lavas on these volcanoes indicate a decrease in the extent of crustal contamination of the mantle-derived magmas from north (Lautaro, etc.) to south (Cook Island) where the lavas approach the composition of MORBs.

4 2 Rhyolites and Ignimbrites of Central America and Mexico

Active volcanic activity in Central America occurs in two separate belts identified in Fig. 4.13. The Mexican volcanic belt extends from the volcano Ceboruco on the west coast across the country to the volcano San Martin on the east coast of Mexico. San Martin is actually located at the southern end of the Eastern Cordillera which is characterized by alkali-rich lavas (Verma 1983).

The Central American volcanic belt is located along the west coast and includes southern Mexico, Guatemala, El Salvador, Nicaragua, and Costa Rica. The volcanic activity in Mexico and Central America is related to the subduction of oceanic lithosphere of the Cocos Plate eastward into the Middle America Trench. Whereas the crust underlying Mexico, Guatemala, and El Salvador includes igneous and metamorphic rocks of Precambrian and early Paleozoic age, the crust in Nicaragua and Costa Rica consists of Cretaceous ultramafic complexes overlain by volcanic and sedimentary rocks of Cenozoic age (Thorpe et al. 1979b)

The Quaternary lavas of Mexico and Central America are underlain by very large deposits of rhyolite ignimbrites of Tertiary age. For example, the Sierra Madre Occidental, which extends for about 1 200 km along the west coast of Mexico is capped by rhyolite ignimbrites (27 to 34 Ma) whose average thickness is close to 1 000 m. According to information cited by Lanphere et al. (1980b), the volume of these ignimbrites is about 2×10^5 km^3 and exceeds that of similar deposits in Central America ($>0.1 \times 10^5$ km^3), the Central Andes ($>0.7 \times 10^5$ km^3, Sect. 4.1.3) and on the North Island of New Zealand ($>0.15 \times 10^5$ km^3, Sect. 3.6.3). The Tertiary ignimbrites of Central America extend for about 900 km from Guatemala, El Salvador, and Honduras into Nicaragua where granitic basement rocks are absent. (Montigny et al. 1969).

The origin of the rhyolite lavas and ignimbrites of Mexico and Central America is constrained by their generally low initial ^{87}Sr/^{86}Sr ratios combined in some cases with low Sr concentrations and unusual enrichment in silica.

4.2.1 Rhyolite Ignimbrites of Central America

The Central American volcanic belt in Fig. 4.13 contains numerous volcanoes that have erupted calc-alkaline basalt, andesite, and dacite lavas as well as rhyolite flows and ignimbrites in Late Tertiary to Recent time. The ^{87}Sr/^{86}Sr ratios of basalts, andesites and dacites in Fig. 4.14 are strongly clustered between 0.7036 and 0.7042. Therefore, these rocks originated from the mantle wedge above the Benioff zone without significant contamination by the continental crust through which the magmas passed on their way to the surface (Walker et al. 1990). However, Grant et al. (1984) observed correlations of ^{87}Sr/^{86}Sr ratios with the concentrations of K$_2$O, Rb, Zr, Ba, and

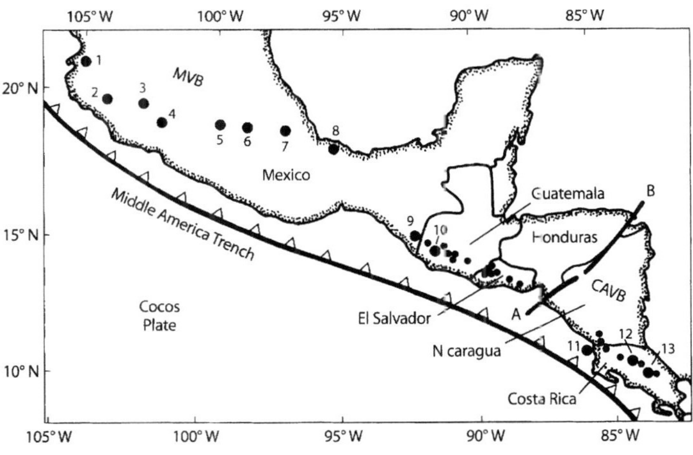

Fig. 4.13.
Map of Central America showing the Mexican volcanic belt (*MVB*) and the Central American volcanic belt (*CAVB*) both of which are associated with subduction of the Cocos Plate eastward into the Middle America Trench. Several prominent volcanoes are identified by number: 1. Ceboruco; 2. Colima; 3. Paricutin; 4. Jorulla; 5. Xitli; 6. Popocatépetl; 7. Orizaba; 8. San Martin; 9. Tacaná; 10. Santa María; 11. Arenal; 12. Poas; 13. Irazu (*Sources: adapted from Moorbath et al. 1978; Thorpe et al. 1979a,b)*

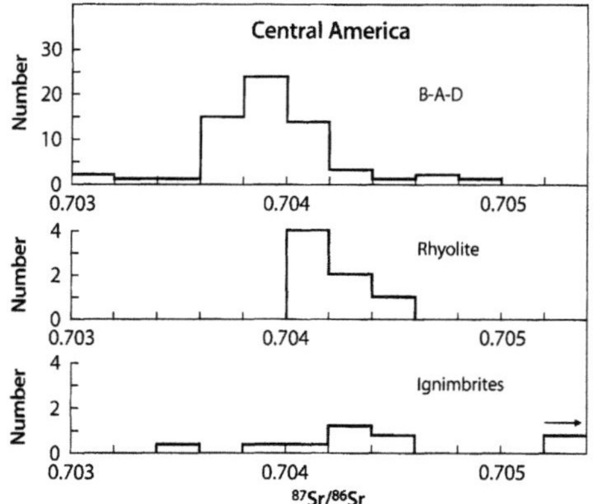

Fig. 4.14. Initial $^{87}Sr/^{86}Sr$ ratios of calc-alkaline lavas and ignimbrites of Mid-Tertiary to Recent age in Central America (Guatemala, El Salvador, Honduras, Nicaragua, and Costa Rica): B = basalt, A = andesite, D = dacite. The $^{87}Sr/^{86}Sr$ ratios of rhyolite ignimbrites in Honduras range up to 0.7170 (*Sources*: Pushkar 1968; Pushkar et al. 1972; Montigny et al. 1969; Rose et al. 1977, 1979; Thorpe et al. 1979a,b; Feigenson and Carr 1986; Walker et al. 1990)

Rb/Sr ratios of basalts and andesites of volcanoes in Guatemala and attributed these correlations to contamination of magmas that ponded within the continental crust for varying periods of time. Similarly, Feigenson and Carr (1986) demonstrated a relationship between the isotope ratios of Sr and Nd of Central American lavas with the thickness of the underlying continental crust. In addition, the volcanic rocks on some of the volcanoes in Costa Rica (Irazu, Pacayas, and Poas) have elevated $\delta^{18}O$ values from +5.30 to +8.83‰ (Montigny et al. 1969). Therefore, the mafic magmas that formed in the mantle wedge under Central America did, in some cases, interact with rocks of the continental crust as they differentiated by fractional crystallization in crustal magma chambers. Clark et al. (1998) investigated the process of magma formation under the volcano Irazu as a result of partial melting in the mantle wedge in response to aqueous and carbonatic fluids released by the downgoing slab.

The $^{87}Sr/^{86}Sr$ ratios of Central American rhyolite lavas in Fig. 4.14 are similar to those of the associated basalts, andesites, and dacites. Therefore, these rhyolites did not form by partial melting of the granitic basement rocks whose $^{87}Sr/^{86}Sr$ ratios range from 0.7037 to 0.7481 (Pushkar et al. 1972). Instead, Pushkar (1968) and Pushkar et al. (1972) considered the rhyolites to be genetically related to the mafic lavas either as a result of magmatic differentiation or by partial melting of basalt and andesite at depth in the crust. The rhyolites of Central America have comparatively high Sr concentrations between 118 to 152 ppm (Pushkar et al. 1972) and from 80 to 423 ppm (Rose et al. 1979), which indicates that the magmas ex-

perienced only limited differentiation while they pooled in magma chambers at shallow depth prior to eruption.

The initial $^{87}Sr/^{86}Sr$ ratios of Miocene and Quaternary rhyolite ignimbrites of Central America in Fig. 4.14 range more widely than those of the rhyolite lavas from 0.7022 to 0.7170. The difference in $^{87}Sr/^{86}Sr$ ratios is an indication that the ignimbrites may have been contaminated by crustal rocks during the eruption or that their magmas formed by partial melting of crustal rocks as result of heating by mafic magmas arising from the mantle wedge. Pushkar et al. (1972) actually demonstrated experimentally that the chemical compositions of glasses formed by partial melting of granitic basement rocks of this region at 760 °C are similar to those of the Tertiary rhyolite ignimbrites of Guatemala.

4.2.2 Ignimbrites of the Sierra Madre Occidental, Mexico

The Sierra Madre Occidental of western Mexico contains the largest deposit of rhyolite ignimbrite in the world. The initial $^{87}Sr/^{86}Sr$ ratios of andesites, dacites, rhyolites, and rhyolite ignimbrites in the Batopilas area (34 to 27 Ma) have a narrow range from 0.70430 to 0.70491 with a mean of 0.70475 ±0.00012 ($2\bar{\sigma}$, N = 13) relative to 0.71025 for NBS 987 (Lanphere et al. 1980b). The silica concentrations of the rhyolite lavas and ignimbrites reach a high value of 76.2% and the average Sr concentration of these rocks is 197 ±69 ppm ($2\bar{\sigma}$, N = 7). Therefore, these rhyolite lavas and ignimbrites occupy a small area in Fig. 4.15 in coordinates of $^{87}Sr/^{86}Sr$ and $(1/Sr) \times 10^3$ ratios.

The initial $^{87}Sr/^{86}Sr$ ratios of rhyolite lavas and ignimbrites in the vicinity of Creel (Lanphere et al. 1980b) and in the provinces of Zacatecas and San Luis Potosi (30 ±8 Ma) range from 0.70499 to 0.71175 relative to 0.7080 for E&A (Verma 1984). The silica concentrations of the ignimbrites in Zacatecas run as high as 83.55%, whereas their average Sr concentration is 53.8 ±14.6 ppm ($2\bar{\sigma}$, N = 12).

The rhyolite lavas and ignimbrites of Creel, Zacatecas, and San Luis Potosi define separate data fields in Fig. 4.15 which distinguish them from the isotopically homogeneous andesites and ignimbrites of Batopilas. The distribution of the data points indicates that the ignimbrites of Creel, Zacatecas, and San Luís Potosi formed by varying degrees of crustal contamination of rhyolite magmas like those that gave rise to the ignimbrites of Batopilas. The most likely explanation for the origin of these magmas is that they formed by partial melting of volcanic rocks at depth in the crust in response to increases in temperature caused by heat transported by mafic magmas from the mantle wedge (Huppert and Sparks 1988). The extremely high concentrations of SiO_2 (up to 83.55%), K_2O (up to 9.18%), and Rb (up to 485 ppm), as well as with the low Sr concentrations (down to 15.5 ppm)

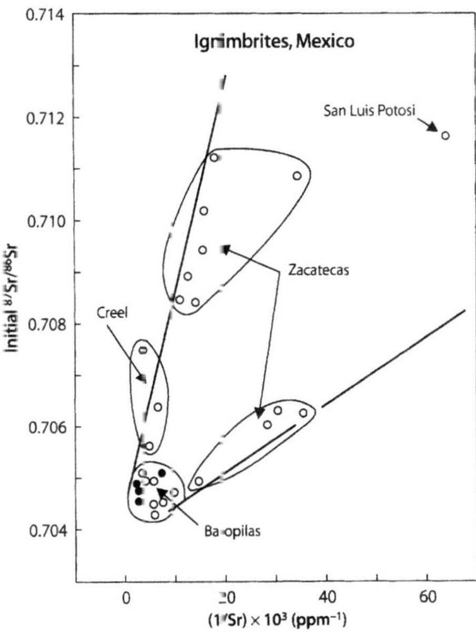

Fig. 4.15. Tertiary rhyolite ignimbrites (*open circles*) and andesites (*solid circles*) in the Sierra Madre Occidental of Mexico. The initial $^{87}Sr/^{86}Sr$ ratios of the ignimbrites of the Batopilas area are indistinguishable from those of the andesites. However, the initial $^{87}Sr/^{86}Sr$ ratios of ignimbrites in Zacatecas and in the Creel area vary widely and are higher than those of andesites. These ignimbrites could not have formed by differentiation of andesite magma but originated from magmas that assimilated crystalline basement rocks having high $^{87}Sr/^{86}Sr$ ratios and low Sr concentrations (*Sources:* Lanphere et al. 1980, Batopilas and Creel; Verma 1984, Zacatecas)

of the ignimbrites are the result of differentiation of the rhyolite magmas at Zacatecas and San Luis Potosi by crystallization of feldspar at shallow depths in crustal magma chambers. The resulting rhyolite magmas were subsequently erupted explosively to form the thick deposits of ignimbrites that characterize the Sierra Madre Occidental of western Mexico as well as the mountains of Central America.

4.2.3 Rhyolites of the Mexican Volcanic Belt

The Mexican volcanic belt (Fig. 4.13) includes several centers of rhyolite lavas of Quaternary to Recent age, including La Primavera at the western end of the belt and Los Humeros at the eastern end. The initial $^{87}Sr/^{86}Sr$ ratios of precaldera rhyolites and rhyodacites at Los Humeros range from 0.70403 to 0.70482 relative to 0.71025 for NBS 987 (Verma 1983). The concentrations of SiO_2 and K_2O of these rocks reach 76.55% and 4.24%, respectively, whereas the Sr concentrations decline to

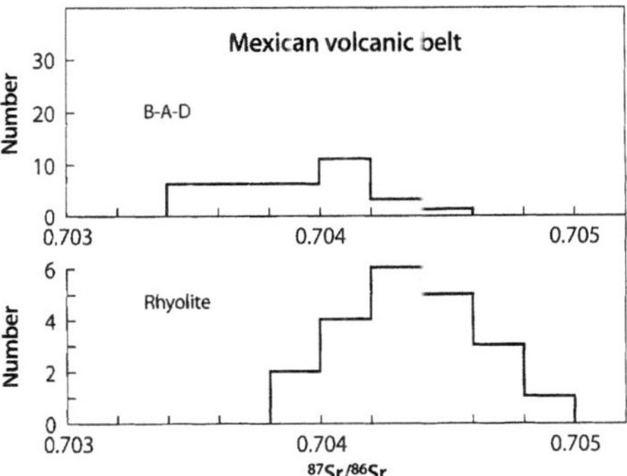

Fig. 4.16. $^{87}Sr/^{86}Sr$ ratios of mafic and rhyolitic volcanic rocks of Quaternary to Recent age in the Mexican volcanic belt. (*B* = basalt, *A* = andesite, *D* = dacite). The mafic lavas are from the volcanoes Nevada de Colima, Colima, Ceboruco, and San Martin as well as from Los Humeros and Sierra La Primavera which are centers of rhyolite volcanism (*Sources:* Moorbath et al. 1978; Verma 1983; Mahood and Halliday 1988)

10.15 ppm. Postcaldera basalts, andesites, and dacites have an average $^{87}Sr/^{86}Sr$ ratio of 0.70415 ±0.00007 ($2\bar{\sigma}$, N = 20) relative to 0.71025 for NBS 987 (Verma 1983).

The isotope ratios of Sr of the basalts, andesites, and dacites at Los Humeros and on other volcanoes of the Mexican volcanic belt are consistent with the standard model for the petrogenesis of volcanic rocks associated with subduction zones. Accordingly, Verma (1983) concluded that the mafic magmas formed in the mantle wedge which had been altered by aqueous fluids derived from the subducted oceanic crust. The resulting magmas subsequently differentiated by fractional crystallization in crustal magma chambers without significant contamination by the granitic basement rocks of this area.

The $^{87}Sr/^{86}Sr$ ratios of rhyolites of Sierra La Primavera at the western end of the Mexican volcanic belt range from 0.70400 to 0.70496 in Fig. 4.16 and are only slightly higher than those of basalts, andesites, and dacites on the volcanoes of this region (Moorbath et al. 1978; Verma 1983). Only one rhyolite specimen from the Southcentral Dome of La Primavera has an elevated initial $^{87}Sr/^{86}Sr$ ratio of 0.708411 (Mahood and Halliday 1988). The average Rb and Sr concentrations of these rhyolites are 168 ppm and 0.97 ppm, respectively, giving them an average Rb/Sr ratio of 173. Consequently, significant in situ decay of ^{87}Rb has occurred in these rocks in less than 10^5 years. In spite of the low Sr concentrations, which make the rhyolites at Sierra La Primavera vulnerable to contamination by crustal Sr, their initial $^{87}Sr/^{86}Sr$ do not correlate with chemical parameters, such as Rb/Sr ratios. In addition, the $\delta^{18}O$ values reported by Mahood and Halliday (1988) are generally low (quartz = +7.4 ±0.1‰; sanidine = +6.6 ±0.3‰; glass = +7.5 ±1.1‰) and pro-

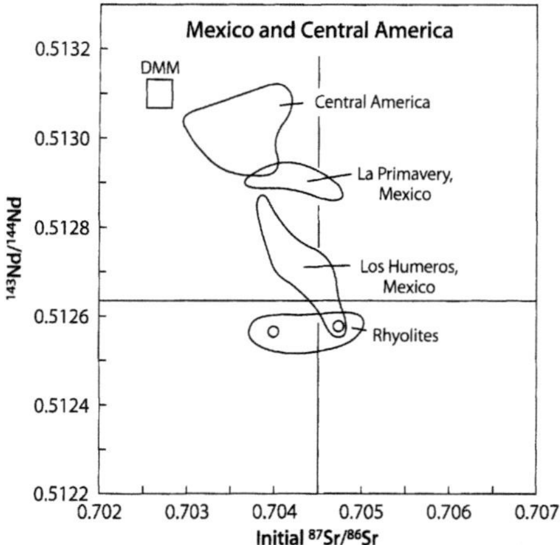

Fig. 4.17. Isotope ratios of Sr and Nd of Quaternary volcanic rocks in Mexico and Central America (*Sources:* Walker et al. 1990; Verma 1983; Mahood and Halliday 1988; Feigenson and Carr 1986)

vide no compelling evidence for extensive contamination of the rhyolite magmas by incorporation of crustal rocks.

The petrogenesis of the rhyolites at La Primavera is constrained by the similarity of their initial ^{87}Sr/^{86}Sr ratios to those of the coexisting mafic lavas, by the extremely low Sr concentrations combined with their evident Rb-enrichment, and by the δ^{18}O values of minerals and glass which do not differ appreciably from those of mantle-derived and uncontaminated basalts and andesites (e.g. Montigny et al. 1969). An additional constraint arises from the facts that the ^{143}Nd/^{144}Nd ratios of the rhyolites at La Primavera in Fig. 4.17 are high compared to the rhyolites at Los Humeros (Verma 1983), that they do not correlate with the initial ^{87}Sr/^{86}Sr ratios, and that they are comparable in value to the ^{143}Nd/^{144}Nd ratios of the basalts and andesites in Central America (Feigenson and Carr 1986; Walker et al. 1990).

These constraints exclude the possibility that the rhyolite magmas formed by partial melting of crystalline basement rocks of central Mexico (Ruiz et al. 1988; Patchett and Ruiz 1987; Verma and Verma 1986; Yañez et al. 1991; Karlstrom et al. 1997). Instead, the similarity of the isotope ratios of Sr and Nd of the rhyolites and the associated mafic lavas in Central America and Mexico permits the existence of a genetic link between them.

4.2.4 Standard Model of Rhyolite Ignimbrite Petrogenesis

The isotopic compositions of Sr, Nd, and Pb of rhyolite lavas and ignimbrites combined with their chemi-

cal compositions and their occurrences in the field can be accounted for by a model that applies not only to the rhyolite ignimbrites of Mexico and Central America but also to those in the Andes of South America and in the Rocky Mountains of North America. This "standard model" includes the following sequence of events:

1. Mafic magmas form by partial melting of rocks in the mantle wedge that was enriched in Sr and Pb (but less so in Nd) by aqueous fluids emanating from the subducted oceanic crust.
2. The mafic magmas may differentiate by fractional crystallization and (to a lesser extent) by assimilation of crustal rocks.
3. The resulting calc-alkaline lavas and associated subsurface plutons may be remelted by heat transported into the crust by magmas arising from the mantle wedge.
4. The resulting secondary magmas are felsic in composition, and their isotope ratios of Sr, Nd, and Pb are similar to those of the young volcanic rocks from which they formed.
5. In some cases, the felsic magmas are erupted to the surface without significant differentiation. In other cases, (e.g. La Primavera) the felsic magmas evolve chemically by crystallization of feldspar causing their Sr concentrations to decrease and allowing their ^{87}Sr/^{86}Sr ratios to rise because of in situ decay of ^{87}Rb depending on the magnitude of the Rb/Sr ratios and the length of residence in the magma chamber.
6. The rhyolite magmas in crustal magma chambers may become chemically stratified with the most highly evolved magma at the top of the chamber where it may be contaminated by incorporation of wallrock.
7. Ultimately, the differentiated rhyolite magma is erupted explosively to form rhyolitic tuff (ignimbrites) at the surface. The tuff may be stratified chemically and isotopically because of the inversion of the original stratification in the magma chamber.
8. The resulting layers of rhyolite tuff may be contaminated during the eruption by mixing with fragments of crustal rocks and later by the action of hydrothermal solutions and groundwater.

The origin of rhyolite lavas and ignimbrites at a particular location may deviate from the standard model in response to local conditions. For example, the rhyolite magmas in the Taupo Volcanic Zone of New Zealand (Sect. 3.6.3) appear to have formed by partial melting of oceanic crust, whereas the rhyolite magmas of Japan originated in large part by melting of granitic basement rocks. The standard model is not only sufficiently flexible to accommodate such local effects, but it also excludes other kinds of explanations for the origin of felsic volcanic rocks, such as:

1. Partial melting of ultramafic rocks in the mantle.
2. Differentiation of basalt or andesite magmas by fractional crystallization (except in small amounts).
3. Derivation of rhyolite magma exclusively by partial melting of granitic basement rocks.

4.2.5 Quaternary Lavas of San Luis Potosí, Mexico

The state of San Luis Potosí of Mexico contains several volcanic centers of Quaternary age where alkali-rich basalts (basanites) have been erupted. The Quaternary lavas at Ventura and Santo Domingo contain inclusions of mantle-derived spinel lherzolites and of granulite from the lower crust (Schaaf et al. 1994; Roberts and Ruiz 1989; Luhr et al. 1989; Pier et al. 1989). The lavas also contain kaersutite xenocrysts (titaniferous amphibole related to arfvedsonite: $Na_3Fe_4Al(OH)_2Si_8O_{22}$) which originated from veins in the lherzolite inclusions. The occurrence of kaersutite veins in mantle-derived inclusions is a sign that the lithospheric mantle under San Luis Potosí was altered by the intrusion of aqueous fluids which caused the formation of kaersutite by reacting with pyroxenes.

The $^{87}Sr/^{86}Sr$ ratios of the Quaternary basanites of San Luis Potosí, Mexico, range from 0.70302 to 0.70389 relative to 0.71025 for NBS 987 and correlate inversely with the $^{143}Nd/^{144}Nd$ ratios (0.51283 to 0.51299) in Fig. 4.18 (Pier et al. 1989). The kaersutite crystals at Santo Domingo have lower $^{87}Sr/^{86}Sr$ ratios (0.70293 to 0.70312) and higher $^{143}Nd/^{144}Nd$ ratios (0.51297 to 0.51301) than the host basanite. According to references cited by Pier et al. (1989), such discordances of isotope compositions have been taken as evidence that the kaersutite veins in the lithospheric mantle formed before the magmatic activity that resulted in the eruption of basanite lavas at the surface. Alternatively, the basanite magmas could have been contaminated by assimilating crustal rocks without affecting the isotope compositions of Sr and Nd in the kaersutite crystals. This alternative is not plausible in the present case because the Quaternary basanites as well as the kaersutites of San Luis Potosí have elevated Sr and Nd concentrations (Pier et al. 1989). This is shown in Table 4.3, where the number of samples in each average is indicated in parentheses. The high concentrations of Sr and Nd of the basanite lavas and kaersutite crystals make their isotope ratios insensitive to contamination by assimilation of granitic basement rocks. The data fields of the basanite lavas and of the kaersutite xenocrysts in Fig. 4.18 lie close to a mixing line between DMM and EM1, meaning that their magma sources in the mantle wedge were contaminated by Sr and Nd derived from subducted oceanic sediment.

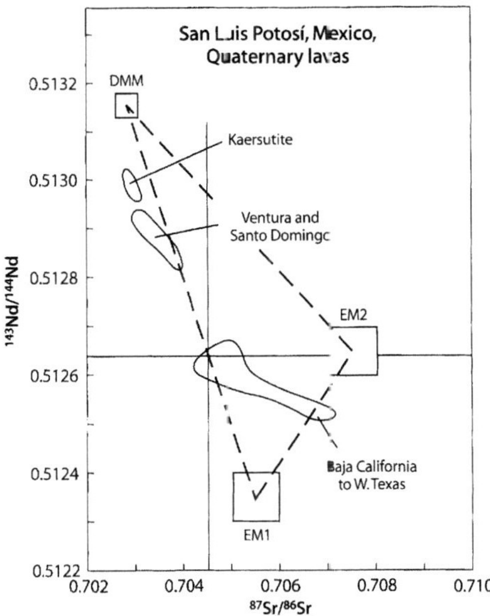

Fig. 4.18. Isotope ratios of Sr and Nd of Quaternary basanite lavas in Santo Domingo and Ventura and kaersutite crystals (megacrysts and xenocrysts) that occur in the volcanic rocks in the Santo Domingo area of San Luis Potosí Mexico (*Source:* Pier et al. 1989). In addition, Middle to Late Tertiary dacites in a belt extending from Baja California, Mexico, to west Texas originated from the mantle wedge without significant contamination by Precambrian crustal rocks (*Source:* Cameron and Cameron 1985)

Table 4.3. Sr and Nd concentrations of the Quaternary basanites and the kaersutites of San Luis Potosí (Pier et al. 1989). The number of samples in each average is indicated in parentheses

	Sr (ppm)	Nd (ppm)
Basanite	853 ±13 (18)	44.0 ±7.0 (18)
Kaersutite	487 ±46 (5)	17.0 ±6.0 (5)

Middle to Late Tertiary dacites in a belt extending from Baja California on the west coast of Mexico to Chihuahua and to west Texas analyzed by Cameron and Cameron (1985) also have uniform $^{87}Sr/^{86}Sr$ and $^{143}Nd/^{144}Nd$ ratios that are characteristic of magma sources in the mantle (Fig. 4.18). The high average Sr and Nd concentrations of these rocks (367 ±80 ppm and 34.7 ±11.6 ppm, respectively), likewise prevent their isotope ratios from being changed by assimilation of crustal rocks by the mantle-derived magmas. Moreover, Cameron and Cameron (1985) demonstrated by AFC modeling that the dacites in this cross-section of the orogenic belt north of San Luis Potosí did not assimilate significant amounts of Precambrian basement rocks.

4.3   Rhyolites and Ignimbrites of California

Large volumes of rhyolite lavas and ignimbrites occur in the Rocky Mountains of the USA, especially in California, Arizona, and Nevada (Hildreth 1981). The prevalence of volcanic rocks of rhyolitic composition and their relation to granitic plutons at depth were the subjects of symposia edited by Barker (1981) and Papike (1987). The tectonic setting of the Late Tertiary to Quaternary volcanic activity of the southwestern USA differs from that of Mexico and Central America because, in this area, the North American Plate has overridden both the East Pacific Rise and the associated northward extension of the Middle America Trench (Fig. 4.13). The subcrustal tectonic activity has manifested itself at the surface by the development of the Basin and Range province (Utah, Nevada, southeastern California), by the uplift of the Colorado Plateau (Colorado, Utah, Arizona, New Mexico), and by the formation of the Rio Grande Rift (New Mexico) identified in Fig. 4.19 (Hamilton and Myers 1966; Leeman and Harry 1993). The volcanic rocks of the southwestern USA range widely in composition and include not only calc-alkaline basalt and rhyolite, but also alkali-rich varieties which are described in Sect. 6.9.1 to 6.9.5.

The rhyolites and ignimbrites of California are associated with isolated volcanic centers located primarily along the eastern slopes of the Sierra Nevada Mountains in central California and in the Mojave Desert in the southeastern part of the state. The principal occurrences of rhyolite lavas and tuffs are associated with the Long Valley Caldera (37°40' N, 118°50' W) and the related Inyo and Mono Craters south of Mono Lake; with the Coso volcanic field (36°03' N, 117°50' W) between Death Valley National Monument in the east and the southern end of the Sierra Nevada Mountains in the west; and with the Woods Mountains volcanic center (35°10' N, 115°18' W) located a short distance west of the Cima volcanic center in the Mojave Desert in southeastern California. The petrogenesis of rhyolites and mafic lava flows at these volcanic centers in California and elsewhere in the southwestern USA is complicated because of the unusual tectonic setting in which the magmas formed and evolved in this region.

4.3.1   Long Valley Caldera and the Bishop Tuff

The Long Valley Caldera (Fig. 4.19) is the site of episodic volcanic activity that has lasted from the Pliocene Epoch to the present time (Bailey et al. 1976; Bailey 1984, 1989). The area is underlain by granitic rocks of Jurassic and Cretaceous age of the Sierra Nevada Batholith and by Paleozoic and Mesozoic metasedimentary and metavolcanic rocks that now form roof pendants in the batholith (Kistler and Peterman 1973).

The oldest lavas of the Long Valley area were erupted between 3.2 and 2.6 Ma and consist of basalts, andesites, and rhyodacites. These rocks were followed between 2.1 and 0.79 Ma by the silica-rich rhyolites of Glass Mountain located along the northeastern edge of the present caldera (Metz and Mahood 1985). The eruption of rhyolite flows, lava domes, and pyroclastics was terminated at 0.725 ±0.015 Ma by the violent explosion that produced the Bishop tuff (Dalrymple 1980). Subsequently, the roof of the magma chamber collapsed to

Fig. 4.19.
Physiographic provinces of the southwestern region of the USA (*Source:* adapted from Mukasa and Wilshire 1997)

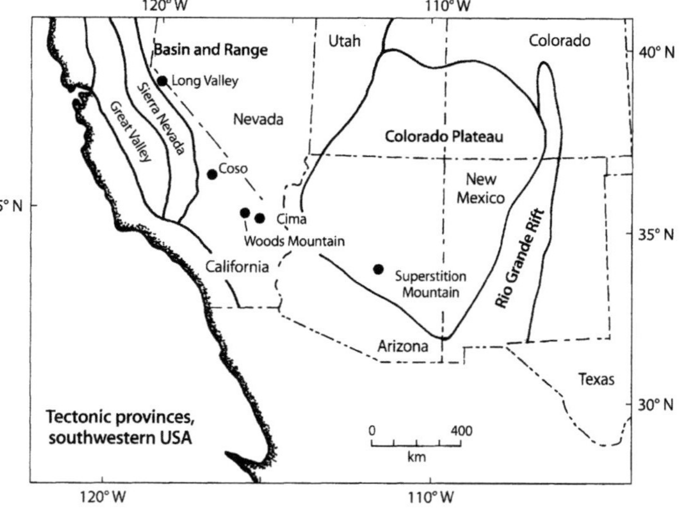

form the present Long Valley Caldera. However, volcanic activity did not cease, but has continued with the eruption of the Resurgent Dome and a number of smaller centers around the periphery of the dome. The Mono Craters and Inyo Craters are the sites of the most recent eruptions of rhyolite lavas in the area (Sieh and Bursik 1986; Vogel et al. 1989; Friedman 1989; Heumann and Davies 1997).

The precaldera rhyolites on Glass Mountain were erupted during two episodes between 2.1 and 1.2 Ma (older lavas) and from 1.1 to 0.70 Ma (younger lavas). The older lavas are highly differentiated with elevated Rb (156 to 311 ppm) but very low Sr concentrations (0.101 to 3.92 ppm). Their initial ^{87}Sr/^{86}Sr ratios in Fig. 4.20 range widely from 0.7068 to 0.7389 relative to 0.71025 for NBS 987 (Halliday et al. 1989). In contrast to the initial ^{87}Sr/^{86}Sr ratios, the isotope ratios of Nd and Pb of the older lavas on Glass Mountain are virtually constant

with average ^{143}Nd/^{144}Nd = 0.512472 ±0.000011 ($2\bar{\sigma}$, $N = 14$) and ^{206}Pb/^{204}Pb = 19.149 ±0.007 ($2\bar{\sigma}$, $N = 9$; Halliday et al. 1989).

The younger precaldera rhyolites on Glass Mountain are less severely differentiated than their precursors (Rb: 158 to 182 ppm; Sr: 1.20 to 2.49 ppm) and their initial ^{87}Sr/^{86}Sr ratios in Fig. 4.20 are more uniform (0.70622 to 0.70722). However, their average ^{143}Nd/^{144}Nd ratio is slightly higher at 0.512588 ±0.000008 ($2\bar{\sigma}$, $N = 8$) than that of the older precaldera rhyolite flows. The ^{206}Pb/^{204}Pb ratios of the younger lavas on Glass Mountain are indistinguishable from those of the older lavas (Halliday et al. 1989).

The Bishop tuff, described by Hildreth (1979, 1981) and Bailey et al. (1976), has high silica concentrations ranging from 77.4% (early) to 75.5% (late) on a water-free basis. The concentrations of other elements (Fe, Mn, Ti, P, Ca, and Mg) vary widely within this narrow range of silica concentrations, implying the existence of steep compositional gradients in the magma chamber. The rhyolite magma of the Bishop tuff, whose volume was about 600 km^3, was expelled in a single eruption culminating its evolution in a magma chamber at shallow depth.

The initial ^{87}Sr/^{86}Sr ratios of whole-rock samples of the Bishop tuff in Fig. 4.20 analyzed by Halliday et al. (1984) range from about 0.7060 to 0.7092 and generally rise with decreasing Sr content and increasing Rb/Sr ratios. The initial ^{87}Sr/^{86}Sr ratios of sanidine phenocrysts (0.7061 to 0.7069) are lower than or equal to the ^{87}Sr/^{86}Sr ratios of the rocks in which they occur. Halliday et al. (1984) concluded from this and other evidence that:

1. The rhyolite magma was contaminated by addition of Sr from the wallrock.
2. The sanidine crystals had formed prior to the contamination of the magma.
3. The uncontaminated rhyolite had an initial ^{87}Sr/^{86}Sr ratio of about 0.7061.

The postcaldera rhyolites analyzed by Heumann and Davies (1997) have higher Sr and lower Rb concentrations than the precaldera rhyolites (Halliday et al. 1989) and the Bishop tuff (Halliday et al. 1984; Noble and Hedge 1969). These and other aspects of their chemical composition indicate that the postcaldera magmas are less severely differentiated than the precaldera lavas presumably because of shorter residence times in crustal magma chambers. The isotope ratios of Sr and Nd of the rhyolites of the Resurgent Dome in Fig. 4.20 are remarkably uniform with average values of 0.706613 ±0.000029 ($2\bar{\sigma}$, $N = 14$) and 0.512531 ±0.000009 ($2\bar{\sigma}$, $N = 7$), respectively. The ^{206}Pb/^{204}Pb ratios of the rhyolite of the Resurgent Dome also vary only between narrow limits, have a mean of 19.252 ±0.012 ($2\bar{\sigma}$, $N = 14$), and are only slightly higher than those of the precaldera rhyolites (Heumann and Davies 1997).

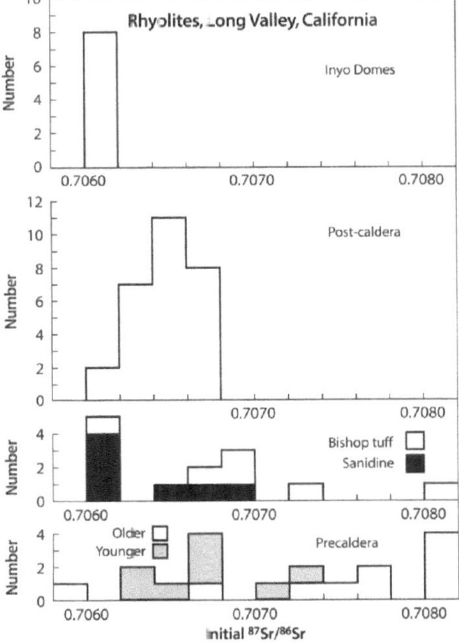

Fig. 4.20. Initial ^{87}Sr/^{86}Sr ratios of Late Tertiary and Quaternary rhyolites at Long Valley, California. The diagram illustrates the fact that the initial ^{87}Sr/^{86}Sr ratios of the rhyolites have decreased with time and that the rhyolite magmas at Long Valley had an ^{87}Sr/^{86}Sr ratios of about 0.7060. The initial ^{87}Sr/^{86}Sr ratios of the older precaldera rhyolites range up to 0.7389 and 75% of the samples analyzed (18 of 24) have ^{87}Sr/^{86}Sr ratio higher than 0.7080. Sanidine crystals in the Bishop tuff have lower ^{87}Sr/^{86}Sr ratios than the rocks in which they occur (*Sources:* Davies et al. 1994; Halliday et al. 1984, 1989; Heumann and Davies 1997)

4.3.2   Petrogenesis of the Rhyolites at Long Valley

The record of eruptions of the Long Valley Volcano raises questions about the origin of the rhyolite magmas and their subsequent differentiation by fractional crystallization, wallrock assimilation, and magma mixing while residing in crustal magma chambers (Christensen and DePaolo 1993; Halliday et al. 1991). Although the initial $^{87}Sr/^{86}Sr$ ratios of the rhyolite lavas at Long Valley range widely, the lowest values of about 0.706 are similar to the $^{87}Sr/^{86}Sr$ ratios of Late Tertiary basaltic rocks in the area. Therefore, the rhyolite magmas could have formed by remelting of mafic igneous rocks or of other kinds of rocks in the lower continental crust. The heat required for melting was presumably provided by basaltic magmas that originated in the underlying mantle in accordance with the standard model of petrogenesis in subduction zones (Huppert and Sparks 1988).

The low Sr concentrations of the precaldera lavas indicate that the rhyolite magmas differentiated by crystallization of feldspar while they resided in one or several large chambers under the original Long Valley Volcano (Lu et al. 1992). These magma chambers existed for about 1.5 million years until the catastrophic eruption that produced the Bishop tuff at 0.7 Ma. The highly variable initial $^{87}Sr/^{86}Sr$ ratios (0.7074 to 0.7389) of the older suite of precaldera lavas can be attributed to assimilation of crustal rocks by the magma at the top of the magma chamber. The comparatively low initial $^{87}Sr/^{86}Sr$ ratios (0.70622 to 0.70722) of the younger precaldera lavas are an indication that these flows formed from magma that was derived from lower levels in the chamber where wallrock assimilation was not significant. In addition, the eruption of the younger precaldera lavas implies that at 1.1 Ma the magma chamber had refilled with rhyolitic magma from sources at depth in the crust.

At the time of the eruption of the Bishop tuff, the magma chamber was highly stratified chemically (Halliday et al. 1984; Hildreth 1979, 1981; Noble and Hedge 1969). The vertical zonation of the Bishop tuff is inverted because the most highly differentiated rocks at the bottom of the Bishop tuff originated from the top of the magma chamber, whereas the less differentiated magma at depth in the chamber was erupted last and formed the top layers of the Bishop tuff.

The initial $^{87}Sr/^{86}Sr$ ratios of sanidine crystals in the Bishop tuff (Fig. 4.20) range from 0.70606 to 0.70687 and are, in most cases, lower than the $^{87}Sr/^{86}Sr$ ratios of the whole-rock samples in which they occur. Therefore, the sanidine crystallized before the magma or the resulting rocks were contaminated. The sanidine crystals also have lower $\delta^{18}O$ values (+6.72 to +7.94‰) than whole-rock samples (+5.92 to +10.30‰), but the differences are not large in most cases (Halliday et al. 1984) Since the sanidines crystallized prior to the addition of crustal Sr, their average $\delta^{18}O$ value of +7.0 ±0.2% probably represents the oxygen isotope composition in the rhyolite magma prior to contamination. The comparatively low $\delta^{18}O$ value of the magma indicates that it originated from mantle-derived rocks rather than by melting of granitic rocks of the underlying crust because the $\delta^{18}O$ values of such rocks are higher than +7.0‰ in many cases.

The addition of varying amounts of Sr to rocks or magmas having a range of Sr concentrations can be illustrated by means of a mixing diagram devised by Lutz et al. (1988). The contaminant Sr in Fig. 4.21 is assumed

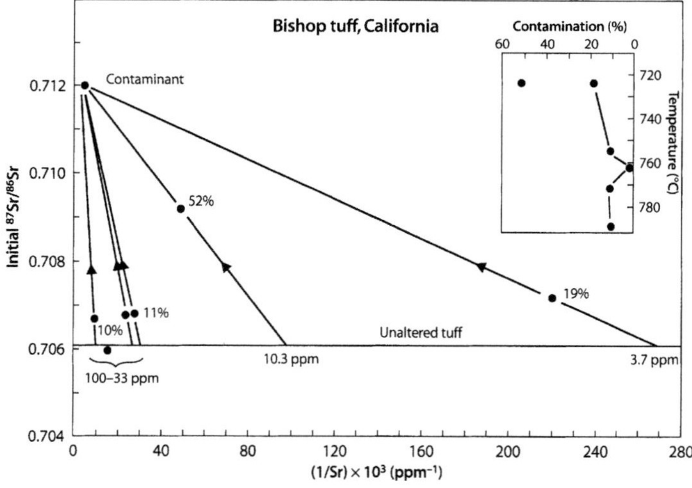

Fig. 4.21.
Contamination of the Bishop tuff with Sr having an elevated $^{87}Sr/^{86}Sr$ of 0.712. The construction of this diagram is explained in the text following Lutz et al. (1988), who invented it. The results summarized in the inset indicate that the extent of contamination of Sr increased toward the top of the magma chamber (decreasing Fe-Ti temperatures). The contamination of the tuff that originated from the top of the chamber may have occurred by interactions of the magma with wallrocks, by incorporation of xenocrystic material during the explosive eruption, or by subsequent alteration of the tuff by groundwater (*Source:* Halliday et al. 1984)

:c have an $^{87}Sr/^{86}Sr$ ratio of 0.712, whereas the unaltered Bishop tuff has $^{87}Sr/^{86}Sr = 0.7061$. The whole-rock samples are plotted in coordinates of their initial $^{87}Sr/^{86}Sr$ and $(1/Sr) \times 10^3$ ratios. Mixing lines are then drawn from the point representing the contaminant through each data point to intersect the line at $^{87}Sr/^{86}Sr = 0.7061$. The x-coordinates of the points of intersection yield the Sr concentration of each rock prior to contamination. The fraction of contaminant Sr added to each rock sample is determined by the lever rule. The resulting abundances of contaminant Sr in the inset to Fig. 4.21 decrease with increasing crystallization temperature and hence with increasing depth in the magma chamber. The addition of contaminant Sr may have been caused by wallrock interaction at the top of the magma chamber, by incorporation of xenocrystic material during the explosive eruption, or by circulating groundwater after deposition of the Bishop tuff as proposed by Halliday et al. (1984). In all cases, the low Sr concentrations of the rocks facilitated the contamination process.

The petrogenesis of the rhyolites at Long Valley was discussed in a comment by Sparks et al. (1990) with replies by Halliday (1990) and Mahood (1990). The principal question at issue was the evolution of rhyolite magma in crustal reservoirs. Sparks et al. (1990) proposed that the different suites of rhyolite that were erupted at Long Valley formed as separate melts of granitic basement rocks rather than by fractional crystallization in a magma chamber that was refilled episodically (Halliday et al. 1989). However, age determinations by the Rb-Sr isochron method of minerals and melt inclusions subsequently confirmed the timescale of the evolution of rhyolite magma at Long Valley (Davies et al. 1994; Christensen and Halliday 1996. Davies and Halliday 1998).

The wide range of initial $^{87}Sr/^{86}Sr$ ratios of the rhyolites of Long Valley contrasts sharply in Fig. 4.22 with the much smaller range of the $^{143}Nd/^{144}Nd$ ratios. The elevated $^{87}Sr/^{86}Sr$ and comparatively low $^{143}Nd/^{144}Nd$ ratios of these rocks are both expressions of the involvement of crustal rocks in the petrogenesis of these rhyolites.

4.3.3 Rhyolites of the Coso Range

The Coso Range of California (Fig. 4.19) consists of more than 30 rhyolite domes and flows most of which are less than 300 000 years old. The lava domes are aligned along north-south trending faults of the Basin and Range tectonic province. These rocks are part of a complex of Late Tertiary volcanic rocks that were erupted in two episodes between 4.0 to 2.5 Ma and from about 1.1 to less than 0.2 Ma (Lanphere et al. 1975; Duffield et al. 1980). The older suite of lavas consists of about 31 km^3 of calc-alkaline rocks (basalt to rhyolite), whereas the younger suite is composed of basalt and rhyolite in about equal proportions. The volcanic activity in the Coso Range is coeval with the eruptions of the Long Valley Volcano located about 225 km northwest.

The volcanic rocks are underlain by granitic basement rocks of the Sierra Nevada Batholith. Two specimens of these basement rocks (granodiorite and diorite) of Early Jurassic age have present-day $^{87}Sr/^{86}Sr$ ratios of 0.7057 and 0.7062, respectively, relative to 0.7080 for E&A (Bacon et al. 1984). Kistler and Peterman (1973) reported that the initial $^{87}Sr/^{86}Sr$ ratios of the Mesozoic granitoids of the nearby Sierra Nevada Mountains range primarily from 0.7031 to 0.7082 and that the 0.7060 contour line defines a region of elevated initial $^{87}Sr/^{86}Sr$ ratios in the southern part of this mountain range. In addition, they observed that the overlying Late Tertiary basalts and andesites have similar $^{87}Sr/^{86}Sr$ ratios as the granitic basement rocks.

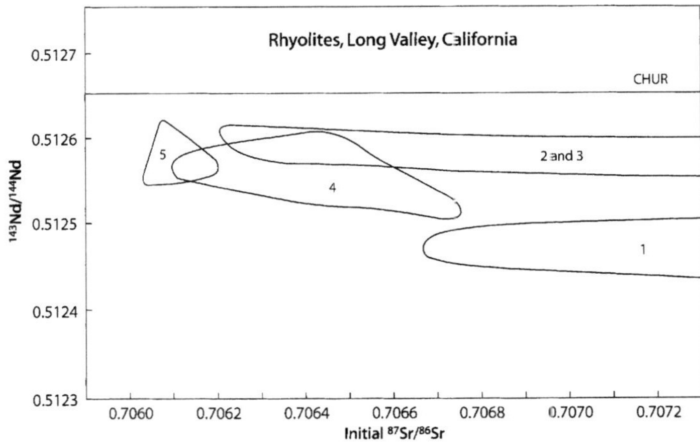

Fig. 4.22.
Isotope ratios of Sr and Nd of rhyolites at Long Valley, California. The data clusters are identified by number in the sequence in which they were erupted: 1. Precaldera lavas, older suite; 2. Precaldera lavas, younger suite; 3. Bishop tuff; 4. Postcaldera lavas; 5. Inyo Craters. The initial $^{87}Sr/^{86}Sr$ ratios vary widely, whereas the $^{143}Nd/^{144}Nd$ ratios are restricted to a narrow range. Both features reveal that the rhyolites are strongly affected by assimilation of crustal rocks (*Sources:* Halliday et al. 1984, 1989; Heumann and Davies 1997)

The granitic basement rocks and overlying basalt lavas in the Coso Range have high Sr concentrations between 600 and 800 ppm, whereas the Quaternary rhyolites contain only 9.3 ±5.3 ppm Sr, but have high Rb concentrations with a mean of 255 ±74 ppm ($2\bar{\sigma}$, $N = 3$; Bacon et al. 1984). The initial $^{87}Sr/^{86}Sr$ ratios of Quaternary rhyolite obsidian in the Coso Range (0.70530 to 0.70585) are similar to the present $^{87}Sr/^{86}Sr$ ratio of the Early Jurassic basement rocks (0.7057 to 0.7062), but differ from the $^{87}Sr/^{86}Sr$ ratios of the associated basalts which range from 0.70364 to 0.70537 with a mean of 0.7043 ±0.0007 ($2\bar{\sigma}$, $N = 4$).

These data suggest that the rhyolite magma formed by melting of granitic basement rocks like those of the Sierra Nevada Batholith rather than by differentiation of basalt magma or by melting of Tertiary gabbro at depth in the crust. In addition, the low Sr concentrations of the rhyolite reveal that their magmas differentiated by crystallization of feldspar prior to eruption to the surface. The heat required for melting the granitic basement rocks was transported into the continental crust by basalt and andesite magmas which originated in the underlying mantle having an $^{87}Sr/^{86}Sr$ ratio of about 0.7036 or less. The variation of the initial $^{87}Sr/^{86}Sr$ ratios of the basalts (0.70364 to 0.70537) is probably the result of assimilation of crustal rocks (Glazner and O'Neil 1989).

The isotope ratios of Sr and Pb of the basalt in Fig. 4.23 define a mixing hyperbola that extrapolates to a magma source composed of depleted MORB mantle (DMM). Accordingly, the basalt magma originated in the mantle wedge to which Sr and Pb derived from the subducted 'sediment had been added. The basalt magma ($^{87}Sr/^{86}Sr = 0.70364$; $^{206}Pb/^{204}Pb = 19.047$) subsequently assimilated rocks of the continental crust, causing its $^{87}Sr/^{86}Sr$ ratio to rise to 0.70537 and its $^{206}Pb/^{204}Pb$ ratio to 19.199.

The rhyolites in Fig. 4.23 are displaced from the basalt trajectory because they have higher $^{87}Sr/^{86}Sr$ and $^{206}Pb/^{204}Pb$ ratios than the basalts. Two inclusions also deviate from the basalt trajectory and may be related to the rhyolites. The isotope ratios of Sr and Pb are consistent with the hypothesis that the rhyolite magma originally formed by partial melting of granitic basement rocks having present $^{87}Sr/^{86}Sr$ ratios between 0.7050 and 0.7060.

4.3.4 Woods Mountains Volcanic Center

The effect of crustal contamination on the isotope ratios of Sr and Nd in mantle-derived magmas is especially well represented by the lava flows and ignimbrites of the Woods Mountains volcanic center located in the southeastern part of the state (Fig. 4.19). The basement rocks in this area are approximately 2-billion-year old granitoids. The volcanic rocks of Miocene age consist primarily of the Wild Horse Mesa tuff (15.8 Ma) and overlying basalt and andesite flows (10.0 Ma). The Wild Horse Mesa tuff is a rhyolite ignimbrite that was erupted in three closely spaced episodes followed by the collapse of the magma chamber to form a caldera with a diameter of about 10 km. The younger mafic flows were erupted through fissures along the western side of the caldera (Musselwhite et al. 1989).

The silica concentrations of the Wild Horse Mesa tuff range from 73.7 to 75.5% with Sr concentrations between 9.8 and 28.3 ppm. In addition, the Sr is notably enriched in ^{87}Sr with initial $^{87}Sr/^{86}Sr$ ratios that range from 0.70940 to 0.71407 relative to 0.71025 for NBS 987. The overlying basalt and andesite have initial $^{87}Sr/^{86}Sr$ ratios of about 0.7058 (Musselwhite et al. 1989).

The initial $^{87}Sr/^{86}Sr$ ratios of the Wild Horse Mesa tuff (0.70949 to 0.71407) are higher than those of the rhyolites in the Coso Range and at Long Valley, except for the older precaldera rhyolites at the latter site. In addition, the $^{143}Nd/^{144}Nd$ ratios (renormalized to $^{146}Nd/^{144}Nd = 0.7219$) of the volcanic rocks at Woods Mountains are lower than those of the Coso Range and Long Valley, California, with an average of 0.51229 ±0.00002 ($2\bar{\sigma}$, $N = 6$, Musselwhite et al. 1989). These isotopic data demonstrate a strong crustal imprint in the rhyolite ignimbrites of the Wild Horse Mesa tuff.

The tectonic setting of the Woods Mountains, like that of the Coso Range, is characterized by rifting of the continental crust and of the underlying lithospheric man-

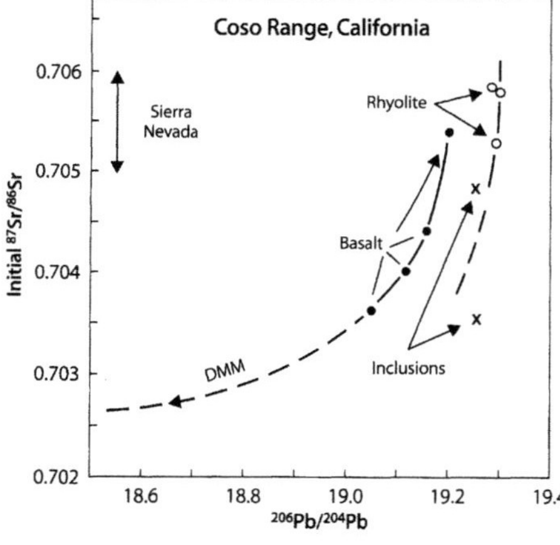

Fig. 4.23. Isotope ratios of Sr and Pb of Quaternary basalt (*solid circles*), rhyolite (*open circles*), and inclusions (*crosses*) in the Coso Range, Mojave Desert, California. The basalts form a mixing hyperbola that extrapolates to the DMM component (off scale). The rhyolites and inclusions are displaced from the basalt trajectory suggesting that they are not directly related to the basalt (*Source: data from Bacon et al. 1984*)

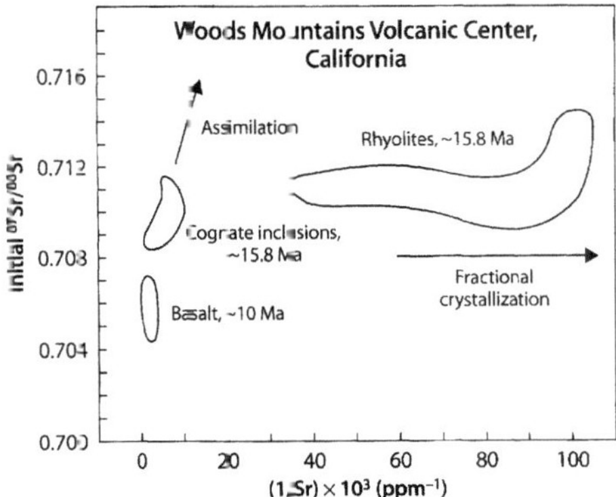

Fig. 4.24. Strontium-isotope mixing diagram of rhyolite, cognate inclusions, and younger basalt at the Woods Mountains volcanic center in southeastern California. The *arrows* indicate the changes in the concentration and isotope composition of Sr in a magma as a result of fractional crystallization and wallrock assimilation (*Source:* data from Musselwhite et al. 1989)

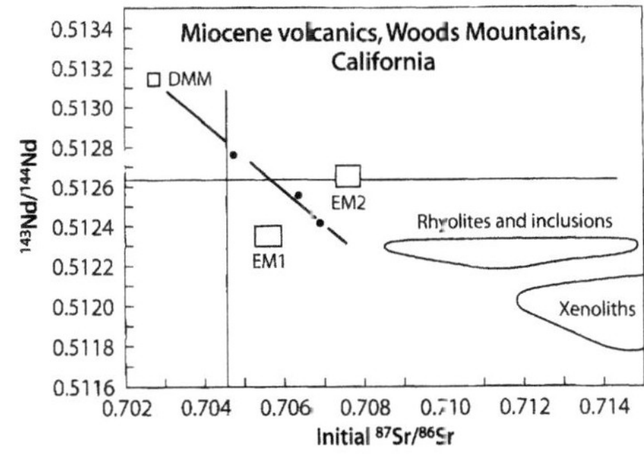

Fig. 4.25. Volcanic rocks of Miocene age at the Woods Mountains volcanic center in southeastern California (Figure 4.19). The rhyolite ignimbrites and associated cognate inclusions of the Wild Horse Mesa tuff (15.8 Ma) formed by assimilation of crustal rocks and simultaneous fractional crystallization of magmas that originated by decompression melting in the subcrustal mantle (*Source:* data from Musselwhite et al. 1989; Hart 1988)

le during the formation of the Basin and Range province. Consequently, basaltic magmas, formed by decompression melting in the subcrustal mantle, could have risen into the continental crust where they differentiated by assimilation of Precambrian granitoids and by simultaneous fractional crystallization. This scenario is consistent with the distribution of data points in Fig. 4.24, which demonstrates that the initial $^{87}Sr/^{86}Sr$ ratios of the rhyolites of the Wild Horse Mesa tuff are, for the most part, restricted to values between 0.710 and 0.712 and do not correlate with their Sr concentrations which vary by a factor of about three. Various kinds of cognate inclusions in the tuff (trachyte, trachydacite, quartz monzonite) represent intermediate stages in the differentiation of the magma. The younger basalt, alkali basalt, and basaltic andesite at Woods Mountains have much lower $^{87}Sr/^{86}Sr$ ratios than the cognate inclusions and rhyolite tuff. These rocks represent mantle-derived magmas with only limited contamination by crustal rocks. Two xenoliths (not shown in Fig. 4.24) composed of biotite-orthopyroxene gneiss and garnet biotite-gneiss have high $^{87}Sr/^{86}Sr$ ratios (at 15.8 Ma) of 0.7575 and 0.7823, respectively (Musselwhite et al. 1989).

The $^{143}Nd/^{144}Nd$ ratios of the basaltic rocks in Fig. 4.25 form an array that lies between EM1 and EM2 and extrapolates to the DMM component representing depleted mantle. According to the standard model of petrogenesis in subduction zones, magmas having these characteristics originate from sources in the mantle wedge altered by aqueous fluids expelled by subducted oceanic crust. Therefore, the origin of the rhyolite ignimbrites at this site can be explained by assimilation of crustal rocks and fractional crystallization of mantle-derived magmas.

4.3.5 Magma Sources, Cima Volcanic Field

Xenoliths derived from the lithospheric mantle and lower crust occur also in the hawaiite basalts of the Cima volcanic field in California (Fig. 4.19) located only about 30 km east of the Woods Mountains in the Mojave Desert. The volcanic activity at this site started at 7.6 Ma and continued episodically to about 15 000 years ago (Dohrenwend et al. 1984). Consequently, the eruptions postdate the activity of the Woods Mountains volcanic center and the main phase of extensional Basin and Range tectonics which ended at about 12 Ma.

The xenoliths in the Cima volcanic field include peridotite and pyroxenite derived from the underlying mantle as well as mafic igneous rocks from the lower continental crust (Wilshire et al. 1991). Several Cr-diopside-bearing ultramafic inclusions analyzed by Mukasa and Wilshire (1997) contain amphiboles, plagioclase, and even phlogopite of metasomatic origin. The $^{87}Sr/^{86}Sr$ ratios of these minerals range from 0.70514 to 0.70727 (relative to 0.71025 ±0.000010 for NBS 987) and are identical in each case to the $^{87}Sr/^{86}Sr$ ratios of coexisting clinopyroxene. Amphibole-plagioclase veins also have elevated $^{87}Sr/^{86}Sr$ ratios similar to those of the Cr-diopside-bearing ultramafic inclusions.

The gabbroic inclusions have surprisingly low $^{87}Sr/^{86}Sr$ ratios between 0.70267 and 0.70345. These inclusions represent basaltic magmas that originated by melting of depleted mantle and crystallized in the lower crust and upper lithospheric mantle. The $^{87}Sr/^{86}Sr$ ratios of the basalt lavas at Cima are similar to those of the gabbro inclusions, but differ markedly from those of the ultramafic inclusions. The gabbro inclusions and host basalt of the Cima volcanic field define two over-

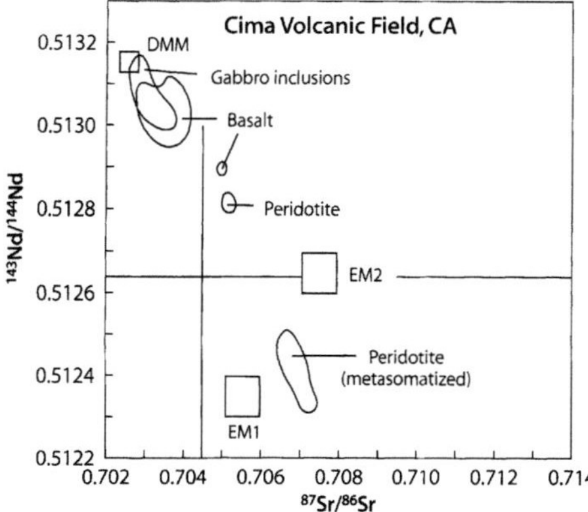

Fig. 4.26. Isotope ratios of Sr and Nd of basalt lava flows and their entrained mafic and ultramafic inclusions of the Cima volcanic field in southeastern California. The gabbro inclusions and basalt flows have similar isotope compositions of Sr and Nd and therefore may have formed from magma derived from asthenospheric mantle that intruded the lithospheric mantle (represented by the ultramafic inclusions) during the development of the Basins and Range province. The DMM, EM1, and EM2 mantle components serve as points of reference on the diagram (*Sources:* Mukasa and Wilshire; Farmer et al. 1995; Hart 1988)

lapping data fields in Fig. 4.26 located close to the DMM component. Therefore, the basalt magmas of the Cima volcanic field originated from depleted mantle which intruded rifts in the subcrustal lithosphere that formed by extension during the development of the San Andreas transform fault in southern California (Leeman and Harry 1993; Farmer et al. 1995).

The metasomatized ultramafic inclusions presumably represent the lithospheric mantle that may have been enriched in alkali metals, Sr, Pb, and other elements by aqueous fluids derived from the subducted Pacific Plate. A comparison of the isotope ratios of Sr and Nd reveals that the basalt flows erupted at Cima and at Woods Mountains had quite different magma sources. The Cima basalts originated from depleted asthenospheric mantle (low $^{87}Sr/^{86}Sr$, high $^{143}Nd/^{144}Nd$), whereas the basalts in the Woods Mountains formed by melting of metasomatized lithospheric mantle (high $^{87}Sr/^{86}Sr$, low $^{143}Nd/^{144}Nd$) represented by the ultramafic inclusions of the Cima volcanic field.

4.4   Silicic Volcanic Rocks of the Southwestern USA

Rhyolite ignimbrites of Late Tertiary age occur not only near the leading edge of North American Plate but formed also in the Basin and Range province of Arizona and Nevada identified in Fig. 4.19. At these sites, magmas formed in metasomatically altered lithospheric

mantle and subsequently evolved by assimilation of Precambrian basement rocks and by fractional crystallization in crustal magma chambers. The resulting volcanic rocks have elevated initial $^{87}Sr/^{86}Sr$ ratios similar to those of the Coso Range and Woods Mountains in California. The ages of silicic volcanic rocks in the Basin and Range province of Nevada and Utah were measured by Armstrong et al. (1969), Armstrong (1970), Lee et al. (1970), and Marvin et al. (1970) based on the K-Ar method. Deposits of silicic volcanic rocks occur in the Superstition Mountains near Globe, Arizona, and in Nye County of southern Nevada. Other occurrences of silicic volcanic rocks in the Basin and Range province have been described by Lipman (1966), Noble and Hedge (1969), Sheridan et al. (1970), Moyer and Nealey (1989), and Moyer and Esperanza (1989).

4.4.1   Superstition Mountains, Arizona

The volcanic rocks of the Superstition Mountains near Globe, Arizona, (Fig. 4.19) consist of Oligocene basalt (30 Ma) overlain by dacite flows and quartz-latite-rhyolite ashflow tuffs of the Miocene Geronimo Head, Superstition, and Apache Leap Formations. The Superstition tuff contains, in order of decreasing age, the Siphon Draw Member (24.4 Ma), the Dogie Spring Member (18.4 Ma), and the Canyon Lake Member (15.0 Ma). Stuckless and O'Neil (1973) reported that the initial $^{87}Sr/^{86}Sr$ ratios of these ashflow tuffs range from 0.7068 to 0.7135 relative to 0.7080 for E&A.

The initial $^{87}Sr/^{86}Sr$ ratios of the Superstition tuff and of the older dacite and basalt flows record the systematic evolution of the magma that produced these rocks. The data of Stuckless and O'Neil (1973) indicate that the $^{87}Sr/^{86}Sr$ ratios of the rocks increased with time from 0.7053 (basalt, 30 Ma) to 0.7090 (Canyon Lake Member, 15 Ma) relative to 0.7080 for E&A and that, in many cases, phenocrysts of plagioclase and sanidine have lower $^{87}Sr/^{86}Sr$ ratios than the rocks in which they occur. Moreover, the $^{87}Sr/^{86}Sr$ ratios of the Canyon Lake and Dogie Spring Members decrease up-section, whereas those of the older Siphon Draw Member do not vary stratigraphically. The $\delta^{18}O$ values of whole-rock samples of the Superstition tuff, the Apache Leap tuff, and of various glassy dikes range from +7.68‰ to +20.45‰ compared to only +7.80‰ for one sample of Precambrian basement gneiss.

The elevated initial $^{87}Sr/^{86}Sr$ ratio of the Superstition tuff demonstrates that the magma which produced these rocks either formed by melting of rocks in the crust, or assimilated such rocks. Since the volcanic rocks were deposited on Precambrian basement rocks, Stuckless and O'Neil (1973) analyzed a suite of these rocks and reported present-day $^{87}Sr/^{86}Sr$ ratios ranging from 0.7228 (diorite, 1 640 Ma) to 1.0903 (granite, 1 395 Ma). The weighted average $^{87}Sr/^{86}Sr$ ratio of the basement rocks is 0.7890 with an average Sr concentration of 183 ±83 ($2\bar{\sigma}$, $N = 6$). Since all of the basement rocks analyzed by Stuckless and O'Neil (1973) have much higher $^{87}Sr/^{86}Sr$ ratios than any of the Cenozoic volcanic rocks in the area, the magma of the Superstition tuff did not form by partial melting of the basement rocks. However, these granitic basement rocks could have been assimilated by a magma whose $^{87}Sr/^{86}Sr$ ratio was 0.7054 as recorded by the older dacite and basalt. The distribution of data points in the Sr-isotope mixing diagram in Fig. 4.27 indicates that both the $^{87}Sr/^{86}Sr$ ratio and the Sr concentrations of the Superstition tuff increased with time, and confirms that the local basement rocks are potential contaminants of the rhyolite magmas.

According to Stuckless and O'Neil (1973), magma formed by partial melting of amphibolite in the lower crust. The silicic magma so produced evolved by fractional crystallization and was pooled in a magma chamber at shallow depth where it assimilated radiogenic ^{87}Sr from the Precambrian basement rocks. Magma near the roof of the chamber was more strongly contaminated than magma at lower levels in the chamber. In some cases, the addition of ^{87}Sr occurred after plagioclase and sanidine phenocrysts had crystallized. Isotope re-equi-

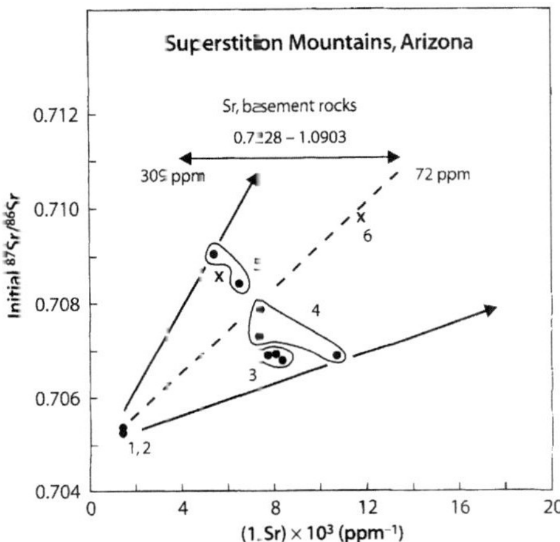

Fig. 4.27. Sr-isotope mixing diagram (without AFC) for ashflow tuffs of the Superstition-Superior area of Arizona. (*1*, *2*) older basalt (30 Ma) and dacite (29.0 Ma); (*3*) Siphon Draw Member, Superstition tuff (24.1 Ma); (*4*) Dogie Spring Member, Superstition tuff (18.4 Ma); (*5*) Canyon Lake Member, Superstition tuff (15.0 Ma); (*6*) Apache Leap tuff (19.8 Ma). The distribution of data points permits the rocks of the Canyon Lake Member and the Apache Leap tuff to have formed by contamination of basalt or dacite magma ($^{87}Sr/^{86}Sr = 0.7054$) by assimilation of local Precambrian basement rocks. The rocks of the Siphon Draw and Dogie Spring Members require contaminants with Sr concentrations <72 ppm, and may be magmatic differentiates of contaminated magmas (*Source:* Stuckless and O'Neil 1973)

libration was prevented by the rapid eruption of the magma to the surface. The magma near the top of the chamber was erupted first thus causing the basal part of the Dogie Spring and Canyon Lake Members to have higher $^{87}Sr/^{86}Sr$ ratios than later deposits of rhyolite tuff.

4.4.2 Nuclear Test Site, Nye County, Nevada

The testing of nuclear devices that started in 1957 in Nye County required detailed geological studies that included work on the extensive Tertiary ashflow tuffs whose ages range from 26.5 to 7.5 Ma and which include the Timber Mountain tuff (2 080 km³) and the Paintbrush tuff (500 km³) (Marvin et al. 1970). These volcanic rocks were deposited on Paleozoic and Precambrian basement rocks and lie within the Basin and Range tectonic province (Fig. 4.19) in southern Nevada (Ekren 1968; Noble et al. 1968).

The ashflow layers of the Timber Mountain and Paintbrush tuff units are vertically zoned with rhyolite tuff at the base, grading upward into quartz latite. The groundmass of the upper quartz latite tuffs generally contains less SiO_2 and more Al_2O_3, FeO/Fe_2O_3, MgO, CaO, TiO_2, MnO, P_2O_5 and Sr than the matrix of the rhyolite tuffs. The chemical zonation of these ignimbrite layers reflects the original zonation of the magma prior to its eruption with rhyolite near the top and quartz latite near the bottom of the chamber (Noble and Hedge 1969).

The initial $^{87}Sr/^{86}Sr$ ratios of whole-rock samples of tuffs in this area range from 0.7065 (Rainier Mesa Member, Timber Mountain tuff, 11.3 Ma) to 0.7123 (Topopah Spring Member, Paintbrush tuff, 13.2 Ma). The basal rhyolite tuffs have significantly higher $^{87}Sr/^{86}Sr$ ratios than the overlying latites, implying that the rhyolite in the upper part of the magma chamber was more strongly contaminated with radiogenic ^{87}Sr from the country rock than the latite magma. In addition, the $^{87}Sr/^{86}Sr$ ratios of sanidine phenocrysts are higher than those of the matrix or of whole-rock samples. Noble and Hedge (1969) suggested that the sanidine phenocrysts had formed in the upper, more contaminated, part of the magma chamber and had subsequently sunk towards the bottom of the chamber prior to eruption. The $^{87}Sr/^{86}Sr$ ratios of the whole-rock samples generally rise with decreasing Sr concentrations as expected for contaminated magmas, but the data points scatter on the isotope-mixing diagram indicating that the magmas may have interacted with a variety of crustal rocks.

The stratigraphic zonation of isotope ratios of Sr and Nd of individual layers of rhyolite ignimbrites of southwestern Nevada were confirmed by Farmer et al. (1991) based on a study of felsic volcanic rocks at the Timber Mountain/Oasis Valley Caldera. These rocks were erupted between 13.9 and 9.8 Ma and include lava flows, pumice, tuff, and vitrophyre most of which contain sanidine phe-

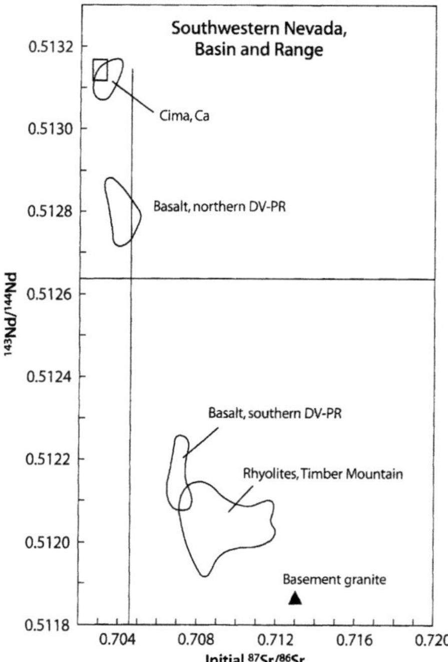

Fig. 4.28. Isotope ratios of basalt (<10 Ma) and rhyolites (13.9 to 0.9 Ma) in the Basin and Range province of southwestern Nevada and neighboring California. The rhyolites of the Timber Mountain/Oasis Valley Caldera (Nevada) have a strong crustal imprint, whereas the isotope ratios of Sr and Nd of the basalts (<10 Ma) from the Death Valley/Pancake Range (*DV-PR*) area have a bimodal distribution which is discussed in the text. The ^{143}Nd/^{144}Nd ratios of the basalts were recalculated to 0.512638 for CHUR (*Sources:* Farmer et al. 1989, 1991)

nocrypts. The isotope ratios of Sr and Nd of these rhyolites place them far from the DMM component in Fig. 4.28 into the quadrant characterized by granitic rocks of the continental crust. In fact, a granite xenolith in the Topopah Spring Member has Sr = 347 ppm, Nd = 30.8 ppm, ^{87}Sr/^{86}Sr = 0.71300 at 13.4 Ma, and ^{143}Nd/^{144}Nd = 0.511860. The isotope ratios of the rhyolites at Timber Mountain are explainable by the incorporation of 20 to 40% of rocks like this into magmas that intruded the upper crust (Farmer et al. 1991).

Basalt (<10 Ma) in the area of the Death Valley-Pancake Range of southern Nevada and adjacent California (including the Timber Mountain/Oasis Valley Caldera) form three separate data fields in Fig. 4.28. The basalt flows of the Cima volcanic field in southeastern California (Figs. 4.19 and 4.26) plot close to DMM and were derived from depleted lithospheric mantle without crustal contamination. The basalt flows of the northern part of the region in Fig. 4.28 originated from upwelling

asthenospheric mantle in response to extension of the lithospheric mantle (Farmer et al. 1989; Lum et al. 1989; Leeman and Harry 1993). The basalts in the southern region have high ^{87}Sr/^{86}Sr ratios (0.70671 to 0.70747) and low ^{143}Nd/^{144}Nd ratios (0.51209 to 0.51256, relative to 0.512638 for CHUR) which are attributable to assimilation of granitic basement rocks of the upper crust by the basalt magmas. However, large amounts of crustal rocks are required to change the isotope ratios of Sr and Nd of the magmas because the basalts of the southern region have high average concentrations of Sr (1 097 ±175 ppm) and Nd (62.6 ±9.3 ppm). In addition, the narrow range of variation of the isotope ratios of the southern basalts is not consistent with large-scale assimilation of crustal rocks, regardless of whether the assimilation was selective or whether it occurred in bulk. Therefore, Farmer et al. (1989) preferred the explanation that the magmas originated from old subcrustal lithospheric mantle in agreement with Menzies et al. (1983).

4.4.3 Bimodal Basalt-Rhyolite Suites

A series of volcanic centers in west-central Arizona define a lineament that strikes northeast and is collinear with the Bright Angel Fault of the Grand Canyon in northern Arizona (Moyer and Nealey 1989). The volcanic centers, listed sequentially from southwest to northeast, are the Castaneda Hills, Kaiser Spring, Mt. Hope, and Mt. Floyd. The ages of volcanic rocks decrease in a northeasterly direction from 15.1 to 10.3 Ma at the Castaneda Hills to Mt. Floyd which was active from 9.8 to 2.7 Ma. The volcanic rocks at these sites consist of basalt (alkali olivine basalt, tholeiite, and basaltic andesites) and rhyolite. The silica concentrations of the mafic lavas range from >44 to <59%, compared to >68% for all of the rhyolites. Moyer and Nealey (1989) distinguished two kinds of rhyolite based on their silica concentrations. High silica: >74 to <78%; low silica: >68 to <74%).

Moyer and Esperanza (1989) determined that the high-silica rhyolites at the Kaiser Spring volcanic field had formed by fractional crystallization of low-silica rhyolite magma. In addition, the high-silica rhyolites were extruded before the eruption of the low-silica rhyolite, thereby implying that the high-silica rhyolites originated from the tops of the low-silica magma chambers. The two types of rhyolite at Kaiser Spring have similar isotope ratios of Sr, Nd, and Pb (s. Table 4.4) that differ from those of the alkali olivine basalts (Moyer and Esperanza 1989).

Moyer and Esperanza (1989) considered that the alkali-olivine basalt (listed above) originated from the subcontinental lithospheric mantle, whereas the rhyolites formed by melting in the lower crust with possible input of mantle-derived magma. In addition, quartz-bearing mafic andesites and tholeiite basalts at Kaiser

Table 4.4. Isotope ratios of Sr, Nd, and Pb for the two types of rhyolite at Kaiser Spring and alkali olivine basalts (Moyer and Esperanza 1989)

	$(^{87}Sr/^{86}Sr)_i$	$^{143}Nd/^{144}Nd$	$^{206}Pb/^{204}Pb$
Rhyolite	0.7056–0.7086	0.51202–0.51233	16.96–17.32
Basalt	0.7042	0.51270	18.18

Spring also crystallized from mixed magmas or were derived by melting of mafic rocks in the lower crust.

Several of the large volcanic centers in the western USA (e.g. the San Juan Mountains of Colorado and the Indian Peak volcanic field of Utah and Nevada) extruded large volumes (>3 000 km^3) of Oligocene and Miocene rhyolitic and dacitic ash-flow tuffs interbedded with much smaller volumes (<300 km^3) of andesite lavas. Askren et al. (1997) explained the association of felsic and mafic volcanic rocks at these and other sites in the western USA by the following process: Mantle-derived basaltic and andesitic lavas rising through the continental crust were trapped by pools of rhyolitic or dacitic magmas because of the difference in their densities. While trapped, the mafic magmas evolved by fractional crystallization and by mixing with the overlying felsic magmas to form hornblende andesites. When the overlying rhyolitic and dacitic magmas were erupted to the surface, the underlying andesite magmas were discharged into the caldera that had formed by collapse of the rhyolite-magma chamber. Some mafic magmas rose to the surface without encountering pools of felsic magma and were erupted around the periphery of the calderas.

4.5 Mafic Volcanic Rocks, Southwestern USA

The magmatic activity in the southwestern USA was discussed in a series of papers introduced by Leeman and Fitton (1989). In addition, Wes Hildreth, Tim Grove, and Michael Dungan organized a conference on open magmatic systems held in August 1984 in Taos, New Mexico, during which many aspects of the petrogenesis of volcanic rocks in subduction zones were discussed. Some of the papers presented at that conference were later published in the "Journal of Geophysical Research" with an introduction by Hildreth et al. (1986). The same issue of the journal also contains a section on the tectonics and magmatic activity of the Rio Grande Rift of New Mexico (Fig. 4 19).

The alkali-rich rocks of Wyoming, California, Arizona, and New Mexico are discussed in Sect. 6.8. In addition, the large igneous provinces of the northwestern USA (e.g. the Columbia River basalt, Snake River Plain, Yellowstone Caldera, and the Absaroka volcanic field) are presented in Sect. 5 2 and 5.3.

4.5.1 Basin and Range Province

The lithospheric mantle underlying the Basin and Range province in Fig. 4.19 was thinned and fragmented by extension allowing asthenospheric mantle to well up in some places (Leeman and Harry 1993). As a result of the disruption of the lithospheric mantle and continental crust, the mafic volcanic rocks of Tertiary to Recent age in the western USA are heterogeneous in both chemical and isotope composition (Menzies 1989; Lipman and Glazner 1991).

The isotope composition of Sr of the Tertiary calcalkaline volcanic rocks in the Basin and Range province in Fig. 4.29 was measured and interpreted by Leeman (1970, 1982), Hedge and Noble (1971), and by Scott et al. (1971), whereas Doe (1968) and Zartman (1974) discussed the isotope composition of Pb in the Cenozoic igneous rocks and associated ore deposits. Leeman (1982) noted the wide variation of $^{87}Sr/^{86}Sr$ ratios of the Late Tertiary mafic lavas of the western region of the USA (including the Colorado Plateau) and concluded that the magmas originated from an isotopically heterogeneous lithospheric mantle. The mantle-derived magmas subse-

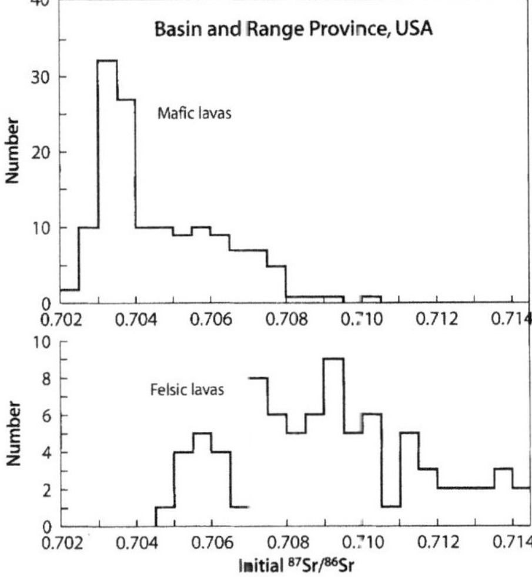

Fig. 4.29. Initial $^{87}Sr/^{86}Sr$ ratios of mafic and felsic volcanic rocks of Tertiary and Quaternary age in the Basin and Range province including Nevada, Utah, Arizona, and the Coso Range, Woods Mountains, and the Cima volcanic center of California. The $^{87}Sr/^{86}Sr$ ratios of the mafic lavas are unimodal (0.7030 to 0.7040) but skewed toward higher values, whereas the $^{87}Sr/^{86}Sr$ ratios are polymodal and significantly displaced to higher values than the mafic rocks (*Sources:* Noble and Hedge 1969; Leeman 1970, 1974; Scott et al. 1971; Hedge and Noble 1971; Stuckless and O'Neil 1973; Pushkar and Condie 1973; Pushkar and Stoeser 1975; Bacon et al. 1984; Musselwhite et al. 1989; Farmer et al. 1995)

quently assimilated crustal rocks as they differentiated by fractional crystallization and caused partial melting by releasing heat they had transported into the crust. Leeman (1982) also pointed out that Early Cenozoic volcanic rocks of the western USA generally have higher and more variable initial $^{87}Sr/^{86}Sr$ ratios than the Late Cenozoic lavas. This difference was caused by the change from compressive tectonics in the subduction zone along the west coast of North America in Early Cenozoic time to extensional tectonics after the North American Plate had overridden the subduction zone and the associated spreading ridge.

Alternatively, the chemical and isotope compositions of Tertiary volcanic rocks in settings of extensional tectonics may also be related to the difference in the densities of the mafic magmas and of the overlying continental crust (Glazner and Ussler 1989). Such density differences determine the buoyancy of the magmas and control the level to which they can rise in the crust as a result of hydrostatic pressure alone. Basalt magmas that stagnate in the crust are exposed to crustal contamination which causes their densities to decrease because of the segregation of olivine and pyroxene and thus allow them to resume their upward progress towards the surface. In this way, Glazner and Ussler (1989) related the $^{87}Sr/^{86}Sr$ ratios of mafic volcanic rocks to the bulk density of the crust rather than to extension and compression of the American Plate.

4.5.2 Colorado Plateau

The Colorado Plateau in Fig. 4.19 occupies southwestern Colorado, northwestern New Mexico, northeastern Arizona, and southeastern Utah. In contrast to the Basin and Range province, the Colorado Plateau has been a stable part of the American Plate since the Cambrian Period and has a thick continental crust. Nevertheless, volcanic activity of Tertiary age has occurred along its southern border with the Basin and Range province as well as at several other locations (Leat et al. 1989). The volcanic fields along the southern margin of the Colorado Plateau include, from west to east (Condit et al. 1989):

- Western Grand Canyon, northwestern Arizona (Leeman 1974);
- San Francisco Mountains, west-central Arizona (Pushkar and Stoeser 1975);
- Springerville, east-central Arizona (Condit et al. 1989);
- Zuni-Bandera, west-central New Mexico (Laughlin et al. 1971);
- Mount Taylor, west-central New Mexico (Duffield and Ruiz 1992);
- Jemez, north-central New Mexico (Stix and Gorton 1990).

Other noteworthy volcanic centers associated with the Colorado Plateau include the Mogollon-Datil field of New Mexico (Davis and Hawkesworth 1995, 1994, 1993; Davis et al. 1993), the Henry Mountains of southeastern Utah (Nelson and Davidson 1993), and the San Juan Mountains of southwestern Colorado.

4.5.3 San Juan Mountains, Colorado Plateau

The San Juan volcanic field is located in southwestern Colorado in the tectonic province of the Colorado Plateau. This area is more than 1 700 km east of the leading edge of the North American Plate. The volcanic rocks of the San Juan Mountains include andesites that were erupted during the Oligocene Epoch (34.7 to 26.4 Ma) followed by late basalts and rhyolites (26.4 to 3.6 Ma) (Lipman et al. 1970). The rocks that formed during the main phase of the volcanic activity consist primarily of andesites, quartz latites, and extensive deposits of ashflow tuffs of intermediate chemical composition which were erupted from several centers now represented by calderas at Silverton, Lake City, San Luis, Creede, Mount Hope, Summitville, Cochetopa Park, and Bonanza (Lipman et al. 1978).

The initial $^{87}Sr/^{86}Sr$ ratios of the main body of the volcanic rocks (34.7 to 18.5 Ma) range from 0.7048 to 0.7075, whereas the younger basalt and rhyolite flows (26.8 to 5.0 Ma) have yielded lower values between 0.7037 and 0.7053. The Sr concentrations of the rocks range from 2 030 ppm (rhyodacite flow, Cimarron Ridge, 32.1 Ma) to 40.5 ppm (Sunshine Peak tuff, $SiO_2 = 76\%$, 22.5 Ma). In general, the silicic lavas and tuffs having $SiO_2 > 72\%$ have low Sr concentrations between 157 and 40.5 ppm, whereas their Rb concentrations lie between 281 and 132 ppm. The $^{206}Pb/^{204}Pb$ ratios of the lavas in the San Juan Mountains range from 17.33 to 18.87 (Lipman et al. 1978).

The time-averaged concentrations of SiO_2 and the isotope ratios of Sr and Pb in Fig. 4.30 increased to maximum values at about 25 Ma and then declined. The correlation between the average silica concentrations and the isotope compositions of the lavas in the San Juan volcanic field implies that changes occurred in the sources of the magma and/or in their subsequent evolution. Lipman et al. (1978) emphasized the prevalence of andesites among the Oligocene lavas and interpreted the more silicic ashflow tuffs as differentiates of the andesitic magmas. The parent magmas of the lavas erupted in the San Juan Mountains presumably originated in the lithospheric mantle that may have been altered by aqueous fluids emanating from subducted oceanic crust of the Pacific Plate. These magmas then pooled in the lower crust where they differentiated by fractional crystallization and assimilation of wallrock. In addition, the heat released by the mantle-derived magmas may

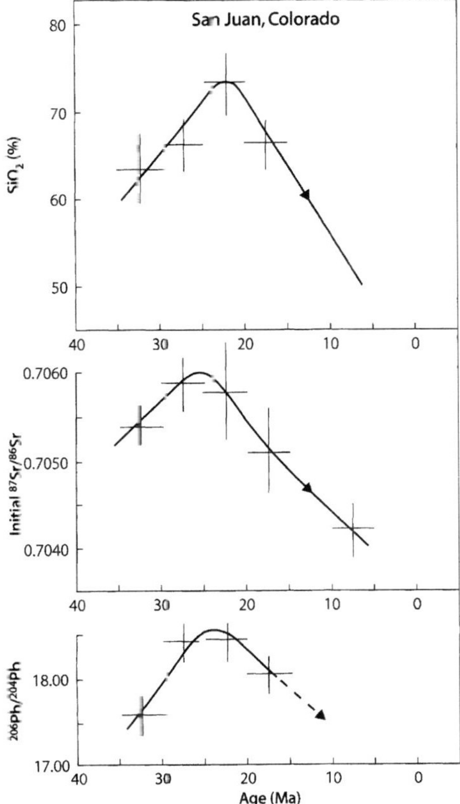

Fig. 4.30. Variation of average SiO_2 concentrations, initial $^{87}Sr/^{86}Sr$, and $^{206}Pb/^{204}Pb$ ratios of the volcanic rocks of the San Juan Mountains of southwestern Colorado as a function of their ages. The evident correlation between the silica concentrations and the isotope ratios of Sr and Pb implies that changes occurred in the sources and evolution of the magmas (*Sources:* Lipman et al. 1978; Doe et al. 1969a,b; Hedge 1966)

have caused local melting of crustal rocks. The silicic magmas so generated, accumulated in the upper crust between 30 to about 25 Ma and were erupted explosively to form very large deposits of ashflow tuff (ignimbrites) of intermediate chemical composition.

The Fish Canyon tuff (27.8 Ma) is one of these ashflow deposits with a total volume of about 5 000 km³ (Whitney and Stormer 1985). It is composed of homogeneous phenocryst-rich dacite indicating that the magma was not stratified in its chemical and isotope composition (Lipman et al. 1997). The Fish Canyon tuff has an initial $^{87}Sr/^{86}Sr$ ratio of 0.7059 (relative to 0.7080 for E&A), $Sr = 1 050$ ppm, and $Rb = 2.6$ ppm. In addition, Lipman et al. (1978) reported that its average $^{206}Pb/^{204}Pb$ ratio is 18.40 ±0.02.

The more recent basalts and rhyolites of the San Juan Mountains have been assigned to the Servilleta Formation (basalt, 4.5 to 3.6 Ma), the Hindsdale Formation (basalt and rhyolite, 26.4 to 4.7 Ma), and the Sunshine Peak tuff (22.5 Ma). The Servilleta basalt extends into the Taos Plateau within the Rio Grande Rift of northern New Mexico (Dungan et al. 1986). These rocks formed in a setting of extensional tectonics in contrast to the andesitic lavas erupted prior to about 25 Ma which originated during a period of plate convergence along the west coast of North America. The mafic rocks consist of alkali-rich olivine basalts and of basaltic andesites containing abundant xenocrysts. Lipman et al. (1978) attributed the alkali basalts to melting in the lithospheric mantle without significant crustal contamination, whereas the xenocrystic basaltic andesites contain a contaminant from the lower crust.

In retrospect, the petrogenesis of the lavas of the San Juan volcanic field of Colorado can be explained as a consequence of the tectonic history of the western margin of the North American Plate. During the main phase of the igneous activity between 35 and 25 Ma, large volumes of magma that originated from the lithospheric mantle intruded the overlying crust and differentiated by wallrock assimilation and fractional crystallization. After about 25 Ma, extension of the crust allowed the mantle-derived magmas to reach the surface without being contaminated by the continental crust. The shallow crustal magma chambers from which the ashflow tuffs of the San Juan Mountains were erupted are now represented by a large composite batholith whose presence is indicated by a gravity anomaly. The igneous activity that occurred in this area mobilized large volumes of water which deposited quartz veins containing native gold and sulfide minerals of Ag, Pb, Zn and Cu. The resulting ore deposits have been mined at many places in the San Juan Mountains, including Silverton, Creed, Ouray, and Telluride.

4.6 The Cascade Range of the Northwestern USA

The Cascade Range in Fig. 4.31 is located about 300 km east of the west coast of North America and extends from northern Washington south through Oregon into northern California (McBirney 1978). The principal stratovolcanoes in the Cascade Range in Fig. 4.31 include, from north to south: Mt. Baker, Glacier Peak, Mt. Rainier, Mt. St. Helens, and Mt. Adams in Washington; Mt. Hood, Mt. Jefferson, Mt. Washington, Three Sisters, Crater Lake, and Mt. Loughlin in Oregon; as well as Mt. Shasta and Lassen Peak in California. A belt of high-alumina basalt volcanoes east of the Cascade Range contains the Simcoe volcanic complex of Washington, the Newberry Volcano in Oregon, and the Medicine Lake Highland Volcano in California. Both volcanic provinces postdate

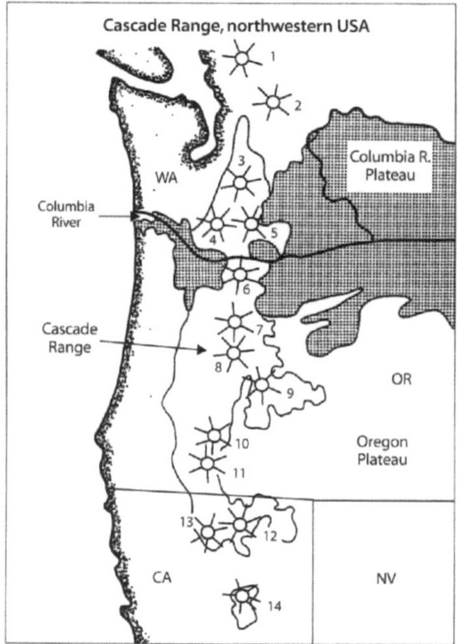

Fig. 4.31. Cascade Range, Columbia River Plateau, and the Oregon Plateau of the northwestern USA. The volcanoes of the Cascade Range are identified by number: *1.* Mt. Baker; *2.* Glacier Peak; *3.* Mt. Rainier, *4.* Mt. St. Helens; *5.* Mt. Adams; *6.* Mt. Hood; *7.* Mt. Jefferson; *8.* Three Sisters; *9.* Newberry Caldera; *10.* Crater Lake; *11.* Mt. Loughlin (Formerly Mt. Pitt); *12.* Medicine Lake volcanic center; *13.* Mt. Shasta; *14.* Lassen Peak (*Source:* Church and Tilton 1973)

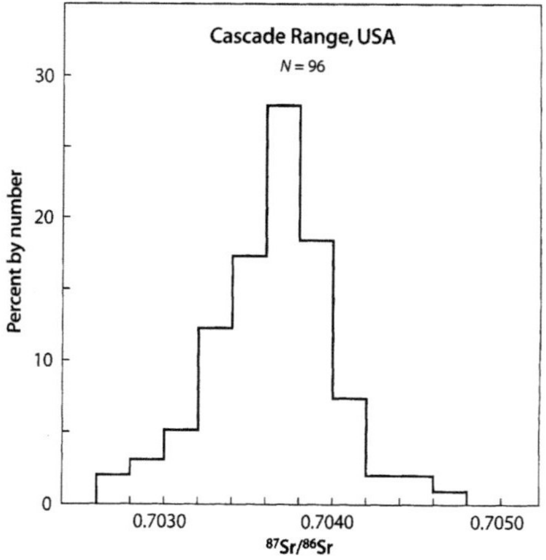

Fig. 4.32. Range of $^{87}Sr/^{86}Sr$ ratios of calc-alkaline lavas (basalt to rhyolite) of the stratovolcanoes of the Cascade Mountains of Washington, Oregon, and California (*Sources:* Peterman et al. 1970c; Hedge et al. 1970; Church and Tilton 1973; Halliday et al. 1983; Goles and Lambert 1990)

and in part overlie the Columbia River basalt of southern Washington, northern Oregon, and western Idaho (Higgins 1973).

The Cascade Range volcanic arc is underlain by the continental crust of North America including Precambrian metamorphic rocks in northern Washington and younger accreted terranes of Paleozoic and Mesozoic age in Oregon and southwestern Washington (Church et al. 1986). K-Ar dating of lavas in the central Cascade Range by McBirney et al. (1974) and by Armstrong (1975) indicates that volcanic activity in this area was episodic with peak intensities at about 16 to 14 Ma (Columbia episode), 11 to 9 Ma, 6 to 4 Ma, and 1.5. to 0 Ma. The apparent contradiction between continuous plate motions and the discontinuous character of local volcanic activity was discussed by Gilluly (1973). The volcanic activity of the Cascade Range is attributable to the subduction of the Juan de Fuca Plate under the North American continent. The history of igneous and tectonic activity of this region during the Cenozoic Era was reviewed and dis-

cussed by Hamilton and Myers (1966), Atwater (1970), Stewart and Carlson (1978), Armstrong (1978b), White and McBirney (1978), and McBirney (1978). Only the most recent volcanic activity associated with the stratovolcanoes of the Cascade Range is included in this section. A comprehensive review of the isotopic composition of Sr in Cenozoic lavas of the western USA was published by Leeman (1982).

The Quaternary lavas of the stratovolcanoes of the High Cascade Range are calc-alkaline in composition and consist primarily of basalt and andesite with subsidiary aomounts of dacite and rhyolite. The $^{87}Sr/^{86}Sr$ ratios of these rocks in Fig. 4.32 range from about 0.7027 to 0.7048, but more than 76% of the measured ratios lie between 0.7032 and 0.7040 relative to 0.7080 for E&A or 0.71025 for NBS 987. Therefore, the early work of Peterman et al. (1970b), Hedge et al. (1970), and Church and Tilton (1973) led to the conclusion that assimilation of crustal rocks had not played an important part in the petrogenesis of these rocks.

However, the more recent data of Halliday et al. (1983) in Fig. 4.33 demonstrate that the $^{87}Sr/^{86}Sr$ ratios of the lavas on Mount St. Helens do vary with the concentrations of SiO_2 (50 to 69%) and with Rb/Sr ratios (0.034 to 0.136). These correlations of isotopic and chemical parameters indicate that the magma on Mount St. Helens assimilated varying amounts of crustal rocks having $^{87}Sr/^{86}Sr$ ratios greater than about 0.704. Church and Tilton (1973) actually reported measurements of $^{87}Sr/^{86}Sr$

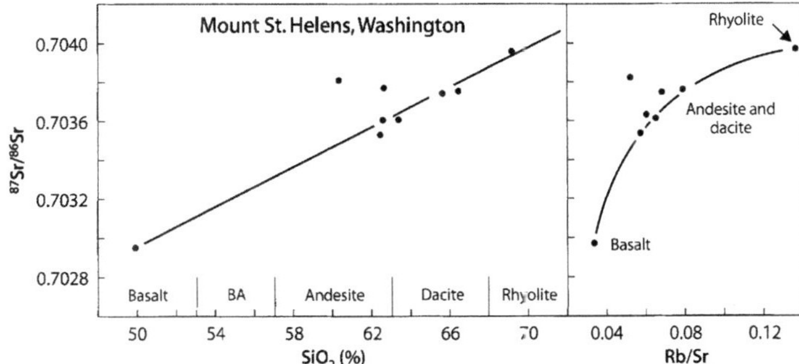

Fig. 4.33.
Relation between $^{87}Sr/^{86}Sr$ ratios and the chemical composition in the lavas of Mount St. Helens, Cascade Range, Washington, USA. The evident correlation of $^{87}Sr/^{86}Sr$ ratios with SiO_2 and Rb/Sr ratios indicates that these lavas formed from a basaltic parent magma by assimilation of varying amounts of crustal rocks having $^{87}Sr/^{86}Sr$ ratios > 0.7040 (*Source:* Halliday et al. 1983)

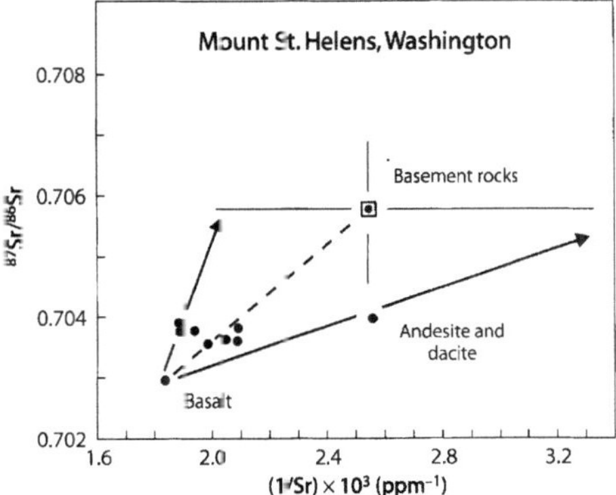

Fig. 4.34. Strontium-isotope mixing diagram (without AFC) for lavas from Mount St. Helens, Cascade Range, Washington, USA. The distribution of data points indicates that the andesites and dacites are mixtures of basalt magma with a diversified suite of basement rocks in northern Washington. The *dashed line* is a hypothetical mixing Lne for one specimen of andesite discussed in the text (*Sources:* Halliday et al. 1983, lavas; Church and Tilton 1973, basement rocks)

ratios of basement rocks in northern Washington ranging from 0.7042 to 0.7119 with a weighted mean of 0.70575 ±0.0011 ($2\bar{\sigma}$, $N = 19$) and an average Sr content of 383 ±102 ppm ($2\bar{\sigma}$, $N = 19$).

The distribution of data points in Fig. 4.34 representing lavas from Mount St. Helens and average basement rocks in northern Washington indicates that the $^{87}Sr/^{86}Sr$ ratios and Sr concentrations of the andesites and dacites can be accounted for by the incorporation of basement rocks into mantle-derived basalt magma having $^{87}Sr/^{86}Sr$ = 0.70295 and Sr = 577 ppm. For example, specimen SH16 (andesite, $^{87}Sr/^{86}Sr$ = 0.703535, Sr = 531 ppm) can be treated as a two-component mixture of basalt and average basement rock as indicated by the dashed line in Fig. 4.34. The weight fraction of basement rock in SH16 (calculated from Eq. 1.29) is 0.25. Halliday et al.

(1983) used chemical and isotopic data to outline a history of magmatic activity at Mount St. Helens involving interactions between mantle-derived basalt magma and andesite-dacite magma that formed by partial melting at the base of the crust.

Contamination of mantle-derived magma by the overlying crust was also demonstrated by Goles and Lambert (1990) for the lavas of Newberry Volcano in Oregon. In addition, the lavas at Medicine Lake Highland in California display strong correlations between $^{87}Sr/^{86}Sr$ and Rb/Sr ratios. Such data arrays may be interpreted as pseudo-isochrons reflecting the age of heterogeneous magma sources in the mantle (Mertzman 1979). Alternatively, such arrays may also be attributed to assimilation of crustal rocks by mantle-derived magmas or to mixing of crustal and mantle-derived magmas. Moreover, Church (1976) concluded from the isotope compositions of Pb that silicic lavas in the Cascade Range contain a crustal component.

In summary, the most precise measurements of $^{87}Sr/^{86}Sr$ ratios reveal that the mantle-derived magmas of the Cascade Range did assimilate some of the crustal rocks through which they were erupted. However, the $^{87}Sr/^{86}Sr$ ratios of the resulting rocks were not greatly changed because the continental crust in Oregon and southern Washington consists of mafic igneous and associated sedimentary rocks having low $^{87}Sr/^{86}Sr$ ratios.

4.7 Cenozoic Volcanic Rocks of the Oregon Plateau

The Oregon Plateau in Fig. 4.35 is located west of the Cascade Range in the northernmost part of the Basin and Range tectonic province in southern Oregon where it occupies an area of about 1.3×10^5 km^2. The northern border of the plateau is defined in part by the Olympic-Wallowa lineament which extends from the northeastern corner of Oregon to Puget Sound in the State of Washington. This lineament represents the southern extent of old continental crust to the north and the

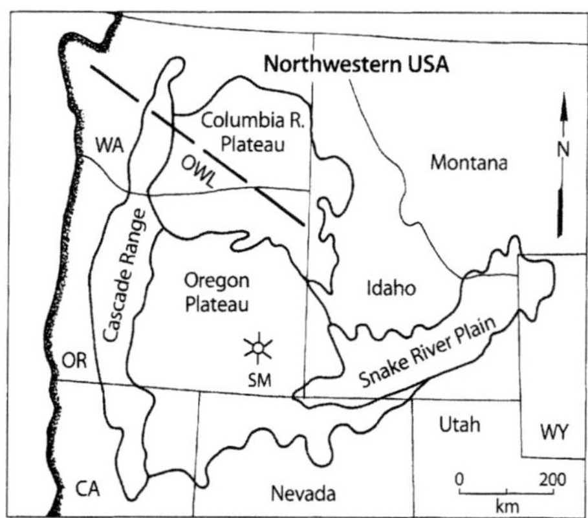

Fig. 4.35. Physiographic regions of the northwestern USA including the Oregon Plateau, the Cascade Range, the Columbia River Plateau, and the Snake River Plain. *OWL* is the Olympic-Wallowa lineament and *SM* is Steens Mountain (*Source:* adapted from Carlson and Hart 1987)

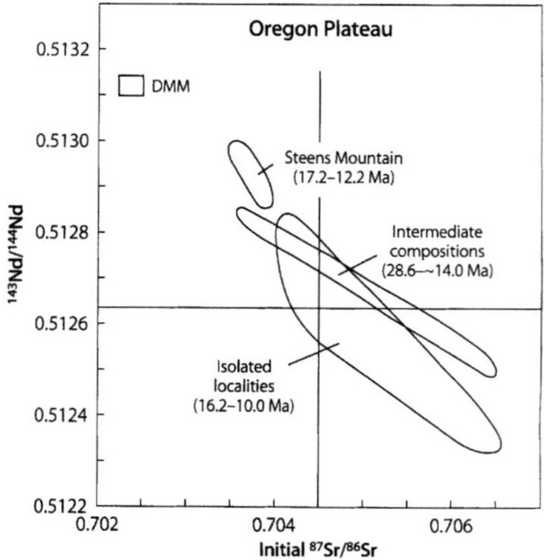

Fig. 4.36. Isotope ratios of Sr and Nd of Cenozoic volcanic rocks on the Oregon Plateau in the northernmost part of the Basin and Range tectonic province. The basalt flows on Steens Mountain originated from enriched mantle rocks without significant crustal contamination, whereas the magmas of intermediate compositions and from isolated localities in south-central Oregon show signs of having assimilated crustal rocks (*Source:* Carlson and Hart 1987)

younger rocks that underlie the Oregon Plateau in the south (Carlson and Hart 1987).

The volcanic activity of this region started in late Eocene time and continued intermittently to the present. The oldest volcanic rocks consist of intermediate to silicic lavas that were extruded from latest Eocene to early Miocene time. At about 18 to 17 Ma, the composition of the lavas changed to Fe-rich basalt which is best preserved on Steens Mountain (Gunn and Watkins 1970; Helmke and Haskin 1973). The eruption of voluminous basalt flows continued until about 11 Ma and was coeval with the extrusion of silicic lavas in southeastern Oregon. A third suite of high-Al olivine-normative tholeiites began to be erupted at about 10.5 Ma (Hart et al. 1984). The volcanic activity that formed the Oregon Plateau overlapped in time the extrusion of the Columbia River basalt (16.5 to 6.0 Ma) in northern Oregon and Washington (Carlson et al. 1981; Carlson 1984).

The initial ^{87}Sr/^{86}Sr and ^{143}Nd/^{144}Nd ratios of basalt and intermediate rocks on the Oregon Plateau in Fig. 4.36 form three separate data fields. The basalt flows of Steens Mountain and related localities plot closest to the DMM component and therefore originated from magma sources in the subcontinental lithospheric mantle without significant contamination by rocks of the continental crust. The lavas of intermediate compositions, which were erupted before the Steens Mountain basalt, have a wide range of initial ^{87}Sr/^{86}Sr and ^{143}Nd/^{144}Nd ratios implying that their mantle-derived magmas assimilated crustal rocks as they differentiated by fractional crystallization in crustal magma chambers (AFC). The vol-

canic rocks at isolated sites on the Oregon Plateau likewise show evidence of petrogenesis by AFC.

The work of Carlson and Hart (1987) highlights the difficulty in reconstructing the petrogenesis of volcanic rocks that form along the margins and in the interiors of continents. In such cases, the range and absolute values of the isotope ratios of Sr, Nd, and Pb are attributable either to the heterogeneity of the magma sources in the mantle, or to subsequent contamination of mantle-derived magmas by assimilation of crustal rocks, or both. Contamination at the source in the mantle is especially likely for magmas that form in the mantle wedge above subduction zones or in back-arc basins along continental margins. However, such tectonic settings also provide opportunities for magmas to assimilate rocks from the walls of crustal magma chambers where they differentiate by fractional crystallization. In some cases, trace element ratios can be used to constrain the various options. However, a realistic assessment of the petrogenesis of volcanic rocks that form along the margins and in the interiors of continents is that all options must be considered. In other words, the isotope ratios and chemical compositions of volcanic rocks that form in these tectonic settings are affected by the heterogeneity of the magma sources, by subsequent assimilation of heterogeneous crustal rocks, and by the segregation of crystals that form while magmas pool within the continental crust or are in transit to the surface.

4.8 Summary: The Tertiary Ignimbrites of the Americas

Subduction of oceanic lithosphere along the west coast of the Americas has produced a bimodal suite of mafic and felsic volcanic rocks of Tertiary age. The mafic lavas, consisting of basalt and andesite, originated by partial melting in the mantle wedge which had been altered by aqueous fluids emanating from the subducted oceanic sediment. The felsic rocks include voluminous ashflow tuffs (ignimbrites) as well as lava flows of rhyolitic and dacitic compositions. The felsic magmas formed by partial melting of mafic igneous and metamorphic rocks in the lower continental crust and subsequently pooled in shallow crustal magma chambers where they differentiated by fractional crystallization and by assimilation of wallrocks. Crystallization of sanidine by the rhyolite magmas reduced their Sr concentrations, which caused the $^{87}Sr/^{86}Sr$ ratios of the magmas to rise in response to interactions with granitic basement rocks of the upper continental crust. As a result, the rhyolite magmas became stratified such that the magma in the upper parts of the chamber were enriched in radiogenic ^{87}Sr, SiO_2, and alkali metals compared to the magma at depth.

The incubation of rhyolite magmas typically ended with explosive eruptions causing the felsic magmas to disintegrate into glowing volcanic ashflows that buried the landscape surrounding the volcanic centers. The empty magma chambers were refilled, in many cases, by fresh magma rising from depth, thus allowing several ignimbrite layers to be deposited by successive eruptions from the same volcanic center. Many of these magma chambers ultimately collapsed to form calderas which have been preserved to the present time. The original stratification of the rhyolite magmas was inverted during the eruption because the highly differentiated magmas from the top of the chamber were erupted first and formed the basal layer of the resulting ignimbrite deposit.

The isotope ratios of Nd and Pb of the rhyolite magmas also varied with depth in the magma chambers. The $^{143}Nd/^{144}Nd$ ratios of the most highly contaminated parts of the magma near the top of the chamber were lowered by the introduction of Nd from rocks of the upper crust. As a result, the present isotope ratios of Sr and Nd of the felsic volcanic rocks are inversely correlated and, in some cases, define data fields in the "crustal" quadrant of the Sr-Nd isotope mixing diagram. The isotope ratios of Pb of the felsic rocks depend on the isotope compositions of Pb in the magma sources in the lower crust and on the Pb the magmas assimilated from the rocks of the upper crust. The resulting correlations between the $^{87}Sr/^{86}Sr$ and $^{206}Pb/^{204}Pb$ ratios are generally compatible with the interpretations derived from the Sr and Nd isotope ratios.

The petrogenetic scenario described above applies primarily to the volcanic rocks of the Andes of South America and to the Cordillera of Central America and Mexico where subduction of oceanic lithosphere is still occurring. The tectonic setting in the western USA is more complex because the North American Plate has overridden the subduction zone as well as the spreading ridge that existed during the Mesozoic Era west of the advancing North American Plate. As a result, the western region of the North American Plate was first compressed and then stretched as the plate first overrode the subduction zone followed by the spreading ridge. These events have caused the development of different physiographic provinces which reflect the tectonic processes that prevailed in each. The disruption of the American Plate has resulted in the eruption of a wide variety of volcanic rocks of Tertiary to Recent age including not only a bimodal suite of mafic lavas and felsic ignimbrites but also alkali-rich lavas and plutons to be discussed in Chap. 6. The tectonic disturbance of the American Plate extends to areas that are now located more than 1 000 km east of the coast of California.

The mafic volcanic rocks of the Cascade Range and of the Columbia River Plateau in Oregon and Washington record the variety of tectonic and volcanic activity associated with subduction of oceanic lithosphere under the North American Plate. In this tectonic setting we find not only the stratovolcanoes of the Cascade Range, but also the flood basalts of the Columbia River Plateau which were erupted through fissures in a back-arc basin.

Chapter 5

Continental Flood Basalt Provinces

Since the start of the Mesozoic Era about 230 million years ago, large volumes of tholeiite basalt have been erupted through fissures on all of the continents. In several cases, the extrusion of these lavas occurred in conjunction with the break-up of the supercontinents of Gondwana and Pangea. Therefore, these basalt plateaus are now situated on the passive margins of the present continents on opposite sides of ocean basins that have opened between them. Typical examples of this phenomenon are the complementary basalts of Scotland and Greenland, as well as the basalt plateaus in Namibia (southwest Africa) and in the state of Paraná (Brazil). Fissures can also form in back-arc basins behind subduction zones located off the continental coast, causing basalt magma to form by decompression melting of lithospheric mantle. The standard model of flood-basalt petrogenesis developed by White and McKenzie (1989) and by McKenzie and Bickle (1988) is based on the conclusion that flood basalts are erupted thorough continental rifts caused by the interaction of asthenospheric plumes with the lithospheric mantle and with the overlying continental crust (Milanovskiy 1976). The arrival of an asthenospheric plume at the base of the continental lithosphere causes extension and rifting before large-scale decompression melting in the head of the plume can occur (Kent et al. 1992a).

5.1 Petrogenesis of Flood Basalts

The flood basalts of the North and South Atlantic regions are genetically related to large mantle plumes that are located along the Mid-Atlantic Ridge and were described in Sect. 2.3.3 (Iceland) and 2.5.6 (Tristan da Cunha). These and other plumes originated in the asthenospheric mantle as a result of buoyancy caused by a localized increase in temperature from heat released by radioactive decay of U, Th, and K in previously subducted oceanic crust (Hofmann and White 1982). The temperature difference between the head of a plume and the surrounding asthenospheric mantle does not have to be large. For example, the temperature of the plume that underlies the Cape Verde Plateau (Sect. 2.5.3) is only about 100 to 300 °C higher than that of the surrounding asthenospheric mantle (Courtney and White 1986).

The impact of mantle plumes against the underside of the lithospheric mantle beneath Gondwana and Pangea (Cox 1978, 1980) caused uplift, rifting, volcanic activity, crustal thinning, and ultimately the formation of an ocean basin containing a spreading ridge. According to White and McKenzie (1989), even a modest temperature differential of 100 to 200 °C can cause the formation of large volumes of Mg-rich basalt magma by decompression melting in the plume and in the overlying lithospheric mantle. The magmas subsequently rise through the crustal fractures and are erupted at the surface at a rapid rate to form thick layers of tholeiite basalt, which literally flood the landscape and ultimately form plateaus that may be several thousand meters thick. In addition, basalt magma is intruded into the crust to form sills, dikes, and plutons that can reach batholithic proportions (Hill et al. 1991). The sequence of events outlined above is exemplified by some of the flood basalt provinces of the Mesozoic and Cenozoic age in Table 5.1.

Table 5.1. Locations, ages, and approximate volumes of continental basalt plateaus of Cenozoic and Mesozoic age (Wilson 1989; Rampino and Stothers 1988; White et al. 1987; Baksi et al. 1987)

Province or location	Age (Ma)	Volume (km^3)
North America, Columbia Plateau	17 – 6	0.3 × 10^6
North America, Snake River Plain	17 – 0	
Ethiopia	35	
North Atlantic, British Islands	65 – 50	~10.0 × 10^6
India, Deccan Traps	~65	1.0–2.5 × 10^6
India, Rajmahal	110	~0.12 × 10^6
South America, Paraná, Brazil	140 – 110	2.2 × 10^6
South Africa, Etendeka, Namibia	114 – 110	0.07 × 10^6
South Africa, Karoo basalt	206 – 166	1.3 × 10^6
Antarctica, Ferrar Supergroup	179 ±7	0.007 × 10^6
Eastern North America	200	
Russia, Siberian Platform	248 – 216	5.3 × 10^6

Whatever the sequence of events may have been in any given region, floods of basalt have been erupted through fissures in the continental crust since the Archean Eon, which started at 3.8 Ga (Palmer 1983). The ancient deposits of this type are now represented in most cases by large differentiated mafic intrusions (Chap. 7), because the basalt layers on the surface have long since been eroded. The Keweenawan basalts in the midcontinent area of North America (Sect. 7.1.1) and the Coppermine River basalt in the Northwest Territories of Canada (Sect. 7.1.2) are important exceptions to this generalization.

The isotope compositions of Sr, Nd, and Pb of the basalt flows depend on the proportions of plume rocks and lithospheric mantle in the magma sources, as well as on the amount and composition of crustal rocks the magmas assimilated while they differentiated by fractional crystallization in crustal magma chambers (Campbell and Turner 1987; Marsh 1989). The amount of crustal rocks the magmas can assimilate increases with the temperature of the crust (Fodor et al. 1985b). The contamination of mantle-derived magma by assimilation of crustal rocks is difficult to distinguish from contamination resulting from magma formation by melting of heterogeneous source rocks in the heads of plumes and in the mantle wedge above subduction zones along compressive continental margins (Chap. 4). In addition, magmas derived from different sources may mix to form hybrid magmas whose chemical and isotopic compositions depend on the proportions of mixing and on the compositions of the component magmas. The literature addressing these problems is extensive and includes consideration of both isotopic and trace-element data starting with the work of Gast (1968), Shaw (1970), and Harris (1989).

The standard petrogenetic model for the origin of flood basalts has been endorsed in general outline by many authors. However, some authors have found it necessary to modify the theory in order to explain certain aspects of specific deposits (e.g. Menzies and Cox 1988; Storey et al. 1993), or have proposed an alternative explanation (e.g. King and Anderson 1995). Gallagher and Hawkesworth (1992) and Hawkesworth et al. (1990) attributed the origin of flood basalt to decompression melting in the hydrated subcontinental lithosphere, whereas Arndt and Christiansen (1972) and Arndt et al. (1997) concluded that most magmas are generated by melting in the plume head rather than in the lithospheric mantle, because the transfer of heat from the plume to the overlying lithosphere is too slow to allow large-scale melting during periods of only 2 to 3 million years (White et al. 1987). In addition, Cordery et al. (1997) emphasized that the plume heads must contain eclogite derived from subducted oceanic crust in order to yield large volumes of basalt magma.

In spite of these and other disagreements, the eruption of large volumes (e.g. $10 \times 10^6 \, \text{km}^3$, North Atlantic) of basalt magma virtually requires the existence of crustal fractures. In addition, the transfer of heat to the lithospheric mantle and overlying continental crust takes place primarily by the upward flow of basalt magma rather than by conduction (Huppert and Sparks 1988). Consequently, partial melting can occur in the lithospheric mantle and even in the continental crust. Therefore, magmas derived by melting in the asthenospheric mantle may be contaminated by mixing with lithospheric or crustal magmas and/or by assimilation of rocks in the crust or lithospheric mantle (Hart 1985; Hooper 1985; Watson 1982). The formation of basalt plateaus and the associated mafic plutons on the continents (Table 5.1 and Chap. 7) has contributed significantly to the growth of the continental crust, and therefore merits the attention of geochemists, petrologists, and geophysicists whose work was recorded in books edited by Hawkesworth and Norry (1983), Macdougall (1988), Morton and Parson (1988), Puffer and Ragland (1992), and Storey et al. (1993), among others.

Continental rifts are also the centers of volcanic activity of alkali-rich lavas and associated plutons. In some cases, the eruption of alkali-rich lavas is a precursor to the subsequent outpouring of floods of tholeiite basalt. The petrogenesis of the alkali-rich lavas of the East African Rift valleys, of the Rhine Graben of Germany, and the Oslo Graben in Norway are discussed in Chap. 6.

The presentations of plateau basalt provinces in this chapter are arranged in approximate chronological order, starting with the most recent deposits. The emphasis is on recognition of the cause for the magmatic activity, on evidence of crustal contamination of the basalt magma, and on the origin of associated rhyolites and granites.

5.2 Columbia River Basalt, USA

The Columbia Plateau of Washington and Oregon (Fig. 5.1) consists of about $0.3 \times 10^6 \, \text{km}^3$ of basaltic rocks that were erupted from fissures between 16.5 and 6 Ma to form thick layers of tholeiite basalt. The basalt layers are continuous for distances of up to about 600 km in an east-west direction and have been subdivided stratigraphically into five major mappable units proposed by Swanson et al. (1979) based in part on the previous work of Waters (1961), Bingham and Grolier (1966), and Wright et al. (1973). In the new stratigraphic terminology, Swanson et al. (1979) elevated the Yakima basalt to a subgroup composed of three formations and fourteen members. These are underlain by two additional formations, which occur locally (Table 5.2).

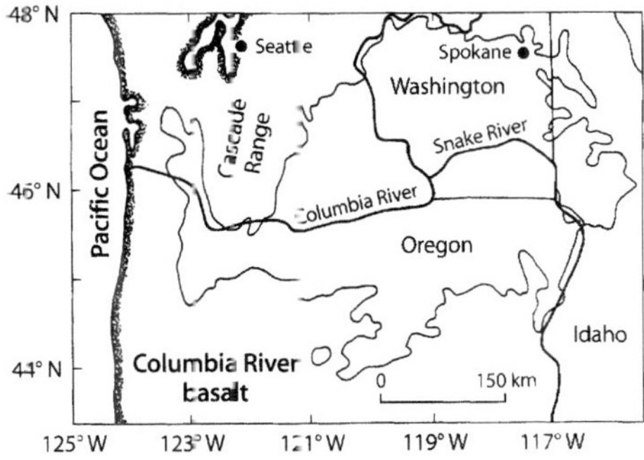

Fig. 5.1. Outcrop area of the Columbia River basalt in the states of Washington and Oregon, USA (*Source:* adapted from Wright et al. 1973)

Table 5.2. Additional formations of the Columbia River Plateau, which occur locally

Subgroup	Formation	Volume (%)
Yakima	Saddle Mountain (youngest)	1
	Wanapum	5
	Grande Ronde	75
	Picture Gorge	9
	Imnaha (oldest)	10

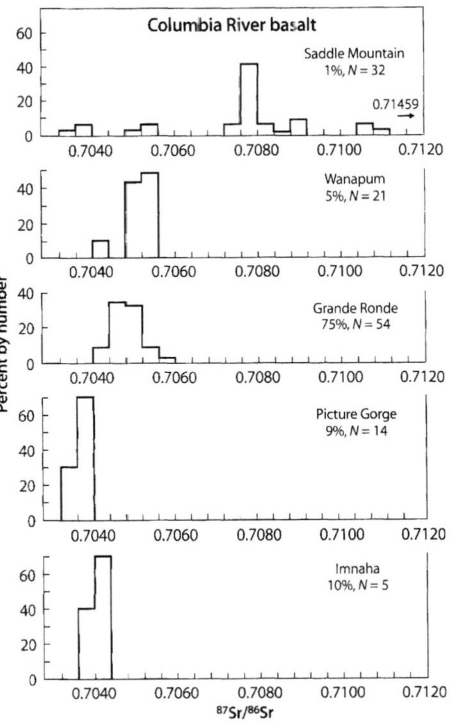

Fig. 5.2. Range of $^{87}Sr/^{86}Sr$ ratios of the Columbia River basalt in Washington and Oregon, USA. The general increase of the $^{87}Sr/^{86}Sr$ ratios up-section implies progressive contamination of the mantle-derived magmas by assimilation of a variety of crustal rocks. The $^{87}Sr/^{86}Sr$ ratios of the Saddle Mountain Formation range widely from 0.70354 to 0.71459 relative to 0.71025 for NBS 987 (*Sources:* McDougall 1976; Carlson et al. 1981; Carlson 1984)

The Picture Gorge Formation occurs only in north central Oregon, whereas the Imnaha Formation is restricted to the tri-state area of Oregon, Washington, and Idaho (Fig 5.1). The volumes of basalt assigned to the formations in Table 5.2 by Swanson et al. (1979) range from 75% for the Grande Ronde to only 1% for the Saddle Mountain Formations.

The chemical compositions of the basalt layers vary significantly but within narrow limits. For example, Wright et al. (1973) defined eleven chemical types of basalt based primarily on their concentrations of SiO_2 (47.09 to 54.98%) and MgO (2.69 to 6.82%). The range of $^{87}Sr/^{86}Sr$ ratios and chemical compositions of the Columbia River basalts has been variously attributed to the heterogeneity of magma sources in the mantle (Hooper 1982, 1984, 1988; Nelson 1983), to fractional crystallization of the magmas, to the assimilation of crustal rocks, and to a combination of all three processes (Carlson et al. 1981; Carlson 1984; Carlson and Hart 1988).

Most authors agree that the large volumes of basalt magma represented by the Columbia River basalt formed by decompression melting in the mantle in a setting of extensional tectonics. The magmas were erupted through fissures that are now represented by

sets of feeder dikes known from the southeastern region of the basalt plateau (Carlson et al. 1981; Hooper 1982, 1984). The dikes strike northwest-southeast parallel to the subduction zone that originally occurred offshore before it was overridden by the North American Plate. The formation of these fissures is therefore related to the subduction of the Juan de Fuca Plate under the North American continent and to the resulting development of a back-arc basin in this region.

The $^{87}Sr/^{86}Sr$ ratios of the Columbia River basalt reported by Hedge et al. (1970), McDougall (1976), Carlson et al. (1981), and Carlson (1984) increase up-section from average values of 0.70398 (Imnaha) and 0.70357 (Picture Gorge) to 0.70478 in the Grande Ronde Formation and 0.70799 for the Saddle Mountains basalt relative to 0.7080 for E&A. The data in Fig. 5.2 reveal that the $^{87}Sr/^{86}Sr$ ratios of the Saddle Mountain

Table 5.3.
Isotope ratios of Sr, Nd, Pb, and O of the Columbia River basalt

	$^{87}Sr/^{86}Sr$ [a]	$^{143}Nd/^{144}Nd$ [b]	$^{206}Pb/^{204}Pb$	$\delta^{18}O$ (‰)
Saddle Mountain (youngest)	0.70354 to 0.71459	0.51177 to 0.51308	17.736 to 19.751	+5.81 to +8.03
Wanapum	0.70435 to 0.70546	0.51259 to 0.51277	18.729 to 19.009	+6.36 to +6.82
Grande Ronde	0.70439 to 0.70558	0.51264 to 0.51288	18.785 to 18.978	+4.94 to +7.66
Picture Gorge	0.70351 to 0.70380	0.51302 to 0.51303	–	+5.64 to +6.24
Imnaha	0.70360 to 0.70406	0.51294 to 0.51304	18.769 to 19.086	+5.60 to +6.48

[a] Relative to 0.71025 for NBS 987.
[b] Corrected for isotope fractionation to $^{146}Nd/^{144}Nd = 0.7219$.

Formation, which comprises only about 1% of the total volume, actually vary from about 0.7034 to 0.7145. The isotope ratios of Sr, Nd, Pb, and O of the Columbia River basalt reported by the authors cited above, in most cases, range only within narrow limits (Table 5.3).

The explanation of the petrogenesis of the Columbia River basalt is complicated by the occurrence of geographical variations in the $^{87}Sr/^{86}Sr$ ratios of flows extruded within the same interval of time. These geographic patterns have been attributed to the existence of a crustal suture represented by the Olympic-Wallowa lineament (Fig. 4.35). Carlson (1984) observed that basalts of the Grande Ronde and Saddle Mountain Formations north and east of lineament have higher $^{87}Sr/^{86}Sr$ ratios than those erupted south and west of it.

The isotope ratios of Sr and Nd in Fig. 5.3 reveal that the Columbia River basalt forms several discrete data fields, suggesting significant differences in the petrogenesis of the lava flows extruded in the course of the formation of these rocks. The oldest flows assigned to the Imnaha and Picture Gorge Formations, which together constitute about 19% of the total volume, have comparatively low $^{87}Sr/^{86}Sr$ and high $^{143}Nd/^{144}Nd$ ratios that approach those of the DMM component in the mantle. The magma sources of the Grande Ronde and Wanapum basalts (about 80% of the total volume) have either been enriched in Sr and Nd derived from the subducted Juan de Fuca Plate; or these flows could have originated from previously enriched subcrustal lithospheric mantle.

The basalt flows of the Saddle Mountain Formation (about 1% of the total) have high $^{87}Sr/^{86}Sr$ and low $^{143}Nd/^{144}Nd$ ratios characteristic of crustal rocks. Therefore, they formed by large-scale assimilation of crustal rocks by mantle-derived magmas and/or by mixing of basalt magmas with crustal melts (Hooper 1985). However, the lava flows of the Saddle Mountain Formation that are exposed along the eastern edge of the Grande

Ronde Graben near LaGrande, Oregon, actually have low $^{87}Sr/^{86}Sr$ and high $^{143}Nd/^{144}Nd$ ratios that average 0.70361 ±0.00007 and 0.512957 ±0.00005 relative to NBS 987 = 0.71025 for Sr and 0.511929 for the La Jolla Nd Standard, respectively (Carlson et al. 1981). The isotope ratios of Sr and Nd of these flows are similar to those of the Picture Gorge/Imnaha basalts and approach the composition of the DMM component in the mantle (Hart 1988).

The basement rocks that underlie the Columbia River basalt, represented by xenoliths analyzed by Carlson et al. (1981), Carlson (1984), and Nelson (1983), have elevated $^{87}Sr/^{86}Sr$ and low $^{143}Nd/^{144}Nd$ ratios, as expected, but their $^{206}Pb/^{204}Pb$ ratios are similar to those of the Columbia River basalt. However, the oxygen of the xenoliths is enriched in ^{18}O ($\delta^{18}O = +9.35$ to +12.37‰). Therefore, the isotope composition of Pb in the basalt layers is not sensitive to contamination by crustal Pb, whereas their $\delta^{18}O$ values should have responded to the assimilation of crustal rocks by the basalt magmas, as they did in the flows of the Saddle Mountain Formation ($\delta^{18}O = 5.81$ to 8.03‰). Nelson (1983) concluded from the lack of correlation of $^{87}Sr/^{86}Sr$ ratios and $\delta^{18}O$ values that only certain basalt layers in the Grande Ronde and Saddle Mountain Formations were contaminated and that the isotopic heterogeneity of the basalt layers was inherited from their heterogeneous magma sources. Although Hooper (1984) agreed that the basalt magmas assimilated only minor amounts of upper-crustal rocks in most cases, he emphasized that the magmas differentiated by fractional crystallization in magma chambers near the base of the crust prior to eruption.

In summary, the evidence concerning the petrogenesis of the Columbia River basalt indicates that magmas formed by decompression melting in the tectonic setting of a back-arc basin. The magma sources included old subcontinental lithosphere and mantle rocks that had been altered by aqueous solutions emanating from subducted sediment of the Juan de Fuca

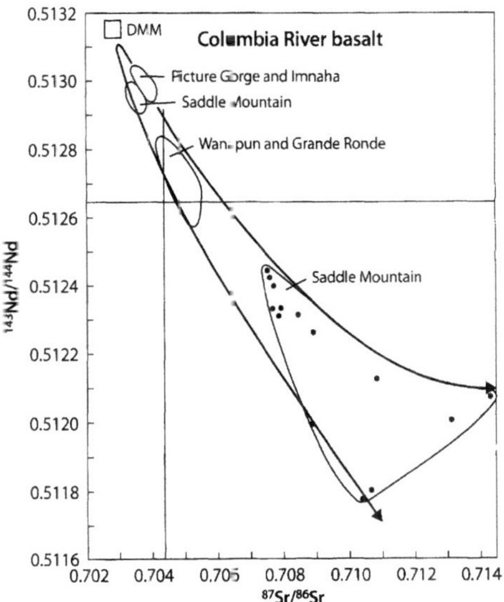

Fig. 5.3. Isotope ratios of the Columbia River basalt of Oregon and Washington. The grouping of data clusters along hypothetical mixing hyperbolas is an indication that the mantle-derived magmas evolved by assimilating varying amounts of isotopically heterogeneous rocks of the continental crust. In addition, the magmas differentiated by fractional crystallization (*Sources:* Carlson et al. 1981; Carlson 1984; Hart 1998)

Plate. The magmas differentiated to a limited extent by fractional crystallization in magma chambers near the base of the crust. In addition, some of the magmas of the Grande Ronde and the Saddle Mountain Formations assimilated rocks from the upper crust having elevated $^{87}Sr/^{86}Sr$ and low $^{143}Nd/^{144}Nd$ ratios as well as being enriched in ^{18}O (DePaolo 1983a; Carlson et al. 1983; Schmincke 1967; Nathan and Fruchter 1974).

The Columbia River Plateau was evaluated in the 1980s as a potential site for storage of radioactive waste. For this reason, the basalt of the Grande Ronde Formation on the Hanford Reservation of Washington was drilled to determine the permeability of the rocks and to obtain evidence for migration of groundwater. Brookins et al. (1986) measured the $^{87}Sr/^{86}Sr$ ratios of fracture minerals (zeolites, quartz, cristobalite, and illite) and associated basalt in one of the cores (DC-6). The results indicate that the fracture minerals have different $^{87}Sr/^{86}Sr$ ratios than the basalt. In six out of ten samples, the $^{87}Sr/^{86}Sr$ ratios of the fracture minerals are higher than those of the associated basalt for an average difference of +0.0029. This evidence indicates that the fracture minerals were deposited by aqueous solutions containing Sr derived in part from sources outside the Grande Ronde basalt.

5.3 The Snake River Plain and Yellowstone Caldera

The basalt plateau of the Snake River Plain in Fig. 5.4 forms a broad belt in southern Idaho with a northeast-southwest strike. The volcanic rocks in this belt consist primarily of tholeiite basalt flows that were erupted through fissures in the underlying continental crust. Rhyolites are much less abundant than basalts, and volcanic rocks of intermediate composition occur only locally; for example, at Magic Reservoir along the northern edge of the central region of the Snake River Plain (Honjo and Leeman 1987).

Rhyolites are much more abundant in the Yellowstone Caldera located at the northeastern end of the Snake River Plain. The lavas at this locality were erupted episodically from a shallow crustal magma chamber starting at about 1.9 Ma. Geophysical data indicate that rhyolite magma may still exist beneath the present caldera and may be the source of heat for the continuing hydrothermal activity (Eaton et al. 1975).

The orientation of the belt of volcanic rocks of the Snake River Plain and the bimodal composition of the lavas, including the rhyolites of the Yellowstone Caldera, suggest a different tectonic setting than existed at the time of eruption of the Columbia River basalt. The strike of the Snake River Plain is approximately at right angles to the edge of the North American Plate, and therefore the volcanic activity in this region is not related to the subduction of the Juan de Fuca Plate. Instead, Menzies et al. (1983) attributed the volcanic activity of the Snake River Plain to a mantle plume, which

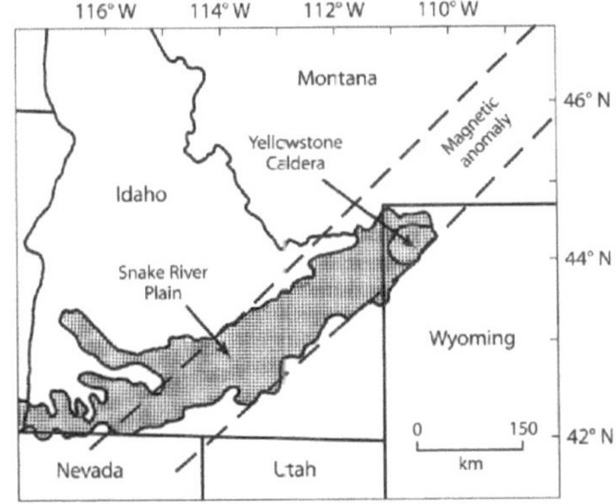

Fig. 5.4. The Snake River Plain (Idaho) and the Yellowstone Caldera (Wyoming) in the northeastern USA. The *dashed parallel lines* define the boundaries of a magnetic anomaly with which both volcanic provinces are associated (*Source:* adapted from Eaton et al. 1975)

is presently located under the Yellowstone Caldera. In that case, the petrogenesis of the basalts is a consequence of decompression melting in the head of the plume and in the overlying lithospheric mantle, whereas the rhyolites formed either by melting of crystalline basement rocks of the continental crust or by partial melting of previously emplaced mantle-derived mafic rocks in the lower continental crust.

5.3.1 The Snake River Plain, Idaho

The lavas of the Snake River Plain have been subdivided into the Idavada volcanics (early to middle Pliocene), the Idaho Group (late Pliocene to middle Pleistocene), and the Snake River Group (Pleistocene to Holocene). The Idavada volcanics consist primarily of felsic lavas, whereas the overlying rocks of the Idaho and Snake River Groups are composed of interbedded olivine tholeiite lavas and sedimentary rocks of continental origin. The olivine tholeiites have uniform chemical compositions and $^{87}Sr/^{86}Sr$ ratios between 0.7056 and 0.7076 relative to 0.7080 for E&A (Leeman and Manton 1971; Lum et al. 1989).

Lavas of more variable chemical composition than the olivine tholeiites occur at Craters of the Moon National Monument, King Hill, Magic Reservoir, and the canyon of the Blackfoot River. Some of the volcanic rocks at these locations are enriched in alkali elements and Fe and include latites containing 62.09% of SiO_2 on average. The Idavada volcanics have even higher SiO_2 concentrations with an average of 74.55%. The $^{87}Sr/^{86}Sr$ ratios of the lavas at Craters of the Moon are higher than those of the olivine tholeiites and range from 0.7086 to 0.7104, whereas the $^{87}Sr/^{86}Sr$ ratios of

King Hill have even more extreme values between 0.7103 and 0.7138 relative to 0.7080 for E&A. Silicic lavas of Pliocene age are enriched in Rb (164 ±20 ppm, $N = 5$) and depleted in Sr (60 ±22 ppm, $N = 5$) with high initial ratios between 0.7103 and 0.7125 at 5 Ma relative to 0.7080 for E&A (Leeman and Manton 1971).

The $^{87}Sr/^{86}Sr$ ratios of the whole compositional spectrum of the lavas of the Snake River Plain in Fig. 5.5 range widely from about 0.7056 to 0.7179 relative to 0.7080 for E&A and 0.71025 for NBS 987 (Leeman and Manton 1971; Leeman et al. 1976; Menzies et al. 1983; Honjo and Leeman 1987) and are significantly higher than those of lavas in the Cascade Range (Fig. 4.32) and in most of the basalts of the Columbia River Plateau (Fig. 5.2), except the Saddle Mountain Formation. Nearly 50% of the analyzed basalt samples from the Snake River Plain have $^{87}Sr/^{86}Sr$ ratios between 0.7056 and 0.7076 relative to 0.7080 for E&A and 0.71025 for NBS 987. Rhyolites and latites of the Snake River Plain have significantly higher $^{87}Sr/^{86}Sr$ ratios than the basalts.

The lavas at the Craters of the Moon National Monument, located along the northern edge of the Snake River Plain in south-central Idaho, range in composition from ferrobasalt to ferrolatite ($SiO_2 = 44.4$ to 72.2%). The $^{87}Sr/^{86}Sr$ ratios of these rocks also vary widely from 0.7079 to 0.7124 (Leeman et al. 1976) and correlate positively with SiO_2 concentrations in Fig. 5.6.

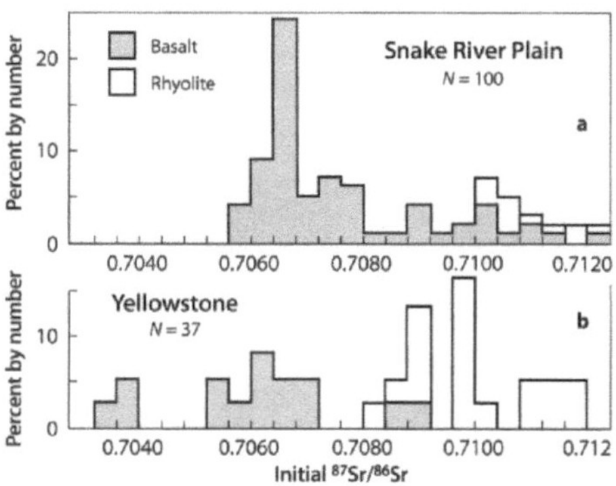

Fig. 5.5. Range of initial $^{87}Sr/^{86}Sr$ ratios of basalt and rhyolite relative to 0.7080 for E&A; **a** Snake River Plain; **b** Yellowstone Caldera (*Sources:* Leeman and Manton 1971; Leeman et al. 1976, 1985; Doe et al. 1982; Menzies et al. 1983; Honjo and Leeman 1987)

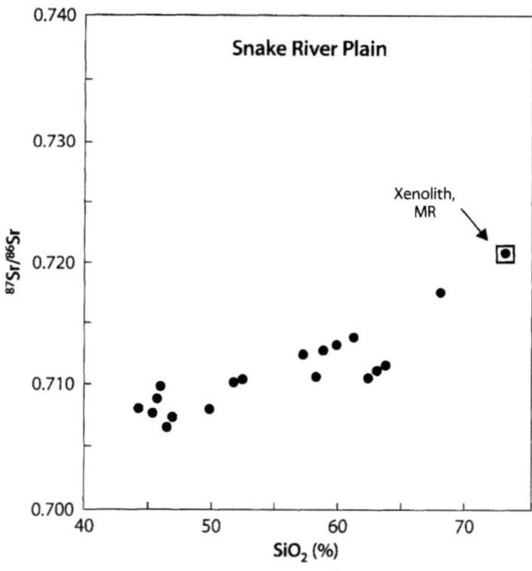

Fig. 5.6. Correlation of $^{87}Sr/^{86}Sr$ ratios and concentrations of SiO_2 in lavas of the Snake River Plain at Craters of the Moon and Magic Reservoir (*MR*), Idaho. The correlation between $^{87}Sr/^{86}Sr$ ratios and SiO_2 indicates contamination of the magmas by assimilation of silica-rich crustal rocks with elevated $^{87}Sr/^{86}Sr$ ratios (*Sources:* Leeman and Manton 1971; Leeman et al. 1976; Honjo and Leeman 1987)

Ferrolatites at the nearby locality of Magic Reservoir vary similarly and confirm the correlation between $^{37}Sr/^{86}Sr$ ratios and SiO_2 concentrations (Honjo and Leeman 1987).

The high initial $^{37}Sr/^{86}Sr$ ratios of the basalts in the Snake River Plain (lowest value 0.7056) reflect the isotope composition of Sr in the magma sources in the mantle (i.e. the plume head and/or subcontinental lithospheric mantle). In addition, the wide range of $^{37}Sr/^{86}Sr$ ratios of the lavas at the Craters of the Moon and at the Magic Reservoir, together with their positive correlation with SiO_2 concentrations in Fig. 5.6, are evidence that the mantle-derived basalt magmas at these locations assimilated crustal rocks represented by the xenoliths at Magic Reservoir. All of the rhyolites of the Snake River Plain included in Fig. 5.5 have elevated $^{87}Sr/^{86}Sr$ ratios greater than about 0.7100, because they formed by partial melting of crustal rocks as a result of heat given off by the basalt magmas (Huppert and Sparks 1938).

The evidence for assimilation of crustal rocks by basalt magmas at Magic Reservoir and elsewhere in the Snake River Plain is strengthened by the negative correlation of $^{87}Sr/^{85}Sr$ and $^{143}Nd/^{144}Nd$ ratios in Fig. 5.7, based on data from Honjo and Leeman (1987). All of the data points plot in the fourth quadrant, which is typically occupied by granitic basement rocks of the continental crust and by sedimentary rocks derived from them.

The lavas of the Snake River Plain contain a variety of xenoliths, consisting primarily of granulites and

metasediments that yield a Sm-Nd whole-rock isochron date of 2.75 ±0.18 Ga (Leeman et al. 1985). The $^{87}Sr/^{86}Sr$ ratios of the xenoliths in the lavas at Craters of the Moon range from 0.71522 to 0.81725 relative to 0.71025 for NBS 987 and have an average Sr concentration of 176 ±117 ppm ($2\bar{\sigma}$, N = 4). The $^{87}Sr/^{86}Sr$ ratios of xenoliths in lava flows at other localities within the Snake River Plain range even more widely from 0.7023 to 0.835 and generally have high Sr concentrations with an average of 304 ±86 ($2\bar{\sigma}$, N = 18). The average $^{87}Sr/^{86}Sr$ ratio of the xenoliths, weighted by their Sr concentrations, is 0.7316. The high average Sr concentration and $^{87}Sr/^{86}Sr$ ratio of these xenoliths make these rocks effective contaminants of the Sr in a magma into which they are assimilated. Therefore, these results support the conclusion by Leeman et al. (1976) that the isotope composition of Sr in the lavas of the Snake River Plain was altered by assimilation of Precambrian basement rocks, whereas the concentrations of the major and trace elements were not appreciably affected.

5.3.2 Yellowstone Caldera, Wyoming

The volcanic center at Yellowstone, Wyoming, is located at the eastern end of the Snake River Plain (Fig. 5.4). The volcanic activity was concentrated in three major episodes at 1.9, 1.2, and 0.6 Ma during which about 6×10^3 km^3 of lavas were extruded, consisting primarily of rhyolite (95%) and subsidiary amounts of basalt (5%). The magmas originated from large shallow chambers that subsequently collapsed to form the Yellowstone Caldera, whose long axis trends NE for more than 70 km (Eaton et al. 1975; Leeman et al. 1977; Christiansen and Blank 1972). The most recent eruptions occurred about 70 000 years ago along ring fractures within the main caldera.

The rhyolites of the Yellowstone Caldera in Fig. 5.5 have high initial $^{87}Sr/^{86}Sr$ ratios ranging from 0.7084 to 0.7268 (Doe et al. 1982; Leeman et al.1977). The initial $^{87}Sr/^{86}Sr$ ratios of associated basalts are generally lower than those of the rhyolites, with values between 0.7035 and 0.7089. These measurements leave no doubt that the rhyolites did not form by differentiation of mantle-derived basalt magma which, in this case, had a low $^{87}Sr/^{86}Sr$ ratio of 0.7035. The distribution of data points in Fig. 5.8 indicates that the basalts and some of the rhyolites are mixtures of basalt magma ($^{87}Sr/^{86}Sr$ = 0.7035) with granitic basement rocks represented by the Lamar Canyon gneiss for which Doe et al. (1982) reported $^{87}Sr/^{86}Sr$ = 0.7427, Rb = 94.5 ppm, and Sr = 276 ppm. However, most of the rhyolite data points in Fig. 5.8 scatter widely across the diagram, because of their low Sr concentrations caused by fractional crystallization of contaminated magma. The iso-

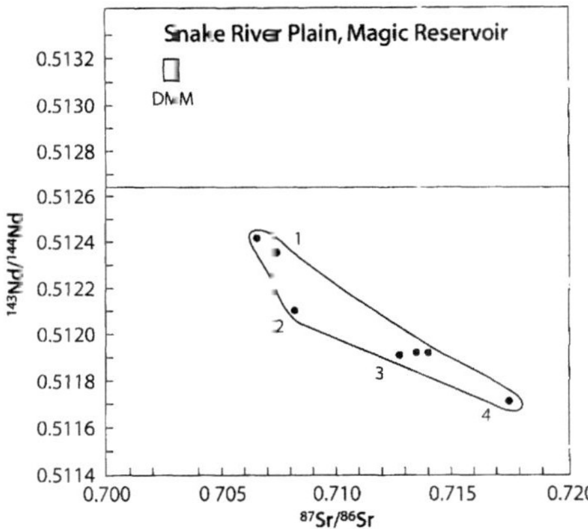

Fig. 5.7. Range of variation of the isotope ratios of Sr and Nd in the volcanic rocks at Magic Reservoir in the Snake River Plain of Idaho. The rock types are identified by number: *1*. Olivine tholeiites; *2*. McHan basalt, Havada volcanics; *3*. Square Mountain ferrolatite; *4*. Rhyolite. All of these rocks plot in the fourth quadrant of the diagram where granitic basement rocks of the continental crust reside (*Source:* Honjo and Leeman 1987)

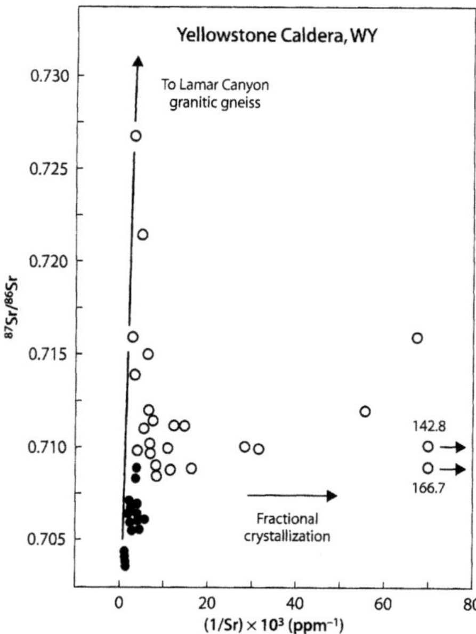

Fig. 5.8. Sr-isotope mixing diagram (without AFC) for basalt and rhyolite from the Yellowstone Caldera, Wyoming. The initial $^{87}Sr/^{86}Sr$ ratios of the basalt (*solid circles*) range from 0.7035 (lower flow, the Narrows) to 0.7089 (Falls River), whereas those of the rhyolites (*open circles*) vary even more widely from 0.7084 (Mesa Falls tuff) to 0.7268 (Huckleberry Ridge tuff). The basalts and some of the rhyolites lie close to a hypothetical mixing line between the basalts and Precambrian granitic gneiss of Lamar Canyon. However, most of the rhyolites scatter to the right of this mixing line, because their Sr concentrations (down to 6 ppm, Lava Creek tuff) were reduced by fractional crystallization of the rhyolite magma (*Source:* Doe et al. 1982)

topic data permit the conclusion that the rhyolite magmas erupted at Yellowstone formed by melting of crustal rocks and subsequently evolved by fractional crystallization and by interaction with hydrothermal fluids containing radiogenic ^{87}Sr derived from the Precambrian basement rocks. The water discharged by Mammoth Springs and travertine deposits have average $^{87}Sr/^{86}Sr$ ratios of 0.7103 ($N = 1$) and 0.7106 ±0.0004 ($2\bar{\sigma}, N = 4$), respectively (Leeman et al. 1977).

5.3.3   Absaroka Volcanic Field, Wyoming

The Absaroka Mountains of northwestern Wyoming and southwestern Montana are located east of and adjacent to the Yellowstone Caldera. The volcanic rocks of this area consist of andesite and basalt of Eocene to Oligocene age and thus predate the lavas of Yellowstone and the Snake River Plain. In addition to the calc-alkaline suite of volcanic rocks (andesite to

rhyolite), the Absaroka Mountains contain alkali-rich rocks (shoshonites and absarokites) whose K_2O concentrations are about twice those of the calc-alkaline suite. The rocks of both suites are primarily volcaniclastic in origin and were erupted through vents at Electric Peak, Mt. Washburn, and the Sunlight volcanic center (Peterman et al. 1970c).

The initial $^{87}Sr/^{86}Sr$ ratios (at 50 Ma) of the volcanic rocks in the Absaroka Mountains range from 0.7042 to 0.7090 relative to 0.7080 for E&A and have high Sr concentrations between 630 and 2 480 ppm (Peterman et al. 1970c). The heterogeneity of the isotope compositions of Sr (and Pb) cannot be explained by assimilation of crustal rocks, because the $^{87}Sr/^{86}Sr$ and $^{206}Pb/^{204}Pb$ ratios of the rocks are not correlated as expected if mantle-derived magmas had been contaminated with Sr and Pb derived from crustal rocks. Therefore, Peterman et al. (1970c) concluded that the calc-alkaline and alkali-rich lavas formed by different degrees of partial melting of isotopically heterogeneous rocks in the lower crust or upper mantle without becoming homogenized in the process. Subsequently, magmas differentiated in magma chambers at shallow depth before being erupted explosively to form the ashflow tuffs.

5.4   North Atlantic Igneous Province

The areas bordering on the North Atlantic Ocean in Fig. 5.9 contain an extensive igneous province of Early Tertiary age. This province encompasses the Inner Hebrides Islands of northwest Scotland (Skye, Mull, Eigg, and others), as well as the Antrim Plateau in Northern Ireland, the Faeroe Islands, Rockall Bank in the Atlantic Ocean, parts of East and West Greenland, and Cape Dyer on the east coast of Baffin Island. The volcanic centers in East Greenland include the Skaergaard Intrusion made famous by the work of Wager and Deer (1939), Wager (1953), and many other petrologists who have visited the area or carried out specialized studies on the rocks. The North Atlantic igneous province also includes Iceland and Jan Mayen, discussed in Sect. 2.3 in connection with the Late Tertiary to Recent volcanic activity associated with the Mid-Atlantic Ridge. In addition, the Late Cretaceous diabase and basalt of Spitsbergen may represent an early stage of development of igneous activity in this region (Turner and Verhoogen 1960).

The magmatic activity of the North Atlantic igneous province between 70 and 50 Ma was caused by the thermal effects of a large mantle plume which is presently located under Iceland (Sect. 2.3.3). When the plume first interacted with the subcrustal lithosphere of Pangea, it was located in what later became East Greenland (Fig. 5.9). The head of the plume expanded

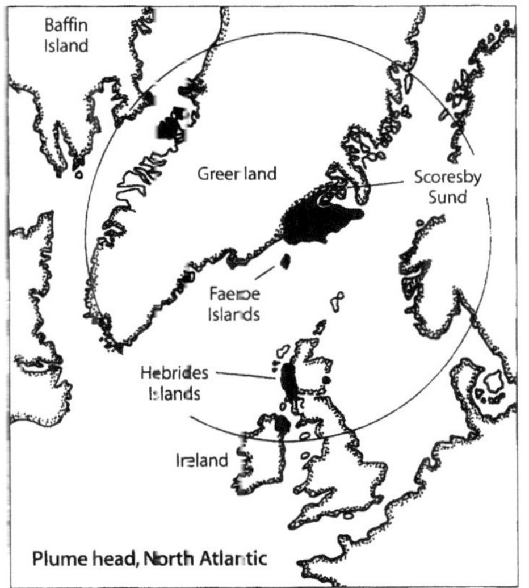

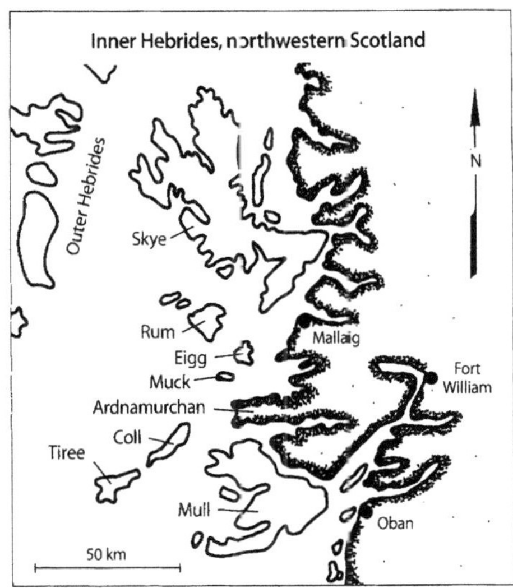

Fig. 5.9. The North Atlantic Igneous Province at about 60 Ma when magmatic activity may have peaked in intensity. The principal igneous centers (volcanic and intrusive) were in (1) the Hebrides Islands (Skye, Mull, Eigg, etc.) and Northern Ireland; (2) the Faeroe Islands located off the coast of Greenland at that time; (3) the Skaergaard-Kangerdlugssuaq Intrusions south of Scoresby Sund in Eastern Greenland; (4) the Svartenhuk-Disko-Cape Dyer area of West Greenland and Baffin Island. All of these sites are located along rifts that were active during the opening of the North Atlantic Ocean and of Davis Strait between Baffin Island and Greenland. The circle defines the approximate area that was affected by the expanded head of the plume, which was located south of Scoresby Sund along the present east coast of Greenland (Sources: adapted from Carter et al. 1979 and White and McKenzie 1989)

Fig. 5.10. Map of the Inner Hebrides of northwest Scotland including the islands of Skye, Rum, Eigg, Muck, Mull, Coll, and Tiree as well as the Ardnamurchan Peninsula (Source: adapted from Moorbath and Bell 1965a)

by lateral streaming and thereby caused rifting of the overlying lithospheric mantle and continental crust leading ultimately to the opening of the North Atlantic Ocean (Taylor et al. 1937; Thirlwall et al. 1994; White and Morton 1995; White and McKenzie 1989).

The process started with the formation of large-scale basalt plateaus along the rifted margins of the proto-Atlantic Ocean and continued with the intrusion of gabbroic and granitic magmas. In most but not all cases, the mantle-derived magmas interacted with rocks of the continental crust through which they were extruded (Carter et al. 1979).

5.4.1 Skye, Inner Hebrides, Scotland

The geology of Skye in Fig. 5.10 has been studied by geologists for more than a century (Harker 1904; Richey 1932; Anderson and Dunham 1966; Bell 1966, 1976; Gass and Thorpe 1976; Meighan 1979; Sutherland 1981; Thompson 1932). The rocks consist of a Precam-

brian basement complex composed of the Lewisian gneisses, Moine and Dalradian schists, and Torridonian sandstone overlain by Cambrian limestone and Mesozoic sandstones (Giletti 1959; Moorbath et al. 1969, 1975a; Moorbath and Park 1971; van Breemen et al. 1971; Whitehouse 1990). These rocks were intruded by granites and gabbros of Early Tertiary age and are overlain in the northern part of the island by a sequence of alkali olivine basalt flows including hawaiites, mugearites, and trachytes. The lava flows approach a total thickness of 2 300 m and have been divided into several groups:

- Osdale Group, 490 m (youngest)
- Bracadale Group, 120 m
- Beinn Totaig Group, 610 m
- Ramascaig Group, 760 m
- Beinn Edra Group, 300 m (oldest)

The plutonic rocks were emplaced after the extrusion of the lava flows and occur primarily in the central part of the island. They consist of the layered mafic intrusion in the Cuillin Hills (Wager and Brown 1968), the arcuate granitic ring dikes of the Western Redhills Centre (including hybrid rocks termed "marscoite"), and the Eastern Redhills Centre including the Beinn an Dubhaich and other granites (Wager et al. 1953, 1965; Lambert 1969; Moorbath and Welke 1969a; Bell 1984; Vogel et al. 1984).

The origin of the granitic rocks on Skye and at other centers of this petrologic province has been variously attributed to:

1. Fractional crystallization of mantle-derived basalt magmas;
2. Partial melting of the Precambrian basement rocks;
3. Assimilation of the Precambrian basement rocks by mantle-derived basalt magmas;
4. Mixing of magmas of basaltic and granitic composition.

The bimodal composition of the igneous rocks on Skye and the large volume of granitic rocks relative to basalt and gabbro cannot be explained exclusively by fractional crystallization of basalt magma. Therefore, the origin of the granitic rocks has been attributed to partial melting of the basement rocks, based in part on the experimental work of Brown (1963).

The origin of the granitic rocks on Skye was first investigated by means of isotopic data by Moorbath and Bell (1965a,b) who measured $^{87}Sr/^{86}Sr$ ratios of basaltic and granitic rocks. Their work evolved from the conclusion of Faure and Hurley (1963) that granitic rocks that had formed by partial melting of old crustal rocks have higher initial $^{87}Sr/^{86}Sr$ ratios than granitic differentiates of basalt magma derived from the upper mantle. Moorbath and Bell (1965a) demonstrated that the granites, marscoites, and ferrodiorites of Skye have consistently higher initial $^{87}Sr/^{86}Sr$ ratios than the basalts, dolerites, gabbros, mugearites, and peridotites. Therefore, they concluded that the granitic magmas had formed by partial melting of the Archean Lewisian gneisses, which constitute the basement of the central part of Skye. In addition, Moorbath and Bell (1965a) noted that the initial $^{87}Sr/^{86}Sr$ ratios of the granitic rocks vary widely, indicating that the Sr in the source rocks of the granitic magmas was isotopically heterogeneous and that the granitic rocks of Skye had not formed from one large isotopically homogeneous body of magma.

Subsequent studies by other investigators including Carter et al. (1978b), Moorbath and Thompson (1980), Dickin (1981), Dickin and Exley (1981), and Thompson et al. (1982, 1986) have generally confirmed the conclusion that the granitic rocks of Skye are derivatives of the underlying basement rocks rather than products of fractional crystallization of basalt magmas. Their data in Fig. 5.11 demonstrate that the $^{87}Sr/^{86}Sr$ ratios of the granitic rocks are, in most cases, higher than those of the basalts and gabbros and range widely from 0.70589 (Dickin 1981) to 0.7340 (Dickin and Exley 1981) relative to 0.7080 for E&A.

Most of the igneous rocks on Skye contain evidence of hydrothermal alteration manifested by mineralogical criteria and low $\delta^{18}O$ values. Therefore, Dickin et al. (1980) investigated whether the $^{87}Sr/^{86}Sr$ ratios of the Coire Uaigneich Granophyre of Skye had been affected by hydrothermal alteration. The results indicated that the initial $^{87}Sr/^{86}Sr$ ratios of the rocks range only from 0.7298 to 0.7320 relative to 0.7080 for E&A and do not correlate with $\delta^{18}O$ values, even though the rocks are strongly depleted in ^{18}O ($\delta^{18}O$ between +1.9 and –1.8‰). However, the initial $^{87}Sr/^{86}Sr$ ratios of one whole-rock sample, taken 0 to 1 mm from the center of a shear zone, had apparently been lowered by about 0.5% from 0.73191 ±0.00004 ($2\bar{\sigma}$, $N = 7$) to 0.72838, whereas all the rocks in this shear zone are strongly depleted in ^{18}O and have an average $\delta^{18}O$ value of –1.3 ±0.2‰ ($2\bar{\sigma}$, $N = 9$). Evidently, the hydrothermal fluids pervasively altered the isotope composition of oxygen in the rocks, but had only a very small effect on their $^{87}Sr/^{86}Sr$ ra-

Fig. 5.11. Range of initial $^{87}Sr/^{86}Sr$ ratios of mafic and felsic igneous rocks on the Isle of Skye, northwest Scotland. The wide range and elevated initial $^{87}Sr/^{86}Sr$ ratios of the granitic rocks indicate that the granitic magmas originated by partial melting of heterogeneous basement rocks consisting primarily of the Lewisian gneisses of Archean age. The basaltic magmas that formed the mafic rocks on Skye originated from the mantle, but also assimilated varying amounts of crustal rocks as indicated by their initial $^{87}Sr/^{86}Sr$ ratios that range from less than 0.7030 to greater than 0.7060 relative to 0.7080 for E&A (*Sources:* Moorbath and Bell 1965a; Carter et al. 1978b; Moorbath and Thompson 1980; Pankhurst 1969; Dickin 1981; Dickin and Exley 1981; Thompson et al. 1982)

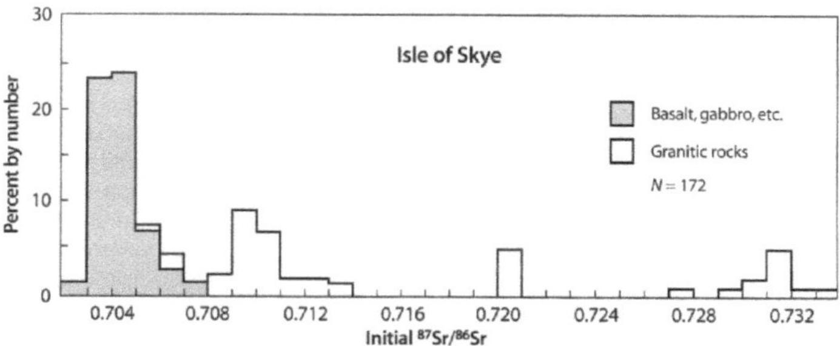

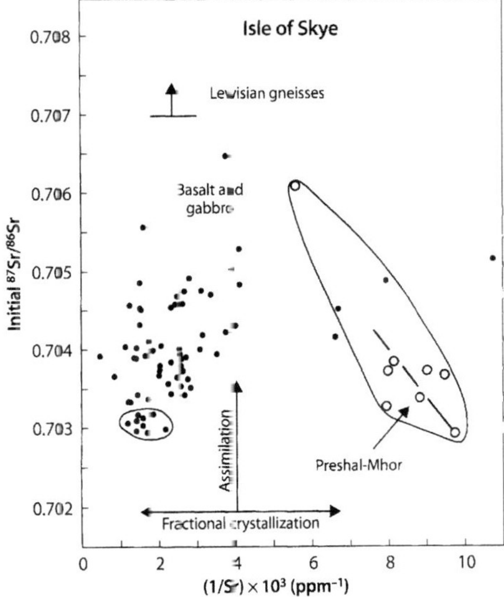

Fig. 5.12. Sr isotope mixing diagram (without AFC) for basalt and gabbro on the Isle of Skye, northwest Scotland. The evident scatter of the data points is consistent with the conclusion that basalt magmas from the mantle ($^{87}Sr/^{86}Sr < 0.7030$, $Sr \approx 900$ ppm) assimilated varying amounts of Sr from different metamorphic facies of the Lewisian gneisses that form the basement complex of Skye. In addition, the basalt magmas differentiated by fractional crystallization, which caused the Sr concentrations to vary widely from less than 100 to more than 2000 ppm. The Preshal-Mhor type lavas and tholeiite dikes have lower Sr concentrations than the Main lava Series of Skye. This group of flows and dikes may have formed from low-Sr mantle-derived magmas that assimilated varying amounts of Lewisian basement rocks. Preshal-Mhor type flows occur also on Mull and Arisaig (*Sources:* Moorbath and Thompson 1980; Thompson et al. 1982; Carter et al. 1978b; Dickin 1981).

tios, presumably because the aqueous solutions had low Sr concentrations. In addition, the data of Dickin et al. (1980) suggest that the hydrothermal solutions that altered the Coire Uaigneich Granophyre had lower $^{87}Sr/^{86}Sr$ ratios than the rocks they interacted with. The significance of these results is that the $^{87}Sr/^{86}Sr$ ratios of granitic rocks on Skye were not appreciably changed by hydrothermal alteration and therefore are reliable indicators of the petrogenesis of these rocks.

The initial $^{87}Sr/^{86}Sr$ ratios of the mafic rocks of Skye also vary significantly from 0.70295 (Preshal-Mhor type basalt, Talisker Bay) to 0.70609 (Preshal-Mhor type tholeiite, South Sleat Coast) relative to 0.7080 for E&A (Moorbath and Thompson 1980). The data points representing rocks of the Main lava Series of Skye form a fan-shaped array on the Sr-isotope mixing diagram in Fig. 5.12. This evidence is consistent with the postulate of Moorbath and Thompson (1980) that the magmas from which they formed were selectively contami-

nated with Sr derived from crustal rocks such as the amphibolite facies of the Lewisian basement rocks, for which Dickin (1981) reported average $^{87}Sr/^{86}Sr = 0.7174$ and Sr = 493 ppm.

The Preshal-Mhor type lavas and dikes on Skye, Mull, and Arisaig have low Sr concentrations (Sr = 124 ±15 ppm, $2\bar{\sigma}$, N = 9; Moorbath and Thompson 1980) compared to the flows of the Main lava Series, but their $^{87}Sr/^{86}Sr$ ratios encompass the full range of variation observed in the mafic rocks of Skye. Consequently, the Preshal-Mhor type rocks form a distinctive array in Fig. 5.12 that is consistent with assimilation of Lewisian basement rocks by low-Sr, mantle-derived magmas. The low Sr content of the Preshal-Mhor magmas may have been caused by crystallization and segregation of Sr-rich minerals (e.g. plagioclase) prior to contamination of the residual magma by Sr-rich Lewisian basement rocks. Alternatively, the low-Sr Preshal-Mhor magmas may also have formed as a result of fractional melting of depleted spinel-bearing lherzolite in the upper mantle (Thompson et al. 1972).

The contamination of mantle-derived magmas by assimilation of Precambrian basement rocks is strongly indicated by the inverse correlation with the isotope ratios of Sr and Nd of the basalts on the islands of Mull, Skye, Muck and Eigg in Fig. 5.13. The data of Carter et al. (1978b) demonstrate that the initial $^{87}Sr/^{86}Sr$ ratios of the basalt flows on these islands range widely from a MORB-like value of 0.70277 on Mull to 0.71261 for a felsite on Eigg relative to 0.7080 for E&A. The corresponding $^{143}Nd/^{144}Nd$ ratios range from 0.51307 to 0.51146. The isotope ratios of Sr and Nd of Tertiary granite and related intrusives on Skye in Fig. 15.13 are typical of granitic gneisses (amphibolite facies) of the Lewisian basement rocks and thereby identify them as the source of the granitic magmas on this island (Moorbath and Bell 1965a).

The evidence for contamination of the Tertiary basalts and granites of Skye by Precambrian basement rocks is further strengthened by the isotope ratios of Pb published by Moorbath and Welke (1969a). Their data in Fig. 5.14 demonstrate that the $^{206}Pb/^{204}Pb$ and $^{207}Pb/^{204}Pb$ ratios of these rocks at 60 Ma define a mixing line whose end-members are leads that separated at 3.1 Ga and 60 Ma from a U-Pb system whose present $^{238}U/^{204}Pb$ ratio is 8.92. The distribution of data points along the mixing line implies that up to 80% of the Pb in some granites on Skye originated from granitic gneisses in the Precambrian basement, whereas less than about 10% of the Pb in some of the basalts and gabbros originated from that source (Moorbath and Welke 1969a). These results therefore support the interpretation that the granites on Skye originated by melting of the Lewisian gneisses followed by mixing with felsic differentiates of the voluminous basalt magmas (Dickin et al. 1987).

Fig. 5.13.
Isotope ratios of Sr and Nd of Tertiary basalt on the islands of Mull (*crosses*), Skye (*solid circles*), Muck (*open triangles*), and Eigg (*solid triangles*), as well as granitic rocks on Skye (*open circles*). The distribution of data points is an indication that the mantle-derived basalt magmas assimilated varying amounts of Precambrian granitic basement rocks (*Source:* Carter et al. 1978b)

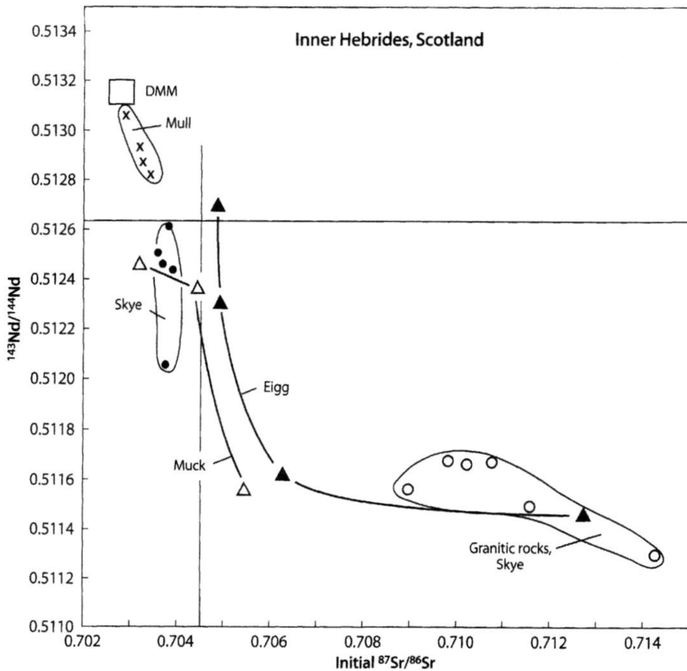

Fig. 5.14.
Isotope ratios of Pb in the Tertiary basalt (*solid circles*) and granite (*open circles*) on the Isle of Skye, Inner Hebrides. The data points define a linear mixing array whose endmembers are the Pb in granite gneiss and basalt magma. The Pb in the granite gneiss separated at 3.1 Ga from a U-Pb system whose present $^{238}U/^{204}Pb$ ratio is 8.92. The Pb in the basalt magma originated from the same source at 60 Ma. All of the Pb ratios were corrected for in situ decay of U based on measured concentrations of U and Pb and assuming an age of 60 Ma for all of the rocks (*Source:* data and interpretation by Moorbath and Welke 1969a)

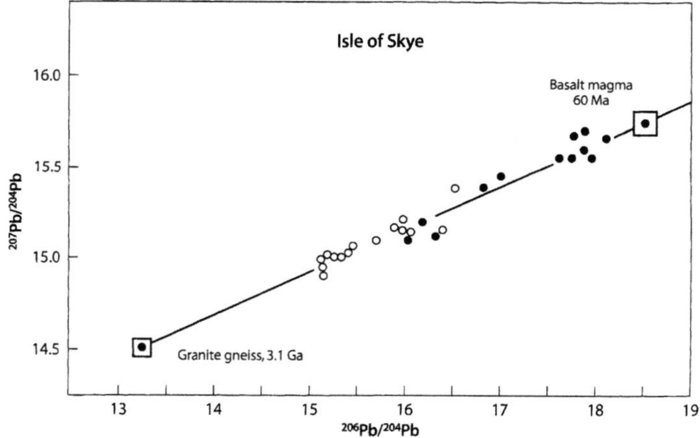

5.4.2 Mull, Inner Hebrides, Scotland

The volcanic rocks on the island of Mull (Fig. 5.10) have been divided into three groups based on their chemical compositions and $^{87}Sr/^{86}Sr$ ratios. The data of Beckinsale et al. (1978) in Fig. 5.15 demonstrate that the alkali-rich lavas of Group I (basalt-hawaiite-mugearite) have a low average initial $^{87}Sr/^{86}Sr$ ratio of 0.70288 ±0.00007 (2$\bar{\sigma}$, N = 9) relative to 0.7080 for E&A. The $^{87}Sr/^{86}Sr$ ratios of these flows are identical to those of plume MORBs extruded along mid-ocean ridges (Fig. 2.7).

The flows of Group II on Mull are composed of tholeiitic basalt and trachyte similar to the Main lava Series on Skye. The initial $^{87}Sr/^{86}Sr$ ratios range from 0.70514 to 0.70616 with a mean of 0.70550 ±0.00027

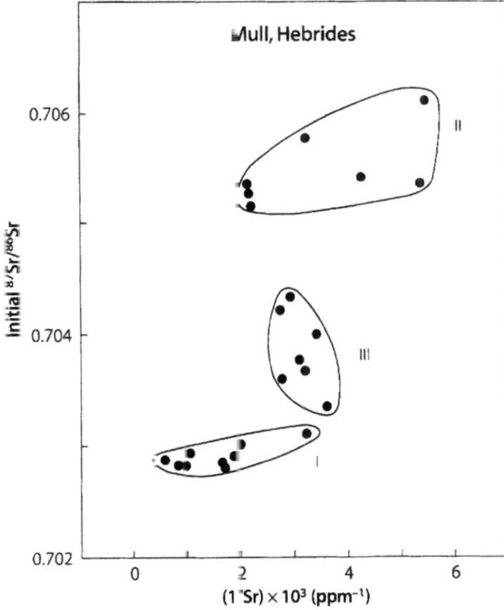

Fig. 5.15. Sr isotope mixing diagram (without AFC) for the three groups of Tertiary lava flows on the Isle of Mull, Inner Hebrides, northwest Scotland. The average initial ^{87}Sr/^{86}Sr ratio of the Group I flows is identical to that of plume MORBs extruded along mid-ocean ridges (Figure 2.7). The flows of Group II have higher ^{87}Sr/^{86}Sr ratios than those of Group I and originated from a Rb-enriched source presumably located in the mantle. The Group III lavas are intermediate in isotope composition and either formed by mixing of Group I and II magmas or by contamination of Group I magma with Lewisian gneisses (Source: data from Beckinsale et al. 1978)

$2\bar{\sigma}$, $N = 7$) compared to 0.7080 for E&A. These rocks also have higher Rb concentrations, but contain less Sr than the lavas of Group I. These differences suggest that the Group II lavas originated from different sources than the flows of Group I. Beckinsale et al. (1978) suggested that the Group I magmas formed by varying degrees of partial melting of a garnet lherzolite; whereas those of Group II originated from Rb-enriched plagioclase lherzolite.

The lavas of Group I I are intermediate between those of Groups I and II in terms of their ^{87}Sr/^{86}Sr ratios which range from 0.70332 to 0.70432. Beckinsale et al. (1978) attributed the origin of these flows to (1) mixing of Group I and Group II magmas; or (2) deri-

vation of Group III magma from mixed sources; or (3) contamination of Group I magma by Lewisian gneisses. Evidence for contamination of basalt magma by the Lewisian basement rocks was presented by Kerr et al. (1995).

The Glen Cannel granophyre of Mull has an elevated initial ^{87}Sr/^{86}Sr ratio of 0.7094 ±0.0003 based on a whole-rock Rb-Sr isochron obtained by Beckinsale (1974). The rocks on Mull and on the Ardnamurchan Peninsula are depleted in ^{18}O, and their average δ^{18}O value is −2.3‰ (Taylor and Forester 1971; Forester and Taylor 1976). Beckinsale (1974) concluded that the Glen Cannel granophyre was enriched in radiogenic ^{87}Sr by hydrothermal alteration and thus could be a magmatic differentiate of tholeiite basalt. However, the δ^{18}O values do not correlate with the ^{87}Rb/^{86}Sr or ^{87}Sr/^{86}Sr ratios of the rocks. In addition, Dickin et al. (1980) subsequently demonstrated that the isotope composition of Sr in the Coire Uaigneich Granophyre on Skye was not significantly changed by hydrothermal alteration that had radically altered the isotope composition of oxygen of the rocks. Therefore, the magma of the Glen Cannel granophyre assimilated crustal rocks and is not a product of fractional crystallization of basalt magma.

Other studies of granitic rocks on the Isle of Mull are by Beckinsale and Obradovich (1973) and by Pankhurst et al. (1978), whereas Skelhorn and Elwell (1966) discussed the structure of a granophyric quartz dolerite on Ardnamurchan. The petrogenesis of the lava flows and of a related layered mafic intrusion on the island of Rhum is discussed in Chap. 7.

5.4.3 The Faeroe Islands

The Tertiary basalts of the Faeroe Islands in Fig. 5.9 have been divided into three-stratigraphic series whose total thickness reaches 3 000 m (Noe-Nygaard and Rasmussen 1968). Gariépy et al. (1983) recognized that the initial ^{87}Sr/^{86}Sr ratios and concentrations of REE and LIL elements of most of the flows of the Upper Series differ significantly from those of the Middle and Lower Series. The average initial ^{87}Sr/^{86}Sr ratios and concentrations of Rb and Sr of the two groups of flows are shown in Table 5.4 (excluding two anomalous samples), where all errors are two standard deviations of the mean.

Table 5.4.
Average initial ^{87}Sr/^{86}Sr ratios and concentrations of Rb and Sr of the two groups of flows of the Tertiary basalts of the Faeroe Islands (All errors are two standard deviations of the mean)

Series	Group 1			Group 2		
	Lower and Middle			Upper		
^{87}Sr/^{86}Sr	0.70362	±0.00023	$N = 12$	0.70310	±0.00027	$N = 9$
Sr (ppm)	262	±21	$N = 15$	88.9	±5.5	$N = 9$
Rb (ppm)	5.8	±3.0	$N = 15$	1.95	±1.19	$N = 9$

The distribution of data points in Fig. 5.16 illustrates the division of the Faeroe lavas into the two groups mentioned above. In addition, the data confirm the conclusion of Gariépy et al. (1983) that both magma types assimilated crustal rocks and suggest that sample K-1 is highly contaminated with radiogenic ^{87}Sr. The isotopic and chemical differences between the two groups of flows require that they formed from different magmas derived from different sources in the mantle. The data of Gariépy et al. (1983) indicate that the magma of Group 1 flows had an $^{87}Sr/^{86}Sr$ ratio of about 0.7032 and was enriched in the light REE compared to chondrite meteorites, whereas the magma of the Group 2 flows had a low $^{87}Sr/^{86}Sr$ ratio of 0.7026 and was depleted in the light REEs. These characteristics are consistent with the conclusion of Schilling and Noe-Nygaard (1974) and Gariépy et al. (1983) that the Group 1 magma originated from the mantle plume, whereas the Group 2 flows on the Faeroe Islands have the chemical and isotopic characteristics of normal MORBs (Fig. 2.7). The subsequent contamination of both magma types requires the presence of old crustal rocks under the Faeroe Islands.

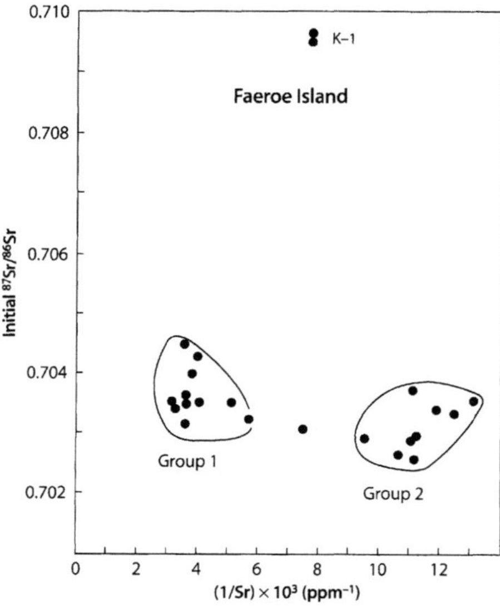

Fig. 5.16. Sr-isotope mixing diagram (without AFC) of the Early Tertiary basalt flows on the Faeroe Islands. The flows form two groups based on their initial $^{87}Sr/^{86}Sr$ ratios, Sr concentrations, and other geochemical criteria. The Group *1* flows originated from the mantle plume, whereas the magma of the Group *2* flows formed from depleted MORB-like source rocks in the mantle. Both magma types were subsequently contaminated by Sr derived from old crustal rocks under the Faeroe Islands. Flow *K-1* is highly contaminated with radiogenic ^{87}Sr (*Source:* Gariépy et al. 1983)

5.4.4 Rockall Bank, North Atlantic Ocean

The seafloor west of the Outer Hebrides of northwestern Scotland includes several small submerged basalt plateaus including Rockall Bank (1.02×10^5 km^2 at a depth of 1 000 m) as well as the smaller Rosemary and Lousy banks identified in Fig. 5.17 (Féraud et al. 1986; Morton and Taylor 1987). A very small part of the Rockall Bank is exposed above sealevel to form Rockall Islet (24 to 30 m wide), which is composed of aegirine granite described by Sabine (1960). The $^{87}Sr/^{86}Sr$ ratio of one specimen of this granite (recalculated to 60 Ma and $\lambda(^{87}Rb) = 1.42 \times 10^{-11}$ yr^{-1}) is 0.7068 ±0.0005 ($2\bar{\sigma}$) based on five replicate analyses and relative to 0.7080 for E&A (Moorbath and Welke 1969b). This value is intermediate between the $^{87}Sr/^{86}Sr$ ratios of rhyolites on Iceland and the granitic rocks on Skye reported by Moorbath and Bell (1965a). In addition, the isotope ratios of Pb of the Rockall granite define a point that lies close to the Pb-Pb mixing line between mantle-derived basalt on Skye and Precambrian basement rocks (Moorbath and Welke 1969a).

These isotopic data support the conclusion of Moorbath and Welke (1969b) that the Rockall granite has continental affinities. In other words, the magma from which it formed assimilated substantial amounts of granitic rocks of Precambrian age. The location of the Rockall Bank at 60 Ma was much closer to the present coast of East Greenland than it was to northwest Scotland. Therefore, the Rockall granite is related to the alkali-rich intrusives of Greenland rather than to the Tertiary granites of the Inner Hebrides of Scotland.

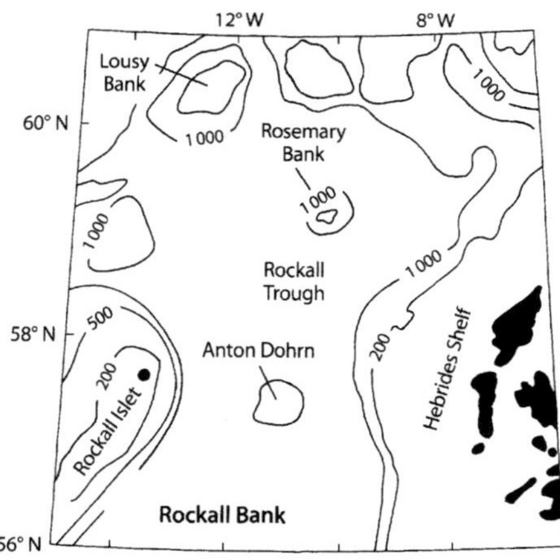

Fig. 5.17. Location of Rockall Islet on Rockall Bank in the North Atlantic Ocean west of the Outer Hebrides Islands of northwest Scotland (*Source:* adapted from Morton et al. 1995)

The Late Cretaceous age of the basalt on Rosemary Bank was later confirmed by Morton et al. (1995) based on biostratigraphic evidence from a drill core. The average isotope ratios of Sr, Nd, and Pb of three basalt specimens recovered in this core are: initial $^{87}Sr/^{86}Sr$ = 0.70296 ±0.00002 at 60 Ma, $^{143}Nd/^{144}Nd$ = 0.513054 ±0.00001, and $^{206}Pb/^{204}Pb$ = 18.448 ±0.120. These values are similar to those of plume MORBs extruded along the present Mid-Atlantic Ridge (Sect. 2.2.1) and therefore permit the conclusion that the basalt of Rosemary Bank originated from the head of the Iceland Plume, which was located along the present east coast of Greenland at about 60 Ma (Fig. 5.9).

5.4.5 Antrim Plateau, Northern Ireland

The Tertiary basalts of the Antrim Plateau of Northern Ireland in Fig. 5.18 are part of the British Igneous Province (White and McKenzie 1989). The basalt lavas were extruded at about 60 Ma through fissures in the continental crust, which is about 30 km thick. The upper part of the crust is composed of Mesozoic carbonates and Paleozoic sediments underlain by late Precambrian metasedimentary rocks (Dalradian) and by Caledonian granites.

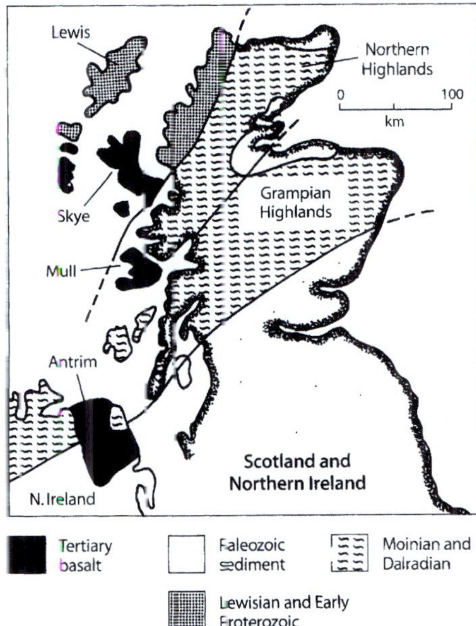

Fig. 5.18. The Antrim Plateau of Northern Ireland in relation to the Tertiary volcanic rocks of the Inner Hebrides Islands of northwest Scotland (*Source:* adapted from Barrat and Nesbitt 1996)

The Tertiary lavas flows of the Antrim Plateau (about 300 m) have been subdivided into three formations (Old 1975; Barrat and Nesbitt 1996):

- Upper Basalt
- Middle Series, including the Causeway tholeiite Member and the Tardree rhyolite complex
- Lower Basalt

The volcanic activity was episodic and occurred between 62 and 58 Ma (Mussett 1988). These dates are in good agreement with Rb-Sr dates obtained by Meighan et al. (1988) for the central complexes of the Mourne Mountains, the Slieve Gullion complex, and with K-Ar dates of Tertiary dolerites in West Connacht (Mitchell and Mohr 1986).

The origin of the basalts of the Antrim Plateau has been the subject of numerous studies based on petrographic, geochemical, and isotopic data (Gamble 1979; Moorbath and Thompson 1980; Kitchen 1984; Thompson et al. 1986; Wallace et al. 1994; Kerr et al. 1995; Barrat and Nesbitt 1996). The consensus of opinions is that the magmas that produced the majority of the basalt layers of the Antrim Plateau assimilated rocks of the Dalradian Supergroup (Late Proterozoic to Cambrian) prior to their eruption at the surface. However, it is not clear whether they also assimilated Proterozoic basement rocks that have been identified on the islands of Islay and Inishtrahull by Marcantonio et al. (1988), Muir et al. (1992), and Dickin and Bowes (1991). These Proterozoic basement rocks (1.7 to 1.9 Ga) appear to underlie the rocks of the Dalradian Supergroup in Northern Ireland (Daly et al. 1991). The composition of the lower continental crust of the British Igneous Province has been investigated by studies of xenoliths included in the Tertiary lava flows (Upton et al. 1983; Dickin and Bowes 1991; Halliday et al. 1993). The results suggest that the deep crustal rocks of Northern Ireland differ from the Lewisian gneisses that underlie the Inner Hebrides of northwest Scotland (Wallace et al. 1994).

The Middle Series represents an interval of quiescence of sufficient duration to permit the formation of laterites by the chemical weathering of basalt. The lava flows that were erupted locally during this interval include the rhyolites and obsidians of the Tardree rhyolite in the central region of the Antrim Plateau (Meighan et al. 1984). In addition, up to seven quartz tholeiite flows that form the Giant's Causeway were extruded along the northern coast of Antrim. These flows are unusually thick (exceeding 30 m in some cases) and have well-developed columnar jointing. The silica concentrations of the Causeway flows range from 49 to 55% indicating that some quartz andesites occur in this sequence (Wallace et al. 1994).

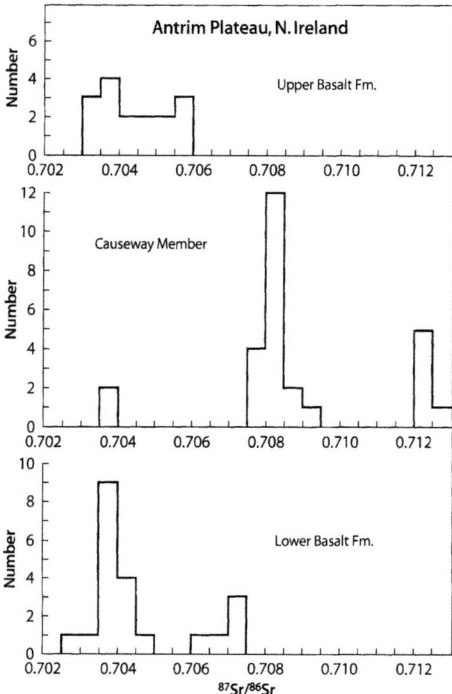

Fig. 5.19. Range of ^{87}Sr/^{86}Sr ratios of the Upper Basalt, Causeway Member of the Middle Basalt, and the Lower Basalt of the Tertiary Antrim Plateau, Northern Ireland (*Sources:* Wallace et al. 1994; Barrat and Nesbitt 1996)

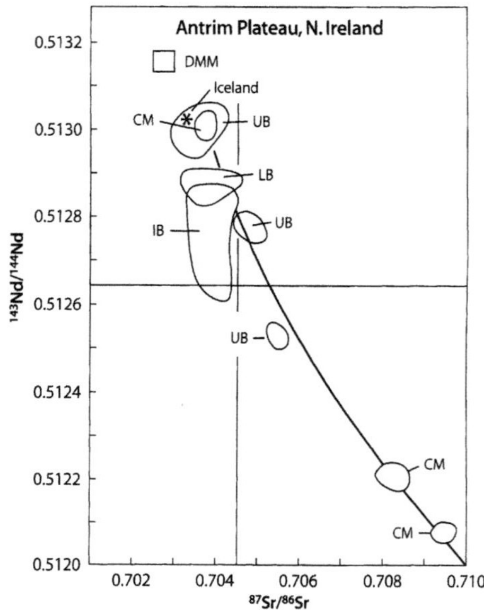

Fig. 5.20. Isotope ratios of Sr and Nd of the Tertiary basalt of the Antrim Plateau, Northern Ireland: *UB* = Upper Basalt Formation; *CM* = Causeway Member of the Middle Basalt Formation; *LB* = Lower Basalt Formation; *IB* = Intrusive basalt. The star identifies the Iceland Plume, which was the source of the basalt magmas of the North Atlantic igneous province (*Source:* Barrat and Nesbitt 1996; Hémond et al. 1993; Hart 1988)

The initial ^{87}Sr/^{86}Sr ratios of the basalt flows of the Antrim Plateau at 60 Ma vary widely from 0.702911 (Lower Basalt Formation) to 0.715292 (Causeway tholeiite) relative to 0.71025 for NBS 987 (Wallace et al. 1994). Meighan et al. (1988) reported 0.707210 for the initial ^{87}Sr/^{86}Sr ratio of one specimen of the Tardree rhyolite (Rb/Sr = 3.7840). The data of Wallace et al. (1994) and Barrat and Nesbitt (1996) in Fig. 5.19 indicate that the ^{87}Sr/^{86}Sr ratios of the Upper and Lower Basalt Formations vary primarily between 0.7030 and 0.7060, whereas most of the ^{87}Sr/^{86}Sr ratios of the Causeway tholeiite lie between 0.7075 and 0.7095, with some exceptionally high values between 0.7120 and 0.7130. The highest value (0.715292) reported by Wallace et al. (1994) is not shown in Fig. 5.19. The elevated and wide-ranging initial ^{87}Sr/^{86}Sr ratios of the basalt flows of the Causeway Member are evidence that the magmas from which they formed assimilated rocks of the continental crust of Northern Ireland.

The evidence in favor of crustal contamination of the magmas that formed the Antrim Plateau is confirmed by the wide range of ^{143}Nd/^{144}Nd ratios and

their inverse correlation with the ^{87}Sr/^{86}Sr ratios in Fig. 5.20. Barrat and Nesbitt (1996) concluded that the mantle-derived basalt magmas of the Causeway tholeiite Member assimilated primarily Dalradian metasedimentary rocks which are characterized by the following geochemical and isotopic parameters: Sr = 244 (81–515) ppm, ^{87}Sr/^{86}Sr (at 60 Ma) = 0.7258 (0.7178–0.7663), Nd = 35 (7–65) ppm, and ^{143}Nd/^{144}Nd (60 Ma) = 0.51159 (0.51116 –0.51200). However, other kinds of crustal rocks probably also contributed to the contamination of the Causeway tholeiite.

The isotope ratios of Sr and Nd of some of the basalt flows in the Upper Basalt Formation and in the Causeway Member in Fig. 5.20 are indistinguishable from the average ^{87}Sr/^{86}Sr and ^{143}Nd/^{144}Nd ratios of basalts on Iceland for which Hémond et al. (1993) reported average values of 0.703318 ±0.00003 and 0.51302 ±0.00001, respectively, relative to 0.71025 for NBS 987 and 0.511837 for the La Jolla Nd standard. Therefore, the lavas of the Antrim Plateau originated from the Iceland Plume. In some cases, the magmas assimilated crustal rocks while in transit to the surface or while they differentiated in crustal magma chambers (e.g. the Causeway flows). The extent of crustal contamination of most of the flows in the Upper and Lower Basalt Formations is small. Their isotope ratios of Sr and Nd were affected by differences in the magma sources, which included the heterogeneous assemblage of rocks in the plume head as well as by melting of previously altered rocks of the lithospheric mantle.

5.4.6 Skaergaard Intrusion, East Greenland

The early Tertiary volcanic and plutonic rocks of East Greenland in Fig. 5.9 occur primarily in the vicinity of Scoresby Sund (also known as Ittoqqortoormiitt) at about 70° N and include not only the Skaergaard Intrusion in Fig. 5.21 but also the basaltic lava flows at Mikis Fjord (Carter et al. 1979; Holm 1988; Hansen and Nielsen 1999), the Kangerdlugssuaq alkaline intrusion (Pankhurst et al. 1976), and the Nualik complex at Ersingerseq (Rex et al. 1979). In addition, Brown et al. (1977b) described the late Tertiary Kialineq complex (about 35 Ma) and Rex et al. (1979) dated several alkali-rich intrusives exposed in Kong Oscar Fjord at 72° N. Seismic surveys reviewed by White and McKenzie (1989) reveal the presence of voluminous submarine basalt flows off the coast of East Greenland from its southern tip north to about 78° N. The basalt plateau located onshore at Mikis Fjord (Fig. 5.21) reaches a thickness of 6 to 7 km but thins landward, indicating that the source of the eruptions was located off the present coast in the Greenland Sea.

The Skaergaard Intrusion (outcrop area about 50 km²) was originally described by Wager and Deer (1939), Wager (1953, 1960), Wager and Brown (1969), and Chayes (1970). These studies have been continued primarily by Professor A. R. McBirney of the University of Oregon (e.g. McBirney 1975, 1979, 1989b, 1995, 1996; McBirney and Noyes 1979; McBirney and Nicolas 1997; Boudreau and McBirney 1997; Sonnenthal and McBirney 1998) as well as by Naslund (1984), Hunter and Sparks (1987), and Hoover (1989). The importance of the Skaergaard Intrusion to our understanding of magma differentiation by fractional crystallization in a closed magma chamber was recognized by the publication of a Commemorative Issue of the "Journal of Petrology" (vol. 30, no. 2, 1989).

The very detailed studies of Wager and Deer (1939) demonstrated that the Skaergaard Intrusion has the shape of a steep-sided funnel whose diameter decreases with depth and whose axis is inclined to the south at about 45°. The basalt magma that originally filled this chamber differentiated as it cooled to form a sequence of layered gabbros.

The resulting stratigraphy of the Skaergaard Intrusion includes the following units from top to bottom:

- Marginal Border Group: is composed of fine grained gabbros within about 30 m of the contact grading into coarse gabbro with flow texture parallel to the walls of the original magma chamber.
- Upper Border Group: consists of late-stage differentiates including hedenbergite granophyre and acid granophyre.
- Layered Series: includes the main exposed mass of the intrusion.
- Hidden Layered Series: comprises about 60% of the total thickness of the Skaergaard Intrusion which underlies the Layered Series but is not exposed.

The layering of the rocks in the Layered Series arises from variations in the proportions of the principal minerals (plagioclase, clinopyroxene, orthopyro-

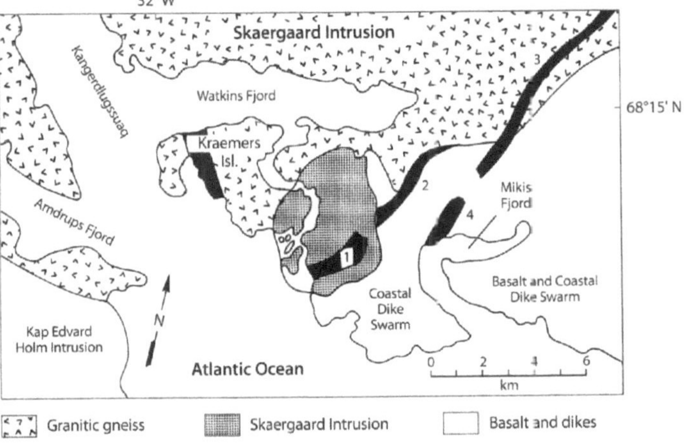

Fig. 5.21.
Map of the Kangerdlugssuaq area of East Greenland including the Skaergaard Intrusion, the East Greenland basalt plateau, and related dike swarm. The macrodikes and sills (*black*) are identified by number. *1.* Basistoppen Sill; *2.* Vandfaldsdalen macrodike; *3.* Soedalen segment; *4.* Eskimonaes segment (*Sources:* adapted from Taylor and Forester 1979; with modification of the macrodikes from Blichert-Toft et al. 1992. The direction of north is approximate)

xene, and olivine) as well as from "cryptic layering" caused by progressive changes in the chemical compositions of each of the principal minerals, which become enriched in the low-temperature end-members (i.e. Fe and Na). In addition, there are abrupt changes in the mineral assemblages, because Mg-rich olivine and orthopyroxene drop out as other minerals such as magnetite, iron-rich apatite, and quartz appear in the rocks. The entire sequence of rocks in the Skaergaard Intrusion formed by fractional crystallization of a tholeiite basalt magma and by the sedimentation of crystals on the floor of the magma chamber. The crystals, formed near the top of the magma chamber, were transported to the floor of the magma chamber by convection currents and by gravitational settling. The chemical compositions of the different rock types contained within the Skaergaard Intrusion vary widely (e.g. SiO_2 = 44.81 to 75.03%; FeO = 0.58 to 22.89‰; MgO = 0.15 to 9.61%; CaO = 0.69 to 11.29%; Na_2O = 2.44 to 4.24%; and K_2O = 0.14 to 3.85%) (Wager and Deer 1939).

The work of Wager and Mitchell (1951) originally supported the view that the concentrations of certain trace elements in the Skaergaard Intrusion depend on the minerals that make up the rocks and on the chemical composition of the residual magma from which the minerals crystallized. Their data showed that the Sr concentration initially increased from about 100 to 600 ppm until more than 50% of the magma had crystallized and then declined to about 300 ppm, when less than about 1% of the magma remained. After an initial increase to more than 400 ppm, the Sr concentration of the late-stage granophyres decreased to near 0 ppm in the acid granophyre (SiO_2 = 75.03%, Na_2O = 4.24%, K_2O = 3.85%).

Subsequent work by others (e.g. Paster et al. 1974; Naslund 1984; Hoover 1989) has greatly increased the number of available trace-element analyses and has forced a reconsideration of the attractively simple model proposed by Wager and Mitchell (1951). Recent work by McBirney (1998) indicates that the concentrations of trace elements (e.g. Ni, Cr, and V) in general decrease up-section with progressive crystallization of the magma but are not well correlated with the modal mineral compositions of the rocks. Therefore, the trace-element concentrations of the minerals are not governed by crystal-liquid partition coefficients (Sect. 1.9) but appear to reflect the mobility of the trace elements during metasomatic alteration of the rocks.

The presence of gold in the rocks of the Skaergaard Intrusion was first detected by Vincent and Crocket (1960), who reported concentrations ranging from 3.5 ppb for gabbro of the Marginal Border Group to 73 ppb in a late-stage granophyre. They also noted that gold is strongly concentrated in copper sulfide minerals. More detailed studies later showed that sulfide-rich rocks in the Skaergaard Intrusion contain mineable concentrations of gold (Bird et al. 1986, 1988).

The magma of the Skaergaard Intrusion was emplaced during the main stage of the eruption of flood basalt in East Greenland and crystallized at 55.65 ±0.30 Ma (Hirschmann et al. 1997; Brooks and Gleadow 1977). The magma presumably originated form the head of the Iceland Plume, which at that time was located near the present site of the Skaergaard Intrusion. The crustal rocks into which the magma was intruded consist of Precambrian granitic gneisses overlain by early Tertiary flood basalt which are cut by the basalt dikes of the Coastal Dike Swarm. These dikes were formed at about 58 Ma prior to the emplacement of the Skaergaard magma and are aligned parallel to the present coast. The spacing of the dikes increases inland, suggesting that they represent an early stage of the opening of the North Atlantic Ocean. After the emplacement of the Skaergaard Intrusion, a second set of basalt dikes formed between 36 and 41 Ma parallel to the preceding dike swarm. All of the basalt dikes and the lava flows they intruded are intensely jointed, and the older dikes contain evidence of hydrothermal alteration by the presence of chlorite, epidote, actinolite, calcite, prehnite, and quartz.

Taylor and Forester (1979) demonstrated that the oxygen in the basalt flows and dikes was strongly depleted in ^{18}O such that their present $\delta^{18}O$ values decrease in the direction of the coast from about +4 to –1‰. They attributed this phenomenon to the interaction of circulating heated groundwater (consisting of local meteoric water) with the basalt flows and dikes. Since most unaltered basalts elsewhere in the world have $\delta^{18}O$ values of about +5.7‰, Taylor and Forester (1979) concluded that all of the basalts of the East Greenland Plateau were hydrothermally altered.

The $\delta^{18}O$ values of plagioclase in the gabbros of the Skaergaard Intrusion itself also decrease up-section from about +6.4‰ in the Lower Zone of the Layered Series to –2.4‰ in the Upper Border Group (Taylor and Epstein 1963; Taylor and Forester 1979). Clinopyroxene, which is more resistant to alteration of the oxygen isotope composition, was also depleted in ^{18}O and had its $\delta^{18}O$ values lowered from +5.2 to +3.5‰. Taylor and Forester (1979) estimated from their data that the isotope composition of hydrogen and oxygen in the circulating heated groundwater was $\delta D \approx$ –100‰ and $\delta^{18}O \approx$ –14‰, respectively.

The hydrothermal circulation was initiated soon after the emplacement of the Skaergaard magma and caused oxygen-isotope exchange with the rocks at high temperatures exceeding 400 to 500 °C. The large-scale circulation of groundwater through the rocks of the Skaergaard Intrusion was modeled quantitatively by Norton and Taylor (1979) based on assumed permeabilities on the order of 10^{-13} cm^2 for the intrusion, 10^{-11} cm^2 for the basalt plateau, and 10^{-16} cm^2 for the Precambrian gneisses. The hydrothermal alteration of layered

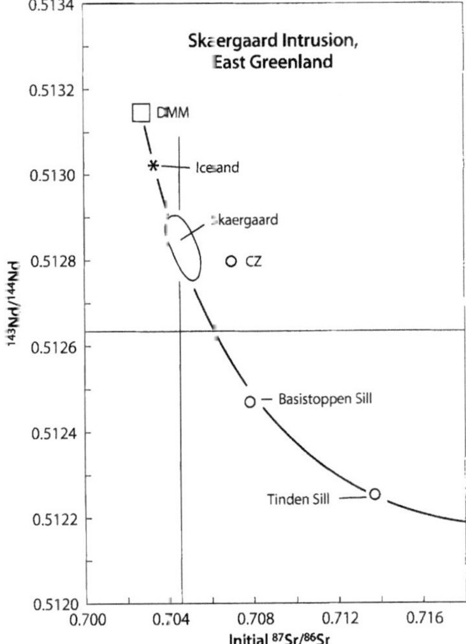

Fig. 5.22. Isotope ratios of Sr and Nd in rocks of the Skaergaard Intrusion of East Greenland. The rocks of the Layered Series and of the Upper Border Zone define a small data field, whose position on the diagram is distinctly different from the average isotope ratios of basalts on Iceland. The rocks of the Basistoppen and Tinden Sills have been contaminated with Sr and Nd derived from the Late Archean gneisses of this area. CZ is the chilled zone of the Skaergaard Intrusion. The $^{143}Nd/^{144}Nd$ ratios were renormalized to a value of 0.7219 for the $^{146}Nd/^{144}Nd$ ratio (Sources: Stewart and DePaolo 1991; Hémond et al. 1993; Hart 1988)

gabbros in East Greenland was also discussed by Bird et al. (1988).

The $^{87}Sr/^{86}Sr$ ratios of rocks in the Skaergaard Intrusion were first measured by Hamilton (1963), who reported that the rocks of the Tinden granophyre sill (Upper Border Group) are enriched in radiogenic ^{87}Sr compared to the main body of layered gabbros. The initial $^{87}Sr/^{86}Sr$ ratios of the Skaergaard gabbros analyzed by Leeman and Dasch (1978) have a mean of 0.70426 ±0.00029 ($2\bar{\sigma}$, $N = 3$) compared to 0.70364 for five samples of Tertiary basalt (Carter et al. 1979) from East Greenland, all relative to 0.7080 for E&A. The initial $^{87}Sr/^{86}Sr$ ratios of concordant melanophyres (0.7043) are similar to those of the gabbros, indicating that the melanophyres may be differentiates of the Skaergaard magma. However, the transgressive acid granophyres (e.g. Tinden and Sydtoppen Sills) and the felsic series of the Vandfalcsdalen Macrodike (White et al. 1989; Hirschmann 1992) have elevated initial $^{87}Sr/^{86}Sr$ ratios

ranging from 0.7067 to 0.7287 and presumably formed by partial melting of the Archean gneisses that constitute the basement rocks of this area (2.86 ±0.04 Ga, Kays et al. 1989). The $^{87}Sr/^{86}Sr$ ratios of these gneisses on Kraemers Island (Fig. 5.21) range from 0.7105 to 0.7158. However, Pankhurst et al. (1976) reported higher $^{87}Sr/^{86}Sr$ ratios for gneiss, metasediment, and amphibolite from the Kangerdlugssuaq area with a weighted average of 0.7552 and an average Sr content of about 160 ppm.

The isotope ratios of Sr and Nd of gabbros in the Layered Series and upper Border Zone in Fig. 5.22 vary within narrow limits, indicating that the main body of Skaergaard magma assimilated only small amounts of the Late Archean gneisses of the area (Stewart and DePaolo 1991). In contrast, the magmas of the Basistoppen and Tinden Sills analyzed by Stewart and DePaolo (1991) were significantly contaminated by assimilation of the gneisses in agreement with the early work of Hamilton (1963).

The contamination of the macrodikes was especially well demonstrated by Blichert-Toft et al. (1992) who analyzed samples of gabbros, diabase, hybrid rocks, granophyre, and gneisses of the Soedalen segment of the Mikis Fjord Macrodike and from the Vandfaldsdalen Dike (Fig. 5.21). Their data in Fig. 5.23 demonstrate that the isotope ratios of Sr and Nd of hybrid rocks of these dikes range from values that are close to those of the gabbros of the Skaergaard Intrusion (Stewart and DePaolo 1991; Fig. 5.21) to those of quartz-feldspar gneisses of the Precambrian basement rocks. The initial $^{87}Sr/^{86}Sr$ and $^{143}Nd/^{144}Nd$ ratios of granophyres of the Soedalen segment are similar to the isotope ratios of the gneisses, suggesting that the magmas of these granophyres formed by melting of the local granitic basement rocks.

Blichert-Toft et al. (1992) concluded that the isotope ratios as well as the chemical compositions of the hybrid rocks in the two dikes are not compatible with mixing of basaltic and felsic magmas nor with assimilation and simultaneous fractional crystallization of basalt magma (AFC). Instead, they proposed that the hybrid rocks had formed by interdiffusion between the two silicate liquids as the less dense granophyric magma rose toward the top of the respective magma chambers. The importance of diffusion during mingling of melts of differing chemical and isotope composition has been demonstrated by Rosing et al. (1989) for Tertiary magmas in East Greenland, by Grove et al. (1988) at Medicine Lake Volcano in California, and by Frost and Mahood (1987) in the Lamarck granodiorite of the Sierra Nevada Mountains, California. In addition, the theoretical and experimental basis for interdiffusion of specific elements between two silicate melts has been presented by Sparks and Marshall (1986), Lesher (1986, 1990), and Koyaguchi (1989).

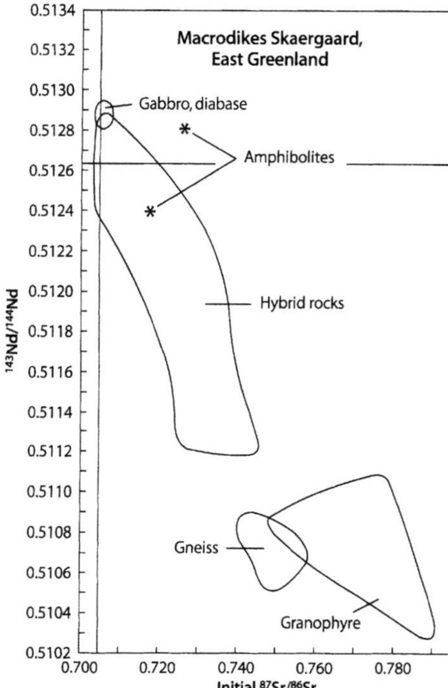

Fig. 5.23. Isotope ratios of Sr and Nd of gabbro and diabase, hybrid rocks, and granophyres of the Soedalen segment (Mikis Fjord) and the Vandfaldsdalen Macrodike, Skaergaard Intrusion, East Greenland (Figure 5.21). In addition, the diagram includes data for amphibolite and quartz-feldspar gneiss of the Precambrian basement complex (*Sources:* Blichert-Toft et al. 1992 and Stewart and DePaolo 1991)

5.4.7 Kangerdlugssuaq Syenite Complex, East Greenland

The Kangerdlugssuaq alkaline intrusion is located only about 20 km south of the Skaergaard Intrusion along the south side of Kangerdlugssuaq Fjord (Wager 1965; Kempe et al. 1970). The intrusion has an outcrop area of about 800 km^2 and is concentrically zoned with foyaite and pulaskite in the center and syenite and quartz syenite on the periphery. This intrusion represents the late stage of Tertiary igneous activity in East Greenland. The age of the intrusion is 50 ±1 Ma based on age determinations by the whole-rock Rb-Sr method (Pankhurst et al. 1976). Small masses of intrusive rock also occur both within the body of the alkaline intrusion (earlier Tertiary intrusions) and along the contact with the basement rocks (Snout Series). In addition, tholeiite and alkali basalts of Tertiary age overlie the basement gneisses and occur as inclusions

in the alkaline intrusion. The basalts as well as the basement gneisses (Kays et al. 1989) are cut by the Coastal Dike Swarm whose petrogenesis has been discussed by Brooks and Nielsen (1978) based on samples from Kraemers Island.

The initial $^{87}Sr/^{86}Sr$ ratios of the alkaline intrusion, reported by Pankhurst et al. (1978), range from 0.70434 (foidite, center of the intrusion) to 0.70949 (quartz nordmarkite, near the margin) The rocks of the Snout Series have about the same range of initial $^{87}Sr/^{86}Sr$ ratios (0.70427 to 0.71000), but tholeiite basalts from Scoresby Sund reveal their derivation from sources in the mantle with initial $^{87}Sr/^{86}Sr$ ratios between 0.70313 to 0.70323. Pankhurst et al. (1978) concluded that the magma of the Kangerdlugssuaq alkaline intrusion had an initial $^{87}Sr/^{86}Sr$ ratio of 0.7043 similar to the magma that had previously formed the layered gabbros of the nearby Skaergaard Intrusion. The elevated initial $^{87}Sr/^{86}Sr$ ratios of the quartz syenite and syenite are attributable to contamination of these rocks by Sr derived from the Archean gneisses.

5.4.8 Tertiary Volcanics of Northeast Greenland

Remnants of the Tertiary basalt plateau of East Greenland occur at the Hold with Hope Peninsula and on Wollaston Forland located north of Scoresby Sund at about 74° N latitude (Hansen and Nielsen 1999). The lava plateau in this area has a thickness of about 1 km and occupies an area of about 16 000 km^2. The basalt flows (~51 Ma) at Hold with Hope include the Lower Series which is composed of quartz tholeiites of uniform chemical composition. These flows are conformably overlain by the lavas of the Upper Series, which consists of a highly diverse suite of volcanic rocks ranging in composition from porphyritic quartz tholeiites to basanites. The volcanic rocks of the Upper Series are absent on the Wollaston Forland. Field evidence indicates that the basalt flows of the Lower Series were erupted through fissures located to the east of the present coast of Greenland, whereas the lavas of the Upper Series were erupted by a shield volcano (Upton et al. 1984; Thirlwall et al. 1994).

The isotope ratios of Sr and Nd of the tholeiites of the Lower Series in Fig. 5.24 are restricted to a narrow range and closely resemble those of the basalt on Iceland. Therefore, Holm (1988) and Thirlwall et al. (1994) concluded that the lavas of the Lower Series at Hold with Hope and Wollaston Forland originated from the head of the Iceland Plume during the opening of the North Atlantic Ocean. The rocks of the Upper Series in Fig. 5.24 have a wide range of initial $^{87}Sr/^{86}Sr$ and $^{143}Nd/^{144}Nd$ ratios, caused by extensive assimilation of crustal rocks by magmas whose isotope compositions prior to contamination were similar to those of the Lower Series.

The Tertiary lavas at Hold with Hope and Wollaston Forland are underlain by a thick section of late Paleozoic and Mesozoic sedimentary rocks that were deposited on a basement of Caledonian granitic rocks. The only sample of the underlying crust available for analysis by Thirlwall et al. (1994) is a granulite xenolith (Fig. 5.24) whose isotope ratios of Sr, Nd, and Pb are 0.71562, 0.512242, and 17.745, respectively. This specimen appears to be a suitable contaminant of the magmas of the Upper Series.

Volcanic rocks of Late Cretaceous to Paleocene age (63.7 ±3 Ma) have also been found at Kap Washington and Kap Kane in J. V. Jensen Land at about 33°30' S and 38 to 40° W along the north coast of Greenland (Larsen 1982). The rocks consist of basalt and trachyte, and include rhyolite flows and ignimbrites. The average initial $^{87}Sr/^{86}Sr$ ratio of rhyolites is 0.7070 ±0.0004 based on a whole-rock Rb-Sr isochron. The Rb concentrations of these rocks range from 20 to 259 ppm, whereas Sr varies between 5 and 225 ppm. Larsen (1982) cited suggestions in the literature that the volcanic activity in northern Greenland is related to the opening of Makarov Basin in the Arctic Ocean in Late Cretaceous time. If these suggestions are correct, then the Kap Washington Group of volcanic rocks are not related to the Iceland Plume and do not belong to the North Atlantic igneous province.

5.4.9 Tertiary Volcanic Rocks of West Greenland

The Tertiary basalt lavas on the Svartenhuk Peninsula, on Ubekendt and Disko Islands, and on the Nugssuaq Peninsula of West Greenland in Fig. 5.25 are part of the North Atlantic petrologic province (Fig. 5.9). In addition, the Dyer Peninsula of the east coast of Baffin

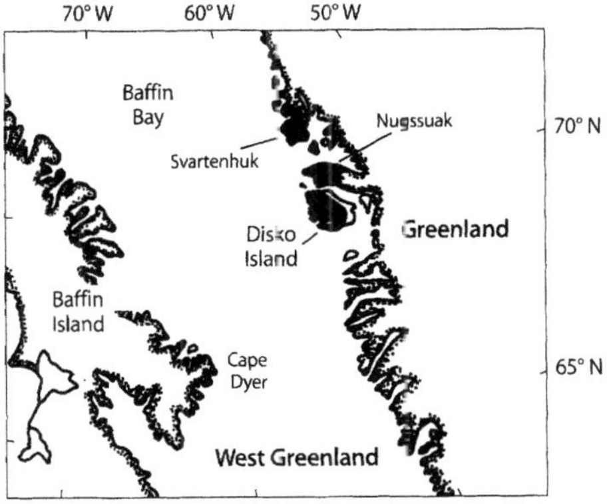

Fig. 5.25. Map of Baffin Bay showing the exposures of Early Tertiary volcanic rocks of West Greenland and on the east coast of Baffin Island. The volcanic and plutonic rocks belong to the North Atlantic petrologic province outlined in Fig. 5.9 (*Source:* adapted from O'Nions and Clarke 1972)

Fig. 5.24. Isotope ratios of Sr and Nd of Early Tertiary lava flows of Hold with Hope Peninsula and Wollaston Forland on the coast of East Greenland at about 76° N. The isotope ratios of quartz tholeiites of the Lower Series resemble closely those of the basalt on Iceland, whereas the lavas of the Upper Series at Hold with Hope have a wide range of isotope ratios of Sr and Nd overlapping those of the Skaergaard Intrusion. These data indicate that the magmas of the Upper Series assimilated granitic basement rocks represented by a crustal xenolith (*Sources:* Thirlwall et al. 1994; Stewart and DePaolo 1991; Hémond et al. 1988; Hart 1988)

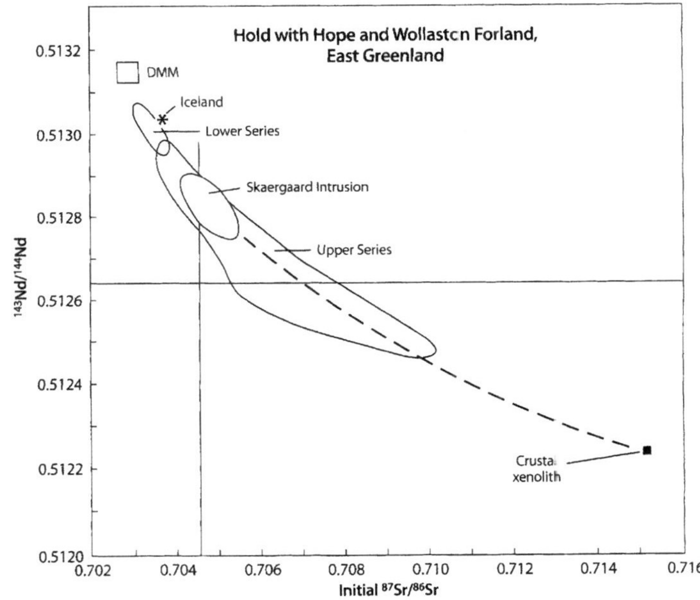

Island also contains Tertiary lava flows that are part of the North Atlantic suite. The volcanic and plutonic rocks at these sites originated from the Iceland Plume in the course of rifting of the continental crust and the resulting opening of the North Atlantic Ocean.

The $^{87}Sr/^{86}Sr$ ratios of basalt flows on the Svartenhuk Peninsula of West Greenland and at Cape Dyer on Baffin Island (Clarke 1970; O'Nions and Clarke 1972; Carter et al. 1979; Holm et al. 1993) in Fig. 5.26 range from about 0.7030 to 0.7040 relative to 0.7080 for E&A. The low and relatively constant $^{87}Sr/^{86}Sr$ ratios of the lava flows at Svartenhuk and Cape Dyer imply that the basalt magmas at these locations were not contaminated with Sr from the rocks of the continental crust. Therefore, Carter et al. (1979) concluded that crustal contamination of basalt magmas in the North Atlantic Igneous Province was a local phenomenon whose intensity varied with the temperature of the rocks in the lower crust.

The volcanic rocks on Disko Island (55 Ma) are of special interest because they assimilated varying amounts of carbonaceous shale and sandstone of Mesozoic age and contain an unusual assemblage of minerals, including native iron, armalcolite ($FeMgTi_2O_5$), wustite (FeO), and troilite (FeS) (Pedersen 1981; Goodrich 1984; Ulff-Möller 1990).

The initial $^{87}Sr/^{86}Sr$ ratios of the volcanic rocks on Disko Island (basalt, andesite, dacite) in Fig. 5.27 range from 0.7033 to 0.7152 relative to 0.7080 for E&A (Pedersen and Pedersen 1987). Two rhyolite specimens have even higher initial $^{87}Sr/^{86}Sr$ ratios of 0.7176 and 0.7203,

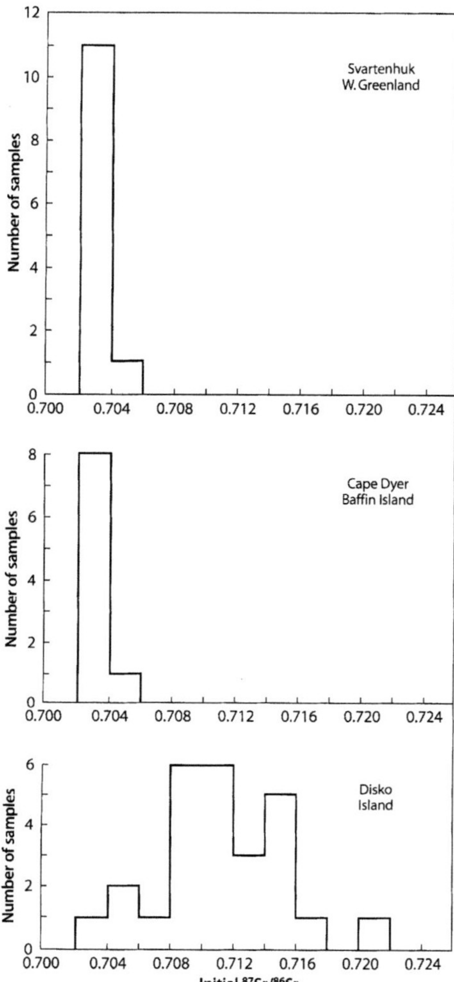

Fig. 5.26. Range of initial $^{87}Sr/^{86}Sr$ ratios of Early Tertiary lava flows on the Svartenhuk Peninsula, Cape Dyer, and Disko Island in the West Greenland sector of the North Atlantic petrologic province. The low initial $^{87}Sr/^{86}Sr$ ratios of basalt at Svartenhuk and Cape Dyer indicate that their magmas did not assimilate basement gneisses. The volcanic rocks on Disko Island are variably contaminated with carbonaceous shale and sandstone of Mesozoic age and contain native iron and other unusual minerals (*Sources:* Pedersen and Pedersen 1987; Carter et al. 1979; O'Nions and Clarke 1972)

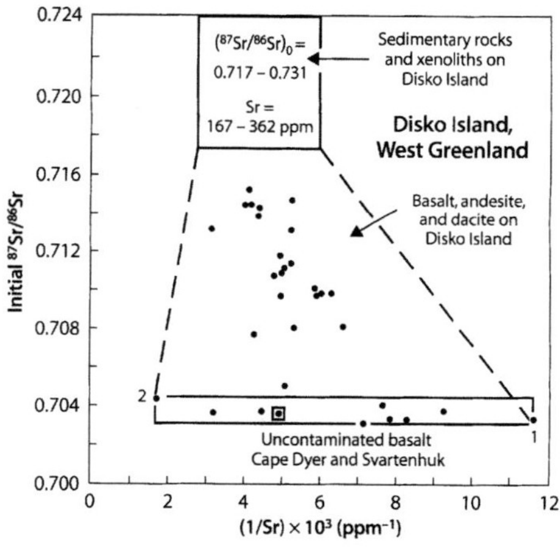

Fig. 5.27. Sr-isotope mixing diagram (without AFC) for the hybrid volcanic rocks on Disko Island of West Greenland. Most of the volcanic rocks on Disko Island scatter in the area between uncontaminated basalt at Svartenhuk and Cape Dyer, and the local Mesozoic to Early Tertiary sandstone and shale. Samples *1* and *2* are apparently uncontaminated basalts on Disko Island. The volcanic rocks on Disko Island are remarkable because of the occurrence of metallic iron and unusual minerals including armalcolite ($Fe, MgTi_2O_5$) wustite (FeO), and troilite (FeS) found also in meteorites and lunar rocks. Specimens of iron-bearing rocks are displayed in the Geological Museum in Copenhagen, Denmark (*Sources:* Carter et al. 1979; Pedersen and Pedersen 1987)

whereas the $^{87}Sr/^{86}Sr$ ratios (at 55 Ma) of sedimentary rocks of Mesozoic to Early Tertiary age are 0.7183 (sandstone) and 0.7269 (shale composite).

The distribution of data points in the Sr-isotope mixing diagram in Fig. 5.27 indicates that most of the basalts, andesites, and dacites on Disko Island originated by contamination of mantle-derived basalt magma ($^{87}Sr/^{86}Sr$ = 0.70359 ±0.00028, Sr = 203 ±105 ppm; $2\bar{\sigma}$, N = 9) with Sr derived from the sedimentary rocks on Disko Island. The primary magmas are represented by the basalt flows at Svartenhuk and Cape Dyer as well as by Specimens 1 and 2 on Disko Island identified on Fig. 5.27. The scatter of data points representing the contaminated volcanic rocks on Disko Island is consistent with the wide range of Sr concentrations of the uncontaminated basalt in West Greenland.

5.4.10 Sarqâta qáqâ Complex of Ubekendt Island

Ubekendt Ejland (Island) is a small island located south of Svartenhuk Peninsula and north of the Nugssuaq Peninsula on the coast of West Greenland (Fig. 5.25). The island consists of a sequence of basaltic and picritic lava flows, including flows of intermediate and felsic composition. The total thickness of these flows exceeds 8 000 m (Beckinsale et al. 1974). The lava flows in the southern part of Ubekendt Island were intruded by several small plutons, which are referred to collectively as the Sarqâta qáqâ complex whose outcrop area is approximately 15 km². The complex consists of a layered mafic intrusion overlain by a sill-like sheet of granite. The exposed part of the layered complex (about 300 m) ranges in composition from eucrites at the base

to diorite near the top. The relationship between the granite and the underlying mafic complex is not clear, although Beckinsale et al. (1974) considered that the intrusion of the granite magma postdated the intrusion of the mafic magma.

The coexistence of granitic and mafic plutonic rocks in the Sarqâta qáqâ complex raises the question whether the granite magma formed by melting of Precambrian basement gneisses (as at Skye; Moorbath and Bell 1965a) or whether the granite is comagmatic with the mafic rocks of the complex. Beckinsale et al. (1974) reported that the acid rocks define an errorchron in Fig. 5.28, which extrapolates to the data points of the mafic rocks. The slope of this errorchron yields a date of 67 ±5 Ma (recalculated to $\lambda(^{87}Rb)$ = 1.42 × 10^{-11} yr^{-1}). The initial $^{87}Sr/^{86}Sr$ ratio derived from the isochron is 0.7045 ±0.0003 relative to 0.7080 for E&A. This value is indistinguishable from the initial $^{87}Sr/^{86}Sr$ ratio of gabbros and related rocks of the Skaergaard Intrusion for which Stewart and DePaolo (1991) reported 0.70453 ±0.00009 ($2\bar{\sigma}$, N = 21).

Biotite concentrates derived from two gabbros on Ubekendt Island yielded K-Ar dates of 54.0 ±1.5 and 55.4 ±1.5 Ma. In addition, Beckinsale et al. (1974) reported that the $\delta^{18}O$ values of four samples of the Sarqâta granite range from +6.30 to +7.18‰ and are similar to those of two mafic rocks (+5.50 and +5.56‰). The authors concluded that the Sarqâta granite did not form by fusion of Precambrian basement rocks because its initial $^{87}Sr/^{86}Sr$ ratio is characteristic of Sr derived from sources in the mantle. In addition, there is no evidence for assimilation of crustal rocks by mantle-derived magmas on Ubekendt Island or for hydrothermal alteration of the plutonic rocks, because the isotope compositions of oxygen are within the range of $\delta^{18}O$ values observed in mantle-derived rocks occurring elsewhere in the world. Beckinsale et al. (1974) attributed the discordance of the K-Ar biotite dates (55 Ma) and the whole-rock Rb-Sr isochron date (65 Ma) to slow cooling of the intrusive after it crystallized at about 65 Ma. This interpretation of the dates implies that the basalt lavas on Ubekendt Island were extruded in Late Cretaceous time (i.e. prior to 65 Ma).

A question could be raised concerning the Rb-Sr date because it was derived from a combination of data for granites and gabbros, which presupposes a comagmatic relationship between them. However, an unweighted least-squares regression of the granite data alone yields a date of 63.4 Ma and an initial $^{87}Sr/^{86}Sr$ ratio of 0.70477, both of which are similar to the results derived from the combined data. The gabbros and related basalts analyzed by Beckinsale et al. (1974) cannot be used to calculate a Rb-Sr date because these rocks have low Rb/Sr ratios that vary only between narrow limits (0.014 to 0.039). However, these rocks do yield an average initial $^{87}Sr/^{86}Sr$ ratio at 65 Ma of

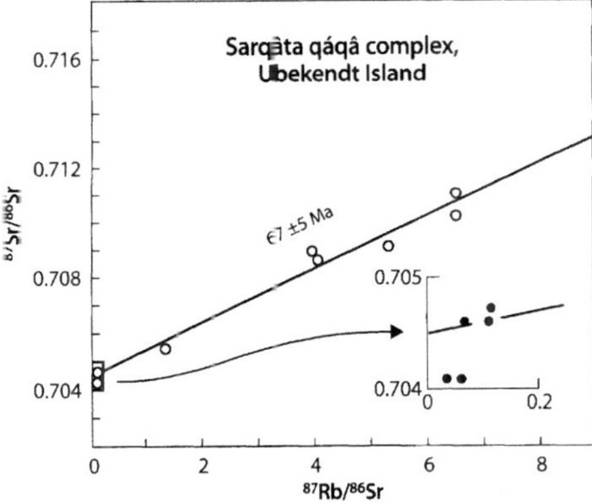

Fig. 5.28. Whole-rock Rb-Sr isochron diagram for granite (*open circles*) and gabbros (*solid circles*) of the Sarqâta qáqâ igneous complex on Ubekendt Island of West Greenland (*Source:* Beckinsale et al. 1974)

0.70434 ±0.00024. Therefore, the age of the Sarqâta complex and its initial $^{87}Sr/^{86}Sr$ ratio are not significantly altered when the granite and gabbro are interpreted separately.

The evidence that the Sarqâta granite on Ubekendt Island is comagmatic with the underlying gabbros sets it apart from the Tertiary granites on the island of Skye, which originated in large part by fusion of the underlying Lewisian gneisses. The data presented by Beckinsale et al. (1974) indicate that the Sarqâta granite is either a product of fractional crystallization of mantle-derived basalt magma, or that it formed by partial melting of basalt or gabbro at the base of the lava plateau.

5.5 Deccan Plateau, India

The flood basalts of the Deccan Plateau in Fig. 5.29 occupy an area of 5×10^5 km^2 in west-central India. The flows were extruded in Late Cretaceous time and reach a thickness of 1.2 km in the Mahabaleshwar District, where 48 flows have been identified. Along the southern border of the plateau only six flows (50 m) overlie Precambrian basement rocks (Lightfoot and Hawkesworth 1988). The basalt layers of the Deccan Plateau were extruded through fissures that formed while the Indian Plate was located above a large mantle plume whose remnant is presently close to Réunion Island at the southern end of the Mascarene Ridge in the Indian Ocean (Richards et al. 1989; Sect. 2.16.3).

According to the reconstruction of the tectonic history by White and McKenzie (1989) in Fig. 5.30, the Deccan basalt flows were extruded during the separation of the Seychelles crustal block from India. The subsequent movement of the Seychelles block caused volcanic activity that resulted in the formation of the Chagos-Laccadive submarine ridge and the Mascarene submarine plateau. The ages of the volcanic rocks on the islands of the Mascarene Plateau decrease in a southerly direction toward Réunion. Similarly, the northward motion of the Indian Plate caused the center of volcanic activity under the Deccan Plateau to shift south. The episode of volcanic activity that produced the Deccan Plateau lasted only a short time. Estimates of its duration reviewed by White and McKenzie (1989) range from <3 to only 0.4 million years.

The lavas of the western Deccan Plateau near Mahabaleshwar (Fig. 5.29) have been subdivided into three subgroups and eleven formations (Table 5.5) based on their chemical compositions and field evidence (Cox and Hawkesworth 1984, 1985; Beane et al. 1986; Lightfoot and Hawkesworth 1988; Peng et al. 1998).

Some of the formations (e.g. the Khandala, Poladpur, Ambenali Formations) have been traced for more than 300 km from their type sections (Subbarao et al.

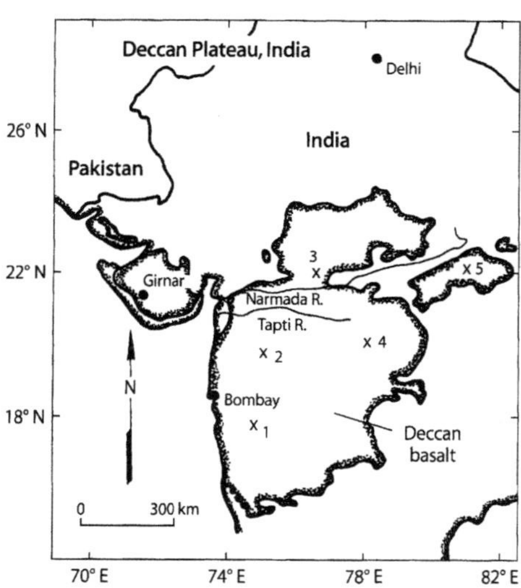

Fig. 5.29. The Deccan Plateau of western India, composed primarily of tholeiite basalt flows and lesser amounts of rhyolite extruded between 60 and 70 Ma. Localities mentioned in the text are identified by number: *1.* Mahabaleshwar; *2.* Igatpuri; *3.* Mhow; *4.* Chikaldara; *5.* Jabalpur (*Sources:* adapted from Paul et al. 1977 and Peng et al.1998)

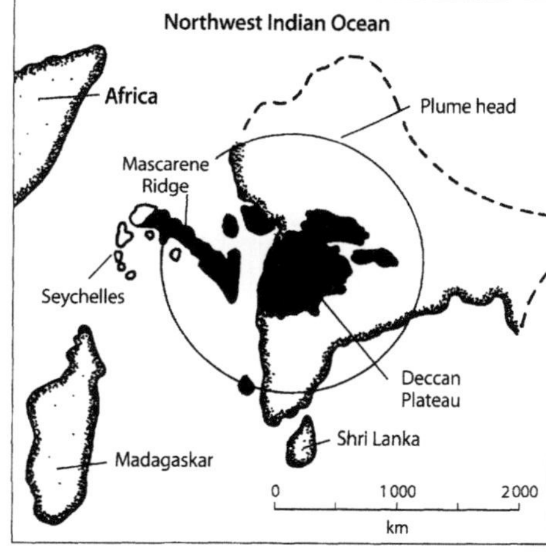

Fig. 5.30. Initial separation of the Seychelles block from India at about 65 Ma as a result of continental rifting caused by a mantle plume, whose area of influence is outlined by the circle. The eruption of tholeiite basalt through a system of crustal fissures led to the formation of the Deccan Plateau and the Mascarene Ridge associated with the Seychelles Islands. The Chagos-Laccadive Ridge formed later as the Indian Plate continued to move north (*Source:* adapted from White and McKenzie 1989)

Table 5.5. Subgroups and formations of the lavas of the western Deccan Plateau near Mahabaleshwar based on their chemical compositions and field evidence (Cox and Hawkesworth 1984, 1985; Beane et al. 1986; Lightfoot and Hawkesworth 1988; Peng et al. 1998)

Subgroup	Formation	Thickness (m)
Wai	Panhala	>175
	Mahabaleshwar	280
	Ambenali	500
	Poladpur	375
Lonavala	Bushe	325
	Khandala	140
Kalsubai	Bhimashankar	140
	Thakurvadi	650
	Neral	100
	Igatpur	
	Jawhar	>700
Total thickness		>3375

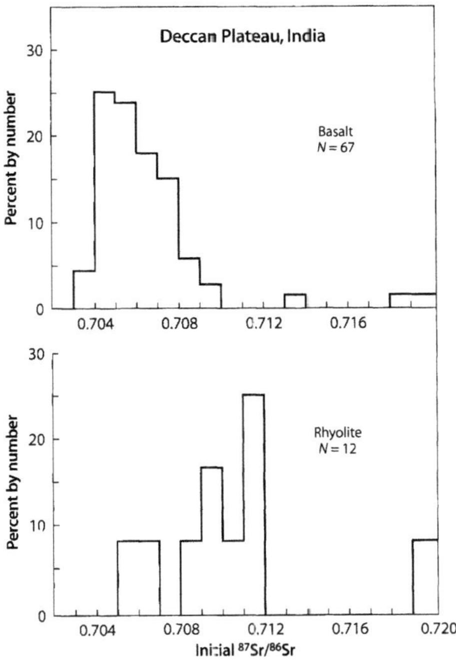

Fig. 5.31. Range of initial ^{87}S-/^{86}Sr ratios of basalt and rhyolite of the Deccan Plateau, India. The $^{87}Sr/^{86}Sr$ ratios of the basalt range widely from 0.7039 to 0.7197, presumably because mantle-derived magmas ($^{87}Sr/^{86}Sr < 0.7039$) assimilated varying amounts of crustal rocks having a range of Sr concentrations and $^{87}Sr/^{86}Sr$ ratios. Rhyolites have higher $^{87}Sr/^{86}Sr$ ratios than the associated tholeiites and formed from magmas that originated as partial melts of mafic igneous and metamorphic rocks in the lower crust (*Sources:* Mahoney et al. 1982, 1985; Lightfoot et al. 1987; Alexander 1981; Paul et al. 1977; Alexander and Paul 1977)

1994; Mitchell and Widdowson 1991). However, the cumulative thickness of >3 385 m is not realized in any part of the present Deccan Plateau.

Age determinations of the Deccan basalt by the K-Ar method have yielded dates between about 70 and 50 Ma (Courtillot et al. 1986), which include the K-T boundary. Therefore, the climatic consequences of the eruption of the Deccan basalt have been cited as a possible cause for the extinction of dinosaurs and other animals and plants at 65 Ma (Officer and Drake 1985). However, Baksi (1987) concluded that the available age determinations do not constrain the time of eruption well enough to link the volcanic activity of the Deccan Plateau to the extinction event at the K-T boundary.

Courtillot et al. (1988) subsequently narrowed the range of dates for the eruption of the Deccan basalt from 69 to 65 Ma, while Duncan and Pyle (1988) reported dates between 68.5 and 66.7 Ma. Precise ^{40}Ar-^{39}Ar dates by Venkatesan et al. (1993) later demonstrated that the basalt flows in the lower 2 km-section of the Deccan Plateau erupted within one million years close to 67 Ma. This date is directly comparable to the ages of tektites on Haiti (65.2 ±0.1 Ma) and to impact melt in the Chicxulub Crater (64.98 ±0.10 Ma), because all of them were determined by the ^{40}Ar-^{39}Ar method using the same standard (MMhb-1) whose age was assumed to be 520.4 Ma (Faure 1986). Therefore, Venkatesan et al. (1993) concluded that the main body of the Deccan basalt (reversely magnetized) was erupted more than one million years before the meteoroid impact at Chicxulub on the Yukatan Peninsula of Mexico and hence before the K-T boundary.

The petrology and chemical composition of selected layers of basalt of the Deccan Plateau have been discussed by many authors, including: Chandrasekharam and Parthasarathy (1978), Alexander (1980a), Sen (1983), Paul et al. (1984), Sethna et al. (1987), Devey and Cox (1987) as well as by authors whose reports were published in Memoir 10 of the Geological Society of India edited by Subbarao (1988). The rocks of the Deccan Plateau consist primarily of tholeiite basalt, but also include differentiated igneous intrusives at Girnar (Paul et al. 1977), as well as rhyolites and trachytes along the Narmada River (Fig. 5.29), on Salsette Island (Lightfoot et al. 1987), and at Pavagarh Hill (Alexander 1980b). In addition, K-rich lavas occur in the northern region of the plateau (Mahoney et al. 1985).

The initial $^{87}Sr/^{86}Sr$ ratios of tholeiites at Mahabaleshwar vary widely from about 0.70391 to 0.71972 relative to 0.71025 for NBS 987 (Mahoney et al. 1982). Nevertheless, the data in Fig. 5.31 indicate that 82% of the analyzed tholeiites have initial $^{87}Sr/^{86}Sr$ ratios between 0.7040 and 0.7080. Devey and Cox (1987) reported that the initial $^{87}Sr/^{86}Sr$ ratios and the Mg numbers of basalt flows in the Mahabaleshwar region are correlated and pointed out, in agreement with Huppert and Sparks (1985), that this relation is opposite to trends observed in other flood basalt provinces where the evolved flows having low Mg numbers have elevated $^{87}Sr/^{86}Sr$ ratios indicating that they are more extensively contaminated. Devey and Cox (1987) concluded from this evidence that the mafic magmas that formed the Deccan Plateau were contaminated during their ascent through fissures rather than while being pooled in crustal magma chambers.

The initial $^{87}Sr/^{86}Sr$ ratios of the rhyolites in Fig. 5.31 at Rajpipla on the Narmada River (Mahoney et al. 1985), on Salsette Island near Bombay (Lightfoot et al. 1987), and on Pavagarh Hill (Alexander 1980b) range from 0.70534 to 0.71912 and are significantly higher on average than those of the tholeiites. Silicic porphyries of the Girnar intrusive complex (Fig. 5.29) have even higher initial $^{87}Sr/^{86}Sr$ ratios ranging up to 0.7281 at 60 Ma (not included in Fig. 5.31). Paul et al. (1977) attributed the origin of these silicic porphyries to partial melting of Precambrian granitic basement rocks and concluded that the gabbro-syenite suite of the Girnar intrusive formed by contamination of basalt magma with rocks of the continental crust.

The rhyolites exposed along the Narmada River (Fig. 5.29) and on Salsette Island have low Sr concentrations (8 to 120 ppm) with a mean of about 32 ppm. The Rb concentrations (193 to 369 ppm) average about 284 ppm. Consequently, the rhyolites have high Rb/Sr ratios (7.7 to 34.1) and their present $^{87}Sr/^{86}Sr$ ratios range up to 0.80067 compared to the average initial $^{87}Sr/^{86}Sr$ ratio of 0.7095. The rhyolites on Salsette Island actually form a Rb-Sr whole-rock isochron yielding a date of 61.5 ±1.9 Ma (Lightfoot et al. 1987). The initial $^{87}Sr/^{86}Sr$ ratios of the rhyolites on Salsette Island are not as high as those of the silicic porphyry of the Girnar complex; they do not correlate with reciprocal Sr concentrations, and intermediate rock types are missing. Therefore, these rhyolites did not form by partial melting of granitic basement rocks (because their initial $^{87}Sr/^{86}Sr$ ratio is too low), nor by contamination of basalt magma, nor by fractional crystallization of basalt magma. Instead, Lightfoot et al. (1987) concluded that the rhyolites on Salsette Island formed by partial melting of mafic rocks in the lower crust or by remelting of basalt deep in the lava pile. In this regard, the petrogenesis of these rhyolites is similar to that of the rhyolites of Iceland (Sect. 2.3.2).

The initial $^{87}Sr/^{86}Sr$ and $^{143}Nd/^{144}Nd$ ratios of tholeiite basalts in the western and northeastern regions of the Deccan Plateau in Fig. 5.32 vary widely and present clear evidence that the basalt magmas assimi-

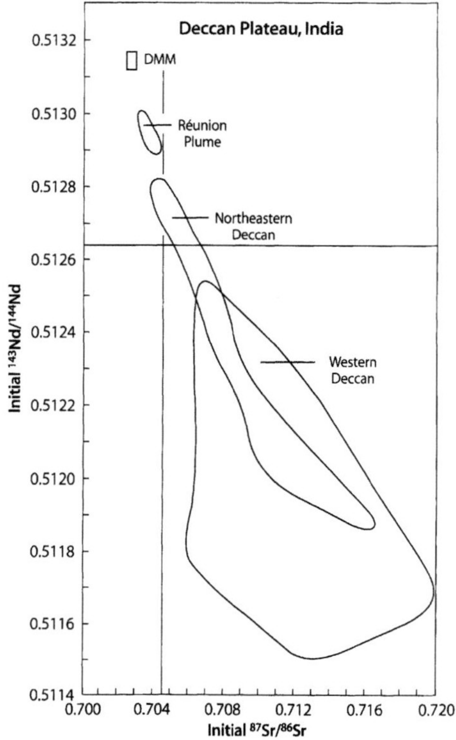

Fig. 5.32. Initial isotope ratios of Sr and Nd of basalt on the western and northeastern regions of the Deccan Plateau in relation to the isotope ratios of the Réunion Plume. The wide range of elevated $^{87}Sr/^{86}Sr$ and low $^{143}Nd/^{144}Nd$ ratios is an indication that the basalt magmas assimilated Precambrian basement rocks prior to their eruption (*Sources:* Peng et al. 1994; Peng et al. 1998; White et al. 1990; Hart 1988)

lated Precambrian basement rocks. The isotope ratios of Sr and Nd in the rocks of the two regions overlap extensively, but the magmas of the western Deccan assimilated a more diverse suite of basement rocks than those of the northeastern region. Nevertheless, the lava flows in both regions could have originated from the Réunion Plume represented in Fig. 5.32 by the isotope ratios of basalt on the Chagos-Laccadive and the Mascarene Plateaus (White et al. 1990). The overlap of the isotope ratios of Sr and Nd in Fig. 5.32 implies that some of the formations of the western Deccan extend to the northeastern region where the stratigraphy has not yet been worked out (Peng et al. 1998). The results included in Fig. 5.32, as well as those of Allègre et al. (1982) and Mahoney et al. (1982), provide one of the most spectacular examples of the effect the assimilation of Precambrian basement rocks has on the isotope ratios of Sr and Nd of mantle-derived basalt magmas.

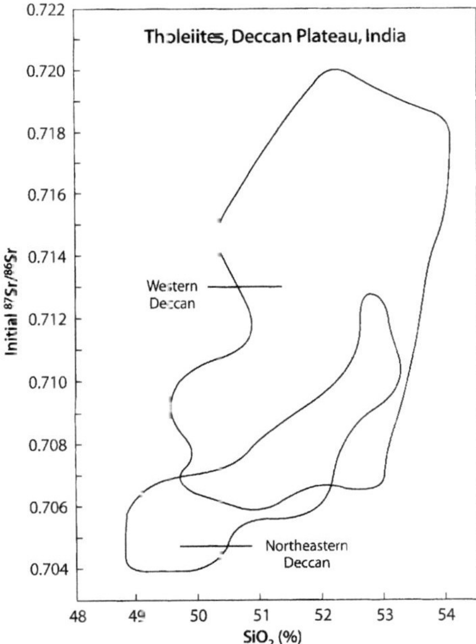

Fig. 5.33. Initial ^{87}Sr/^{86}Sr ratios and silica concentrations of tholeiites of the western and northeastern regions of the Deccan Plateau of India. The data points representing the western region are more widely scattered than those of the northeastern Deccan but positive correlations of these two parameters are evident for basalts of both regions (*Sources:* Peng et al. 1994, 1998)

The evidence in favor of crustal contamination of basalt magmas is strengthened by the positive correlation of SiO$_2$ concentrations (48.88 to 54.04%) and initial ^{87}Sr/^{86}Sr ratios (c.70413 to 0.71996) for suites of basalt of the western and northeastern Deccan Plateau in Fig. 5.33. In addition, Peng et al. (1994) demonstrated a positive correlation between the initial ^{87}Sr/^{86}Sr ratios and δ^{18}O values (+5.9 to +7.6‰) of the Deccan magmas.

The heterogeneity of isotopic compositions of the assimilated continental crust is evident in Fig. 5.34, which combines the initial ^{87}Sr/^{86}Sr ratios and the present ^{206}Pb/^{204}Pb ratios of basalts analyzed by Peng et al. (1994, 1998). The ^{206}Pb/^{204}Pb ratios range unusually widely from 16.768 (Khandala Formation) to 22.426 (Neral Formation). The Pb in the lavas of the western Deccan is much more heterogeneous in its isotopic composition than the Pb in the basalts of the northeastern region. The lavas at Jabalpur (northeastern Deccan) and two samples of the Bhimashankar Formation (Western Deccan) in Fig. 3.54 define a mixing trajectory between two components (E1 and E2) characterized by their ^{87}Sr/^{86}Sr and ^{206}Pb/^{204}Pb ratios. In addition, a third component (E3) must be present to

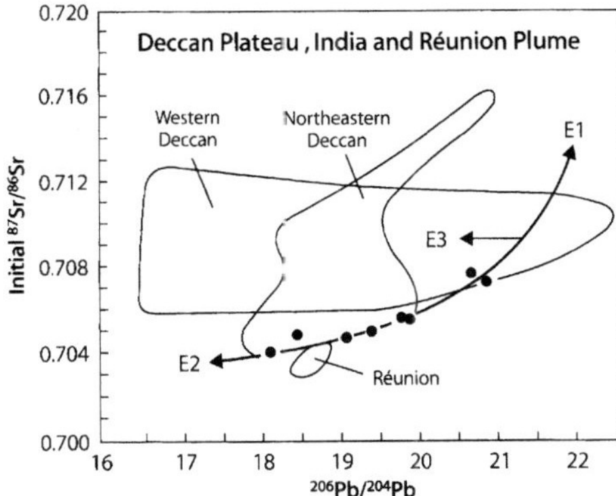

Fig. 5.34. Initial ^{87}Sr/^{86}Sr and present ^{206}Pb/^{204}Pb ratios of tholeiite basalt in the western and northeastern Deccan Plateau, India. The evident scatter of data points requires that the magmas assimilated at least three different types of Precambrian basement rocks identified as *E1*, *E2*, and *E3*. The mixing line between *E1* and *E2* is defined primarily by the lavas at Jabalpur (northeastern Deccan) and the Bhimashankar Formation of the western Deccan. The scatter of data points requires at least a two-stage process of contamination of magmas that originated from the Réunion Plume (*Sources:* Peng et al. 1994, 1998)

account for the way the isotope ratios of Sr and Pb scatter to the left of the hypothetical E1-E2 mixing line.

Peng et al. (1994) identified the E1 component as amphibolite of Archean age dated by Gopalan et al. (1990) by the Sm-Nd method. The E2 component resembles granulitic gneisses of Archean age exemplified by the Lewisian gneisses of northwest Scotland (Sect. 5.4.1), whereas E3 has low ^{206}Pb/^{204}Pb but elevated ^{87}Sr/^{86}Sr ratios. Peng et al. (1994) concluded that the isotope compositions of Pb of the Deccan basalt require a two-stage process of contamination of magmas first by E1 (high ^{206}Pb/^{204}Pb, high ^{87}Sr/^{86}Sr) and E2 (low ^{206}Pb/^{204}Pb, low ^{87}Sr/^{86}Sr) and then by E3 (low ^{206}Pb/^{204}Pb, high ^{87}Sr/^{86}Sr). However, a complete description of this process is not yet available.

5.6 Rajmahal Plateau, Northeastern India

Volcanic rocks of Cretaceous age in Bengal, Bihar, and Assam in northeastern India, known as the Rajmahal, Sylhet, and Bengal Traps are composed primarily of tholeiite basalt and basaltic andesite with some alkali basalt (Baksi et al. 1987). The lava flows of the Rajmahal Plateau in Fig. 5.35 were extruded in about two million years at 117 Ma (Baksi et al. 1987, 1995; Class et al. 1993; McDougall and McElhinny 1970) as a result of magma formation in the head of a mantle plume presently located near Heard Island on the Kerguelen Plateau (Sect. 2.15; Mahoney et al. 1983). The Indian Plate separated from Antarctica at about 125 Ma and moved north

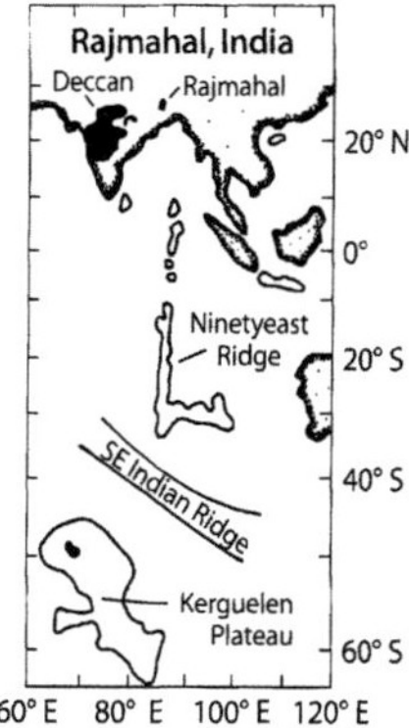

Fig. 5.35.
Location of the Rajmahal Plateau of eastern India relative to the Ninetyeast Ridge and the submarine Kerguelen Plateau (*Source:* adapted from Mahoney et al. 1983)

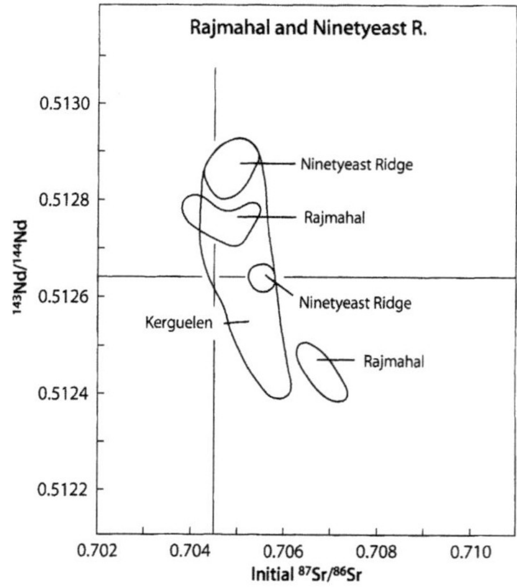

Fig. 5.36. Isotope ratios of Sr and Nd of the Cretaceous flood basalts in the Rajmahal Plateau, India, compared to data for basalt of the Ninetyeast Ridge and the Kerguelen Plateau (*Sources:* Mahoney et al. 1983; Salters et al. 1992)

resulting in the formation of a chain of submarine volcanoes that define the Ninetyeast Ridge in the eastern basin of the Indian Ocean (Sect. 2.15.3, Storey et al. 1992).

The volume of the Rajmahal Traps in Bengal is about 2.4×10^3 km^3 (area ~4 000 km^2, thickness ~600 m), whereas the Sylhet Traps in Assam amount to about 100 km^3 (area ~200 km^2, thickness ~500 m). The Bengal Traps are covered by sedimentary rocks and are known only from drill cores. Baksi et al. (1987) reported whole-rock K-Ar dates between 107 ±3 and 128 ±7 Ma for five specimens of the Bengal Traps, thus demonstrating that they are coeval with the Rajmahal and Sylhet Traps. Consequently, the combined volume of volcanic rocks of this petrologic province is about 1.2×10^5 km^3 (area ~2×10^5 km^2, thickness ~600 m), making it comparable in volume to the Columbia River basalt of North America.

The isotope ratios of Sr and Nd of basalt in the Rajmahal Plateau and in the Ninetyeast Ridge (Mahoney et al. 1983) as well as the isotope ratios of basalt on the Kerguelen submarine plateau (Salters et al. 1992) are negatively correlated in Fig. 5.36. The initial ^{87}Sr/^{86}Sr ratios of the Rajmahal basalt range from 0.70412 to 0.70710 relative to 0.71025 for NBS 987. The ^{143}Nd/^{144}Nd ratios also have a large range from 0.512391 to 0.51276 compared to 0.511859 for the La Jolla Nd standard (Mahoney et al. 1983). These results support the hypothesis that the magmas that formed the Rajmahal basalts originated from the Kerguelen Plume at 117 Ma (Baksi 1995; Class et al. 1993) during the initial rifting and subsequent separation of the Indian Plate from Antarctica in the context of the fragmentation of Gondwana. The elevated initial ^{87}Sr/^{86}Sr and low ^{143}Nd/^{144}Nd ratios of some of the Rajmahal lavas indicate that the basalt magmas assimilated basement rocks of Precambrian age. Therefore, the basalts of the Rajmahal Plateau conform to the standard model of flood-basalt petrogenesis as a consequence of rifting caused by the stress applied by a large asthenospheric plume

on the overlying lithospheric plate composed of continental crust and associated upper mantle (White and McKenzie 1989).

The genetic link of the Rajmahal basalts to the Kerguelen Plume was challenged by Curray and Munasinghe (1991) who suggested that the Crozet Plume (Sect. 2.16.6) was the source of the Rajmahal magmas. However, the ^{87}Sr/^{86}Sr and 1/Sr ratios of volcanic rocks of Crozet Island reported by Hedge et al. (1973) are not collinear with those of the Kerguelen Island and the Rajmahal Traps (not shown). In addition, Kent et al. (1992b) presented tectonic arguments to support the conclusion that magmas of the Rajmahal basalt originated from the same plume as the volcanic rocks of the Kerguelen Islands rather than those of Crozet. This interpretation was also supported by Saunders et al. (1991), Storey et al. (1992), and Weis and Frey (1991).

5.7 The South Atlantic Igneous Province

The continental flood basalts of the Paraná Plateau in southern Brazil (Fig. 5.37) are related to the basalt lavas of the Etendeka Group of Namibia because both originated from the Tristan da Cunha Plume in Early Cretaceous time (Peate et al. 1992a; and Sect. 2.5.6 and 2.6). The complementary basalt plateaus of eastern Brazil and western Namibia in Fig. 5.38 are one of the prime examples of the association of flood basalts with

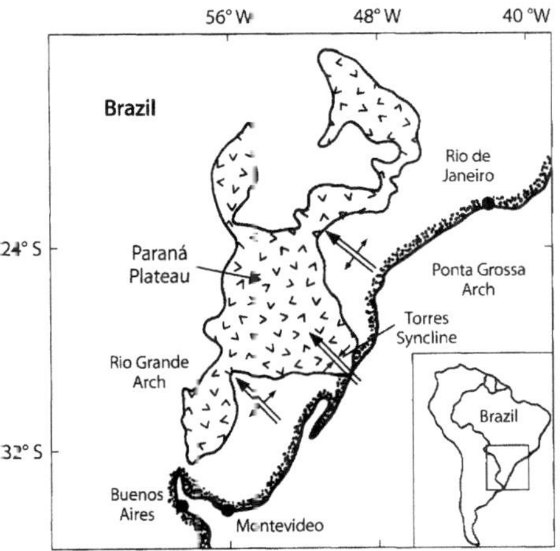

Fig. 5.37. Map of southern Brazil showing the extent of the Paraná basalt plateau of Cretaceous age (*Source:* adapted from Fodor 1987)

Fig. 5.38. Complementary flood-basalt plateaus in the state of Paraná Brazil, and Namibia, SW Africa at about 120 Ma (*Source:* adapted from White and McKenzie 1989)

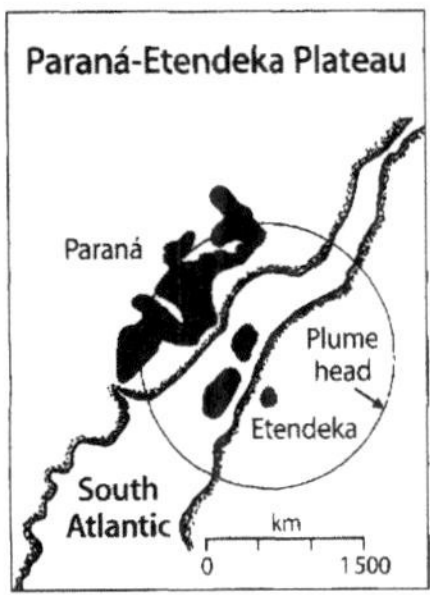

plume-related rifting of continental crust and subsequent seafloor spreading to form an ocean basin (White and McKenzie 1989). Under these conditions, the magmas that form by decompression melting in the plume head and adjacent lithospheric mantle are contaminated by assimilation of crustal rocks either during their ascent to the surface or while they are pooled in magma chambers in the lower crust, or both. In addition, the heat transported into the crust by the basalt magmas causes melting of Precambrian basement rocks in the upper crust and of magic igneous and metamorphic rocks in the lower crust to form felsic magmas. As a result, the flood basalts may be interbedded with or intruded by felsic volcanic rocks.

After crystallization, the lavas and associated plutonic rocks may have been altered by convecting groundwater that was energized by the heat that emanated from the volcanic and plutonic rocks as they cooled. The late-stage hydrothermal alteration of the lavas and associated plutons caused selective changes

in their chemical compositions and in the isotope ratios of Sr, Nd, Pb, and O (Iacumin et al. 1991; Murata et al. 1987).

5.7.1 Paraná Plateau of Brazil

The basalts of the Paraná Plateau (Sera Geral Formation) formed in Early Cretaceous time (110–140 Ma), occupy an area of about 1.2×10^6 km^2, and are among the most voluminous outpourings of continental basalt with an estimated volume of about 2.2×10^6 km^3 (Wilson 1989). Most of the rocks are basaltic in composition, but felsic lavas occur in the upper part of the sequence especially in the eastern and south-central part of this igneous province (Mantovani et al. 1985a). The petrogenesis of these rocks has been investigated by means of isotopic, geochemcial, and mineralogical data by Cordani et al. (1980, 1988), Bellieni et al. (1984), Hughes et al. (1986), Piccirillo and Melfi (1988), Fodor et al. (1985a,b, 1990), Peate et al. (1990), and others referred to in this section.

The volcanic rocks of the Paraná Plateau have been subdivided into four suites defined in Fig. 5.39a by

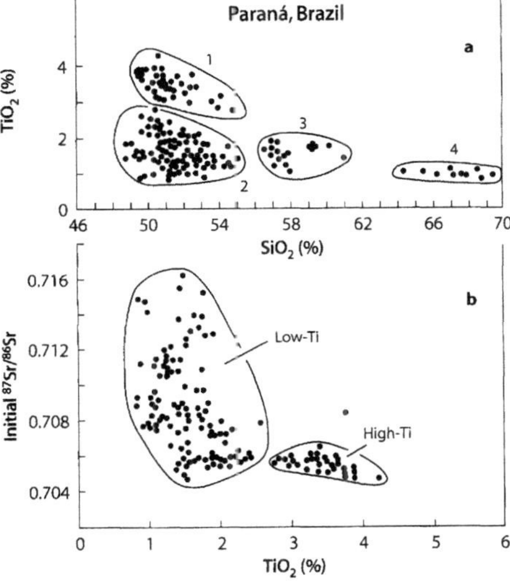

Fig. 5.39. a Variation of TiO$_2$ and SiO$_2$ concentrations of volcanic rocks (Early Cretaceous), in the Paraná Plateau of southern Brazil. The rocks form four distinctive suites: *1.* High-Ti basalts; *2.* Low-Ti basalts; *3.* Intermediate tholeiites; *4.* Felsic rocks; **b** Range of initial ^{87}Sr/^{86}Sr ratios and TiO$_2$ concentrations. Most of the high-Ti rocks (TiO$_2 > 2.75\%$) have low initial ^{87}Sr/^{86}Sr ratios (0.7047 to 0.7065), whereas the low-Ti rocks (TiO$_2 < 2.50\%$) have a wide range of initial ^{87}Sr/^{86}Sr ratios (0.7047 to 0.7163 and higher) (*Sources:* Mantovani et al. 1985b; Hawkesworth et al. 1986; Petrini et al. 1987; Piccirillo et al. 1987, 1989)

Table 5.6. Subdivision of the lavas of the Paraná Plateau of Brazil

Ti content	Magma-type
Low	Gramado (Peate and Hawkesworth 1996) Esmeralda (Peate and Hawkesworth 1996) Ribeira
High	Urubici (Peate et al. 1999) Pitanga Paranapanema

their concentrations of TiO_2 and SiO_2. Suite 1 consists of high-Ti basalts that occur in the northern part of the Paraná Plateau, whereas the low-Ti basalts of Suite 2 occur primarily in the southern region. Suite 3 includes intermediate low-Ti tholeiites identified by Piccirillo et al. (1989) from both northern and southern parts of the region, and Suite 4 consists of felsic volcanics which have $SiO_2 > 64\%$ but $TiO_2 < 1.5\%$. The lavas of the Paraná Plateau were further subdivided by Peate et al. (1992b) based on a large body of chemical data pertaining to these rocks (Table 5.6). In addition, Bellieni et al. (1986) referred the rhyolites to Palmas-type magmas.

The lavas of the Paraná Plateau were extruded through fissures; some of which have been preserved as Early Cretaceous dikes in the Precambrian basement and Paleozoic cover rocks of the Ponta Grossa arch along the east coast of Brazil (Fig. 5.37). The data of Piccirillo et al. (1990) indicate that these dikes range in composition from tholeiite to basaltic andesite and include minor amounts of latites, dacites, and rhyolites. The initial $^{87}Sr/^{86}Sr$ ratios (at 135 Ma) of these rocks vary from 0.70516 to 70792, but do not correlate with concentrations of SiO_2 (50.70 to 66.91%) or TiO_2 (0.55 to 3.45%). The average initial $^{87}Sr/^{86}Sr$ ratio of the Ponta Grossa Dikes is 0.70598 ±0.00027 ($2\bar{\sigma}, N = 21$) relative to 0.71025 for NBS 987.

In addition to the flood basalts and rhyolites of the Paraná Plateau and to the feeder dikes in the Ponta Grossa region, K-rich mafic igneous rocks were intruded in Early and Late Cretaceous time into the sedimentary rocks on the periphery of the Paraná Basin. Gibson et al. (1996) recognized both high-Ti and low-Ti varieties of these rocks and related their regional distribution to the occurrence of high-Ti and low-Ti basalts of the Paraná Plateau.

The initial $^{87}Sr/^{86}Sr$ ratios of the volcanic rocks of the Paraná Plateau in Fig. 5.39b range widely from about 0.7047 to a high value of 0.72855 reported by Fodor et al. (1985a) for a rhyolite from the state of Rio Grande do Sul in southern Brazil. This sample was characterized by 73.01% SiO_2, Mg number = 17.5%, (100 Mg/Mg+Fe, molar), and $\delta^{18}O = +15.11‰$ all of which indicate that this rhyolite formed either by melting of rocks in the continental crust or by differentiation of a mantle-derived magma that had assimilated large amounts of such rocks. The display of data in Fig. 5.39b indicates that the lowest value of the initial $^{87}Sr/^{86}Sr$ ratios of both low-Ti and high-Ti basalts is 0.7047 but that the high-Ti rocks have much more uniform initial $^{87}Sr/^{86}Sr$ ratios than the low-Ti basalts. Therefore, both suites originated from similar source rocks in the mantle, but the low-Ti magmas assimilated larger amounts of crustal rocks prior to eruption. The only initial $^{87}Sr/^{86}Sr$ ratios lower than 0.7047 were reported by Halpern et al. (1974) for basalt specimens from the state of Santa Catarina, which yielded an average value of 0.7038 ±0.00025 recalculated to 0.71025 for NBS 987 and 1.42×10^{-11} yr^{-1} for $\lambda(^{87}Rb)$.

The petrogenesis of the high-Ti and low-Ti basalts in the Paraná Plateau of Brazil is constrained by four basic facts:

1. The initial $^{87}Sr/^{86}Sr$ ratios of the two magma types were identical (<0.7047).
2. The high-Ti suite of rocks is enriched in LIL elements including Rb and Sr.
3. The initial $^{87}Sr/^{86}Sr$ ratios of the high-Ti basalt suite have a smaller range than those of the low-Ti suite.
4. The high-Ti basalts are restricted to the northern part of the Paraná Plateau.

These facts were explained by Fodor et al. (1985b) and Fodor (1987) based on the hypothesis that the high-Ti magmas are the products of lower degrees of partial melting than the low-Ti magmas, but that both suites originated from the same source rocks in the mantle. Subsequently, the high-Ti magmas assimilated smaller amounts of crustal rocks than the low-Ti magmas, because the crust under the northern part of the Paraná Plateau was cooler than the crust under the central and southern parts of the plateau.

Calculations by Fodor (1987) indicated that the low-Ti basalts of the Paraná Plateau could have formed from a 25% partial melt of source rocks in the mantle followed by crystallization of 63% of olivine, clinopyroxene, and plagioclase. The low-Ti basalts produced by this process would have $TiO_2 = 1.4\%$, $P_2O_5 = 0.16\%$, Rb = 17.8 ppm, and Sr = 328 ppm. The high-Ti basalt evolved from 11% partial melts of the same source rocks, followed by crystallization of 72% olivine, clinopyroxene, and plagioclase to produce basalt containing $TiO_2 = 3.6\%$, $P_2O_5 = 0.45\%$, Rb = 52 ppm, and Sr = 495 ppm.

The heat required to drive this process was provided by the large Tristan Plume which domed up and fractured the continental crust of Gondwana. The degree of partial melting within the plume decreased from the center outward and resulted in the formation of picritic melts, as originally proposed by Cox

(1980). The plume also increased the temperature of the overlying lithospheric mantle and continental crust and thereby facilitated the assimilation of "warm" crust by low-Ti (25% melt fraction) magmas over the center of the plume, which caused their $^{87}Sr/^{86}Sr$ ratios to increase. Around the margin of the plume, where temperatures were lower, high-Ti (11% melt fraction) magmas could assimilate only a small amount of crustal rocks. Therefore, their $^{87}Sr/^{86}Sr$ ratios increased only moderately.

The volcanic activity ended when South America and Africa drifted apart leaving a trail of submarine volcanoes in their wake. These are now represented by the Rio Grande Rise located in the South Atlantic between Tristan da Cunha and southern Brazil (Sect. 2.2.3). The other half of the Paraná Plateau is represented by the Tafelberg and other exposures of basalt and latites of the Etendeka Group in Namibia, southwest Africa. The trail of the Etendeka volcanics from their place of origin on the Mid-Atlantic Ridge to their present residence is marked by the Walvis Ridge, studied by Richardson et al. (1982, 1984) as described in Sect. 2.2.4.

The volcanic rocks on Tristan da Cunha are alkali-rich (alkali basalt, ankaramite, trachybasalt) presumably because magma is presently forming by small degrees of partial melting in the center of the plume. Studies by O'Nions and Pankhurst (1974) and White and Hofmann (1982) yielded an average $^{87}Sr/^{86}Sr$ ratio of 0.70508 ±0.00004 (2σ) for 11 specimens relative to 0.7080 for E&A. This value is higher than the lowest initial $^{87}Sr/^{86}Sr$ ratios of basalt in the Paraná Plateau. Nevertheless, a genetic relationship of the two suites of volcanic rocks in Paraná and Namibia to the Tristan Plume is well established by reconstructions of the tectonic history of the South Atlantic Ocean (Morgan 1981, 1983). These conclusions imply that the plume under Tristan da Cunha has been active for up to about 140 million years and that it contributed significantly to the rifting of Gondwana and the opening of the Atlantic Ocean.

The isotope ratios of Sr and Nd of Low-Ti basalts from the southern part of the Paraná Plateau in Fig. 5.40 represent the Gramado and Esmeralda magma types. The MgO concentrations of both magma types range from about 3.0 to 8.0%; however, the Gramado flows have higher concentrations of SiO_2, Sr, and Ba but lower concentrations of total Fe, TiO_2, and Y than the Esmeralda flows. In addition, the initial $^{87}Sr/^{86}Sr$ ratios of the Gramado flows in Fig. 5.40 are higher than those of the Esmeralda type, whereas their $^{143}Nd/^{144}Nd$ ratios are lower.

The isotope ratios of Sr and Nd of the Esmeralda flows in Fig. 5.40 partly overlap those of the alkali-rich rocks of Tristan da Cunha (Cliff et al. 1991; LeRoex et al. 1990). Therefore, the Esmeralda-type magmas of the Paraná Plateau could have originated by partial melt-

ing in the head of the Tristan Plume. Stratigraphic information summarized by Peate and Hawkesworth (1996) supports their conclusion that the Gramado flows were erupted first and were followed by the Esmeralda flows. This is significant because it means that the distribution of data fields in Fig. 5.40 cannot be explained by concluding that the Gramado-type

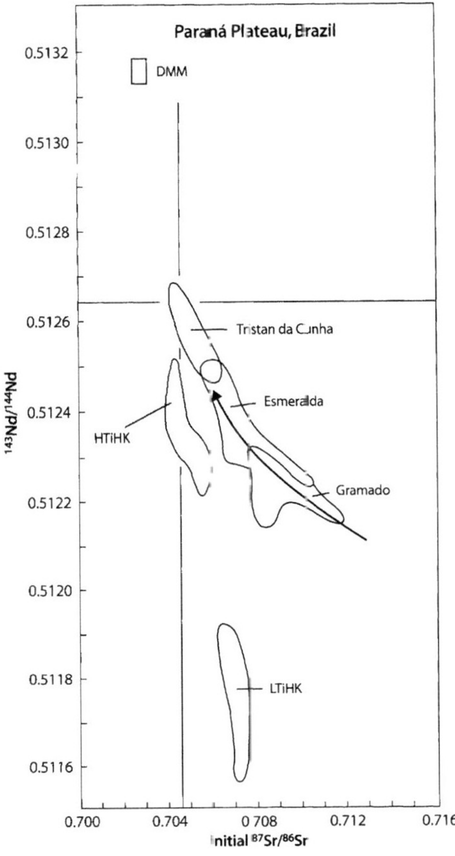

Fig. 5.40. Isotope ratios of Sr and Nd of basalt on the Paraná Plateau of Brazil compared to the alkali-rich lavas derived from the mantle plume beneath Tristan da Cunha. The interpretation of the data fields hinges on the fact that the Esmeralda flows overlie and are, in some cases, interbedded with the older Gramado flows. Therefore, the Gramado flows are thought to have originated from the subcrustal lithospheric mantle, whereas the Esmeralda flows are mixtures of magma derived from the Tristan Plume and the overlying lithosphere. The high-K lavas that occur along the periphery of the Paraná Plateau are composed of the high-Ti-high-K (*HTiHK*) and low-Ti-high-K (*LTiHK*) subgroups, both of which differ from the low-K tholeiites of the Paraná Plateau (*Sources:* Paraná basalt: Peate and Hawkesworth 1996; K-rich lavas: Gibson et al. 1996; Tristan da Cunha: Cliff et al. 1991 and LeRoex et al. 1990; DMM: Hart 1988)

magmas evolved from the Esmeralda-type magmas by assimilating more crustal rocks.

Instead, Peate and Hawkesworth (1996) interpreted the subtle differences in the chemical compositions of the Gramado and Esmeralda flows and their Sr and Nd isotope ratios as evidence that these magmas originated from different sources. The Gramado-type magma formed primarily by melting in the subcontinental lithospheric mantle, whereas the later Esmeralda-type magmas are mixtures in varying proportions of a plume-derived component with Gramado-type magmas.

The isotope ratios of Sr and Pb of the Gramado and Esmeralda flows in Fig. 5.41 are aligned relative to the data field of Tristan da Cunha exactly as the Sr and Nd ratios in Fig. 5.40. The $^{206}Pb/^{204}Pb$ of the Paraná Plateau lavas have a narrow range 18.366 to 18.904 with an average of 18.678 ±0.054 ($2\bar{\sigma}$, N = 24) compared to 18.596 ±0.048 ($2\bar{\sigma}$, N = 27) for the alkali-rich lavas of Tristan da Cunha (LeRoex et al. 1990; Cliff et al. 1991).

The homogeneity of the $^{206}Pb/^{204}Pb$ ratios of the Gramado and Esmeralda flows in the southern Paraná Plateau reported by Peate and Hawkesworth (1996) is in sharp contrast to the heterogeneity of the isotope composition of Pb in the basalt of the Deccan Plateau (Fig. 5.34). Since the wide variation of $^{206}Pb/^{204}Pb$ ratios of the Deccan basalt is attributable to the assimi-

lation of heterogeneous crustal rocks by the magmas, the evident uniformity of $^{206}Pb/^{204}Pb$ ratios of the Gramado and Esmeralda basalts implies that crustal assimilation was not an important factor in the petrogenesis of these rocks. Consequently, this comparison reinforces the interpretation by Peate and Hawkesworth (1996) that the Gramado lavas originated from the subcrustal lithospheric mantle rather than by crustal contamination of Esmeralda-type magmas which originated from the head of the Tristan Plume.

The scenario favored by Peate and Hawkesworth (1996) implies that the magma sources of the Paraná basalt changed with time from the subcrustal lithospheric mantle to the asthenospheric rocks of the Tristan Plume and reinforces the view expressed by Professor Hawkesworth and his collaborators that flood basalts may originate to a significant extent from the subcontinental lithospheric mantle that is being stretched and heated by an asthenospheric plume (Hawkesworth et al. 1988).

5.7.2 Etendeka Group, Namibia

The lavas of the Etendeka Group of Namibia (Milner et al. 1995a) consist of basalt and interbedded silicic latite flows which were erupted in Late Jurassic to Early Cretaceous time prior to the separation of southern Africa from South America (Erlank et al. 1984). The lava flows occur in the form of several erosional remnants and are primarily exposed in the Tafelberg identified in Fig. 5.42. The Etendeka lavas of the Tafelberg (about 850 m) consist of basalt (70%), as well as of

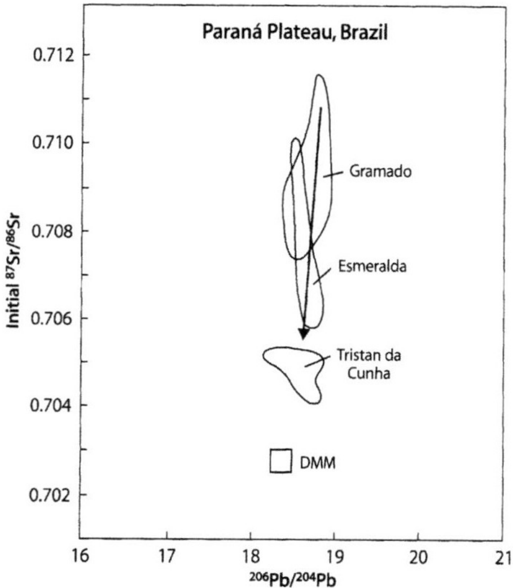

Fig. 5.41. Isotope ratios of Sr and Pb of the Gramado and Esmeralda basalt flows of the southern Paraná Plateau, Brazil. The Sr-Pb isotopic data confirm that the Esmeralda magmas are mixtures of a plume component and Gramado-type magma (*Source:* Peate and Hawkesworth 1996; LeRoex et al. 1990; Cliff et al. 1991; Hart 1988)

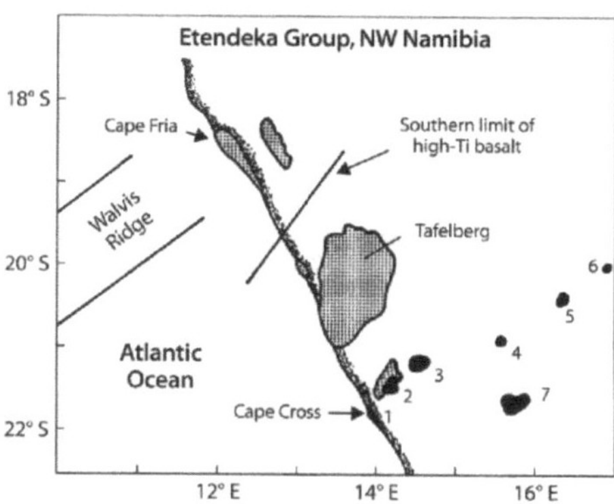

Fig. 5.42. Flood basalt of the Etendeka Group and the Mesozoic Damaraland igneous complexes of northwestern Namibia: *1.* Cape Cross; *2.* Messum; *3.* Brandberg; *4.* Okenyenya; *5.* Paresis; *6.* Okorusu; *7.* Erongo (*Source:* adapted from Milner and LeRoex 1996)

interbedded latites (5%), and quartz latites (25%) whose silica concentrations increase from about 60 to nearly 70% toward the top of the sequence. The lavas overlie shale and sandstones of the Karoo Group which was deposited on a basement complex of Precambrian (Middle to Early Proterozoic) schists and gneisses.

The lava flows of the Etendeka Group of the Tafelberg consist of the basal Awahab Formation and the disconformably overlying Tafelberg Formation, both of which include basalt and silicic quartz latite layers. Milner et al. (1995b) correlated the silicic lavas of the Awahab Formation in Namibia with rhyolite flows of the Palmas magma type in the southern part of the Paraná Plateau of Brazil. They also correlated quartz latites of the Tafelberg Formation of Namibia with similar silicic units in the upper Palmas sequence of Paraná. These correlations are based on discriminant-function analysis of the chemical compositions of the rock units and are supported by comparisons of phenocryst assemblages, initial ^{87}Sr/^{86}Sr ratios, and petrographic data.

Exposures of Etendeka lavas also occur at scattered sites north of Tafelberg along the coast up to Cape Fria, as well as in the vicinity of the Mesozoic Messum and Brandberg igneous complexes of the Goboboseb Mountains south of the Tafelberg (Fig. 5.42). The basalts of the Etendeka Group north of the Tafelberg have elevated concentrations of Ti, P, Zr and other high-field-strength elements compared to the basalts of the Tafelberg and those that are exposed in the southern part of the region (Fig. 5.42). The regional distribution of high-Ti basalts in the north and low-Ti basalts farther south mirrors the distribution of these rock types in the Paraná Plateau.

The initial ^{87}Sr/^{86}Sr ratios at 121 Ma of the tholeiite basalts of the Etendeka Group in Fig. 5.43 range widely

from about 0.7078 to 0.7134 and are strongly enriched in radiogenic ^{87}Sr compared to the lavas of Tristan da Cunha (0.70508 ±0.00004, 2$\bar{\sigma}$, N = 11, Sect. 5.7.1). However, the initial ^{87}Sr/^{86}Sr ratios of the basalts on Tafelberg in Namibia are similar to those of the low-Ti Gramado and Esmeralda flows of the Paraná Plateau. In addition, the latites of the Etendeka Group and the Palmas rhyolites of the southern Paraná Plateau have similar initial ^{87}Sr/^{86}Sr ratios. These comparisons suggest that the basalt magmas of Namibia originated from the subcrustal lithospheric mantle rather than directly from the Tristan Plume and that the latite flows contain a significant component of Sr from the rocks of the continental crust.

The isotope ratios of Sr and Nd of the low-Ti basalts of the Tafelberg reported by Hawkesworth et al. (1984a) define a data field in Fig. 5.44 that overlaps the field of the low-Ti Gramado flows of the Paraná Plateau and that is considerably displaced from the data field of Tristan da Cunha. Consequently, the low-Ti magmas of the Tafelberg in Namibia originated from the subcrustal lithospheric mantle much like the Gramado flows of Paraná (Peate and Hawkesworth 1996). The latites and quartz latites evolved either from basalt magmas that assimilated Precambrian basement rocks or the latite magmas originated as crustal melts.

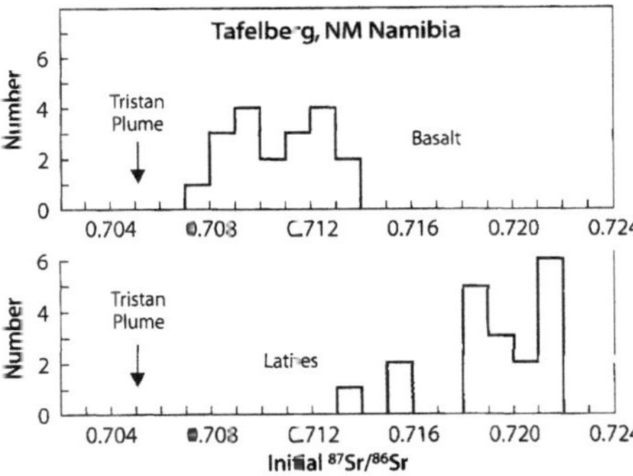

Fig. 5.43. Initial ^{87}Sr/^{86}Sr ratios at 121 Ma of tholeiite basalts and latites of the Etendeka Group, Tafelberg, northwestern Namibia (Source: Compilation by Bristow et al. 1984)

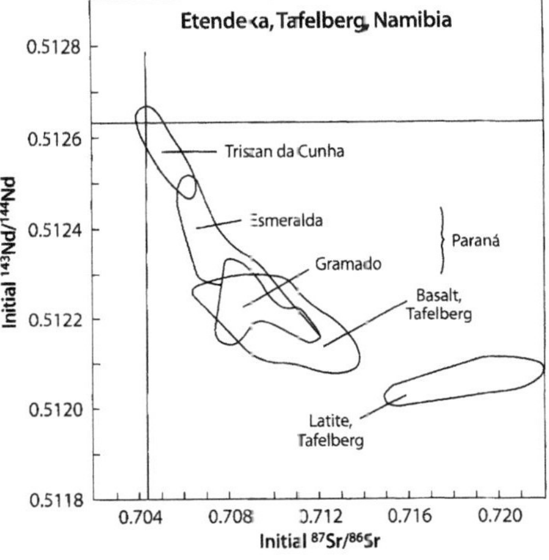

Fig. 5.44. Isotope ratios of Sr and Nd of basalt and latite flows of the Etendeka Group on Tafelberg of northwestern Namibia. In addition, the data fields representing the lavas on Tristan da Cunha and Paraná, Brazil, are shown for comparison. The isotope ratios of Sr and Nd of the basalts on Tafelberg of Namibia resemble those of the Gramado flows in the southern region of the Paraná Plateau of Brazil (Sources: Tafelberg: Hawkesworth et al. 1984; Paraná: Peate and Hawkesworth 1996; Tristan da Cunha: LeRoex et al. 1990 and Cliff et al. 1991)

5.7.3 Damaraland Igneous Complexes, Namibia

The Damaraland igneous complexes in Fig. 5.42 (Cape Cross, Messum, Brandberg, Okenyenya, Paresis, and Okorusu) range in age from 137 to 124 Ma (Milner et al. 1995c) and define a lineament that extends from the coast in a northeasterly direction for a distance of about 350 km. Martin et al. (1960) classified the complexes as being mafic (Cape Cross, Messum), granitic (Brandberg), and alkali-rich (Paresis) in composition. The Messum complex of the Goboboseb Mountains was the source of the quartz latites that occur in the Awahab Formation. Milner et al. (1992) identified these quartz latite layers as ash-flow tuffs (rheo-ignimbrites). Their correlation with the Palmas succession of the Paraná Plateau implies that the ash flows originally extended more than 340 km from their source prior to the opening of the South Atlantic Ocean and that the Messum complex erupted a total of about 8 600 km^3 of volcanic material (Milner et al. 1995b).

The Messum complex in the Goboboseb Mountains south of the Tafelberg in Namibia was an important volcanic center where a bimodal suite of volcanic rocks was erupted (Ewart et al. 1998a,b). The lavas associated with the Messum Volcano belong to the Awahab Formation of the Etendeka Group and consist of a basal suite of low-Ti and low-Zr basalts overlain by the Goboboseb quartz latite, the Messum Mountains basalt, and the Springbok quartz latite. These lava flows were intruded by dikes, sills, and small plutons composed primarily of dolerite, but including rhyolite and carbonatite in some cases.

The Messum complex itself has a diameter of about 20 km and consists of a variety of rock types, including volcanic breccia and rhyolite in the core, with gabbro, granite, and syenite arranged around it. All of these rocks types contain an abundance of xenoliths of granite and, in some cases, Karoo sandstone.

Age determinations of these rocks by Milner et al. (1995c) and Renne et al. (1996), indicate that the Messum complex and the Goboboseb and Springbok quartz latites formed during a short interval of time at 132 Ma. Age determinations of the Brandberg and Okenyenya complexes have yielded dates between 124 to 137 Ma, indicating that they formed at about the same time as the Messum complex during the episode of rift-related volcanic activity of the Etendeka Group (Watkins et al. 1994; Milner et al. 1993, 1995a).

The Okenyenya complex contains an even wider range of rock types than the Messum complex. Watkins and LeRoex (1994), assigned these rocks to an older tholeiite suite (picritic gabbro, olivine gabbro, quartz monzodiorite, and syenite) and a younger suite of alkali-rich rocks (gabbro, essexite, nepheline syenite, and late-stage lamprophyric rocks).

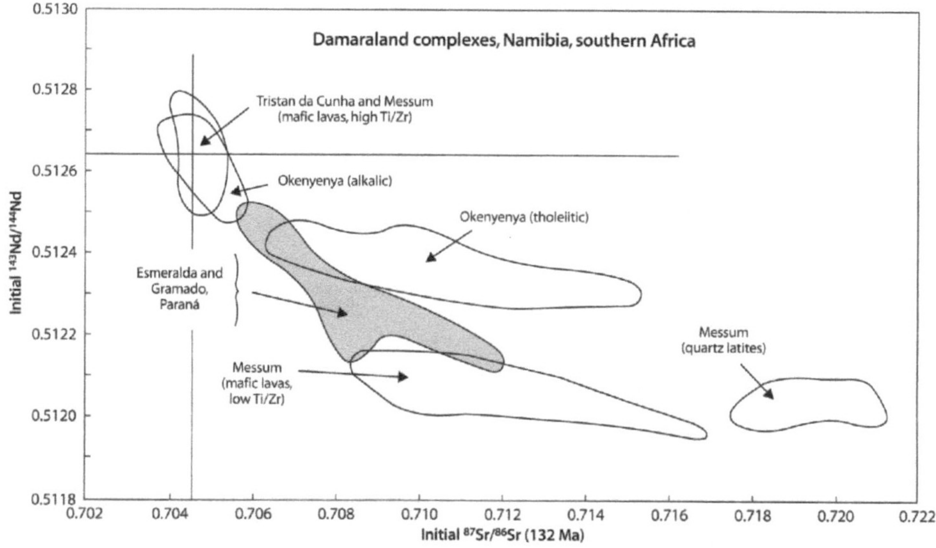

Fig. 5.45. Initial isotope ratios of Sr and Nd (at 132 Ma) of the volcanic rocks erupted by the Messum Volcano and of the plutonic rocks of the Okenyenya igneous complex of Namibia. These and other Mesozoic Damaraland complexes of Namibia are synchronous with the eruption of the flood basalt and quartz latite ignimbrites that comprise to Etendeka Group (*Sources:* Milner and LeRoex 1996; Ewart et al. 1998a,b; Cliff et al. 1991; LeRoex et al. 1990)

The $^{87}Sr/^{86}Sr$ and $^{143}Nd/^{144}Nd$ ratios of the lavas erupted by the Messum Volcano define three distinct data fields in Fig. 5.45 (Ewart et al. 1998a,b). The low-Ti and low-Zr (high Ti/Zr) mafic lavas have low initial $^{87}Sr/^{86}Sr$ (0.70420 to 0.70501) and high initial $^{143}Nd/^{144}Nd$ (0 512510 to 0.51277) ratios relative to 0.71025 for NBS 987 and 0.51264 for BCR-1, respectively. These values overlap the isotope ratios of Sr and Nd in the alkali-rich lavas of Tristan da Cunha (Cliff et al. 1991; LeRoex et al. 1990). Therefore, the low-Ti mafic lavas having high Ti/Zr ratios erupted by the Messum Volcano originated from the Tristan Plume with only limited contributions from the lithospheric mantle.

A second suite of low-Ti and low-Zr basalts (low Ti/Zr) is characterized by high and variable initial $^{87}Sr/^{86}Sr$ (0.70867 to 0.71686) and nearly constant but low $^{143}Nd/^{144}Nd$ (0.511974 to 0.512144) ratios. Ewart et al. (1998a) attributed the isotope compositions and trace-element concentrations of these rocks to assimilation of crustal rocks by plume-derived magmas accompanied by fractional crystallization.

The quartz latites erupted by the Messum Volcano define a third data field in Fig. 5.45. These rocks have extremely high initial $^{87}Sr/^{86}Sr$ (0.71748 to 0.72117) ratios relative to 0.71025 for NBS 987, whereas their initial $^{143}Nd/^{144}Nd$ ratios are similar to those of the mafic lavas with low Ti/Zr ratios. Ewart et al. (1998b) explained the extreme ^{87}Sr-enrichment of the quartz latites as a result of large-scale assimilation of crustal rocks by hybridized basalt magmas (low Ti/Zr) with simultaneous fractional crystallization. Two samples of breccia in the core of the Messum complex have initial $^{87}Sr/^{86}Sr$ and $^{143}Nd/^{144}Nd$ ratios that are intermediate between those of the two types of mafic rocks (e.g. 0.70604 to 0.70887 and 0.512446 to 0.512553, respectively).

The intrusive rocks of the Okenyenya complex define two data fields in Fig. 5.45 depending on their chemical compositions. The isotope ratios of Sr and Nd of the alkali-rich rocks (gabbro, essexite, syenite) as well as picritic gabbros overlap those of the lavas of Tristan da Cunha and of the low-Ti (high Ti/Zr basalts erupted by the Messum Volcano. Therefore, the alkali-rich magmas of the Okenyenya complex originated from the Tristan Plume. The rocks of the tholeiite suite (olivine gabbro, quartz monzodiorite, and quartz syenite) have a wide range of initial $^{87}Sr/^{86}Sr$ ratios (0.70634 to 0.71309) relative to 0.71025 for NBS 937, but their initial $^{143}Nd/^{144}Nd$ ratios vary only from 0.51227 to 0.51259 relative to 0.511816 for the La Jolla Nd standard. These results leave no doubt that the tholeiitic suite of rocks at Okenyenya formed from magmas that had assimilated crustal rocks. Milner and LeRoex (1996) confirmed the contamination of the tholeiitic magmas of the Okenyenya complex and of the quartz latites of the Etendeka Group by demonstrating a strong correlation of initial $^{87}Sr/^{86}Sr$ ratios and $\delta^{18}O$ values, which range from about +8 to nearly +12‰.

The similarity of isotope compositions of Sr and Nd demonstrated in Fig. 5.45 clearly establishes a genetic link between the Damaraland igneous complexes (Messum and Okenyenya) with the Tristan Plume, which was located in their vicinity at about 130 Ma. However, the data fields of the tholeiitic suite of Okenyenya and of the low-Ti (low Ti/Zr) basalts of the Messum complex exceed the limits of the Gramado flows of the Paraná Plateau. Therefore, the subcrustal lithospheric magmas of both complexes assimilated more crustal rocks than seems to be the case for the low-Ti basalts of the Tafelberg.

5.8 The Brazil-West Africa Connection

Evidence of the igneous activity that was associated in space and time with the opening of the South Atlantic Ocean is preserved in the form of Mesozoic flood basalts of the Maranhão Plateau on the Brazilian mainland and by high-Ti basalts of Eocene age in the Abrolhos Archipelago at about 18° S latitude on the continental shelf of northeastern Brazil. In addition, Fodor and McKee (1986) described tholeiite basalts of early Middle Jurassic (184 Ma) and Early Cretaceous (127 Ma) ages from a drill core recovered at 3°45′ N and 50°00′ W in the Amapá Basin about 100 km off the north coast of Brazil. Basaltic lavas of Cretaceous age including both high-Ti and low-Ti varieties occur also south of the Abrolhos Islands between 18 and 24° S along the east coast of Brazil. Fodor and Vetter (1984) reported that the chemical compositions of these rocks are transitional between continental flood basalts and MORBs. However, their isotope ratios of Sr and Nd have not yet been determined.

The evidence on the African side consists of the Cretaceous sedimentary and volcanic rocks of the Benue Trough (Cameroon) which formed in Early Cretaceous time as a failed branch of a triple junction that resulted in the separation of central Africa from Brazil. The African Plate subsequently moved north, causing a deep crustal fracture to form south of the Benue Trough, along which alkali-rich lavas continue to be erupted in a volcanic province known as the Cameroon Line. In addition, basaltic dikes and lavas flows of Jurassic age in Liberia and Sierra Leone as well as the differentiated Freetown igneous complex (Sierra Leone) originated from to a mantle plume presently located near Ascension Island (Morgan 1983).

Rifting between central Africa and northeastern Brazil implies the presence of asthenospheric plumes that caused doming of the continental crust and

Table 5.7. Average initial $^{87}Sr/^{86}Sr$ ratios of the St. Helena Plume and the Cameroon Line relative to 0.7080 for E&A (Halliday 1988, 1990; Grant et al. 1976)

Cameroon Line	Offshore islands	0.70318 ±0.00006 (2$\bar{\sigma}$, N = 30)
	On the mainland	0.70326 ±0.00009 (2$\bar{\sigma}$, N = 33)
St. Helena	Lavas (Sr > 100 ppm)	0.70316 ±0.00007 (2$\bar{\sigma}$, N = 56)

subcrustal lithospheric mantle in this area leading to rifting and ultimately to separation of the opposite sides of the Atlantic Rift. A tectonic reconstruction by Morgan (1983) indicates that the plumes presently located near the islands of St. Helena and Ascension Island close to the Mid-Atlantic Ridge (Sect. 2.5.5) contributed significantly to the formation of igneous rocks on the African Plate. The relation of the St. Helena Plume to the Neogene volcanic activity along the Cameroon Line is supported by a comparison of the average initial $^{87}Sr/^{86}Sr$ ratios reported by Halliday (1988, 1990) for the lavas of the Cameroon Line and by Grant et al. (1976) for St. Helena relative to 0.7080 for E&A (Table 5.7).

5.8.1 Abrolhos Archipelago, Brazil

The Abrolhos Islands are located on the continental shelf of Brazil at about 18° S and 38°30' W. The islands, which may represent a deeply dissected volcano, include Santa Barbara, Redonda, Sueste, Siriba, and Guarita in order of decreasing size. The largest island, Santa Barbara, is barely 1.5 km long and less than 0.5 km wide.

The Abrolhos Islands may be related to a series of seamounts, including Hotspur, located about 500 km east of the town of Caravelas on the east coast of Brazil. A second chain of seamounts south of the Abrolhos Islands is located at about 20° S latitude and includes the Victoria and Jasseur Seamounts as well as the islands of Trinidade and Martin Vaz. The northwestward motion of the South American Plate has left a track in Fig. 5.46 marked by the presence of volcanic centers whose ages decrease from about 200 Ma on the north coast of Colombia to near 0 Ma on the island of Trinidad (Morgan 1983).

The Abrolhos Islands contain high-Ti basalt (average TiO_2 = 4.77%) and diabase (average TiO_2 = 4.52%) of Eocene age (40 to 50 Ma; Fodor and Asmus 1983). The high Ti concentrations of the basalt flows were explained by Fodor et al. (1989) as the result of 90% fractional crystallization of a picritic parent magma which formed by a low degree of partial melting (10 to 15%) of source rocks in the mantle (Fodor 1987).

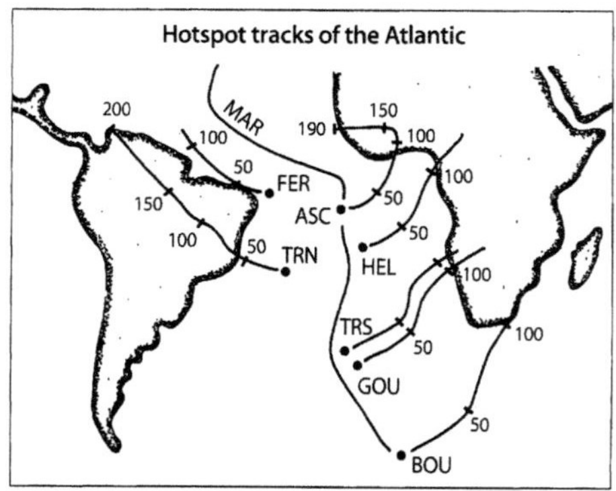

Fig. 5.46. Selected hotspot tracks between South America and Africa. The plumes are identified by letters: *FER* = Fernando de Noronha; *ASC* = Ascension Island; *TRN* = Trinidade; *HEL* = St. Helena; *TRS* = Tristan da Cunha; *GOU* = Gough; *BOU* = Bouvet. *MAR* is the Mid-Atlantic Ridge. Numbers along the tracks are the ages in Ma of volcanic rocks that were erupted when their present location was over the stationary plumes (*Source:* adapted from Morgan 1983)

The initial $^{87}Sr/^{86}Sr$ ratios of the basalts and diabases on the Abrolhos Islands at 45 Ma range from 0.70362 to 0.70378 relative to 0.71025 for NBS 987 (Fodor et al. 1989). The diabases have elevated initial $^{87}Sr/^{86}Sr$ ratios between 0.70381 and 0.70461 at 45 Ma. In general, high-Ti basalt and diabase on the Abrolhos Islands are not only significantly younger than the high-Ti basalts of the Paraná Plateau but also have lower initial $^{87}Sr/^{86}Sr$ ratios (Hawkesworth et al. 1986).

The isotope ratios of Sr and Nd of basalt and diabase on the Abrolhos Islands form an array in Fig. 5.47 that closely approaches the data field of MORBs along the Mid-Atlantic Ridge between 10° S and 40° S latitude, as reported by Fontignie and Schilling (1996). Both arrays lie between the DMM and the EM1 components of magma sources in the mantle identified by Hart (1988). According to Weaver (1991), the EM1 component consists of subducted oceanic crust containing a small amount of old pelagic marine sediment. Fodor et al. (1989) concluded that the Abrolhos magmas originated from mixed sources consisting of an asthenospheric plume and adjacent lithospheric mantle. The magmas subsequently differentiated by crystallization of olivine, clinopyroxene, and plagioclase at the base of the continental crust. The residual liquid then rose toward the surface and pooled in shallow magma chambers under the Abrolhos Platform. In spite of the evolution of the Abrolhos magmas within the continental crust, they did not become contaminated by crustal rocks, presumably because the temperature of the crust was too low to permit large-scale assimilation. In accordance with the reconstruction of plume tracks in the Atlantic Ocean by Morgan

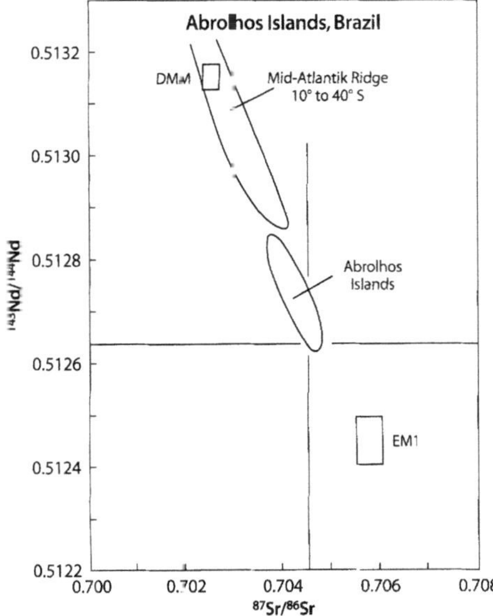

Fig. 5.47. Isotope ratios of Sr and Nd of basalt and diabase Eocene, 45 Ma) on the Abrolhos Islands located at 18° S latitude or the continental shelf of the east coast of Brazil (*Sources:* Abrolhos Islands: Focor et al. 1989; Mid-Atlantic Ridge: Fontignie and Schilling 1996; EM1 and DMM: Hart 1988 and Weaver 1991)

(1983) in Fig. 5.46, the Abrolhos basalts could have formed when the area was located above the Trinidade Plume.

5.8.2 Cameroon Line, West Africa

The connection of the St. Helena Plume to the volcanoes of the Cameroon Line is well established based on the work of Morgan (1983) in Fig. 5.46 and on $^{40}Ar/^{39}Ar$ dates of basalts dredged from the St. Helena Seamount chain (O'Connor and LeRoex 1992). The crystallization ages of the volcanic rocks along this chain increase from about 7 Ma on St. Helena (Baker et al. 1967) to about 81 Ma for a seamount located a short distance southwest of Pagalu Island in Fig. 5.48 (O'Connor and LeRoex 1992). Although the offshore islands and the volcanoes of the Cameroon Line appear to be an extension of the St. Helena chain of seamounts, the volcanic islands of Pagalu, São Tomé, Principe, and Bioko in the Gulf of Guinea and the volcanoes of the Cameroon Line constitute a separate volcanic province. The ages of the volcanic rocks in the Cameroon Line range from 30 to 0 Ma (e.g. Mt. Cameroon identified in Fig. 5.48).

The rocks of the Cameroon Line include both intrusive plutons as well as lava flows and are generally

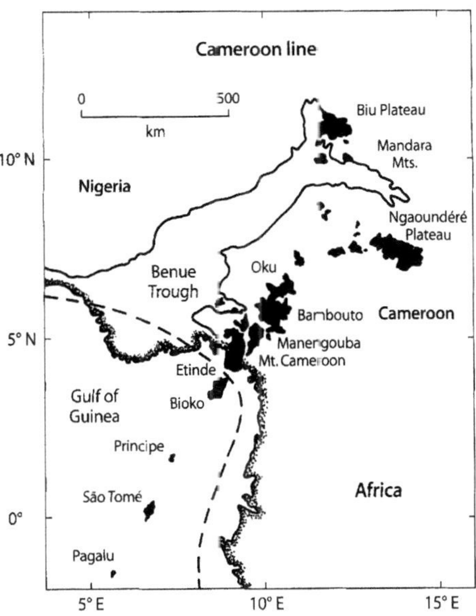

Fig. 5.48. Map of the volcanic centers of the Cameroon Line (*black*) and of the Benue Trough in equatorial West Africa. The Late Cenozoic volcanic centers on the off-shore islands are underlain by oceanic crust (*dashed line*), whereas those on the mainland of Africa formed on continental crust. However, the initial $^{87}Sr/^{86}Sr$ ratios of the alkali-rich lavas along the Cameroon Line are uniformly low, indicating that the magmas did not interact appreciably with rocks of the oceanic or continental crust (*Source:* adapted from Halliday et al. 1988)

alkalic in composition, consisting of alkali basalts, basanites, trachytes, phonolites, tephrites, and nephelinites (Fitton and Dunlop 1985; Fitton 1987; Halliday et al. 1990).

The line of volcanic centers in Cameroon runs parallel to and south of the Benue Trough, which is filled with folded Cretaceous sediment and some volcanic rocks. The Benue Trough in Fig. 5.48 splits into two branches at its northeastern end. One branch extends north toward Lake Chad, while the other runs due east and dies out about 250 km east of the split. The sedimentary/volcanic rocks of the Benue Trough were intruded by small plutons of alkali-rich rock (2.8 to 23 Ma, Grant et al. 1972). The igneous activity they record was synchronous with the volcanic eruptions along the Cameroon Line and is probably related to it.

The initial $^{87}Sr/^{86}Sr$ ratios of the volcanic rocks extruded along the Cameroon Line range only from 0.70291 to 0.70414 (Halliday et al. 1988, 1990) relative to 0.71025 ±0.00002 for NBS 987. Even the intrusive and extrusive igneous rocks from the Benue Trough have primarily low initial $^{87}Sr/^{86}Sr$ ratios (Grant et al. 1972). Although the islands Pagalu, São Tomé, Principe, and Bioko in the Gulf of Guinea (Fig. 5.48) are underlain by oceanic crust, the initial $^{87}Sr/^{86}Sr$ ratios of the volcanic rocks on these islands do not differ significantly from those of the volcanic centers of the Cameroon Line that formed on Precambrian basement rocks on the continent.

The island of Principe described by Fitton and Hughes (1977) contains four suites of volcanic and intrusive rocks dated by Dunlop and Fitton (1979):

1. Intrusive tristanite (4.89 ±0.15 Ma), phonolite (5.32 ±0.17 Ma and 5.48 ±0.19 Ma), and trachyphonolite plutons, (6.93 ±0.68 Ma);
2. Younger lava series composed of nephelinite (5.6 ±0.32 Ma) and bassanite (3.51 ±0.15 Ma);
3. Older lava series including basalt (23.6 ±0.7 Ma) and hawaiite (19.1 ±0.5 Ma);
4. Basal palagonite breccia (30.6 ±2.1 Ma).

The phonolite plugs as well as the nephelinite and basanite flows of the younger lava series form a Rb-Sr isochron that yields a date of 5.9 ±0.3 Ma with an initial $^{87}Sr/^{86}Sr$ ratio of 0.70299 ±0.00007 (2$\bar{\sigma}$, $N = 19$) relative to 0.71025 for NBS 987. Therefore, Dunlop and Fitton (1979) supported the suggestion that the phonolites, tristanites, and trachyphonolites, whose Sr concentrations vary widely from 9.37 ppm to 1 230 ppm, could have formed by fractional crystallization of the basanite magma. However, the older lavas (basalt and hawaiite), as well as the tholeiite of the basal palagonite breccia, have a significantly higher average initial $^{87}Sr/^{86}Sr$ ratio of 0.70329 ±0.00009 (2$\bar{\sigma}$, $N = 14$) relative to 0.71025 for NBS 987. Therefore, the older lavas formed from magma that originated from different kinds of source rocks than the younger lavas and the associated phonolites/tristanite/trachyphonolite suite.

The isotope ratios of Sr and Nd of the rock suites representing the offshore islands and the Cameroon Line in Fig. 5.49 (Halliday et al. 1988, 1990) define two overlapping data fields. All of the data points lie in the quadrant occupied by mantle-derived rocks, whereas two granulite-facies xenoliths in lava flows at Bambouto on the Cameroon Line have elevated $^{87}Sr/^{86}Sr$ ratios at 18 Ma (0.71041 and 0.72097) relative to 0.71025 for NBS 987 and low $^{143}Nd/^{144}Nd$ ratios (0.512168 and 0.511644) compared to 0.51185 for the La Jolla Nd standard (Halliday et al. 1988).

The isotope ratios of the high-Ti basalt and diabase on the Abrolhos Islands of Brazil define a data field in Fig. 5.49 that overlaps that of the Cameroon volcanoes (Fodor et al. 1989). Therefore, all three rock suites could have originated from magmas that formed by mixing of melts derived from mantle plumes at St. Helena and Trinidade and from associated subcrustal lithospheric mantle represented by the EM1 component. The virtual absence of crustal contamination is confirmed by the

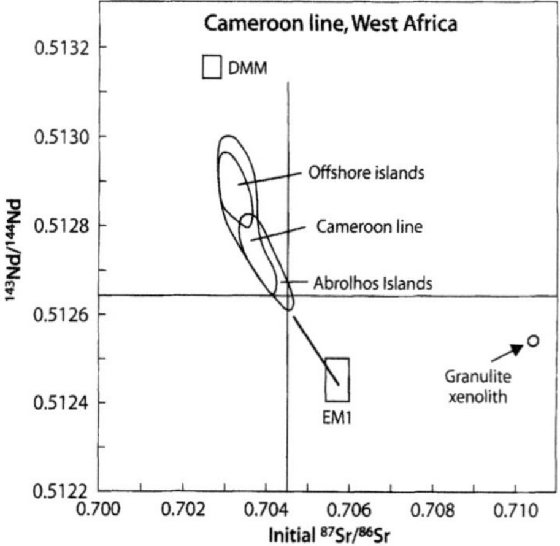

Fig. 5.49. Isotope ratios of Sr and Nd of the alkali-rich volcanic rocks (Oligocene to Holocene) of the Cameroon Line and the associated offshore islands off the coast of West Africa. The magmas of the Cameroon Line (Oligocene to Holocene) and of the Abrolhos Islands (Brazil) can be viewed as mixtures of melts derived from asthenospheric plumes and the EM1 components (*Sources:* Cameroon Line: Halliday et al. 1988, 1990; Abrolhos Islands: Fodor et al. 1989)

$\delta^{18}O$ values of the volcanic rocks of the Cameroon Line, which range from +4.53 to +7.60‰ with an average of +5.82 ±0.28 (2$\bar{\sigma}$, $N = 32$) (Halliday et al. 1988).

The similarity of $^{87}Sr/^{86}Sr$ ratios of the volcanic rocks on the Cameroon Line and on St. Helena permits the conclusion that the source of the Cameroon magmas is a remnant of the St. Helena Plume whose head originally spread both east and west under the subcrustal lithospheric mantle at the time when the plume defined the position of the Atlantic Rift.

The origin of the Cameroon magmas is further constrained by the high $^{206}Pb/^{204}Pb$ ratios of the lavas, which range from 18.327 (rhyolite) to 20.522 (nephelinite, Etinde, <0.5 Ma). Therefore, the volcanic rocks of the Cameroon Line have the characteristics of the HIMU mantle component which was defined partly by the high $^{206}Pb/^{204}Pb$ ratios of the basalt on St. Helena (Gast 1968).

The isotope ratios of Pb and Sr of the volcanic rocks along the Cameroon Line of West Africa define a data field that lies between the HIMU and EM1 mantle components in Fig. 5.50 (Hart 1988). The Eocene basalt and diabase of the Abrolhos Islands partly overlap the data field of the Cameroon lavas indicating that the magmas at both locations may have formed from mixed EM1-HIMU sources in the mantle. Alternatively, the magmas may be mixtures of melts derived separately from EM1 and HIMU sources. In either case, the data in Fig. 5.50 (Pb-Sr) are consistent with the evidence in

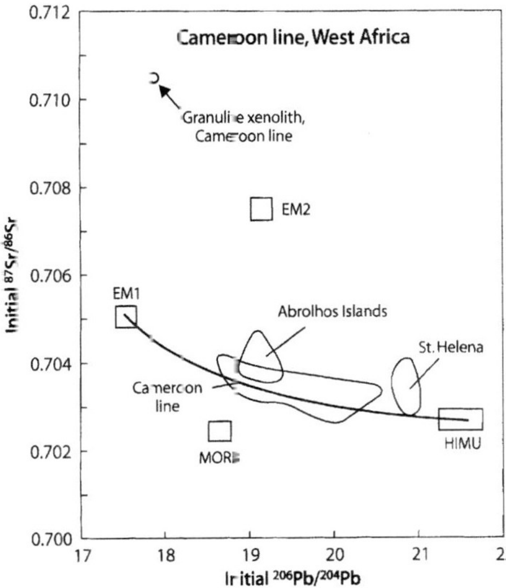

Fig. 5.50. Initial isotope ratios of Pb and Sr of the volcanic rocks (Oligocene to Holocene) of the Cameroon Line. The data field of the Eocene basalt and diabase on the Abrolhos Island partly overlaps that of the Cameroon Line. Both suites have elevated $^{206}Pb/^{204}Pb$ ratios and thereby resemble the lavas on St. Helena which were used in part to define the HIMU mantle component. The isotope compositions of Pb and Sr indicate that the magmas of the Cameroon Line and of the Abrolhos Islands formed either from mixed EM1-HIMU sources, or are mixtures of magmas derived separately from the EM1 and HIMU components (*Sources:* Halliday et al. 1988–1990; Grant et al. 1976; Gast 1968; Fodor et al. 1989; Hart 1988)

Fig. 5.49 (Sr-Nd) concerning the origin of the magmas and strengthens the hypothesis that the volcanic rocks of the Abrolhos Islands and of the Cameroon Line are both related to plumes that are characterized by having high $^{206}Pb/^{204}Pb$ ratios.

The $^{87}Sr/^{86}Sr$ ratios of the Late Tertiary lava flows and small intrusives within the Benue Trough vary more widely than those of the Cameroon Line (i.e. 0.7025 to 0.7129, Grant et al. 1972). Therefore, some of the magmas that intruded the rocks of the Benue Trough did assimilate crustal rocks on a local scale. Nevertheless, the Late Tertiary rocks of the Benue Trough originated from the same magma sources as the lavas of the Cameroon Line (Grant et al. 1972).

5.8.3 Anorogenic Plutons of Cameroon and Nigeria

The geology of Cameroon and neighboring Nigeria is characterized by the presence of small anorogenic plutons of Mesozoic to early Tertiary ages (Turner and Bowden 1979; van Breemen et al. 1975; Bowden et al.

1976; Bowden 1985; Vidal et al. 1979). The Cameroon Line itself includes about 40 volcano-plutonic complexes whose ages range from 65 to 35 Ma and which, therefore, predate the more recent volcanic activity of this petrologic province (Jacquemin et al. 1982).

The Golda Zuelva and Mboutou complexes occur at the northeastern end of the Cameroon Line and have diameters of 6 to 8 km. The Golda Zuelva complex consists of concentric cone sheets of gabbro and monzonite surrounding a core of riebeckite-arfvedsonite bearing biotite granite. The Mboutou complex, which has two centers, is primarily composed of gabbro that encloses syenite in the northern core and monzonite in the southern core.

The alkali-rich granites of the Golda Zuelva complex have low Sr (4.96 to 160 ppm) but high Rb concentrations (45.1 to 212 ppm). A whole-rock Rb-Sr errorchron of these granites and associated rhyolites yielded a date of 66 ±3 Ma ($\lambda = 1.42 \times 10^{-11}$ yr^{-1}, $MSWD = 13.8$) and an initial $^{87}Sr/^{86}Sr$ ratio of 0.7025 ±0.0012 relative to 0.71025 for NBS 987 (Jacquemin et al. 1982). Olivine gabbro, micro-gabbro, and riebeckite diorite have higher initial $^{87}Sr/^{86}Sr$ ratios (0.70311 to 0.70352) than the granites, as do various rhyolites (0.70375 to 0.70456). The initial $^{87}Sr/^{86}Sr$ ratios of the Mboutou complex range from 0.70358 to 0.70406 (northern center) and from 0.70459 to 0.70603 (southern center) relative to 0.71025 for NBS 987 and are higher than those of the Golda Zuelda complex located about 175 km to the northeast along the Cameroon Line (Jacquemin et al. 1982).

The isotope ratios of Sr and Pb of the Golda Zuelva and Mboutou complexes in Fig. 5.51 define a data field

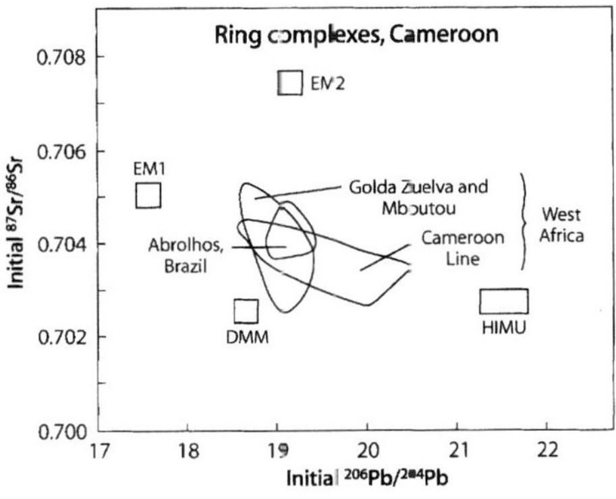

Fig. 5.51. Isotope ratios of Pb and Sr of granitic and gabbroic rocks of the Golda Zuelva and Mboutou complexes located at the northeastern end of the Cameroon Line. The data field defined by these complexes partly overlaps the data field of the volcanic rocks of the Cameroon Line (*Sources:* Golda Zuelva and Mboutou: Jacquemin et al. 1982; Cameroon Line; Halliday et al. 1988, 1990; Abrolhos Islands: Fodor et al. 1989; mantle components: Hart 1988)

that partly overlaps the data fields of the volcanic rocks of the Cameroon Line. However, the $^{206}Pb/^{204}Pb$ ratios of the two complexes rise only to a value of 19.59 (syenite, Mboutou complex, northern center) which means that the HIMU component did not contribute appreciably to the formation of their magmas. In addition, the position of the data field of the Golda Zuelva and Mboutou complexes within the EM1-EM2-HIMU-DMM polygon in Fig. 5.51 implies that the magmas originated in the mantle and did not assimilate significant amounts of crustal rocks, even though they were intruded into Precambrian granitic basement rocks. The only exceptions to this generalization are the rocks associated with the southern center of the Mboutou complex whose elevated initial $^{87}Sr/^{86}Sr$ ratios reflect crustal contamination of magmas on a local scale.

Jacquemin et al. (1982) noted the occurrence of low initial $^{87}Sr/^{86}Sr$ ratios in both granitic and gabbroic rocks in these two complexes and attributed their chemical diversity to the effects of fractional crystallization of mantle-derived magmas in crustal magma chambers. This interpretation is strengthened by the $\delta^{18}O$ values of the constituent minerals, which are clustered between values of +5 and +8‰. However, an alternative explanation for the $\delta^{18}O$ values and low $^{87}Sr/^{86}Sr$ ratios of the granites is that the granitic magmas originated by partial melting of previously formed mafic igneous rocks in the crust.

Seven anorogenic complexes in Nigeria range in age from 171 ±5 to 151 ±4 Ma ($\lambda = 1.42 \times 10^{-11}$ yr^{-1}; van Breemen et al. 1975). The ages decrease systematically from north to south, consistent with the northward movement of the African Plate. The initial $^{87}Sr/^{86}Sr$ ratios determined from whole-rock Rb-Sr isochrons range from 0.7065 ±0.0013 (Pankshin complex) to 0.752 ±0.021 (albite-riebeckite granite, Liruei complex) relative to 0.7080 for E&A. The elevated initial $^{87}Sr/^{86}Sr$ ratios of most of the complexes analyzed by van Breemen et al. (1975) are the result of crustal contamination of magmas that had low Sr concentrations. For example, nineteen specimens of granitic rocks of the Liruei complex contain only about 7.4 ppm Sr on average (0.948 to 26.3 ppm), whereas their average Rb concentration is 809 ppm (126 to 1570 ppm). Such extreme chemical differentiation is the result of fractional crystallization of felsic magmas in crustal magma chambers.

The Nigerian ring complexes dated by van Breemen et al. (1975) are all older than the Golda Zuelva and Mboutou complexes of Cameroon. In addition, the magmas of the Nigerian complexes were strongly contaminated by Sr derived from crustal rocks while they differentiated by extensive fractional crystallization causing the rocks to become depleted in Sr and enriched in Rb. Although the granites of the Golda Zuelva

complex of Cameroon also have low Sr and high Rb concentrations, they were not significantly contaminated by crustal Sr as indicated by their low initial $^{87}Sr/^{86}Sr$ ratio of 0.7025 ±0.0012.

Neither the basaltic rocks on the Abrolhos Islands of Brazil nor the plutonic and volcanic rocks of the Cameroon Line are classifiable as flood basalts. However, these rocks are products of magmatic activity associated with rifting of continental crust in response to stresses caused by asthenospheric plumes. The absence of tholeiitic flood basalts indicates that large-scale melting of the lithospheric mantle did not occur here, perhaps because of insufficient release of pressure.

5.9   The Northeastern Brazil-Liberia Connection

Reconstructions of the central Atlantic region by White and McKenzie (1989) and Morgan (1983) indicate that the Mesozoic flood basalts of northeastern Brazil formed in close proximity to the diabase dikes of Liberia and the Freetown gabbro complex of Sierra Leone. In addition, Schilling et al. (1994) drew attention to the fact that the Ceará and Sierra Leone Rises in the central Atlantic Ocean are located symmetrically on opposite sides of the present Mid-Atlantic Ridge, and considered the hypothesis that they may have formed by the volcanic activity associated with the Sierra Leone Plume located close to St. Peter and St. Paul's Rocks in the central Atlantic Ocean (Sect. 2.2.2). The reconstructions of the central Atlantic Ocean presented by Schilling et al. (1994) also place northeastern Brazil south of Liberia and Sierra Leone of West Africa prior to about 100 Ma.

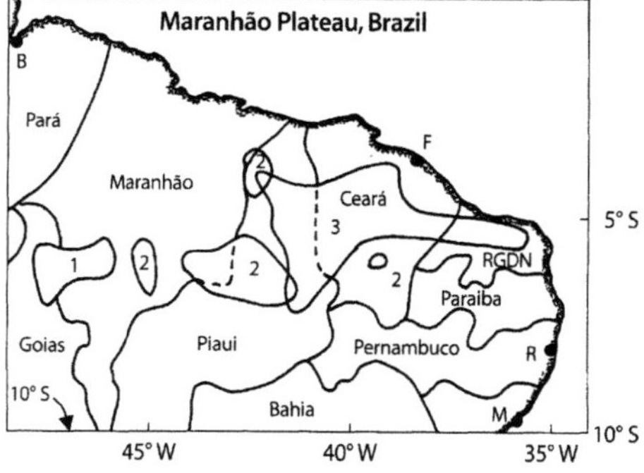

Fig. 5.52. Distribution of tholeiite and alkali basalt of Mesozoic and Cenozoic age in northeastern Brazil. Remnants of the Maranhão flood basalt plateau are identified by number: 1. Low-Ti tholeiite; 2. High-Ti tholeiite; 3. Dikes and alkali basalt. Major cities are identified by letter: B = Belem; F = Fortaleza; R = Recife; M = Maceio. The state of Rio Grande do Norte is abbreviated *RGDN* (*Sources:* Fodor et al. 1990; Bellieni et al. 1992)

5.9.1 Maranhão Plateau, Northeastern Brazil

Remnants of a continental flood basalt plateau of Mesozoic age in the states of Maranhão, Goias, Piaui, and Ceará in northeastern Brazil (Fig. 5.52) are collectively referred to as the Maranhão Plateau (Fodor et al. 1990; Bellieni et al. 1992). The work of Fodor et al. (1990) indicates that low-Ti basalts occur primarily west of the 45° W merician, whereas high-Ti basalts occur in the eastern part of the plateau. Whole-rock K-Ar dates of the low-Ti basalts from the western region range from 189 to 154 Ma (Early to Late Jurassic), whereas the lavas from the eastern region (both high- and low-Ti basalts) yielded younger dates from 144 to

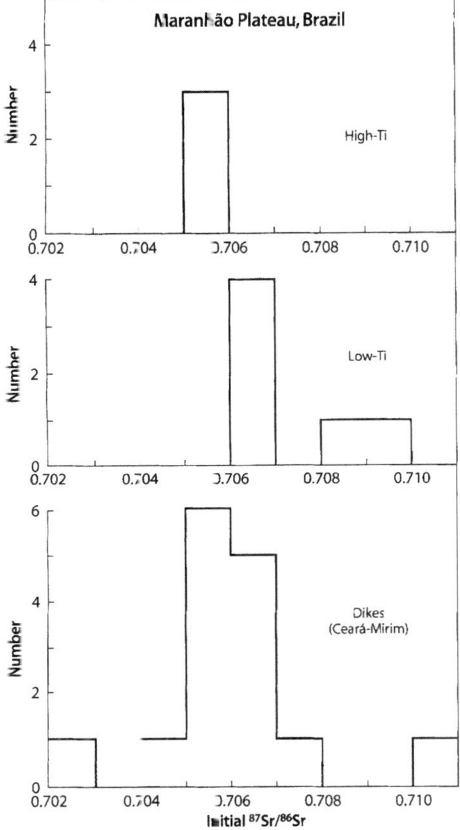

Fig. 5.53. Initial ^{87}Sr/^{86}Sr ratios of the low-Ti and high-Ti basalt layers of the Maranhão Plateau of northeastern Brazil compared to the initial Sr-isotope ratios of the Ceará-Mirim basaltic dikes of northeastern Brazil. Most of the Ceará-Mirim Dikes have high TiO_2 concentrations greater than 2.50% (Bellieni et al. 1992)

115 Ma (Early Cretaceous). These results by Fodor et al. (1990) imply that the centers of volcanic activity shifted from west to east across northern Brazil, and that the composition of the basalt changed from low-Ti (early) to high-Ti (late).

The initial ^{87}Sr/^{86}Sr ratios of the low-Ti basalts in Fig. 5.53 range from 0.70532 to 0.70545 and are higher than those of the high-Ti basalts, which range only from 0.70565 to 0.70583 relative to 0.71025 for NBS 987. The $\delta^{18}O$ values of these rocks (+6.50 to +12.64‰) correlate positively with their initial ^{87}Sr/^{86}Sr ratios (not shown), presumably because the magmas assimilated varying amounts of crustal rocks (Fodor et al. 1990). In other words, the older low-Ti magmas (high ^{87}Sr/^{86}Sr, high $\delta^{18}O$) assimilated more crustal rocks than the younger high-Ti magmas (low ^{87}Sr/^{86}Sr, low $\delta^{18}O$).

The basaltic dikes of the Ceará-Mirim region of northeastern Brazil (Fig. 5.52) formed during two episodes of magmatic activity between 175 to 160 Ma (Middle Jurassic) and 145 to 125 Ma (Early Cretaceous). Bellieni et al. (1992) pointed out that the ring complexes of Nigeria (van Breemen 1975; Bowden et al. 1984; Rahaman et al. 1984) and the tholeiite dikes of Liberia (Dalrymple et al. 1975) were also intruded during the Jurassic Period.

The initial ^{87}Sr/^{86}Sr ratios of the Ceará-Mirim Dikes in Fig. 5.53 range from 0.70293 to 0.71012 relative to 0.71025 for NBS 987 (Bellieni et al. 1992). However, the initial ^{87}Sr/^{86}Sr ratios of nearly 75% of the dikes vary only from 0.7050 to 0.7070 and thus resemble both types of basalt flows of the Maranhão Plateau (Fig. 5.53), even though most of the dikes have high TiO_2 concentrations of 2.50% or higher. As in the case of the Paraná Plateau, the initial ^{87}Sr/^{86}Sr ratios of the few low-Ti dikes vary more widely than those of the more numerous high-Ti variety.

The isotope ratios of Sr and Nd of the Maranhão basalts and Ceará-Mirim Dikes in Fig. 5.54 cluster in the quadrant that is populated by crustal rocks. The data fields labeled 1 and 2 represent the low-Ti and high-Ti basalts, respectively. Both varieties have similar $^{143}Nd/^{144}Nd$ ratios of about 0.51245. The $^{206}Pb/^{204}Pb$ ratios of these rocks (not shown) range from 18.223 to 18.783, meaning that the Maranhão basalts lack the HIMU signature of the St. Helena Plume. Most of the Ceará-Mirim Dikes define a third data field in Fig. 5.54, that includes the field of the high-Ti basalts of the Maranhão Plateau. However, several dikes analyzed by Bellieni et al. (1992) scatter widely in Fig. 5.54; perhaps because they originated from different magma sources or because they were contaminated by local basement rocks, or both.

The relatively small data fields in Fig. 5.54 indicate that the magmas of the Maranhão basalts originated from similar magma sources. The low-Ti basalts rep-

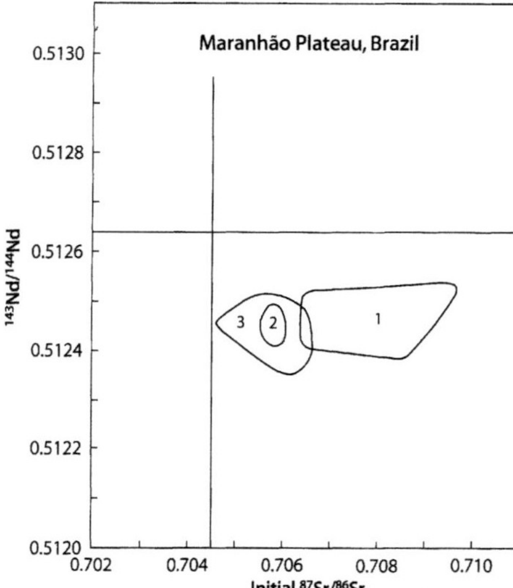

Fig. 5.54. Isotope ratios of Sr and Nd of Mesozoic flood basalts of the Maranhão Plateau and of tholeiite dikes of northeastern Brazil identified by number: *1.* Low-Ti basalts, east and west Maranhão; *2.* High-Ti basalts, eastern Maranhão; *3.* Basaltic dikes, Ceará-Mirim (*Sources:* Fodor et al. 1990; Bellieni et al. 1992)

resent large-scale melts (~25%) having low Sr concentrations of about 270 ppm, whereas the high-Ti basalts formed by a smaller degree of melting (~15%), yielding high Sr concentrations with an average of about 460 ppm. The low-Sr, low-Ti magmas were more susceptible to crustal contamination than the high-Sr, high-Ti magmas, thus explaining the elevated and variable $^{87}Sr/^{86}Sr$ ratios of the low-Ti basalts. The age differences of the different types of basalt indicate that the low-Ti basalts (Early to Late Jurassic) are related to the initial rifting between South America and West Africa, whereas the high-Ti basalts formed about 60 million years later in Early Cretaceous time during rifting in the equatorial Atlantic Ocean. The apparent isotopic homogeneity of both types of basalts suggests that their magmas originated by decompression melting of the subcrustal lithospheric mantle.

5.9.2 Late-stage Magmatic Activity of NE Brazil

Late-stage igneous activity along the trailing edge of South American Plate takes the form of the anorogenic Cabo de Santo Agostinho granite pluton located along the coast of northeastern Brazil about 30 km south of Recife (Fig. 5.52). The rocks of this pluton resemble the ring complexes of Nigeria by having high concentra-

tions of SiO_2 (70.44 to 76.15%), K_2O (5.07 to 6.08%), and Na_2O (3.54 to 4.52%), but very low concentrations of FeO (0.00 to 1.06%), MgO (0.00 to 0.06%), and CaO (0.09 to 1.40%). In addition, the rocks are depleted in Sr (2.26 to 20.9 ppm) and enriched in Rb (210 to 265 ppm). Long et al. (1986) obtained a whole-rock Rb-Sr isochron date of 104.8 ±1.8 Ma with an initial $^{87}Sr/^{86}Sr$ ratio of 0.7084 ±0.0011. The $\delta^{18}O$ values of the granites range narrowly from +8.2 to +8.8‰ with a mean of +8.5 ±0.3‰.

Long et al. (1986) pointed out that in predrift reconstructions the Cabo de Santo Agostinho granite of Brazil is the southernmost and youngest of the Nigerian plutons that were dated by van Breemen et al. (1975). In addition, they endorsed a previous proposal by Sial (1976) that the Agostinho Pluton is located at the point where the trace of the Ascension Plume intersects the coast of Brazil. These circumstances permit several alternative explanations for the origin of the Agostinho granite: (*1*) Partial melting of mantle rocks without contributions from the continental crust; (*2*) Large-scale assimilation of crustal rocks by a mantle-derived magma; and (*3*) Partial melting of crustal rocks by heat transported into the crust by mantle-derived basalt magma. Long et al. (1986) used the chemical and mineralogical compositions as well as the isotope ratios of Sr and O of the granites to demonstrate that the Agostinho magma formed by partial melting (about 20%) of feldspathic arenite in the continental crust at pressures of 5 to 6 kbar. In addition, tectonic reconstructions support the hypothesis that the basalt magma originated from the Ascension Plume.

A second example of late-stage igneous activity associated with the opening of the central Atlantic Ocean is provided by the alkali basalts in the states Rio Grande do Norte and Pernambuco of northeastern Brazil (Fig. 5.52). These volcanic rocks were erupted between 30 to 13 Ma, when the area was located above the Fernando de Noronha hotspot (Sect. 2.5.3). The rocks have elevated concentrations of MgO (12.27 to 17.11%) with SiO_2 between 41.07 and 48.44% (Fodor et al. 1998). Their isotope ratios of Sr and Nd define a small data field (not shown) that extensively overlaps the field of the volcanic rocks on Fernando de Noronha (Gerlach et al. 1987).

In addition, the $^{206}Pb/^{204}Pb$ and $^{87}Sr/^{86}Sr$ ratios of these rocks in Fig. 5.55 are similar to those of the lavas on Fernando de Noronha (19.132 to 19.565), thereby indicating that both suites carry the HIMU signature that characterizes the Noronha Plume. Therefore, the isotope ratios of Sr, Nd, and Pb of the Tertiary alkali basalts of the states of Rio Grande do Norte and Pernambuco establish the existence of a close genetic relationship to the lavas on Fernando de Noronha and hence to the Noronha Plume located about 250 km off the coast of northeastern Brazil.

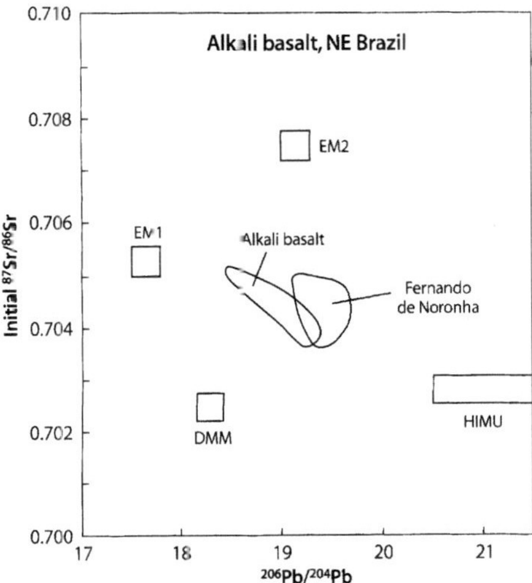

Fig. 5.55. Isotope ratios of Pb and Sr in alkali basalts (30 to 12 Ma) of northeastern Brazil. The alignment of the data fields between the HIMU and EM1 components supports the view that the alkali basalts of northeastern Brazil as well as the lavas of Fernando de Noronha originated from mixtures of melts formed in the Noronha Plume (HIMU component) and the lithospheric mantle (EM1 component) (*Sources:* Fodor et al. 1998; Gerlach et al. 1987; Hart 1938)

Fodor et al. (1993) concluded from these results and from the $\delta^{18}O$ values (+7.06 to +7.89‰) that the magmas originated by mixing of melts derived from the Noronha Plume (HIMU component) and from the overlying lithospheric mantel (EM1 component). The magmas subsequently underwent limited differentiation by fractional crystalization and were not significantly contaminated by assimilation of crustal rocks.

5.9.3 Mesozoic Diabase Dikes of West Africa

The opening of the Atlantic Ocean also left a record of igneous activity in West Africa in the form of diabase dikes and rare basalt flows in Liberia, Sierra Leone, and Ivory Coast. In addition, a differentiated gabbro complex of Jurassic age occurs near Freetown in Sierra Leone. The Mesozoic dikes intrude the Precambrian gneisses of the Liberian age province (2700 Ma) dated by Hurley et al. (1971), Hedge et al. (1975), Beckinsale et al. (1980), and Onstott and Dorbor (1987). Parts of the Liberian age province were rejuvenated during the Eburnian (2000 Ma) and Pan-African (550 Ma) events. In addition, the Mesozoic dikes intrude sedimentary rocks of the early to middle Paleozoic Paynesville Formation which overlies the basement complex along the coast of Liberia and Sierra Leone (White 1972).

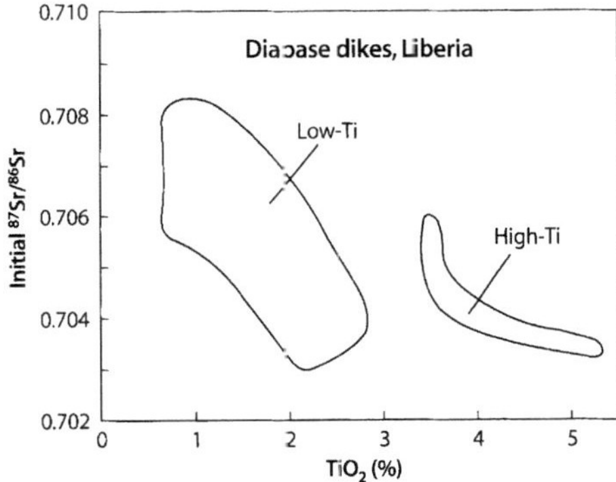

Fig. 5.56. Relation of the concentration of TiO_2 and the initial $^{87}Sr/^{86}Sr$ ratios of Jurassic diabase dikes of Liberia. The low-Ti rocks ($TiO_2 > 2.5\%$) have a wider range of initial $^{87}Sr/^{86}Sr$ ratios than the high-Ti variety and in this respect resemble the Mesozoic flood basalts of the Paraná and Maranhão Plateaus (*Sources:* Mauche et al. 1989; Dupuy et al. 1988b)

The diabase dikes strike NW-SE, range in width from 15 to 45 m, and can be traced for up to 50 km. Dalrymple et al. (1975b) and Mauche et al. (1989) reported that dikes intruding crystalline basement rocks contain excess ^{40}Ar and yield dates ranging up to about 1700 Ma, whereas dikes that are in contact with the Paynesville sandstone yield Early to Middle Jurassic dates (200 to 173 Ma) in agreement with paleomagnetic pole positions derivable from the dikes. Therefore, all of the diabase dikes of West Africa are considered to be of Jurassic age and are believed to have formed during the initial rifting of Pangea (Behrendt and Wotorson 1970; May 1971).

The dikes are composed of tholeiite basalt with silica concentrations ranging from 46.6 to 51.88% (Mauche et al. 1989; Dupuy et al. 1988b). The rocks have been subdivided in Fig. 5.56, based on their concentrations of TiO_2. As in the basalts of the Paraná (Fig. 5.39) and Maranhão Plateaus of Brazil, the low-Ti diabase dikes of Liberia have a wide range of initial $^{87}Sr/^{86}Sr$ ratios (0.7032 to 0.7080), whereas the initial $^{87}Sr/^{86}Sr$ ratios of the high-Ti basalts have a significantly narrower range (0.7034 to 0.7058). The $\delta^{18}O$ values of the Liberian dikes range from +5.6 to +9.1‰ and correlate positively with the initial $^{87}Sr/^{86}Sr$ ratios (Mauche et al. 1989).

The isotope ratios of Sr and Nd of the Liberian diabase dikes are negatively correlated in Fig. 5.57, leading to the interpretation that the magmas were mixtures of melts derived from an asthenospheric plume ($^{87}Sr/^{86}Sr \approx 0.7032$, $^{143}Nd/^{144}Nd \approx 0.51290$) and from lithospheric mantle represented by EM1 (Dupuy et al. 1988b). These magmas differentiated by fractional crystallization prior to their intrusion, causing the Mg numbers of the dikes to vary from 0.67 to 0.24.

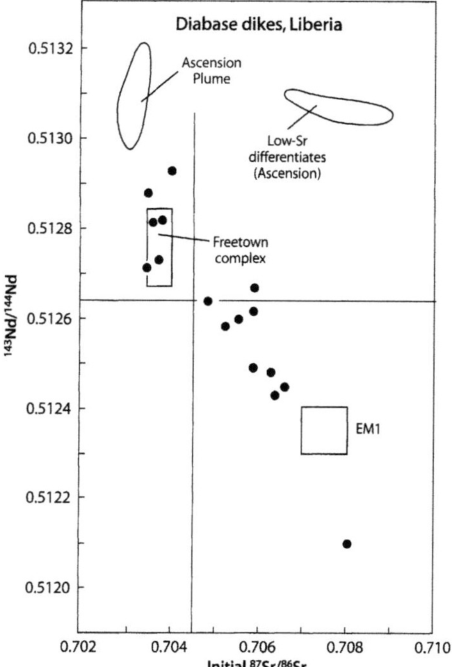

Fig. 5.57. Isotope ratios of Sr and Nd of Jurassic diabase dikes (*black circles*) in Liberia, West Africa, compared to those of Ascension Island. The position of the data field for the lavas of Ascension Island relative to the Liberian diabase dikes supports the hypothesis that the Ascension Plume is one of the source components of the Liberian diabase magmas. The data field of the Freetown layered gabbro complex establishes the close genetic relationship of this complex to the Liberian diabase dikes and the Ascension Plume (*Sources:* Weis et al. 1987; Dupuy et al. 1988b; Hart 1988; Morgan 1983; Hattori and Chalokwu 1995)

In addition, Mauche et al. (1989) used the correlation of initial ^{87}Sr/^{86}Sr ratios and δ^{18}O values to demonstrate that some of the magmas assimilated crustal rocks as they differentiated (AFC) and that the subsequent conversion of pyroxene to amphibole in the presence of heated groundwater caused the δ^{18}O values of some diabase samples to decrease.

The wide range of ^{87}Sr/^{86}Sr and ^{143}Nd/^{144}Nd ratios of the Liberian diabase dikes (Fig. 5.57) contrasts with the more homogeneous isotope compositions of the flood basalts of the Maranhão Plateau in northeastern Brazil (Fig. 5.54). Therefore, the isotopic data suggest that the flood basalts of the Maranhão Plateau originated primarily from subcrustal lithospheric mantle, whereas the Liberian diabase dikes represent mixtures of magmas derived from two components, including an asthenospheric plume and the subcontinental lithospheric mantle.

The reconstruction of plate motions by Morgan (1983) in Fig. 5.46 indicates that at about 190 Ma the Ascension Plume was located near the coast of West Africa. The subsequent drift of the African Plate caused the hotspot track first to move east for about 50 million years and then to turn southwest towards the present location of the Ascension Plume near the Mid-Atlantic Ridge (Sect. 2.5.4). This interpretation is strengthened by the fact that the isotope ratios of Sr and Nd of rocks on Ascension Island in Fig. 5.57 define a data field that fits the plume component of the Liberian diabase magmas (Weis et al. 1987a).

An even more compelling association of the Liberian diabase dikes with the Ascension Plume could be based on the isotope compositions of Pb. Harris et al. (1982) and Gast et al. (1964) reported that the ^{206}Pb/^{204}Pb ratios of the lava flows on Ascension Island exceed 19.00 in most cases and range up to 19.72. Consequently, the rocks that make up the Ascension Plume include the HIMU component defined partly by the nearby St. Helena Plume. Unfortunately, the ^{206}Pb/^{204}Pb ratios of the Liberian diabase dikes have not yet been measured.

5.9.4 Freetown Layered Gabbro Complex, Sierra Leone

The Freetown complex is a differentiated gabbroic intrusive located on the coast of West Africa south of Freetown, Sierra Leone. It extends for about 60 km along strike and is about 12 km wide (Umeji 1983, 1985). The complex is composed of layers of troctolite, gabbro, and anorthosite dipping southwest into the Atlantic Ocean with an exposed thickness of about 7 km. Beckinsale et al. (1977) dated granitic veins cutting the mafic rocks and obtained an Early Jurassic date of 193 ±3 Ma ($\lambda = 1.42 \times 10^{-11}$ yr^{-1}) with an initial ^{87}Sr/^{86}Sr ratio of 0.70389 ±0.00006 relative to 0.7080 for E&A. The granitic veins appear to have formed by in situ differentiation of the mafic magma, because the veins and the mafic rocks have the same initial ^{87}Sr/^{86}Sr ratios.

The Freetown complex formed by multiple intrusions of basaltic magma into Precambrian basement rocks followed by cooling and fractional crystallization. Wells (1962) originally identified four thick zones based on the development of characteristic topographic features and on the stratigraphic repetition of certain rock types. Chalokwu and Seney (1995) demonstrated cryptic layering in Zone 3 expressed by changes in the Mg concentration of olivine and the Ca concentration of plagioclase. These variations resulted from the repeated introduction of magma into the chamber, followed by cooling and fractional crystallization.

The initial $^{87}Sr/^{86}Sr$ ratios of clinopyroxene and plagioclase in Zones 3 and 4 decrease up-section from 0.70394 (cyclically layered subzone of Zone 3) to 0.70339 (gabbro-norite in Zone 4). Hattori and Chalokwu (1995) concluded from these results that the magmas originated from two sources in the mantle, having different $^{87}Sr/^{86}Sr$ ratios and that the source with the lower $^{87}Sr/^{86}Sr$ became dominant as a function of time. The comparatively low values of $^{87}Sr/^{86}Sr$ ratios exclude significant contamination of the magmas by Sr derived from the Archean basement rocks of West Africa, even though Hattori et al. (1991) reported the presence of crustal Os in these rocks.

The reconstruction of plume tracks by Morgan (1983) in Fig. 5.46 indicates that the Ascension Plume was located in the vicinity of the Freetown complex at about 190 Ma and therefore was one of the sources from which the magmas originated. The initial isotope ratios of Sr and Nd of the Freetown complex in Fig. 5.57 define an area that overlaps the data field of the Liberian diabase dikes (Hattori and Chalokwu 1995). The position of the Freetown data indicates that the magmas that formed this complex were two-component mixtures derived from the Ascension Plume and from the EM1 component represented by the subcrustal lithospheric mantle. The stratigraphic decrease of initial $^{87}Sr/^{86}Sr$ ratios implies that the magma source shifted from the subcontinental lithospheric mantle (high $^{87}Sr/^{86}Sr$ ratio) to the asthenospheric plume (low $^{87}Sr/^{86}Sr$ ratio). This interpretation differs from that of Hattori and Chalokwu (1995) who considered that the plume-derived magmas dominated initially, followed later by magmas that originated from depleted mantle. The Liberian diabase dikes and the Freetown complex represent the plumbing system through which magmas passed before being erupted at the surface. The complementary flood basalt plateau of West Africa has been eroded except for a few remaining flows exposed along the coast of Liberia.

5.10 Magmatic Activity Along the Atlantic Coast of North America

The Mesozoic diabase sills and basalt flows that occur along the east coast of North America record the igneous activity that was associated with the opening of the North Atlantic Ocean. These rocks were subdivided by McHone and Butler (1984) into four suites:

1. Early Cretaceous plutons of the New England-Quebec province including the Monteregian Hills;
2. Plutons of the White Mountain Magma Series (mainly Early Jurassic) in New Hampshire, Vermont, and Maine;

3. Diabase sills and dikes of Early Jurassic age extending from Alabama to Newfoundland, including the North Mountain basalts of Nova Scotia (Jones and Mossman 1988);
4. Alkalic complexes of Permian to Triassic age.

The plutons of the White Mountain Magma Series (110 to 120 Ma) in New Hampshire, Vermont, and Main are associated in Fig. 5.53 with the track of the Great Meteor Plume, which includes the New England Seamounts and merges with the track of the Verde Plume (Morgan 1983). An earlier episode of magmatic activity in New England between 170 to 130 Ma was caused by the Verde Plume, which preceded the Great Meteor Plume by about 40 million years. A third occurrence of igneous activity in New England at about 230 Ma (Triassic) is not accounted for by the model of Morgan (1983). A third plume track, caused by the Bermuda Plume, which is now located about 300 km east of the island of Bermuda, crossed the east coast of North America in North Carolina and Tennessee causing an increase in the uplift of the Appalachian Mountains in this area.

The Triassic igneous rocks of New England are alkalic in composition and include plutons at Agamenticus, Abbott, and Litchfield in the state of Maine. According to age determinations by Hoefs (1967), Burke et al. (1969), and Foland and Faul (1977), these intrusives crystallized at about 210 to 240 Ma and are the oldest Mesozoic plutons in New England.

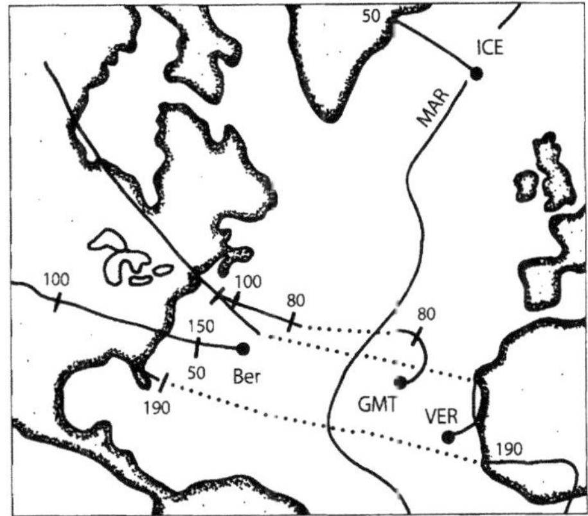

Fig. 5.58. Selected plume tracks in the Atlantic Ocean. The plumes are assumed to be stationary and are identified by the letters: *BER* = Bermuda, *VER* = Verde, *GMT* = Great Meteor Tablemount, *ICE* = Iceland, and *MAR* = Mid-Atlantic Ridge. The east coast of North America passed over the Verde Plume at 180 Ma and over the Meteor Plume about 40 million years later (*Source:* Morgan 1983)

5.10.1 Basalt and Diabase of the East Coast of North America

An extensive literature exists concerning the occurrence of tectonic basins and associated basalt flows and diabase sills on the east coast of North America, extending from Newfoundland to Florida where Mesozoic tholeiites and diabase occur beneath the Cenozoic sedimentary rocks of the Coastal Plain of southern Georgia and Florida (Heatherington and Mueller 1999, 1991). In addition, Mesozoic basalts in Tunisia, Morocco, Algeria, Spain, and Mauritania are probably correlative with the diabase sills and basalt flows along the east coast of North America (Kurtz 1983; Bertrand et al. 1982; Manspeizer et al. 1978). The literature on the Mesozoic basalt and diabase of eastern North America includes books edited by Manspeizer (1988) and by Robinson and Froelich (1985), as well as papers by Weigand and Ragland (1970), McHone (1978), Philpotts and Reichenbach (1985), Jones and Mossman (1988), and many others listed by McHone and Butler (1984) and Pegram (1990).

Age determinations by the K-Ar and ^{40}Ar/^{39}Ar methods published by Armstrong and Besancon (1970), Dallmeyer (1975), Sutter and Smith (1979), Seidemann et al. (1984), and Seidemann (1988) have yielded a spectrum of dates. For example, Seidemann (1988) reported dates ranging from 167 ±7 to 959 ±154 Ma for basalt flows in the Hartford Basin of Connecticut and attributed this range of dates to the presence of varying amounts of excess radiogenic ^{40}Ar. Accordingly, Seidemann (1988) preferred a date of 187 ±3 Ma (Early Jurassic) for the Talcott flow of the Hartford Basin in Connecticut. Sutter and Smith (1979) reported an even younger date of 175.0 ±3.2 Ma for several sills in Connecticut and Maryland, and obtained a date of 191.1 ±4.2 Ma for the Mt. Carmel Sill that is similar to the dates obtained by Dallmeyer (1975) for the well-known Palisade Sill of New Jersey (192.5 ±9 Ma). The available evidence indicates that the basalt flows and diabase sills along the east coast of the USA crystallized in Early Jurassic time and thus record the onset of rifting leading to the opening of the North Atlantic Ocean. In this respect, the Early Jurassic basalt flows and sills of the east coast of North America are equivalent to the diabase dikes of Liberia, which crystallized at about 185 Ma (Dalrymple et al. 1975b).

The first measurements of initial ^{87}Sr/^{86}Sr ratios by Faure and Hurley (1963) yielded values of 0.7034 and 0.7045 for the Palisade Sill (New Jersey) and 0.7049 for the Mt. Carmel Sill (Connecticut) relative to 0.7080 for E&A and assuming an age of 180 Ma. In addition, Barker and Long (1969) reported an initial ^{87}Sr/^{86}Sr ratio of 0.7057 for a specimen of unaltered quartz diabase at Brookville, New Jersey, based on

$\lambda = 1.42 \times 10^{-11}$ yr^{-1} and an age of 180 Ma. These authors also demonstrated that a granophyre and syenite associated with the Brookville Sill had been contaminated by assimilation of the Lockatong Argillite, which is the local host of the sill.

The initial isotope ratios of Sr and Nd of diabase sills in North and South Carolina, Pennsylvania, New Jersey, and Connecticut define a data field in Fig. 5.59 that overlaps the EM1 mantle component (Pegram 1990). The diagram also contains the data field of lavas on the Cape Verde Islands (Gerlach et al. 1988a; Sect. 2.5.2), because at about 180 Ma the east coast of North America was passing over to Verde Plume (Morgan 1983), which provided heat to the subcrustal lithospheric mantle of the North American Plate thus causing basalt magma to form by partial melting.

The lavas on the Cape Verde Islands define an elongated data field in Fig. 5.59 that extends from the DMM component towards EM1 (Gerlach et al. 1998a). The diabase sills of the east coast of North America extend

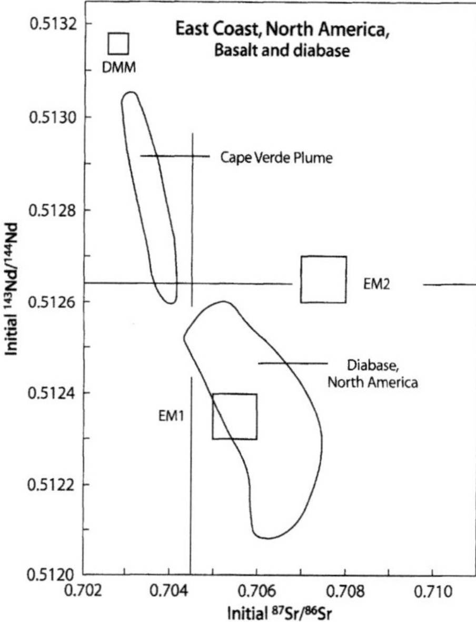

Fig. 5.59. Initial isotope ratios of Sr and Nd at 180 Ma of diabase sills of Early Jurassic age along the east coast of North America from Massachusetts to South Carolina. The diagram also contains the data field of lavas on the Cape Verde Islands, because the trailing edge of the North American Plate was passing over the Verde Plume in Early Jurassic time. The diabase magmas originated primarily by decompression melting in the subcrustal lithospheric mantle, which contained a component of previously subducted continental sediment (*Sources:* Pegram 1990; Gerlach et al. 1988a; Hart 1988; Morgan 1983)

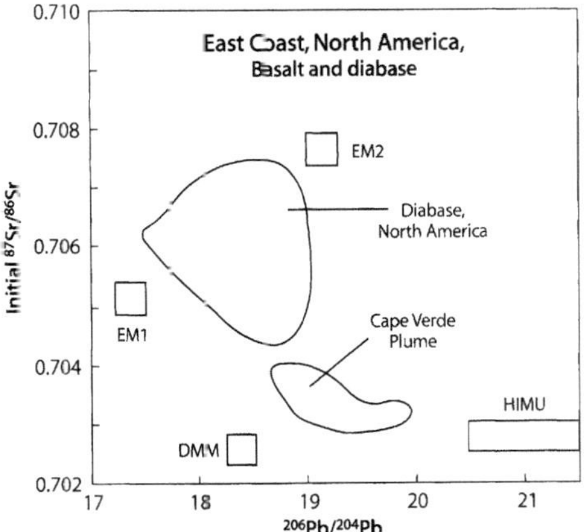

Fig. 5.60. Isotope ratios of Sr and Pb of Early Jurassic diabase along the east coast of North America and on the Cape Verde Islands. The data fields indicate that the Cape Verde magmas contain Pb derived from the HIMU source and that the North American diabase magmas derived Pb from the EM2 component, which contains previously subducted sediment of continental origin (*Sources:* Pegram 1990; Gerlach et al. 1988a; Hart 1988; Morgan 1983)

this trend beyond the EM1 component in some cases. These data indicate that the Early Jurassic diabase magmas of North America formed primarily by partial melting of subcrustal lithospheric mantle, which contained previously subducted sediment derived from continental sources (Pegram 1990). In some cases, the magmas may have contained a component derived from the Verde Plume which itself included a heterogeneous assemblage of mantle rocks (Gerlach et al. 1988a).

The ^{206}Pb/^{204}Pb ratios of the Cape Verde lavas in Fig. 5.60 range up to 19.881, indicating that the HIMU component contributed to the magmas that were erupted on these islands (Gerlach et al. 1988a). In addition, the combination of Sr and Pb isotope compositions emphasizes the importance of subducted continental sediment (EM2) among the magma sources of the diabase sills along the east coast of North America. A comparison of the data fields of the Liberian diabase dikes in Fig. 5.57 and of the diabase sills of the eastern margin of the North American Plate in Fig. 5.59 reinforces the interpretation that both originated from magmas that formed primarily by decompression melting of subcrustal lithospheric mantle and (to a lesser extent) rocks in the respective asthenospheric plumes. The plumes caused extension and rifting of the overlying lithospheric plate and provided the heat that facilitated the formation of large volumes of basalt magma as the Atlantic rift widened and the plate fragments drifted apart.

5.10.2 Anorogenic Plutons of New England

The granitic and syenitic plutons in New Hampshire, Vermont, and Maine are commonly referred to as the White Mountain Magma Series. However, McHone and Butler (1984) restricted that term to the Early Jurassic plutons in the White Mountains of New Hampshire and excluded the Early Cretaceous intrusives that also occur in this region. The geology of the two suites of intrusives was reviewed by Eby (1987), whereas Zartman (1988) summarized the available age determinations.

The plume tracks in Fig. 5.58 indicate that in Early Jurassic time the New England Coast was passing over the Verde Plume (Morgan 1983). About 40 million years later, the same area passed over the Meteor Plume, which caused the intrusion of a second suite of plutons of Early Cretaceous age and resulted in the development of the chain of the New England Seamounts in the North Atlantic Ocean.

The Early Jurassic plutons of New Hampshire and Vermont are predominantly composed of alkali-rich granites containing abundant biotite, as well as riebeckite and hastingsite. In addition, some of these plutons include syenite, monzonite, diorite, and gabbro. The plutonic rocks were originally associated with voluminous deposits of rhyolite, andesite, basalt, and pyroclastics; most of which have been eroded. Field relations indicate that the plutons formed by sequential intrusion of magmas ranging in composition from mafic to granitic such that the alkali-rich biotite granites were the last rocks to be intruded (Turner and Verhoogen 1960).

The initial ^{87}Sr/^{86}Sr ratios of the Early Jurassic plutons of New England range from 0.70310 to 0.70880 relative to 0.71025 ±0.00015 for NBS 987 (Foland and Allen 1991). The Early Cretaceous plutons have similar initial ^{87}Sr/^{86}Sr ratios between 0.70338 and 0.70650 (Eby 1985a; Foland et al. 1988). Foland and Allen (1991) demonstrated by modeling in Nd-Sr isotope coordinates that the Jurassic and Cretaceous granites in New England originated from mantle-derived magmas, which assimilated crustal rocks and differentiated by fractional crystallization. The syenite magmas had a similar origin but experienced less differentiation by fractional crystallization than the granite magmas. If the granitic magmas had formed by extensive differentiation of basaltic magmas, then complementary mafic rocks should exist at depth, but the geophysical evidence does not support that expectation in all cases (Foland and Allen 1991).

The complementary relationship between mafic rocks (gabbro/diorite) and granite is illustrated in by data from Foland et al. (1985) for the Early Cretaceous composite pluton at Mount Ascutney, Vermont. The granites and some of the gabbros form a Rb-Sr isochron yielding a date of 122.2 ±1.2 Ma for both suites of rocks with an initial ^{87}Sr/^{86}Sr ratio of 0.70400 ±0.00006 relative to 0.71025 for NBS 987. Therefore, the granites could have formed by fractional crystallization of basalt magmas derived from the subcrustal lithospheric mantle and/or the Meteor Plume. The variation of initial ^{87}Sr/^{86}Sr ratios of some of the mafic rocks on Mt. Ascutney was caused by assimilation of local Precambrian gneisses by the mafic magmas. Alternatively, the granitic magmas could have formed by partial melting of mafic rocks previously intruded into the lower continental crust.

The significance of the Sr isotope data derived from the Mesozoic plutons of New England is that the highly evolved granite plutons such as Mt. Ascutney (Foland et al. 1985), Merrymeeting Lake (Eby 1985a; Foland and Allen 1991), and Pawtuckaway (Eby 1985a) have low initial ^{87}Sr/^{86}Sr ratios similar to those of the associated gabbros.

5.10.3 New England Seamounts

The Cretaceous plutons of New England are related to the New England Seamounts, because both originated from the Meteor Plume as the North American Plate passed over it (Morgan 1983). Duncan (1984) demonstrated by means of ^{40}Ar/^{39}Ar dating that the ages of the New England Seamounts increase in a northwesterly direction from 82 Ma (Nashville; 35°18.4' N, 57°33.6' W) to 103 Ma (Bear; 39°49.2' N, 67°26.5' W). These results confirm the relation of these seamounts to the Meteor Plume and hence support the association of the seamounts to the Cretaceous plutons of New Hampshire (Duncan 1984; Foland et al. 1986, 1988).

The initial ^{87}Sr/^{86}Sr ratios of volcanic rocks dredged from the New England Seamounts (Fig. 5.61) range only from 0.70311 to 0.70348 relative to 0.7080 for E&A (Taras and Hart 1987). These data were derived from separated minerals (hornblende and clinopyroxene) and from whole-rock samples after leaching the latter with 6.2 N HCl for 10 to 12 h at 125 °C. Hand-picked mineral grains were cleaned twice in 2.5 N HCl for 30 min at 125 °C, once in 5% HF for 10 min, and then rinsed with water. The initial ^{87}Sr/^{86}Sr ratios so obtained are higher than those of MORBs (Fig. 2.5) and represent the magma sources in the head of the Meteor Plume and/or of the overlying sub-oceanic lithospheric mantle.

The initial ^{87}Sr/^{86}Sr ratios of the fine-grained monzonites of the Pawtuckaway Pluton (129 ±5Ma) (Eby 1985a) and of quartz syenite in the Mascoma Quadrangle of New Hampshire (Foland and Allen 1991) lie within the range of initial ^{87}Sr/^{86}Sr ratios of the New England Seamounts. This result is consistent with the hypothesis that the magmas that formed these plutons origi-

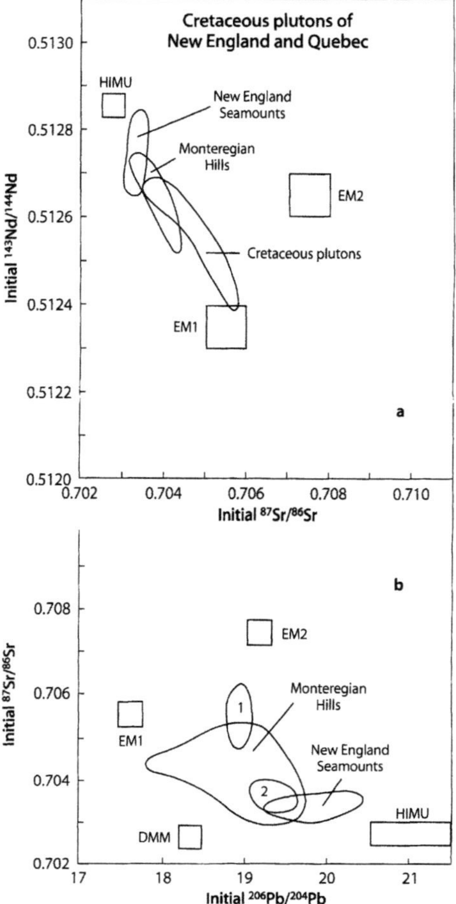

Fig. 5.61. a Initial isotope ratios of Sr and Nd of basalts of the New England Seamounts, of the Cretaceous anorogenic plutons of Quebec (Monteregian Hills), and of the New England states in the USA (*Sources:* Taras and Hart 1987; Foland et al. 1988; Hart 1988); b Initial isotope ratios of Pb and Sr of the New England Seamounts and the Cretaceous anorogenic plutons of Quebec and the New England states. The data fields identified by number are: *1.* Merrymeeting and *2.* Pawtuckaway. The extensive overlap of the data fields in Parts *a* and *b* strongly supports the hypothesis of Morgan (1983) and others that all three suites are related to the Meteor Plume (*Sources:* Taras and Hart 1987; Eby 1985a; Hart 1988)

nated from the Meteor Plume without significant crustal contamination. However, most of the plutons examined by Foland and Allen (1991) have higher initial ^{87}Sr/^{86}Sr ratios than the New England Seamounts, either because their magmas originated from the subcrustal lithospheric mantle, or because they were contaminated by assimilating crustal rocks, or both.

5.10.4 Monteregian Hills, Quebec

The trend of the New England Seamounts and Cretaceous plutons of New Hampshire is continued by the plutons of the Montereg an Hills in Quebec, Canada, which include (from southeast to northeast): Megantic, Brome, Shefford, Yamaska, Johnson, Rougemont, St. Hilaire, St. Bruno, and Mt. Royal, as well as a carbonatite at Oka located about 30 km west of Montreal (Sect. 6.15). These plutons are composed of multiple intrusions of different alkali-rich rock types. Following the early work by Faure and Hurley (1963) and Fairbairn et al. (1963), Eby (1984a) reported that slightly undersaturated to saturated rocks have ages of about 136 Ma, whereas moderately to strongly undersaturated rocks cluster around an age of 118 Ma. However, Foland et al. (1986) demonstrated that $^{40}Ar/^{39}Ar$ dates of biotite and amphibole separated from the rocks of seven Monteregian plutons range only from 123.1 ±1.2 Ma (Brome) to 127.4 ±1.5 Ma (Bruno). The latter result may be too old due to the presence of excess ^{40}Ar identified by the elevated $^{40}Ar/^{39}Ar$ ratios of Ar released at low temperature during the analysis. Therefore, Foland et al. (1986) concluded that the plutons they studied had been intruded during a short time interval at 124 ±1 Ma and cooled rapidly. The carbonatite at Oka appears to be younger than the other plutons based on K-Ar dates of biotite that average 116 ±4 Ma (Shafiqullah et al. 1970).

The initial $^{87}Sr/^{86}Sr$ ratios of the rocks of the Monteregian plutons vary internally and among each other, but do not correlate with Sr concentrations. The data of Eby (1985a) indicate that the strongly undersaturated rocks of Mt. Johnson have the lowest average initial $^{87}Sr/^{86}Sr$ ratio of 0.70332 ±0.00004 (2$\bar{\sigma}$, N = 5, at 119 Ma), whereas the granites of Mt. Megantic have 0.70518 ±0.00000 (2$\bar{\sigma}$, N = 4, at 128 Ma). The initial $^{87}Sr/^{86}Sr$ ratios of the Monteregian Hills give evidence that the magmas originated from the Meteor Plume and were variously contaminated by assimilating crustal rocks (Foland and Allen 1991). In addition, the magmas differentiated by fractional crystallization to produce the wide range of chemical compositions represented by the rocks of the Monteregian Hills (Eby 1984b, 1985b; Bedard 1994).

The carbonatite at Oka (Sect. 6.15) is located near the intersection of the Ottawa and Champlain Grabens with the St. Lawrence Rift system (Gold 1969). It was intruded at about 120 Ma (Fairbairn et al. 1963; Shafiqullah et al. 1970) into Precambrian gneisses and anorthosites of the Grenville structural province. The intrusive consists of arcuate sheets of alkali-rich silicate rocks that surround cores of soevite (calcite carbonatite). The intrusion of carbonatitic magma is the most recent manifestation of igneous activity in the New England-Quebec igneous province.

Calcite of the carbonatite at Oka has a low average initial $^{87}Sr/^{86}Sr$ ratio of 0.70335 ±0.00002 (2$\bar{\sigma}$, N = 6) relative to 0.71025 for NBS 987 (Grünenfelder et al. 1986). Subsequently, Wen et al. (1987) reported 0.70329 ±0.00001 (2$\bar{\sigma}$, N = 25) for a variety of rocks and minerals at Oka. The average Sr concentration of the calcite is 18 240 (11 100 to 24 640) ppm making it quite insensitive to contamination with Sr from crustal rocks. Powell et al. (1962, 1965a,b) and Powell (1965a,b, 1966) first drew attention to the low $^{87}Sr/^{86}Sr$ ratios of the Oka carbonatite and used this fact to discredit the idea that carbonatites are recrystallized limestone xenoliths. Powell (1966) concluded that carbonatites are comagmatic with the associated alkali-rich rocks and that their magmas originated from subcrustal sources in the mantle. Bell et al. (1982) and Wen et al. (1987) subsequently emphasized that the isotope compositions of Sr in carbonatites represents the magma sources located in the mantle more than 100 km below the surface.

5.11 The South Africa-Antarctica Connection

The breakup of Gondwana started with the separation of southeast Africa from Queen Maud Land of Antarctica (Cox 1978; Storey et al. 1992b; Burke 1996). Rifting may have begun during the Permian Period, but full-scale separation was delayed until about 178 Ma (White and McKenzie 1989). The onset of rifting by seafloor spreading between southeast Africa and Queen Maud Land of Antarctica in Fig. 5.62 was preceded by the eruption of a very large volume of tholeiite basalt at 193 ±5 Ma (Early Jurassic) on the African continent and off-shore along the Andenes-Explora Escarpment (Kristoffersen and Haughland 1986; Fitch and Miller 1984). A second outpouring of basalt lava at 178 ±5 Ma (Early to Middle Jurassic) is recorded both in South Africa and in Queen Maud Land of Antarctica (White and McKenzie 1989). Both episodes of volcanic activity occurred during relatively short intervals of time, implying very rapid rates of eruption (Fitch and Miller 1984).

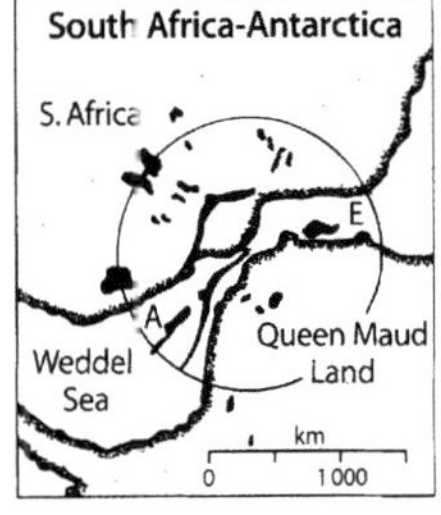

Fig. 5.62.
Rifting of Gondwana between southeast Africa and Queen Maud Land of East Antarctica. The black areas are remnants of the former flood basalt plateau. The rift between the two continents contains basalt plateaus of the Andenes Escarpment (A) and the Explora Wedge (E) (*Source:* adapted from White and McKenzie 1989)

The total volume of lava erupted at the surface and intruded in the form of sills and dikes in southeast Africa and Queen Maud Land is difficult to estimate because only erosional remnants are preserved in some areas and because the basalt flows are partly covered by younger rocks in South Africa or by the Antarctic ice sheet in Queen Maud Land. Cox (1970, 1972) estimated that the Karoo lavas in southeast Africa alone may have covered an area of about 2×10^6 km^2. According to Eales et al. (1984), the thickness of lavas buried under the coastal plain of Mozambique is between 6 and 13 km. Therefore, the volume of Karoo basalt in southeast Africa alone is probably greater than 10×10^6 km^3.

The eruption of the volcanic rocks of Early Jurassic age in southeast Africa and Queen Maud Land occurred in a setting of extensional tectonics that developed when the head of an asthenospheric plume impinged against the bottom of the subcontinental lithosphere of Gondwana and spread out, reaching a diameter of about 2 000 km (Fig. 5.62; White and McKenzie 1989). The Karoo basalts that resulted from this tectonic activity are older than the lavas of the Etendeka Group of Namibia, which, in addition, are related to the Tristan Plume in the South Atlantic Ocean.

5.11.1 The Karoo Volcanic Province, Southeastern Africa

Dolerite sills and dikes as well as basalt and rhyolite flows of Early to Middle Jurassic age occur widely throughout southeastern Africa in Fig. 5.63. The principal exposures of the erosional remnants of the Karoo basalt are located in Lesotho, in eastern Botswana, and in the Nuanetsi-Lebombo Monocline that extends from Swaziland on the east coast north into Zimbabwe and Mozambique (Eales et al. 1984). The extent of the Karoo igneous province is significantly increased by the presence of basalt flows, dolerite sill and dikes below post-Karoo rocks in the Kalahari Desert of Botswana and northeastern Namibia, as well as in Barotseland of western Zambia. In addition, the basalts and rhyolites of the Nuanetsi-Lebombo monocline in southeastern Africa dip under Cretaceous to Tertiary sediment and may extend in subsurface more than 300 km east to northern Madagaskar (Melluso et al. 1997). The Karoo igneous province also encompasses ring complexes of highly differentiated basaltic and alkalic rocks, including carbonatites. These complexes occur in linear belts in Damaraland of Namibia (Sect. 5.7.3), in Nuanetsi of southeast Zimbabwe, and in Angola (Eales et al. 1984).

The lavas erupted in southeastern Africa are composed primarily of tholeiite basalt, but do encompass a range of compositions from early-formed picrites in

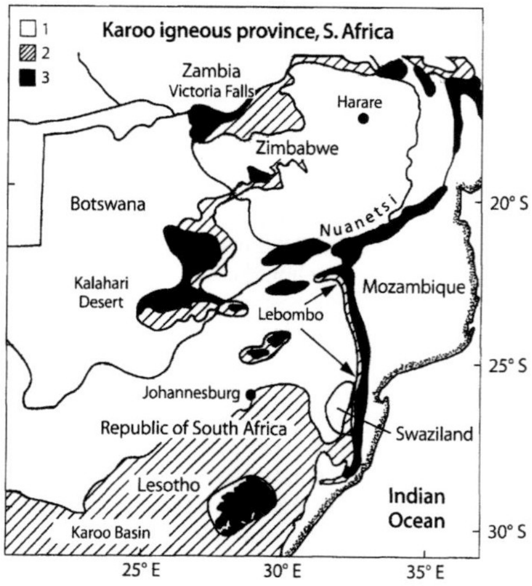

Fig. 5.63. Outcrop areas of Karoo basalt and sediment in southeastern Africa. The northeastern region discussed in the text includes primarily Nuanetsi (Zimbabwe) and Lebombo (South Africa), whereas the southern region consists of Lesotho, southern Lebombo, and the eastern and western Cape Province in the Karoo Basin of South Africa. The major rock units are identified by number: *1.* Precambrian basement and post-Karoo cover; *2.* Karoo sediments; *3.* Karoo basalt (*Source:* adapted from Eales et al. 1984)

the Nuanetsi area to late-stage rhyolites in the Lebombo monocline (Eales et al. 1984). The petrogenesis of the igneous rocks of the Karoo igneous province was discussed in books edited by Clifford and Gass (1970), Dingle et al. (1983), Erlank (1984), Weaver and Johnson (1987), and Macdougall (1988). Additional contributions to the study of these rocks have been made by Manton (1968), LeRoex and Reid (1978), Betton (1979), Richardson (1979), Cox (1988), and by many others referred to by them.

The history of igneous activity in southern Africa during the Mesozoic Era includes the events listed in Table 5.8 (Fitch and Miller 1984).

The volcanic activity started in the Cape Province of South Africa (DeWit and Ransome 1992; Veevers et al. 1994) and in Lesotho at about 193 ±5 Ma, and moved north into Botswana and Zimbabwe. The voluminous rhyolites of the Nuanetsi area also formed at 191 ±9 Ma (Allsopp et al. 1984) and were succeeded at 178 ±5 Ma by basalt and rhyolite along the Lebombo monocline. Late-stage ignimbrites were extruded at 133 ±4 Ma (e.g. the Bumbeni complex, southern Lebombo; Allsopp et al. 1984).

The Lebombo monocline in Fig. 5.63 extends from the Nuanetsi area of Zimbabwe in the north for a distance of more than 1 000 km south to Natal in South

Table 5.8. Events in the history of igneous activity in southern Africa during the Mesozoic Era (Fitch and Miller 1984)

Event (Ma)	Manifestation
204±5	Alkaline intrusive complexes
193±5	Major eruption of basalt and intrusion of sills and dikes
186±3	Minor event Marangudzi alkaline complex, Zimbabwe
178±5	Major eruption of basalt and intrusion of sills and dikes
165±5	Intrusion of sills and dikes
150±5	Intrusion of dikes and of the Messum complex in Namibia
137±5	Various dolerite sills and rhyolite flows
121±1	Extrusion of Etendeka formation in western Namibia
85–90	Intrusion of kimberlite pipes

Africa. The volcanic and sedimentary rocks of the Karoo Group in this belt dip to the east under the rocks of the coastal plain of Mozambique. The volcanic rocks consist primarily of tholeiite basalt (up to 5 km thick) overlain by a similar thickness of rhyolite flows, which are overlain by another sequence of basalt flows in the central part of the Lebombo monocline (Manton 1968; Cleverly et al. 1984).

The major stratigraphic units of the volcanic rocks in the Lebombo monocline (Fig. 5.63), starting with the most recent, include (Cleverly et al. 1984):

- Movene basalt (youngest)
- Bumbeni complex
- Mbuluzi rhyolite
- Jozini rhyolite
- Sabie River basalt
- Letaba River basalt
- Mashikiri nephelinite (oldest)

The initial ^{87}Sr/^{86}Sr ratios of the Karoo basalt in the Nuanetsi-Lebombo Belt in Fig. 5.64 vary widely, but are concentrated primarily between 0.7040 and 0.7060. The associated rhyolites of this area have similar initial ^{87}Sr/^{86}Sr ratios to those of the basalt. The rhyolites of the Nuanetsi Formation are an exception with initial ^{87}Sr/^{86}Sr ratios between 0.7066 and 0.7089. In addition, the Mkutshane rhyolites of Swaziland and southern Lebombo have very high initial ^{87}Sr/^{86}Sr ratios, ranging from 0.7150 to 0.7383 (Bristow et al. 1984; Hawkesworth et al. 1984a; Betton 1979).

In Swaziland and southern Lebombo the Sabie River basalt constitutes the basal volcanic sequence, because the underlying Letaba River basalt and the Mashikiri nephelinites are absent in this area. Therefore, the Sabie River basalt was deposited directly on the sedimentary rocks of the Karoo Group, which are underlain by granitic gneisses of the Archean Kaapvaal Craton. Betton (1979) reported that the different lithologic units of the Kaapvaal Craton are variously enriched in radiogenic ^{87}Sr depending on their ages and Rb/Sr ratios. These units also had ^{87}Sr/^{86}Sr ratios ranging from 0.73 to 4.7 at 197 Ma. In addition, Betton (1979) demonstrated that the Mkutshane rhyolites de-

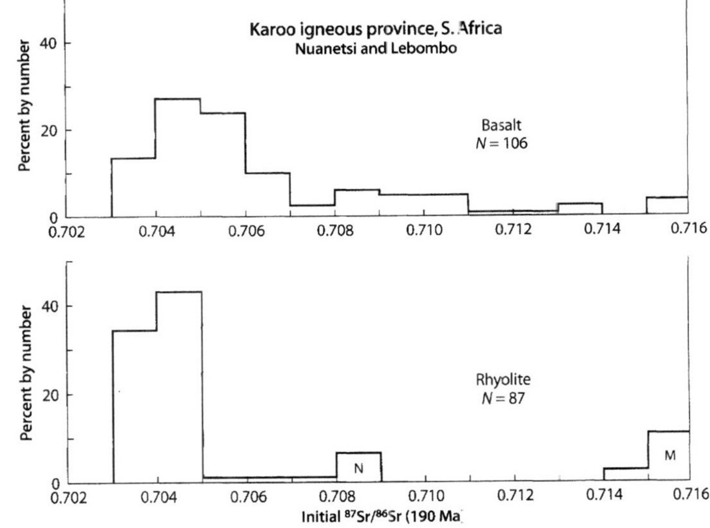

Fig. 5.64. Initial ^{87}Sr/^{86}Sr ratios of Karoo basalt and rhyolite at 190 Ma in the Nuanetsi and Lebombo areas of Zimbabwe and South Africa identified in Figure 5.63. The rhyolites of Nuanetsi (N) and the Mkutshane rhyolite of Swaziland (M) are specifically identified. The principal feature of these data is that basalts and associated rhyolites have similar initial ^{87}Sr/^{86}Sr ratios (*Sources:* data from Bristow et al. 1984; Hawkesworth et al. 1984; Betton 1979)

fine a straight line on the Rb-Sr isochron diagram, yielding a fictitious date of 1 294 Ma, whereas Manton (1968) reported a date of 198 Ma for Karoo rhyolites and granophyres in Swaziland. Based on this and other evidence, Betton (1979) concluded that the Mkutshane rhyolite contains Sr derived from the granitic basement rocks of the Kaapvaal Craton that underlies the sedimentary and volcanic rocks of the Karoo Group in Swaziland.

The contamination of the magmas with crustal Sr recognized by Betton (1979) in the case of the Mkutshane rhyolites is also evident in the average initial $^{87}Sr/^{86}Sr$ ratios of the Sabie River basalt (Cox and Bristow 1984), which range from 0.70605 ±0.0078 (2$\bar{\sigma}$, $N = 5$) in Nuanetsi in the north to 0.70481 in the Olifants River area of northern Lebombo, to 0.70578 ±0.00110 (2$\bar{\sigma}$, $N = 5$) in central Lebombo and Swaziland, and to 0.70829 ±0.00149 (2$\bar{\sigma}$, $N = 13$) in southern Lebombo. Actually, both the lowest (0.7031) and the highest (0.71343) initial $^{87}Sr/^{86}Sr$ ratios of the Sabie River basalt have been reported from southern Lebombo, implying considerable isotopic heterogeneity of the magmas either in time or place or both. The Sabie River basalts of Swaziland and southern Lebombo were subsequently intruded by the Rooi Rand dolerite dikes, which have comparatively low initial $^{87}Sr/^{86}Sr$ ratios ranging from 0.70350 to 0.70416 with a mean of only 0.70382 ±0.00009 (2$\bar{\sigma}$, $N = 13$) (Bristow et al. 1984; Hawkesworth et al. 1984a; Armstrong et al. 1984).

The Karoo basalts and rhyolites of Nuanetsi have higher $^{87}Sr/^{86}Sr$ ratios on average than those of the Lebombo in the south (Manton 1968). This phenomenon is exemplified by the rhyolites of the Nuanetsi Formation and by the basalts of the Sabie River Formation, both of which were mentioned above. The $^{87}Sr/^{86}Sr$ ratios of the rhyolites in Lebombo and Swaziland (Jozini, Twin Ridge, Mbuluzi, and undifferentiated) have consistently low initial $^{87}Sr/^{86}Sr$ ratios similar to those of most of the basalts in Fig. 5.64. In addition, age determinations by Allsopp et al. (1984), using the whole-rock Rb-Sr method, indicate that the rhyolites of Nuanetsi are older (191 ±9 Ma) than those of northern Lebombo (177 ±6 Ma) and of Swaziland (southern Lebombo) (179 ±4 Ma). These authors also reported whole-rock Rb-Sr dates and initial $^{87}Sr/^{86}Sr$ ratios for the Kuleni rhyolite (145 ±3.2 Ma; 0.7114 ±0.0103) and the Bumbeni complex (132.9 ±4.3; 0.70547 ±0.00022). These are the last lavas that were erupted along the Nuanetsi-Lebombo Belt.

The $^{206}Pb/^{204}Pb$ ratios of basalts and rhyolites in the Nuanetsi-Lebombo Belt are generally low (17.0 to 18.2) and not distinguishable from each other (Betton et al. 1984). The rhyolites of the Mkutshane rhyolites in Swaziland are a major exception to this generalization with an average $^{206}Pb/^{204}Pb$ ratio of 19.48 ±0.09 (2$\bar{\sigma}$, $N = 9$). The elevated average $^{206}Pb/^{204}Pb$ ratio of the

Mkutshane rhyolite presumably reflects the presence of Pb derived from the granitic gneisses of the underlying Kaapvaal Craton. Betton et al. (1984) also observed small differences in the isotope ratios of Pb in the basalt and rhyolite of the Nuanetsi area, which they attributed to contamination of the magmas with Pb, derived from the underlying basement rocks of the Lebombo Belt.

The measured $\delta^{18}O$ values of rhyolites in the Lebombo monocline range from +5.0 to +10.3‰, whereas those of pyroxene phenocrysts (augite) range only from +3.7 to +6.1‰ (Harris and Erlank 1992). The measured $\delta^{18}O$ values of the rhyolites appear to correlate negatively with the initial $^{87}Sr/^{86}Sr$ ratios of the rocks, although two of the nine samples analyzed by Harris and Erlank (1992) deviate from this trend. This relationship (not shown) could be interpreted to mean that the rhyolite magmas formed from source rocks in the crust that had interacted with meteoric water containing Sr having elevated $^{87}Sr/^{86}Sr$ ratios. Harris and Erlank (1992) calculated the $\delta^{18}O$ values of the rhyolite magmas based on the measured oxygen isotope ratios of pyroxenes. These calculated $\delta^{18}O$ of the rhyolites range only from +4.4 to +6.8‰ and are lower than expected for magmas that formed by partial melting of rocks in the crust or lithospheric mantle. On the basis of this evidence, Harris and Erlank (1992) concluded that the rocks of the magma sources, believed to be Karoo-age gabbros of the Rooi Rand suite, had interacted with meteoric water that percolated into the crust through fractures that formed as a result of extension during rifting between southern Africa and Antarctica.

The most compelling evidence for extensive crustal contamination of the basalt and rhyolite magmas of the Karoo igneous province is provided in Fig. 5.65 by the initial isotope ratios of Sr and Nd. The data fields of both basalt and rhyolite in the different areas diverge from the isotope composition of the Rooi Rand Dikes and the most primitive Sabie River basalt of southern Lebombo. The divergence of the data fields reveals the existence of significant differences in the isotope ratios of Sr and Nd of the granitic basement gneisses that underlie each of the three regions identified in Fig. 5.65. The Mkutshane rhyolites (M) have the lowest initial $^{143}Nd/^{144}Nd$ (0.51152) and the highest $^{87}Sr/^{86}Sr$ (0.7150) ratios of any of the Karoo lavas analyzed by Hawkesworth et al. (1984a).

To summarize, the eruption of the Karoo basalts and rhyolites occurred in a setting of extensional tectonics associated with rifting between southeastern Africa and Queen Maud Land of East Antarctica, which started at about 190 Ma and resulted in the opening of the Indian Ocean at about 178 Ma. White and McKenzie (1989) postulated that the rifting and eventual separation of Africa from Antarctica was caused by the presence of an asthenospheric plume; however, the present location of this plume is not known with certainty. The initial isotope ratios of Sr and Nd of the Karoo basalts are compatible with magma formation by decompression melting in the subcrustal lithospheric mantle, which was heterogeneous on a regional scale. The resulting magmas subsequently differentiated in crustal magma chambers (Stolper and Walker 1980) and, in some cases, assimilated heterogeneous wallrocks (e.g. the Sabie River basalt). Most of the rhyolites that are interbedded with basalt originated by remelting of Karoo-age mafic rocks near the base of the crust as a result of local temperature increases, caused by heat emanating from large volumes of basalt magmas rising toward the surface. The Mkutshane rhyolite of Swaziland and southern Lebombo is exceptional because it originated by partial melting of granitic basement rocks of the Kaapvaal Craton.

The basalt and quartz latite flows of the Etendeka Group in northwestern Namibia are not only younger than the Karoo basalt, but they also formed as a result of tectono-magmatic activity caused by the plume now located beneath Tristan da Cunha in the South Atlantic Ocean. Any similarities of the Etendeka basalts with

those of the Karoo are therefore attributable to the fact that both suites are the products of the same petrogenetic process operating in similar tectonic settings at different times.

5.11.2 Basalts of Queen Maud Land, Antarctica

The Mesozoic basalt flows and dikes of Queen Maud Land occur in Vestfjella (73°20' S, 14°10' W), Heimefrontfjella (74°35' S, 11°00' W), Bjornnutane (74°00' S, 10°00' W), Sembberget (74°35' S, ~8° W), and in the Kirwan Escarpment (73°25' S, 5°30' W), all of which consist of groups of nunataks projecting through the ice sheet of Queen Maud Land in Fig. 5.66 (Brewer and Brook 1991). The nunataks of Vestfjella consist of altered flows of pyroxene andesite interbedded with tuffaceous sandstone and dolerite intrusions (Juckes 1968). Two specimens of dolerite from Vestfjella yielded whole-rock K-Ar dates of 168 ±6 and 172 ±6 Ma, suggesting a Middle Jurassic age for these flows (Rex 1967; Peters et al. 1991). Additional age determinations by Rex (1972) later confirmed the Jurassic age for basalt lavas and dikes of Milorgfjella (74°29' S, 9°20' W) but also indicated the occurrence of pre-Permian basalt dikes at this location and at Mannefallknausane. The initial $^{87}Sr/^{86}Sr$ ratios of basalt and dolerite from Queen Maud Land reported by Faure and Elliot (1971) and by Faure et al. (1972) demonstrated that the Mesozoic basalts and dolerites of Queen Maud Land have significantly lower initial $^{87}Sr/^{86}Sr$ ratios than the age-equivalent rocks in the Transantarctic Mountains and of the Dufek Intrusion of Antarctica, and resemble those of the Karoo igneous province of southern Africa.

The geology of Queen Maud Land has been described by Hjelle and Winsnes (1972), Furnes and

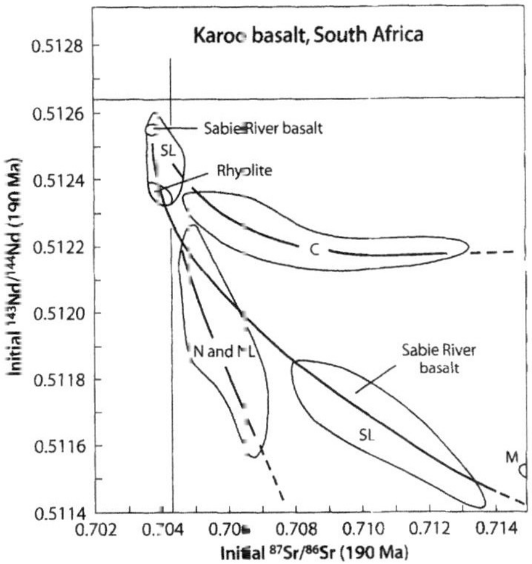

Fig. 5.65. Initial isotope ratios of Sr and Nd at 190 Ma of basalt and associated rhyolites of the Karoo Group in southern Africa. C = central region; N = Nuanetsi region; NL = northern Lebombo; SL = southern Lebombo; M = Mkutshane rhyolite, southern Lebombo. The reference lines representing CHUR and UR have been adjusted to an age of 190 Ma. The mixing hyperbolas converge to the isotope ratios of the Rooi Rand Dikes and the most primitive Sabie River basalt in southern Lebombo. These results indicate that the basalt and rhyolite magmas originated from source rocks having similar isotope compositions and were subsequently contaminated by Sr and Nd derived from heterogeneous basement rocks (Source: Hawkesworth et al. 1984)

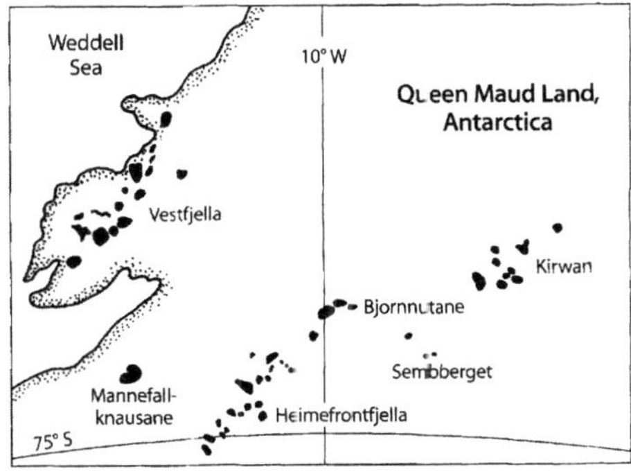

Fig. 5.66. Location of the principal rock exposures in Queen Maud Land. The relation of Queen Maud Land to southeastern Africa is apparent in Figure 5.62 (adapted from Furnes et al. 1987)

Mitchell (1978), and Luttinen et al. (1998) (Vestfjella), Aucamp et al. (1972) and Harris et al. (1990), (Kirwan), and Juckes (1968) (Mannefallknausane). In general, the Mesozoic basalt flows of Queen Maud Land were deposited on flat-lying sedimentary rocks of late Paleozoic age which presumably correlate with the sedimentary rocks of the Karoo Group in southern Africa. The late Paleozoic sandstones and shales are unconformably underlain by highly deformed volcano-sedimentary rocks and granitic intrusives of Middle to Late Proterozoic age (Eastin et al. 1970; Eastin and Faure 1970; Allsopp and Neethling 1970). In addition, Juckes (1968) described granitic gneisses that form the basement of the Mannefallknausane nunataks.

The chemical composition of the Jurassic basalt lavas and dikes of Queen Maud Land (Table 5.9) vary only within narrow limits consistent with low-Ti tholeiite basalts, although some dikes are composed of high-Ti basalt with alkaline affinity (Furnes and Mitchell 1978).

Rhyolites like those of the Nuanetsi-Lebombo Belt in southern Africa have not been reported from Queen Maud Land of Antarctica. The major and trace-element compositions of the lava flows and dikes of Vestfjella were later discussed by Furnes et al. (1982, 1987).

The initial $^{87}Sr/^{86}Sr$ ratios of lava flows and dikes from all parts of Queen Maud Land identified in Fig. 5.66 are strongly clustered between 0.7030 and 0.7060 in Fig. 5.67 based on data by Faure and Elliot (1972), Faure et al. (1979), Furnes et al. (1982, 1987), Harris et al. (1990), and Luttinen et al. (1998) after all $^{87}Sr/^{86}Sr$ ratios were recalculated to $t = 170$ Ma, $\lambda = 1.42 \times 10^{-11}$ yr^{-1}, and E&A = 0.7080 as necessary. The range of initial $^{87}Sr/^{86}Sr$ ratios of the Jurassic basalts of Queen Maud Land closely matches that of the basalts of the Nuanetsi-Lebombo Belt of southern Africa. Therefore, these data support the interpretation that the Jurassic basalts of southern Africa and Queen Maud Land both formed by decompression melting in the subcontinental lithospheric mantle of Gondwana prior to and during the initiation of rifting.

The basalt lavas and dikes of Vestfjella were subdivided by Luttinen et al. (1998) into four groups iden-

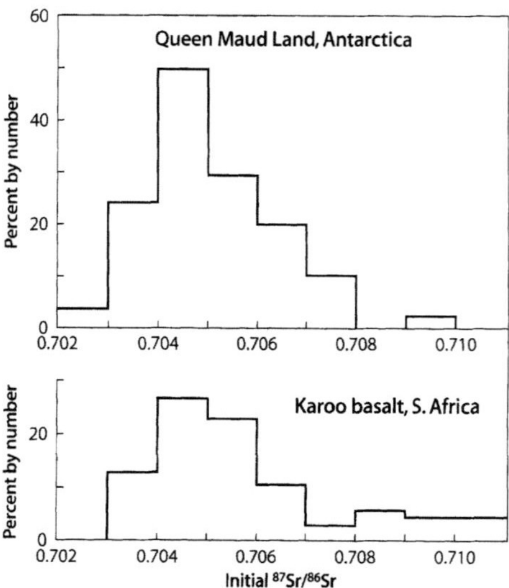

Fig. 5.67. Range of initial $^{87}Sr/^{86}Sr$ ratios of the Jurassic basalt lavas and dikes of the various clusters of nunataks of Queen Maud Land, Antarctica. All isotope ratios were corrected for decay of ^{87}Rb to an age of 170 Ma and a decay constant of $\lambda = 1.42 \times 10^{-11}$ yr^{-1} for ^{87}Rb. The histogram of Karoo basalts in southern Africa is from Figure 5.64 (*Sources:* Faure and Elliot 1971; Faure et al. 1979; Furnes et al. 1982, 1987; Harris et al. 1990; Luttinen et al. 1998)

tified as CT1, CT2, CT3, and CT4 based on their concentration of TiO_2, Mg numbers, as well as on their Ti/Zr and Ti/P ratios. The Sr and Nd isotope ratios of CT2 and 3 at 170 Ma define a data field in Fig. 5.68 which overlaps a small data field of the basalt lavas on the Kirwan Escarpment analyzed by Harris et al. (1990). The initial $^{87}Sr/^{86}Sr$ and $^{143}Nd/^{144}Nd$ ratios of these Vestfjella-Kirwan lavas plot close to a hypothetical mixing line between the asthenospheric plume (represented by CT4) and EM1 composed of subducted oceanic crust and marine sediment which underplated the lithospheric mantle of Gondwana prior to its break-up. The CT1 flows in Vestfjella have exceptionally low initial $^{143}Nd/^{144}Nd$ ratios, placing these rocks well within the "crustal" quadrant in Fig. 5.68.

The tectonic setting of Queen Maud Land during the Mesozoic Era can accommodate several petrogenetic scenarios for the origin of the CT1 lavas (Luttinen et al. (1998):

1. Contamination of plume-derived magma by assimilation of Precambrian basement rocks from the continental crust;
2. Magma formation in subcrustal lithospheric mantle of Archean age;

Table 5.9. Chemical composition of the Jurassic basalt lavas and dikes of Queen Maud Land (Furnes and Mitchell 1978)

		Lava flows	Dikes
SiO_2	(%)	49.13 (46.11–50.47)	48.49 (45.35–52.72)
TiO_2	(%)	1.59 (1.19–2.32)	2.30 (0.87–4.21)
Na_2O	(%)	2.10 (1.10–2.66)	2.20 (0.78–3.14)
K_2O	(%)	0.44 (0.09–1.27)	0.69 (0.04–1.43)

3. Contamination of plume-derived magma by mixing with alkali-rich melts derived from Archean lithospheric mantle.

Luttinen et al. (1998) preferred the second alternative based on modeling of isotopic and trace-element data. Alternatively, the hybrid (plume + lithosphere) magmas, which produced the Vestfjella-Kirwan suite of flows, could also have been contaminated by Sr and Nd derived from the rocks of the continental crust.

The initial $^{87}Sr/^{86}Sr$ and $^{143}Nd/^{144}Nd$ ratios of basalt lavas at Vestfjella and Kirwan in Queen Maud Land of Antarctica in Fig. 5.68 are similar to those of the basalt and rhyolite flows of the Karoo Group in south-

eastern Africa in Fig. 5.65. The similarity of the isotope compositions is consistent with the theory that these rocks formed under similar circumstances that accompanied the separation of southern Africa and Queen Maud Land as the result of plume-induced rifting of Gondwana.

5.12 Jurassic Tholeiites of the Transantarctic Mountains

Mesozoic dolerite sills of the Ferrar Group are widely distributed throughout the length of the Transantarctic Mountains in Fig. 5.69 where they intrude flat-lying sandstones, shales, and coal measures of the Beacon Supergroup of middle to early Mesozoic ages. The dolerite sills are comagmatic with the Kirkpatrick basalt flows, which are preserved as erosional remnants of large flood basalt plateaus. The principal occurrences of tholeiite basalt flows in the Transantarctic Mountains are identified in Fig. 5.69:

1. Mesa Range of northern Victoria Land.
2. Brimstone-Peak-Griffin Nunatak area (Prince Albert Mountains) of southern Victoria Land.
3. Queen Alexandra Range of the Beardmore Glacier area.

Another manifestation of Mesozoic magmatic activity in the Transantarctic Mountains is the Dufek Intrusion, located at 82°36′ S and 52°30′ W in the Pensacola Mountains. The Dufek Intrusion is a large differentiated body of ultramafic rocks, gabbro, and grano-

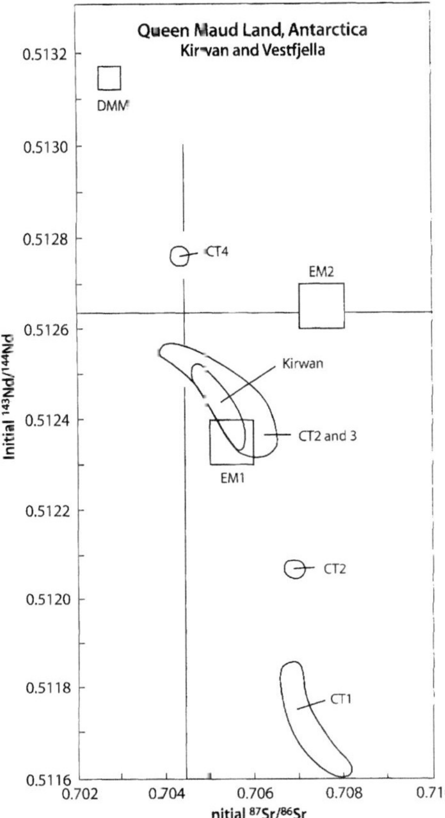

Fig. 5.68. Isotope ratios of Sr and Nd of basalt flows and dikes of the Kirwan Escarpment and of Vestfjella in Queen Maud Land, Antarctica. The volcanic rocks of Vestfjella from four groups labeled CT1, CT2, CT3, and CT4. The isotope ratios of Sr and Nd were corrected for decay to 170 Ma and the $^{143}Nd/^{144}Nd$ ratios are relative to 0.7219 for $^{146}Nd/^{144}Nd$ (Sources: Harris et al. 1990; Luttinen et al. 1998)

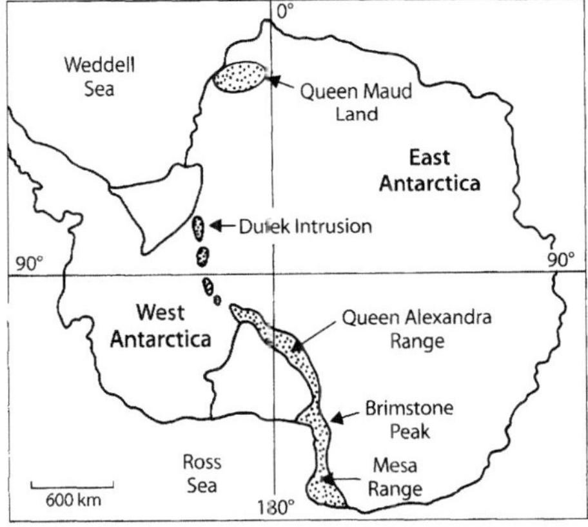

Fig. 5.69. Locations of the principal centers of Jurassic basalt and dolerite in Antarctica. The stippled areas represent rock exposures of the Transantarctic Mountains and Queen Maud Land (adapted from Furnes et al. 1987)

phyre whose age coincides with that of the Ferrar dolerites and Kirkpatrick basalt.

The dolerite sills in southern Victoria Land were first mapped by Ferrar (1907) who was a member of Robert F. Scott's "Discovery Expedition" (1901–1904). Rock specimens collected during that expedition were later described by Prior (1907), Benson (1916), and Smith (1924). The study of the geology of the Transantarctic Mountains resumed during the International Geophysical Year (1957–1958) with the work of McKelvey and Webb (1959, 1962), Webb and McKelvey (1959), Gunn (1962), and others listed by Hamilton (1965). Age determinations by McDougall (1963b) based on the K-Ar method first suggested a Middle Jurassic age for the dolerite sills in the Transantarctic Mountains.

The unusual chemical composition and high initial $^{87}Sr/^{86}Sr$ ratios of the dolerites in the Transantarctic Mountains were originally revealed by Compston et al. (1968) who demonstrated that the different types of dolerite classified by Gunn (1962) all have initial $^{87}Sr/^{86}Sr$ ratios of about 0.711 similar to those of Jurassic dolerites in Tasmania studied by Heier et al. (1965). In addition, the Kirwans dolerite on the South Island of New Zealand has been identified as belonging to the Ferrar Group of Antarctica (Mortimer et al. 1995). The high initial $^{87}Sr/^{86}Sr$ ratios of the basalts and dolerites of the Transantarctic Mountains, Tasmania, and New Zealand distinguish these rocks from the basalts and dolerites of all other continental basalt provinces described in this chapter. The origin of these rocks has been attributed to:

1. Partial melting in the lithospheric mantle which had been enriched in radiogenic ^{87}Sr after a prior increase of the Rb/Sr ratio;
2. Large-scale contamination of basalt magma derived from lithospheric mantle by assimilation of crustal rocks;
3. Partial melting of rocks in the continental crust with only minor subsequent assimilation of granitic basement rocks.

These and other proposals to explain the petrogenesis of the Jurassic dolerites and basalts in the Transantarctic Mountains must take into consideration the remarkable uniformity of initial $^{87}Sr/^{86}Sr$ ratios and the size of this petrologic province that extends for more than 2 500 km from the Rennick Glacier of northern Victoria Land (70°30' S, 160°45' E) to the Theron Mountains (79°05' S, 28°15' W). Although the tholeiite basalts and dolerites of the Transantarctic Mountains formed in a setting of extensional tectonics, the cause for the rifting is still somewhat uncertain. In addition, the basalts and dolerites are not restricted to the Transantarctic Mountains but occur also under the ice sheet in some parts of East Antarctica

(Faure et al. 1993). The Transantarctic Mountains themselves consist of crustal blocks that were uplifted during the Mesozoic Era and were subsequently dissected by streams and more recently by the East Antarctic ice sheet and its outlet glaciers (Behrendt et al. 1991).

5.12.1 Mesa Range, Northern Victoria Land

The Mesa Range of northern Victoria Land occurs within the Rennick Graben, which has a width of about 50 km and extends for about 500 km in a north-south direction parallel to the coast of northern Victoria Land in the Transantarctic Mountains (Gair 1967). It consists of numerous mesas composed of flat-lying flows of the Kirkpatrick basalt and interbedded nonmarine sedimentary rocks of Mesozoic age. The principal mesas in Fig. 5.70 are Pain, Tobin, Gair, and Sheehan, as well as Sculpture Mountain, Solo Nunatak, and Agate Peak. The Rennick Graben is occupied by the Rennick Glacier which flows north towards the coast of Antarctica.

The Kirkpatrick basalt in the Mesa Range consists of low-Ti tholeiite whose combined thickness exceeds 600 m (Mensing et al. 1991). The low-Ti flows are overlain by six high-Ti flows, which also have elevat-

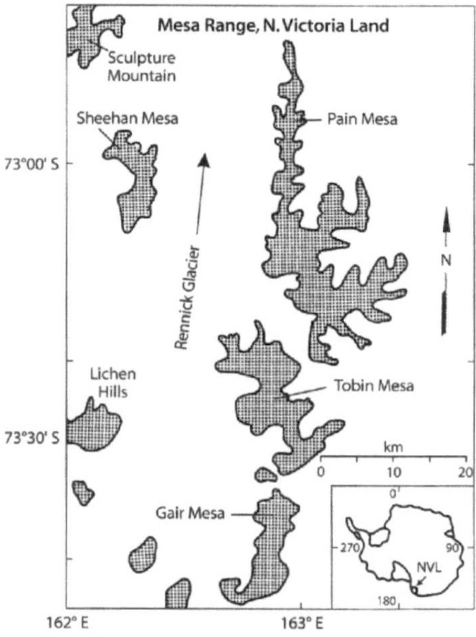

Fig. 5.70. The Mesa Range of northern Victoria Land, Antarctica (*Source:* Mensing et al. 1991)

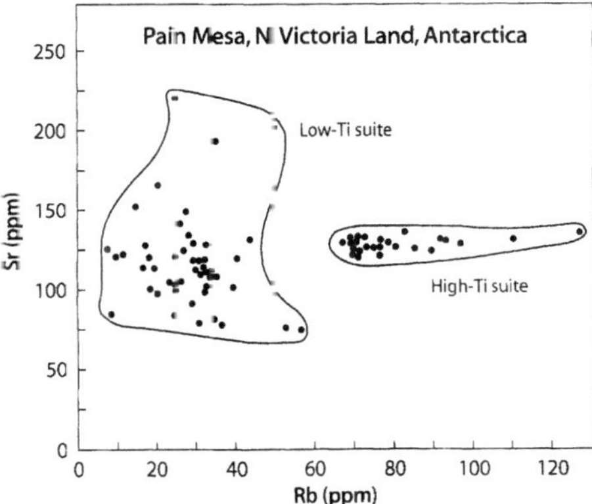

Fig. 5.71. Variation of Rb and Sr concentrations of basaltic lava flows on Pain Mesa, northern Victoria Land, Antarctica. The differences in the Rb and Sr concentrations of the two types of basalt were caused by differences in the degree of partial melting of similar kinds of source rocks (*Source:* Mensing 1987; Mensing et al. 1991)

ed concentrations of P_2O_5 and total Fe but lower concentrations of Al_2O_3, CaO, and MgO than the low-Ti flows. (Siders and Elliot 1985; Mensing et al. 1991). In addition, the two types of basalt on Pain Mesa have distinctive concentrations of Rb and Sr in Fig. 5.71 reminiscent of the Paraná basalt of Brazil in Fig. 5.39 and the diabase dikes of Liberia in Fig. 5.56. However, the so-called high-Ti flows of the Mesa Range contain only about 2% of TiO_2, whereas the low-Ti flows contain less than 1%.

The difference in the Rb and Sr concentrations of the two types of basalt on Pain Mesa can be attributed to differences in the degree of partial melting of magma sources in the mantle, such that the low-Ti magmas represent larger melt fractions than the high-Ti magmas (Focor 1987). In addition, their wide range of Sr concentrations indicates that the low-Ti magmas experienced more extensive differentiation by fractional crystallization and/or assimilation of crustal rocks than the high-Ti flows, which have a narrower range of Sr concentrations. The elevated Rb concentrations of the high-Ti flows that cap the sequence in the Mesa Range can also be attributed, at least in part, to the effects of hydrothermal alteration of the high-Ti flows at about 100 Ma (Early Cretaceous) (Mensing and Faure 1996; Fleming et al. 1992).

The crystallization age of both types of basalt in the Mesa Range and elsewhere in the Transantarctic Mountains is 176.6 ±1.8 Ma (Middle Jurassic) based on $^{40}Ar/^{39}Ar$ dating of feldspar concentrates by Heimann et al. (1994), although Encarnacíon et al. (1996) reported U-Pb dates of zircon and baddeleyite of 183.4 and 183.8 Ma for two dolerites in the Transantarctic

Mountains. The work of Heimann et al. (1994) confirmed the $^{40}Ar/^{39}Ar$ dates reported previously by McIntosh et al. (1986) and later by Fleming et al. (1993), whereas the conventional whole-rock K-Ar method has yielded discordant dates ranging from 90 to about 200 Ma (Faure and Mensing 1993). The wide range of the whole-rock K-Ar dates of the Mesozoic basalt in the Transantarctic Mountains was attributed by Fleck et al. (1977) and Foland et al. (1993) to loss (or gain) of radiogenic ^{40}Ar from (or by) the glassy or fine-grained matrix of the basalt. The short duration of the volcanic activity in the Transantarctic Mountains reported by Heimann et al. (1994) is consistent with evidence from other flood basalt provinces mentioned previously in this chapter.

The distribution of Rb-Sr data points in Fig. 5.72 demonstrates that the low-Ti flows of Pain Mesa scatter above and below a reference isochron corresponding to a date of 180 Ma. The high-Ti flows have higher $^{87}Rb/^{86}Sr$ ratios than the low-Ti flows and define an errorchron yielding a date of only 100 ±15 Ma (Mensing and Faure 1996). However, since both basalt suites in the Mesa Range were erupted within one million years of each other at 176.6 Ma (Heimann et al. 1994), the Rb-Sr date of the high-Ti flows on Pain Mesa records an episode of hydrothermal alteration that caused an increase of their Rb concentrations. The high-Ti flows of Pain Mesa responded more readily to the hydrothermal alteration than the underlying low-Ti flows, because the former contain 55% of matrix composed of devitrified opaque glass, whereas the low-Ti flows contain only about 20% of matrix (Mensing and Faure 1996). Therefore, the preferred interpretation is that the high-Ti flows of the Mesa Range are only slightly younger (e.g. less than about one million years) than the low-Ti flows which underlie them. Subsequently, all of the rocks of this area came in contact with heated groundwater at about 100 Ma as a consequence of tectonic activity associated with the separation of Australia from northern Victoria Land and the concurrent displacement of West Antarctica along major transform faults (Schmidt and Rowley 1986; Behrendt et al. 1991).

The initial isotope ratios of Sr and Nd of the lava flows and dolerite sills on Tobin Mesa in Fig. 5.73 indicate that both rock suites contain significant proportions of Sr and Nd derived from rocks of crustal origin. Since these rocks formed in a setting of extensional tectonics, their magmas presumably originated by decompression melting in the subcrustal lithospheric mantle. The crustal character of the isotope ratios of Sr and Nd was attributed by Mensing et al. (1984) to assimilation of crustal rocks by mantle-derived magmas. Alternatively, the magmas were contaminated at the source by partial melting of a mechanical mixture of previously subducted ocean crust

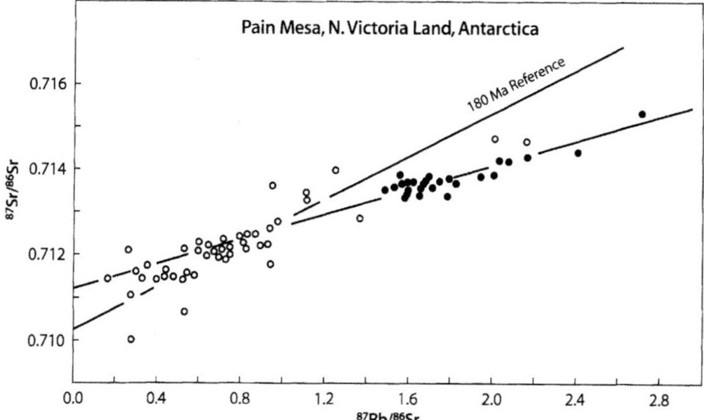

Fig. 5.72. Whole-rock Rb-Sr isochron diagram of the low-Ti basalts (*open circles*) and high-Ti basalts (*solid circles*) on Pain Mesa (73°08' S, 163°00' E) in northern Victoria Land, Antarctica. The data points representing low-Ti flows scatter widely above and below the 180 Ma reference isochron. The high-Ti flows define an errorchron that yields an anomalously low date of 100 ±15 Ma and an initial ^{87}Sr/^{86}Sr ratio of 0.71129. Since the age of the high-Ti flows is known to be 176 ±1.8 Ma based on ^{40}Ar/^{39}Ar dating of plagioclase concentrates by Fleming et al. (1993) and Heimann et al. (1994), the errorchron date of these flows was attributed to aqueous alteration of the high Ti flows, which increased the Rb/Sr ratios of these glass-rich rocks but did not consistently change the Rb/Sr ratios of the low-Ti flows (*Sources:* Mensing 1987; Mensing et al. 1991; Mensing 1994; Mensing and Faure 1996)

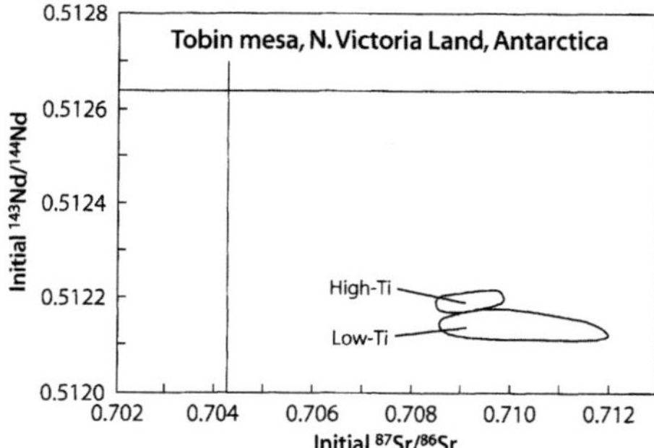

Fig. 5.73. Initial isotope ratios of Sr and Nd at 180 Ma of basalt flows and dolerite sills on Tobin Mesa, Mesa Range, northern Victoria Range, Antarctica. The location of the data fields reveals that a significant fraction of Sr and Nd in these volcanic rocks originated from rocks of the continental crust (*Source:* data from Fleming et al. 1995; Elliot et al. 1999)

containing sediment and rocks of the subcrustal lithosphere (Hergt et al. 1989b).

The hydrothermal alteration of high-Ti flows on Pain Mesa (Fig. 5.72; Mensing 1987) and Tobin Mesa (Fleming et al. 1992) obscures their initial ^{87}Sr/^{86}Sr ratios, which cannot be calculated accurately because of the increase of their Rb/Sr ratios at about 100 Ma. Therefore, the apparent uniformity of the initial ^{87}Sr/^{86}Sr ratios of the high-Ti basalt and dolerite in the Transantarctic Mountains is problematical even when precisely determined ^{40}Ar/^{39}Ar dates of plagioclase are used in the calculation. Nevertheless, Elliot et al. (1999) demonstrated that high-Ti basalt flows, occurring at scattered locations in the Transantarctic Mountains, have nearly constant chemical compositions. One possible explanation for this phenomenon proposed by Elliot et al. (1999) is that the magma originated from a single reservoir and was distributed to the eruption sites through crustal fractures for distances of several hundred kilometers. Alternatively, the chemical and isotopic homogeneity of the flood basalts in the Transantarctic Mountains and elsewhere in the world can be attributed to large-scale decompression melting of subcrustal lithospheric mantle.

5.12.2 Brimstone Volcanic Center, Southern Victoria Land

The Brimstone volcanic center encompasses the nunataks of the Prince Albert Mountains exposed in an area of about 7500 km² on the polar plateau at the heads of the Mawson, Harbord, and David Glaciers of southern Victoria Land. Some of the prominent topographic features identified in Fig. 5.74 are the Griffin Nunataks with Ambalada Peak (75°57' S, 158°23' E), Brimstone Peak (75°38' S, 158°33' E), the Ricker Hills (75°41' S, 159°10' E), Mt. Joyce (75°36' S, 160°51' E), and Mt. Armytage (76°02' S, 160°45' E). These and many smaller nunataks consist of Ferrar dolerite sills that intruded the flat-lying sedimentary rocks of the Beacon Supergroup. In addition, Kirkpatrick basalt flows occur on the Griffin Nunataks and on Brimstone Peak. The

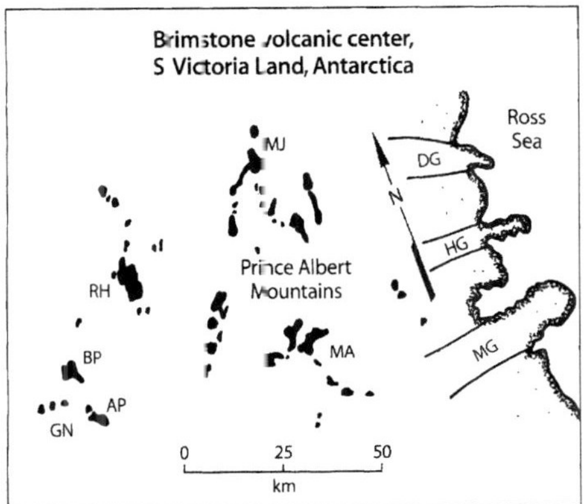

Fig. 5.74. Brimstone volcanic center (75°38' S, 158°33E) in the Prince Albert Mountains, southern Victoria Land, Antarctica. The localities mentioned in the text are abbreviated as follows: GN = Griffin Nunataks; AP = Ambalada Peak; BP = Brimstone Peak; RH = Ricker Hills; MJ = Mt. Joyce; MA = Mt. Armytage; MG = Mawson Glacier; HG = Harbord Glacier; DG = David Glacier (*Source:* adapted from Molzahn et al. 1996)

basalts and dolerites are generally tholeiites with andesitic tendencies and belong predominantly to the low-Ti suite. However, high-Ti flows do occur on Brimstone Peak (Kyle et al. 1930; Wörner 1992; Hornig 1993; Molzahn et al. 1994, 1996). High-Ti basalt and dolerite clasts were also reported by Mensing (1991) from the Elephant Moraine on the polar plateau.

Conventional whole-rock K-Ar dates of the basalts and dolerites in the Brimstone volcanic center and in the ice-free valleys of southern Victoria Land range widely. Foland et al. (1993) reported a $^{40}Ar/^{39}Ar$ date of 176.7 ±0.5 Ma for basalt at the base of Carapace Nunatak (76°54 S, 159°27' E) and Kyle et al. (1981) obtained 175.8 ±3 o Ma for a basalt flow on Ambalada Peak of the Griffin Nunatacks. In addition, Molzahn et al. (1996) published a Re-Os isochron date of 172 ±5 Ma ($MSWD$ = 8.6) for basalts and dolerites of the Brimstone volcanic center. These age determinations are in agreement with the results of Heimann et al. (1994) and Elliot et al. (1999), which indicate that the volcanic activity of the Brimstone center occurred at the start of the Middle Jurassic Epoch at about 177 ±1 Ma.

The chemical and isotope compositions of basalt flows on Brimstone Peak having high or low Ti concentrations (Table 5.10) vary in the same sense as observed elsewhere (Molzahn et al. 1996).

These results demonstrate that the low-Ti basalts have a higher average initial $^{87}Sr/^{86}Sr$ ratio than the high-Ti basalt and that their initial $^{143}Nd/^{144}Nd$ ratios are indistinguishable. In addition, the low-Ti basalts are enriched in ^{18}O but have lower Rb concentrations than the high-Ti basalts.

Table 5.10. Chemical and isotope compositions of basalt flows on Brimstone Peak (Molzahn et al. 1996)

	Low-Ti	High-Ti
Initial $^{87}Sr/^{86}Sr$	0.71017	0.70956
Rb (ppm)	19.7	64.0
Sr (ppm)	138.1	128.2
Initial $^{143}Nd/^{144}Nd$	0.51224	0.51223
$\delta^{18}O$ (‰)	+8.03	+6.01

Dolerite sills in the Prince Albert Mountains analyzed by Molzahn et al. (1996) have low Ti concentrations and correspondingly high initial $^{87}Sr/^{86}Sr$ ratios ranging from 0.71137 to 0.71205 with a mean of 0.71168 ±0.00021 ($2\bar{\sigma}$, N = 5) at 177 Ma. Their initial $^{143}Nd/^{144}Nd$ ratios are virtually constant at 0.51213 ±0.00001 ($2\bar{\sigma}$, N = 5), but their whole-rock $\delta^{18}O$ values range from +4.79 to +7.01‰. The $\delta^{18}O$ values of the basalt flows and dolerite sills in the Brimstone volcanic center are not well correlated with their initial $^{87}Sr/^{86}Sr$ ratios (not shown) (Molzahn et al. 1996).

Surprisingly, the constituent minerals of the basalts have different initial $^{87}Sr/^{86}Sr$ ratios at 177 Ma from the whole rock samples. A data set (Table 5.11) representing a low-Ti basalt flow on Brimstone Peak illustrates this phenomenon (Molzahn et al. 1996).

The concentrations of Rb and Sr of the minerals do not account for the concentrations of these elements in the rock as a whole. Therefore, a third phase must be present that contains most of the Rb and Sr. This phase is presumably the devitrified matrix, which must also be enriched in ^{87}Sr in order to explain the elevated initial $^{87}Sr/^{86}Sr$ ratio of the whole rock. Therefore, the phenocrysts of plagioclase and clinopyroxene could not have crystallized from this matrix, meaning that the matrix was altered after the phenocrysts had crystallized from it.

The dolerite samples analyzed by Molzahn et al. (1996) are aligned on the Rb-Sr isochron diagram in Fig. 5.75 and yield a date of 244 ±12 Ma (λ = 1.42 × 10^{-11} yr^{-1}) with

Table 5.11. Ratios of $^{87}Rb/^{86}Sr$ and initial $^{87}Sr/^{86}Sr$, concentrations of Sr and Rb, and $\delta^{18}O$ values of a low-Ti basalt flow on Brimstone Peak (Molzahn et al. 1996)

	Whole rock	Plagioclase	Clinopyroxene
Initial $^{87}Sr/^{86}Sr$	0.71017	0.70990	0.70960
Rb (ppm)	19.7	5.1	0.4
Sr (ppm)	138.1	44.1	23.0
$^{87}Rb/^{86}Sr$	0.413	0.335	0.457
$\delta^{18}O$ (‰)	+8.03	+13.33	+6.18

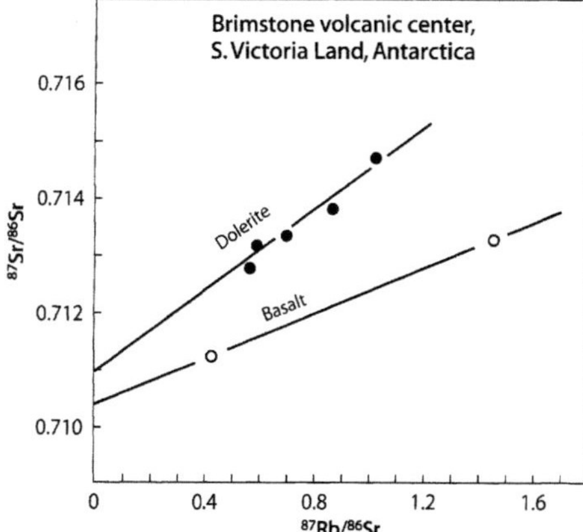

Fig. 5.75. Whole-rock Rb-Sr isochron diagram of Ferrar dolerite sills and Kirkpatrick basalt flows in the Brimstone volcanic center, Prince Albert Mountains, southern Victoria Land, Antarctica. The dates derivable from these data are 244 ±12 Ma for the dolerites and 137 ±3 Ma for the basalts (*Source:* data from Molzahn et al. 1996)

an initial $^{87}Sr/^{86}Sr$ ratio of 0.71099 ±0.00013 resulting from an unweighted least-squares regression of the data. The reliability of this date is questionable because of the possible alteration of these rocks seen more clearly in the basalts on Brimstone Peak. Two basalt samples on Brimstone Peak (high-Ti and low-Ti) define a second errorchron whose slope corresponds to a date of 137 ±3 Ma (Early Cretaceous) with an initial $^{87}Sr/^{86}Sr$ ratio of 0.71041 ±0.00002. This anomalously low date is attributable to alteration of these basalts by aqueous fluids, which increased the Rb concentration of the devitrified matrix, and also added Sr enriched in radiogenic ^{87}Sr. The large difference of 7.15‰ ($\Delta^{18}O$) between the $\delta^{18}O$ values of plagioclase and clinopyroxene of the low-Ti basalt cited above implies a low temperature of alteration. The $\Delta^{18}O$ values of two plagioclase-clinopyroxene pairs of dolerite samples analyzed by Molzahn et al. (1996) are much lower than those of the basalt (2.03 and 0.55‰) and can probably be reconciled with the effects of isotope fractionation during initial crystallization and subsequent cooling of the rocks.

The conclusions derivable from the data of Molzahn et al. (1996) are that the volcanic activity in the Brimstone volcanic center was synchronous with that of the Mesa Range in northern Victoria Land. The age of the dolerite sills in this area remains to be confirmed. The Rb and Sr concentrations of the Brimstone basalts were changed by aqueous solutions at comparatively low temperature in Early Cretaceous time. The high initial $^{87}Sr/^{86}Sr$ and low $^{143}Nd/^{144}Nd$ ratios indicate the presence of substantial amounts of crustal Sr and Nd

in the magmas either as a result of assimilation of crustal rocks or because of contamination at the source. Molzahn et al. (1996) concluded that the magmas formed by melting of mixed sources in the subcrustal lithospheric mantle containing about 3% of previously subducted terrigenous sediment. In the absence of evidence for the existence of a plume either in the Mesa Range or at Brimstone Peak, the magmas formed by decompression melting in a setting of extensional tectonics.

5.12.3 Queen Alexandra Range, Central Transantarctic Mountains

The Queen Alexandra Range is located at about 84° S and 168° E in the Beardmore Glacier area and contains extensive exposures of the Kirkpatrick basalt as well as numerous Ferrar dolerite sills. According to Elliot (1972), the lava flows are composed of tholeiite basalt having high silica concentrations between 53.14 and 58.09% and low MgO concentrations ranging from 7.58 to 2.19%. The TiO_2 concentrations of these rocks range widely from 0.42 to 2.33%. High-Ti flows ($TiO_2 > 1.40\%$) occur on Storm Peak and on Mt. Bumstead (Elliot 1972) as well as on Mt. Falla (Faure et al. 1982). These high-Ti flows also have elevated silica concentrations between 56.20 and 58.20%. Elliot (1972) suggested that the basalt flows in the Queen Alexandra Range are the acidic differentiates of more mafic dolerite sills, which were intruded into the granitic basement and into the sedimentary rocks of the Beacon Supergroup.

Age determinations by McDougall (1963b) yielded a Middle Jurassic K-Ar date for plagioclase and pyroxene of dolerites of the Beardmore area. This result was confirmed and refined by Fleck et al. (1977) who obtained a $^{40}Ar/^{39}Ar$ date of 170.4 ±1.4 Ma for a basalt flow on Mt. Bumstead, whereas Elliot et al. (1985) published a wide range of discrepant whole-rock K-Ar dates between 205 and 114 Ma. Six lava flows on Mt. Falla of the Queen Alexandra Range define a straight line on the Rb-Sr isochron diagram from which Faure et al. (1982) calculated a date of 173 ±6 Ma and an initial $^{87}Sr/^{86}Sr$ ratio of 0.7128 ±0.0001. Kyle (1980) obtained a similar Rb-Sr date of 170 Ma by regressing a widely scattered data set representing basalts and dolerites of the Transantarctic Mountains and Tasmania. The dates reported by Fleck et al. (1977) and Faure et al. (1982) were subsequently confirmed by Heimann et al. (1994) who reported three concordant $^{40}Ar/^{39}Ar$ dates for feldspar separates of two flows on Storm Peak with a mean of 176.8 ±0.4 Ma. Elliot et al. (1999) later reported a $^{40}Ar/^{39}Ar$ date of 177.2 ±0.5 Ma for the Grosvenor Mountains.

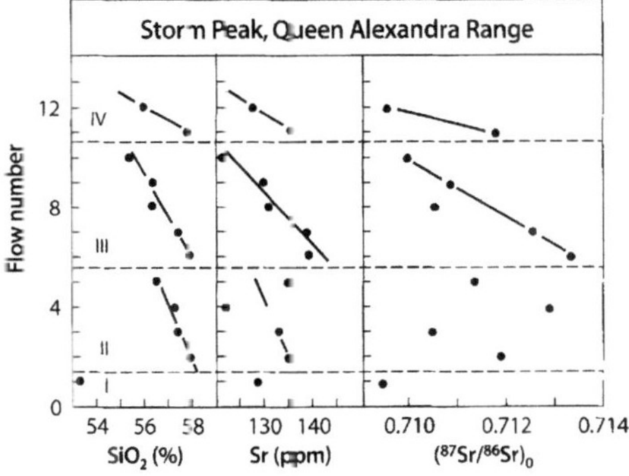

Fig. 5.76. Evidence for the existence of four eruptive cycles based on the stratigraphic variation of average concentrations of SiO_2 and Sr, and initial $^{87}Sr/^{86}Sr$ ratios of basalt flows on Storm Peak (84°35′ S, 163°55′ E) in the Queen Alexandra Range, Transantarctic Mountains. The stratigraphic variation of these and other chemical parameters was caused by the stratification of the chemical composition and $^{87}Sr/^{86}Sr$ ratios of the magma in a magma chamber from which lava was erupted periodically, starting with the most highly contaminated magma at the top of the chamber (*Source:* Faure et al. 1974)

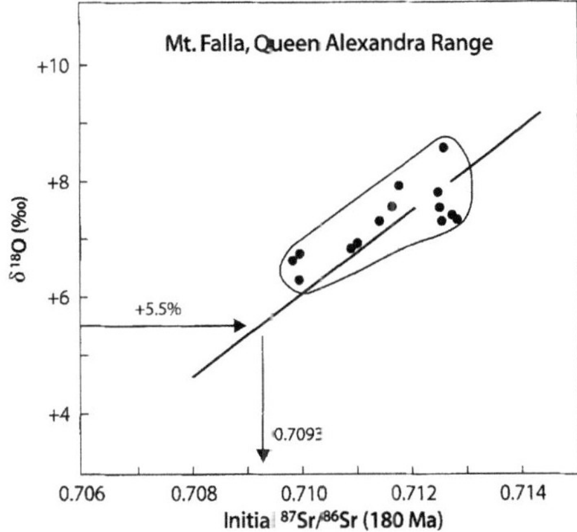

Fig. 5.77. Positive correlation of initial $^{87}Sr/^{86}Sr$ ratios at 178 Ma and the $\delta^{18}O$ values of basalt flows on Mt. Falla, Queen Alexandra Range. The data indicate that mantle-derived magmas ($\delta^{18}O = +5.5‰$) had an $^{87}Sr/^{86}Sr$ ratio of 0.7093 at 180 Ma (*Sources:* Hoefs et al. 1980; Faure et al. 1982)

The major-element concentrations and the initial $^{87}Sr/^{86}Sr$ ratios of lava flows on Storm Peak (84°35′ S, 163°55′ E) in the Queen Alexandra Range vary systematically up-section and define four eruptive cycles (Faure et al. 1974). The data in Fig. 5.76 demonstrate that the average initial $^{87}Sr/^{86}Sr$ ratios as well as the concentrations of SiO_2 and Sr of the flows decrease in ascending stratigraphic sequence in each cycle. In addition, the concentrations of TiO_2, Al_2O_3, MgO, CaO and MnO increase up-section, whereas those of total iron, Na_2O, K_2O, and P_2O_5 decrease (not shown). Faure et al. (1974) attributed the stratigraphic variation of the isotopic and chemical compositions of the flows to the effects of simultaneous fractional crystallization of basaltic magma and assimilation of granitic basement rocks (AFC), which caused the magma to become stratified. During periodic eruptions, the magma chamber was emptied from the top downward in such a way that the first flow in each cycle was more highly evolved and had higher initial $^{87}Sr/^{86}Sr$ ratios than flows erupted subsequently. Similar stratigraphic variations of lava flows were reported from volcanoes in Japan (Sect. 3.5.3) and for the Bishop tuff in California (Sect. 4.3.1). In addition, Faure et al. (1984) reported stratigraphic variations of the initial $^{87}Sr/^{86}Sr$ ratios of lava flows on Mount Falla (84°22′ S, 164°55′ E) in the Queen Alexandra Range

The assimilation of crustal rocks by magmas derived from the subcrustal lithospheric mantle of Gondwana is supported by the correlation of the initial $\delta^{18}O$ values and the initial $^{87}Sr/^{86}Sr$ ratios of the lava flows on Mt. Falla in Fig. 5.77. Hoefs et al. (1980)

fitted a straight line to these data and determined that the mantle-derived magma at $\delta^{18}O = +5.5‰$ had an initial $^{87}Sr/^{86}Sr$ ratio of c.7093.

The extrapolation of initial $^{87}Sr/^{86}Sr$ ratios of the basalt flows on Mt. Falla to 0.7093 at $\delta^{18}O = +5.5‰$ reported by Hoefs et al. (1980) emphasizes the point that the magmas had high initial $^{87}Sr/^{86}Sr$ ratios even before they assimilated granitic basement rocks. The cause for the high initial $^{87}Sr/^{86}Sr$ ratios of the Kirkpatrick basalt and Ferrar dolerite in the Transantarctic Mountains has been discussed by many other authors including Cox (1978), Kyle (1980), Kyle et al. (1983, 1987) and Thompson et al. (1983). The consensus is that the "crustal" characteristics of the magmas are associated with the magma sources in the mantle. Cox (1978) and Kyle et al. (1987) especially emphasized the possibility that the magmas may have been contaminated at the source by oceanic crust and pelagic sediment subducted under the Pacific margin of Gondwana in pre-Jurassic time.

The possibility of magma contamination at the source by partial melting of subducted sediment was subsequently explored by Hergt et al. (1989a) based on a suite of samples from a sill of Ferrar dolerite exposed on Portal Peak located at 83°50′ S and 165°40′ E in the Queen Alexandra Range. This sill is 150 m thick and is composed of low-Ti tholeiite basalt (SiO_2: 51.94 to 54.41%; TiO_2: 0.55 to 0.78%). The initial $^{87}Sr/^{86}Sr$ ratios reported by Hergt (1989a) range from 0.70910 to 0.71086 at 175 Ma and vary systematically through the body of the sill. However, the $^{143}Nd/^{144}Nd$ ratios are virtually constant and have an average value of

0.51235 ±0.00001 ($2\bar{\sigma}$, $N = 6$) corrected for fractionation to $^{146}Nd/^{144}Nd = 0.7219$. The elevated initial $^{87}Sr/^{86}Sr$ ratio (0.70910) and the low $^{143}Nd/^{144}Nd$ ratio of the Portal Peak Sill, together with the correlation of $\delta^{18}O$ and initial $^{87}Sr/^{86}Sr$ ratios of the basalt flows on Mt. Falla (Hoefs et al. 1980), lead to the conclusion that the basalt magmas of the Queen Alexandra Range acquired their "crustal" characteristics at the source in the subcrustal lithospheric mantle. Hergt et al. (1989a) demonstrated by modeling calculations that both are attributable to magma formation by partial melting of a mixture of depleted lithospheric mantle and less than about 3% of previously subducted shale-like sediment. The sedimentary rocks contributed a large proportion of the incompatible elements and radiogenic ^{87}Sr, whereas the mantle rocks controlled the oxygen isotope composition of the magma.

Therefore, the Ferrar and Kirkpatrick magmas in the Queen Alexandra Range formed from a hybrid magma having an $^{87}Sr/^{86}Sr$ ratio of about 0.7093 and $\delta^{18}O \approx +5.5‰$ (Hoefs et al. 1980), which subsequently assimilated crustal rocks thereby increasing the $^{87}Sr/^{86}Sr$ ratio of some lava flows up to about 0.7143 and $\delta^{18}O$ to +8.6‰. In addition, the chemical and isotopic compositions of the flows as well as sills were modified in some cases by hydrothermal fluids either at the time of crystallization or later during the Cretaceous Period.

5.12.4 Ferrar Dolerite Sills

The flat-lying sedimentary rocks of the Beacon Supergroup of the Transantarctic Mountains were intruded in Middle Jurassic time by thick dolerite sills of the Ferrar dolerite composed of low-Ti tholeiite. The sills in the ice-free valley region of southern Victoria Land were described by Hamilton (1965) whose report contains excellent aerial views of the topography of the Transantarctic Mountains. He recognized the so-called Basement Sill and the Peneplain Sill on the basis of their stratigraphic positions within the granitic basement rocks and at or close to the nonconformity between the basement rocks and the overlying Beacon rocks, respectively. The Basement Sill is composed primarily of gabbro and includes a layer of granophyre (3 to 5 m thick) near the top. Hamilton (1965) concluded that the granophyre formed both by fractional crystallization of magma in the sill and by recrystallization of wallrock inclusions. The silica concentrations of the Basement Sill range from 53.09% (gabbro) to 69.22% (granophyre), whereas the concentrations of TiO_2 are less than 0.70% except for dioritic pegmatoids which contain up to 1.3% TiO_2. One specimen of the Basement Sill in Wright Valley (GFBS3) was later analyzed by Thompson et al. (1983).

The sills of Ferrar dolerite in southern Victoria Land were subdivided by Gunn (1962, 1966) into three types: hypersthene tholeiites, pigeonite tholeiites, and olivine tholeiites based on their chemical and mineralogical compositions. The Basement Sill of the ice-free valley region is a hypersthene tholeiite, whereas the Peneplain Sill is a pigeonite tholeiite. The geology of the prominent Peneplain Sill on Terra Cotta Mountain along the upper Taylor Glacier was described by Morrison and Reay (1995) who presented evidence for hydrothermal alteration of the dolerite sill and of the associated basement granitoids and Beacon sandstones in this area.

Compston et al. (1968) discovered that Jurassic dolerites of the Transantarctic Mountains have initial $^{87}Sr/^{86}Sr$ ratios of about 0.711 and concluded that their magmas had originated in the mantle either by partial melting of source rocks having unusual chemical compositions or by assimilation of 10 to 30% of crustal rocks by magmas whose initial composition approached that of oceanic tholeiites. They also recognized that the low K/Rb ratios of the Antarctic and Tasmanian tholeiites (~200) are difficult to reconcile with crustal contamination, because the average K/Rb ratio of granitic rocks is about 230 and that of oceanic tholeiites is about 1400. Consequently, Compston et al. (1968) considered the possibility that the tholeiite magmas were contaminated by selective diffusion of certain elements or by mixing with partial melts of crustal rocks. In addition, they recognized that the Jurassic basalts and dolerites of the Transantarctic Mountains constitute a separate petrologic province that is distinguishable by their high initial $^{87}Sr/^{86}Sr$ ratios from the Karoo igneous province of southeastern Africa and from the Paraná basalt plateau of Brazil.

The initial $^{87}Sr/^{86}Sr$ ratios of dolerite and pegmatoids of the Ferrar Sills of southern Victoria Land in Fig. 5.78 illustrate an important point. Some of the pegmatoid samples of the Basement Sill at Solitary Rocks and the Peneplain Sill at New Mountain in southern Victoria Land have anomalously high initial $^{87}Sr/^{86}Sr$ ratios ranging up to 0.7178, indicating that they contain Sr derived from the Irizar granite, which is the local basement rock. Compston et al. (1968) agreed with Hamilton (1965) that the pegmatoids are not only magmatic differentiates of the dolerite but also contain Sr derived from reconstituted and assimilated inclusions of granitic basement rocks.

Another significant feature of the data set published by Compston et al. (1968) is that the initial $^{87}Sr/^{86}Sr$ ratios of the dolerites are not constant but vary just as much as the initial $^{87}Sr/^{86}Sr$ ratios of the Kirkpatrick basalt flows in both the Mesa and Queen Alexandra Ranges. Moreover, the lowest initial $^{87}Sr/^{86}Sr$ ratios are between 0.7090 and 0.7100 in agreement with the data

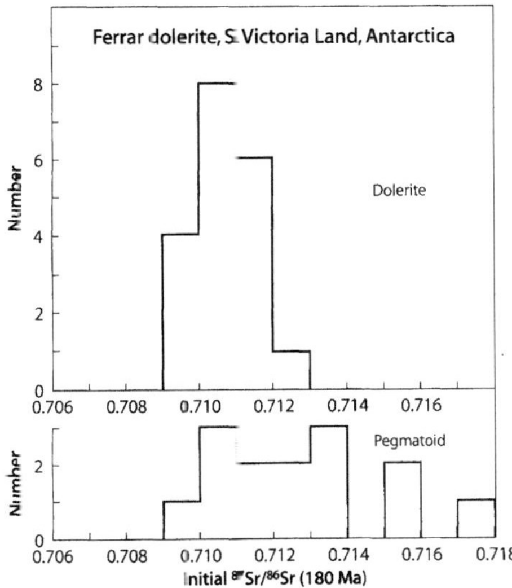

Fig. 5.78. Initial $^{87}Sr/^{84}Sr$ ratios of dolerites and pegmatoids in sills of the Transantarctic Mountains. The elevated initial $^{87}Sr/^{86}Sr$ ratios of some pegmatoid samples in the Basement and Peneplain Sills of southern Victoria Land indicate that they contain Sr derived from the local granitic country rocks (*Source:* Compston et al. 1968; Jones et al. 1973)

by Hergt et al. (1989a) for the Portal Peak Sill in the Queen Alexandra Range.

In spite of the prominence of the sills of Ferrar dolerite sills in the Transantarctic Mountains, they have not received the same attention as the Kirkpatrick basalt. Following the pioneering work of Compston et al. (1968), the only detailed isotopic studies of dolerite sills are by Hergt et al. (1989a) of a sill at Portal Peak in the Queen Alexandra Range and by Faure et al. (1991) who analyzed suites of samples from three sills at Mt. Achernar (84°11' S, 160°59' E; MacAlpine Hills) and two sills at Roadend Nunatak (79°48' S, 158°01' E; Darwin Glacier).

One of the sills at Roadend Nunatak intrudes the local granitic basement consisting of the medium grained, equigranular Hope granite (Cambro-Ordovician), whereas the other occurs at the unconformity between the granite and the overlying sandstone and conglomerate of the Beacon Supergroup. All of the sills at Mt. Achernar as well as the sill at Portal Peak (Hergt et al. 1989a) are located within the flat-lying sedimentary rocks of the Beacon Supergroup. The sills range in thickness from about 60 m (lowest sill, Mt. Achernar) to more than 150 m (Basement Sill, Roadend Nunatak).

The chemical compositions of these sills indicate that they represent different stages in the fractional crystallization of magma reservoirs at depth and that additional fractional crystallization occurred after intrusion. The top sill at Mt. Achernar represents a lower stage of differentiation (SiO_2 = 52.5%, K_2O = 0.67%; 12 samples) than the lowest sill (SiO_2 = 54.7%, K_2O = 0.96%; 8 samples) suggesting that the magma reservoir from which they formed was compositionally stratified with the most silica-rich and K-rich magma at the top of the chamber and more primitive magma near the bottom; and that the sills were intruded sequentially in increasing stratigraphic order. The average chemical compositions of the two sills at Roadend Nunatak also differ significantly in that the Basement Sill (SiO_2 = 52.2%, K_2O = 0.36%; 1 sample) appears to be less evolved than the Peneplain Sill (SiO_2 = 55.3%, K_2O = 1.10%; 3 samples).

The whole-rock samples of the sills at Roadend Nunatak form two parallel errorchrons yielding dates of 182 ±13 (Peneplain Sill) and 188 ±23 Ma (Basement Sill). These dates are consistent with the Middle Jurassic age of 176.6 ±1.8 Ma of the Kirkpatrick basalt (Heimann et al. 1994) and with the U-Pb date of 183.4 Ma of dolerite on Dawson Peak (Encarnacíon et al. 1996). The two sills at Roadend Nunatak have significantly different initial $^{87}Sr/^{86}Sr$ ratios of 0.71047 ±0.00016 (Basement Sill) and 0.71167 ±0.00014 (Peneplain Sill) relative to 0.7080 for E&A and 0.71025 for NBS 987 (Faure et al. 1991).

The sills at Mt. Achernar also form errorchrons, but with more scatter of the data points than occurs in the sills at Roadend Nunatak. The pooled date derived from the three sets of samples at Mt. Achernar is 186 ±5 Ma, but each of the three sills has a different initial $^{87}Sr/^{86}Sr$ ratio 0.708517 ±0.00017 (lowest sill), 0.70808 ±0.00009 (middle sill), and 0.71012 ±0.00008 (top sill) (Faure et al. 1991). Hergt et al. (1989a) reported an initial $^{87}Sr/^{86}Sr$ ratio of 0.7097 ±0.00063 for the Portal Peak Sill. These results suggest that sills in a given area formed from different batches of magma that originated from a crustal magma chamber in which magma was differentiating by simultaneous assimilation and fractional crystallization.

Profiles of $\delta^{18}O$ values, concentrations of Rb and Sr, and initial $^{87}Sr/^{86}Sr$ ratios (at 178 Ma) in Fig. 5.79 reveal that the oxygen isotope composition and Rb concentrations of whole-rock samples of the lowest sill on Mount Achernar are lower at the contacts than those in the center of the sill. Therefore, Faure et al. (1991) concluded that the $\delta^{18}O$ values and Rb concentrations of the rocks at the contact of the sill had been altered by aqueous solutions derived from the country rock. The concentrations of Sr and the initial $^{87}Sr/^{86}Sr$ ratios of the rocks at the contacts are not significantly changed, and both increase up-section and vary in-step. The apparent absence of contact effects implies that the change of Rb concentrations occurred

Fig. 5.79.
Profiles of the lowest Ferrar dolerite sill exposed on Mt. Achernar (84°12' S, 160°56' E) near the Queen Alexandra Range, Transantarctic Mountains. The $\delta^{18}O$ values and Rb concentrations of whole-rock samples decrease near the top and bottom of the sill, suggesting aqueous alteration of both parameters. The concentrations of Sr and initial $^{87}Sr/^{86}Sr$ ratios at 178 Ma both increase up-section and display similar stratigraphic variations. The absence of variation of the initial $^{87}Sr/^{86}Sr$ ratios at the contacts indicates that the alteration of the Rb concentration occurred a short time after intrusion and crystallization of the sill (*Source:* Faure et al. 1991)

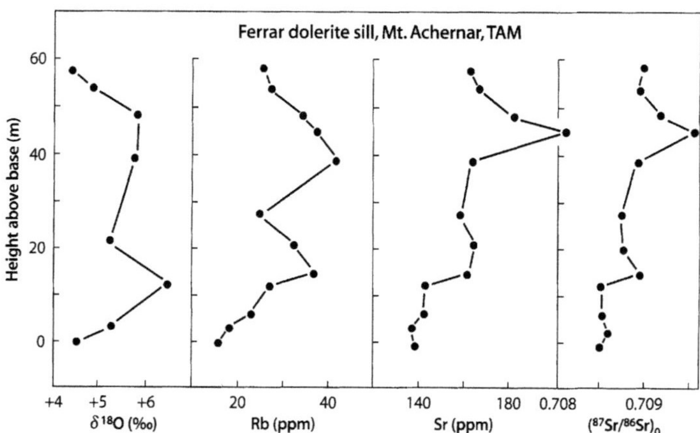

soon after intrusion of the sill, because it is not reflected by variations of the calculated initial $^{87}Sr/^{86}Sr$ ratios.

The information concerning sills of Ferrar dolerite in the Transantarctic Mountains can be summarized as follows:

1. The initial $^{87}Sr/^{86}Sr$ ratios of the sills vary stratigraphically, indicating that the dolerite magmas were not isotopically homogeneous, or that the sills formed by multiple injections of batches of magma having different $^{87}Sr/^{86}Sr$ ratios, or that the magmas assimilated wallrock inclusions.

2. Even the lowest initial $^{87}Sr/^{86}Sr$ ratio of the Ferrar dolerite (0.70808 ±0.00007, Rb-Sr errorchron, middle sill, Mt. Achernar, Faure et al. 1991) is anomalously high and requires that the magmas of the Ferrar dolerite and Kirkpatrick basalt originated from sources in the subcrustal lithospheric mantle that contained previously subducted terrigenous sediment as demonstrated by Hergt et al. (1989a,b).

3. The magmas that formed the sills differentiated in situ by fractional crystallization and, in some cases, by recrystallization and assimilation of inclusions of the country rocks.

4. The $\delta^{18}O$ values of rocks at the contacts of sills that intruded Beacon sandstones were lowered, which is attributable to isotope exchange of oxygen in the rocks with heated groundwater of meteoric origin.

5. The Rb concentrations of the contact rocks of dolerite sills were also affected by the aqueous alteration in some cases, depending on local conditions (e.g. concentrations of Rb, flux of the hydrothermal solutions, presence of fractures or joints in the sill, cooling rate of the rocks, etc.).

5.12.5 Dufek Intrusion, Pensacola Mountains

The Dufek Intrusion is a large body of stratified gabbros, anorthosites, and pyroxenite in the Dufek Massif (82°36' S, 52°30' W) capped by a layer of granophyre in the Forrestal Range (83°00' S, 49°30' W) which is located about 50 km southeast of the Dufek Massif in Fig. 5.80. The area between the two mountain ranges is covered by ice of the West Antarctic ice sheet. In addition, the base of the intrusion is not exposed, making it difficult to determine the true size of this body. Ford and Boyd (1968) originally estimated that the Dufek Intrusion is more than 7 km thick and extends over an area of about 8 000 km², giving it a volume of 5.6×10^4 km³. However, airborne geophysical surveys (gravity and magnetism) by Behrendt et al. (1966, 1980) greatly increased the estimated area of the Dufek Intrusion to more than 50 000 km², and Ford (1976) determined that it is between 8 and 9 km thick, thus increasing its volume to more than 4×10^5 km³. More recently, Ferris et al. (1998) reduced the extent of the Dufek Intrusion to only 6 600 km² based on a reexamination of aeromagnetic data. This revision reduces the estimate of the volume of the Dufek Intrusion from 4×10^5 km³ to only about 0.6×10^5 km³. Nevertheless, Ford (1990) discussed the possible occurrence of platinum group elements (PGE) in the Dufek Intrusion by comparison with the PGE deposits of the Bushveld Complex (South Africa), the Great Dyke (Zimbabwe), and the Stillwater Complex (Montana).

Conventional K-Ar age determinations of whole-rock samples and concentrates of pyroxene and plagioclase published by Ford and Kistler (1980) yielded a spectrum of dates from 97.5 to 174.1 Ma for the Dufek Intrusion. The oldest date 174.1 ±4.4 Ma for a plagioclase is consistent with the age of the Kirkpartick ba-

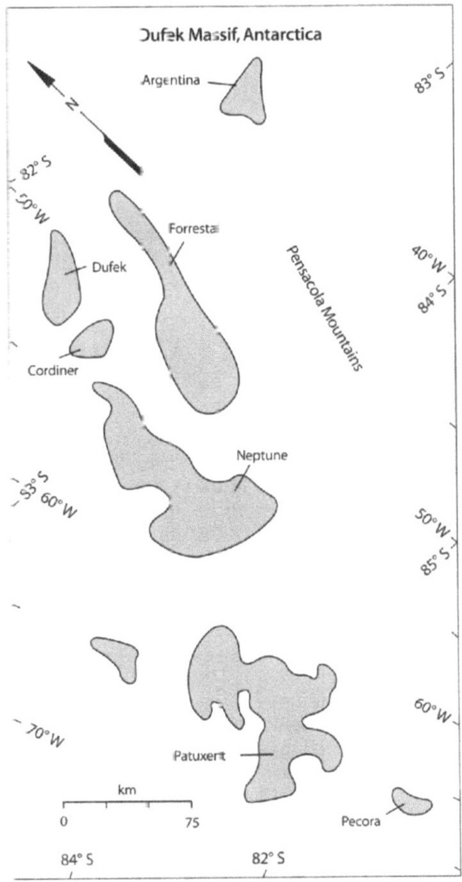

Fig. 5.80. Outline map of the Pensacola Mountains, Transantarctic Mountains. The Dufek Intrusion forms the Dufek Massif and part of the Forrestal Range. The individual nunataks (rock exposures) that form these ranges are not shown. The *white areas* outside of the mountain ranges represent the continental ice sheet of Antarctica (*Source:* adapted from Ford and Kistler 1980)

salt and Ferrar dolerite in the Transantarctic Mountains (176.6 ±1.8 Ma) reported by Heimann et al. (1994). Therefore, the Dufek Intrusion is coeval with the magmatic activity that occurred in the Transantarctic Mountains in Middle Jurassic time. The low K-Ar dates obtained by Ford and Kistler (1980) for the Dufek Intrusion are attributable to loss of radiogenic ^{40}Ar. However, a suite of whole-rock samples, pyroxenes, and plagioclase of a Ferrar dolerite sill in the Pecora Escarpment at 85°37' S and 68°44' W in Fig. 5.80 analyzed by Ford and Kistler (1980) have a range of older dates from 177.7 ±0.4 Ma (plagioclase) to 223.1 ±5.6 Ma

(whole rock), indicating variable gain of ^{40}Ar, primarily by pyroxene. The initial ^{87}Sr/^{86}Sr ratio of the sill in the Pecora Escarpment is 0.71036 (180 Ma, $\lambda = 1.42 \times 10^{-11}$ yr^{-1}). Ford and Kistler (1980) also reported an initial ^{87}Sr/^{86}Sr ratio of 0.71249 for a dolerite dike in the Cordiner Peaks at 82°46' S and 53°21' W. Both values are compatible with initial ^{87}Sr/^{86}Sr ratios of Ferrar dolerites elsewhere in the Transantarctic Mountains.

A comparatively small set of data indicates that the rocks of the Dufek Intrusion have elevated initial ^{87}Sr/^{86}Sr ratios similar to those of the Ferrar dolerite and the Kirkpatrick basalt in the Transantarctic Mountains. Faure et al. (1972) reported initial ^{87}Sr/^{86}Sr ratios of 0.7084 and 0.7095 (relative to 0.7080 for E&A) for pyroxenites collected and described by Walker (1958). Subsequently, Ford et al. (1986) obtained initial ^{87}Sr/^{86}Sr ratios between 0.70765 and 0.70836 for three pyroxene concentrates and 0.70920 for a plagioclase sample. These results confirm that the Dufek Intrusion and the sills and dikes of the entire Pensacola Range are part of the Ferrar dolerite igneous province of the Transantarctic Mountains, even though they are located close to Queen Maud Land.

A noteworthy feature of the Dufek Intrusion is the wide range of δ^{18}O values from about 0 to +6.9‰. The δ^{18}O values of the rocks near the base of the intrusion in the Dufek Massif range from about +5 to +7‰, whereas those at the top of the intrusion in the Forrestal Range vary from nearly 0 to +5.8‰. Ford et al. (1986) attributed the variations in the oxygen isotope composition to subsolidus exchange of oxygen in the rocks with meteoric/hydrothermal solutions near the top of the Dufek Intrusion. A layer of granophyre (300 m thick) that caps the intrusion (Ford 1970) has comparatively low δ^{18}O values between about +3.8 and +5.8‰.

The Dufek Intrusion formed by progressive fractional crystallization of basalt magma in a large magma chamber. Early formed plagioclase, pyroxene, and titaniferous magnetite accumulated in layers at the bottom of the chamber. There is no evidence to indicate whether any parts of the magma were erupted at the surface because volcanic rocks that may have existed in the area are either covered by the ice sheet or have been removed by erosion. The low δ^{18}O values of the upper part of the Dufek Intrusion indicate that the rocks interacted with convecting groundwater of meteoric origin.

5.12.6 Dolerites of Tasmania

Dolerites sills and basalt flows of Middle Jurassic age occur in Tasmania as well as on Kangaroo Island at the mouth of the Gulf of St. Vincent near Adelaide in southern Australia (Milnes et al. 1982) and on the

South Island of New Zealand (Mortimer et al. 1995). These rocks are related to the Ferrar dolerite of Antarctica by virtue of their characteristic chemical compositions and elevated initial $^{87}Sr/^{86}Sr$ ratios, and because reconstructions of the break-up of Gondwana demonstrate that both Tasmania and New Zealand originated from the vicinity of northern Victoria Land in Antarctica.

The unusual enrichment of the Tasmanian dolerites in radiogenic ^{87}Sr and other large-ion lithophile elements was originally recognized by Heier et al. (1965) following a study of magmatic differentiation of these rocks by McDougall (1962). Heier et al. (1965) demonstrated that the Jurassic dolerite and granophyre at Red Hill and Great Lake in Tasmania had an average initial $^{87}Sr/^{86}Sr$ ratio of 0.71153 ±0.00005 ($2\bar{\sigma}, N = 7$) at 165 Ma. However, the initial $^{87}Sr/^{86}Sr$ ratios of the dolerites and granophyres are not appreciably correlated with the concentrations of SiO_2 (53.91 to 68.94%), K_2O, Rb, and Sr, which confirms the conclusion of McDougall (1962) that the dolerites and granophyres formed by fractional crystallization of a homogeneous magma. Heier et al. (1965) concluded that the "crustal" characteristics of these rocks (e.g. their Th/K, U/K, K/Rb ratios) must be attributed to the magma sources in the mantle and that contamination of the magma by assimilation of crustal rocks was negligible. The same conclusions presumably apply to the Jurassic basalts on Kangaroo Island near Adelaide in South Australia for which Milnes et al. (1982) reported an average initial $^{87}Sr/^{86}Sr$ ratio of 0.7112 ±0.0006 ($2\bar{\sigma}, N = 3$) at 170 Ma.

The petrogenesis of the Tasmanian dolerite sills was reconsidered by Hergt et al. (1989b) who used geochemical and isotopic data to support the proposal that the dolerite magmas had formed in the mantle by mixing of partial melts derived from previously subducted shale-like sediment and depleted mantle rocks. The isotope composition of Sr in these hybrid magmas was strongly influenced by the sedimentary source rocks, which had become enriched in radiogenic ^{87}Sr by decay of ^{87}Rb. The mixed magma subsequently differentiated by fractional crystallization without additional assimilation of crustal rocks. Accordingly, the range of initial $^{87}Sr/^{86}Sr$ ratios reported by Hergt et al. (1989b) (0.70937 to 0.71281) relative to 0.71025 for NBS 987 and 0.70801 for E&A was caused by imperfect mixing of magmas at the source rather than by contamination of magma by crustal rocks.

The petrogenesis of the Tasmanian dolerites and granophyres has a direct bearing on the origin on the Jurassic basalt flows and dolerite sills in the Transantarctic Mountains of Antarctica because Tasmania was connected to Antarctica at the time these rocks formed (Hergt et al. 1989a). However, the Ferrar dolerite sills and Kirkpatrick basalt flows of the Transantarctic Mountains exhibit a variety of features that resulted from interactions between the hybrid mantle-derived magmas and the continental crust of Gondwana. These include the recrystallization of granite inclusions in the Basement Sill of southern Victoria Land (Compston et al. 1968), the progressive change of the initial $^{87}Sr/^{86}Sr$ ratios of sequentially erupted basalt flows on Storm Peak for the Queen Alexandra Range (Faure et al. 1974), and the correlation of $\delta^{18}O$ values and initial $^{87}Sr/^{86}Sr$ ratios of basalt flows on Mt. Falla in the same area (Hoefs et al. 1980). Even the 300 m thick layer of granophyre at the top of the Dufek Intrusion provides evidence for assimilation of granitic basement rocks by the Dufek magmas. In addition, the Jurassic basalts and dolerites of the Transantarctic Mountains exemplify the important effect of water-rock interactions on the trace-element composition of these rocks and on their $^{87}Sr/^{86}Sr$ ratios, which were affected not only by the addition of Sr from the country rocks but also by the increase of their Rb/Sr ratios. The hydrothermal alteration of all of the rocks in the Transantarctic Mountains was locally intensified by the development or reactivation of deep crustal rifts during the Cretaceous Period associated with the separation of Australia, Tasmania, and New Zealand from Antarctica. The remaining crustal fragments of Gondwana then moved relative to each other and reassembled into East and West Antarctica as we know them today.

5.12.7 Tectonic History of Gondwana

The petrogenesis of the continental tholeiites of Gondwana has been discussed in terms of the heterogeneity of the underlying lithospheric mantle manifested by the initial $^{87}Sr/^{86}Sr$ and Rb/Sr ratios of the rocks (e.g. Brooks and Hart 1978; Kyle 1980) and in terms of the timing of crustal separation (Scrutton 1973). However, the most revealing approach has been to view the igneous activity as a consequence of tectonic processes that acted on the lithospheric plate of Gondwana.

Reconstructions of the fragments of Gondwana (e.g. Smith and Hallam 1970) indicate that Antarctica was located along the Pacific margin of Gondwana where the Pacific Plate was being subducted during the Paleozoic Era. Therefore, the formation of basalt and dolerite along the Pacific Coast of Gondwana can be viewed in terms of the tectonics of subduction zones along continental margins. In fact, subduction of the Pacific and related plates under the American continents is still going on at the present time as documented in Chap. 4. Since continental flood basalts may form by decompression melting in a setting of extensional tectonics, Elliot (1974) postulated that the basalts and dolerites of the Transantarctic Mountains and Tasmania formed in a back-arc basin that developed along the continental margin of Gondwana as a result of subduction of the Pacific Plate. Elliot (1992) subsequently elaborated this scenario by identifying a magmatic arc along the plate margin, a belt of silicic igneous activity inboard of the arc, and the low-Ti basalt province in the back-arc

The evidence that supports this hypothesis has been obscured by lateral displacements of crustal fragments (e.g. Tasmania), by movement along the Jurassic rift system (Schmidt and Rowley 1986; Behrendt et al. 1991), and by the extensive ice cover over West Antarctica and the Ross embayment. Nevertheless, Elliot (1992) assembled the fragmented evidence to support his hypothesis, which provides a realistic framework within which local variations of the magmatic activity can be considered. The geologic evolution of Antarctica, including the fragmentation of Gondwana and the associated igneous activity, has also been discussed in books edited by Thompson et al. (1991) and Storey et al. (1992). The relation between the Jurassic basalts of Antarctica and similar volcanic rocks in South America and southeastern Australia was demonstrated by Storey and Alabaster (1991) and Hergt et al. (1991), respectively.

5.13 The Siberian Flood-Basalt Province

The flood basalts on the Siberian Precambrian shield of Russia are located north of the Arctic Circle in Fig. 5.81 and range in age from Late Permian to early Triassic. The so-called Siberian Traps occur in three contiguous areas: (1) Putorana; (2) Noril'sk-Talnakh; and (3) Meimecha-Kotui (Sharma et al. 1991). The basalt lavas and associated mafic intrusives of the Noril'sk area have been studied most intensively because of the presence of Ni-Cu ore deposits associated with some of the intrusives (Al'Mukhamedov and Zolotukhin 1988; Czamanske et al. 1992; Naldrett et al. 1992; Brügmann et al. 1993; Horan et al. 1995). The city of Noril'sk located at about 65°30' N and 88°00' E was founded in 1930 and is now the second-largest city north of the Arctic Circle, with a population of 183 000 inhabitants.

The Siberian Traps were extruded during a short interval of time between 248 and 250 Ma, which coincides with the Permo-Triassic boundary (Dalrymple et al. 1991; Renne and Basu 1991; Campbell et al. 1992). The original volume of magma extruded in this area was greater than 1.5×10^6 km^3 (Zolotukhin and Al'Mukhamedov 1988) making the Siberian Traps one of the largest eruptions of flood basalt on record. The igneous activity is attributable to the interaction of a large asthenospheric plume with the overlying subcontinental lithospheric mantle and continental crust of the Asian Plate. As a result, large volumes of basaltic mag-

ma formed by decompression melting in the head of the plume and in the adjacent lithospheric mantle in a setting of extensional tectonics. However, this plume did not permanently split the Asian lithospheric plate unlike the plumes that caused the break-up of Gondwana.

The basement rocks of the Noril'sk area in Fig. 5.81 consist of gneisses and schists of Archean and Early Proterozoic age overlain by marine carbonates, argillites, and sulfate-bearing evaporite rocks of Paleozoic age. These rocks are overlain by non-marine clastic sediments of the Tungusskaya Series (Carboniferous to Late Permian), which includes coal measures (Hawkesworth et al. 1995; Wooden et al. 1993).

The apparent similarity of the Ni-Cu ore deposits of Noril'sk and Sudbury, Ontario, was the topic of a symposium the proceedings of which were edited by Lightfoot and Naldrett (1994). The sulfide ores of Noril'sk are significant not only because of their economic value, but also because the ore-bearing intrusives are associated with basaltic lava flows and volcanic ash that were deposited at the surface. The pres-

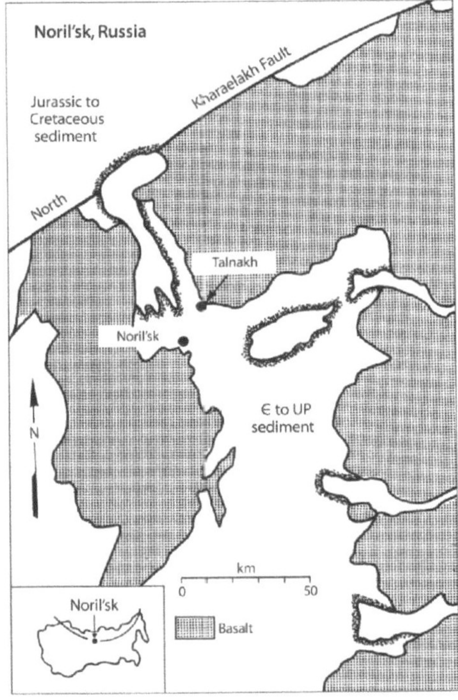

Fig. 5.81. Flood basalt plateau (250 Ma) in the Noril'sk area north of the Arctic Circle in Siberia, Russia (*Sources:* adapted from Lightfoot et al. 1993; Hawkesworth et al. 1995)

ence of tuff interbedded with the lava flows indicates that the volcanic eruptions were in part explosive, in contrast to other continental basalt provinces. In addition, the lava flows have amygdaloidal tops that constitute about 40% by volume of each flow. Since the age of the Siberian flood basalts coincides with the Permo-Triassic boundary, Campbell et al. (1992) proposed that the intense volcanic eruptions on the Siberian Platform contributed to the extinction of fauna and flora at the end of the Permian Period.

The volcanic rocks of the Noril'sk area have been subdivided into eleven formations (shown in Table 5.12) based on petrologic and geochemical criteria (Fedorenko 1981, 1994; Lightfoot et al. 1994).

The chemical compositions and isotope ratios of Sr, Nd, and Pb of the Putorana region have been reported by Sharma et al. (1991, 1992). The initial $^{87}Sr/^{86}Sr$ ratios (at 250 Ma) of the volcanic rocks in the Noril'sk area (SiO_2: 47.13 to 54.90%) range from 0.703246 (tholeiitic picrite basalt, Sharma et al. 1991) to 0.70872 (porphyritic basalt and tholeiite, Nadezhdinsky Formation; Wooden et al. 1993) relative to 0.71025 for NBS 987. In addition, Fig. 5.82 reveals that the initial $^{87}Sr/^{86}Sr$ ratios and the Rb concentrations vary stratigraphically such that the older formations (Ivakinsky to Nadezhdinsky) of Assemblages I and II have higher $^{87}Sr/^{86}Sr$ ratios and Rb concentrations than the younger formations. Table 5.13 shows the average Rb and Sr concentrations and $^{87}Sr/^{86}Sr$ ratios of the three assemblages of volcanic rocks (Wooden et al. 1993).

The number of samples included in each average is indicated in parentheses. These data demonstrate that the rocks of the first and second assemblages have significantly higher average concentrations of Rb and Sr

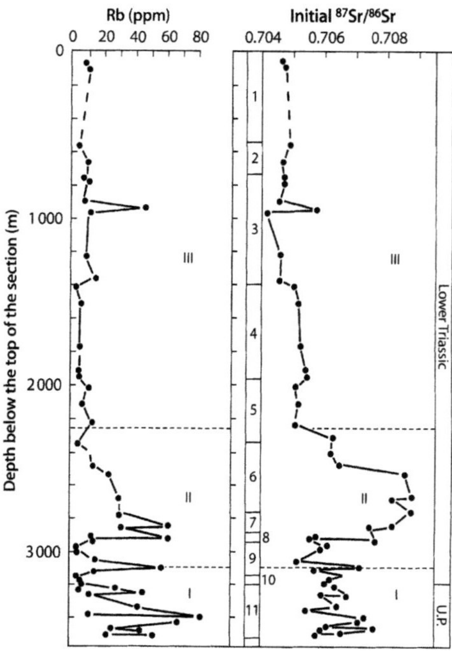

Fig. 5.82. Stratigraphic variation of the initial $^{87}Sr/^{86}Sr$ ratios (at 250 Ma) and Rb concentrations of basaltic lava flows and tuff in the Noril'sk area of Siberia, Russia. The formations are identified by number (Fedorenko 1981, 1994): 1. Samoedsky; 2. Kumginsky; 3. Karaelakhsky; 4. Mokulaevsky; 5. Morongovsky; 6. Nadezhdinsky; 7. Tuklonsky; 8. Khakanchansky; 9. Gudchikinsky; 10. Syverminsky; 11. Ivakinsky. The data suggest that the initial $^{87}Sr/^{86}Sr$ ratios and Rb concentrations of the rocks generally decreased as the volcanic activity progressed. The Roman numerals refer to suites of lava formations. The rocks of assemblages *I* and *II* have higher average concentrations of Rb and Sr and initial $^{87}Sr/^{86}Sr$ ratios than the rocks of assemblage *I* (*Source:* Wooden et al. 1993)

Table 5.12. Formations of the volcanic rocks of the Noril'sk area (Fedorenko 1981, 1994; Lightfoot et al. 1994)

	Formation	Thickness (m)	Assemblage
1.	Samoedsky (youngest)	575	
2.	Kumginsky	175	
3.	Kharaelakhsky	600	III
4.	Mokulaevsky	575	
5.	Morongovsky	360	
6.	Nadezhdinsky	400	
7.	Tuklonsky	120	II
8.	Khakanchansky	60	
9.	Gudchikhinsky	200	
10.	Syverminsky	50	I
11.	Ivakinsky (oldest)	260	
	Total thickness:	3 375	

as well as higher initial $^{87}Sr/^{86}Sr$ ratios than the rocks of the third assemblage. Evidently, the magmas that were erupted at the start of the volcanic activity were alkali-rich and contained more radiogenic Sr than the voluminous tholeiite basalts that followed.

The data summarized in Fig. 5.83 demonstrate that the basalt flows and mafic intrusives of the Noril'sk area have similar initial $^{87}Sr/^{86}Sr$ ratios ranging from >0.7030 to <0.7090. This result supports the assumption that the basalt flows on the surface are genetically related to the ore-bearing mafic intrusives below the surface. In addition, the wide range of the initial $^{87}Sr/^{86}Sr$ ratios confirms that the basalt flows of the Noril'sk area exemplify volcanic activity caused by the interaction of an asthenospheric mantle plume with the overlying lithospheric mantle and continental crust. (Campbell and Griffiths 1990; Griffiths and Campbell 1990, 1991).

Table 5.13.
Average Rb and Sr concentrations and $^{87}Sr/^{86}Sr$ ratios of the three assemblages of volcanic rocks in the Noril'sk area (Wooden et al. 1993)

Assemblage	Volume (%)	Formations	Rb (ppm)	Sr (ppm)	$(^{87}Sr/^{86}Sr)_0$
III (youngest)	79	1 – 5 2 – 46	9.5 (19) 2 – 46	207 (19) 16.4 – 394	0.70491 (19) 0.70459–0.70541
II	14	6 – 8 3 – 88	28.2 (17) 3 – 88	267 (17) 138 – 337	0.70707 (17) 0.70528–0.70872
I (oldest)	7	9 – 11 4 – 82	28.7 (17) 4 – 82	371 (17) 140 – 937	0.70618 (17) 0.70546–0.70710

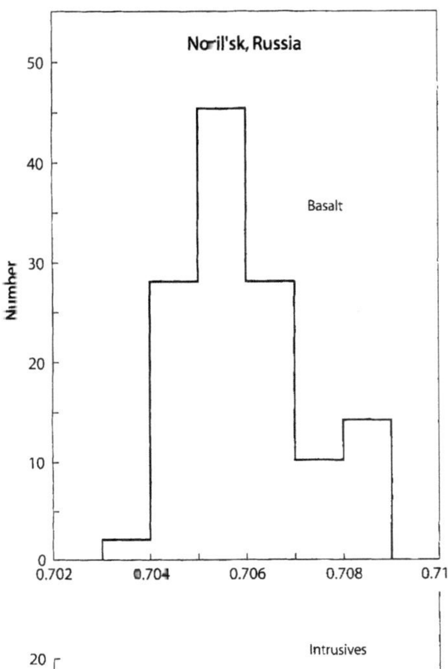

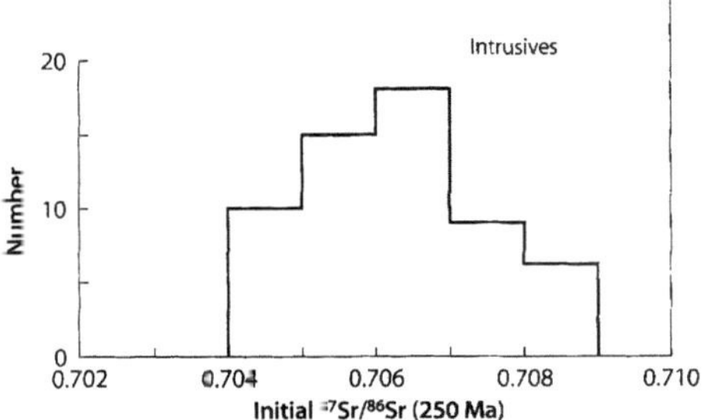

Fig. 5.83. Initial $^{87}Sr/^{86}Sr$ ratios of basalt flows and mafic intrusives at 250 Ma of the Noril'sk area in Siberia, Russia (*Sources:* DePaolo and Wasserburg 1979; Lightfoot et al. 1990, 1993; Hawkesworth et al. 199)

Table 5.14. Measured and modeled (Wooden et al. 1993) concentrations of SiO_2 and Sr and isotope ratios of Sr for Assemblage III and the Nadezhdinsky Formation in the Noril'sk area

Component	SiO_2 (%)	Sr (ppm)	$^{87}Sr/^{86}Sr$
Assemblage III			
Parent magma	47.0	60	0.7030
Contaminant	66.0	1 000	0.7070
Model basalt	49.9	194	0.7052
Measured basalt	49.4	200	0.7051
Nadezhdinsky formation			
Parent magma	47.0	60	0.7030
Contaminant	66.0	1 200	0.7100
Model basalt	52.8	341	0.7084
Measured basalt	53.0	310	0.7085

Wooden et al. (1993) modeled the formation of basalts of assemblage III and of the Nadezhdinsky Formation in crustal magma chambers, assuming episodic replenishment and tapping of magma with continuous fractionation and assimilation of granitic wallrock. The evolution of magmas by this process was discussed by O'Hara (1977, 1980) and by O'Hara and Matthews (1981). The modeling by Wooden et al. (1993) achieved excellent agreement with the measured concentrations of both major and trace elements as well as with isotope ratios of Sr based on assimilation of granitic basement rocks by basaltic magmas (Table 5.14).

The agreement of the model results with the measured values of the parameters listed in Table 5.14 supports this explanation of the petrogenesis of the Siberian basalts. The high Sr concentrations and relatively low $^{87}Sr/^{86}Sr$ of the granitic contaminant are noteworthy. The amount of this granitic contaminant added to the basalt magma decreased from 15% (Nadezhdinsky Formation, Assemblage II) to only 6% in Assemblage III.

The evidence for crustal contamination of the basalt flows of the Noril'sk area is strengthened by the wide range of their initial $^{143}Nd/^{144}Nd$ and $^{206}Pb/^{204}Pb$ ratios in Figs. 5.84 and 5.86. The data fields in Fig. 5.84 demonstrate a strong negative correlation of initial $^{87}Sr/^{86}Sr$ and $^{143}Nd/^{144}Nd$ ratios consistent with the results of mixing of several different crustal components with magmas that may have been derived by de-

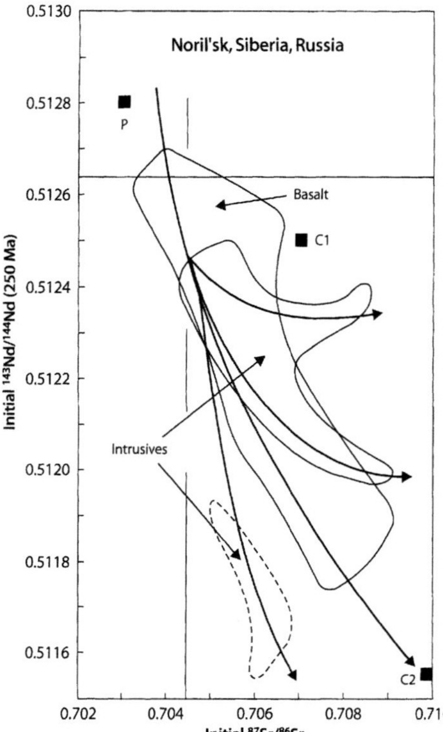

Fig. 5.84. Initial isotope ratios of Sr and Nd (at 250 Ma) in basalt and intrusives of the Noril'sk area, Siberia, Russia. The hypothetical mixing trajectories suggest the range of isotope compositions of Sr and Nd in the crustal contaminants of the mantle-derived magmas. *C1* and *C2* are crustal contaminants whereas *P* is the parent magma used by Wooden et al. (1993) in modeling calculations (*Sources:* Lightfoot et al. 1990, 1993; Hawkesworth et al. 1995; Wooden et al. 1993; Sharma et al. 1992)

Fig. 5.85. Initial isotope ratios of Sr and Pb (at 250 Ma) in basalts of the Noril'sk area of Siberia, Russia. *C1* and *C2* are crustal contaminants, and *P* is the parent magma of model calculations by Wooden et al. (1993). The scatter of data points requires more than two types of contaminants (*Sources:* Wooden et al. 1993; Sharma et al. 1992)

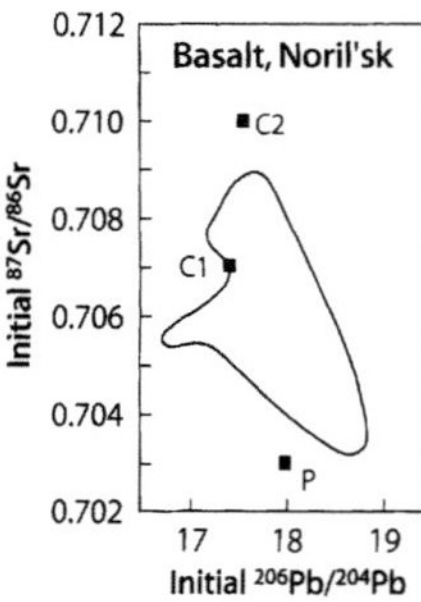

The Siberian Traps (Permo-Triassic) are the oldest Mesozoic flood basalts to be considered in this chapter. Their origin can be understood as the result of the interaction of an asthenospheric plume with the overlying lithospheric mantle and continental crust of the Asian Plate. The mantle-derived magmas assimilated varying amounts of different kinds of crustal rocks as they differentiated by fractional crystallization in crustal magma chambers. The wide range of initial isotope ratios of Sr, Nd, and Pb implies considerable heterogeneity of the lithospheric mantle and continental crust of the Siberian Precambrian shield.

5.14 Cenozoic Basalts of Eastern China

Early Tertiary to Quaternary basalt lavas occur at many locations in Fig. 5.86 in the Marginal Pacific Tectonic Domain (MPTD) of eastern China, extending from Mongolia and Manchuria in the north to the Luichow Peninsula and the island of Hainan in the south (Zhang et al. 1984). This region includes the Sino-Korean (>1.7 Ga) and Yangtze (0.7 Ga) Cratons composed of metamorphic and igneous rocks of Precambrian age and is characterized by having a crustal thickness of less than 40 km. In addition, the Marginal Pacific Tectonic Domain contains large transcurrent faults (including the Tanscheng-Lujiang fault) which strike NNE parallel to the eastern margin of the Asian Plate. The Tanscheng-Lujiang fault extends for 2 400 km along strike and cuts the continental crust to the Mohorovičić discontinuity. In addition, the Marginal Pacific Tectonic Domain is crossed by major east-west trending thrust faults that reflect tectonic compression from the south. For example, the Tsinling thrust fault occurs within the Phanerozoic fold belt that separates the Sino-Korean Craton from the Yangtze Craton. This thrust fault has been traced west for about 2 500 km to the Tarim Craton (Zhang et al. 1984; Zhou and Armstrong 1982).

compression melting of plume rocks and of the lithospheric mantle. The hypothetical mixing hyperbolas in Fig. 5.84 imply that the crustal assimilants had a wide range of isotope ratios and concentrations of Sr and Nd.

The initial ^{206}Pb/^{204}Pb ratios (at 250 Ma) of the basalt flows at Noril'sk range from 16.815 (Wooden et al. 1993) to 18.719 (Sharma et al. 1992). The distribution of data points in Fig. 5.85 indicates that more than two crustal contaminants are required to account for the full range of isotope ratios of Sr and Pb. A basalt sample of the Gudchikinsky Formation analyzed by Sharma et al. (1992) has ^{87}Sr/^{86}Sr = 0.703238, ^{143}Nd/^{144}Nd = 0.51261, and ^{206}Pb/^{204}Pb = 18.719 at 248 Ma. These isotope ratios represent mantle-derived magmas that contained the smallest amount of Sr, Nd, and Pb crustal origin.

The Marginal Pacific Tectonic Domain in northern China is characterized by basin-and-range topography

Table 5.15.
Concentrations of SiO_2, Rb and Sr and isotope ratios for basalts in the Beijing area of northern China and in the surrounding Hebei Province averaged on the basis of age reported by Zhou and Armstrong (1982) (SA = subalkaline; A = alkaline)

Age	Type	SiO_2 (%)	Rb (ppm)	Sr (ppm)	$(^{87}Sr/^{86}Sr)_0$
Early T.	SA	49.4 (4)	4.80	514	0.70429
Early T.	A	49.2 (1)	22.0	1154	0.70390
Neogene	SA	50.8 (3)	11.4	402	0.70429
Neogene	A	45.4 (7)	34.5	1124	0.70417
Quaternary	A	48.8 (6)	23.9	764	0.70370

Table 5.16. Effect of crustal thickness on the average initial $^{87}Sr/^{86}Sr$ ratios illustrated by the early Tertiary subalkaline basalt flows from Bohai Bay located off the coast of Hebei Province (Zhou and Armstrong 1982)

Region	Crust (km)	$(^{87}Sr/^{86}Sr)_0$
Bohai Bay	<34	0.70332
Hebei and Beijing	~40	0.70429

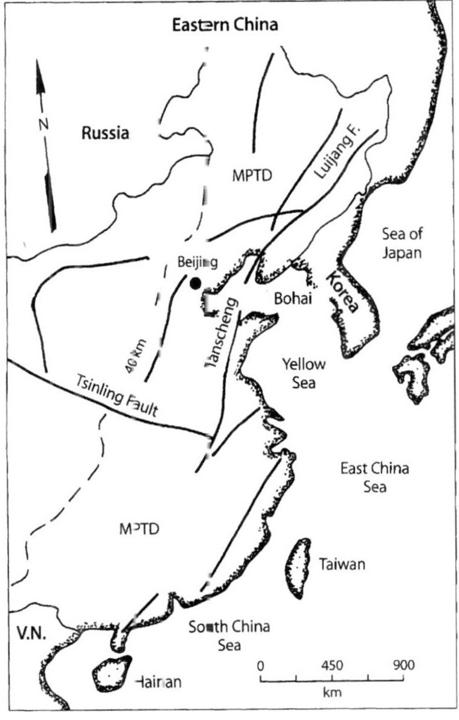

Fig. 5.86. The Early Tertiary to Quarternary basalt of eastern China in relation to the major fault systems (*Source:* adapted from Zhou and Armstrong 1982)

caused by extensional stresses that accompanied the separation of the Japanese Islands from the mainland of Asia in Late Cretaceous time. The subduction of oceanic crust beneath the Japanese Islands subsequently led to the formation of a back-arc basin presently occupied by the Sea of Japan. The extensional forces also caused rifting in northern China, which resulted in the eruption of tholeiitic and alkalic basalts ranging in age from Paleogene (Early Tertiary) to Quaternary. The Tertiary lavas were erupted primarily through fissures associated with the major faults rather than through volcanic vents and are known primarily from dia-

mond-drill cores. The Quaternary flows were erupted explosively through vents and are exposed at the surface (Zhou and Armstrong 1982).

The initial $^{87}Sr/^{86}Sr$ ratios of Tertiary and Quaternary basalt flows in northern and southern China range from 0.70294 (Hainan Island, South China) to 0.70478 (Henan Province, North China) relative to 0.70800 for E&A (Zhou and Armstrong 1982). These authors also observed that the $^{87}Sr/^{86}Sr$ ratios of Early Tertiary basalts decreased after leaching with hydrochloric acid that removed from 64 to 72% of the Sr. Therefore, all of the rock samples they analyzed were leached for 20 min in 2 N HCl with ultrasonic vibration followed by 8 h of leaching in 6 N HCl at 50 °C in order to remove carbonate and clay minerals prior to the final dissolution and measurement of the $^{87}Sr/^{86}Sr$ ratios.

The initial $^{87}Sr/^{86}Sr$ ratios of the Tertiary and Quaternary basalts of the Marginal Pacific Tectonic Domain of China vary with age, chemical composition, and crustal thickness. For example, Zhou and Armstrong (1982) reported the data in Table 5.15 for basalts in the Beijing area of northern China and in the surrounding Hebei Province averaged on the basis of age.

These results demonstrate that the alkali-rich basalts have lower average concentrations of SiO_2 but higher concentrations of Rb and Sr than the subalkaline basalts. In addition, the alkaline lavas have lower initial $^{87}Sr/^{86}Sr$ ratios than subalkaline basalts in the same age bracket. The effect of crustal thickness on the average initial $^{87}Sr/^{85}Sr$ ratios is illustrated by the early Tertiary subalkaline basalt flows from Bohai Bay located off the coast of Hebei Province (Table 5.16; Zhou and Armstrong 1982).

5.14.1 Northern China

The volcanic centers along the Tanscheng-Lujiang fault system in northern China include the Wudalianchi, Changbaishan, and Kuandian areas, all of which are located north of the 40th parallel east of Beijing; whereas the Datong, Hannuoba, and Chifeng volcanic fields lie west of Beijing (Fan and Hooper 1991; Liu et al. 1994). The petrology and petrogenesis of the Hannuoba basalts was discussed by Zhi et al. (1990), Song et al. (1990), and Basu et al. (1991). In addition, Song and Frey (1989) and Chen et al. (1995) described the xenoliths from Hannuoba, whereas Xu et al. (1996) derived evidence for mantle metasomatism from the presence of K-rich glass in wehrlite inclusions in northeastern China.

The initial $^{87}Sr/^{86}Sr$ ratios of Early Tertiary to Quaternary lavas of subalkaline and alkaline compositions north of the 40th parallel in northern China (Hebei, Liaoning, Jilin, and Heilongjiang provinces, as well as Bohai Bay) are compiled in the Fig. 5.87. Based on the relatively narrow range and low values of the $^{87}Sr/^{86}Sr$ ratios (>0.7030 to <0.7060) and on the basis of trace-element data, Zhou and Armstrong (1982) concluded that the tholeiite magmas originated from the subcontinental lithospheric mantle with only limited contamination by crustal rocks. The alkali-basalts, having somewhat lower $^{87}Sr/^{86}Sr$ ratios than the tholeiites, may have originated from deeper sources in the mantle, possibly from plume-type rocks of the asthenosphere. However, the similarity of $^{87}Sr/^{86}Sr$ ratios of the basalts in northern China and those of oceanic islands is probably a coincidence because of the difference in the tectonic setting.

Additional evidence about the petrogenesis of the basalts of northern China is provided in Fig. 5.88a and b by the combination of the isotope ratios of Sr, Nd, and Pb (Tatsumoto et al. 1992; Peng et al. 1986). The data fields in Fig. 5.88a demonstrate the wide range of the $^{143}Nd/^{144}Nd$ ratios from 0.51294 to 0.51238 and their negative correlation with the $^{87}Sr/^{86}Sr$ ratios. In addition, the data fields form an apparent two-component mixing array between the DMM and the EM1 mantle components (Hart 1988). The peridotite (spinel lherzolite) inclusions in the basanite (alkali basalt) lavas at Hannuoba also have a wide range of $^{87}Sr/^{86}Sr$ and $^{143}Nd/^{144}Nd$ ratios (0.702156; 0.513585 to 0.704300; 0.512491), confirming that the mantle under northern China contains both depleted and enriched components represented by DMM and EM1 (Song and Frey

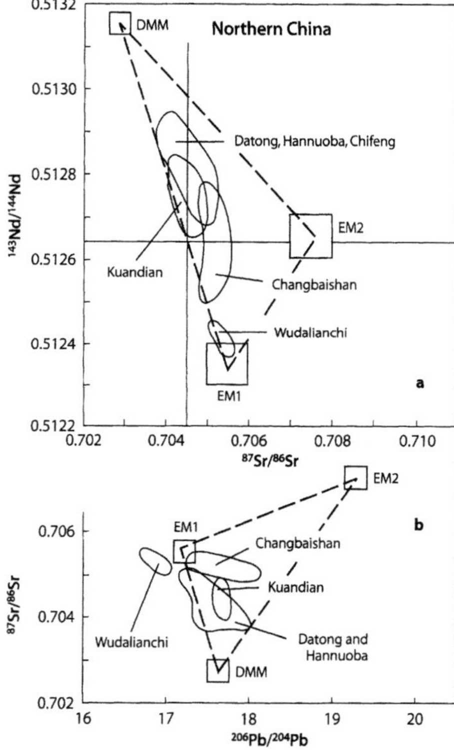

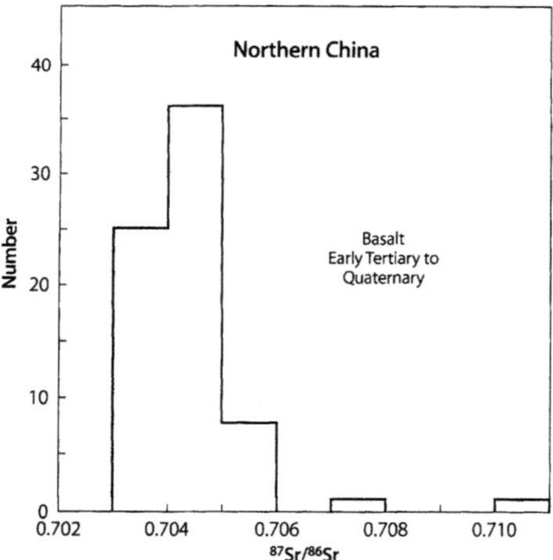

Fig. 5.87. Alkali-rich and subalkaline basalt of Early Tertiary to Quaternary basalt in the northern provinces of China (Hebei, Beijing, Jilin, Heilongjiang) and in Bohai Bay of the Yellow Sea (*Sources:* Zhou and Armstrong 1982; Basu et al. 1991; Han et al. 1999)

Fig. 5.88 a,b. Isotope ratios of Sr, Nd, and Pb of Late Tertiary to Quaternary basalt of northern China including the provinces of Hebei, Beijing, Liaoning, Jilian, and Heilongjiang north of latitude 40° N (*Sources:* Basu et al. 1991; Han et al. 1999; Hart 1988)

1989). The $^{206}Pb/^{204}Pb$ and $^{87}Sr/^{86}Sr$ ratios in Fig. 5.88b are likewise negatively correlated between the isotope ratios of MORB sources and the EM1 mantle component. The Quaternary lavas at Wudalianchi overlap the isotope compositions of EM1 in both Fig. 5.88a and b.

Basu et al. (1991) suggested that the EM1 component is the subcontinental lithospheric mantle, whereas the DMM component is the asthenospheric mantle that has invaded the lithospheric mantle as a result of the tectonic processes in the subduction zone associated with the Japanese Islands. Although the resulting magmas differentiated by fractional crystallization, assimilation of crustal rocks was of minor importance, perhaps because the temperature of the continental crust did not rise sufficiently to facilitate assimilation.

In conclusion, the Early Tertiary to Quaternary volcanic rocks of northern China are associated with deep transcurrent faults that run parallel to the eastern edge of the Asian Plate. These faults developed in Late Cretaceous to Early Tertiary time during the separation of the Japanese Islands from the mainland and as a result of the subsequent development of a back-arc basin in the Sea of Japan. Therefore, the volcanic rocks of northern China formed in a setting of extensional tectonics by decompression melting in the subcontinental lithospheric mantle, which was altered, in some cases, by the addition of Sr, Nd, and Pb derived from the subducted oceanic crust. The magmas were predominantly alkalic in composition and did not assimilate significant amounts of crustal rocks.

5.14.2 Hainan Island and the South China Sea

The chemical compositions of Cenozoic volcanic rocks along the east coast of China in the provinces of Shandong, Anhui, and Jiangsu between latitudes of 38° and 30° N were described by Zhou and Armstrong as well as by Dostal et al. (1988, 1991). These occurrences, like those of northeastern China, are associated with the extensional tectonics of the Tanscheng-Lugjiang Fault (Fig. 5.86). Farther south, the South China Sea formed as a result of the fragmentation of continental crust by rifting, crustal attenuation, and sea-floor spreading following the collision of India with the Eurasian Plate (Hayes 1980, 1983; Tu et al. 1992).

Hainan Island (Fig. 5.86) is a continental fragment that separated from the mainland in Paleocene time. It consists of the Lei-Qiong Graben in the north and a horst composed of sialic rocks in the south. Most of the volcanic rocks are located in the northern part of Hainan Island and on the Luichow Peninsula, which is separated from the island by a major strike-slip fault (Flower et al. 1992).

The lava flows of northern Hainan Island overlie and are interbedded with petroliferous sedimentary

rocks of Tertiary age. Evidence from drill cores indicates that the lava flows have a composite thickness of about 1 000 m. In addition, older sedimentary rocks (Cretaceous to early Oligocene) were intruded by sills of unknown age down to a depth of about 3 000 m. Most of the volcanic rocks consist of tholeiite basalts that were erupted through fissures, whereas less abundant alkali basalts were extruded through central volcanic vents. Some of the alkali basalt flows contain spinel-lherzolite and harzburgite inclusions. Flower et al. (1992) estimated that about 5 000 km^3 of magma were erupted on Hainan Island in less than 50 000 years, yielding a rate of more than 0.1 km^3 yr^{-1}.

The $^{87}Sr/^{86}Sr$ ratios of the Quaternary basalts on Hainan Island and in Guangdong Province on the adjacent mainland range from 0.70294 relative to 0.70800

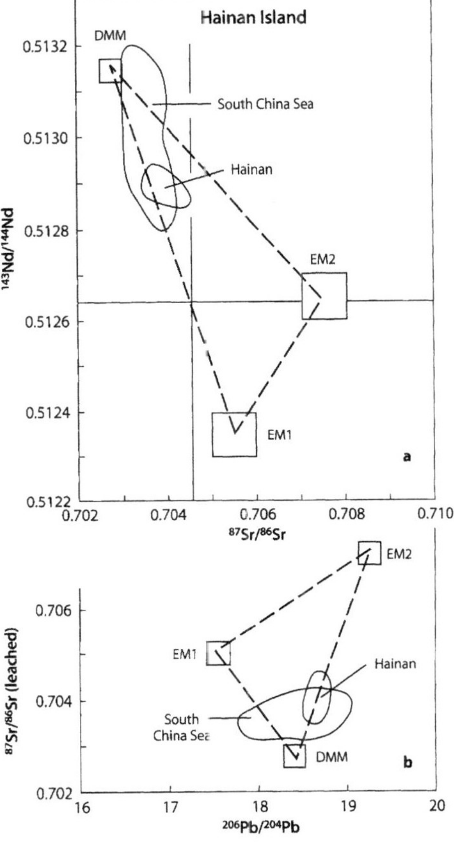

Fig. 5.89 a,b. Isotope ratios of Sr, Nd, and Pb of Late Tertiary basalts on Hainan Island, southern China, and in the South China Sea (*Sources:* Tu et al. 1991, 1992; Hart 1988)

for E&A (Zhou and Armstrong 1982) to 0.704474 relative to 0.71025 for NBS 987 (Tu et al. 1991). In addition, Zhou and Armstrong (1982) reported a value of 0.70291 for a hawaiite on Xisha Island (16°35' N, 112°59' E) and 0.70326 for a trachybasalt from 14°58' N and 116°30' E on the floor of the South China Sea.

The isotope ratios of Sr, Nd, and Pb of tholeiites and alkali basalts on Hainan Island (Tu et al. 1991) and of sea-floor basalts in the South China Sea (Tu et al. 1992) are presented in Fig. 5.89a and b. In general, both data fields are more closely associated with the DMM component than the basalts of northern China in Fig. 5.88a and b. Although fragments of continental crust occur in the South China Sea, they played an insignificant role in the petrogenesis of these volcanic rocks. The lavas that erupted within the South China Sea are mixtures derived from depleted mantle rocks similar to the DMM component and less depleted or enriched source rocks represented by the EM1 component. The basalt flows on Hainan Island contain some Sr, Nd, and Pb derived from the EM1 and EM2 components that presumably reside in the lithospheric mantle underlying the China Sea.

5.14.3 Khorat Plateau, Thailand

The crustal extension of the Marginal Pacific Tectonic Domain of eastern China during the Tertiary Period also affected the continental crust of southeast Asia (Barr and Macdonald 1981). As a result, lavas of Late Tertiary to Quaternary age were erupted along the western margin of the Khorat Plateau (Lop Buri basalt) and at scattered locations in its southern region. The Khorat Plateau in Fig. 5.90 is part of the Indochina Terrane which is separated from the Sukhothai and Shan-Thai terranes of western Thailand by the Nan suture (Zhou and Mukasa 1997).

The volcanic rocks along the southern margin of the Khorat Plateau are composed of alkali olivine basalt and hawaiite with less abundant basanite, tholeiite, and other rock types. Zhou and Mukasa (1997) cited a whole-rock K-Ar date of 0.9 Ma for a basalt from one of the exposures on the Khorat Plateau. The Indochina Terrane is composed primarily of sedimentary rocks of late Paleozoic age, overlain by continental beds of the Mesozoic Khorat Group and by Permo-Triassic lavas. These rocks were later intruded by a few granitic plutons (Barr and Macdonald 1991).

The isotope ratios of Sr and Nd of the alkali basalts on the Khorat Plateau divide these rocks into two groups in Fig. 5.91. Both are collinear between the depleted MORB mantle (DMM) and enriched lithospheric mantle (EM2). Zhou and Mukasa (1997) concluded from this and other evidence that the magmas contain varying proportions of two components that originated from the asthenospheric and the lithospheric mantle. There is no evidence for assimilation of significant amounts of Sr and Nd from the continental crust. The $^{206}Pb/^{204}Pb$ ratios of the rocks in Group I (18.239 to 18.469) are slightly lower than those of Group II (18.494 to 18.638), thereby confirming that

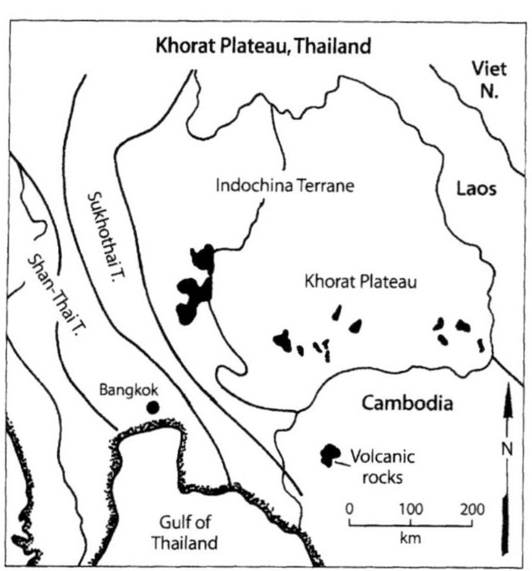

Fig. 5.90. Map of Thailand identifying the location of the volcanic centers in the southern part of the Khorat Plateau (*Source:* Zhou and Mukasa 1997)

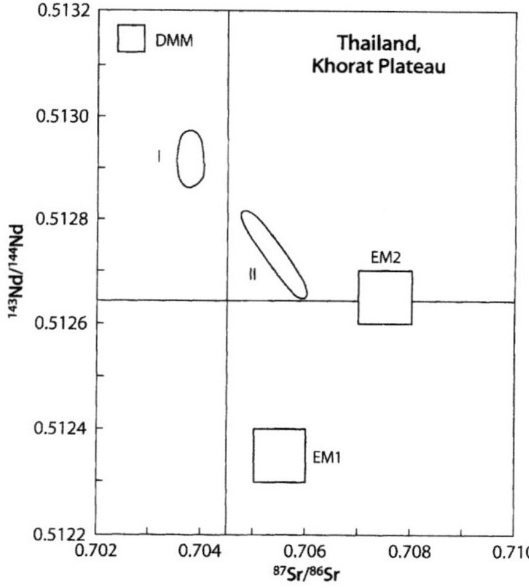

Fig. 5.91. Isotope ratios of Sr and Nd of alkali basalt flows of the Khorat Plateau of Thailand (*Source:* Zhou and Mukasa 1997)

the magmas are mixtures of two components (i.e. DMM and EM2).

The volcanic rocks of the Khorat Plateau resemble those of southern China in Fig. 5.89 in several ways. Both formed under conditions of extensional tectonics in Late Tertiary to Quaternary time from magma sources in the asthenospheric and lithospheric mantle. The magmas differentiated by fractional crystallization without significant assimilation of crustal rocks. In addition, the magmas are distinctly alkali-rich in composition, presumably because the extent of partial melting was less than required for the formation of quartz-normative tholeiite basalt. The Late Tertiary to Quaternary basalts of eastern China and of southeast Asia exemplify the close association of quartz-normative tholeiites with undersaturated alkali-rich basalts. This association arises because both types of lavas form in settings of extensional tectonics by different degrees of partial melting of the same kinds of source rocks in the asthenospheric and lithospheric mantle.

5.15 Flood Basalts of Ethiopia and Yemen

The basalt plateaus of Ethiopia, Djibouti, and western Yemen in Fig. 5.92 form a single petrologic province whose origin is directly related to rifting in this area that started during the Tertiary Period and continues to the present. According to information reviewed by White and McKenzie (1989), the Tertiary basalt plateaus of Ethiopia and Yemen formed after a large asthenospheric plume reached the base of the lithospheric mantle in the Afar region of Ethiopia at about 30 Ma. The heat released by this plume increased the temperature of the overlying lithosphere and caused significant regional uplift, which resulted in rifting of the continental crust and thinning of the lithospheric mantle. The expansion of the head of the so-called Afar Plume caused rifting of the Afro-Arabian Plate. The rifts subsequently widened, thereby forming the Red Sea, the Gulf of Aden, and the Main Ethiopian Rift, which together define the Afar triple junction. The head of the Afar Plume underlies this triple junction and has a present diameter of about 2 000 km.

As a consequence of this tectonic activity, large volumes of flood basalt were erupted through fissures in Ethiopia as well as in Yemen and Saudi Arabia. The number of lava flows and their thickness decrease with increasing distance from the major rifts. The Ethiopian basalt plateau originally covered an area of about 750 000 km², whereas the basalts in Yemen and Saudi Arabia had an area of about 130 000 km².

The main episode of volcanic activity started at 30 Ma (McDougall et al. 1975) and lasted for about ten million years. However, volcanic activity in the area

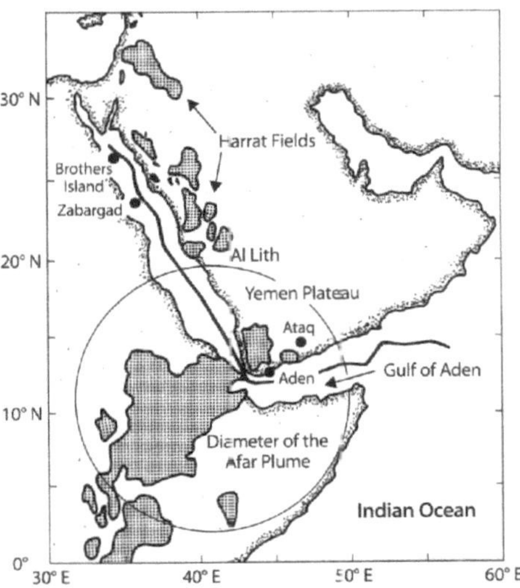

Fig. 5.92. The Afro-Arabian Plate including the continental flood basalts of Ethiopia, Yemen, and Saudi Arabia. The *circle* defines the head of the Afar Plume, which is still active at the present time (*Source:* Schilling et al. 1992)

of the Main Ethiopian Rift started about 15 million years earlier at about 45 Ma (Stewart and Rogers 1996; Berhe et al. 1987). About 20 million years ago, volcanic eruptions ceased in Yemen and became localized in the main rift valley of Ethiopia. A second phase of volcanism started at 4.5 Ma throughout the region and coincided with renewed widening of the Red Sea.

The Tertiary basalts of Ethiopia are primarily alkali-rich tholeiites and picrites, which compose about 85% of the lavas. The chemical compositions of these lavas differ from the alkali-rich lavas of the East African Rift valleys, which are not connected with the Main Ethiopian Rift.

5.15.1 Ethiopian Plateau

The Tertiary basalts of Ethiopia form two large plateaus on both sides of the Main Rift in Fig. 5.92 (Brotzu et al. 1974; Woldegabriel et al. 1990). The plateaus consist of the basalt flows of the Trap Series (Oligocene), which are unconformably overlain in southwestern Ethiopia by welded tuffs composed of rhyolite (pantellerite) of probable Pliocene age (Mohr 1968). The rhyolite tuffs were erupted through vents located within the Main Rift and aligned along fissures parallel to the Rift. Their maximum thickness along the western margin of the Main Rift is about 500 m. More recently, welded tuffs of Quaternary age were erupted at centers of silicic volcanic activity along the Wonji fault belt in the center of the Main Ethiopian Rift.

In northern Ethiopia, the eruption of the basalts of the Trap Series occurred during the Eocene Epoch, whereas in southern Ethiopia the Trap basalts were erupted during the Oligocene (Mohr 1968). Subsequently, large shield volcanoes composed primarily of alkali olivine basalt (Shield Group) formed during early Miocene time. At about the same time, alkali basalts including mugearites were erupted in the Afar region to the north. The main phase of basalt volcanism in Ethiopia was followed in late Pliocene time by the eruption of the rhyolite tuff mentioned above. At about the same time, small volumes of highly alkali-rich trachytes, phonolites, and comendites were erupted through vents in south-central Ethiopia. After a short interval of erosion in late Pliocene-early Pleistocene time, tectonic activity resumed in the mid-Pleistocene with major faulting that formed the present scarps of the Main Ethiopian Rift and in the Afar region. The most recent volcanic eruptions produced scoriaceous alkali-olivine basalt that covers the floor of the Main Rift in Ethiopia and the Afar region. Silica-rich volcanic rocks of late Pleistocene age occur only in a few places in the Wonji fault belt within the Main Rift (Mohr 1968).

The bimodal distribution of silica concentrations of the Tertiary volcanic rocks of Ethiopia and Yemen was demonstrated graphically by Mohr (1971), based on a catalog of chemical analyses published by Mohr (1970). The data reveal a gap in the silica concentrations between about 47% for alkaline and subalkaline basalts and 72% for alkali-rich rhyolites. Moreover, the average alkali concentrations of the dominant alkali-rich basalts are $Na_2O = 3.1\%$ and $K_2O = 1.3\%$ compared to 2.0% and 0.6%, respectively, for the calc-alkaline basalt.

The rhyolite tuffs ($Na_2O = 4.8\%$; $K_2O = 4.5\%$) make up less than 15% of the estimated volume of 335 000 km³ of the Tertiary volcanic rocks in Ethiopia and Afar (Mohr 1968). These rocks do not appear to be magmatic differentiates of basalts magmas, because rhyolites and basalts are rarely interbedded, and intermediate rock types are uncommon. Consequently, Mohr (1968) favored the explanation that the silicic magmas originated by melting of rocks in the continental crust. However, the elevated concentrations of Na_2O and K_2O of the rhyolites are not consistent with magma formation by crustal fusion. However, the rhyolite magmas could have formed by partial melting of alkali-rich gabbros in the continental crust like the rhyolites of Iceland (Sect. 2.3.2).

The initial $^{87}Sr/^{86}Sr$ ratios of the basalt lavas that were erupted in the Main Rift of Ethiopia and that form the adjacent plateaus range from 0.70318 (Stewart and Rogers 1996) to 0.70665 (Hart et al. 1989) relative to 0.71025 for NBS 987 and 0.70800 for E&A, respectively. Actually, 72% of the 68 basalt specimens from

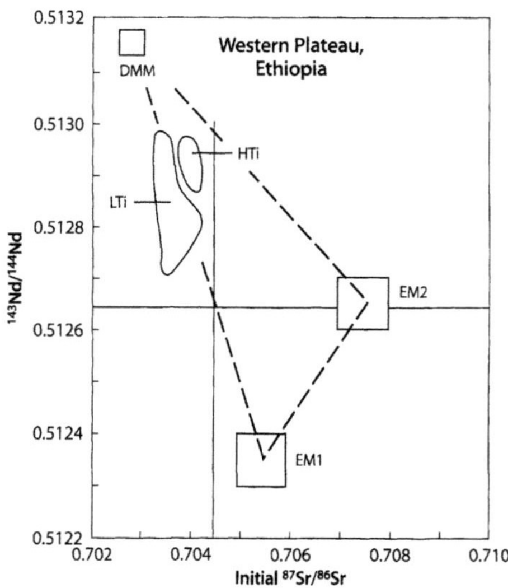

Fig. 5.93. Isotope ratios of Sr and Nd of basalt (Oligocene) of the lava plateau in northwestern Ethiopia. *HTi* = high Ti; *LTi* = low Ti. A group of mixed HTi-LTi flows are not shown because they overlap the HTi and LTi data fields (*Source:* Pik et al. 1999)

the Main Rift and the adjacent plateaus analyzed by Barbieri et al. (1976), Hart et al. (1989), and Stewart and Rogers (1996) have initial $^{87}Sr/^{86}Sr$ ratios that are less than 0.70400. The Oligocene flood basalts of northwestern Ethiopia have been subdivided into three suites based primarily on their Ti concentrations. Pik et al. (1999) reported that the low-Ti basalts located around Lake Tana near the city of Gondar have low initial $^{87}Sr/^{86}Sr$ (0.70337 to 0.70420 relative to 0.71025 for NBS 987) and high $^{143}Nd/^{144}Nd$ (0.51271 to 0.51295) ratios consistent with their derivation from the lithospheric mantle. The isotope ratios of some of the high-Ti basalts in Fig. 5.93 indicate that they originated from the Afar Plume, whereas a second group of high-Ti basalt (not shown in Fig. 5.93) formed from mixed plume-lithosphere magmas. Pik et al. (1999) considered that both components may have occurred in the head of the Afar Plume. Pan-African metavolcanic basement rocks that underlie the basalt plateau are potential contaminants of the mantle-derived magmas. Pik et al. (1999) concluded that the correlations of trace element concentrations and isotope ratios of Sr and Nd of the low-Ti basalts are evidence that the magmas did interact with wallrocks while they differentiated in crustal magma chambers.

The initial $^{87}Sr/^{86}Sr$ ratios of the basalts in southern Ethiopia reported by Stewart and Rogers (1996) decrease with age in Fig. 5.94 from about 0.7043 at 45 Ma to about 0.7033 at 20 Ma (Miocene) and then

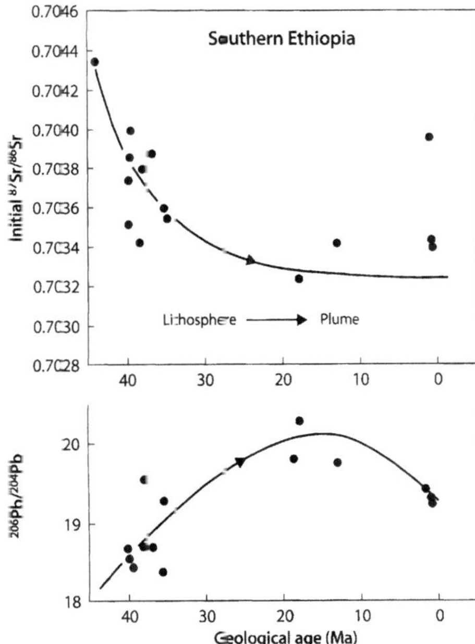

Fig. 5.94. Variation of the isotope ratios of Sr and Pb with age of the basalt flows in southern Ethiopia (northern transect of Stewart and Rogers 1996). The ^{143}Nd/^{144}Nd ratios also increased with time. The change in the isotope compositions of Sr, Nd, and Pb indicates that, prior to about 30 Ma, the magmas originated from the lithospheric mantle, whereas after about 20 Ma they originated from the head of the Afar Plume (*Source:* Stewart and Rogers 1996)

remained at this value until 0.68 Ma (Pleistocene). Similarly, the ^{206}Pb/^{204}Pb ratios of these rocks increased to about 19.5 at 30 Ma, reached 20.0 from 20 to 10 Ma, and then declined to about 19.3. The ^{143}Nd/^{144}Nd ratios (not shown) also increased with time from about 0.51261 at 45 Ma to 0.51284 at 1.34 Ma. Stewart and Rogers (1996) attributed the systematic time-dependent variation of the isotope ratios of Sr, Nd, and Pb to changes in the origin of the magmas such that magmas derived from the head of the Afar Plume have low ^{87}Sr/^{86}Sr (~0.7033), high ^{143}Nd/^{144}Nd (~0.51284), and high ^{206}Pb/^{204}Pb (~20.3) ratios, whereas lavas that were erupted prior to about 30 Ma originated primarily from the lithospheric mantle.

The eruption of silicic lavas in the Main Rift of Ethiopia during the Quaternary Period is illustrated by the volcano Fantale, which is composed primarily of peralkaline trachytes comendite, pantellerite, and some basalt. Major and trace-element data cited by Dickinson and Gibson (1972) permit the conclusion that these rocks formed by differentiation of magma in a chamber beneath the volcano. The ^{87}Sr/^{86}Sr ratios of the lavas on Fantale range widely from 0.7038 (ba-

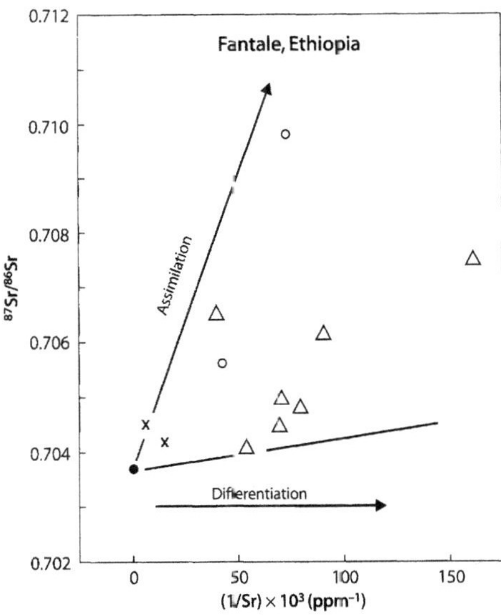

Fig. 5.95. Sr-isotope mixing diagram for lavas on the volcano Fantale in the Main Ethiopian Rift. Basalt: *solid circles*; trachyte: *crosses*; rhyolite: *open circles*; pitchstone: *open triangles* (*Source:* Dickinson and Gibson 1972)

salt) to 0.7099 (rhyolite) relative to 0.7080 for E&A and increase with rising Rb/Sr ratios. The measured ^{87}Sr/^{86}Sr and 1/Sr ratios of whole-rock samples on Fantale scatter widely on Fig. 5.95 consistent with the effects of simultaneous fractional crystallization of magma and simultaneous assimilation of crustal rocks. One basalt sample analyzed by Dickinson and Gibson (1972) (^{87}Sr/^{86}Sr = 0.7038; Sr = 520 ppm) is located at the apex of the fan-shaped data array and therefore represents the parent magma from which the other rock types on Fantale evolved.

In addition, the data of Dickinson and Gibson (1972) demonstrate that anorthoclase phenocrysts have consistently higher ^{87}Sr/^{86}Sr ratios, higher Sr concentrations, and lower Rb concentrations than the rhyolite glass in which they occur. This evidence indicates that the magma was stratified in its chemical and Sr-isotope compositions because of the assimilation of wall-rocks from the roof of the chamber. Feldspar crystals that formed near the top of the chamber from contaminated magma sank to deeper levels in the chamber where the magma had a lower ^{87}Sr/^{86}Sr ratio, because it was less contaminated than the magma at the top. The removal of anorthoclase from the upper regions of the chamber depleted the magma in Sr but enriched it in Rb so that the pitchstones and rhyolites on Fantale have a high average Rb concentration (120 ±16 ppm) and a low Sr concentration (15.5 ±4.1 ppm).

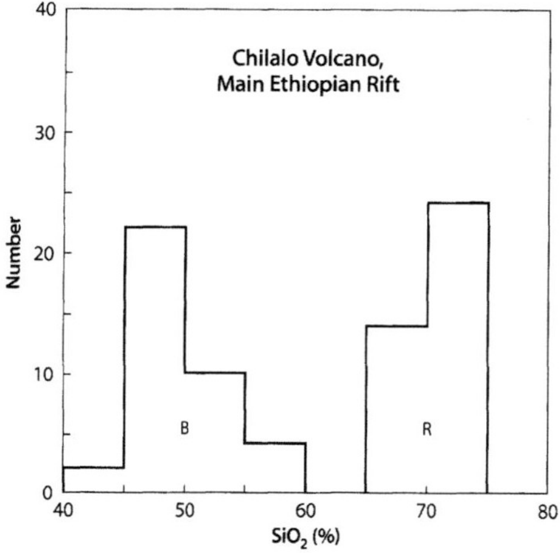

Fig. 5.96. Bimodal silica concentrations of volcanic rocks of Pliocene to Recent age erupted by Chilalo Volcano in the Main Ethiopian Rift about 100 km southeast of Addis Ababa in Ethiopia. The basalts (*B*) consist of transitional and alkali basalt, hawaiite, mugearite, and benmoreite. The rhyolites (*R*) are composed of pantellerite ignimbrites and comendite flows (*Source:* Trua et al. 1999)

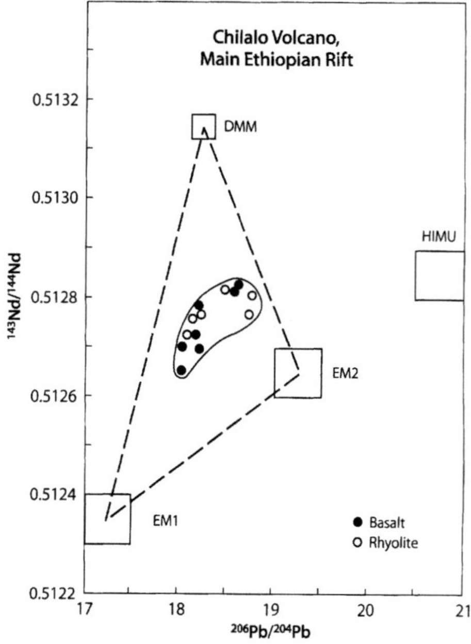

Fig. 5.97. Isotope ratios of Nd and Pb of alkali-rich basalts and pantellerite ignimbrites on the volcano Chilalo located along the eastern wall of the Main Ethiopian Rift about 100 km southeast of Addis Ababa, Ethiopia (*Sources:* Trua et al. 1999; Hart 1998)

The volcano Chilalo is another example of the eruption of bimodal volcanic rocks from Pliocene to Recent in the Main Ethiopian Rift. Chilalo is located along the eastern margin of the rift about 100 km southeast of Addis Ababa. The lavas erupted by this volcano include basalt (Pliocene) followed by felsic volcanic rocks (<1.7 Ma) consisting largely of pantellerite ignimbrites (Na-rich rhyolites) and obsidian flows. The bimodal chemical compositions of the volcanic rocks erupted by the volcano Chilalo are illustrated by the SiO_2 concentrations in Fig. 5.96 based on data by Trua et al. (1999). Most of the basaltic rocks contain from 45 to 50% SiO_2 whereas most of the felsic rocks contain between 70 and 75% silica. The basaltic rocks also have a high average Sr concentration (617 ±60 ppm, 2$\bar{\sigma}$, N = 38) compared to the felsic rocks (13.3 ±3.4 ppm, 2$\bar{\sigma}$, N = 38).

Contrary to their bimodal chemical compositions, the alkali-rich basaltic and felsic volcanic rocks on Chilalo have similar isotope ratios of Nd and Pb. The $^{87}Sr/^{86}Sr$ ratios of the basaltic rocks range from 0.70395 to 0.70513 relative to 0.71025 for NBS 987 as expected for magmas derived from the Afar Plume and associated lithospheric mantle. Trua et al. (1999) did not report Sr isotope ratios for the pantellerite ignimbrites. However, the $^{143}Nd/^{144}Nd$ and $^{206}Pb/^{204}Pb$ ratios of the alkali-rich basalts and pantellerite together define a small data field in Fig. 5.97 inside a mixing triangle formed by DMM, EM1, and EM2. Consequently, both suites originated from the Afar Plume and associated

subcontinental lithospheric mantle, which may contain subducted pelagic and continental sediment. In addition, the data demonstrate that the pantellerite magmas did not form by partial melting of upper crustal rocks. However, the felsic magmas could have formed by partial melting of mafic igneous rocks derived from the Afar Plume during an earlier episode of magmatic activity, or by fractional crystallization of basalt magma as proposed by Gibson (1972), Weaver et al. (1972), Barberi et al. (1975), and Deniel et al. (1994). However, an origin by fractional crystallization of basaltic magma is difficult to reconcile with the observation by Mohr (1992) that the pantellerites constitute about 90% of the volcanic rocks erupted within the Main Ethiopian Rift.

Model calculations by Trua et al. (1999) supported the hypothesis that the felsic magmas erupted by Chilalo originated as 10% partial melts of basaltic rocks in the lower crust. The magmas subsequently differentiated by crystallizing alkali feldspar, plagioclase, titanomagnetite, and apatite at shallow depth prior to eruption. The felsic magmas may also have assimilated wallrocks while they differentiated in crustal magma chambers, but this process did not alter the chemical and isotopic composition appreciably.

5.15.2 Afar Region and Djibouti

The Afar region of Ethiopia and the state of Djibouti in Fig. 5.92 are located above the Afar Plume, which means that the complicated history of tectonic activity and volcanism of this region is directly attributable to the interactions between the plume and the overlying lithospheric mantle (Schilling 1973b; Pilger and Rösler 1975; Markis et al. 1991; Fuchs et al. 1997). The volcanic rocks of Afar and Djibouti are not underlain by continental crust except along the margins of the Afar region. Therefore, the compositional variety of the volcanic rocks has resulted from the heterogeneity of the magma sources, from the extent of melting during magma formation, by mixing of magmas derived from different mantle sources, and by fractional crystallization of magmas at shallow depth below the surface. Assimilation of Precambrian basement rocks of granitic composition is not an option in the Afar region of Ethiopia nor in Djibouti.

Most of the volcanic rocks exposed at the surface in the Afar region were erupted after about 3.5 Ma. A more complete section in the Gulf of Tadjoura of Djibouti reveals the extensive history of volcanic activity in this region starting at about 30 Ma (Deniel et al. 1994; Barrat et al. 1993; Vidal et al. 1991):

1. Asal Rift and active volcanoes (present time)
2. Initial Series, basalt, 2 to 1 Ma, opening of the Gulf of Tadjoura
3. Stratoid Series, basalt, 4 to 1 Ma
4. Dalha Series, basalt, 9 to 4 Ma
5. Mablas Series, alkali basalt and rhyolite, 15 to 10 Ma
6. Ali Adde Series, rhyolite, 25 to 19 Ma
7. Adolei Series, basalt, 28 to 15 Ma

The basaltic lava flows of the Adolei Series are age-equivalent with the flood basalts of Ethiopia and western Yemen.

The lavas of the Afar region and Djibouti include tholeiites, alkali basalt, rhyolite, and pantellerites. The compositional gap between basalts and rhyolites in the Afar region is bridged in some cases (e.g. the Boina volcanic center, Barberi et al. 1975) by what are described as andesine basalts, ferrobasalts, and dark trachytes. The pantellerites at Boina are strongly enriched in Rb (150 ±3 ppm) and depleted in Sr (2.7 ±0.5 ppm), and their silica concentrations range up to 72.58%. Barberi et al. (1975) demonstrated that pantellerite magmas could have formed by sequential crystallization of olivine, plagioclase, clinopyroxene, Fe-Ti oxides, and alkali feldspar from alkali-basalt magmas. Weaver et al. (1972) likewise concluded that pantellerites in the East African Rift valleys formed by progressive fractional crystallization of alkali-basalt magmas. However, par-

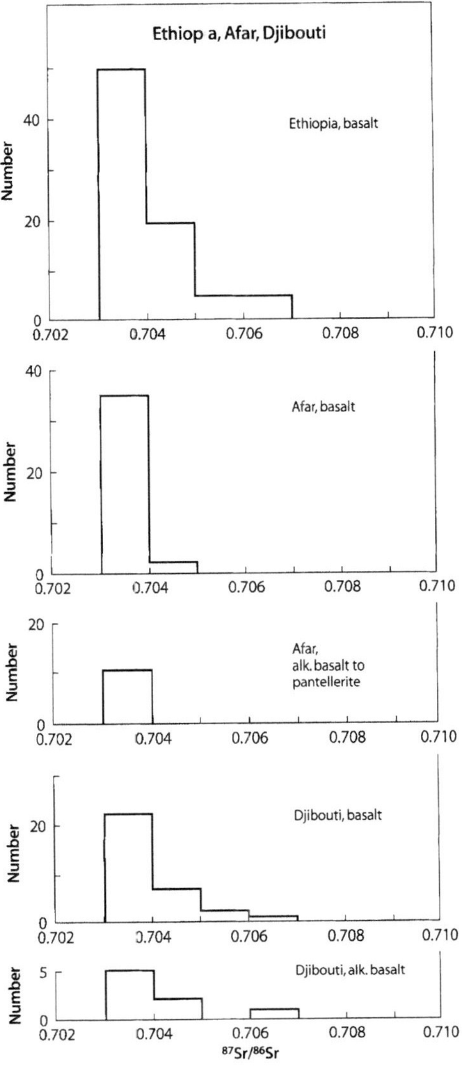

Fig. 5.98. Range of $^{87}Sr/^{86}Sr$ ratios of subalkaline basalts, alkali basalts, and various alkali-rich lavas including pantellerites in the Main Rift and adjacent lava plateaus of Ethiopia, in the Afar region, and in Djibouti (*Sources:* Barberi et al. 1975, 1980; Hart et al. 1989; Stewart and Rogers 1996; Vidal et al. 1991)

tial remelting of alkali basalt or gabbro at depth in the crust also results in the formation of felsic magmas, which can differentiate by fractional crystallization to form rhyolite (pantellerite) magmas (Trua et al. 1999).

The $^{87}Sr/^{86}Sr$ ratios of the volcanic rocks of the Afar region in Fig. 5.98 have a limited range between 0.7030

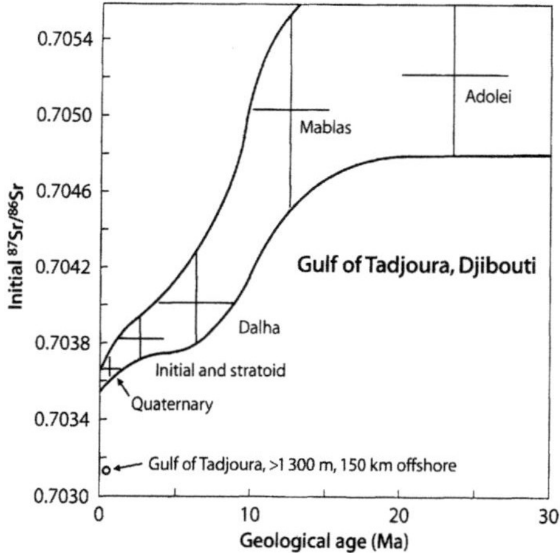

Fig. 5.99. Average ^{87}Sr/^{86}Sr ratios of basalt extruded prior to and after the opening of the Gulf of Tadjoura, Republic of Djibouti. The magmas initially originated from the lithospheric mantle beneath the continental crust of Africa. As the rifts widened, magma formed increasingly in the Afar Plume. The eruption of the Initial and Stratoid Series accompanied the opening of the Gulf of Tadjoura. The Quaternary lavas of this region have an average ^{87}Sr/^{86}Sr ratio of 0.70366 ±0.00007 (2$\bar{\sigma}$, N = 37), which presumably represents Sr in the Afar Plume. Basalt extruded in the axial valley of the Gulf of Tadjoura at a depth of more than 1 300 m and about 155 km offshore has a low average ^{87}Sr/^{86}Sr ratio of 0.70314 ±0.00003 (2$\bar{\sigma}$, N = 5) similar to basalts in the median valley of the Red Sea at 18° N (Eissen et al. 1989) (*Sources:* Vidal et al. 1991; Barrat et al. 1993)

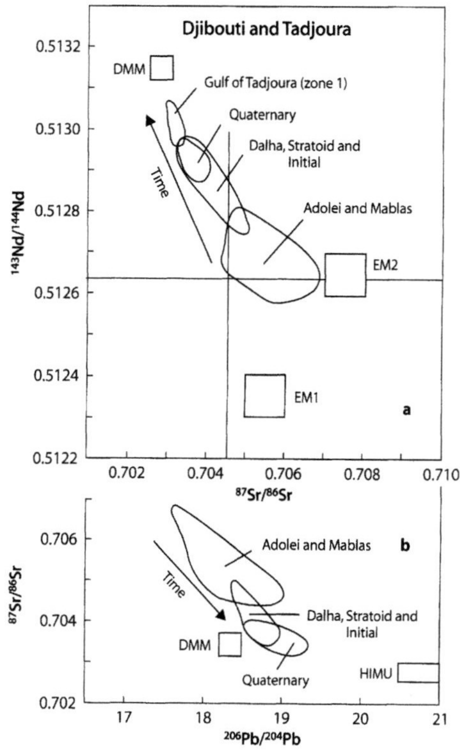

Fig. 5.100. Isotope ratios of Sr, Nd, and Pb in the volcanic rocks of Djibouti and the Gulf of Tadjoura. The data demonstrate a strong time-dependent change of the magma sources from subcrustal lithospheric mantle (Adolei and Mablas Series) to the head of the Afar Plume (*Sources:* Vidal et al.1991; Barrat et al. 1993)

and 0.7040, much like the lavas of the plateaus and the Main Rift of Ethiopia. The data of Barberi et al. (1980) demonstrate that the Late Tertiary alkali basalt-pantellerite lavas in the Afar region have similar ^{87}Sr/^{86}Sr ratios to the subalkaline basalts. However, the ^{87}Sr/^{86}Sr ratios of the basalts of Djibouti range widely from 0.703438 to 0.706705 relative to 0.71025 for NBS 987 (Vidal et al. 1991). The reason for this is that the lavas of the Adolei, Mablas, and Dalha Series (Oligocene to Miocene) of the Gulf of Tadjoura have higher initial ^{87}Sr/^{86}Sr ratios than the Pliocene to Quaternary lavas.

The time-dependent decrease of the ^{87}Sr/^{86}Sr ratios of the volcanic rocks in the Gulf of Tadjoura in Fig. 5.99 was attributed by Vidal et al. (1991) to the initial formation of magmas in old subcontinental lithospheric mantle starting at about 30 Ma. Subsequently, continuing stretching and thinning of the lithospheric mantle permitted magmas from the head of the plume to reach the surface in Pliocene to Quaternary time. Thus the history of volcanic activity of Djibouti is similar to that of southern Ethiopia as revealed by Stewart and Rogers (1996) in Fig. 5.94.

Basalts in the axial valley of the Gulf of Tadjoura analyzed by Barrat et al. (1993) have low ^{87}Sr/^{86}Sr ra-

tios ranging down to 0.70309. The ^{87}Sr/^{86}Sr ratios of these basalts are similar to those of MORBs in the axial valley of the Red Sea (Volker et al. 1993; Altherr et al. 1988, 1999; Eissen et al. 1989) and those of the Gulf of Aden (Schilling et al. 1992).

The time-progressive variation of the isotope composition of Sr, Nd, and Pb is illustrated in Fig. 5.100a and b based on the data of Vidal et al. (1991) and Barrat et al. (1993). The lavas of the Adolei and Mablas Series (28 to 10 Ma) have comparatively high ^{87}Sr/^{86}Sr and low ^{143}Nd/^{144}Nd ratios that approach the isotope composition of EM2 in Fig. 5.100a. The isotope ratios of Sr and Nd of the younger Dalha, Stratoid, and Initial (or Tadjoura) Series change toward those of the DMM component and overlap the data field of the subaerially erupted Quaternary lavas. The MORB-like lavas in the axial valley in the Gulf of Tadjoura (Zone 1 of Barrat et al. 1993) have Sr and Nd isotope ratios that are closest to those of DMM. The data of Betton and

Civetta (1984) indicate that the lavas of the Afar region plot in the field of the Quaternary lavas of Djibouti.

The ^{206}Pb/^{204}Pb ratios of the lavas of Djibouti in Fig. 5.100b also change with time from an average of 18.36 ±0.31 (Adolei and Mablas) to 18.90 ±0.09 (Quaternary). The youngest (plume-derived) lavas in Fig. 5.100b contain HIMU Pb, which presumably resides in the head of the Afar Plume and may represent in situ growth of ^{206}Pb by decay of ^{238}U in the entrained subducted oceanic crust

All of the isotopic data considered here indicate that the volcanic rocks of the Afar region and of the Gulf of Tadjoura in Djibouti are transitional between the flood basalt in the plateaus of Ethiopia and the MORBs in the axial valley of the Red Sea and the Gulf of Aden. These rocks began to form after the Afro-Arabian Plate had been split and the opposite sides had begun to drift apart, as revealed by a study of fission tracks by Omar and Steckler (1995). The Afar Plume is still active and causes continuing volcanic activity in the Afar region and in Djibouti.

5.15.3 Plateau Basalts of Western Yemen

The lava plateau of western Yemen in Fig. 5.92 is composed of interbedded basalt, rhyolitic pyroclastics, and pantellerites of Oligocene age (Yemen Trap Series) intruded locally by Miocene dikes and granite plutons (Mohr 1991). The stratigraphic thickness of these rocks decreases with distance from the Red Sea Rift and ranges from about 1 500 m in west Yemen to about 500 m farther east. The basalt flows are thick, tabular units composed of olivine and clinopyroxene with varying amounts of plagioclase and oxides of Fe and Ti. The basaltic rocks range in composition from subalkaline basalt to alkali basalt, hawaiite, and basanite. Their concentrations of Na$_2$O and K$_2$O vary between 1.66 and 5.29% and between 0.255 and 2.025%, respectively (Manetti et al. 1991; Menzies et al. 1990; Chiesa et al. 1989). Rifting of the continental crust in the Red Sea area probably started at about 34 Ma (Omar and Steckler 1995) and resulted in the extrusion of about 350 000 km^3 of basalt and rhyolite pyroclastics prior to the oceanization of the Red Sea Rift (Civetta et al. 1978; Chiesa et al. 1989). Age determinations by the ^{40}Ar/^{39}Ar method of basalts and rhyolites in western Yemen indicate that the volcanic eruptions lasted only about 4.4 million years from 30.9 to 26.5 Ma (Baker et al. 1996a). These authors also demonstrated that conventional whole-rock K-Ar dates of basalt in this area are unreliable because of loss as well as gain of radiogenic ^{40}Ar.

The initial ^{87}Sr/^{86}Sr ratios of the Oligocene plateau basalts of western Yemen in Fig. 5.101 are strongly clus-

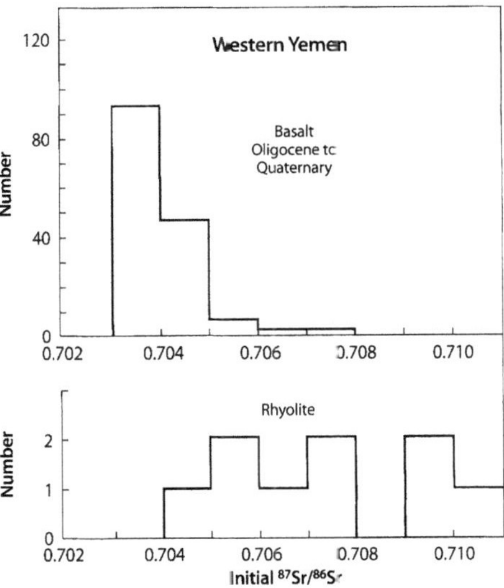

Fig. 5.101. Initial ^{87}Sr/^{86}Sr ratios of the flood basalts (Oligocene to Quaternary) of western Yemen and the rhyolites (pantellerites) associated with them (*Sources:* Chiesa et al. 1989; Manetti et al. 1991; Chazot and Bertrand 1993; Baker et al. 1996b, 1997)

tered in the interval between 0.7030 to 0.7040 but range upward to values of greater than 0.7070. Baker et al. (1996b) reported ^{87}Sr/^{86}Sr ratios between 0.70365 and 0.70555 relative to 0.71025 for NBS 987 for the Oligocene basalts. The ^{87}Sr/^{86}Sr ratios of most of the plateau basalts of western Yemen are higher than those of the magma sources in the mantle of this region including the Afar Plume, the lithospheric mantle, and the sources of MORBs in the Red Sea. Therefore, Baker et al. (1996b) concluded that the magmas assimilated a heterogeneous assemblage of crustal rocks accompanied in some cases by crystallization of plagioclase. The ^{87}Sr/^{86}Sr ratios of Pliocene to Quaternary lavas of the Sana'a-Amran area of western Yemen analyzed by Baker et al. (1997) range similarly from 0.70525 (SiO$_2$ = 47.2%, Sr = 700 ppm) to 0.70463 (SiO$_2$ = 58.9%, Sr = 322 ppm) relative to 0.71025 for NBS 987 and therefore are included with the Oligocene flood basalts in Fig. 5.101.

The distribution of initial ^{87}Sr/^{86}Sr and ^{143}Nd/^{144}Nd ratios of the basalts of western Yemen in Fig. 5.102b indicates that the magma component derived from the Afar Plume is somewhat more abundant in the Quaternary basalts than in the Oligocene basalts. In addition, the felsic rocks (rhyolites, comendites, and pantellerites) formed by partial melting of heterogeneous crustal rocks or by fractional crystallization and simultaneous assimilation of crustal rocks by mantle-derived basalt magma (Chazot and Bertrand 1993) rather than by remelting of Oligocene plateau basalt.

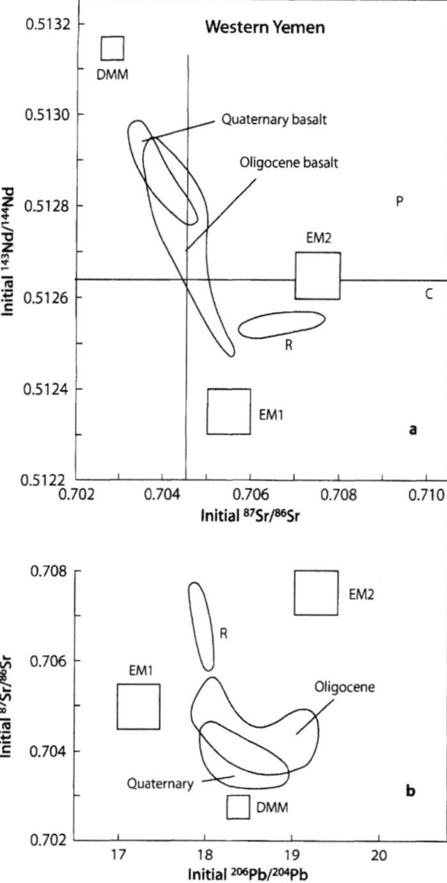

Fig. 5.102. Initial isotope ratios of Sr, Nd, and Pb of Oligocene and Pliocene-Quaternary basalt in western Yemen. R = rhyolite; P = pantellerite; C = comendite (*Sources:* Baker et al. 1996b, 1997, basalts; Chazot and Bertrand 1993, felsic rocks only; Hart 1988, mantle components)

The initial ^{206}Pb/^{204}Pb ratios of the Oligocene flood basalts of west Yemen range from 17.920 (SiO_2 = 46.68%) to 19.258 (SiO_2 = 50.53%) and overlap the ^{206}Pb/^{204}Pb ratios of the Quaternary basalts in Fig. 5.102b. However, the data field of the Quaternary basalts lies closer to the DMM component because of their lower initial ^{87}Sr/^{86}Sr (Baker et al. 1996b, 1997). The rhyolites, comendites, and pantellerites analyzed by Chazot and Bertrand (1993) are separated from the data fields of the basalt in Fig. 5.102b, because of their elevated initial ^{87}Sr/^{86}Sr ratios.

Some indication of the diversity of the isotope ratios of Sr and Nd of the lithospheric mantle underly-

ing this region was provided by the ultramafic inclusions and megacrysts that occur in Recent alkali basalt lavas at Ataq in southern Yemen (Fig. 5.92). Menzies and Murthy (1980a) reported that whole-rock samples of amphibole-bearing lherzolite inclusions at this location have a range of ^{87}Sr/^{86}Sr ratios from 0.70363 to 0.70540 relative to 0.71025 for NBS 987, which fits well with the range of initial ^{87}Sr/^{86}Sr ratios of Oligocene and Quaternary basalts of western Yemen in Fig. 5.101. The ^{143}Nd/^{144}Nd ratios of pargasite and diopside in these inclusions (0.51301 to 0.51242) exceed the range of Nd isotope ratios of the basalts in Fig. 5.102a. The presence of hydrous phases in the lherzolite inclusions (e.g. the amphibole pargasite) indicates that the alkali-rich basalt magmas formed by low degrees of partial melting of subcontinental lithospheric mantle that had been previously altered by metasomatism. The evidence for metasomatic alteration of the subcontinental lithospheric mantle provided by the amphibole-bearing inclusions in the alkali basalt at Ataq is an important clue to the origin of alkali-rich rocks that are the subject of Chap. 6 (Menzies and Murthy 1980a,b).

5.15.4 The Aden Volcano of Southern Yemen

Six volcanoes of Upper Miocene to Pliocene age along the southern coast of Yemen extend from the city of Aden (Fig. 5.92) westward to the straits of Bab el Mandeb at the entrance to the Red Sea. The lavas extruded by these volcanoes rest unconformably on the Yemen Trap Series, which was deposited on Upper Jurassic limestones overlying granitic gneisses and biotite schist of the basement complex. The lavas extruded by each volcano range in composition from mildly alkaline hawaiite to trachyandesites, trachytes, and peralkaline rhyolites (pantellerites).

Dickinson et al. (1969) reported that the initial ^{87}Sr/^{86}Sr ratios of lavas from three of the six volcanoes are positively correlated with their ^{87}Rb/^{86}Sr ratios and define two lines (not shown) whose slopes yield Rb-Sr dates of 30 ±5 and 24.5 ±7 Ma (λ = 1.42 × 10^{-11} yr^{-1}). The authors recognized that these dates are not consistent with the K-Ar dates of these rocks and concluded that the Rb-Sr dates reflect an event in the mantle that may have been related to magma formation associated with the extrusion of the Yemen Trap Series. Such positive correlations between the ^{87}Sr/^{86}Sr and ^{87}Rb/^{86}Sr ratios can also originate by assimilation of crustal rocks by mantle-derived basalt magmas. However, Dickinson et al. (1969) pointed out that in this case assimilation of granitic basement rocks would have produced a mixing line having a steeper slope controlled by the 600 Ma age of the basement. Carter and Norry (1976) confirmed that lavas of the Main Cone and Shamsan Caldera Series on the Aden

Volcano form pseudo-isochrons corresponding to dates of 33.8 ±5.0 Ma (Main Cone) and 11.3 ±3.0 Ma (Shamsan Caldera) ($\lambda = 1.42 \times 10^{-11}$ yr^{-1}), even though other lavas on the Aden Volcano scatter on the isochron diagram. The authors concluded that the lavas formed by remelting of volcanic or plutonic rocks that had previously crystallized at about 34 Ma with a range of Rb/Sr ratios from 0.04 to more than 1.0.

The data of Carter and Norry (1976) for the Aden Volcano are re-examined in Fig. 5.103 in coordinates of $^{87}Sr/^{86}Sr$ and 1/Sr ratios. The lavas of the Main Cone Series (solid circles) and the Shamsan Caldera (open circles) define two mixing lines with different slopes, whereas eight lavas from the Aden Volcano form a tight cluster ($^{87}Sr/^{86}Sr$: c.70350 to 0.70371 for E&A = 0.7080; Sr = 540 to 410 ppm). The lavas that form the cluster in Fig. 5.103 are all hawaiites or basalt and therefore represent the most uncontaminated and undifferentiated magmas erupted by the Aden Volcano. The Main Cone line is defined by lavas that were variously contaminated by rocks having a present age of about 34 Ma or more, as proposed by Carter and Norry (1976). Although the apparent age of the contaminant fits the Yemen Trap Series, rocks of basaltic composition are excluded because of their high Sr concentration. However, the data permit the conclusion that the mantle-derived magmas of the Aden Volcano could have assimilated felsic differentiates (e.g rhyolites) of the Yemen Trap Series to produce the lavas of the Main Cone Series. The magma of the Shamsan Caldera Series also assimilated felsic rocks of the Yemen Trap Series, but was later contaminated by assimilating younger felsic lavas at the base of the Aden Volcano. The contaminated magmas of the Main Cone and Shamsan Caldera Series were erupted before the Sr was isotopically homogenized and without significant fractional crystallization which would have altered the Rb and Sr concentrations.

5.16 Continental Basalts of Arabia

Centers of Late Tertiary (Oligocene to Quaternary) volcanic activity extend from western Yemen northwest all the way to Israel along the eastern Mediterranean Sea. The lavas that were extruded in this region are alkali-rich basalts, basanites, hawaiites, phonolites, and tephrites. The volcanic activity coincides with the extensional tectonics that led to the opening of the Red Sea, but is primarily confined to the eastern side of the Red Sea Rift (Bosworth and Strecker 1997). However, alkali basalt lavas of Late Cretaceous age were erupted at Wadi Natash in the Eastern Desert of Egypt (Hubbard et al. 1987).

5.16.1 Al Lith Area, Saudi Arabia

Volcanic rocks on the coastal plain of Al Lith about 100 km south of Jiddah on the west coast of Saudi Arabia (Fig. 5.92) record the volcanic activity that accompanied the opening of the Red Sea Rift. The basement rocks of the Al Lith area consist of volcanic, plutonic, and sedimentary rocks of Proterozoic age that were metamorphosed at about 800 Ma during the Pan-African event to biotite-granite gneiss, muscovite schist, and kyanite-muscovite-quartzite. These basement rocks are unconformably overlain by clastic marine sediment and volcanic tuff deposited prior to about 30 Ma. Rifting, accompanied by intrusion of the Damm Dike complex and volcanic activity (Sita Formation) occurred between about 30 and 20 Ma. At about 20 Ma the volcano-sedimentary complex of the Sita Formation was intruded by dikes of gabbro and monzo-gabbro of the Gumayqah complex. The last episode of volcanic activity occurred between 11 and 3 Ma, when lavas of the Harrat flood basalt covered large areas along the western margin of the Arabian Shield (Pallister 1987; Hegner and Pallister 1989).

The Al Lith area of Saudi Arabia is located at the extreme northern extent of the head of the Afar Plume. Therefore, the volcanic activity of this area cannot

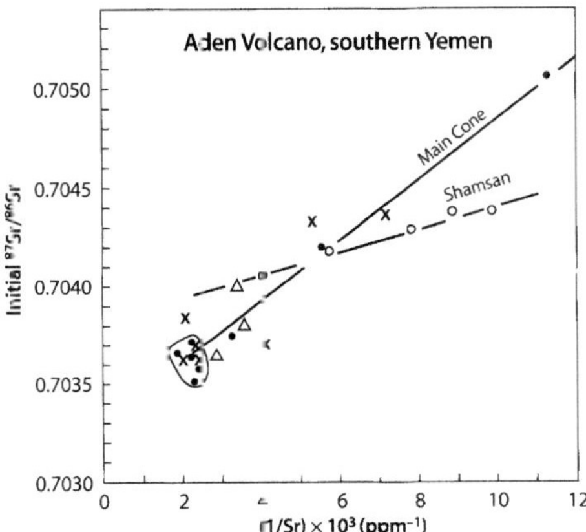

Fig. 5.103. $^{87}Sr/^{86}Sr$ ratios and reciprocal Sr concentrations of hawaiites-mugearites-trachytes-rhyolites of the Main Cone Series (*solid circles*) of the Aden Volcano form a linear array of data points. Trachyandesite and trachytes (*open circles*) from the Shamsan Caldera and overflow lavas (*open triangles*) form a second array, although some specimens lie close to the Main Cone array. In addition, basalt, trachyandesite, trachyte and hawaiite (*crosses*) from several other vents (Amen Khal, Ma'Alla, and Tawahi) scatter above and below the line formed by the Main Cone lavas. These linear data arrays can be explained by postulating that the basalt magmas assimilated felsic differentiates of the Yemen Trap Series (*Main Cone*) and felsic volcanic rocks at the base of Aden Volcano (*Shamsan*) (*Source:* data by Carter and Norry 1976)

be attributed with certainty to the presence of the Afar Plume. Instead, the centers of Tertiary volcanic activity of northern Saudi Arabia, Jordan, Israel, and Syria imply the existence of one or more additional asthenospheric plumes which contributed to the opening of the Red Sea and caused the eruption of alkali basalt flows at numerous volcanic centers in this region.

Preliminary measurements of initial ^{87}Sr/^{86}Sr ratios by Pallister (1987) of plutonic and volcanic rocks in the Al Lith area range from 0.7047 (tholeiite basalt dike, >20 Ma) to 0.7031 (hawaiite, 3.25 Ma) and suggested a progressive change of the magma sources with time. The ^{87}Sr/^{86}Sr ratios of the lavas of the Sita Formation (30 Ma) and of the rocks of the Damm Dike complex (30 Ma) reported by Hegner and Pallister (1989) range widely from 0.70297 to 0.70446, excluding felsic differentiates whose low Sr concentrations (6.1 to 9.8 ppm) and high Rb/Sr ratios (12.9 to 30.2) add uncertainty to the calculated values of their initial ^{87}Sr/^{86}Sr ratios. Both suites of rocks contain alkali-rich varieties (hawaiites-trachybasalt-trachytes) as well as calc-alkaline members (tholeiite basalt-andesite-dacite), but the initial ^{87}Sr/^{86}Sr ratios of the alkaline and calc-alkaline suites overlap. In addition, average ^{87}Sr/^{86}Sr ratios based on age fail to show a consistent time-dependent trend, probably because of alteration of the rocks (Hegner and Pallister 1989).

The initial ^{87}Sr/^{86}Sr ratios and reciprocal Sr concentrations reported by Hegner and Pallister (1989) in Fig. 5.104 form a fan-shaped array. Two tholeiite basalts of the Damm Dike complex actually plot close to the apex of this array with ^{87}Sr/^{86}Sr ratios of 0.70297 and Sr = 505 ppm. The Miocene-Pliocene flood basalts also have a low average initial ^{87}Sr/^{86}Sr ratio of 0.70310 ±0.00002 and Sr = 660 ppm. These rocks originated from the asthenospheric mantle underlying the Red Sea (Eissen et al. 1989). However, other specimens of the Sita Formation, the Damm Dike complex, and the Gumayqah Dike complex have elevated initial ^{87}Sr/^{86}Sr ratios for several possible reasons:

1. Magmas that originated in the asthenosphere assimilated rocks of the lithospheric mantle having higher ^{87}Sr/^{86}Sr ratios, or the magmas actually formed by partial melting in the lithospheric mantle;
2. The magmas assimilated crustal rocks while they pooled at the base of the continental crust during early stages of rifting;
3. The rocks were altered by deposition of secondary minerals after initial crystallization.

Since there is no compelling evidence in favor of crustal contamination of the mafic volcanic rocks, Hegner and Pallister (1989) concluded that the range of initial ^{87}Sr/^{86}Sr ratios at Al Lith reflects the hetero-

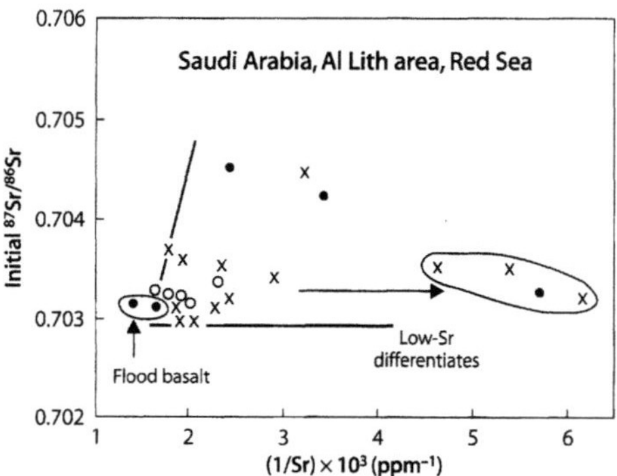

Fig. 5.104. Initial ^{87}Sr/^{86}Sr ratios and reciprocal Sr concentrations of lava flows and dikes at Al Lith, Saudi Arabia, during the opening of the Red Sea starting at about 30 Ma. *Open circles:* lavas, Sita Formation, 30 Ma; *crosses:* Damm Dike complex, 30 Ma; *solid circles:* Gumayqah Dike complex and flood basalts (11 to 3.0 Ma). Most of the data points form a fan-shaped array pointing to the composition of a hypothetical basaltic parent magma that originated from the asthenospheric mantle. Some of the rocks that formed between 30 and 20 Ma may have assimilated rocks in the lithospheric mantle having an elevated ^{87}Sr/^{86}Sr of about 0.7050. Rocks having low ^{87}Sr/^{86}Sr and low Sr concentrations are products of differentiation by fractional crystallization of magma (*Source:* Hegner and Pallister 1989)

geneity of magma sources in the mantle. In particular, the Sr isotope ratios permit the conclusion that basaltic magmas, formed by partial melting in the upwelling asthenospheric mantle, subsequently assimilated rocks of the lithospheric mantle and thereby raised their ^{87}Sr/^{86}Sr ratios from about 0.7030 up to about 0.7045. However, the distribution of data points in the fan-shaped array in Fig. 5.104 indicates that the magmas of the most recently erupted flood basalt and some of the tholeiites of the older Damm Dike complex and Sita Formation were not significantly contaminated and represent almost pure asthenospheric melts.

The importance of fractional crystallization of magmas is suggested by a group of rocks in Fig. 5.104 that have relatively low ^{87}Sr/^{86}Sr ratios (0.7032 to 0.7035), but which are displaced from the main array because they have low Sr concentrations (215 to 161 ppm). The felsic rocks mentioned above are also part of this array, but their low Sr concentration causes them to plot off scale.

5.16.2 Volcanic Centers of Saudi Arabia

The lava flows in the numerous other Late Tertiary volcanic centers of Saudi Arabia are primarily alkali-rich in composition (alkali basalt, basanite, hawaiite, and phonotephrite) and range in age from Lower Miocene

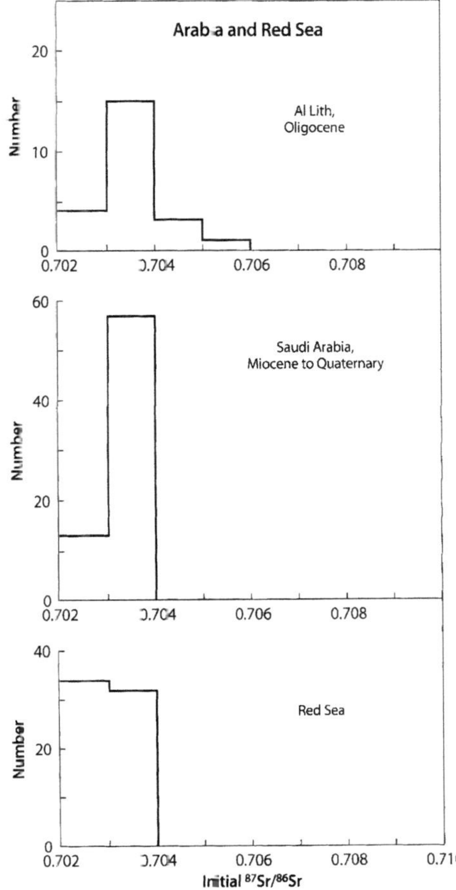

F g. 5.105. Initial ^{87}Sr/^{86}Sr ratios of the primarily alkali-rich basalts of Saudi Arabia and of MORBs in the Red Sea. These basalts have lower initial ^{87}Sr/^{86}Sr on average than those of western Yemen (Figure 5.101) and Ethiopia (Figure 5.98) (*Sources:* Altherr et al. 1988, 1990; Hegner and Pallister 1989; Eissen et al. 1989; Volker et al. 1993)

to Quarternary (Altherr et al. 1990). The ^{87}Sr/^{86}Sr ratios of these rocks in Fig. 5.105 are all less than 0.7040 and range to low values of 0.70278 (hawaiite, Nawasif-Al Buqum, Quaternary) relative to 0.7080 for E&A. The comparatively low ^{87}Sr/^{86}Sr ratios indicate that these and other Quaternary lavas of this region originated from depleted magma sources similar to those of MORBs extruded along the axial valley of the Red Sea included in Fig. 5.105. Lavas having ^{87}Sr/^{86}Sr ratios between 0.7030 and 0.7040 originated from the subcontinental lithospheric mantle along the eastern rift shoulder or formed from magmas that assimilated rocks from that source.

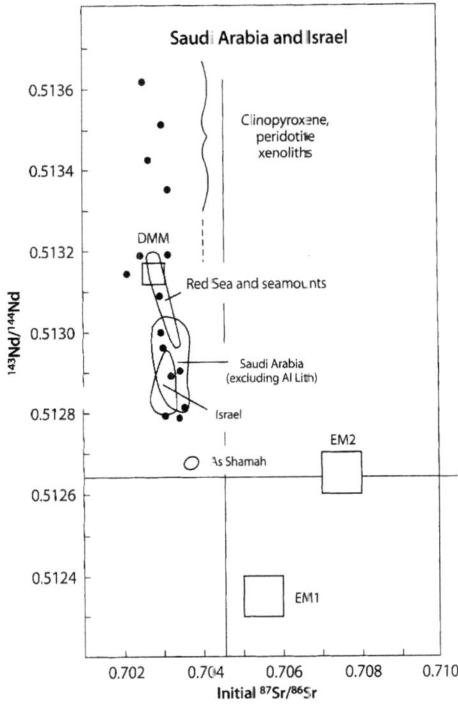

Fig. 5.106. Initial ^{87}Sr/^{86}Sr and ^{143}Nd/^{144}Nd ratios of alkali-rich basalts of Miocene to Quaternary age of Saudi Arabia, Jordan, and Israel as well as MORBs in the Red Sea. The *solid circles* represent clinopyroxenes in lherzolite inclusions of the basalt in Saudi Arabia (*Sources:* Altherr et al. 1990; Stein and Hofmann 1992; Volker et al. 1993; Hart 1988)

The isotope ratios of Sr and Nd of the Late Tertiary lavas of the Red Sea, Saudi Arabia, and Israel define overlapping data fields, all of which are located within the mantle quadrant in Fig. 5.106 (Altherr et al. 1990; Volker et al. 1993; Stein and Hofmann 1992). The data fields are aligned along a hypothetical mixing line connecting depleted MORB mantle (DMM) to the EM1 component that resides in the lithospheric mantle. Therefore, the origin of these volcanic rocks can be attributed to the formation of magmas in mixed asthenospheric-lithospheric sources or to assimilation of rocks of the subcontinental lithosphere by magmas derived from upwelling asthenospheric mantle. In other words, the Late Tertiary lavas of Saudi Arabia formed from mantle-derived magmas without significant contributions from the continental crust.

Direct evidence for the heterogeneity of the subcontinental lithospheric mantle underlying Saudi Arabia was provided by spinel-bearing lherzolite and harzburgite inclusions which occur in the Pliocene to

Pleistocene lavas at many of the volcanic centers of Saudi Arabia. Blusztajn et al. (1995) reported that clinopyroxenes in these xenoliths have wide-ranging trace element concentrations (e.g. Sr = 7 to 1040 ppm; Ti = 190 to 3800 ppm; Zr = 3 to 160 ppm). In addition, the ^{143}Nd/^{144}Nd ratios also vary widely from 0.512792 up to 0.513614 relative to 0.51264 for BCR-1 of the U.S. Geological Survey. However, the ^{87}Sr/^{86}Sr ratios are restricted to 0.70208 to 0.70356 compared to 0.7080 for E&A. The data points representing the clinopyroxenes scatter widely in Fig. 5.106 and actually exceed the range of isotope ratios of Sr and Nd in the host basalts analyzed by Altherr et al. (1990).

The highest ^{143}Nd/^{144}Nd ratios occur in spinel lherzolites, which contain Cr-diopside that is depleted in the light rare earths. The exceptional enrichment in radiogenic ^{143}Nd of these rocks means that their Sm/Nd ratios were increased as a result of previous melt events in the mantle. Subsequently, the decay of ^{147}Sm in these melt residues increased their ^{143}Nd/^{144}Nd ratios to high values that exceed the ^{143}Nd/^{144}Nd ratios of the DMM component in the sub-oceanic mantle. Blusztajn et al. (1995) concluded from trace-element data that the Cr-diopside-bearing xenoliths originated from the subcontinental lithospheric mantle under Arabia, and that they are not the source rocks of the MORB magmas. Although the Late Tertiary volcanic activity at Al Lith and elsewhere in Saudi Arabia occurred in a setting of extensional tectonics related to the widening of the Red Sea Rift, there is no direct evidence for the existence of plumes in this area.

5.16.3 Evidence for a Fossil Plume

The volcanic centers on the eastern side of the Red Sea in Saudi Arabia are aligned along crustal fractures that run parallel to the Red Sea (Stein and Hofmann 1992). This system of fractures intersects another set of fractures, which extend in a north-south direction parallel to the Mediterranean coast of Syria, Lebanon, and Israel. These fracture systems, which intersect in the vicinity of the Sinai Peninsula in Fig. 5.107, are two branches of a triple junction that formed in response to extensional stress applied to the base of the Afro-Arabian Plate in this area.

The first manifestations of plume-related volcanic activity in the eastern Mediterranean region are the Late Triassic basalts (up to 2.5 km thick) which occur in the subsurface near the Mediterranean coast of Israel. Lang and Steinitz (1989) reported K-Ar dates for these rocks ranging from 200 to 180 Ma. The second episode of volcanic activity occurred in Late Jurassic to Early Cretaceous time (130 to 95 Ma) when basalt lavas were erupted in an area now occupied by Israel, Lebanon, Egypt, and Syria. Volcanism resumed during

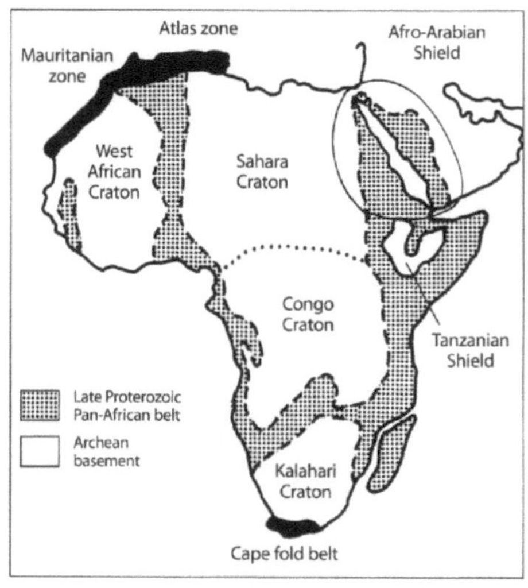

Fig. 5.107. Outline of the Archean cratons and associated Late Proterozoic and Pan-African orogenic belts of Africa. The basement rocks of the Afro-Arabian Shield underlie the alkali-rich volcanic and plutonic rocks of Cenozoic age of Saudi Arabia, Jordan, Israel, and Egypt (*Source:* adapted from Abdel-Rahman and Doig 1987)

the Oligocene Epoch (30 Ma) and continued until the Quaternary Period. Volcanic rocks dating from this episode constitute the large petrologic province in Saudi Arabia, Jordan, Israel, Lebanon and Syria. In Israel, four pulses of volcanic activity have been defined (Stein and Hofmann 1992):

- Young basalts, Quaternary
- Cover basalts, Pliocene, 5.5 to 1.7 Ma
- Intermediate basalts, Late Miocene, 8.3 to 5.5 Ma
- Lower basalts, Miocene, 17 to 9 Ma.

The isotope ratios of Sr and Nd of the volcanic rocks from Israel (200 to 0.14 Ma) define a small data field in Fig. 5.106 that extensively overlaps the field of the Late Tertiary lavas of Saudi Arabia (Stein and Hofmann 1992; Altherr et al. 1990). The ^{206}Pb/^{204}Pb ratios of these rocks range from 19.181 (Golan Heights, 0.99 to 0.74 Ma) to 20.406 (Ramon, 115 Ma). The apparent enrichment in radiogenic ^{206}Pb is characteristic of the continental basalts of this entire region including Saudi Arabia (Altherr et al. 1990).

In addition, Stein and Hofmann (1992) reported that the basalts of Israel have elevated Nb/U ratios between 40 and 60 similar to those of oceanic basalts, whereas the Nb/U ratios of most crustal rocks are less than 12 (Hofmann et al. 1986). Therefore, they concluded that the magmas originated from sources in the mantle without contamination by crustal rocks. The isotope compositions of Sr, Nd, and Pb of the volcanic rocks in Israel, Saudi Arabia, and adjacent countries are certainly consistent with the derivation of magmas from sources in the mantle. The only question is whether the magmas originated from the lithospheric mantle or from a plume of asthenospheric rocks like the Afar Plume of Ethiopia, Djibouti, and western Yemen.

Stein and Hofmann (1992) advanced the hypothesis that the cause for the volcanic activity of Arabia is a "weak" plume, which impacted the underside of the lithosphere but was unable to penetrate it. Therefore, the head of this plume flattened itself against the underside of the lithosphere and remained there as a "fossil plume." Rifting of the overlying continental crust and lithosphere later caused decompression melting in the fossil plume, thereby producing basalt magmas with isotopic compositions characteristic of plumes: $^{87}Sr/^{86}Sr = 0.7032$ and $^{143}Nd/^{144}Nd = 0.5130$. In fact, the $^{206}Pb/^{204}Pb$ ratios of the plume head may have been increased by decay of ^{238}U to ^{206}Pb during its initial ascent and subsequent "fossilization."

The hypothesis of Stein and Hofmann (1992) is an extension of the work of McNutt and Fischer (1987) on the plumes of the South Pacific (Sect. 2.10) and of Halliday et al. (1988) on the volcanic rocks of the Cameroon Line (Sect. 5.8.2). If only the "strong" plumes are able to penetrate continental lithospheric plates, a large number of "weak" plumes may have become fossilized at the base of the lithospheric mantle. The arrival of such weak plumes may cause regional uplift accompanied by a limited amount of extension and thinning of the continental crust and by episodic small-scale volcanic activity.

According to this hypothesis, the Late Tertiary lavas erupted in the volcanic centers of Saudi Arabia along the eastern rim of the Red Sea Rift and in Israel, originated by decompression melting of a fossil plume rather than of the overlying lithospheric mantle. As the Red Sea widened, the fossil plume under the Red Sea became exhausted, causing magmas to form by melting of depleted asthenospheric mantle and resulting in the extrusion of normal MORBs within the median valley of the Red Sea (Altherr et al. 1988, 1990; Eissen et al. 1989; Volker et al. 1993).

5.16.4 Origin of the Arabian-Nubian Shield

The formation of the continental crust of the Arabian-Nubian Shield began with the eruption of oceanic tholeiites between 900 and 870 Ma to form an oceanic basalt plateau that extended eastward from Egypt to Israel and Jordan and south of Sudan and Saudi Arabia. According to Stein and Goldstein (1996), such oce-

anic basalt plateaus can evolve into continental nuclei, because their thickness prevents them from being subducted back into the mantle. Consequently, oceanic basalt plateaus may be enlarged by subduction of oceanic crust along their margins and by the associated eruption of andesites and other volcanic rocks of intermediate composition.

Stein and Goldstein (1996) suggested that oceanic basalt plateaus form above plume heads that have risen to the base of the oceanic lithosphere. They cited evidence from the literature that the origin of the Birrimian lithotectonic province of Africa, the Superior province of North America, as well as the Afro-Arabian Shield in Fig. 5.107 all evolved from oceanic basalt plateaus which formed by volcanic activity associated with plumes in the mantle (Hill et al. 1991). Apparently, plumes have affected geological activity on the surface of the Earth since the Archean Eon and are not restricted to the Mesozoic and Cenozoic Eras that have been the focus of this chapter.

In the case of the Afro-Arabian Shield, the formation of the oceanic basalt plateau at 900 to 870 Ma was followed by volcanic activity associated with island arcs at about 650 Ma. The resulting volcanic-sedimentary complexes of the growing continental nucleus were severely deformed and metamorphosed resulting in the intrusion of granitic batholiths between 640 to 600 Ma, followed by anorogenic alkali-rich granites and dolerite dikes, which were intruded until about 540 Ma (Bentor 1985; Roobol et al. 1983; Duyverman et al. 1982; Duyverman and Harris 1982; Engel et al. 1980; Greenwood et al. 1976).

5.16.5 The Pan-African Orogeny in Saudi Arabia

The results of the extensive fieldwork combined with age determinations of the rocks indicate that the Afro-Arabian Shield was joined to the African Plate by closure of a marine basin and by compression of the volcano-sedimentary complexes that had formed at subduction zones. The basin closure and resulting orogeny occurred during the Pan-African event in Late Proterozoic to Early Paleozoic time about 650 to 500 million years ago. The orogenic belts that formed during this period elsewhere in Africa are identified in Fig. 5.107. Their geographic extent and origin are the subject of an extensive literature with notable contributions by Clifford and Gass (1970); Hurley and Rand (1969), Hurley (1972, 1974), Kröner et al. (1979, 1987, 1991), Kröner (1981), Stern et al. (1984), Cahen et al. (1984), Thorweihe and Schandelmeier (1993), and many others referred to by them.

Histograms of whole-rock Rb-Sr isochron dates of the metavolcanic basement complexes and granitic plutons of Arabia in Fig. 5.108 record the history of

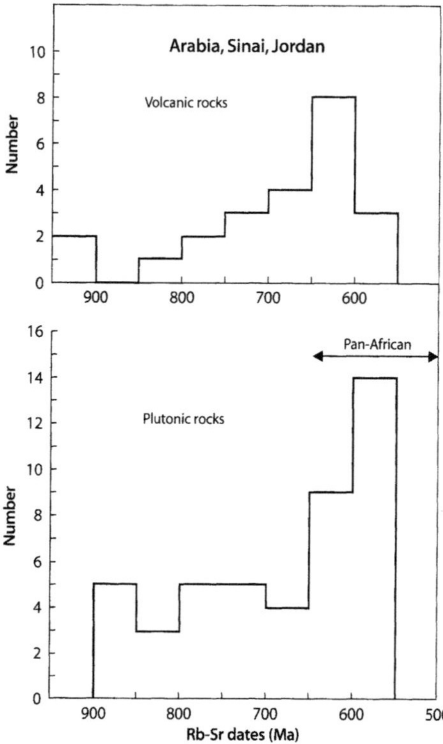

Fig. 5.108. Histograms of whole-rock Rb-Sr isochron dates of metavolcanic and plutonic rocks from the Arabian Shield, including the Sinai Peninsula and Jordan. The volcanic rocks formed in island arcs primarily between 900 and 650 Ma (Hijaz tectonic cycle). The chemical composition of the intrusive rocks evolved with time from diorite and trondhjemite (900 to 800 Ma) to quartz diorite and granodiorite (750 to 650 Ma), and finally to late-stage granites (650 to 550 Ma) that were intruded during the Pan-African orogeny (*Sources:* Shimron and Brookins 1974; Fleck et al. 1979, 1980; Kröner et al. 1979; Bokhari and Kramers 1981; Bielski et al. 1979; Halpern and Tristan 1981; Duyverman et al. 1982; Marzouki et al. 1982; Darbyshire et al. 1983; Jarrar et al. 1983; Al-Shanti et al. 1984; Stacey and Hedge 1984; Radain et al. 1987; Stern and Manton 1987; Ayalon et al. 1987)

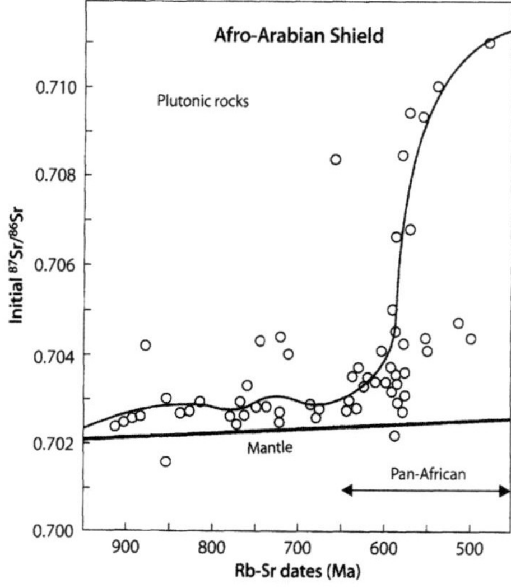

Fig. 5.109. Evolution of the isotope composition of Sr in the magma sources of plutonic rocks in the Afro-Arabian Shield; including Saudi Arabia, Sudan, Egypt, the Sinai Peninsula, and Jordan. The line representing the mantle extends from 0.699 at 4.5 Ga to 0.7030 at the present time. The initial ^{87}Sr/^{86}Sr ratios of most igneous intrusives between 950 and 650 Ma are less than 0.7030. However, between 650 and 450 Ma, the initial ^{87}Sr/^{86}Sr ratios of plutons composed predominantly of granite rise up to 0.7110. This increase of the ^{87}Sr/^{86}Sr reflects the progressive stabilization of the Afro-Arabian Shield and the derivation of magma from rocks with increasing crustal residence ages. The Sr-isotope evolution line for rocks older than 600 Ma was drawn by averaging 64 initial ^{87}Sr/^{86}Sr ratios derived from whole-rock Rb-Sr isochrons ($\lambda = 1.42 \times 10^{-11}$ yr^{-1}) of plutonic igneous rocks in 50 million-year increments. The ^{87}Sr/^{86}Sr ratios were adjusted to 0.7080 for E&A or 0.71025 for NBS 987 where possible (*Sources* for Saudi Arabia: Duyverman et al. 1982; Radain et al. 1987; Marzouki et al. 1982; Kröner et al. 1979; Fleck et al. 1979; Al-Shanti et al. 1984. Sudan: Kröner et al. 1987, 1991; Ries et al. 1985; Klemenic 1985; Abdel-Rahman and Doig 1987. Egypt: Stern and Hedge 1985. Sinai Peninsula: Bielski et al. 1979; Stern and Manton 1987; Halpern and Tristan 1981; Ayalon et al. 1987. Jordan: Jarrar et al.1983)

magmatic and tectonic activity of the Arabian Shield. Volcanic activity associated with island arcs occurred for more than 300 million years from about 900 to 600 Ma. During this time interval, the volcano-sedimentary complexes were intruded by cogenetic plutons. Intrusion of granite plutons continued during the Pan-African orogeny (650 to 500 Ma) when the Arabian Basin was closed and the volcano-sedimentary complexes were joined to the African Plate. These time-dependent trends in the evolution of the Afro-Arabian Shield were accompanied by increases in the concentrations of alkali metals noted by Greenwood et al. (1976). The significance of such patterns was challenged on the basis of local exceptions or subjective selection of data (Stern et al. 1982). In addition, questions arise about the interpretation of whole-rock Rb-Sr isochron dates of igneous rocks in a region known to have been affected by pervasive regional metamorphism during the Pan-African event.

The initial ^{87}Sr/^{86}Sr ratios of plutonic rocks in Saudi Arabia, Sudan, Egypt, Jordan, and the Sinai Peninsula in and Fig. 5.109 are less than 0.7040 between 950 and 650 Ma. The magmas that formed these rocks originated by partial melting of mantle-derived rocks with short crustal residence ages. Some plutons that formed

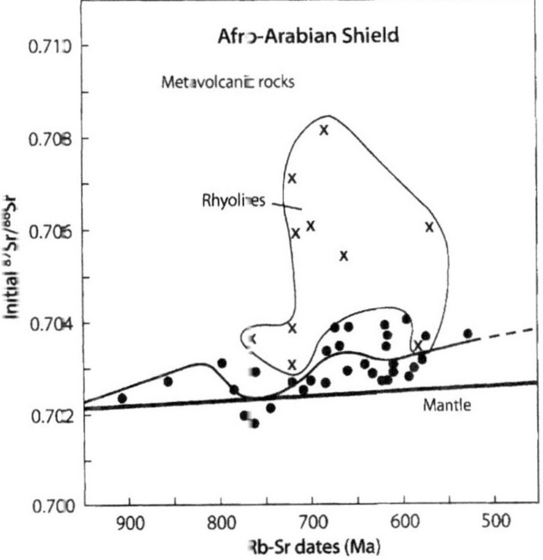

Fig. 5.110. Evolution of the isotope composition of Sr in andesites and basalts (*solid circles*) and rhyolites (*crosses*) of the Afro-Arabian Shield including Saudi Arabia, Sudan, and Egypt. The initial ^{87}Sr/^{86}Sr ratios of the mafic lavas are only marginally higher than those of the upper mantle represented here by a straight line drawn between 0.699 at 4.5 Ga and 0.7030 at the present time. Therefore, the isotope composition of Sr of the mafic lavas is consistent with their origin in subduction zones in marine basins. Some of the rhyolites have higher initial ^{87}Sr/^{86}Sr ratios than the associated mafic lavas. Possible reasons for the apparent enrichment of the rhyolites in radiogenic ^{87}Sr are discussed in the text (*Sources* for Saudi Arabia: Fleck et al. 1979, 1980; Bokhari and Kramers 1981; Duyverman et al. 1982; Darbyshire et al. 1983. Sudan: Kröner et al. 1991; Ries et al. 1985; Kemenic 1985, 1987; Abdel-Rahman and Doig 1987; Fitches et al. 1983. Egypt: Stern and Hedge 1985; Willis et al. 1988)

during the Pan-African orogeny between about 650 and 450 Ma have elevated initial ^{87}Sr/^{86}Sr ratios caused by partial melting of rocks in the previously-formed continental crust of the Afro-Arabian Shield. This evidence is significant, because the increase of the initial ^{87}Sr/^{86}Sr ratios of granitic plutons of Pan-African age in the Afro-Arabian Shield records the increasing recycling of crustal rocks in the formation of magma and hence the progressive stabilization of the newly-formed continental crust.

The initial ^{87}Sr/^{86}Sr ratios of mafic volcanic rocks (basalt and andesite) in Saudi Arabia, Sudan, and Egypt (averaged in 50 million-years increments of their whole-rock Rb-Sr isochron dates) in Fig. 5.110 demonstrate that all of the measured initial ^{87}Sr/^{86}Sr ratios of mafic lavas (solid circles) are less than 0.7040. The line based on the incremental averages lies close to the mantle-Sr isotope evolution line and rises only slightly with time. Therefore, these results are consistent with the derivation of the mafic volcanic rocks in oceanic subduction zones with only minimal contami-

nation of magmas, with Sr derived from seawater or terrestrial sediment of the subducted oceanic crust.

The data in Fig. 5.110 also demonstrate that the initial ^{87}Sr/^{86}Sr ratios of Late Proterozoic rhyolites vary widely between about 0.7030 and 0.7080. The apparent enrichment of some of the rhyolites in radiogenic ^{87}Sr is attributable to one of several causes:

1. Magma originated by partial melting of previously-formed rocks of island-arc complexes that had sufficiently high Rb/Sr ratios to increase their ^{87}Sr/^{86}Sr ratios up to about 0.7080
2. Contamination of rhyolitic magmas by interaction with wallrocks of the island-arc complexes or by assimilation of marine carbonates and/or connate water derived from seawater.
3. Partial melting of sialic rocks of continental crust in the Afro-Arabian Plate.

In summary, the continental crust of the Afro-Arabian Shield consists of a compositionally diverse suite of volcanic and plutonic rocks that formed between 900 and 500 Ma in island arcs located in marine basins. The process climaxed with the closure of the basins and with the intrusion of large granitic plutons during the Pan-African orogeny.

5.17 Summary: Plumes Dominate the Mantle

The formation of continental plateau basalts discussed in this chapter is a manifestation of the activity of plumes which rise from the base of the upper asthenospheric mantle at 670 km and/or from the base of the lower asthenospheric mantle at 2 900 km until their heads impinge against the underside of the rigid lithospheric mantle. Strong plumes intrude the subcontinental lithosphere and cause large-scale eruptions of tholeiite basalt magma through crustal fissures. Weak plumes do not penetrate the lithospheric mantle but do cause uplift and extension of the continental crust, resulting in episodic volcanic activity of limited scope. In addition, weak plumes can become "fossilized" as a layer beneath the lithospheric mantle and may later become the source of magma when the overlying lithospheric plate is rifted by the arrival of a strong plume. In general, weak plumes are probably more common than strong plumes.

The heads of strong as well as weak plumes consist of blocks of previously subducted oceanic crust and of entrained undepleted asthenospheric mantle. When the head of a large plume reaches the underside of the lithospheric mantle, its diameter increases and may reach 2 000 km, as in the case of the plumes that underlie Iceland, Tristan da Cunha, and Afar. The outward flow of the plume head applies extensional forces on the overlying lithosphere and causes fractures to form, which may propagate to the surface as rifts of the continental crust. In addition, the head of the plume raises the temperature of the lithospheric mantle and continental crust by conduction of heat as well as by the upward movement of basalt magma formed by decompression melting at the base of the lithospheric mantle, or in the head of the plume, or both. As the rifts widen, very large volumes of tholeiite basalt magma can form in a short interval of time (1 to 4 million years) by comparatively large degrees of partial melting (20 to 30%) of the magma sources in the mantle.

The composition of the magma sources may change with time from mainly lithospheric mantle to primarily plume rocks and, in some cases, to depleted asthenospheric mantle underlying the lithospheric mantle. Partial melting of rocks in the lithospheric mantle may be facilitated by previous episodes of aqueous metasomatism and/or by the formation of carbonatite veins. In most cases, the rate of magma production is high, allowing large flood basalt plateaus to form in only a few million years.

The basalt plateaus that form above strong plumes on both sides of a continental rift may ultimately become separated from each other by an ocean that opens between them. Such paired basalt plateaus exist on both sides of the Atlantic Ocean (e.g. in northwestern Scotland and East Greenland; in the state of Paraná, Brazil, and in northwestern Namibia), and in the Indian Ocean (e.g. the Karoo basalt of southeastern Africa and the basalts of Queen Maud Land, Antarctica). In the case of the Deccan Plateau of India, most of the lava flows accumulated on the Indian subcontinent and comparatively smaller volumes of basalt occur on the Seychelles Islands and on the floor of the West Indian Ocean that now separates these continental fragments from the Indian Plate. In some cases, even a strong plume may not be able to break up a continent, but can still cause very large volumes of basalt to be erupted in a continental setting (e.g. Siberia).

Although flood basalt plateaus are invariably associated with deep crustal rifts, plumes are not the only cause for the formation of crustal rifts. In some cases, rifting was caused by subduction of oceanic crust along continental margins (e.g. Columbia River basalt, Transantarctic Mountains, eastern China). The magmas in this kind of tectonic setting form not only by decompression melting, but also as a result of fluxing by water or carbon dioxide released by the down-going slab. In addition, previously subducted oceanic crust with associated terrigenous or marine sediment in the magma sources can significantly affect the chemical and isotopic composition of the magmas that are generated in such tectonic settings.

In many cases (e.g. Paraná, Brazil, and Deccan, India) the tholeiite basalt magmas pool in the lower crust where they evolve by simultaneous fractional crystallization and assimilation of granitic or granulitic crustal rocks. The ease with which basalt magmas can assimilate crustal rocks depends on the temperature of the crust. In cases where large volumes of basalt magma are produced, the temperature of the crust through which these magmas pass is increased significantly, allowing extensive crustal contamination of the magmas.

Crustal contamination of basalt magmas by assimilation of granitic rocks in most cases causes an increase in the $^{87}Sr/^{86}Sr$ and $^{18}O/^{16}O$ ratios and a corresponding decrease of $^{143}Nd/^{144}Nd$ ratio of the magma. The effect of crustal assimilation on the isotope composition of Pb in volcanic rocks depends on the ages and U/Pb ratios of assimilated rocks. In addition, assimilation of crustal rocks causes isotope ratios of volcanic rocks to correlate with the concentrations of certain major or trace elements (e.g. $^{87}Sr/^{86}Sr$ and SiO_2). In addition, the $\delta^{18}O$ values of volcanic rocks may be altered after crystallization by exchange of oxygen isotopes between silicate minerals and heated meteoric water (or seawater).

A special point of interest is the association of felsic lavas with flood basalts. The felsic lavas in some cases formed by remelting of recently crystallized basalt or gabbro at depth in the crust. In other cases, the felsic magmas originated by remelting of older sedimentary, volcanic, or plutonic rocks in the upper continental crust as a result of heating by large volumes of mantle-derived basalt magma. In such cases, the rhyolites have elevated $^{87}Sr/^{86}Sr$ and low $^{143}Nd/^{144}Nd$ ratios, and their $\delta^{18}O$ may be higher than those of the coexisting tholeiites. Rhyolitic magmas differentiate in shallow crustal magma chambers and are erupted explosively, resulting in the deposition of large volumes of rhyolite ignimbrite.

Basalt plateaus in the oceans can serve as continental nuclei because they resist subduction into the mantle. After they have become stuck above a subduction zone, submarine basalt plateaus grow in size by the deposition of volcano-sedimentary complexes associated with subduction zones that develop along their margins. The addition of volcano-sedimentary rocks of intermediate to felsic composition gradually converts submarine tholeiite plateaus into continental crust. The evolution of submarine basalt plateaus into continental crust exemplifies the far-reaching consequences mantle plumes have on the geology of the Earth and implies that plumes have played a dominant role in the interaction of the mantle and the crust since earliest Precambrian time.

The flood basalts of Ethiopia, Djibouti, Yemen, Saudi Arabia, Jordan, and Israel are, in many cases, alkali-rich. The enrichment of these lavas in alkali metals cannot be attributed solely to magma formation by small degrees of partial melting in the lithospheric mantle, nor to assimilation of alkali-rich crustal rocks, nor to fractional crystallization of tholeiite basalt magma. Instead, the enrichment of the Tertiary basalts of this region in alkali metals appears to be a characteristic property of the lithospheric mantle. This insight sets the stage for a comprehensive examination of the isotopic composition of the alkali-rich igneous rocks in Chap. 6.

Chapter 6

Alkalic Igneous Rocks on the Continents

Alkalic igneous rocks are characterized by elevated concentrations of Na_2O and K_2O and typically have low concentrations of SiO_2, CaO, MgO, and FeO compared to silica-saturated granitic rocks. They are also characterized by high abundances of relatively rare elements including P, F, Cl, Zr, Ti, Nb, Ta, and REEs. Alkali-rich plutons tend to be elliptical or circular in plan and are smaller in volume than plutons of granitic rocks. They are composed of alkali feldspar, feldspathoids (e.g. nepheline and leucite), and alkalic pyroxenes or amphiboles. Above all, the alkali-rich volcanic and plutonic rocks are characterized by a wide range of chemical and mineral compositions requiring elaborate classification schemes (e.g. Streckeisen 1967; Middlemost 1975, 1980; LeBas et al. 1986; Wilkinson 1986). Cox et al. (1979) proposed to classify alkali-rich volcanic rocks on the basis of $Na_2O + K_2O$ and SiO_2 (Fig. 2.2, Sect. 2.1). In addition, kimberlites and carbonatites are included among the alkalic rocks because kimberlites have high concentrations of Na_2O and K_2O, whereas carbonatites are commonly associated with differentiated alkali-rich plutons (Mitchell 1995; Harmer 1997; Bell 1989; Heinrich 1966). Alkali-rich igneous rocks are considered in all textbooks on igneous petrology (e.g. Turner and Verhoogen 1960; Cox et al. 1979; Best 1982; Wilson 1989). In addition, numerous monographs have been published on alkali-rich rocks (e.g. Sørensen 1974; Palmason 1982; Fitton and Upton 1987).

Alkali-rich igneous rocks occur in all of the tectonic settings in the ocean basins and on the continents considered in the previous chapters: along mid-ocean ridges, on oceanic islands, in subduction zones in the oceans and along continental margins, as well as along rift zones and above plume tracks on the continents. In addition, alkali-rich rocks are not restricted to the Mesozoic or Cenozoic Eras but formed throughout Precambrian and Phanerozoic time. Therefore, a discussion of the petrogenesis of alkali-rich rocks requires an understanding of all of the tectonic settings in which igneous rocks form and an expansion of the time frame to encompass the entire history of the Earth. Many occurrences of alkali-rich igneous rocks have already been described in this book and others will be mentioned in subsequent chapters.

Questions about the origin of alkali-rich igneous rocks on the continents are concerned not only with the tectonic settings in which they occur, but also with the causes of the alkali-enrichment of the magmas, including the remarkable enrichment of some alkali-rich igneous rocks in potassium. The occurrence of phlogopite and kaersutite in some ultramafic inclusions of alkali-rich lavas demonstrates that the lithospheric mantle underlying the continental crust in some areas was enriched in alkali metals as well as in water. Consequently, alkali-rich lavas formed in some areas by small degrees of partial melting within hydrated domains of the lithospheric mantle. The resulting magmas also have elevated $^{87}Sr/^{86}Sr$ ratios due to the decay of ^{87}Rb and/or the addition of ^{87}Sr by the aqueous fluids that caused the alkali-enrichment.

The metasomatic alteration of the lithospheric mantle can be caused by water-rich fluids that emanated from subducted oceanic basalt and sediment or from plume heads rising through the asthenospheric mantle. In either case, the ultimate source of the alkali metals is subducted seawater, while the subsequent formation of alkali-rich magmas is a consequence of the dynamics of mantle plumes. The occurrence of alkali-rich rocks along the rift valleys of East Africa and Europe is closely related to the extensional tectonics associated with plumes, whereas the K-rich lavas of Italy and Spain have formed as a consequence of the prior subduction of K-rich oceanic crust.

The enrichment of some alkali-rich rocks in ^{87}Sr can also be attributed to assimilation of granitic rocks of the continental crust by magmas that originated in the lithospheric mantle or in plume heads (Maury and Bizouard 1974). Although crustal contamination has been considered in many cases, it is not an attractive option because alkali-rich magmas typically have high Sr concentrations that protect their $^{87}Sr/^{86}Sr$ ratios from being changed by the addition of Sr derived from old granitic rocks of the continental crust. Nevertheless, the basement rocks that underlie alkali-rich lavas in some cases are a potential source of alkali metals (e.g. evaporite rocks) and radiogenic ^{87}Sr (e.g. Rb-rich metamorphic complexes). Therefore, the geology of the crust does play a role in the petrogenesis of alkali-rich rocks in some areas.

6.1   East African Rift System

The East African Rift valleys in Fig. 6.1 consist of the eastern and western branches. The eastern branch (Gregory Rift) extends from Kenya into Tanzania (Baker 1987). The western branch is marked by a chain of lakes that starts with Lake Albert and Lake Edward on the border between Uganda and Upper Zaire and extends south through Lake Kivu (Rwanda) into Lake Tanganyika (Burundi, Tanzania, and Zambia) and Lake Rukwa (Tanzania). The western Rift valley continues south to Lake Malawi (Tanzania, Malawi, and Mozambique) for a total distance of more than 2300 km.

The Gregory Rift of East Africa also contains a number of lakes including Lake Turkana and Lake Magadi in Kenya, as well as Lake Natron and Lake Manyara in Tanzania. The two branches of the East African Rift system outline the Precambrian shield of Tanzania, which includes Lake Victoria and contains the Kavirondo Rift extending from Lake Victoria eastward to the Gregory Rift. The resulting triple junction was the site of volcanic activity from about 20 Ma (middle Miocene) to 7 Ma when the activity shifted into the Gregory Rift and the volcanoes in the Kavirondo Rift became inactive (K-Ar dates, Jones and Lippard 1979).

The granitic rocks of the Precambrian shield of East Africa are potential contaminants of mantle-derived alkali-rich magmas that have been extruded through the rift system of this region. An examination of the lithologic composition and geological history of the Precambrian rocks of Tanzania will enable us to evaluate the

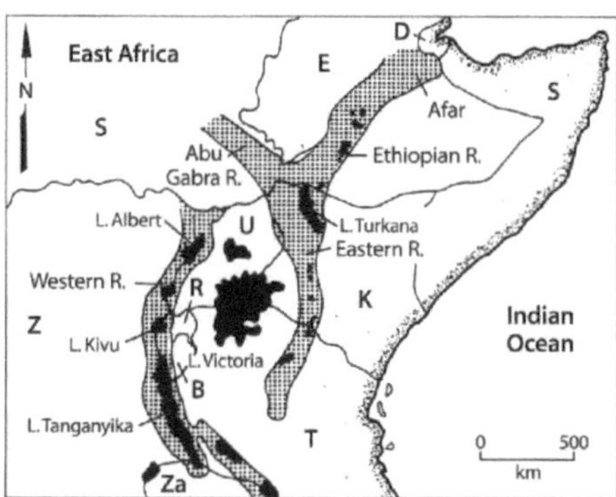

Fig. 6.1. The continental rift system of East Africa including the eastern and western rifts, as well as the Abu Gabra and Ethiopian Rifts. The political boundaries define the territories of the following countries: *T* = Tanzania; *K* = Kenya; *U* = Uganda; *E* = Ethiopia; *S* = Somalia; *D* = Djibouti; *S* = Sudan; *Z* = Zaire; *R* = Rwanda; *B* = Burundi; *Za* = Zambia (*Source:* adapted from Cohen et al. 1984)

possibility that the alkali enrichment of the lavas in the rift valleys of East Africa is the result of crustal assimilation of mantle-derived basalt magmas.

6.1.1   Precambrian Shield of East Africa

The Precambrian shield of East Africa is located in Tanzania where it extends from the vicinity of Lake Rukwa (Tanzania) in the south to the area north of Lake Victoria (Kenya and Uganda) (Fig. 6.1). It is bordered by the Mozambique fold belt on the east, the Ubendian-Usagaran Belt in the south, and the Karakwe-Ankolan Belt in the west. The Tanzanian Shield is composed of granitic gneisses associated in some localities with folded supracrustal metavolcanic and metasedimentary rocks of the Dodoman, Nyanzian, and Kavirondian Greenstone Belts (Clifford 1970; Bell and Dodson 1981). A summary of K-Ar and Rb-Sr dates of minerals by Cahen and Snelling (1966) suggested that the Dodoman and Kavirondian Greenstone Belts are older than about 2.5 Ga. The whole-rock Rb-Sr isochron dates ($\lambda = 1.42 \times 10^{-11}$ yr^{-1}) of granitic rocks from Tanzania published by Bell and Dodson (1981) range from about 2.8 to 2.3 Ga, although Old and Rex (1971) reported an older date of 3.0 Ga for the Masaba granite in southeastern Uganda.

The Rb-Sr dates of the Tanzanian Shield in Fig. 6.2 are clustered between 2.8 and 2.4 Ga and thus straddle the boundary between the Late Archean and Early Proterozoic Eras. Rocks from the Ubendian-Usagaran Belt in Malawi and Tanzania crystallized at about 1750 Ma, whereas the granites in the Karakwe-Ankolan Belt of western Uganda and Zaire have Rb-Sr dates between about 1560 and 900 Ma. The Pan-African event (650 to 500 Ma) is recognizable primarily in Rb-Sr and K-Ar dates of biotite as well as in U-Pb zircon concordia dates of the rocks of this area (Harper et al. 1972; Maboko et al. 1985).

The initial ^{87}Sr/^{86}Sr ratios derived from whole-rock Rb-Sr isochrons and errorchrons by Bell and Dodson (1981) are generally low indicating that the granitic magmas formed by partial melting of protoliths with short prior residence times in the crust. Therefore, the granitic crust underlying Tanzania and parts of Kenya and Uganda formed at about the time indicated by their Rb-Sr isochron dates in Fig. 6.2 and does not consist of rejuvenated older rocks.

This brief review of the relevant literature indicates that the Late Archean craton of Tanzania is surrounded by Middle to Late Proterozoic orogenic belts including syntectonic gneisses and posttectonic granite plutons. The granitic rocks in these orogenic belts have elevated initial ^{87}Sr/^{86}Sr ratios indicating that they originated primarily from rocks of the continental crust. The presence of the Tanzanian Shield and the marginal orogenic

Fig. 6.2.
Histogram of whole-rock Rb-Sr isochron/errorchron dates ($\lambda = 1.42 \times 10^{-11}$ yr^{-1}) of granitic plutons and gneisses of the Precambrian basement rocks of East Africa, including Tanzania, Kenya, Uganda, Upper Zaïre, Rwanda, and Malawi. The Tanzanian Shield is surrounded by the Mozambique Belt in the east, the Ubendian-Usagaran in the south, and the Karakwe-Ankolan in the west. The K-Ar and Rb-Sr dates of biotite in these rocks contain an imprint of the Pan-African orogenic event (*Sources:* Bell and Dodson 1981; Spooner et al. 1970; Leggo 1974; Wendt et al. 1977; Priem et al. 1979; Dodson et al. 1975; Gerards and Ledent 1975; Cahen et al. 1972; Vernon-Chamberlain and Snelling 1974)

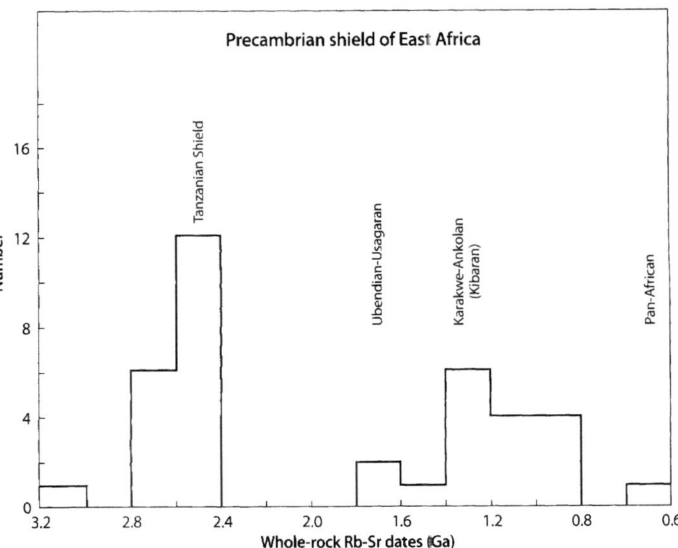

belts that surround it on at least three sides affected the location of deep crustal rifts that formed in East Africa during the Cenozoic Era. However, the Precambrian basement rocks of Tanzania are not particularly enriched in alkali metals and therefore do not appear to be responsible for the alkali enrichment of the volcanic rocks in the East African Rift valleys.

6.1.2 Gregory Rift, Kenya and Tanzania

The eastern branch of the East African Rift system (Gregory Rift) has been active for about 45 million years with episodes of volcanic activity between 44 to 38 Ma (Eocene), 16 to 11 Ma (middle Miocene), and 5 to 0 Ma (Pliocene to Recent) (Baker et al. 1971; Wilkinson et al. 1986). The Gregory Rift contains some of the best-known volcanoes of East Africa including Mt. Kenya (Kenya), as well as Kilimanjaro, Meru, Oldoinyo Lengai, and Ngorongoro (Tanzania). The origin of the alkali-rich volcanic rocks extruded by these and other volcanoes of the Gregory Rift was discussed by Goles (1976). Baker et al. (1977) concluded from a review of the literature and on the basis of their own work north of the volcano Olorgesailie (1°40′ S, 36°25′ E) in Kenya that the basalts, benmoreites, and trachytes in their study area are products of fractional crystallization of basaltic parent magmas without significant assimilation of wallrock, magma mixing, partial melting of crustal rocks, or transfer of volatile elements. The apparent absence of crustal contamination is consistent with the generally low ^{87}Sr/^{86}Sr ratios reported by Bell

and Powell (1969, 1970) for alkali-rich volcanic rocks and carbonatites in the eastern and western branches of the East African Rift system.

The summit of Mt. Kenya (5 200 m) exposes lavas of phonolitic and trachytic compositions that were erupted in Pliocene to Pleistocene time. These rocks overlie older tuffs and basalts that were deposited on Precambrian gneisses and schists (Baker 1967). The initial ^{87}Sr/^{86}Sr ratios of the phonolites, trachytes, and associated mafic rocks in Fig. 6.3 are strongly clustered between 0.70339 and 0.70380 and have a mean of 0.70354 (Price et al. 1985; NBS 987 = 0.71025). The narrow range of variation of the initial ^{87}Sr/^{86}Sr ratios of the rocks on the summit of Mt. Kenya tends to confirm the conclusions of Baker et al. (1977) that the alkali-rich rocks of the Gregory Rift formed from mantle-derived magmas without significant assimilation of wallrocks, magma mixing, or partial melting of crustal rocks.

The isotope ratios of Sr, Nd, and Pb of Late Tertiary lavas (8 to 0.09 Ma) in northern Tanzania (south of Mt. Kenya) define data fields in Fig. 6.4a,b which are positioned between the HIMU and EM1 components (Paslick et al. 1995). Therefore, the lithospheric mantle from which the magmas originated is not homogeneous but does contain at least two components having different Rb/Sr, Sm/Nd, and U/Pb ratios. The ^{87}Sr/^{86}Sr ratios of the lavas range from 0.7035 to 0.7058 relative to 0.71025 for NBS 987 and do not correlate with the Sr concentrations (444 to 3 650 ppm) of the silicate lavas (not shown). Therefore, the alkali-rich lavas of northern Tanzania were not significantly contaminated by assimilation of crustal rocks. However, the wide range

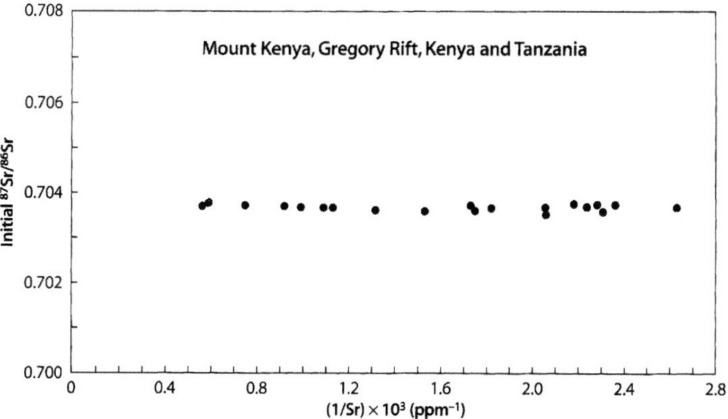

Fig. 6.3.
Initial $^{87}Sr/^{86}Sr$ ratios and reciprocal Sr concentrations of lavas on Mt. Kenya, in the Gregory Rift of Kenya and Tanzania. The lavas on Mt. Kenya have low and virtually constant initial $^{87}Sr/^{86}Sr$ ratios but a wide range of Sr concentrations. These rocks therefore formed by fractional crystallization of magmas derived from undepleted sources in the mantle (Price et al. 1985)

Fig. 6.4 a,b. Isotope ratios of Sr, Nd, and Pb in Late Tertiary (8 to 0.09 Ma) alkali-rich lavas on volcanoes of northern Tanzania. The carbonatites of the volcano Kerimasi lie close to the HIMU component in Parts *a* and *b* (*K* = Kerimasi). The lavas originated from magma sources in the lithospheric mantle containing the HIMU and EM1 components (*Sources:* Paslick et al. 1995; Hart 1988)

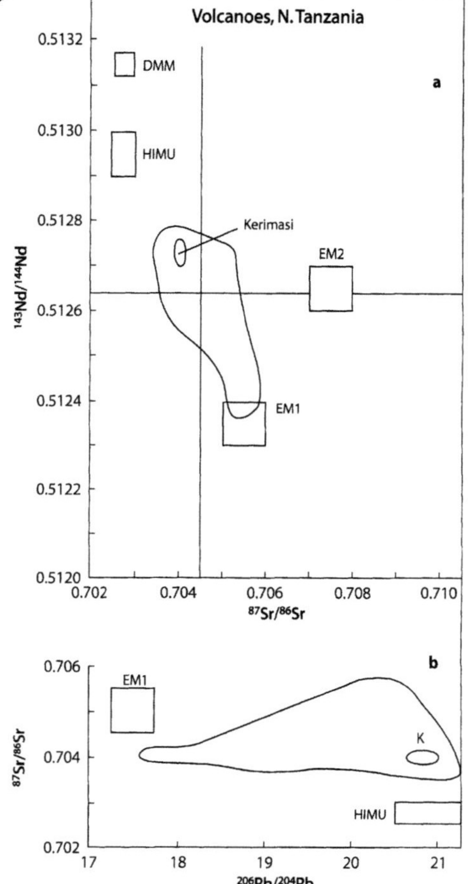

of Sr and silica concentrations (37.42 to 57.10%) is evidence for fractional crystallization presumably in a magma chamber in the lower crust. Three samples of carbonatite on the volcano Kerimasi analyzed by Paslick et al. (1995) have an average $^{87}Sr/^{86}Sr$ ratio of 0.703984 ±0.000021 ($2\bar{\sigma}$, $N = 3$; 0.710250 for NBS 987) and an average Sr concentration of 4050 ±150 ppm. These carbonatites plot near the HIMU end of the data fields defined by the alkali-rich silicate lavas of northern Tanzania in Fig. 6.4a,b.

The elevated $^{206}Pb/^{204}Pb$ ratios (17.639 to 21.220) are attributable to the presence of magma sources having an elevated U/Pb ratio. Paslick et al. (1995, 1996) proposed that this component consists either of an old plume head (fossil plume) that had underplated the lithospheric mantle, or that it resulted from metasomatic alteration of the lithosphere by melts that originated in the asthenosphere. In either case, the magma-source rocks remained isolated for about 2 billion years before rifting permitted magma formation by decompression melting.

6.1.3   Natrocarbonatite of Oldoinyo Lengai, Tanzania

Oldoinyo Lengai, located about 15 km south of Lake Natron in Tanzania, is the only active volcano of the Eastern Rift that has erupted carbonatite lavas (Hay 1989; Bell and Keller 1995). The most recent eruption in

June 1993 was described by Dawson et al. (1994). A color photograph of its crater was published by Dawson et al. (1990) following an earlier eruption in 1988. The volcano rises about 2 000 m above the surrounding plain, and its summit has an elevation of 2 878 m above sea-level. Oldoinyo Lengai is composed primarily of ijolitic pyroclastics interbedded with relatively thin lava flows of phonolite and nephelinite. The first description of the geology by Uhlig (1907) mentioned the occurrence of white sodium salts on the upper slopes of Oldoinyo Lengai. Following a minor eruption in 1960, Dawson (1962a,b) reported that sodium carbonate lava had been extruded at the summit of the mountain. He also mentioned that during an eruption in 1917 several streams containing warm salt water appeared on the northeast side of Oldoinyo Lengai.

During an eruption from mid-August to early October 1966, Oldoinyo Lengai ejected volcanic ash containing crystals of melanite garnet, pyroxene, Na-melilite, nepheline, Ti-magnetite, and wollastonite with lesser amounts of combeite ($Na_{1.3}Ca_{1.7}Si_3O_9$), larnite (Ca_2SiO_4), sylvite (KCl), barite ($BaSO_4$), and other rare minerals including Na-Ca-K phosphates and sulfides of K, Na, Fe, Mn, Si, P, and Cl. A chemical analysis of alkali carbonate from the summit of Oldoinyo Lengai indicated the presence of 54.00% Na_2O, 0.18% K_2O, and only trace amounts of SiO_2, Al_2O_3, MgO, CaO, FeO, and MnO (Dawson et al. 1992). The authors attributed the texture and mineral composition of the pyroclastic deposits to chemical reactions caused by mixing of carbonatite and silicate (ijolite) magmas during the explosive eruption of 1966.

During a more recent eruption in June 1988, Krafft and Keller (1989) measured the temperatures of carbonatite lava flows and lakes in the summit caldera of Oldoinyo Lengai. The highest temperature they recorded was 544 °C in a carbonatite lava lake. Rapidly moving carbonatite lava flows had somewhat lower temperatures between 502 and 519 °C. Dawson et al. (1990) reported similar values. These temperatures are several hundred degrees lower than those of silicate lavas. Nevertheless, the viscosity of the carbonatite lavas was lower than that of basaltic lavas whose temperatures exceed 1 000 °C. The natrocarbonatite lavas that erupted in June 1988 contained about 31 to 32% Na_2O, 8 to 9% K_2O, 16 to 17% CaO + SrO + BaO, and 27 to 32% CO_2 (see also Dawson et al. 1990). Fresh samples of lava that were collected hot and sealed before cooling were found to contain 2.5 to 4.5% F, 3.5 to 5% Cl, and 4.0 to 5.5% SO_3, but virtually no water (Krafft and Keller 1989). The principal Na-bearing minerals in the natrocarbonatite lavas of Oldoinyo Lengai are nyerereite [($Na_{0.88}K_{0.12}$)$_2$ Ca(CO_3)$_2$] and gregoryite [($Na_{0.78}K_{0.05}$)$_2$ ($Ca_{0.17}CO_3$)]. Both minerals are soluble in water leaving a residue of calcite. The petrology and

origin of natrocarbonatite on Oldoinyo Lengai was discussed by Petersen (1990), Pyle et al. (1995), and Gittins and Jago (1998), whereas Kapustin and Polyakov (1985) reported chemical analyses of carbonatite lavas not only on Oldoinyo Lengai but also on the volcano Kerimasi and other volcanoes of East Africa.

The $^{87}Sr/^{86}Sr$ ratios of lavas and blocks of plutonic rocks ejected by Oldoinyo Lengai vary widely from 0.7030 to 0.7093 relative to 0.7080 for E&A (Bell et al. 1973). The sodium carbonatite lavas also have an elevated average $^{87}Sr/^{86}Sr$ ratio of 0.7057 that is similar to that of many of the silicate lavas. Most of the nephelinite and phonolite lavas (solid circles) in Fig. 6.5 form a linear array in coordinates of $^{87}Sr/^{86}Sr$ and 1/Sr ratios. This array may be a result of mixing of magmas having different $^{87}Sr/^{86}Sr$ ratios and Sr concentrations, or it may have formed by contamination of a mantle-derived nephelinite magma ($^{87}Sr/^{86}Sr = 0.7031$) that assimilated crustal rocks enriched in radiogenic ^{87}Sr. The fact that some lavas deviate from the array indicates that the contaminated magmas differentiated by fractional crystallization, thereby forming rocks that were, in most cases, depleted in Sr and therefore plot to the right of the mixing array in Fig. 6.5.

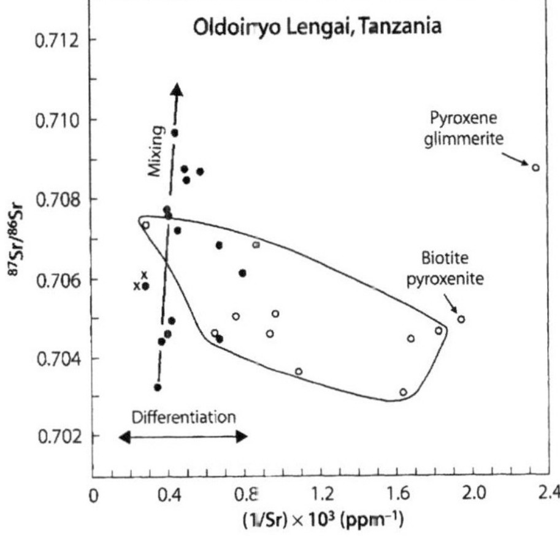

Fig. 6.5. Strontium-isotope mixing diagram for silicate lavas (*solid circles*), natrocarbonatite (*crosses*), and inclusions of alkali-rich plutonic rocks (*open circles*) on the volcano Oldoinyo Lengai in Tanzania. The silicate lavas form a linear array which permits the interpretation that nephelinite magma of mantle origin was contaminated by varying amounts of radiogenic ^{87}Sr derived from the continental crust or from trona deposits and Na-rich brines in subsurface. Inclusions of plutonic rocks in deposits of pyroclastics, composed of nepheline syenite and ijolite-melteigite-jacupirangite, have similar $^{87}Sr/^{86}Sr$ ratios as the lavas and could have formed by fractional crystallization of the same magmas as the flows. Two mica-bearing inclusions are identified in the diagram, but their origin is uncertain (*Source:* Bell et al. 1973)

The blocks of plutonic rocks (open circles in Fig. 6.5) included in the pyroclastic deposits consist primarily of nepheline syenite, ijolite, melteigite, and jacupirangite. The three last-mentioned rock types are composed of nepheline and pyroxene with increasing abundances of mafic minerals causing the rocks to become progressively darker in color from ijolite (30 to 70% pyroxene), to melteigite (70 to 95% pyroxene), and jacupirangite (>95% pyroxene) (Williams et al. 1955). The range of $^{87}Sr/^{86}Sr$ ratios and Sr concentrations of the inclusions of plutonic rocks permits the conclusion that they could have formed by fractional crystallization of the same magmas that erupted at the surface.

The two samples of sodium carbonatite lava (crosses in Fig. 6.5) from the summit of Oldoinyo Lengai may also be genetically related to the alkali-rich silicate magmas, since their average $^{87}Sr/^{86}Sr$ ratio (0.7057 relative to 0.7080 for E&A) is well within the range of $^{87}Sr/^{86}Sr$ ratios of the lavas. Bell et al. (1973) pointed out that the $^{87}Sr/^{86}Sr$ ratios of the carbonatite lavas on Oldoinyo Lengai are higher than those of typical carbonatites (Powell 1966; Bell 1989) including the carbonatites of eastern Uganda (Bell and Powell 1970). However, carbonated lavas of the Toro and Ankole volcanic centers in the Western Rift of southwestern Uganda also have elevated $^{87}Sr/^{86}Sr$ ratios between 0.7039 and 0.7055 (Bell and Powell 1969).

Weathering of natrocarbonatite ash erupted by Oldoinyo Lengai may have resulted in the formation of trona deposits in Lake Magadi (Kenya) and caused the Na-enrichment of Lake Natron on the border between Tanzania and Kenya. However, Milton (1968) and Eugster (1970, 1986) took the opposite view by suggesting that the natrocarbonatite lavas of Oldoinyo Lengai are the products of reactions between nephelinite magma and subsurface trona deposits or Na-rich groundwater. In order to test this hypothesis, Bell et al. (1973) measured the $^{87}Sr/^{86}Sr$ ratios of Na carbonate deposits in Lake Magadi and Lake Natron. The $^{87}Sr/^{86}Sr$ ratios of trona ($Na_2CO_3 \cdot NaHCO_3 \cdot 2H_2O$) in Lake Magadi range from 0.7071 to 0.7111, whereas a mixture of trona and halite from Lake Natron has an $^{87}Sr/^{86}Sr$ ratio of 0.7045 relative to 0.7080 for E&A. Therefore, the isotope compositions of Sr in the Na salts of the lakes are compatible with, but do not prove, the hypothesis of Milton and Eugster. The discharge of hot brines at several sites on the slopes of Oldoinyo Lengai during the eruption in 1917, mentioned by Dawson (1962a,b), also does not prove that these brines originated from Lake Natron or that they contributed to the petrogenesis of the natrocarbonatite lava.

Although Oldoinyo Lengai is the only active carbonatite volcano in the world, Kerimasi, located about 10 km south of it, also emitted natrocarbonatite tephra during its final eruptions between 0.6 to 0.4 Ma (Hay 1976, 1983). The occurrence of natrocarbonatite lavas on Kerimasi proves that magmas of this composition are not restricted to Oldoinyo Lengai. Hay (1983) pro-

posed that alkali-rich carbonatite magmas may be depleted in Na and K by hydrothermal fluids that cause fenitization of the wallrock, leaving a Ca-rich carbonatite rock. An example of alkali depletion of carbonatite by fenitization of the country rock was described by Bailey (1989) from southeast Zambia. The actual conversion of natrocarbonatite to a Ca-rich carbonatite as a result of leaching of alkali metals and the addition of Ca was described by Dawson et al. (1987) from a flow exposed in the active northern crater of Oldoinyo Lengai. Similarly, Hay and O'Neil (1983) cited petrographic and geochemical evidence to suggest that Ca-rich carbonatite pyroclastic deposits in the Laetolil beds of Tanzania and at Kaiserstuhl in Germany were originally composed of alkali carbonatite and that the alkali metals were leached after deposition, leaving a calcite residue. Other occurrences of Ca-rich carbonatite lavas in East Africa include the Tinderet Foothills of western Kenya (Deans and Roberts 1984), the confluence of the Rufunsa and Luangwa Rivers in southeast Zambia (Bailey 1989), and the Fort Portal area of western Uganda (von Knorring and Dubois 1961).

6.1.4 Carbonatites of East Africa

The East African Rift valleys contain many examples of carbonatites that occur as small plutons and veins at the numerous volcanic centers of this region. The $^{87}Sr/^{86}Sr$ ratios of some of these carbonatites were first measured by Powell et al. (1962, 1965b), Hamilton and Deans (1963), Powell (1966), and Powell and Bell (1974) in order to demonstrate that the carbonatites of East Africa and elsewhere in the world are igneous rocks and not recrystallized marine limestones. More recently, all aspects of carbonatites were reviewed in a book edited by Bell (1989). It includes summaries of the isotope ratios of Sr and Nd of carbonatites worldwide by Bell and Blenkinsop (1989) and of Pb by Kwon et al. (1989). In addition, the book edited by Bell (1989) contains a chapter by Deines (1989) on the isotope compositions of C, O, and S in carbonatites.

The $^{87}Sr/^{86}Sr$ ratios of alkali-rich lavas reported by Rock (1976) from Mt. Kisingiri, Homa Mountain, the Ruri Hills, and the Wasaki Peninsula along the eastern shore of Lake Victoria in Kenya range widely, but these carbonatites have a high Sr concentration of 3 232 ppm, which assures that they were not contaminated by radiogenic ^{87}Sr from the rocks of the Precambrian basement of this area. Therefore, the carbonatites of Miocene to Quaternary age which occur along the eastern shore of Lake Victoria (Fig. 6.1) can provide information about the isotope compositions of Sr, Nd, and Pb in the underlying lithospheric mantle from which the alkali-rich lavas and associated carbonatites of this area originated. Kalt et al. (1997) postulated that these carbonatites should be immune from crustal contamination because of:

1. High concentrations of Sr and Nd (Bell and Blenkinsop 1989);
2. Low crystallization temperatures (Wyllie 1989; Krafft and Keller 1989, Dawson et al. 1990);
3. Low viscosity (Treiman 1989; Krafft and Keller 1989; Dawson et al. 1990);
4. Rapid rate of ascent through the crust (Williams et al. 1986).

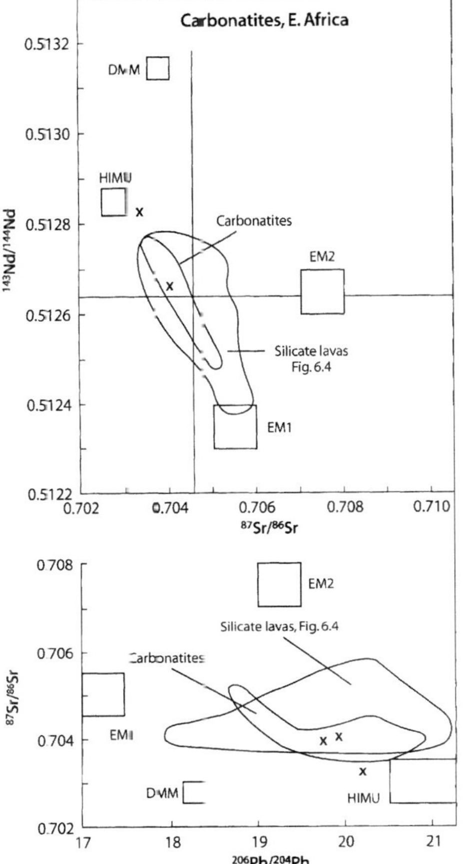

Fig. 6.5. Isotope ratios of Sr, Nd, and Pb in carbonatites of East Africa Homa Mountains and Wasaki Peninsula along the eastern shore of Lake Victoria in Kenya and on Mt. Kerimasi in Tanzania. Associated alkali-rich melilitites and ijolites are represented by crosses. The data fields reveal that the carbonatites originated from mixed sources in the lithospheric mantle composed of the HIMU and EM1 components, just like the silicate lavas of northern Tanzania in Figure 6.4. (*Source:* Kalt et al. 1997)

The carbonatites in the Homa Mountains and on the Wasaki Peninsula of Kenya and on Mt. Kerimasi of Tanzania in Fig. 6.6 originated from mixed magma sources in the lithospheric mantle consisting of the HIMU and EM1 components. In this regard, the isotope ratios of Sr, Nd, and Pb of the carbonatites analyzed by Kalt et al. (1995) are similar to those of alkali-rich lavas in northern Tanzania analyzed by Paslick et al. (1995). The isotopic evidence for the heterogeneity of the lithospheric mantle underlying the East African Rift valleys was also noted by Bell and Blenkinsop (1987a) and Bell and Dawson (1995). In addition, Simonetti and Bell (1994, 1995) and Bell and Simonetti (1996) demonstrated that nephelinites of Napak Volcano in Uganda and on Mt. Elgon along the border of Uganda and Kenya originated from isotopically heterogeneous magma sources in the mantle. A mantle source was also indicated by Harmer et al. (1998) for nephelinites and carbonatites in the Buhera District of Zimbabwe.

The similarity of isotope ratios of alkali-rich silicate lavas and the associated carbonatites along the Eastern Rift of Tanzania and Kenya in Fig. 6.5 supports the widely held view that carbonatite magmas originate from alkali-rich silicate magmas by liquid immiscibility as originally demonstrated experimentally by Wyllie and Tuttle (1960, 1962) and, most recently, by Lee and Wyllie (1997), Kramm et al. (1997), and by Ray (1998).

The ^{87}Sr/^{86}Sr ratios of carbonatites along the eastern and western branches of the East African Rift valleys compiled by Kalt et al. (1997) appear to have a wide range of values from 0.703016 (Kangank, Malawi, 126 Ma) to 0.707738 (Chasweta, Zambia, 113 Ma). Actually, the data in Fig. 6.7 demonstrate that 29 of 46 carbonatites in this region have ^{87}Sr/^{86}Sr ratios between 0.7030 and 0.7040 and that only a few carbonatites exceed 0.7050. The ^{143}Nd/^{144}Nd ratios range from 0.51283 to 0.512228 and vary inversely as the ^{87}Sr/^{86}Sr ratios. In addition, the East African carbonatites have elevat-

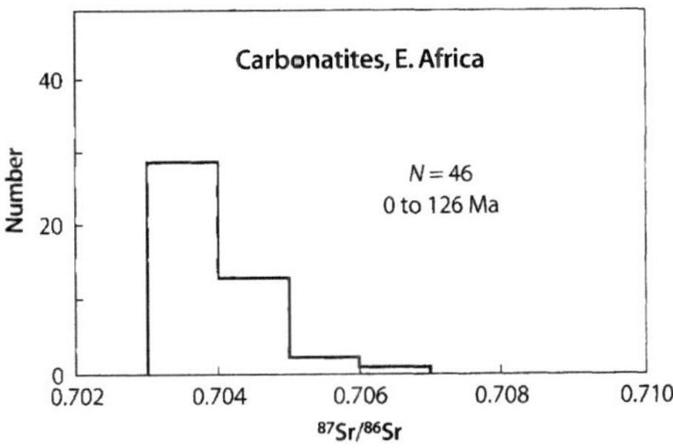

Fig. 6.7. Isotope ratios of ratios Sr of carbonatites (126 to 0 Ma) along the eastern and western rifts of East Africa including Tanzania, Kenya, Uganda, Zambia, and Malawi (*Source:* Kalt et al. 1997)

ed $^{206}Pb/^{204}Pb$ ratios ranging up to 21.154 (e.g. Sukulu, eastern Uganda).

The enrichment in ^{206}Pb is a characteristic property of the alkali-rich lavas and carbonatites along both the eastern and western rifts of East Africa. This feature also characterizes the alkali-rich lavas of Arabia and was tentatively attributed by Stein and Hofmann (1992) to the presence of a fossil plume, which underplated the lithospheric mantle of that region (Sect. 5.16.3). A similar explanation probably applies to the alkali-rich lavas and carbonatites of East Africa.

6.1.5 Xenoliths of Northern Tanzania

The extrusion of alkali-rich and carbonatitic lavas by volcanoes in the East African Rift valleys raises the question whether the magma sources in the mantle underlying this region were enriched by alkali metals before rifting and volcanic activity commenced, or whether the composition of the magmas is attributable primarily to low degrees of partial melting of ultramafic rocks in the lithospheric mantle. Information that may help to answer this question can be obtained by the study of xenoliths that occur in the lava flows of the volcano Lashaine located southwest of the city of Arusha in Tanzania. Lashaine consists entirely of pyroclastics composed of ankaramite (augite-rich olivine basalt) and carbonatite. The xenoliths include basalt, granulites from the underlying Precambrian basement, and ultramafic rocks: garnet lherzolite, spinel lherzolite, wehrlite, and mica dunite. The $^{87}Sr/^{86}Sr$ ratios and concentrations of Rb and Sr of these xenoliths measured by Hutchison and Dawson (1970) and Cohen et al. (1984) reveal that the history of the mantle underlying Lashaine is more complex than expected. The mineralogy, petrology, and chemical composition of these xenoliths were discussed by Dawson et al. (1970), Reid et al. (1975), Rhodes and Dawson (1975), Ridley and Dawson (1975), and Jones et al. (1983).

The $^{87}Sr/^{86}Sr$ ratios of a spinel lherzolite and a garnet lherzolite inclusion from Lashaine Volcano analyzed by Hutchison and Dawson (1970) are higher than those of the host rocks which means that the inclusions are not of cognate origin but may have originated from the subcrustal lithospheric mantle. In addition, these inclusions have higher Rb/Sr ratios (0.143 and 0.189) than the ankaramite whose Rb/Sr ratio is only about 0.038. The apparent enrichment in Rb of these inclusions is reflected by the presence of small amounts of primary mica in the spinel lherzolite and in other inclusions at Lashaine Volcano (Dawson and Powell 1969). The addition of Rb to the two lherzolite specimens presumably occurred at some time in the past, thus causing their $^{87}Sr/^{86}Sr$ ratios to increase to about 0.706 as recorded by the clinopyroxenes of both inclusions.

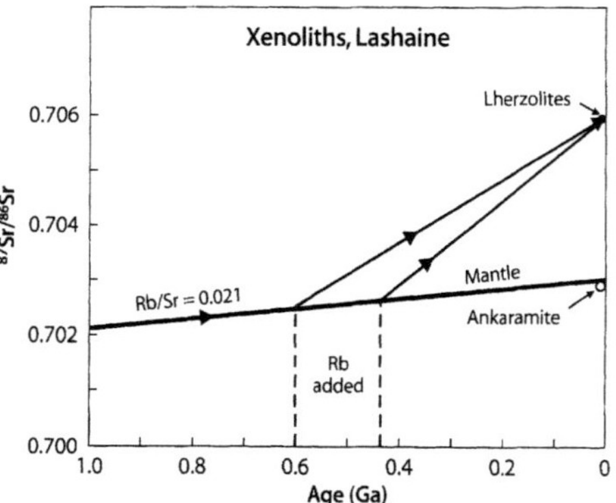

Fig. 6.8. Hypothetical scenario to explain the elevated $^{87}Sr/^{86}Sr$ ratio of lherzolite xenoliths in ankaramite pyroclastics of Lashaine Volcano, Tanzania. The ankaramite was derived from a source in the mantle whose $^{87}Sr/^{86}Sr$ ratio increased from 0.699 at 4.5 Ga to 0.7030 at the present time. The lherzolite xenoliths could have originated from mantle rocks, whose Rb/Sr ratios were increased from 0.021 to 0.143 and 0.188 between 0.60 and 0.43 Ga such that the present $^{87}Sr/^{86}Sr$ ratio of this region in the mantle is 0.7060. A few tens of thousands of years ago, ankaramite magma formed by partial melting of unenriched mantle rocks at depth below the Rb-enriched region of the mantle. The resulting magma then transported blocks of the overlying Rb-enriched lherzolite to the surface during explosive volcanic eruptions (*Source:* Hutchison and Dawson 1970)

The evolution of the $^{87}Sr/^{86}Sr$ ratios of the lithospheric mantle beneath the Lashain Volcano has been reconstructed in Fig. 6.8 based on the work of Hutchison and Dawson (1970). The $^{87}Sr/^{86}Sr$ ratio of the lithospheric mantle is assumed to have evolved from 0.699 at 4.5 Ga to 0.703 at the present time, implying that it had a Rb/Sr ratio of 0.021. At some time in the past, the rocks in a part of the lithospheric mantle in northern Tanzania were enriched in alkali metals by water-rich silicate melts that emanated either from an asthenospheric plume at depth or from subducted oceanic crust that had underplated the lithospheric mantle of East Africa. The alkali-rich fluid permeated the rocks and crystallized phlogopite mica both in the form of disseminated flakes and in the form of veins (e.g. Pello Hill, Tanzania; Cohen et al. 1984).

The Rb/Sr ratios of the two lherzolite inclusions from Lashaine Volcano were used in Fig. 6.8 to plot Sr-isotope evolution lines backwards in time from 0.706 at the present time to points of intersection with the hypothetical Sr-isotope evolution line of the lithospheric mantle. The results indicate that the Rb enrichment of the region in the mantle, from where the lherzolite inclusions originated, occurred between about 0.6 and 0.4 Ga, or at about 500 Ma as stated by Hutchison and Dawson (1970).

The chemical and isotope compositions of inclusions on Lashaine Volcano and at nearby Pello Hill reveal an

Table 6.1. Isotope ratios of Sr, Nd, and Pb and concentration of the parent and daughter elements in the minerals of a garnet lherzolite inclusion (BD-738) collected at Lashaine (Cohen et al. 1984)

	Clinopyroxene	Garnet	Phlogopite
$^{87}Sr/^{86}Sr$	0.83604	0.83453	0.81960
Rb (ppm)	1.182	0.085	583.0
Sr (ppm)	564.3	1.427	50.14
$^{87}Rb/^{86}Sr$	0.006	0.174	33.84
$^{143}Nd/^{144}Nd^a$	0.51127	0.51167	–
Sm (ppm)	2.839	0.9396	–
Nd (ppm)	8.83	1.223	–
$^{147}Sm/^{144}Nd$	0.093	0.466	–
$^{206}Pb/^{204}Pb$	15.55	–	–
$^{207}Pb/^{204}Pb$	15.39	–	–
$^{208}Pb/^{204}Pb$	36.00	–	–
U (ppm)	0.022	–	–
Pb (ppm)	0.659	–	–
$^{238}U/^{204}Pb$	1.96	–	–

[a] Corrected for fractionation to $^{146}Nd/^{144}Nd = 0.7219$.

even more astonishing degree of complexity. Cohen et al. (1984) reported isotope ratios (Table 6.1) of Sr, Nd, and Pb and concentration of the parent and daughter elements in the minerals of a garnet lherzolite inclusion (BD-738) collected at Lashaine.

The $^{87}Rb/^{86}Sr$ ratios of the clinopyroxene (0.006) and garnet (0.174) in this inclusion are not compatible with their elevated $^{87}Sr/^{86}Sr$ ratios. For example, the Rb-Sr age of the garnet indicated by these data is about 55 Ga. Apparently, the radiogenic ^{87}Sr was produced by decay of ^{87}Rb in the phlogopite and then diffused into the garnet and clinopyroxene which retained it quantitatively. Therefore, the age of Rb enrichment of this rock can be estimated by using the $^{87}Sr/^{86}Sr$ ratio of the clinopyroxene (0.83604) and the $^{87}Rb/^{86}Sr$ ratio of the coexisting phlogopite (33.94). The resulting date of metasomatism is about 300 Ma, which is of the same order of magnitude as the crystallization age of the lherzolite inclusions at Lashaine estimated by Hutchison and Dawson (1970). However, an amphibole-phlogopite-olivine vein in another spinel lherzolite inclusions collected at Pello Hill about 125 km northwest of Lashaine has an $^{87}Sr/^{86}Sr$ ratio of only 0.70344 similar to that of the coexisting clinopyroxene (0.70364; Cohen et al. 1984). The age of metasomatism estimated from these data is about 200 Ma based on the assumptions that the clinopyroxene quantitatively retained the radiogenic ^{87}Sr formed by decay of ^{87}Rb in the vein minerals ($^{87}Rb/^{86}Sr = 0.531$) and that the initial $^{87}Sr/^{86}Sr$ ratio was 0.7020.

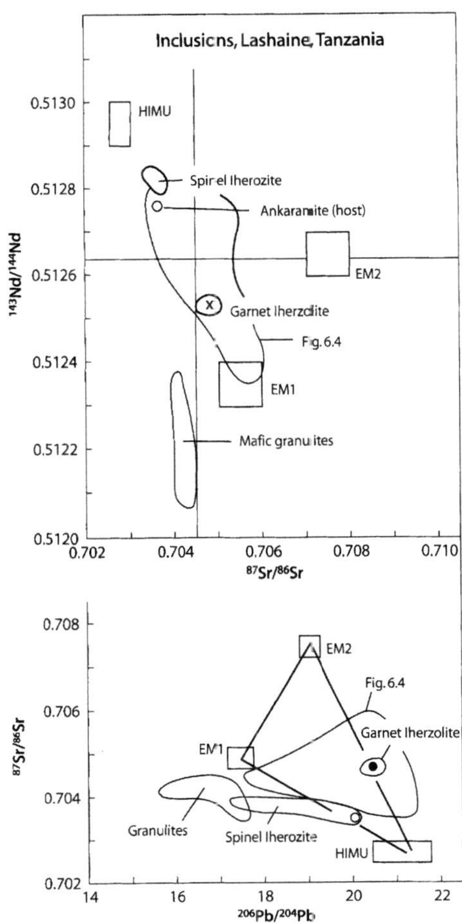

Fig. 6.9 a,b. Isotope ratios of Sr, Nd, and Pb of lherzolite and mafic granulite inclusions in ankaramite pyroclastics at Lashaine Volcano in northern Tanzania. The isotope ratios of these inclusions of mantle rocks are similar to those of alkali-rich lavas of northern Tanzania in Figure 6.4 analyzed by Paslick et al. (1995) (*Sources:* Hutchison and Dawson 1970; Paslick et al. 1995; Hart 1988)

The isotope ratios of Sr and Nd of clinopyroxene and vein minerals (amphibole-phlogopite-olivine) of the other lherzolite and pyroxenite inclusions from Lashaine and Pello Hill analyzed by Cohen et al. (1984) in Fig. 6.9a are aligned between the HIMU and EM1 components. In this respect, the isotope ratios of Sr and Nd of these mantle-derived inclusions are consistent with those of alkali-rich silicate lavas of northern Tanzania analyzed by Paslick et al. (1995). However, the mafic granulite inclusions form a separate data field indicating that they are not related to the lherzolite inclusions and did not contribute Sr and Nd to the magmas that were erupted in northern Tanzania.

The $^{206}Pb/^{204}Pb$ ratios of the constituent minerals of the spinel lherzolite inclusions in Fig. 6.9b range from 19.26 to 20.44 and appear to be mixtures of Pb from the HIMU and EM1 components. The elevated $^{87}Sr/^{86}Sr$ ratio of clinopyroxene in the garnet lherzolite causes this mineral to lie on a mixing line between EM2 and HIMU and the clinopyroxene of the pyroxenite has a low $^{206}Pb/^{204}Pb$ ratio (17.36) that places it in the data field of the mafic granulites.

The data fields in Fig. 6.9b demonstrate that the lherzolite inclusions at the Lashaine Volcano have high $^{206}Pb/^{204}Pb$ ratios just like the alkali-rich lavas and carbonatites of this region in Figs. 6.4 and 6.6. In addition, the range of isotope ratios of Sr and Nd of the inclusions is similar to that of the silicate lavas and carbonatites. Therefore, the isotope compositions of Sr, Nd, and Pb lead to the conclusion that the alkali-rich lavas and carbonatites of the Eastern Rift originated by partial melting of magma sources in the lithospheric mantle like those represented by the ultramafic inclusions. The data also demonstrate that the mafic granulites of the Precambrian shield of Tanzania did not contribute significantly to the formation of alkali-rich magmas.

Another point worth emphasizing is that the presence of amphiboles and phlogopites in some of the lherzolite inclusions at Lashaine and Pello Hill clearly demonstrates that the lithospheric mantle under the Eastern Rift was enriched in alkali metals by metasomatism caused by water-rich fluids. The alkali metasomatism of the lithospheric mantle and its enrichment in radiogenic ^{206}Pb is circumstantial evidence of the presence of previously subducted oceanic crust, which had been enriched in alkali metals and water by interaction with seawater. The subducted oceanic crust may be present in the head of a strong asthenospheric plume which has been exerting stress on the lithospheric mantle since about 45 Ma (Baker et al. 1971). Alternatively, the head of a weak plume may have underplated the lithospheric mantle prior to about 500 Ma and caused the alteration of the overlying lithosphere. In either case, the cause for the alkalic composition of the volcanic rocks of the Eastern Rift appears to be the alkali-enrichment of the magma sources in the underlying lithospheric mantle.

6.1.6　Alkalic Rocks of Eastern Uganda

The alkalic complexes of eastern Uganda occur along the Uganda-Kenya border north of Lake Victoria (Fig. 6.1) and are located along fractures that extend northwest from the Gregory Rift. The igneous complexes intrude Precambrian granitic gneisses and represent

the cores of deeply dissected volcanoes. The complexes were classified by Davies (1956) into an older post-Karoo but pre-Miocene suite, and a younger post-Miocene suite. The older complexes are circular or elliptical in cross-section and have central cores of carbonatite surrounded by alkali-rich ultramafic rocks (e.g. urtite and ijolite) as well as by fenite formed by alkali metasomatism of the country rocks. The younger complexes consist of variously dissected volcanoes exemplified by Mt. Elgon (4 321 m) composed largely of agglomerates and tuff interbedded with nephelinite, phonolite, and trachyte flows. The $^{87}Sr/^{86}Sr$ ratios of rock specimens from both the older and younger suites were reported by Bell and Powell (1970), whereas Rock (1976) only published data for Mount Elgon.

The initial $^{87}Sr/^{86}Sr$ ratios (at 100 Ma) of the ijolites, nepheline syenites, carbonatites, and soevites of the older (pre-Miocene) complexes range between limits of 0.7029 and 0.7043 (Bell and Powell 1970) and are independent of the Sr concentrations of the rocks (not shown). These observations permit the conclusion that the older complexes formed from mantle-derived magmas, which differentiated by fractional crystallization without significant contamination with radiogenic ^{87}Sr from crustal rocks. The average $^{87}Sr/^{86}Sr$ ratio of all rock specimens from the older complexes analyzed by Bell and Powell (1970) is 0.7037 ±0.0003 ($2\bar{\sigma}$, $N = 13$). This value is indistinguishable from the average $^{87}Sr/^{86}Sr$ ratios of the lavas on the summit of on Mt. Kenya (0.70354 ±0.00003 ($2\bar{\sigma}$, $N = 26$; relative to 0.71025 for NBS 987; Price et al. 1985).

The younger dissected volcanoes of eastern Uganda present quite a different picture. The initial $^{87}Sr/^{86}Sr$ ratios of these rocks (at 30 Ma) range widely from 0.7032 to 0.7061 (Bell and Powell 1970). Therefore, the lavas on the younger volcanoes (Elgon, Moroto, Napak, and Toror) formed from mantle-derived magmas that were contaminated during fractional crystallization prior to eruption.

The $^{87}Sr/^{86}Sr$ ratios of the carbonatites in eastern Uganda are not affected by crustal contamination because of their high Sr concentrations (2 031 to 5 075 ppm). Therefore, they should have preserved the $^{87}Sr/^{86}Sr$ ratio of their sources in the mantle. The four carbonatites analyzed by Bell and Powell (1970) from Toror, Budeda, Sukulu, and Tororo have an average $^{87}Sr/^{86}Sr$ ratio of 0.7035 ±0.0002 ($2\bar{\sigma}$) that is indistinguishable from the average $^{87}Sr/^{86}Sr$ ratios of the lavas on Mt. Kenya (Price et al. 1985). The origin of the carbonatite at Sukulu, Uganda, was the subject of a paper by Ting et al. (1994).

The intrusion of corrosive alkali-rich magmas into deep crustal fractures in eastern Uganda was accompanied by extensive metasomatic alteration of the granitic country rocks. The resulting fenites resemble the magmatic rocks of the intrusives associated with them. However, Bell and Powell (1970) demonstrated that the

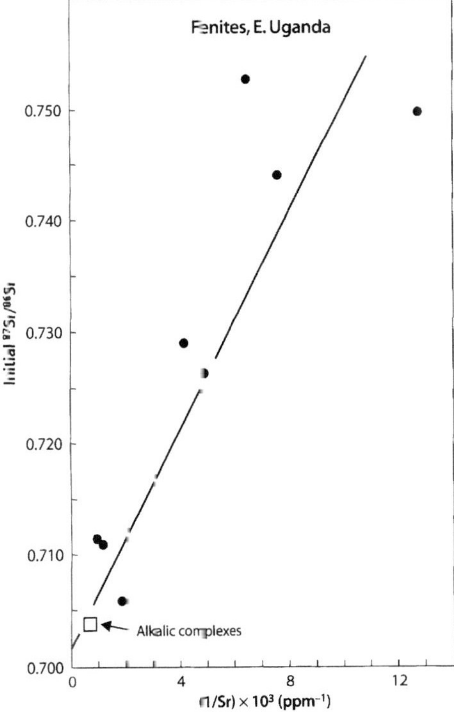

Fig. 6.10. Correlation of the initial ^{87}Sr/^{86}Sr ratios (at 100 Ma) of fenites in eastern Uganda and the reciprocals of their Sr concentrations. The collinearity of these data points indicates that the fenitizing solutions added Sr to the granitic country rocks, thereby increasing their Sr concentrations and lowering their ^{87}Sr/^{86}Sr ratios (*Source:* data and interpretation by Bell and Powell 1970)

^{87}Sr/^{86}Sr ratios of the fenites of eastern Uganda have preserved a record of their Precambrian protolith. The initial ^{87}Sr/^{86}Sr ratios of the fenites in Fig. 6.10 are positively correlated with the reciprocals of their Sr concentrations. This result indicates that the fenitizing solutions added varying amounts of Sr to the granitic country rocks, thereby increasing their Sr concentrations and lowering their ^{87}Sr/^{86}Sr ratios. This evidence confirms that fenites are metasomatized basement rocks as originally recognized by Brögger (1921) at Fen in Norway.

6.2 Potassic Rocks of the Western Rift of East Africa

The Western Rift of East Africa (Fig. 6.1) includes the Toro-Ankole area between Lake Albert (north) and Lake Edward (south) in southwestern Uganda, the

Virunga volcanic field located at the junction between Zaire, northwest Rwanda, and southwest Uganda north of Lake Kivu, and the south Kiva field (Pasteels et al. 1989). The Virunga area contains eight volcanoes including Mt. Nyamuragira (2 972 m) whose most recent eruption occurred on December 2, 1996 (Aoki and Kurasawa 1984). Most of the other major volcanoes and numerous subsidiary vents have also been active since the Pliocene Epoch. The volcanoes of the Western Rift are underlain by folded Precambrian argillites and rare dolomitic limestones of the Karakwe-Ankolan Belt that wraps around the western side of the Late Archean gneisses of the Tanzanian Shield (Sect. 6.1.1). The argillites were intruded by granite magma along southeast-trending anticlines. The geology of the Western Rift was originally described by Arthur Holmes in a series of papers between 1932 and 1956 referenced by Bell and Powell (1969).

The volcanic rocks of the Western Rift are potassic to ultrapotassic in composition and share this distinction with only a few other volcanic centers on the Earth: the Roman province of Italy, Jumilla in southeastern Spain, Laacher See of Germany, the Leucite Hills of Wyoming, several localities in Brazil (Carlson et al. 1996; Ferreira et al. 1997), and the West Kimberley area of Western Australia. These rocks are characterized by high concentrations of K_2O (up to 12.0%), by $K_2O > Na_2O$ or even $K_2O > Al_2O_3$, low SiO_2 resulting in the presence of feldspathoids (e.g. leucite), high TiO_2, and elevated concentrations of CaO, MgO, BaO, and SrO, as well as of Rb, Zr, Nb, La, and Y (Bell and Powell 1969). As a result of their extreme enrichment in K_2O and certain other elements, the ultrapotassic rocks have unusual mineral associations, such as augite and leucite (ugandite), augite and kalsilite (mafurite), and melilite with K-rich glass (katungite).

Bell and Powell (1969) reported ^{87}Sr/^{86}Sr ratios for 117 rock samples from the Virunga and Toro-Ankole volcanic fields. The authors demonstrated that the ^{87}Sr/^{86}Sr ratios (averaged by rock type) correlate positively with the Rb/Sr ratio of the rocks and proposed that this relation is the result of mixing of two components having different ^{87}Sr/^{86}Sr and Rb/Sr ratios. The evidence in favor of mixing during the petrogenesis of the volcanic rocks in the Virunga volcanic province is examined in Fig. 6.11. Each volcano is presumably tapping its own magma chamber whose products of fractional crystallization and assimilation of wallrock are preserved on the surface. Nevertheless, the 117 rock samples analyzed by Bell and Powell (1969) and the additional samples of Vollmer and Norry (1983b) form a single coherent pattern of data points in coordinates of ^{87}Sr/^{86}Sr ratios and reciprocal Sr concentrations.

A melilite nephelinite lava flow labeled (1) in Fig. 6.11 on Mt. Nyiragongo in the Virunga field has the lowest ^{87}Sr/^{86}Sr ratio of 0.7036 which is indistinguishable from

Fig. 6.11.
$^{87}Sr/^{86}Sr$ ratios and reciprocal Sr concentrations of K-rich lavas in the Birunga volcanic field of the Western Rift of East Africa. The data field extends from carbonated lava flows (low $^{87}Sr/^{86}Sr$, high Sr concentration) to the latite flows (high $^{87}Sr/^{86}Sr$, low Sr concentration) of Mt. Sabinyo. The Sr-isotope geochemistry of these rocks is generally compatible with mixing of two components one of which was a mantle-derived magma whose $^{87}Sr/^{86}Sr$ ratios was similar to that of the phonolites and trachytes on Mt. Kenya. The other component originated in the continental crust. In addition, the distribution of data points indicates that some of the lavas in the Birunga area were enriched or depleted in Sr as a result of fractional crystallization. However, the data do not reveal by what process the two components were mixed and they do not explain why the lavas of the Western Rift have elevated concentrations of K₂O (*Sources:* Bell and Powell 1969; Vollmer and Norry 1983b)

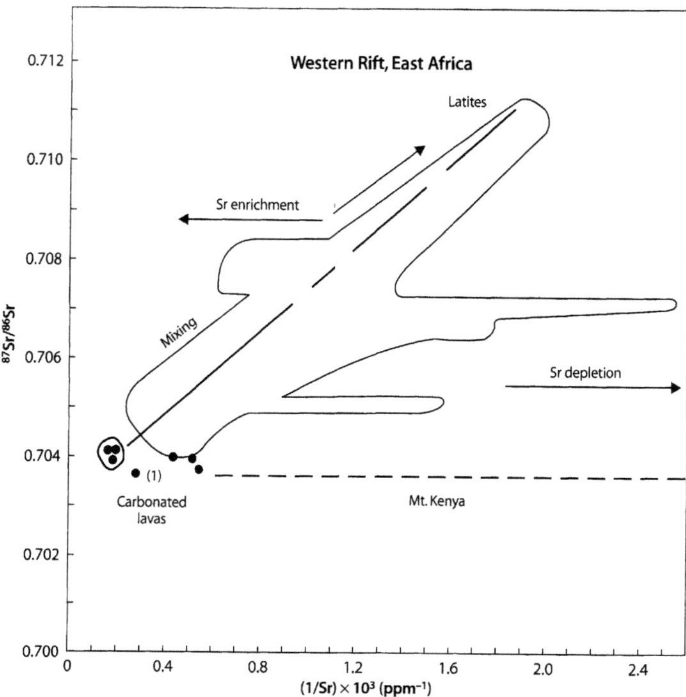

the average $^{87}Sr/^{86}Sr$ ratio of phonolites and trachytes on the summit Mt. Kenya (Sect. 6.1.2). Therefore, one of the components in the petrogenesis of the volcanic rocks of the Western Rift was magma derived either directly from the lithospheric mantle or by partial melting of mantle-derived rocks with short prior crustal residence ages. The mantle-derived magma had an average $^{87}Sr/^{86}Sr$ ratio of 0.7038 ±0.0001 ($2\bar{\sigma}, N = 5$) based on five rock samples whose $^{87}Sr/^{86}Sr$ ratios are equal to or less than 0.7040 (Bell and Powell 1969).

The $^{87}Sr/^{86}Sr$ ratios of most of the lavas in the Western Rift range up to about 0.710, which characterizes the latites on the summit of Mt. Sabinyo in the Virunga field. According to the mixing model proposed by Bell and Powell (1969), the $^{87}Sr/^{86}Sr$ ratios of the magmas increased, whereas their Sr concentrations decreased, as the magmas assimilated Precambrian basement rocks from the walls of their magma chambers. The correlation of $^{87}Sr/^{86}Sr$ ratios and reciprocal Sr concentrations is consistent with the hypothesis of Bell and Powell (1969) that, on a regional scale, the lavas are two-component mixtures. The linear correlation is well developed in the data for Mt. Sabinyo, whereas those of the other volcanoes in the area cluster along the mixing line in Fig. 6.11 (e.g. Mt. Nyiragongo). The deviation of some of the data points from the main trend in

Fig. 6.11 indicates that the magmas were also differentiating by fractional crystallization. The collinearity of the carbonated lavas in the Western Rift with the mixing line, including the latites on Mt. Sabinyo having the highest $^{87}Sr/^{86}Sr$ ratios of the flows in this area, is consistent with the immiscibility of carbonatic and silicate melts reported by Van Groos and Wyllie (1966) and others mentioned in Sect. 6.1.4. Since the carbonatic lavas have low $^{87}Sr/^{86}Sr$ ratios, they presumably separated from the mantle-derived silicate magmas before these magmas had assimilated substantial amounts of crustal rocks.

Although these insights advance our understanding of the petrogenesis of the alkalic lavas along the Western Rift, neither the concentrations of Rb and Sr of the rocks nor their $^{87}Sr/^{86}Sr$ ratios explain their unusual K enrichment. Partial melting of silicate rocks and fractional crystallization of basalt magma in a closed system is capable of forming products enriched both in Na and K, but it does not cause enrichment of K over Na. Therefore, the origin of K-rich rocks has been attributed either to prior K-enrichment of the magma sources in the mantle or to crustal contamination of magmas. Many authors have supported the concept of mantle enrichment (e.g. Wass and Rogers 1980; Edgar and Arima 1981; Bailey 1982; Hawkesworth et al. 1984b;

Thompson 1985). However, crustal contamination must also be considered and has been invoked in some cases (e.g. Downes 1984).

The case for prior metasomatic enrichment of the mantle sources of K-rich magmas in the Western Rift was investigated directly by Lloyd et al. (1985) who melted clinopyroxenite nodules from the Katwe-Kikorongo and Bunyaruguru volcanic fields (southwestern Uganda) containing 37% phlogopite and containing 4.96% K$_2$O. They reported that 20 to 30% partial melts, formed at about 1 250 °C and 30 kbar, contained 3.07 to 5.05% K$_2$O, 35.0 to 39.2% SiO$_2$, K / K + Na = 0.54 to 0.71, K + Na / Al = 0.99 to 1.08, and Mg / Mg + Fe (total) = 0.59 to 0.62. These results permit the conclusion that the K-rich lavas of the Western Rift could have formed by relatively high degrees of partial melting of phlogopite-rich peridotites in the mantle under the Western Rift. However, the high modal abundance of phlogopite in the nodules studied by Lloyd et al. (1985) may not be typical of the source rocks of these K-rich magmas.

The role of crustal contamination in the petrogenesis of the K-rich volcanic rocks of Mt. Karisimbi (Virunga field) is evaluated in Fig. 6.12 based on a study of DeMulder et al. (1986). In Fig. 6.12a the Sr concentrations of the rocks initially rise with increasing Rb concentration and then decline abruptly. This pattern of variation is attributable to the preferential removal of Sr by crystallization of feldspar from a cooling magma (Sect. 1.10.3). Therefore, the concentrations of Sr of these rocks appear to be controlled by fractional crystallization of the magma rather than by assimilation of country rocks. However, the ^{87}Sr/^{86}Sr ratios of the lavas on Karisimbi are not constant as expected for a magma crystallizing in a closed system, but range significantly from 0.70539 to 0.70799 causing the data points to scatter in Fig. 6.12b. Therefore, DeMulder et al. (1986) suggested that the magmas not only differentiated by fractional crystallization but also assimilated crustal rocks. In addition, the lowest ^{87}Sr/^{86}Sr ratio on Karisimbi (0.70539, sample (1) in Fig. 6.12b) is significantly higher than the ^{87}Sr/^{86}Sr ratio of nephelinite on nearby Mt. Nyiragongo (0.7036, Bell and Powell 1969). This difference means either that the magma sources in the mantle underlying Karisimbi and Nyiragongo have different ^{87}Sr/^{86}Sr ratios over a distance of only 20 km, or that the magma erupted at Karisimbi was more extensively contaminated than at Nyiragongo.

The crustal-contamination hypothesis is presented in Fig. 6.12b by means of mixing lines drawn from ^{87}Sr/^{86}Sr = 0.7036, Sr = 3 543 ppm (Nyiragongo) to outline a fan-shaped array of data points representing the lavas of Karisimbi. This interpretation implies that Sr in the magma sources of Nyiragongo and Karisimbi was isotopically homogeneous, but that the Sr in the rocks in the overlying crust is isotopically heterogeneous. As

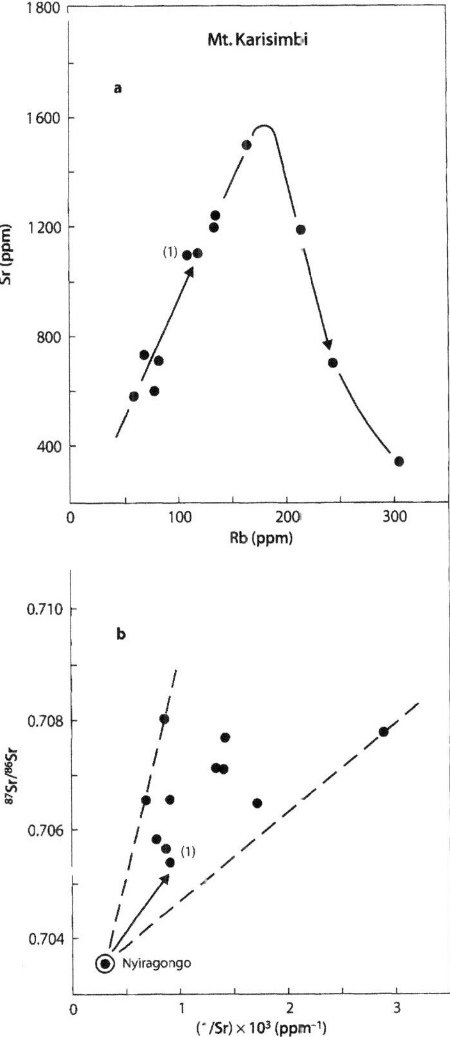

Fig. 6.12. **a** Evidence for the effects of fractional crystallization of magma on the Rb and Sr concentrations of K-rich lavas on Mt. Karisimbi of the Birunga field in Uganda north of Lake Kivu; **b** Hypothetical interpretation of the origin of the lavas on Karisimbi by contamination of mantle-derived magma represented by nephelinite at nearby Mt. Nyiragongo (0.7036) with heterogeneous rocks of the continental crust. Sample (1) in *a* and *b* has the lowest ^{87}Sr/^{86}Sr ratio of all Karisimbi lavas (*Sources:* DeMulder et al. 1986; Bell and Powell 1969)

a result, the ^{87}Sr/^{86}Sr ratios of the lava flows on Karisimbi are controlled by assimilation of heterogeneous crustal rocks, whereas the Rb and Sr concentrations reflect primarily the progress of fractional crystallization of the magma. A similar explanation may apply to the lavas of Nyiragongo whose $^{87}Sr/^{86}Sr$ ratios range from 0.7036 to 07058 (Bell and Powell 1969).

The crustal assimilation model of the petrogenesis of the K-rich lavas of Karisimbi is strengthened by the fact that their $^{143}Nd/^{144}Nd$ ratios also vary. DeMulder et al. (1986) were able to model the $^{87}Sr/^{86}Sr$ and $^{143}Nd/^{144}Nd$ ratios of most of the Karisimbi lavas by means of AFC mixing hyperbolas. However, they noted that ankaramitic and picritic basanites deviate from the AFC model and concluded that these rocks had originated from a different source than the majority of the rocks.

The evidence therefore permits the conclusion that the K-enrichment of the alkalic lavas of the Virunga field is attributable either to assimilation of crustal rocks by alkali-rich mantle-derived magmas, or to the formation of magma in K-rich source rocks in the mantle. Accordingly, the K-enrichment can be attributed hypothetically to assimilation of Precambrian argillites of the Karakwe-Ankolan volcano-sedimentary complex. These rocks may contain sufficient illite, sericite, or muscovite to cause the K-enrichment of the Virunga lavas. Alternatively, the K-enrichment may be the result of partial melting of metasomatically-altered magma sources in the lithospheric mantle or melting of source rocks consisting of a mixture of subcontinental lithosphere and previously subducted oceanic crust and marine sediment. Such mixed source rocks may occur in the head of a plume which underplated the lithospheric mantle long before melting occurred, or they may occur in the head of a plume which arrived in Tertiary time and caused uplift, extension, and volcanic activity in East Africa.

The difficulty in explaining the high K concentrations of the Virunga lavas arises because the isotopic data strongly support crustal contamination (Fig. 6.10b), but crustal contamination is not linked to K-enrichment. On the other hand, metasomatic alteration of the lithospheric mantle under East Africa and elsewhere in the world is indicated by the occurrence of veined ultramafic inclusions containing phlogopite and kaersutite.

6.3   Volcanic Activity of North-Central Africa

Middle Tertiary to Recent volcanic activity in north-central Africa in Fig. 6.13 is associated with the Darfur Dome of western Sudan, the Tibesti Massif on the border between Libya and Chad, and the Hoggar Mountains of southern Algeria. The relation of these volcanic centers to the East African Rift valleys and to each other is somewhat uncertain. However, the Darfur Dome is located at the western end of the Abu Gabra Rift, which extends in a northwesterly direction from Kenya and

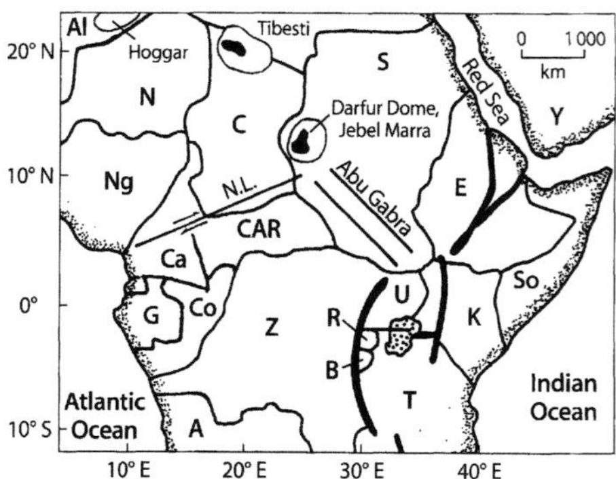

Fig. 6.13. Map of central Africa showing the Abu Gabra Rift in relation to the East African Rift system and the alkali-rich volcanic centers of Jebel Marra (Sudan), Tibesti (Chad and Libya), and Hoggar (Algeria). The capital letters identify countries: *A* = Angola; *Al* = Algeria; *B* = Burundi; *C* = Chad; *Ca* = Cameroon; *CAR* = Central African Republic; *E* = Ethiopia; *G* = Gabon; *K* = Kenya; *N* = Niger; *Ng* = Nigeria, *R* = Rwanda; *S* = Sudan; *So* = Somalia; *T* = Tanzania; *U* = Uganda; *Y* = Yemen. The straight line labeled *N.L.* is the Ngaoundere lineament (*Source:* adapted from Davidson and Wilson 1989)

Uganda across southern Sudan. In addition, the Darfur Dome is linked to the Tibesti Massif and the Hoggar Mountains by gravity anomalies (Fairhead 1978; Davidson and Wilson 1989).

6.3.1   Darfur Volcanic Center of Western Sudan

The Darfur Dome in Fig. 6.14 is located at about 14° N and 24° E. The dome began forming during the Cretaceous Period and became the site of extensive volcanic activity which started at 36 Ma in the Marra Mountains (or Jebel Marra) and migrated in a northeasterly direction to the Tagabo Hills and the Meidob Hills for a distance of about 350 km. The volcanic activity of the Meidob Hills started at 7 Ma and lasted until the Holocene.

The Darfur Dome is the surface expression of a weak asthenospheric plume that reached the base of the local lithospheric mantle in Early Cretaceous time. The head of the plume expanded causing fractures to form in the lithospheric mantle, which led to magma formation by decompression melting and volcanic activity at the surface.

The volcanic rocks of the Jebel Marra volcanic complex range in composition from basalt to hawaiite, mugearite, benmoreite, trachyte, and phonolite. The $^{87}Sr/^{86}Sr$ ratios of these rocks extend from an average value of 0.70357 ±0.00004 ($2\bar{\sigma}$, $N = 4$) for basalt to 0.706664 for a trachyte reported by Davidson and Wilson (1989) relative to 0.710250 for NBS 987. The av-

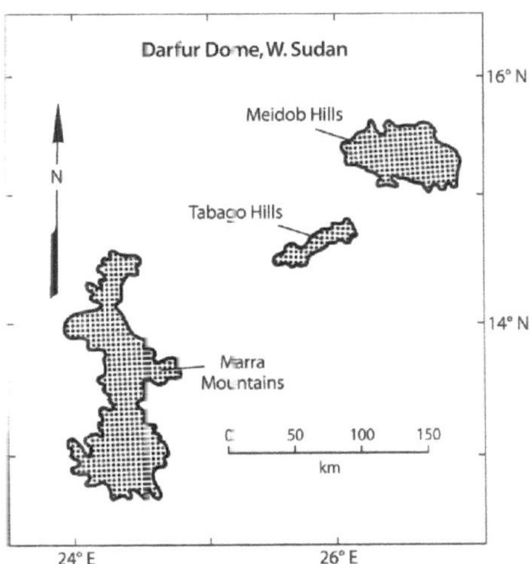

Fig. 6.14. Map of the Darfur Dome in western Sudan; including the Jebel Marra, Tagabo Hills, and Meidob Hills volcanic centers (36 Ma to Recent) (*Source:* adapted from Franz et al. 1999)

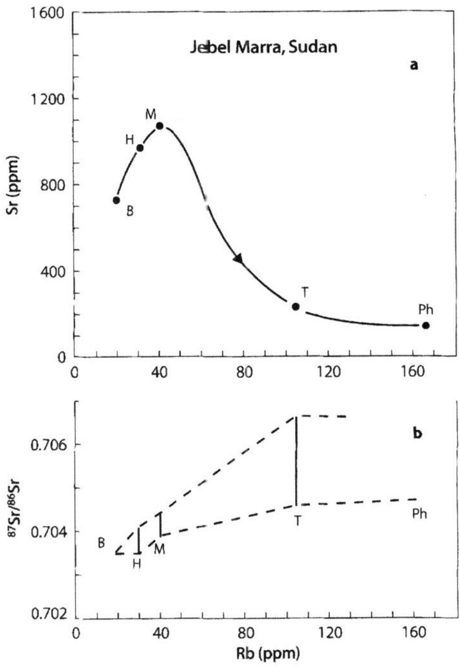

Fig. 6.15. a Variation of the average concentrations of Rb and Sr in the alkali-rich lavas of Jebel Marra, Sudan. B = basalt; H = hawaiite; M = mugearite; T = trachyte; Ph = phonolite. The systematic variation of the Sr concentration is the result of fractional crystallization of basalt magma; **b** Range of $^{87}Sr/^{86}Sr$ ratios of the lavas on Jebel Marra with increasing degree of differentiation recorded by the Rb concentration. The increase of the $^{87}Sr/^{86}Sr$ ratios is evidence for assimilation of granitic basement rocks by magmas in crustal magma chambers underlying the volcanic complex (*Source:* Davidson and Wilson 1989)

erage Sr concentrations of the different rock types at Jebel Marra initially rise (basalt to mugearite) and then decline (mugearite to phonolite) with increasing Rb concentrations. This relationship, shown graphically in Fig. 6.15a, is evidence of extensive differentiation of basalt magma by fractional crystallization. The initial increase of the Sr concentration (basalt to mugearite) was caused by crystallization of minerals that exclude Sr (e.g. clinopyroxene, olivine, and oxides of Fe and Ti) followed by minerals that incorporate Sr (e.g. plagioclase and apatite) thereby reducing the Sr concentration of the residual magma (mugearite to phonolite). The increase of the $^{87}Sr/^{86}Sr$ ratios from basalt to trachyte and their range of variation (e.g. trachyte) demonstrated in Fig. 6.15a,b reveals that the magmas also assimilated crustal rocks containing Sr with elevated $^{87}Sr/^{86}Sr$ ratios. Thus Fig. 6.15a and b are consistent with the conclusion of Davidson and Wilson (1989) that the petrogenesis of the lavas of Jebel Marra involved AFC. The basalt samples analyzed by Davidson and Wilson (1989) probably approximate the composition of the magma, which originated from the plume that caused the uplift of the Darfur Dome. The average $^{87}Sr/^{86}Sr$ ratio of the basalt on Jebel Marra (0.70357 ±0.00003) is identical to that of lavas on Mt. Kenya for which Price et al. (1985) reported an average $^{87}Sr/^{86}Sr$ ratio of 0.70355 ±0.0003 (2σ, N = 26) relative to 0.71025 for NBS 987.

The basement rocks of the Darfur Dome consist of foliated granite and leucocratic gneiss (upper crust) as well as of granite and biotite gneiss (lower crust). A sample of foliated upper-crustal granite (Rb/Sr = 2.13, $^{87}Sr/^{86}Sr$ = 0.76448) yields a model Rb-Sr date of about 700 Ma relative to an initial $^{87}Sr/^{86}Sr$ ratio of 0.7030. This date coincides with the Pan-African event. However, the lower-crustal biotite gneiss (Rb/Sr = 0.297, $^{87}Sr/^{86}Sr$ = 0.73063) has a significantly older model Rb-Sr date of about 2 300 Ma for an assumed initial $^{87}Sr/^{86}Sr$ ratio of 0.7025. These results suggest that the crystalline basement of the Darfur Dome consists of granitic gneisses of Late Archean to Early Proterozoic age that were later intruded by granites during the Pan-African event (Davidson and Wilson 1989).

The lavas of the Meidob Hills range in composition from basanite (oldest) to phonolite and trachyte pyroclastics. The basanites have low $^{87}Sr/^{86}Sr$ ratios (0.70309), whereas more differentiated rocks (basalt, hawaiites, benmoreites, phonolites, and trachytes) have elevated ratios ranging up to 0.70503 (benmoreite) relative to 0.71025 for NBS 987 (Franz et al. 1999).

The isotope ratios of these lavas define straight lines in Fig. 6.16a and b, which means that the magmas primarily contain two components. One of these was the magma that originated from the head of the Darfur Plume whose isotope compositions of Sr, Nd, and Pb are recorded in the early-formed basanite lavas. The high $^{206}Pb/^{204}Pb$ ratio (20.071) of the basanites reveals that the plume contained the HIMU component in the form of subducted oceanic crust that had been incubated sufficiently to allow the $^{206}Pb/^{204}Pb$ ratio to increase to values in excess of 20.0.

The second component was enriched in ^{87}Sr but depleted in ^{143}Nd and ^{206}Pb compared to the HIMU component and could be the basal lithospheric mantle above the head of the Darfur Plume. However, that hypothesis does not explain the correlation between the chemical composition of the lavas and the isotope ratios of Sr, Nd, and Pb evident in Fig. 6.16a and b. There-

fore, it is more likely that the basanite magma differentiated by assimilating small amounts of wallrocks during fractional crystallization in crustal magma chambers. According to this scenario, the basanite magmas differentiated by AFC to form the observed range of chemical and isotope compositions of the lavas at Meidob Hills in agreement with the petrogenesis of the lavas at Jebel Marra (Davidson and Wilson 1989).

6.3.2 Tibesti Massif, Libya and Chad

The Ben Ghnema Batholith of southern Libya is located in the northwest region of the Tibesti Massif (Fig. 6.13). The batholith is composed of diorite, granodiorite, and gabbro in its eastern parts, and consists of quartz monzonite and feldspathic granites in the west. The metamorphic country rocks include a variety of clastic metasediments, but the details of the geology are not well known because of the unconsolidated sand that covers large areas (Pegram et al. 1976). The Tibesti area also contains volcanic rocks of Tertiary age described by Vincent (1970).

Whole-rock Rb-Sr isochron dates of different parts of the Ben Ghnema Batholith reported by Pegram et al. (1976) range from 574 ±27 Ma to 522 ±5 Ma ($\lambda = 1.42 \times 10^{-11}$ yr^{-1}). The initial $^{87}Sr/^{86}Sr$ ratios of these rocks range from 0.7053 to 0.7066 relative to 0.7080 for E&A. Therefore, the authors considered the possibility that these dates represent a time of metamorphic rehomogenization of Sr in the Ben Ghnema Batholith during the Pan-African orogeny. However, they demonstrated that the prior crustal residence age of the granites was relatively short and that the Rb-Sr date of the Ben Ghnema Batholith is well within the time frame of the Pan-African orogeny (650 to 500 Ma).

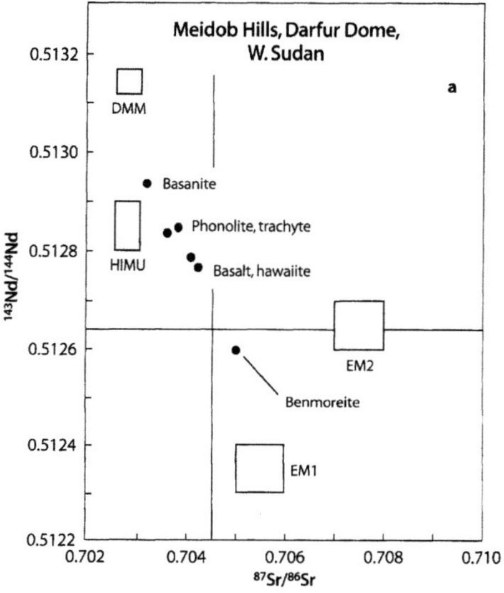

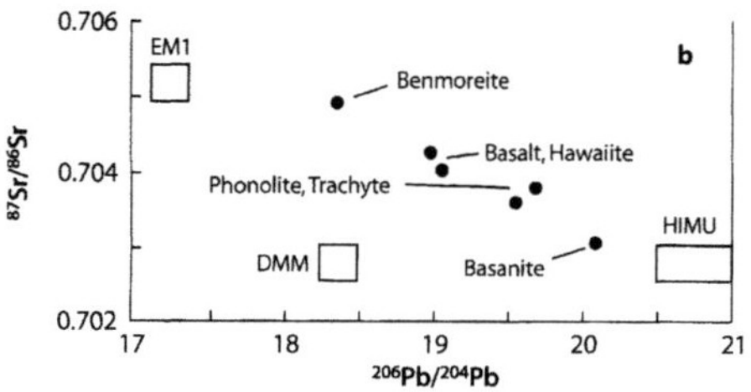

Fig. 6.16 a,b. Isotope ratios of Sr, Nd and Pb of a suite of alkali-rich lavas from middle Tertiary to Holocene in the Meidob Hills of the Darfur Dome in western Sudan (*Sources:* Franz et al. 1999; Hart 1988)

6.3.3 Tahalra Volcanic Center, Hoggar, Algeria

The Hoggar (or Ahaggar) Mountains of southern Algeria (Fig. 6.13) expose Archean basement rocks (2 960 Ma, $\lambda = 1.42 \times 10^{-11}$ yr^{-1}) that were extensively recrystallized at about 1 800 Ma (Ferrara and Gravelle 1966; Bertrand and Lassere 1976) and again at about 600 Ma during the Pan-African event (Hurley and Rand 1969). Alkali basalts and nephelinites were erupted in this area during the Tertiary Period at several volcanic centers associated with a system of north-south trending faults. Some of the basalt flows contain lherzolite xenoliths, which appear to be unaltered suggesting that the basalt magmas rose rapidly to the surface without pooling in crustal magma chambers.

Allègre et al. (1981) reported that alkali basalt and nephelinite at three volcanic centers in the Hoggar area have similar $^{87}Sr/^{86}Sr$ ratios averaging 0.70337 ±0.00010

($2\bar\sigma, N = 6$) relative to NBS 987=0.71025. These lavas also have a high average Sr concentration of 993 ± 203 ($2\bar\sigma$, $N = 6$) which is typical of alkali-rich magmas that formed by small degrees of partial melting in the mantle (e.g. 3 to 5%). The $^{87}Sr/^{86}Sr$ ratios of a nephelinite analyzed by Dautria et al. (1987) and the lavas studied by Allègre et al. (1981) range from 0.70323 to 0.70356 and thus could have formed from the basanite magma by assimilation of crustal rocks and simultaneous fractional crystallization.

The xenoliths in the lavas and pyroclastites of the Tahalra volcanic center in the Hoggar Mountains contain megacrysts of amphibole (up to 15 cm in length), clinopyroxene (5 cm) and, more rarely, plagioclase, ilmenite, and zircon. The amphibole megacrysts are enriched in alkali metals (Na_2O = 1.80 to 2.46%; K_2O = 2.04 to 2.07%; Rb = 8–14 ppm) and thus have the composition of kaersutite (Dautria et al. 1988). Measurements by Dautria et al. (1987) indicate that amphibole (kaersutite) megacrysts, amphibole-rich veins in lherzolite xenoliths, and basanite lava flow all have similar $^{87}Sr/^{86}Sr$ ratios with an average of 0.70306 ±0.00001 ($2\bar\sigma$, $N = 3$). The presence of lherzolite xenoliths containing kaersutite-rich veins in the lavas and pyroclastic rocks at Tahalra indicates that the lithospheric mantle beneath the Hoggar Mountains of Algeria was hydrated and enriched in alkali metals in the relatively recent past prior to the onset of volcanic activity.

6.4 Intraplate Alkaline Plutons of Africa

Alkali-rich plutons of Phanerozoic age occur widely in the Sudan and in Egypt, as well as in other parts of Africa, including Zimbabwe (Foland and Henderson 1976), Nigeria (Jacobsen et al. 1958; Dickin et al. 1991), Mali (Weis et al. 1986, 1987b), and Algeria (Boissonnas et al. 1969). Almost 100 alkaline ring complexes have been described in the Sudan and Egypt by Vail (1985), Harris (1982), and Stern and Gottfried (1986).

The ring complexes of the Sudan occur primarily in the Red Sea Hills (e.g. Jebel Sha and Jebel Mindara), in the Bayuda Desert north of the city of Khartoum (e.g. Silitat es Sufr), and in a relatively small area in the valley of the Nile south of Khartoum. Most of the ring complexes of Egypt consist of multiple intrusive phases ranging in composition from gabbro to nepheline syenite and are associated with trachytes, rhyolites, and phonolites. Age determinations by Serencsits et al. (1981) based on the K-Ar method revealed that the ring complexes of Egypt range in age from Cambrian to Cretaceous and formed during a time interval of about 450 million years. Consequently, they are not directly related to continental rifting and the opening of the Red Sea.

6.4.1 Abu Khruq Complex, Egypt

The Abu Khruq complex is one of the largest composite intrusives in the Eastern Desert of Egypt with a diameter of about 7 km and a roughly circular cross-section. It formed by the intrusion of alkali gabbro followed by several incomplete cone sheets of nepheline syenite and quartz syenite, which intruded cogenetic trachyte, and rhyolite lavas that were extruded on the surface. The field relations as well as K-Ar dating by Serencsits et al. (1981) indicate that the syenites of the Abu Khruq complex were intruded at shallow depth at 89 Ma and cooled rapidly. However, whole-rock Rb-Sr isochron dates of the syenite reported by Lutz et al. (1988) varied widely up to 100 Ma with a range of initial $^{87}Sr/^{86}Sr$ ratios from 0.70354 to 0.70746 relative to 0.7080 for E&A, whereas the gabbros yielded a concordant Rb-Sr isochron date of 89 ±2 Ma with a low initial $^{87}Sr/^{86}Sr$ ratio of 0.702932 ±0.00015 ($\lambda = 1.42 \times 10^{-11}$ yr^{-1}).

Lutz et al. (1988) noted that the syenites have low Sr concentrations ranging from 6.70 to 223.1 ppm with a mean of 43.0 ppm and cited petrographic evidence indicating that the syenites had been hydrothermally altered. The gabbros have much higher Sr concentrations than the syenites ranging from 668 to 919 ppm which made the isotope composition of the Sr in the gabbros insensitive to change by hydrothermal alteration. Since the syenites do not contain secondary calcite, Lutz et al. (1988) concluded that the Sr derived from the hydrothermal solutions was sorbed on mineral surfaces and microfractures. Therefore, the effect of the hydrothermal alteration on the isotope composition of Sr in the syenites depended on the minerals they contained and on the concentration of Sr before alteration. The data suggested that nepheline and analcime were more susceptible to hydrothermal alteration than feldspar and pyroxene.

The amount of Sr added to minerals by hydrothermal fluids can be estimated by a procedure devised by Lutz et al. (1988). The $^{87}Sr/^{86}Sr$ ratio of the hydrothermal Sr at Abu Khruq was arbitrarily fixed at 0.725 because the Sr originated from the rocks of the continental crust (Faure et al. 1963a). Therefore, the hydrothermal Sr is represented in Fig. 6.17 by point P whose coordinates are: (1/Sr) × 10^3 = 0.0, $^{87}Sr/^{86}Sr$ = 0.725. In addition, the $^{87}Sr/^{86}Sr$ ratio of the minerals prior to the addition of hydrothermal Sr was assumed to be 0.7030, based on the initial $^{87}Sr/^{86}Sr$ ratio of the Sr-rich gabbro at Abu Khruq. The Sr concentration of any sample of a hydrothermally altered rock or mineral of this complex is determined by drawing a straight line from P through the selected data point to intersect the line representing the original $^{87}Sr/^{86}Sr$ ratio (0.7030). The desired Sr concentration is then calculated from the (1/Sr) × 10^3 coordinate of the point of intersection.

Fig. 6.17.
Initial $^{87}Sr/^{86}Sr$ ratios (at
89.5 Ma) and reciprocal Sr
concentrations of separated
minerals of the nepheline
syenites and quartz syenites
of the Abu Khruq ring complex, Eastern Desert, Egypt.
Feldspar: *solid circles*; nepheline and analcime: *open circles*; pyroxene: *crosses*. The
line defined by the data points
was drawn from 0.7030 (intercept) to the nepheline with
the highest $^{87}Sr/^{86}Sr$ ratio of
0.7163. Point *P* represents Sr in
a hydrothermal fluid (heated
groundwater) which added
about 2.2 µg of Sr per gram to
the minerals that plot on or
are close to the hydrothermal
alteration line. The resulting
increase in the $^{87}Sr/^{86}Sr$ ratios
of the minerals is inversely
proportional to their Sr concentrations prior to the alteration (*Source:* Lutz et al. 1988)

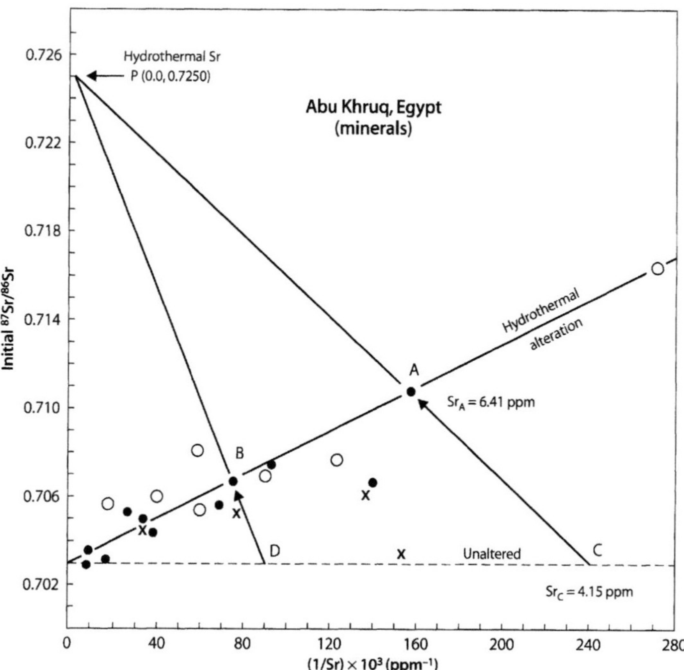

For example, the amount of hydrothermal Sr added
to Feldspar A in Fig. 6.17 is 2.26 µg g^{-1} or about 54% of
the Sr present originally. Similarly, 2.16 µg g^{-1} of hydrothermal Sr were added to Feldspar B, thus increasing
its Sr concentration by about 19%. These results reveal
that the amount of Sr added to the feldspar by the hydrothermal alteration was not large and that the effect
on the $^{87}Sr/^{86}Sr$ ratio increases as the original Sr concentration decreases. This procedure invented by Lutz
et al. (1988) can be used to calculate the amount of Sr
added not only to the constituent minerals of a rock
sample, but also to whole rock samples.

The hydrothermal fluids associated with the intrusion of plutons at shallow depth in the crust consist of
local groundwater that is heated by the intrusion and
hence flow by convection through the country rocks as
well as through the intrusion. Convection of heated
groundwater also changes the isotope composition of
oxygen in the minerals of the intrusion and of the surrounding country rock. In addition, the circulation of
heated groundwater in the vicinity of igneous intrusions can cause the formation of ore deposits within
the body of the intrusion and in the adjacent country
rock (Taylor 1974).

Hydrothermal alteration of igneous intrusions and
of the adjacent country rocks by heated groundwater is
a common occurrence and is a form of contact metamorphism. The effect of such alteration on the isotope
composition of Sr in the rocks of the intrusion depends
on several factors; including the Sr concentrations of
the intrusive rocks, the permeability of the country
rocks, the $^{87}Sr/^{86}Sr$ ratio and Sr concentration of the
hydrothermal fluids, and the rate of cooling of the intrusive rocks because that determines the duration of
the process. The effects of hydrothermal alteration on
the isotope composition of Sr in igneous intrusives may
be overlooked in cases where the rocks have a high Sr
concentration, where the $^{87}Sr/^{86}Sr$ ratio of the hydrothermal Sr is similar to that of the igneous rocks, or
where the rocks cooled rapidly. Nevertheless, the addition of Sr by hydrothermal solutions can lead to an
overestimated age of igneous intrusions by the Rb-Sr
method because of the increase of their $^{87}Sr/^{86}Sr$ ratios
and the concurrent decrease of the $^{87}Rb/^{86}Sr$ ratios
(Lutz and Srogi 1986).

6.4.2 Alkali-rich Ring Complexes of Sudan

The Jebel Sha and Jebel Mindara complexes in the Red
Sea Hills in Sudan intrude volcanic and plutonic rocks
of Late Proterozoic age (Kröner et al. 1991). The intrusive rocks range in composition from plagioclase-
rich monzonite to alkali-rich syenite and quartz syenite.

The Mindara complex consists of five concentric cone sheets of quartz syenite and biotite syenite, and is associated with trachytic lava flows. The center of the Jebel Sha ring complex moved along a curved path causing eight overlapping complexes to form over a distance of about 11 km.

Whole-rock Rb-Sr isochrons ($\lambda = 1.42 \times 10^{-11}$ yr^{-1}) indicate that the Jebel Sha and Mindara complexes crystallized at 135 ±3 and 139 ±3 Ma, respectively, with nearly identical $^{87}Sr/^{86}Sr$ ratios of 0.7030 ±0.0001 (Klemenic 1987). Both isochrons are based on nine or more whole-rock samples without excessive scatter ($MSWD < 2.6$). Since the rocks contain evidence of late-stage hydrothermal alteration (arfvedsonite, albite, and calcite), their Rb-Sr dates may have been increased by the addition of hydrothermal Sr having a high $^{87}Sr/^{86}Sr$ ratio. Nevertheless, Klemenic (1987) cited the good fit of the data points on the isochrons to support the conclusion that the Rb-Sr dates represent the time of intrusion of these complexes. Therefore, the Sha and Mindara complexes in the Red Sea Hills of Sudan are of Cretaceous age and are older than the Abu Khruq complex of Egypt.

The Silitat es Sufr complex near Khartoum has a Rb-Sr date of 169 ±2 Ma (5 data points, $MSWD = 4.5$) with a high initial $^{87}Sr/^{86}Sr$ rat o of 0.7091 ±0.0025 (Klemenic 1987). These rocks contain calcite, fluorite, and aegirine deposited by late-stage hydrothermal solutions. Two whole-rock samples have low Sr concentrations of 3 and 4 ppm making these rocks vulnerable to contamination by hydrothermal Sr. Therefore, the Rb-Sr date of the Silitat es Sufr complex may overestimate the age of this intrusion. Nevertheless, Klemenic (1987) cited a K-Ar date of 165 ±4 Ma for the Silitat es Sufr complex and concluded that the magma had assimilated Archean gneiss like that at Uweinat for which Klerkx and Deutsch (1977) reported a whole-rock Rb-Sr isochron date of 2 679 ±21 Ma.

The other ring complex near Khartoum (Jebel Abu Tulleih) dated by Klemenic (1987) yielded an early Paleozoic Rb-Sr date of 459 ±8 Ma (5 data points, $MSWD = 0.5$) with a low initial $^{87}Sr/^{86}Sr$ ratio 0.7038 ±0.0002, thus confirming a Rb-Sr isochron date of 465 ±14 Ma reported by Harris et al. (1983). The rocks of this complex have relatively high Sr concentrations (184 to 202 ppm) making the isotope composition of Sr insensitive to the addition of hydrothermal Sr. However, even these rocks contain late-stage aegirine-rimming arfvedsonite crystals.

6.4.3 Marangudzi Complex, Zimbabwe

The alkali-rich ring complex of Mesozoic age located near Marangudzi in Zimbabwe has a diameter of about 10 km and is composed of cone sheets of quartz syenites, nepheline syenites, and several other alkalic rock types including juvite (coarse grained nepheline-rich rocks containing K-rich alkali feldspar) in the core. Foland and Henderson (1976) reported a whole-rock Rb-Sr isochron date of 186.5 ±1.3 Ma ($\lambda = 1.42 \times 10^{-1}$ yr^{-1}) with an initial $^{87}Sr/^{86}Sr$ ratio of 0.70760 ±0.00003 (0.71025 for NBS 987) based on 24 samples from all of the major rock types.

The Rb-Sr date (186.5 Ma) and its geographic location associate the Marangudzi complex with the Lebombo-Nuanetsi igneous province of South Africa (Manton 1968, Sect. 5.11.1). Additional evidence for this association is provided by the high initial $^{87}Sr/^{86}Sr$ ratio of the alkali-rich rocks at Marangudzi (0.70760). Manton (1968) obtained similar initial $^{87}Sr/^{86}Sr$ ratio of 0.7083 ±0.0007 for granophyres and granites of the Dembe-Divula and Masukwe Intrusions with a whole-rock Rb-Sr isochron date of 173 ±7 Ma ($\lambda = 1.42 \times 10^{-11}$ yr^{-1}). The Mesozoic plateau basalt flows of the Karoo volcanic province and the felsic rocks of the related Lebombo-Nuanetsi igneous province in South Africa are well known for having anomalously high initial $^{87}Sr/^{86}Sr$ ratios (Sect. 5.11.1).

6.5 Mediterranean Region

The volcanic rocks on the Hellenic and Eolian Island arcs of the Mediterranean Sea (Sect. 3.12.1 and 3.12.2) as well as in Italy and southern Spain are strongly enriched in alkali elements but the tectonic setting of this region is quite different from that of the rift valleys of East Africa and of the plume-related volcanic and plutonic complexes of Sudan and Egypt. The outstanding properties of the alkali-rich lavas of Italy are their high concentration of K_2O and their elevated $^{87}Sr/^{86}Sr$ ratios. These features continue to be investigated as indicated by the "Comments" by Peccerillo (1990) and the "Reply" by Vollmer (1990) and by recent contributions of Conticelli (1998), Conticelli et al. (1991), and by Spera et al. (1998).

6.5.1 Italy

The volcanic rocks of Italy are of Late Tertiary to Recent age and occur in five regions of southern and central Italy identified in Fig. 6.18: (1) Sicily; (2) the Campanian region; (3) the Roccamonfina volcanic complex; (4) the Roman region; and (5) Tuscany. Much has been written about the origin of the potassic alkali-rich lavas which characterize the volcanic province of southern and central Italy. This relatively small area along the west coast of Italy contains volcanic rocks that have remarkably high concentrations of K_2O, Rb, and Sr. In addition, these rocks have elevated $^{87}Sr/^{86}Sr$ ra-

Fig. 6.18. Map of the volcanic fields of Italy including Mt. Etna on the island of Sicily, the Campanian region around Naples, the Roccamonfina volcanic complex, the Roman region, and Tuscany (*Source:* adapted from Hawkesworth and Vollmer 1979)

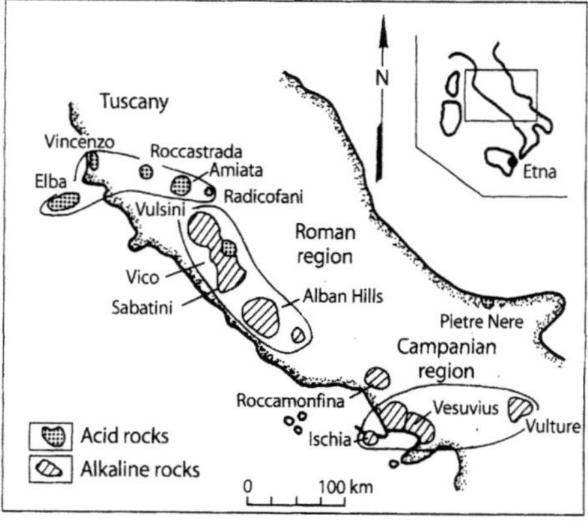

Fig. 6.19.
Histograms of ^{87}Sr/^{86}Sr ratios of volcanic rocks of Tertiary to Recent age in Italy. The data are presented by region from Mt. Etna on the island of Sicily north to Tuscany in central Italy. The volcanic rocks in the Campanian region near Naples, on Roccamonfina Volcano, and in the area around Rome are strongly undersaturated and K-rich, whereas those on Mt. Etna and in Tuscany are saturated calc-alkaline lavas. The diagrams demonstrate that the alkali-rich lavas have elevated ^{87}Sr/^{86}Sr ratios and that these ratios increase progressively from south to north (*Sources:* Carter et al. 1978; Hawkesworth and Vollmer 1979; Cortini and Hermes 1981; Vollmer 1976; Hoefs and Wedepohl 1968; Hurley et al. 1966; Poli et al. 1989; Rogers et al. 1985; Barbieri et al. 1975, 1988; Ferrara et al. 1986; Cox et al. 1976; Carter and Civetta 1977)

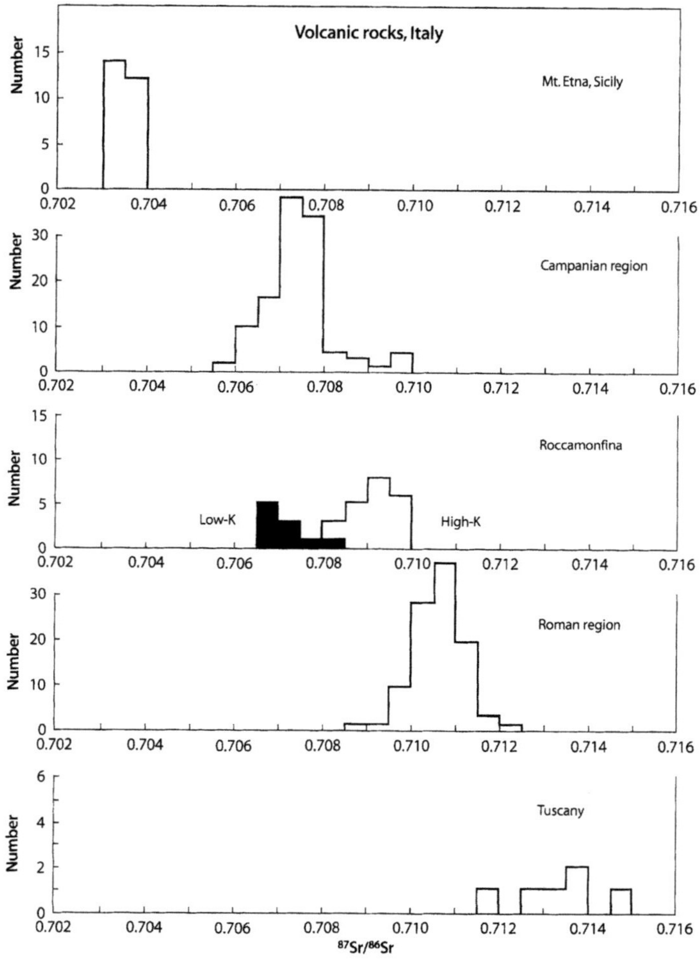

ios, generally ranging from 0.7060 to 0.7120. Consequently, a debate has arisen about the origin of these rocks based on the following alternatives (Ellam and Rogers 1988):

1. Alkali-rich magmas originated from regions in the lithospheric mantle having elevated $^{87}Sr/^{86}Sr$ ratios because of prior enrichment in K and Rb.
2. Magmas formed by small degrees of partial melting of unenriched source rocks in the mantle and subsequently assimilated K-rich sedimentary rocks of the continental crust.
3. Partial melting of mixed source rocks including lithospheric mantle and subducted sediment.

In the third scenario, the contamination of the magmas occurs at the source rather than in a crustal magma chamber.

The assimilation of sedimentary rocks by mantle-derived magmas was strongly advocated by Taylor et al. (1979) based on the ^{18}O-enrichment of the potassic lavas of the Roccamonfina area in central Italy. Ellam and Rogers (1988) supported the sediment-assimilation model by demonstrating that the isotope composition of Sr and O in the lavas at Vulsini, Roccamonfina, and Stromboli (Eolian Island arc) fit an isotopic mixing hyperbola joining sediment ($\delta^{18}O$ = +12.0‰, $^{87}Sr/^{86}Sr$ = 0.710) and mantle rocks ($\delta^{18}O$ = +6.0‰, $^{87}Sr/^{86}Sr$ = 0.7030). These results support the contamination hypothesis but do not indicate whether the contamination occurred in crustal magma chambers (Hypothesis 2, above) or at the source (Hypothesis 3).

The $^{87}Sr/^{86}Sr$ ratios of the volcanic rocks of Italy are presented as histograms in Fig. 6.19 in order to illustrate the differences between the various volcanic centers. Mount Etna on the island of Sicily (Figs. 3.49 and 6.18) is located on a rift associated with the subduction of oceanic crust in the Eolian Island arc north of Sicily (Sect. 3.12.2). It is composed of olivine tholeiite overlain by alkali-rich lavas (hawaiite, mugearite, and benmoreite). The Rb concentrations of the alkali-rich rocks range from 25 to 71 ppm and are modest by comparison with the Rb concentrations of alkali-rich lavas on the Italian mainland which range from 26 to 674 ppm with a mean of about 600 ppm. The $^{87}Sr/^{86}Sr$ ratios of the olivine tholeiites on Mt. Etna reported by Carter et al. (1978a) and Carter and Civetta (1977) average 0.70325, indicating their derivation from mantle sources without significant crustal contamination.

Mount Vesuvius near Naples in the Campanian region on the mainland of southern Italy (Fig. 6.18) has extruded strongly alkali-enriched lavas even though it is located only about 300 km northwest of Mount Etna (Spera et al. 1998). The lavas on Mt. Vesuvius and Mt. Somma contain leucite ($KAlSi_2O_6$) and have high average concentrations of Rb (273 ppm, N = 64), Sr

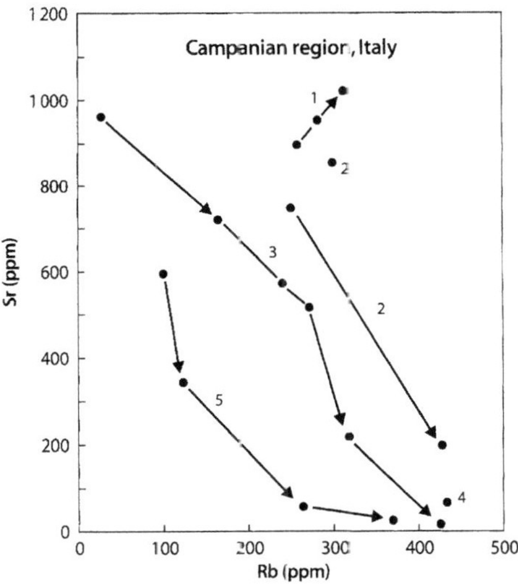

Fig. 6.20. Evidence for magmatic differentiation of the K-rich lavas of southern Italy. 1. Vesuvius; 2. Somma; 3. Phlegrean fields (ejecta); 4. Campanian ignimbrite; 5. Ischia Island. The data points represent averages of different rock types at each locality. The decrease of the average Sr concentrations with increasing Rb concentrations is characteristic of magmatic differentiation by fractional crystallization. The lavas on Mt. Vesuvius included in this diagram are more uniform in chemical composition than those at the other locations and thereby deviate from the general trend (Source: Cortini and Hermes 1981; Hoefs and Wedepohl 1968)

(870 ppm, N = 64), and K_2O (6.62% N = 59). (Cortini and Hermes 1981; Hoefs and Wedepohl 1968). The K_2O concentrations of these and other alkali-rich lavas of Italy are more than five times the crustal average for igneous rocks (Taylor and McLennan 1985) making them potential ore deposits of K (Washington 1918).

Another noteworthy feature of the lavas of Mount Vesuvius reported by Cortini and Hermes (1981) is the systematic decrease of the $^{87}Sr/^{86}Sr$ ratios with the date of the eruption between 1754 and 1944. The authors identified two parallel linear trends for lavas erupted between 1754 to 1881 and between 1861 to 1944. The temporal evolution of the $^{87}Sr/^{86}Sr$ ratios implies that the lavas erupted by Vesuvius originated from two reservoirs containing magmas that were zoned with respect to their $^{87}Sr/^{86}Sr$ ratios, either because of mixing of magmas or as a result of wallrock assimilation. In either case, the lavas erupted from these chambers had progressively lower $^{87}Sr/^{86}Sr$ ratios that approach a value of about 0.7072 in both suites.

The $^{87}Sr/^{86}Sr$ ratios of the alkali-rich lavas on the Italian mainland increase in a northerly direction from values between 0.7060 and 0.7080 on Mt. Vesuvius and in the surrounding Campanian region to 0.7100 for the high-K lavas of the Roccamonfina volcanic complex.

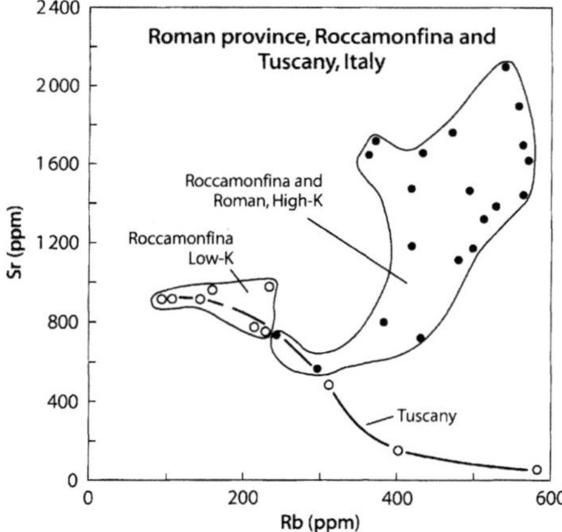

Fig. 6.21. Average Rb and Sr concentrations of different types of the alkalic (high-K) volcanic rocks of the Roman province and Roccamonfina (*solid circles*) Italy. The low-K lavas of Roccamonfina and the saturated lavas of Tuscany are represented by *open circles*. The irregular distribution of data points of alkalic rocks in Roccamonfina and the Roman province contrasts with the volcanic rocks of the Campanian region and suggests that magma differentiation by fractional crystallization was not important in this region. However, the saturated rocks of Tuscany and the low-K lavas of Roccamonfina do define a fractional crystallization trajectory (*Sources:* Cortini and Hermes 1981; Hawkesworth and Vollmer 1979; Poli et al. 1989; Carter et al. 1978; Rogers et al. 1985; Cox et al. 1976; Barbieri et al. 1988)

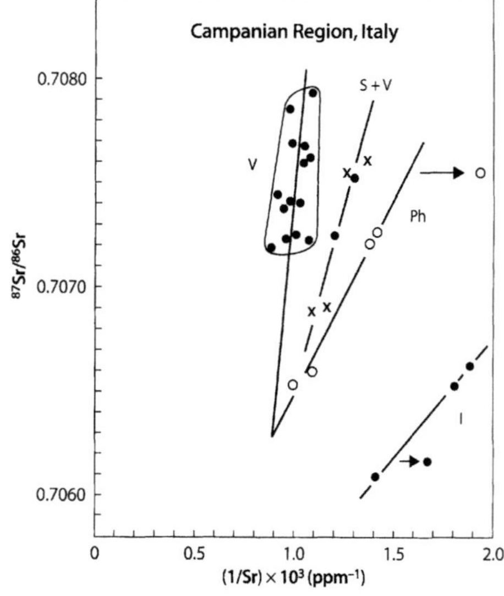

Fig. 6.22. Sr-isotope mixing diagram for K-rich lavas of southern Italy: V = Vesuvius; S = Somma; Ph = Phlegrean fields; I = Ischia Island. The lavas of these volcanic centers (except Ischia), form linear arrays that converge to $^{87}Sr/^{86}Sr \approx 0.70630$ and Sr $\approx$ 1100 ppm. Highly differentiated lavas with Sr<500 ppm are displaced from the arrays and are not shown (*Source:* Cortini and Hermes 1981; Poli et al. 1989)

Still farther north in the vicinity of Rome, the $^{87}Sr/^{86}Sr$ ratios of alkali-rich lavas rise to 0.7120. The lavas of Tuscany, north of the Roman region, are silica saturated and have $^{87}Sr/^{86}Sr$ ratios between 0.71184 and 0.72548.

The alkali-rich lavas in the Campanian region differ from those on Roccamonfina and in the Roman province because the average Sr concentrations of different rock types in Fig. 6.20 decrease with increasing average Rb concentrations. Such correlations are caused by the crystallization of Sr-rich minerals (e.g. plagioclase and apatite) from cooling magmas and indicate that the alkali-rich magmas differentiated in magma chambers underlying the Campanian volcanic region. The relationship between the average concentrations of Rb and Sr of different rock types erupted at Roccamonfina and in the Roman province in Fig. 6.21 differs greatly from that shown in Fig. 6.20 for the Campanian region. At Roccamonfina and in the Roman region the average Sr concentrations of the lavas increase with rising Rb concentrations, although the data points scatter considerably. The low-K lavas of Roccamonfina cluster in a separate part of the diagram and may be related to the silica-saturated lavas of Tuscany because their average Sr concentration decrease with rising Rb concentrations, perhaps because both suites formed by fractional crystallization of magma.

The K-rich lavas of Roccamonfina have unusually high average concentrations of Rb (530 ppm) and Sr (1755 ppm) giving them an average Rb/Sr ratio of 0.30. This ratio would cause their average $^{87}Sr/^{86}Sr$ ratio to increase by only 0.00012 in one million years. At this rate, the $^{87}Sr/^{86}Sr$ ratio of magma would increase from 0.7030 to 0.7120 in about 730 million years. This result therefore demonstrates that the elevated $^{87}Sr/^{86}Sr$ ratios of the K-rich lavas in Italy cannot be explained by decay of ^{87}Rb in the magmas even though their Rb concentrations were very high.

The $^{87}Sr/^{86}Sr$ ratios of the lavas in the Campanian region define straight lines in Fig. 6.22 that may be the result of assimilation of different kinds of country rocks by magmas that were crystallizing in crustal magma chambers. The apparent convergence of the mixing lines of Vesuvius, Somma, and the Phlegrean fields indicates the existence of a common parent magma with $^{87}Sr/^{86}Sr \approx 0.7063$ and Sr $\approx$ 1100 ppm. The lavas of the island of Ischia are highly differentiated, meaning that their Sr concentrations have been reduced to values as low as 6.0 ppm (Poli et al. 1989). For this reason, the slope of the Ischia array in Fig. 6.22 is uncertain, and many specimens actually plot off scale to the right. However, the trachytes on Ischia have the lowest $^{87}Sr/^{87}Sr$ ratio (0.70607 relative to E&A = 0.7080) of

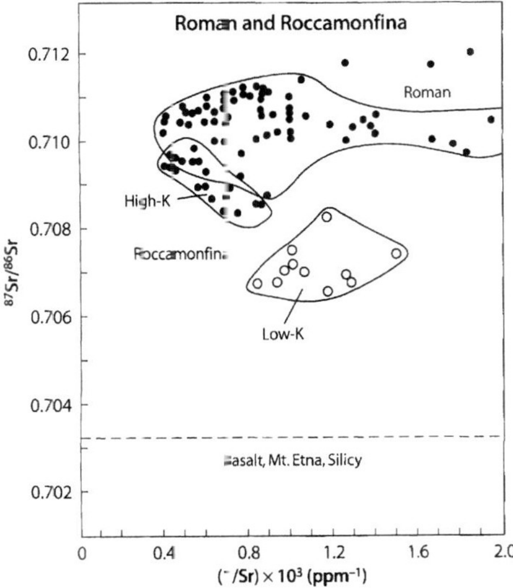

Fig. 6.23. Strontium isotope mixing diagram for alkali-rich volcanic rocks of Roccamonfina and the Roman province of Italy. The average $^{87}Sr/^{86}Sr$ ratios are exceptionally high and contrast sharply with the $^{87}Sr/^{86}Sr$ ratios (0.70328) of olivine tholeiites on Mt. Etna, Sicily. The high-K lavas at Roccamonfina are tightly clustered in the diagram, but overlap the Sr concentrations and $^{87}Sr/^{86}Sr$ ratios of the alkalic lavas of the Roman province. The Sr concentrations and $^{87}Sr/^{86}Sr$ ratios distinguish these lavas from the volcanic rocks of the Campanian region which form linear mixing arrays that converge towards a common point. The low-K lavas of Roccamonfina separate from the high-K lavas without defining mixing arrays (Sources: Cortini and Hermes 1981; Hawkesworth and Vollmer 1979; Poli et al. 1989; Carter et al. 1978; Rogers et al. 1985; Cox et al. 1976; Barbieri et al. 1988; Ferrara et al. 1986)

any of the lavas in the Campanian region (Hurley et al. 1966; Hawkesworth and Vollmer 1979; Cortini and Hermes 1981). Therefore, the hypothetical parent magma of the lavas on the island of Ischia had an $^{87}Sr/^{86}Sr$ ratio of 0.70607 or less. However, even that value is unusually high for Sr derived from the lithospheric mantle, suggesting that crustal rocks contributed to the formation of the magma either at the source or by later assimilation.

In contrast to the Campanian region, the $^{87}Sr/^{86}Sr$ ratios of the volcanic rocks on Roccamonfina and in the Roman region are not correlated with their reciprocal Sr concentrations, but scatter widely in Fig. 6.23. Because of the high average Sr concentration of these rocks (about 1 200 ppm) the $^{87}Sr/^{86}Sr$ ratios of the magmas were insensitive to change by assimilation of crustal rocks. In addition, the lack of correlation between average Rb and Sr concentrations (Fig. 6.21) indicates that fractional crystallization did not play an important role in the petrogenesis of the high-K lavas of Roccamonfina. Instead, Cox et al. (1976) concluded

that the K-rich lavas of Roccamonfina are undifferentiated and that their chemical compositions and $^{87}Sr/^{86}Sr$ ratios are similar to those of the magmas from which they formed. Hawkesworth and Vollmer (1979) attributed the apparent heterogeneity of the mantle under Roccamonfina to metasomatism by a fluid having a high $^{87}Sr/^{86}Sr$ ratio and high concentrations of K and Rb, and Rogers et al. (1985) later proposed that the hydrous silicic fluids were released by subducted sediment.

In this way, the chemical and isotopic data relate the high $^{87}Sr/^{86}Sr$ ratios of the alkalic lavas of Italy to the presence of subducted sediment in the lithospheric mantle. Accordingly, magmas formed by partial melting of metasomatically altered lithospheric mantle or from mechanical mixtures of oceanic sediment and lithospheric mantle. In some areas, the resulting magmas differentiated in crustal magma chambers (e.g. the Campanian region), whereas in others they were erupted rapidly with little time for fractional crystallization and assimilation of crustal rocks (e.g. Roccamonfina and Roman region). In either case, the K-enrichment and elevated $^{87}Sr/^{86}Sr$ ratios of the volcanic rocks of Italy are a consequence of the presence of subducted oceanic sediment, which contaminated the lithospheric mantle in this region.

The silica-saturated lavas of Tuscany range in composition from shoshonitic basaltic andesites, latites, rhyodacites, to rhyolites (Hawkesworth and Vollmer 1979; Vollmer 1976; Ferrara et al. 1986, 1989). These rocks differ from the undersaturated K-rich lavas of the Roman region by having low concentrations of Sr (330 ppm) and high $^{87}Sr/^{86}Sr$ ratios (0.71184 to 0.72548). Consequently, the lavas of Tuscany are considered to have formed from crustal magmas (Hawkesworth and Vollmer 1979).

The most direct demonstration that the alkali-rich magmas of Italy are mixtures of components derived from the mantle and crust is provided in Fig. 6.24 based on the $^{87}Sr/^{86}Sr$ and $^{143}Nd/^{144}Nd$ ratios of the rocks in all of the major volcanic fields (Hawkesworth and Vollmer 1979). The lavas of Pietre Nere in the Campanian region on the east coast of Italy (Fig. 6.18) actually plot in the mantle quadrant of Fig. 6.24, whereas the lavas at other volcanic centers (e.g. Somma/Vesuvius, Ischia, etc.) contain varying proportions of Sr and Nd of crustal origin. The volcanic rocks of Campania are transitional between melts of the lithospheric mantle and the magmas of Roccamonfina, the Roman region, and Tuscany which contain increasing proportions of Sr and Nd derived from the continental crust (Ferrara et al. 1985). The well-established occurrence of subduction within the Mediterranean Basin favors the conclusion that the magmas formed by partial melting of mixtures of subducted sediment and ultramafic rocks of the lithospheric mantle. The geographic vari-

Fig. 6.24.
^{87}Sr/^{86}Sr and ^{143}Nd/^{144}Nd ratios of the alkali-rich lavas of Italy. The isotope ratios of Sr and Nd of the alkali-rich rocks of Italy form a series of clusters that extend from lithospheric melts (Campanian region) into Quadrant IV thereby demonstrating the presence of increasing proportions of Sr and Nd derived from crustal rocks (*Source:* Hawkesworth and Vollmer 1979)

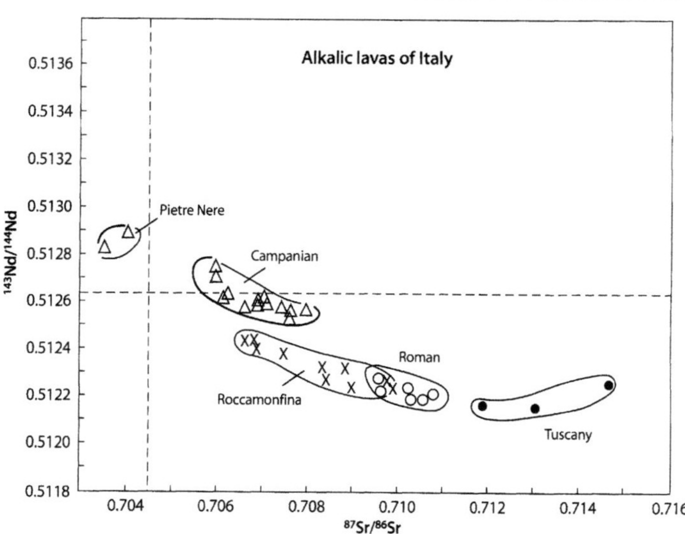

ation of ^{87}Sr/^{86}Sr and ^{143}Nd/^{144}Nd ratios suggests that the proportion of melt derived from the sedimentary component increased systematically from south to north.

The igneous rocks at Pietre Nere (Fig. 6.18) play a key role in the explanation of the origin of the K-rich lavas of Italy. The rocks were brought to the surface by a salt dome, but are now covered by the foundation of a hotel (Vollmer 1990). Two samples of alkali syenite collected at Pietre Nere yielded the lowest ^{87}Sr/^{86}Sr and highest ^{143}Nd/^{144}Nd ratios among all of the Tertiary volcanic rocks of Italy in Fig. 6.24 (Hawkesworth and Vollmer (1979). The values they reported are: ^{87}Sr/^{86}Sr = 0.70401 ±0.00003 and 0.70355 ±0.00005 relative to 0.7080 for E&A; ^{143}Nd/^{144}Nd = 0.51289 ±0.00003 and 0.51283 ±0.00003 normalized to ^{146}Nd/^{144}Nd = 0.7219, respectively. In addition, these rocks crystallized at 56 Ma and are much older than any of the Late Tertiary to Recent alkali-rich lavas of Italy (Vollmer 1976, 1977). Nevertheless, Vollmer (1989) proposed that Pietre Nere is the product of igneous activity related to an asthenospheric plume and that the volcanic rocks of the Campanian and Roman regions are located along the track of this plume that resulted from the counterclockwise rotation of Italy. As a result, the hypothetical Pietre Nere Plume is presently located in Tuscany.

Based on this hypothesis, Vollmer (1989, 1990) proposed that the plume contributed significantly to the petrogenesis of the alkali-rich lavas of the Campanian and Roman regions in addition to the subducted marine sediment which is present in the lithospheric mantle. Peccerillo (1990) questioned the existence of an asthenospheric plume and maintained that the rocks at Pietre Nere are unrelated to the volcanic rocks of the

Campanian and Roman regions. However, Vollmer (1990) cited evidence from the literature that a large heat-flow anomaly does exist in Tuscany that can be attributed to the presence of a plume beneath the lithospheric mantle of this region.

On balance, most workers agree that the volcanic activity in Italy and elsewhere in the Mediterranean region is caused by subduction of oceanic crust and terrigenous sediment resulting from the northward movement of the African Plate. Both the Hellenic and Eolian Island arcs (Sect. 3.12.1 and 3.12.2) are manifestations of this tectonic process, which will ultimately consume the Mediterranean Basin. The complex magmatic processes that occur in this setting are exemplified by the volcanic activity of Stromboli in the Eolian Island arc. Francalanci et al. (1989, 1999) reported evidence for source heterogeneity, magma mixing, crustal contamination, and fractional crystallization in the lavas of this volcano. In addition, Civetta et al. (1991) described the evolution of volcanic activity on the island of Ischia in the past 55 000 years in terms of episodes of magma mixing and assimilation of wallrocks by magmas during fractional crystallization in isotopically zoned magma chambers. The volcanic rocks on these islands and on the Italian mainland owe more to the magmatic processes of subduction zones than they do to asthenospheric plumes.

6.5.2 Southeastern Spain (Alkali-rich Rocks)

Ultrapotassic igneous rocks in the form of small plutons and dikes occur along the eastern side of the Betic Cordillera between the cities of Malaga and Va-

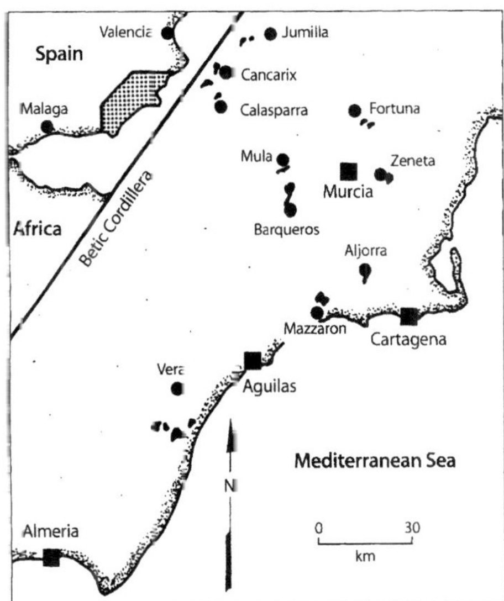

Fig. 6.25. Outcrops of Miocene ultrapotassic dikes and small plutons (*black*) in southeastern Spain (*Source:* adapted from Venturelli et al. 1984)

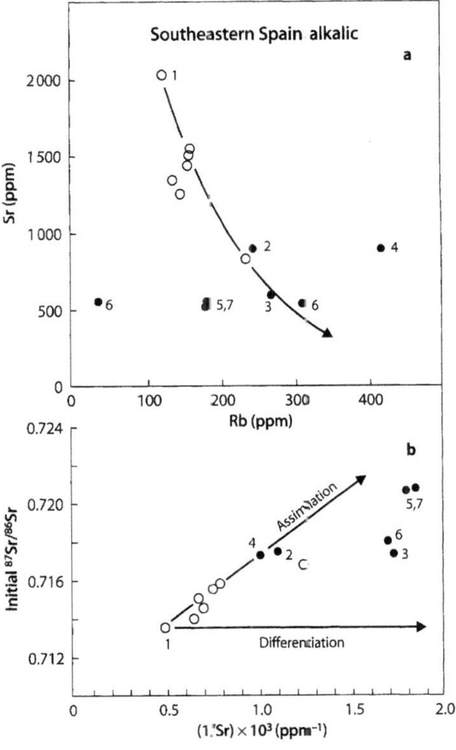

Fig. 6.26. a Rb and Sr concentrations of lamproites in southeastern Spain: *1.* Jumilla; *2.* Cancarix; *3.* Calasparra; *4.* Mula; *5.* Zeneta; *6.* Fortuna; *7.* Aljorra. The curved trajectory is attributable to fractional crystallization of a common magma that produced the rocks at Jumilla, Cancarix, Calasparra, and Fortuna. The relation of the rocks at Mula, Zeneta, and Aljorra to the group that form the Rb-Sr trajectory is not clear; **b** Strontium-isotope mixing diagram for the same rock specimens shown in Part *a*. The distribution of data points in this diagram, combined with the evidence for magmatic differentiation in Part *a*, suggests that these rocks could have formed either from magmas that assimilated crustal rocks as they differentiated or from magmas that originated by partial melting of subducted sedimentary rocks in the lithospheric mantle (*Sources:* data from Powell and Bell 1970; Nelson et al. 1986)

encia along the Mediterranean coast of southeastern Spain (Fig. 6.25). These alkali-rich rocks intruded sedimentary rocks of Late Tertiary age and crystallized between 8.6 and 7.2 Ma, based on K-Ar dates of phlogopite and K-feldspar reported by Nobel et al. (1981). The Betic Cordillera is known to have formed as a result of tectonic activity associated with the collision of the African and Eurasian Plates. However, the orientation of the subduction zone in this area is the subject of conflicting interpretations by Araña and Vegas (1974) and by Torres-Roldan (1979).

The alkali-rich igneous rocks occur at different places in a fairly large area along the eastern side of the Betic Cordillera near the towns of Jumilla, Cancarix, Calasparra, Fortuna, Zeneta, Mula, Aljorra, Mazzaron, and Vera identified in Fig. 6.25. The mineral composition of the dikes and plutons varies from place to place, but olivine, phlogopite, apatite, and sanidine are always present (Venturelli et al. 1984). In addition, the rocks contain a variety of xenoliths of crustal rocks (e.g. mica schist, granitic rocks, limestone, and xenocrysts of quartz, K-feldspar, plagioclase, kyanite, and andalusite).

The $^{87}Sr/^{86}Sr$ ratios of the K-rich rocks at Jumilla were first determined by Powell and Bell (1970) who reported very high values between 0.7134 and 0.7156 (relative to 0.7080 for E&A) with Sr concentrations between 1 278 and 2 039 ppm. Therefore, the alkali-rich rocks at Jumilla in Spain have even higher $^{87}Sr/^{86}Sr$ ra-

tios than the lavas in the Roman region of Italy. The high $^{87}Sr/^{86}Sr$ ratios of the alkali-rich rocks at Jumilla were later confirmed by Nelson et al. (1986) who also analyzed rocks from other locations in this area.

The scattered occurrence and relatively small volume of the alkali-rich rocks in southeastern Spain raises questions about their origin from a common source and about evidence of magmatic differentiation by fractional crystallization. The data in Fig. 6.26a demonstrate that rock specimens from Jumilla (1), Cancarix (2), Calasparra (3), and Fortuna (6) form an array in coordinates of Rb and Sr concentrations that is attributable to fractional crystallization of a single volume of magma from which the rocks at these sites evolved. The rocks at Mula (4), Zeneta (5), and Aljorra (7) deviate from the Rb-Sr array and may have originated from separate parent magmas.

The wide range and elevated values of the $^{87}Sr/^{86}Sr$ ratios reported by Powell and Bell (1970) and by Nelson et al. (1986) require that the magmas either assimilated crustal rocks as they differentiated, or that the magmas formed by partial melting of heterogeneous mixtures of subducted sediment and ultramafic rocks in the lithospheric mantle. The similarity of the tectonic setting of southeastern Spain to that of southern Italy favors the conclusion that the alkali-rich rocks of southern Spain originated by partial melting of subducted sediment in the lithospheric mantle. Alternatively, the distribution of data points in Fig. 6.26b is also explainable by alkali-rich magma of Jumilla assimilating crustal rocks having elevated $^{87}Sr/^{86}Sr$ ratios (>0.720) and low Sr concentrations accompanied by fractional crystallization (AFC).

6.5.3 Southeastern Spain (Calc-alkaline Lavas)

The alkali-rich igneous rocks of southeastern Spain are closely associated in time and space with calc-alkaline lavas many of which contain cordierite. The calc-alkaline rocks (15.5 to 6.6 Ma) consist primarily of dacite and andesite lavas and occur at locations close to the Mediterranean coast of southeastern Spain.

The $^{87}Sr/^{86}Sr$ ratios of the lavas range widely from 0.7095 to 0.7171 relative to 0.7080 for E&A (Munksgaard 1984). The Sr isotope ratios of the lavas at each of four localities in Fig. 6.27 are positively correlated with their reciprocal Sr concentrations indicating that the magmas either assimilated a variety of crustal rocks or that they mixed with crustal magmas. However, even the lowest recorded $^{87}Sr/^{86}Sr$ ratio (0.7095 at Vera) is far above values associated with the lithospheric mantle. In this regard, the calc-alkaline lavas of southeastern Spain resemble those of Tuscany in Italy whose $^{87}Sr/^{86}Sr$ ratios range from 0.71184 to 0.72548 relative to E&A = 0.70805 (Hawkesworth and Vollmer 1979).

Munksgaard (1984) also reported that the $\delta^{18}O$ values of the calc-alkaline lavas of southeastern Spain range from +12.2 to +16.0‰. These data demonstrate that the calc-alkaline lavas of southeastern Spain are enriched in ^{18}O relative to mantle-derived basalts whose $\delta^{18}O$ values are strongly clustered between +5.0 and +6.0‰. The elevated $\delta^{18}O$ values and $^{87}Sr/^{86}Sr$ ratios of the calc-alkaline lavas indicate that the magmas from which they formed contained a significant amount of a crustal component. The involvement of crustal rocks in the petrogenesis of the calc-alkaline lavas is also indicated by the correlations of their $^{87}Sr/^{86}Sr$ and

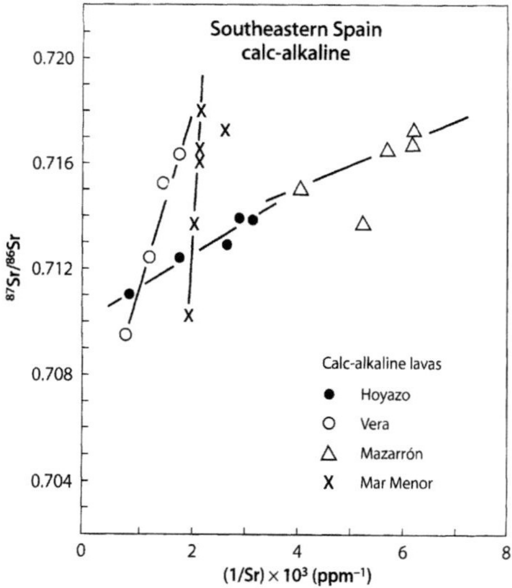

Fig. 6.27. Linear correlations of $^{87}Sr/^{86}Sr$ and 1/Sr ratios of calc-alkaline lavas (15.5 to 6.6 Ma) of southeastern Spain. The high $^{87}Sr/^{86}Sr$ ratios (0.7095 to 0.7171) of these rocks as well as their elevated $\delta^{18}O$ values (+12.2 to +16.0‰) indicate that these rocks formed from mixtures of crustal magmas (*Source:* data from Munksgaard 1984)

$^{87}Rb/^{86}Sr$ ratios reported by Munksgaard (1984). The resulting mixing lines yielded fictitious Rb-Sr dates between 173 ±23 Ma (Mar Menor) and 535 ±22 Ma (Vera).

6.5.4 Monchique Complex, Southern Portugal

In contrast to the small plutons and dikes of alkali-rich rocks of southeastern Spain, the Monchique complex of southern Portugal is a large differentiated intrusive composed of alkali-rich gabbros (berondrites), nepheline syenites, pulaskites, and peralkaline trachytes. In addition, the Monchique complex is significantly older than the alkali-rich rocks of southeastern Spain and Italy, judging on the basis of a whole-rock Rb-Sr date of 76 Ma reported by Rock (1976). The Monchique complex is the type locality for monchiquite, a lamprophyre composed of alkali pyroxene or amphibole, olivine, and biotite, but lacking feldspar. In addition, foyaites consisting of alkalic pyroxene (aegirine) and nepheline enclosed in Na-orthoclase or perthite were first described from the Monchique complex (Williams et al. 1955).

The Rb and Sr concentrations of the rocks in the Monchique complex range widely and are negatively correlated as expected for a differentiated alkali gabbro intrusive (not shown). The most Rb-poor and Sr-rich rocks are the "berondrites" (Rb = 60 ppm,

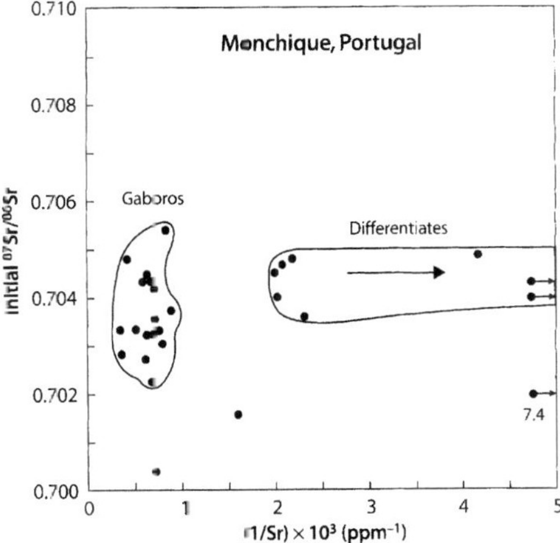

Fig. 6.28. Initial ^{8}Sr/^{86}Sr ratios (at 76 Ma) and reciprocal Sr concentrations of the alkali gabbros and related magmatic differentiates of the Monchique complex in southern Portugal. The gabbroic rocks have higher Sr concentrations (Sr > 1 000 ppm) than the differentiates (Sr < 500 ppm), but the two suites have similar initial ^{87}Sr/^{86}Sr ratios ranging primarily from 0.7030 to 0.7045. These results permit the conclusion that the alkali-rich gabbroic magma originated from the lithospheric mantle and differentiated by fractional crystallization without significant input from crustal rocks (*Source:* Rock 1976)

Sr = 1 840 ppm, whereas nordmarkites (quartz-bearing alkali syenite) are most enriched in Rb (293 ppm) and most depleted in Sr (23 ppm) among the rock types of this complex (Rock 1976). The country rock surrounding the Monchique complex was fenitized by alkali-rich fluids and has high Rb concentrations. One specimen of fenite analyzed by Rock (1976) contained 314 ppm Rb and 176 ppm Sr, but had preserved a high initial ^{87}Sr/^{86}Sr ratio of 0.7161 at 75 Ma. In this regard, the fenites in the Monchique area resemble those of eastern Uganda (Sect. 6.2.6) described by Bell and Powell (1970).

The alkali gabbros with Sr concentrations of more than 1 000 ppm have a range of initial ^{87}Sr/^{86}Sr ratios from 0.7022 to 0.7054 and form a cluster of data points in Fig. 6.28. The magmatic differentiates (Sr < 500 ppm) form a separate cluster with initial ^{87}Sr/^{86}Sr ratios between 0.7035 and 0.7049. The conclusions to be drawn from these data are:

1. The initial ^{87}Sr/^{86}Sr ratios of the Monchique complex are significantly lower than those of the alkali-rich rocks of southeastern Spain.
2. More than 60% of the analyzed samples of gabbro and their magmatic differentiates have ^{87}Sr/^{86}Sr ratios between 0.7030 and 0.7045.
3. The alkali-rich magmas of the Monchique complex originated by partial melting in the lithospheric mantle without significant assimilation of crustal rocks.

6.6 Alkali-rich Rocks of Central Europe

The continental crust of western and central Europe in Fig. 6.29 contains three terranes which are separated from each other by large east-west trending thrust faults that dip south. The principal terranes that form the crust of Europe are (from north to south): (1) The Rhenohercynian; (2) the Saxothuringian; and (3) the Moldanubian. The thrust faults, which define these terranes, cut several large massifs of igneous rocks of Variscan age between 360 and 260 Ma. These Variscan massifs of Europe include (from west to east): (1) the Armorican Massif of Brittany; (2) the Massif Central of central France; (3) the Rhenish Massif of western Germany, and the Bohemian Massif of the Czech Republic and southern Poland (Wilson and Downes 1991).

The subsequent formation of the Alps during the Tertiary Period was accompanied by crustal rifting in the Rhone and Bresse Depressions of France and in the Upper Rhine Valley in Germany. The tectonic activity was accompanied by extensive intraplate volcanic activity in the Massif Central of France, in central Germany, in the Czech Republic, and in southwest Poland. In addition, calc-alkaline volcanic rocks of Eocene and Miocene age were erupted in the Carpathian Mountains of Hungary.

The volcanic activity of central Europe is attributable to a variety of causes resulting from the complex tectonic history of this region. In many cases, magma formed by decompression melting of the lithospheric mantle which had been variously altered by alkali-rich

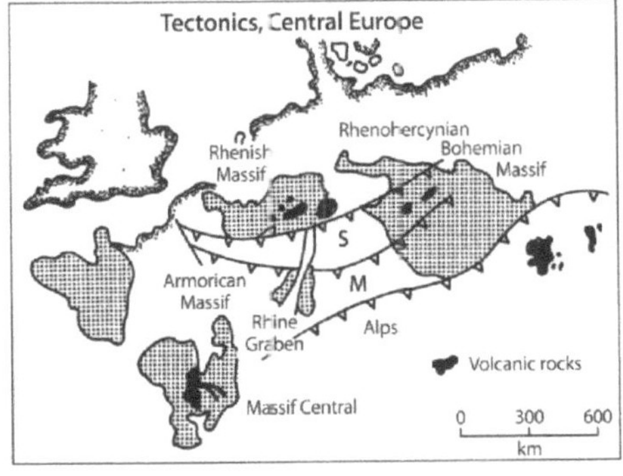

Fig. 6.29. Simplified tectonic map of western and central Europe identifying the major terranes: *RH* = Rhenohercynian; *S* = Saxothuringian; *M* = Moldanubian. The Variscan Massifs (360 to 260 Ma) are identified by name: *1.* Armorican; *2.* Massif Central; *3.* Rhenish; *4.* Bohemian. The principal volcanic centers (Tertiary to Quaternary) are shown in *black* (*Source:* adapted from Wilson and Downes 1991)

aqueous fluids emanating from previously subducted oceanic crust. In addition, weak asthenospheric plumes may have contributed magmas by decompression melting either at the base of the lithospheric mantle or after intruding along deep fractures. The magmas derived from such mantle sources mixed with each other in some cases and/or assimilated wallrocks as they differentiated by fractional crystallization in crustal magma chambers.

6.6.1 Massif Central, France

The close association of alkali-rich volcanic rocks with silica-saturated lavas, described previously from southeastern Spain and Tuscany in Italy, also occurs in the Tertiary volcanoes of the Massif Central of France. The volcano Cantal, located about 100 km southwest of the city of Clermont Ferrand, erupted from 10 to 2.5 Ma. The lavas consist of silica-saturated alkali olivine basalts, trachyandesites, trachytes, and rhyolites and strongly alkali-rich nepheline-normative basanites, tephrites, and phonolites. In addition, the Tertiary lavas of Cantal and elsewhere in the Massif Central contain xenoliths derived from the mantle (peridotite) and the continental crust (granulite, granite gneiss, and granite).

The lavas of both suites on Cantal in Fig. 6.30 are highly differentiated with Sr concentrations initially rising to about 1 000 ppm and then decreasing to <100 ppm as the Rb concentrations increase from <50 to >250 ppm. The $^{87}Sr/^{86}Sr$ ratios of the lavas on Cantal Volcano reported by Downes (1984) range from 0.70345 to 0.70804 relative to 0.71025 for NBS 987 and are in agreement with the results of Stettler and Allègre (1979). The latter demonstrated that the $^{87}Sr/^{86}Sr$ and 1/Sr ratios of the lavas on Cantal form a wedge-shaped array on the Sr-isotope mixing diagram (not shown). Stettler and Allègre (1979) concluded from this and other evidence that both fractional crystallization and assimilation of crustal rocks contributed to the formation of the lavas of Cantal.

The petrogenesis of these rocks is most effectively represented on the Sr-Nd isotope mixing diagram in Fig. 6.31 based on the data of Downes (1984) and Chauvel and Jahn (1984). On this diagram, the data points representing saturated lavas (solid circles) and the alkali-rich rocks (open circles) form a linear array that extends from the mantle quadrant toward the crustal quadrant, occupied in this case by lower-crustal granulite xenoliths (crosses) having average $^{87}Sr/^{86}Sr$ and $^{143}Nd/^{144}Nd$ ratios of 0.70975 and 0.51229, respectively (Downes 1984). The coordinates of the mantle-derived component (basalt and basanite) are 0.70350 and 0.51293. The good fit of the data points to the same mixing line indicates that the calc-alkaline and alkalic magmas originated from the same magma sources in the lithospheric mantle and assimilated

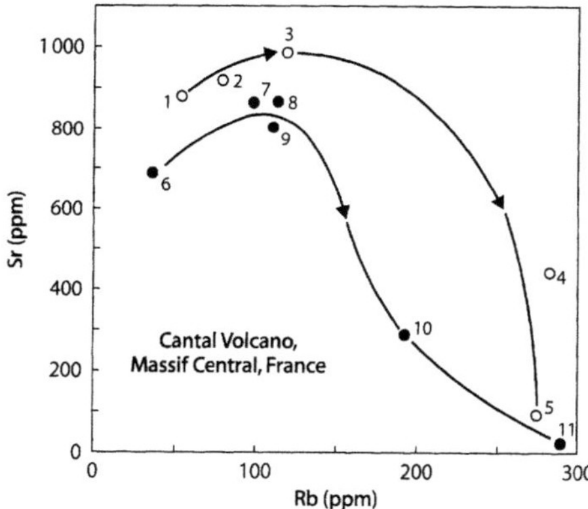

Fig. 6.30. Variation of average Rb and Sr concentrations of the undersaturated alkali-rich lavas (*open circles*) and the saturated lavas (*solid circles*) on the shield volcano Cantal of the Massif Central, France. The alkali-rich suite of lavas includes: *1*. basanite; *2*. nephelinite; *3*. tephrite; *4*. trachyte; *5*. phonolite. The mildly alkaline saturated suite contains: *6*. basalt; *7*. gabbro; *8*. andesite; *9*. trachyandesite; *10*. latite; *11*. rhyolite. The pattern of variation of the Rb and Sr concentrations is caused by fractional crystallization of the respective magmas (*Sources:* Downes 1984; Chauvel and Jahn 1984; Stettler and Allègre 1979)

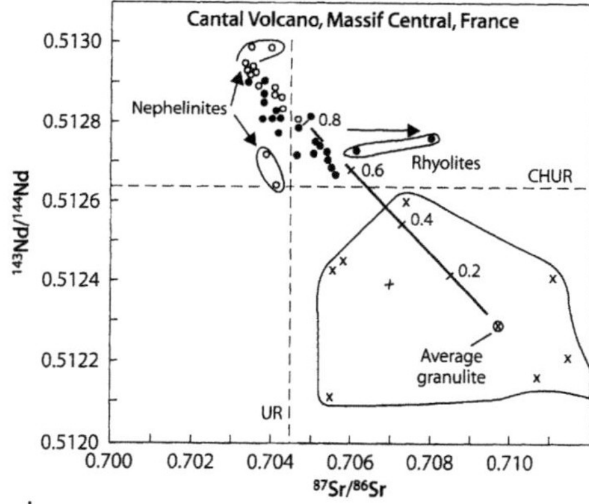

Fig. 6.31. Sr-Nd isotope mixing diagram for silica saturated volcanic rocks (*solid circles*) and alkali-rich lavas (*open circles*) on the volcano Cantal in the Massif Central of France. The data points representing both suites of samples form a linear array and could have formed by mixing of mantle-derived basalt and basanite magma with average lower-crustal granulite (*crosses*) represented by xenoliths in the lava flows. The $^{87}Sr/^{86}Sr$ ratios of two rhyolites were increased by addition of contaminant Sr from aqueous solutions (*Source:* Downes 1984; Chauvel and Jahn 1984)

similar rocks of the continental crust. According to the mixing model in Fig. 6.31, the saturated magmas assimilated up to about 40% of deep-crustal granulite, whereas the alkali-rich magmas assimilated less than about 20%. In addition, both magmas differentiated by fractional crystallization causing the systematic variation of Rb and Sr concentrations of the lavas in Fig. 6.30 Two rhyolites identified in Fig. 6.31 deviate from the mixing array because they have elevated $^{87}Sr/^{86}Sr$ ratios (0.70614 to 0.70804), but their $^{143}Nd/^{144}Nd$ ratios appear to be unaffected. The increase of the $^{87}Sr/^{86}Sr$ ratios was caused by the alteration of these rocks by heated groundwater containing Sr derived from the granitic basement rocks (Stettler 1977).

Nephelinites analyzed by Downes (1984) and Chauvel and Jahn (1984) also deviate from the mixing array. Downes (1984) concluded that the nephelinites originated from a different magma source in the underlying mantle. The evidence for heterogeneity of the mantle under the Massif Central is in agreement with chemical, mineralogical, and isotopic studies of ultramafic xenoliths from the Massif Central and elsewhere in France by Leggo and Hutchison (1968), Hutchison et al. (1975), Brown et al. (1980), Dostal et al. (1980), Downes, (1984) and Zangana et al. (1999).

6.6.2 Central Germany

Alkali-rich volcanic rocks were erupted during the Tertiary Period in a broad belt that extends across Germany from the West Eifel Mountains eastward across the Rhine Valley to the Siebengebirge, Westerwald, Vogelsberg, Rhön, and Heldburger Gangschar, and continues through Doubrovske Hory and Ceske Stredhori (Czech Republic) all of which are identified in Fig. 6.32. In addition, volcanic activity occurred in the northern Hessian Depression and in the Upper Rhine Valley (Harmon et al. 1987). The ages of the lavas that erupted at these centers range from Eocene (e.g. Pfalz, 50 Ma) to Miocene (e.g. Siebengebirge, Westerwald, Vogelsberg, northern Hessian Depression, Rhön, and Heldburger Gangschar) and to Pleistocene (e.g. Eifel and Westerwald) (Lippolt 1983). The lavas are strongly silica undersaturated and contain leucite, nepheline, melilite, and hauyne ($Na_{5\,to\,8}\,Ca_{0\,to\,2}\,K_{0\,to\,1}[(AlO_2)_6\,(SiO_2)_6]\,(SO_4)_{1\,to\,2}$). In addition, tholeiite basalts have been reported from Vogelsberg and the northern Hessian Depression.

The $^{87}Sr/^{86}Sr$ and $^{143}Nd/^{144}Nd$ ratios reported by Wörner et al. (1986) range from 0.7033 to 0.7047 and from 0.51260 to 0.51290, respectively. The distribution of data points in Fig. 6.33 reveals that the Sr and Nd in these rocks originated from two or more sources and that one of these was the EM1 component. Consequently, most of the alkali-rich lavas of central Germany originated from sources in the subcrustal lithospheric mantle without significant contamination by crustal rocks. Two granulite xenoliths from the Eifel Mountains analyzed by Wörner et al (1985) have low $^{143}Nd/^{144}Nd$ and elevated $^{87}Sr/^{86}Sr$ ratios compared to most of the lavas except for the phonolite tephra of the Laacher See in the East Eifel Mountains which have elevated $^{87}Sr/^{86}Sr$ ratios that range up to 0 71122 relative to 0.7080 for E&A.

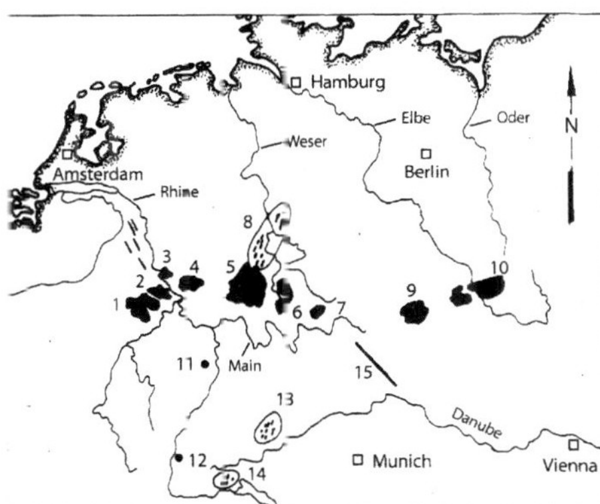

Fig. 6.32. Alkali-rich volcanic rocks of Tertiary age in central Germany and the Czech Republic. *1.* West Eifel; *2.* East Eifel (Laacher See); *3.* Siebengebirge; *4.* Westerwald; *5.* Vogelsberg; *6.* Rhön; *7.* Heldburger Gangschar; *8.* Northern Hessian Depression; *9.* Doubrovske Hory; *10.* Ceske Stredhori; *11.* Pfalz; *12.* Kaiserstuhl; *13.* Urach; *14.* Fegau; *15.* Oberpfalz (*Source:* adapted from Wörner et al. 1986)

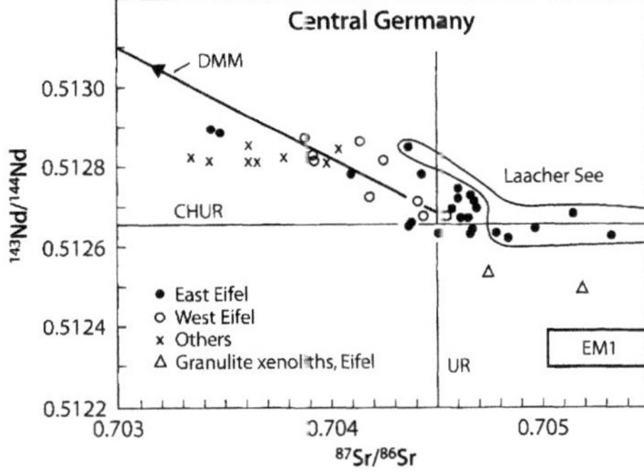

Fig. 6.33. Sr-Nd isotope mixing diagram for alkali-rich volcanic rocks of Tertiary age in the East and West Eifel (including Laacher See) and at other locations in central Germany. The magmas originated from the subcrustal lithospheric mantle and were not significantly contaminated by assimilating crustal rocks, except for highly differentiated lavas and pyroclastic deposits at Laacher See in the East Eifel Mountains (*Source:* data from Wörner et al. 1985, 1986)

6.6.3 Laacher See, East Eifel

The East Eifel Mountains contain several centers of volcanic activity including primarily the Laacher See, the Rieden complex, Schellkopf (Fuhrman and Lippolt 1985), the Dreiser Weiher (Paul 1971), and the Wehr Volcano (Wörner et al. 1982). The phonolite tephra deposits of the Laacher See volcanic field were studied by Wörner et al. (1985), Wörner and Schmincke (1984a,b), Wörner and Wright (1984), and Wörner et al. (1983), among others.

The evidence indicates that the alkali-rich rocks in the Laacher See volcanic field were erupted from a crustal magma chamber, which was zoned such that the most highly differentiated magma was at the top of the chamber with the denser and less differentiated magma near the bottom. The zonation of the magma chamber was inverted during eruption because magma from the top of the chamber was erupted first, followed by magma from deeper levels, and ending with the least differentiated magma which formed the tephra layers at the top of the section.

The Sr concentrations of the phonolite tephra at Laacher See reported by Wörner et al. (1985) range from about 300 ppm at the bottom of the magma chamber to only 4 ppm near the top. The low Sr concentration of the differentiated magma caused the isotopic composition of Sr in the magma to be susceptible to alteration by the introduction of Sr from the country rocks. As a result, the $^{87}Sr/^{86}Sr$ ratios of the magma increased from 0.70436 near the bottom to 0.71122 at the top of the magma chamber relative to 0.7080 for E&A, whereas the $^{143}Nd/^{144}Nd$ ratios of the phonolite tephra were largely unchanged (Fig. 6.33).

The $^{87}Sr/^{86}Sr$ ratios of the phenocryst minerals in the phonolite tephra in Fig. 6.34 indicate the sequence

in which they crystallized from a magma that was assimilating Sr from the country rock. Sanidine started to form when the $^{87}Sr/^{86}Sr$ ratio of the magma was 0.70472 followed closely by hauyne at 0.70475 and plagioclase at 0.70476. The crystallization of "feldspar", clinopyroxene, and sphene was confined to the period when the $^{87}Sr/^{86}Sr$ ratio of the magma increased from 0.70481 to 0.70498, whereas sanidine and hauyne continued to crystallize until the $^{87}Sr/^{86}Sr$ ratio of the magma had increased to 0.70514. The remaining magma ultimately formed the matrix as the $^{87}Sr/^{86}Sr$ rose from 0.70486 to 0.70754.

The addition of crustal Sr to the parental magma of the Laacher See Volcano is demonstrated in Fig. 6.35 by the method of Lutz et al. (1988) based on the assumption that the initial $^{87}Sr/^{86}Sr$ ratio of the mantle-derived magma was 0.70436. Xenoliths of Devonian slate and mica schist analyzed by Wörner et al. (1985) have an average $^{87}Sr/^{86}Sr$ ratio of 0.747 ±0.002 which is taken to be representative of the Sr added to the magma. Fig. 6.35 indicates that the Sr concentration of sample 1 017 increased by only 1.7 µg g⁻¹ from 33.3 to 35.0 µg g⁻¹, which raised its $^{87}Sr/^{86}Sr$ ratio from 0.70436 to 0.70673 ±0.00004. The crustal Sr apparently resides on grain boundaries because the matrix and phenocrysts of this rock all have lower $^{87}Sr/^{86}Sr$ ratios than the whole-rock sample (0.70486 to 0.70533). Wörner et al. (1985) proposed a three-stage petrogenesis process for the alkalic rocks at Laacher See, including frac-

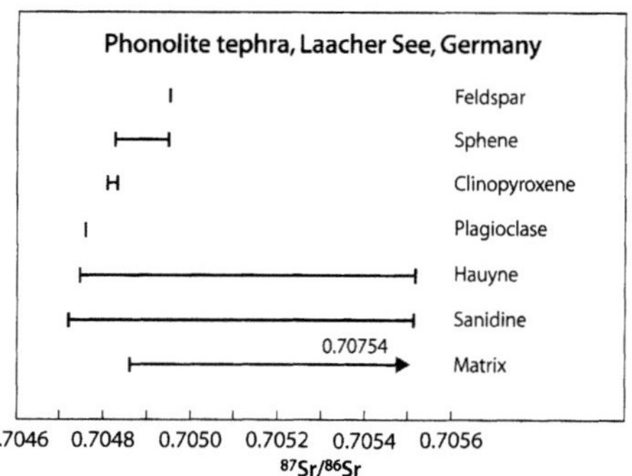

Fig. 6.34. Sequence and duration of crystallization of phenocryst minerals in the phonolite tephra deposits, Laacher See, Germany (*Source:* Wörner et al. 1985)

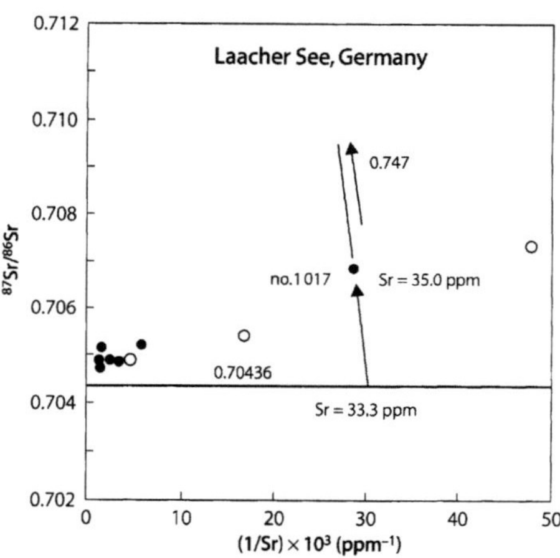

Fig. 6.35. $^{87}Sr/^{86}Sr$ ratios and reciprocal Sr concentrations of whole-rock samples (*solid circles*) and matrix (*open circles*) of phonolite tephra at Laacher See, Eifel Mountains, Germany. The diagram illustrates the effect of adding Sr from the country rock ($^{87}Sr/^{86}Sr = 0.747$ ±0.002) to the magma of sample 1 017 whose Sr concentration increased by 1.7 µg g⁻¹ from 33.3 to 35.0 µg g⁻¹ while its $^{87}Sr/^{86}Sr$ ratio rose from 0.70436 to 0.70673 (*Source:* Wörner et al. 1985)

tional crystallization of a mantle-derived basanite magma accompanied by assimilation of crustal rocks on a small scale.

6.6.4 Xenoliths, West Eifel Mountains

The lavas in the Eifel Mountains of Germany contain ultramafic inclusions of different origin such as: (1) Crystal cumulates derived from the host lava; (2) Residual mantle rocks after partial melting to form magma; and (3) Accidental inclusions entrained in the magma during its ascent to the surface. The isotope ratios of Sr, Nd, and Pb of crystal cumulates and residues of partial melting should be identical to those of the host lava, unless the inclusions were contaminated during its ascent to the surface or after eruption. Ultramafic inclusions generally have low Sr concentrations regardless of their origin, making their $^{87}Sr/^{86}Sr$ ratios susceptible to change by addition of Sr from other sources.

Ultramafic inclusions at Dreiser Weiher (Pleistocene) analyzed by Paul (1971) have a wide range of $^{87}Sr/^{86}Sr$ ratios from 0.7028 to 0.7067 compared to 0.7035 for the host basalt relative to 0.7080 for E&A. The average $^{87}Sr/^{86}Sr$ ratio of wehrlite and clinopyroxenite inclusions (0.70355 ±0.00054, $2\bar{\sigma}$, N = 4) is similar to that of the host basalt (0.70355) which means that they are of cognate origins. The $^{87}Sr/^{86}Sr$ ratios of lherzolite inclusions are higher than those of the host

basalt and range widely from 0.7039 to 0.7067. Therefore, the lherzolites are accidental inclusions. The inclusion featured in Fig. 6.36 is noteworthy because it has a high whole-rock $^{87}Sr/^{86}Sr$ ratio of 0.7067 relative to 0.7080 for E&A. The constituent minerals of this inclusion have widely different $^{87}Sr/^{86}Sr$ ratios ranging from 0.7022 (clinopyroxene) to 0.7095 (olivine). The isotope composition of Sr in DW15 and its constituent minerals is explainable by the addition of Sr derived from crustal rocks. The process is illustrated in Fig. 6.36 assuming that the contaminant Sr had an $^{87}Sr/^{86}Sr$ ratio of 0.7200 and that the original $^{87}Sr/^{86}Sr$ of this inclusion was preserved by the clinopyroxene (0.7022; Sr = 69.88 ppm). The results are summarized in Table 6.2.

The contaminant Sr could have been added to this lherzolite xenolith by an aqueous fluid either before or after it was transported to the surface by the ascending basalt magma.

The work of Paul (1971) on the mantle xenoliths of Dreiser Weiher was later extended by Stosch et al. (1980), Stosch and Seck (1980), and Stosch and Lugmair (1986). The latter reported that the $^{87}Sr/^{86}Sr$ ratios of clinopyroxenes in anhydrous lherzolite xenoliths range from 0.702297 to 0.704371, thereby confirming the low $^{87}Sr/^{86}Sr$ ratio of the clinopyroxene (0.7022) in sample DW15 reported by Paul (1971). A second suite of hydrous lherzolite and wehrlite inclusions analyzed by Stosch and Lugmair (1986) is characterized by the presence of amphiboles and phlogopite. These minerals are strongly enriched in Rb as well as Sr compared to clinopyroxene in the anhydrous inclusions. Nevertheless, the $^{87}Sr/^{86}Sr$ ratios of amphiboles and phlogopite range only from 0.703518 (amphibole) to 0.703704 (phlogopite) and are in general agreement with the $^{87}Sr/^{86}Sr$ ratios of the host basalt (Paul 1971) and basanite Stosch and Lugmair (1986). The occurrence of Rb-rich minerals in the hydrous xenoliths indicates that some parts of the mantle under the Eifel Mountains were altered by alkali-rich aqueous fluids (Witt-Eickschen et al. 1998). The presence of Rb-rich amphibole and phlogopite in the magma sources may have contributed to the alkali-enrichment of the magmas that were erupted in this area.

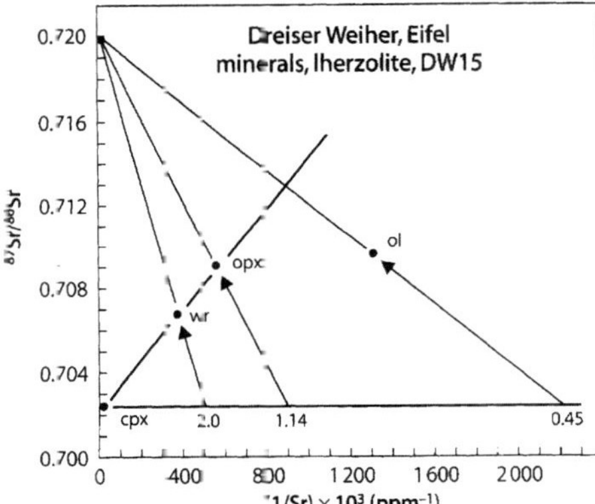

Fig. 6.36. Reconstruction of the contamination of a lherzolite inclusion (DW15) in basalt lava at Dreiser Weiher, and Eifel by Sr derived from crustal sources based on the method of Lutz et al. (1988). The $^{87}Sr/^{86}Sr$ ratio of the contaminant Sr is assumed to be 0.720 and the clinopyroxene is assumed to have preserved the original $^{87}Sr/^{86}Sr$ ratio of this rock (0.7022). The reconstruction yields the following estimates of the original Sr concentrations of the whole rock and of its constituent minerals: whole rock: 2.0 ppm; orthopyroxene: 1.14 ppm; olivine 0.45 ppm (Source: Paul 1971)

Table 6.2. Measured and original Sr concentration in Lherzolite DW15

Lherzolite DW15	Sr (ppm) measured	original	Increase (%)
Whole rock	2.64	2.0	32
Olivine	0.76	0.45	69
Orthopyroxene	1.79	1.14	57
Clinopyroxene	69.88	69.88	0

6.6.5 Northern Hessian Depression

The igneous rocks of the northern Hessian Depression occur in the form of dikes and lava flows at several hundred scattered locations. The rocks were classified by Wedepohl (1985) as: (1) quartz tholeiites; (2) alkali olivine basalts; and (3) nepheline basanites, limburgites, and olivine nephelinites. The alkali olivine basalts predominate with 73% of all outcrops studied by Wedepohl (1985). The volcanic activity started at 20 Ma with the extrusion of the quartz tholeiites, reached a climax between 13 and 12 Ma, and ended at 8 Ma with the eruption of the nepheline-bearing volcanic rocks. The isotope compositions of O, H, and S of the volcanic rocks in the northern Hessian Depression were determined and interpreted by Harmon et al. (1987).

The $^{87}Sr/^{86}Sr$ ratios of the tholeiites and alkalic rocks of this area range from 0.70330 (nepheline basanite) to 0.70432 (tholeiite) relative to 0.71025 for NBS 987. The tholeiites form a linear array in coordinates of $^{87}Sr/^{86}Sr$ and 1/Sr ratios of Fig. 6.37, whereas the alkali-rich rocks form a separate cluster of data points. These results clearly indicate that the tholeiite magmas assimilated crustal rocks whereas the alkali-rich magmas did not. Therefore, the data in Fig. 6.37 are in agreement with conclusions of Wedepohl (1985), Mengel et al. (1984), and Harmon et al. (1987) that the alkali-rich magmas formed by partial melting of metasomatically altered spinel peridotites and were erupted rapidly without opportunity for fractional crystallization and assimilation of crustal rocks.

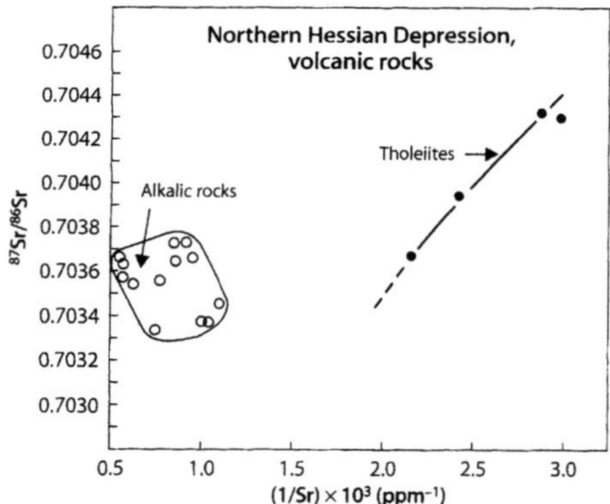

Fig. 6.37. Sr-isotope mixing diagram of tholeiites and alkali-rich rocks of Tertiary volcanic rocks at scattered localities in the northern Hessian Depression of Germany. The collinearity of the tholeiites is a sign of crustal contamination of the tholeiite magma. The alkali-rich rocks show little evidence of crustal contamination or fractional crystallization and formed by partial melting of metasomatized mantle rocks (*Source:* Mengel et al. 1984)

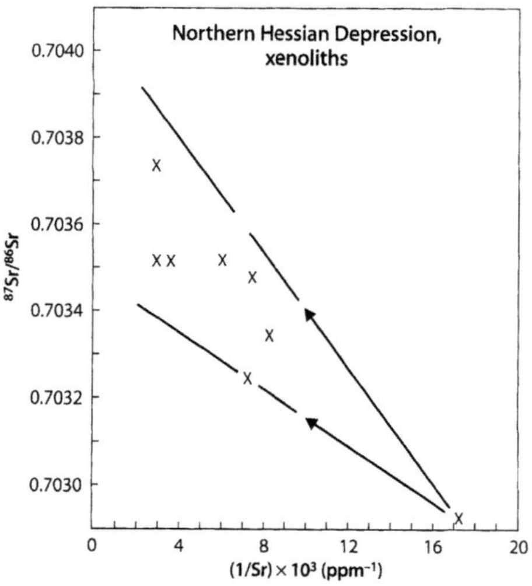

Fig. 6.38. $^{87}Sr/^{86}Sr$ ratios and reciprocal Sr concentrations of acid-leached clinopyroxene separates of peridotite xenoliths that occur in the Tertiary alkali-rich lavas of the northern Hessian Depression of Germany. The interpretation of the data is that both the $^{87}Sr/^{86}Sr$ ratios and Sr concentrations of the clinopyroxenes were increased from 0.70292 (Sr = 58 ppm) by Sr-rich metasomatic fluids with $^{87}Sr/^{86}Sr$ ratios between about 0.7034 and 0.7039 relative to 0.71025 for NBS 987 (*Source:* Mengel et al.1984)

The alkali-rich lava flows of the northern Hessian Depression contain xenoliths composed of spinel lherzolites and spinel harzburgites. The $^{87}Sr/^{86}Sr$ ratios of whole-rock xenoliths range from 0.70343 to 0.70402 relative to 0.71025 for NBS 987 (Mengel et al. 1984) and therefore are similar to those of the alkali-rich lavas in which they occur. The $^{87}Sr/^{86}Sr$ ratios of acid-leached clinopyroxene separates increase from 0.70292 (Sr = 58 ppm) to 0.70373 (Sr = 363 ppm), indicating that the metasomatic fluid raised both their Sr concentrations and $^{87}Sr/^{86}Sr$ ratios. The data points representing the acid-washed clinopyroxenes in Fig. 6.38 form a fan-shaped array in agreement with the conclusion of Mengel et al. (1984) that the metasomatic fluid consisted of two components characterized by different $^{87}Sr/^{86}Sr$ ratios of about 0.7034 and 0.7039. In addition, the clinopyroxenes had an $^{87}Sr/^{86}Sr$ ratio of 0.70292 ±0.00004 prior to metasomatism relative to 0.71025 for NBS 987.

6.6.6 Upper Rhine Graben

The Upper Rhine Graben encompasses an area that is about 35 km wide and 300 km long located between the cities of Mainz in the north and Basel in the south. It is flanked by the Vosges Mountains and the Pfalz in the

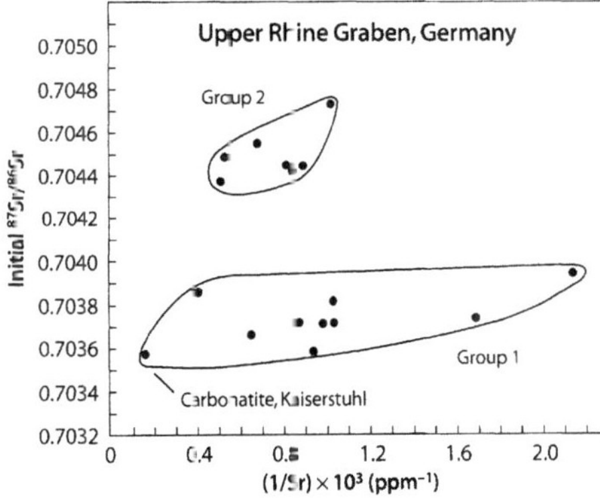

Fig. 6.39. Average initial $^{87}Sr/^{86}Sr$ ratios and reciprocal Sr concentrations of alkali-rich volcanic rocks from different locations within the Upper Rhine Graben. The Group 1 lavas (tholeiites, alkali olivine basalt, basanite, olivine nephelinite, olivine melilitite, limburgite, plagioclase basalt, shonkinite, and alkali syenite) are characterized by having $^{87}Sr/^{86}Sr < 0.7040$ averaged by rock type from different localities. The Group 2 lavas (melilite ankaratrite, trachyte, shonkinite porphyry, essexite, tephrite, phonolite, sanidine nephelinite, and nepheline mica porphyry) have average $^{87}Sr/^{86}Sr$ ratios > 0.70430. The distribution of data points in Group 2 is indicative of crustal contamination of the alkali-rich magmas, whereas the Group 1 lavas could have originated from heterogeneous magma sources in the subcrustal mantle without significant crustal interactions (*Sources:* Calvez and Lippolt 1980; Wörner et al. 1986; Hegner et al. 1995; Hegner and Vennemann 1997)

west and by the Black Forest and the Odenwald in the east. The faults that define this rift were reactivated during the Eocene Epoch causing an increase in the intensity of volcanic activity that had started in Late Cretaceous time, peaked during the Eocene, and reached its largest volume during Miocene Epoch. The lavas are alkali-rich and were erupted at several centers including the Kaiserstuhl, Urach, Hegau, and Odenwald, among others. The volcanic rocks include olivine nephelinites, melilite ankaratrites, tephrites, phonolites, and trachytes (Calvez and Lippolt 1980).

The Rb and Sr concentrations of the volcanic and plutonic rocks in this region vary widely and provide evidence for magmatic differentiation by fractional crystallization of the magmas. In general, the olivine nephelinites, alkali olivine basalts, and tholeiites have comparatively low Rb (20–40 ppm) and high Sr (1 200 to 1 400 ppm) concentrations, whereas some phonolites, trachytes, and tephrites are enriched in Rb (150 to 300 ppm) and depleted in Sr (50 to 500 ppm). However, no simple patterns of variation emerge from the data of Calvez and Lippolt (1980) and Wörner et al. (1986), perhaps because of the considerable range in the ages of these rocks (100 to 9 Ma), their occurrence at many localities scattered throughout this region (Fig. 6.45), and the possible heterogeneity of the subcrustal magma sources.

The $^{87}Sr/^{86}Sr$ ratios of the volcanic rocks in the Upper Rhine Graben range from 0.70329 (olivine nephelinite, Kaiserstuhl) to 0.7054 (phonolite, Spessart) relative to 0.7080 for E&A (Calvez and Lippolt 1980; Wörner et al. 1986; Hegner et al. 1995; Hegner and Vennemann 1997). However, the data points representing average $^{87}Sr/^{86}Sr$ ratios and Sr concentrations of the same rock type from different volcanic centers form two clusters in Fig. 6.39. The $^{87}Sr/^{86}Sr$ ratios of the lavas in Group 1 (tholeiites, alkali olivine basalts, basanites, olivine nephelinites, melilitites, limburgites, plagioclase basalts, Na-shonkinites, and alkali syenites) are all less than 0.7040, whereas those of Group 2 (ankaratrites, essexites, tephrites, phonolites, trachytes, sanidine nephelinites, and porphyries) are greater than 0.7043. Calvez and Lippolt (1980) as well as Schleicher et al. (1984) considered that the $^{87}Sr/^{86}Sr$ ratios of the Group 1 lavas reflect the heterogeneity of the isotope composition of Sr in the mantle underlying the Upper Rhine Valley, whereas the elevated $^{87}Sr/^{86}Sr$ ratios of the Group 2 lavas were caused by crustal contamination of differentiated magmas (e.g. tephrites, trachytes, etc.)

6.6.7 Kaiserstuhl

The Kaiserstuhl in Fig. 6.32 is the most widely known volcanic center in the Upper Rhine Graben because of the presence of an intrusive carbonatite in the core of the volcano in addition to evidence for the eruption of carbonatite lavas. The volcanic activity at this location occurred at 13.5 Ma (Miocene) toward the end of the lengthy period of volcanism in the Upper Rhine Graben (100 to 9 Ma).

The lavas include tephrites and phonolites as well as olivine nephelinites and olivine melilitites. The intrusive mass of carbonatite in the center of the volcanic complex is composed of coarse-grained calcite with accessory apatite, magnetite, pyrochlore, and Nb-bearing perovskite. In addition, fine-grained carbonatite dikes are concentrated in the center of the complex. The intrusive carbonatite body has an outcrop area of about one square kilometer and formed near the end of the magmatic evolution of the Kaiserstuhl complex (Keller 1981).

The eruption of alkali-rich volcanic rocks at the Kaiserstuhl was accompanied by the deposition of carbonatite pyroclastics composed of lapilli that range in diameter from 0.5 to 15 mm. The existence of the carbonatite tuff beds indicates that carbonatite lavas were erupted at the Kaiserstuhl, much like those of Oldoinyo Lengai and Kerimasi in Tanzania (Sect. 6.1.3). The lapilli are composed of calcite crystals with magnetite that has elevated concentrations of Mn, Mg, and Al. The Sr concentration of the calcite is more than 4 000 ppm, whereas Rb has concentrations of less than 8 ppm (Keller 1981).

The $^{87}Sr/^{86}Sr$ ratios of the silicate rocks reported by Calvez and Lippolt (1980) Schleicher et al. (1984), Wörner et al. (1986) range from 0.70329 (olivine nephelinite) to 0.70507 (tephrite) relative to E&A = 0.7080 and encompass both Group 1 and Group 2 in Fig. 6.39. The $^{87}Sr/^{86}Sr$ ratios of essexites, phonolites, and tephrites of Group 2 range from 0.7040 to values in excess of 0.7050. Schleicher et al. (1984) reported that the Sr isotope ratios of constituent minerals of these rock suites increase in the order of their crystallization and concluded that these magmas were assimilating crustal rocks as they crystallized. However, the authors were unable to demonstrate significant chemical contamination, suggesting that Sr was selectively transferred from the country rocks into the essexite, phonolite, and tephrite magmas.

The $^{87}Sr/^{86}Sr$ ratios of the carbonatites measured by Calvez and Lippolt (1980) and by Schleicher et al. (1984) have a narrow range from 0.70356 to 0.70405 indicating that the carbonatites belong to Group 1 in Fig. 6.39. A group of rocks known as "bergalites" (calcite-sodalite-melilite-nepheline-perovskite) have similar $^{87}Sr/^{86}Sr$ ratios between 0.7038 to 0.7041. The bergalites appear to be transitional between the silicate and carbonatite rocks and thereby provide support for the formation of the carbonatite magma by liquid immiscibility with the alkali-rich silicate magmas.

6.6.8 Lower Silesia, Southwest Poland

The belt of alkali-rich volcanic rocks of central Germany extends eastward into Lower Silesia in southwest Poland. The rocks occur in the form of lava flows and small intrusives at numerous locations south of the city of Wroclaw and are composed of olivine basalts, basanites, nephelinites, and trachytes. The ages of the volcanic rocks range from Late Cretaceous to late Miocene and indicate that the intensity of the igneous activity peaked during the Eocene and Miocene Epochs.

The variation of the Sr concentrations of the basanites and alkali olivine basalts with rising Rb concentrations in Fig. 6.40a is consistent with the effects of fractional crystallization of the parent magmas (Blusztajn and Hart 1989). The $^{87}Sr/^{86}Sr$ and $^{143}Nd/^{144}Nd$ ratios of all of the rock types form only one array that lies within the mantle quadrant of the Sr-Nd isotope-mixing diagram in Fig. 6.40b. Accordingly, Blusztajn and Hart (1989) attributed the range of isotope ratios of the lavas to mixing of magmas derived from two or more components in the lithospheric mantle (DMM and EM1). The $^{206}Pb/^{204}Pb$ ratios of these rocks range from 19.419 to 19.944 because the Pb originated in part from the HIMU component that takes the form of subducted oceanic crust that underplated the litho-

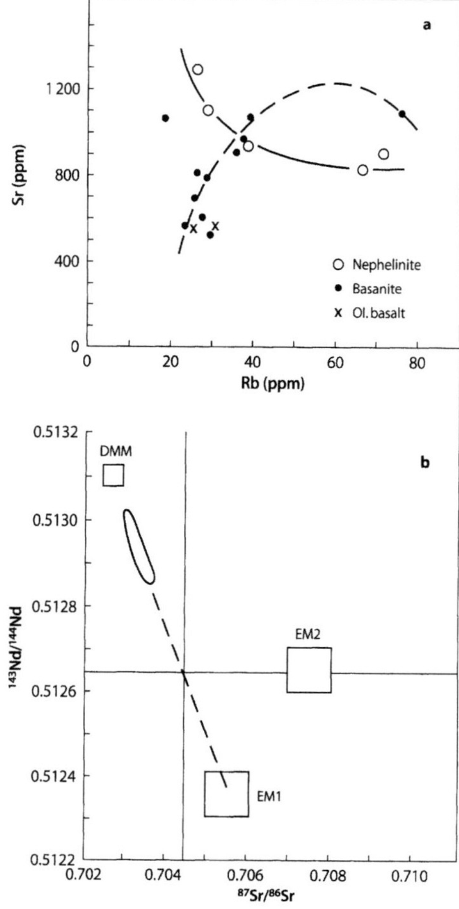

Fig. 6.40. a Variation of the concentrations of Rb and Sr in the alkali-rich volcanic rocks of Tertiary age in Lower Silesia, Poland, caused by fractional crystallization of the basanite and nephelinite magmas; **b** The isotope ratios of Sr and Nd of the basanites and nephelinites form a single two-component mixing array between the DMM and the EM1 components (*Source:* data from Blusztajn and Hart 1989; Hart 1988)

spheric mantle (not shown). There is no evidence for crustal contamination of the magmas even though the Rb and Sr concentrations of the rocks show the effects of fractional crystallization. The absence of crustal contamination of the magmas implies that the crust was "cold" and that magmas rose to the surface through fractures and did not pool in magma chambers in the upper crust. Alibert et al. (1987) reached similar conclusions on the basis of $^{87}Sr/^{86}Sr$ and $^{143}Nd/^{144}Nd$ ratios of a different suite of lavas from Lower Silesia in Poland.

6.6.9 Carpathian Mountains, Hungary

The Carpathian Mountains are an eastern extension of the Alps and formed in a subduction zone that was active during the Eocene and Miocene Epochs. The volcanic rocks that were erupted during these episodes are calc-alkaline in composition and were followed in Pliocene to Pleistocene time by alkali-rich lavas. The latter formed in a setting of extensional tectonics and contain lherzolite xenoliths implying that the alkali-rich magmas originated from the underlying mantle (Burchfield and Royden 1982).

The Sr concentrations of the calc-alkaline and alkalic lavas reported by Salters et al. (1988) tend to decrease with increasing degree of differentiation reflected by their Rb concentrations, but the data points do not form coherent patterns. Similarly, the $^{87}Sr/^{86}Sr$ ratios are not well correlated with their reciprocal Sr concentrations. Nevertheless, the $^{87}Sr/^{85}Sr$ ratios of the calc-alkaline rocks increase with decreasing Sr concentrations from 0.7042 to 0.71125 relative to 0.7080 for E&A, whereas the $^{87}Sr/^{86}Sr$ ratios of the alkali-rich rock vary from 0.703199 to 0.70555 as the Sr concentrations decrease from 937 to 375 ppm.

The most unequivocal insight into the petrogenesis of the Tertiary lavas of the Carpathian Mountains is provided by their initial $^{87}Sr/^{86}Sr$ and $^{143}Nd/^{144}Nd$ ratios in Fig. 6.41 (Salters et al. 1988). The data points representing the calc-alkaline and alkali-rich rocks form a single mixing hyperbola even though the two suites of lavas were erupted at different times. All but one of the calc-alkaline volcanic rocks plot in the quadrant that characterizes rocks of the continental crust and their isotope ratios of Sr and Nd exceed those of the EM1 and EM2 components. Therefore, these rocks formed by progressive contamination of a mantle-derived magma ($^{87}Sr/^{86}Sr = 0.7042$) with a variety of crustal rocks having low $^{143}Nd/^{144}Nd$ ratios and high $^{87}Sr/^{86}Sr$ ratios greater than 0.7110. Salters et al. (1988) determined that the most silicic rocks of the calc-alkaline suite contain up to 67% of the crustal contaminant.

The alkali-rich rocks have lower $^{87}Sr/^{86}Sr$ and higher $^{143}Nd/^{144}Nd$ ratios than the calc-alkaline suite and plot in the mantle quadrant. Therefore, the magmas that formed these rocks originated from sources in the lithospheric mantle and were much less contaminated by crustal rocks than the calc-alkaline rocks. Nevertheless, the range of Sr and Nd isotope ratios of the alkali-rich rocks indicates that the magmas originated from the lithospheric mantle and that they contain Sr and Nd derived from a mixture of enriched EM1 and EM2 components.

The $^{206}Pb/^{204}Pb$ ratios of both suites of rocks in Fig. 6.41b cluster close to 19.0 (18.694 to 19.426) and

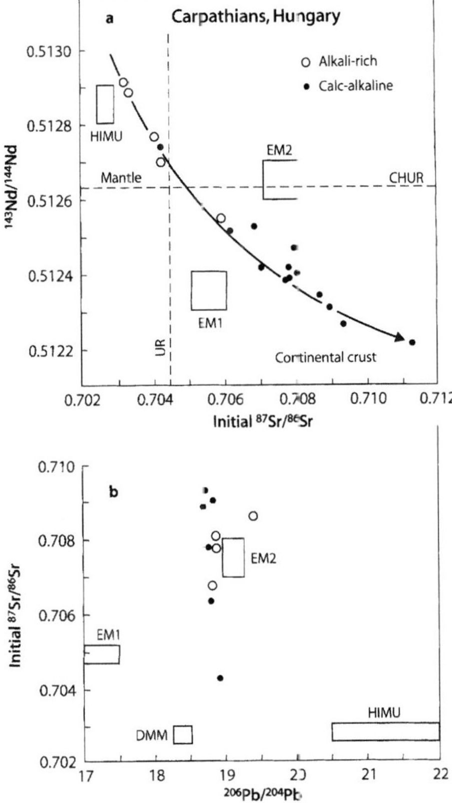

Fig. 6.41. a Sr-Nd isotope mixing diagram for alkali-rich (Pliocene-Pleistocene) and calc-alkaline (Eocene and Miocene) volcanic rocks, Carpathian Mountains, Hungary. The initial $^{87}Sr/^{86}Sr$ and $^{144}Nd/^{143}Nd$ ratios of these rocks form a single well-defined mixing hyperbola. The distribution of data points indicates that the alkali-rich rocks originated from magma sources in the lithospheric mantle that contained Sr and Nd derived from a mixture of EM1 and EM2 components. The calc-alkaline rocks contain varying amounts of crustal Sr and Nd, either as a result of assimilation of crustal rocks or by mixing with a crustal magma; b Isotope ratios of Sr and Pb of the alkali-rich and calc-alkaline lavas, Carpathian Mountains, Hungary. The $^{206}Pb/^{204}Pb$ ratios of the magmas are virtually independent of the $^{87}Sr/^{86}Sr$ ratios because the $^{206}Pb/^{204}Pb$ ratio of the crustal Pb was only slightly lower than that of the mantle Pb (*Source:* Salters et al. 1988; Hart 1988)

generally confirm that the Pb in these rocks originated from the EM2 and HIMU components of the lithospheric mantle. The $^{206}Pb/^{204}Pb$ ratio of the crustal Pb was only slightly lower than that of the mantle Pb which caused the $^{206}Pb/^{204}Pb$ ratio of the contaminated magmas to remain virtually independent of the extent of crustal contamination.

6.6.10 Lamprophyre Dikes

The igneous and metamorphic basement rocks of Hercynian age (Upper Carboniferous) of western Europe are cut by swarms of lamprophyre dikes that occur from southwest England to France and to the central Bohemian Pluton of the Czech Republic. In France, minettes occur in Normandy, on the Channel Islands, and in the Vosges Mountains, whereas kersantites are found in the Armorican Massif and in the Massif Central (Fig. 6.29). The minettes are distinguished from kersantites by the presence of phlogopite and alkali feldspar whereas kersantites contain biotite, plagioclase and, in some cases, quartz and alkali feldspar. Both minette and kersantite carry altered olivine.

Lamprophyres occur widely on all of the continents in the form of dikes in settings of extensional tectonics (Rock 1987, 1991). In addition, lava flows having the chemical composition of minette were described by Wallace and Carmichael (1989) from the Los Volcanes volcanic center at the western end of the Mexican volcanic belt (Sect. 4.2.3). In addition, Bernard-Griffiths et al. (1991) used the isotope compositions of Sr, Nd, O, and C to document the extensive crustal contamination of Tertiary lamprophyre dikes at Tamazert in the Atlas Mountains of Morocco.

The lamprophyres analyzed by Turpin et al. (1988) in the Hercynian Mountains of western Europe are K-rich alkalic rocks containing elevated concentrations of Rb and Sr. The minettes have a higher average Rb concentration (253 ± 49 ppm; $2\bar{\sigma}$, $N = 7$) than the kersantites (119 ± 30 ppm; $2\bar{\sigma}$, $N = 9$), but their average Sr concentrations are similar at about 745 ppm. The initial ^{87}Sr/^{86}Sr and 1/Sr ratios are scattered and fail to shed light on the petrogenesis of these rocks (not shown), but the kersantites do have consistently higher initial ^{87}Sr/^{86}Sr ratios (0.70694 to 0.70794) than the minettes (0.70545 to 0.70637) relative to 0.71025 for NBS 987. The best perspective on the origin of these lamprophyres is provided by Sr-Nd isotope-ratio mixing diagram in Fig. 6.42. All of the lamprophyres from the Hercynian orogen of Europe plot in the quadrant normally occupied by rocks of the continental crust having elevated initial ^{87}Sr/^{86}Sr ratios and low initial ^{143}Nd/^{144}Nd ratios. Turpin et al. (1988) favored the interpretation that the mantle-derived lamprophyre magmas had been contaminated at the source by subducted pelagic sediment. The alignment of data points in Fig. 6.42 between hypothetical mixing hyperbolas is consistent with that interpretation. Stille et al. (1989) reached a similar conclusion concerning the petrogenesis of calc-alkaline lamprophyres of central Europe based on a study of the isotope composition of Hf and Nd.

The Tertiary lamprophyre dikes associated with the Tamazert alkaline complex (Mesozoic) in the High At-

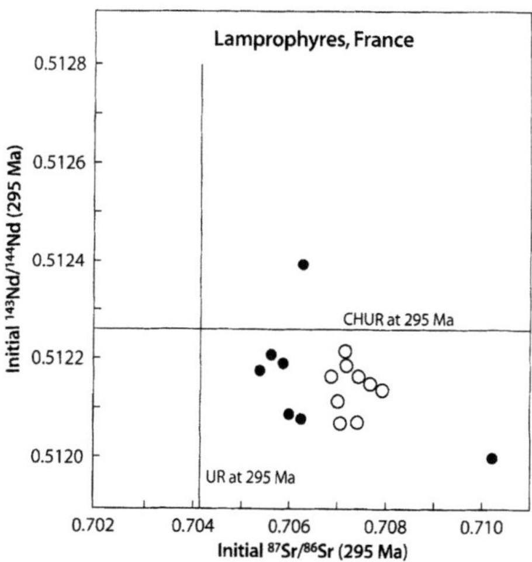

Fig. 6.42. Lamprophyre dikes in Hercynian granitic basement rocks in the Armorican Massif of Normandy, the Channel Islands, Massif Central, and the Vosges Mountains of France (including one sample from the Schwarzwald of Germany). *Solid circles*: minette; *open circles*: kersantite. The high initial ^{87}Sr/^{86}Sr and low ^{143}Nd/^{144}Nd ratios at 295 Ma demonstrate that these lamprophyres formed from mantle-derived magmas that contained a significant proportion of a crustal component. The contamination probably occurred at the source by partial melting of subducted pelagic sediment (*Source:* Turpin et al. 1988; Hart 1988)

las Mountains of Morocco present a quite different picture. The isotope ratios of Sr and Nd of these lamprophyre dikes in Fig. 6.43 define a data field that lies on a hypothetical mixing trajectory between the DMM and the EM1 components. The ^{87}Sr/^{86}Sr ratios of the lamprophyres that define this data field range from 0.70328 to 0.70407 relative to 0.71025 for NBS 987, whereas the ^{143}Nd/^{144}Nd of these rocks are restricted to values between 0.512921 and 0.512681 relative to 0.511878 for the La Jolla Nd standard. The ^{87}Sr/^{86}Sr ratios of lamprophyre dikes which intruded Early Jurassic marine limestone (Sr = 2 300 ppm, ^{87}Sr/^{86}Sr = 0.7075) were increased up to 0.70665, whereas their ^{143}Nd/^{144}Nd ratios remained virtually unchanged because the Nd concentration of the limestone is less than 1 ppm and its ^{143}Nd/^{144}Nd ratio is approximately 0.51245 (Bernard-Griffiths et al. 1991).

The isotope compositions of Sr and Nd of the uncontaminated lamprophyre dikes associated with the Tamazert Intrusion of Morocco in Fig. 6.43 relate their magmas to sources in the subcontinental lithospheric mantle. In addition, the magma sources were enriched in alkali metals and radiogenic ^{87}Sr by aqueous fluids derived from subducted oceanic crust and marine sediment of the EM1 component. Therefore, the Moroccan lamprophyre dikes in Fig. 6.43 were contaminated only during and subsequent to their intrusion in marked

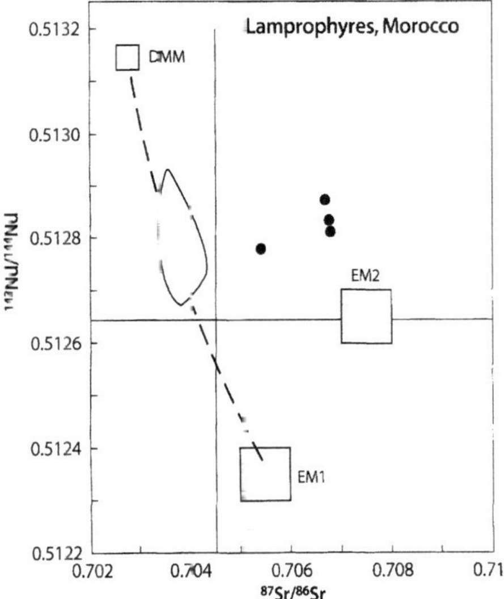

Fig. 6.43. Isotope ratios of Sr and Nd of Tertiary lamprophyre dikes associated with the Tamazert alkali-rich igneous complex in the High Atlas Mountains of Morocco. Several dikes which intruded Early Jurassic limestone were contaminated with Sr having an $^{87}Sr/^{86}Sr$ ratio of about 0.7075, but the $^{143}Nd/^{144}Nd$ ratios of these dikes were not significantly changed (*Sources:* Bernard-Griffiths et al. 1991; Hart 1988)

contrast to the Hercynian lamprophyres of France in Fig. 6.42, which were contaminated at the source during the formation of their magmas.

6.7 Scandinavia

The alkali-rich rocks of Scandinavia occur primarily in the Oslo area and at Fen in Norway, on Alnö Island in Sweden, and on the Kola Peninsula of Russia. The alkalic igneous rocks associated with the Oslo Graben were originally studied by Brögger (1921, 1933), Barth (1945, 1954), and more recently by Heier and Compston (1969), Neumann (1978, 1980), and Neumann et al. (1985, 1986, 1990, 1992). The alkali-rich intrusives and carbonatite of Alnö were described by von Eckermann (1948), while the Khibina and Lovozero Massifs on the Kola Peninsula were studied by Zak et al. (1972) and by Gerasimovsky et al. (1966a,b), respectively. As a result of these pioneering studies, the petrology, mineralogy, and geochemistry of the alkali-rich rocks of Scandinavia have become the standard to which alkalic rocks of other areas are compared.

Doig (1970) pointed out that alkali-rich igneous rocks were intruded at about 565 Ma along a rift system that extended from central Canada to eastern Swe-

den. Parts of this rift system were later reactivated and caused local magmatic activity in Norway (e.g. the Oslo Graben) and formation of calcite veins in the St. Lawrence Rift system of Canada (Carignan et al. 1997). However, Vartiainen and Woolley (1974) related the Cambrian alkali complexes of Scandinavia to the Caledonian orogeny, which affected the rocks of western Norway, eastern Greenland, as well as some parts of eastern North America.

6.7.1 Oslo Graben, Norway

The mildly alkaline igneous rocks of Permo-Carboniferous age of southern Norway are closely associated with the Oslo Graben which extends north from the south coast of Norway for about 200 km to Lake Mjösa. The rift has been eroded to a depth of 2 to 3 km and exposes a cross-section of plutonic igneous rocks, whereas only small exposures of the associated volcanic rocks have been preserved. Age determinations by Sundvoll and Larsen (1990) based mainly on the whole-rock Rb-Sr method suggest that igneous activity started at 294 ±6 Ma ($\lambda = 1.42 \times 10^{-11}$ yr^{-1}) in the southern part of the rift (Vestfold segment) and migrated northward, ending at 241 ±3 Ma in the Akershus segment.

The volcanic rocks consist of three suites composed of basaltic lavas, intermediate rhomb-porphyries, and highly differentiated trachytes and rhyolites. The plutonic rocks of the Oslo Rift form four composite plutons identified by geographic place names (Table 6.3).

The Rb and Sr concentrations of the igneous rocks in the Oslo Graben vary widely and thereby provide evidence for magmatic differentiation by fractional crystallization. In general, the Sr concentrations of the rocks decline with increasing Rb concentrations from basalt (Rb = 90 ppm, Sr = 680 ppm) to rhyolite (Rb = 165 ppm, Sr = 45 ppm) and granite (Rb = 230 ppm, Sr = 115 ppm). The alkali-rich syenites contain 140 ppm Rb and 170 ppm Sr on the average. Barth (1945) proposed a "family tree" for the plutonic igneous rocks of the Oslo Graben based on the fractional crystallization of a syenitic magma leading to the formation of two series of rocks characterized by increasing concentra-

Table 6.3. Plutonic rocks of the Oslo Rift form four composite plutons identified by geographic place names

Pluton	Principal rock type
Tønsberg-Larvik	Monzonites (larvikites)
Skrim-Eikeren	Monzonites (larvikites)
Hurum-Finnemarka	Granitic rocks
Nordmarka-Hurdal	Syenites

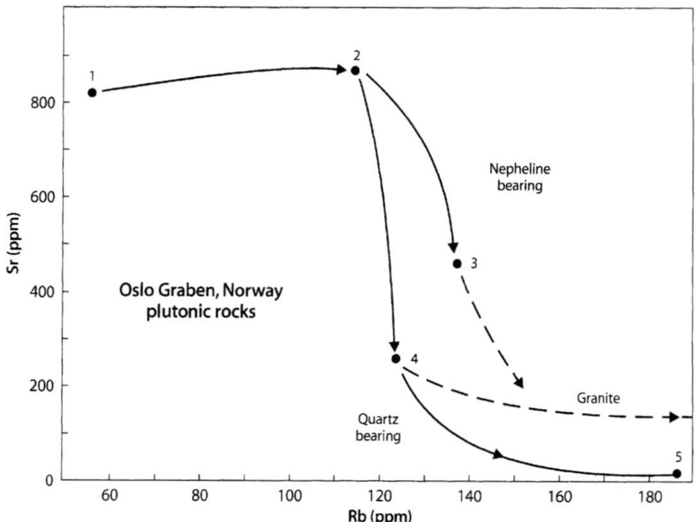

Fig. 6.44.
Average Rb and Sr concentrations of two series of differentiated plutonic rocks in the Oslo Graben of Norway. The rocks are characterized by the presence of quartz and nepheline, respectively, and are arranged sequentially by decreasing concentrations of anorthite. The nepheline-bearing series includes: *1.* kjelsåsite; *2.* larvikite; *3.* lardalite; nepheline pegmatite (no data). The quartz-bearing suite separates from larvikite and leads to *4.* nordmarkite and *5.* ekerite or granite (*Source:* Barth 1945; Heier and Compston 1969)

tions of nepheline (kjelsåsite-larvikite-lardarite-nepheline pegmatite) and of quartz (kjelsåsite-larvikite-nordmarkite-ekerite, and granite), respectively. The Rb and Sr concentrations of these rock series are illustrated in Fig. 6.44 based on the data of Heier and Compston (1969). The relation of the granites to the other plutonic rocks in the Oslo Graben is uncertain.

The initial $^{87}Sr/^{86}Sr$ ratios of the igneous rocks of the Oslo Graben range from 0.70387 (rhomb porphyry, Krokskogen, Akershus segment) to 0.71009 (trachyte, Vestfold segment) relative to 0.71025 for NBS 987 (Sundvoll and Larsen 1990). The initial $^{87}Sr/^{86}Sr$ ratios reported by Neumann et al. (1988) range even more widely from 0.70309 (larvikite at Larvik, Vestfold segment) to 0.71965 (granite, Hurdal, Akershus segment) relative to 0.71025 for NBS 987. The initial $^{87}Sr/^{86}Sr$ ratios of Neumann et al. (1988), as well as those of Anthony et al. (1989), are based on single samples, whereas those of Sundvoll and Larsen (1990) are based, for the most part, on whole-rock Rb-Sr isochrons.

The initial $^{87}Sr/^{86}Sr$ ratios and average reciprocal Sr concentrations reported by Sundvoll and Larsen (1990) in Fig. 6.45 demonstrate that the parent magma had an $^{87}Sr/^{86}Sr$ ratio of 0.70387 ±0.00004) (relative to 0.71025 for NBS 987) derived from the rhomb porphyry at Krokskogen in the Akershus segment, and that these magmas differentiated by fractional crystallization and by assimilation of crustal rocks.

The Sr-Nd isotope mixing diagram in Fig. 6.46 based on the data of Neumann et al. (1988) and Anthony et al. (1989) demonstrates that the basalts, rhomb porphyries, and larvikites (monzonites) originated from sources in the lithospheric mantle without significant contamina-

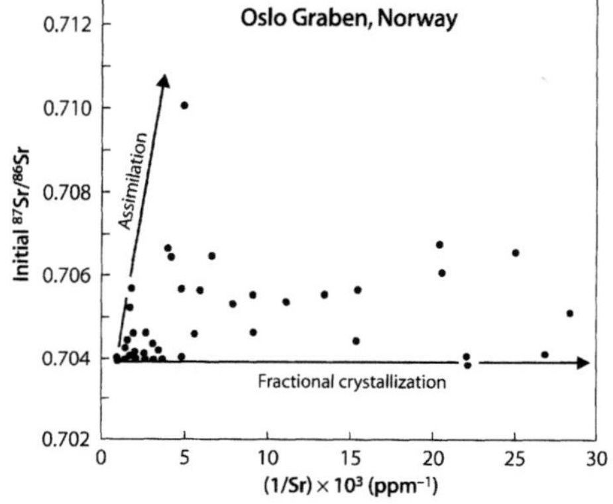

Fig. 6.45. Average initial $^{87}Sr/^{86}Sr$ and 1/Sr ratios of volcanic and plutonic igneous rocks of the Oslo Graben in Norway. The initial $^{87}Sr/^{86}Sr$ ratios were determined from Rb-Sr isochrons based primarily on whole-rock samples, but including minerals of the plateau lavas (rhomb porphyries). The distribution of data points indicates that the magmas differentiated both by fractional crystallization and by assimilation of crustal rocks having elevated $^{87}Sr/^{86}Sr$ ratios (*Source:* data from Sundvoll and Larsen 1990)

tion by crustal rocks. However, the wide range of initial $^{87}Sr/^{86}Sr$ ratios of the syenites and granites was caused by the assimilation of varying amounts of crustal rocks having elevated $^{87}Sr/^{86}Sr$ and low $^{143}Nd/^{144}Nd$ ratios. Mixing models by Neumann et al. (1988) indicated that some syenites and granites may contain more than 50% of pre-Permian upper crust of southern Norway. Nevertheless, the isotope ratios of these rocks converge to-

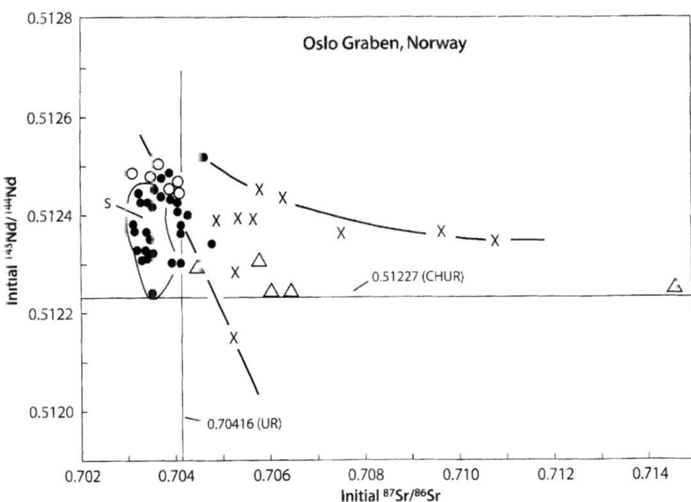

Fig. 6.46. Initial ^{87}Sr/^{86}Sr and ^{143}Nd/^{144}Nd ratios of igneous rocks in the Oslo Graben, Norway. Basalt and rhomb porphyry: *solid circles*; larvikites (monzonites) *open circles*; syenites: *crosses*; granites: *open triangles*. The isotope ratios of UR and CHUR have been recalculated for $t = 290 \times 10^6$ years starting with present values of 0.7045 for ^{87}Sr/^{86}Sr and ^{87}Rb/^{86}Sr = 0.0816 (UR) and 0.512638 for ^{143}Nd/^{144}Nd and ^{147}Sm/^{144}Nd = 0.1967 (CHUR). The basalts, rhomb porphyries and larvikites originated from sources in the lithospheric mantle with only minor contamination by crustal rocks. However, the magmas that formed the syenites and granites assimilated varying amounts of crustal rocks having elevated ^{87}Sr/^{86}Sr and low ^{143}Nd/^{144}Nd ratios. The nephelinites, basanites, and alkali basalts at Skien (S) have low ^{143}Nd/^{144}Nd ratios and contain contributions from several different mantle reservoirs (*Sources:* Neumann et al. 1988; Anthony et al. 1989)

wards those of the basaltic and monzonitic rocks thereby implying that the magmatic activity in the Oslo Graben originated in the subcontinental lithospheric mantle by decompression melting caused by the development of the Oslo Graben.

The nephelinites, basanites, and alkali basalts at Skien (Vestfold segment) studied by Segalstad (1979) and analyzed by Anthony et al. (1989) form a separate array labeled S in Fig. 6.46. The nephelinites, which were erupted first, originated from a source in the mantle having a low ^{87}Sr/^{86}Sr ratio (0.70307) as well as a low ^{143}Nd/^{144}Nd ratio (0.512302). The initial isotope ratios of the basanites and alkali basalt, extruded later, both increase and approach those of the monzonites (larvikites) and alkali basalt in the Vestfold segment of the Oslo Graben (Neumann et al. 1988). These results demonstrate the isotopic heterogeneity of the mantle underlying the Oslo Graben. The results of modeling by Anthony et al. (1989) support the hypothesis that parts of the underlying lithospheric mantle were enriched in Nd and other light rare earths at about 550 Ma during the magmatic event at nearby Fen.

6.7.2 Fen, Southwestern Norway

The carbonatite complex at Fen is located in the province of Telemark west of the Oslo Graben. It is roughly circular in plan with a diameter of about 3 km and intruded granitic gneisses of Middle Proterozoic age. The Fen complex contains a variety of rocks including ijolites (nepheline and pyroxene), fenites, nepheline syenites, as well as carbonatites composed of calcite, dolomite, and ankerite. A large part of the Fen complex consists of "tuffisitic" ferrocarbonatite (ankerite with calcite and chlorite) that was emplaced as a fluidized carbonatite tuff. The ijolitic rocks are surrounded by a large mass of fenite intruded by dikes of calcite carbonatite (sövite), silicosövite, and nepheline syenite. The fenite as well as the ferrocarbonatite were intruded by a porphyritic lamprophyre locally called damtjernite consisting of titaniferous augite, amphibole, nepheline, perthite, and calcite with phenocrysts of biotite and olivine (Andersen 1986, 1987; Mitchell and Crocket 1972).

Age determinations reviewed by Mitchell and Crocket (1972) included a study by Faul et al. (1959) who reported a K-Ar date of 565 Ma for biotite in sövite. More recent work by Verschure et al. (1983) confirmed the Cambrian age of the Fen complex. According to Andersen (1987), a Pb-Pb isochron of carbonatites and associated rocks also yielded at date of 539 ±14 Ma.

The Fen complex is the type locality for fenites, which form by metasomatic alteration of the country rock surrounding alkali-rich intrusives. The fenites at Fen are composed of alkali feldspar and pyroxene (aegirine) and formed by metasomatic alteration of Precambrian gneisses by fluids that emanated from an ijolitic magma. The fenites were subsequently intruded by nepheline syenite and by a variety of carbonatite dikes that caused pervasive phlogopitization of the fenite. The fenites of eastern Uganda (Sect. 6.1.6) and around the Monchique complex of southeastern Portugal (Sect. 6.5.4) were previously mentioned.

The sequence of events and the processes that contributed to the formation of the Fen complex were originally studied by Brögger (1921), and subsequently by Saether (1957), Barth and Ramberg (1967), and many others identified by Andersen (1987). The formation of the fenites is no longer in doubt, but the origin of the carbonatites has been the subject of a debate. Brögger (1921) originally concluded that the carbonatites crystallized from carbonate magma that had formed by the fusion of Precambrian limestone at depth. However, Bowen (1924, 1926) proposed that the carbonatite had formed by metasomatic replacement of ijolites and fenites.

The physical chemistry of carbonate melts and their relation to silicate melts was illuminated by the experimental work of Wyllie and Tuttle (1960, 1962) who demonstrated that carbonate and silicate melts are immiscible under certain conditions, allowing carbonatite magmas to form from alkali-rich magmas and be erupted as lava flows. The validity of this conclusion was amply confirmed when Dawson (1962b) reported carbonatite lavas from the summit of Oldoinyo Lengai in Tanzania (Sect. 6.1.3). Subsequent trace-element studies by Mitchell and Brunfelt (1975) and Mitchell (1980) confirmed that the pyroxene sövite at Fen crystallized from carbonatite melts related to ijolitic and nepheline syenitic magma. However, biotite-amphibole sövite, the dolomite carbonatite, and apatite cumulates

originated from a Mg-rich calcite carbonatite melt that coexisted with a mafic silicate magma (Andersen 1986).

The relationship between the silicate and carbonate rocks of Fen was investigated by Mitchell and Crocket (1972) by means of initial $^{87}Sr/^{86}Sr$ ratios. They reported that the silicate and carbonate rocks have a wide range of initial $^{87}Sr/^{86}Sr$ ratios (at 550 Ma) from 0.7025 for one specimen of calcite carbonatite (sövite) to 0.7101 for a specimen of rauhaugite (ankerite carbonatite) relative to 0.7080 for E&A. Andersen (1987) later demonstrated overwhelmingly that the sövite at Fen has an average initial $^{87}Sr/^{86}Sr$ ratio (at 540 Ma) of 0.703035 ±0.000042 ($2\bar{\sigma}$, $N = 8$) relative to 0.71025 for NBS 987. In addition, the initial $^{87}Sr/^{86}Sr$ ratios of the dolomite and ferrocarbonatites (dike facies) are indistinguishable from that of the sövite. However, the apatite cumulate has a slightly lower initial $^{87}Sr/^{86}Sr$ ratio than the sövite, whereas the tuffisitic ferrocarbonatite and the hema-

Table 6.4. Initial $^{87}Sr/^{86}Sr$ ratios of the silicate and carbonate rocks of Fen (Mitchell and Crocket 1972)

Rock type	Initial $^{87}Sr/^{86}Sr$ ($2\bar{\sigma}$, N)	
Calcite carbonatite (sövite)	0.703025 ±0.000042	(8)
Dolomite carbonatite	0.703097 ±0.000040	(20)
Pyroxene sövite and silicosövite	0.703060 ±0.000018	(2)
Ferrocarbonatite dike (ankerite)	0.702967 ±0.000029	(2)
Apatite cumulate	0.702869 ±0.000050	(2)
Tuffisitic ferrocarbonatite	0.703333 ±0.000014	(1)
Hematitic ferro-carbonate (rödberg)	0.707257 ±0.000236	(2)

Fig. 6.47.
Carbonatites (*solid circles*) and ijolite-melteigite (*open circles*) of the Fen carbonatite complex, Telemark, southern Norway (540 Ma). The different types of carbonatite composed of calcite (*C*), dolomite (*D*), and ankerite (*F1*) all have low initial $^{87}Sr/^{86}Sr$ ratios of about 0.7030. However, the abundant tuffisitic ferrocarbonatite (*F2*) and hematitic carbonatite (*R*), derived from them by hydrothermal alteration have elevated initial $^{87}Sr/^{86}Sr$ ratios. The fenites (*crosses*) that surround the ijolite-carbonatite core of the complex have low $^{87}Sr/^{86}Sr$ ratios compared to the Precambrian granite gneisses (*G*) of the country rocks. Silicocarbonatites (*open triangles*) are composed of mixtures of carbonate and silicate minerals (*Sources:* Mitchell and Crocket 1972; Andersen 1987)

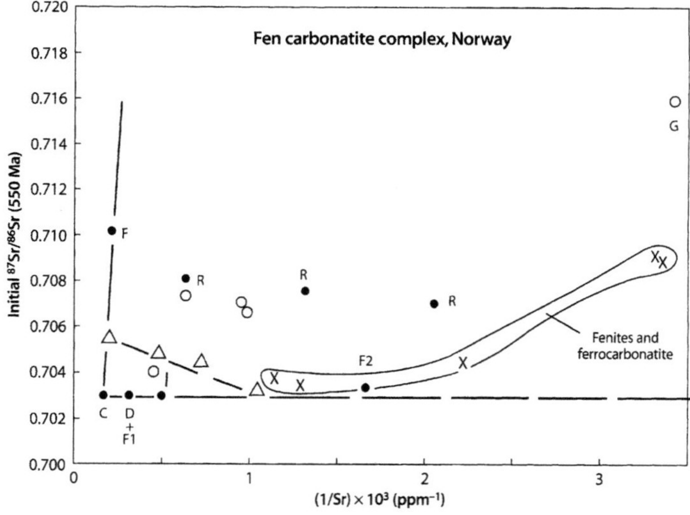

titic carbonatite (rödberg) are significantly enriched in ^{87}Sr (Table 6.4).

The data sets of Mitchell and Crocket (1972) and Andersen (1987) have been merged in Fig. 6.47 in order to shed light on the petrogenesis of the carbonate and silicate rocks at Fen.

The results demonstrate that: (1) The calcite, dolomite, and ankerite (dike) carbonatites have identical initial ^{87}Sr/^{86}Sr ratios with an average value of 0.70307 and therefore are differentiates of the same carbonatite magma. (2) The tuffisitic ferrocarbonatite (F2), which is one of the major rock units at Fen, and the hematitic carbonatites (R) have elevated initial ^{87}Sr/^{86}Sr ratios compared to the calcite-dolomite-ankerite carbonatites and were hydrothermally altered. (3) The ^{87}Sr/^{86}Sr ratios of the fenites decrease with increasing Sr concentration in the same way as the fenites of eastern Uganda in Fig. 6.10.

The low ^{87}Sr/^{86}Sr ratios of the carbonatites are consistent with their derivation from sources in the lithospheric mantle (Griffin 1973). Their intimate association with silicate rocks in the form of silicocarbonatites further indicates that the carbonatites are genetically related to silicate magmas represented by the ijolite-melteigite suite. The silicate magmas were variably contaminated by Sr derived from the continental crust, as was the large mass of tuffisitic ferrocarbonatite. The latter was locally altered by heated groundwater to form hematitic carbonatite characterized by having a wide range of Sr concentrations and elevated ^{87}Sr/^{86}Sr ratios. The intrusion of the silicate and carbonatite magmas was apparently preceded by the formation of fenites as a result of metasomatic alteration of the Precambrian gneisses by corrosive alkali-rich fluids emanating from magma chambers at depth.

6.7.3 Alnö, Sweden

The Alnö alkaline complex is located on an island at about 62°30' N latitude in the Gulf of Bothnia off the east coast of Sweden (Woolley 1989). The complex was brought to the attention of the geological community by the work of von Eckermann (1948). More recently, Brueckner and Rex (1980) determined an age of 553 ±6 Ma for the alkali-rich rocks, which include alkaline pyroxenite, ijolite-melteigite rocks, malignite, and nepheline syenite. In addition, the complex and fenitized country rocks contain numerous dikes of phonolite, alkali trachyte, tinguaite, nephelinite, and dolomite carbonatite. The complex also contains major bodies of calcite carbonatite. These diversified carbonate and alkali-rich silicate rocks formed during a complex sequence of events. The present understanding of the origin of the Alnö complex differs from that proposed by von Eckermann (1948) who concluded that most of the

silicate rocks are fenites that were remobilized during the intrusion of the carbonatites.

The initial ^{87}Sr/^{86}Sr and 1/Sr ratios of silicate and carbonate rocks of the Alnö complex in Fig. 6.48 indicate that both fractional crystallization and assimilation of crustal rocks contributed to the formation of the rocks. The initial ^{87}Sr/^{86}Sr ratios of two calcite carbonatite are 0.70344 and 0.70399, relative to 0.71025 for NBS 987 (Brueckner and Rex 1980). Some of the peripheral phonolite and nephelinite dikes and the foyaite in the core of the complex also have low initial ^{87}Sr/^{86}Sr ratios between 0.70394 and 0.70400. Therefore, the carbonatite and silicate magmas had similar initial ^{87}Sr/^{86}Sr ratios between 0.70390 and 0.70400 indicative of a subcrustal magma source.

The Alnö complex of Sweden intruded migmatites of Precambrian age much like the coeval Fen complex of Norway. One specimen of this migmatite has a Rb/Sr ratio of 0.836 (Rb = 174 ppm, Sr = 208 ppm) and a present ^{87}Sr/^{86}Sr ratio of 0.75819 ±0.00009 relative to 0.71025 for NBS 987 (Brueckner and Rex 1980). The ^{87}Sr/^{86}Sr ratio of this migmatite at 553 Ma (0.73910) was much higher than that of two samples of fenites whose initial ^{87}Sr/^{86}Sr ratios are 0.70547 and 0.70603. Therefore, fenitization of the migmatite caused its ^{87}Sr/^{86}Sr ratio to be lowered either by addition of Sr from the silicate and/or carbonatite magmas or by selective migration of ^{87}Sr from the migmatite into the mag-

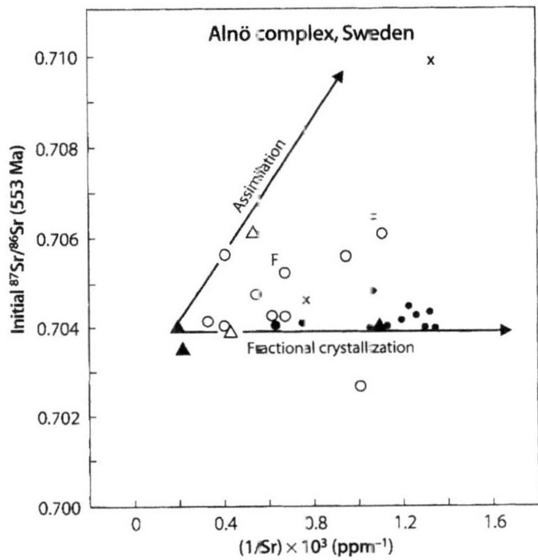

Fig. 6.48. Initial ^{87}Sr/^{86}Sr and 1/Sr ratios of alkali-rich silicate rocks and carbonatites on Alnö Island, Sweden. Phonolite: *solid circles*; alkali trachytes: *crosses*; nephelinites: *open triangles*; plutonic silicate rocks including alnöites: *open circles*; calcite and dolomite carbonatites: *solid triangles*; fenites: *open circles* labeled "F". The distribution of data points indicates that both fractional crystallization of mantle-derived magmas and assimilation of crustal rocks contributed to the formation of this complex (*Source:* data from Brueckner and Rex 1980)

mas. The model age of the migmatite, based on the Rb-Sr data of Brueckner and Rex (1980) and assuming an initial $^{87}Sr/^{86}Sr$ ratio of 0.7030, is about 1580 Ma ($\lambda = 1.42 \times 10^{-11}$ yr^{-1}).

6.8   Kola Peninsula, Russia

The Kola Peninsula is a major province of alkali-rich igneous rocks and carbonatites. Although Precambrian alkali-rich igneous complexes do occur here, most of the alkalic rocks are of Paleozoic age (Gerasimovsky et al. 1974; Kogarko 1987; Kogarko et al. 1981, 1983; Woolley 1989; Dudkin and Mitrofanov 1994). The two largest are the Khibina and Lovosero Massifs both of which intruded Archean granitic gneisses and Proterozoic metavolcanics and metasediments of the Baltic Shield at about 365 Ma (Kogarko 1987).

6.8.1   Muntsche-Tundra Pluton

The geochronology of the Kola Peninsula was first studied by E. K. Gerling and his colleagues at the Institute of Precambrian Geology in St. Petersburg. Gerling (1984) described how in the early 1960s the K-Ar dates he measured for pyroxenites and other mafic rocks in the Muntsche-Tundra Pluton on the Kola Peninsula ranged from 3.8 to 10.0 Ga indicating that these rocks contained excess ^{40}Ar (Gerling et al. 1962). The discovery by Gerling and his colleagues of excess ^{40}Ar in the mafic rocks of Precambrian age in the Kola Peninsula was later confirmed by Kirsten and Müller (1967), Lippolt and Wasserburg (1973), Kaneoka (1974), and by Ashkinadze et al. (1978).

The presence of excess ^{40}Ar in the rocks of the Muntsche-Tundra Pluton was a challenge to geochronologists because the excessively old K-Ar dates that had been reported cast doubt on the reliability of all isotopic age determinations. For this reason, Birck and Allègre (1973) dated a small suite of whole-rock specimens from the Muntsche-Tundra Pluton by the Rb-Sr method. Their results yielded a whole-rock isochron date of 1.98 ±0.15 Ga ($\lambda = 1.42 \times 10^{-11}$ yr^{-1}) and an initial $^{87}Sr/^{86}Sr$ ratio of 0.70205 ±0.00009 relative to 0.70800 for E&A. The authors interpreted this date as the time of initial crystallization of Muntsche-Tundra intrusive and concluded that K-Ar dates of this complex that are older than about 2 billion years indicate the presence of excess ^{40}Ar. The initial $^{87}Sr/^{86}Sr$ ratios of minerals separated from norite and anorthosite of the Muntsche-Tundra Pluton range from 0.70011 (clinopyroxene, norite) to 0.70291 (clinopyroxene, anorthosite) calculated at 1.98 Ga and $\lambda = 1.42 \times 10^{-11}$ yr^{-1} (Birck and Allègre 1973). This discrepancy reveals that the Muntsche-Tundra intrusive was slightly altered

after crystallization, even though the texture of the rocks does not record such an event. The norite and anorthite together with their orthopyroxene and plagioclase separates yield a Rb-Sr date of about 1900 Ma with an initial $^{87}Sr/^{86}Sr$ ratio of 0.70206±0.00008 ($\lambda = 1.42 \times 10^{-11}$ yr^{-1}, E&A = 0.70800). However, the clinopyroxenes in these rocks do not fit this mineral-whole rock isochron.

Lippolt and Wasserburg (1973) also reported Rb-Sr dates based on minerals of a gabbro and a peridotite from the Nittis mining area of the Monche Range: 1.63 ±0.22 Ga for the gabbro and 3.27 ±0.68 Ga for the peridotite ($\lambda = 1.42 \times 10^{-11}$ yr^{-1}). The K-Ar dates of the same rock specimens reported by Kirsten and Müller (1967) were 3.4 Ga for the gabbro and 3.9 Ga for the peridotite. The K-Ar dates of the gabbro and peridotite are significantly older than the Rb-Sr mineral dates and thereby confirm the presence of excess ^{40}Ar in these rocks. More recently, Tolstikhin et al. (1992) reported Sm-Nd dates of 2.492 ±0.048 Ga for the Sopcha pyroxenite and 2.492 ±0.031 Ga for the Nud gabbro from the Monche Range. The initial $^{87}Sr/^{86}Sr$ ratios are 0.70227 (pyroxenite) and 0.70189 (gabbro) relative to 0.71025 for NBS 987.

The Kola Peninsula also contains alkali-rich complexes of Precambrian age at Yelet' ozero and Gremiakha-Vyrmes in Russia and at Almunge and Norra-Karr in Sweden. The ages of these complexes range from 1900 to 1600 Ma (Kogarko 1987). The Yelet' ozero (50 km²) and Gremiakha-Vyrmes (130 km²) complexes are composite intrusions containing both mafic and alkalic rocks.

6.8.2   Khibina and Lovozero Intrusives

The Khibina Massif on the Kola Peninsula (1 327 km²) is the largest composite intrusive of alkali-rich rocks in the world (Zak et al. 1972). It is concentrically zoned with a core of foyaite surrounded by cone sheets of ijolite, urtite, and malignites, rischorrite, khibinite, and nepheline syenites (rischorrite consists of microcline, orthoclase, nepheline, aegirine, augite, arfvedsonite, and biotite, whereas khibinite contains microcline, nepheline, aegirine, and arfvedsonite; Gerasimovsky et al. 1974). The contact zone between the Khibina intrusive and the Precambrian country rocks is several hundred meters wide and is composed of fenite and other products of alkali metasomatism.

The Lovozero Massif (650 km²) is likewise a composite intrusion studied in detail by Gerasimovsky et al. (1966a,b). It is composed primarily of eudialyte lajuvrite (microcline, aegirine, eudialyte, arfvedsonite, nepheline), sodalite and nosean syenites, and other rocks of lesser abundance. (The formula of eudialyte is: $(Na, Ca, Fe)_6 Zr (OH, Cl) (SiO_3)_6$). The rocks of the

Table 6.5. Age determinations by the whole-rock Rb-Sr isochron method of the Lovozero and Khibina Massifs (Kogarko et al. 1981; Kogarko et al. 1983; Kramm and Kogarko 1994). The dates and initial $^{87}Sr/^{86}Sr$ ratios of the Khibina Massif are for two different intrusions of the composite massif

Date (Ma)	$^{87}Sr/^{86}Sr)_0$	Reference
Khibina Massif		
365 ±13	0.70333 ±15	Kogarko et al. (1981)
366 ±19.8	0.70399 ±8	Kramm and Kogarko (1994)
367.5 ±5.5	0.70323 ±5	Kramm and Kogarko (1994)
Lovozero Massif		
362 ±17	0.70363 ±24	Kogarko et al. (1983)
370 ±6.7	0.70362 ±6	Kramm and Kogarko (1994)

Lovozero Massif are Na-rich with an excess of alkali metals over alumina ($Na_2O + K_2O / Al_2O_3 = 1.40$). They also include highly differentiated pegmatites containing rare minerals (Gerasimovsky et al. 1974; Khomyakov 1994; Semenov 1994).

Age determinations by the whole-rock Rb-Sr isochron method of the Lovozero and Khibina Massifs published by Kogarko et al. (1981), Kogarko et al. (1983), and by Kramm and Kogarko (1994) are in good agreement and indicate a Late Devonian age (Table 6.5). The dates and initial $^{87}Sr/^{86}Sr$ ratios of the Khibina Massif in Table 6.5 are for two different intrusions of the composite massif.

The Khibina Massif includes a large carbonatite stock with a diameter of about 800 m. Carbonatite veins and carbonatization of foyaite have also been reported (e.g. Pyatenko and Saprykina 1976). An age determination by U. Kramm indicated a date of 363 Ma for the carbonatite and an elevated initial $^{87}Sr/^{86}Sr$ ratio of 0.70410 (Kogarko 1987; Kramm 1993). In addition, Dudkin and Mitrofanov (1994) reported initial $^{87}Sr/^{86}Sr$ ratios for apatites and sphene from the Khibina Massif between 0.7030 and 0.7040. The Khibina Massif and other alkalic intrusives of the Kola Peninsula also contain the largest deposit of magmatic apatite in the world originally discovered under the leadership of Alexander Fersman. The exploitation of the Khibina deposit has stimulated the industrial development of this region, including the construction of the city of Apatity. The origin of the apatite ore bodies of the Khibina Massif was discussed by Kogarko (1987).

The initial isotope ratios of Sr and Nd at 370 Ma of the Khibina Massif form two separate clusters labeled I and II in Fig. 6.49. The rocks in Group I include a variety of alkali-rich silicate rocks (nepheline syenite, urtite, and ijolite) as well as rocks composed predominantly of apatite and sphene. Group II includes carbonatite, tinguaite, pyroxenite, and ijolite all of which are

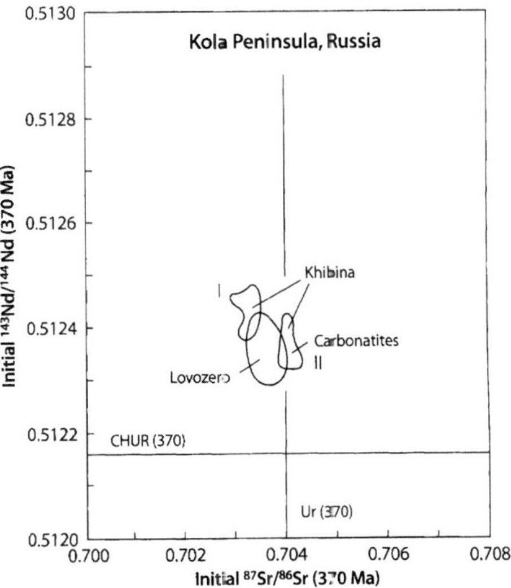

Fig. 6.49. Isotope ratios of Sr and Nd of the Khibina and Lovozero Massifs on the Kola Peninsula of Russia. The alkali-rich rocks of the Khibina Massif form two separate data fields, labeled *I* and *II*. The rocks defining these fields have identical ages (Late Devonian) but those of Group *II* have a higher average initial $^{87}Sr/^{86}Sr$ ratio (0.70398) than those in Group *I* (0.70329) (*Source:* Kramm and Kogarko 1994)

part of a separate intrusion within the Khibina Massif. The rocks of Group II have a higher average initial $^{87}Sr/^{86}Sr$ ratio of 0.70393 ±8 than the rocks of Group I for which Kramm and Kogarko (1994) reported 0.70328 ±5 relative to 0.71025 for NBS 987. The rocks of the Lovozero Massif (including such exotic rock types as eudialyte lajuvrite and apatite-sphene lajuvrite) have an intermediate average initial $^{87}Sr/^{86}Sr$ ratio of 0.70362 ±6.

The values of CHUR and UR in Fig. 6.49 were recalculated to an age of 370×10^6 years based on the following constants (Sect. 1.6):

- $(^{87}Sr/^{86}Sr)^0_{UR}$ = 0.7045
- $(^{87}Rb/^{86}Sr)_{UR}$ = 0.0816
- $(^{143}Nd/^{144}Nd)^0_{CHUR}$ = 0.512638
- $(^{147}Sm/^{144}Nd)^0_{CHUR}$ = 0.1967

The results are:

- $(^{87}Sr/^{86}Sr)^{370}_{UR}$ = 0.70407
- $(^{143}Nd/^{144}Nd)^{370}_{CHUR}$ = 0.512162

These values place the Khibina and Lovozero Massifs in Fig. 6.49 into the quadrant populated by igneous rocks derived from the subcontinental lithospheric mantle without significant assimilation of crustal rocks.

However, the familiar components of magma sources in the mantle cannot be translated from the present to the Late Devonian Epoch because their Rb/Sr and Sm/Nd ratios are unknown. The intrusion of the alkali-rich plutons on the Kola Peninsula coincided with subsidence of the Kontozero Graben which strikes NE-SW and is part of the Late Caledonian Arctic-North Atlantic Megashear System (Kramm et al. 1993). Consequently, the alkali-rich magmas probably originated by decompression melting in the lithospheric mantle, rose into the upper continental crust composed of Archean and Proterozoic metamorphics, and cooled rapidly. The wide range of Sr concentrations from 31 940 ppm (apatite rock) to 254.9 ppm (nepheline syenite) in the Khibina Massif indicates that the magmas differentiated by fractional crystallization. In addition, the carbonatites and apatite rocks may have formed by liquid segregation (Ryabchikov and Hamilton 1994).

Large intrusives composed of alkali-rich rocks also occur elsewhere in Russia as for example the Synnyr complex north of Lake Baykal (Orlova et al. 1994), in the Chulym-Yenesey Basin (Zubkov et al. 1991), and in the central Aldan region (Bogatikov et al. 1994; Mitchell et al. 1994; Pokrovskiy and Vinogradov 1991; Apt et al. 1998). Brief summaries of the alkali-rich rocks of Siberia including a bibliography were published by Butakova (1974).

6.9 Western North America

The continent of North America contains a large number of alkali-rich plutons and volcanic centers ranging in age from Proterozoic to Pleistocene. Barker (1974) listed 127 locations in Canada and the USA where such rocks occur. Some important examples of alkali-rich rocks in North America are discussed elsewhere in this book: (1) Syenite plutons of New England (Sect. 5.10.2); (2) Monteregian Hills (Sect. 5.10.4; and (3) Coldwell complex and related alkalic intrusives on the north shore of Lake Superior (Sect. 7.1.1). Other well-studied examples of alkali-rich volcanic rocks in North America occur in Montana and Wyoming, in California, and in the Navajo petrologic province of Arizona and New Mexico.

The first comprehensive study of the Sr-isotope geochemistry of Tertiary alkali-rich rocks in the western USA by Powell and Bell (1970) included suites of specimens from the Leucite Hills of Wyoming, the Bearpaw Mountains, Highwood Mountains, and diatremes of Montana, and from the Hopi Buttes and Navajo Lands of Arizona and New Mexico. The results in Fig. 6.50 demonstrate that the alkali-rich rocks at the localities in Montana and Wyoming have a wide range of initial $^{87}Sr/^{86}Sr$ ratios from 0.7034 (Diatremes) to 0.7087 (Highwood Mountains) relative to 0.7080 for E&A. The elevated $^{87}Sr/^{86}Sr$ ratios of the alkali-rich

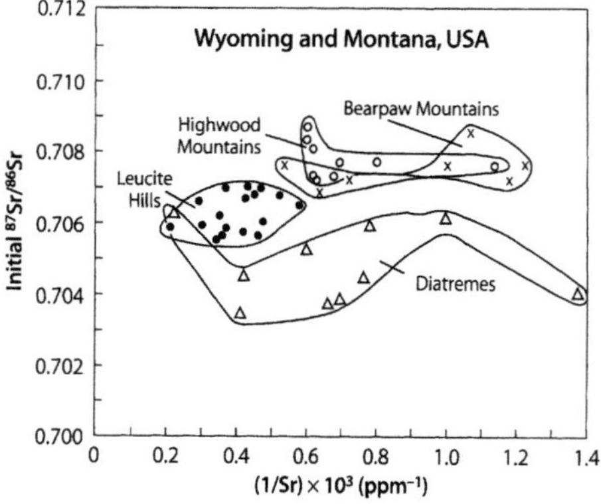

Fig. 6.50. Initial $^{87}Sr/^{86}Sr$ ratios and reciprocal Sr concentrations of alkali-rich igneous rocks in the western USA: Leucite Hills, Wyoming; Highwood Mountains, Bearpaw Mountains, and diatremes, Montana. The $^{87}Sr/^{86}Sr$ ratios of these rocks range widely from 0.7034 (Montana diatremes) to 0.7087 (Highwood Mountains, Montana). The absence of linear data arrays indicates that these rocks did not form by crustal contamination of mantle-derived magmas or by mixing of magmas derived from sources in the crust and mantle (*Source:* data from Powell and Bell 1970)

rocks of Montana and Wyoming resemble those of the continental crust. However, the alkalic rocks are not associated with large volumes of basalt that could have provided the necessary heat for melting of rocks in the crust. Alternatively, alkali-rich magmas derived from the mantle could have assimilated crustal rocks having elevated $^{87}Sr/^{86}Sr$ ratios. However, the high Sr concentrations of the alkali-rich rocks (450 to 6 100 ppm) made their $^{87}Sr/^{86}Sr$ ratios insensitive to change. In addition, the data points in Fig. 6.50 do not form straight mixing lines as expected, if assimilation of crustal rocks or mixing of magmas derived from sources in the crust and mantle had occurred. The clustering of data points in Fig. 6.50 suggests instead that the alkali-rich magmas of Montana and Wyoming originated from sources in the subcontinental lithospheric mantle that contained subducted sediment. The resulting magmas differentiated by fractional crystallization but without significant assimilation of crustal rocks.

The volcanic activity that occurred in Montana and Wyoming during the Tertiary Period is difficult to explain because the mantle-derived magmas had to penetrate the Wyoming Craton composed of granitic gneisses of Archean age. The evidence originally presented by Powell and Bell (1970) indicates that the magmas did not assimilate significant amounts of Sr from the crust on their way to the surface and that the heterogeneity of the $^{87}Sr/^{86}Sr$ ratios of the volcanic rocks was inherited from the magma sources in the subcontinental lithospheric mantle. These insights raise additional questions about the cause for the chemical and isotopic heterogeneity of the subcrustal magma sources and why magma formed in this setting. In spite of the apparently unfavorable circumstance, O'Brien et al. (1995) identified 14 volcanic centers in Montana and three in Wyoming which were active during the Tertiary Period and all of which erupted alkali-rich rocks. The lavas erupted at two of these: the Leucite Hills of Wyoming and the Highwood Mountains of Montana are potassium-rich with K_2O concentrations of 5 to 6%.

6.9.1 Leucite Hills, Wyoming

The Pleistocene lavas of the Leucite Hills in Fig. 6.51 are exposed in a series of mesas about 40 km north-east of the town of Rock Springs on Interstate 80 in Wyoming. These lavas are ultrapotassic in composition with average K_2O concentrations of 6.9 ±1.3% in madupites, 10.2 ±2.4% in wyomingites, and 10.9 ±1.0% in orendites, compared to low average concentrations of Na_2O of 0.9 ±0.1, 1.52 ±1 1, and 1.2 ±0.2%, respectively (Vollmer et al. 1984). These rock types are defined as follows (Williams et al. 1935):

- Madupite: Glass-rich leucitite containing about 50% diopside, 20% phlogopite, and about 10% perovskite, apatite, and iron oxides.

- Orendite: Leucite phonolite composed 40% leucite, 30% sanidine, 15% diopside, 10% phlogopite phenocrysts, about 5% apatite and brookite.
- Wyomingite: Similar to orendite, but having a glassy matrix and lacking sanidine.

The K-rich lavas of the Leucite Hills were erupted between 3.1 and 1.4 Ma, which makes them significantly younger than the volcanic and plutonic rocks at the other volcanic centers in Montana and Wyoming most of which were active between 54 and 45 Ma. The only exceptions are the lamproites of Smokey Butte in north-central Montana (28 Ma) and the rhyolites of Yellowstone Park in northwestern Wyoming (1.9 to 0.07 Ma, Sect. 5.3.2). The ultrapotassic lavas of the Leucite Hills were erupted through sedimentary rocks (Paleozoic to Eocene) that are underlain by granitic basement rocks of the Wyoming Craton which crystallized in Late Archean time between 2810 and 2408 Ma (Wooden et al. 1982; Armstrong and Hills 1967). The origin of K-rich magmas of the Leucite Hills and their subsequent crystallization has been investigated by Carmichael (1967), Barton (1979), Barton and van Bergen (1981); Barton and Hamilton (1978, 1979, 1982), and Kuehner et al. (1981).

The $^{87}Sr/^{86}Sr$ and 1/Sr ratios of the madupites in Fig. 6.52a define a straight line with a positive slope along which the $^{87}Sr/^{86}Sr$ ratios increase from 0.70539 ±0.00003 to 0.70563 ±0.00005 relative to 0.7080 for E&A. Vollmer et al. (1984) concluded that assimilation of Archean basement rocks does not explain the enrichment in light REEs of the madupite and orendite/wyomingite lavas. Therefore, magma mixing is the more likely explanation for the good fit of the madupites to a straight line. The data points of the orendites and wyomingites form a cluster in Fig. 6.52a consistent with the results of Powell and Bell (1970) in Fig. 6.50. The isotopic data of Vollmer et al. (1984) therefore strengthen the evidence that the orendite and wyomingite magmas at Leucite Hills originated from source rocks that had a range of elevated $^{87}Sr/^{86}Sr$ ratios between 0.70530 and 0.70609 relative to 0.7080 for E&A and that their magmas were not significantly differentiated by fractional crystallization or by assimilation of crustal rocks.

The $^{87}Sr/^{86}Sr$ and $^{143}Nd/^{143}Nd$ ratios of the ultrapotassic lavas of the Leucite Hills form two clusters of data points in the "crustal" quadrant of Fig. 6.52b. These clusters appear to define a mixing line consistent with the hypothesis that the magmas of the madupites and orendite/wyomingites formed from mixtures of lithospheric mantle and subducted sediment analogous to the origin of the K-rich lavas of Italy (Sect. 6 5.1). Alternatively, Vollmer et al. (1984) favored the explanation that one of the magma sources in the mantle had been altered by metasomatism prior to the formation of magmas.

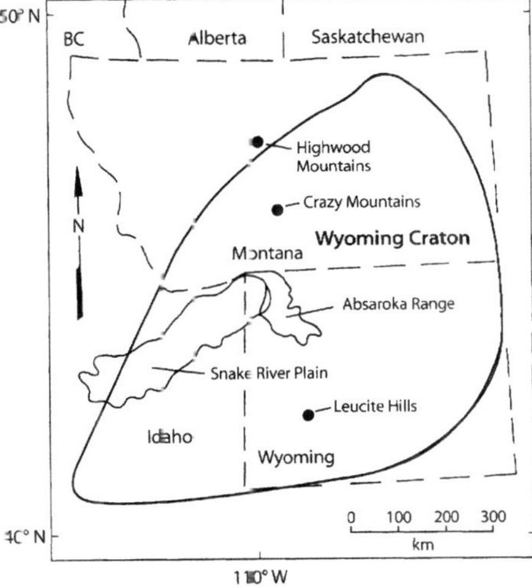

Fig. 6.51. Location of the Leucite Hills of Wyoming and of the Highwood Mountains, Montana, USA. The Leucite Hills are underlain by the Archean gneisses of the Wyoming Craton, whereas the Highwood Mountains are located in the Great Falls Tectonic Zone along the northwestern margin of the craton (*Source:* O'Brien et al. 1995)

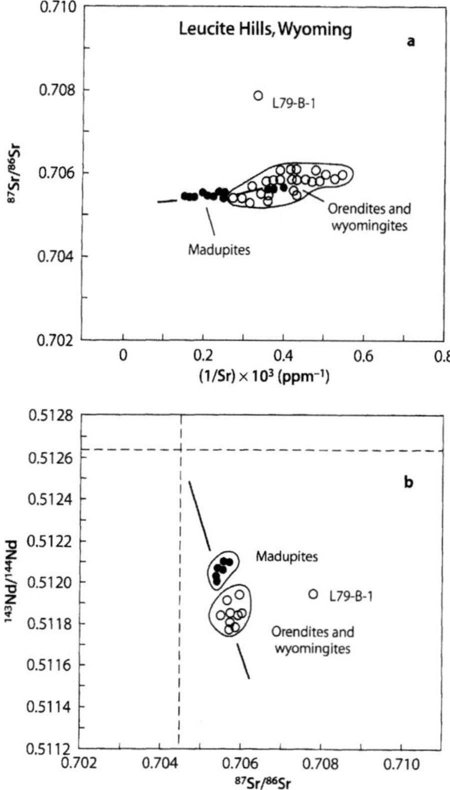

(1991). This area was a major center of volcanic activity at 52 Ma (Eocene), which resulted in the intrusion of several laccoliths into Cretaceous sedimentary rocks and the eruption of leucitite and minette lava flows. These rocks have high concentrations of K_2O ranging from 4.42 to 7.61% (Burgess 1941; O'Brien et al. 1995) much like the rocks of the Leucite Hills in Wyoming (Fig. 6.51).

The initial $^{87}Sr/^{86}Sr$ and $^{143}Nd/^{144}Nd$ ratios of the minettes and phonolites of the Highwood Mountains range from 0.706512 to 0.708806 and from 0.511636 to 0.512018, respectively. The $^{206}Pb/^{204}Pb$ ratios lie between 16.109 and 17.905 (O'Brien et al. 1995). Clearly, the isotope ratios of Sr, Nd, and Pb (like those of the lavas of the Leucite Hills) appear to have a strong crustal signature that is reinforced in Fig. 6.53 by the location of the Highwood data field well within the "crustal" quadrant.

However, the alkali-rich magmas of the Highwood Mountains could not have formed by partial melting of granitic basement rocks. In addition, trace-element data interpreted by O'Brien et al. (1995) also do not support the expectation that the magmas assimilated

Fig. 6.52. a Strontium-isotope mixing diagram of madupites (*solid circles*) and orendite/wyomingites (*open circles*), Leucite Hills, Wyoming, USA. The madupites form a linear array explainable either by contamination of the madupite magma by assimilation of crustal rocks or by mixing of magmas having slightly different $^{87}Sr/^{86}Sr$ ratios. The data points of the orendites/wyomingites scatter as reported by Powell and Bell (1970) and shown in Figure 6.50. Sample *L79-B-1* (Boar's Tusk Mesa) has an anomalously high $^{87}Sr/^{86}Sr$ ratio (0.70779 ±0.00005) and a low Rb concentration (130 ppm); **b** $^{87}Sr/^{86}Sr$ and $^{143}Nd/^{144}Nd$ ratios of K-rich lavas in the Leucite Hills, Wyoming. The data points representing madupites and orendites and wyomingites form two clusters in the quadrant normally occupied by rocks of the continental crust. These data are explainable by magmas that formed in the mantle by mixing of melts that formed from subducted sediment (high $^{87}Sr/^{86}Sr$, low $^{143}Nd/^{144}Nd$) and ultra-mafic rocks (low $^{87}Sr/^{86}Sr$, high $^{143}Nd/^{144}Nd$) (*Source:* data by Vollmer et al. 1984)

6.9.2 Highwood and Crazy Mountains, Montana

The Highwood Mountains of Montana in Fig. 6.51 are located near the town of Geraldine, which lies about 80 km east of the city of Great Falls on the Missouri River. The geology of the Highwood Mountains was described by Larsen et al. (1941) and, more recently, by O'Brien et al.

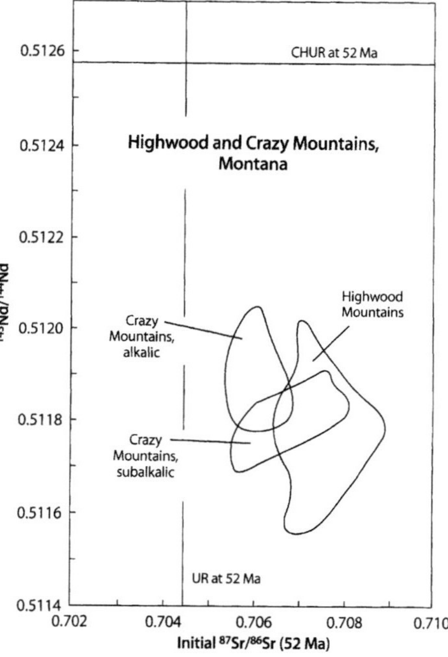

Fig. 6.53. Isotope ratios of Sr and Nd of the alkali-rich volcanic rocks of the Highwood Mountains and of the Crazy Mountains in Montana. The lavas of the Highwood Mountains are K-rich whereas those of the Crazy Mountains are Na-rich. The isotope ratios of CHUR and UR were recalculated to 52 Ma (*Sources:* O'Brien et al. 1995; Dudas et al. 1987)

significant amounts of crustal rocks. Therefore, the isotope compositions of Sr Nd, and Pb as well as the high K_2O concentrations of the rocks reflect the isotopic and chemical composition of the magma sources in the mantle.

O'Brien et al. (1995) proposed that the magmas originated in the asthenospheric mantle and assimilated metasomatically altered lithospheric mantle represented by inclusions of harzburgite containing phlogopite veins. According to this hypothesis, the presence of phlogopite in the lithospheric mantle underlying the Wyoming Craton is a consequence of alkali metasomatism caused by subduction of oceanic crust of the Farallon Plate. Melting of phlogopite accounts for the high K concentration of the resulting magma.

The problem of explaining the eruption of alkali-rich lavas in Montana and Wyoming during the Tertiary Period was also investigated by Dudas et al. (1987) in the Crazy Mountains of Montana (Fig. 6.51). The volcanic activity at this location was synchronous with that of the Highwood Mountains, but the rocks are Na-rich rather than K-rich. The lavas of the Crazy Mountains form a data field in Fig. 6.53 that partly overlaps the data field of the K-rich lavas of the Highwood Mountains. In addition, the isotope ratios of Sr, Nd, and Pb of monticellite peridotite at Haystack Butte (48 Ma), located about 5 km northeast of Geraldine, are similar to those of the Highwood Mountains (O'Brien et al. 1995). Therefore, the volcanic rocks at the three centers considered here are characterized by having elevated initial $^{87}Sr/^{86}Sr$ ratios as well as low $^{143}Nd/^{144}Nd$ and $^{206}Pb/^{204}Pb$ ratios (Table 6.6).

The regional similarity of isotopic and chemical compositions of the Tertiary alkali-rich lavas of Montana and Wyoming implies that the magma sources in the subcrustal lithospheric mantle have certain characteristics on the same regional scale. These may have been provided by the head of a weak asthenospheric plume which flattened itself against the underside of the lithospheric mantle or by subducted oceanic crust. In addition, the low $^{143}Nd/^{144}Nd$ and $^{206}Pb/^{204}Pb$ ratios of the lavas in this petrologic province require an old component that was depleted in Sm and U. This component may be the subcontinental lithospheric mantle that underlies the Wyoming Craton and whose Sm/Nd and U/Pb ratios were reduced at about 1.5 ±0.1 Ga as indicated by model Nd-dates calculated by Dudas et al.

(1987). These dates were calculated by relating the $^{143}Nd/^{144}Nd$ ratios of the lavas to the isotopic evolution of Nd in CHUR. The formation of magmas at about 50 Ma was triggered either by decompression caused by fracturing of the overlying crust or by heat provided by the hypothetical plume, or both. The unusual circumstances surrounding the petrogenesis of the Tertiary alkali-rich lavas that were extruded through the thick crust of the Archean Wyoming Craton have contributed to the importance of studies of mantle-derived inclusions at volcanic centers throughout the western states of the USA.

6.9.3 Sierra Nevada Mountains, California

Ultrapotassic lavas and small intrusives also occur at scattered locations in the central Sierra Nevada Mountains of California (Bateman et al. 1963) southeast of Yosemite National Park and in the area defined by latitudes 37°20' to 37°38' N and from longitudes 119°00' to 119°38' W (Van Kooten 1980). The lavas, which were erupted through the Sierra Nevada Batholith between 3.6 and 3.4 Ma, consist of interbedded high-K and low-K basalts and leucitites. The latter occur in Deep Springs Valley at 37°24' N and 118°03' W in the Blanco Mountain Quadrangle of California. The K_2O concentration of the basaltic lavas ranges from 1.9% to 8.4%.

The granitic rocks of the Sierra Nevada Batholith and the eruption of the associated calc-alkaline lavas between 210 and 79 Ma was attributed by Hay (1976) to subduction of oceanic crust along the western margin of North America. The igneous activity in the central sector of the Sierra Nevada Mountains ended at about 79 Ma when a decrease of the dip of the subducted slab caused the overlying lithospheric mantle and continental crust to be shielded from the heat flow of the underlying asthenosphere. After subduction ceased at about 10 Ma, heat from the asthenosphere increased the temperature of the lithospheric mantle and thus caused the Late Cenozoic uplift of the Sierra Nevada Mountains (Crough and Thompson 1977).

The initial $^{87}Sr/^{86}Sr$ and 1/Sr ratios of the suite of K-rich basalts (basanites and olivine basalts) and low-K nodule-bearing basalt in the Sierra Nevada Mountains of California form separate clusters of data points in Fig. 6.54. The initial $^{87}Sr/^{86}Sr$ ratios of

Table 6.6.
Initial $^{87}Sr/^{86}Sr$, $^{143}Nd/^{144}Nd$ and $^{206}Pb/^{204}Pb$ ratios for Leucite Hills, Highwood Mountains, Haystack Butte and Crazy Mountains

Location	$^{87}Sr/^{86}Sr$	$^{143}Nd/^{144}Nd$	$^{206}Pb/^{204}Pb$	Reference
Leucite Hills	0.7058	0.51193	–	Vollmer et al. (1984)
Highwood Mts.	0.7078	0.51173	17.7	O'Brien et al. (1995)
Haystack Butte	0.7050	0.51250	17.87	O'Brien et al. (1995)
Crazy Mts.	0.7060	0.51190	16.94	Dudas et al. (1987)

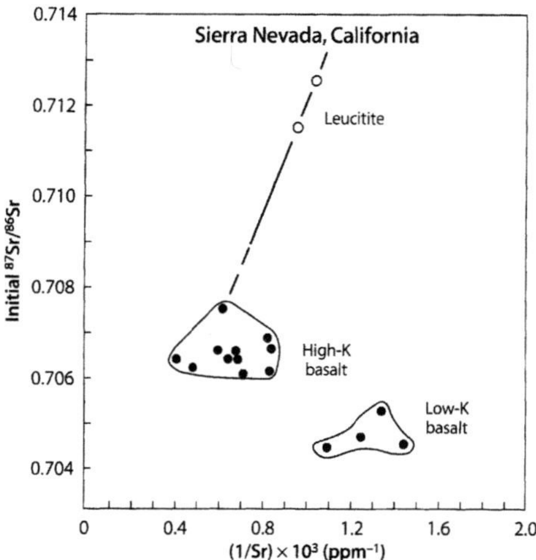

Fig. 6.54. Initial ^{87}Sr/^{86}Sr and 1/Sr ratios of basaltic rocks of Pliocene age in the Sierra Nevada Mountains of central California. The high-K and low-K lavas form separate clusters indicating that their magmas originated from different sources. The leucitites of Deep Springs Valley have high ^{87}Sr/^{86}Sr ratios and may have formed from K-rich basaltic magma by assimilation of crustal rocks or by partial melting of K-rich source rocks in the mantle (*Source:* Van Kooten 1981)

the K-rich basalt (0.70608 to 0.70754) are consistently higher than those of the low-K basalt (0.70441 to 0.70532) relative to 0.71025 for NBS 987 which means that the two magma types formed from different kinds of source rocks.

Van Kooten (1980, 1981) proposed that the dehydration of the subducted oceanic crust under the western edge of the North American continent released alkalirich aqueous fluids that permeated the overlying lithospheric mantle and enriched it in K and ^{87}Sr. The subsequent increase in temperature and the presence of water caused melting in the metasomatized lithospheric mantle represented by the phlogopite-bearing pyroxenite and peridotite inclusions in the low-K basalt flows. Accordingly, the K-rich magmas formed by melting of the metasomatized lithospheric mantle, whereas the low-K lavas originated by partial melting of the subducted oceanic crust.

The two leucitites from Deep Springs Valley in Fig. 6.54 have high ^{87}Sr/^{86}Sr ratios of 0.71152 and 0.71251 relative to 0.71025 for NBS 987. Their data points form a straight line that extrapolates to the cluster of high-K basalts. This interpretation permits the leucitites to have evolved from high-K basalt magma by assimilation of crustal rocks (e.g. granulites) whose ^{87}Sr/^{86}Sr ratio was greater than 0.713, or even by melting of K-rich sedimentary rocks.

The study by Van Kooten (1981) provides a plausible explanation for the K-rich lavas of central California as a consequence of subduction of oceanic crust under the North American Plate. This tectonic setting permits magma formation from several different kinds of source rocks, including: (*1*) Metasomatically altered rocks of the lithospheric mantle wedge; (2) Hydrated basalt and/or marine sediment of the subducted oceanic crust; and (*3*) Igneous or metamorphic rocks of the basal crust. In addition, magmas derived from these kinds of sources can assimilate rocks from other sources, as exemplified by the leucitites of Deep Springs Valley.

6.9.4 Navajo Petrologic Province, Arizona and New Mexico

A third example of ultrapotassic lavas and dikes occurs in the land of the Navajo, south of the Four Corners area where Utah, Arizona, New Mexico, and Colorado join at 37°00' N and 109°03' W. The Oligocene volcanic rocks of this region include minettes, trachybasalts, and monchiquites. The K_2O concentrations of the minettes analyzed by Rogers et al. (1982) range from 4.57 to 6.93% and their K_2O/Na_2O ratio is greater than 2.5. The presence of crustal as well as ultramafic inclusions in the minettes indicates that their magmas originated in the mantle. Powell and Bell (1970) reported that the volcanic rocks of the Navajo petrologic province have elevated initial ^{87}Sr/^{86}Sr ratios ranging from 0.7050 to 0.7079 relative to E&A = 0.7080. The presence of ultramafic inclusions excludes a crustal origin for the magmas, while the high Sr concentration of these rocks (average Sr = 1 306 ±145 ppm, 2$\bar{\sigma}$, N = 11) makes their ^{87}Sr/^{86}Sr ratios insensitive to change by assimilation of crustal rocks. The initial ^{87}Sr/^{86}Sr and 1/Sr ratios of the rocks analyzed by Powell and Bell (1970) and by Roden (1981) form a cluster of data points in Fig. 6.55 somewhat like the K-rich lavas of the Sierra Nevada (Fig. 6.54). The clustering of the data points limits the extent of fractional crystallization in the petrogenesis of the Navajo volcanics. Several lines of evidence, therefore, point to the conclusion that the K-rich magmas were relatively unfractionated and uncontaminated melts that formed either by partial melting of subducted oceanic sediment or of lithospheric mantle that was enriched in alkali metals by metasomatism, or both.

Six samples of minette and one trachybasalt analyzed by Powell and Bell (1970) and Roden (1981) define a straight line in Fig. 6.55. The collinearity of some of the data points supports the hypothesis that a mantle-derived magma was contaminated at the source by mixing with varying amounts of magma

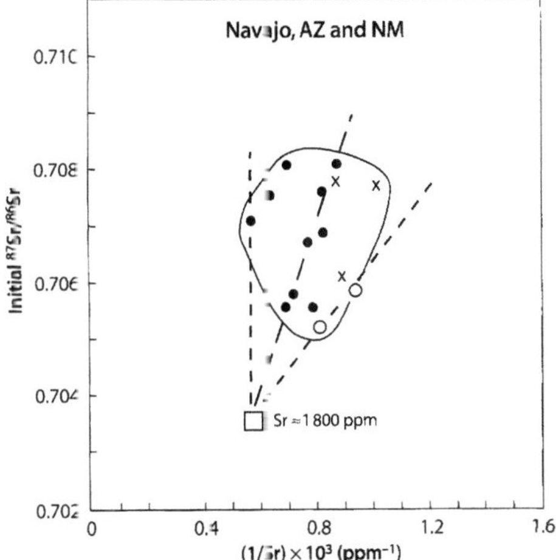

Fig. 6.55. Strontium-isotope mixing diagram for minettes (*solid circles*), trachybasalts (*crosses*), and monchiquites (*open circles*) in the Navajo petrologic province of the Four Corners area of Arizona and New Mexico. The scatter of data points is tentatively explained here by mixing of a hypothetical magma (^{87}Sr/^{86}Sr ≈ 0.7035, Sr = 1800 ppm) with a variety of partial melts derived from subducted sediment and/or Rb-enriched ultramafic rocks in the mantle (*Sources:* data from Powell and Bell 1970; Roden 1981)

derived from Rb-enriched ultramafic rocks or from subducted sediment. Extrapolation of the main trend in Fig. 6.55 to ^{87}Sr/^{86}Sr = 0.7035 yields an estimated Sr concentration of 1800 ppm for the primary magma component.

6.9.5 Hopi Buttes, Arizona

The Pliocene alkalic rocks of the Hopi Buttes at about 35°30' N and 110°00 W in north-central Arizona may be genetically related to those of the Navajo petrologic province located approximately 120 km to the north-east (Williams 1936). The alkali-rich igneous rocks of this area consist of monchiquites, teschenites, syenites, and tuff. Powell and Bell (1970) originally reported that the initial ^{87}Sr/^{86}Sr ratios of these rocks range from 0.7036 to 0.7092 relative to 0.7080 for E&A. The data points representing these rocks in Fig. 6.56 scatter like those of the Navajo petrologic province. The low initial ^{87}Sr/^{86}Sr ratio of the teschenite (0.7036) indicates that it formed from an uncontaminated mantle-derived parent magma. The initial ^{87}Sr/^{86}Sr ratios and Sr concentrations of the other rock types in the Navajo and Hopi Buttes areas are derivable from the teschenite magma either by assimilation of heterogeneous crustal rocks or by mixing with partial melts of subducted sediment

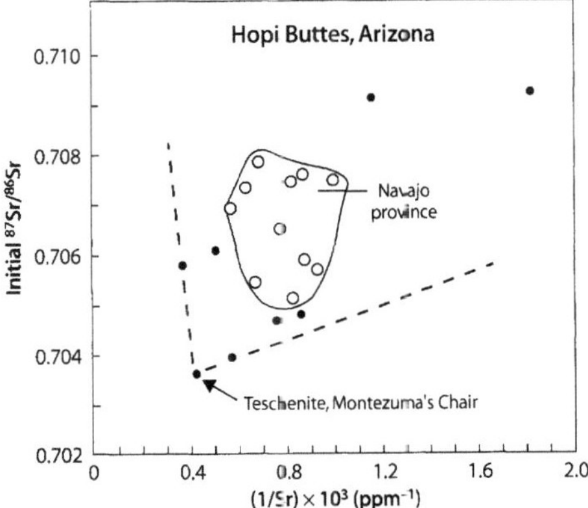

Fig. 6.56. Initial ^{87}Sr/^{86}Sr and 1/Sr ratios of alkali-rich volcanic rocks of Pliocene age in the Hopi Buttes area (*solid circles*) and in the Navajo petrologic province (*open circles*), Arizona. A teschenite from Montezuma's Chair has the lowest initial ^{87}Sr/^{86}Sr ratio (0.7036) and represents the uncontaminated mantle-derived parent magma of the alkalic rocks in both areas. The scatter of data points within the wedge-shaped field is attributable either to assimilation of heterogeneous crustal rocks or to mixing with partial melts derived from subducted sediment or from Rb-enriched ultramafic rocks (*Source:* Powell and Bell 1970)

or Rb-enriched ultramafic rocks. Fractional crystallization played only a minor role in the petrogenesis of the rocks in both areas.

6.10 Ultramafic Inclusions, Western North America

The K-rich alkalic rocks in western North America, Italy, southeastern Spain, the Laacher See in Germany, and in the Western Rift of East Africa all have elevated and variable ^{87}Sr/^{86}Sr ratios, perhaps because the magma sources in the mantle at these locations were enriched in Rb by alkali metasomatism long before the onset of volcanic activity. This explanation is supported by the presence of phlogopite and alkali-rich amphiboles in the ultramafic nodules that were brought to the surface by the alkali-rich lava flows. The apparent isotopic heterogeneity of the mantle underlying western North America was noted by Menzies et al. (1983) and has motivated a large number of studies of ultramafic nodules in the alkali-rich lava flows in this region (e.g. Best 1970; Zartman and Tera 1973 (Pb); Leeman 1979 (Pb); Irving 1980; Wilshire et al. 1980; Bergman et al. 1981; Mathez et al. 1984 (C); Othman et al. 1990; Condie et al. 1999). Ultramafic nodules have been described primarily from New Mexico, Arizona, Nevada, and California, as well as in Texas and in northern Mexico (Baja California) at sites identified in Fig. 6.57.

Fig. 6.57. Locations of Tertiary volcanic fields containing ultramafic inclusions in the southwestern USA: *1.* Bandera Crater; *2.* Puerco Necks; *3.* Kilbourne Hole; *4.* San Carlos; *5.* Hoover Dam; *6.* Grand Canyon; *7.* San Francisco volcanic field; *8.* Dish Hill; *9.* San Quintin; *10.* Black Rock; *11.* Knippa (*Source:* adapted from Frey and Prinz 1978)

6.10.1 New Mexico

The principal occurrences of ultramafic nodules in New Mexico are at Bandera Crater (Laughlin et al. 1971), the Puerco Necks (Kudo et al. 1972), and at Kilbourne Hole and Potrillo (Stueber and Ikramuddin 1974; Basu 1978). The data of Laughlin et al. (1971) demonstrate that the spinel lherzolite inclusions in basalt at Bandera Crater (no. 1, Fig. 6.57) have a range of $^{87}Sr/^{86}Sr$ ratios from 0.7020 to 0.7052 (relative to 0.7080 for E&A) and are not genetically related to the host basalt (0.7025 to 0.7031).

The inclusions at the Puerco Necks (no. 2, Fig. 6.57) located only about 30 km northeast of the Bandera Crater have even higher $^{87}Sr/^{86}Sr$ ratios than those at Bandera Crater. Kudo et al. (1972) reported $^{87}Sr/^{86}Sr$ ratios between 0.7068 and 0.7089 for lherzolite and websterite nodules from this location, compared to only 0.7040 ±0.0001 ($2\bar{\sigma}$, $N = 3$) for the host basalt, relative to 0.7080 for E&A. The basalts at the Puerco Necks and at the Bandera Crater have a high average Sr concentration (470 ±100 ppm) compared to the ultramafic inclusion which contains only 47 ±27 ppm Sr on the average. Nevertheless, there is no evidence at either locality that the inclusions were contaminated by Sr from the host basalt.

The elevated and wide-ranging $^{87}Sr/^{86}Sr$ ratios of the ultramafic nodules at Bandera Crater and Puerco Necks can be interpreted in quite different ways:

1. The source regions of ultramafic nodules have a range of Rb/Sr ratios as a result of alkali metasomatism that occurred long before the nodules were transported to the surface by magma.

2. The alkali-rich aqueous fluids also increased the $^{87}Sr/^{86}Sr$ ratios of the rocks in the source regions of the inclusions.

3. The Sr-poor nodules were contaminated by groundwater (or even rain) containing Sr derived locally from the rocks of the continental crusts.

In the second case, the aqueous fluid originated from subducted oceanic crust that underplated the lithospheric mantle or from the head of a weak asthenospheric plume that contained blocks of subducted oceanic crust. In either case, the Sr was of marine origin and had an elevated $^{87}Sr/^{86}Sr$ ratio compared to the Sr in the lithospheric mantle. Therefore, the $^{87}Sr/^{86}Sr$ ratio increased at the time of the metasomatism instead of by in situ decay of ^{87}Rb as required in Case 1.

In Alternatives 1 and 2, the $^{87}Sr/^{86}Sr$ ratios of ultramafic inclusions are assumed to be identical to those of the domain in the lithospheric mantle (or basal crust) from which they originated. However, the low Sr concentrations of ultramafic inclusions makes their $^{87}Sr/^{86}Sr$ ratios vulnerable to change by the addition of Sr from external sources such as the magma that transported them to the surface and groundwater or even rain water after their eruption.

Although alkali metasomatism of ultramafic rocks in inclusions may cause the formation of phlogopite and kaersutite, the presence of these minerals is not necessarily proof that they formed as a result of metasomatic alteration of the lithospheric mantle. For example, Laughlin et al. (1971) reported that phlogopite and kaersutite in some of the lherzolite inclusions at Bandera Crater occur in reaction rims at contacts between the basalt magma and the inclusions.

Another noteworthy feature of lherzolite inclusions at Puerco Necks in New Mexico (no. 2, Fig. 6.57) is that olivine and enstatite are not in isotopic equilibrium with the Sr in the whole-rock sample. Kudo et al. (1972) reported an $^{87}Sr/^{86}Sr$ ratio of 0.7068 for a lherzolite inclusion, 0.7083 for olivine in that inclusion, and 0.7084 for enstatite. Similarly, the constituent minerals of a lherzolite inclusion at the Kilbourne Hole in New Mexico (no. 3, Fig. 6.57) analyzed by Stueber and Ikramuddin (1974) also have discordant $^{87}Sr/^{86}Sr$ ratios whose values increase with decreasing Sr concentrations. Their data in Fig. 6.58 reveal a wide range of $^{87}Sr/^{86}Sr$ ratios in clinopyroxene (0.7026), orthopyroxene (0.7059), and olivine (0.7061) relative to 0.7080 for E&A. In this case, isotopic disequilibrium exists not only among the minerals of the lherzolite inclusion but also between these minerals and the host basalt (0.70325). Stueber and Ikramuddin (1974) reported that the nodule in Fig. 6.58 has xenomorphic granular texture, is extremely fresh, and does not contain secondary minerals or glass. Nevertheless, the minerals in this

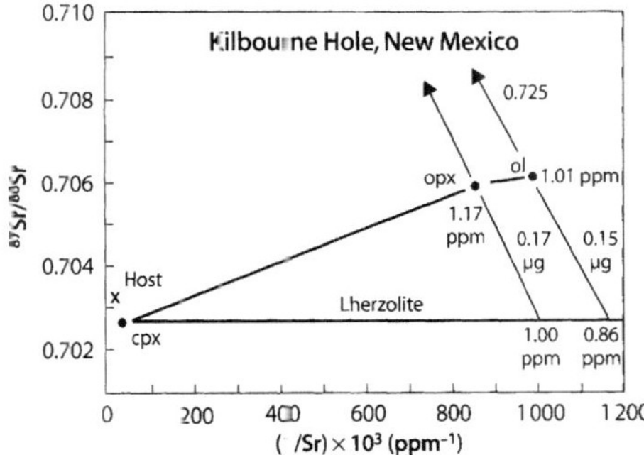

Fig. 6.58. Explanation for the elevated ^{87}Sr/^{86}Sr ratios of olivine and orthopyroxene in an ultramafic inclusion collected at Kilbourne Hole, New Mexico. The reconstruction is based on the method of Lutz et al. (1988) and assumes that small amounts of Sr having an ^{87}Sr/^{86}Sr ratio of 0.725 were added to all of the minerals in this inclusion. The ^{87}Sr/^{86}Sr ratios of olivine and orthopyroxene increased as shown, because their Sr concentrations were low (0.86 ppm for olivine and 1.0 ppm for orthopyroxene). The ^{87}Sr/^{86}Sr ratio of clinopyroxene was not increased significantly because of its comparatively high Sr concentration of 43.7 ppm (*Source:* data from Stueber and Ikramuddin 1974)

nodule form a linear array in coordinates of ^{87}Sr/^{86}Sr and ^{87}Rb/^{86}Sr (not shown), which the authors interpreted as an isochron yielding a date of 1 243 ±230 Ma ($\lambda = 1.42 \times 10^{-11}$ yr^{-1}). According to this interpretation, the isotope composition of Sr in the constituent minerals of this inclusion was not homogenized by diffusion, even though the rock resided in the mantle for more than one billion years before a fragment of it was entrained in the basalt magma erupted at Kilbourne Hole in Late Tertiary time.

The alternative explanation is that the minerals were contaminated after the eruption with Sr derived from the country rocks by groundwater. The reconstruction illustrated in Fig. 6.58 is based on the method of Lutz et al. (1988) and assumes that the Sr added to the minerals had an ^{87}Sr/^{86}Sr ratio of 0.725. The results demonstrate that the present ^{87}Sr/^{86}Sr ratios of orthopyroxene and olivine are explainable by the addition of about 0.16 ±0.01 µg of the Sr contaminant per gram of mineral. The resulting change in the ^{87}Sr/^{86}Sr ratio and concentration of Sr in the minerals depends on the magnitude of the initial Sr concentrations of the minerals. The construction in Fig. 6.58 demonstrates that the Sr concentration of olivine and orthopyroxene increased by about 17%, whereas the addition of 0.16 µg of Sr to one gram of clinopyroxene (Sr = 43.7 ppm) increased its Sr concentration by only about 0.4% and therefore did not increase its ^{87}Sr/^{86}Sr ratio appreciably. According to this interpretation, the internal Rb-Sr isochron date derived from the minerals of this inclusion is fictitious.

The small amounts of Sr required to explain the discordance of ^{87}Sr/^{86}Sr ratios of low-Sr minerals may be deposited on grain surfaces and in fractures not only by groundwater but also by rain. Graustein and Armstrong (1983) reported that rain collected at the airport in Santa Fe, New Mexico, had an ^{87}Sr/^{86}Sr ratio of 0.7088 and a Sr concentration of 8.4 µg L^{-1}. Rain in the Santa Fe Range was found to have higher ^{87}Sr/^{86}Sr ratios (0.7100 to 0.7104) but a lower average Sr concentration of 1.34 µg L^{-1}. These results demonstrate that only a small volume of rainwater in New Mexico can provide sufficient Sr to alter the ^{87}Sr/^{86}Sr ratios of low-Sr minerals such as olivine and enstatite. The contaminant Sr on the surfaces of these minerals can be removed by leaching with dilute acid as demonstrated by Basu and Murthy (1977b).

According to the hypothesis of groundwater alteration, all of the minerals of utramafic inclusions in lava flows have the same ^{87}Sr/^{86}Sr ratio at the time of eruption because of diffusion of ^{87}Sr at the elevated temperatures that prevail in the lithospheric mantle. This assumption was confirmed by Basu (1978) who demonstrated that phlogopite and Cr-diopside crystals in a lherzolite inclusion at Kilbourne Hole have identical ^{87}Sr/^{86}Sr ratios of 0.70332 ±0.0007 (phlogopite) and 0.70335 ±0.0009 (Cr-diopside) relative to 0.71025 for NBS 987. These minerals were not significantly contaminated with Sr from external sources because of their high Sr concentrations of 150.1 ppm (phlogopite) and 132.22 ppm (Cr-diopside). In addition, the phlogopite and Cr-diopside grains maintained the same ^{87}Sr/^{86}Sr ratio while they resided in their mantle source even though the Rb/Sr ratio of the phlogopite was 56 times higher than that of the Cr-diopside.

The study of ultramafic inclusions in the Tertiary volcanic centers of New Mexico therefore supports the conclusions that the low-Sr minerals of ultramafic nodules are prone to alteration by the addition of small amounts of Sr having elevated ^{87}Sr/^{86}Sr ratios by exposure to groundwater or even rainwater.

6.10.2 Arizona

Evidence for contamination of ultramafic inclusions in Tertiary lavas flows is not restricted to sites in New Mexico but has been reported also from volcanic centers in Arizona including Peridot Mesa in the San Carlos Indian Territory.

Peridot Mesa is located about 30 km east of the town of Globe, Arizona (no. 4 in Fig. 6.67). The mesa is about 3 km in diameter and is capped by a basalt flow (3 to 6 m thick) that originated from a volcanic cone in the southwest corner of Peridot Mesa. Ultramafic nodules occur at this locality within basalt bombs and as inclusions in the basalt flow and associated agglomerate.

Frey and Prinz (1978) distinguished two types of ultramafic inclusions characterized by the presence of Cr-diopside (Group I) and Al-augite (Group II).

The minerals of a lherzolite inclusions from San Carlos analyzed by Stueber and Ikramuddin (1974) have discordant $^{87}Sr/^{86}Sr$ ratios ranging from 0.7026 (clinopyroxene, Sr = 24.6 ppm) to 0.7097 (olivine, Sr = 1.65 ppm), whereas the host basalt has an $^{87}Sr/^{86}Sr$ ratio of 0.7029 relative to 0.7080 for E&A. These minerals do not form a straight line in the Rb-Sr isochron diagram, but their $^{87}Sr/^{86}Sr$ ratios do increase as their Sr concentrations decrease. The authors reported that the ultramafic nodules at San Carlos contain evidence of weathering and that they are "extremely porous" making them susceptible to addition of Rb and Sr by groundwater.

The addition of crustal Sr ($^{87}Sr/^{86}Sr$ = 0.725) to the minerals of the lherzolite inclusion analyzed by Stueber and Ikramuddin (1974) caused their $^{87}Sr/^{86}Sr$ ratios to increase as shown in Fig. 6.59. The olivine gained 0.54 µg g^{-1} of crustal Sr causing its Sr concentration to increase by about 47%. The Sr concentration of the orthopyroxene increased about 31% by the addition of 0.71 µg g^{-1} of crustal Sr. As a result, the $^{87}Sr/^{86}Sr$ ratios of both minerals increased, whereas the $^{87}Sr/^{86}Sr$ ratio of the clinopyroxene was virtually unaffected because of its relatively high Sr concentration of 24.6 ppm.

The resistance of Sr-rich minerals in ultramafic inclusions to contamination is further illustrated by the work of Basu (1978) who reported an $^{87}Sr/^{86}Sr$ ratio of 0.70278 ±0.00005 (relative to 0.71025 for NBS 987) and a high Sr concentration of 711.4 ppm for kaersutite in a San Carlos nodule. In addition, a kaersutite from a lherzolite nodule at Hoover Dam, Arizona, (no. 5, Fig. 6.57) with a high Sr concentration (577 ppm) also has a low $^{87}Sr/^{86}Sr$ ratio of 0.70282 ±0.00005 that is indistinguishable from the $^{87}Sr/^{86}Sr$ ratio of the San Carlos kaersutite.

A similar case was reported by Bergman et al. (1981) for a wehrlite nodule from the Lunar Crater volcanic field in Nevada. This nodule contains a vein or dike composed of kaersutite, feldspar, and ilmenite. The kaersutite and feldspar have high Sr concentrations of about 600 and 2 663 ppm, respectively. Their $^{87}Sr/^{86}Sr$ ratios are identical at 0.70323. The isotope ratios of Sr of the two minerals in this nodule exemplify the pattern seen elsewhere: the Sr-rich minerals of the inclusion have identical low $^{87}Sr/^{86}Sr$ ratios that are unaffected by contamination by groundwater or rain.

The inverse relation between the $^{87}Sr/^{86}Sr$ and the Sr concentration of minerals in ultramafic inclusions applies even for cognate inclusions. For example, a wehrlite inclusion at Crater 160 in the San Francisco volcanic field of Arizona analyzed by Stueber and Ikramuddin (1974) is a cumulate that formed from the same magma as the host basalt because the $^{87}Sr/^{86}Sr$

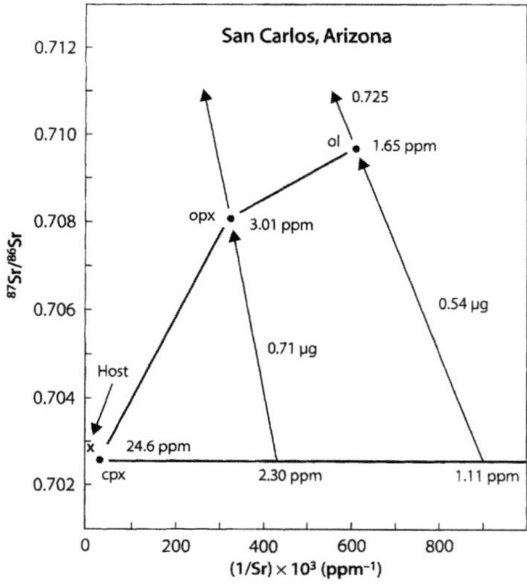

Fig. 6.59. Alteration of the $^{87}Sr/^{86}Sr$ ratios of the minerals of a lherzolite inclusion in the San Carlos Indian Territory, Arizona, as a result of additions of Sr derived from the country rocks ($^{87}Sr/^{86}Sr$ = 0.725) by circulating groundwater. The resulting increase of the $^{87}Sr/^{86}Sr$ ratios of the minerals is inversely proportional to their Sr concentrations. The $^{87}Sr/^{86}Sr$ ratio of the host basalt (0.7029) differs from that of the clinopyroxene (0.7026) indicating that this inclusion is unrelated to the basalt host (*Source:* data from Stueber and Ikramuddin 1974)

ratio of the clinopyroxene of the inclusion (0.7034, Sr = 53.1 ppm) is similar to that of the host basalt (0.7032, Sr = 785 ppm). Even in this case, the olivine (Sr = 1.62 ppm) in this inclusion has an elevated $^{87}Sr/^{86}Sr$ ratio (0.7044), either because of contamination with crustal Sr, or because of in situ decay of ^{87}Rb after the Rb/Sr ratio of the olivine increased at about 1 280 Ma. The choice between these alternative interpretations of these data hinges on the question whether minerals in the mantle can develop and maintain different $^{87}Sr/^{86}Sr$ ratios at high temperatures and over long periods of geologic time.

The scale of isotopic and chemical heterogeneity of ultramafic rocks in the mantle was considered by Hofmann and Hart (1975, 1978) and by Nelson and Dasch (1976), among others. Hofmann and Hart (1975) concluded that the homogeneity of isotopic compositions of Sr and Pb in mantle rocks is maintained by diffusion on a local scale. For example, they determined that a phlogopite sphere with a diameter of one centimeter equilibrates Sr with a melt at 1 000 °C in about 1 200 years. Consequently, phlogopite crystals in magma sources in the mantle cannot accumulate radiogenic ^{87}Sr by in situ decay of ^{87}Rb and therefore must have the same $^{87}Sr/^{86}Sr$ ratios as adjacent mineral grains having low Rb/Sr ratios.

6.10.3 California

Ultramafic inclusions in Quaternary basalt at Dish Hill in southeastern California (no. 8, Fig. 6.57) display the same kind of heterogeneity of $^{87}Sr/^{86}Sr$ ratios reported from occurrences in New Mexico and Arizona. The inclusions occur in trachybasalts and nepheline-normative basanites that were erupted in an area extending from Barstow in San Francisco County to the Nevada state line (Wise 1969; Wilshire et al. 1980). Trachybasalts on Mt. Pisgah have $^{87}Sr/^{86}Sr$ ratios between c.7037 and 0.7043, whereas a basanite sample from Dish Hill yielded 0.7031 relative to 0.7080 for E&A (Peterman et al. 1970d). A specimen of alkali basalt at Dish Hill analyzed by Basu (1978) has an $^{87}Sr/^{86}Sr$ ratio of c.70383 ±0.00005 relative to 0.71025 for NBS 987.

The $^{87}Sr/^{86}Sr$ ratios of the constituent minerals of a lherzolite inclusion from Dish Hill reported by Peterman et al. (1970e) range from 0.7016 ±0.0002 in Cr-diopside (Sr = 74 ppm) to 0.7087 ±0.0002 in olivine (Sr = 11 ppm), and to 0.708 ±0.001 in orthopyroxene (Sr = 2.4 ppm). These results are similar to those reported from other localities in the southwestern USA where the $^{87}Sr/^{86}Sr$ ratios of the constituent minerals of ultramafic inclusions increase with decreasing Sr concentrations and the nodules themselves are not in isotopic equilibrium with their host basalt.

Kaersutite megacrysts from Deadman Lake (Pliocene, about 10 km south of Dish Hill) and from Dish Hill analyzed by Basu (1978) have high Sr concentrations (577 ±103 ppm, 2σ, N = 5) and a narrow range of $^{87}Sr/^{86}Sr$ ratios from 0.70271 ±0.00008 to 0.70295 ±0.00006 with a mean of 0.70279 ±0.00009 (2σ, N = 5). In fact, all of the kaersutite megacrysts from ultramafic inclusions in Arizona (Hoover Dam and San Carlos) and California (Dish Hill and Deadman Lake) analyzed by Basu (1978) have the same average $^{87}Sr/^{86}Sr$ ratio of 0.70279 ±0.00006 (2σ, N = 7), whereas high-Sr phlogopite and Cr-diopside at Kilbourne Hole, New Mexico, have an average $^{87}Sr/^{86}Sr$ ratio of 0.70333 ±0.00003 (2σ, N = 2). These regional average $^{87}Sr/^{86}Sr$ ratios of Sr-rich minerals represent the isotope composition of Sr in the rocks of the upper mantle underlying the southwestern USA.

6.10.4 Baja California, Mexico

Additional support for the contamination of ultramafic nodules by crustal Sr is contained in a study by Basu and Murthy (1977b) of an anhydrous spinel lherzolite inclusion in the San Quintin volcanic field along the west coast of Baja California in Mexico (no. 9, Fig. 6.57). The authors leached mineral separates with cold 2 N HCl for three minutes followed by washing with dis-

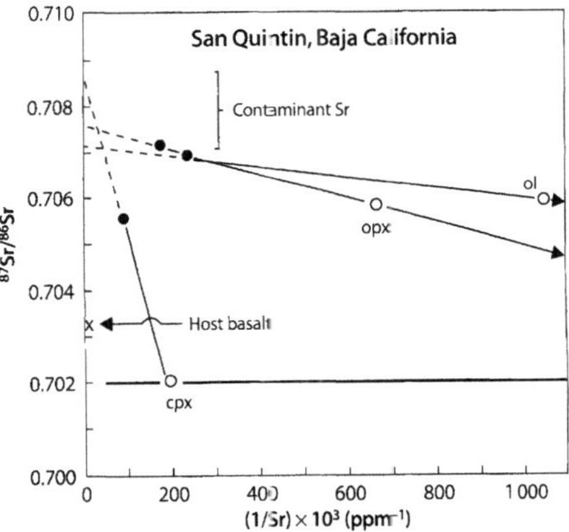

Fig. 6.60. Leaching of the constituent minerals of a lherzolite inclusion in alkali basalt at San Quintin, west coast of Baja California, Mexico. The results show that the leached minerals (*open circles*) have lower Sr concentrations and lower $^{87}Sr/^{86}Sr$ ratios than the unleached minerals (*solid circles*). The contaminant Sr had $^{87}Sr/^{86}Sr$ ratios between c.7072 and 0.7087 and could have originated from seawater and the host basalt. The authors of this study demonstrated that the leached minerals define a Rb-Sr isochron that yields a date of 3.4 Ga. However, the line is a mixing line rather than an isochron. An alternative explanation for the elevated $^{87}Sr/^{86}Sr$ ratios of the olivine and orthopyroxene is that the leaching procedure did not remove all of the contaminant Sr from these minerals. Extrapolation of the leaching lines to $^{87}Sr/^{86}Sr = 0.70203$ (clinopyroxene leached) yields estimates of the original Sr concentrations of clinopyroxene (0.46 ppm) and olivine (0.23 ppm) (*Source:* Basu and Murthy 1977b)

tilled water and acetone prior to analysis. The leaching caused significant decreases of the $^{87}Sr/^{86}Sr$ ratios and the concentrations of Rb and Sr of olivine, orthopyroxene, and clinopyroxene. The relationship between the washed and the unwashed grains in Fig. 6.60 suggests that the Sr removed by washing had $^{87}Sr/^{86}Sr$ ratios between 0.7072 and 0.7087 and may have been a mixture of Sr derived from seawater (0.7091) and the host basalt (0.70318 ±0.00008).

The washed clinopyroxene has a low $^{87}Sr/^{86}Sr$ ratio of 0.70203 ±0.00011 relative to 0.71025 for NBS 987, but the washed olivine and orthopyroxene have surprisingly high $^{87}Sr/^{86}Sr$ ratios of 0.70592 ±0.00018 and 0.70586 ±0.00007, respectively. Basu and Murthy (1977b) demonstrated that the washed mineral grains define a straight line in coordinates of $^{87}Sr/^{86}Sr$ and $^{87}Rb/^{86}Sr$ ratios (not shown) that yields a date of 3.4 ±0.3 Ga (λ = 1.42 × 10^{-11} yr^{-1}) and an initial $^{87}Sr/^{86}Sr$ ratio of 0.70064 ±0.00004. Taken at face value, this date means that the minerals were closed systems for 3.4 billion years while this ultramafic rock resided in the upper mantle. In that case, the effective diffusion coefficient at 1 100 °C was about 10^{-17} cm^2 sec^{-1}, or four orders of magnitude smaller than the diffusion coefficient of phlogopite assumed by Hofmann and Hart (1975, 1978).

Alternatively, the line defined by the minerals of the lherzolite inclusion from San Quintin is a mixing line drawn between the clinopyroxene and the points representing olivine and orthopyroxene. The slope of such a mixing line has no time significance. Although this alternative discredits the date, it does not explain the elevated $^{87}Sr/^{86}Sr$ ratios of the washed olivine and orthopyroxene grains. Basu and Murthy (1977b) favored the explanation that the Sr in these minerals was not isotopically homogenized while the rock resided in the mantle because of the absence of a melt phase in this anhydrous rock even though the minerals indicate an equilibration temperature of about 1100 °C.

A second explanation for the anomalous $^{87}Sr/^{86}Sr$ ratios of the washed olivine and orthopyroxene is that the leaching procedure did not completely remove all of the contaminant Sr from the surfaces and fractures of olivine and orthopyroxene grains. This alternative is investigated in Fig. 6.60. The equation of the leaching line for the orthopyroxene is:

$$^{87}Sr/^{86}Sr = 0.70752 - 0.00000252\,(1/Sr) \times 10^3$$

If the $^{87}Sr/^{86}Sr$ ratio of the orthopyroxene was identical to that of the clinopyroxene (0.70203) at the time of eruption, its Sr concentration was 0.46 ppm instead 1.51 ppm as measured. A similar procedure yields a Sr concentration of 0.23 ppm for olivine prior to eruption. These Sr concentrations are not unreasonably low because the leached whole-rock nodule itself contains only 1.591 ppm Sr. In addition, the extrapolated Sr concentration of the orthopyroxene (0.46 ppm) is higher than the extrapolated Sr concentration of olivine (0.23 ppm) as observed also in Fig. 6.58 (Kilbourne Hole, New Mexico) and Fig. 6.59 (San Carlos, Arizona). Basu and Murthy (1977b) concluded from the low concentrations of large-ion lithophile elements (e.g. K, Rb, Sr, etc.) of this nodule that it originated from a depleted region of the mantle. This conjecture is consistent with the comparatively low $^{87}Sr/^{86}Sr$ ratio of the washed clinopyroxene (0.70203), which represents the Sr in this nodule prior to its contamination by marine Sr.

The isotopic disequilibrium among the constituent minerals of ultramafic inclusions is plausibly explained by groundwater contamination of low-Sr minerals. In that case, the elevated $^{87}Sr/^{86}Sr$ ratios of olivine and orthopyroxene, and of the nodules themselves are not the cause of the high $^{87}Sr/^{86}Sr$ ratios of alkali-rich rocks in the southwestern USA (e.g. the Navajo petrologic province and the Hopi Buttes, Arizona; and the Sierra Nevada Mountains, California). In other words, the ultramafic inclusions do not explain why K-rich volcanic rocks of this region have such high $^{87}Sr/^{86}Sr$ ratios.

6.11 Australia

The Great Dividing Range of southeastern Australia contains numerous volcanic centers primarily of Tertiary age composed of basalt and alkali-rich volcanic rocks. The volcanic centers of this region are scattered along the strike of the mountains over a distance of about 1 300 km from southern Queensland across New South Wales and Victoria and formed in a setting of extensional tectonics associated with the opening of the Tasman Sea and the separation of Tasmania from the Australian mainland. The volcanic rocks of southeastern Australia were dated by Wellman and McDougall (1974a,b) and have been described by Kesson (1973), Cundari (1973), Frey et al. (1978), and others referenced in this section. The Newer basalt province of southern Victoria near Melbourne is composed primarily of alkali-rich rocks containing megacrysts and ultramafic inclusions studied by Irving (1974), and Irving and Green (1976). In western Australia, alkali-rich lamproites and kimberlites (Bergman 1987) occur in the Fitzroy Trough, the Lennard Shelf, and the Halls Creek Mobile belt which together define the boundaries of the Precambrian Kimberley block (Atkinson et al. (1984). These rocks are not only K-rich, but have remarkably high initial $^{87}Sr/^{86}Sr$ ratios suggesting that their formation was preceded by extensive Rb-enrichment of the magma sources in the mantle.

6.11.1 New South Wales

The Tertiary basalts of New South Wales in Fig. 6.61 are closely associated with alkali-rich volcanic rocks and both rock types contain abundant ultramafic inclusions (Wass 1980; Kesson 1973). The basalts studied by O'Reilly and Griffin (1984) have $^{87}Sr/^{86}Sr$ ratios between about 0.7030 and 0.7040, although values ranging up to 0.70544 (relative to NBS 987 = 0.71025) occur in the Monaro volcanic province in southern New South Wales (Fig. 6.61). The basalts form a cluster of data points in Fig. 6.62, confirming petrologic and geochemical evidence cited by O'Reilly and Griffin (1984) that the basalt magmas did not assimilate crustal rocks but differentiated by fractional crystallization.

The basalts of the Monaro province are noteworthy because the $^{87}Sr/^{86}Sr$ ratios vary from 0.703837 to 0.705440 and correlate positively with the Rb/Sr ratios (O'Reilly and Griffin 1984). The slope of the line fitted to the data points yielded a date of 929 ±238 Ma (not shown). O'Reilly and Griffin (1984) doubted the significance of this date, but the discordant $^{87}Sr/^{86}Sr$ ratios of the lavas indicate that the Sr in the magma sources of the basalt was isotopically heterogeneous.

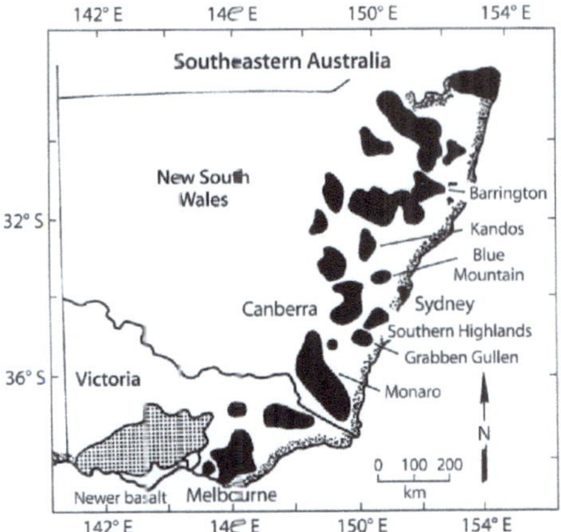

Fig. 6.61. Tertiary volcanic provinces in southeastern Australia (*Source:* adapted from O'Reilly and Griffin 1984)

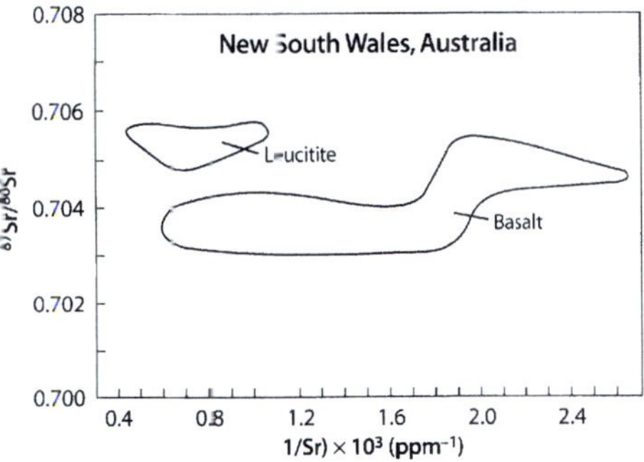

Fig. 6.62. ^{87}Sr/^{86}Sr and reciprocal Sr concentrations of Tertiary basalts and leucitites in the Great Dividing Range of New South Wales, Australia. The leucitites have higher ^{87}Sr/^{86}Sr ratios than the basalts and therefore originated from source regions in the mantle that had been enriched in alkali metals and radiogenic ^{87}Sr prior to magma formation. The clustering of the data points confirms petrologic and geochemical evidence that the basalt and leucitite magmas did not assimilate crustal rocks. The basalt samples originated from the Barrington, Kandos, Blue Mountain, Grabben Gullen, Monaro, and Southern Highland volcanic provinces (*Sources:* data from Nelson et al. 1986; O'Reilly and Griffin 1984; Menzies and Wass 1983; Wass 1980)

The explanation for the range of ^{87}Sr/^{86}Sr ratios of the volcanic rocks in southeastern Australia favored by many investigators is that the lithospheric mantle under this region was enriched in alkali metals and hence in radiogenic ^{87}Sr prior to magma formation (e.g. Wass and Rogers 1980; Menzies and Wass 1983; O'Reilly and Griffin 1984; Nelson et al. 1986). The enrichment pre-

sumably occurred by fluids that originated from deeper levels in the mantle and caused veining of the magma sources that produced the basalt magmas. Similar explanations have been proposed for the formation of alkali-rich volcanic rocks elsewhere in the world. The presence of phlogopite and alkalic amphibole in ultramafic inclusions entrained in the lava flows is unmistakable evidence for the presence of aqueous fluids in some parts of the lithospheric mantle.

The potassium-rich olivine leucitites and analcimites of southeastern Australia in Fig. 6.62 occur in a belt that extends from the western margin of the Lachlan fold belt in central New South Wales to the town of Cosgrove in northern Victoria (Cundari 1973). The leucitites are strongly enriched in Rb (154 ±29 ppm) and Sr (1 468 ±179 ppm) and their ^{87}Sr/^{86}Sr ratios range from 0.70496 to 0.70572 relative to E&A = 0.7080 (Nelson et al. 1986). The K-rich lavas form a separate cluster of data points in Fig. 6.62, implying that the leucitite magmas originated from different sources in the mantle than the basalt magmas and rose to the surface without appreciable differentiation by fractional crystallization or assimilation of crustal rocks.

6.11.2 Ultramafic inclusions, Kiama, New South Wales

The ultramafic inclusions in the Tertiary basalts of New South Wales described by Wilshire and Binns (1961) are composed predominantly of peridotite and pyroxenite, but gabbro inclusion occurs as well. A dike near the town of Kiama close to the Southern Highlands volcanic province contains an unusual suite of apatite-rich inclusions described by Wass (1980) and Wass and Rogers (1980). The ^{87}Sr/^{86}Sr ratios of the inclusions and most of their constituent minerals vary only within narrow limits relative to 0.71025 to NBS 987:

- whole-rock inclusion: 0.70379 to 0.70406,
  average: 0.70396 ±0.00009 ($2\bar{\sigma}$, $N = 9$)
- apatite: 0.70364 to 0.70373
- diopside: 0.70370
- amphibole: 0.70372 to 0.70377
- spinel: 0.70424
- mica: 0.70756 to 0.70805

Only the low-Sr phases (spinel and mica) have elevated ^{87}Sr/^{86}Sr ratios in contrast to the Sr-rich minerals (apatite, diopside, and amphibole) whose average ^{87}Sr/^{86}Sr ratio is 0.70370 ±0.0004 ($2\bar{\sigma}$, $N = 7$). The host rock of the dike in which the apatite-bearing nodules occur has a higher ^{87}Sr/^{86}Sr ratio (0.70420) than the Sr-rich minerals of the inclusions (0.70370) relative to 0.71025 for NBS 987.

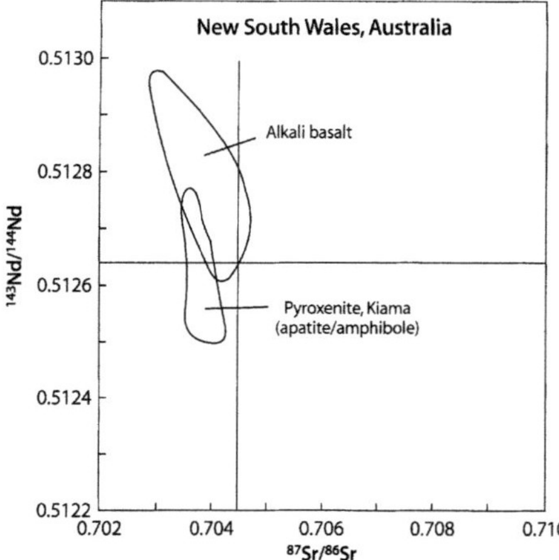

Fig. 6.63. $^{87}Sr/^{86}Sr$ and $^{143}Nd/^{144}Nd$ ratios of Early Tertiary alkali basalt lavas in the Southern Highlands of New South Wales and apatite/amphibole-bearing pyroxenite inclusions at Kiama (*Source:* Menzies and Wass 1983)

The isotope ratios of Sr and Nd of the pyroxenite inclusions at Kiama define a small data field in Fig. 6.63 that partly overlaps that of the alkali basalts of the nearby Southern Highland volcanic province identified on Fig. 6.61. Menzies and Wass (1983) concluded that the evidence favors magma formation in the lithospheric mantle that was enriched in alkali metals by CO_2-rich metasomatic fluids. The metasomatic alteration transformed lherzolite of the lithospheric mantle into pyroxenite containing veins composed of apatite, kaersutite, mica, and diopside. The resulting changes in the Rb/Sr and Sm/Nd ratios of the metasomatic pyroxenite caused its $^{87}Sr/^{86}Sr$ and $^{143}Nd/^{144}Nd$ ratios to deviate from the data field of the local alkali basalts and to generate the near-vertical data array defined by the Kiama pyroxenite inclusions.

The significance of the work by Menzies and Wass (1983) arises from their conclusions that carbon dioxide derived from the mantle contributed to the metasomatic alteration of the lithospheric mantle underlying the Southern Highland volcanic province of southeastern Australia. The presence of carbon dioxide in the source regions of alkali-rich magmas is consistent with the common association of carbonatites and alkali-rich igneous rocks. In addition, carbon dioxide, water, and other volatiles (e.g. F_2 and Cl_2) can contribute to the explosive eruption of alkali-rich magmas that pooled at the base of the continental crust. These insights are relevant to the petrogenesis of the alkali-rich lavas of the Montana and Wyoming (e.g. Leucite Hills and Highwood Mountains and to the intrusion of alnoite plutons on the island of Malaita) (Sect. 3.9.2).

In addition, the presence of volatiles can explain the intrusion of recently discovered kimberlite pipes in the Lac de Gras area into the Archean basement rocks of the Slave Craton in the Northwest Territories of Canada (Davis and Kjarsgaard 1997).

6.11.3 Newer Basalts, Victoria

The $^{87}Sr/^{86}Sr$ ratios of the alkali-rich basalts of the Newer basalt volcanic province (Fig. 6.61) were measured by Dasch and Green (1975), Burwell (1975), Stuckless and Irving (1976), and by McDonough et al. (1985). Several of the specimens were analyzed by more than one group of investigators. For this reason, only the data of McDonough et al. (1985) are presented in Fig. 6.64. The data points of the alkali basalt of Victoria are clustered like those of the basalts in New South Wales, but the $^{87}Sr/^{86}Sr$ ratios of the Newer basalts (0.7038 to 0.7045) are higher on average than those of the alkali-rich basalts in New South Wales. Therefore, the rocks of the Newer basalts also formed from mantle-derived magmas without significant assimilation of crustal rocks.

The alkali-rich lavas of Victoria, New South Wales, and Queensland contain megacrysts of clinopyroxene, orthopyroxene, anorthoclase, and kaersutite (Stuckless and Irving 1976; Basu 1978). The $^{87}Sr/^{86}Sr$ ratios of these minerals differ from those of their host lavas in some cases, but not in others. For example, the data of Stuckless and Irving (1976) indicate that the $^{87}Sr/^{86}Sr$

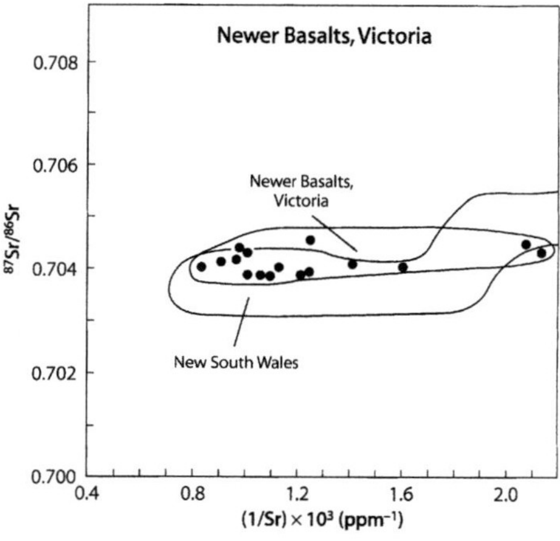

Fig. 6.64. $^{87}Sr/^{86}Sr$ and 1/Sr ratios of the alkali-rich volcanic rocks of the Newer basalts (Pliocene to Recent), Victoria, southeast Australia. The data cluster of the Newer basalts partly overlaps that of New South Wales. The irregular range of $^{87}Sr/^{86}Sr$ ratios and Sr concentrations of the Newer basalts confirms petrologic and geochemical evidence that the magmas did not assimilate crustal rocks (*Source:* data from McDonough et al. 1985)

ratios of clinopyroxene megacrysts (Sr: 100 to 122 ppm) are concordant with the $^{87}Sr/^{86}Sr$ ratios of a nepheline hawaiite flow on Mt. Gambier and with the $^{87}Sr/^{86}Sr$ ratio of a nepheline mugearite on Mt. Franklin in Victoria. However, anorthoclase megacrysts (Sr: 6750 to 8300 ppm) on Mt. Franklin have higher $^{87}Sr/^{86}Sr$ ratios than their host. The situation is reversed on Anakies (east) where anorthoclase (Sr: 1340 ppm) is isotopically concordant, but kaersutite (Sr: 653 to 823 ppm) is discordant with the Sr in their host. One low-Sr orthopyroxene megacryst (Sr: 1.02 ppm) in a hawaiite flow at Anakies (west) analyzed by Stuckless and Irving (1976) has a high $^{87}Sr/^{86}Sr$ ratio of 0.70453 compared to c.70363 in the lava flow relative to 0.71025 for NBS 987. The evident enrichment of this orthopyroxene megacryst in ^{87}Sr may have been caused by the addition of crustal Sr by groundwater or rain. Kaersutite megacrysts at Spring Mountains and Wee Jasper in New South Wales analyzed by Basu (1978) have low $^{87}Sr/^{86}Sr$ ratios (0.70253 to 0.70254), and a kaersutite from Queensland has $^{87}Sr/^{86}Sr$ = 0.70292 relative to 0.71025 for NBS 987.

These results are consistent with the conclusion of Irving (1974) that, in many cases, megacrysts did not crystallize from the same magma as their present host. In this regard, the megacrysts resemble the abundant lherzolite inclusions that occur within the Tertiary alkali basalts of Queensland, New South Wales, and Victoria. Many of these inclusions contain Sr, Nd, and Pb whose isotope compositions differ from those of their hosts.

6.11.4 Ultramafic Inclusions, Newer Volcanics, Victoria

The abundant lherzolite inclusions of the Newer volcanics in Victoria are composed of olivine, enstatite, and Cr-diopside with accessory spinel and, in some cases, hornblende, phlogopite, and apatite. Cooper and Green (1969) demonstrated that the Pb in these inclusions has different isotopic compositions than the basanite lavas and concluded that the ultramafic inclusions are not genetically related to their host. This conclusion was confirmed by Frey and Green (1974) based on geochemical and mineralogical data. In addition, these authors demonstrated that the inclusions contain a highly fractionated melt phase enriched in K, Ti, P, LREE, Th, and U that was added to a refractory residue left over from an episode of partial melting. Griffin et al. (1984) subsequently reported that spinel lherzolite inclusions at the Bullenmerri and Gnotuk Maars in Victoria contain amphiboles, apatite and, in some cases, phlogopite, that was introduced into anhydrous lherzolite.

The first study of Sr isotope compositions by Stueber (1969) of a peridotite inclusion from Mt. Leura near

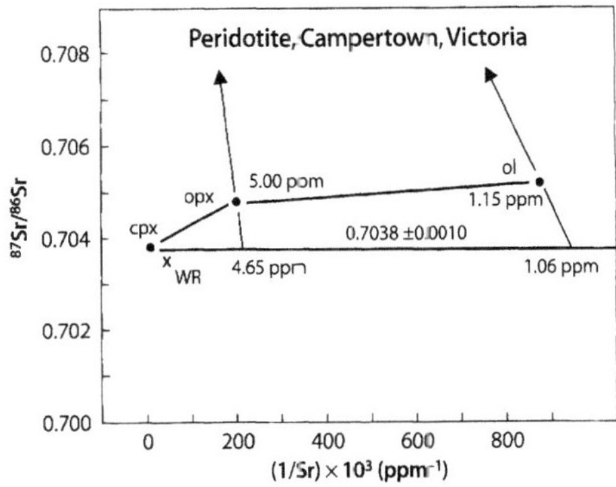

Fig. 6.65. $^{87}Sr/^{86}Sr$ and 1/Sr ratios of the minerals in a peridotite inclusion on Mt. Leura near Camperdown, Victoria, Australia. The $^{87}Sr/^{86}Sr$ ratios of enstatite (*opx*) and olivine (*ol*) increase with decreasing Sr concentration, which is attributed in this diagram to surface contamination of these minerals with crustal Sr ($^{87}Sr/^{86}Sr$ = 0.725) deposited by groundwater and rain. The $^{87}Sr/^{86}Sr$ ratio of clinopyroxene (0.7038 ±0.0010) is the best estimate of the isotope composition of Sr in this nodule at the time of its eruption (*Source:* data from Stueber 1969)

Camperdown is especially significant because he measured concentrations and $^{87}Sr/^{86}Sr$ ratios of the constituent minerals. The results indicate that the $^{87}Sr/^{86}Sr$ ratios of the minerals increase with decreasing Sr concentrations, just like in the nodules at Kilbourne Hole in New Mexico and at San Carlos, Arizona (Sect. 6.10.1 and 6.10.2). The presentation of Stueber's data in Fig. 6.65 makes the point that the $^{87}Sr/^{86}Sr$ ratios of the minerals are explainable by small additions of crustal Sr ($^{87}Sr/^{86}Sr$ = 0.725). Olivine gained only 0.09 µg g^{-1}, whereas enstatite gained 0.35 µg g^{-1}, presumably by deposition from groundwater and rain. The addition of 0.35 µg of crustal Sr to diopside (Sr = 194 ppm) would have raised its Sr content by less than 0.2% and therefore did not increase its $^{87}Sr/^{86}Sr$ ratio appreciably.

The information about the ultramafic inclusions on Mt. Leura was greatly increased by Burwell (1975) who analyzed seven inclusions and their separated minerals. Concentrations of Rb and Sr were measured at the University of Leeds by isotope dilution using spikes of nearly pure ^{87}Rb and ^{84}Sr. All Sr determinations were duplicated with good reproducibility, and the E&A isotope standard yielded a weighted average $^{87}Sr/^{86}Sr$ ratio of 0.70814 ±0.00002. These details are relevant because the results of these analyses are remarkable.

The $^{87}Sr/^{86}Sr$ ratios of the nodules analyzed by Burwell (1975) range from 0.7034 ±0.0002 to 0.7073 ±0004, compared to 0.7040 ±0.0001 for the host basanite relative to 0.7080 for E&A. The $^{87}Sr/^{86}Sr$ ratio of the host basanite (0.7040) reported by Burwell (1975) is similar to the value of 0.7039 measured by Stueber (1969) and to 0.70381 determined by McDonough et al.

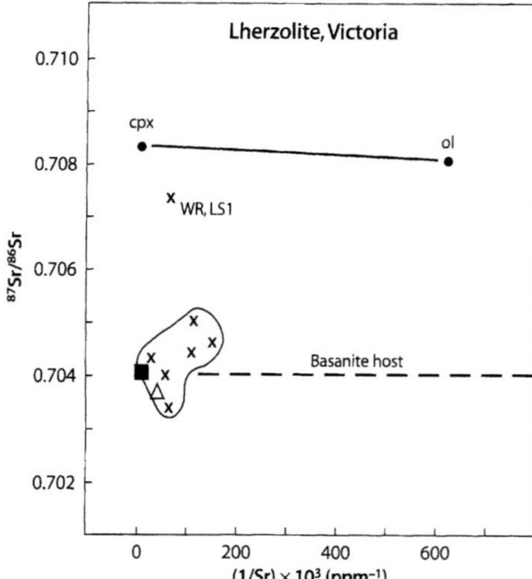

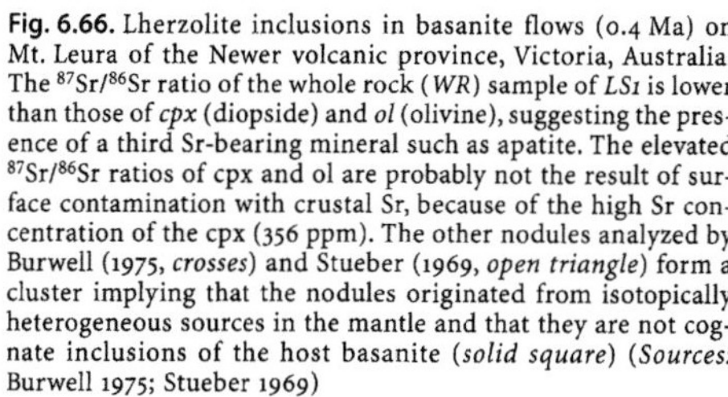

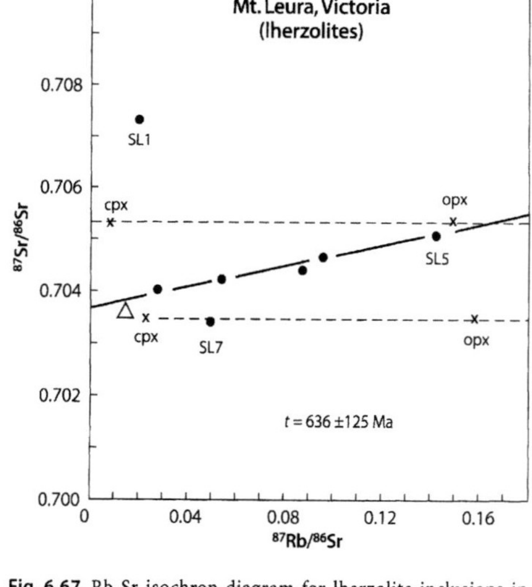

Fig. 6.66. Lherzolite inclusions in basanite flows (0.4 Ma) on Mt. Leura of the Newer volcanic province, Victoria, Australia. The $^{87}Sr/^{86}Sr$ ratio of the whole rock (*WR*) sample of *LS1* is lower than those of *cpx* (diopside) and *ol* (olivine), suggesting the presence of a third Sr-bearing mineral such as apatite. The elevated $^{87}Sr/^{86}Sr$ ratios of cpx and ol are probably not the result of surface contamination with crustal Sr, because of the high Sr concentration of the cpx (356 ppm). The other nodules analyzed by Burwell (1975, *crosses*) and Stueber (1969, *open triangle*) form a cluster implying that the nodules originated from isotopically heterogeneous sources in the mantle and that they are not cognate inclusions of the host basanite (*solid square*) (*Sources:* Burwell 1975; Stueber 1969)

Fig. 6.67. Rb-Sr isochron diagram for lherzolite inclusions in basanite lava flows on Mt. Leura, Newer volcanics, Victoria. Five of seven nodules analyzed by Burwell (1975, *solid circles*) define a straight line that yields a date of 636 ±125 Ma ($\lambda = 1.42 \times 10^{-11}$ yr^{-1}) and an initial $^{87}Sr/^{86}Sr$ ratio of 0.7037 ±0.0001 relative to 0.7080 for E&A. The peridotite nodule from this location analyzed by Stueber (1969, *open triangle*) is compatible with the line; however, the nodules from Mt. Leura analyzed by Dasch and Green (1975) scatter widely. Inclusions *SL1* and *SL7* do not fit the line but their minerals (*crosses*) demonstrate the internal isotopic equilibrium of most of the inclusions analyzed by Burwell (1975) (*Source:* data from Burwell 1975; Stueber 1969)

(1985) for other samples of the basanite flows on Mt. Leura. Most of the Sr of the inclusions analyzed by Burwell (1975) resides in clinopyroxene (diopside) which has an average Sr concentration of 207 ±65 ppm ($2\bar{\sigma}$, $N = 7$) compared to only 14.7 ±3.4 ppm in the whole-rock lherzolites. Orthopyroxene (enstatite) and olivine (Fo$_{87}$ to Fo$_{92}$) contain only 2.10 ±1.19 ppm and 1.14 ±0.89 ppm of Sr, respectively. Nevertheless, the $^{87}Sr/^{86}Sr$ ratios of these Sr-poor minerals are, in most cases, indistinguishable from those of the Sr-rich diopsides.

The Rb-Sr data of inclusion SL1 analyzed by Burwell (1975) are illustrated in Fig. 6.66. This inclusion differs from all others on Mt. Leura by having a high $^{87}Sr/^{86}Sr$ ratio of 0.7073, far above the Sr isotope ratio of the basanite host. The $^{87}Sr/^{86}Sr$ ratios of other nodules included in the cluster in Fig. 6.66 (0.7034 to 0.7051) vary beyond the analytical errors (±0.0003 or less), indicating that all of the inclusions originated from source regions in the mantle that were heterogeneous with respect to the isotope composition of Sr. The general concordance of the $^{87}Sr/^{86}Sr$ ratios of the constituent minerals excludes the possibility that the elevated $^{87}Sr/^{86}Sr$

ratios of most of the nodules analyzed by Burwell (1975) are the result of contamination with crustal Sr deposited on the surfaces and fractures of mineral grains by groundwater or rain.

Burwell (1975) demonstrated that five of the seven nodules he analyzed define a straight line in the Rb-Sr isochron diagram shown in Fig. 6.67. If this line is an isochron, its slope yields a date of 636 ±125 Ma ($\lambda = 1.42 \times 10^{-11}$ yr^{-1}) and an initial $^{87}Sr/^{86}Sr$ ratio of 0.7037 ±0.0001 relative to 0.7080 for E&A. After crystallization at 636 Ma, isotopic equilibrium among the constituent minerals of each nodule was maintained, as expected at the elevated temperatures of the upper mantle, where the rocks resided before being transported to the surface at about 0.4 Ma.

A plausible explanation for the Rb-Sr systematics of the nodules analyzed by Burwell (1975) is that they are fragments of large masses of rocks that crystallized at 636 ±125 Ma from magmas in the lithospheric mantle under Victoria. These rocks formed domains whose $^{87}Sr/^{86}Sr$ ratios increased with time by decay of ^{87}Rb. Accidental inclusions from these domains were subsecuently transported to the surface in basanite magma formed by an unrelated melting event in regions of the mantle below the earlier-formed domains. In this case, the metasomatic alteration of the mantle is the result of partial melting and subsequent fractional crystallization of the resulting magma (i.e. zone refining, Harris 1957).

There are, however, inconsistencies to this interpretation because two of the lherzolite specimens (SL1 and SL7) from Mt. Leura analyzed by Burwell (1975) do not fit the straight line in Fig. 6.67, and neither do the four inclusions from Mt Leura analyzed by Dasch and Green (1975). The validity of this interpretation is, therefore, open to question. Nevertheless, the data of Burwell (1975) demonstrate that accidental inclusions in lavas extruded at a specific site could form Rb-Sr isochrons that date episodes of partial melting and fractional crystallization of magma in the lithospheric mantle. Mantle isochrons based on ultramafic inclusions are more likely to date real events in the mantle than the mantle isochrons that are based on the Rb-Sr systematics of lava flows (e.g. French Polynesia, Sect. 2.10.7).

The $^{87}Sr/^{86}Sr$ ratios of the lherzolite inclusions on Mt. Leura analyzed by Dasch and Green (1975) range from 0.7027 to 0.7071 and thus confirm the conclusion of Burwell (1975) that the lithospheric mantle from which they originated is heterogeneous with respect to the isotope composition of Sr. Dasch and Green (1975) also reported that the $^{87}Sr/^{86}Sr$ ratios of lherzolite inclusions at Mt. Gambier in Victoria rise to high values between 0.7077 and 0.7107 relative to 0.7080 for E&A. The heterogeneity of Sr isotope ratios of the Mt. Gambier inclusions contrasts with the more uniform $^{87}Sr/^{86}Sr$ ratios of the pyroxenite inclusions at Kiama located 800 km northeast of Mt. Leura.

The complexity of the Rb-Sr systematics of the lherzolite inclusions in the Newer volcanics of Victoria goes beyond their heterogeneous isotope compositions of Sr. Dasch and Green (1975) reported that $^{87}Sr/^{86}Sr$ ratios of the minerals of a lherzolite inclusion from Mt. Noorat in Fig. 6.68 range from 0.7029 (diopside) to 0.7130 (olivine), whereas the $^{87}Sr/^{86}Sr$ ratio of the whole nodule is 0.7040 relative to 0.7080 for E&A. An acid wash of the olivine reduced its Sr concentrations by less than 0.5% and did not change its $^{87}Sr/^{86}Sr$ ratio. However, the acid wash reduced the Rb concentration of the olivine by about 30%, indicating that a significant fraction of the Rb in the olivine (and presumably in the other minerals) resided on the surfaces of the mineral grains. In addition, Dasch and Green (1975) observed that the abundances of the principal minerals (olivine, 58%; enstatite, 34%; diopside, 5%; spinel, 2%) and their Rb and Sr concentrations are inconsistent with the measured concentrations of these elements

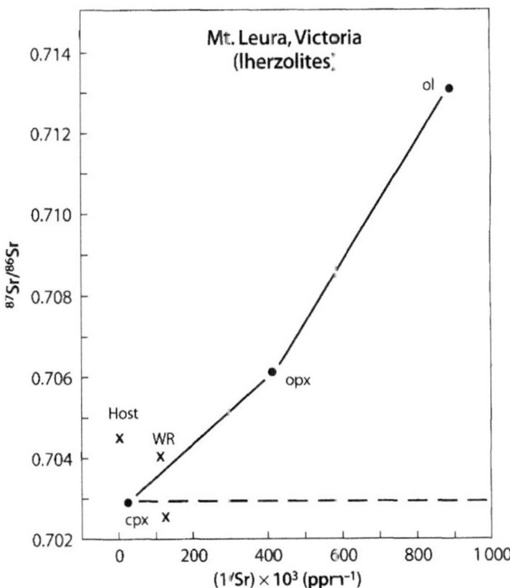

Fig. 6.68. $^{87}Sr/^{86}Sr$ and 1/Sr ratios of the whole rock (*cross*, *WR*) and constituent minerals (*solid circles*) of a lherzolite nodule in the Newer volcanics, Mt. Noorat, Victoria. Australia. The elevated $^{87}Sr/^{86}Sr$ ratios of olivine (*ol*) and enstatite (*opx*) are attributed in this case to the decay of ^{87}Rb on grain boundaries and to the diffusion of the resulting ^{87}Sr into olivine and enstatite at high temperatures in the mantle. The original $^{87}Sr/^{86}Sr$ ratio of the ultramafic inclusion (0.7029) is preserved by the diopside (*cpx*) and not by the whole-rock (*WR*) sample of the nodule. Consequently, the $^{87}Sr/^{86}Sr$ ratio of the inclusion differs significantly from that of the host basanite (*Source:* data from Dasch and Green 1975)

in the whole rock. A mass balance for Sr can be achieved by increasing the abundance of diopside (Sr = 42.11 ppm) from 5% (reported by Dasch and Green 1975) to 18%.

However, even this arbitrary adjustment of the data leaves about 60% of the Rb in this lherzolite nodule unaccounted for. Therefore, the authors concluded that the excess Rb resided in grain boundaries and in fractures of the minerals. The radiogenic ^{87}Sr produced by decay of the excess ^{87}Rb was transferred from the grain boundaries into olivine and enstatite crystals by diffusion at high temperature. As a result the radiogenic ^{87}Sr in olivine is not removable by leaching.

A troublesome feature of this mechanism is that the minerals appear to have acquired different $^{87}Sr/^{86}Sr$ ratios in spite of the high temperature in the mantle where this process presumably occurred. Internal isotope equilibration of Sr in this nodule would have yielded an $^{87}Sr/^{86}Sr$ ratio of 0.7054 in the nodule as a whole, whereas Dasch and Green (1975) actually measured 0.7040.

In retrospect, the Rb-Sr systematics of ultramafic inclusions associated with Tertiary volcanic activity in western North America and southeastern Australia are remarkably complex. The observed range of variation of $^{87}Sr/^{86}Sr$ ratios of whole-rock specimens is attributable to several alternative causes:

1. Contamination of the nodules after eruption by deposition of crustal Sr on the surfaces and in fractures of mineral grains by groundwater or rain. Evidence:
   a The $^{87}Sr/^{86}Sr$ ratios of minerals increase with decreasing Sr concentrations.
   b The Sr is removable by leaching with dilute acid and the $^{87}Sr/^{86}Sr$ ratios of the minerals are changed by this procedure.
   c Other elements that occur in groundwater or rain are also leachable from the nodules (e.g. Na, K, Rb, Mg, Ca, Cl, etc.).
2. Partial melting and fractional crystallization of magma (zone refining) in the lithospheric mantle resulting in the formation of domains with elevated Rb/Sr ratios and hence high $^{87}Sr/^{86}Sr$ ratios by in situ decay of ^{87}Rb. Evidence:
   a Whole-rock samples of nodules define Rb-Sr isochrons that date the time of zone refining.
   b The $^{87}Sr/^{86}Sr$ ratios of the minerals are concordant within individual nodules, because of continuous isotopic re-equilibration of Sr isotopes among the minerals at mantle temperatures.
   c Neither Rb nor Sr are removable from the minerals to a significant extent by leaching with acid.
3. Metasomatic alteration of mantle domains by aqueous or CO_2-rich fluids arising from mantle plumes in the asthenosphere and from subducted slabs of oceanic crust. Evidence:
   a Presence of veins containing alkali-rich amphiboles (kaersutite) and mica (phlogopite) in the ultramafic inclusions.
   b The $^{87}Sr/^{86}Sr$ ratios of the constituent minerals are concordant, provided that contaminant Sr is first removed by leaching, especially from the low-Sr minerals.
4. Contamination of the nodules with radiogenic ^{87}Sr produced by the decay of ^{87}Rb deposited on grain boundaries by aqueous fluids in the mantle. Evidence:
   a The $^{87}Sr/^{86}Sr$ ratios of minerals increase with decreasing Sr concentrations as in Case 1 above.
   b The Sr is not removable by leaching with dilute acid and the $^{87}Sr/^{86}Sr$ ratios of the minerals are not lowered by this procedure.
   c The resulting disequilibrium of the isotope compositions of Sr in the minerals is a troublesome feature that requires confirmation.

A great deal of analytical information is required before the Rb-Sr systematics of ultramafic nodules can be attributed to one of the processes listed above:

1. Abundances of the major and accessory minerals.
2. The $^{87}Sr/^{86}Sr$ ratios as well as Rb and Sr concentrations of the whole-rock sample of the nodule and of all of its constituent minerals.
3. The same, after leaching the whole-rock nodule and all of the minerals with dilute acid.

This information is rarely provided because of the large amount of analytical labor required to obtain it and because the amount of material available for analysis may be limited.

In the final analysis, metasomatism of the lithospheric mantle is indicated by the presence of hydrous minerals (kaersutite and phlogopite) in ultramafic nodules. However, since the nodules are not genetically related to the lavas flows in many cases, the derivation of alkali-rich lavas from metasomatized mantle rocks is indicated by their elevated $^{87}Sr/^{86}Sr$ ratios and by the absence of evidence for assimilation of crustal rocks.

6.11.5 Western Australia: Origin of K-rich Lavas

The K-rich lamproites of early Miocene age in Western Australia occur primarily in the mobile belts that surround the Precambrian Kimberley block in Fig. 6.69. The Fitzroy Trough along the southern margin of the Kimberley block contains leucite lamproites and olivine lamproites near the towns of Ellendale, Calwynyardah, and Noonkanbah. Olivine lamproites also occur in the Halls Creek Mobile zone that forms the east-

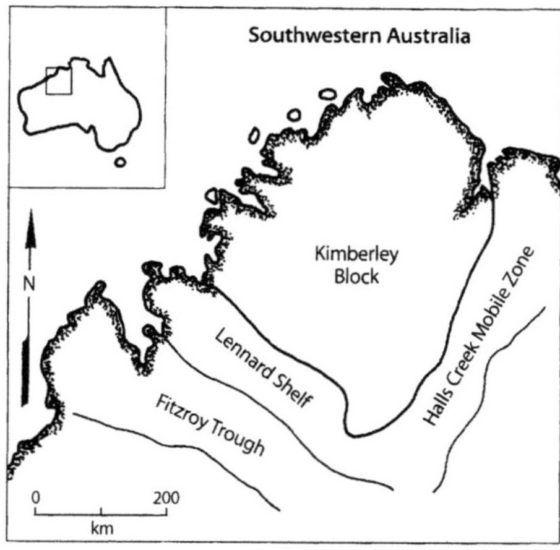

Fig. 6.69. Occurrence of ultrapotassic lamproites of Tertiary age in the mobile belts surrounding the Kimberley block composed of igneous and metamorphic rocks of Precambrian age in Western Australia (*Source:* adapted from Nelson et al. 1986)

ern boundary of the Kimberley block and along the north coast within the Kimberley block itself.

Lamproites are ultrapotassic (Mg-rich) igneous rocks with $K_2O / Na_2O > 3$ and comparatively low SiO_2 concentrations between 45 and 55%. They occur in many different forms including dikes, lava flows, diatremes, and pipes. Lamproites range in age from Proterozoic to Quaternary and tend to occur in mobile belts around the margins of cratons, in contrast to kimberlites, which occur in the interiors of cratons. Lamproites have also been identified as absarokites, leucite basalt, leucite basanite, and leucitite (Wilson 1989).

The olivine-rich lamproites of Western Australia contain diamonds (Scott-Smith and Skinner 1984) and therefore originated from similar kinds of source regions in the mantle as kimberlites. The lamproites of Western Australia are strongly enriched in K and have K_2O/Na_2O ratios that typically exceed 8.0. The $^{87}Sr/^{86}Sr$ ratios of these rocks reported by Powell and Bell (1970) range from 0.7124 to 0.7214 relative to 0.7080 for E&A. Values of this magnitude are typical of Precambrian

granitoids and had never before been reported for K-rich volcanic rocks of Tertiary age. In addition, Powell and Bell (1970) observed that the $^{87}Sr/^{86}Sr$ ratios of different rock specimens from the same locality vary far beyond the analytical error (e.g. Mt. North: 0.7126 to 0.7152; Mamilu Hill: 0.7189 to 0.7214). The high values and the variability of the $^{87}Sr/^{86}Sr$ ratios are not attributable to contamination of the magmas by crustal rocks because of the high Sr concentrations of the rocks (1 106 to 1 801 ppm). Instead, Powell and Bell (1970) concluded that the K-rich magmas originated from source regions in the mantle that had high $^{87}Sr/^{86}Sr$ ratios because of prior enrichment in Rb and K as a result of zone refining (Harris 1957).

Subsequent work by McCulloch et al. (1983a), Fraser et al. (1985/86), and Nelson et al. (1986) confirmed the high and variable $^{87}Sr/^{86}Sr$ ratios of the lamproites in the Fitzroy Trough reported by Powell and Bell (1970). The $^{87}Sr/^{86}Sr$ and 1/Sr ratios of these rocks form a large cluster in Fig. 6.70, confirming that the magmas were not differentiated by assimilation of crustal rocks or by fractional crystallization. The presence of diamonds

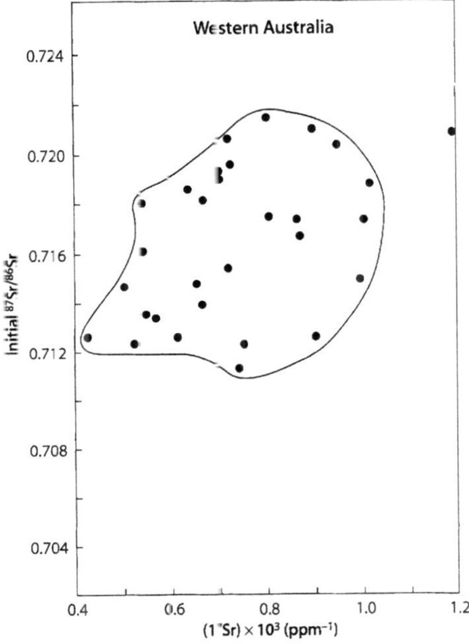

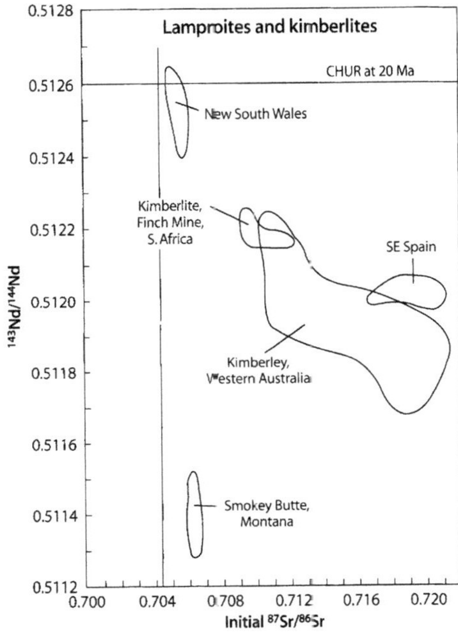

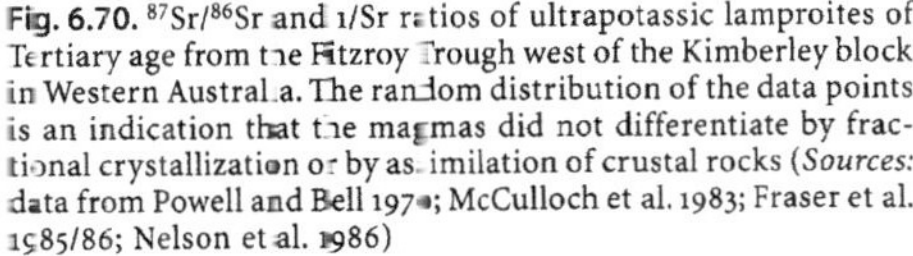
Fig. 6.70. $^{87}Sr/^{86}Sr$ and 1/Sr ratios of ultrapotassic lamproites of Tertiary age from the Fitzroy Trough west of the Kimberley block in Western Australia. The random distribution of the data points is an indication that the magmas did not differentiate by fractional crystallization or by assimilation of crustal rocks (Sources: data from Powell and Bell 1970; McCulloch et al. 1983; Fraser et al. 1985/86; Nelson et al. 1986)

Fig. 6.71. Isotope ratios of Sr and Nd of ultrapotassic lamproites and kimberlites of the Kimberley block, Western Australia; New South Wales, southeastern Australia; Finch Mine, South Africa; southeastern Spain, and Smokey Butte, Montana (Sources: McCulloch et al. 1983a; Fraser et al. 1985/86; Nelson et al. 1986. The $^{143}Nd/^{144}Nd$ ratios of McCulloch et al. (1983a) were renormalized to $^{146}Nd/^{144}Nd = 0.7219$ by multiplying them by 1.00154)

in some of the olivine lamproites of Western Australia is strong evidence that the magmas originated directly from highly alkali-enriched regions in the lithospheric mantle underlying this region.

The isotope ratios of Sr and Nd of lamproites in the Kimberley block of Western Australia form a large data field in the "crustal" quadrant of Fig. 6.71 because they have high initial $^{87}Sr/^{86}Sr$ ratios (0.710 to 0.721) as well as low $^{143}Nd/^{144}Nd$ ratios (0.5116 to 0.5122) (Nelson et al. 1986; Fraser et al. 1985/86; McCulloch et al. 1983a). Lamproites in New South Wales, Australia; Smokey Butte, Montana; southeastern Spain; and kimberlites in the Finch Mine of South Africa in Fig. 6.71 also have high $^{87}Sr/^{86}Sr$ and low $^{143}Nd/^{144}Nd$ ratios. Therefore, the magmas at all of these sites originated from heterogeneous source regions in the lithospheric mantle which had been enriched in Rb and depleted in Sm during a prior episode (Boettcher and O'Neil 1980). McCulloch et al. (1983) determined by means of Nd model-dates relative to CHUR that this episode occurred at 1.14 ±0.07 Ga.

The scatter of the isotope ratios of Sr and Nd of the Kimberley lamproites is attributable to incomplete mixing of magmas derived from different source components in the lithospheric mantle or in upwelling asthenospheric rocks. The presence of the Kimberley block prevented magma formation in the underlying lithospheric mantle for more than one billion years which allowed the $^{87}Sr/^{86}Sr$ ratios to increase with time by decay of ^{87}Rb. At the same time, the isotope evolution of Nd was retarded because the lithospheric mantle under the Kimberley block had a low Sm/Nd ratio (less than that of CHUR) since the rocks were enriched in Nd rather than Sm. The resulting isotope evolution of Sr and Nd is a direct consequence of the incubation time of the metasomatically altered magma sources beneath the thick continental crust of the Kimberley Craton. Such isotopic evolution does not occur in the mantle wedge of subduction zones because in that case, magma formation occurs shortly after the metasomatic alteration.

The $^{206}Pb/^{204}Pb$ ratios of the lamproites in Western Australia are uniformly low with values between 17.228 and 17.882 (Nelson et al. 1986; Fraser et al. 1985/86). Lamproites on Smokey Butte in Montana on the Wyoming Craton have even lower $^{206}Pb/^{204}Pb$ ratios ranging from 16.025 to 16.643, whereas those of kimberlite in the Finch Mine of South Africa vary between 17.739 and 18.245 (Fraser et al. 1985/86). These data indicate that the magma sources at these centers of igneous activity had low U/Pb ratios for long periods of time, thereby preventing the $^{206}Pb/^{204}Pb$ ratios from increasing significantly.

The alkali-rich lavas of Western Australia confirm and reinforce the evidence seen elsewhere that they originated from magma sources in the lithospheric mantle, which had been previously enriched in alkali metals and volatile compounds. The metasomatizing fluids take the form of silica-rich melts containing water, carbon dioxide, alkali metals, and other elements that are concentrated in the melt fraction. These fluids can originate from the heads of asthenospheric plumes and from subducted oceanic crust that underplated the lithospheric mantle. Metasomatic alteration of the lithospheric mantle takes the form of veins composed of amphibole and mica, both of which are hosts for alkali metals and water. These altered regions of the lithospheric mantle beneath thick Precambrian cratons are prevented from venting to the surface and can incubate for one or even two billion years. When partial melting finally does occur, the magmas may be trapped temporarily beneath the continental crust and thus form megacrysts until they finally break through the overlying crust and are propelled to the surface by the pressure of gases they contained. This hypothesis also accounts for the explosive intrusion of the K-rich and diamond-bearing kimberlite pipes which typically are much younger than the rocks of the Precambrian cratons where they occur.

6.12 Antarctica

The alkali-rich lavas of Antarctica occur primarily in the Transantarctic Mountains including Ross Island and the Balleny Islands, in Marie Byrd Land of West Antarctica, and on Gaussberg (66°48' S, 89°12' E) on the coast of East Antarctica. The alkalic volcanic rocks of the Transantarctic Mountains range in age from about 20 Ma to Recent and are referred to the McMurdo Volcanic Group, because of the presence of such lavas near McMurdo on Ross Island shown in Fig. 6.72 (Kyle et al. 1979a). The volcanic activity on Ross Island and at scattered locations in the Transantarctic Mountains is related to the presence of deep crustal rifts (Kyle and Cole 1974). The Tertiary to Recent volcanic activity at isolated centers along the Transantarctic Mountains to be discussed here is unrelated to the earlier intrusion of sills of Ferrar dolerite and the eruption of tholeiite flows of the Kirkpatrick basalt (Middle Jurassic) in the same area.

The alkali basalts of Marie Byrd Land occur within the West Antarctic Rift system on continental crust composed of Paleozoic metasedimentary rocks intruded by granitic plutons that range in age from Paleozoic to Late Mesozoic (LeMasurier and Wade 1977). The alkali basalts occur at the base of the sequence of lavas and in parasitic cones on the flanks of large volcanic mountains.

Gaussberg is a small and isolated volcano 370 m above sealevel on the coast of East Antarctica. It consists of pillow lavas composed of leucitite containing 11.61% K_2O, compared to only 1.65% Na_2O (Sheraton

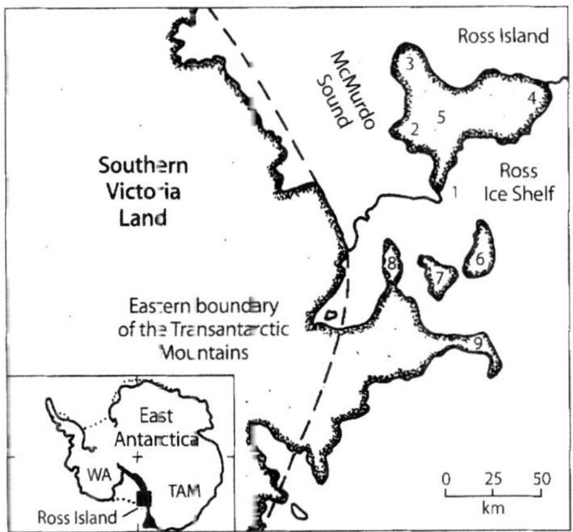

Fig. 6.72. Ross Island and southern Victoria Land of Antarctica. The following localities are identified by number: *1*. Hut Point Peninsula and McMurdo Station; *2*. Cape Royds; *3*. Cape Bird; *4*. Cape Crozier, *5*. Mount Erebus; *6*. White Island; *7*. Black Island; *8*. Brown Peninsula; *9*. Minna Bluff Peninsula. The Cenozoic alkali-rich lavas of this area constitute the Erebus volcanic province. *TAM* = Transantarctic Mountains; *WA* = West Antarctica (*Source:* adapted from Kalamarides et al. 1987)

and Cundari 1980). These ultrapotassic lavas are probably Pleistocene in age and are not related to any known exposures of K-rich lavas in East Antarctica or elsewhere. The Gaussberg was discovered in March 1902 after the ship of the German Antarctic Expedition, the Gauss, was trapped in the ice nearby on February 2, 1902, forcing the members of the expedition to winter over (Drygalski 1912; Philippi 1912; Rheinisch 1912).

The alkali-rich volcanic rocks of Antarctica formed in different tectonic settings that encompass most of the sites discussed in this chapter.

6.12.1 McMurdo Volcanic Group

The lavas of the McMurdo Volcanic Group were extruded in four volcanic centers identified by Kyle and Cole (1974) as the Balleny, Hallett, Melbourne, and Erebus provinces. A fifth volcanic center at Mt. Early-Sheridan Bluff (87° S and 153° W) erupted alkali-olivine basalt in Early Miocene time between about 20 and 15 Ma (Treves 1967; Stump et al. 1980). Most of the alkali-rich lavas of the McMurdo volcanics in the Erebus province on Ross Island and in the Transantarctic Mountains of southern Victoria Land were extruded less than 4 million years ago based on K-Ar dates measured by Armstrong et al. (1968), Armstrong (1978a), and Kyle et al. (1979b). Lavas of the Hallett and Melbourne volcanic provinces in the northern Victoria Land also formed in this time interval.

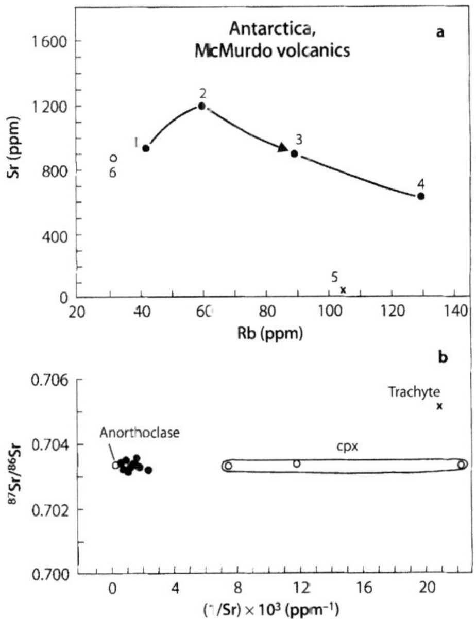

Fig. 6.73. a Average Rb and Sr concentrations of alkali-rich volcanic rocks (about 20 to 0 Ma) of the McMurdo volcanics on Ross Island and in southern Victoria Land, Antarctica. *1*. Basanite; *2*. Trachybasalt; *3*. Anorthoclase phonolite (kenyte); *4*. Pyroxene and hornblende phonolite; *5*. Trachyte (one sample only); *6*. Younger basanites. The variation of average Rb and Sr concentrations is consistent with fractional crystallization of basanite magmas (*Source:* Goldich et al. 1975); b $^{87}Sr/^{86}Sr$ and 1/Sr ratios of alkali-rich lavas (*solid circles*) and separated minerals (*open circles*) on Ross Island and southern Victoria Land, Antarctica. The uniformly low $^{87}Sr/^{86}Sr$ ratios of the lava flows (0.70317 to 0.70351) and separated minerals (0.70324 to 0.70338) support the conclusion that the basanite magmas differentiated primarily by fractional crystallization and not by assimilation of crustal rocks. The only exception is the inclusion-rich trachyte (*Sources:* Part *a*: Goldich et al. (1975); Part *b*: Stuckless and Ericksen 1976)

Volcanic eruptions of Mt. Erebus on Ross Island were first reported by Sir James Ross in January 1841 and have occurred intermittently to the present time (Giggenbach et al. 1973; Kyle et al. 1982). The lava flows on the summit of Mt. Erebus are composed of anorthoclase phonolite porphyry (also referred to as Antarctic kenyte) whose chemical composition is similar to that of the volcanic rocks on Mt. Kenya and Mt. Kilimanjaro in East Africa (Sect. 6.1.2). The petrography of lava flows on the Hut Point Peninsula and at other sites on Ross Island was described by Treves (1962) and by Kyle (1981). Other notable contributions to the occurrence and petrography of lava flows in southern Victoria Land were published by Cole and Ewart (1968), McIver and Gevers (1970), and by Kyle et al. (1979a). The geology of the Hallett volcanic province in northern Victoria Land was discussed by Hamilton (1972) and by Hart and Kyle (1993).

The alkali-rich lavas on Ross Island range in composition from basanite to trachybasalt and phonolite. Trachytes are present only in small amounts. Goldich et al. (1975) concluded from geochemical data that the volcanic rocks of Ross Island formed by fractional crystallization of basanite magmas. The average Rb and Sr concentrations of these rocks in Fig. 6.73a are consistent with the expected results of fractional crystallization.

The $^{87}Sr/^{86}Sr$ ratios of the Erebus volcanic center on Ross Island and adjacent areas of southern Victoria Land were first reported by Jones and Walker (1972) and by Halpern (1969). The $^{87}Sr/^{86}Sr$ ratios of the Erebus volcanic center in Fig. 6.73b are tightly clustered between 0.70317 and 0.70351 relative to 0.71025 for NBS 987 (Stuckless and Erickson 1976). The limited range of the $^{87}Sr/^{86}Sr$ ratios of the Erebus volcanic province and the evidence for fractional crystallization of basanite magma confirms the conclusions of Goldich et al. (1975), Sun and Hanson (1975b), and Stuckless and Ericksen (1976) that the alkali-rich lavas of the Erebus volcanic center are primarily the result of fractional crystallization of mantle-derived basanite magmas without significant assimilation of crustal rocks.

6.12.2 Inclusions, McMurdo Volcanic Group, Ross Island

The lavas of the McMurdo Volcanic Group of the Erebus province contain a variety of crustal and subcrustal inclusions. The inclusions on Ross Island consist of plagioclase-pyroxene granulites, wehrlites, dunites, sandstones, harzburgites, and lherzolites in order of decreasing abundance. Inclusions of calc-silicate and granitic rocks are very rare (Stuckless and Ericksen 1976). Some of the granulite inclusions are banded and many contain kaersutite, both in veins and as interstitial grains. The ultramafic nodules contain clinopyroxene, olivine, orthopyroxene, and opaque oxides but lack kaersutite.

The $^{87}Sr/^{86}Sr$ ratios of the granulite nodules range from 0.70290 to 0.70390 relative to 0.71025 for NBS 987 and are independent of the Sr concentrations. Stuckless and Ericksen (1976) observed that the $^{87}Sr/^{86}Sr$ ratios of several granulite nodules are positively correlated with their Rb/Sr ratios and define a straight line on the Rb-Sr isochron diagram in Fig. 6.74. If this line is an isochron, its slope yields a date of 155 ±22 Ma ($\lambda = 1.42 \times 10^{-11}$ yr^{-1}). The authors suggested that this date represents an episode of isotopic re-equilibration of granulites in the lower continental crust at the time of the intrusion of the Ferrar dolerites at 151 ±18 Ma (Compston et al. 1968). However, $^{40}Ar/^{39}Ar$ dates by Heimann et al. (1994) later yielded an older date of 176.6 ±1.8 Ma for basalts of the Ferrar Supergroup (Sect. 5.12.1). In addition, several other granulite inclu-

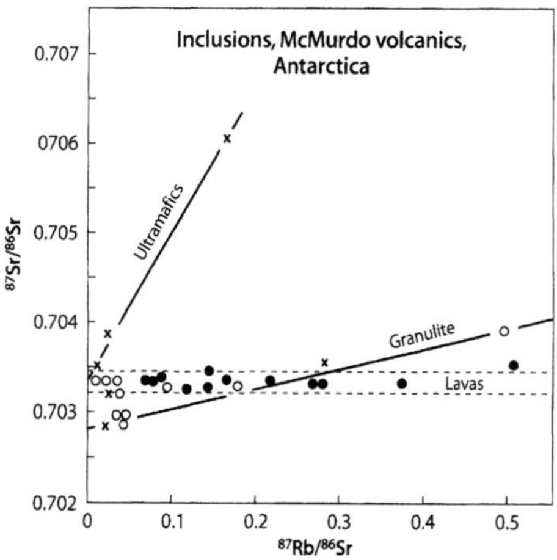

Fig. 6.74. Rb-Sr isochron diagram for granulite (*open circles*) and ultramafic (*crosses*) inclusions in the alkali-rich volcanic rocks on Ross Island and the offshore islands of Antarctica. The *straight lines* defined by five granulites and three ultramafic inclusions probably are not isochrons. The $^{87}Sr/^{86}Sr$ ratios of volcanic rocks on Ross Island (*solid circles*) have a small range between about 0.7032 and 0.7035. The isotope composition of Sr in several granulite inclusions appears to have been re-equilibrated with Sr in the volcanic rocks (*Source:* Stuckless and Erickson 1976)

sions in Fig. 6.74 do not fit the hypothetical granulate isochron. Therefore, the significance of the Rb-Sr date of the granulite nodules is open to question (Kalamarides et al. 1987). Nevertheless, the distribution of data points in Fig. 6.74 indicates that the $^{87}Sr/^{86}Sr$ ratios of some of the granulite nodules differ from those of the alkali-rich lavas on Ross Island and, in several cases, are lower than the $^{87}Sr/^{86}Sr$ ratios of their host rocks. In addition, Kalamarides et al. (1987) demonstrated the existence of differences in the initial $^{87}Sr/^{86}Sr$ ratios (at 900 Ma) and $\delta^{18}O$ values of granulite inclusions in the Transantarctic Mountains of southern Victoria Land and those on Ross Island and related volcanic centers. They concluded that the offshore islands are separated from the Transantarctic Mountains on the mainland of Antarctica by a major tectonic boundary.

Three ultramafic nodules analyzed by Stuckless and Ericksen (1976) constrain a second line in Fig. 6.74 that yields a date of about 1 150 Ma ($\lambda = 1.42 \times 10^{-11}$ yr^{-1}). The magnitude of this date is strongly affected by the high $^{87}Sr/^{86}Sr$ ratio (0.70605) of a harzburgite nodule whose whole-rock Sr concentration is only 3.01 ppm. This specimen contains 80% olivine, 19% orthopyroxene, 1% Cr diopside, and trace amounts of oxides. There is no evidence that this nodule was enriched in Rb at the indicated date because its whole-rock Rb concentration is only 0.17 ppm and because Rb-bearing phases such as kaersutite and phlogopite are absent.

Consequently, the $^{87}Sr/^{86}Sr$ ratio of this nodule may have been increased after eruption by contamination with marine Sr present in snow.

Faure and Jones (1989) demonstrated that soil salts (thenardite, Na_2SO_4) on Ross Island contain Sr and S derived both from the volcanic rocks and from marine sources. The $^{87}Sr/^{86}Sr$ ratios of thenardite and other soil salts at Cape Bird, Cape Royds, Hut Point Peninsula and on the summit of Mt. Erebus range from 0.70912 (Cape Bird) to 0.70339 (Fang Glacier, Mt. Erebus) relative to 0.7080 for E&A.

6.12.3 Northern Victoria Land and the Balleny Islands

The existence of a major tectonic boundary along the coast of Victoria Land, inferred by Kalamarides et al. (1987) from a study of granulite inclusions in the lavas of the Erebus province, was supported by Hart and Kyle (1993) who demonstrated that the alkali-rich lavas of the McMurdo Volcanic Group on the Balleny Islands, Scott Island (Wörner and Orsi 1992), Cape Adare, and

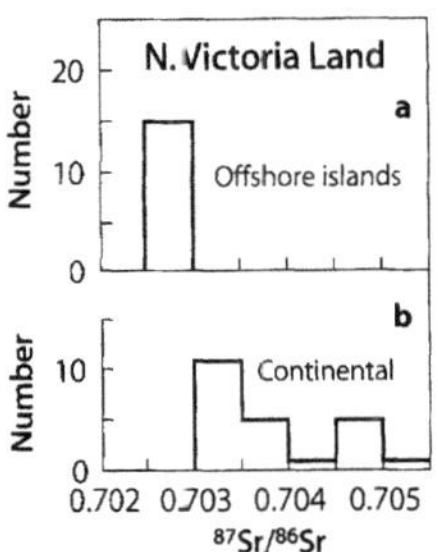

Fig. 6.76. Histograms of $^{87}Sr/^{86}Sr$ ratios of alkali basalts of the McMurdo Volcanic Group in Northern Victoria Land, Antarctica; a Lavas on the Balleny Islands, Scott Island, Cape Adare, Possession Island, Coulman Island, Franklin Island, and Mt. Melbourne (alkali basalt only); b Lavas of the McMurdo Volcanic Group on the mainland of northern Victoria Land including Mt. Melbourne, Mt. Overlord, and the Daniell and Hallett Peninsulas. The $^{87}Sr/^{86}Sr$ ratios of the alkali-rich volcanic rocks in this region define two petrologic domains separated by a tectonic boundary located near the coast of Victoria Land (*Source:* Hart and Kyle 1993; Wörner et al. 1989)

the offshore islands (Possession, Coulman, and Franklin) in Fig. 6.75 have low $^{87}Sr/^{86}Sr$ ratios compared to similar rocks on the Antarctic mainland. The differences are clearly expressed in Fig. 6.76, which also contains the data of Wörner et al. (1989) for the lavas on Mt. Melbourne. The $^{87}Sr/^{86}Sr$ ratios of the Cenozoic volcanic rocks on Coulman Island and Mt. Melbourne include both high and low values suggesting that they straddle the tectonic boundary. The geochemical evolution of magmas associated with the Ross-Sea Rift of northern Victoria Land was discussed by Rocholl et al. (1995).

The Balleny Islands in Fig. 6.75 are aligned along a northwest trending lineament which suggests that they formed by partial melting in a stationary asthenospheric plume as a result of movement of the Antarctic Plate during the opening of the Tasman Sea (Wright and Kyle 1990; Lanyon et al. 1993). The plume trace extends from the Balleney Islands to the vicinity of Soela Seamount on the East Tasman Plateau about 300 km southeast of Tasmania (Duncan and McDougall 1989). Lanyon et al. (1993) suggested that the Tertiary alkali-rich lavas on Tasmania are also related to the Balleny Plume, consistent with their low $^{87}Sr/^{86}Sr$ ratios (average; 0.70299 ±0.0003, 2$\overline{\sigma}$, $N = 5$) reported by McDonough et al. (1985) compared to those on the Australian mainland (Sect. 6.11, Fig. 6.64).

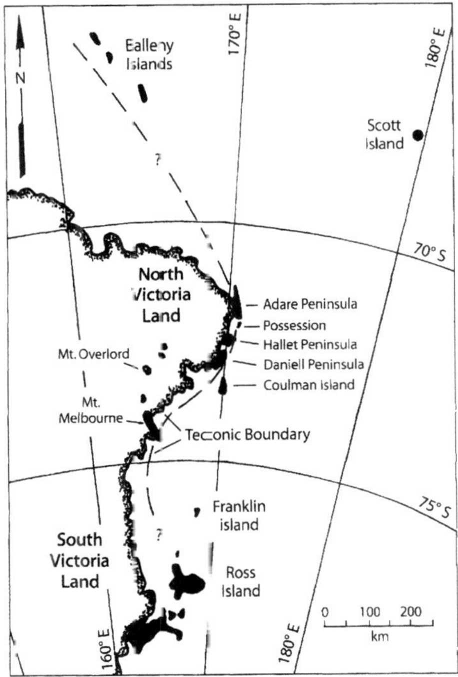

Fig. 6.75. Centers of alkali-rich volcanic rocks of the McMurdo volcanics in Victoria Land, Antarctica, and on the offshore islands (*Source:* adapted from Hart and Kyle 1993)

6.12.4 Marie Byrd and Ellsworth Land, West Antarctica

The Cenozoic volcanic rocks of West Antarctica were deposited on a volcano-sedimentary complex of Devonian to Cretaceous age that formed as a result of pro-

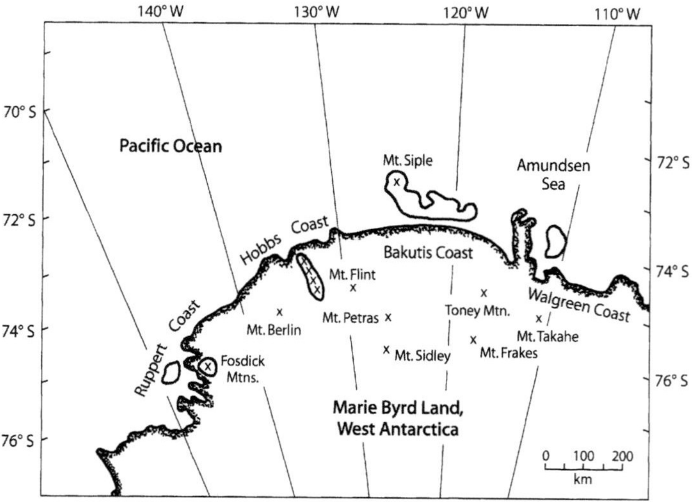

Fig. 6.77.
Marie Byrd Land of *West Antarctica* including the Ruppert Coast, Hobbs Coast, and Bakutis Coast (adapted from Hart et al. 1997)

longed subduction of oceanic crust along the Pacific margin of Gondwana (LeMasurier and Thompson 1990). The Cenozoic volcanic activity started about 30 million years ago in a large dome that encompasses the area adjacent to the Hobbs and Bakutis Coasts of Marie Byrd Land in Fig. 6.77 (LeMasurier 1990a,b). The Cenozoic lavas include alkali-rich basalt, basanites, mugearites, hawaiites, and trachytes.

Measurements of the $^{87}Sr/^{86}Sr$ ratios of volcanic rocks from West Antarctica were first reported by Jones and Walker (1972). Subsequent measurements by Futa and LeMasurier (1983) indicated a low average value of 0.70279 ±0.00008 (2$\bar{\sigma}$, $N = 11$) relative to 0.7080 for E&A. These authors related the Cenozoic volcanic activity to the West Antarctic Rift which extends from the coast of Victoria Land along the front of the Transantarctic Mountains to the Bellingshausen Sea at the base of the Antarctic Peninsula.

The $^{87}Sr/^{86}Sr$ and 1/Sr ratios of alkali-rich basalts from volcanic centers along a lineament near the Hobbs Coast in Marie Byrd Land and from the Jones Mountains in Ellsworth Land define clusters of data points in Fig. 6.78 (Hart et al. 1995, 1997). The volcanic rocks from the Hobbs lineament range in age from 11.7 ±1.0 to 2.34 ±0.11 Ma and consist of basalt, basanite, hawaiite, and mugearite. Their $^{87}Sr/^{86}Sr$ ratios are all less than 0.7030 and have an average value of 0.70280 ±0.00005 (2$\bar{\sigma}$, $N = 13$) relative to 0.71025 for NBS 987 (Hart et al. 1997). Therefore, the magma that produced these rocks originated in the lithospheric mantle and did not assimilate rocks from the continental crust. The positive correlation of the Rb and Sr concentrations of these rocks is attributable either to fractional crystallization of magma or to variations in the degree of partial melting.

The $^{87}Sr/^{86}Sr$ ratios of the volcanic rocks from Marie Byrd Land analyzed by Futa and LeMasurier (1983) and by Hart et al. (1997) are similar to those of the Balleny Islands (Sect. 6.12.3) but contrast with the elevated $^{87}Sr/^{86}Sr$ ratios of alkalic volcanic rocks at other Antarctic locations described in this chapter. Evidently, the alkali-rich magmas extruded along the Hobbs Coast and elsewhere in Marie Byrd Land originated from sources in the mantle having a low time-integrated Rb/Sr ratio of only 0.020. The alkali-enrichment of the Tertiary lavas in Marie Byrd Land was caused by low degrees of partial melting rather than by prior metasomatic alteration of the lithospheric mantle. Hole and LeMasurier (1994) attributed the Cenozoic volcanic activity in Marie Byrd Land to the presence of a large plume that underlies the West Antarctic Rift system. The extension in the plume head, and hence the degree of partial melting, may have been restricted by the fact that the Antarctic Plate is encircled by spreading ridges. Other contributions to the tectonic history of Marie Byrd Land are by Weaver et al. (1994) and by Pankhurst et al. (1993).

The data points representing the Jones Mountains of Ellsworth Land in Fig. 6.78 are displaced from the Hobbs cluster because the basalts in the Jones Mountains have higher $^{87}Sr/^{86}Sr$ ratios and lower Sr concentrations. The average $^{87}Sr/^{86}Sr$ ratio of these rocks analyzed by Hart et al. (1995) is 0.70348 ±0.00021 (2$\bar{\sigma}$, $N = 6$) relative to 0.71025 for NBS 987. Hole et al. (1994) who analyzed two basalt specimens from the Jones Mountains (0.70311 and 0.70337) concluded that the magma had originated from the asthenospheric mantle, whereas Hart et al. (1995) proposed a lithospheric source that had been modified as a result of prior sub-

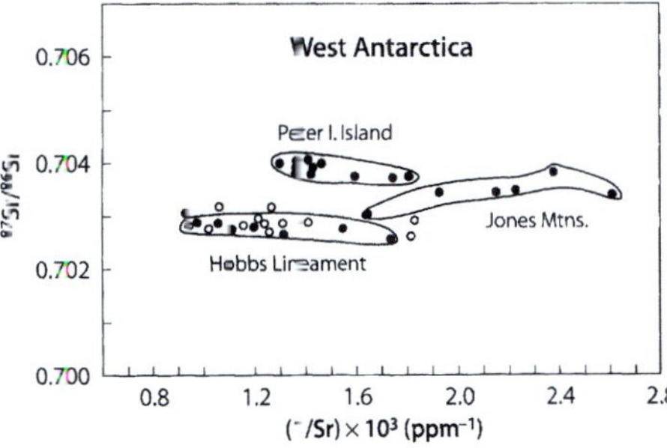

Fig. 6.78. ^{87}Sr/^{86}Sr and 1/Sr ratios of Cenozoic alkali-rich volcanic rocks of the Hobbs lineament in Marie Byrd Land and the Jones Mountains of Ellsworth Land in West Antarctica. Peter I. Island is located at 68.8° S and 98.6° W off the coast of Ellsworth Land. The *open circles* represent rock samples from various localities in West Antarctica whose ^{87}Sr/^{86}Sr and 1/Sr ratios overlap those of rocks collected along the Hobbs lineament. The distribution of data points indicates that the magmas originated from the mantle and did not assimilate crustal rocks. Instead, the magmas differentiated primarily by fractional crystallization or as a result of variations in the extent of partial melting. The differences in the ^{87}Sr/^{86}Sr ratios of the rocks at the localities represented here reveal that the magmas tapped different source regions in the underlying mantle (*Sources:* Futa and LeMasurier 1983; Prestvik et a. 1990; Hart et al. 1995; Hart et al. 1997)

duction of oceanic crust. In either case, the Rb-Sr systematics of these rocks are not attributable to crustal contamination of mantle-derived magmas.

6.12.5 Ultramafic Inclusions, Marie Byrd Land

Volcanic rocks of Tertiary age in the Executive Committee Range (76°30' S, 126° W) and on Mt. Aldaz (76°03' S, 124°25' W) in Marie Byrd Land contain lherzolite inclusions two of which were analyzed by Stueber and Ikramuddin (1974). In addition, Halpern (1969) reported ^{87}Sr/^{86}Sr ratios of 0.7032 and 0.7030 relative to 0.7080 for E&A for two ultramafic nodules from Mt. Perkins (75°32' S, 144°08' W) in the Ford Ranges (77°30' S, 145° W) of Marie Byrd Land. The nodule from the Executive Committee Range analyzed by Stueber and Ikramuddin (1974) has a low Sr concentration of 1.06 ppm and a high ^{87}Sr/^{86}Sr ratio of 0.7052 compared to 0.7026 for the host basalt relative to 0.7080 for E&A. The discordance of the ^{87}Sr/^{86}Sr ratios identifies this nodule as an accidental inclusion in the basalt, although the high ^{87}Sr/^{86}Sr ratio of the nodule could have been caused by contamination with Sr derived from melting snow. The low ^{87}Sr/^{86}Sr ratio of the basalt is consistent with the ^{87}Sr/^{86}Sr ratios of volcanic rocks in Marie Byrd Land reported later by Futa and LeMasurier (1983) and by Hart et al. (1997).

Mt. Aldaz is a volcano located at 76°03' S and 124°25' W, about 60 km northeast of the northern end of the Ex-

ecutive Committee Range. The ^{87}Sr/^{86}Sr ratios of the constituent minerals of a lherzolite inclusion from this location analyzed by Stueber and Ikramuddin (1974) increase with decreasing Sr concentrations from 0.7023 (clinopyroxene, Sr = 16.0 ppm) to 0.7059 (orthopyroxene, Sr = 0.97 ppm), and to 0.7078 (olivine, Sr = 3.05 ppm). This pattern of variation is attributable to contamination of the minerals by marine Sr in melting snow. However, the ^{87}Sr/^{86}Sr ratios also correlate with the Rb/Sr ratios of the minerals and yield a nominal Rb-Sr date of about 600 ±110 Ma ($\lambda = 1.42 \times 10^{-11}$ yr^{-1}). Stueber and Ikramuddin (1974) suggested that this date refers to the last time the lherzolite was partially melted, but also considered the possibility that the minerals were contaminated after eruption of the nodule.

6.12.6 Peter I. Island

Peter I. Island is a volcano of Pleistocene age that rests on oceanic crust and rises to a height of 1 640 m above sealevel. It is located at 68.8° S and 90.6° W, about 400 km off the Eights Coast of Ellsworth Land and appears to be the most southern oceanic island in the world. The volcanic rocks exposed on Peter I. Island range in composition from basalt to hawaiite and include benmoreite and trachyte (Prestvik and Duncan 1991).

The ^{87}Sr/^{86}Sr ratios of basalts and hawaiites on Peter I. Island reported by Prestvik et al. (1990) range from 0.70378 to 0.70404 and average 0.70386 ±0.00006 ($2\bar{\sigma}$, $N = 11$). One specimen of benmoreite has a low ^{87}Sr/^{86}Sr ratio of 0.70369. Hart et al. (1995) measured ^{87}Sr/^{86}Sr ratios of five specimens from Peter I. Island including two (PI7 and PI8) that had been previously analyzed by Prestvik et al. (1990). The differences in the reported ^{87}Sr/^{86}Sr ratios appear to be random, and the average ^{87}Sr/^{86}Sr ratio reported by Hart et al. (1995) is 0.70394 ±0.00008 ($2\bar{\sigma}$, $N = 4$) which is indistinguishable from the average of Prestvik et al. (1990). However, a trachyte analyzed by Hart et al. (1995) has an anomalously low ^{87}Sr/^{86}Sr ratio of 0.703010. The ^{87}Sr/^{86}Sr and 1/Sr ratios of the volcanic rocks on Peter I. Island form a cluster in Fig. 6.76 similar to that of the Jones Mountains in Ellsworth Land. Hart et al. (1995) proposed that the magma extruded on Peter I. Island originated from a mantel plume without significant contamination by rocks of the oceanic crust.

6.12.7 Antarctic Peninsula

Alkali basalts of the Antarctic Peninsula (Pankhurst 1982) postdate the end of subduction along this part of the Pacific margin and were attributed by Hole and

LeMasurier (1994) to the formation of small melt fractions in the underlying asthenospheric mantle. This late-stage magmatic activity was facilitated by the development of slab-free windows that permitted upwelling of the asthenospheric mantle and subsequent decompression melting (Hole et al. 1991a, 1993). The origin of the alkali-rich basalts of the Antarctic Peninsula, therefore, differs from that of the volcanic rocks in Marie Byrd Land which formed by melting in the head of a large plume underlying that area. The petrogenesis of the alkali-rich rocks of the Antarctic Peninsula was discussed by Storey and Garretts (1985), Smellie (1987), Smellie et al. (1988), Hole (1988, 1990), Hole et al. (1991b,c), and by Hole and Larter (1993). The volcanic activity related to subduction in the South Shetland Islands (Pankhurst et al. 1980 and Tanner et al. 1982) west of the northern tip of the Antarctic Peninsula was reviewed in Sect. 3.9.5.

6.12.8 Gaussberg

The leucitite lavas of late Pleistocene age on Gaussberg, located at 66°48' S and 89°12' E on the coast of East Antarctica, have high $^{87}Sr/^{86}Sr$ ratios, and their concentrations of K_2O range from 4.7 to 12.4% (Collerson and McCulloch 1983; Sheraton and Cundari 1980). The Gaussberg is collinear with a submarine ridge that extends for about 2000 km from Kerguelen Island to McDonald and Heard Islands in the Indian Ocean to about 63° S (Sect. 2.15). The volcanic activity on Gaussberg at 0.056 ±0.005 Ma coincided with eruptions on Heard and McDonald Islands but postdates those on Kerguelen Island which was active from 27 to 8 Ma (Collerson and McCulloch 1983).

The leucitites on Gaussberg have high concentrations of Rb and Sr with averages of 310 ±11 ppm and 1760 ±80 ppm ($2\bar{\sigma}, N = 7$), respectively. The $^{87}Sr/^{86}Sr$ ratios of these rock have a mean of 0.70955 ±0.0019 ($2\bar{\sigma}, N = 7$) relative to 0.71025 for NBS 987 (Collerson and McCulloch 1983). These results distinguish the leucitites of Gaussberg from the alkali basalt on Heard Island in the Indian Ocean for which Storey et al. (1988) reported Rb = 44.2 (34–57) ppm, Sr = 735 (622–845) ppm, and $^{87}Sr/^{86}Sr$ = 0.70565 relative to 0.71025 for NBS 987.

Two leucite-clinopyroxene-magnetite inclusions also have high Rb and Sr concentrations as well as high $^{87}Sr/^{86}Sr$ ratios (0.71033) relative to 0.71025 for NBS 987 (Collerson and McCulloch 1983). These inclusions therefore are cognate cumulates derived from the leucitite lavas extruded at Gaussberg. However, two crustal xenoliths and one sample of clinopyroxene of crustal origin have elevated $^{87}Sr/^{86}Sr$ ratios ranging from 0.72741 to 0.76255.

The elevated $^{87}Sr/^{86}Sr$ ratio of the leucitites can be attributed to one of several possible causes:

1. Contamination of a mantle-derived basalt magma by assimilation of rocks of the continental crust.
2. Partial melting within the continental crust.
3. Derivation from source regions in the mantle that had been enriched in Rb (and K) long enough to increase the $^{87}Sr/^{86}Sr$ ratios to the observed values (0.70918 to 0.71085).

Collerson and McCulloch (1983) discredited crustal contamination of mantle-derived magma because the high Sr concentration (1582 to 1890 ppm) makes the leucitites insensitive to contamination by crustal rocks having low Sr concentrations (158 to 303 ppm). They also discredited a crustal origin of the leucitite magma because their $^{87}Sr/^{86}Sr$ and $^{147}Nd/^{144}Nd$ ratios in Fig. 6.79 form a tight cluster along the extension of the magma sources in the lithospheric mantle. Therefore, Collerson and McCulloch (1983) agreed with Sheraton and Cundari (1980) that the leucitites originated from Rb-enriched sources in the mantle. The enrichment of the rocks in K_2O (11.61%) relative to Na_2O (1.65%) is strong evidence that phlogopite was present in the source rocks. The high $^{87}Sr/^{86}Sr$ and low $^{143}Nd/^{144}Nd$ ratios of the leucitites on Gaussberg are similar to those of K-rich lavas elsewhere in the world (e.g. the Roman prov-

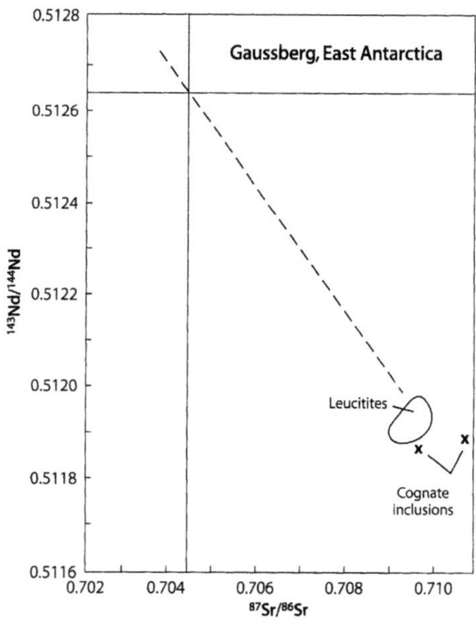

Fig. 6.79. Isotope ratios of Sr and Nd of leucitite lavas and cognate inclusions (X) on Gaussberg on the coast of East Antarctica. The $^{143}Nd/^{144}Nd$ ratios were recalculated based on 0.512638 for the $^{143}Nd/^{144}Nd$ ratio of CHUR (*Source:* Collerson and McCulloch 1983)

ince of Italy, the Leucite Hills and Highwood Mountains of Wyoming and Montana, and the Fitzroy Trough-Kimberley block of Western Australia). The age of the alkali-enriched magma sources in the mantle under Gaussberg can be estimated from the average $^{87}Sr/^{86}Sr$ (0.70955) and $^{87}Rb/^{86}Sr$ (0.473) ratios of the leucitites based on the model for Sr isotope evolution in the mantle discussed in Sect. 1.9. The resulting date of 1.10 Ga is an underestimate, because the leucitite magma was probably enriched in Rb during partial melting of the source rocks. Collerson and McCulloch (1983) calculated similar dates of 1.22 to 1.28 Ga from the isotope composition of Nd in the leucitites at Gaussberg.

6.13 Summary: The Case for Mantle Metasomatism

Alkali-rich igneous rocks form under a wide variety of tectonic conditions and therefore are widely distributed all over the Earth. In most cases, the alkali-enrichment of the rocks is inherited from the magma sources in the mantle and is not the result of crustal contamination of calc-alkaline magmas. Although alkali-rich volcanic rocks have a wide range of mineral compositions, fractional crystallization of magma in crustal reservoirs is not an important part of their petrogenesis. Instead, the evidence indicates that alkali-rich magmas are erupted rapidly and forcefully from magma sources located within the lithospheric mantle.

The isotope compositions of Sr, Nd, and Pb of alkali-rich igneous rocks range widely, indicating that their magma sources are isotopically heterogeneous. The magnitudes of the isotope ratios of Sr, Nd, and Pb as well as their range of variation in alkali-rich igneous rocks are related to the physical and chemical properties of their magma sources in the mantle, to the extent of partial melting, and to subsequent mixing of magmas derived from different components in the lithospheric mantle.

The alkali-enrichment of the magmas can be explained by low degrees of partial melting (<5%) of the source rocks. In addition, the prior alkali-enrichment of the magma sources is indicated by the presence of kaersutite (an alkali-rich amphibole) and phlogopite (mica), which occur in ultramafic inclusions at many volcanic centers. These inclusions are unmetamorphosed in many cases, indicating that the magma in which they were entrained erupted rapidly. Kaersutite and phlogopite form either veins or occur in disseminated form in the inclusions. Both minerals are not only carriers of alkali metals, but also contain water. Therefore, the presence of water-bearing minerals in mantle-derived inclusions is unmistakable evidence that the lithospheric mantle was altered by water-rich silicate melts or aqueous fluids prior to magma formation.

The intrusions of such aqueous fluids enrich the affected region of the lithospheric mantle in alkali metals and water as well as in other volatiles such as carbon dioxide and fluorine. The presence of volatile compounds (H_2O and CO_2) in some parts of the lithospheric mantle lowers the melting temperature of the rocks and predisposes these domains to become sources of alkali-rich magmas. Partial melting may be initiated by rifts which propagate from the base of the lithospheric mantle towards the surface or by heat emanating from the heads of asthenospheric plumes that apply pressure against the base of the lithosphere, or both. Consequently, alkali-rich lavas are erupted along rifts in the continental crust (e.g. the East African Rift valleys, the Rhine Graben in central Europe, the Oslo Graben of Scandinavia, as well as in the Basin and Range province of the southwestern USA). Alkali-rich rocks are also erupted along plume tracks in the Pacific Ocean (e.g. the Hawaiian Islands and in Polynesia) and on many other oceanic islands that developed above mantle plumes. In addition, alkali-rich volcanic rocks have been erupted in island arcs (e.g. the Mediterranean Basin, Indonesia, Japan, etc.) where alkali-rich magmas form by low degrees of partial melting of the mantle wedge.

The aqueous fluids that cause alkali metasomatism of the lithospheric mantle originate from sources located at the base of the lithospheric mantle in the form of subducted oceanic crust and associated marine and terrigenous sediment. Blocks of subducted oceanic crust either underplate the subcontinental lithospheric mantle, or sink to deeper levels in the asthenosphere where they cause the development of plumes because of buoyancy in response to radioactive heating. In either case, the ultimate source of the water and of the alkali metals appears to be seawater that was trapped in the subducted oceanic crust or was incorporated into secondary minerals such as clay minerals, oxyhydroxides of Mn and Fe, silica gel, serpentine, chlorite, zeolites, and others.

The metasomatizing fluids also contain carbon dioxide that originates from subducted organic compounds, marine carbonate minerals, and from sources in the asthenospheric mantle. The importance of carbon dioxide in the petrogenesis of alkali-rich rocks is indicated by the association of carbonatites with alkali-rich volcanic and plutonic rocks as well as by the occurrence of diamonds in K-rich kimberlite pipes and lamproites in the continental cratons of South Africa, Siberia, South America, North America, and Australia. In fact, Antarctica is the only continent in which diamond-bearing kimberlites have not yet been found, although intrusive kimberlites could be associated with the K-rich lamproite lavas of Gaussberg on the margin of the East Antarctic Craton.

The isotope ratios of Sr, Nd, and Pb of alkali-rich igneous rocks depend on the Rb/Sr, Sm/Nd, and U/Pb ratios of their magma sources and on the time that passed between the alteration of the magma sources and the onset of partial melting. In cases where the incubation time is short (e.g. in the mantle wedge of subduction zones), the abundances of radiogenic isotopes of the magma sources do not increase appreciably by decay of their parents, but may be changed by mixing with the Sr, Nd, and Pb derived from the subducted oceanic crust. If magma formation is delayed for several hundred million or even one to two billion years (e.g. beneath Precambrian cratons), the isotope compositions of Sr, Nd, and Pb in the resulting alkali-rich rocks are controlled by in situ decay or by the lack thereof. For example, the K-rich volcanic and plutonic rocks at several sites on the Earth have high $^{87}Sr/^{86}Sr$ but low $^{143}Nd/^{144}Nd$ and $^{206}Pb/^{204}Pb$ ratios (e.g. Leucite Hills, Wyoming; Highwood Mountains, Montana). The explanation is that the magma sources had high Rb/Sr ratios but low Sm/Nd and U/Pb ratios. Consequently, the $^{87}Sr/^{86}Sr$ increased significantly by decay of ^{87}Rb during the incubation time, whereas the $^{143}Nd/^{144}Nd$ and $^{206}Pb/^{204}Pb$ ratios increased only slightly and still have the low values that existed prior to the episode of metasomatism. Consequently, volcanic and plutonic rocks derived from such source regions plot in the "crustal" quadrant of the Sr-Nd mixing diagram, even though their magmas did not assimilate crustal rocks.

The mineral composition of mantle-derived inclusions has provided important information about the metasomatic alteration of the magma sources of alkali-rich lavas in which they occur. However, the isotope ratios of Sr of these inclusions are sensitive to alteration during transport by the magma and by exposure to aqueous solutions at the surface of the Earth. Olivine and orthopyroxene are especially vulnerable, because of their low Sr concentrations, which are less than 2 ppm in most cases. Consequently, the $^{87}Sr/^{86}Sr$ ratios of whole-rock samples of ultramafic inclusions depend on the extent of contamination of the constituent minerals. In addition, the constituent minerals of ultramafic inclusions, in some cases, have different $^{87}Sr/^{86}Sr$ ratios, contrary to expectations for rocks that originated from the lithospheric mantle where the ambient temperature exceeds 1 000 °C. In some cases, the contaminant Sr is removable by leaching the rocks with dilute acid, because it was deposited on the surfaces of mineral grains. The isotope composition of Pb of mantle-derived inclusions is also prone to alteration because of deposition of Pb by aqueous solutions at the surface of the Earth.

The petrogenetic processes that result in the formation of igneous rocks in different tectonic settings in the ocean basins and along the margins and in the interiors of continents have been active throughout geologic time. Remnants of these igneous rocks and of the sedimentary rocks that were derived from them still exist on the continents of the Earth. These rocks therefore contain a record of the tectono-magmatic processes that have contributed to the growth and evolution of the continental crust.

Chapter 7

Differentiated Gabbro Intrusives

The extrusion of large volumes of tholeiite basalt on the continents discussed in Chap. 5 has been an important phenomenon throughout most of geologic time. Remnants of Precambrian basalt plateaus still exist in scattered locations on the Precambrian shields of the major continents (e.g. the Coppermine River basalt, Northwest Territories, Canada). However, the basalt flows have been eroded in most cases, leaving only swarms of feeder dikes or large differentiated gabbroic intrusives which formed in the magma reservoirs that supported the volcanic activity at the surface. The Skaergaard Intrusion of East Greenland (Sect. 5.4.6), the Freetown Complex of Liberia (Sect. 5.9.4), and the Dufek Intrusion of Antarctica Sect. 5.12.5) have already been mentioned. The differentiated gabbroic intrusives to be considered in this chapter range in age from Archean to Cenozoic and will be examined to detect evidence of crustal contamination of the magmas, to draw attention to mineral deposits associated with these kinds of rocks, and to understand the origin and evolution of the magmas in large crustal magma chambers.

Large bodies of stratified ultramafic rocks, layered gabbros, and granophyre occur on virtually every continent. Some of these have been studied intensively as examples of fractional crystallization of large volumes of basalt magma derived from sources in the mantle. The results have illuminated the processes that occur in magma chambers such as fractional crystallization, assimilation of wallrocks, injection of new magma into magma chambers from sources at depth, and mixing of magmas (Wyllie 1967; Irvine 1970a; O'Hara 1977; Hargraves 1980; O'Hara and Matthews 1981; DePaolo 1985).

The rocks that are produced consist of early-formed cumulus minerals; (e.g. olivine, pyroxene, plagioclase, etc.) and late-forming intercumulus minerals (e.g. pyroxene, plagioclase, magnetite, etc.) that crystallize from the residual intercumulus liquid. This liquid is mobile and therefore is not in thermodynamic equilibrium with the crystals it comes in contact with (Irvine 1980). In addition, intercumulus liquids can have different $^{87}Sr/^{86}Sr$ ratios than the cumulus crystals. As a result, intercumulus mineral phases in the resulting rocks have different $^{87}Sr/^{86}Sr$ ratios than the cumulus minerals, and the isotope composition of Sr in whole-rock samples depends on both the concentration and isotope composition of Sr and on the relative abundances of intercumulus and cumulus phases. Thus the $^{87}Sr/^{86}Sr$ ratios of whole-rock samples of mafic plutonic rocks may be controlled in part by the porosity of the mush of cumulus crystals.

When crustal magma chambers are replenished by the injection of magma from sources at depth (Huppert and Sparks 1980; O'Hara 1977), the new magmas may differ in density, temperature, and chemical as well as isotopic composition from the magma already in the chamber. In cases where the new magma is denser than the resident magma, it remains at the bottom of the chamber, but the injection of new magma may cause the chamber to overflow and thereby causes eruption of lava at the surface.

The chemical and isotopic compositions of mafic magmas in crustal reservoirs may also be altered by mixing with partial melts of the wallrocks (DePaolo 1985), by assimilation of wallrocks, and by selective transfer of certain elements or specific isotopes from the country rock into the magma (Pankhurst 1969). In addition, significant alteration of minerals resulting in chemical and isotopic changes may also occur after crystallization by circulating aqueous fluids whose movement is controlled by temperature gradients and by the permeability of the rocks (Taylor and Forester 1979).

The study of layered mafic intrusions has been stimulated by the occurrence of valuable deposits of platinum group elements (PGE), sulfide minerals of Ni and Cu which in some cases contain gold, and by the presence of layers of chromite and magnetite. In addition, layered mafic intrusives of different ages have preserved a record of the isotopic evolution of Sr in the mantle sources of mafic magmas. The results of these studies have revealed an unexpected complexity in the evolution of mafic magmas in large crustal reservoirs. Multiple injections of magma, mixing of magmas, assimilation of wallrocks, fractional crystallization, and hydrothermal alteration all contribute to the petrogenesis of layered mafic intrusives. The rocks produced by these processes have a wide range of chemical compositions and isotope ratios of Sr, Nd, and Pb, which make it difficult to determine the isotope ratios of these elements in the magma sources in the mantle.

7.1   Midcontinental Rift System of North America

The volcanic province associated with the Midcontinental Rift system in Fig. 7.1 is centered on Lake Superior in North America and formed during a relatively short interval of time at about 1100 Ma (Green 1982). The rocks in this province are composed primarily of tholeiite basalt that formed a flood-basalt plateau, but include also rhyolite, quartz-feldspar porphyry intrusions, and thick deposits of conglomerate, sandstone, and shale. In addition, the Duluth gabbro of Minnesota and smaller gabbroic intrusives in Wisconsin formed during this period of igneous activity. The lava flows and interbedded sedimentary rocks are exposed primarily at Mamainse Point on the eastern shore of Lake Superior, on the Keweenaw Peninsula of Michigan, in northern Wisconsin, along the western shore of Lake Superior in Minnesota, and on Isle Royal.

The volcanic rocks that were erupted along the Midcontinental Rift system of North America are similar to these of younger flood basalts discussed in Chap. 5. For example, they include tholeiite basalt whose total stratigraphic thickness exceeds 5 km. In addition, the flows are interbedded with terrigenous sedimentary rocks and with rhyolites. The volcano-sedimentary complex of the Midcontinental Rift system was intruded by late-stage anorogenic granitic plutons and contains large differentiated bodies of gabbro (e.g. the Duluth gabbro of Minnesota). These bodies are genetically related to the tholeiite basalt flows and probably represent former magma reservoirs in the continental crust about one billion years ago.

The volcano-sedimentary rocks of the Lake Superior region differ from younger basalt plateaus only because they were structurally deformed and now have regional dips of 30 to 40°, because they were buried by younger sedimentary rocks of latest Precambrian to Cambrian age, and because they were altered by hydrothermal solutions that moved within the sedimentary interbeds and through the highly vesicular flow tops (Nicholson et al. 1992). These solutions deposited native copper in the basalt flows and interbedded conglomerates of the Keweenaw Peninsula of Michigan, which was once an important mining district based on the extensive deposits of native copper in the Copper Harbor Conglomerate and in the vesicular flow tops of the Portage Lake Lava Series (Butler and Burbank 1929; White 1968). In addition, chalcocite (Cu_2S) and minor native copper occur near White Pine, Michigan, in the Nonesuch Shale, which overlies the Copper Harbor Conglomerate (White and Wright 1954, 1966).

The Midcontinental Rift system of North America in Fig. 7.1 extends from Lake Superior in a southerly direction into Iowa, Nebraska, and Kansas. In addition, the volcanic rocks of this igneous province have been traced in subsurface into the lower peninsula of Michigan (Van Schmus and Hinze 1985; Cannon et al. 1989). The igneous activity in the Lake Superior area was caused by magma formation in a plume of astheno-spheric mantle rocks (Hutchinson et al. 1990; Klewin and Shirey 1992).

7.1.1   Mamainse Point Formation, Ontario

A stratigraphic section consisting of about 350 lava flows and interbedded conglomerates at Mamainse Point on the eastern shore of Lake Superior has a thickness of about 5250 m and contains the most complete record of igneous activity associated with the Midcontinental Rift system. The lava flows strike northwest and dip southwest under Lake Superior at about 20°. Age determinations cited by Shirey et al. (1994) indicate that these rocks formed between 1108 and 1088 Ma. The geology of this area was described by Thomson (1953), Giblin (1969), and Annells (1973), whereas Van Schmus (1971a) reported Rb-Sr age determinations of felsites and basalt. The origin of the basalt flows at Mamainse Point has been discussed by Massey (1983), Berg and Klewin (1988), and Klewin and Berg (1990) based on geochemical and petrologic data. Richards and Spooner (1989) attributed the origin Cu-Ag fissure veins in the area to mixing of hydrothermal solutions of meteoric and magmatic origin.

The data of Shirey et al. (1994) in Fig. 7.2 indicate that the average concentrations of Rb and Sr in groups of lava flows at Mamainse Point vary systematically

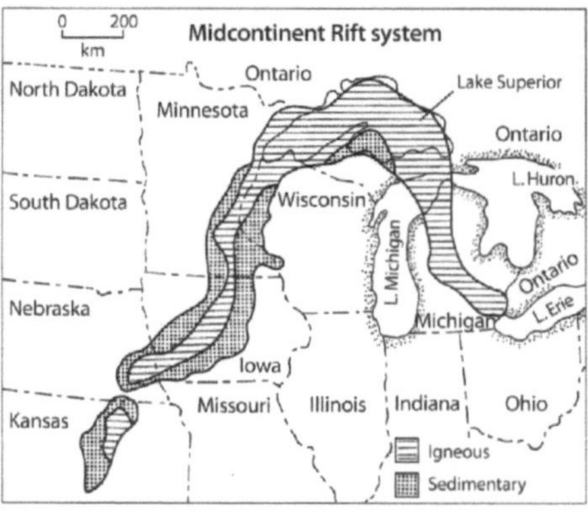

Fig. 7.1. Midcontinental Rift system of North America containing tholeiite lava flows with minor rhyolite, interbedded conglomerates, sandstones, and shales as well as differentiated gabbroic and alkalic intrusives. These rocks were deposited within a failed rift system at about 1100 Ma and are exposed in the Lake Superior area (*Source:* adapted from Van Schmus et al. 1987 and Nicholson et al. 1992)

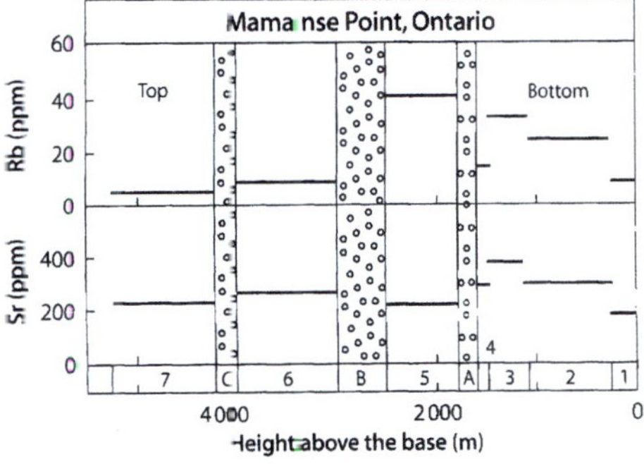

Fig. 7.2. Stratigraphic variation of average Rb and Sr concentrations of groups of basalt flows in the Mamainse Point Formation exposed along the eastern shore of Lake Superior, Ontario. These flows are part of the Midcontinental Rift system that was caused by plumes rising from the asthenosphere. Magmas, formed by partial melting in a plume or in the overlying lithospheric mantle, were subsequently modified chemically and isotopically by interactions with the upper lithospheric mantle and continental crust, as well as by fractional crystallization. The groups of lava flows (1 through 7) are interbedded with conglomerate beds (A through C). The average Rb and Sr concentrations of the early-formed basalt groups increase up-section. However, after deposition of the Great Conglomerate (B), the average concentrations of both elements stabilized at relatively low values in Groups 6 and 7 (*Source:* Shirey et al. 1994)

up-section. The average Rb concentrations rise from 9.3 ±2.3 ppm in Group 1 (250 m) to 40.9 ±12.4 ppm in Group 5 (700 m) with a decrease to 15.5 ±4.9 in Group 4 (125 m). The average Sr concentrations also rise from 185 ±17 ppm in Group 1 to 378 ±88 ppm in Group 3 and then decline in Groups 4 (298 ±22 ppm) and 5 (232 ±33 ppm). The flows in Groups 6 and 7 above the Great Conglomerate (250 m) have uniformly low average concentrations of Rb and Sr. Shirey et al. (1994) reported that the flows below the Great Conglomerate have low $^{143}Nd/^{144}Nd$ ratios (at 1 100 Ma) compared to those above the Great Conglomerate. In addition, all basalt flows are enriched in radiogenic ^{206}Pb and ^{207}Pb, probably as a result of assimilation of Archean basement rocks (1 to 3%) by the basalt magmas. Shirey et al. (1994) concluded that the magmas of Groups 1 through 5 below the Great Conglomerate formed by partial melting in an asthenospheric plume followed by assimilation of varying amounts of subcrustal lithospheric mantle and continental crust. The lavas in Groups 6 and 7 above the Great Conglomerate have uniform trace-element concentrations and isotope ratios, and originated from mixtures of magmas derived from the plume and from depleted rocks in the mantle.

The rocks exposed on Michipicoten Island in Lake Superior include olivine tholeiite flows of the Mamainse Point Formation which were intruded by quartz porphyry and more mafic rocks of basaltic andesite

composition. The intrusive rocks are unconformably overlain by lava flows of the Michipicoten Formation ranging in composition from andesite to rhyolite. Van Schmus (1971) reported that a felsite sample from Michipicoten Island has an initial $^{87}Sr/^{86}Sr$ ratio of 0.7041 at 953 Ma. Subsequently, Wanless and Loveridge (1978) obtained a Rb-Sr isochron date of 918 ±78 Ma ($\lambda = 1.42 \times 10^{-11}$ yr^{-1}) and an initial $^{87}Sr/^{86}Sr$ ratio of 0.7068 ±0.0010 for volcanic rocks in the Channel Lake Member of the Michipicoten Formation (E&A = 0.7080). In addition, they measured a Rb-Sr date of 1 004 ±64 Ma for the quartz porphyry intrusive with an elevated initial $^{87}Sr/^{86}Sr$ ratio of 0.711 ±0.024. A quartz-feldspar porphyry at Jodran Mine (10 km east of Mamainse Point on the mainland) also has a high initial $^{87}Sr/^{86}Sr$ ratio of 0.7088 at 1 047 Ma (Van Schmus 1971). The initial $^{87}Sr/^{86}Sr$ ratios of the late-stage porphyry intrusives at the Jodran Mine (Van Schmus 1971) and on Michipicoten Island (Wanless and Loveridge 1978) are higher than those of the volcanic rocks, suggesting a crustal origin of the magma or more extensive contamination of basalt magmas that originated from the Midcontinent Plume.

7.1.2 Keweenaw Peninsula, Michigan

The Portage Lake volcanics on the Keweenaw Peninsula in Fig. 7.3 consist of about 200 tholeiite basalt flows whose cumulative thickness approaches 5 km. Interflow sediments and felsic volcanic rocks constitute about 3 to 4% by volume (Paces and Bell 1989). The flows strike northeast and dip to the northwest at about 60° under the western basin of Lake Superior (White 1966).

The work of Chaudhuri and Faure (1967) established a range of ages (recalculated to $\lambda = 1.42 \times 10^{-11}$ yr^{-1})

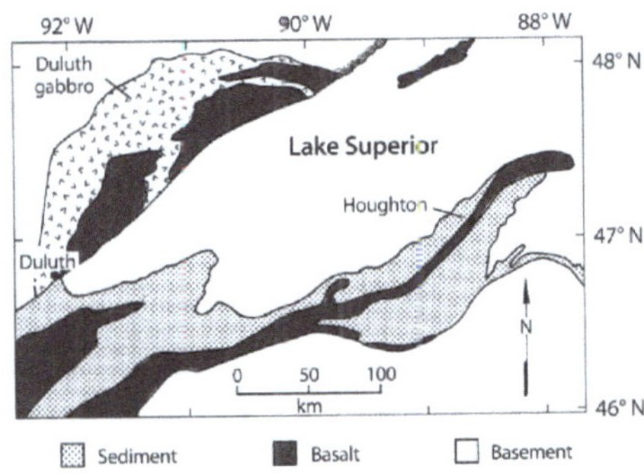

Fig. 7.3. Geology of the western basin of Lake Superior including tholeiite basalt flows and continental sedimentary rocks as well as the Duluth gabbro of Middle Proterozoic age (*Source:* White 1966)

from about 960 to 1 080 Ma for rhyolite and quartz-feld-spar porphyry near Bergland and White Pine on the Keweenaw Peninsula. The initial ^{87}Sr/^{86}Sr ratios relative to E&A = 0.7080 range from 0.7039 (rhyolite, Bergland) to 0.7178 (quartz-feldspar porphyry, 8 miles south of White Pine). The relatively low initial ^{87}Sr/^{86}Sr ratio of the rhyolites is remarkable, because the magmas that produced them cannot have formed by melting of rocks in the continental crust. A possible explanation is that the rhyolite magmas originated by remelting of young basalt at depth in the crust. However, the quartz-feldspar porphyry originated from a crustal melt that formed at 957 ±40 Ma with an initial ^{87}Sr/^{86}Sr ratio of 0.7178 relative to 0.7080 for E&A.

More recent work by Paces and Bell (1989) and Nicholson and Shirey (1990) indicates that the initial ^{87}Sr/^{86}Sr ratios of basalt of the Portage Lake volcanics range from 0.70235 to 0.70635 at 1 095 Ma. These values depend on the assumption that the Rb/Sr ratios of the rocks have remained constant and that the rocks actually crystallized at 1 095 Ma. In addition, the inverse correlation between the initial ^{87}Sr/^{86}Sr ratios and the reciprocal Sr concentrations of basalt in Fig. 7.4 is attributable to assimilation by mantle-derived magmas (^{87}Sr/^{86}Sr ≈ 0.7025, Sr ≈ 235 ppm) of crustal rocks (^{87}Sr/^{86}Sr ≈ 0.7065, Sr ≈ 1 000 ppm). However, the petrogenetic significance of the data in Fig. 7.4 is question-

able in this case because of evidence for aqueous alteration of the lava flows of the Portage volcanics presented by Jolly and Smith (1972).

The hydrothermal alteration of the rhyolites analyzed by Nicholson and Shirey (1990) is indicated by inconsistencies in the slope and intercept values of a linear data array in coordinates of ^{87}Sr/^{86}Sr and ^{87}Rb/^{86}Sr ratios. Five rhyolite specimens, collected from the basal section of the Portage Lake volcanics near the eastern tip of the Keweenaw Peninsula, form a straight line (correlation coefficient 0.99913) that yields a date of 1 022 Ma and an initial ^{87}Sr/^{86}Sr ratio of 0.6994 (not shown). This date is lower than the age of the Copper City flow near the base of the Portage Lake volcanics for which Davis and Paces (1990) reported a U-Pb zircon date of 1 096.2 ±1.8 Ma. In addition, the initial ^{87}Sr/^{86}Sr ratio of the rhyolites is too low to be compatible with the isotopic evolution of Sr in the Earth (Sect. 1.12). Therefore, Nicholson and Shirey (1990) suggested that the rhyolite samples they analyzed have not remained closed systems.

The age of the native copper deposits in the amygdaloidal flows of the Portage Lake volcanics and interbedded conglomerates was determined by Bornhorst et al. (1988) by analyzing minerals in amygdules. Their data yielded Rb-Sr isochron dates of 1 047 ±33 Ma (chlorite and K-feldspar) and 1 060 ±20 Ma (calcite, epidote, and K-feldspar) with initial ^{87}Sr/^{86}Sr ratios of 0.7147 ±0.0167 and 0.7047 ±0.0091, respectively. These results favor the interpretation that the deposition of the native copper occurred after the eruption of the lava flows and was caused by hydrothermal solutions that were generated by burial metamorphism at depth in the basin where heat was provided by cooling basalt dikes. The low initial ^{87}Sr/^{86}Sr ratios of calcite (0.7035) and epidote (0.7047) indicate that these minerals formed from solutions that had interacted with the basalts of the Portage Lake volcanics. The K-feldspar and chlorite, which have high initial ratios (0.713 to 0.715), were deposited by different solutions that contained radiogenic ^{87}Sr derived either from older basement rocks or from the detrital sediment in the conglomerates. The range of initial ^{87}Sr/^{86}Sr ratios of sequentially deposited minerals implies that the ^{87}Sr/^{86}Sr ratios of the solutions flowing through the vesicular lava flows decreased with time as the interactive volume of rocks decreased from basement plus lava flows to only the flows of the Portage Lake volcanics.

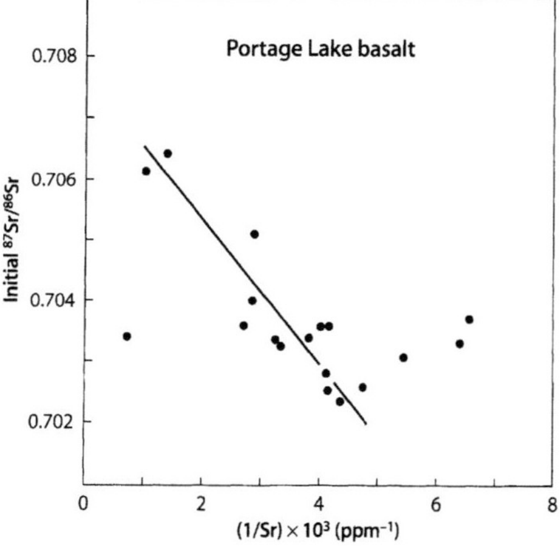

Fig. 7.4. Sr-isotope mixing diagram for basalt of the Portage Lake Lava Series on the Keweenaw Peninsula of Michigan. The apparent correlation of initial ^{87}Sr/^{86}Sr ratios (at 1 095 Ma) and reciprocal Sr concentrations is the result of assimilation of rocks (^{87}Sr/^{86}Sr ≈ 0.7065, Sr ≈ 1 000 ppm) by mantle-derived magma (^{87}Sr/^{86}Sr ≈ 0.7025, Sr ≈ 235 ppm). The deviation of four data points from the array was caused by decreases of the Sr concentrations resulting from fractional crystallization or hydrothermal alteration during cooling of the flows (*Sources:* Paces and Bell 1989; Nicholson and Shirey 1990)

7.1.3 Nonesuch Shale, White Pine, Michigan

The Nonesuch Shale near White Pine on the Keweenaw Peninsula of Michigan is composed of black shales and siltstones noted for containing disseminated chalcocite as well as organic matter including petroleum. The discovery of organic compounds of biological origin in one-billion-year old sedimentary rocks was a milestone in the study of the evolution of life on Earth that continues to attract attention (Meinschein et al. 1964; Eglinton et al. 1964; Barghoorn et al. 1965; Ho et al. 1990). The origin of the copper mineralization was discussed by White and Wright (1954, 1966), White (1971), Burnie et al. (1972), and Brown (1974).

Chaudhuri and Faure (1967) reported that whole-rock samples of the Nonesuch Shale form a linear array on the Rb-Sr isochron diagram indicating a date of 1052 ± 50 Ma (recalculated to $\lambda = 1.42 \times 10^{-11}$ yr^{-1}) and an initial ^{87}Sr/^{86}Sr ratio of 0.7077 ± 0.0008 relative to c.7080 for E&A. The authors considered this date to be an overestimate of the age of deposition of the Nonesuch Shale because the rocks are probably mixtures of sediment components derived from the Portage Lake Lava Series and from the underlying basement rocks.

Although the distribution of copper in the Nonesuch Shale is stratigraphically controlled, detailed studies have demonstrated that the ore at the White Pine mine is of epigenetic origin (e.g. Brown 1974). Therefore, Ruiz et al. (1984) measured the age of the ore by comparing the average ^{87}Sr/^{86}Sr ratio of the host rocks to the average ^{87}Sr/^{86}Sr ratio of late-stage calcite veins based on the assumption that the Sr in the calcite veins was derived from the wallrock. The results indicate an age of 1047 ± 35 Ma in good agreement with the whole-rock Rb-Sr date of 1052 ± 50 Ma reported by Chaudhuri and Faure (1967) and with the ages of minerals in amygdules of the Portage Lake volcanics obtained by Bornhorst et al. (1988).

7.1.4 Sibley Group, Ontario

The sedimentary rocks of the Sibley Group were deposited in a continental basin that extended north from the Midcontinent Rift system in the Lake Superior area of Ontario. The present outcrop belt of these rocks reaches from the Sibley Peninsula and the north-shore of Lake Superior north to the south shore of Lake Nipigon. J. M. Franklin (n Wanless and Loveridge 1978) reported that the Sibley Group rests unconformably on the granitic gneisses of the Superior Craton north of Lake Superior. The rocks of the Sibley Group were intruded by the Logan diabase sills related to the magmatic activity of the Midcontinental Rift system. Although these sills have caused local contact metamorphism, the rocks of the Sibley Group are otherwise undeformed and unaltered.

The rocks of the Sibley Group are overlain by the Osler Group composed of lava flows and interbedded red-bed sedimentary rocks. The relation of the Sibley and Osler Groups to the basalt flows and interflow sedimentary rocks of the Keweenaw Peninsula of Michigan

and Mamainse Point in Ontario is not clear. Wanless and Loveridge (1978) reported a whole-rock Rb-Sr isochron date of 1336 ± 33 Ma ($\lambda = 1.42 \times 10^{-11}$ yr^{-1}) and initial ^{87}Sr/^{86}Sr ratio of c.7056 ± 0.0016 (E&A = 0.7080) for specimens of hematitic dolomite of the Sibley Group. The authors interpreted this date to be the age of diagenesis which is only slightly less than the time of deposition in this case. Therefore, the Sibley Group was deposited in a failed branch of the Midcontinental Rift system and predates the main phase of igneous activity represented by the basalt flows and gabbro intrusives in and around Lake Superior.

7.1.5 Duluth Gabbro and Related Intrusives

The Duluth gabbro in Fig. 7.3 is a sill-like body of differentiated mafic rocks that was intruded at the base of the North Shore volcanic rocks along the western shore of Lake Superior in Minnesota (Goldich et al. 1961). The geology of the Duluth gabbro was described by Grout (1918a,b,c) and by Taylor (1964). Several sills of diabase and granophyre within the North Shore volcanic rocks (Endion Sill, Northland Sill, and Lester River Sills) are probably related to the Duluth gabbro (Schwartz and Sandberg 1940). In addition, the Logan Sills near Lakehead City, Ontario (Hanson 1975), and small gabbroic intrusives in Michigan (e.g. Mt. Bohemia intrusive, Chaudhuri and Faure 1968) and Wisconsin (e.g. Potato River Intrusion, Klewin and Berg 1990) are also related to the Duluth gabbro.

The lava flows of the North Shore volcanics as well as the Duluth gabbro and the associated sills strike northeast and dip to the southeast under the western basin of Lake Superior (White 1966). A petrographic study by Ernst (1960) indicated that the granophyre at the top of the Endion Sill either formed by fractional crystallization of basalt magma or is a separate intrusion unrelated to the underlying diabase.

An age determination by the Rb-Sr method based on whole-rock samples of gabbro and granophyre of the Duluth gabbro yielded a date of 1091 ± 14 Ma (recalculated to $\lambda = 1.42 \times 10^{-11}$ yr^{-1}) and an initial ^{87}Sr/^{86}Sr ratio of 0.7052 ± 0.0003 relative to 0.7080 for E&A (Faure et al. 1969). The same authors reported an age of 1070 ± 15 Ma for diabase and granophyre of the Endion Sill (recalculated) and an initial ^{87}Sr/^{86}Sr ratio of 0.7043 ± 0.0006. These results confirmed that the Duluth gabbro and the Endion Sill formed during the episode of igneous activity related to the Midcontinental Rift system from magma that had assimilated varying amounts of crustal rocks (Ripley et al. 1998). Subsequently, Grant and Molling (1981) reported a Rb-Sr isochron date of 1050 ± 105 Ma for troctolite near the base of the Duluth gabbro and an initial ^{87}Sr/^{86}Sr ratio of 0.70500 ± 0.00018 relative to 0.7080 for E&A. The

Mount Bohemia intrusive of Keweenaw County, Michigan is composed of syenodiorite that formed at 1106 ±35 Ma with an initial $^{87}Sr/^{86}Sr$ ratio of 0.7042 (Chaudhuri and Faure 1968).

7.1.6 Coldwell Complex and Related Alkalic Intrusives

The igneous rocks that formed in the Lake Superior area during the episode of magmatic activity at about 1100 Ma also include the Coldwell Complex on the north shore of Lake Superior and several smaller plutons that intruded rocks of the Superior structural province. The Coldwell Complex consists of gabbro, ferroaugite syenite, nepheline syenite, and quartz syenite (Fairbairn et al. 1959). A whole-rock Rb-Sr isochron published by Platt and Mitchell (1982) yielded a date of 1044.5 ±6.2 Ma and an initial $^{87}Sr/^{86}Sr$ ratio of 0.70354 ±0.00016 relative to 0.7080 for E&A ($\lambda = 1.42 \times 10^{-11}$ yr^{-1}) which were subsequently discussed by Blenkinsop and Bell (1983). These results indicate that the Coldwell Complex also formed during the magmatic activity associated with the Midcontinental Rift from alkali-rich magma that originated in the mantle. A second age determination of the Coldwell Complex by Turek et al. (1985) based on U-Pb dating of zircons yielded a date of 1188 ±56 Ma which is older than the Rb-Sr date of Platt and Mitchell (1982). Although this result subsequently provoked a discussion from Thorpe (1986), it did not alter the evidence that the Coldwell Complex is related to the magmatic activity that resulted in the extrusion of a large volume of tholeiite basalt.

The occurrence of alkali-rich plutons of Late Precambrian age in northwestern Ontario was documented by Gittins et al. (1967) who associated them with the Kapuskasing High, which is an area of gravity and magnetic anomalies extending from Lake Superior towards James Bay (Percival and Card 1983). Gittins et al. (1967) used the K-Ar method to demonstrate that the ages of the alkali-rich plutons along the Kapuskasing High cluster between 1000 to 1100 Ma and from 1655 to 1740 Ma. These plutons have attracted attention because they contain carbonatites (Bell 1989). In addition, the Lackner Lake Complex contains magmatic deposits of apatite, magnetite, and pyrochlore ($CaNb_2O_6 \cdot CaTiO_3$). Owen and Faure (1977) reported a Rb-Sr isochron date of 1055 ±7 Ma for this pluton ($\lambda = 1.42 \times 10^{-11}$ yr^{-1}) with an initial $^{87}Sr/^{86}Sr$ ratio of 0.70285 ±0.00011 relative to 0.70800 ± 0.00014 for E&A.

The $^{87}Sr/^{86}Sr$ ratios of carbonatites associated with the alkali-rich plutons north of Lake Superior were measured by Powell (1965a, 1966) and by Bell et al. (1982, 1987). Bell (1989) listed initial $^{87}Sr/^{86}Sr$ ratios for seven plutons with ages between 1000 and 1100 Ma (Table 7.1).

Table 7.1. Initial $^{87}Sr/^{86}Sr$ ratios for seven plutons with ages between 1000 and 1100 Ma (Bell 1989) located north of Lake Superior

Pluton	Age (Ma)	$(^{87}Sr/^{86}Sr)_0$
Big Beaver House	1060	0.70244
Nemegosenda Lake	1015	0.70371
Prairie Lake	1170	0.70260
Firesand River	1060	0.70242
Clay-Howells	1075	0.70372
Schryburt Lake	1140	0.70234
Seabrook Lake	1100	0.70256

These and other alkali-rich plutons and carbonatites have low $^{87}Sr/^{86}Sr$ ratios as pointed out originally by Powell et al. (1962) who reported an average $^{87}Sr/^{86}Sr$ ratio of 0.7025 ±0.0003 for 21 carbonatites relative to 0.7080 for E&A. Bell et al. (1982) concluded that the alkali-rich plutons and associated carbonatites listed above form two groups having average initial $^{87}Sr/^{86}Sr$ ratios of 0.70255 ±0.00016 ($2\bar{\sigma}$, $N = 6$) and 0.70367 ±0.00015 ($2\bar{\sigma}$, $N = 4$) relative to 0.7080 for E&A. The difference in the initial $^{87}Sr/^{86}Sr$ ratios indicates that magmas were either derived from two different source regions in the mantle or that some of the magmas were contaminated by assimilation of crustal rocks. Bell et al. (1982) considered crustal assimilation to be unlikely because the high Sr concentrations of carbonatites and the associated syenites make the isotope composition of Sr in their magmas insensitive to alteration by assimilation of crustal rocks. Therefore, Bell et al. (1982) and Bell and Blenkinsop (1987b) used the initial $^{87}Sr/^{86}Sr$ ratios of carbonatites ranging in age from 110 to 2700 Ma to define a Sr-development line for the source regions of carbonatites in the mantle under the Lake Superior district indicating a Rb/Sr ratio of 0.020 ±0.002 and a present $^{87}Sr/^{86}Sr$ ratio of 0.70342.

7.2 Muskox Intrusion and the Coppermine River Basalt, Northwest Territories, Canada

The Muskox Intrusion described by Irvine and Smith (1967) is a large stratified body composed of ultramafic rocks, gabbro, and granophyre. It is located in the Canadian Arctic about 145 km east of Great Bear Lake and about 80 km south of the settlement of Coppermine on Coronation Gulf shown in Fig. 7.5. The Muskox Intrusion was injected into igneous and sedimentary rocks of Proterozoic age at 1270 ±4 Ma, based on U-Pb dating by LeCheminant and Heaman (1989). It is tabular in plan but funnel-shaped in cross section. The Muskox Intrusion extends in outcrop for about 120 km and probably continues in subsurface for an additional dis-

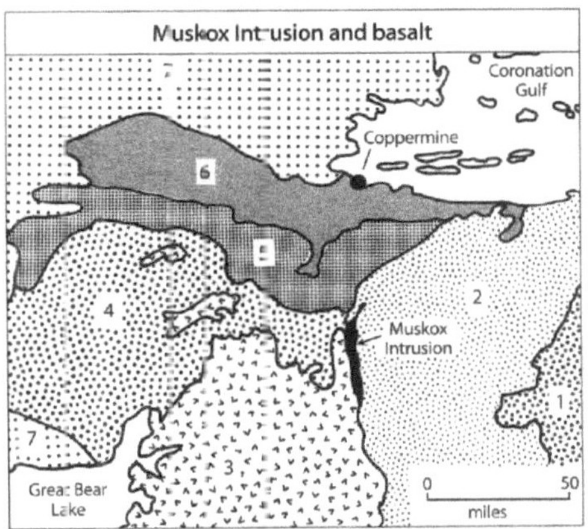

Fig. 7.5. Regional geology of the Coppermine River area, Northwest Territories, Canada. The Coppermine River basalt flows formed by eruption of magma originating from the magma chamber now filled by the Muskox Intrusion. *1.* Archean granitic rocks and volcanics; *2.* Early Proterozoic metasediments; *3.* Proterozoic granitic igneous and metamorphic rocks; *4.* Middle Proterozoic metasediments; *5.* Coppermine basalt and Muskox Intrusion; *6.* Late Proterozoic metasedimentary rocks; *7.* Paleozoic sedimentary rocks (*Source:* adapted from Irvine and Smith 1967)

tance of 120 km as suggested by a positive gravity anomaly. The roughly tabular shape of the body suggests that it occupies a former rift in the Precambrian basement.

The relation of the Coppermine basalt to the Muskox Intrusion has been uncertain because the two units are not in contact and because age determinations have not been definitive. The age of the Coppermine basalt was determined by Wanless and Loveridge (1972) by the whole-rock Rb-Sr method. Since the basalt flows have low Rb/Sr ratios, the authors analyzed samples of flow tops containing amygdules of orthoclase. The slope of the isochron yielded a date of 1257 ±45 Ma ($\lambda = 1.42 \times 10^{-11}$ yr^{-1}) and an initial ^{87}Sr/^{86}Sr ratio of 0.7045 ±0.0011 relative to 0.7080 for E&A. In addition, Baragar (1972) reviewed previously published K-Ar dates of the Coppermine basalt ranging from 740 to 1200 Ma and concluded that the clustering of K-Ar dates around 740 Ma reflects loss of ^{40}Ar caused by thermal effects associated with the intrusion of gabbro sills at 604 to 718 Ma into sedimentary rocks that unconformably overlie the Coppermine basalt. K-Ar dates of the Muskox Intrusion range from 1095 to 1155 Ma, whereas the Mackenzie diabase dikes of the area have yielded dates between 1100 to 1200 Ma. Therefore, the K-Ar dates do not prove that the Muskox Intrusion and the Coppermine basalt belong to the same magmatic system. The problem was ultimately resolved by Kerans (1983), who demonstrated by means of stratigraphic

and structural information that the Coppermine basalt and the Muskox Intrusion are coeval and comagmatic and that both are manifestations of the Mackenzie igneous event that occurred in this area at about 1200 Ma. Therefore, the Muskox Intrusion does appear to occupy the magma chamber from which the Coppermine basalt flows were derived.

7.2.1 Muskox Intrusion

The Muskox Intrusion formed by multiple injections of basalt magma into a chamber at shallow depth below the surface (Irvine 1970b, 1980). Each injection is now represented by a layer of cumulates between 150 and 200 m thick. The basalt magmas contained sufficient heat to melt the country rocks, thus forming a layer of low-density silica-rich magma that floated on the basalt magma and is now represented by a layer of granophyre whose thickness ranges from 0 to 70 m (Irvine and Smith 1967).

Stewart and DePaolo (1992) measured the initial ^{87}Sr/^{86}Sr ratios (at 1257 Ma) of rocks collected at increasing height above the base of Unit 25. The results in Fig. 7.6a demonstrate that the initial ^{87}Sr/^{86}Sr ratios increase up-section from 0.70501 to 0.70968 (relative to 0.71025 for NBS 987) compared to 0.73758 for the granophyre at the top of the magma chamber. The pattern of stratigraphic variation of initial ^{87}Sr/^{86}Sr ratios has been extrapolated in Fig. 7.6a to the base of Unit 25 in order to estimate the value of the initial ^{87}Sr/^{86}Sr ra-

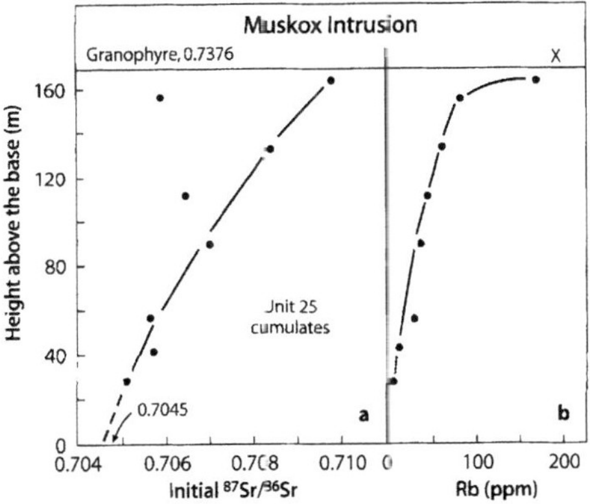

Fig. 7.6. a Variation of the initial ^{87}Sr/^{86}Sr ratios at 1257 Ma of cumulates in Unit 25 of the Muskox Intrusion, Northwest Territories, Canada. This unit is overlain by a xenolith-rich granophyre that formed by melting of country rocks in the roof of the magma chamber; b Stratigraphic variation of the Rb concentrations of the mafic rocks in Unit 25. The increase of the Rb concentration up-section is consistent with fractional crystallization of a layer of mafic magma (*Source:* Stewart and DePaolo 1992)

tio at the time of intrusion of the mafic magma. The value so obtained (0.7045) is identical to the initial $^{87}Sr/^{86}Sr$ ratio obtained by Wanless and Loveridge (1972) for the Coppermine basalt. In contrast to the isotope composition of Sr, the $^{143}Nd/^{144}Nd$ ratios of the cumulate rocks in Unit 25 are virtually constant. The wallrocks (granite, orthogneiss, hornfels) adjacent to the Muskox Intrusion had elevated $^{87}Sr/^{86}Sr$ ratios between 0.70537 and 0.79154 at 1 257 Ma with a mean of 0.7132 weighted in accordance with their Sr concentrations.

The stratigraphic increase of the initial $^{87}Sr/^{86}Sr$ ratios of Unit 25 in Fig. 7.6a cannot be modeled by AFC because assimilation of wallrock by the magma cannot be reconciled with the constant initial $^{143}Nd/^{144}Nd$ ratios. Therefore, Stewart and DePaolo (1992) proposed that radiogenic ^{87}Sr diffused from the granophyre magma into the underlying mafic magma, while the latter was being differentiated chemically by fractional crystallization. The results of numerical modeling indicate that the process of diffusional exchange combined with fractional crystallization can account for the isotope ratios of both Sr and Nd in the rocks of Unit 25. In addition, the results of the modeling indicate that this unit crystallized in about 8 000 years and that the Muskox Intrusion as a whole crystallized in about 50 000 to 100 000 years.

The data of Stewart and DePaolo (1992) in Fig. 7.6b also demonstrate that the Rb concentrations of the rocks in Unit 25 increase up-section from 5.0 ppm at 29 m above the base to 166 ppm at the top. This evidence is consistent with the effects of fractional crystallization of basaltic magma in this layer. The Sr concentrations of the rocks likewise rise up-section from 111 ppm at 29 m to 333 ppm at 111 m but then decline to 155 ppm at the top of the section. The Sr concentrations of the rocks depend primarily on the relative proportions of orthopyroxene, Sr = 25.8 ppm; clinopyroxene, Sr = 2.4 ppm; and plagioclase, Sr = 567 ppm.

The available evidence indicates that the Muskox Intrusion represents the magma chamber from which the Coppermine River basalt flows originated at about 1 270 Ma. The magmas formed by partial melting in the mantle and rose toward the surface in a rift within the continental crust. After the magmas had entered the magma chamber, they deposited layers of cumulates before being expelled to the surface thus building up a basalt plateau (Irvine 1977).

7.2.2 Coppermine River Basalt

The Coppermine basalt flows described by Baragar (1969) crop out only a few kilometers north of the exposed part of the Muskox Intrusion, but are not in contact with it. The flows have a total thickness of about 3 000 m, dip north at about 3 to 10°, and are uncon-

formably overlain by red sandstones and interbedded basalt flows of late Precambrian age. The chemical compositions of the Coppermine basalt flows were determined by Baragar (1969) and by Dostal et al. (1983b). The results indicate that the basalts are typical continental tholeiites containing evidence of extensive differentiation by fractional crystallization and interaction with rocks from the continental crust. Dostal et al. (1983b) attributed wide variations in the concentrations of K, Rb, and other mobile elements to selective diffusion rather than to bulk assimilation of crustal rocks by the basalt magmas.

The Coppermine River basalt include about 150 flows ranging in thickness from 4 to about 100 m, which were erupted in a time interval of less than 5 million years. These flows have been subdivided into two formations (Griselin et al. 1997):

- Husky Creek (1 200 m)
  - basalt flows and interbedded
  - red fluvial sandstones
- Copper Creek (200–3 500 m)
  - basalt flows only

The initial isotope ratios of Sr and Nd at 1 270 Ma in Fig. 7.7 provide evidence that the magmas of the Coppermine River basalt originated from sources in the mantle and assimilated crustal rocks before they were erupted. Dupuy et al. (1992) concluded that the magmas formed by partial melting in the head of an asthenospheric plume and in an adjacent lithospheric mantle. The crustal rocks that were assimilated by these

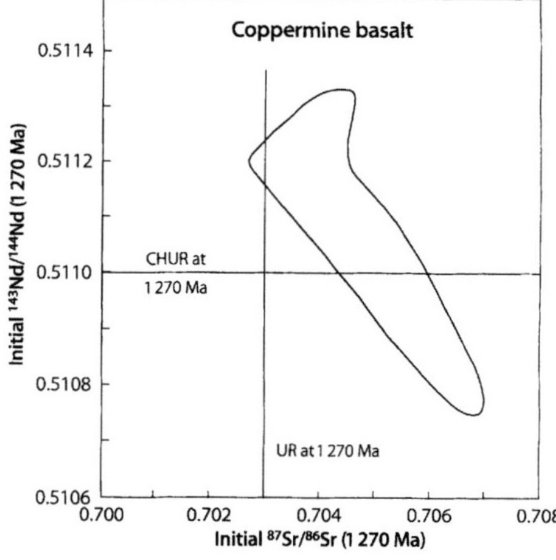

Fig. 7.7. Initial isotope ratios of Sr and Nd of the Coppermine River basalt in the Northwest Territories of Canada at 1 270 Ma (*Source:* Dupuy et al. 1992)

magmas originated from the Precambrian basement which, in this area, consists of the Wopmay Orogen (2.2 to 1.8 Ga; Bowring and Podosek 1989) overlain by the Hornby Bay Group of quartz-rich metasedimentary rocks (1 663 ±8 Ma; Bowring and Ross 1985). Griselin et al. (1997) subsequently reported that the initial $^{143}Nd/^{144}Nd$ ratios of the Coppermine River basalt increase up-section, indicating that the extent of crustal contamination of the mantle-derived magmas decreased with time during the eruption.

7.3 Diabase Dikes and Sills of the Canadian Precambrian Shield

The Precambrian shield of Canada is cut by swarms of diabase dikes and sills, some of which can be traced for hundreds of miles (Halls and Fahrig 1987). The dikes were subdivided into thirteen sets based on their strike and K-Ar-dates ranging from 2 485 to 640 Ma (Fahrig and Wanless 1963). In addition, a large diabase sill known as the Nipissing Diabase intrudes the Huronian sedimentary rocks north of Lake Huron in Ontario (Van Schmus et al. 1963; Fairbairn et al. 1960a,b, 1969). The dikes formed at different times as a result of extension of the continental crust of North America and may have been associated with complementary basalt plateaus at the surface that have since been eroded.

7.3.1 Mackenzie Dikes

The Mackenzie Dikes in Fig. 7.8 converge on the location of the Muskox Intrusion and the Coppermine River basalt. Gates and Hurley (1973) dated the diabase dikes on the Canadian Shield by the Rb-Sr method using pyroxene and plagioclase separated from the diabase. The results ranged widely from about 2.7 to 1.3 Ga ($\lambda = 1.42 \times 10^{-11}$ yr^{-1}) and yielded initial $^{87}Sr/^{86}Sr$ ratios between 0.700 and 0.704 (relative to 0.7080 for E&A). The dates have large uncertainties (up to ±9.1%) not only because of the low Rb/Sr ratios of the minerals used for dating, but also because of alteration of the magmas at the time of intrusion. Gates and Hurley (1973) reported that the Rb and Sr concentrations of the dikes vary depending on the composition of the adjacent country rocks (Table 7.2).

The apparent addition of Rb and Sr from the wall-rock to the diabase magma at the time of intrusion changed both the Rb/Sr and the $^{87}Sr/^{86}Sr$ ratios of the magmas. Changes of the Rb/Sr ratio of the magma at the time of intrusion do not affect the dates derivable from these rocks. However, changes of the $^{87}Sr/^{86}Sr$ ratios cause the data points to scatter above and below the Rb-Sr isochron and therefore do affect the dates derivable from the rocks by this method. Alteration of

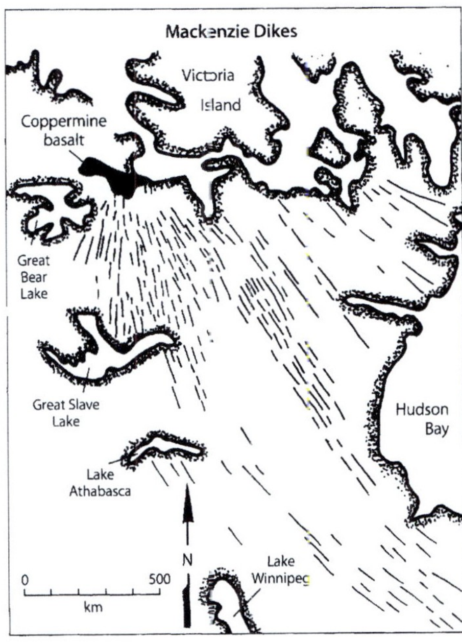

Fig. 7.8. Map of the Mackenzie dike swarm in the Northwest Territories of Canada in relation to the Coppermine River basalt and the related Muskox Intrusion (*Source:* adapted from Griselin et al. 1997)

Table 7.2. Rb and Sr concentrations of the diabase dikes in relation to composition of the adjacent country rocks (Gates and Hurley 1973)

Country rock	Diabase dike Rb (ppm)	Sr (ppm)
Basalt (greenstone)	26	144
Metasediment	59	277
Syenite	64	314

the isotope composition of Sr at the time of intrusion of magma can be recognized on a Sr-isotope mixing diagram, provided the age of the rocks is known. Since that information was not available to them, Gates and Hurley (1973) assumed that the scatter of data points was random rather than systematic and therefore did not affect the calculated dates appreciably.

The interaction of the magma of Mackenzie Dikes with the country rock was later confirmed by Fratta and Shaw (1974) and by Dostal and Fratta (1977). The former sampled one of the Mackenzie Dikes in the northern Lake of the Woods area of northwestern Ontario where it is in contact with granite at one site and mafic volcanic rocks at another. Their results demonstrated that the diabase was enriched in K, Li, Rb, and Tl in contact with granite compared to where it is adjacent to mafic volcanics. Fratta and Shaw (1974) concluded that these elements had diffused from the wallrock into the diabase magma before it crystallized. However, Sr was not detectably mobilized in this case.

The alteration of the diabase dikes on the Canadian Precambrian shield makes them unsuitable for dating by the K-Ar and Rb-Sr methods. Therefore these rocks were subsequently reliably dated by the U-Pb method based on analyses of zircon and baddeleyite (Krogh et al. 1987).

7.3.2 Nipissing Diabase Sill

The Nipissing diabase sill intrudes the Huronian sedimentary rocks exposed along the north shore of Lake Huron and extends north to the former mining town of Cobalt (Hriskevich 1968). It consists of a complex of sills and dikes composed of tholeiite diabase or gabbro and includes granophyre near the top of the principal sills. The sills are generally concordant with the bedding of the Huronian sedimentary rocks and are up to 600 m thick, but only erosional remnants remain of the original volume of this intrusive.

Following the early pioneering study of Fairbairn et al. (1960b) on the geochronology of the Sudbury-Blind River area, Van Schmus et al. (1963) and Van Schmus (1965) dated the Nipissing diabase and other rocks from the north shore of Lake Huron by the Rb-Sr method. The latter refined the initial estimate of the age of the Nipissing sills to 2 109 ±80 Ma ($\lambda = 1.42 \times 10^{-11}$ yr^{-1}) from the slope of a whole-rock Rb-Sr isochron with an initial ^{87}Sr/^{86}Sr ratio of 0.7067 relative to 0.7080 for E&A.

The age of the Nipissing diabase is important because it places a lower limit on the age of deposition of the Huronian rocks made famous by the studies of Collins (1925) and of Young and Church (1966). The age of the Huronian rocks was determined by Fairbairn et al. (1969) who dated a Nipissing sill at Gowganda, Ontario, as well as the rocks of the Gowganda Formation which were intruded by the sill. Whole-rock samples from a drill core through the sill yielded a date of 2 116 ±27 Ma ($\lambda = 1.42 \times 10^{-11}$ yr^{-1}) in good agreement with the date obtained by Van Schmus (1965) of 2 109 ±80 Ma for whole-rock samples of granophyre from different localities. These results constrained the age of the Huronian System to less than 2 500 Ma (the age of the granitic basement rocks) and greater than 2 116 ±27 Ma. Fairbairn et al. (1969) refined this estimate by dating whole-rock samples of the Gowganda Formation (quartzose greywacke and argillite) which yielded

a satisfactory isochron indicating a date of 2 240 ±87 Ma and an initial ^{87}Sr/^{86}Sr ratio of 0.7063 ±0.0019. Since the rocks of the Gowganda Formation in the Gowganda-Cobalt area were not structurally deformed or significantly metamorphosed, Fairbairn et al. (1969) concluded that the age determinations of the Nipissing diabase and the Gowganda Formation effectively constrain the age of the Huronian System.

7.4 Stillwater Complex, Montana

The Stillwater igneous complex of southern Montana in Fig. 7.9 is widely known to petrologists as an important example of a stratified gabbroic intrusion. (Turner and Verhoogen 1960; Hess 1960; Wyllie 1967). It intrudes Precambrian schists and gneisses of the Beartooth Plateau, dips steeply to the north, has an outcrop length of nearly 50 km, and an exposed thickness of about 4 900 m. The upper section of the intrusion, amounting to about 40% of the total thickness, is not exposed and may have been eroded before deposition of sedimentary rocks of Paleozoic and Mesozoic ages. The magma apparently cooled slowly, allowing early-formed crystals to be deposited in layers on the floor of the magma chamber. The base of the complex consists of chilled diabasic norite (90 m) followed up-section by banded ultramafic rocks (760 m) composed of pyroxenite, harzburgite, and some dunite. The ultramafic rocks are overlain by layers of norite, gabbro, and anorthosite. No basalt flows derived from the Stillwater magma chamber have been identified either because of the absence of volcanic activity or because they were eroded. In addition, no granophyres are exposed near the top of the Stillwater Complex either because they did not form, or because they were eroded, or because

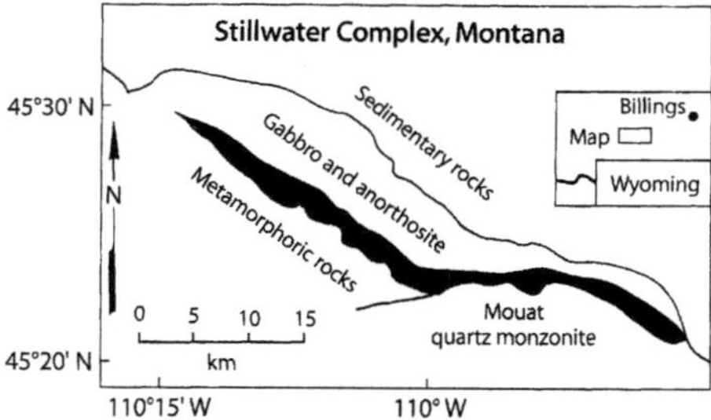

Fig. 7.9. Stillwater Complex (2.7 Ga) southwest of Billings in southern Montana, USA. The complex consists of layers of ultramafic cumulates (*black*) overlain by stratified gabbro and anorthosites. It is underlain by metamorphic rocks and by the Mouat Quartz Monzonite. The sedimentary rocks north of the complex are of Paleozoic and Mesozoic age (*Source:* Jackson 1961)

they are covered by the sedimentary rocks that unconformably overlie the Complex.

The age of the Stillwater Complex has been difficult to determine by the Rb-Sr method because the ultramafic rocks have low concentrations of Rb and Sr making these rocks sensitive to alteration by groundwater or regional metamorphism. In addition, the ultramafic rocks and gabbros have low Rb/Sr ratios with a limited range of variation, causing large uncertainties in the slopes of isochrons. Attempts to date the Stillwater Complex by the K-Ar method (Kistler et al. 1969) were not successful because the pyroxenes contain excess ^{40}Ar (Schwartzman and Giletti 1968). Kistler et al. (1969) reported Rb-Sr dates on phlogopite and biotite from the Stillwater Complex ranging from 2 210 to 2 530 Ma ($\lambda = 1.42 \times 10^{-11}$ yr^{-1}). Powell et al. (1969) analyzed samples of metasedimentary rocks collected between 0.64 and 9.6 km from the base of the intrusion. The samples (hornfelses and schists) formed a linear array on the Rb-Sr isochron diagram that yielded a date of 2 670 ±150 Ma ($\lambda = 1.42 \times 10^{-11}$ yr^{-1}) and an initial ^{87}Sr/^{86}Sr ratio of 0.705 ±0.012 (not shown). Powell et al. (1969) concluded that this date represents either the age of sedimentation, the age of regional metamorphism, or the time of intrusion of the Stillwater Complex. Several years later, Mueller and Wooden (1976) repeated the work of Powell et al. (1969) by dating hornfels samples collected less than one kilometer from the lower contact of the Stillwater Complex. A suite of twelve specimens yielded a whole-rock Rb-Sr date of 2 690 ±45 Ma ($\lambda = 1.42 \times 10^{-11}$ yr^{-1}) and an initial ^{87}Sr/^{86}Sr ratio of 0.705 ±0.003 relative to 0.7080 for E&A. The satisfactory fit of data points to the isochron is demonstrated in Fig. 7.10. Mueller and Wooden (1976) interpreted the Rb-Sr date of the hornfels to be an underestimate of the age of the Stillwater Complex.

The age determinations of the Stillwater Complex by Powell et al. (1969) and Mueller and Wooden (1976) were confirmed by DePaolo and Wasserburg (1979) who reported a Sm-Nd isochron date of 2 701 ±8 Ma for the constituent minerals (orthopyroxene, clinopyroxene, and plagioclase) and the associated whole-rock sample of gabbro. Additional whole-rock samples also fit the internal mineral isochron. However, the same minerals yielded a date of only 2.2 Ga by the Rb-Sr method. Therefore, DePaolo and Wasserburg (1979) concluded that the Rb-Sr systematics of the Stillwater Complex had been disturbed after its crystallization at 2 701 ±8 Ma. The age of the Stillwater Complex was redetermined by Lambert et al. (1989) based in part on the Re-Os method.

The importance of the Stillwater Complex lies in the fact that it is the product of crystallization of a very large volume of basaltic magma that originated from sources in the mantle at an early time in the history of

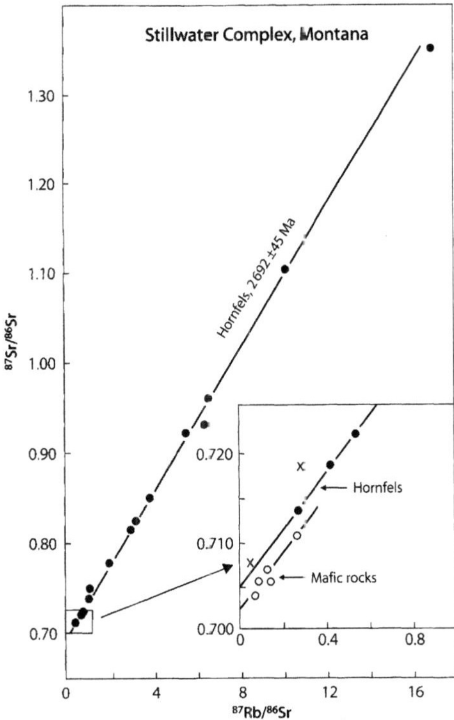

Fig. 7.10. Whole-rock Rb-Sr isochron of hornfels (*solid circles*) underlying the Stillwater Complex, Montana. The initial ^{87}Sr/^{86}Sr ratio at 2 692 ±45 Ma ($\lambda = 1.42 \times 10^{-11}$ yr^{-1}) is 0.705 ±0.003 relative to 0.7080 for E&A. The Rb-Sr systematics of serpentinized harzburgites (*crosses*) deviate significantly from those of apparently unaltered igneous rocks of the complex (*open circles*) (*Sources:* Mueller and Wooden 1976; Fenton and Faure 1969)

the Earth. Therefore, the initial ^{87}Sr/^{86}Sr ratio of these rocks is a milestone in the isotopic evolution of Sr in magma sources of the mantle at 2.7 Ga. The value of this ratio, based on the data of Fenton and Faure (1969) and recalculated to an age of 2.7 Ga, is 0.7021 ±0.0005 ($2\bar{\sigma}$, $N = 5$) relative to 0.7080 for E&A. In addition, the Stillwater Complex is an important host of significant accumulations of ore minerals including chromite, sulfides of Ni and Cu, and platinum group elements (PGE). These minerals are concentrated primarily in the ultramafic zone (layers of chromitite and platinum group elements) and in the basal zone and underlying hornfels (sulfides). (Jackson 1967, 1968; Page 1971; Page et al. 1976; Todd et al. 1982; Irvine et al. 1983). The origin of the anorthosite layers of the Stillwater Complex was discussed by Czamanske and Zientek (1985), Czamanske and Bohlen (1990), and Haskin and Salpas (1992).

7.5   Kiglapait Intrusion, Labrador

The Kiglapait Intrusion (57°00' N, 61°30' E) on the coast of Labrador in Fig. 7.11 is a large differentiated body of mafic igneous rocks (Morse 1969) mentioned briefly in Sect. 1.10.2. It is circular in plan view with a diameter of about 30 km and bowl-shaped in cross-section with a thickness of about 8.6 km. The geochemistry of Rb and Sr in the rocks of this intrusion was discussed by Morse (1981, 1982), respectively. The Kiglapait magmas were intruded into Archean gneisses and into the Nain anorthosite at 1 305 ±22 Ma based on a Sm-Nd mineral isochron reported by DePaolo (1985). Subsequently, the Kiglapait Complex was intruded by the Manvers granite at 1 250±26 Ma (Barton 1977).

Morse (1969) demonstrated that the Kiglapait Intrusion formed by extreme fractional crystallization of a troctolitic magma resulting in the development of an iron-rich syenite liquid that contained 15 to 22% $FeO + Fe_2O_3$, 6.5 to 8.0% alkalies, but only 0.1% MgO. The stratigraphy of the Kiglapait Intrusion includes a Lower Zone (troctolite, 5.0 km, 0 to 84% solidified), an Upper Zone (olivine gabbro, 3.2 km, 84 to ~100% solidified), and an Upper Border Zone (ferrosyenite, ~0.6 km, ~100% solidified).

Isotopic compositions of Sr of rocks from the Kiglapait Intrusion were reported by Simmons and Lambert (1981), Morse (1983), and DePaolo (1985). Three whole-rock samples analyzed by Morse (1983) were reanalyzed by DePaolo (1985). Therefore, in this presentation the initial $^{87}Sr/^{86}Sr$ ratios reported by Morse

(1983) were first recalculated for decay of ^{87}Rb to an age of 1 305 Ma and then increased by +0.00022 in order to make them consistent with those of DePaolo (1985). Subsequently, the initial $^{87}Sr/^{86}Sr$ ratios of both data sets were reduced by 0.00005 to make them compatible with 0.71025 for NBS 987 for which DePaolo reported a value of 0.71030. The data in Fig. 7.12 reveal that the initial $^{87}Sr/^{86}Sr$ ratios of the rocks in the Lower Zone comprising 84% of the intrusion vary only from 0.70403 to 0.70428 relative to 0.71025 for NBS 987 (DePaolo 1985). In the Upper Zone (16% of the intrusion) the initial $^{87}Sr/^{86}Sr$ ratios rise steeply from 0.70411 (89.4% solidified), to 0.70726 (99.986% solidified, Morse 1983). The data suggest a decrease in the initial $^{87}Sr/^{86}Sr$ ratio from 0.70429 at the top of the Lower Zone to 0.70411 near the base of the Upper Zone. A second step-change occurs between 95.5 and 98.6% consolidated where the ratio increases from 0.70514 to 0.70576.

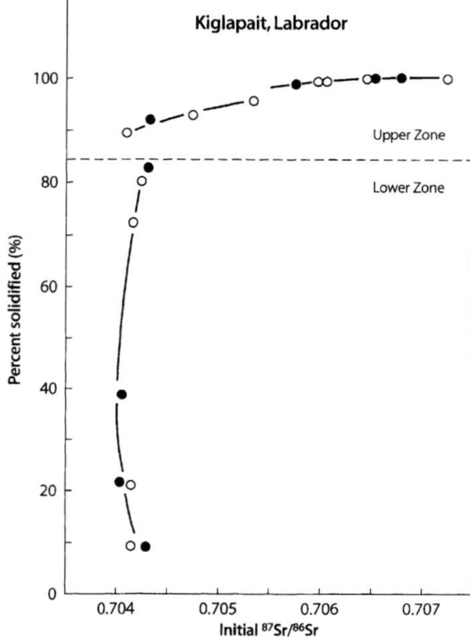

Fig. 7.12. Variation of the initial $^{87}Sr/^{86}Sr$ ratios (at 1 305 ±22 Ma) of the Kiglapait Intrusion (57°00' N, 61°30' E) of Labrador with increasing extent of solidification. The initial $^{87}Sr/^{86}Sr$ ratios of the Lower Zone vary only slightly from 0.70403 to 0.70428 relative to 0.71025 for NBS 987. However, the initial $^{87}Sr/^{86}Sr$ ratios in the Upper Zone increased rapidly during progressive crystallization from 0.70411 or less to 0.70726 or higher. The increase of the initial $^{87}Sr/^{86}Sr$ ratios of the magma was caused by assimilation of Archean gneisses which formed the wallrock along the northern side of the magma chamber (*Sources:* Morse 1983; DePaolo 1985)

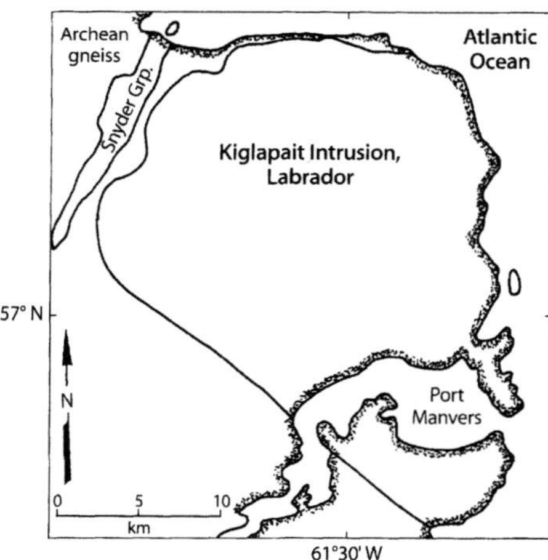

Fig. 7.11. Kiglapait Intrusion on the coast of Labrador in the Province of Newfoundland, Canada (*Source:* adapted from Kalamarides 1984)

The variation of initial ^{87}Sr/^{86}Sr ratios in the rocks of the Kiglapait Intrusion can be understood in terms of magma evolution by fractional crystallization, assimilation, replenishment, and venting (O'Hara 1977). During the crystallization of the rocks of the Lower Zone, assimilation played a minor role compared to crystallization, replenishment, and venting of the magma chamber (DePaolo 1985). Replenishment probably stopped during the crystallization of the Upper Zone making the remaining liquid increasingly sensitive to contamination by the Archean gneisses that formed part of the roof of the magma chamber. The apparent discontinuity of initial ^{87}Sr/^{86}Sr ratios at the base of the Upper Zone indicates that the chamber was replenished at that time with magma having an ^{87}Sr/^{86}Sr ratio of 0.7041 or less. Subsequently, the magma evolved by assimilation and fractional crystallization (AFC) without additional replenishment. The small step-like increase of the ^{87}Sr/^{86}Sr ratio when 95.5 to 98.6% of the intrusion had solidified could have been caused by an increase in the ^{87}Sr/^{86}Sr ratio of the contaminant or by increased efficiency of assimilation. The isotope compositions of Sr and Nd reported by Morse (1983) and DePaolo (1985) reveal that wallrock assimilation, fractional crystallization, and replenishment of magma played an important part in the formation of this intrusion, especially during the final stages of crystallization of the Upper Zone.

7.6 Caledonian Gabbros of Northeast Scotland

The so-called "Newer gabbros" of northeast Scotland exemplify the process of magma contamination by crustal rocks to such an extent that the ^{87}Sr/^{86}Sr ratios of some magmas became equal to that of the clay-rich sedimentary rocks (Dalradian metasediments) into which the basalt magmas were intruded (Pankhurst 1969). The gabbro intrusions include peridotites, troctolites, gabbros, and norites. There are eight separate bodies (including Insch Belhelvie, Haddo House, and Arnage), all of which may have formed from the same magma. The Insch Intrusion formed by extreme fractional crystallization of a single batch of tholeiite basalt magma producing basal dunites and peridotites that are overlain by gabbro. The sequence ends with ferrogabbro, syenogabbro, and syenite. The final differentiates, representing less than 0.15% of the total volume, are strongly enriched in incompatible elements (K = 6.2%, Ba = 18 000 ppm, Zr = 3 000 ppm, Rb = 100 ppm)

A Rb-Sr whole-rock isochron reported by Pankhurst (1969) based on the most highly differentiated rocks of the Insch Intrusion (ferrogabbro to syenite) yielded a date of 498 ±13 Ma ($\lambda = 1.42 \times 10^{-11}$ yr^{-1}) and an initial ^{87}Sr/^{86}Sr ratio of 0.7119 ±0.0001 (0.7080 for E&A). The

hornfels from the Dalradian country rock surrounding the Haddo House Intrusion yielded a similar Rb-Sr whole-rock isochron date of 488 ±13 Ma with an initial ^{87}Sr/^{86}Sr ratio of 0.7185 ±0.0002. This date is indistinguishable from the age of the gabbro intrusions and indicates that the Sr in the hornfelsed country rock around the Haddo House Intrusion was isotopically homogenized by thermal metamorphism caused by the intrusion of the gabbro.

The initial ^{87}Sr/^{86}Sr ratios of the gabbros in the Insch Intrusion in Fig. 7.13 range widely from 0.7032 to 0.7114 relative to 0.7080 for E&A but are not well correlated with the Sr concentrations of these rocks. Therefore, Pankhurst (1969) concluded that the ^{87}Sr/^{86}Sr ratios of the gabbroic magma had increased by selective addition of ^{87}Sr from the country rock during crystallization. The peridotites also have a range of initial ^{87}Sr/^{86}Sr ratios (0.7045 to 0.7082) and are cumulates that formed by fractional crystallization of the gabbro

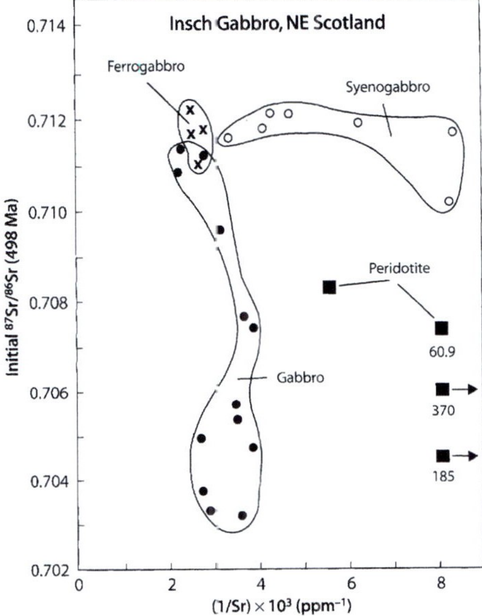

Fig. 7.13. Strontium-isotope mixing diagram for mafic rocks of the Insch Intrusion, northeast Scotland. The initial ^{87}Sr/^{86}Sr ratios of the gabbros (*solid circles*) range from 0.7032 to 0.7113, but are not well correlated with their reciprocal Sr concentrations. The ferrogabbros (*crosses*) are closely related to the gabbros but have an elevated average initial ^{87}Sr/^{86}Sr ratio of 0.7117 ±0.0005 ($2\bar{\sigma}$, N = 4, E&A = 0.7080). The distribution of data points indicates that the syenogabbros (*open circles*) and peridotites (solid squares) are magmatic differentiates of the ferrogabbro and gabbro magmas, respectively. The lack of correlation of ^{87}Sr/^{86}Sr ratios and Sr concentrations of the gabbros was caused by selective assimilation of ^{87}Sr by the magma from the country rock consisting of Dalradian metasediments (*Source:* Pankhurst 1969)

magma while its $^{87}Sr/^{86}Sr$ ratio was increasing. The syenogabbros have elevated initial $^{87}Sr/^{86}Sr$ ratios similar to those of the ferrogabbro and some of the gabbros. Therefore, the syenogabbros are products of fractional crystallization of late-stage gabbroic and ferrogabbroic magmas.

The evidence for selective migration of ^{87}Sr from the country rocks into the mafic magma at Insch is supported by the apparent lowering of the $^{87}Sr/^{86}Sr$ ratios in the contact metamorphic aureole surrounding the Haddo House Intrusion. According to Pankhurst (1969), the $^{87}Sr/^{86}Sr$ ratios in these rocks at the time of intrusion of the gabbros at about 500 Ma (0.7179 to 0.7187) are lower than those of Dalradian rocks elsewhere in Scotland because of isotope equilibration with Sr in the gabbroic magma over distances of 2 to 3 km. The consequences of the equilibration of isotope compositions of Sr between mafic magma and country rock without large-scale assimilation of major elements are (Pankhurst 1969):

1. Mafic magmas with elevated initial $^{87}Sr/^{86}Sr$ ratios may have originated from sources in the mantle having low $^{87}Sr/^{86}Sr$ ratios;
2. Igneous rocks having different $^{87}Sr/^{86}Sr$ ratios may, nevertheless, be products of the same magma.

Evidence for increasing $^{87}Sr/^{86}Sr$ ratios during the crystallization of mafic magmas has been found in other intrusions (e.g. Kiglapait, Labrador; Bushveld Complex, South Africa). In these and other cases, such variations of $^{87}Sr/^{86}Sr$ ratios have been explained by mixing of mafic magmas having different $^{87}Sr/^{86}Sr$ ratios (Bushveld Complex) or with partial melts derived from the country rocks along the roof of the magma chamber (Kiglapait).

The initial $^{87}Sr/^{86}Sr$ ratios of the gabbros and ultramafic rocks at Belhelvie in Scotland range only from 0.7041 to 0.7084 and are similar to those of the gabbros in the Insch Intrusion. The extreme differentiates (syenogabbro and syenite) that occur at Insch are absent at this location. The quartz norites at Haddo House and Arnage have a wide range of elevated initial $^{87}Sr/^{86}Sr$ ratios: 0.7093 to 0.7152 at Haddo House and 0.7087 to 0.7144 at Arnage (relative to 0.7080 for E&A, $\lambda = 1.42 \times 10^{-11}$ yr^{-1}, and 498 ±13 Ma). The reciprocal concentrations of Sr of these intrusives are not well correlated with the isotope ratios, implying selective additions of ^{87}Sr from the country rock to the gabbroic magmas during crystallization (not shown). A quartz-cordierite norite at Haddo House has extremely high initial $^{87}Sr/^{86}Sr$ ratios between 0.7202 and 0.7224 exceeded only by quartz-cordierite-garnet wall rocks with $^{87}Sr/^{86}Sr$ ratios of 0.7215 to 0.7261.

The initial $^{87}Sr/^{86}Sr$ ratios of the cordierite-bearing igneous rocks at Haddo House overlap those of the associated schists and gneisses of the Dalradian country rocks (0.716 to 0.730) at about 500 Ma. Pankhurst (1969) therefore concluded that the cordierite-bearing "igneous" rocks at Haddo House formed by partial melting of the local country rock. The gneisses at Arnage (initial $^{87}Sr/^{86}Sr$ ratio: 0.7135 to 0.7334) are interpreted to be Caledonian granites derived from deep crustal sources.

7.7 Rhum Intrusion, Inner Hebrides, Scotland

The North Atlantic igneous province of Cenozoic age (Chap. 5) includes the island of Rhum (Fig. 5.10) which contains a differentiated igneous complex consisting of alternating layers of peridotite and gabbro composed of olivine and anorthite (allivalite), as well as of olivine and labradorite/bytownite (troctolite) (Mussett 1984). These layers formed by a complex process of fractional crystallization of mafic magmas and intercumulus liquids in a chamber that was vented to the surface and periodically replenished (Palacz 1984). However, the basaltic lava flows that originated from this chamber were eroded prior to uplift and exposure of the ultramafic rocks. The geology of the Rhum Complex was discussed in papers published in the "Geological Magazine" (vol. 122(5), 1985).

Measurements by Palacz and Tait (1985) indicate that the $^{87}Sr/^{86}Sr$ ratios in stratigraphic Unit 10 in Fig. 7.14 increase up-section from 0.70381 (basal peridotite) to 0.70649 near the center of the overlying allivalite layer relative to 0.71025 for NBS 987. The Sr concentrations of these rocks also vary stratigraphically and depend on the abundance of plagioclase which contains about 500 ppm Sr on average, whereas olivine has less than 1.0 ppm, and clinopyroxene contains about 20 ppm (Palacz and Tait 1985). The intercumulus plagioclase and clinopyroxene in the peridotites crystallized from a residual liquid derived from contaminated basaltic magma ($^{87}Sr/^{86}Sr \approx 0.7065$) that is now represented by the allivalite. Therefore, the $^{87}Sr/^{86}Sr$ ratios and Sr concentrations of the peridotites are controlled primarily by the abundance of intercumulus plagioclase and clinopyroxene.

The difference in the $^{87}Sr/^{86}Sr$ ratios of the peridotite and the allivalites cannot be explained by fractional crystallization in a closed magma chamber and requires the presence of a contaminant (high $^{87}Sr/^{86}Sr$ ratio) in the basaltic magma that produced the allivalite. Palacz and Tait (1985) postulated that a dense, uncontaminated picritic magma was injected beneath a less dense, contaminated, and evolved basaltic magma modeled previously by Huppert and Sparks (1980). The

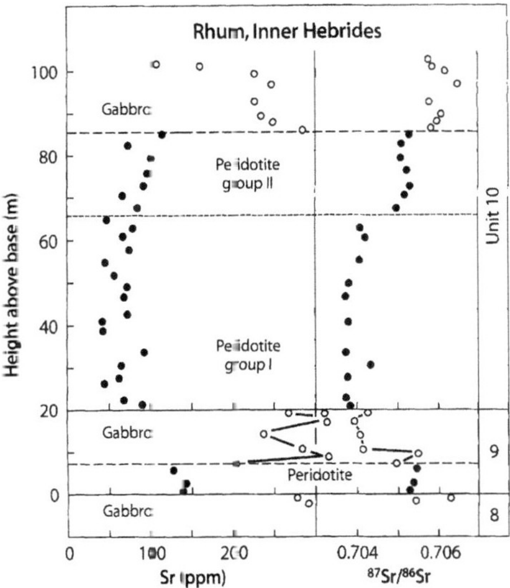

Fig. 7.14. Stratigraphic profiles of Sr concentrations and $^{87}Sr/^{86}Sr$ ratios in Units 9 and 10 of the Rhum Intrusion, Inner Hebrides of Scotland. The peridotite layers (*solid circles*) formed from uncontaminated, dense picrite magma injected along the bottom of a magma chamber filled with a contaminated, evolved, basaltic magma which formed the gabbro at the top of this layer (*open circle*). The picritic magma in Unit 10 displaced the contaminated magma without mixing with the resident magma (*Sources:* Palacz 1984; Palacz and Tait 1985)

picrite magma crystallized olivine and the basalt magma formed plagioclase, olivine, and pyroxene. The $^{87}Sr/^{86}Sr$ ratio of the peridotite depends on the amount of intercumulus plagioclase and pyroxene (high $^{87}Sr/^{86}Sr$ ratio) and hence on the porosity of the mush of cumulus crystals. The increase of the $^{87}Sr/^{86}Sr$ ratio of the peridotites in Group II in Fig. 7.14 suggests that the upper part of the cumulate pile had a higher porosity than the basal section represented by Group I. Palacz and Tait (1985) determined that the $^{87}Sr/^{86}Sr$ ratios of the peridotites in Group I require the addition of only 20% of an intercumulus liquid ($^{87}Sr/^{86}Sr \approx 0.706$; $Sr \approx 250$ ppm).

The variation of $^{87}Sr/^{86}Sr$ ratios and Sr concentrations of the allivalite layer in Fig. 7.14 indicates either that the magma was not isotopically homogeneous or that the mush of cumulus crystals of the allivalite was invaded by an intercumulus liquid that had a lower $^{87}Sr/^{86}Sr$ ratio than the allivalite cumulate. This liquid had a high Mg concentration as indicated by the observation of Palacz and Tait (1985) that the forsterite content of olivine near the top of the allivalite increases up-section from 77 to 84% whereas the Sr concentrations and $^{87}Sr/^{86}Sr$ ratios decrease from 244 to 109 ppm

and from 0.70649 to 0.70581, respectively (relative to 0.71025 for NBS 987).

The stratigraphic variation of initial $^{87}Sr/^{86}Sr$ ratios in the underlying Unit 9 differs from that shown in Fig. 7.14 for Unit 10. Unit 9 is 21 m thick and consists of a basal peridotite layer (6 m) grading upward into troctolite (4 m) and then into gabbro (11 m). The $^{87}Sr/^{86}Sr$ ratios reported by Palacz (1984) decrease up-section from the basal peridotite (0.70525 to 0.70633) to the troctolite (0.70496 to 0.70550), and to the gabbro layer (0.70404 to 0.70425) with an anomalously high value of 0.70537 for gabbro at the top of the section. The basal peridotite of Unit 10 directly above this gabbro has a low $^{87}Sr/^{86}Sr$ ratio of 0.70381. Palacz (1984) also reported that clinopyroxene and whole-rock samples have the same $^{87}Sr/^{86}Sr$ ratios, indicating that migrating intercumulus liquids were not a factor in the formation of Unit 9.

The gabbro in Unit 8 directly below the basal peridotite of Unit 9 also has a high $^{87}Sr/^{86}Sr$ ratio (0.70543 to 0.70633). Therefore, the basal peridotite of Unit 9 may have formed from a mafic magma that was injected along the bottom of the magma chamber and mixed with contaminated magma that had formed the gabbro of Unit 8. The sharp decrease of $^{87}Sr/^{86}Sr$ ratios from the gabbro of Unit 9 to the peridotite of Unit 10 indicates that the contaminated magma that formed the gabbro layer of Unit 9 was displaced by the injection of the mafic magma (low $^{87}Sr/^{86}Sr$) from which the peridotite of Unit 10 was subsequently deposited.

The Rhum Intrusion therefore exemplifies the results of several processes that occurred in the magma chamber:

1. Fractional crystallization of magma contaminated with Sr derived from crustal rocks (gabbro, units 9 and 10);
2. Mixing of dense picritic magma and of evolved and contaminated basalt magma (basal peridotite, Unit 9);
3. Displacement of the resident magma by a dense picritic magma injected along the bottom of the chamber without mixing (basal peridotite, Unit 10); and
4. Crystallization of intercumulus minerals by a migrating intercumulus liquid, causing alteration of $^{87}Sr/^{86}Sr$ ratios and olivine compositions (upper peridotite, Unit 10).

In addition, Greenwood et al. (1992) presented evidence of fluid flow along the contact of the Rhum Intrusion. The complex magma-chamber processes recorded in the Rhum Intrusion are probably typical of the petrogenesis of other layered mafic complexes, including the Bushveld Complex of South Africa.

7.8 Bushveld Complex, South Africa

Southern Africa includes at least six layered gabbroic intrusives of Precambrian age, four of which are aligned along a north-south lineament extending about 1700 km in Fig. 7.15. These intrusives, starting at the northern end of the lineament, are: Great Dyke (Zimbabwe), Bushveld, Losberg, and Trompsburg (South Africa). The Usushwana Complex is located east of the others along the border between Transvaal and Swaziland, and the Modipe gabbro straddles the border between South Africa and Botswana, (McElhinny 1966). These igneous intrusives consist of basal ultramafic rocks (dunite, pyroxenite, harzburgite) followed upwards by layered gabbro and, in some cases, by granophyre and granite. Although many different explanations for the origin of these complexes have been proposed in the past, they formed from mantle-derived magmas that pooled in large crustal reservoirs and differentiated by fractional crystallization.

After the early work of Nicolaysen et al. (1958) and Schreiner (1958) on the age of the Bushveld Complex and by Faure et al. (1963b) and Allsopp (1965) on the Great Dyke, Davies et al. (1970) dated several of the layered complexes by the Rb-Sr method. Although the alignment of four large mafic intrusions of the Precambrian age may indicate the existence of a deep fracture in the continental crust of Africa, the wide range of ages from 2480 ±90 Ma (Great Dyke) to 1340 ±142 Ma (Trompsburg Complex) makes it unlikely that these intrusives are related to each other except in a very general way. However, Coetzee and Kruger (1989) proposed that the Losberg Complex (1961 ±151 Ma) is a southern extension of the Bushveld Complex (2050 ±24 Ma, Hamilton 1977).

7.8.1 Mafic Rocks of the Layered Suite

The Bushveld igneous complex in Fig. 7.16 was intruded into the Transvaal Supergroup of sedimentary and volcanic rocks of Early Proterozoic age whose combined maximum thickness is about 21 km (Jackson 1967; Visser and von Gruenewaldt 1970; von Gruenewaldt et al. 1985; Hall and Hughes 1989). The magmas were injected along the contact between the Pretoria Group and the overlying Rooiberg Group. The contact metamorphic aureole at the base of the intrusion is up to 5 km thick and consists of a thick andalusite zone that grades upward into a sillimanite-microcline zone. The uppermost sedimentary rocks of the Pretoria Group closest to the base of Bushveld Complex were partially melted and migmatized. Uken and Watkeys (1997) reported that the intrusion of the Bushveld magmas caused the formation of diapirs in the rocks of the Pretoria Group.

The Bushveld Complex is about 9 km thick (Sharpe 1985) and has been subdivided into four "zones" which,

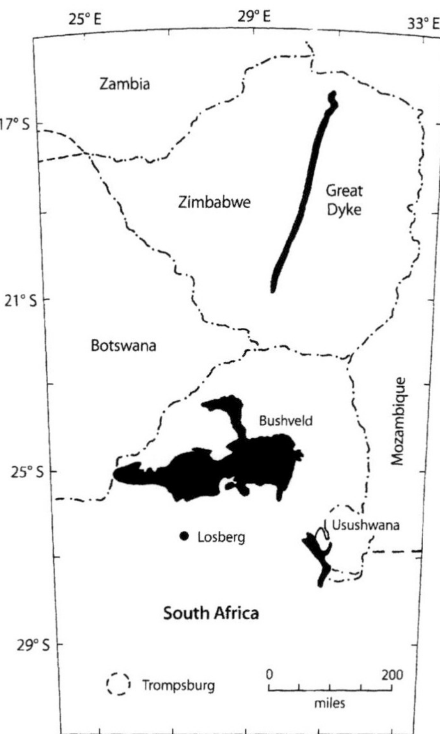

Fig. 7.15. Locations of the five major mafic differentiated intrusives of Precambrian age in southern Africa: Great Dyke (Zimbabwe), and Bushveld, Losberg, Trompsburg, and Usushwana complexes (South Africa) (*Source:* adapted from Davies et al. 1970)

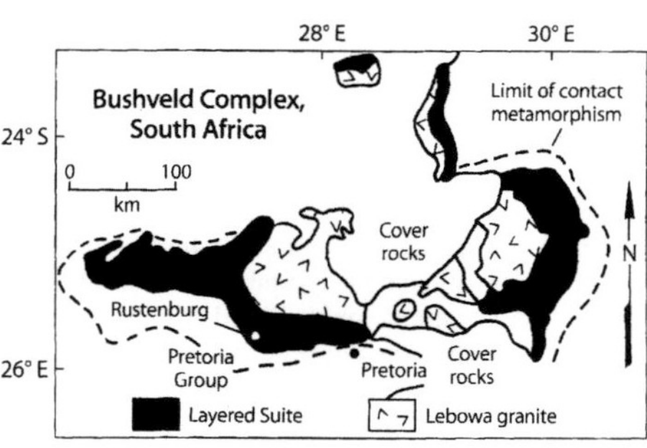

Fig. 7.16. Outcrop area of the Rustenburg layered suite of the Bushveld complex (*black*) and of the related Lebowa granite (*checks*) of South Africa (*Source:* adapted from Uken and Watkeys 1997)

Table 7.3. Generalized stratigraphic section of the Transvaal Supergroup including the intrusive Bushveld igenous complex and the younger Lebowa granite suite. *Wavy line:* unconformity; *dashed line:* intrusive contact *Source:* von Gruenewaldt et al. 1985)

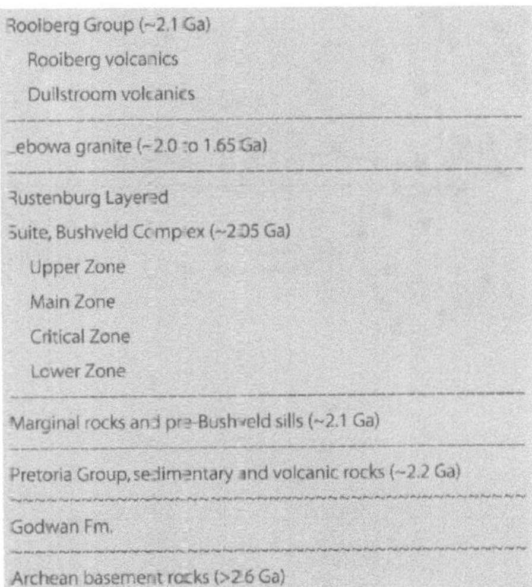

together with the associated sills and marginal rocks, are included in a generalized stratigraphic column of the Transvaal Supergroup in Table 7.3. The mafic rocks of the Bushveld Complex were formed by multiple injections of two distinctly different magmas identified as "ultramafic" (U-type) and "anorthositic" (A-type). According to von Gruenewaldt et al. (1985), the U magmas had the chemical composition of olivine boninites (Sect. 3.2.6) with elevated concentrations of SiO_2 (52–56%), MgO (12–16%), Cr (800–2000 ppm), Zr (150–400 ppm), and Rb (20–50 ppm) but relatively low initial $^{87}Sr/^{86}Sr$ ratios (0.703 to 0.705). The A magmas were tholeiitic in composition with low concentrations of SiO_2 (48 to 50%), low MgO (8 to 10%), and low concentrations of incompatible elements (e.g. Cr, Zr, Rb, etc.) but high initial $^{87}Sr/^{86}Sr$ ratios (0.707 to 0.708). The compositional variations of rocks in the Bushveld Complex are caused in part by mixing of the A and U magma types. The decrease of the $^{87}Sr/^{86}Sr$ ratio at the base of the Upper Zone may require the existence of a third magma type composed of an iron-rich quartz tholeiite with an intermediate $^{87}Sr/^{86}Sr$ ratio of about 0.7067 (Kruger et al. 1987).

The Bushveld Complex grew by sequential additions of magma of progressively changing composition from primarily U-type in the Lower Zone to A-type in the Main Zone (Sharpe 1985). The Critical Zone was formed by mixing of both magma types which contributed to the formation of chromite layers and precious metal

deposits in this zone. The last major input of magma occurred during the crystallization of the Upper Zone as indicated by the presence of the "Pyroxenite Marker" at the base of that zone (von Gruenewaldt et al. 1985; Kruger et al. 1987; Cawthorn et al. 1991).

The mafic rocks of the Bushveld Complex formed in four chambers that were probably connected to each other and to a larger master magma chamber at depth (i.e. the western, far-western, eastern (Rustenburg Layered Suite), and northern compartments). The stratigraphy of the rocks in these lobes of the Bushveld Complex differs significantly in spite of broad similarities. However, the lithologic composition and layering of the Critical Zone and the lower Main Zone in the eastern and western Bushveld are virtually identical indicating that these two chambers were closely connected.

The upper part of the Critical Zone contains economic deposits of platinum group elements (PGE) in two layers known as the UG-2 chromitite layer and the Merensky Reef located about 500 m above it (e.g. the Rustenburg platinum mine). The mineral composition and distribution of PGEs of these layers were described in three series of papers published in "Economic Geology" (vol. 77, 1982; vol. 80, 1985; and vol. 81, 1986).

7.8.2 Age and Strontium-Isotope Stratigraphy

The age of the mafic rocks of the Bushveld Complex was determined by Hamilton (1977) who demonstrated that these rocks formed at 2050 ±24 Ma ($\lambda = 1.42 \times 10^{-11}$ yr^{-1}). More recently, Walraven et al. (1990) reported an age of 2061 ±27 Ma. In addition, Hamilton (1977) demonstrated that the average initial $^{87}Sr/^{86}Sr$ ratios of the Rustenburg Layered Suite increase stratigraphically from 0.70561 ±0.00004 in the Lower Zone to 0.70956 ±0.00007 in the Upper Zone relative to 0.7080 for E&A. The increases of the initial $^{87}Sr/^{86}Sr$ ratio occur stepwise in places where the mineral composition of the rock layers changes (Sharpe 1985).

The data in Fig. 7.17 demonstrate that the rocks of the Merensky Reef, the Critical Zone, and the Lower Zone have numerically distinct but constant initial $^{87}Sr/^{86}Sr$ ratios in contrast to the rocks of the Main Zone and the Upper Zone which vary widely and have higher average values. In addition, there is evidence of fractional crystallization of the magmas that formed the Merensky Reef, the Critical Zone, and the Lower Zone.

Kruger and Marsh (1982) reported that the initial $^{87}Sr/^{86}Sr$ ratios of the Merensky Reef at the top of the Critical Zone in the western lobe of the Bushveld increase by 0.15% from 0.70636 at the base of the Reef to 0.70739 at the top ($\lambda = 1.42 \times 10^{-11}$ yr^{-1}, $t = 2100$ Ma, NBS 987 = 0.71025) over a stratigraphic thickness of only about 10 m. The abrupt change of the initial $^{87}Sr/^{86}Sr$ ratio within the Merensky Reef contrasts with

Fig. 7.17.
Sr isotope mixing diagram of mafic igneous rocks of the Rustenburg Layered Suite of the Bushveld Complex, South Africa. The initial $^{87}Sr/^{86}Sr$ ratios of the Upper Zone (*crosses*) vary widely and overlap those of the Main Zone (*solid circles*). The initial $^{87}Sr/^{86}Sr$ ratios of the Critical Zone, Merensky Reef, and Lower Zone vary only within narrow limits and are lower than those of the Upper Zone and the Main Zone (*Source:* Hamilton 1977; Sharpe 1985)

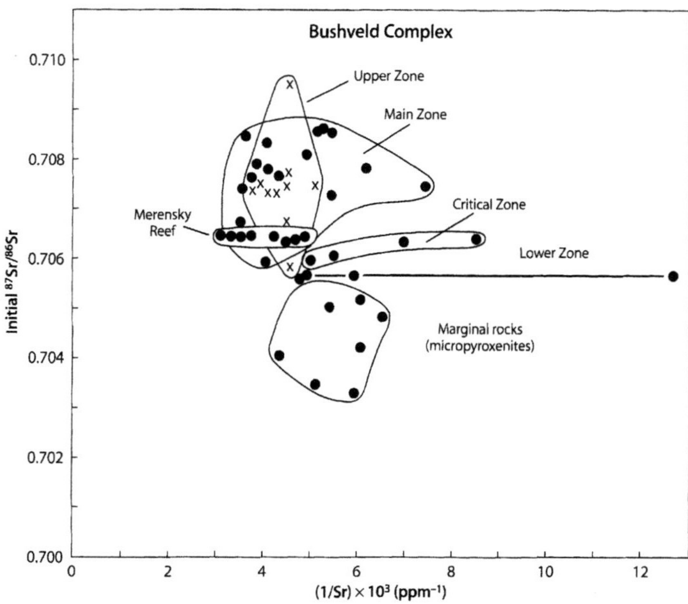

Table 7.4. Initial $^{87}Sr/^{86}Sr$ ratio of the Merensky Reef and the Main and Critical Zone (Kruger and Marsh 1982)

Unit	Average initial $^{87}Sr/^{86}Sr$
Main Zone (60 m)	0.70764 ±0.00016 (2σ, N = 6)
Merensky Reef	0.70636 to 0.70739
Critical Zone	0.70636 ±0.00002 (2σ, N = 5)

the much more gradual changes within the zones above and below the Reef (Table 7.4; Kruger and Marsh 1982).

Kruger and Marsh (1982) considered whether the magma of the Merensky Reef was contaminated by assimilation of wallrocks, but favored the widely held view that the change in the isotope composition of Sr at the base of the Main Zone was caused by the intrusion of a gabbroic magma having an elevated $^{87}Sr/^{86}Sr$ ratio. In this scenario, the Merensky Reef was formed by the accumulation of minerals that crystallized as the new magma (high $^{87}Sr/^{86}Sr$) mixed with the existing magma (low $^{87}Sr/^{86}Sr$). The final $^{87}Sr/^{86}Sr$ ratio of the rocks of the Merensky Reef was subsequently affected by the addition of intercumulus liquid from the underlying Critical Zone. Another step-change in the initial $^{87}Sr/^{86}Sr$ ratio occurs at the Pyroxenite Marker which is located near the top of the Main Zone (Cawthorn et al. (1991).

Kruger et al. (1987) demonstrated that the magnetite gabbro, norite, and ferrodiorite of the Upper Zone

in a stratigraphic section of 2.1 km form a Rb-Sr isochron yielding an age of 2 066 ±58 Ma and an initial $^{87}Sr/^{86}Sr$ ratio of 0.7075 ±0.0001 ($\lambda = 1.42 \times 10^{-11}$ yr⁻¹, 0.71025 for NBS 987). The collinearity of the data points indicates that these rocks formed by fractional crystallization of a mafic magma that was homogeneous in terms of the isotope composition of Sr. The Upper Zone contains 28 magnetite layers whose origin has been attributed to magma mixing by analogy with the Critical Zone which is characterized by numerous chromite layers (Irvine 1977). The isotopic homogeneity of Sr in the Upper Zone indicates that the rocks of this zone could have originated by fractional crystallization of one large batch of magma that formed by thorough mixing of the Main-Zone magma ($^{87}Sr/^{86}Sr \approx 0.7085$) with another magma ($^{87}Sr/^{86}Sr \approx 0.7067$) to produce an isotopically homogeneous mixture ($^{87}Sr/^{86}Sr = 0.7075$) prior to the onset of crystallization. Therefore, the numerous magnetite layers in the Upper Zone cannot be attributed to multiple intrusions of magma unless all such magmas had the same $^{87}Sr/^{86}Sr$ ratio of 0.7073 (Kruger et al. 1987).

The chemical and isotopic evolution of Bushveld magmas is also recorded in the Marginal Rocks and the associated sills exposed in the eastern lobe. Harmer and Sharpe (1985) subdivided these rocks into two principal groups in order of increasing stratigraphic level. The first group includes pyroxenites, peridotites, and norites that form a shell around the base of the complex up to about the middle of the Critical Zone. Above

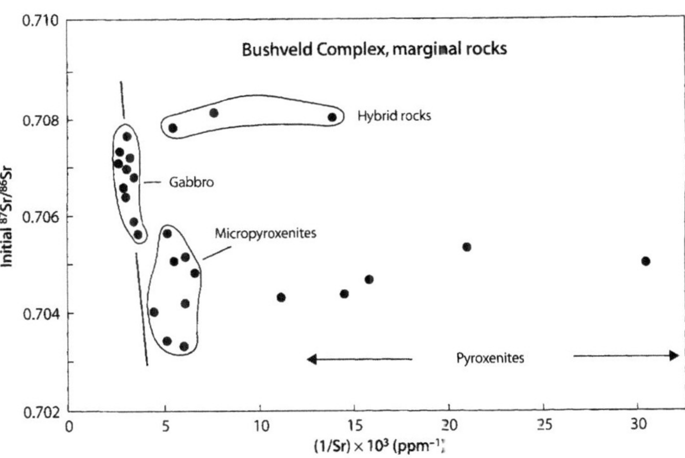

Fig. 7.18.
Sr-isotope mixing diagram of the marginal rocks of the Bushveld Complex. The apparent collinearity of the gabbros and micropyroxenites is the result either of magma mixing or of assimilation of crustal rocks. The pyroxenites and hybrid rocks are products of strong fractional crystallization of magmas with quite different ^{87}Sr/^{86}Sr ratios (*Source:* Harmer and Sharpe 1985)

that level, gabbroic rocks make up the border of the upper Critical Zone and of the Main Zone. Harmer and Sharpe (1985) reported that the initial ^{87}Sr/^{86}Sr ratios of the lower part of the Marginal Rocks (at 2 050 Ma, $\lambda = 1.42 \times 10^{-11}$ yr^{-1}) range from 0.7034 to 0.7059 relative to 0.71025 for NBS 987. The gabbros of the upper Marginal Rocks have distinctly higher initial ^{87}Sr/^{86}Sr ratios between 0.7059 and 0.7081. The distribution of data points in Fig. 7.18 suggests that the gabbros are products of magma mixing or assimilation of crustal rocks. The displacement of the pyroxenites, micropyroxenites, and hybrid rocks from the hypothetical mixing line of the gabbros is evidence for fractional crystallization of magmas having different ^{87}Sr/^{86}Sr ratios. The micropyroxenites of the Marginal Rocks have a low average initial ^{87}Sr/^{86}Sr ratio of 0.7042 ±0.0006 ($2\bar{\sigma}$, $N = 7$) relative to 0.71025 for NBS 987. This value presumably represents the isotope composition of Sr in one of the magma sources of the Bushveld Complex.

Although the mafic magmas of the Bushveld Complex originated in the subcrustal mantle, their chemical compositions and ^{87}Sr/^{86}Sr were altered before the magmas entered the crustal compartments where they differentiated by fractional crystallization during slow cooling. Since the magmas were not contaminated by crustal rocks from the walls of the magma chambers, except locally (see below), Harmer and Sharpe (1985) concluded that the marginal rocks (and presumably the layered rocks as well) formed by mixing of magmas derived from different regions in the mantle. However, the long history of magmatic activity in the Transvaal Basin of South Africa as well as the large volume and compositional diversity of the volcanic and plutonic rocks is also consistent with the existence of a giant mantle plume interacting with the overlying lithospheric mantle and continental crust (Hatton 1995).

7.8.3 Platreef, Northern Bushveld

The interaction of the mafic magmas of the Bushveld Complex with the wallrock of the northern magma compartment was studied by Cawthorn et al. (1985) and Barton et al. (1986) in the area north of Potgietersrus. Here a platinum-bearing layer of pyroxenite is in direct contact with Archean granitoids and with dolomite of the Transvaal Supergroup. The pyroxenite body, known as the Platreef, has been subdivided into three stratigraphic layers in descending order based on differences in mineral composition and texture:

- C Reef: feldspathic pyroxenite overlain by gabbros of the Main Zone;
- B Reef: medium to coarse grained pyroxenite including chromite;
- A Reef: heterogeneous feldspathic pyroxenite intruding rocks of the Transvaal Supergroup and the Archean basement.

Cawthorn et al. (1985) demonstrated that the feldspathic pyroxenite (A Reef) of the Platreef at Farm Overysel about 30 km north of Potgietersrus has elevated initial ^{87}Sr/^{86}Sr ratios ranging from 0.7079 to 0.7227. These rocks also contain Na-rich plagioclase, biotite, and quartz suggesting assimilation of a partial melt derived from the granitic basement rocks by the Bushveld magma. In general, where the Platreef is underlain by Archean granite (Farm Overysel), it is enriched in Rb (3.84 to 31.3 ppm) and has high initial ^{87}Sr/^{86}Sr ratios (at 2.05 Ga) that vary irregularly relative to 0.71025 for NBS 987 (Table 7.5; Barton et al. 1986).

The gabbros of the Main Zone that overlie the Platreef at Farm Overysel have an average initial ^{87}Sr/^{86}Sr

Table 7.5. Initial $^{87}Sr/^{86}Sr$ ratios of the Main Zone and the Platreef underlain by Archean granite (Barton et al. 1986)

Unit	$(^{87}Sr/^{86}Sr)_0$
Main Zone (343.8 m)	0.7082 – 0.7095
Platreef (113.8m), underlain by Archean granitoids	
C Reef (12.3m)	0.7087 – 0.7132 (top)
B Reef (12.3m)	0.7123 – 0.7203
A Reef (89.2m)	0.7135 – 0.7225 (bottom)

Table 7.6. Initial $^{87}Sr/^{86}Sr$ ratios of the Main Zone and the Platreef underlain by Transvaal dolomite (Barton et al. 1986)

Unit	$^{87}Sr/^{86}Sr_0$
Main Zone (50.5 m)	0.7068 – 0.7092
Platreef (326.5 m), underlain by Transvaal dolomite	
C Reef (7.4 m)	0.7085 – 0.7100 (top)
B Reef (6.9 m)	0.7089 – 0.7146
A Reef (312.2 m)	0.7053 – 0.7139 (bottom)

ratio of 0.70855 ±0.00017 ($2\bar{\sigma}$, $N = 5$, omitting one sample) that is similar to the initial $^{87}Sr/^{86}Sr$ ratios of the Main Zone elsewhere in the Bushveld Complex based on measurements by Hamilton (1977), Kruger and Marsh (1982), and Sharpe (1985).

An orthopyroxene in the C Reef has an initial $^{87}Sr/^{86}Sr$ ratio of 0.7079 which is significantly lower than the average initial $^{87}Sr/^{86}Sr$ ratio of the overlying Main Zone, whereas plagioclase from the same rock has a high initial ratio of 0.7122 relative to 0.71025 for NBS 987. Therefore, Barton et al. (1986) concluded that the magma of the C Reef was contaminated and that the orthopyroxene had formed prior to the plagioclase and recorded the $^{87}Sr/^{86}Sr$ ratio of the magma before contamination. The contaminant originated from the Archean basement and permeated the crystal mush of the Platreef from which plagioclase was crystallizing. The formation of intercumulus plagioclase depleted the contaminant in Sr and thereby reduced its ability to affect the isotope composition of Sr in the rocks of the overlying Main Zone. In addition, the rocks of the Main Zone were insensitive to contamination because they contain a high proportion of Sr-rich cumulus plagioclase. However, the rocks of the Main Zone above the Platreef do have anomalously high Rb concentrations (13.4 ±4.4 ppm, $2\bar{\sigma}$, $N = 6$) compared to rocks of the Main Zone elsewhere which contain only about 4.0 ppm Rb. Therefore, the Rb concentrations of the rocks in the Main Zone above the Platreef were increased by the contaminant, but the $^{87}Sr/^{86}Sr$ ratios of the rocks were not changed significantly.

Barton et al. (1986) also analyzed rocks of the Platreef at a site (Farm Sandsloot) where it overlies dolomite of the Transvaal Supergroup. The dolomite at this location has a very high initial $^{87}Sr/^{86}Sr$ ratio of 0.7234 (at 2.05 Ga, Cawthorn et al. 1985). Other carbonate rocks within 10 km of the contact with the Platreef have a wide range of initial $^{87}Sr/^{86}Sr$ ratios from 0.7032 to 0.7231. The Sr-weighted average initial $^{87}Sr/^{86}Sr$ ratio of all carbonate rocks of the Transvaal Supergroup analyzed by Barton et al. (1986) is only 0.7039. Although the overlying rocks of the Platreef at this location have also been contaminated with ^{87}Sr, the initial $^{87}Sr/^{86}Sr$ ratios re-

Table 7.7. Average initial $^{87}Sr/^{86}Sr$ ratios (Main Zone) for Archean granite and Transvaal dolomite (Barton et al. 1986; 0.71025 for NBS 987)

Country rock	Average initial $^{87}Sr/^{86}Sr$ (Main Zone)
Archean granite	0.70856 ±0.00017 ($2\bar{\sigma}$, $N = 5$)
Transvaal dolomite	0.70776 ±0.0009 ($2\bar{\sigma}$, $N = 3$)

ported by Barton et al. (1986) relative to 0.71025 for NBS 987 are not as high as those where the Platreef is in contact with Archean granitoids (Table 7.6).

The initial $^{87}Sr/^{86}Sr$ ratios of all three layers of the Platreef where it is in contact with Transvaal dolomite are significantly lower than where the Platreef intruded Archean granite. Even the rocks of the Main Zone have slightly lower average initial $^{87}Sr/^{86}Sr$ ratio (Table 7.7; Barton et al. 1986, 0.71025 for NBS 987).

In contrast to the average initial $^{87}Sr/^{86}Sr$ ratios, the rocks of the Main Zone adjacent to the Transvaal dolomite have a high average Rb concentration of 32.9 ±19.1 ppm compared to only 13.4 ±4.4 ppm Rb where the Main Zone is in contact with Archean granite. Even the rocks of the Platreef adjacent to the dolomite have high Rb concentrations (C Reef: 15.2 ±1.1 ppm; B Reef: 16.3 ±3.7 ppm; A Reef: 9.1 ±5.7 ppm). Therefore, both the Platreef and the Main Zone at this location were contaminated both with ^{87}Sr and Rb, but the dolomite contributed less ^{87}Sr than the Archean granite to the rocks of the Platreef at the base of the Bushveld Complex.

7.8.4 Rooiberg Felsites

The volcanic rocks of the Rooiberg Group were erupted before the intrusion of the mafic magmas of the Bushveld Complex and the younger associated Lebowa granite. According to Twist (1985), the volcanic rocks reach a thickness of 3 520 m at the Loskop Dam in the southern region of the eastern lobe of the Bushveld Complex and originally had a volume of about

300 000 km³. Therefore, the Rooiberg felsites are the most voluminous deposit of rhyolitic lava of Precambrian age on Earth. The Rooiberg Group consists of silicic volcanic rocks (felsites) ranging in composition from silicic andesites to dacites, rhyodacites, and rhyolites erupted at about 2150 Ma. Twist (1985) divided the stratigraphic succession of flows at the Loskop Dam into nine units numbered in stratigraphic sequence from the bottom towards the top based on the occurrence of interbedded sandstones, shales, and tuff.

The average Rb and Sr concentrations of these units vary systematically up-section in Fig. 7.19, thereby revealing extensive differentiation of the magma in the course of time. Twist (1985) recognized three different magma types on the basis of their concentrations of MgO and FeO that were derived from at least two different chambers (Table 7.8).

The overall stratigraphic variation of the average Rb and Sr concentrations of the Rooiberg felsites results from the presence or absence of different lava types in the stratigraphic units and from the differentiation of the different types of magma.

The high-Mg lavas are interbedded with low-Mg lavas in Units 1 and 2, becoming dominant in Unit 2. The Mg concentrations of the high-Mg lavas in Unit 2 increase stratigraphically, suggesting that successive lava flows were withdrawn from increasingly deeper levels of a compositionally stratified magma chamber. Similar chemical stratification has been observed in much younger ignimbrite and rhyolite deposits (e.g. Hildreth 1979,1981). The eruption of high-Mg lavas was abruptly terminated by the volcaniclastic rocks that cap Unit 2.

Table 7.8. Rb and Sr concentrations and occurrence of three different magma types, recognized on the basis of their concentrations of MgO and FeO by Twist (1985)

Magma type	Occurrence	Rb (ppm)	Sr (ppm)
1. High MgO (>1.7%)	Lower 1 400 m	135 ±25	187 ±67
2. Low MgO (<1.0%)	In all units	178 ±46	100 ±56
3. High FeO (~10%)	Units 1 and 3 only	110 ±39	159 ±16

The low-Mg lavas remained uniform in composition from Unit 1 to 6, but evolved significantly in the three uppermost Units (7, 8, and 9, Fig. 7.19) in which the concentrations of MgO, TiO_2, P_2O_5, and Sc decrease stepwise up-section, whereas those of Nb, Zr, and Y increase. Twist (1985) pointed out that these changes in chemical composition of the low-Mg lavas imply that they were erupted from a stratified magma chamber with reverse zonation because the least differentiated magma seems to have been at the top of the chamber while more highly differentiated magma existed below it. Actually, apparent reverse zonation of stratigraphic sequences of lava flows can be caused by withdrawal of magma through conduits located below the tops of normally stratified magma chambers (Elston 1984). Therefore the low-Mg magma chamber, which was much larger than the high-Mg chamber, was probably stratified in the normal manner with a Rb-rich and Sr-poor liquid floating on a more homogeneous dacitic to rhyodacitic magma. However, the low-Mg magma chamber was tapped such that the less differentiated homogeneous magma was expelled first, followed by the upper layer as the chamber was gradually emptied.

The Rooiberg felsites must be older than about 2 050 Ma because the mafic rocks of the Bushveld Complex were intruded after the extrusion of the lava flows of the Rooiberg Group. Walraven (1985) attempted to date whole-rock samples of the Rooiberg felsite by the Rb-Sr method but obtained an isochron date of only 1 604 ±37 Ma and an initial $^{87}Sr/^{85}Sr$ ratio of 0.7321 ±0.0029. The date is about 500 million years younger than required by the known age of the Bushveld Complex. Therefore, the Rb-Sr systematics of the Rooiberg felsites were disturbed after eruption, presumably by aqueous fluids. Nicholson and Shirey (1990) came to the same conclusion when they attempted to date Precambrian felsite on the Keweenaw Peninsula of Michigan.

Several authors (e.g. Hamilton 1970; Rhodes 1975; Twist and French 1983; Elston and Twist 1989) have considered the hypothesis that the Bushveld Complex formed as a result of the impact of an asteroid and that the rocks of the Rooiberg Group are related to such an event. The impact origin of the Bushveld Complex is not as far-fetched as it may appear, because of the analogy

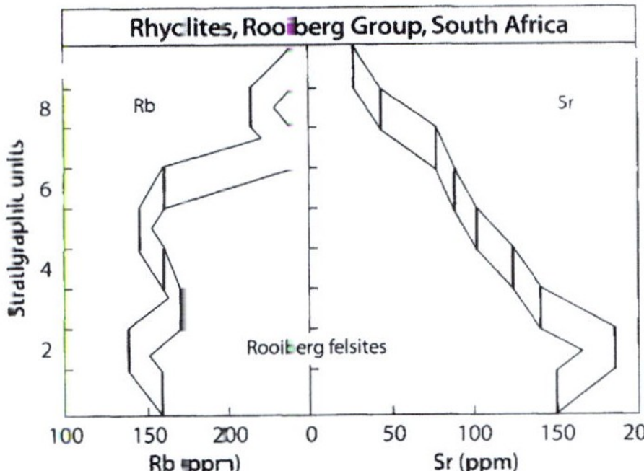

Fig. 7.19. Stratigraphic variation of average concentrations of Rb and Sr in the principal stratigraphic units of the Rooiberg felsites (2150 Ma). The observed changes in the chemical composition of these rocks were caused primarily by the evolution of low-Mg magma in a large crustal reservoir which was emptied so that the layer of differentiated magma in the upper part of the chamber was erupted last (*Source:* Twist 1985)

to the norite and associated rocks at Sudbury, Ontario. The origin of the Sudbury "nickel irruptive" as a result of the impact of an asteroid was first recognized by Dietz (1964) based in part on the occurrence of the intrusive breccia and shatter cones. Subsequent studies by French (1967, 1970), French and Short (1968), and Brocoum and Dalziel (1974) strongly support the impact origin of the Sudbury igneous complex. As a result, recent papers on the geologic history of this complex by Lowman (1992) and Fueten and Redmond (1997), on the petrogenesis by Lightfoot et al. (1997), and geochronology by Corfu and Lightfoot (1996) all accept the impact origin.

In the case of the Bushveld Complex, a recent study by Buchanan and Reimold (1998) failed to detect any features in the felsites of Rooiberg Group that support an impact origin for these rocks. In addition, the rocks of the older Pretoria Group contain no evidence of shock metamorphism, even through they would have been the target rocks if an asteroid impact had occurred.

7.8.5 Granophyres

The abundant granophyres of the Bushveld Complex are collectively referred to as the Rashoop granophyre suite. These rocks differ from the rocks of the Lebowa granite suite which form the so-called Bushveld granite. The origin of the Bushveld granophyres has been attributed to magmatic differentiation, but Walraven (1985) demonstrated that several different types of granophyre are present and that none of them formed by direct magmatic differentiation of the mafic magmas of the Bushveld Complex.

The Rashoop granophyre suite includes several different varieties identified by name as:

- Zwartbank pseudogranophyre
- Diepkloof granophyre
- Stavoren granophyre
- Rooikop porphyry

These granophyres occur as disconnected bodies that occupy different stratigraphic positions: (1) Between the Upper Zone of the mafic Layered Suite and the overlying Lebowa granite suite; (2) Between the Layered Suite and the Rooiberg Group; (3) Along the contact between the Lebowa granites and the Rooiberg felsites; and (4) In the Transvaal Supergroup near the base of the Layered Suite at Pretoria. In view of the diversity of stratigraphic settings occupied by the granophyre, it is hardly surprising that they do not have a common origin.

Walraven (1985) demonstrated that the concentrations of Rb, Sr, and Ba of the Stavoren and Diepkloof

granophyres are similar to those of the Rooiberg felsites, but differ from those of the granites of the younger Lebowa suite. However, age determinations of the Stavoren granophyre by the whole-rock Rb-Sr method yielded an errorchron date of only 1 859 ±135 Ma and an initial $^{87}Sr/^{86}Sr$ ratio of 0.7001 ±0.0162. The date is inaccurate because it is younger than the known age of the Nebo granite of the Lebowa suite which intrudes the Stavoren granophyre. Additional Rb-Sr age determinations of the granodiorite roof rock above the mafic rocks of the Bushveld and of the Klipkloof granite (aplite phase of the Nebo granite) also yielded dates known to be too young. Walraven (1985) attributed these anomalous Rb-Sr dates of the felsite and granophyres to loss of ^{87}Sr caused by circulating groundwater at moderate temperatures.

On the basis of field geology, chemical composition, and thermal modeling, Walraven (1985) concluded that the Zwartbank pseudogranophyre formed by metamorphism of sedimentary rocks through heat provided from the mafic intrusive, that the Diepkloof granophyre originated similarly by melting of Rooiberg felsite, and that the Stavoren granophyre and Rooikop porphyry crystallized from magmas related to the Rooiberg felsite that were intruded into the volcanic rocks and the underlying sedimentary rocks of the Transvaal Supergroup. Therefore, the Stavoren granophyre and Rooikop porphyry predate the intrusion of the mafic magmas, whereas the Zwartbank and Diepkloof granophyres are products of contact metamorphism and partial melting caused by heat emanating from the magma compartments of the Layered Suite of the Bushveld complex.

7.8.6 Lebowa Granite

The intrusion of a sheet-like body of granite magma along the contact between the Upper Zone of the mafic rocks of the Bushveld Complex and the felsite roof rocks of the Rooiberg Group was the last major igneous event of this area. The resulting Lebowa granite suite (Bushveld granite) includes the Nebo granite and its marginal variants such as the Klipkloof granite and the Boobejaankop granite. Some of the Bushveld granites contain deposits of fluorite and tin (e.g. the Zaaiplaats tin mine in the northern part of the Bushveld Complex; von Gruenewaldt et al. 1985; Gulson and Jones 1992).

According to age determinations by Davies et al. (1970), Schreiner (1958), and Nicolaysen et al. (1958), the Lebowa granite is not only younger at 1910 ±30 Ma than the mafic rocks of the Bushveld Complex, but also has a higher initial $^{87}Sr/^{86}Sr$ ratio of 0.715 ±0.0088 compared to 0.7080 for E&A. The elevated initial $^{87}Sr/^{86}Sr$ ratio of the Lebowa granite indicates that it is the product of a crustal magma that formed by melting of the Archean basement rocks underlying the Transvaal Basin.

7 8.7 Archean Granitoids

Granitic rocks of Archean age form the floor of the northern lobe of the Bushveld Complex and are accessible in the area north of Potgietersrus discussed in Sect. 7.8.3. At this location, the pyroxene cumulate rocks of the Platreef are in contact with Archean granitic basement rocks. Cawthorn et al. (1985) described several textural facies of the granitic rocks including the Utrecht granite close to the contact and the Lunsklip granite about 5 km below the base of the Bushveld Complex. In addition, xenoliths of banded gneisses occur close to the contact in association with the Utrecht granite. The age of the Lunsklip granite is 2.504 ±0.109 Ga and its initial ^{87}Sr/^{86}Sr ratio is 0.7033 ±0.0006 (NBS 987: 0.71025 ±0.00006; $\lambda = 1.42 \times 10^{-11}$ yr^{-1}; Cawthorn et al. 1985). However, samples of the Utrecht granite scatter widely on the isochron diagram because their Rb-Sr systematics were disturbed by the intrusion of the Bushveld magmas at 2.05 Ga. Cawthorn et al. (1985) calculated initial ^{87}Sr/^{86}Sr ratios of the granitic basement rocks at 2.05 Ga. The Sr-weighted averages and ranges of these ratios are shown in Table 7.9.

These results demonstrate that the Archean granitic rocks that underlie the Transvaal Basin had elevated ^{87}Sr/^{86}Sr ratios at the time of the intrusion of the mafic magmas of the Bushveld Complex and were capable of forming granitic magmas containing Sr with an ^{87}Sr/^{86}Sr ratio of about 0.715 recorded by the Lebowa granite. However, the magmas of the Lebowa granite probably originated from greater depth in the crust and did not necessarily form by partial melting of the granitic basement rocks exposed near Potgietersrus.

The Rb-Sr systematics of the granitic basement rocks underlying the northern lobe of the Bushveld Complex reveal an interesting anomaly illustrated in Fig. 7.20 by means of a Sr-isotope evolution diagram based on Eq. 1.16 in Sect. 1.2 and demonstrated in

Table 7.9. Initial ^{87}Sr/^{86}Sr ratios of the granitic basement rocks at 2.05 Ga calculated by Cawthorn et al. (1985)

Unit	Initial ^{87}Sr/^{86}Sr at 2.05 Ga
Banded gneiss	0.7084 (N = 11) 0.7024 – 0.7258
Leucocratic granite veins	0.7191 (N = 6) 0.7164 – 0.7225
Lunsklip granite	0.7336 (N = 10) 0.7294 – 0.7373
Utrecht granite	0.7622 (N = 6) 0.7316 – 0.7906
All of the above	0.7673 (N = 33) 0.7024 – 0.7906

Fig. 1.2. The straight lines in Fig. 7.20 represent the banded gneiss and the associated leucocratic granite veins as well as the Lunsklip granite. In addition, the time-dependent change of the ^{87}Sr/^{86}Sr ratio of magma sources in the mantle is represented by a line that extends from 0.699 (BABI) at 4.5 Ga to 0.7035 at the present time (Sect. 1.12). BABI is the ^{87}Sr/^{86}Sr ratio of primordial Sr that was incorporated into the Earth at the time of its formation.

The phenomenon illustrated in Fig. 7.20 is that the lines representing the banded gneisses and leucocratic-granite veins do not intersect the Sr-isotope evolution line of the mantle within the time allowed by the age of the Earth, whereas the evolution line of the Lunsklip granite does intersect the mantle line at about 2.5 Ga. Cawthorn et al. (1985) recognized this situation and concluded that the Rb/Sr ratios of the banded gneisses and leucocratic granite had been lowered by the removal of Rb relative to Sr. Consequently, the age of these rocks can only be determined by other isotopic systems such as the Sm-Nd method based on whole-rock samples or by the U-Pb method applied to zircons, both of which are less susceptible to alteration than the Rb-Sr system. Since the necessary information about the his-

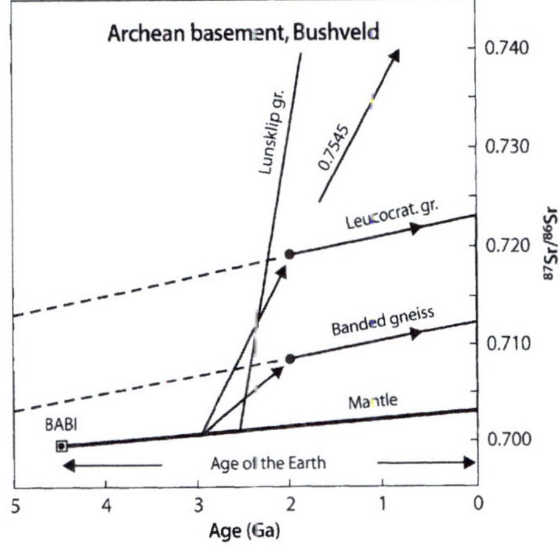

Fig. 7.20. Sr-isotope evolution diagram for banded gneisses and associated veins of leucocratic granite and for the Lunsklip granite, all of which compose the Archean basement near Potgietersrus at the base of the northern lobe of the Bushveld Complex. The lines were drawn from Sr-weighted average ^{87}Sr/^{86}Sr ratios of each rock type as reported by Cawthorn et al. (1985). The line representing magma sources in the mantle extends from 0.699 (Basaltic Achondrite Best Initial) at 4.5 Ga to 0.7035 at present. The diagram illustrates a hypothetical geologic history of the banded gneiss and leucocratic veins. The arrows in the diagram suggest that these rocks originated from sources in the mantle at about 3.0 Ga, evolved isotopically until about 2.05 Ga, and then lost Rb as a result of partial melting associated with the intrusion of the Bushveld magmas (*Source:* Cawthorn et al. 1985)

tory of these rocks is lacking, Fig. 7.20 is based on the assumption that the rocks originated from mantle-derived magma at 3.0 Ga, evolved isotopically until 2.05 Ga as indicated by the arrows, and lost Rb as a consequence of metamorphism or partial melting caused by the intrusion of the mafic magmas of the Bushveld. The Rb removed from these granitic basement rocks may have been transferred to the mafic rocks of the Main Zone and the Platreef of this area (Sect. 7.8.3).

The decrease of the Rb/Sr ratio illustrated in Fig. 7.20 means that the leucocratic granite and banded gneisses at Potgietersrus are restites from which a Rb-rich partial melt was removed. The reduction in slope of the isotope-evolution lines explains why these restites fail to intersect the Sr-evolution line of source rocks in the mantle.

7.9   Great Dyke, Zimbabwe

The Great Dyke (Fig. 7.15) is a remnant of a much larger body that was intruded into Precambrian basement rocks along a rift extending more than 530 km across Zimbabwe in a north-south direction. Worst (1958, 1960) identified four feeders which now form the Musengezi (Darwendale), Hartley (Sebakwe), Selukwe, and Wedza Complexes. The Great Dyke is a remnant of these complexes which have been preserved in a graben. The rocks are layered and consist of dunite, harzburgite, and pyroxenite overlain by gabbros that define the feeder complexes. The layers dip inward toward the center of the intrusive whose width ranges from about 5 to 11 km. The main body of the Great Dyke is flanked on both sides by satellite dikes which run parallel to it at distances of up to 25 km.

The age of the Great Dyke has been difficult to determine because the rocks are not well suited for dating by the K-Ar and Rb-Sr methods (e.g. Faure et al. 1963b; Allsopp 1965). Davies et al. (1970) reported a whole-rock Rb-Sr isochron date of 2 479 ±89 Ma ($\lambda = 1.42 \times 10^{-11}$ yr^{-1}), whereas Robertson and van Breemen (1970) obtained 2 545 ±120 Ma and initial ^{87}Sr/^{86}Sr = 0.7038 ±0.0012 for the Southern Satellite Dykes. The age of the Great Dyke was redetermined by Hamilton (1977) who reported 2 460 ±16 Ma ($\lambda = 1.42 \times 10^{-11}$ yr^{-1}) and an initial ^{87}Sr/^{86}Sr ratio of 0.70261 ±0.0004 relative to 0.7080 for E&A based on a Rb-Sr isochron of whole-rocks and minerals from the Hartley and Wedza Complexes. Mukasa et al. (1998) reported a very similar Rb-Sr date of 2 467 ±85 Ma with an initial ^{87}Sr/^{86}Sr ratio of 0.7026 ±0.0004 for rocks and minerals of the Darwendale sub-chamber.

All of the Rb-Sr dates cited above are about 100 million years younger than the Sm-Nd isochron date of 2 586 ±16 Ma reported by Mukasa et al. (1998). This date is identical to a U-Pb date of rutiles in a feldspathic pyroxenite of the Selukwe component of the Great Dyke. In addition, Pb extracted from whole-rock samples and minerals in the Darwendale sub-chamber yielded a Pb-Pb date (^{207}Pb/^{204}Pb vs. ^{206}Pb/^{204}Pb) of 2 596 ±14 Ma. However, zircon in the same rocks yielded discordant U-Pb dates indicating loss of radiogenic Pb at about 830 Ma during the Pan-African tectonothermal event.

The difference between the Rb-Sr dates and those obtained by the Sm-Nd, U-Pb and Pb-Pb methods means that Rb-Sr dates are not representative of the age of the Great Dyke. Mukasa et al. (1998) considered several possible explanations for the alteration of the Rb-Sr systematics of the rocks (e.g. slow cooling and/or low-grade metamorphism), but concluded that the alteration was probably caused by ion-exchange reactions between the minerals of the cooling intrusive and an aqueous fluid. This process did not affect the Sm-Nd systematics of silicate minerals and the U-Pb systematics of rutile. The apparent alteration of U-Pb date of zircons was caused by radiation damage resulting from their high U concentrations (42.6 to 63.1 ppm) compared to only 2.46 to 2.86 ppm U in the rutile.

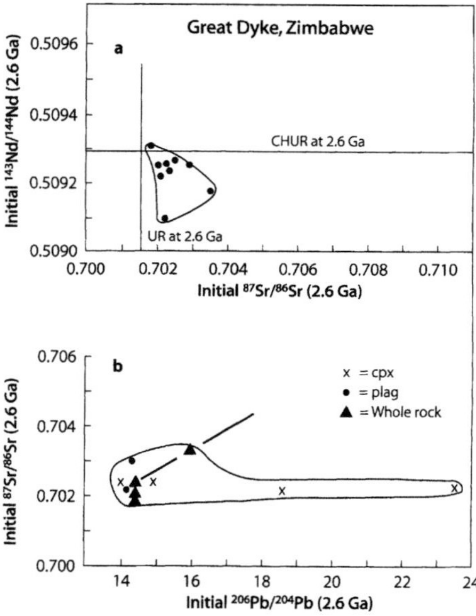

Fig. 7.21 a,b. Initial isotope ratios of Sr, Nd, and Pb of core samples (whole rocks, clinopyroxene, and plagioclase) of the Darwendale component of the Great Dyke, Zimbabwe. All measured isotope ratios were corrected for in situ decay of the parent elements by using the Rb/Sr, Sm/Nd, and U/Pb ratios or the samples based on the crystallization age of 2.6 Ga (*Source:* Mukasa et al. 1998)

Mukasa et al (1998) associated the intrusion of the Great Dyke with the merger of the Kaapvaal and Zimbabwe Cratons by basin closure. Consequently, the magma sources of the Great Dyke were altered by aqueous fluids derived from subducted oceanic crust. The Great Dyke is about 500 million years older than the Bushveld Complex (Sect. 7.7) and has significantly lower initial $^{87}Sr/^{86}Sr$ ratios. Therefore, the Great Dyke and Bushveld Complex are not closely related by petrogenesis or age. However, the Modipe gabbro on the border between South Africa and Botswana (Fig. 7.15) has yielded an imprecise whole-rock Rb-Sr date of 2559 ± 376 Ma ($\lambda = 1.42 \times 10^{-11}$ yr^{-1}).

The initial $^{87}Sr/^{86}Sr$ and $^{143}Nd/^{144}Nd$ ratios (at 2.6 Ga) of gabbros and their constituent minerals of the Darwendale component of the Great Dyke form a data field in Fig. 7.21a which extends into the "crustal" quadrant of the diagram. These isotopic data reveal that the magma of the Darwendale component of the Great Dyke originated in the mantle and could have assimilated wallrocks while it pooled in a crustal magma chamber. However, Mukasa et al. (1998) concluded that the magma sources had been altered by aqueous fluids as a result of subduction of oceanic crust. Both alternatives are consistent with the origin of the continental flood basalt plateaus of Mesozoic age discussed in Chap. 5 (e.g. Wanapum and Grande Ronde Formations of the Columbia River basalt in Fig. 5.3). Contamination of the magma at the source, preferred by Mukasa et al. (1998), is more likely than assimilation of crustal rocks because of the narrow range of variation of the initial isotope ratios of Sr and Nd of the gabbros throughout all parts of the Great Dyke.

The initial $^{206}Pb/^{204}Pb$ ratios of the gabbros cluster between 14.0 and 15.0 in Fig. 7.21b and demonstrate only a weak correlation between the initial $^{87}Sr/^{86}Sr$ and $^{206}Pb/^{204}Pb$ ratios. Two clinopyroxene separates were enriched in ^{206}Pb, presumably by diffusion from U-rich minerals. The concentrations of Pb of the clinopyroxenes (0.322 ppm) are consistently lower than those of plagioclase (1.67 ppm) and the whole-rock samples (1.433 ppm) causing the $^{206}Pb/^{204}Pb$ ratio of the clinopyroxenes to be more susceptible to change by addition of small amounts of ^{206}Pb than the $^{206}Pb/^{204}Pb$ ratios of plagioclase and whole-rock samples.

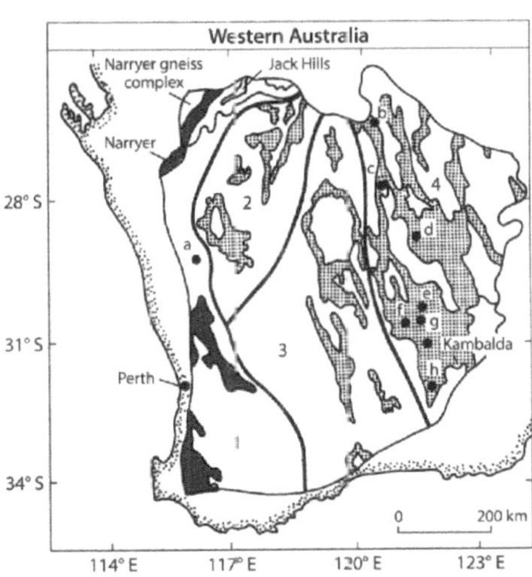

Fig. 7.22. Greenstone belts of the Yilgarn block of Western Australia and its subdivisions: *1* Western gneiss terrain; *2.* Murchison Province; *3.* Southern Cross Province; *4.* Eastern Goldfields Province. In addition, the map shows the location of the Narryer metasediments and the Narryer gneiss complex at the northern end of the Western gneiss terrain (*Source:* adapted from Pidgeon and Wilde 1990)

7.10 Mafic Intrusives of Australia

Layered gabbroic complexes occur primarily in the Yilgarn and Pilbara blocks of Western Australia and in the Musgrave block in central Australia. The Yilgarn block has been subdivided into the Western gneiss terrain, the Murchison province, the Southern Cross province and the Eastern Goldfields province outlined in Fig. 7.22. The geology of the mafic intrusives in the Yilgarn block has been described by Hockley (1971), McCall (1971), Williams and Hallberg (1973), and Jaques (1976) among others.

The Pilbara block includes the Millindinna Complex (Korsch and Gulson 1986), the Munni Complex, the Mount Hall-Carlow Castle Complex, and the Sherlock Bay Complex. The Millindinna Complex is Late Archean in age (about 2.9 Ga, Korsch and Gulson 1986) and contains platinum group metals (Fitton et al. 1975).

The Musgrave block of central Australia lies along the boundary between South Australia and the Northern Territory and extends about 250 km into Western Australia. The granulite gneisses of the Musgrave block were intruded by several layered gabbro bodies of the Giles Complex, including the Kalka, Ewarara, and Gosse Pile Intrusions (Nesbitt et al. 1970; Goode and Moore 1975).

7.10.1 Windimurra Gabbro, Western Australia

The Late Archean Windimurra gabbro in the northern part of the Southern Cross province of the Yilgarn block of Western Australia is a part of a differentiated sill whose total volume is similar to that of the Stillwater Complex of Montana (Hockley 1971; Ahmat and DeLaeter 1982). It is composed of anorthositic gabbro containing from 60 to 90% plagioclase and includes

Table 7.10. Whole-rock Rb-Sr isochron dates of granitic country rock, Windimurra gabbro, pegmatite, various felsic and mafic dikes

Unit	Rb-Sr isochron date (Ma)
Granitic country rock	2574 ±74
Windimurra gabbro	1496 ±189
Pegmatites	2472 ±46
Porphyry dikes	2669 ±135
Quartz dolerite dikes (hornblende bearing)	2426 – 2485

Table 7.11. Average initial $^{87}Sr/^{86}Sr$ ratios for Windimurra gabbro, porphyry dikes, quartz dolerite and pegmatite

Unit	Average ($^{87}Sr/^{86}Sr)_0$
Windimurra gabbro	0.70122 ±0.00010
Porphyry dikes	0.70204 ±0.0007
Quartz dolerite	0.70264 ±0.0009
Pegmatite	0.75100 ±0.093

layers of pyroxenite, serpentinized dunites, and magnetite that dip inward and whose total thickness approaches 7 km. The Windimurra gabbro is cut by a variety of dikes composed of dolerite, feldspar porphyries, pegmatite, and quartz veins.

Ahmat and DeLaeter (1982) reported an internal Rb-Sr mineral isochron date of 1496±189 Ma and an initial $^{87}Sr/^{86}Sr$ ratio of 0.7014 ±0.0006 for a specimen of gabbro ($\lambda = 1.42 \times 10^{-11}$ yr^{-1}, 0.7080 for E&A and 0.71029 for NBS 987). However, the whole-rock Rb-Sr isochron dates of the pegmatite and various felsic and mafic dikes that intrude the Windimurra gabbro are all significantly older than the Rb-Sr date of the gabbro (Table 7.10).

The field relations require the conclusion that the Windimurra gabbro is older than the pegmatites and cross-cutting dikes and younger than the granitic country rocks. Therefore, Ahmat and DeLaeter (1982) concluded that the Windimurra gabbro is older than 2 600 Ma and attributed the anomalously young mineral-isochron date to alteration of the mafic minerals which have low Rb and Sr concentrations. The mineral isochron may record a time of low-temperature alteration of the rocks by heated saline groundwater which occurred about 1.1×10^6 years after the crystallization of these rocks. The effect of low-grade metamorphism on the Rb-Sr systematics of dolerite dikes in North America and Scandinavia has also been documented by Patchett and Bylund (1977), Patchett (1978), Patchett et al. (1979) and Armstrong et al. (1982).

The initial $^{87}Sr/^{86}Sr$ ratios of the Windimurra gabbro (recalculated to 2 670 Ma and corrected to 0.71025 for NBS 987) range only from 0.70106 to 0.70137 and have a mean of 0.70122 ±0.00010 (2$\bar{\sigma}$, $N = 5$). Therefore, these rocks formed from magma that originated from sources in the mantle without contamination by rocks of the continental crust. The dikes have higher initial $^{87}Sr/^{86}Sr$ ratios than the Windimurra gabbro (Table 7.11).

Although the porphyry and dolerite dikes formed from magmas derived either from the mantle or the lower crust, the pegmatite is clearly the product of a crustal melt. However, the granitoids that form the country rock have a surprisingly low initial $^{87}Sr/^{86}Sr$ ratio of 0.70494 ±0.0031 at 2 574 ±74 Ma relative to 0.71025 for NBS 987 (Ahmat and DeLaeter 1982). This result requires that the pegmatite magma was contaminated by selective assimilation of radiogenic ^{87}Sr from the granitic basement rocks.

7.10.2 Kalka Intrusion, Giles Complex, Central Australia

The Kalka Intrusion is one of fourteen components of the Middle Proterozoic Giles Complex in the Musgrave block of central Australia. It is composed of a basal pyroxenite overlain by layers of norite, olivine gabbro, and anorthosite whose total thickness is about 5.0 km (Goode 1977). The Kalka Intrusion formed between 1.1 and 1.2 Ga and was assigned an age of 1.15 Ga by Gray and Goode (1981). It consists of about 4 000 m of basal pyroxenite and gabbro/norite capped by more than 800 m of olivine-bearing anorthosite (Gray 1987). The anorthosite layer is separated from the underlying gabbro/norite layer by a transition zone that is about 500 m thick. Gray (1987) demonstrated an inverse correlation between the anorthite content of plagioclase and the initial $^{87}Sr/^{86}Sr$ ratios of the rocks in the Kalka Intrusion (Table 7.12).

In addition, the gabbro/norite zone contains a layer of olivine gabbro that is characterized by variable initial $^{87}Sr/^{86}Sr$ ratios between 0.70568 and 0.7091 (Gray and Goode 1981; Gray et al. 1981; Gray 1987) compared

Table 7.12. Inverse correlation between the anorthite content of plagioclase and the initial $^{87}Sr/^{86}Sr$ ratios of the rocks in the Kalka Intrusion (Gray 1987)

Section	($^{87}Sr/^{86}Sr)_0$	An content of plagioclase (%)
Anorthosite	0.7049	76
Transition	0.7055–0.7077	60–72
Gabbro/norite	0.7075	60–65

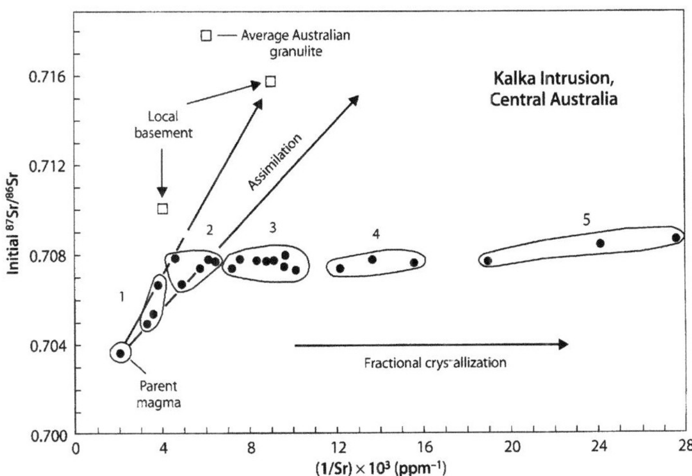

Fig. 7.23. Sr-isotope mixing diagram for the rocks of the Kalka intrusion, Giles Complex, central Australia: 1. Anorthosites; 2. Norites; 3. Olivine gabbro; 4. Melagabbro; 5. Pyroxenites. The norites and anorthosites formed from magma that had assimilated varying amounts of granitic basement rocks. The olivine gabbro and melagabbro as well as the basal pyroxenites formed by fractional crystallization of contaminated magmas (*Sources:* Gray and Goode 1981; Gray et al. 1981)

to uniformly high values in the norite layers above and below the olivine gabbro. The norites have an average initial ^{87}Sr/^{86}Sr ratio of 0.70780 ±0.00008 (2$\bar{\sigma}$, $N = 6$; Gray and Goode 1981; Gray et al. 1981) compared to 0.71025 for NBS 987.

The anorthosite layer of the Kalka Intrusion is composed of cumulate plagioclase with lesser amounts of olivine-bearing leucotroctolites, orthopyroxene-bearing leuconorites, and rare layers of magnetite and ilmenite (Goode 1977). The Kalka Intrusion is the only part of the Giles Complex that contains a complete succession of rock types from pyroxenite to anorthosite. Goode (1977) attributed the origin of the anorthosites of the Kalka Intrusion to fractional crystallization of a tholeiite basalt parent magma.

The elevated and variable initial ^{87}Sr/^{86}Sr ratios of the rocks in the Kalka Intrusion suggest that the rocks formed from magmas that had assimilated crustal rocks. This hypothesis is supported by Fig. 7.23 which confirms that the anorthosite (1) and norite (2) define mixing lines between mantle-derived magma (^{87}Sr/^{86}Sr ≈ 0.7034) and granulites of the local continental crust (Gray 1977, 1978, 1987; Gray and Compston 1978). In addition, Gray et al. (1981) reported ^{87}Sr/^{86}Sr = 0.7180 (at 1 150 Ma) and Sr = 165 ppm for high-grade metamorphic rocks from 14 localities in Australia. The distribution of data points in Fig. 7.23 suggests that olivine gabbro (3), melagabbro (4), and pyroxenites (5) formed by fractional crystallization of the contaminated norite magma (2).

The initial isotope ratios of Sr and Nd (at 1 150 Ma) of the Kalka Intrusion in Fig. 7.24 define a mixing array consisting of mantle-derived magma and granulites. Therefore, these data reinforce the evidence that the magmas of the Kalka Intrusion assimilated significant

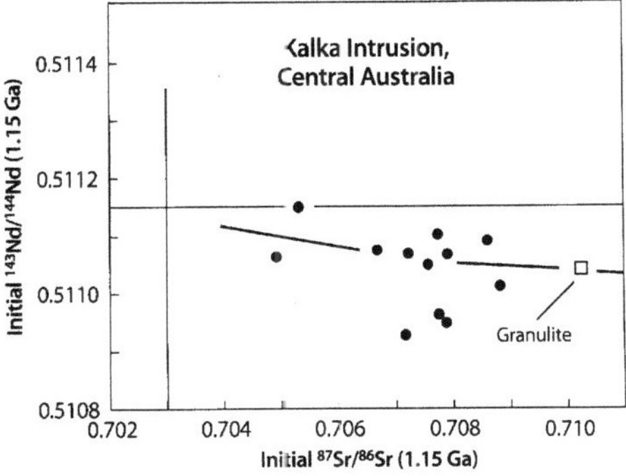

Fig. 7.24. Initial isotope ratios of gabbro, pyroxenite (websterite), and anorthosite of the Kalka Intrusion at 1 150 Ma in central Australia. The data points define a mixing line which indicates that mantle-derived basalt magmas assimilated significant amounts of the granulite and granitic gneisses of the local continental crust (*Source:* Gray et al. 1981)

quantities of crustal rocks having elevated ^{87}Sr/^{86}Sr but low ^{143}Nd/^{144}Nd ratios (Gray et al. 1981). Korsch and Gulson (1986) reached a similar conclusion from a study of the isotope ratios of Nd and Pb of the Millindinna Complex in the Pilbara block of Western Australia. They reported a Sm-Nd isochron date of 2 830 ±20 Ma (Late Archean) for this intrusion with an initial ^{143}Nd/^{144}Nd ratio of 0.50892 ±0.00002. This value is lower than the ^{143}Nd/^{144}Nd ratio of CHUR at 2.83 Ga (0.50900) which thereby supports the hypothesis that the mantle-derived magma of the Millindinna Complex also assimilated crustal rocks having low ^{143}Nd/^{144}Nd ratios.

7.11 Summary: Petrogenesis in the Precambrian

The large stratified intrusives composed of gabbros and related anorthosites and ultramafic rocks occupy former magma chambers in the continental crust. The ages of these bodies range widely from early Cenozoic (e.g. Rhum and Skaergaard Intrusions) to Late Archean (e.g. Stillwater Complex and Great Dyke). In some cases, remnants of tholeiite basalt plateaus occur nearby suggesting that the lavas originated from the magma chambers that are now represented by the stratified gabbro intrusives (e.g. Skaergaard, Rhum, Freetown, Dufek, Duluth, and Muskox Intrusions). In other cases, plateau basalts that may have been erupted at the time the intrusives were forming have been eroded (e.g. Stillwater, Bushveld, Great Dyke, Kalka, and Millindinna).

The stratification of the rocks is the result of fractional crystallization of mafic magmas and the deposition of early-formed crystals of olivine and pyroxene. In addition, the magma chambers were periodically refilled by magmas that originated from reservoirs at depth. The intrusion of new magma caused the reservoirs to overflow by eruption of lava at the surface. In other words, episodic volcanic activity at the surface was caused by intrusion of batches of new magma into magma chambers at depth in the crust.

The new magmas differed from the resident magmas in density, chemical composition, isotope composition, and temperature. In cases where the new magma was denser than the resident magma, it spread along the bottom of the magma chamber without mixing extensively with the resident magma. In other cases, the two magmas mixed to varying degrees thereby forming hybrid magmas which ranged from being homogeneous in chemical and isotope composition to being heterogeneous. In any case, the intrusion of new magma into a partially consolidated chamber caused a discontinuity in the progress of fractional crystallization.

The isotope compositions of Sr, Nd, and Pb of the gabbros as well as their trace element compositions support the view that the magmas originated in the subcontinental mantle. In most cases, the magmas formed by decompression melting associated with rifting of the continental crust and underlying lithospheric mantle (e.g. Skaergaard Intrusion, Duluth gabbro, Muskox Intrusion and Mackenzie Dikes, Great Dyke, and others). Such rifts can form in continental back-arc basins associated with subduction zones and by the stresses applied to the base of the lithospheric mantle by plumes that originated from the asthenospheric mantle. The plume heads not only applied divergent stresses on the lithosphere but also added heat by conduction and by advection in the form of magma that originated by partial melting in the plume head.

The rocks of stratified gabbro intrusives have a range of initial isotope ratios of Sr, Nd, and Pb because of contamination of their magmas under a variety of circumstances:

1. Contamination at the source by partial melting of mixed source rocks consisting of asthenospheric mantle and metasomatically altered lower lithospheric mantle.
2. Assimilation of old granulites and granitic gneisses of the lower continental crust before the magmas entered upper-crustal chambers where they solidified by fractional crystallization.
3. Interaction of the cooling magma with the wallrock of the magma chamber by melting and mixing and by diffusion of ions from the crustal melt into Sr-depleted residual magmas at or near the top of the magma chamber.
4. Migration of contaminated residual magma through the mush of early-formed crystals leading to crystallization of interstitial minerals whose isotope compositions of Sr, Nd, and Pb differed from those of the early-formed crystals.
5. Transfer of alkali metals, Sr, and Pb from the local country rocks into the border zone of stratified gabbro intrusives by convection of heated groundwater.
6. Granophyres which may occur near the tops of stratified gabbro intrusives are, in most cases, crustal melts that may have mixed partly with the underlying mafic magma.

The processes listed above have contributed to the origin and evolution of stratified gabbro intrusives of all ages ranging in age from Cenozoic to Archean. The only difference is that the isotope compositions of Sr and Pb in some Archean intrusives were altered by regional metamorphism long after they crystallized. Under these circumstances, the isotope composition of Nd is more resistant to alteration than those of Sr and Pb. Therefore, the Sm-Nd method of dating in some cases yields older (and more reliable) age determinations of mafic rocks of Precambrian age than the K-Ar, Rb-Sr, or the U-Pb method, except when the latter is applied to U-rich accessory minerals which are resistant to alteration (e.g. zircon, rutile, baddeleyite, etc.).

The contamination of large volumes of basalt magma at the source or by assimilation of crustal rocks, or both, limits the reliability of the conclusions that can be derived from the initial isotope ratios of Sr, Nd, and Pb of stratified gabbro intrusives about the evolution of their magma sources in the subcrustal mantle. The petrogenesis of stratified gabbro intrusives reinforces the conclusion that the mantle has always been the driving force of the tectonic and magmatic activity to which the continental crust of the Earth has been subjected.

Chapter 8

Archean Greenstone Belts and Granitic Gneisses of North America

The Precambrian shields of all continents contain belts of folded metavolcanic and metasedimentary rocks surrounded by granitic gneisses. These so-called greenstone belts are the remnants of large volcano-sedimentary complexes of Archean and Proterozoic age. The volcanic rocks range in composition from komatiites and low-K tholeiites to dacites and rhyolites and consist of lava flows, sills, and stratified tuffs. The sedimentary rocks are typically composed of greywacke, quartzite, argillites, and cherty iron formations with uncommon carbonate units (de Wit and Ashwal 1997; Goodwin 1996).

The granitic rocks associated with greenstone belts are composed of quartz diorites (tonalites and trondhjemites) and granodiorites (Barker 1979; Glickson 1979). Tonalites and trondhjemites are typically deficient in K-feldspar, but their concentrations of SiO_2 identify them as acidic rather than intermediate in composition. According to Larsen (1948), the tonalites in the batholith of southern California are composed of andesine (55 to 60%, quartz 20 to 25%), hornblende (10%), biotite (10%), and orthoclase (<5%). Trondhjemites are biotite diorites composed of Na-rich plagioclase (oligoclase), biotite, and quartz with minor hornblende, Fe-Ti oxides, and are virtually lacking in K-feldspar.

The volcano-sedimentary complexes and associated granitic rocks which formed during successive orogenic cycles during the Precambrian contributed to the growth of the continental crust as we know it today. The process of crustal growth continues to be the subject of a vigorous debate. A sampling of the extensive literature includes contributions by Hurley et al. (1962b), Goodwin (1968a,b), Windley (1976, 1977), Moorbath (1975a,b, 1976, 1978), Allègre and Ben Othman (1980), DePaolo (1983b), Taylor et al. (1984), Shaw et al. (1986), Armstrong (1991), Whitford and Kinny (1991), Glickson (1995), Rudnick and Fountain (1995), and Nelson (1998).

Geological mapping of the Precambrian shield areas of North America progressed slowly during the 20th century and in a piecemeal fashion. This work revealed a structural complexity that initially defied explanation until the geologic maps of individual townships and counties were combined to represent large regions that had experienced similar tectonic processes related to closure of marine basins and collisions of microcontinents. The ultimate success in the reconstruction of the tectonic histories and associated magmatic activity of Precambrian shield areas is attributable to precise age determinations based primarily on U-Pb concordia dating of U-bearing refractory minerals (e. g. zircon, rutile, sphene, baddeleyite, etc.) and to the applications of the principles of plate tectonics.

The first step in the study of greenstone belts is to determine the ages and initial $^{87}Sr/^{86}Sr$ ratios of the metavolcanic and granitic rocks (Hart et al. 1970a; Brooks et al. 1969a; Hart and Brocks 1977). This has been a difficult task because the Rb-Sr systematics of the rocks were disturbed during structural deformation and metamorphism. Therefore, much information about the magma sources of Precambrian volcanic rocks has been lost. In addition, the differences in the isotope ratios of Sr, Nd, and Pb in the lithospheric mantle and in the continental crust were smaller during the Archean Eon than they are at present. Therefore, assimilation of crustal rocks by mantle-derived magmas in Archean time did not change their isotope ratios of Sr, Nd, and Pb as much as it did during Phanerozoic time. As a result, the initial $^{87}Sr/^{86}Sr$ and $^{143}Nd/^{144}Nd$ ratios of granitic rocks of Archean age are similar in many cases to those of the associated metavolcanic rocks.

8.1 Greenstone Belts of the Superior Craton

The greenstone belts of the Canadian Precambrian shield (Card 1990) have been studied for many decades not only because they were thought to contain ancient rocks that formed at the beginning of geologic time, but also because they contain large deposits of metals (Cu, Ni, Pb, and Zn) as well as gold-bearing quartz veins. The ore deposits of the greenstone belts have been and continue to be one of the principal natural assets of Canada. Therefore, the economic importance of greenstone belts has justified the effort to understand their origin beginning with the classical studies of A. C. Lawson of the greenstone belt of the Rainy Lake district of Ontario.

8.1.1 Rainy Lake, Northwestern Ontario

The region along the border between northeastern Minnesota and northwestern Ontario was studied intensively by Lawson (1885, 1888, 1913) in an effort to identify the granite basement or "fundamental gneiss" on which subsequent volcanic and sedimentary rocks were thought to have been deposited. He recognized the Keewatin metavolcanics and the associated (but older) metasedimentary rocks of the Coutchiching Series, as well as the younger Knife Lake (Seine River) Series. Lawson concluded that the "Laurentian granites" upon which the sedimentary and volcanic rocks were supposed to have been deposited actually intrude the Keewatin Series and are therefore younger than the metavolcanic and metasedimentary rocks. In addition, Lawson demonstrated that the metasedimentary rocks of the Knife Lake Series were deposited unconformably on the Keewatin volcanics, that a conglomerate at the base of the Knife Lake Series contains boulders of Laurentian granite, and that the metasedimentary rocks of the Knife Lake Series were themselves intruded by younger "Algoman" granites.

Therefore, the oldest Precambrian rocks known at the time appeared to record a lengthy history of geological activity consisting of two tectono-magmatic cycles illustrated in Fig. 8.1. The first commenced with deposition of the sedimentary rocks of the Coutchiching Series followed by the volcanic rocks (basalt to rhyolite) of the Keewatin Series. The Coutchiching and Keewatin rocks were folded, metamorphosed, and intruded by granite plutons during the Laurentian orogeny. After uplift and erosion of the Laurentian complex, sedimentary rocks of the Knife Lake (Seine River) Series were deposited, folded, metamorphosed, and in-

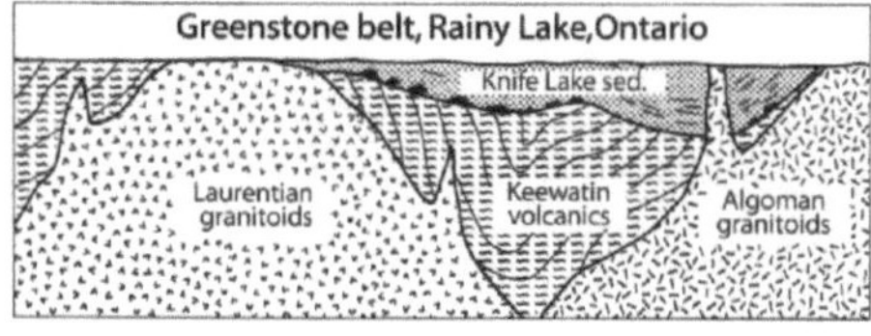

Fig. 8.1. Schematic representation of the geologic history of the Rainy Lake area, northwestern Ontario. The Keewatin metavolcanics and underlying Coutchiching metasediments were folded, metamorphosed, and intruded by the Laurentian "granite." After uplift and erosion, the sedimentary rocks of the Knife Lake series were deposited unconformably on the erosion surface with a basal conglomerate containing pebbles of Laurentian granite. The Knife Lake metasediments were subsequently folded and intruded by the younger suite of Algoman granites. Age determinations by Hart and Davis (1969) of zircons from the major rock units indicate that both geological cycles were completed in a surprisingly short interval of time of about 50 million years at about 2.70 Ga (adapted from Dunbar 1949)

truded by granite plutons during the Algoman orogeny. The Algoman orogeny of Lawson (1913) was later renamed the Kenoran orogeny by Stockwell (1982).

The controversies that arose from Lawson's work in the Rainy Lake area of northwestern Ontario later helped to motivate an extensive program of age determinations led by S. S. Goldich at the University of Minnesota (Goldich et al. 1961, 1970; Goldich 1968; Hanson et al. 1971; Sims and Morey 1972; Peterman et al. 1972) and subsequently continued by Jahn et al. (1974), Arth and Hanson (1975), and Jahn and Murthy (1975). At about the same time, R. K. Wanless of the Geological Survey of Canada started a research program to date the Precambrian rocks of the Canadian Shield (e.g. Lowdon 1960; Wanless and Loveridge 1972; Stockwell 1982).

The results of early geochronological studies in the Minnesota-Ontario border area were reviewed and reinterpreted by Hart and Davis (1969) who also dated zircons extracted from most of the major rock units in the Rainy Lake area. They reported that all of these zircons define a single discordia line on the concordia diagram (Sect. 1.8.1) yielding a date of 2.75 ±0.03 Ga. This result indicates that the zircons from different rock units, thought to represent two cycles of geologic activity, apparently formed within a time interval that is too short to resolve by the U-Pb method of dating. Alternatively, the zircons may have suffered complete loss of radiogenic Pb at 2750 Ma, or the younger rocks may have inherited zircons from the older rocks. Although Hart and Davis (1969) did not entirely rule out these alternatives, they favored the straightforward interpretation that all of the rocks in the Rainy Lake area had formed within a short time interval.

Whole-rock Rb-Sr isochron dates reported by Hart and Davis (1969) for rocks of the Laurentian cycle ($\lambda = 1.42 \times 10^{-11}$ yr^{-1}) are younger than the U-Pb concordia date of the zircons and yielded low initial ^{87}Sr/^{86}Sr ratios relative to 0.7080 for E&A. The results of this study confirmed earlier conclusions of Goldich et al. (1961) who had reduced the significance of the "Laurentian orogeny" and relegated the Coutchiching to a minor sedimentary sequence interbedded with the Keewatin lavas. Apparently, the Laurentian granites (Bad Vermilion and Saganaga Lakes in northwestern Ontario) were intruded into the sequence of Keewatin lava flows and Coutchiching interflow sediments soon after their deposition. In addition, Hart and Davis (1969) concluded that the Coutchiching sediments had a local volcanic source and contain no evidence for the existence of an older sialic crust in this area. Arth and Hanson (1975) later came to a similar conclusion based on a trace-element study of the major rock units.

Peterman et al. (1972) subsequently confirmed that zircons from the major stratigraphic units (Coutchiching to Algoman) crystallized at 2730 ±30 Ma and that the Rb-Sr whole-rock isochron dates ($\lambda = 1.42 \times 10^{-11}$ yr^{-1})

of the Coutchiching metasediments (2560 ±50 Ma), of the Keewatin metavolcanics (2540 ±45 Ma), and of the Algoman granites (2486 ±90 Ma) in the Rainy Lake area of Ontario are all younger than the U-Pb age of the zircons in these rocks. The lowering of the Rb-Sr dates of all of the rocks in the Rainy Lake area of Ontario was attributed to loss of ^{87}Sr during metamorphism and subsequent uplift of these rocks. In contrast to the Rb-Sr dates of Hart and Davis (1969) and Peterman et al. (1972), the metavolcanic, metasedimentary, and plutonic rocks of the Rainy Lake area analyzed by Shirey and Hanson (1984) yielded a Sm-Nd isochron date of 2737 ±42 Ma, thereby confirming the U-Pb zircon dates previously reported for rocks of this area (Goldich and Fischer 1986). The concordance of U-Pb and Sm-Nd dates is good evidence that the Sm-Nd isochron date of Shirey and Hanson (1984) represents the time of crystalization of these rocks rather than the time of subsequent alteration during the Kenoran (Algoman) orogeny.

A second demonstration of the chronology of the so-called Laurentian and Algoman granites was provided by the granitic plutons along the Ontario-Minnesota border. The Saganaga tonalite of this area is one of the Laurentian "granites" because it intrudes folded Keewatin metavolcanic rocks and the associated Northern Light gneiss. The Saganaga tonalite as well as the Northern Light gneiss are unconformably overlain by metasedimetary rocks of the Knife Lake Group. The younger Algoman granites of this area are represented by the Icarus Pluton which intruded both the Saganaga tonalite and the Northern Light gneiss.

In spite of the fact that the Saganaga tonalite and the Icarus Pluton intruded rocks that formed in separate cycles of volcanic activity and sedimentation followed by folding and uplift, their Rb-Sr dates and initial ^{87}Sr/^{86}Sr ratios are indistinguishable (Hanson et al. 1971): Northern Light: 2682 ±100 Ma, Saganaga: 2653 ±560 Ma, and Icarus: 2633 ±480 Ma. The Rb-Sr dates have large uncertainties in this case because the Rb/Sr ratios of these rocks are low and have only a narrow range, but the initial ^{87}Sr/^{86}Sr ratios are precisely determined by the data of Hanson et al. (1971).

The geologic history of the Archean rocks of the Rainy Lake district was most clearly revealed by the U-Pb zircon dates of Davis et al. (1989). Their results demonstrate that the volcanic and associated plutonic rocks formed in less than 10 million years between 2728 and 2725 Ma. They were succeeded by the sedimentary rocks of the Coutchiching Group deposited between 2704 ±3 and 2692 ±5 Ma. Since the Coutchiching metasedimentary rocks are overlain by Keewatin greenstones, Davis et al. (1989) concluded that the Coutchiching rocks are allochthonous and were tectonically emplaced beneath the Keewatin Greenstones during compressive tectonic activity associated with the colli-

sion of plates. Since the U-Pb dates of detrital zircons of the Coutchiching Group range up to 3059 ±3 Ma, these zircons originated from crustal rocks that predated the eruption of the Keewatin volcanics and were a source of sediment for the Coutchiching Series. The low initial ^{87}Sr/^{86}Sr ratios of Coutchiching rocks reported by Hart and Davis (1969) and Peterman et al. (1972) did not reveal the existence of such a source.

The sedimentary rocks of the Seine River (Knife Lake) Group were deposited between 2696 and 2686 Ma. Although this time interval coincides partly with the deposition of the Coutchiching sediments, the Seine River Group is younger than the Coutchiching Series on structural grounds.

The significance of these results arises from the recognition that the rocks along the Minnesota-Ontario border formed during a relatively short interval of time by vigorous geological activity involving eruption of volcanic rocks (basalt to rhyolite), erosion and sedimentation, folding, and intrusion of granitic magmas. The rocks formed in a continuous outburst of activity that ended with the intrusion of the Algoman granites. The apparent vigor of the geological activity and the close

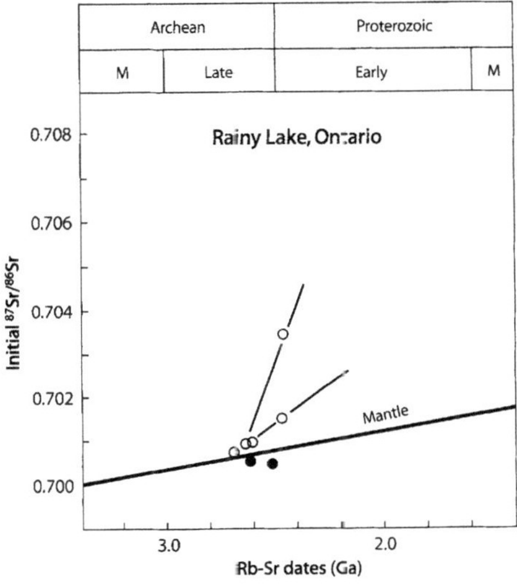

Fig. 8.2. Initial ^{87}Sr/^{86}Sr ratios of Keewatin Greenstones (*solid circles*) and of granitic intrusives (*open circles*) in the Rainy Lake area of northwestern Ontario, Canada. Error bars were omitted to avoid crowding. The mantle-Sr evolution line extends from 0.699 at 4.5 Ga to 0.7030 at 0 Ga. The elevated initial ^{87}Sr/^{86}Sr ratios of the Bad Vermilion and Saganaga intrusives (0.7034 ±0.0019; Hart and Davis 1969) and of Algoman granites (0.7015 ±0.0009; Peterman et al. 1972) were caused by metamorphic re-equilibration of Sr isotope compositions. Therefore, the initial ratios of the greenstones and associated granites are consistent with the isotope composition of Sr in magma sources in the mantle (*Sources:* Hart and Davis 1969; Peterman et al. 1972; Hanson et al. 1971)

association of surficial rocks (volcanic and sedimentary) with granitic plutons is a general characteristic of Archean greenstone belts.

In conclusion, since the Rb-Sr dates of these rocks were reset during low-grade regional metamorphism, the initial $^{87}Sr/^{86}Sr$ ratios derived from whole-rock Rb-Sr isochrons are higher than they were at the time of crystallization. In spite of this prediction, the initial $^{87}Sr/^{86}Sr$ of the Keewatin metavolcanic rocks reported by Hart and Davis (1969) and Peterman et al. (1972) fit the mantle-Sr evolution line (0.699 at 4.5 Ga to 0.7030 at 0 Ga) in Fig. 8.2 within the uncertainties of the measurements. In addition, the granitic rocks analyzed by Hanson et al. (1971) (e.g. Northern Light, Saganaga, and Icarus) all have low initial $^{87}Sr/^{86}Sr$ ratios and also plot on the mantle-Sr evolution line. However, the Bad Vermilion and Saganaga intrusives dated by Hart and Davis (1969) and the Algoman granites dated by Peterman et al. (1972) do have anomalously high initial $^{87}Sr/^{86}Sr$ ratios. The Rb-Sr dates of these granitic rocks were reset by rehomogenization of the isotope composition of Sr. The key evidence requiring this conclusion is that the U-Pb dates of the granitic intrusives (about 2.75 Ga) are older than the Rb-Sr dates, which means that the $^{87}Sr/^{86}Sr$ ratios of the granitic rocks increased by in situ decay of ^{87}Rb between the time of original crystallization (U-Pb date) and subsequent metamorphism (Rb-Sr date).

extends from Saganaga Lake on the Minnesota-Ontario border for more than 150 km in a southwesterly direction to Lake Vermilion in Minnesota. It is bordered on the north by the Vermilion granite (dated by Peterman et al. 1972) and in the south by the Giants Range granite. The stratified rocks of the Vermilion Greenstone Belt of Minnesota include the Ely Greenstone, Soudan iron formation, Knife Lake Group, Lake Vermilion Formation, Newton Lake Formation, and the Northern Light gneiss. Jahn et al. (1974) concluded on the basis of trace-element concentrations that the metavolcanic rocks of the Vermilion Greenstone Belt formed in an island arc setting.

Age determinations by the whole-rock Rb-Sr method show the same similarity of initial $^{87}Sr/^{86}Sr$ ratios of mafic volcanic and granitic plutonic rocks already noted in the Rainy Lake area. Therefore, both were presumably affected by metamorphism at the time indicated by the most precisely determined Rb-Sr date (e.g. Vermilion granite, 2 640 ±50 Ma, Jahn and Murthy 1975). In spite of the probable increases of the initial $^{87}Sr/^{86}Sr$ ratios caused by the resetting of Rb-Sr dates, the data points representing these rocks fit the mantle-Sr evolution line in Fig. 8.4. The Rb-Sr dates and initial $^{87}Sr/^{86}Sr$ dates of the volcano-sedimentary rocks and of the associated granitic intrusives in the Vermilion Greenstone Belt of Minnesota as reported by (1) Jahn and

8.1.2 Vermilion Greenstone Belt, Minnesota

The metavolcanic rocks of the Vermilion Greenstone Belt of northeastern Minnesota are equivalent in age and chemical composition to the Keewatin metavolcanic rocks of northwestern Ontario. The belt in Fig. 8.3

Fig. 8.3. Major units of the Precambrian rocks of northeastern Minnesota, USA. *1.* Saganaga tonalite (Laurentian); *2.* Vermilion granite (Algoman); *3.* Giants Range granite (Algoman); *4.* Duluth gabbro complex (1.05 Ga, Sect. 7.2); *5.* Biwabik iron Formation. The Vermilion Greenstone Belt includes the Ely Greenstone, Soudan iron formation, Knife Lake group, Lake Vermilion Formation, and the Newton Lake Formation (adapted from Jahn et al. 1974)

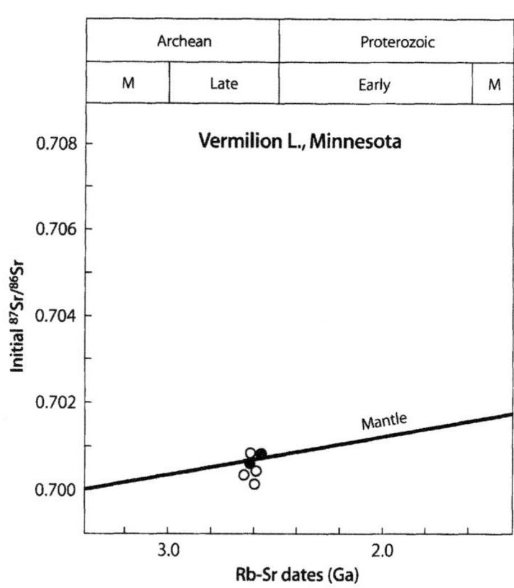

Fig. 8.4. Initial $^{87}Sr/^{86}Sr$ ratios and Rb-Sr whole-rock isochron dates of greenstones (*solid circles*) and intrusive granites in the Vermilion district of northeastern Minnesota, USA. The Sr in these rocks is consistent with the $^{87}Sr/^{86}Sr$ ratios of magma sources in the mantle (*Sources:* Peterman et al. (1972); Prince and Hanson 1972; Jahn et al. 1974; Jahn and Murthy 1975)

Table 8.1.
Rb-Sr dates and initial $^{87}Sr/^{86}Sr$ ratios of the volcano-sedimentary rocks and of the associated granitic intrusives in the Vermilion Greenstone Belt of Minnesota relative to 0.71025 for NBS 987 and 0.7080 for E&A: (1) Jahr and Murthy 1975, (2) Peterman et al. 1972, (3) Prince and Hanson 1972

Unit	Rb-Sr date (Ma)	Initial $^{87}Sr/^{86}Sr$
Volcano-sedimentary complex		
Ely greenstone (1)[a]	2 630 ±80	0.70056 ±0.00026
Newton Lake (1)	2 590 ±110	0.70086 ±0.00024
Granitic intrusives		
Vermilion (2)	2 623 ±95	0.70050 ±0.00012
Vermilion (1)	2 640 ±50	0.70040 ±0.0003
Giants Range (3)	2 614 ±65	0.70020 ±0.0019
Linden syenite (3)	2 682	0.70090 ±0.0004
Granite pebbles (1), Ely and Newton Lake	2 630 ±280	0.70080 ±0.0006

[a] The reference is indicated in parentheses.

Murthy (1975), (2) Peterman et al. (1972), and (3) Prince and Hanson (1972) relative to 0.71025 for NBS 987 and c.7080 for E&A are shown in Table 8.1.

The low initial $^{87}Sr/^{86}Sr$ ratios of the granitic magmas of the Vermilion Greenstone Belt indicate that these rocks did not form by partial melting of continental crust having a lengthy history preceding the eruption of the mafic lavas. Instead, the isotopic data indicate that the felsic volcanic rocks and the granitic intrusives both formed from partial melts of quartz eclogite or garnetiferous amphibolite representing basalt and gabbro at the base of the volcanic pile (Hanson and Goldich 1972; Arth and Hanson 1975; Shirey and Hanson 1984; Shirey 1986).

One of the questions remaining to be answered concerns the existence of Early Archean continental crust in this area. Although there is no evidence in the rocks for the existence of an older continental crust, the Morton gneiss in southwestern Minnesota does contain zircons that crystallized at 3 550 Ma (Goldich et al. 1970). These rocks were subsequently metamorphosed at 2 650 Ma during the Kenoran (Algoman) orogeny when the rocks lost radiogenic ^{87}Sr, and again at 1 850 Ma when the K-Ar and Rb-Sr systematics of biotite were reset. Therefore, the Morton gneiss in the Minnesota River valley represents the continental crust at the time the greenstone sequences of the Minnesota-Ontario border were deposited. Goldich et al. (1970) concluded that the Morton and Montevideo gneisses are not some kind of protocrust, but actually formed as a result of an earlier cycle of geological activity that ended with the intrusion of the gneisses during the "Mortonian event" at 3 550 Ma. Additional evidence for "pre-Laurentian" geological activity comes from the Lumby Lake Greenstone Belt in the Wabigoon subprovince of northwestern Ontario where Davis and Jackson (1988) reported a concordant U-Pb date of 2 999 ±1 Ma for felsic metavolcanic rocks and 3 003 ±5 Ma for the underlying Marmion Lake Batholith.

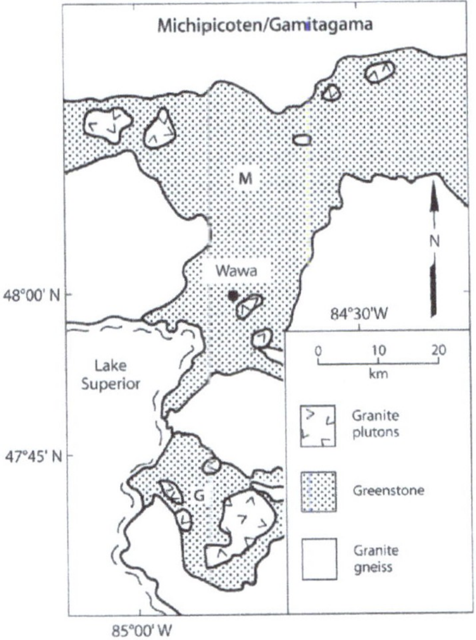

Fig. 8.5. Michipicoten (*M*) and Gamitagama (*G*) Greenstone Belts along the eastern shore of Lake Superior, Ontario, Canada (adapted from Turek et al. 1984)

8.1.3 Michipicoten and Gamitagama Greenstone Belts, Ontario

The Michipicoten Greenstone Belt in Fig. 8.5 extends from the town of Wawa on the eastern shore of Lake Superior toward the northeast for at least 120 km. The Gamitagama Greenstone Belt is located about 50 km

Table 8.2.
Whole-rock Rb-Sr dates and initial $^{87}Sr/^{86}Sr$ ratios of the volcano-sedimentary rocks and of the associated granitic plutons of the Michipicoten and Gamitagama Greenstone Belts relative to 0.71025 for NBS 987 and 0.7080 for E&A

Unit	Rb-Sr (Ma)	$(^{87}Sr/^{86}Sr)_0$	U-Pb (zircon) (Ma)
Metavolcanic rocks			
Lower Series	2618 ±90 (3)[a]	0.7019 ±0.0014 (4)	2747 ±3 (2)
Middle Series	–	–	2713 ±3 (1)
Upper Series	2496 ±100 (1)	0.7022 ±0.0003 (1)	2695 ±3 (2)
Iron formation	–	–	2744 (Sm-Nd)
Granite rocks			
Basement	–	–	2888 ±2 (1)
Intrusives	2597 ±39 (6)	0.7025 ±0.0013 (6)	2717 ±38 (4)

[a] The number of determinations included in the averages is indicated in parentheses.

south of Wawa and is separated from the Michipicoten Greenstone belt only by about 5 km of granitic rocks. Therefore, the metavolcanic rocks of the Gamitagama Belt may be correlative with those of the Michipicoten Belt (Turek et al. 1984). The metavolcanic and metasedimentary rocks (including iron formations) of the Michipicoten Greenstone Belt were originally mapped and described by Goodwin (1962) who divided the metavolcanic rocks into Upper, Middle, and Lower Series. The volcano-sedimentary complexes were intruded by "internal" granitic plutons and are in contact with "external" granitic gneisses upon which the volcanic and sedimentary rocks of the greenstone belts may have been deposited. The geological history of the Michipicoten Greenstone Belt was worked out in detail by Turek et al. (1982, 1984) based on U-Pb zircon dates because whole-rock Rb-Sr dates of the metavolcanic and granitic rocks are lower in all cases than the U-Pb zircon dates of the same rocks.

The whole-rock Rb-Sr dates and initial $^{87}Sr/^{86}Sr$ ratios of the volcano-sedimentary rocks and of the associated granitic plutons of the Michipicoten and Gamitagama Greenstone Belts were reported by Brooks et al. (1969b), Hart and Brooks (1977), Turek et al. (1981), Turek et al. (1982), and Jacobsen and Pimentel-Klose (1988) relative to 0.71025 for NBS 987 and 0.7080 for E&A. The results yield the average values shown in Table 8.2.

The U-Pb zircon dates of the external granitic basement rocks reported by Turek et al. (1984) range from 2 888 ±2 Ma (biotite granite) to 2 662 ±5 Ma (granite gneiss). The older of these dates is the earliest recorded evidence of plutonic activity in the area. Therefore, the biotite granite (2 888 Ma) is part of the basement upon which the volcanic rocks were subsequently deposited. The Lower Metavolcanic Series was extruded at about 2 749 ±2 Ma based on a U-Pb date of zircons extracted from dacite in the Helen Mine near Wawa (Turek et al. 1982). A second U-Pb date of 2 744 ±10 Ma for zircons in a quartz-feldspar tuff is consistent with that date.

The Middle Series of metavolcanic rocks was only dated in the Gamitagama Lake area where Krogh and Turek (1982) obtained a U-Pb date of 2 713 ±3 Ma for zircons extracted from felsic volcanics. The Upper Series was extruded at about 2 696 ±2 Ma (U-Pb, zircon, Catfish Lake) as reported by Turek et al. (1982). This date was subsequently confirmed by Turek et al. (1984) with a U-Pb zircon date of 2 698 ±11 Ma for fragmental volcanic rocks.

According to the U-Pb zircon dates, the volcanic eruptions in the Michipicoten and Gamitagama Greenstone Belts occurred during an interval of about 60 million years from 2 749 to 2 696. The iron formation, which is an important economic resource of the Michipicoten area, was also deposited within this time interval at about 2 744 Ma based on a whole-rock Sm-Nd date reported by Jacobsen and Pimentel-Klose (1988). In addition, several internal plutons of the Michipicoten area dated by the U-Pb zircon method (Turek et al. 1982, 1984) were intruded while volcanic activity was still in progress. Even some of the external granites dated by Turek et al. (1982, 1984) with U-Pb zircon ages between 2 747 ±6 Ma and 2 662 ±5 Ma were intruded during and after the period of active volcanism and therefore are not part of the granitic basement rocks.

The initial $^{87}Sr/^{86}Sr$ ratios of the granitic rocks of the Michipicoten/Gamitagama area have an average value of 0.7025 ±0.0013 ($2\bar{\sigma}$, $N = 6$). The initial $^{87}Sr/^{86}Sr$ ratios of the Lower and Upper Series of metavolcanic rocks at Michipicoten reach an extreme value of 0.72765 ±0.0052 in the Wawa tuff. However, the average initial $^{87}Sr/^{86}Sr$ ratio of the Lower Series of metavolcanic rocks of the Michipicoten Greenstone Belt is 0.7019 ±0.0014 ($2\bar{\sigma}$, $N = 3$) and their average Rb-Sr isochron date is 2 618 ±90 Ma ($2\bar{\sigma}$, $N = 4$). The initial $^{87}Sr/^{86}Sr$ ratio of the Upper Series is 0.7022 ±0.0003 and its Rb-Sr age is 2 496 ±100 Ma. The U-Pb zircon dates of the metavolcanic rocks in the Michipicoten area are 2 747 ±3 Ma (Lower Series), 2 713 ±3 Ma (Middle Series) and 2 695 ±3 Ma (Upper Series). In each case, the whole-

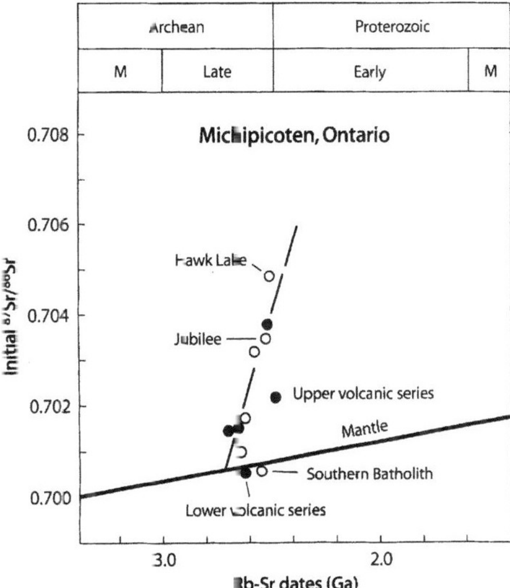

Fig. 8.6. Initial $^{87}Sr/^{86}Sr$ ratios and Rb-Sr whole-rock isochron dates of metavolcanic and granitic rocks in the Michipicoten (*solid circles*) and Gamitagama Greenstone Belt (*open circles*) near Wawa northern Ontario, Canada. The data points define a Sr-isotope evolution line that intersects the mantle-Sr development line at about 2740 Ma. The slope of the evolution line corresponds to a Rb/Sr ratio of 0.48 (*Sources:* Brooks et al. 1969b; Hart and Brooks 1977; Turek et al. 1981, 1982)

rock Rb-Sr isochron dates of the metavolcanic and granitic rocks are less than the U-Pb concordia dates of zircons extracted from the same rocks. The average difference is about 122 million years.

The initial $^{87}Sr/^{86}Sr$ ratios and Rb-Sr isochron dates of the metavolcanic and granitic rocks in the Michipicoten/Gamitagama Greenstone Belts in Fig. 8.6 define a Sr-isotope evolution line whose slope yields an average Rb/Sr ratio of 0.48 and which intersects the mantle-Sr development line at an age of 2740 Ma. The history of magmatic activity and regional metamorphism indicated by these results can be summarized as follows:

1. Basalt magma, formed in the lithospheric mantle at about 2740 Ma (Late Archean), was erupted at the surface to form a large volume of mafic volcanic rocks.

2. The resulting volcano-sedimentary complex also contained rocks of intermediate and felsic composition which originated by partial melting of the mafic mantle-derived igneous rocks at depth.

3. During the time period between 2740 and 2550 Ma, the $^{87}Sr/^{86}Sr$ ratios of volcano-sedimentary complex increased with time at a rate controlled by its average Rb/Sr ratio (0.48).

4. The heat emanating from the voluminous mafic magmas caused partial melting within the volcano-sedimentary complex to form felsic magmas which crystallized as granitic plutons within the complex.

The sequence of events outlined above is consistent with the tectonic evolution of a subduction zone along the compressive margin of a microcontinent. Therefore, the greenstone belt of the Michipicoten/Gamitagama area probably formed during subduction of an oceanic plate leading to structural deformation and regional metamorphism of the volcano-sedimentary complex. The resulting greenstone belt and associated granitic plutons were fused to the older continental nucleus, thereby causing the continental crust to increase in volume and surface area.

The rocks of the Michipicoten/Gamitagama area have remained stable since uplift and cooling in Late Archean-Early Proterozoic time. However, this part of the Superior Craton was later rifted during the Late Proterozoic Era resulting in the eruption of tholeiite basalt along the Midcontinental Rift system mentioned in Sect. 7.1. Similar tectonic histories apply to the metavolcanic rocks and associated sedimentary rocks near Rainy Lake, Ontario, and Vermilion Lake, Minnesota.

8.1.4 Komatiites

The Late Archean greenstone belts of the Superior tectonic province of Canada contain a variety of Mg-rich basalt called komatiite. This type of basalt occurs almost exclusively in volcano-sedimentary complexes of Archean age all over the Earth, but has also been discovered on Gorgona Island off the west coast of Colombia (Sect. 3.11.1) and on northeastern Newfoundland (Gale 1973).

Komatiites were first identified in the Komati Formation at the base of the Early Archean Onverwacht Group in the Barberton Greenstone Belt of southern Africa by Viljoen and Viljoen (1969) who proposed the name komatiite for these rocks and recognized peridotitic and basaltic varieties. The characteristic chemical composition of basaltic komatiite includes:

- SiO_2 = 50.77 to 53.90%
- Na_2O = 0.42 to 2.71%
- MgO = 10.33 to 21.86%
- K_2O = 0.06 to 0.47%
- Al_2O_3 = 5.49 to 10.05%
- CaO/Al_2O_3 = 1.02 to 2.38
- CaO = 8.91 to 13.07%

In some cases, the high CaO/Al_2O_3 ratios may be the result of Al_2O_3 loss during metamorphism and therefore not all komatiites share this feature (Nesbitt and

Sun 1976). Rocks fitting the description of komatiites are known from the Precambrian shield areas of Australia, (Williams 1972), Canada, (Arndt et al. 1977), India (Viswanathan 1974), and Zimbabwe (Bickle et al. 1975; Hawkesworth and O'Nions 1977). The occurrence and composition of komatiites was the subject of a book edited by Arndt and Nisbet (1982).

Ultramafic and mafic komatiite lava flows characteristically have spinifex texture in which "large plates, needles or complex skeletal grains of olivine or clinopyroxene are oriented randomly or parallel to one another in matrices of fine skeletal grains of clinopyroxene, and devitrified glass" (Arndt et al. 1977, p. 324). This texture results from quenching of Mg-rich silicate liquids (Donaldson 1982).

Brooks and Hart (1972) reported the occurrence of komatiite in the Michipicoten volcanic belt near Wawa, Ontario. In this area, the komatiite lava flows are pillowed and are associated with basalt, andesite, dacite, and rhyolite of Late Archean age that are similar in chemical composition to modern island-arc volcanic sequences (Chap. 3). Therefore, Brooks and Hart (1972) concluded that basaltic and peridotitic komatiite flows of Archean age were extruded in island arcs. However, komatiites are also known from other tectonic settings including continental rifts and mantle plumes.

The major and trace-element compositions of komatiites indicate that they formed by high degrees of partial melting in the mantle (Ohtani 1990) and were extruded at a temperature of about 1600 °C (Nisbet et al. 1993). The high eruption temperature of komatiite flows has been attributed to magma formation in plumes that ascended from the core-mantle boundary (Campbell et al. 1990; Herzberg 1992). The large degrees of partial melting implied by the chemical composition of some kinds of komatiites (Ohtani 1990; Jochum et al. 1991) and their elevated eruption temperatures are consistent with the high geothermal gradients in the mantle during the Archean Eon and may explain why these kinds of rocks rarely occur in lava flows of Phanerozoic age.

Recent work by Herzberg (1992) indicates that high-Al komatiite magmas formed by partial melting at depths greater than 300 km, whereas low-Al komatiite magmas formed at shallower depths. Similarly, Xie et al. (1993) discussed the origin of komatiite-basalt sequences in the Abitibi Greenstone Belt of Ontario and Quebec in terms of partial melting at different depths within rising mantle plumes. The intensive study of komatiites has been motivated by the expectation that their chemical composition approaches that of their source rocks in the mantle because of the high degree of melting required to produce them. However, the isotope composition of Sr in komatiites is susceptible to alteration because of their low average Sr concentrations ranging from about 10 to 100 ppm. As a result,

whole-rock Rb-Sr dates of Archean komatiites and the associated greenstones and felsites in most cases reflect the time of alteration rather than the time of extrusion. The alteration of the Rb-Sr couple in komatiites and greenstones may involve changes in the Rb/Sr ratios of the rocks caused by gain or loss of Rb and Sr as well as changes in the $^{87}Sr/^{86}Sr$ ratio. As a result, data points may scatter erratically on the isochron diagram making accurate determination of the initial $^{87}Sr/^{86}Sr$ ratios impossible.

In addition to Mg-rich komatiites, mafic lava flows with high Fe concentrations have been reported from several Archean greenstone belts (e.g. Boston Creek flow, south of Kirkland Lake, Abitibi Greenstone Belt, Ontario, Stone et al. 1995). The chilled top of the Boston Creek flow has spinifex texture and contains 15% MgO suggesting that it is a komatiite. However, it also contains about 16% FeO, but only about 5% Al_2O_3. Therefore, Stone et al. (1995) concluded that this flow is a ferropicrite and is related to, but distinct from, the komatiite series of volcanic rocks. Mafic lavas with these characteristics have been reported from Minnesota, (Arndt and Brooks 1978), India (Rajamani et al. 1985), and elsewhere. Stone et al. (1995) attributed the high Fe content of ferropicrites to partial melting at a depth of <100 km of depleted mantle rocks that were metasomatically enriched by a fluid that had formed by very small degrees of partial melting in a plume rising through the garnet-stability zone in the mantle.

8.1.5 Abitibi Greenstone Belt, Ontario and Quebec

The Abitibi Greenstone Belt in Fig. 8.7 extends in a northeasterly direction between the Kapuskasing gneiss belt in the west and the Grenville Province in the east. It includes the important mining areas (primarily gold and copper) in the vicinity of Timmins and Kirkland Lake, Ontario, and around Noranda and Val d'Or in Quebec. In addition, it includes the towns of Matagami and Chibougamau in Quebec. The Abitibi Greenstone Belt is about 300 km wide, measured at right angles to its strike, and is about 400 km long giving it an area of close to 120 000 km^2. The geology of this important part of the Precambrian shield of Canada was reviewed by Goodwin and Ridler (1970), Baragar and McGlynn (1976), Goodwin (1977, 1982), Dimroth et al. (1982, 1983), and Ludden et al. (1986).

The literature concerning the geology of the Abitibi Belt is extensive partly because of the occurrence of large deposits of gold in quartz veins (e.g. Kirkland Lake and Timmins; Fryer et al. 1979; Kerrich et al. 1984) and of base-metal sulfide mineralization in the metavolcanic rocks (e.g. Noranda and Kidd Creek; Maas et al. 1986). The close association of base-metal sulfide deposits with the volcano-sedimentary complexes of

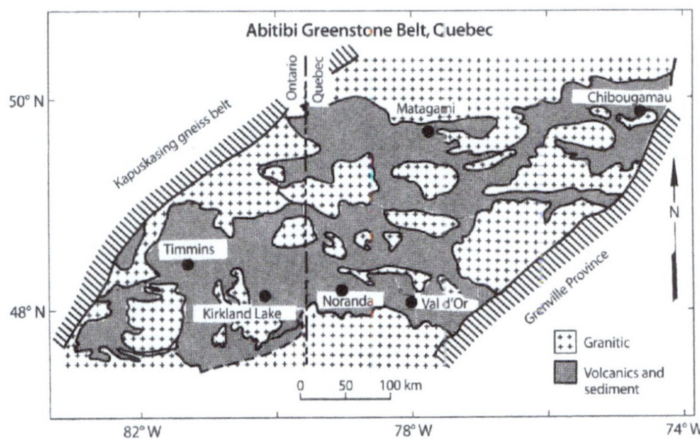

Fig. 8.7.
The Abitibi Greenstone Belt of northern Ontario and Quebec including the principal cities which serve as points of reference (*Source:* adapted from Gariépy and Allègre 1985)

the Abitibi Greenstone Belt has stimulated research to determine the reason for this association in the hope that the resulting geotectonic models may have predictive value (Dimroth et al. 1982; Kerrich et al. 1987). The origin of ore deposits in the Abitibi Greenstone Belt was the subject of a symposium organized by Spooner and Barrie (1993) that included a synthesis by Corfu (1993) of the evolution of the region based on U-Pb dates of zircon and baddeleyite (ZrO_2).

Arndt et al. (1977) provided a detailed description of the occurrence, petrography, and chemical composition of the komatiite flows and sills in Munro Township about 50 km north of the city of Kirkland Lake. The komatiites at that locality have a thickness of about 1 000 m and are overlain and underlain by even greater thicknesses of flows and sills ranging in composition from tholeiite basalt to dacite. Arndt et al. (1977) subdivided the komatiite series of rocks of Munro Township into three types based on their concentrations of MgO and SiO_2:

1. Ultramafic: MgO ≈ 30%, SiO_2 ≈ 44%
2. Mafic: MgO ≈ 8%, SiO_2 ≈ 52%
3. Intermediate: MgO ≈ 12%, SiO_2 ≈ 56%

The average Sr concentrations of these komatiites increase from only 10.4 ppm (periodotitic komatiites), to 58 ppm (pyroxenitic komatiites), and 96 ppm (basaltic komatiites). Cattell and Arndt (1987) later recognized six chemically distinct komatiite magma types based on the concentrations of Al and REEs of komatiite lavas in Newton Township of the Swayze Greenstone Belt in Ontario adjacent to the Abitibi Belt.

The time of extrusion of the komatiites and associated calc-alkaline metavolcanic rocks of the Abitibi Greenstone Belt has been determined by the Sm-Nd,

U-Pb, Pb-Pb, and Re-Os methods of dating because the Rb-Sr systematics of these rocks were compromised during greenschist-facies metamorphism. The very precise U-Pb zircon dates published by Corfu et al. (1989) and Corfu (1993) reveal that the volcano-sedimentary complexes of the Abitibi Greenstone Belt near Timmins and Kirkland Lake formed in about 30 million years primarily between 2 730 and 2 700 Ma. The volcanic rocks were later intruded by porphyry stocks at 2 691 to 2 688 Ma, and (at 2 673 Ma) by an albitite dike which predates the formation of gold-bearing quartz veins. (Fryer et al. 1979; Spooner and Barrie 1993).

The alteration of the Rb-Sr systematics of komatiites in the Abitibi Greenstone Belt is illustrated in Fig. 8.8 by comparison to dates determined by other methods. The crystallization age of these rocks is 2 726 ±93 Ma (R-Os method, Walker et al. 1988) and 2 738 ±26 Ma (Pb-Pb method, Dupré and Arndt 1990). However, the data points scatter widely on the Rb-Sr isochron diagram in Fig. 8.8 and yield a poorly constrained date of 1 028 Ma with an initial $^{87}Sr/^{86}Sr$ ratio of 0.70415. Walker et al. (1988) suggested that this date reflects the time of alteration of these rocks during the Grenville event. Consequently, the initial $^{87}Sr/^{86}Sr$ ratio of komatiites (0.70415) increased by the addition of radiogenic ^{87}Sr that had formed by in situ decay of ^{87}Rb in these rocks during a time interval of 1 700 million years between the dates of 2 726 and 1 028 Ma.

The wide-spread alteration of Rb-Sr systematics of the metavoclanic rocks of greenstone belts and of the associated granitic intrusives in the Superior tectonic province of Canada obscures information concerning their magma sources. In addition, Nd-Sr isotope mixing diagrams, which are very effective in the study of igneous rocks of Phanerozoic age, are not applicable to the rocks of Archean greenstone belts. Even when the

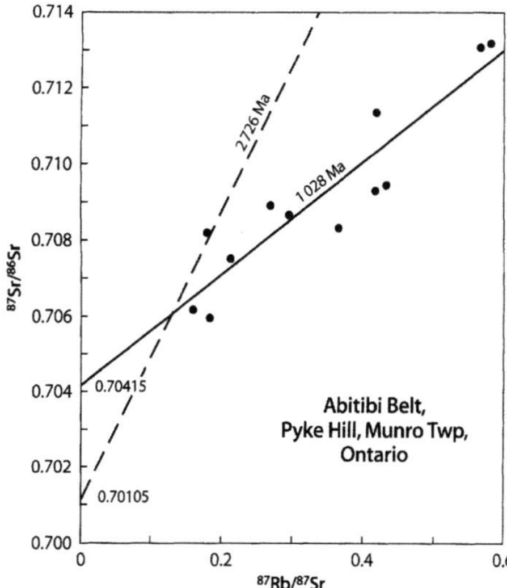

Fig. 8.8. Rb-Sr errorchron for komatiites, Pyke Hill, Munro Township, Ontario (Abitibi Greenstone Belt). The data points scatter widely and yield a poorly constrained errorchron date of 1 028 Ma with initial ^{87}Sr/^{86}Sr = 0.70415. The crystallization age of these rocks of 2 726 ±93 Ma was determined by the Re-Os method (*Source:* Walker et al. 1988)

crystallization age is known, the initial ^{87}Sr/^{86}Sr ratios cannot be calculated for rocks whose ^{87}Sr/^{86}Sr and/or ^{87}Rb/^{86}Sr ratios were changed during an episode of alteration.

The sensitivity of the Rb-Sr decay scheme to alteration means that dates determined by this method record postcrystallization events involving thermal metamorphism and/or circulation of aqueous solutions. This point was illustrated by Barrie and Shirey (1991) in the Kamiskotia-Montcalm area located about 25 km west of Timmins, Ontario, in the westernmost Abitibi Subprovince (Fig. 8.7) where the authors reported a sevenpoint whole rock and mineral Sm-Nd isochron age of 2 710 ±30 Ma for the Kamiskotia suite (basalt and rhyolite). This date agrees with U-Pb dates for zircons in a synvolcanic gabbro complex (2 707 ±2 Ma) and in a rhyolite (2 705 ±2 Ma) analyzed by Barrie and Davis (1990). However, a Rb-Sr isochron formed by a wholerock ferroan gabbro, and by clinopyroxene and zircon separates of this rock, yielded a date of only 2 450 ±35 Ma ($\lambda = 1.42 \times 10^{-11}$ yr^{-1}) and an initial ^{87}Sr/^{86}Sr ratio of 0.70085 ±0.00004. (The zircon contained 0.081 ppm Rb and 0.87 ppm Sr and had an ^{87}Sr/^{86}Sr ratio of 0.710378 ±0.000110). In addition, Barrie and Shirey (1991) reported a Rb-Sr whole-rock and mineral isochron date of 2 530 ±35 Ma ($\lambda = 1.42 \times 10^{-11}$ yr^{-1}) for the Ground-

hog River tonalite that intrudes the Kamiskotia metavolcanics, whereas U-Pb dates of zircon (2 696 ±2 Ma) and sphene (2 692 ±5 Ma) of the same rocks are 164 million years older than the Rb-Sr date. The evidence for isotopic re-equilibration of Sr in the Groundhog River tonalite means that its measured initial ^{87}Sr/^{86}Sr ratio of 0.70124 ±0.0002 is higher than the value of this ratio at the time of crystallization.

In order to avoid the effects of alteration of the Rb-Sr systematics, Hart and Brooks (1977) analyzed clinopyroxenes separated from mafic and ultramafic rocks in the Abitibi Greenstone Belt. The unaltered clinopyroxene grains have an average initial ^{87}Sr/^{86}Sr ratio of 0.70114 ±13 at 2.7 Ga relative to 0.7080 for E&A. However, two composite samples of basalt from the Abitibi Greenstone Belt have a higher (and probably incorrect) initial ^{87}Sr/^{86}Sr ratio of 0.70162 ±0.00006, weighted by the number of samples in each composite. The use of unaltered minerals to determine the initial ^{87}Sr/^{86}Sr ratios of Archean Greenstones (and hence of the magma sources in the mantle) was later extended by Machado et al. (1986) who obtained an average initial ^{87}Sr/^{86}Sr ratio of 0.70105 ±0.00006 ($t = 2\,702$ Ma; $\lambda = 1.42 \times 10^{-11}$ yr^{-1}; E&A = 0.7080) for six clinopyroxene and one plagioclase concentrates derived from basalt flows of the Abitibi Belt. Since the initial ^{87}Sr/^{86}Sr ratios have only a narrow range (0.70092 to 0.70129) and since the samples originated from widely spaced locations in the Abitibi Greenstone Belt, the authors concluded that the Sr of the magma sources in the mantle at 2.7 Ga was isotopically homogeneous. The initial ^{87}Sr/^{86}Sr ratio of the metavolcanic rocks determined by Machado et al. (1986) is similar to that of the Chibougamau Pluton for which Brooks (1980) and Jones et al. (1974) reported a Rb-Sr date of 2 730 ±100 Ma and 0.7009 ±0.0004 relative to 0.7080 for E&A.

The granitic rocks of the Abitibi Belt yield dates that are similar to those of the volcanic rocks. For example, Krogh and Davis (1971) reported a U-Pb zircon date of 2 755 Ma for a granodiorite pluton that intrudes the volcanic rocks. The ages of six other granitic plutons in the Abitibi Greenstone Belt were later determined by Gariépy and Allègre (1985) by the Pb-Pb method which yielded dates between 2 616 ±19 Ma (Waswanipi Pluton) and 2 718 ±12 Ma (Renault Pluton).

The low initial ^{87}Sr/^{86}Sr ratio of many Archean granitic plutons and felsic volcanic rocks associated with mafic lavas cannot be explained by:

1. Derivation of mafic and felsic magmas from the same source rocks in the mantle;
2. Formation of felsic magmas by fractional crystallization of mafic magmas;
3. Partial melting of old continental crust to produce magmas of felsic composition.

The explanation for this important problem in igneous petrology is that the granitic magmas associated with large-scale eruption of mafic lavas of Archean age formed by partial melting of mantle-derived basalt or gabbro (or their metamorphic equivalents) at the base of the volcanic pile (Martin 1987).

A question concerning the origin of Archean greenstone belts is whether they were deposited on pre-existing granitic crust. In the Abitibi Greenstone Belt, there is no direct evidence for the existence of granitic rocks older than about 2 720 Ma (Pb-Pb date, Renault Pluton, Gariépy and Allègre 1985). Therefore, Gariépy et al. (1984) tried to find evidence for the existence of granitic basement by dating individual detrital zircon grains extracted from the Pontiac turbidites about 8 km south of Noranda, Quebec. Three morphologically distinct types of zircon grains revealed a spectrum of dates ranging from 2 925 Ma (thick prisms) to 2 713 Ma (translucent needles), and 2 686 Ma (rounded prisms). In addition, the thick prisms provided a date of 2 095 Ma from the lower intercept of the discordia line with concordia. The oldest date (2 925 Ma) exceeds the age of any of the known granitic rocks in the Abitibi Belt thereby implying that the thick zircon prisms originated from igneous or sedimentary basement rocks that formed before the extrusion of the basal komatiites.

The geologic history of the entire Abitibi Greenstone Belt has been revealed most clearly by U-Pb dates of zircon and baddeleyite. According to the compilation of dates by Corfu (1993) the rocks in this belt formed in about 80 million years from 2 750 to 2 670 Ma. The belt consists of several overlapping volcano-sedimentary complexes containing coeval komatiites, tholeiites, and calc-alkaline volcanic and plutonic rocks that formed in paired island arc-back-arc systems in an oceanic setting. Continental crust may have been present some distance to the west. The volcanic activity was terminated by compression, folding, emplacement of granitic plutons between 2 700 and 2 688 Ma, and deposition of turbidites. The late-stage magmatic rocks between 2 681 and 2 676 Ma are predominantly alkalic in composition. At about the same time, alluvial-fluvial sediment was deposited and subsequently folded and thrust-faulted between 2 672 and 2 670 Ma. The final orogenic activity was accompanied by intrusion of lamprophyre dikes. Hydrothermal activity followed until about 2 580 Ma and resulted locally in the formation of quartz-albite-gold veins, presumably derived from the newly formed underlying crust.

The initial $^{87}Sr/^{86}Sr$ ratios of the metavolcanic rocks of the Abitibi Greenstone Belt range from 0.70105 ±0.00006 at 2.70 Ga (clinopyroxenes, Machado et al. 1986 and Hart and Brooks 1977) to 0.7028 ±0.0003 at 2.28 Ga (Roy Group, Jones et al. 1974) and scatter widely in Fig. 8.9. Only the initial $^{87}Sr/^{86}Sr$ ratios of the clino-

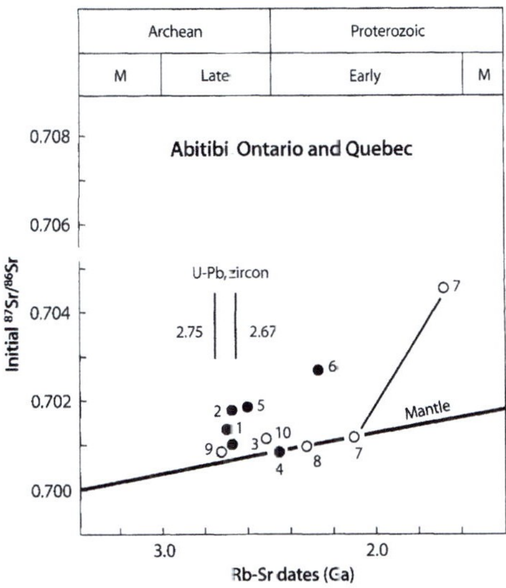

Fig. 8.9. Strontium-isotope evolution of metavolcanic (*solid circles*) and granitic (*open circles*) rocks in the Abitibi Greenstone Belt of Ontario and Quebec. The data points identified by number are: *1.* Kinojevis Group; *2.* Blake River Group; *3.* clinopyroxene; *4.* ferroan gabbro, Kamiskotia; *5.* Dore Lake complex; *6.* Roy Group; *7.* Otto stock; *8.* Round Lake Batholith; *9.* Chibougamau Pluton; *10.* Groundhog River tonalite. The mantle-Sr evolution line extends from 0.699 at 4.5 Ga to 0.7030 at 0 Ga. The U-Pb zircon dates between 2.75 and 2.67 Ga encompass the period of igneous activity in this area (*Sources:* Purdy and York 1968; Wanless et al. 1970; Jones et al. 1974; Bell and Blenkinsop 1976; Hart and Brooks 1977; Brooks 1980; Machado et al. 1986; Barrie and Shirey 1991; Corfu 1993)

pyroxenes (Item 3) analyzed by Machado et al. (1986) and the ferroan gabbro (Item 4) at Kamiskotia dated by Barrie and Shirey (1991) fit the mantle line. All of the other mafic rocks in Fig. 8.9 have higher initial $^{87}Sr/^{86}Sr$ ratios than magma sources in the mantle during the Late Archean Era. The most likely explanation is that the Rb-Sr systematics of these rocks were disturbed during regional metamorphism after the time of crystallization between 2.75 and 2.67 Ga. Consequently, the elevated initial $^{87}Sr/^{86}Sr$ ratios cannot be used to conclude that the mafic magmas were derived from different kinds of source rocks in the mantle or that the magmas assimilated Sr from the rocks of the continental crust.

The initial $^{87}Sr/^{86}Sr$ ratios of the granitic rocks of the Abitibi Belt are compatible with the mantle-Sr evolution line, except for the Otto stock south of Kirkland Lake (Item 7, Fig. 8.9). This body crystallized at 2 680 ±1 Ma (U-Pb, zircon and titanite, Corfu et al. 1989), but yielded highly discordant whole-rock Rb-Sr isochron dates of 1 693 ±50 Ma (Purdy and York 1968) and 2 114

±80 Ma (Bell and Blenkinsop 1976; $\lambda = 1.42 \times 10^{-11}$ yr^{-1}; 0.7080 for E&A). The initial ^{87}Sr/^{86}Sr ratio of the Otto stock reported by Bell and Blenkinsop (1976) actually lies on the mantle-Sr evolution line and therefore implies that the granitic magma formed by partial melting of very young mantle-derived rocks at 2 114 ±80 Ma. However, this date is in conflict with the U-Pb date reported by Corfu et al. (1989). The nearby Round Lake Batholith (Item 8) also fits the mantle-Sr evolution line even though its Rb-Sr systematics may also have been reset after initial crystallization. The Chibougamau Pluton (Item 9, Fig. 8.8) was studied both by Jones et al. (1974) and by Brooks (1980). The latter applied petrographic and geochemical "filters" to the data to select the least-altered set of rocks which yielded 2 730 ±100 Ma ($\lambda = 1.42 \times 10^{-11}$ yr^{-1}) and 0.7009 ±0.0004 relative to 0.7080 for E&A.

Therefore, the metavolcanic and granitic rocks of the Abitibi Greenstone Belt display the same characteristic properties of the Rb-Sr system noted in other greenstone belts discussed in this section:

1. Whole-rock Rb-Sr dates of greenstones and granites are lower than U-Pb zircon and Sm-Nd whole-rock dates of the same rocks;
2. The most reliably determined initial ^{87}Sr/^{86}Sr ratios of the two rock types are compatible with the mantle-Sr evolution line, except when the Sr in the rocks was isotopically rehomogenized after initial crystallization.

8.1.6 Duxbury Massif, Quebec

The Archean gneisses of the Superior Craton have been affected by tectonic deformation, regional metamorphism, and igneous activity over long periods of geologic time. The compositional diversity and tectono-thermal history of these rocks are exemplified by the tonalitic gneisses that form an easterly plunging antiform centered on Duxbury Lake in the Eastmain River area of Quebec, located north of the Abitibi Greenstone Belt and east of James Bay at about 52°25' N and 77°30' W.

Verpaelst et al. (1980) reported that the tonalitic and granodioritic gneisses of the Duxbury Massif have a combined whole-rock Rb-Sr errorchron date of 2 915 ±180 Ma ($\lambda = 1.42 \times 10^{-11}$ yr^{-1}) and an initial ^{87}Sr/^{86}Sr ratio of 0.7013 ±0.0002 (E&A = 0.7080). However, the granodiorites, regressed separately, yielded a date of only 2 500 ±85 Ma and initial ^{87}Sr/^{86}Sr = 0.7023 ±0003, whereas selected specimens and slabs of tonalite formed an errorchron with a date of 3 080 ±180 Ma and an intercept of 0.7014 ±0.0003. Although the granodiorites are a distinct border facies of the Duxbury Massif, they grade into the tonalites that constitute the main

body of the massif. Therefore, the apparent difference in the Rb-Sr dates of these two rock types cannot be the result of separate intrusion, but was caused by alteration after crystallization. However, even some suites of tonalite samples yielded low Rb-Sr dates ranging from 2 500 to 1 220 Ma.

The range of discordant Rb-Sr dates of the Duxburg Massif is attributable to several possible causes (Verpaelst et al. 1980):

1. Successive intrusion of magmas of different ages.
2. Variable contamination of the magmas during intrusion.
3. Subsequent metamorphism and metasomatic alteration of the rocks.
4. Differences in the ages and Rb-Sr systematics of the protoliths.
5. Formation of the Duxbury Massif by in situ remobilization of older crustal rocks.

The authors favored the last alternative on the grounds that the average ^{87}Sr/^{86}Sr ratio of the tonalites would have increased from 0.7014 at 3.1 Ga to 0.7028 at 2.5 Ga, thereby coming close to matching the initial ^{87}Sr/^{86}Sr ratio (0.7023 ±0.0003) of the granodiorite. This scenario is illustrated in Fig. 8.10 where Sr isotope evolution lines for the tonalite (selected total rocks and slabs) have been drawn from 0.7014 ±0.0003 for an average ^{87}Rb/^{86}Sr ratio of 0.1425 ±0.0704 ($2\bar{\sigma}, N = 10$). The Sr-isotope evolution lines define a band that includes not only the granodiorite (symbolized by x), but also all of the suites of tonalites that yielded dates between 2.50 and 1.22 Ga. Therefore, the granodiorite and all of the "younger" tonalites could have been derived by remobilization of pre-existing sialic rocks represented by the tonalites in the core zone of the antiform. In this process, the average Rb/Sr ratio of the rocks increased from 0.049 ±0.024 (selected tonalites) to 0.088 ±0.037 (granodiorite).

Verpaelst et al. (1980) concluded that the tonalites of the Duxbury Massif formed by recrystallization of a protolith composed of calc-alkaline volcanic rocks and related sedimentary rocks at about 3.1 Ga near the base of Archean crust at a depth of 15 to 20 km and at a temperature of about 630 °C. The protolith had resided in the crust for a least 400 × 10^6 years after being removed from sources in the mantle between 3.5 and 3.6 Ga (Verpaelst et al. 1980). The Sr in these rocks was subsequently isotopically homogenized during recrystallization at about 3.1 Ga, and again between 2.5 and 2.3 Ga. The lengthy crustal history of the protolith and the evidence for remobilization of the rocks accompanied by isotopic homogenization of Sr is typical of Archean gneisses elsewhere in the Superior Craton. In contrast to these Late Archean rocks, the granitic gneisses of Early Archean age in the Godthåb area of West Green-

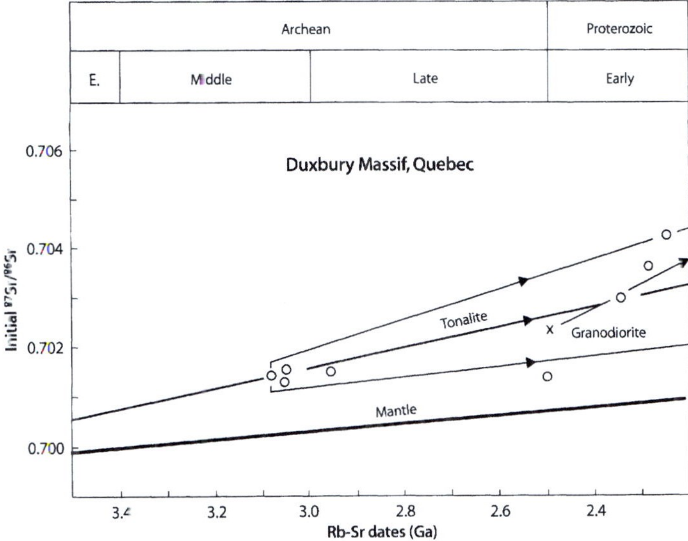

Fig. 8.10. Isotope evolution of Sr in the tonalites of the Duxbury Massif, Quebec. A selected suite of tonalites and tonalitic gneisses yielded a Rb-Sr errorchron date of 3 080 ±180 Ma ($\lambda = 1.42 \times 10^{-11}$ yr^{-1}) and an initial ^{87}Sr/^{86}Sr ratio of 0.7014 ±0.0003 (B&A = 0.7080). The slopes of the Sr development lines emanating from that point are based on the average ^{87}Rb/^{86}Sr ratio of these tonalites of 0.1425 ±0.0704 ($2\bar{\sigma}$, $N = 10$) and on the resulting limiting values of c.2129 (*upper*) and 0.0721 (*lower*). The granodiorite at the margins of the massif and the younger tonalite suites lie within the band that defines the isotope evolution of the Sr in the tonalite at the core of the massif. Therefore, all of the rocks of the massif could have formed by recrystallization and isotopic rehomogenization of Sr of a protolith at about 3.1 Ga. The protolith originated from sources in the mantle several hundred million years before it was recrystallized to form the tonalite (*Source:* Verpaelst et al. 1980)

land and in southern Africa originated from volcanic rocks that were derived from sources in the mantle only a short time before they were remelted or recrystallized. (Moorbath 1975a, 1975b, 1976).

8.2   Mantle-Separation Dates

In several cases considered in the preceding sections of this chapter, the whole-rock Rb-Sr isochron dates of granitic gneisses were found to be lower than U-Pb concordia dates of zircons in the same rocks. (Beakhouse et al. 1983). In addition, the Rb-Sr isochron dates and initial ^{87}Sr/^{86}Sr ratios of the granitic intrusives associated with some greenstone belts define straight lines in the Sr-isotope evolution diagram (e.g. Michipicoten). The slopes of such straight lines can be used to calculate the average Rb/Sr ratios of the protoliths from which the granitic gneisses formed at various times. In

addition, these Sr-isotope evolution lines can be extrapolated to the Sr-isotope evolution line of magma sources in the mantle, assumed to be a straight line extending from ^{87}Sr/^{86}Sr = 0.699 at 4.5 Ga to 0.7030 at the present time. The time-coordinate of the point of intersection of a Sr-isotope evolution line of crustal rocks with the evolution line of mantle-Sr is the "mantle-separation" date of the protoliths of the igneous or high-grade metamorphic rocks. Mantle-separation dates can also be calculated for suites of samples that define a Rb-Sr isochron or errorchron. In this case, the average measured ^{87}Sr/^{86}Sr and ^{87}Rb/^{86}Sr ratios can be used to plot the Sr-isotope evolution line. The validity of the mantle-separation dates obtained by this procedure is limited by the assumptions on which they are based:

1. The average measured Rb/Sr ratio of the crustal rocks is identical to the average Rb/Sr ratio of the protoliths from which they formed.
2. The isotope composition of Sr of the protoliths was homogenized in the course of their transformation thereby allowing specimens of the resulting granitic rocks to form an isochron with an elevated initial ^{87}Sr/^{86}Sr ratio.
3. The Sr-isotope evolution line of the mantle is an accurate representation of time-dependent changes in the ^{87}Sr/^{86}Sr ratio of magma sources in the mantle.

Strictly speaking, none of these assumptions are valid or even plausible, suggesting that mantle-separation dates are not reliable indicators of the ages of petrogenetic events. Nevertheless, in some cases, such dates help to reduce the difference between U-Pb zir-

Fig. 8.11.
a Formation of granitic orthogneiss by partial melting of protoliths in the crust; **b** Formation of paragneiss by recrystallization of sedimentary rocks without melting

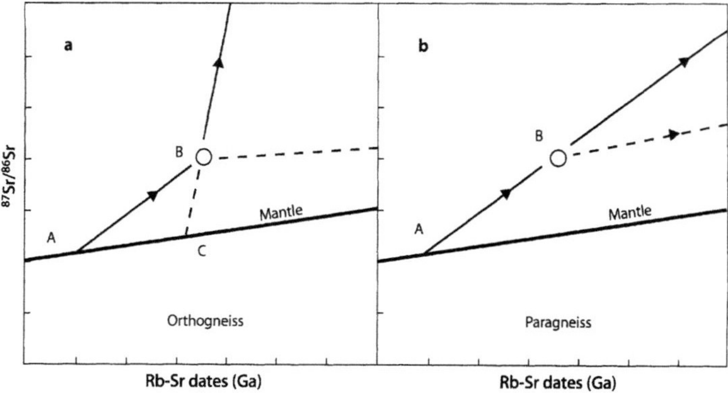

con and whole-rock Rb-Sr dates and their evaluation focuses attention on the petrogenetic processes that are involved in the formation of granitic rocks having elevated initial $^{87}Sr/^{86}Sr$ ratios. The extent to which the assumptions on which mantle-separation dates are based are actually satisfied depends on the type of rock under consideration and on their petrogenesis as illustrated in Fig. 8.11a and b.

Volcanic and plutonic rocks at Point A in Fig. 8.11a form by crystallization of mantle-derived magma. The average $^{87}Sr/^{86}Sr$ ratio of the resulting igneous rocks increases with time at a rate depending on their Rb/Sr ratio. At Point B partial melting of these rocks (protoliths) causes a granitic magma to form which is enriched in Rb relative to the protoliths and has an $^{87}Sr/^{86}Sr$ ratio which is greater than mantle-Sr at that time. The granitic magma subsequently intrudes the rocks of the upper crust and crystallizes within a relatively short time after its formation. The $^{87}Sr/^{86}Sr$ ratio of the resulting granitic rocks increases with time more rapidly than that of the protoliths before magma formation. The residue (restite) remains in the lower continental crust and its $^{87}Sr/^{86}Sr$ ratio evolves slowly with time (dashed line) because of its low Rb/Sr ratio.

The time-coordinate of Point C in Fig. 8.11a is the mantle-separation date obtained by extrapolating the Sr-evolution line of the granitic orthogneisses from the present to its point of intersection with the hypothetical mantle evolution line. This "mantle-separation" date underestimates the age of the protoliths and overestimates the crystallization age of the granitic orthogneiss. The restite which formed at Point B has a low Rb/Sr ratio but a high initial $^{87}Sr/^{86}Sr$ ratio. Such restites are exemplified by the granitic gneisses that underlie the northern lobe of the Bushveld Complex at Potgietersrus (Sect. 7.8.4).

The petrogenesis of paragneisses in Fig. 8.10b starts at Point A where sediment is produced by erosion of young mantle-derived rocks and is enriched in Rb dur-

ing early-stage diagenesis. Subsequently, the average $^{87}Sr/^{86}Sr$ ratio of the sedimentary and volcanic rocks (protoliths) increases with time in accordance with its Rb/Sr ratio. At Point B the protoliths are subjected to regional metamorphism of amphibolite grade and are recrystallized to form a granitic paragneiss. The Sr is isotopically homogenized by diffusion and by the movement of an aqueous fluid during regional metamorphism and tectonic deformation. If the average Rb/Sr ratio of the protoliths remains unchanged, the mantle-separation date at Point A approaches the time of deposition and diagenesis of the protoliths, but underestimates the age of the provenance of the sediment. If the Rb/Sr ratio of the protolith decreases at Point B as a result of granulite-facies metamorphism causing loss of Rb, the mantle-separation date of the paragneiss may exceed the age of the Earth and hence has no real meaning. An increase of the Rb/Sr of the protoliths during amphibolite-facies metamorphism at Point A is possible but unlikely because it requires input of Rb from an external source, or loss of Sr, or both. Therefore, Rb-Sr mantle-separation dates are most appropriate for metamorphosed sedimentary rocks that formed by erosion of mantle-derived igneous rocks and were later metamorphosed without melting.

8.3   Isotopic Homogenization of Granitic Gneisses

The apparent isotopic homogenization of Sr in granitic gneisses is indicated in cases where the rocks yield satisfactory Rb-Sr isochron dates that are lower than the U-Pb concordia dates of zircons in the same rocks. In many cases, the U-Pb dates reflect the time of crystallization of the rocks from magma (assuming that the zircons lack inherited cores), whereas the Rb-Sr dates record a time of isotopic re-equilibration of Sr in the rocks. Moorbath (1975b) expressed doubt that the isotope composition of Sr and Pb in granitic gneisses can be homogenized by regional metamorphism and cited evidence reported by Krogh and Davis (1973) that differences in the $^{87}Sr/^{86}Sr$ ratios over distances of a few centimeters had survived amphibolite-grade metamorphism during the Grenville metamorphic event in Ontario. However, evidence for isotopic homogenization of Sr during regional metamorphism of granitic rocks has been reported by van Breemen and Dallmeyer (1984), Roddick and Compston (1977), Brooks (1966, 1968), Arriens et al. (1966), Pidgeon and Compstons, and Wasserburg et al. (1964). Several possible mechanisms have been mentioned in the literature for the equalization of $^{87}Sr/^{86}Sr$ ratios of paragneisses either at the time of their formation during metamorphic recrystallization of a protolith or during subsequent metamorphic episodes:

1. The protolith had uniform $^{87}Sr/^{86}Sr$ ratios at the time of recrystallization;
2. The radiogenic ^{87}Sr that had accumulated within the protolith by in situ decay of ^{87}Rb was selectively removed from the rocks during recrystallization;
3. The $^{87}Sr/^{86}Sr$ ratios of isotopically heterogeneous protoliths were equalized by redistribution of Sr by aqueous fluids released by dehydration of the rocks;
4. The rocks remained open to Rb and Sr during slow cooling after crystallization meaning that the Rb-Sr clock did not start until the rocks had cooled through the "closure temperature," whereas the U-Pb clock in zircon started at the time of crystallization.

Although all of these scenarios seem contrived, they are at least worth considering to explain specific instances of apparent isotopic re-equilibration of Sr within large volumes of solid rock.

The foregoing discussion highlights the importance of the origin of the granitic gneisses in the interpretation of their whole-rock Rb-Sr and U-Pb zircon dates. Paragneisses that formed by recrystallization of pre-existing sedimentary and volcanic rocks can contain zircon crystals that predate the formation of the gneiss. In such cases, the U-Pb concordia dates indicate the time of original crystallization of the zircons rather than the time of formation of the gneiss in which they now occur. Geochronologists who use the U-Pb method of dating are well aware of this possibility and therefore present evidence for or against the existence of detrital zircon cores with younger overgrowths. In addition, the cores and overgrowths can be dated separately by means of ion-probe mass spectrometers such as SHRIMP designed and operated by Dr. Compston's research team at the Australian National University in Canberra. In cases where the presence of old cores in zircons is recognized and dealt with appropriately, the U-Pb zircon dates are more reliable indicators of the

crystallization age of the rocks than whole-rock Rb-Sr isochron dates because the loss of Pb and gain or loss of U by the zircons can be corrected on the concordia diagram, whereas disturbed Rb-Sr systematics cannot be restored. Open-system behavior of the minerals of polymetamorphic granitic rocks is a well known phenomenon. In many cases, the constituent minerals of igneous and high-grade metamorphic rocks form "internal mineral isochrons" whose slopes record the time elapsed since the last isotopic re-equilibration of Sr among the minerals of a handspecimen of rock. This phenomenon was recognized and discussed by Fairbairn et al. (1961), Pidgeon and Hopgood (1975), and Faure (1986).

8.4 The Wyoming Craton

The North American continent includes the Superior, Wyoming, and Slave Cratons of Archean greenstone belts and associated gneisses. The greenstone belts discussed in this chapter exemplify the Archean rocks of the Superior Craton where the initial $^{87}Sr/^{86}Sr$ ratios of some of the granitic gneisses are low enough to be compatible with mantle Sr of Late Archean age. Therefore, these rocks formed by partial melting of mantle-derived rocks with short crustal residence times. The granitic gneisses and anorthosites (Geist et al. 1990) of the Wyoming Craton in Fig. 8.12 do not conform to this generalization because their initial $^{87}Sr/^{86}Sr$ ratios are too high to be compatible with mantle Sr of Archean age. The Archean craton of Wyoming, Montana, and Idaho contains a number of mountain ranges identified in Fig. 8.12 that extend from southeastern Wyoming into southwestern Montana and southern Idaho. The geology of this area consists of metasedimentary and metavolcanic rocks that are embedded in and intruded by granites and granitic gneisses (e.g. Owl Creek and Beartooth Mountains).

The granitic gneisses of the Wyoming Craton were among the first rocks to be dated as isotopic methods were developed in the 1950s and 1960s (Giletti 1968). The U-Pb zircon dates of some granitic rocks in the Wyoming Craton exceed their whole-rock Rb-Sr isochron dates as noted above for Archean gneisses of the Superior Cratons. For example, Naylor et al. (1970) reported a U-Pb zircon date of 2687 ± 15 Ma for the Louis Lake Pluton of the Wind River Range, whereas the Rb-Sr whole-rock isochron date is only 2574 ± 20 Ma ($\lambda = 1.42 \times 10^{-11}$ yr^{-1}), initial $^{87}Sr/^{86}Sr = 0.702 \pm 0.001$. The Rb-Sr dates of minerals of the Louis Lake Pluton range from 2535 ± 50 to 1958 Ma. In addition, the initial $^{87}Sr/^{86}Sr$ ratios determined from mineral-rock isochrons increase with decreasing Rb-Sr dates from 0.703 ± 0.001 at 2535 ± 50 Ma to 0.780 at 1958 Ma relative to 0.7091 for seawater. The discordance of whole-

Fig. 8.12. Exposures of Archean granitoids of the Wyoming Craton of North America (adapted from Mueller et al. 1983)

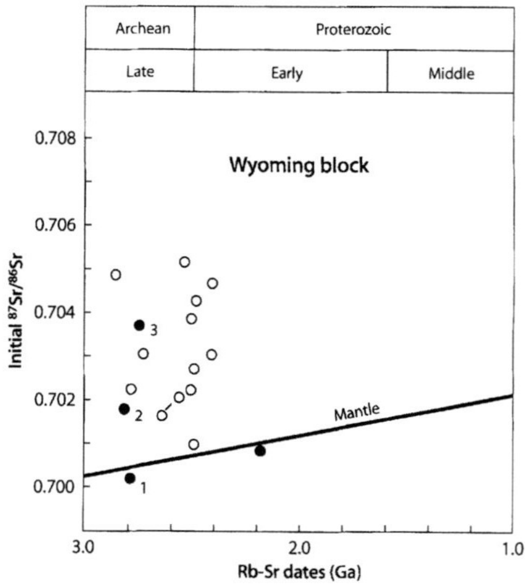

Fig. 8.13. Initial $^{87}Sr/^{86}Sr$ ratios and whole-rock Rb-Sr dates of granitic rocks (*open circles*) and metavolcanic rocks (*solid circles*) of the Wyoming Craton. Most of the granitic gneisses in this craton have higher initial $^{87}Sr/^{86}Sr$ ratios than mantle-Sr of Late Archean age. In addition, their Rb-Sr dates range widely from 2.86 to 2.40 Ga. The metavolcanic rocks are identified by number: *1.* Rendezvous metagabbro, Teton Range; *2.* Andesite amphibolite, Beartooth Mountains; *3.* Metadacites, Wind River Canyon, Owl Creek Mountains. The mantle-Sr evolution line extends from 0.699 at 4.5 Ga to 0.7030 at 0 Ga (*Sources:* Armstrong and Hills 1967; Hills et al. 1968; Naylor et al. 1970; Reed and Zartman 1973; Hills and Armstrong 1974; Johnson and Hills 1976; Divis 1976; Peterman and Hildreth 1978; Wooden et al. 1982; Stuckless et al. 1985; Wooden and Mueller 1988)

rock Rb-Sr and U-Pb zircon dates of the Louis Lake Pluton was later confirmed by Stuckless et al. (1985).

In contrast to the Louis Lake Pluton of the Wind River Range, granites in the Granite Mountains (Lankin Dome and Tincup Mountain) have concordant whole-rock Rb-Sr and U-Pb zircon dates. Peterman and Hildreth (1978) measured a Rb-Sr whole rock date of $2\,540 \pm 30$ Ma ($\lambda = 1.42 \times 10^{-11}$ yr^{-1}) for granites in the Granite Mountains and reported a U-Pb zircon concordia date of $2\,594 \pm 40$ Ma determined by Ludwig and Stuckless (written communication 1976). The apparent concordance of whole-rock Rb-Sr and U-Pb zircon dates contrasts with the discordance of their Rb-Sr mineral dates (microcline and biotite) which range from about 1 450 to 2 300 Ma.

Most of the granitic rocks of the Wyoming Craton in Fig. 8.13 have high initial $^{87}Sr/^{86}Sr$ ratios compared to mantle-Sr during the Late Archean Era. Therefore, Peterman and Hildreth (1978) concluded that crustal rocks with significant crustal residence ages had contributed to the formation of granitic magmas in the Wyoming Craton in contrast to the granitic rocks of the Rainy Lake and Vermilion Lake areas of Ontario and Minnesota where the initial $^{87}Sr/^{86}Sr$ ratios of the Late Archean granitic rocks are compatible with the $^{87}Sr/^{86}Sr$ ratios of mantle Sr.

The Wyoming Craton also contains remnants of supracrustal rocks (greenstones) that were intruded by the granitic gneisses. For example, Henry et al. (1982) described granulite-grade supracrustals in the Quad Creek area of the Beartooth Mountains. Andesitic amphibolites from this area were dated by Wooden et al. (1982) who found that most amphibolite specimens plot along the whole-rock Rb-Sr isochron of the enclosing granite indicating a date of $2\,810 \pm 40$ Ma with an initial $^{87}Sr/^{86}Sr$ ratio of 0.7018 ±0.0002. Metavolcanic rocks also occur in the Owl Creek Mountains, southeast of the Beartooth Mountains in north-central Wyoming (Granath 1975). The Rb-Sr whole-rock isochron date of meta-dacites from the southern end of Wind River Canyon in the Owl Creek Mountains is $2\,755 \pm 101$ Ma with an initial $^{87}Sr/^{86}Sr$ ratio of 0.7037 ±0.0015 (Mueller et al. 1981). In addition, iron formations and interbedded supracrustal rocks, metamorphosed to the almandine amphibolite facies, in the southern Wind River Range near Atlantic City were described by Pride and Hagner (1972) who concluded that the iron formation is Late Archean in age based on a U-Pb zircon date of 2680 ± 20 reported by Naylor et al. (1970) for the Louis Lake granodiorite which intrudes the metasedimentary rocks.

8.5 Amîtsoq Gneiss and Isua Supracrustals, West Greenland

The North American greenstone belts described in the previous sections all formed in the Late Archean Era between about 2.75 and 2.65 Ga. This period in the geologic history of the Earth was a time of widespread volcanic activity followed by orogenies caused by compression during basin closure and continental collisions. These processes resulted in the formation of microplates which were subsequently welded together during the Proterozoic Eon to form larger units of continental crust. The North American continent is a fairly typical example of this process. The Archean gneisses and associated supracrustal volcano-sedimentary rocks of Greenland and the east coast of Labrador form part of the North Atlantic Craton which includes also the Archean granulites of Scotland and northern Scandinavia. The southern part of West Greenland includes a block of Archean gneisses bounded by the Ketilidian Belt in the south (Blaxland et al. 1976) and by the Nagssugtoquidian Belt in the north (Kalsbeek et al. 1987). Both belts contain granitic gneisses and volcano-sedimentary complexes that were severely deformed during the Proterozoic Eon. The Archean rocks of Greenland and Labrador in Fig. 8.14 merit special attention because they have preserved a record of geological activity during Early Archean time and because they have been very thoroughly studied since the time their exceptional age was first revealed by the work of McGregor (1973), Black et al. (1971), and Moorbath et al. (1972).

8.5.1 Amîtsoq Gneisses, Godthåbsfjord

The antiquity of the Amîtsoq gneisses in the Godthåb area of West Greenland in Fig. 8.15 was first indicated by Black et al. (1971) who reported a whole-rock Rb-Sr isochron date of 3896 ± 170 Ma ($\lambda = 1.42 \times 10^{-11}$ yr^{-1}) and an initial $^{87}Sr/^{86}Sr$ ratio of 0.6989 ± 0.0010 for these rocks relative to 0.7080 for E&A. These values were, at the time, the oldest date and the lowest initial $^{87}Sr/^{86}Sr$ ratio that had ever been measured in any terrestrial rocks. Additional whole-rock Rb-Sr dating by Moorbath et al. (1972) subsequently confirmed the Early Archean age of the Amîtsoq gneisses, but reduced the original age estimate of Black et al. (1971) to about 3640 ± 30 Ma.

The Rb-Sr date of the Amîtsoq gneisses reported by Moorbath et al (1972) is in agreement with U-Pb and Th-Pb concordia intercepts of zircons at 3650 ± 50 Ma and 3648 ± 85 Ma, respectively, reported by Baadsgaard (1973). Moorbath et al. (1975b) later reported a Pb-Pb date of 3800 ± 120 Ma and Michard-Vitrac et al. (1977) obtained a U-Pb zircon date of 3.65 Ga for Amîtsoq

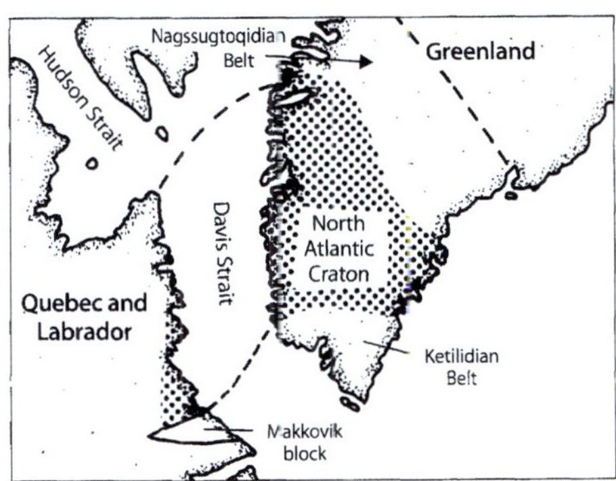

Fig. 8.14. Outline of the North Atlantic Craton composed of granitic gneisses and volcano-sedimentary complexes of Archean age in southern Greenland and along the east coast of Labrador (adapted from Wardle et al. 1990)

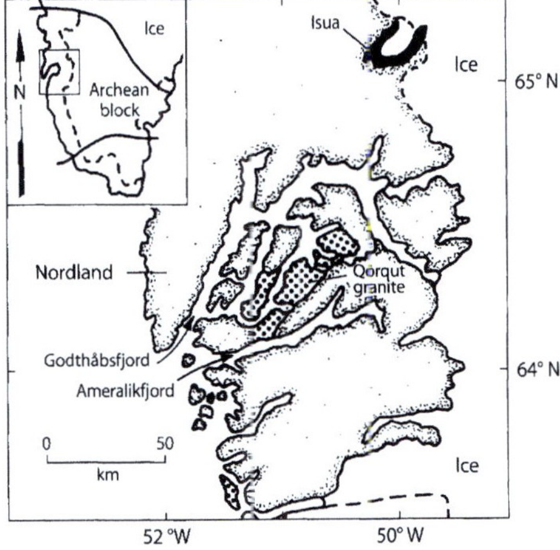

Fig. 8.15. Godthåbsfjord of southern West Greenland and the Isua supracrustals (adapted from Baadsgaard et al. 1984)

gneisses in the Godthåb area. In addition, Kinny (1986) demonstrated that the zircons in the Amîtsoq gneiss have a lengthy geologic history that started with their crystallization from magma at 3822 ± 5 Ma and was followed by the deposition of overgrowths at about 3630 Ma during an episode of regional metamorphism. Metamorphic episodes were also uncovered by the work of Baadsgaard et al. (1976) who analyzed minerals of the Amîtsoq gneisses by means of a variety of dating methods which yielded a spectrum of dates with major events at 3.6, 2.5, and 1.52 Ga.

Although the metamorphic grade of the Amîtsoq gneisses in the Godthåbsfjord area is the amphibolite facies, the gneisses on the islands south of Godthåbsfjord were metamorphosed to the granulite facies at 3 650 ±140 Ma (Early Archean) indicated by a whole-rock Rb-Sr isochron date published by Griffin et al. (1980). However, Black et al. (1973) obtained a whole-rock Pb-Pb date of only 2 890 ±60 Ma (Late Archean) for granulites from Sukkertoppen and Nordland north of Godthåbsfjord, whereas the granulites at Fiskenaesset yielded a Pb-Pb date of 2 850 ±100 Ma.

The age determinations of the Amîtsoq gneiss are remarkable not only because these rocks crystallized during the Early Archean Era but also because, in this case, whole-rock Rb-Sr isochron dates, U-Pb concordia dates of zircon, and whole-rock Pb-Pb dates are in good agreement. The concordance is even more remarkable because of evidence that the Amîtsoq gneiss experienced several episodes of regional metamorphism. The data of Moorbath et al. (1972, 1975b, 1977) and Griffin et al. (1980) yield an average initial $^{87}Sr/^{86}Sr$ ratio of 0.70096 for the Amîtsoq gneiss at an age of 3.6 Ga.

8.5.2 Isua Supracrustals

The Isua supracrustals are located about 150 km northeast of Godthåb near the edge of the Greenland ice sheet (Fig. 8.15). They include mafic volcanic rocks, iron formation, greywacke, quartzite, and conglomerate all of which were metamorphosed to the amphibolite facies. The Isua supracrustals form a refolded syncline and were intruded by granitic gneisses whose Rb-Sr date of 3 622 ±140 Ma is identical to that of the Amîtsoq gneisses near Godthåb (Moorbath et al. 1972). Moorbath et al. (1977) later refined the Rb-Sr date of the gneisses at Isua to 3 641 ±60 Ma ($\lambda = 1.42 \times 10^{-11}$ yr^{-1}). This date is in good agreement with a U-Pb concordia date of 3.56 Ga of sphene in the gneisses at Isua (Baadsgaard 1983).

The age of the Isua supracrustals was first measured by Moorbath et al. (1973) who reported a whole-rock Pb-Pb date of 3 760 ±70 Ma for samples of metamorphosed iron formation. Age determinations of the metavolcanic and metasedimentary rocks at Isua by the Sm-Nd method confirmed that these rocks crystallized at 3 770 ±42 Ma (Hamilton et al. 1978). In addition, a whole-rock Rb-Sr date of muscovite-bearing schist and associated boulders composed of altered felsic volcanic rocks is marginally lower at 3 661 ±60 Ma (Moorbath et al. 1977) than the Sm-Nd date. The age determinations of the Isua supracrustals demonstrate that even in Early Archean time large bodies of water were present on the surface of the Earth.

Metavolcanic and metasedimentary rocks also occur as enclaves within the Amîtsoq gneisses at the mouth of Ameralik Fjord (Fig. 8.15) and on the islands southwest of it. These rocks are collectively referred to as the Akilia association (McGregor and Mason 1977)

and include amphibolites, iron formation, quartz-rich and calc-silicate rocks, as well as felsic gneisses, mica schists, and mafic/ultramafic plutonic rocks. These kinds of inclusions also occur in the Amîtsoq gneisses at Isua where the basal section of the supracrustal rocks appears to grade into Akilia-type lithologies. Therefore, Baadsgaard et al. (1984) concluded that the rocks of the Akilia association and of the Isua supracrustals are parts of the same volcano-sedimentary complex that was disrupted by the intrusion of the Amîtsoq gneisses. Zircons of the Akilia association collected near the mouth of Ameralik Fjord yielded a date of 3 587 ±38 Ma compared to 3 812 Ma for zircons of the Isua supracrustals (Baadsgaard et al. 1984). The lower U-Pb date of the Akilia zircons was caused by alteration during metamorphism and tectonic deformation of the rocks along the coast of West Greenland at about 3.6 Ga, whereas the rocks at Isua were not significantly affected by this event.

In conclusion, the age of the Isua supracrustals is constrained to be less than 3.80 Ga (U-Pb, detrital zircon, Baadsgaard et al. 1984) and greater than 3.73 Ga (Sm-Nd, Hamilton et al. 1978). Evidently, only about 70 million years elapsed between the crystallization of zircon in granitic rocks that became the source of sediment deposited with mafic and felsic volcanic rocks in the Isua Basin and the subsequent folding of the volcano-sedimentary complex and its intrusion by granitic rocks. The U-Pb zircon date of 3 650 ±50 Ma reported by Baadsgaard (1973) restricts the age of the Amîtsoq gneisses to the interval 3.70 to 3.60 Ga. All of the whole-rock Rb-Sr dates of the Amîtsoq gneisses reported by Moorbath et al. (1972, 1975b) and Griffin et al. (1980) are consistent with these limits.

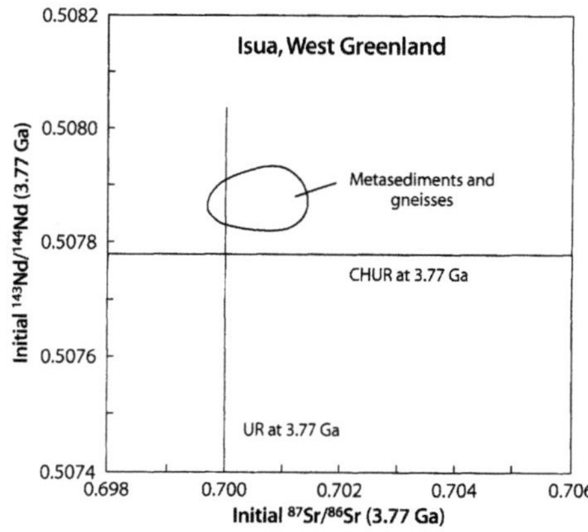

Fig. 8.16. Initial isotope ratios of Sr and Nd at 3.77 Ga in the metasedimentary rocks and gneisses at Isua, West Greenland (*Sources:* Moorbath et al. 1975b, 1977; Hamilton et al. 1983b)

The initial $^{87}Sr/^{86}Sr$ and $^{143}Nd/^{144}Nd$ ratios of the metasedimentary rocks and gneisses at Isua at 3.77 Ga range from 0.699 to 0.7015 and from 0.50783 to 0.50795, respectively (Moorbath et al. 1975b, 1977; Hamilton et al. 1983b). These values place the resulting data field in Fig. 8.16 close to the intersection between CHUR and UR at that time. This result is consistent with the idea that in Early Archean time the volcano-sedimentary rocks originated from magma sources in the mantle whose isotope ratios of Sr and Nd were similar to those of bulk Earth. In other words, the Earth was much more homogeneous at that time than it was later in its history when the continental crust and lithospheric mantle developed distinctly different $^{87}Sr/^{86}Sr$ and $^{143}Nd/^{144}Nd$ isotope ratios.

8.5.3 Malene Supracrustals and Nûk Gneisses

The Godthâb area of West Greenland also contains volcano-sedimentary complexes of Late Archean age known as the Malene supracrustals. These rocks were apparently deposited on the eroded weathering surface of the Amîtsoq gneisses and are preserved on some of the small coastal islands south of the mouth of Godthâbsfjord (McGregor 1973; Nutman and Bridgwater 1983; Dymek et al. 1983). The rocks consist of amphibolites and metasedimentary rocks that could have formed by deposition of sediment derived from the Amîtsoq gneisses. The origin of the Malene supracrustals and their relation to the Amîtsoq gneisses is obscured by subsequent tectonic activity which resulted in the intercalation of the Malene supracrustals with the Amîtsoq gneisses followed by the intrusion of dikes, sills, and plutons that were subsequently metamorphosed to the granulite and amphibolite facies to form the Nûk gneisses (Black et al. 1973). Therefore, the Nûk gneisses are significantly younger than the Amîtsoq gneisses, and they are distinguished from the latter in the field by the absence of mafic Ameralik dikes which were intruded into the Amîtsoq gneisses prior to the deposition of the Malene supracrustals (Gill and Bridgwater 1976).

The Nûk gneisses were originally dated by Pankhurst et al. (1973a) who reported a Rb-Sr whole-rock date of 2 976 ±50 ($\lambda = 1.42 \times 10^{-11}$ yr^{-1}) and an initial $^{87}Sr/^{86}Sr$ ratio of 0.7016 ±0.0004. Subsequent age determinations of the Nûk gneisses by the Rb-Sr method have revealed a range of dates from 3 076 ±27 Ma (Baadsgaard and McGregor 1981) to 2 676 ±65 Ma (Kalsbeek and Pidgeon 1980). The U-Pb dates of zircon extracted from Nûk gneisses by Baadsgaard (1976) and Baadsgaard and McGregor (1981) also range from 2 820 to 3 065 Ma and are consistent with the whole-rock Rb-Sr dates. The range of U-Pb dates of the Nûk gneisses was attributed by Baadsgaard and McGregor (1981) to

varying Pb-loss caused by metamorphism of the rocks. A study of Nûk gneisses in the Godthâb area by Taylor et al. (1980) indicated that these rocks contain mixtures of mantle-Pb with unradiogenic Pb derived from the Amîtsoq gneisses into which the magmatic precursors of the Nûk gneisses were intruded.

The distribution of data points in Fig. 8.17 demonstrates that the Nûk gneisses have higher initial $^{87}Sr/^{86}Sr$ ratios than the Amîtsoq gneisses. However, since the latter have a high average Rb/Sr ratio of about 0.36, the average $^{87}Sr/^{86}Sr$ ratio of the Amîtsoq gneisses increased rapidly with time and had reached an average value of about 0.7105 at 3.0 Ga. Therefore, the Nûk magmas could not have formed by partial melting of Amîtsoq gneisses, although they probably did assimilate Sr from the Amîtsoq gneisses at the time of intrusion. Taylor et al. (1980) proposed that trace elements and radiogenic Pb may have been transferred to the younger gneisses by means of aqueous fluids produced by dehydration of the Early Archean basement rocks which were heated by burial and regional metamor-

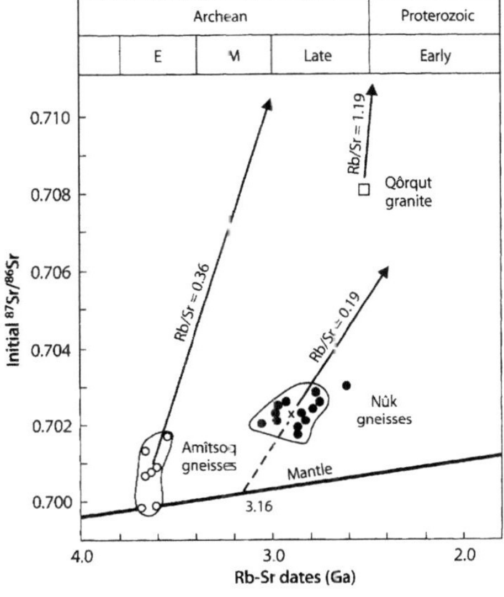

Fig. 8.17. Strontium-isotope evolution diagram for Amîtsoq gneisses (*open circles*) Nûk gneisses (*solid circles*), and the Qôrqut granite (*open square*). The average initial $^{87}Sr/^{86}Sr$ ratio of the Amîtsoq is 0.7008 ±0.0005 ($2\bar{\sigma}, N = 7$) at 3 633 ±34 Ma, whereas that of the Nûk gneisses is 0.7024 ±0.0002 ($2\bar{\sigma}, N = 13$) at 2 868 ±69 Ma. The mantle-separation date of the Nûk gneisses is 3.16 Ga indicating that they are not derivatives of the Amîtsoq gneisses. However, the magma of the younger Qôrqut granite (2 530 Ma, initial $^{87}Sr/^{86}Sr = 0.7081$ ±0.0008) could have formed from a mixture of Amîtsoq and Nûk gneisses (*Sources:* Moorbath et al. 1972, 1975b, 1981; Griffin et al. 1980; Pankhurst et al. 1973a,b; Taylor et al. 1980; Moorbath and Pankhurst 1976; Baadsgaard and McGregor 1981; Pidgeon and Hopgood 1975; Kalsbeek and Pidgeon 1980)

phism associated with an episode of igneous activity that led to the formation of the Late Archean gneisses. Elsewhere in southern West Greenland where Early Archean gneisses are absent (Grant and Hickman 1984), the younger gneisses contain no evidence of having been affected by older crustal rocks.

The Amîtsoq and Nûk gneisses can be regarded as products of major orogenies that terminated the deposition of volcano-sedimentary complexes represented by the Isua supracrustals (Akilia association) and the Malene supracrustals, respectively. Therefore, the ages of the gneisses restrict the time of deposition of the Malene supracrustals to be less than 3.6 Ga (Amîtsoq gneisses) but greater than 3.07 (Nûk gneisses). However, age determinations of metasedimentary rocks of the Malene supracrustals have not fulfilled this expectation in all cases. Although Stecher et al. (1986) did report a whole-rock Sm-Nd date of 3 190 Ma for metavolcanic rocks from the Ameralik Fjord, other investigators obtained dates that are less than 2.9 Ga. For example, Baadsgaard (1976) reported U-Pb dates between 2.6 and 2.9 Ga for zircons from the Malene supracrustals, whereas Hamilton et al. (1983b) obtained Sm-Nd model dates relative to CHUR (Sect. 1.5) of 2 660 ±20 and 2 820 ±30 Ma. Evidently, the dates of the Malene supracrustals determined by Baadsgaard (1976) and by Hamilton et al. (1983b) are younger than the oldest reported Rb-Sr date of 3 076 ±27 Ma for the Nûk gneisses (Baadsgaard and McGregor 1981).

Additional evidence that some parts of the Malene supracrustals are younger than some of the Nûk gneisses was provided by U-Pb dates of single detrital zircon grains in Malene metasediments on the islands of Rypeø and Simiutâ south of Godthâb. Schiøtte et al. (1988) obtained a date of about 2 900 Ma for these zircon grains and 2 650 Ma for metamorphic overgrowths. Therefore, these authors concluded that the Malene supracrustals in part postdate the Nûk magmatic event, whereas other parts predate the formation of Nûk gneisses.

8.5.4 Qôrqut Granite

The last major rock forming event in the Godthâb area was the intrusion of the Qôrqut granite consisting of leucocratic granite, grey biotite granite, and pegmatitic granite (Fig. 8.15). The latter was dated by Pankhurst et al. (1973b) who reported a mineral Rb-Sr isochron date of 2 525 ±30 Ma ($\lambda = 1.42 \times 10^{-11}$ yr^{-1}) and an initial ^{87}Sr/^{86}Sr ratio of 1.0 ±0.3. This date was confirmed by Baadsgaard (1976) with a U-Pb zircon date of 2 530 ±30 Ma. Moorbath et al. (1981) carried out a comprehensive study of the different rock types collected throughout the outcrop area of the Qôrqut complex and found that all of the specimens fit one Rb-Sr iso-

chron from which they derived a date of 2 530 ±30 Ma ($\lambda = 1.42 \times 10^{-11}$ yr^{-1}) and an initial ^{87}Sr/^{86}Sr ratio of 0.7081 ±0.0008. The same samples yielded a Pb-Pb date of 2 580 ±80 Ma, in good agreement with the Rb-Sr date.

The elevated initial ^{87}Sr/^{86}Sr ratio of the Qôrqut granite complex (0.7081 ±0.0008) clearly indicates that it is a derivative of crustal rocks. Figure 8.16 illustrates the petrogenetic model for the Qôrqut granite complex considered by Moorbath et al. (1981). The slope of the Sr-evolution line of the Amîtsoq gneisses corresponds to an average Rb/Sr ratio of 0.36 ±0.07 ($2\bar{\sigma}$, $N = 71$), whereas the Nûk gneisses have an average Rb/Sr ratio of only 0.19 ±0.04 ($2\bar{\sigma}$, $N = 109$) (Moorbath et al. 1972, 1977; Moorbath and Pankhurst 1976; Taylor et al. 1980). The diagram demonstrates that at 2 530 Ma, when the Qôrqut magma crystallized, its ^{87}Sr/^{86}Sr ratio (0.7081 ±0.0008) was intermediate between that of the Amîtsoq gneisses (0.7174) and that of the Nûk gneisses (0.7051). Therefore, the Qôrqut magma could have formed by partial melting of a mixture of Amîtsoq and Nûk gneisses.

8.6 Summary: The Origin of the Oldest Rocks

The Archean greenstone belts consist of volcano-sedimentary complexes that formed in island arcs along the margins of microcontinents. The microcontinents consisted of basalt plateaus and granitic intrusives formed from them in previous orogenic cycles. Age determinations of zircons by the U-Pb concordia method indicate that during the Archean Eon geological activity proceeded at a rapid pace such that a geological cycle was completed in only a few tens of millions of years. As a result, rock units separated by angular unconformities formed in rapid succession and were incorporated into the growing continental crust. The resulting microcontinents, in some cases, remained undisturbed by the geological activity which occurred around them later in the history of the Earth.

Remnants of Early Archean volcano-sedimentary complexes and associated granitic gneisses were first recognized in the Godthâb area of southern West Greenland because they yielded concordant dates of about 3.7 Ga by the whole-rock Rb-Sr and Sm-Nd isochron methods, and by the U-Pb concordia method. The sedimentary rocks of the Isua area near Godthâb were deposited at about 3.8 Ga when the Earth was only about 700 million years old. Nevertheless, even at this early age, sediment was derived by erosion of previously erupted volcanic rocks of basaltic and felsic composition. In addition, the presence of a sedimentary iron formation at Isua reveals the existence of standing bodies of water on the surface of the Earth in which chemical precipitates did accumulate. The Early Archean Amîtsoq gneiss which formed by remelting of mafic to felsic volcanic rocks did not reveal its exceptional age until it was dated by isotopic methods. Consequently, the origin of even the oldest igneous and sedimentary rocks is explainable by geologic processes that have continued to the present time. The apparent uniformity of igneous petrogenesis reaffirms the Principle of Uniformitarianism.

In addition, the evidence derived from trace-element abundances of the metavolcanic rocks and the occurrence of these rocks in roughly linear belts reveals that they formed in subduction zones. Therefore, plates of lithospheric mantle and overlying basalt crust were already being subducted by Late Archean time and are exemplified by the volcano-sedimentary complexes preserved in northern Ontario and Minnesota. The subduction of plates during the Late Archean Era implies that the asthenospheric mantle was convecting at that time and that basalt crust was forming along spreading ridges. Convection was initially caused by the instability of the asthenospheric mantle resulting from heat released during core formation and by radioactive decay of isotopes of U, Th, and K. The subduction of oceanic crust subsequently replenished the radioactive heat sources of the asthenospheric mantle and allowed convection to continue in the form of plumes. Therefore, plumes rising from the asthenospheric mantle continue to provide the driving force for most of the geological activity that occurs on the surface of the Earth. The only other major forces acting on the Earth are of extraterrestrial origin and are related to the rotation of the Earth, the gravitational interactions between the Earth and the Moon, the impacts of asteroids and comets, as well as the electromagnetic radiation and nuclear particles emitted by the Sun.

Although the petrogenesis of the oldest igneous rocks is plausibly explainable by the geophysical phenomena that characterized the early history of the Earth, the information that is derivable from the abundances of trace elements and radiogenic isotopes has been degraded partly by alteration of the rocks and partly because of the properties of isotopic systems. The time-dependent evolution of the isotopic compositions of Sr, Nd, and Pb was affected by the chemical differentiation of the Earth which caused the Rb/Sr, Sm/Nd, and U/Pb ratios of magma sources to diversify. Consequently, the isotope ratios of Sr, Nd, and Pb in modern volcanic rocks are the result of mixing of these elements derived from different reservoirs. During the Archean Eon, the range of isotope compositions of these elements was much smaller than it is today. Therefore, the initial isotopic compositions of Sr, Nd, and Pb of igneous rocks of Archean age are not sensitive indicators of the assimilation of crustal rocks by magma derived from the mantle wedge above subduction zones. At the time the Amîtsoq gneiss and the Isua supracrustals formed during the Early Archean Era, all rocks on the Earth were young and their $^{87}Sr/^{86}Sr$, $^{143}Nd/^{144}Nd$, and $^{206}Pb/^{204}Pb$ ratios were much more uniform than they are at the present time.

The study of the origin of igneous rocks has provided insights into the inner workings of the Earth and how these processes have evolved through time. The most remarkable conclusions are that igneous rocks have always formed as a consequence of tectonic processes that occur in the mantle of the Earth and that plumes of hot rocks, rising from the asthenosphere, continue to energize the geological activity of the Earth. These plumes move the lithospheric plates, open and close ocean basins, break up continents and later reassemble the fragments, cause mountains to rise and continents to sink beneath the surface of the oceans. This is how the Earth has worked for more than four billion years, thereby giving life a chance to evolve and to diversify on its surface. The plumes that drive the internal activity of the Earth cause U, Th, and K to be recycled into the asthenosphere in the form of subducted oceanic crust. The heat generated by these elements ultimately causes new plumes to rise. In this way, the Earth continually rejuvenates itself by recycling the heat-producing elements.

References

Abdel-Monem A, Gast PW (1967) Age of volcanism on St. Helena. Earth Planet. Sci. Letters 2:415–418

Abdel-Monem A, Fernandez LA, Boone GM (1975) K-Ar ages from the eastern Azores Group (Santa Maria, São Miguel and the Formigas Islands). Lithos 8:247–254

Abdel-Rahman A-FM, Doig F (1987) The Rb-Sr geochronlogical evolution of the Ras Gharib segment of the northern Nubian Shield. J Geol Soc 144:577–586

Åberg G, Aguirre L, Levi, Nyström JO (1984) Spreading-subsidence and generation of ensialic marginal basins: an example from the Early Cretaceous of central Chile. In: Kokelaar BP, Howells MF (eds) Marginal Basin Geology. Geol Soc, London, Spec Pub 16, pp 185–193

Åberg G, Bollmark B, MacIntyre RM (1987) Age of the Austurhorn intrusion; a net-veined complex in southeastern Iceland. Geol Förening Stockholm Förhand 109(4):291–293

Adams CJD (1975) Discovery of Precambrian rocks in New Zealand: Age relations of the Greenland Group and Constant gneiss, West Coast, South Island. Earth Planet Sci Letters 28:98–104

Ahmat AL, De Laeter JL (1982) Rb-Sr isotopic evidence for Archaean-Proterozoic crustal evolution of part of the central Yilgarn Block, Western Australia: Constraints on the age and source of the anorthositic Windimurra gabbroic. J Geol Soc Australia 29:177–190

Albarède F, Bottinga Y (1972) Kinetic disequilibrium in trace element partitioning between phenocrysts and host lava. Geochim Cosmochim Acta 36:141–146

Aldrich LT, Herzog LF, Doak JB, Davis GL (1953) Variations in strontium isotope abundance in minerals. Part I. Mass spectrometric analysis of mineral sources of strontium. Trans Amer Geophys Union 34:457–460

Alexander PO (1980a) Petrogenesis of low potassic quartz normative Deccan tholeiites. J Geol Soc India 21:261–272

Alexander PO (1980b) The Pavagarh rhyolite. J Geol Soc India 21:453–457

Alexander PO (1981) Strontium-isotopic composition of Dhandhuka basalts, western India. Chem Geol 32:129–138

Alexander PO, Paul DK (1977) Geochemistry and strontium isotopic composition of basalts from the eastern Deccan volcanic province. Indian Mineral Mag 41:165–172

Alexander RT, Macdonald KC (1996) Small off-axis volcanoes on the East Pacific Rise. Earth Planet Sci Letters 139:387–394

Alibert C (1991) Mineralogy and geochemistry of basalt from site 738: Implications for the tectonic history of the southernmost part of the Kerguelen Plateau. Proc Ocean Drilling Program, Sci Res 119:293–298

Alibert C, Leterrier J, Panasiuk M, Zimmerman J (1987) Trace and isotope geochemistry of the alkaline Tertiary volcanism in southwestern Poland. Lithos 20:311–321

Allan JF, Batiza R, Perfit MR, Fornari DJ, Sack RO (1989) Petrology of lavas from the Lamont Seamount Chain and adjacent East Pacific Rise, 10° N. J Petrol 30:1245–1298

Allan JF, Chase RL, Cousens B, Michael PJ, Gorton MP, Scott SD (1993) The Tuzo Wilson volcanic field, NE Pacific: Alkaline volcanism at a complex, diffuse, transform-trench-ridge triple junction. J Geophys Res 98(B12):22367–22387

Allègre CJ, Condomines M (1982) Basalt genesis and mantle structure through Th isotopic geochemistry. Nature 299:21–24

Allègre CJ, Hart SR (eds) (1978) Trace Elements in Igneous Petrology. Elsevier, New York

Allègre CJ, Othman DB (1980) Nd-Sr isotopic relationship in granitoid rocks and continental crust development: a chemical approach to orogenesis. Nature 286:335–342

Allègre CJ, Turcotte DL (1985) Geodynamic mixing in the mesosphere boundary layer and the origin of oceanic islands. Geophys Res Letters 12:207–210

Allègre CJ, Pineau F, Bernat M, Javoy M (1971) Evidence for the occurrence of carbonatites on the Cape Verde and Canary islands. Nature (Phys Sci) 233:103–104

Allègre CJ, Treuil M, Minster JF, Minster B, Albarède F (1977) Systematic use of trace elements in ig

Allègre CJ, Dupré B, Lambret B, Richard P (1981) The subcontinental versus suboceanic debate. I. Lead-neodymium-strontium isotopes in primary alkaline basalts from a shield area: The Ahaggar volcanic suite. Earth Planet Sci Letters 52:85–92

Allègre CJ, Dupré B, Richard P, Rousseau D, Brooks C (1982) Subcontinental versus suboceanic mantle. II. Nd-Sr-Pb isotopic comparison of continental tholeiites with mid-ocean ridge tholeiites, and the structure of the continental lithosphere. Earth Planet Sci Letters 57:25–34

Allègre CJ, Hart SR, Minster J-F (1983a) Chemical structure and evolution of the mantle and continents determined by inversion of Nd and Sr isotopic data. I. Theoretical methods. Earth Planet Sci Letters 66:177–.90

Allègre CJ, Hart SR, Minster J-F (1983b) Chemical structure and evolution of Nd and Sr isotopic data. II. Numerical experiments and discussion. Earth Planet Sci Letters 66:191–213

Allègre CJ, Hamelin B, Dupré B (1984) Statistical analysis of isotopic ratios in MORB: The mantle blob cluster model and the convective regime of the mantle. Earth Planet Sci Letters 71: 71–84

Allègre CJ, Hamelin B, Provost A, Dupré B (1986) Topology is isotopic multispace and origin of mantle heterogeneities. Earth Planet Sci Letters 81:319–337

Allsopp HL (1965) Rb-Sr and K-Ar age measurements on the Great Dyke of Southern Rhodesia. J Geophys Res 70:977–984

Allsopp HL, Neethling DC (1970) Rb-Sr isotopic ages of Precambrian intrusives from Queen Maud Land, Antarctica. Earth Planet Sci Letters 8:66–70

Allsopp HL, Manton WI, Bristow JW, Erlank AJ (1984) Rb-Sr geochronology of the Karoo felsic volcanics. In: Erlank AJ (ed) Petrogenesis of the Volcanic Rocks of the Karoo Province. Spec Publ Geol Soc South Africa 13:273–280

Al'Mukhamedov AI, Zolotukhin VV (1988) Geochemistry of low-potassic basalts from Siberian and Deccan traps. In: Subbarao KV (ed) Deccan Flood Basalts. Geol Soc India, pp 341–351

Al-Rawi Y, Carmichael ISE (1976) A note on the natural fusion of granite. Amer Mineralog 52:1806–1814

Al-Shanti AMS, Abdel-Monem AA, Marzouki FH (1984) Geochemistry, petrology, and Rb-Sr dating of trondhjemite and granophyre associated with Jabal Tays ophiolite, Idsas area, Saudi Arabia. Precamb Res 24:321–334

Alt JC, Shanks III WC, Jackson MC (1993) Cycling of sulfur in subduction zones: The geochemistry of sulfur in the Mariana Island Arc and back-arc trough. Earth Planet Sci Letters 119: 477–494

Altherr R, Henjes-Kunst F, Puchelt H, Baumann A (1988) Volcanism in the Red Sea axial trough-Evidence for a large mantle diapir? Tectonophys 150:121–133

Altherr R, Henjes-Kunst F, Baumann A (1990) Asthenosphere versus lithosphere as possible sources for basaltic magmas erupted during formation of the Red Sea: Constraints from strontium, lead, and neodymium isotopes. Earth Planet Sci Letters 96:269–286

Alvarado GE, Denyer P, Sinton CW (1997) The 89 Ma Tortugal komatiite suite, Costa Rica: Implications for a common geological origin in the Caribbean and Eastern Pacific region from a mantle plume. Geology 25:439–442

Amari S, Lewis RS, Anders E (1994) Interstellar grains in meteorites. I. Isolation of SiC, graphite, and diamond. Size distributions of SiC and graphite. Geochim Cosmochim Acta 58: 459–470

Andersen T (1986) Magmatic fluids in the Fen carbonatite complex, S.E. Norway: Evidence of mid-crustal fractionation from solid and fluid inclusions in apatite. Contrib Mineral Petrol 93:491–503

Andersen T (1987) Mantle and crustal components in a carbonatite complex, and the evolution of a carbonatite magma: REE and isotopic evidence from the Fen complex, southeast Norway. Chem Geol 65:147–166

Anderson DL (1996) Enriched asthenosphere and depleted plumes. International Geology Review 38:1–21

Anderson DL (1992) The Earth's interior. In: Brown G, Hawkesworth C, Wilson C (eds) Understanding the Earth. Cambridge University Press, Cambridge, pp 44–66

Anderson FW, Dunham KC (1966) The geology of northern Skye. Mem Geol Surv Scotland

Annells RN (1973) Proterozoic flood basalts of eastern Lake Superior: The Keweenawan volcanic rocks of the Mamainse Point area, Ontario. Canadian Geol Surv, Paper 72–10

Anthony EY, Segalstad TV, Neumann E-R (1989) An unusual mantle source region for nephelinites from the Oslo Rift, Norway. Geochim Cosmochim Acta 53:1067–1076

Aoki K-I, Kurasawa H (1984) Sr isotope study of the tephrite series from Nyamuragira volcano, Zaire. Geochem J 18:95–100

Apt YE, Akinin VV, Wright JE (1998) Sr, Nd, and Pb isotopes in Neogene melanephelinites and deep-seated xenoliths from northeast Russia. Geochem Internal 36(1):24–34

Arakawa Y (1984) Rb-Sr ages of the gneiss and metamorphosed intrusive rocks of the Hida metamorphic belt in the Urushiyama area, Gifu Prefecture, central Japan. J Japanese Assoc Mineralogists Petrologists and Econ Geologists 79:431–442

Arakawa Y, Shinmura T (1995) Nd-Sr isotopic and geochemical characteristics of two contrasting types of calc-alkaline plutons in the Hida belt, Japan. Chem Geol 124:217–232

Araña V, Vegas R (1974) Plate tectonics and volcanism in the Gibraltar arc. Tectonophysics 24:197–212

Araña V, Marti J, Aparicio A, Garcia-Cacho L, Garcia-Garcia R (1994) Magma mixing in alkaline magmas: an example from Tenerife, Canary Island. Lithos 32:1–19

Arculus RJ (1976) The geology and geochemistry of the alkali basalt, andesite association of Grenada, Lesser Antilles island arc. Geol Soc Amer Bull 87:612–624

Arculus RJ (1978) Mineralogy and petrology of Grenada, Lesser Antilles island arc. Contrib Mineral Petrol 65:413–424

Arculus RJ, Johnson RW (1981) Island arc magma sources: A geochemical assessment of the roles of slab-derived components and crustal contamination. Geochem J 15:109–134

Arculus RJ, DeLong SE, Kay RW, Brooks C, Sun S-S (1977) The alkalic rock suite of Bogoslof Island, eastern Aleutian arc, Alaska. J Geol 85:177–186

Armstrong RA, Bristow JW, Cox KG (1984) The Rooi Rand dyke swarm, southern Lebombo. In: Erlank AJ (ed) Petrogenesis of the Volcanic Rocks of the Karoo Province. Spec Publ Geol Soc South Africa, 13:77–86

Armstrong RL (1970) Geochronology of Tertiary igneous rocks, eastern Basin and Range province, western Utah, eastern Nevada, and vicinity, U.S.A. Geochim Cosmochim Acta 34: 203–232

Armstrong RL (1971) Isotopic and chemical constraints on models of magma genesis in volcanic arcs. Earth Planet Sci Letters 12:137–144

Armstrong RL (1975) Episodic volcanism in the central Oregon Cascade Range: Confirmation and correlation with the Snake River Plain. Geology 3(8):356

Armstrong RL (1978a) K-Ar dating: Late Cenozoic McMurdo Volcanic Group and dry valley glacial history, Victoria Land, Antarctica. New Zealand J Geol Geophys 21:685–698

Armstrong RL (1978b) Cenozoic igneous history of the U.S. Cordillera from lat 42° to 49° N. Geol Soc Amer Mem 152: 265–282

Armstrong RL (1991) The persistent myth of crustal growth. Australian J Earth Sci 38:613–630

Armstrong RL, Besancon J (1970) A Triassic time-scale dilemma: K-Ar dating of Upper Triassic igneous rocks, eastern U.S.A. and Canada and post-Upper Triassic plutons, western Idaho, U.S.A. Eclogae geol Helveticae 63:15–28

Armstrong RL, Hills FA (1967) Rb-Sr and K-Ar geochronologic studies of mantled gneiss domes, Albion range, southern Idaho, USA. Earth Planet Sci Letters 3:114–124

Armstrong RL, Hamilton W, Denton GH (1968) Glaciation in Taylor Valley, Antarctica, older than 2.7 million years. Science 159:187–188

Armstrong RL, Erken EB, McKee EH, Noble DC (1969) Space-time relations of Cenozoic silicic volcanism in the Great Basin of the western United States. Amer J Sci 267:478–490

Armstrong RL, Eisbacher GH, Evans P (1982) Age and stratigraphic-tectonic significance of Proterozoic diabase sheets, Mackenzie Mountains, northwestern Canada. Can J Earth Sci 19(2):316–323

Arndt NT, Brooks C (1978) Iron-rich basaltic komatiites in the early Precambrian Vermilion District: Discussion. Can J Earth Sci 15:856–857

Arnst NT, Christensen U (1992) The role of lithospheric mantle in continental flood volcanism. J Geophys Res 97:10967–10981

Arndt NT, Nisbet EG (eds) (1982) Komatiites. Allen and Unwin, London

Arndt NT, Naldrett AJ, Pyke DR (1977) Komatiitic and iron-rich tholeiitic lavas of Munro Township, northeast Ontario. J Petrol 18:319–369

Arndt NT, Kerr AC, Tarney J (1997) Dynamic melting in plume heads: the formation of Gorgona komatiites and basalts. Earth Planet Sci Letters 146:289–301

Arriens PA, Brooks C, Bofinger VM, Compston W (1966) The discordance of mineral ages in granitic rocks resulting from the redistribution of rubidium and strontium. J Geophys Res 71(20):4981–4994

Arth JG (1976) Behaviour of trace elements during magmatic processes-a summary of theoretical models and their applications. J Res US Geol Surv 4:41–47

Arth JG, Hanson GN (1975) Geochemistry and origin of the early Precambrian crust of northeastern Minnesota. Geochim Cosmochim Acta 39:325–362

Ashkinadze GSh, Gorokhovskiy BM, Shukolyukov YA (1978) ^{40}Ar/^{39}Ar dating of biotite containing excess ^{40}Ar. Geochem Internat 1977:172–176

Askren DR, Roden MF, Whitney JA (1997) Petrogenesis of Tertiary andesite lava flows interlayered with large-volume felsic ash-flow tuffs of the western USA. J Petrol 38: 1021–1046

Atkinson WF, Hughes FE, Smith CB (1984) A review of the kimberlitic rocks of Western Australia. In: Kornprobst J (ed) Kimberlites and Related Rocks. Elsevier, New York (Porceedings of the Third International Kimberlite Conference, vol I, pp 195–224)

Atwater T (1970) Implications of plate tectonics for the Cenozoic tectonic evolution of western North America. Geol Soc Amer Bull 81:3513–3536

Aucamp APH, Welmarans LG, Neethling DC (1972) The Urfjell Group, a deformed (?) early Paleozoic sedimentary sequence, Kirwanveggen, western Dronning Maud Land. In: Adie RJ (ed) Antarctic geology and geophysics. Universitetsforlaget, Oslo, pp 557–562

Austrheim H (1983) Comment on "The statistical analysis and interpretation of imperfectly-fitted Rb-Sr isochrons from polymetamorphic terrains" by M. Cameron et al. Geochim Cosmochim Acta 47:655–658

Ayalon A, Steinitz G, Starinsky A (1987) K-Ar and Rb-Sr whole-rock ages reset during Pan African event in the Sinai Peninsula (Ataqua area). Precamb Res 37:191–198

Baadsgaard H (1973) U-Th-Pb dates on zircons from the early Precambrian Amîtsoq gneisses, Godthaab district, West Greenland. Earth Planet Sci Letters 19:22–28

Baadsgaard H (1976) Further U-Pb dates on zircons from the early Precambrian rocks of the Godthåbsfjord area, West Greenland. Earth Planet Sci Letters 33:261–267

Baadsgaard H (1983) U-Pb isotope systematics on minerals from the gneiss complex at Isukasia, West Greenland. Rapp Grønlands geol Uncers 112:35–42

Baadsgaard H, McGregor VR (1981) The U-Th-Pb systematics of zircons from the type Núk gneisses, Godthåbsfjord, West Greenland. Geochim Cosmochim Acta 45:1099–1110

Baadsgaard H, Lambert RSJ, Krupicka J (1976) Mineral isotopic age relationships in the polymetamorphic Amîtsoq gneiss, Godthaab district, West Greenland. Geochim Cosmochim Acta 40:513–529

Baadsgaard H, Nutman AP, Bridgwater D, Rosing M, McGregor VR, Allaart JH (1984) The zircon geochronology of the Akilia association and Isua supracrustal belt, West Greenland. Earth Planet Sci Letters 68:221–228

Bacon CR, Kurasawa H, Delevaux MH, Kistler RW, Doe BR (1984) Lead and strontium isotopic evidence for crustal interaction and compositional zonation in the source regions of Pleistocene basaltic and rhyolitic magmas of the Coso volcanic field, California. Contrib Mineral Petrol 85:366–375

Bailey DK (1982) Mantle metasomatism-continuing chemical change within the Earth. Nature 296:525–530

Bailey DK (1989) Carbonate melt from the mantle in the volcanoes of south-east Zambia. Nature 338:415–417

Bailey JC, Larsen O, Frolova TI (1987) Strontium isotope variations in Lower Tertiary–Quaternary volcanic rocks from the Kurile island arc. Contrib Mineral Petrol 95:155–165

Bailey RA (1984) Chemical evolution and current state of the Long Valley magma chamber. US Geol Surv Open File Rept 84-939:24–40

Bailey RA (1989) Geologic map of the Long Valley Caldera, Mono-Inyo craters volcanic chain, and vicinity, eastern California. US Geol Surv Map I-1933 (1:62500)

Bailey RA, Dalrymple GB, Lanphere MA (1976) Volcanism, structure, and geochronology of Long Valley Caldera, Mono County, California. J Geophys Res 81:725–744

Baker BH (1967) Geology of the Mount Kenya area. Geol Surv Kenya Rept 79

Baker BH (1987) Outline of the petrology of the Kenya Rift alkaline province. In: Fitton JG, Upton BGJ (eds) Alkaline igneous rocks. Geol Soc London Spec Pub 30:293–311

Baker BH, Miller JH (1963) Geology and geochronology of the Seychelles Islands and the structure of the floor of the Arabian Sea. Nature 199:346–348

Baker BH, Williams LAJ, Miller JA, Fitch FJ (1971) Sequence and geochronology of the Kenya Rift volcanics. Tectonophysics 11:191–215

Baker BH, Goles GG, Leeman WP, Lindstrom MM (1977) Geochemistry and petrogenesis of a basalt-benmoreite-trachyte suite from the southern part of the Gregory Rift, Kenya. Contrib Mineral Petrol 64:303–332

Baker I (1969) Petrology of the volcanic rocks of Saint Helena Island, South Atlantic. Geol Soc Amer Bull 80:1283–1310

Baker I, Gale NH, Simons J (1967) Geochronology of the St. Helena volcanoes. Nature 215:1451–1456

Baker JA, Snee L, Menzies MA (1996a) A brief period of flood volcanism in Yemen: Implications for the duration and rate of continental flood volcanism at the Afro-Arabian triple junction. Earth Planet Sci Letters 138:39–55

Baker JA, Thirlwall MF, Menzies MA (1996b) Sr-Nd-Pb isotopic and trace element evidence for crustal contamination of plume-derived flood basalts: Oligocene flood volcanism in western Yemen. Geochim Cosmochim Acta 60(14):2559–2583

Baker JA, Menzies MA, Thirlwall MF, MacPherson CG (1997) Petrogenesis of Quaternary intraplate volcanism, Sana'a, Yemen: Implications for plume-lithosphere interaction and polybaric melt hybridization. J Petrol 38(10):1359–1390

Baker PE (1973) Volcanism at destructive plate margins. J Earth Sci Leeds 8:183–195

Baker PE (1982) Evolution and classification of orogenic volcanic rocks. In: Thorpe S (ed) Andesites; Orogenic Andesites and Related Rocks. Wiley, Chichester, UK, pp 11–23

Baker PE, Tomblin JF (1964) A recent volcanic eruption on Bouvetoya, South Atlantic Ocean. Nature 303:1055–1056

Baker PE, Gass IG, Harris PG, LeMaitre RW (1964) The volcanological report of the Royal Society Expedition to Tristan da Cunha, 1962. Phil Trans Roy Soc A256:439–578

Baker PE, Buckley F, Holland JG (1974) Petrology and geochemistry of Easter Island. Contrib Mineral Petrol 44:85–100

Baksi AK (1987) Critical evaluation of the age of the Deccan Traps, India: Implications for flood-basalt volcanism and faunal extinctions. Geol 15:147–150

Baksi AK (1995) Petrogenesis and timing of volcanism in the Rajmahal flood basalt province, northeastern India. Chem Geol 121:73–90

Baksi AK (1999) Reevaluation of plate motion based on hotspot tracks in the Atlantic and Indian Oceans. J Geol 107(1):13–26

Baksi AK, Barman TR, Paul DK, Farrar E (1987) Widespread Early Cretaceous flood basalt volcanism in eastern India: Geochemical data from the Rajmahal-Bengal-Sylhet Traps. Chem Geol 63:133–141

Balashov YA (1979) Evolution of isotopic nonhomogeneity of the upper mantle. Geokhimia 1979(12):1767–1783

Baragar WRA (1969) The geochemistry of Coppermine River basalt. Geol Surv Canada, Paper 69-44

Baragar WRA (1972) Coppermine River basalts: geological setting and interpretation. In: Wanless RK, Loveridge WD (eds) Rubidium-Strontium isochron age studies report 1. Geol Surv Canada, Paper 72-23, pp 21–24

Baragar WRA, McGlynn JC (1976) Early Archean basement in the Canadian Shield: A review of the evidence. Geol Surv Canada, Paper 76-14

Barberi F, Borsi S, Ferrara G, Innocenti F (1969) Strontium isotopic composition of some recent basic volcanites in the southern Tyrrhenian Sea and Sicily Channel. Contrib Mineral Petrol 23:157–172

Barberi F, Gasparini P, Innocenti F, Villari L (1973) Volcanism of the southern Tyrrhenian Sea and its geodynamic implications. J Geophys Res 78:5221–5232

Barberi F, Ferrara G, Innocenti F, Keller J, Villari L (1974) Evolution of Eolian arc volcanism (southern Tyrrhenian Sea). Earth Planet Sci Letters 21:269–309

Barberi F, Treuil M, Varet J (1975) A transitional basalt-pantellerite sequence of fractional crystallization, the Boina Centre (Afar Rift, Ethiopia). J Petrol 16:22–56

Barberi F, Civetta L, Varet J (1980) Sr isotopic composition of Afar volcanics and its implication for mantle evolution. Earth and Planet Sci Letters 50:247–259

Barbieri M, Penta A, Turi B (1975) Oxygen and strontium isotope ratios in some ejecta from the Alban Hills volcanic area, Roman comagmatic region. Contrib Mineral Petrol 51:127–134

Barbieri M, Brotzu P, Morbidelli L, Penta A, Piccirillo EM, Traversa G (1976) Trace elements and $^{87}Sr/^{86}Sr$ ratios of the basic stratoid vulcanism in the south-eastern Ethiopian plateau. Periodico di Mineralogia 45:129–146

Barbieri M, Masi U, Tolomeo L (1979) Origin and distribution of strontium in the travertines of Latium (central Italy). Chem Geol 24:181–188

Barbieri M, Peccerillo A, Poli G, Tolomeo L (1988) Major, trace element and Sr isotopic composition of lavas from Vico volcano (central Italy) and their evolution in an open system. Contrib Mineral Petrol 99:485–497

Barca D, Crisci GM, Ranieri GA (1988) Further developments of the Rayleigh equation for fractional crystallization. Earth Planet Sci Letters 89(2):170–172

Bargar KE, Jackson ED (1974) Calculated volumes of individual shield volcanoes along the Hawaiian-Emperor chain. J Research US Geol Survey 2:545–550

Barghoorn ES, Tyler SA (1965) Microfossils from the Gunflint chert. Science 147:563–577

Barker DS (1974) Alkaline rocks of North America. In: Sörensen H (ed) The Alkaline Rocks. Wiley, London, pp 160–171

Barker DS (1983) Igneous rocks. Prentice-Hall, Englewood Cliffs, N.J.

Barker DS, Long LE (1969) Feldspathoidal syenite in a quartz diabase sill, Brookville, New Jersey. J Petrol 10(2):201–221

Barker F (ed) (1979) Trondhjemites, dacites and related rocks. Elsevier, Amsterdam

Barker F (1981) Introduction to special issue on granites and rhyolites: A commentary for the nonspecialist. J Geophys Res 86(B11):10131–10135

Barley ME, Weaver SD, de Laeter JR (1988) Strontium isotope composition and geochronology of intermediate-silicic volcanics, Mt. Somers and Banks Peninsula, New Zealand. New Zealand J Geol Geophys 31:197–206

Barling J, Goldstein SL (1990) Extreme isotopic variations in Heard Island lavas and the nature of mantle reservoirs. Nature 348:59–62

Barling J, Goldstein SL, Nicholls IA (1994) Geochemistry of Heard Island (southern Indian Ocean): Characterization of an enriched mantle component and implications for enrichment of the sub-Indian mantle. J Petrol 35:1017–1053

Barr SM, Macdonald AS (1981) Geochemistry and geochronology of Late Cenozoic basalts of Southeast Asia: Summary. Geol Soc Amer Bull 92:508–512

Barr SM, Macdonald AS (1991) Toward a late Paleozoic-early Mesozoic tectonic model for Thailand. J Thai Geosci 1:11–22

Barrat JA, Nesbitt RW (1996) Geochemistry of the Tertiary volcanism of Northern Ireland. Chem Geol 129:15–38

Barrat JA, Jahn BM, Fourcade S, Joron JS (1993) Magma genesis in an ongoing rifting zone: The Tadjoura Gulf (Afar area). Geochim Cosmochim Acta 57:2291–2302

Barreiro B (1983) Lead isotopic compositions of South Sandwich Island volcanic rocks and their bearing on magmagenesis in intra-oceanic island arcs. Geochim Cosmochim Acta 47:817–822

Barrett TJ, Friedrichsen H (1987) Oxygen-isotopic composition of basalts from young spreading axes in the eastern Pacific. Can J Earth Sci 24:2105–2118

Barrie CT, Davis DW (1990) Timing of magmatism and deformation in the Kamiskotia-Kidd Creek area, western Abitibi subprovince. Precamb Research 46:217–240

Barrie CT, Shirey SB (1991) Nd- and Sr-isotope systematics for the Kamiskotia-Montcalm area: Implications for the formation of Late Archean crust in the western Abitibi subprovince, Canada. Can J Earth Sci 28:58–76

Barth TFW (1945) Studies on the igneous rock complex of the Oslo region. II. Systematic petrography of the plutonic rocks. Skrifter Det Norske Vid.-Akad. I Oslo. I. Mat.-Naturv. Kl. 1944, No. 9

Barth TFW (1954) Studies on the igneous rock complex of the Oslo region. XIV Provenance of the Oslo magmas. Skrifter Det Norske Vid.-Akad. I Oslo. I. Mat.-Naturv. Kl. 1954, No. 4

Barth TFW, Ramberg IB (1967) The Fen circular complex. In: Tuttle OF, Gittins J (eds) The carbonatites. Interscience, New York, pp 225–257

Barton JM Jr (1977) Rb-Sr ages and tectonic setting of some granite intrusions, coastal Labrador. Can J Earth Sci 14:1635–1646

Barton JM Jr, Cawthorn RG, White J (1986) The role of contamination in the evolution of the Platreef of the Bushveld complex. Econ Geol 81:1096–1104

Barton M (1979) A comparative study of some minerals occurring in the potassium-rich alkaline rocks of the Leucite Hills, Wyoming, the Vico volcano, Western Italy, and the Toro-Ankole region, Uganda. Neues Jahrbuch Mineral Abhandl 137:114–134

Barton M, Hamilton DL (1978). Water-saturated melting relations to 5 kilobars of three Leucite Hills lavas. Contrib Mineral Petrol 66:41–49

Barton M, Hamilton DL (1979). The melting relationships of a madupite from the Leucite Hills, Wyoming, to 30 kb. Contrib Mineral Petrol 69:133–142

Barton M, Hamilton DL (1982) Water-undersaturated melting experiments bearing upon the origin of potassium-rich magma. Mineral Mag 45:267–278

Barton M, Bergren MJ van (1981) Green clinopyroxenes and associated phases from the Leucite Hills, Wyoming, to 30 kb. Contrib Mineral Petrol 69:133–142

Barton M, Salters VJM, Huijsmans JPP (1983) Sr isotope and trace element evidence for the role of continental crust in calc-alkaline volcanism on Santorini and Milos, Aegean Sea, Greece. Earth Planet Sci Letters 63:273–291

Basaltic Volcanism Study Project (1981) Basaltic volcanism on the terrestrial planets. Pergamon Press, New York

Basu AR (1978) Trace elements and Sr-isotopes in some mantle-derived hydrous minerals and their significance. Geochim Cosmochim Acta 42:659–668

Basu AR, Murthy VR (1977a). Kaersutites, suboceanic low-velocity zone, and the origin of mid-oceanic ridge basalts. Geol 5:365–368

Basu AR, Murthy VR (1977b) Ancient lithospheric xenoliths in alkali basalt from Baja California. Earth Planet Sci Letters 35:239–246

Basu AR, Wang J, Huang W, Xie G, Tatsumoto M (1991) Major element, REE, and Pb, Nd, and Sr isotopic geochemistry of Cenozoic volcanic rocks of eastern China: Implications for their origin from suboceanic-type mantle reservoirs. Earth Planet Sci Letters 105:149–169

Bateman PC, Clark LS, Huber NK, Moore JG, Rinehart CD (1963) The Sierra Nevada batholith-a synthesis of recent work across the central part. US Geol Surv, Prof Paper 44-D, pp 1–46

Bau M, Knittel U (1993) Significance of slab-derived partial melts and aqueous fluids for the genesis of tholeiitic and calc-alkaline island-arc basalts: Evidence from Mt. Arayat, Philippines. Chem Geol 105:233–251

Baxter AN (1975) Petrology of the Older Series of lavas from Mauritius, Indian Ocean. Geol Soc Amer Bull 86:1449–1458

Baxter AN (1976) Geochemistry and petrogenesis of primitive alkali basalt from Mauritius, Indian Ocean. Geol Soc Amer Bull 87:1028–1034

Baxter AN (1978) Ultramafic and mafic nodule suites in shield-forming lavas from Mauritius, Indian Ocean. J Geol Soc London 135:565–581

Baxter AN, Upton BGJ, White WM (1985) Petrology and geochemistry of Rodrigues Island, Indian Ocean. Contrib Mineral Petrol 89:90–101

Beakhouse GP, McNutt RH, Krogh TE (1988) Comparative Rb-Sr and U-Pb zircon geochronology of late- to post-tectonic plutons in the Winnipeg River belt, northwestern Ontario, Canada. Chem Geol 72:337–351

Beane JE, Turner CA, Hooper PR, Subbarao KV, Wash JN (1986) Stratigraphy, composition, and form of the Deccan basalts, western Ghats, India. Bull Volcanol 48:61–83

Beattie P (1993a) The generation of U-series disequilibria by partial melting of spinel peridotite: constraints from partitioning studies. Earth Planet Sci Letters 117:379–391

Beattie P (1993b) U-Th disequilibria and partitioning on melting of garnet peridotite. Nature 363:63–65

Beckinsale RD (1974) Rb-Sr and K-Ar age determinations, and oxygen isotopic data for the Glen Cannel Granophyre, Isle of Mull, Argyllshire, Scotland. Earth Planet Sci Letters 22:267–274

Beckinsale RD, Obradovich JD (1973) Potassium-argon ages for minerals from the Ross of Mull, Argyllshire, Scotland. Scot J Geol 9(2):147–156

Beckinsale RD, Thompson FN, Durham JJ (1974) Petrogenetic significance of initial $^{87}Sr/^{86}Sr$ ratios in the North Atlantic Tertiary Igneous Province in the light of Rb-Sr, K-Ar, and ^{18}O-abundance studies of the Sarqata qaqa intrusive complex, Ubekendt Ejlard, West Greenland. J Petrol 15:525–538

Beckinsale RD, Bowles JFW, Pankhurst RJ, Wells MK (1977) Rubidium-strontium age studies and geochemistry of acid veins in the Freetown complex, Sierra Leone. Mineral Magazine 41:501–511

Beckinsale RD, Pankhurst RJ, Skelhorn RR, Walsh JN (1978) Geochemistry and petrogenesis of the Early Tertiary lava pile of the Isle of Mull, Scotland. Contrib Mineral Petrol 66:415–427

Beckinsale RD, Gale NH, Pankhurst RJ, Macfarlane A, Crow MJ, Arthurs JW, Wilkirson A (1980) Discordant Rb-Sr and Pb-Pb whole-rock isochron ages for the Archean basement of Sierra Leone. Precambrian Res 13:63–76

Bedard JH (1994) Mesozoic east North American alkaline magmatism: Part 1. Evolution of Monteregian lamprophyres, Quebec, Canada. Geochim Cosmochim Acta 58:95–112

Behrendt JC, Wotorson CS (1970) Aeromagnetic and gravity investigations of the coastal area and continental shelf of Liberia, West Africa, and their relation to continental drift. Geol Soc Amer Bull 81:3563–3574

Behrendt JC, Meiser L, Henderson JR (1966) Airborn geophysical study in the Pensacola Mountains of Antarctica. Science 153:1373–1376

Behrendt JC, Drewry DJ, Jankowski E, Grim MS (1980) Aeromagnetic and radio echo ice-sounding measurements show much greater area of the Dufek intrusion, Antarctica. Science 209:1014–1017

Behrendt JC, LeMasurier WE, Cooper AK, Tessensohn F, Tréhu A, Damaske D (1991) Geophysical studies of the West Antarctic Rift System. Tectonics 10(6):1257–1273

Beiersdorf H, Bach W, Duncan R, Erzinger J, Weiss W (1995) New evidence for the production of EM-type ocean island basalts and large volumes of hyaloclastites during the early history of the Manihiki Plateau. Marine Geol 122:181–205

Bell BR (1984) The geochemistry of Lower Tertiary basic dykes in the eastern Red Hill district, Isle of Skye, and their significance for the proposed magmatic evolution of the Skye centre. Mineral Magazine 34:365–372

Bell JD (1966) Granites and associated rocks of the eastern part of the Western Redhills Complex, Isle of Skye. Trans Roy Soc Edinburgh 66:307–343

Bell JD (1976) The Tertiary intrusive complex on the Isle of Skye. Proceed Geol Assoc 87:247–271

Bell K (ed) (1989) Carbonatites; Genesis and Evolution. Unwin Hyman, London, U.K.

Bell K, Blenkinsop J (1976) A Rb-Sr whole-rock isochron from the Otto stock. Can J Earth Sci 13:998–1002

Bell K, Blenkinsop J (1987a) Nd and Sr isotopic compositions of East African carbonatites: implications for mantle heterogeneity. Geol 15:99–102

Bell K, Blenkinsop J (1987b) Archean depleted mantle: evidence from Nd and Sr initial isotopic ratios of carbonatites. Geochim Cosmochim Acta 51:291–298

Bell K, Blenkinsop J (1989) Neodymium and strontium isotope geochemistry of carbonatites. In: Bell K (ed) Carbonatites. Unwin Hyman London, pp 278–300

Bell K, Dawson JB (1995) Nd and Sr isotope systematics of active carbonatite volcano, Oldoinyo Lengai. In: Bell K, Keller J (eds) Carbonatite volcanism. Springer Verlag, Berlin, pp 100–112

Bell K, Dodson MH (1981) The geochronology of the Tanzanian Shield. J Geol 89:109–128

Bell K, Keller J (eds) (1995) Carbonatite volcanism: Oldoinyo Lengai and the petrogenesis of natrocarbonatites. Springer Verlag, Berlin (IAVCEI Proceedings in Volcanology, 4)

Bell K, Powell JL (1969) Strontium isotopic studies of alkalic rocks: The potassium-rich lavas of the Birunga and Toro-Ankole regions, east and central equatorial Africa. J Petrol 10:536–572

Bell K, Powell JL (1970) Strontium isotopic studies of alkalic rocks: The alkalic complexes of eastern Uganda. Geol Soc Amer Bull 81:3481–3490

Bell K, Simonetti A (1996) Carbonatite magmatism and plume activity: Implications from Nd, Pb and Sr isotope systematics of Oldoinyo Lengai. J Petrol 37:1321–1340

Bell K, Dawson JB, Farquhar RM (1973) Sr isotopic studies of alkali rocks: the active carbonatite volcano Oldoinyo Lengai. Geol Soc Amer Bull 84:1019–1030

Bell K, Blenkinsop J, Cole TJS, Menagh DP (1982) Evidence from Sr isotopes for long-lived heterogeneities in the upper mantle. Nature 298:251–253

Bell K, Blenkinsop J, Kwon ST, Tilton GR, Sage RP (1987) Age and radiogenic isotope systematics of the Borden carbonatite complex, Ontario, Canada. Can J Earth Sci 24:24–30

Bellieni G, Brotzu P, Comin-Chiaramonti F, Ernesto M, Melfi AJ, Pacca IG, Piccirillo EM (1984) Flood basalt to rhyolite suites in the southern Paraná Plateau (Brazil): Paleomagnetism, petrogenesis, and geodynamic implications. J Petrol 25:579–618

Bellieni G, Comin-Chiaramonti P, Marques LS, Melfi AF, Nardy AJR, Papatrechas C, Piccirillo EM, Roisenberg A, Stolfa D (1986) Petrogenetic aspects of acid and basaltic lavas from the Paraná Plateau (Brazil): Geological, mineralogical, and petrological relationships. J Petrol 27:915–944

Bellieni G, Macedo MHF, Petrini R, Piccirillo EM, Cavazzini G, Comin-Chiaramonti P, Ernesto M, Macedo JWP, Martins G, Melfi AJ, Pacca IG, De Min A (1992) Evidence of magmatic activity related to Middle Jurassic and Lower Cretaceous rifting from northeastern Brazil (Ceará-Mirim): K/Ar age, paleomagnetism, petrology, and Sr-Nd isotope characteristics. Chem Geol 97:9–32

Bemmelen RW van (1949) The geology of Indonesia. Martinus Nijhoff, The Hague

Ben Othman D, White WM, Patchett PJ (1989) The geochemistry of marine sediment, island arc magma genesis, and crust-mantle recycling. Earth Planet Sci Letters 94:1–21

Bender JF, Langmuir CH, Hanson GN (1984) Petrogenesis of basalt glasses from the Tamayo region, East Pacific Rise. J Petrol 25:213–254

Bennett VC, Esat TM, Norman MD (1996) Two mantle-plume components in Hawaiian picrites inferred from correlated Os-Pb isotopes. Nature 381:221–223

Benson W (1916) Report on the petrology of the dolerites collected by the British Antarctic expedition, 1907–1909. British Antarctic Exped. 1907–1909, Sci Invs Repts, Geology 2(9):153–160

Bentor YK (1985) The crustal evolution of the Arabo-Nubian massif with special reference to the Sinai Peninsula. Precambrian Res 28:1–74

Berg JH, Klewin KW (1988) High-MgO lavas from the Keweenawan Midcontinent Rift near Mamainse Point, Ontario. Geol 16:1003–1006

Bergman SC (1987) Lamproites and other potassium-rich igneous rocks: a review of their occurrence mineralogy and geochemistry. In: Fitton JG, Upton BGJ (eds) Alkaline igneous rocks. Geol Soc Spec Publ 30:103–189

Berman SC, Foland KA, Spera FJ (1981) On the origin of an amphibole-rich vein in a peridotite inclusion from the Lunar Crater volcanic field, Nevada, USA. Earth Planet Sci Letters 56:343–361

Bergman SC, Coffield DQ, Talbot JP, Garrard RA (1996) Tertiary tectonic and magmatic evolution of western Sulawesi and Makassar Strait, Indonesia: evidence for a Miocene continent-continent collision. In: Hall R, Blundell D (eds) Tectonic evolution of southeast Asia. Geol Soc London, Spec Paper 106: 465–481

Berhe SM, Desta F, Nicoletti M, Teferra M (1987) Geology, geochronology, and geodynamic implications of Cenozoic magmatic province in W and SE Ethiopia. J Geol Soc 144(2): 213–226

Berlin R, Henderson CMB (1969) The distribution of Sr and Ba between the alkali feldspar, plagioclase, and groundmass phases of porphyritic trachytes and phonolites. Geochim Cosmochim Acta 33:247–255

Bernard-Griffiths J, Fourcade S, Dupuy C (1991) Isotopic study (Sr, Nd, O, and C) of lamprophyres and associated dykes from Tamazert (Morroco): Crustal contamination processes and source characteristics. Earth Planet Sci Letters 103:190–199

Bertrand H, Dostal J, Dupuy C (1982) Geochemistry of Early Mesozoic tholeiites from Morocco. Earth Planet Sci Letters 58:225–239

Bertrand JM, Lasserre M (1976) Pan-African and pre-Pan-African history of the Hoggar (Algerian Sahara) in light of new geochronological data from the Aleksod area. Precamb Res 3:342–362

Best MG (1970) Kaersutite-periodotite inclusions and kindred megacrysts in basanite lavas, Grand Canyon, Arizona. Contrib Mineral Petrol 27:25–44

Best MG (1982) Igneous and metamorphic petrology. Freeman, San Francisco

Betton PJ (1979) Isotopic evidence for crustal contamination in the Karroo rhyolites of Swaziland. Earth Planet Sci Letters 45:263–274

Betton PJ, Civetta L (1984) Strontium and neodymium isotopic evidence for the heterogeneous nature and development of the mantle beneath Afar (Ethiopia). Earth Planet Sci Letters 71:59–70

Betton PJ, Armstrong RA, Manton WI (1984) Variations in the lead isotopic composition of Karoo magmas. In: Erlank AJ (ed) Petrogenesis of the volcanic rocks of the Karoo Province. Geol Soc S Africa, Spec Pub 13:331–339

Bibikova EV, Kirnozova TI, Maximov AP, Makarov VA (1983) Study of lead isotope compositions of andesites of the Bezymyanny volcano, Kamchatka. Geokhimia 1983(2):163–171

Bickle MS, Martin A, Nisbet EG (1975) Basaltic and peridotitic komatiites and stromatolites above a basal unconformity in the Belingwe greenstone belt, Rhodesia. Earth Planet Sci Letters 27:155–162

Bielski M, Jäger E, Steinitz G (1979) The geochronology of Iqna Granite (Wadi Kid pluton), southern Sinai. Contrib Mineral Petrol 70:159–165

Bielski-Zyskind M, Wasserburg GJ, Nixon PH (1984) Sm-Nd and Rb-Sr systematics in volcanics and ultramafic xenoliths from Malaita, Solomon Islands, and the nature of the Ontong Java Plateau. J Geophys Res 89:2415–2424

Bingham JW, Grolier MJ (1966) The Yakima Basalt and Ellensburg Formation of south-central Washington. US Geol Survey Bull 1224-G

Birck JL (1986) Precision K-Rb-Sr isotopic analysis: application to Rb-Sr chronology. Chem Geol 56:73–83

Birck JL, Allègre CJ (1973) ^{87}Rb-^{87}Sr systematics of Muntsche Tundra mafic pluton (Kola Peninsula, USSR). Earth Planet Sci Letters 20:266–274

Bird DK, Rogers RD, Manning CE (1986) Mineralized fracture systems of the Skaergaard intrusion. Medd Grönland Geosci 16:1–68

Bird DK, Manning CE, Rose NM (1988) Hydrothermal alteration of Tertiary layered gabbros, East Greenland. Amer J Sci 288:405–407

Black LP, Gale NH, Moorbath S, Pankhurst RJ, McGregor VR (1971) Isotopic dating of very early Precambrian amphibolite facies gneisses from the Godthaab district, West Greenland. Earth Planet Sci Letters 12:245–259

Black LP, Moorbath S, Pankhurst RJ, Windley BF (1973) ^{207}Pb/^{206}Pb whole rock age of the Archaean granulite facies metamorphic event in West Greenland. Nature 244:50–53

Blattner P, Reid F (1982) The origin of lavas and ignimbrites of the Taupo volcanic zone, New Zealand, in the light of oxygen isotope data. Geochim Cosmochim Acta 46:1417–1429

Blaxland AB, Breemen O van, Steenfelt A (1976) Age and origin of agpaitic magmatism at Ilimaussaq, South Greenland: Rb-Sr study. Lithos 9:31–38

Blenkinsop J, Bell K (1983) Rb-Sr geochronology of the Coldwell Complex, northwestern Ontario, Canada: Discussion. Can J Earth Sci 20:1499–1500

Blichert-Toft J, Lesher CE, Rosing MT (1992) Selectively contaminated magmas of the Tertiary East Greenland macrodike complex. Contrib Mineral Petrol 110:154–172

Bloomer SH, Stern RJ, Fisk E, Geschwind CH (1989) Shoshonitic volcanism in the northern Mariana arc. 1. Mineralogic and major and trace element characteristics. J Geophys Res 94(B4):4469–4496

Blundy JD, Wood BJ (1991) Crystal-chemical controls on the partitioning of Sr and Ba between plagioclase feldspar, silicate melt, and hydrothermal solutions. Geochim Cosmochim Acta 55:193–209

Blusztajn J, Hart SR (1989) Sr, Nd, and Pb isotopic character of Tertiary basalts from southwest Poland. Geochim Cosmochim Acta 53:2689–2696

Blusztajn J, Hart SR, Shimizu N, McGuire AV (1995) Trace-element and isotopic characteristics of spinel peridotite xenoliths from Saudi Arabia. Chem Geol 123:53–65

Boettcher AL, O'Neil JR (1980) Stable isotope, chemical, and petrographic studies of high pressure amphiboles and micas: Evidence for metasomatism in the mantle source regions of alkali basalt and kimberlites. Amer J Sci 280A:594–621

Bogatikov OA, Kononova VA, Pervov VA, Zhuravlev DZ (1994) Petrogenesis of Mesozoic potassic magmatism of the central Aldan: A Sr-Nd isotopic and geodynamic model. Internat Geol Review 36(7):629–644

Boissonnas J, Borsi S, Ferrara G, Fabre J, Fabriès J, Gravelle M (1969) On the early Cambrian age of two late orogenic granites from west-central Ahaggar. Can J Earth Sci 6:25–37

Bokhari FY, Kramers JD (1981) Island arc character and late Precambrian age of volcanics at Wadi Shwas, Hijaz, Saudi Arabia: Geochemical and Sr and Nd isotopic evidence. Earth Planet Sci Letters 54:409–422

Bonatti E (1990) Subcontinental mantle exposed in the Atlantic Ocean on St. Peter-Paul islets. Nature 345:800–801

Bonatti E, Harrison CGA, Fisher DE, Honnorez J, Schilling J-G, Stipp JJ, Zentilli M (1977) Easter volcanic chain (southeast Pacific) a mantle hot line. J Geophys Res 82:2457–2478

Bonatti E, Ottonello G, Hamlyn PR (1986) Peridotites from the island of Zabargad (St. John's), Red Sea: Petrology and geochemistry. J Geophys Res 91:599–631

Bonini WE, Hargraves RB, Shagam R (eds) (1984) The Caribbean-South American plate boundary and regional tectonics. Geol Soc Amer, Mem 16

Bornhorst TJ, Paces JP, Grant NK, Obradovic JD, Huber NK (1988) Age of native copper mineralization, Keweenaw Peninsula, Michigan. Econ Geol 83:619–625

Borsuk AM, Tsvetkov AA, Chernyshev IV, Gol'tsman YV, Bairova ED (1983) Strontium-isotope distribution in igneous rocks of the Komandor Islands. Doklady Akad Nauk SSSR 268:100–103

Bosworth W, Strecker MR (1997) Stress field changes in the Afro-Arabian rift system during the Miocene to Recent period. Tectonophysics 278:47–62

Boudreau AE, McBirney AR (1997) The Skaergaard Layered Series. Part III: Non-dynamic layering. J Petrol 38:1003–1020

Bougault H, Dmitriev L, Schilling J-G, Sobolev A, Joron JL, Needham HD (1988) Mantle heterogeneity from trace elements, MAR triple junction near 14° N. Earth Planet Sci Letters 88:27–36

Bowden P (1985) The geochemistry and mineralization of alkaline ring complexes in Africa (a review). J African Earth Sci 3:17–39

Bowden P, Breemen O van, Hutchinson J, Turner DC (1976) Paleozoic and Mesozoic age trends for some ring complexes in Niger and Nigeria. Nature 259:297–299

Bowden P, Kinnaird JA, Abba SI, Ike EC, Turaki U (1984) Geology and mineralization of the Nigerian anorogenic ring complexes. Geol Jb (Hannover) B56:1–65

Bowen NL (1924) The Fen area of Telemark. Amer J Sci 8:1–11

Bowen NL (1926) The carbonate rocks of the Fen area in Norway. Amer J Sci 12:499–502

Bowen NL (1928) The evolution of igneous rocks. Princeton Un. Press and Dover Publications (1956)

Bowring SA, Podosek FA (1989) Nd isotopic evidence from Wopmay orogen for 2.0–2.4 Ga crust in western North American. Earth Planet Sci Letters 94:217–230

Bowring SA, Ross GM (1985) Geochronology of the Narakay volcanic complex: implications for the age of the Coppermine homocline and Mackenzie igneous events. Can J Earth Sci 22:774–780

Box SE, Flower MFJ (1989) Introduction to special section on alkaline arc magmatism. J Geophys Res 94(B4):4467–4468

Braitsch O (1971) Salt deposits: Their origin and composition. Springer Verlag, New York

Breemen O van, Aftalion M, Pidgeon RT (1971) The age of the granitic injection complex of Harris, Outer Hebrides. Scottish J Geol 7:139–152

Breemen O van, Hutchinson , Bowden P (1975) Age and origin of the Nigerian Mesozoic granites: A Rb-Sr isotopic study. Contrib Mineral Petrol 50:157–172

Breemen O von, Dallmeyer RD (1984) The scale of Sr isotopic diffusion during post-metamorphic cooling of gneisses in the Inner Piedmont of Georgia, southern Appalachians. Earth Planet Sci Letters 68:141–150

Brett R (1987) Introduction to Juan de Fuca Ridge special section. J Geophys Res 92(B 1):11281–11399

Brewer TS, Brook D (1991) The geochemistry of Mesozoic tholeiites from Coats Land and Dronning Maud Land. In: Thomson MRA et al. (eds) Geological evolution of Antarctica. Cambridge Un. Press, pp 569–572

Briqueu L, Lancelot JR. (1979) Rb-Sr systematics and crustal contamination models for calc-alkaline igneous rocks. Earth Planet Sci Letters 43:385–396

Briqueu L, Lancelot JR. (1983) Sr isotopes and K, Rb, Sr balance in sediments and igneous rocks from the subducted plate of the Vanuatu (New Hebrides) active margin. Geochim Cosmochim Acta 47:191–200

Briqueu L, Lancelot JR, Tatsumoto M (1982) Origin and evolution of Vanuatu island arc magmas: evidence by trace elements and Sr-Nr isotope. 5th Internat. Conf. Geochem. Cosmochem. Isotope Geol., Japan

Briqueu L, Javoy M, Lancelot JR, Tatsumoto M (1986). Isotope geochemistry of recent magmatism in the Aegean arc: Sr, Nd, Hf and O isotopic ratios in the lavas of Milos and Santorini – geodynamic implications. Earth Planet Sci Letters 80:41–54

Bristow JW, Allsopp HL, Erlank AJ, Marsh JS, Armstrong RA (1984) Strontium isotope characterization of Karoo volcanic rocks. In: Erlank AJ (ed) Petrogenesis of the volcanic rocks of the Karoo Province. Spec Publ Geol Soc S Africa 13:295–329

Brögger WC (1921) Die Eruptvgesteine des Kristianiagebietes. IV. Das Fengebiet in Telemark, Norwegen. Vidensk. Skr., I Mat.-Nat. Kl. 1920, 1. Kristiania

Brögger WC (1933) Die Eruptivgesteine des Oslogebietes. VII. Die chemische Zusammensetzung der Eruptivgesteine des Oslogebietes. Skr. Nor. Vidensk.-Akad. 1, Oslo Mat. Naturv. Kl. 1933(1)

Broccum SJ, Dalz el IWD (1974) The Sudbury Basin, the Southern Province, the Grenville Front, and the Penokean Orogeny. Geol Soc Amer Bull 85:1571–1580

Brookins DG, Murphy MT, Matheney RK (1986) Strontium isotopic analysis of Grande Ronde basalt and fracture filling minerals, drill hole DC-6, Hanford Reservation, Washington. Isochron West, 4C:19

Brooks C (1966) The effect of mineral age discordancies on total-rock Rb-Sr isochrons of the Heemskirk granite, Western Tasmania. J Geophys Res 71(22):5447–5458

Brooks C (1968) Relationship between feldspar alteration and the precise post-crystallization movement of rubidium and strontium isotopes in a granite. J Geophys Res 73(14):4751–4757

Brooks C (1980) The Rb/Sr geochronology of the Archean Chibougamau pluton, Quebec. Can J Earth Sci 17:776–783

Brooks C, Hart SR (1972) An extrusive basaltic komatiite from a Canadian Archean metavolcanic belt. Can J Earth Sci 9:1250–1253

Brooks C, Hart SR (1978) Rb-Sr mantle isochrons and variations in the chemistry of Gondwana-lands's lithosphere. Nature 271:220–223

Brooks C, Wendt I, Harre W (1968) A two-error regression treatment and its implication to Rb-Sr and initial $^{87}Sr/^{86}Sr$ ratios of younger Variscan granitic rocks from the Schwarzwald Massif, southwest Germany. J Geophys Res 73(18):6071–6084

Brooks C, Hart SR, Krogh TE, Davis GL (1969a) Carbonate contents and $^{87}Sr/^{86}Sr$ ratios of calcites from Archaean metavolcanics. Earth Planet Sci Letters 6:35–38

Brooks C, Krogh TE, Hart SR, Davis GL (1969b) The initial $^{87}Sr/^{86}Sr$ ratios of the Upper Metavolcanics, Ontario, Canada, and Lower Series, Michipicoten. Ann. Rept., Dept. Terrestrial Magnetism, Carnegie Inst. Washington, 422pp

Brooks C, Hart SR, Wendt I (1972) Realistic use of two-error regression treatments as applied to rubidium-strontium data. Reviews of Geophysics and Space Physics 10(2):551–577

Brooks C, Hart SR, Hofmann A, James DE (1976a) Rb-Sr mantle isochrons from oceanic regions. Earth Planet Sci Letters 32:51–61

Brooks C, James DE, Hart SR (1976b) Ancient lithosphere: its role in young continental volcanism. Science 193:1086

Brooks CK, Gleadow AJW (1977) A fission-track age for the Skaergaard intrusion and the age of the East Greenland basalts. Geol 5:539–540

Brooks CK, Nielsen TFD (1978) Early stages in the differentiation of the Skaergaard magma as revealed by a closely related suite of dike rocks. Lithos 11:1–14

Brooks CK, Jakobsson SP, Campsie J (1974) Dredged basaltic rocks from the seaward extensions of the Reykjanes and Snaefellsnes volcanic zones, Iceland. Earth Planet Sci Letters 22:320–327

Brotzu P, Morbidelli L, Piccirillo EM, Traversa GM (1974) Transitional basalts of alkaline or tholeiitic affinity in the Somali Trap Series (southeastern margin of the Main Ethiopian Rift from 8°10' to 8°70' Lat. N). Bull Volcanol 38:254–269

Brown AC (1974) An epigenetic origin for stratiform Cd-Pb-Zn sulfides in the lower Nonesuch Shale, White Pine, Michigan. Econ Geol 69:271–274

Brown GM (1963) Melting relations of Tertiary granitic rocks in Skye and Rhum. Mineral Mag (London) 33:533

Brown GM, Holland JG, Sigurdsson HG, Tomblin JF, Arculus RJ (1977a) Geochemistry of the Lesser Antilles volcanic arc. Geochim Cosmochim Acta 41:785–801

Brown GM, Pinsent RH, Croisy P (1980) The petrology of spinel-peridotite xenoliths from the Massif Central, France. Amer J Sci 280A:471–498

Brown M, Piccoli PM (1995) The Origin of granites and related rocks. US Geol Survey Circular 1129

Brown PE, Breemen O van, Nobel RH, McIntyre RM (1977b) Mid-Tertiary igneous activity in East Greenland – the Kialineq complex. Contrib Mineral Petrol 64:109–122

Brueckner HK, Rex DC (1980) K-Ar and Rb-Sr geochronology and Sr isotopic study of the Alnö alkaline complex, northeastern Sweden. Lithos 13:171–180

Brueckner HK, Zindler A, Seyler M, Bonatti E (1988) Zabargad and the isotopic evolution of the sub-Red Sea mantle and crust. Tectonophysics 150:163–176

Brügmann GE, Naldrett AJ, Asif M, Lightfoot PC, Gorbachev NS, Fedorenko VA (1993) Siderophile and chalcophile metals as tracers of the evolution of the Siberian Trap in the Noril'sk region, Russia. Geochim Cosmochim Acta 57:2001–2018

Bryan WB, Thompson G, Frey FA, Dickey JS (1976) Inferred settings and differentiation in basalts from the Deep Sea Drilling Project. J Geophys Res 81:4285–4304

Buchanan PC, Reimold WU (1998) Studies of the Rooiberg Group, Bushveld Complex, South Africa: No evidence for an impact origin. Earth Planet Sci Letters 155:149–166

Bunsen RW (1851) Über die Prozesse der vulkanischen Gesteinsbildung Islands. Ann Phys Chem 83:197–272

Burchfield BC, Royden L (1982) Carpathian Foreland Fold and thrust belt and its relation to Pannonian and other basins. Amer Assoc Petrol Geol Bull 66:1179–1195

Burgess CH (1941) Igneous rocks of the Highwood Mountains, Montana. Part IV: The Stocks. Geol Soc Amer Bull 52:1809–1828

Burke K (1988) Tectonic evolution of the Caribbean. Ann Reviews Earth Planet Sci 16:201–230

Burke K (1996) The African plate. S African J Geol 99:341–409

Burke WH, Otto JB, Denison RE (1969) Potassium-argon dating of basaltic rocks. J Geophys Res 74:1082–1086

Burnie SW, Schwartz HP, Crockett JH (1972) A sulfur isotopic study of the White Pine mine, Michigan. Econ Geol 67:895–914

Burns RG, Fyfe WS (1967) Trace element distribution rules and their significance. Chem Geol 2:89–104

Burton KW, O'Nions RK (1990) Fe-Ti oxide chronometry: with applications of granulite formation. Geochim Cosmochim Acta 54:2593–2602

Burwell ADM (1975) Rb-Sr isotope geochemistry of lherzolites and their constituent minerals from Victoria, Australia. Earth Planet Sci Letters 28:69–78

Butakova EL (1974) Regional distribution and tectonic relations of the alkaline rocks of Siberia. In: Sörensen H (ed) The alkaline rocks. Wiley, London, pp 172–188

Butler BS, Burbank WS (1929) The copper deposits of Michigan. US Geol Survey Prof Paper 144

Butler JC (1982) Artificial isochrons. Lithos 15:212–214

Cahen L, Snelling NJ (1966) The geochronology of equatorial Africa. North-Holland, Amsterdam

Cahen L, Delhal J, Deutsch S (1972) A comparison of the ages of granites of S.W. Uganda with those of the Kibaran of central Shaba (Katanga), Rep. Zaire. Musée Roy De l'Afrique Centrale, Terouren, Ser. 1N8 73:51–67

Cahen L, Snelling NJ, Delhal J, Vail JR, Bonhomme M, Ledent D (1984) The geochronology and evolution of Africa. Clarendon Press, Oxford

Calvez JY, Lippolt HJ (1980) Strontium isotope constraints to the Rhinegraben volcanism. Neues Jahrbuch Mineral Abhandl 139:59–81

Cameron KL, Cameron M (1985) Rare earth element, $^{87}Sr/^{86}Sr$, and $^{143}Nd/^{144}Nd$ compositions of Cenozoic orogenic dacites from Baja California, northwestern Mexico, and adjacent west Texas: Evidence for the predominance of a subcrustal component. Contrib Mineral Petrol 91:1–11

Cameron M, Collerson KD, Compston W, Morton R (1981) The statistical analysis and interpretation of imperfectly-fitted Rb-Sr isochrons from polymetamorphic terrains. Geochim Cosmochim Acta 45:1087–1097

Cameron WE, McCulloch MT, Walker DA (1983) Boninite petrogenesis: Chemical and Nd-Sr isotopic constraints. Earth Planet Sci Letters 65:75–89

Campbell IH, Griffiths RW (1990) Implications of mantle plume structure for the evolution of flood basalts. Earth Planet Sci Letters 99:79–93

Campbell IH, Turner JS (1987) A laboratory investigation of assimilation at the top of a basaltic magma chamber. J Geol 95:155–172

Campbell IH, Hill RET, Griffiths RW (1990) Melting in an Archaean mantle plume: heads it's basalts, tails it's komatiites. Nature 339:697–699

Campbell IH, Czamanske GK, Fedorenko VA, Hill RI, Stepanov V, Kunilov VE (1992) Synchronism of the Siberian traps and the Permian-Triassic boundary. Science 258:1760–1763

Cannon WF, Green AG, Hutchinson DR, Lee M, Milkereit B, Behrendt JC, Halls HC, Green JC, Dickas AB, Morey GB, Sutcliffe R, Spencer C (1989) The North American Midcontinent Rift beneath Lake Superior from GLIMPCE seismic reflection profiling. Tectonics 8:305–332

Card KD (1990) A review of the Superior Province of the Canadian Shield, a product of Archean accretion. Precamb Res 48:99–156

Carignan J, Gariépy C, Hillaire-Marcel C (1997) Hydrothermal fluids during Mesozoic reactivation of the St. Lawrence rift system, Canada: C, O, Sr and Pb isotopic characterization. Chem Geol 137:1–21

Carlson RW (1984) Isotopic constraints on Columbia River flood basalt genesis and the nature of the subcontinental mantle. Geochim Cosmochim Acta 48:2357–2372

Carlson RW, Hart WK (1987) Crustal genesis on the Oregon Plateau, J Geophys Res 92(B7):6191–6206

Carlson RW, Hart WK (1988) Flood basalt volcanism in the northwestern United States. In: Macdougall JD (ed) Continental flood basalts. Kluwer Academic Press, Dordrecht, pp 35–62

Carlson RW, Lugmair GW, Macdougall JD (1981) Columbia River volcanism: the question of mantle heterogeneity or crustal contamination. Geochim Cosmochim Acta 45:2483–2499

Carlson RW, Lugmair GW, Macdougall JD (1983) Columbia River volcanism: The question of mantle heterogeneity or crustal contamination (reply to a comment by D.J. DePaolo). Geochim Cosmochim Acta 47:845–846

Carlson RW, Esperanza S, Svisero DP (1996) Chemical and Os isotopic study of Cretaceous potassic rocks from southern Brazil. Contrib Mineral Petrol 125(4):393–405

Carmichael ISE (1967) The mineralogy and petrology of the volcanic rocks from the Leucite Hills, Wyoming. Contrib Mineral Petrol 15:24–66

Caroff M, Maury RC, Guille G, Cotten J (1997) Partial melting below Tubuai (Austral Islands, French Polynesia). Contrib Mineral Petrol 127:369–382

Carroll MR, Wyllie PJ (1989) Experimental phase relations in the system tonalite-peridotite-H_2O at 15 kb; implications for assimilation and differentiation processes near the crust-mantle boundary. J Petrol 30:1351–1382

Carter RM (1967) The geology of Pitcairn Island, South Pacific Ocean. B.P. Bishop Mus Bull 231:1–38

Carter SR, Civetta L (1977) Genetic implications of the isotope and trace element variations in the eastern Sicilian volcanics. Earth Planet Sci Letters 36:168–180

Carter SR, Norry MJ (1976) Genetic implications of Sr isotopic data from the Aden Volcano, South Arabia. Earth Planet Sci Letters 31:161–166

Carter SR, Evenson NM, Hamilton PJ, O'Nions RK (1978a) Continental volcanics derived from enriched and depleted source regions: Nd- and Sr-isotope evidence. Earth Planet Sci Letters 37:401–408

Carter SR, Evensen NM, Hamilton PJ, O'Nions RK (1978b) Neodymium and strontium isotope evidence for crustal contamination of continental volcanics. Science 202:743–747

Carter SR, Evensen NM, Hamilton PJ, O'Nions RK (1979) Basalt magma sources during opening of the North Atlantic. Nature 281:41–42

Castillo PR (1988) The Dupal anomaly as a trace of the upwelling lower mantle. Nature 336:667–669

Castillo PR, Batiza R (1989) Strontium, neodymium and lead isotope constraints on near-ridge seamount production beneath the South Atlantic. Nature 342:262–264

Castillo PR, Carlson RW, Batiza R (1991) Origin of Nauru Basin igneous complex: Sr, Na and Pb isotope and REE constraints. Earth Planet Sci Letters 103(1/4):200–213

Castorina F, Riccardo P (1989) Radiometric geochronology: some constraints to the isochron method by an iterative least-squares approach. Geochem J 23:101–109

Cattell A, Arndt N(1987) Low- and high-alumina komatiites from a Late Archaean sequence, Newton Township, Ontario. Contrib Mineral Petrol 87:218–227

Cavazzini G (1988) Linear correlation between pairs of Rb-Sr isochron ages from coexisting metamorphic micas. Isotope Geosci 8:29–36

Cavazzini G (1994) Increase of $^{87}Sr/^{86}Sr$ in residual liquids of high Rb/Sr magmas that evolve by fractional crystallization. Chem Geol (Isotope Geoscience Section) 118:321–326

Cawthorn RG, Curran EB, Arculus RJ (1973) A petrogenetic model for the origin of the calc-alkaline suite of Grenada, Lesser Antilles. J Petrol 14:327–337

Cawthorn RG, Barton JM Jr, Viljoen MJ (1985) Interaction of floor rocks with the Platreef on Overysel, Potgietersrus, northern Transvaal. Econ Geol 80:988–1006

Cawthorn RG, Meyer PS, Kruger FJ (1991) Major addition of magma at the Pyroxenite Marker in the Western Bushveld complex, South Africa. J Petrol 32:739–763

Chabaux F, Hémond C, Allègre CJ (1999) ^{238}U-^{230}Th-^{226}Ra disequilibria in the Lesser Antilles arc: Implications for mantle metasomatism. Chem Geol 153:171–185

Chalokwu CI, Seney PJ (1995) Open-system magma chamber process in the Freetown Complex of Sierra Leone: Evidence from zone 3. Geol Mag 132:261–266

Chandrasekharam D, Parthasarathy A (1978) Geochemical and tectonic studies on the coastal and inland Deccan Trap volcanics and a model for the evolution of Deccan Trap volcanism. Neues Jahrbuch Miner Abhandl 132(2):214–229

Charrier R, Linares E, Niemeyer H, Skarmeta J (1979) K-Ar ages of basalt flows of the Meseta Buenos Aires in southern Chile and their relation to the southeast Pacific triple junction. Geol 7:436–439

Chaudhuri S, Faure G (1967) Geochronology of the Keweenawan rocks, White Pine, Michigan. Econ Geol 62(8):1011–1033

Chaudhuri S, Faure G (1968) Rubidium-strontium age of the Mount Bohemia intrusion in Michigan. J Geol 76:488–490

Chauvel C, Jahn BM (1984) Nd-Sr isotope and REE geochemistry of alkali basalts from the Massif Central, France. Geochim Cosmochim Acta 48:93–110

Chauvel C, Hofmann AW, Vidal P (1992) HIMU-EM: The French Polynesian connection. Earth Planet Sci Letters 110:99–119

Chauvel C, Goldstein SL, Hofmann AW (1995) Hydration and dehydration of oceanic crust controls Pb evolution in the mantle. Chem Geol 126:65–75

Chauvel C, McDonough W, Guille G, Maury R, Duncan RA (1997) Contrasting old and young volcanism on Rurutu Island, Austral chain. Chem Geol 139:125–143

Chayes F (1963) Relative abundance of intermediate members of the oceanic basalt-trachyte association. J Geophys Res 68: 1519–1534

Chayes F (1970) On estimating the magnitude of the Hidden Zone and the compositions of the residual liquids of the Skaergaard Layered Series. J Petrol 11:1–14

Chayes F (1977) Use of correlation statistics with rubidium-strontium systematics. Science 196:1234–1235

Chazot G, Bertrand H (1993) Mantle sources and magma-continental crust interaction during early Red Sea-Aden rifting in southern Yemen: elemental and Sr-Nd-Pb isotope evidence. J Geophys Res 98:1319–1835

Chen C-H, Shieh Y-N, Lee T, Chen C-H, Mertzman SA (1990) Nd-Sr-O isotopic evidence for source contamination and an unusual mantle component under Luzon arc. Geochim Cosmochim Acta 54:2473–2483

Chen C-Y (1987) Lead isotope constraints on the origin of Hawaiian basalts. Nature 327:49–51

Chen C-Y, Frey FA (1983) Origin of Hawaiian tholeiite and alkalic basalt. Nature 302:785–789

Chen C-Y, Frey FA (1985) Trace element and isotopic geochemistry of lavas from the Haleakala volcano, East Maui, Hawaii: Implications for the origin of Hawaiian basalts. J Geophys Res 90:8743–8768

Chen D, Zhi X, Li B (1995) Chemical and isotopic character of gabbroic xenoliths from Hannuoba, China. Chinese J Geochem 14:276–287

Chiesa S, Civetta L, DeFino M, LaVolpe L, Orsi G (1989) The Yemen Trap Series. Genesis and evolution of a continental basalt province. J Volcanol Geotherm Res 36:337–350

Christensen JN, DePaolo DJ (1993) Time scales of large volume silicic magma systems: Sr isotopic systematics of phenocrysts in glass from the Bishop Tuff, Long Valley, California. Contrib Mineral Petrol 113:100–114

Christensen JN, Halliday AN (1996) Rb-Sr ages and Nd isotopic compositions of melt inclusions from the Bishop Tuff and the generation of silicic magma. Earth Planet Sci Letters 144:547–563

Christiansen RL, Elank HR Jr (1972) Volcanic stratigraphy of the Quaternary rhyolite plateau in Yellowstone National Park. US Geol Survey Prof Paper 729B

Church SE (1973) Limits of sediment involvement in the genesis of orogenic volcanic rocks. Contrib Mineral Petrol 39:17–32

Church SE (1976) The Cascade Mountains revisited: A re-evaluation in light of new lead isotopic data. Earth Planet Sci Letters 29:175–188

Church SE, Tatsumoto M (1975) Lead isotope relations in oceanic ridge basalts from the Juan de Fuca-Gorda Ridge area, N.E. Pacific Ocean. Contrib Mineral Petrol 53:253–279

Church SE, Tilton GR (1973) Lead and strontium isotopic studies in the Cascade Mountains: Bearing on andesite genesis. Geol Soc Amer Bull 84:431–454

Church SE, Letturay AP, Grant AR, Delevaux MH, Gray JE (1986) Lead-isotopic data from sulfide minerals from the Cascade Range, Oregon and Washington. Geochim Cosmochim Acta 50:317–328

Civetta L, La Volpe L, Lirer L (1978) K-Ar ages of the Yemen plateau. J Volcano Geotherm Res 4:307–314

Civetta L, Gallo G, Orsi G (1991) Sr- and Nd-isotopic and trace-element constraints on the chemical evolution of the magmatic system of Ischia (Italy). J Volcanol Geotherm Res 46 (3/4): 213–230

Clague DA (1987) Hawaiian alkaline volcanism. In: Fitton JG, Upton BGJ (eds) Alkaline igneous rocks. Geol Soc London Spec Publ 30:227–252

Clague DA, Dalrymple GB (1988) Age and petrology of alkalic post-shield and rejuvenated stage lava from Kauai, Hawaii. Contrib Mineral Petrol 99:202–218

Clague DA, Frey FA (1982) Petrology and trace element geochemistry of the Honolulu Volcanics, Oahu: Implications for the oceanic mantle below Hawaii. J Petrol 23(3):447–504

Clague DA, Chen D-G, Murnane R, Beeson MH, Lanphere MA, Dalrymple GB, Friesen W, Holcomb RT (1982) Age and petrology of the Kalaupapa basalt, Molokai, Hawaii. Pacific Science 36(4):411–420

Clague DA, Holcomb RT, Sito IM, Detrick RS, Terresan ME (1990) Pliocene and Pleistocene alkalic flood basalts on the seafloor north of the Hawaiian Islands. Contrib Mineral Petrol 98: 175–191

Clark JG, Dymond J (1977) Geochronology and petrochemistry of Easter and Sala y Gomez islands: Implications for the origin of the Sala y Gomez Ridge. J Volcanol Geotherm Res 2:29–48

Clark SK, Reagan MK, Plank T (1998) Trace elements and U-series systematics for 1963–1965 tephras from Irazu Volcano, Costa Rica: implications for magma generation processes and transit times. Geochim Cosmochim Acta 62:2689–2700

Clarke DB (1970) Tertiary basalts of Baffin Bay: possible primary magma from the mantle. Contrib Mineral Petrol 25:203–224

Clarke I, McDougall I, Whitford DJ (1983) Volcanic evolution of Heard and McDonald Islands, southern Indian Ocean. In: Oliver RL, James PR, Jago JB (eds) Antarctic Earth Science. Australian Academy of Sciences, Canberra, ACT, pp 631–635

Class C, Goldstein SL, Galer SJG, Weis D (1993) Young formation age of a mantle source. Nature 362:715–721

Clauer N (1979) Relationship between the isotopic composition of strontium in newly formed continental clay minerals and their source materials. Chem Geol 27:115–124

Clauer N, Hoffert M, Karpoff A-M (1982) The Rb-Sr isotope system as an index of origin and diagenetic evolution of southern Pacific red clays. Geochim Cosmochim Acta 46(12): 2659–2664

Clayton RN (1993) Oxygen isotopes in meteorites. Ann Rev Earth Planet Sci 21:115–149

Cleverly RW, Betton PJ, Bristow JW (1984) Geochemistry and petrogenesis of the Lebombo rhyolites. In: Erlank AJ (ed) Petrogenesis of the volcanic rocks of the Karoo Province. Spec Publ Geol Soc S Africa 13:171–194

Cliff RA, Baker PE, Mateer NI (1991) Geochemistry of Inaccessible Island volcanics. Chem Geol 92:251–260

Clifford TN (1970) The structural framework of Africa. In: Clifford TN, Gass IG (eds) African Magmatism and Tectonics. Oliver and Boyd, Edinburgh, pp 449–460

Clifford TN, Gass IG (eds) (1970) African magmatism and tectonics. Oliver and Boyd, Edinburgh

Cobbing EJ, Ozard JM, Snelling NJ (1977) Reconnaissance geochronology of the crystalline basement rocks of the coastal Cordillera of southern Peru. Geol Soc Amer Bull 88:241–246

Coetzee H, Kruger FJ (1989) The geochronology, Sr- and Pb-isotope geochemistry of the Losberg complex, and the southern limit of the Bushveld complex. S African J Geol 92:37–41

Cohen AS, O'Nions RK (1994) Melting rates beneath Hawaii: evidence from uranium-series isotopes in recent lavas. Earth Planet Sci Letters 120:169–175

Cohen RS, O'Nions RK (1982a) The lead, neodymium and strontium isotopic structure of oceanic ridge basalts. J Petrol 23: 299–324

Cohen RS, O'Nions RK (1982b) Identification of recycled continental material from Sr, Nd, and Pb isotope investigations. Earth Planet Sci Letters 61:73–84

Cohen RS, Evensen NM, Hamilton PJ, O'Nions RK (1980) U-Pb, Sm-Nd and Rb-Sr systematics of mid-ocean ridge basalt glasses. Nature 283:149–152

Cohen RS, O'Nions RK, Dawson JB (1984) Isotope geochemistry of xenoliths from East Africa: implications for development of mantle reservoirs and their interaction. Earth Planet Sci Letters 68:209–220

Cole JW, Ewart A (1968) Contributions to the volcanic geology of the Black Island, Brown Peninsula, and Cape Bird area, McMurdo Sound, Antarctica. New Zealand J Geol Geophys 11(4):793–828

Coleman P (ed) (1973) The western Pacific: Island arcs, marginal seas, geochemistry. West Australia Univ Press, Perth

Collerson KD, McCulloch MT (1983) Nd and Sr isotopic geochemistry of leucite-bearing lavas from Gaussberg, East Antarctica. In: Oliver RL, James PR, Jago JB (eds) Antarctic earth science. Australian Academy of Science, Canberra, ACT, pp 676–680

Colley H, Ash RP (1971) The geology of Erromango: New Hebrides. Geol. Surv. Regional Rept.

Colley H, Warden AJ (1974) Petrology of the New Hebrides. Geol Soc Amer Bull 85:1635–1646

Collins WH (1925) North shore of Lake Huron. Canada Dept Mines and Tech Surveys, Geol Survey Canada, Memoir 143

Compston W, McDougall I, Heier KS (1968) Geochemical comparison of the Mesozoic basaltic rocks of Antarctica, South Africa, South America, and Tasmania. Geochim Cosmochim Acta 32:129–149

Condie KC, Latysh N, Schmus WR Van, Kozuch M, Selverstone J (1999) Geochemistry, Nd and Sr, and U/Pb zircon ages of granitoid and metasedimentary xenoliths from the Navajo volcanic field, Four Corners area, southwestern United States. Chem Geol 156:95–134

Condit CD, Crumpler LS, Aubele J, Elston WE (1989) Patterns of volcanism along the southern margin of the Colorado Plateau: The Springerville field. J Geophys Res 94(B6):7975–7986

Condomines M, Bernat M, Allègre CJ (1976) Evidence for contamination of recent Hawaiian lavas from ^{230}Th-^{238}U data. Earth Planet Sci Letters 33:122–125

Condomines M, Morand P, Allègre CJ, Sigvaldason G (1981) ^{230}Th-^{238}U disequilibria in historical lavas from Iceland. Earth Planet Sci Letters 55:393–406

Condomines M, Grönvold K, Hooker PJ, Muehlenbachs K, O'Nions RK, O'Skarson N, Oxburgh ER (1983) Helium, oxygen, strontium, and neodymium isotopic relationships in Icelandic volcanics. Earth Planet Sci Letters 66:125–136

Condomines M, Hémond C, Allègre CJ (1988) U-Th-Ra radioactive disequilibria and magmatic processes. Earth Planet Sci Letters 90:243–262

Conrad WK, Kay RW (1984) Ultramafic and mafic inclusions from Adak Island: Crystallization history and implications for the nature of primary magmas and crustal evolution in the Aleutian arc. J Petrol 25:88–125

Conticelli S (1998) The effect of crustal contamination on ultrapotassic magmas with lamproitic affinity: mineralogical, geochemical and isotopic data from Torre Alfina lavas and xenoliths, central Italy. Chem Geol 149:51–81

Conticelli S, Francalanci L, AP Santo (1991) Petrology of final stage Latera lavas (Vulsini Mountains): mineralogical, geochemical and Sr-isotopic data and their bearing on the genesis of some potassic magmas in central Italy. J Volcanol Geotherm Res 46:187–212

Cooper JA, Green DH (1969) Lead isotope measurements on lherzolite inclusions and host basanites from western Victoria, Australia. Earth Planet Sci Letters 6:69–76

Cordani UG (1967) K-Ar ages from Fernando de Noronha. Proceedings of the Symposium on Continental Drift in the Southern Hemisphere (Montevideo), p 85

Cordani UG, Sartori PLP, Kawashita K (1980) Geoquimica dos isótopos de estroncio e a evolucão de atividade vulcanica na Bacia do Paraná (Sul do Brasil) durante o Cretáceo. An Acad Brasil Cienc 52:811–818

Cordani UG, Civetta L, Mantovani MSM, Petrini R, Kawashita K, Hawkesworth CJ, Taylor PN, Longinelli A, Cavazzini G, Piccirillo EM (1988) Isotope geochemistry of flood volcanics from the Paraná Basin (Brazil). In: Piccirillo EM, Melfi AJ (eds) The Mesozoic Flood Volcanism of the Paraná Basin. Petrogenetic and Geophysical Aspects. IAG-USP Press, Sao Paolo, pp 157–178

Cordery M, Davies GF, Campbell IH (1997) Genesis of flood basalts from eclogite-bearing mantle plumes. J Geophys Res 102(B9):20179–20197

Corfu F, Krogh TE, Kwok YY, Jensen LS (1989) U-Pb zircon geochronology in the southwestern Abitibi greenstone belt, Superior Province. Can J Earth Sci 26:1747–1763

Corfu F (1993) The evolution of the southern Abitibi greenstone belt in the light of precise U-Pb geochronology. Econ Geol 88:1323–1340

Corfu F, Lightfoot PC (1996) U-Pb geochronology of the sublayer environment, Sudbury igneous complex, Ontario. Econ Geol 91(7):1263–1269

Cortini M, Hermes OD (1981) Sr isotopic evidence for a multi-source origin of the potassic magmas in the Neapolitan area (S. Italy). Contrib Mineral Petrol 77:57–55

Courtillot V, Besse J, Vandamme D, Montigny R, Jaeger JJ, Cappetta H (1986) Deccan flood basalts at the Cretaceous/Tertiary boundary? Earth Planet Sci Letters 80:361–374

Courtillot V, Féraud G, Maluski H, Vandamme D, Moreau MG, Besse J (1988) Deccan flood basalts and the Cretaceous/Tertiary boundary. Nature 333:843–846

Courtney RC, White RS (1986) Anomalous heat flow and geoid across the Cape Verde Rise: evidence for dynamic support from a thermal plume in the mantle. Geophys J Roy Astron Soc 87:815–867

Cousens BL (1988) Isotopically depleted alkali lavas from Bowie Seamount, northeast Pacific Ocean. Can J Earth Sci 25:1708–1716

Cousens BL, Chase RL, Schilling J-G (1984) Basalt geochemistry of the Explorer Ridge area, northeast Pacific Ocean. Can J Earth Sci 21:157–170

Cousens BL, Chase RL, Schilling J-G (1985) Geochemistry and origin of volcanic rocks from Tuzo Wilson and Bowie seamounts, northeast Pacific Ocean. Can J Earth Sci 22:1609–1617

Cousens BL, Spera FJ, Tilton GR (1990) Isotopic patterns in silicic ignimbrites and lava flows of the Mogan and lower Fataga Formations, Gran Canaria, Canary Islands: temporal changes in mantle source composition. Earth Planet Sci Letters 96:319–335

Cox KG (1970) Tectonics and vulcanism of the Karoo period and their bearing on the postulated fragmentation of Gondwanaland. In: Clifford TN, Gass IG (eds) Magmatism and tectonics. Oliver and Boyd, Edinburgh, pp 211–235

Cox KG (1972) The Karoo volcanic cycle. J Geol Soc London 128:311–336

Cox KG (1978) Flood basalts, subduction, and the break-up of Gondwanaland. Nature 274:47–49

Cox KG (1980) A model for flood basalt volcanism. J Petrol 21:629–650

Cox KG (1988) The Karoo province. In: Macdougall JD (ed) Flood basalts. Kluwer Academic Pub., Hingham, Mass

Cox KG, Bristow JW (1984) The Sabie River basalt Formation of the Lebombo monocline and south-east Zimbabwe. In: Erlank AJ (ed) Petrogenesis of the volcanic rocks of the Karoo Province. Geol Soc S Africa Spec Pub 13:125–147

Cox KG, Hawkesworth CJ (1984) Relative contribution of crust and mantle to flood basalt volcanism, Mahabaleshwar area, Deccan Traps. Phil Trans Roy Soc London 310A:627–641

Cox KG, Hawkesworth CJ (1985) Geochemical stratigraphy of the Deccan Traps at Mahalabeshwar, Western Ghats, India, with implications for open system magmatic processes. J Petrol 26:355–377

Cox KG, Hawkesworth CJ, O'Nions RK, Appleton JD (1976) Isotopic evidence for the derivation of some Roman region volcanics from anomalously enriched mantle. Contrib Mineral Petrol 56:173–180

Cox KG, Bell JD, Pankhurst RJ (1979) The interpretation of igneous rocks. Allen and Unwin, London

Crawford AJ (1989) Boninites. Unwin Hyman, London

Crough ST, Thompson GA (1977) Upper mantle origin of Sierra Nevada uplift. Geol 5:396–399

Cundari A (1973) Petrology of the leucite-bearing lavas in New South Wales. J Geol Soc Australia 20(4):465–492

Curray JR, Munasinghe T (1991) Origin of the Rajmahal Traps and the 85° E Ridge: Preliminary reconstructions of the trace of the Crozet hotspot. Geol 19:1237–1240

Czamanske GK, Bohlen SR (1990) The Stillwater complex and its anorthosites: an accident of magmatic underplating? Amer Mineral 75:37–45

Czamanske GK, Zientek ML (1985) Characteristics of the Banded Series anorthosites. In: Czamanske GK, Zientek ML (eds) Stillwater complex: Montana: Geology and Guide. Montana Bureau of Mines and Geology, Special Pub. 92, pp 334–345

Czamanske GK, Kunilov VE, Zientek ML, Cabri LJ, Likhachev AP, Calk LC, Oscarsor RL (1992) A proton-microprobe study of sulfide ores from the Noril'sk-Talnakh district, Siberia. Canadian Mineral 30:249-287

Dallmeyer RD (1975) The Palisade Sill: A Jurassic intrusion? Evidence from $^{40}Ar/^{39}Ar$ incremental release ages. Geol 3: 243-245

Dalmayrac B, Lancelot JR, Leyreloup A (1977) Evidence of 2 b.y. granulites in the Late Precambrian metamorphic basement rocks along the Peruvian coast. Science 198:49-51

Dalrymple GB (1980) K-Ar ages of the Friant Pumice Member of the Turlock Lake Formation, the Bishop Tuff, and the tuff of Reds Meadow, central California. Isochron West 28:3-5

Dalrymple GB, Clague DA (1976) Age of the Hawaiian-Emperor bend. Earth Planet Sci Letters 31:313-329

Dalrymple GB, Jarrard RD, Clague DA (1975a) K-Ar ages of some volcanic rocks from the Cook and Austral Islands. Geol Soc Amer Bull 86:1463-1467

Dalrymple GB, Grommé CH, White RW (1975b) Potassium-argon age and paleomagnetism of diabase dikes in Liberia: Initiation of central Atlantic rifting. Geol Soc Amer Bull 86:399-411

Dalrymple GB, Lanphere MA, Clague DA (1980a) Conventional and $^{40}Ar/^{39}Ar$, K-Ar ages of volcanic rocks from Ojin (site 430), Nintoku (site 432) and Suiko (site 433) seamounts and the chronology of volcanic propagation along the Hawaiian-Emperor chain. Init Repts DSDP 55:659-676

Dalrymple GB, Lanphere MA. Natland JH (1980b) K-Ar minimum age for Meiji Guyot, Emperor Seamount Chain. In: Jackson ED, Koisumi I et al. (eds) Init Repts DSDP 55:677-683

Dalrymple GB, Clague DA, Garcia MO, Birght S-W (1981) Petrology and K-Ar ages of dredged samples from Laysan Island and Northampton Bank volcanoes, Hawaiian ridge, and evolution of the Hawaiian-Emperor chain. Geol Soc Amer Bull 92:884-933

Dalrymple GB, Czamanske GK, Lanphere MA, Stepanov V, Fedorenko V (1991) $^{40}Ar/^{39}Ar$ ages of samples from the Noril'sk-Talnakh ore-bearing intrusions and the Siberian flood basalts. Eos 72:570

Daly JS, Muir RJ, Cliff RA (1991) A precise U-Pb zircon age for the Inishtrahull syenitic gneiss, County Donegal, Ireland. J Geol Soc London 148:639-642

Dalziel IWD (1981) Back-arc extension in the southern Andes: A review and critical reappraisal. Phil Trans Roy Soc London A300:319-335

Darbyshire DPF, Jackson NJ, Ramsay CR, Roobol MJ (1983) Rb-Sr isotope study of latest Proterozoic volcano-sedimentary belt in the Central Arabian Shield. J Geol Soc 140:203-214

Dasch EJ (1969) Strontium isotope disequilibrium in a porphyritic alkali basalt and its bearing on magmatic processes. J Geophys Res 74:550-565

Dasch EJ, Green DH (1975) Strontium geochemistry of lherzolite inclusions and host basaltic rocks, Victoria, Australia. Amer J Sci 275:461-469

Dautria JM, Liotard JM, Cabanès N, Girod M, Briqueu L (1987) Amphibole-rich xenoliths and host alkali basalts: Petrogenetic constraints and implications on the recent evolution of the upper mantle beneath Ahaggar (Central Sahara, southern Algeria). Contrib Mineral Petrol 95:133-144

Dautria JM, Dostal J, Dupuy C, Liotard JM (1988) Geochemistry and petrogenesis of alkali basalts from Tahalra (Hoggar, northwest Africa). Chem Geol 69:17-35

Davidson JP (1983) Lesser Antilles isotopic evidence of the role of subducted sediment in island-arc magma genesis. Nature 306:253-256

Davidson JP (1985) Mechanisms of contamination in Lesser Antilles island-arc magmas from radiogenic and oxygen isotope relationships. Earth Planet Sci Letters 72:163-174

Davidson JP (1986) Isotopic and trace element constraints on the petrogenesis of subduction-related lavas from Martinique, Lesser Antilles. J Geophys Res 91(B6):5943-5962

Davidson JP (1987) Crustal contamination versus subduction zone enrichment: examples from the Lesser Antilles and implications for the mantle source composition of island arc volcanic rocks. Geochim Cosmochim Acta 51:2185-2198

Davidson JP, Wilson IR (1989) Evolution of an alkali basalt-trachyte suite from Jebel Marra volcano, Sudan, through assimilation and fractional crystallization. Earth Planet Sci Letters 95:141-160

Davidson JP, Dungan MA, Ferguson KM, Colucci MT (1987) Crust-magma interactions and the evolution of arc magmas: The San Pedro-Pellado volcanic complex, southern Chilean Andes. Geol 15:443-446

Davidson JP, Ferguson KM, Colucci MT, Dungan MA (1988) The origin and evolution of magmas from the San Pedro-Pellado volcanic complex, S. Chile: Multicomponent sources and open system evolution. Contrib Mineral Petrol 100:429-445

Davies AK (1956) The geology of part of southeast Uganda with special reference to the alkaline complexes. Uganda Geol Surv Mem 8

Davies GR, Halliday AN (1998) Development of the Long Valley rhyolitic magma system: strontium and neodymium isotope evidence from glasses and individual phenocrysts. Geochim Cosmochim Acta 62:3561-3574

Davies GR, Norry MJ, Gerlach DC, Cliff RA (1989) A combined chemical and Pb-Sr-Nd isotope study of the Azores and Cape Verde hot-spots: The geodynamic implications. In: Saunders AD, Norry, MJ (eds) Magmatism in the ocean basins. Geol Soc London Pub. 42:231-255

Davies GR, Halliday AN, Mahood GA, Hall CM (1994) Isotopic constraints on the production rates, crystallization histories, and residence times of pre-caldera silicic magmas, Long Valley, California. Earth Planet Sci Letters 125:17-37

Davies HL, Sun S-S, Frey FA, Gautier I, McCulloch MT, Price RC, Basias Y, Klootwijk CT, Leclaire L (1989) Basalt basement from the Kerguelen plateau and the trail of a Dupal plume. Contrib Mineral Petrol 103:457-469

Davies RD, Allsopp HL, Erlank AJ, Mantor WI (1970) Sr isotopic studies on various layered intrusions in southern Africa. Geol Soc S Africa Spec Publ 1:576-593

Davis AS, Pringle MS, Pickthorn L-BG, Clague DA, Schwab WC (1989) Petrology and age of alkalic lava from the Ratak Chain of the Marshall Islands. J Geophys Res 94(B5):5757-5774

Davis DW, Jackson MC (1988) Geochronology of the Lumby Lake greenstone belt; a 3 Ga complex within the Wabigoon subprovince, northwest Ontario. Geo Soc Amer Bull 100: 818-824

Davis DW, Paces JB (1990) Time resolution of geologic events on the Keweenawan Peninsula and implications for development of the Midcontinent Rift System. Earth Planet Sci Letters 97: 54-64

Davis DW, Poulsen KH, Kamo SL (1989) New insights into Archean crust development from geochronology in the Rainy Lake area, Superior Province, Canada. Geol 97:379-398

Davis GL (1978) Zircons from the mantle US Geol Surv Open-file Report 78-701, pp 86-88

Davis JM, Hawkesworth CJ (1993) The petrogenesis of 30-20 Ma basic and intermediate volcanics from the Mogollon-Datil volcanic field, New Mexico, USA. Contrib Mineral Petrol 115: 165-183

Davis JM, Hawkesworth CJ (1994) Early calc-alkaline magmatism in the Mogollon-Datil volcanic field, New Mexico, USA. J Geol Soc London 151:825-843

Davis JM, Hawkesworth CJ (1995) Geochemical and tectonic transitions in the evolution of the Mogollon-Datil volcanic field, New Mexico, U.S.A. Chem Geol 119:31-53

Davis JM, Elston WE, Hawkesworth CJ (1993) Basic and intermediate volcanism of the Mogollon-Datil volcanic field: Implications for mid-Tertiary tectonic transitions in southwestern New Mexico, USA. In: Prichard HM, Alabaster T, Harris NBW, Neary CR (eds) Magma processes and plate tectonics. Geol Soc London Spec Pub 76:469-488

Davis WJ, Kjarsgaard BA (1997) A Rb-Sr isochron age for a kimberlite from the recently discovered Lac de Gras field, Slave Province, Northwest Canada. J Geol 105:503-510

Dawson JB (1962a) Sodium carbonate lavas from Oldoinyo Lengai, Tanganyika. Nature 195:1075-1076

Dawson JB (1962b) The geology of Oldoinyo Lengai. Bull Volcanol 24:349-387

Dawson JB, Powell DG (1969) Mica in the upper mantle. Contrib Mineral Petrol 22:233-237

Dawson JB, Powell DG, Reid AM (1970) Ultrabasic xenoliths and lava from the Lashaine volcano, northern Tanzania. J Petrol 11:519

Dawson JB, Garson MS, Roberts B (1987) Altered former alkalic carbonatite lava from Oldoinyo Lengai, Tanzania: Inferences for calcite carbonatite lavas. Geol 15(8):765-768

Dawson JB, Pinkerton H, Norton GE, Pyle DM (1990) Physico-chemical properties of alkali carbonatite lavas: Data from the 1988 eruption of Oldoinyo Lengai, Tanzania. Geol 18(3): 260-263

Dawson JB, Smith JV, Steele IM (1992) 1966 ash eruption of the carbonatite volcano Oldoinyo Lengai: mineralogy of lapilli and mixing of silicate and carbonate magma. Mineral Mag 56(1):1-16

Dawson JB, Pinkerton H, Pyle DM, Nyamweru C (1994) June 1993 eruption of Oldoinyo Lengai, Tanzania: exceptionally viscous and large carbonatite lava flows and evidence for coexisting silicate and carbonate magmas. Geol 22:799-802

Deans T, Roberts B (1984) Carbonatite tuffs and lava clasts of the Tinderet Foothills, western Kenya; a study of calcified natro-carbonatite. J Geol Soc London 141:563-580

DeBoer J, Odom LA, Ragland PC, Snider FG, Tilford NR (1980) The Bataan orogene: eastward subduction, tectonic rotations, and volcanism in the Western Pacific. Tectonophysics 67: 251-282

Defant JJ, Drummond MS (1990) Derivation of some modern arc magmas by melting of young subducted lithosphere. Nature 347:662-665

Defant JJ, DeBoer JC, Coles D (1988) The western Central Luzon volcanic arc, the Philippines: Two arcs divided by rifting? Tectonophysics 145:305-317

Defant MJ, Drummond MS (1993) Mount St. Helens: Potential example of partial melting of subducted lithosphere in a volcanic arc. Geol 21:547-550

Deines P (1989) Stable isotope variations in carbonatites. In: Bell K (ed) Carbonatites. Unwin Hyman, London, pp 301-359

DeLong SE, Hodges FN, Arculus RJ (1975) Ultramafic and mafic inclusions, Karaga Island, Alaska, and the occurrence of alka-line rocks in island arcs. J Geol 83:721-736

Demaiffe D, Hertogen J, Michot J, Weis D (1985) Alkaline affinity of the Seychelles granitic rocks: petrologic, geochemical, and Sr isotopic evidence. Contrib Mineral Petrol (submitted)

Demovic R, Hoefs J, Wedepohl KH (1972) Geochemische Unter-suchungen an Travertinen der Slovakei. Contrib Mineral Petrol 37:15-28

DeMulder M, Hertogen J, Deutsch S, André L (1986) The role of crustal contamination in the potassic suite of the Kari-simbi Volcano (Virunga, African Rift Valley). Chem Geol 57: 117-136

Dengo G, Case JE (eds) (1990) The Caribbean Region. Geol Soc Amer, Centennial vol H

Deniel C, Vidal P, Coulon C, Vellutini P-J, Piguet P (1994) Tempo-ral evolution of mantle sources during continental rifting. The volcanism of Djibouti (Afar). J Geophys Res 99:2853-2869

Deniel C (1998) Geochemical and isotopic (Sr, Nd, Pb) evidence for plume-lithosphere interactions in the genesis of Grande Comore magmas (Indian Ocean). Chem Geol 144:281-303

DePaolo DJ (1979) Implications of correlated Nd and Sr isotopic variations for the chemical evolution of the crust and mantle. Earth Planet Sci Letters 43:201-211

DePaolo DJ (1981) Trace element and isotope effects of combined wallrock assimilation and fractional crystallization. Earth Planet Sci Letters 53:189-202

DePaolo DJ (1983a). Comment on "Columbia River volcanism: the question of mantle heterogeneity or crustal contamination" by R.W. Carlson, G.W. Lugmair and J.D. Macdougall. Geochim Cosmochim Acta 47:841-844

DePaolo DJ (1983b) Geochemical evolution of the crust and man-tle. Reviews Geophys Space Phys 21(b):1347-1358

DePaolo DJ (1985) Isotopic studies of processes in mafic magma chambers: I. The Kiglapait intrusion, Labrador. J Petrol 26: 925-951

DePaolo DJ (1988) Neodymium isotope geochemistry. Springer Verlag, Berlin

DePaolo DJ, Johnson RW (1979) Magma genesis in the New Brit-ain Island arc: constraints from Nd and Sr isotopes and trace element patterns. Contrib Mineral Petrol 70:367-379

DePaolo DJ, Wasserburg GJ (1976a) Nd isotope variations and petrogenetic models. Geophys Res Letters 3:249-252

DePaolo DJ, Wasserburg GJ (1976b) Inferences about magma sources and mantle structure from variations of $^{143}Nd/^{144}Nd$. Geophys Res Letters 3:743-746

DePaolo DJ, Wasserburg GJ (1977) The sources of island arcs as indicated by Nd and Sr isotopic studies. Geophys Res Letters 4(10):465-468

DePaolo DJ, Wasserburg GJ (1979) Neodymium isotopes in flood basalts from the Siberian Platform and inferences about their mantle sources. Proceed Nat Acad Sci 76(6):3056-3060

Deruelle B, Harmon RS, Moorbath S (1983) Combined Sr-O iso-tope relationships and petrogenesis of Andian volcanics of South America. Nature 302:814-816

Devey CW, Cox KG (1987) Relationship between crustal contami-nation and crystallization in continental flood basalt magmas with special reference to the Deccan Traps of the Western Ghats, India. Earth Planet Sci Letters 84:59-68

Devey CW, Albarède F, Chiminee J-L, Michard A, Mühe R, Stoffers P (1990) Active submarine volcanism on the Society hotspot swell (W. Pacific): A geochemical study. J Geophys Res 95:5049-5066

DeWit MJ, Ashwal LD (eds) (1997) Greenstone belts. Oxford Uni-versity Press, New York (Oxford Monographs on Geology and Geophysics #35)

DeWit MJ, Ransome I (eds) (1992) Inversion tectonics of the Cape Fold Belt, Karoo and Cretaceous basins of southern Africa. Balkema, Rotterdam

Dickey JS Jr, Frey FA, Hart SR, Watson EB, Thompson G (1977) Geochemistry and petrology of dredged basalts from the Bouvet triple junction, South Atlantic. Geochim Cosmochim Acta 41:1105-1118

Dickin AP (1981) Isotope geochemistry of Tertiary igneous rocks from the Isle of Skye, N.W. Scotland. J Petrol 22:155-190

Dickin AP, Bowes DR (1991) Isotopic evidence for the extent of Early Proterozoic basement in Scotland and northwest Ire-land. Geol Mag 128:385-388

Dickin AP, Exley RA (1981) Isotopic and geochemical evidence for magma mixing in the petrogenesis of the Coire Uaigneich Granophyre, Isle of Skye, N.W. Scotland. Contrib Mineral Pet-rol 76:98-108

Dickin AP, Exley RA, Smith BM (1980) Isotopic measurement of Sr- and O-exchange between meteoritic-hydrothermal fluid and the Coire Uaigneich Granophyre, Isle of Skye, NW Scot-land. Earth Planet Sci Letters 51:58-70

Dickin AP, Fallick AE, Halliday AN, Macintyre RM, Stephens WE (1986) An isotopic and geochronological investigation of the younger igneous rocks of the Seychelles microcontinent. Earth Planet Sci Letters 81:46-56

Dickin AP, Jones NW, Thirlwall MF, Thompson RN (1987) A Ce/Nd isotope study of crustal contamination processes affecting Palaeocene magmas in Skye, Northwest Scotland. Contrib Mineral Petrol 96:455-464

Dickin AP, Halliday AN, Bowden P (1991) A Pb, Sr and Nd isotope study of the basement and Mesozoic ring complexes of the Jos Plateau, Nigeria. Chem Geol (Isotope Geoscience Section) 94(1):23-32

Dickinson DR, Gibson IL (1972) Feldspar fractionation and anomalous $^{87}Sr/^{86}Sr$ ratios in a suite of peralkaline silicic rocks. Geol Soc Amer Bull 83:231-240

Dickinson DR, Dodson MH, Gass IG, Rex DC (1969) Correlation of initial $^{87}Sr/^{86}Sr$ with Rb/Sr in some Late Tertiary volcanic rocks of South Arabia. Earth Planet Sci Letters 6:84-90

Dickinson WR (1967) Tectonic development of Fiji. Tectonophys-ics 4:543-553

Dickinson WR, Hatherton T (1967) Andesitic volcanism and seis-micity around the Pacific. Science 157:801-803

Dickinson WR, Rickard MJ, Coulsen FI, Smith JG, Lawrence RL (1968) Late Cenozoic shoshonitic lavas in northwestern Viti Levu, Fiji. Nature 219:148

Dietz ES (1964) Sudbury structure as an astrobleme. J Geol 72:412-434

Dimroth E, Imreh L, Rocheleau M, Goulet N (1982) Evolution of the south-central part of the Archean Abitibi Belt, Quebec. Part I. Stratigraphy and paleogeographic model. Can J Earth Sci 19:1729-1758

Dimroth E, Imreh L, Goulet N, Rocheleau M (1983) Evolution of the south-central segment of the Archean Abitibi Belt, Quebec Part II: Plutonic and metamorphic evolution and geotectonic model. Can J Earth Sci 20:1374-1388

Dingle RV, Seisser WG, Newton AR (1983) Mesozoic and Tertiary geology of southern Africa. Balkema, Rotterdam

Divis AF (1976) Geology and geochemistry of Sierra Madre Range, Wyoming. Colorado School of Mines Quart 71:1-127

Dixon TH, Batiza R (1979) Petrology and chemistry of recent lavas in the northern Marianas: implications for the origin of island arc basalts. Contrib Mineral Petrol 70:167-181

Dixon TH, Stern RJ (1983) Petrology, chemistry, and isotopic composition of submarine volcanoes in the southern Mariana arc. Geol Soc Amer Bull 94:1159-1172

Dodson MH (1970) Simplified equations for double-spiked isotopic analyses. Geochim Cosmochim Acta 34:1241-1244

Dodson MH (1973) Closure temperatures in cooling geochronological and petrological systems. Contrib Mineral Petrol 40:259-274

Dodson MH (1976) Kinetic processes and the thermal history of slowly cooling rocks. Nature 259:551-553

Dodson MH (1979) Theory of cooling ages. In: Jäger E, Hunziker JC (eds) Lectures in isotope geology. Springer Verlag, Berlin, pp 194-202

Dodson MH (1982) On "spurious" correlations in Rb-Sr isochron diagrams. Lithos 15:217-219

Dodson MH, Cavanagh BJ, Thatcher EC, Aftalion M (1975) Age limits for the Ubendian metamorphic episode in northern Malawi. Geol Mag 112:403-410

Doe BR (1968) Lead and strontium isotopic studies of Cenozoic volcanic rocks in the Rocky Mountain region - a summary. Colorado School of Mines Quart 63(3):149-174

Doe BR, Lipman PW, Hedge CE (1969a) Radiogenic tracers and the source for continental andesites; a beginning at the San Juan volcanic field, Colorado. Oregon Dept Geology and Mineral Industries Bull 65:143-149

Doe BR, Lipman PW, Hedge CE, Kurasawa H (1969b). Primitive and contaminated basalts from the southern Rocky Mountains, U.S.A. Contrib Mineral Petrol 21:142-156

Doe BR, Leeman WP, Christiansen RL, Hedge CE (1982) Lead and strontium isotopes and related trace elements as genetic tracers in the Upper Cenozoic rhyolite-basalt association of the Yellowstone plateau volcanic field. J Geophys Res 87:4785-4806

Dohrerwend JC, McFadden LD, Turrin BD, Wells SG (1984) K-Ar dating of the Cima volcanic field, eastern Mojava Desert, California: Late Cenozoic volcanic history and landscape evolution. Geol 12:163-167

Doig R (1970) An alkaline rock province linking Europe and North America. Can J Earth Sci 7:22-28

Donaldson CH (1982) Spinifex-textured komatiites: a review of textures, compositions, and layering. In: Arndt NT, Nisbet EG (eds) Komatiites. Allen and Unwin, London, pp 213-244

Donnelly TW, Melson K, Kay R, Rogers JJW (1973) Basalts and dolerites of Late Cretaceous age from the central Caribbean. In: Edgar NT, Saunders JB (eds) Initial Reports of the Deep Sea Drilling Project 15:989-1012, U.S. Government Printing Office, Washington D.C.

Dosso L, Murthy VR (1980) A Nd isotopic study of the Kerguelen islands: inferences on enriched oceanic mantle sources. Earth Planet Sci Letters 48:268-276

Dosso L, Vidal P, Cantagrel JM, Lameyre J, Marot A, Zimine S (1979) Kerguelen: continental fragments or oceanic island? Petrology and isotope geochemistry evidence. Earth Planet Sci Letters 43:46-60

Dosso L, Bougault H, Beuzart P, Clavez J-Y, Joron JL (1988) The geochemical structure of the South-East Indian Ridge. Earth Planet Sci Letters 88:47-59

Dosso L, Hanan BB, Bougault H, Schilling J-G, Joron JL (1991) Sr-Nd-Pb geochemical morphology between 10 degrees N and 17 degrees N on the Mid-Atlantic Ridge-a new MORB isotope signature. Earth Planet Sci Letters 106:29-43

Dosso L, Bougault H, Joron JL (1993) Geochemical morphology of the North Mid-Atlantic ridge, 10°-24° N: trace element-isotope complementarity. Earth Planet Sci Letters 120:443-462

Dostal J, Fratta M (1977) Trace element geochemistry of a Precambrian diabase dike from western Ontario. Can J Earth Sci 14:2941-2944

Dostal J, Dupuy C, Leyreloup A (1980) Geochemistry and petrology of meta-igneous granulitic xenoliths in Neogene volcanic rocks of the Massif Central, France - Implications for the lower continental crust. Earth Planet Sci Letters 50:31-40

Dostal J, Dupuy C, Liotard JM (1982) Geochemistry and origin of basaltic lavas from Society Island, French Polynesia (South Central Pacific Ocean). Bull Volcanol 45:51-62

Dostal J, Dupuy C, Carron JP, LeGuen de Kerneizon M, Maury RC (1983a) Partition coefficients of trace elements: Application to volcanic rocks of St. Vincent, West Indies. Geochim Cosmochim Acta 47:525-533

Dostal J, Baragar WRA, Dupuy C (1983b) Geochemistry and petrogenesis of basaltic rocks from Coppermine River area, Northwest Territories. Can J Earth Sci 20:684-698

Dostal J, Dupuy G, Zhai M, Zhi X (1988) Geochemistry and origin of Pliocene alkali basaltic lavas from Anhui-Jiangsu, eastern China. Geochem J 22:165-176

Dostal J, Zhi X, Muehlenbachs K, Dupuy G, Zhai M (1991) Geochemistry of Cenozoic alkali basaltic lavas from Shandong Province, eastern China. Geochem J 25:1-16

Downes H (1984) Sr and Nd isotope geochemistry of coexisting alkaline magma series, Cantal, Massif Central, France. Earth Planet Sci Letters 69:321-334

Drach V von, Marsh BD, Wasserburg GJ (1986) Nd and Sr isotopes in the Aleutians: Multicomponent parenthood of island-arc magmas. Contrib Mineral Petrol 92:13-34

Drake MJ, Holloway JR (eds) (1978) Experimental trace element geochemistry. Geochim Cosmochim Acta 52(6A):657-943

Drake MJ, Weill DF (1975) Partition of Sr, Ba, Ca, Y, Eu^{2+}, Eu^{3+}, and other REE between plagioclase feldspar and magmatic study: an experimental study. Geochim Cosmochim Acta 39:689-712

Drake R, Vergara M, Munizaga F, Vicente JC (1982) Geochronology of Mesozoic-Cenozoic magmatism in central Chile, Lat. 31°-36° S. Earth-Sci Rev 18:353-363

Drygalski E von (1912) Der Gaussberg; seine Kartierung und seine Formen. Südpolar-Expedition 1901-1903. Geographie und Geologie 11(1):1-46

Dudas FO, Carlson RW, Eggler DH (1987) Regional Middle Proterozoic enrichment of the subcontinental mantle source of igneous rocks from central Montana. Geol 15:22-25

Dudkin OB, Mitrofanov FP (1994) Features of the Kola alkali province. Geochem International 31(3):1-11

Duffield WA, Ruiz J (1992) Compositional gradients in large reservoirs of silicic magma as evidenced by ignimbrites versus Taylor Creek Rhyolite lava domes. Contrib Mineral Petrol 110:192-210

Duffield WA, Bacon CR, Brent Dalrymple G (1980) Late Cenozoic volcanism, geochronology, and structure of the Coso Range, Inyo County, California. J Geophys Res 85(B5):2381-2404

Dunbar CO (1949) Historical geology. Wiley, New York

Duncan RA (1978) Geochronology of basalts from the Ninetyeast Ridge and continental dispersion of the eastern Indian Ocean. J Volcanol Geotherm Res 4:283-305

Duncan RA (1984) Age progressive volcanism in the New England Seamounts and the opening of the central Atlantic Ocean. J Geophys Res 89:9980-9990

Duncan RA (1985) Radiometric ages from volcanic rocks along the New Hebrides-Samoa lineament. In: Brocher TM (ed) Geological Investigations of the Northern Melanesian Borderland. Circum-Pacific Council for Energy and Resources, Houston, Texas, pp 67-76

Duncan RA (1990) The volcanic record of the Réunion hotspot. Proc Ocean Drill Program, Sci Results 115:3-10

Duncan RA (1991) Age distribution of volcanism along aseismic ridges in the eastern Indian Ocean. Proc Ocean Drilling Prog, Sci Results 121:507–517

Duncan RA, Compston W (1976) Sr-isotope evidence for an old mantle source region for French Polynesian volcanism. Geol 4:728–732

Duncan RA, McDougall I (1974) Migration of volcanism with time in the Marquesas Islands, French Polynesia. Earth Planet Sci Letters 21:414–420

Duncan RA, McDougall I (1976) Linear volcanism in French Polynesia. J Volcanol Geotherm Res 1:197–227

Duncan RA, McDougall I (1989) Volcanic time-space relationships. In: Johnson RW (ed) Intraplate Volcanism in Eastern Australia and New Zealand. Cambridge University Press, Cambridge, England, pp 43–54

Duncan RA, Pyle DG (1988) Rapid eruption of the Deccan flood basalts at the Cretaceous/Tertiary boundary. Nature 333:841–843

Duncan RA, Storey M (1992) The life cycle of Indian Ocean hotspots. In: Duncan RA, Rea DK, Kidd RB, Rad U von, Weissel JK (eds) Synthesis of Results from Scientific Drilling in the Indian Ocean. Amer. Geophys. Union, Washington D.C (Geophys. Monog. 70, pp 91–103)

Duncan RA, McDougall I, Carter RM, Coombs DS (1974) Pitcairn Island – another Pacific hot spot? Nature 251:679–682

Duncan RA, McCulloch MT, Barsczus HG, Nelson DR (1986) Plume versus lithospheric sources for melts at Ua Pou, Marquesas Islands. Nature 322:534–538

Dungan MA, Lindstrom MM, McMillan NJ, Moorbath S, Hoefs J, Haskin LA (1986) Open system magmatic evolution of the Taos Plateau volcanic field, northern New Mexico; the petrology and geochemistry of the Servilleta Basalt. J Geophys Res 91:5999–6028

Dunlop HM, Fitton JG (1979) A K-Ar and Sr-isotopic study of the volcanic island of Principe, West Africa–evidence for mantle heterogeneity beneath the Gulf of Guinea. Contrib Mineral Petrol 71:125–131

Dupré B, Allègre CJ (1980) Pb-Sr-Nd isotope correlation and the chemistry of the North Atlantic mantle. Nature 286:17–22

Dupré B, Allègre CJ (1983) Pb-Sr isotope variations in Indian Ocean basalts and mixing phenomena. Nature 303:142–146

Dupré B, Arndt NT (1990) Pb isotopic compositions of Archean komatiites and sulfides. Chem Geol 85:25–56

Dupré B, Echeverría LM (1984) Pb isotopes of Gorgona Island (Colombia): isotopic variations correlated with magma type. Earth Planet Sci Letters 67:186–190

Dupré B, Lambret B, Rousseau D, Allègre CJ (1981) Limitations on the scale of mantle heterogeneities under oceanic ridges. Nature 294:552–554

Dupré B, Lambret B, Allègre CJ (1982) Isotopic variations within a single oceanic island: the Terceira case. Nature 299:620–622

Dupuy C, McNutt RH, Coulon C (1974) Determination de $^{87}Sr/^{86}Sr$ dans le andésites cenozoiques et les laves associees de Sardaigne Nord occidentale (Italie). Geochim Cosmochim Acta 38:1287–1296

Dupuy C, Dostal J, Marcelot G, Bougault H, Joron JL, Treuil M (1982) Geochemistry of basalts from central and southern New Hebrides arc: implications for their source rock composition. Earth Planet Sci Letters 60:207–225

Dupuy C, Vidal P, Barsczus HG, Chauvel C (1987) Origin of basalts from Marquesas archipelago (south central Pacific Ocean): Isotope and trace element constrains. Earth Planet Sci Letters 82:145–152

Dupuy C, Barsczus HG, Liotard JM, Dostal J (1988a) Trace element evidence for the origin of ocean island basalts: an example from the Austral Islands (French Polynesia). Contrib Mineral Petrol 98:293–302

Dupuy C, Marsh J, Dostal J, Michard A, Testa S (1988b) Asthenospheric and lithospheric sources for Mesozoic dolerites from Liberia (Africa): Trace element and isotopic evidence. Earth Planet Sci Letters 87:100–110

Dupuy C, Barsczus HG, Dostal J, Vidal P, Liotard J-M (1989) Subducted and recycled lithosphere as the mantle source of ocean island basalts from southern Polynesia, central Pacific. Chem Geol 77:1–18

Dupuy C, Michard A, Dostal J, Dautel D, Baragar WRA (1992) Proterozoic flood basalts from the Coppermine River area, Northwest Territories: Isotope and trace element geochemistry. Can J Earth Sci 29:1937–1943

Dupuy C, Vidal P, Maury RC, Guille G (1993) Basalts from Mururoa, Fangataufa, and Gambier islands (French Polynesia): Geochemical dependence on the age of the lithosphere. Earth Planet Sci Letters 117:89–100

Duyverman HJ, Harris NBW (1982) Late Precambrian evolution of Afro-Arabian crust from ocean arc to craton: discussion. Geol Soc Amer Bull 93:174–178

Duyverman HJ, Harris NBW, Hawkesworth CJ (1982) Crustal accretion in the Pan African: Nd and Sr isotope evidence from the Arabian shield. Earth Planet Sci Letters 59:315–326

Dymek RF, Beak JL, Kerr MT (1983) The Malene metasedimentary rocks of Rypeø, and their relationship to Amitsoq gneisses. Rapp Grønlands Geol Unders 112:53–70

Eaby J, Clague DA, Delaney JR (1984) Sr isotopic variations along the Juan de Fuca Ridge. J Geophys Res 89:78830–7890

Eales HV, Marsh JS, Cox KG (1984) The Karoo igneous province: an introduction. In: Erlank AJ (ed) Petrogenesis of the volcanic rocks of the Karoo Province. Spec Publ Geol Soc S Africa 13:1–26

Eastin R, Faure G (1970) The age of the Littlewood Volcanics of Coats Land, Antarctica. J Geol 79:241–245

Eastin R, Faure G, Neethling DC (1970) The age of the Trollkjellrygg Volcanics of western Queen Maud Land. Ant J US 5:157–158

Eaton GP, Christensen RL, Iyer HM, Pitt AM, Mabey DR, Blank HR Jr, Zietz I, Gettings ME (1975) Magma beneath Yellowstone Park. Science 188:787–796

Eby GN (1984a) Geochronology of the Monteregian Hills alkaline igneous province, Quebec. Geol 12:468–470

Eby GN (1984b) Monteregian Hills I. Petrography, major and trace-element geochemistry, and strontium isotope chemistry of the western intrusions: Mount Royal, St. Bruno, and Johnson. J Petrol 25:421–452

Eby GN (1985a) Sr and Pb isotopes, U and Th chemistry of the alkaline Monteregian Hills and White Mountain provinces, eastern North America. Geochim Cosmochim Acta 49:1143–1154

Eby GN (1985b) Monteregian Hills I. Petrography, major and trace-element geochemistry, and strontium isotope chemistry of the western intrusions: Mount Shefford, Brome and Megantic. J Petrol 26:418–448

Eby GN (1987) The Monteregian Hills and White Mountain alkaline igneous provinces, eastern North America. In: Fitton JG, Upton BG (eds) Alkaline igneous rocks. Geol Soc Spec Pub 30:433–447

Echeverria LM (1980) Tertiary or Mesozoic komatiites from Gorgona Island, Colombia: field relations and geochemistry. Contrib Mineral Petrol 73:253–266

Echeverría LM, Aitken B (1986) Pyroclastic rocks: another manifestation of ultramafic volcanism of Gorgona Island, Colombia. Contrib Mineral Petrol 92:428–436

Eckermann H von (1948) The alkaline district of Alnö Island. Sver Geol Unders Ser. Ga 36:1–176

Edgar AD, Arima M (1981) Geochemistry of three potassium-rich ultrabasic lavas from the west branch of the African Rift: Inferences on their genesis. Neues Jahrbuch Mineral Monatshefte 12:539–552

Eggins SM, Green DH, Falloon TJ (1991) Shallow melting and contamination of an EM1 mantle plume. Earth Planet Sci Letters 107:448–462

Eglinton G, Scott PM, Belsky T, Burlingame AL, Calvin M (1964) Hydrocarbons of biological origin from a 1-billion year old sediment. Science 145:262–264

Eiler JM, Farley KA, Stolper EM (1998) Correlated helium and lead isotope variations in Hawaiian lavas. Geochim Cosmochim Acta 62:1977–1984

Eissen JL, Juteau T, Joron JL, Dupré B, Humler E, Al'Mukhamedov AI (1989) Petrology and geochemistry of the basalts from the axial rift of the Red Sea at 18° North. J Petrol 30:791–839

Ekren EB (1968) Geological setting of Nevada Test Site and Nellis Air Force Base. In: Eckel EB (ed) Nevada Test Site. Geol Soc Amer Mem 110:11–19

Elburg M, Foden J (1999) Sources for magmatism in central Sulawesi: Geochemical and Sr-nd-Pb isotopic constraints. Chem Geol 156:67–93

Ellam RM, Hawkesworth CJ (1988) Elemental and isotopic variations in subduction related basalts: evidence for a three component model. Contrib Mineral Petrol 98(1):72–80

Ellam RM, Rogers NW (1988) Comment on "Mantle mixing and crustal contamination as the origin of the high-Sr radiogenic magmatism of Stromboli (Aeolian arc)" by B. Luais. Earth Planet Sci Letters 91:239

Elliot DH (1972) Major oxide chemistry of the Kirkpatrick Basalt, central Transantarct c Mountains. In: Adie RJ (ed) Antarctic Geology and Geophysics. Universitets-Forlaget, Oslo, pp 413–418

Elliot DH (1974) The tectonic setting of the Jurassic Ferrar Group, Antarctica. In Gonzalez-Ferran O (ed) Proceedings of the Symposium on Andean and Antarctic Volcanology Problems. Int. Assoc. Volcan. and Chem. of the Earth's Interior, Special Series, pp 357–372

Elliot DH (1992) Jurassic magmatism and tectonism associated with Gondwanaland break-up: an Antarctic perspective. In: Storey BC, Alabaster T, Pankhurst (eds) Magmatism and the causes of continental break-up. Geol Soc (London) Spec Pub 68

Elliot DH, Fleck RJ, Sutter JF (1985) Potassium-argon age determinations of Ferrar Group rocks, central Transantarctic Mountains. Amer. Geophys. Union, Washington, D.C (Antarctic Res Ser 36:197–224)

Elliot DH, Fleming TH, Kyle PR, Foland KA (1999) Long-distance transport of magmas in the Jurassic Ferrar large igneous province, Antarctica. Earth Planet Sci Letters 167:89–105

Elliott TR, Hawkesworth CJ, Grönvold K (1991) Dynamic melting of the Iceland plume. Nature 351:201–206

Elston WE (1984) Mid-Tertiary ash-flow tuff cauldrons of southwestern New Mexico. J Geophys Res 89:8733–8750

Elston WE, Twist D (1989) Vredefort-Bushveld enigma of South Africa and the recognition of large terrestrial impact structures: mental leaps and mental obstacles. Abstracts Int Geol Congress 28:449

Emerick CM, Duncan RA (1982) Age progressive volcanism in the Comores archipelago, western Indian Ocean and implications for Somali plate tectonics. Earth Planet Sci Letters 60:415–428

Encarnación JP, Mukasa SB Obille EC Jr (1993) Zircon U-Pb geochronology of the Zambales and Angat ophiolites, Luzon, Philippines: evidence for an Eocene arc-back arc pair. J Geophys Res 98(B11):19,991–20,004

Encarnación J, Fleming TH, Elliot DH, Eales HV (1996) Synchronous emplacement of Ferrar and Karoo dolerites and the early breakup of Gondwana. Geol 24:535–538

Encarnación J, Mukasa SB, Evans CA (1999) Subduction components and the generation of arc-like melts in the Zambales ophiolite Philippines: Pb, Sr and Nd isotopic constraints. Chem Geol 156:343–357

Engel AEJ, Engel CG (1964) Igneous rocks of the East Pacific Rise. Science 146:477–485

Engel AEJ, Engel CG, Havens RG (1965) Chemical characteristics of oceanic basalts and the upper mantle. Geol Soc Amer Bull 76:719–734

Engel AEJ, Dixon TH, Stern RJ (1980) Late Precambrian evolution of Afro-Arabian crust from ocean arc to craton. Geol Soc Amer Bull 91:639–706

Epp D, Smoot NC (1989) Distribution of seamounts in the North Atlantic. Nature 337:254–257

Erlank AJ (ed) (1984) Petrogenesis of the volcanic rocks of the Karoo Province. Geol Soc S Africa Spec Publ 13:395

Erlank AJ, Marsh JS, Duncan AR, Miller RMcG, Hawkesworth CJ, Betton PJ, Rex DC (1984) Geochemistry and petrogenesis of the Etendeka volcanic rocks from SWA/Namibia. In: Erlank AJ (ed) Petrogenesis of the volcanic rocks of the Karoo Province. Geol Soc S Africa Spec Publ, 13:195–245

Ernst WG (1960) Diabase-granophyre relations in the Endion Sill, Duluth, Minnesota J Petrol 1:286–303

Ernst WG (1999) Hornblende, the continent maker-evolution of H₂O during circum-Pacific subduction versus continental collision. Geol 27(3):675–678

Eskola P (1932) Origin of granitic magmas. Schweizer Mineral Petrol Mitteilungen 42

Esson J, Flower MFJ, Strong DF, Upton BGJ, Wadsworth WJ (1970) Geology of the Comores Archipelago, western Indian Ocean. Geol Mag 107:549–557

Eugster HP (1970) Chemistry and origin of the brines of Lake Magadi, Kenya. Mineralog Soc Amer Spec Paper 3:215–236

Eugster HP (1986) Lake Magadi, Kenya: a model for rift valley hydrochemistry and sedimentation. In: Frostrick LE et al. (eds) Sedimentation in the African Rifts. Geol Soc (London) Spec Pub 25:177–189

Ewart A, Bryan WB (1972) Petrography and geochemistry of the igneous rocks from Eua, Tongan Islands. Geol Soc Amer Bull 83:3281–3289

Ewart A, Hawkesworth CJ (1987) The Pleistocene-Recent Tonga-Kermadec arc lavas: interpretation of new isotopic and rare earth data in terms of a depleted mantle source model. J Petrol 28:495–530

Ewart A, Stipp JJ (1968) Petrogenesis of the volcanic rocks of the central North Island, New Zealand, as indicated by a study of $^{87}Sr/^{86}Sr$ ratios, and Sr, Rb, K, U and Th abundances. Geochim Cosmochim Acta 32:699–736

Ewart A, Taylor SR (1969) Trace element geochemistry of the rhyolitic volcanic rocks, central North Island, New Zealand; phenocryst data. Contrib Mineral Petrol 22:127

Ewart A, Hildreth WE, Carmichael ISE (1975) Quaternary acid magma in New Zealand. Contrib Mineral Petrol 51:1–27

Ewart A, Brothers RN, Mateen A (1977) An outline of the geology and geochemistry, and the possible petrogenetic evolution of the volcanic rocks of the Tonga-Kermadec-New Zealand Arc. J Volcanol Geotherm Res 2:205–250

Ewart A, Milner SC, Armstrong RA, Duncan AR (1998a) Etendeka volcanism of the Goboboseb Mountains and Messum igneous complex, Namibia. Part I. Geochemical evidence of Early Cretaceous Tristan Plume melts and the role of crustal contamination in the Paraná-Etendeka CFB. J Petrol 39(2):191–226

Ewart A, Milner SC, Armtrong RA, Duncan AR (1998b) Etendeka volcanism of the Goboboseb Mountains and Messum igneous complex, Namibia. Part II. Voluminous quartz latite volcanism of the Awahab magma system. J Petrol 39(2):227–254

Fahrig WF, Wanless RK (1963) Age and significance of diabase dyke swarms of the Canadian Shield. Nature 200:934–937

Fairbairn HW, Bullwinkel JH Pinson WH, Hurley PM (1959) Age investigations of syenites from Coldwell, Ontario. Proc Geol Assoc Canada 11:141–144

Fairbairn HW, Hurley PM, Finson WH, Cormier RF (1960a) A comparison of the ages of coexisting biotite and muscovite in some Paleozoic granite rocks. Geochim Cosmochim Acta 19:7–9

Fairbairn HW, Hurley PM, Pinson WH (1960b) Mineral and rock ages at Sudbury-Blind River, Ontario. Proc Geol Assoc Canada 12:41–66

Fairbairn HW, Hurley PM, Pinson WH (1961) The relation of discordant Rb-Sr mineral and rock ages in an igneous rock to its time of subsequent $^{87}Sr/^{86}Sr$ metamorphism. Geochim Cosmochim Acta 23:135–144

Fairbairn HW, Faure G, Pinson WH, Hurley PM and Powell JL (1963) Initial ratio of strontium 87 to strontium 86, whole-rock age, and discordant biotite in the Monteregian igneous province, Quebec. J Geophys Res 68:6515–6522

Fairbairn HW, Hurley PM, Card KD, Knight CJ (1969) Correlation of radiometric ages of Nipissing diabase and Huronian metasediments with Proterozoic orogenic events in Ontario. Can J Earth Sci 6:489–497

Fairhead JD (1978) A gravity link between the domally uplifted Cainozoic volcanic centres of North Africa and its similarity to the East African system anomaly. Earth Planet Sci Letters 42:109–113

Falloon TJ, Crawford AJ (1991) The petrogenesis of high-calcium boninite lavas dredged from the northern Tonga ridge. Earth Planet Sci Letters 102:375–394

Falloon TJ, Green DH, McCulloch MT (1989) Petrogenesis of high-Mg and associated lavas from the north Tonga Trench. In Crawford AJ (ed) Boninites. Unwin Hyman, London, pp 357–395

Fan Q, Hooper PR (1991) The Cenozoic basaltic rocks of eastern China: petrology and chemical composition. J Petrol 32: 765–810

Farley KA, Natland JH, Craig H (1992) Binary mixing of enriched and undegassed (primitive?) mantle components (He, Sr, Nd, Pb) in Samoan lavas. Earth Planet Sci Letters 111:183–199

Farmer GL, Perry FV, Semken S, Crowe B, Curtis D, DePaolo DJ (1989) Isotopic evidence on the structure and origin of subcontinental lithospheric mantle in southern Nevada. J Geophys Res 94(B6):7885–7898

Farmer GL, Broxton DE, Warren RG, Pickthorn W (1991) Nd, Sr, and O isotopic variations in metaluminous ash-flow tuffs and related volcanic rocks at the Timber Mountain/Oasis Valley Caldera complex, S.W. Nevada: Implications for the origin and evolution of large-volume silicic magma bodies. Contrib Mineral Petrol 109:53–68

Farmer GL, Glazner AF, Wilshire HG, Wooden JL, Pickthorn WJ, Katz M (1995) Origin of late Cenozoic basalts at the Cima volcanic field, Mojave Desert, California. J Geophys Res 109: 53–68

Faul H, Elmore PL, Brannock WW (1959) Age of the Fen carbonatite (Norway) and its relation to the intrusives of the Oslo region. Geochim Cosmochim Acta 17:153–156

Faure G (1961) The $^{87}Sr/^{86}Sr$ ratio in continental and oceanic basalts and the origin of igneous rocks. PhD dissertation, Dept Geol Geophys M.I.T., Cambridge, Mass

Faure G (1986) Principles of isotope geology, 2nd edn. Wiley, New York

Faure G, Elliot DH (1971) Isotope composition of strontium in Mesozoic basalt and dolerite from Dronning Maud Land. British Antarctic Surv Bull 25:23–27

Faure G, Hurley PM (1963) The isotopic composition of strontium in oceanic and continental basalts: application to the origin of igneous rocks. J Petrol 4(1):31–50

Faure G, Jones LM (1989) Distribution of marine salts along the west coast of Ross Island, Antarctica, based on isotopic compositions of strontium and sulfur. In: Augustithis SS (ed) Weathering; its products and deposits, vol I: Processes. Theophrastus Publications, S.A., Athens, pp 369–381

Faure G, Mensing TM (1993) K-Ar dates and paleomagnetic evidence for Cretaceous alteration of Mesozoic basaltic lava flows, Mesa Range, northern Victoria Land, Antarctica. Chem Geol (Isotope Geoscience Section) 109:305–315

Faure G, Powell JL (1972) Strontium isotope geology. Springer Verlag, Berlin

Faure G, Hurley PM, Fairbairn HW (1963a) An estimate of the isotopic composition of Sr in rocks of the Precambrian Shield of North America. J Geophys Res 68:2323–2329

Faure G, Hurley PM, Fairbairn HW, Pinson WH Jr (1963b) Age of the Great Dyke of Southern Rhodesia. Nature 200:769–770

Faure G, Chaudhuri S, Fenton MD (1969) Ages of the Duluth gabbro complex and of the Endion sill, Duluth, Minnesota. J Geophys Res 74(2):720–725

Faure G, Hill RL, Jones LM, Elliot DH (1972) Isotope composition of strontium and silica content of Mesozoic basalt and dolerite from Antarctica. In: Adie RJ (ed) Antarctic geology and geophysics. Universitetsforlaget, Oslo, pp 617–624

Faure G, Bowman JR, Elliot DH, Jones LM (1974) Strontium isotope composition and petrogenesis of the Kirkpatrick Basalt, Queen Alexandra Range, Antarctica. Contrib Mineral Petrol 48:153–169

Faure G, Bowman JR, Elliot DH (1979) The initial $^{87}Sr/^{86}Sr$ ratios of the Kirwan volcanics of Dronning Maud Land; comparison with the Kirkpatrick Basalt, Transantarctic Mountains. Chem Geol 26:77–90

Faure G, Pace KK, Elliot DH (1982) Systematic variations of $^{87}Sr/^{86}Sr$ ratios and major element concentrations in the Kirkpatrick Basalt of Mount Falla, Queen Alexandra Range, Transantarctic Mountains. In: Craddock C (ed) Antarctic Geosciences. University of Wisconsin Press, Madison, pp 715–723

Faure G, Hoefs J, Mensing TM (1984) Effect of oxygen fugacity on sulfur isotope compositions and magnetite concentrations in the Kirkpatrick Basalt, Mount Falla, Queen Alexandra Range, Ant-arctica. Chem Geol (Isotope Geoscience Section) 2:301–311

Faure G, Mensing TM, Jones JM, Hoefs J, Kibler EM (1991) Isotopic and geochemical studies of Ferar Dolerite sills in the Transantarctic Mountains. In: Ulbrich H, Rocha Campos AC (eds) Gondwana Seven Proceedings. Inst. Geociencias, Universidade de Sao Paulo, pp 669–683

Faure G, Wehn KS, Montello JM, Hagen EH, Strobel ML, Johnson KS (1993) Isotope composition of ice and sub-glacial geology near the Allan Hills, Victoria Land, Antarctica. In: Findlay RH, Bands HR, Veevers JJ, Unrug R (eds) Gondwana 8: Assembly, evolution, and dispersal. A.A. Balkema, Rotterdam, pp 485–496

Fedorenko VA (1981) Petrochemical series of extrusive rocks of the Noril'sk region. Soviet Geol Geophys 22(6):66–74

Fedorenko VA (1994) Evolution of magmatism as reflected in the volcanic sequence of the Noril'sk region. In: Lightfoot PC et al. (eds) Proceedings of the Sudbury-Noril'sk Symposium, Ontario Geol Survey Special Report 5:171–184

Feigenson MD, Hofmann AW, Spera FJ (1983) Case studies on the origin of basalt II; the transition from tholeiitic to alkalic volcanism on Kohala volcano, Hawaii. Contrib Mineral Petrol 84:390–405

Feigenson MD (1984) Geochemistry of Kauai volcanics and a mixing model for the origin of Hawaiian alkali basalts. Contrib Mineral Petrol 87:109–119

Feigenson MD, Carr MJ (1986) Positively correlated Nd and Sr isotope ratios of lavas from the Central American volcanic front. Geol 14:79–82

Fenton MD, Faure G (1969) The isotopic evolution of terrestrial strontium. Geol Soc Amer Abstracts with Programs 1(7):64

Féraud G, York D, Mevel C, Cornen G, Hall CM and Auzende JM (1986) Additional ^{40}Ar-^{39}Ar dating of the basement and the alkaline volcanism of Gorringe Bank (Atlantic Ocean). Earth Planet Sci Letters 79:255–269

Ferguson EM, Klein EM (1993) Fresh basalts from the Pacific Antarctic Ridge extend the Pacific geochemical province. Nature 366:330–333

Ferrar HT (1907) Report on the field geology of the region explored during the "Discovery" Antarctic Expedition (1901–1904) London, National Antarctic Exped., Natural History, Geology 1:1–100

Ferrara G, Gravelle M (1966) Radiometric ages from western Ahaggar (Sahara) suggesting an eastern limit for the western craton. Earth Planet Sci Letters 1:319–324

Ferrara G, Laurenzi MA, Taylor HP Jr, Tonarini S, Turi B (1985) Oxygen and strontium isotope studies of K-rich volcanic rocks from the Alban Hills, Italy. Earth Planet Sci Letters 75:13–28

Ferrara G, Preite-Martinez M, Taylor HP Jr, Tonarini S, Turi B (1986) Evidence for crustal assimilation, mixing of magmas, and a ^{87}Sr-rich upper mantle. An oxygen and strontium isotope study of the M. Vulsini volcanic area, Central Italy. Contrib Mineral Petrol 92:269–280

Ferrara G, Petrini R, Serri G, Tonarini S (1989) Petrology and isotope-geochemistry of San Vincenzo rhyolites (Tuscany, Italy). Bull Volcanol 51(5):379–388

Ferreira VP, Sial AN, Long LE, Pin C (1997) Isotopic signatures of Neoproterozoic to Cambrian ultrapotassic syenitic magmas, northeastern Brazil: evidence for an enriched mantle source. Int Geol Rev 39:660–669

Ferris J, Johnson A, Storey B (1998) Form and extent of the Dufek intrusion, Antarctica, from newly compiled aeromagnetic data. Earth Planet Sci Letters 154:185–202

Field D, Råheim A (1979) A geologically meaningless Rb-Sr total rock isochron. Nature 282:497–498

Fisher RL, Hess HH (1963) Trenches. In: Hill MN (ed) The Sea, vol III. Wiley and Sons, New York, pp 411–436

Fisher RL, Norris RM (1960) Bathymetry and geology of Sala y Gomez, southeast Pacific. Geol Soc Amer Bull 71:497–502

Fitch FJ, Grasty RL, Miller JA (1965) Potassium-argon ages of rocks from Jan Mayen and an outline of its volcanic history. Nature 207:1349–1351

Fitch FJ, Miller JA (1984) Dating Karoo igneous rocks by the conventional K-Ar and $^{40}Ar/^{39}Ar$ age spectrum methods. In Erlank AJ (ed) Petrogenesis of the volcanic rocks of the Karoo Province. Geol Soc S Africa Spec Publ 13:247–266

Fitch FJ, Molnar P (1970) Focal mechanisms along inclined earthquake zones in the Indonesia-Philippine region. J Geophys Res 75:1431–1444

Fitches WR, Graham EH, Hussein IM, Ries AC, Shackleton RM, Price RC (1983) The late Precambrian ophiolite of Sol Hamed, NE Sudan. Precamb Res 19:385–411

Fitton JG (1987) The Cameroon line, West Africa: a comparison between oceanic and continental alkaline volcanism. In: Fitton JG, Upton BGJ (eds) Alkaline igneous rocks. Geol Soc London Spec Publ 30:273–291

Fitton JG, Dunlop HM (1985) The Cameroon line, West Africa, and its bearing on the origin of oceanic and continental alkali basalt. Earth Planet Sci Letters 72:23–38

Fitton JG, Hughes DJ (1977) Petrochemistry of the volcanic rocks of the island of Principe, Gulf of Guinea. Contrib Mineral Petrol 64:257–272

Fitton JG, Upton BGJ (eds) (1987) Alkaline igneous rocks. Geol Soc London Spec Publ 30

Fitton MJ, Horwitz RC, Sylvester G (1975) Stratigraphy of the early Precambrian in the West Pilbara, Western Australia. CSIRO, Division of Mineralogy, Rot FP 11

Fleck RJ, Sutter JF, Elliot DH (1977) Interpretation of discordant $^{40}Ar/^{39}Ar$ age spectra of Mesozoic tholeiites from Antarctica. Geochim Cosmochim Acta 41:15–32

Fleck RJ, Greenwood WR, Hadley DG, Anderson RE, Schmidt DL (1979) Rubidium-strontium geochronology and plate tectonic evolution of the southern part of the Arabian Shield. U.S. Geol Survey, Saudi Arabia Project Report 245

Fleck RJ, Greenwood WR, Hadley DG, Anderson RE, Schmidt DL (1980) Rubidium-strontium geochronology and plate tectonic evolution of the southern part of the Arabian Shield. U.S.G.S. Prof Paper 1131

Fleming TH, Elliot DH, Jones LM, Bowman JR, Siders MA (1992) Chemical and isotopic variations in an iron-rich lava flow from the Kirkpatrick Basalt, north Victoria Land, Antarctica: implications for low-temperature alteration. Contrib Mineral Petrol 111:440–457

Fleming TH, Elliot DH, Foland KA, Jones LM, Bowman JR (1993) Disturbance of Rb-Sr and K-Ar isotopic systems in the Kirkpatrick Basalt, north Victoria Land, Antarctica: implications for middle Cretaceous tectonism. In: Findlay RH, Banks HR, Veevers JJ, Unrug B (eds) Gondwana 8: Assembly, evolution, and dispersal. Balkema, Rotterdam, pp 411–424

Fleming TH, Foland KA, Elliot DH (1995) Isotopic and chemical constraints on the crustal evolution and source signature of Ferrar magmas, north Victoria Land, Antarctica. Contrib Mineral Petrol 121:217–236

Flower MFJ, Schmincke H-U, Thompson RN (1975) Phlogopite stability and the $^{87}Sr/^{86}Sr$ step in basalts along the Reykjanes Ridge. Nature 254:5499, 404–405

Flower MFJ, Schmincke H-U, Bowman H (1976) Rare earths and other trace elements in historic Azorean lavas. J Volcanol Geotherm Res 1:127–147

Flower MFJ, Zhang M, Chen C-Y, Tu K, Xie G (1992) Magmatism in the South China Basin. 2. Post-spreading Quaternary basalts from Hainan Island, south China. Chem Geol 97:65–87

Foden JD, Varne R (1980) The petrology and tectonic setting of Quaternary-Recent volcanic centres of Lombok and Sumbawa, Sunda arc. Chem Geol 30:201–226

Fodor RV (1987) Low- and high-TiO_2 flood basalts of southern Brazil: origin from picritic parentage and a common mantle source. Earth Planet Sci Letters 84:423–430

Fodor RV, Asmus HE (1983) K-Ar ages and the opening of the South Atlantic Ocean: basaltic rocks from the Brazilian margin. Marine Geol 54:1–8

Fodor RV, McKee EH (1986) Petrology and K-Ar ages of rift-related basaltic rocks, offshore northern Brazil. J Geol 94:585–593

Fodor RV, Vetter SK (1984) Rift-zone magmatism: petrology of basaltic rocks transitional from CFB to MORB, southeastern Brazil margin. Contrib Mineral Petrol 88:307–321

Fodor RV, Corwin C, Sial AN (1985a) Crustal signatures in the Serra Geral flood-basalt province, southern Brazil: O- and Sr-isotope evidence. Geol 13:763–765

Fodor RV, Corwin C, Roisenberg A (1985b) Petrology of Serra Geral (Paraná) continental flood basalts, southern Brazil: crustal contamination, source material, and South Atlantic magmatism. Contrib Mineral Petrol 91:54–65

Fodor RV, Mukasa SB, Gomes CB, Cordani UG (1989) Ti-rich Eocene rocks, Abrolhos Platform, offshore Brazil, 18° S: Petrology with respect to South Atlantic magmatism. J Petrol 30(3):763–786

Fodor RV, Sial AN, Mukasa SB, McKee EH (1990) Petrology, isotope characteristics, and K-Ar ages of the Maranhão, northern Brazil, Mesozoic basalt province. Contrib Mineral Petrol 104:555–567

Fodor RV, Mukasa SB, Sial AN (1998) Isotopic and trace-element indications of lithospheric and asthenospheric components in Tertiary alkalic basalts, northeastern Brazil. Lithos, 43(4):197–218

Foland KA, Allen JC (1991) Magma sources for Mesozoic anorogenic granites of the White Mountain magma series, New England, USA. Contrib Mineral Petrol 109:195–211

Foland KA, Faul H (1977) Ages of the White Mountain intrusives-New Hampshire, Vermont, and Maine, USA. Amer J Sci 277:888–904

Foland KA, Henderson CMB (1976) Application of age and Sr isotope data to the petrogenesis of the Marangudzi ring complex, Rhodesia. Earth Planet Sci Letters 29:291–301

Foland KA, Henderson CMB, Gleason J (1985) Petrogenesis of the magmatic complex at Mount Ascutney, Vermont, USA. Contrib Mineral Petrol 90:331–345

Foland KA, Gilbert LA, Sebring CA, Chen J-F (1986) $^{40}Ar/^{39}Ar$ ages for plutons of the Monteregian Hills, Quebec: Evidence for a single episode of Cretaceous magmatism. Geol Soc Amer Bull 97:966–974

Foland KA, Chen J-F, Gilbert LA, Hofmann AW (1988) Nd and Sr isotopic signatures of Mesozoic plutons in northeastern North America. Geol 16:684–687

Foland KA, Fleming TH, Heimann A, Elliot DH (1993) Potassium-argon dating of fine-grained basalts with massive Ar loss: Application of the $^{40}Ar/^{39}Ar$ technique to plagioclase and glass from the Kirkpatrick Basalt, Antarctica. Chem Geol (Isotope Geoscience Section) 107:173–190

Foley SF, Laan SR Van der (eds) (1994) Trace-element partitioning with application to magmatic processes. Chem Geol 117(1/4):1–390

Foley SF, Venturelli G, Green DH, Toscari L (1987) The ultrapotassic rocks: characteristics, classification and constraints for petrogenetic models. Earth-Sci Rev 24:81–134

Fontignie D, Schilling J-G (1991) $^{87}Sr/^{86}Sr$ and REE variations along the Easter microplate boundary (South Pacific): application of multivariate statistical analysis to ridge segmentation. Chem Geol 89(3/4):209–242

Fontignie D, Schilling J-G (1996) Mantle heterogeneities beneath the South Atlantic: A Nd-Sr-Pb isotope study along the Mid-Atlantic Ridge (3° S–46° S). Earth Planet Sci Letters 142:209–221

Ford AB (1970) Development of the layered series and capping granophyre of the Dufek intrusion of Antarctica. Geol Soc S Africa Spec Publ 1:492–510

Ford AB (1976) Stratigraphy of the layered gabbroic Dufek intrusion, Antarctica. US Geol Surv Bull 1405-D

Ford AB (1990) The Dufek intrusion of Antarctica. In: Splettstoesser JF (ed) Mineral resources potential of Antarctica. Amer Geophys Union, Washington D.C. (Antarctic Research Ser, vol 51, pp 15–32)

Ford AB, Boyd WW Jr (1968) The Dufek intrusion, a major stratiform gabbroic body in the Pensacola Mountains, Antarctica. XXIII Internat Geol Congress 2:213–228

Ford AB, Kistler RW (1980) K-Ar age, composition, and origin of Mesozoic mafic rocks related to Ferrar Group, Pensacola Mountains, Antarctica. New Zealand J Geol Geophys 23:371–390

Ford AB, Kistler RW, White LD (1986) Strontium and isotopic study of the Dufek intrusion, Antarctica. Antarctic J US w1:63–66

Forester RW, Taylor HP Jr (1976) ^{18}O-depleted igneous rocks from the Tertiary complex of the Isle of Mull, Scotland. Earth Planet Sci Letters 32:11–17

Francalanci L, Manetti P, Peccerillo A (1989) Volcanological and magmatological evolution of Stromboli volcano (Aeolian Islands): The roles of fractional crystallization, magma mixing, crustal contamination, and source heterogeneity. Bull Volcanol 51:355–378

Francalanci L, Tommasini S, Conticelli S, Davies GR (1999) Sr isotope evidence for short magma residence time for the 20th century activity of Stromboli volcano, Italy. Earth Planet Sci Letters 167:61–70

Francis PW, Moorbath S, Thorpe RS (1977) Strontium isotope data in andesites from Ecuador and north Chile. Earth Planet Sci Letters 37:197–207

Francis PW, Thorpe RS, Moorbath S, Kretschmar GA, Hammill M (1980) Strontium isotope evidence for crustal contamination of calc-alkaline volcanic rocks from Cerro Galan, northwest Argentina. Earth Planet Sci Letters 48:257–267

Francis PW, Sparks RSJ, Hawkesworth CJ, Thorpe RS, Pyle DM, Mantovani MS, Tait S, McDermott F (1989) Petrology and geochemistry of volcanic rocks of the Cerro Galan caldera, northwest Argentina. Geol Mag 126:515–547

Franklyn MT, McNutt RH, Kamineni DC, Gascoyne M, Frape SK (1991) Groundwater $^{87}Sr/^{86}Sr$ values in the Eye-Dashwa Lakes pluton, Canada: Evidence for plagioclase-water reaction. Chem Geol (Isotope Geoscience Section) 86:111–122

Franz G, Steiner G, Volker F, Pudlo D, Hammerschmidt K (1999) Plume related alkaline magmatism in central Africa–the Meidob Hills (W Sudan). Chem Geol 157:27–47

Fraser KJ, Hawkesworth CJ, Erlank AJ, Mitchell RH, Scott-Smith BH (1985/86) Sr, Nd, and Pb isoope and minor element geochemistry of lamproites and kimberlites. Earth Planet Sci Letters 76:57–70

Fratta M, Shaw DM (1974) "Residence" contamination of K, Rb, Li and Tl in diabase dikes. Can J Earth Sci 11:422–429

French BM (1967) Sudbury structure, Ontario: some petrographic evidence for origin by meteorite impact. Science 156:1094–1098

French BM (1970) Possible relations between meteorite impact and igneous petrogenesis as indicated by the Sudbury structure. Ontario, Canada Bull Volcanol 34:466–517

French BM, Short NM (1968) Shock metamorphism of natural materials. Mono Book Co., Baltimore

Frey FA (1980) The origin of pyroxenites and garnet pyroxenites from Salt Lake Crater, Oahu, Hawaii: trace element evidence. Amer J Sci 280A:427–449

Frey FA, Green DH (1974) The mineralogy, geochemistry, and origin of lherzolite inclusions in Victorian basanites. Geochim Cosmochim Acta 38:1023–1059

Frey FA, Prinz M (1978) Ultramafic inclusions from San Carlos, Arizona: petrologic and geochemical data bearing on their petrogenesis. Earth Planet Sci Letters 38:129–176

Frey FA, Roden MF (1987) The mantle source for the Hawaiian islands: constraints from lavas and ultramafic inclusions. In: Menzies MA, Hawkesworth CJ (eds) Mantle metasomatism. Academic Press, London, pp 423–463

Frey FA, Rhodes JM (1993) Intershield geochemical differences among Hawaiian volcanoes: implications for source compositions, melting processes, and magma ascent paths. Philos Trans Roy Soc Lodon 342:121–136

Frey FA, Dickey JS Jr, Thompson G, Bryan WB (1977) Eastern Indian Ocean DSDP sites: correlations between petrography, geochemistry, and tectonic setting. In: Heirtzler JR et al. (eds) Indian Ocean geology and biostratigraphy. Amer Geophys Union, Geophys Monograph, pp 189–258

Frey FA, Green DH, Roy SD (1978) Integrated models of basalt petrogenesis: a study of quartz tholeiites to olivine melilitites from southeastern Australia utilizing geochemical and experimental petrological data. J Petrol 19:463–513

Frey FA, Dickey JS, Thompson G, Bryan WB, Davies HL (1979) Evidence for heterogeneous primary MORB and mantle sources, NW Indian Ocean. Contrib Mineral Petrol 74:387–402

Frey FA, Gerlach DC, Hickey RL, Lopez-Escobar L, Munizaga-Villavicencio F (1984) Petrogenesis of the Laguna del Maule volcanic complex, Chile (36° S). Contrib Mineral Petrol 88:133–149

Frey FA, Jones WB, Davies H, Weis D (1991) Geochemical and petrologic data for basalts from sites 756, 757 and 758: Implication for the origin and evolution of the Ninetyeast Ridge. Proc Ocean Drilling Program Sci Results 121:611–659

Frey FA, Garcia MO, Roden MF (1994) Geochemical characteristics of Koolau volcano: implications of intershield geochemical differences among Hawaiian volcanoes. Geochim Cosmochim Acta 58:1441–1462

Friedman I (1989) Are extrusive rhyolites produced from permeable foam eruptions? Bull Volcanol, 51:69–71

Frost CD, Snoke AW (1989) Tobago, West Indies, a fragment of a Mesozoic oceanic island arc: Petrochemical evidence. J Geol Soc London 146:953–964

Frost TP, Mahood GA (1987) Field, chemical, and physical constraints on mafic-felsic magma interaction in the Lamarck Granodiorite, Sierra Nevada, California. Geol Soc Amer Bull 99:272–291

Fryer BJ, Kerrich R, Hutchinson RW, Peirce MG, Rodgers DS (1979) Archean precious metal hydrothermal systems, Dome Mine, Abitibi greenstone belt. Can J Earth Sci 16:421–439

Fuchs K, Altherr R, Müller B, Prodehl C (eds) (1997) Structure and dynamic processes in the lithosphere of the Afro-Arabian rift system. Tectonophysics 278:1–3

Fueten F, Redmond DJ (1997) Documentation of a 1450 Ma contractional orogeny preserved between the 1850 Ma Sudbury impact structure and the 1 Ga Grenville orogenic front, Ontario. Geol Soc Amer Bull 109(3):268–279

Fuhrman U, Lippolt HJ (1985) Excess argon and dating of Quaternary Eifel volcanism: I. The Schellkopf alkali phonolite, East Eifel. Neues Jahrbuch Geol Paläont Monatshefte 1985:484–497

Furman T, Frey FA, Park K-H (1995) The scale of source heterogeneity beneath the Eastern neovolcanic zone, Iceland. J Geol Soc London 152:997–1002

Furnes H, Mitchell JG (1978) Age relationships of Mesozoic basalt lava and dikes in Vestfjella, Dronning Maud Land, Antarctica. In: Results from Norwegian Antarctic Research 1974–77. Norsk Polarinstitutt Skriffer 169:45–68

Furnes H, Neumann E-R, Sundvoll B (1982) Petrology and geochemistry of Jurassic basalt dykes from Vestfjella, Dronning Maud Land, Antarctica. Lithos 15:295–304

Furnes H, Vad E, Austrheim H, Mitchell JG, Garmann LB (1987) Geochemistry of basalt lavas from Vestfjella and adjacent areas, Dronning Maud Land, Antarctica. Lithos 20:337–356

Futa K, LeMasurier WE (1983) Nd and Sr isotopic studies on Cenozoic mafic lavas from West Antarctica: another source for continental alkali basalts. Contrib Mineral Petrol 83:38–44

Gair HS (1967) The geology from the upper Rennick glacier to the coast, northern Victoria Land, Antarctica. New Zealand J Geol Geophys 10(2):309–344

Gale GH (1973) Paleozoic basaltic komatiite and ocean-floor type basalts from northeastern Newfoundland. Earth Planet Sci Letters 18:22–28

Gale NH, Moorbath S, Simons J, Walker GPL (1966) K-Ar ages of acid intrusive rocks from Iceland. Earth Planet Sci Letters 1, 284–288

Galimov EM, Gerasimovsky VI (1978) Isotopic composition of carbon in Icelandic magmatic rocks. Geochem Internat 15:1–6

Gallagher K, Hawkesworth CJ (1992) Dehydration melting and the generation of continental flood basalts. Nature 358:57–59

Gamble JA (1979) The geochemistry and petrogenesis of dolerites and gabbros from the Tertiary central volcanic complex of Slieve Gullion, North East Ireland. Contrib Mineral Petrol 69:5–20

Ganguly J, Ruiz J (1986) Time-temperature relation of mineral isochrons: a thermodynamic model, and illustrative examples for the Rb-Sr system. Earth Planet Sci Letters 81:338–34

Gansser A, Dietrich VJ, Cameron WE (1979) Paleogene komatiites from Gorgona Island. Nature 278:545–546

Gariépy C, Allègre CJ (1985) The lead isotope geochemistry and geochronology of late-kinematic intrusives from the Abitibi greenstone belt, and the implications for Late Archaean crustal evolution. Geochim Cosmochim Acta 49:2371–2383

Gariépy C, Ludder J, Brooks C (1983) Isotopic and trace element constraints on the genesis of the Faeroe lava pile. Earth Planet Sci Letters 63:257–272

Gariépy C, Allègre CJ, LaJoie J (1984) U-Pb systematics in single zircons from the Pontiac sediments, Abitibi greenstone belt. Can J Earth Sc 21:1296–1304

Gass IG, Thorpe FS (1976) Igneous case study: The Tertiary igneous rocks of Skye, NW Scotland. Open University Press

Gast PW (1960) Limitations on the composition of the upper mantle. J Geophys Res 65:1287–1297

Gast PW (1968) Trace element fractionation and the origin of tholeiitic and alkaline magma types. Geochim Cosmochim Acta 32:1057–1086

Gast PW (1969) The isotopic composition of lead from St. Helena and Ascension islands. Earth Planet Sci Letters 5:353–359

Gast PW, Tilton GR, Hedge C (1964) Isotopic composition of lead and strontium from Ascension and Gough Islands. Science 145:1181–1185

Gates TM, Hurley PM (1973) Evaluation of Rb-Sr dating methods applied to the Matachewan, Abitibi, Mackenzie, and Sudbury dike swarms in Canada. Can J Earth Sci 10:900–919

Gautier I, Weis D, Mennessier J-P, Vidal P, Giret A, Loubet M (1990) Petrology and geochemistry of the Kerguelen Archipelago basalts: evolution of the mantle sources from ridge to intraplate position. Earth Planet Sci Letters 100:59–76

Geist DJ, McBirney AR, Duncan RA (1986) Geology and petrogenesis of lavas from San Cristobal Island, Galapagos Archipelago. Geol Soc Amer Bull 97:555–566

Geist DJ, Myers JD, Frost CD (1988a) Megacryst-bulk rock isotopic disequilibrium as an indicator of contamination processes: the Edgecumbe volcanic field, S.E. Alaska. Contrib Mineral Petrol 99(1):105–112

Geist DJ, White WM, McBirney AR (1988b) Plume-asthenosphere mixing beneath the Galapagos archipelago. Nature 333:657–660

Geist DJ, Frost CD, Kolker A (1990) Sr and Nd isotopic constraints on the origin of the Laramie Anorthosite Complex, Wyoming. Amer Mineral 75:13–20

Gerards J, Ledent D (1976) Les rehomogeneisations isotopiques d'age Lufilien dans les granites du Rwanda. Musée Roy Afr Central, Tervuren (Belg.). Rapp Ann 1975:91–103

Gerasimovsky VI (1974) Trace elements in selected groups of alkaline rocks. In: Sørensen H (ed) The alkaline rocks. Wiley, New York, pp 402–412

Gerasimovsky VI, Volkov VP, Kogarko LN, Polyakov AI, Saprykia TV, Balashov YA (1966a) The geochemistry of the Lovozero alkaline massif Part 1. Geology and petrology. Australian Nat. Un. Press, Canberra (translator: D.A. Brown)

Gerasimovsky VI, Volkov VP, Kogarko LN, Polyakov AI, Saprykina TV, Balashov YA (1966b) The geochemistry of the Lovozero alkaline massif. Part 2. Geochemistry. Australian Nat. Un. Press, Canberra (translator: D.A. Brown)

Gerasimovsky VI, Volkov VP, Kogarko LN, Polyakov AI (1974) Kola Peninsula. In: Sørensen H (ed) The alkaline rocks. Wiley, New York, pp 206–221

Gerlach DC (1990) Eruption rates and isotopic systematics of ocean islands: Further evidence for small-scale heterogeneity in the upper mantle. Tectonophysics 172:273–289

Gerlach DC, Hart SR, Morales VWJ, Palacios C (1986) Mantle heterogeneity beneath the Nazca plate: San Felix and Juan Fernandez islands. Nature 322:165–169

Gerlach DC, Stormer JC, Mueller PA (1987) Isotopic geochemistry of Fernando de Noronha. Earth Planet Sci Letters 85:129–144

Gerlach DC, Cliff RA, Davies GR, Norry M, Hodgson N (1988a). Magma sources of the Cape Verde archipelago: isotopic and trace element constraints. Geochim Cosmochim Acta 52(12):2979–2992

Gerlach DC, Frey FA, Roa Moreno H, Lopez-Escobar L (1988b) Recent volcanism in the Puyehue-Cordon Caulle region, southern Andes, Chile (40.5° S): Petrogenesis of evolved lavas. J Petrol 29:333–382

Gerling EK (1984) Reminiscences about some works connected with the study of noble gases, their isotopic composition, and geochronology. Chem Geol Isotope GeoScience 2:271–289

Gerling EK, Shukolyukov YA, Koltzova TY, Matveeva II, Yakovleva SZ (1962) The determination of the age of basic rocks by K-Ar method. Geokhimiya 1:931–938 (in Russian)

Giblin PE (1969) Mamainse Point area. Ontario Dept. Mines, Prelim Geol Map P554

Gibson IL (1972) The chemistry and petrogenesis of a suite of pantellerites from the Ethiopian rift. J Petrol 13:31–44

Gibson SA, Thompson RN, Dickin AP, Leonardos OH (1996) Erratum to "High-Ti and low-Ti potassic magmas: key to plume-lithosphere interactions and continental flood-basalt genesis." Earth Planet Sci Letters 141:325–341

Giggenbach WF, Kyle PR, Lyon GL (1973) Present volcanic activity on Mount Erebus, Ross Island, Antarctica. Geol 1(3):135–136

Giletti BJ (1959) Rubidium-strontium ages of Lewisian rocks from northwest Scotland. Nature 184:1793–1794

Giletti BJ (1968) Isotopic geochronology of Montana and Wyoming. In: Hamilton EI, Farquhar RM (eds) Radiometric Dating for Geologists. Wiley, InterScience New York

Giletti BJ (1974) Diffusion related to geochronology. In: Hoffman AW, Giletti BJ, Yoder HS Jr, Yund RA (eds) Geochemical transport and kinetics. Carnegie Institute of Washington, Washington (Publ 643, 353 pp)

Gill JB (1970) Geochemistry of Viti Levu, Fiji, and its evolution as an island arc. Contrib Mineral Petrol 27:179–203

Gill JB (1976a) Composition and age of Lau Basin and Ridge volcanic rocks: implications for evolution of an interarc basin and remnant arc. Geol Soc Amer Bull 87:1384–1395

Gill JB (1976b) From island arc to oceanic islands: Fiji, southwestern Pacific. Geol 4:123–126

Gill JB (1981) Orogenic andesites and plate tectonics. Springer Verlag, Berlin

Gill JB (1984) Sr-Pb-Nd isotopic evidence that MORB and OIB sources contribute to oceanic arc magmas in Fiji. Earth Planet Sci Letters 68:443–458

Gill JB (1987) Early geochemical evolution of an oceanic island arc and backarc: Fiji and South Fiji Basin. J Geol 95:589–615

Gill JB, Compston W (1973) Strontium isotopes in island arc volcanic rocks. In: Coleman P (ed) The Western Pacific: Island Arcs, Marginal Seas, Geochemistry. Australia Univ. Press, Perth, pp 483–496

Gill JB, Condomines M (1992) Short-lived radioactivity and magma genesis. Science 257:1368–1376

Gill JB, McDougall I (1973) Biostratigraphic and geologic significance of Miocene-Pliocene volcanism in Fiji. Nature 241:176–180

Gill JB, Whelan P (1989a) Early rifting of an oceanic island arc (Fiji) produced shoshonitic to tholeiite basalts. J Geophys Res 94(B4):4561–4578

Gill JB, Whelan P (1989b) Postsubduction ocean island alkali basalt in Fiji. J Geophys Res 94(B4):4579–4588

Gill RCO, Bridgwater D (1976) The Ameralik dykes of West Greenland, the earliest known basaltic rocks intruding stable continental crust. Earth Planet Sci Letters 29:276–282

Gillot P-Y, Nativel P (1982) K-Ar chronology of the ultimate activity of Piton des Neiges volcano, Reunion Island, Indian Ocean. J Volcanol Geotherm Res 13:131–146

Gilluly J (1973) Steady plate motion and episodic orogeny and magmatism. Geol Soc Amer Bull 84:499–514

Giret A, Lameyre J (1983) A study of Kerguelen plutonism: petrology, geochronology, and geological implicaton. In: Oliver RL, James PR, Jago JB (eds) Antarctic earth science. Australian National Un. Press, Canberra, A.C.T., pp 646–651

Girod M, Camus G, Vialette Y (1971) Sur la presence de tholeiites a l ile Saint-Paul (Ocean Indien). Contrib Mineral Petrol 33:108–117

Gittins J, Jago BC (1998) Differentiation of natrocarbonatite magma at Oldoinyo Lengai volcano, Tanzania. Mineral Mag 62:759–768

Gittins J, Macintyre RM, York D (1967) The ages of carbonatite complexes in eastern Canada. Can J Earth Sci 4:651–653

Glazner AF, O'Neil JR (1989) Crustal structure of the Mojave Desert, California: inferences from Sr and O isotope studies of Miocene volcanic rocks. J Geophys Res 94(B6):7861–7870

Glazner AF, Ussler W III (1989) Crustal extension, crustal density, and the evolution of Cenozoic magmatism in the Basin and Range of the western United States. J Geophys Res 94(B6): 7952–7960

Gledhill A, Baker PE (1973) Strontium isotope ratios in volcanic rocks from the South Sandwich Islands. Earth Planet Sci Letters 19:369–372

Glickson AY (1979) Early Precambrian tonalite-trondhjemite sialic nuclei. Earth-Sci Rev 15:1–73

Glickson AY (1995) Asteroid/comet mega-impacts may have triggered major episodes of crustal evolution. Eos 76(6):49, 54–55

Gold DP (1969) The Oka carbonatite and alkaline complex. Canadian Mineralogist 10:134–135

Goldich SS (1968) Geochronology in the Lake Superior region. Can J Earth Sci 5:715–724

Goldich SS, Fischer LB (1986) Air-abrasion experiments in U-Pb dating of zircon. Chem Geol (Isotope Geoscience Section) 58: 195–215

Goldich SS, Nier AO, Baadsgaard H, Hoffman JH, Krueger HW (1961) The Precambrian geology and geochronology of Minnesota. Minnesota Geol Surv Bull 41:193

Goldich SS, Hedge CE, Stern TW (1970) Age of the Morton and Montevideo gneisses and related rocks, southwestern Minnesota. Geol Soc Amer Bull 81:3671–3696

Goldich SS, Treves SB, Suhr NH, Stuckless JS (1975) Geochemistry of the Cenozoic volcanic rocks of Ross Island and vicinity, Antarctica. J Geol 83:415–435

Goldschmidt VM (1937) The principles of distribution of chemical elements in minerals and rocks. J Chem Soc London 1937: 655–673

Goldstein SJ, Murrell MT, Janecky DR, Delaney JR, Clague DA (1991) Geochronology and petrogenesis of MORB from the Juan de Fuca and Gorda ridges by ^{238}U-^{230}Th disequilibrium. Earth Planet Sci Letters 107(1):25–41

Goles GG (1976) Some constraints on the origin of phonolites from the Gregory Rift, Kenya and inferences concerning basaltic magmas in the Rift system. Lithos 9:1–8

Goles GG, Lambert RSJ (1990) A strontium isotopic study of Newberry volcano, central Oregon; structural and thermal implications. J Volcanol Geotherm Res 43(1/4):159–174

Goode ADT (1977) Flotation and remelting of plagioclase in the Kalka intrusion, central Australia: petrologic implications for anorthosite genesis. Earth Planet Sci Letters 34:375–380

Goode ADT, Moore AC (1975) High-pressure crystallization of the Ewarara, Kalka, and Gosse Pile intrusion, Giles complex, central Australia. Contrib Mineral Petrol 51:51–77

Goodrich CA (1984) Phosphoran pyroxene and olivine in silicate inclusions in natural iron-carbon alloys, Disko Island, Greenland. Geochim Cosmochim Acta 48:1115–1126

Goodwin AM (1962) Structure, stratigraphy, and origin of iron formations, Michipicoten area, Algoma district, Ontario, Canada. Geol Soc Amer Bull 73:561–586

Goodwin AM (1968a) Evolution of the Canadian Shield. Geol Assoc Canada Proceedings 19:1–14

Goodwin AM (1968b) Archaean protocontinental growth and early crustal history of the Canadian shield. XXIII Internat Geol Cong 1:69–89

Goodwin AM (1977) Archaean basin-craton complexes and the growth of Precambrian shields. Can J Earth Sci 14:2737–2759

Goodwin AM (1982) Archean volcanoes in southwestern Abitibi Belt, Ontario and Quebec: Form, composition, and development. Can J Earth Sci 19:1140–1155

Goodwin AM (1996) Principles of Precambrian geology. Academic Press, San Diego

Goodwin AM, Ridler RH (1970) The Abitibi orogenic belt. In: Baer AE (ed) Precambrian Basins and Geosynclines of the Canadian Shield. Geol Surv Canada, Paper 70-40, pp 1–30

Gopalan K, Macdougall JD, Roy AB, Murali AV (1990) Sm-Nd evidence for 3.3 Ga old rocks in Rajasthan, northwestern India. Precamb Res 48:287–297

Gorai M (1950) Proposal of twin method for the study of the "Granite Problem." J Geol Soc Japan 56:149–156

Gorai M (1960) Ultimate origin of granite. Earth Science J Assoc Geol Collaboration Japan 52:1–8

Gorai M, Kagami H, Iizumi S (1972) Reexamination of the source material of granitic magmas. J Geol Soc Japan 76(10): 549–559

Gorton M (1977) The geochemistry and origin of Quaternary volcanism in the New Hebrides. Geochim Cosmochim Acta 41: 1257–1270

Graham DW, Zindler A, Kurz MD, WJ Jenkins, Batiza R, Staudigel H (1988) He, Pb, Sr, and Nd isotope constraints on magma genesis and mantle heterogeneity beneath young Pacific seamounts. Contrib Mineral Petrol 99:446–463

Graham IJ, Gulson BL, Hedenquist JW, Mizon K (1992) Petrogenesis of Late Cenozoic volcanic rocks from the Taupo Volcanic Zone, New Zealand, in the light of new lead isotope data. Geochim Cosmochim Acta 56:2797–2819

Granath J (1975) Wind River Canyon: An example of a greenstone belt in the Archean of Wyoming, USA. Precamb Res 2:71–91

Grant NK, Hickman MH (1984) Rb-Sr isotope systematics and contrasting histories of Late Archean gneisses, West Greenland. Geol 12:599–602

Grant NK, Molling PA (1981) A strontium isotope and trace element profile through the Partridge River troctolite, Duluth complex, Minnesota. Contrib Mineral Petrol 77:296–305

Grant NK, Rex DC, Freeth SJ (1972) Potassium-argon ages and strontium isotope ratio measurements from volcanic rocks of northeastern Nigeria. Contrib Mineral Petrol 35:277–292

Grant NK, Powell JL, Burkholder RF, Walther JV, Coleman ML (1976) The isotopic composition of strontium and oxygen in lavas from St. Helena, South Atlantic. Earth Planet Sci Letters 31:209–223

Grant NK, Rose WI Jr, Fultz LA (1984) Correlated Sr isotope and geochemical variations in basalts and basaltic andesites from Guatemala. In: Harmon RS, Barreiro BA (eds) Andean magmatism, chemical and isotopic constraints. Shiva Geology Series, Shiva Publishing Co, pp 139–149

Graustein WC, Armstrong RL (1983) The use of strontium-87/strontium-86 ratios to measure atmospheric transport into forested watersheds. Science 219:289–292

Gray CM (1977) The geochemistry of central Australian granulites in relation to the chemical and isotopic effects of granulite facies metamorphism. Contrib Mineral Petrol 65:79–89

Gray CM (1978) Geochronology of granulite-facies gneisses in the Western Musgrave Block, central Australia. J Geol Soc Australia 25(7):403–414

Gray CM (1987) Strontium isotopic constraints on the origin of Proterozoic anorthosites. Precamb Res 37:173–189

Gray CM, Compston W (1978) A rubidium-strontium chronology of the metamorphism and prehistory of central Australian granulites. Geochim Cosmochim Acta 42:1735–1747

Gray CM, Goode ADT (1981) Strontium isotope resolution of magma dynamics in layered intrusion. Nature 294:155–158

Gray CM, Cliff RA, Goode ADT (1981) Neodymium-strontium isotopic evidence for extreme contamination in a basic layered intrusion. Earth Planet Sci Letters 56:189–198

Green DH, Wallace ME (1988) Mantle metasomatism by ephemeral carbonatite melts. Nature 336:459–461

Green JC (1982) Geology of the Keweenawan extrusive rocks. Geol Soc Amer Mem 156:47–55

Green TH (1994) Experimental studies of trace-element partitioning applicable to igneous petrogenesis-Sedona 16 years later. Chem Geol 117:1–36

Greene HG, Dalrymple GB, Clague DA (1978) Evidence for northward movement of the Emperor Seamounts. Geol 6:70–74

Greenland LP (1970) An equation for trace element distribution during magmatic crystallization. Amer Mineral 55:455–465

Greenwood RC, Fallick AE, Donaldson CH (1992) Oxygen isotope evidence for major fluid flow along the contact zone of the Rhum ultrabasic intrusion. Geol Mag 129(2):243–246

Greenwood WR, Hadley DG, Anderson RE, Fleck RJ, Schmidt DL (1976) Late Proterozoic cratonization in southwestern Saudi Arabia. Phil Trans Roy Soc London 280A:517–527

Gribble RF, Stern RJ, Bloomer SH, Stuben D, O'Hearn T, Newman S (1996) MORB mantle and subduction components interact to generate basalts in the southern Mariana Trough back-arc basin. Geochim Cosmochim Acta 60:2153–2166

Gribble RF, Stern RJ, Newman S, Bloomer SH, O'Hearn T (1998) Chemical and isotopic composition of lavas from the northern Mariana Trough: implications for magmagenesis in back-arc basins. J Pe rol 39(1):125-154

Griffin WL (1973) Lherzolite nodules from the Fen alkaline complex. Contrib Mineral Pet ol 38:135-146

Griffin WL, McGregor VR, Nutman A, Taylor PN, Bridgwater D (1980) Early Archaean granulite facies metamorphism south of Ameralik, West Greenland. Earth Planet Sci Letters 50: 59-74

Griffin WL, Wass SY, Hollis JD (1984) Ultramafic xenoliths from Bullenmerri and Gnotuk maars, Victoria, Australia: petrology of a sub-continental crust-mantle transition. J Petrol 25: 53-87

Griffiths RW (1986) Dynamics of mantle thermals with constant buoyancy or anomalous internal heating. Earth Planet Sci Letters 78:435-446

Griffiths RW, Campbell IH (1990) Stirring and structure in mantle starting plumes. Earth Planet Sci Letters 99:66-78

Griffiths RW, Campbell IH (1991) Interaction of mantle plume heads with the Earth's surface and onset of small-scale convection. J Geophys Res 96 18,295-18,310

Griselin M, Arndt NT, Baragar WRA (1997) Plume-lithosphere interaction and crustal contamination during formation of Coppermine River basalt, Northwest Territories, Canada. Can J Earth Sci 34:958-975

Groos AFK van(1975) The distribution of strontium between coexisting silicate and carbonate liquids at elevated pressures and temperatures. Geochim Cosmochim Acta 39:27-34

Groos AFK van, Wyllie PJ (1966) Liquid immiscibility in the system $Na_2O-Al_2O_3-SiO_2-CO_2$ at pressures of 1 kilobar. Amer J Sci 264:234-155

Grout FF (1918a) The lopolith; an igneous form exemplified by the Duluth gabbro. Amer Sci (4th Ser) 46:516-522

Grout FF (1918b) Internal structures of igneous rocks, their significance and origin; with special reference to the Duluth gabbro. J Geol 26:439-458

Grout FF (1918c) The pegmatites of the Duluth gabbro. J Geol 26: 255-264

Grove TL, Kinzler RJ, Baker MB, Donnelly-Nolan JM, Lesher CE (1988) Assimilation of granite by basaltic magma at Burnt lava flow, Medicine Lake volcano, northern California: decoupling of heat and mass transfer. Contrib Mineral Petrol 99:320-343

Grünenfelder MH, Tilton GR, Bell K, Blenkinsop J (1986) Lead and strontium isotope relationships in the Oka carbonatite complex, Quebec. Geochim Cosmochim Acta 50:461-468

Gruenewaldt G von, Sharpe MR, Hatton CJ (1985) The Bushveld complex: Introduction and review. Econ Geol 80(4):803-812

Grunder AL (1987) Low $\delta^{18}O$ silicic volcanic rocks at the Calabozos Caldera complex, southern Andes. Contrib Mineral Petrol 95:71-81

Grutzeck MS, Kridelbaugh S, Weill D (1974) The distribution of Sr and REE between diopside and silicate liquid. Geophys Res Letters 1(6):273-275

Gulson BL, Jones MT (1992) Cassiterite: potential for direct dating of mineral deposits and a precise age for the Bushveld complex granite. Geol 20:355-358

Gunn BM (1962) Differentiation in Ferrar dolerites, Antarctica. New Zealand J Geol. Geophys 5(5):820-863

Gunn BM (1966) Modal and element variation in Antarctic tholeiites. Geochim Cosmochim Acta 30:881-920

Gunn BM, Watkins ND (1970) Geochemistry of the Steens Mountain basalt, Oregon. Geol Soc Amer Bull 81:1497-1516

Gunn BM, Watkins ND (1976) Geochemistry of the Cape Verde Islands and Fernando de Noronha. Geol Soc Amer Bull 87:1089-1100

Gunn BM, Coy-Ill E, Watkins ND, Abranson CE, Nougier J (1970) Geochemistry of the oceanite-ankaramite basalt suite from East Island, Crozet archipelago. Contrib Mineral Petrol 28:319

Gunn BM, Abranson CE, Nougier J, Watkins ND, Hajash A (1971) Amsterdam Island, an isolated volcano in the southern Indian Ocean. Contrib Mineral Petrol 32:79-92

Haack U (1990) Datierung mit Rb/Sr-Mischungslinien? Ber Dtsch Mineral Ges 1, Beih Eur J Miner 2:86

Haase KM, Devey CW (1996) Geochemistry of lavas from Ahu and Tupa volcanic fields, Easter hotspot, southeast Pacific: implications for intraplate magma genesis near a spreading axis. Earth Planet Sci Letters 137:129-143

Haase KM, Devey CW, Mertz DF, Stoffers P, Garbe-Schonberg D (1996) Geochemistry of lavas from Mohns Ridge, Norwegian-Greenland Sea: implications for melting conditions and magma sources near Jan Mayen. Contrib Mineral Petrol 123: 223-237

Hajash A, Armstrong RL (1972) Paleomagnetic and radiometric evidence for the age of the Comores Islands, west central Indian Ocean. Earth Planet Sci Letters 16:231-236

Hall R (1996) Reconstructing Cenozoic SE Asia. In: Hall R, Blundell DJ (eds) Tectonic evolution of Southeast Asia. Geol Soc London Spec Pub 106 153-184

Hall R, Blundell DJ (eds) (1996) Tectonic evolution of southeast Asia. Geol Soc London Spec Paper 106:465-481

Hall RP, Hughes DJ (eds) (1989) Early Precambrian basic magmatism. Blackie and Son, New York

Halliday AN, Fallick AE, Dick n AP, Mackenzie AB, Stephens WE, Hildreth WE (1983) The isotopic and chemical evolution of Mount St. Helens. Earth Planet Sci Letters 63:241-256

Halliday AN (1990) Reply to comments by R. Stephen, J. Sparks, H.E. Huppert, and C.J.N. Wilson on "Evidence for long residence times of rhyolitic magma in the Long Valley magmatic system: The isotopic record in precaldera lavas of Glass Mountain". Earth Planet Sci Letters 99:390-394

Halliday AN, Fallick AE, Hutchinson J, Hildreth WE (1984) A Nd, Sr, and O isotopic investigation into the causes of chemical and isotopic zonation in the Bishop Tuff, California. Earth Planet Sci Letters 68:379-391

Halliday AN, Dickin AP, Fallick AE, Fitton JG (1988) Mantle dynamics: A Nd, Sr, Pb, and O isotopic study of the Cameroon Line volcanic chain. J Petrol 29(1):181-211

Halliday AN, Mahood GA, Holden P, Metz JM, Dempster TJ, Davidson JP (1989) Evidence for long residence times of rhyolitic magma in the Long Valley magmatic system: The isotopic record in precaldera lavas of Glass Mountain. Earth Planet Sci Letters 94:274-290

Halliday AN, Davidson JP, Holden P, DeWolf C, Lee D-C, Fitton JG (1990) Trace-element fractionation in plumes and the origin of HIMU mantle beneath the Cameroon line. Nature 347:523-528

Halliday AN, Davidson JP, Hildreth WE, Holden P (1991) Modelling the petrogenesis of high Rb/Sr silicic magma. Chem Geol 92:107-114

Halliday AN, Dickin AP, Hunter RN, Davies GR, Dempster TJ, Hamilton PJ, Upton BGJ (1993) Formation and composition of the lower-continental crust: Eevidence from Scottish xenolith suites. J Geophys Res 98:581-607

Halliday AN, Lee D-C, Tommasini S, Davies GR, Paslick CR, Fitton JG, James DE (1995) Incompatible trace elements in OIB and MORB and source enrichment in the sub-oceanic mantle. Earth Planet Sci Letters 133:379-395

Halls HC, Fahrig WF (eds) Diabase dyke swarms. Geol Assoc Canada Spec Paper 34

Halpern M (1969) $^{87}Sr/^{86}Sr$ ratios of ultramafic nodules and host basalt from the McMurdo area and Ford Ranges, Antarctica. Antarctic J US 4(5):206

Halpern M (1973) Regional geochronology of Chile south of 50° latitude. Geol Soc Amer Bull 84:2407-2422

Halpern M (1978) Geological significance of Rb-Sr isotopic data of northern Chile crystalline rocks of the Andean orogen between latitudes 23° and 27° South. Geol Soc Amer Bull 89:522-532

Halpern M, Carlin GM (1971) Radiometric chronology of crystalline rocks from southern Chile. Antarctic J US 6(5):191-193

Halpern M, Fuenzalida R (1978) Rubidium-strontium geochronology of a transect of the Chilean Andes between latitudes 45° and 46° S. Earth Planet Sc Letters 41:60-66

Halpern M, Tristan N (1981) Geochronology of the Arabian-Nubian Shield in southern Israel and eastern Sinai. J Geol 89:639-648

Halpern M, Cordani UG, Berenholc M (1974) Variations in strontium isotopic composition of Paraná Basin volcanic rocks of Brazil. Revista Brasileira de Geociencias 4:223-227

Hamelin B, Allègre CJ (1985) Large-scale regional units in the depleted upper mantle revealed by an isotope study of the southwest Indian Ridge. Nature 315:196–199

Hamelin B, Dupré B, Allègre CJ (1984) Lead-strontium isotopic variations along the East Pacific Rise and the Mid-Atlantic Ridge: A comparative study. Earth Planet Sci Letters 67:340–350

Hamelin B, Dupré B, Allègre CJ (1986) Pb-Sr-Nd isotopic data of Indian Ocean ridges: new evidence of large-scale mapping of mantle heterogeneities. Earth Planet Sci Letters 76:286–296

Hamilton EI (1963) The isotopic composition of strontium in the Skaergaard intrusion, East Greenland. J Petrol 4:383–391

Hamilton EI (1965) Isotopic composition of strontium in a variety of rocks from Réunion Island. Nature 207:1188

Hamilton EI, Deans T (1963) Isotopic composition of strontium in some African carbonatites and limestones and in strontium minerals. Nature 198:776–777

Hamilton PJ (1977) Sr isotope and trace element studies of the Great Dyke and Bushveld mafic phase and their relation to Early Proterozoic magma genesis in southern Africa. J Petrol 18:24–52

Hamilton PJ, O'Nions RK, Evensen NM, Bridgwater DE, Allart JH (1978) Sm-Nd isotopic investigations of Isua supracrustals and implications for mantle evolution. Nature 272:41–43

Hamilton PJ, Johnson RW, Mackenzie DE, O'Nions RK (1983a) Pleistocene volcanic rocks from the Fly-Highlands province of western Papua New Guinea: A note on new Sr and Nd isotopic data and their petrogenetic implications. J Volcanol Geotherm Res 18:449–460

Hamilton PJ, O'Nions RK, Bridgwater D, Nutman AP (1983b) Sm-Nd studies of Archaean metasediments and metavolcanics from West Greenland and their implications for the Earth's early history. Earth Planet Sci Letters 62:263–272

Hamilton W (1965) Diabase sheets of the Taylor Glacier region, Victoria Land, Antarctica. US Geol Surv Prof Paper 456-B: 1–71

Hamilton W (1970) Bushveld complex-product of impacts? Geol Soc S Africa Spec Publ 1:367–379

Hamilton W (1972) The Hallett volcanic province, Antarctica. US Geol Surv Prof Paper 456-C:1–62

Hamilton W (1979) Tectonics of the Indonesian region. US Geol Surv Prof Paper 1078

Hamilton W, Myers WB (1966) Cenozoic tectonics of the western United States. Rev Geophys 4:509–549

Han B-F, Wang S-G, Kagami H (1999) Trace element and Nd-Sr isotopic constraints on origin of the Chifeng flood basalts, North China. Chem Geol 155:187–201

Hanan BB, Graham DW (1996) Lead and helium evidence from oceanic basalts for a common deep source of mantle plumes. Science 272:991–994

Hanan BB, Schilling J-G (1997) The dynamic evolution of the Iceland mantle plume: the lead isotope perspective. Earth Planet Sci Letters 151:43–60

Hanan BB, Kingsley RH, Schilling J-G (1986) Pb isotope evidence in the South Atlantic for migrating ridge-hotspot interactions. Nature 22:137–144

Hansen H, Nielsen TFD (1999) Crustal contamination in Palaeogene East Greenland flood basalts: plumbing system evolution during continental rifting. Chem Geol 157:89–118

Hanson GN (1975) $^{40}Ar/^{39}Ar$ spectrum ages on Logan intrusions, a lower Keeweenawan flow and mafic dykes in northeastern Minnesota-northwestern Ontario. Can J Earth Sci 12:821–835

Hanson GN (1978) The application of trace elements to the petrogenesis of igneous rocks of granitic composition. Earth Planet Sci Letters 38:26–43

Hanson GN, Goldich SS (1972) Early Precambrian rocks in the Saganaga Lake-Northern Light Lake area, Minnesota-Ontario. Part II: Petrogenesis. Geol Soc Amer Bull Mem 135:179–192

Hanson GN, Goldich SS, Arth JG, Yardley DH (1971) Age of the early Precambrian rocks of the Saganaga Lake-Northern Light Lake area, Ontario-Minnesota. Can J Earth Sci 8:1110–1124

Hansteen TH, Andersen T, Neumann E-R, Jelsma H (1991) Fluid and silicate inclusions in ultramafic and mafic xenoliths from Hierro, Canary Islands: Implications for mantle metasomatism. Contrib Mineral Petrol 107:242–254

Hanyu T, Kaneoka I (1997) The uniform and low $^3He/^4He$ ratios of HIMU basalts as evidence for their origin as recycled materials. Nature 390:273–275

Hardarson BS, Fitton JG (1991) Increased mantle melting beneath Snaefellsjökull volcano during Late Pliocene deglaciation. Nature 353:62–64

Hards VL, Kempton PD, Thompson RN (1995) The heterogeneous Iceland plume: new insights from the alkaline basalts fo the Snaefell volcanic center. J Geol Soc London 152:1003–1009

Hargraves RB (ed) (1980) Physics of magmatic processes. Princeton University Press, Princeton, N.J.

Harker A (1904) The Tertiary igneous rocks of Skye. Mem Geol Survey Great Britain 9:1–482

Harmer RE (ed) (1997) Carbonatites from source to surface. J African Earth Sci 25:1–159

Harmer RE, Eglington BM (1990) A review of the statistical principles of geochronometry: towards a more consistent approach for reporting geochronological data. South African J Geol 93:845–856

Harmer RE, Sharpe MR (1985) Field relations and strontium isotope systematics of the marginal rocks of the eastern Bushveld complex. Econ Geol 80:813–837

Harmer RE, Lee CA, Eglington BM (1998) A deep mantle source for carbonatite magmatism: evidence from the nephelinites and carbonatites of the Buhera district, SE Zimbabwe. Earth Planet Sci Letters 158:131–142

Harmon RS, Barreiro BA (eds) (1984) Andean magmatism, chemical and isotopic constraints. Shiva Geology Series, Nantwich, Cheshire, England

Harmon RS, Hoefs J (1984) Oxygen isotope ratios in Late Cenozoic Andean volcanics. In: Harmon RS, Barreiro BA (eds) Andean Magmatism, Chemical and Isotopic Constraints. Shiva Geology Series, Nantwich, Cheshire, England, pp 9–20

Harmon RS, Thorpe RS, Francis PW (1981) Petrogenesis of Andean andesites from combined O-Sr relationships. Nature 290: 396–399

Harmon RS, Barreiro BA, Moorbath S, Hoefs J, Francis PW, Thorpe RS, Deruelle B, McHugh J, Viglino JA (1984) Regional O-, Sr-, and Pb-isotope relations in Late Cenozoic calc-alkaline lavas of the Andean Cordillera. Geol Soc London J 141: 803–822

Harmon RS, Hoefs J, Wedepohl KH (1987) Stable isotope (O, H, S) relationships in Tertiary basalts and their mantle xenoliths from the northern Hessian Depression, Germany. Contrib Mineral Petrol 95:350–369

Harper CT, Weintraub GS, Leggo PJ, Shackleton RM (1972) Potassium-argon ages from the basement complex and associated Precambrian metasedimentary rocks of Uganda. Geol Soc Amer Bull 83:3449–3456

Harris C (1989) Covariance of initial $^{87}Sr/^{86}Sr$ ratios, $\delta^{18}O$, and SiO_2 in continental basalt suites: The role of contamination and alteration. Geol 17(7):634–636

Harris C, Erlank AJ (1992) The production of large-volume, low $\delta^{18}O$ rhyolites during the rifting of Africa and Antarctica: The Lebombo Monocline, southern Africa. Geochim Cosmochim Acta 56:3561–3570

Harris C, Bell JD, Atkins FB (1982) Isotopic composition of lead and strontium in lavas and coarse-grained blocks from Ascension Island, South Atlantic. Earth Planet Sci Letters 60:79–85

Harris C, Bell JD, Atkins FB (1983) Isotopic composition of lead and strontium in lavas and coarse-grained blocks from Ascension Island, South Atlantic-an addendum. Earth Planet Sci Letters 83:139–141

Harris C, Marsh JS, Duncan AR, Erlank AJ (1990) The petrogenesis of the Kirwan Basalts of Dronning Maud Land, Antarctica. J Petrol 31(2):341–370

Harris NBW (1982) The petrogenesis of alkaline intrusives from Arabia and north-east Africa and their implications for within-late magmatism. Tectonophysics 83:243–258

Harris NBW, Duyverman HJ, Almond DC (1983) The trace element and isotope geochemistry of the Sabaloka igneous complex, Sudan. J Geol Soc London 140:245–256

Harris PG (1957) Zone refining and the origin of potassic basalts. Geochim Cosmochim Acta 12:195–208

Harris WB, Fullagar PD (1991) Middle Eocene and Late Oligocene isotopic dates of glauconitic mica from the Santee River area, South Carolina. Southeastern Geol 32(1):1–19

Harrison JE, Peterman ZE (1982) North American Commission on Stratigraphic Nomenclature. Report 9: Adoption of geochronometric units for divisions of Precambrian time. Amer Assoc Petrol Geologists Bull 66:801–802

Hart SR (1971) K, Rb, Cs, Sr, and Ba contents and Sr isotope ratios of ocean floor basalts. Phil Trans Roy Soc 268:573

Hart SR (1984) A large-scale isotopic anomaly in the Southern Hemisphere mantle. Nature 309:753–757

Hart SR (1988) Heterogeneous mantle domains: signatures, genesis, and mixing chronologies. Earth Planet Sci Letters 90: 273–296

Hart SR, Brooks C (1974) Clinopyroxene-matrix partitioning of K, Rb, Cs, Sr, and Ba. Geochim Cosmochim Acta 38:1797–1806

Hart SR, Brooks C (1977) The geochemistry and evolution of the early Precambrian mantle. Contrib Mineral Petrol 61:109–128

Hart SR, Davis GL (1969) Zircon U-Pb and whole-rock Rb-Sr ages and early crustal development near Rainy Lake, Ontario. Geol Soc Amer Bull 80:595–612

Hart SR, Kyle PR (1993) The geochemistry of McMurdo Group volcanic rocks. Antarct J US 28:14–16

Hart SR, Staudigel H (1989) Isotopic characterization and identification of recycled components. In: Hart SR, Gulen L (eds) Crust/Mantle Convergence Zones. Kluwer Academic Publishers, pp 15–28

Hart SR, Zindler A (1986) In search of a bulk-Earth composition. Chem Geol 57:247–267

Hart SR, Zindler A (1989a) Isotope fractionation laws: a test using calcium. Int J Mass Spectrometry and Ion Processes 89: 287–301

Hart SR, Zindler A (1989b) Constraints on the nature and development of chemical heterogeneities in the mantle. In: Peltier WR (ed) Mantle convection. Gordon and Breach Science Publishers, New York, pp 261–382

Hart SR, Brooks C, Krogh TE, Davis GL (1970a) Ancient and modern volcanic rocks. A trace element model. Earth Planet Sci Letters 10:17–28

Hart SR, Nalwalk AJ (1970b) K, Rb, Cs and Sr relationships in submarine basalts from the Puerto Rico Trench. Geochim Cosmochim Acta 34:145–155

Hart SR, Gunn BM, Watkins ND (1971) Intralava variation of alkali elements in Icelandic basalt. Amer J Sci 270:315–318

Hart SR, Glassley WE, Karig DE (1972) Basalts and sea floor spreading behind the Mariana island arc. Earth Planet Sci Letters 15:12–18

Hart SR, Schilling J-G, Powel JL (1973) Basalts from Iceland and along the Reykjanes Ridge: Sr-isotope geochemistry. Nature Phys Sci 246:104–107

Hart SR, Gerlach DC, White WM (1986) A possible new Sr-Nd-Pb mantle array and consequences for mantle mixing. Geochim Cosmochim Acta 50:1551–1557

Hart SR, Hauri EH, Oschmann LA, Whitehead JA (1992) Mantle plumes and entrainment: the isotopic evidence. Science 256: 517–520

Hart SR, Blusztajn J, Craddock C (1995) Cenozoic volcanism in Antarctica: Jones Mountains and Peter I Island. Geochim Cosmochim Acta 59:3379–3388

Hart SR, Blusztajn J, LeMasurier WE, Rex DC (1997) Hobbs Coast Cenozoic volcanism: Implications for the West Antarctic rift system. Chem Geol 139:223–248

Hart WK (1985) Chemical and isotopic evidence for mixing between depleted and enriched mantle, northwestern USA. Geochim Cosmochim Acta 49:131–144

Hart WK, Aronson JL, Mertzman SA (1984) Areal distribution and age of low-K, high alumina olivine tholeiite magmatism in the northwestern Great Basin. Geol Soc Amer Bull 95:186–195

Hart WK, Woldegabriel E, Walter RC, Mertzman SA (1989) Basaltic volcanism in Ethiopia. Constraints on continental rifting and mantle interaction. J Geophys Res 94(Bb):7731–7748

Haskin LA, Salpas PA (1992) Genesis of compositional characteristics of Stillwater AN- and AN-II thick anorthosite units. Geochim Cosmochim Acta 56:1187–1212

Hatherton T, Dickinson WR (1969) The relationship between andesitic volcanism and seismicity in Indonesia, the Lesser Antilles, and other island arcs. J Geophys Res 74, No. 22:5301–5310

Hatton CJ (1995) Mantle plume origin for the Bushveld and Ventersdorp magmatic provinces. J African Earth Sci 21:571–577

Hattori K, Chalokwu CI (1995) Source of early magmatism related to the opening of the Atlantic Ocean: Sr- and Nd-isotopes of mineral separates from the Freetown Complex of Sierra Leone. Amer Geophys Union (Eos, Abstract supplement, Abstract V41D-7, p F687)

Hattori K, Cabri LJ, Hart SR (1991) Osmium isotope ratios of PGM grains associated with the Freetown Layered Complex, Sierra Leone, and their origin. Contrib Mineral Petrol 109:10–18

Hauri EH (1996) Major-element variability in the Hawaiian mantle plume. Nature 382:415–419

Hauri EH, Shimizu N, Dieu JJ, Hart SR (1993) Evidence for hotspot-related carbonatite metasomatism in the oceanic upper mantle. Nature 365:221–227

Hawkesworth CJ (1982) Isotope characteristics of magmas erupted along destructive plate margins. In: Thorpe RS (ed) Andesites. Wiley, Chichester, pp 549–571

Hawkesworth CJ, Norry MJ (1983) Continental flood basalts and mantle xenoliths. Shiva, Nantwich

Hawkesworth CJ, O'Nions RK (1977) The petrogenesis of some Archaean volcanic rocks from southern Africa. J Geophys Res 18:487–520

Hawkesworth CJ, Powell BM (1980) Magma genesis in the Lesser Antilles island arc. Earth Planet Sci Letters 51:297–308

Hawkesworth CJ, Vollmer R (1979) Crustal contamination versus enriched mantle: ^{143}Nd/^{144}Nd and ^{87}Sr/^{86}Sr evidence from the Italian volcanics. Contrib Mineral Petrol 69:151–165

Hawkesworth CJ, O'Nions RK, Pankhurst RJ, Hamilton PJ, Evensen NM (1977) A geochemical study of island-arc and back-arc tholeiites from the Scotia Sea. Earth Planet Sci Letters 36: 253–262

Hawkesworth CJ, Norry MJ, Roddick JC, Vollmer R (1979a) ^{143}Nd/^{144}Nd and ^{87}Sr/^{86}Sr ratios from the Azores and their significance in LIL-element enriched mantle. Nature 280:28–31

Hawkesworth CJ, O'Nions PK, Arculus RJ (1979b) Nd and Sr isotope geochemistry of island arc volcanics. Earth Planet Sci Letters 45:237–248

Hawkesworth CJ, Norry MJ, Roddick JC, Baker PE, Francis PW, Thorpe RS (1979c) ^{143}Nd/^{144}Nd, ^{87}Sr/^{86}Sr, and incompatible trace element variations in calc-alkaline andesitic and plateau lavas from South America. Earth Planet Sci Letters 42: 45–57

Hawkesworth CJ, Hammill M, Gledhill AR, Calsteren PWC van, Rogers G (1982) Isotope and trace element evidence for late-stage intra-crustal melting in the High Andes. Earth Planet Sci Letters 58:240–254

Hawkesworth CJ, Marsh JS, Duncan AR, Erlank AJ, Norry MJ (1984a) The role of continental lithosphere in the generation of the Karoo volcanic rocks: Evidence from combined Nd- and Sr-studies. In: Erlank AJ (ed) Petrogenesis of the volcanic rocks of the Karoo Province. Geol Soc S Africa Spec Publ 13: 341–354

Hawkesworth CJ, Rogers NW, Calsteren PWC van, Menzies MA (1984b) Mantle enrichment processes. Nature 311:331–335

Hawkesworth CJ, Mantovani MSM, Taylor PN, Palacz Z (1986) Coupled crust-mantle systems: evidence fro the Paraná of S. Brazil. Nature 322:356–358

Hawkesworth CJ, Mantovani MSM, Peate D (1988) Lithosphere remobilisation during Paraná CFB magmatism. In: Menzies MA, Cox KG (eds) Oceanic and continental lithosphere: Similarities and differences. University Press Oxford, Oxford (J Petrol Spec Volume, pp 205–223)

Hawkesworth CJ, Kempton PD, Rogers NW, Ellam RM, Calsteren PWC van (1990) Continental mantle lithosphere and shallow level enrichment processes in the Earth's mantle. Earth Planet Sci Letters 96:256–268

Hawkesworth CJ, Lightfoot PC, Fedorenko VA, Blake S, Naldrett AJ, Doherty W, Gorbachev NS (1995) Magma differentiation and mineralization in the Siberian continental flood basalts. Lithos 34:61–88

Hawkins JW, Melchior JT (1985) Petrology of Mariana Trough and Lau Basin basalts. J Geophys Res 90:11431–11468

Hawkins JW, Natland JH (1975) Nephelinites and basanites of the Samoan linear volcanic chain: their possible tectonic significance. Earth Planet Sci Letters 24:427–439

Hawkins JW, Macdougall JD, Volpe AM (1990) Petrology of the axial ridge of the Mariana Trough back-arc spreading center. Earth Planet Sci Letters 100:226–250

Hay EA (1976) Cenozoic uplifting of the Sierra Nevada in isostatic response to north American and Pacific plate interactions. Geol 4:763–766

Hay RL (1976) Geology of Olduvai Gorge. Un. Cal. Press, Berkeley

Hay RL (1983) Natrocarbonatite tephra of Kerimasi volcano, Tanzania. Geol 11:599–602

Hay RL (1989) Holocene carbonatite-nephelinite tephra deposits of Oldoinyo Lengai. J Volcanol Geotherm Res 37:77–91

Hay RL, Iijima A (1968) Nature and origin of palagonite tuffs of the Honolulu Group on Oahu, Hawaii. Geol Soc Amer Mem 116:371

Hay RL, O'Neil JR (1983) Carbonatite tuffs in the Laetolil beds of Tanzania and the Kaiserstuhl in Germany. Contrib Mineral Petrol 82:403–406

Hayase I, Nohda S (1969) Geochronology of the "Oldest Rock" of Japan. Geochem J 3:45–52

Hayes DE (ed) (1980) The tectonic and geologic evolution of Southeast Asian seas and islands. Amer Geophys Union, Washington, D.C. (Monogr Ser 23)

Hayes DE (ed) (1983) The tectonic and geologic evolution of Southeast Asian seas and islands. Amer Geophys Union, Washington, D.C. (Monogr Ser 27)

Healy J (1962) Structure and volcanism in the Taupo Volcanic Zone, New Zealand. Amer Geophys Union, Washington, D.C. (Geophys Monogr 6, pp 151–157)

Heatherington AL, Mueller PA (1991) Geochemical evidence for Triassic rifting in southwestern Florida. Tectonophysics 188: 291–302

Heatherington AL, Mueller PA (1999) Lithospheric sources of north Florida, USA, tholeiites and implications for the origin of the Suwannee terrane. Lithos 46(2):215–234

Hedge CE (1966) Variations in radiogenic strontium found in volcanic rocks. J Geophys Res 71(24):6119–6126

Hedge CE (1978) Strontium isotopes in basalts from the Pacific Ocean basin. Earth Planet Sci Letters 38:88–94

Hedge CE, Gorshkov GS (1977) Strontium-isotopic composition of volcanic rocks from Kamchatka. Doklady Akad Nauk SSSR 233:163–166

Hedge CE, Knight RJ (1969) Lead and strontium isotopes in volcanic rocks from northern Honshu, Japan. Geochem J 3:15–24

Hedge CE, Lewis JF (1971) Isotopic composition of strontium in three basalt-andesite centers along the Lesser Antilles arc. Contrib Mineral Petrol 32:39–47

Hedge CE, Noble DC (1971) Upper Cenozoic basalt with high $^{87}Sr/^{86}Sr$ and Sr/Rb ratios, southern Great Basin, western United States. Geol Soc Amer Bull 82:3503–3510

Hedge CE, Peterman ZE (1970) The strontium isotopic composition of basalts from the Gordo and Juan de Fuca rises, northeastern Pacific Ocean. Contrib Mineral Petrol 27:114–120

Hedge CE, Hildreth RA, Henderson WT (1970) Strontium isotopes in some Cenozoic lavas from Oregon and Washington. Earth Planet Sci Letters 8:434–438

Hedge CE, Peterman ZE, Dickinson WR (1972) Petrogenesis of lavas from Western Samoa. Geol Soc Amer Bull 83:2709–2714

Hedge CE, Watkins ND, Hildreth RA, Doering WP (1973) $^{87}Sr/^{86}Sr$ ratios in basalts from islands in the Indian Ocean. Earth Planet Sci Letters 21:29–34

Hedge CE, Marvin RF, Naeser CW (1975) Age provinces in the basement rocks of Liberia. J Res US Geol Surv 3:425–430

Hedge CE, Futa K, Engel CG, Fisher RL (1979) Rare earth abundances and Rb-Sr systematics of basalts, gabbro, anorthosite, and minor granitic rocks from the Indian Ocean ridge system, western Indian Ocean. Contrib Mineral Petrol 68:373–376

Hegner E, Pallister JS (1989) Pb, Sr, and Nd isotopic characteristics of Tertiary Red Sea rift volcanics from the central Saudi Arabian coastal plain. J Geophys Res 94(B6):7749–7756

Hegner E, Tatsumoto M (1987) Pb, Sr, and Nd isotopes in basalts and sulfides from the Juan de Fuca Ridge. J Geophys Res 92: 11380–11386

Hegner E, Vennemann TW (1997) Role of fluids in the origin of Tertiary European intraplate volcanism: Evidence from O, H. and Sr isotopes in melilites. Geol 25(11):1035–1038

Hegner E, Unruh DM, Tatsumoto M (1986) Sr, Nd, and Pb isotopic evidence for the origin of the West Maui volcanic suite, Hawaii. Nature 319:478–480

Hegner E, Walter HJ, Satir M (1995) Pb-Sr-Nd isotopic compositions and trace element geochemitry of megacrysts and melilitites from the Tertiary Urach volcanic field: source composition of small volume melts under SW Germany. Contrib Mineral Petrol 122:322–335

Heier KS (1962) Trace elements in feldspars – a review. Norsk Geol Tidsskr 42:415–454

Heier KS, Compston W (1969) Rb-Sr isotopic studies of the plutonic rocks of the Oslo region. Lithos 2:133–148

Heier KS, Compston W, McDougall I (1965) Thorium and uranium concentrations, and the isotopic composition of strontium in the differentiated Tasmanian dolerites. Geochim Cosmochim Acta 29:643–659

Heimann A, Fleming TH, Elliot DH, Foland KA (1994) A short interval of Jurassic continental flood basalt volcanism in Antarctica as demonstrated by $^{40}Ar/^{39}Ar$ geochronology. Earth Planet Sci Letters 121:19–41

Heinrich EW (1966) The geology of carbonatites. Rand McNally

Heinrichs H, Schulz-Dobrick B, Wedepohl K-H (1980) Terrestrial geochemistry of Cd, Bi, Tl, Pb, Zn and Rb. Geochim Cosmochim Acta 44(10):1519–1533

Heller PL, Peterman ZE, O'Neil JR, Shafiqullah M (1985) Isotopic provenance of sandstones from the Eocene Tyee Formation, Oregon Coast Range. Geol Soc Amer Bull 96:770–780

Helmke PA, Haskin LA (1973) Rare-earth element, Co, Sc, and Hf in the Steens Mountain basalts. Geochim Cosmochim Acta 37: 1513–1529

Heming RF (1974) Geology and petrology of Rabaul Caldera, Papua New Guinea. Geol Soc Amer Bull 85:1253–1264

Hémond C, Condomines M, Fourcade S, Allègre CJ, Oskarsson N, Javoy M (1988) Thorium, strontium, and oxygen isotopic geochemistry in recent tholeiites from Iceland: Crustal influence on mantle-derived magmas. Earth Planet Sci Letters 87:273–285

Hémond C, Arndt NT, Lichtenstein U, Hofmann AW (1993) The heterogeneous Icelandic plume: Nd-Sr-O isotopes and trace element constraints. J Geophys Res 98(B9):15833–15850

Hémond C, Devey CW, Chauvel C (1994a) Source compositions and melting processes in the Society and Austral plumes (South Pacific Ocean): Element and isotope (Sr, Nd, Pb, Th) geochemistry. Chem Geol 115:7–45

Hémond C, Hofmann AW, Heusser G, Condomines M, Raczek I, Rhodes JM (1994b) U-Th-Ra systematics in Kilauea and Mauna Loa basalts, Hawaii. Chem Geol 116:163–180

Henderson P (1982) Inorganic geochemistry. Pergamon Press, Oxford

Henry DJ, Mueller PA, Wooden JL, Warner JL, Lee-Berman R (1982) Granulite grade supracrustal assemblages of the Quad Creek area, eastern Beartooth Mountains, Montana. In: Mueller P, Wooden J (eds) Precambrian geology of the Beartooth Mountains. Montana Bur Mines Geol Spec Pub 84:147–156

Hergt JM, Chappell BW, Faure G, Mensing TM (1989a) The geochemistry of Jurassic dolerites from Portal Peak, Antarctica. Contrib Mineral Petrol 102:298–305

Hergt JM, Chappell BW, McCulloch TM, McDougall I, Chivas AR (1989b) Geochemical and isotopic constraints on the origin of the Jurassic dolerites of Tasmania. J Petrol 30:841–883

Hergt JM, Peate DW, Hawkesworth CJ (1991) The petrogenesis of Mesozoic Gondwana low-Ti flood basalts. Earth Planet Sci Letters 105:134–148

Herron EM (1972) Sea-floor spreading and the Cenozoic history of the east-central Pacific. Geol Soc Amer Bull 83:1671–1692

Hervé F, Nelson E, Kawashita E, Suárez M (1981) New isotopic ages and the timing of orogenic events in the Cordillera Darwin, southernmost Chilean Andes. Earth Planet Sci Letters 55: 257–265

Hervé F, Kawashita K, Munizaga F, Bassei M (1984) Rb-Sr isotopic ages from late Paleozoic metamorphic rocks of central Chile. J Geol Soc London 141:877–884

Hervé M, Suárez M, Puig A (1984) The Patagonian batholith S of Tierra del Fuego, Chile: Timing and tectonic implications. J Geol Soc London 141:909–917

Herzberg CT (1992) Depth and degree of melting of komatiites. J Geophys Res 97:4521–4540

Herzberg CT (1995) Generation of plume magmas through time: An experimental perspective. Chem Geol 126:1–16

Herzer RH (1971) Bowie Seamount. A recently active, flat topped seamount in the northeast Pacific Ocean. Can J Earth Sci 8:676–687

Herzog LF, Hurley PM (1955) Investigation of isotopic abundances of strontium, calcium, and argon in certain minerals. Annual Progress Report for 1954–5. AEC Contract AT (30-1)-138

Herzog LF, Aldrich LT, Holyk WK, Whiting FB, Ahrens LH (1953) Variations in strontium isotope abundances in minerals. Part II. Radiogenic ^{87}Sr in biotite, feldspar, and celestite. Trans Amer Geophys Union 34:461–470

Herzog LF, Pinson WH Jr, Cormier RF (1958) Sediment age determination by Rb/Sr analysis of glauconite. Amer Assoc Petrol Geol Bull 42(4):717–733

Hess HH (1960) Stillwater igneous complex, Montana. A quantitative mineralogical study. Geol Soc Amer Mem 80:1–230

Hess PC (1989) Origins of igneous rocks. Harvard University Press, Cambridge, Mass.

Heumann A, Davies GR (1997) Isotopic and chemical evolution of the past-caldera rhyolitic system at Long Valley, California. J Petrol 38:1661–1678

Hickey RL, Gerlach DC, Frey FA (1984) Geochemical variations in volcanic rocks from central-south Chile (33°–42° S): Implications for their petrogenesis. In: Harmon RS, Barreiro BA (eds) Andean magmatism Shiva Pub. Co., Cheshire, England, pp 72–79

Hickey RL, Frey FA, Gerlach DC, Lopez-Escobar L (1986) Multiple sources for basaltic arc rocks from the southern volcanic zone of the Andes (34°–41° S): Trace element and isotopic evidence for contributions from subducted oceanic crust, mantle, and continental crust. J Geophys Res 91(B6):5963–5983

Hickey-Vargas RL (1998) Origin of the Indian Ocean-type isotopic signature in basalts from Philippine Sea plate spreading centers: An assessment of local versus large-scale processes. J Geophys Res 103(19):20,963–20,980

Higgins MW (1973) Petrology of Newberry volcano, central Oregon. Geol Soc Amer Bull 84:455–488

Hildreth WE (1979) Evidence for the origin of compositional zonation in silicic magma chambers. Geol Soc Amer Special Paper 180:43–75

Hildreth WE (1981) Gradients in silicic magma chambers: implications for lithospheric magmatism. J Geophys Res 86:10153–10192

Hildreth WE, Grunder AL, Drake RE (1984) The Loma Seca Tuff and the Calabozos caldera: a major ash-flow and caldera complex in the southern Andes of Chile. Geol Soc Amer Bull 95:45–54

Hildreth WE, Grove TL, Dungan MA (1986) Introduction to a special section on open magmatic systems. J Geophys Res 91(B6):5887–5890

Hildreth WE, Moorbath S (1988) Crustal contributions to arc magmatism in the Andes of Central Chile. Contrib Mineral Petrol 98:455–499

Hill RI (1991) Starting plumes and continental break-up. Earth Planet Sci Letters 104:398–416

Hill RI, Campbell IH, Griffiths RW (1991) Plume tectonics and the development of stable continental crust. Explor Geophys 22:185–188

Hills FA, Armstrong RL (1974) Geochronology of Precambrian rocks in the Laramie Range and implications for the tectonic framework of Precambrian southern Wyoming. Precamb Res 1:213–225

Hills FA, Gast PW, Houston RS, Swainbank IG (1968) Precambrian geochronology of the Medicine Bow Mountains, southeastern Wyoming. Geol Soc Amer Bull 79:1757–1784

Hirschmann MM (1992) Origin of the transgressive granophyres from the Layered Series of the Skaergaard intrusion, East Greenland. J Volcanol Geotherm Res 52:185–207

Hirschmann MM, Renne PR, McBirney AR (1997) $^{40}Ar/^{39}Ar$ dating of the Skaergaard intrusion. Earth Planet Sci Letters 146:645–658

Hjelle A, Winsnes T (1972) The sedimentary and volcanic sequence of Vestfjella, Dronning Maud Land. In: Adie RJ (ed) Antarctic geology and geophysics. Universitetsforlaget, Oslo, pp 557–562

Ho ES, Myers PA, Mauk JL (1990) Organic geochemical study of mineralization in the Keweenawan Nonesuch Formation at White Pine, Michigan. Organic Geochem 16:229–234

Hochstaedter AG, Gill JB, Morris JD (1990) Volcanism in the Sumisu Rift. II. Subduction and non-subduction related components. Earth Planet Sci Letters 100:195–200

Hockley JJ (1971) Occurrence and geological significance of layered stratiform intrusion in the Yilgarn Block, Western Australia. Nature 232:252–253

Hoefs J (1967) A Rb-Sr investigation in the southern York County area, Maine. In: Hurley PM (ed) Variations in Isotopic Abundances of Strontium, Calcium, and Argon and Related Topics. Mass. Inst. Tech., Cambridge, MA, pp 127–129

Hoefs J (1978) Oxygen isotope composition of volcanic rocks from Santorini and Christiani. In: Doumas C (ed) Thera and the Aegean World. London

Hoefs J (1996) Stable isotope geochemistry, 4th edn. Springer Verlag, Berlin

Hoefs J, Wedepohl KH (1968) Strontium isotope studies on young volcanic rocks from Germany and Italy. Contrib Mineral Petrol 19:238–338

Hoefs J, Faure G, Elliot DH (1980) Correlation of $\delta^{18}O$ and initial $^{87}Sr/^{86}Sr$ ratios in Kirkpatrick Basalt on Mt. Falla, Transantarctic Mountains. Contrib Mineral Petrol 75:199–203

Hoernle KA, Schmincke H-U (1993a) The role of partial melting in the 15 Ma geochemical evolution of Gran Canaria: a blob model for the Canary hotspot. J Petrol 34:599–626

Hoernle KA, Schmincke H-U (1993b) The petrology of the tholeiites through melilite nephelinites on Gran Canaria, Canary Islands: Crystal fractionation, accumulation, and depth of melting. J Petrol 34(3):573–597

Hoernle KA, Tilton GR (1991) Sr-Nd-Pb isotope data for Fuerteventura (Canary Islands) basal complex and subaerial volcanics: applications to magma genesis and evolution. Schweiz Mineral Petrogr Mitteilungen 71:5–21

Hoernle KA, Tilton GR, Schmincke H-U (1991) Sr-Nd-Pb isotopic evolution of Gran Canaria: Evidence for shallow enriched mantle beneath the Canary Islands. Earth Planet Sci Letters 106:44–63

Hofmann AW (1988) Chemical differentiation of the Earth: The relationship between mantle, continental crust, and oceanic crust. Earth Planet Sci Letters 90:297–314

Hofmann AW (1997) Mantle geochemistry: the message from oceanic volcanism. Nature 385:219–229

Hofmann AW, Feigenson MD (1983) Case studies on the origin of basalt. I. Theory and reassessment of Grenada basalts. Contrib Mineral Petrol 84:382–389

Hofmann AW, Hart SR (1975) An assessment of local and regional isotopic equilibrium in a partially molten mantle. Carnegie Institution Yearbook 74:195–210

Hofmann AW, Hart SR (1978) An assessment of local and regional isotopic equilibrium in the mantle. Earth Planet Sci Letters 38:44–62

Hofmann AW, White WM (1982) Mantle plumes from ancient oceanic crust. Earth Planet Sci Letters 57:421–436

Hofmann AW, Feigenson MD, Raczek I (1984) Case studies on the origin of basalt: III. Petrogenesis of the Mauna Ulu eruption, Kilauea (1969-1971). Contrib Mineral Petrol 88:24–35

Hofmann AW, Jochum KP, Seifert M, White WM (1986) Nb and Pb in oceanic basalts: New constraints on mantle evolution. Earth Planet Sci Letters 79:33–45

Hofmann AW, Feigenson MD, Raczek I (1987) Kohala revisited. Contrib Mineral Petrol 95 114–122

Hole MJ (1988) Post-subduction alkaline volcanism along the Antarctic Peninsula. J Geol Soc London 145:985–988

Hole MJ (1990) Geochemical evolution of Pliocene-Recent post-subduction alkalic basalts from Seal Nunataks, Antarctic Peninsula. J Volcanol Geotherm Res 40:149–167

Hole MJ, Larter ED (1993) Trench proximal volcanism following ridge crest-trench collision along the Antarctic Peninsula. Tectonics 12:897–910

Hole MJ, LeMasurier WE (1994) Tectonic controls on the geochemical composition of Cenozoic, mafic alkaline volcanic rocks from West Antarctica. Contrib Mineral Petrol 117:187–202

Hole MJ, Rogers G, Saunders AD, Storey M (1991a) The relationship between alkalic volcanism and slab-window formation. Geol. 19:657–660

Hole MJ, Pankhurst RJ, Saunders AD (1991b) The geochemical evolution of the Antarctic Peninsula magmatic arc: The importance of mantle-crust interaction during granitoid genesis. In: Thomson MRA, Crame JA, Thomson JW (eds) The geological evolution of Antarctica. Cambridge Un. Press, Cambridge, UK, pp 369–374

Hole MJ, Smellie JL, Marriner GF (1991c) Geochemistry and tectonic setting of Cenozoic alkaline basalts from Alexander Island, southwest Antarctic Peninsula. Thomson MRA, Crame JA, Thomson JW (eds) The geological evolution of Antarctica. Cambridge Un. Press, Cambridge, UK, pp 521–526

Hole, MJ, Kempton PD, Millar IL (1993) Trace-element and isotopic characteristics of small-degree melts of the asthenosphere: Evidence from the alkalic basalts of the Antarctic Peninsula. Chem Geol 109:51–68

Hole MJ, Storey BC, LeMasurier WE (1994) Tectonic setting and geochemistry of Miocene alkalic basalts from the Jones Mountains, West Antarctica. Antarctic Science 6:85–92

Holm PM (1988) Nd, Sr, and Pb isotope geochemistry of the Lower Lavas, E. Greenland Tertiary igneous province. In: Morton AC, Parson LM (eds) Early Tertiary volcanism and the opening of the NE Atlantic. Geol Soc London Special Paper 39:181–195

Holm PM, Gill RCO, Pedersen AK, Larsen JG, Hald N, Nielsen TFD, Thirlwall MF (1993) The Tertiary picrites of West Greenland: Contributions from "Icelandicy" and other sources. Earth Planet Sci Letters 115:227–244

Holmes A (1936) Transfusion of quartz xenoliths in alkali basic and ultrabasic lavas, south-west Uganda. Mineral Mag 24:408–421

Honjo N, Leeman WP (1987) Origin of hybrid ferrolatite lavas from Magic Reservoir eruptive center, Snake River Plain, Idaho. Contrib Mineral Petrol 96:163–177

Honma H, Kusakabe M, Kagami H, Iizumi S, Sakai H, Kodama Y, Kimura M (1991) Major and trace element chemistry and D/H, $^{18}O/^{16}O$, $^{87}Sr/^{86}Sr$ and $^{143}Nd/^{144}Nd$ ratios of rocks from the spreading center of the Okinawa Trough, a marginal back-arc basin.Geochem J 25(2):121–136

Hooper PR (1982) The Columbia River basalts. Science 215:1463–1468

Hooper PR (1984) Physical and chemical constraints on the evolution of the Columbia River basalt. Geol 12:495–499

Hooper PR (1985) A case of simple magma mixing in the Columbia River Basalt Group: The Wilbur Creek, Lapwai, and Asotin flows, Saddle Mountain Formation. Contrib Mineral Petrol 91:66–73

Hooper PR (1988) The Columbia River basalt. In: Macdougall JD (ed) Continental flood basalts. Kluwer Academic Press, Dordrecht, The Netherlands, pp 1–34

Hoover JD (1989) Petrology of the Marginal Border Series of the Skaergaard Intrusion. J Petrol 30:399–439

Hoppe P, Strebel R, Eberhardt P, Amari S, Lewis RS (1996) Small SiC grains and a nitride grain of circumstellar origin from the Murchison meteorite: Implications for stellar evolution and nucleosynthesis. Geochim Cosmochim Acta 60:883–907

Horan MF, Walker RJ, Fedorenko VA, Czamanske GK (1995) Osmium and neodymium isotopic constraints on the temporal and spatial evolution of Siberian flood basalt sources. Geochim Cosmochim Acta 59:5159–5168

Hornig I (1993) High-Ti and low-Ti tholeiites in the Jurassic Ferrar Group, Antarctica. In: Danske D, Fritsch J (eds) German Antarctic North Victoria Land Expedition 1988/89, Ganovex V., 335–369. Geologisches Jahrbuch, Reihe E, vol 47, Hannover, Germany

Hriskevich ME (1968) Petrology of the Nipissing diabase sill of the Cobalt area, Ontario, Canada. Geol Soc Amer Bull 79:1387–1404

Hubbard HB, Wood LF, Rogers JJW (1987) Possible hydration anomaly in the upper mantle prior to Red Sea rifting: Evidence from petrologic modeling of the Wadi Natash alkali basalt sequence of eastern Egypt. Geol Soc Amer Bull 98:92–98

Hughes CW, Turner CC (1977) Upraised Pacific Ocean floor, southern Malaita, Solomon Islands. Geol Soc Amer Bull 88:412–424

Hughes DJ, Brown GC (1972) Basalts from Madeira: A petrochemical contribution to the genesis of oceanic alkali rock series. Contrib Mineral Petrol 37:91–109

Hughes SS, Schmitt RA, Wang YL, Wasserburg GJ (1986) Trace element and Sr-Nd isotopic constraints on the compositions of lithospheric primary sources of Serra Geral continental flood basalts, southern Brazil. Geochem J 20:173–190

Humphris SE, Thompson G, Schilling J-G, Kingsley RH (1985) Petrological and geochemical variations along the Mid-Atlantic Ridge between 46° S and 32° S: Influence of the Tristan da Cunha mantle plume. Geochim Cosmochim Acta 49:1445–1464

Hunter AG, Kempton PD, Greenwood P (1999) Low-temperature fluid-rock interaction–an isotopic and mineralogical perspective of upper crustal evolution, eastern flank of the Juan de Fuca Ridge, ODP Leg 168. Chem Geol 155:3–28

Hunter RH, Sparks RSJ (1987) The differentiation of the Skaergaard intrusion. Contrib Mineral Petrol 95:451–461

Huppert HE, Sparks RSJ (1980) The fluid dynamics of a basaltic magma chamber replenished by influx of hot, dense ultrabasic magma. Contrib Mineral Petrol 75:279–289

Huppert HE, Sparks RSJ (1985) Cooling and contamination of mafic and ultramafic magmas during ascent through continental crust. Earth Planet Sci Letters 74:371–386

Huppert HE, Sparks RSJ (1988) The generation of granitic magmas by intrusion of basalt into continental crust. J Petrol 29:599–624

Hurley PM (1972) Can the subduction process of mountain building be extended to Pan-African and similar orogenic belts? Earth Planet Sci Letters 15:305–314

Hurley PM (1974) Pangeaic orogenic system. Geol 2:373–376

Hurley PM, Rand JR (1969) Pre-drift continental nuclei. Science 164:1229–1242

Hurley PM, Hughes H, Pinson WH Jr, Fairbairn HF (1962a) Radiogenic argon and strontium diffusion parameters in biotite at low temperatures obtained from alpine-fault uplift of New Zealand. Geochim Cosmochim Acta 26:67–80

Hurley PM, Hughes H, Faure G, Fairbairn HW, Pinson WH (1962b) Radiogenic strontium-87 model of continent formation. J Geophys Res 67(13):5315–5334

Hurley PM, Fairbairn HW, Pinson WH Jr (1966) Rb-Sr isotopic evidence in the origin of potash-rich lavas of western Italy. Earth Planet Sci Letters 1:301–306

Hurley PM, Leo GW, White HW, Fairbairn HW (1971) Liberian age province (about 2700 m.y.) And adjacent provinces in Liberia and Sierra Leone. Geol Soc Amer Bull 82:3483–3490

Hussong DM, Wipperman LK, Kroenke LW (1979) The crustal structure of the Ontong Java and Manihiki plateaus. J Geophys Res 84:6003–6010

Hutchinson DR, White RW, Cannon WF, Schulz KJ (1990) Keweenaw hot spot: Geophysical evidence for a 1.1 Ga mantle plume beneath the Midcontinent Rift System. J Geophys Res 85:10869–10884

Hutchison R, Dawson JB (1970) Rb, Sr and $^{87}Sr/^{86}Sr$ in ultrabasic xenoliths and host rocks, Lashaine volcano, Tanazania. Earth Planet Sci Letters 9:87–92

Hutchison R, Chambers AL, Paul DK, Harris PG (1975) Chemical variations among French ultramafic xenoliths: evidence for a heterogeneous upper mantle. Mineral Mag 40:153–170

Iacumin P, Piccirillo EM, Longinelli A (1991) Oxygen isotopic composition of Lower Cretaceous tholeiites and Precambrian basement rocks from the Paraná Basin (Brazil): The role of water-rock interaction. Chem Geol (Isotope Geoscience Section) 83:225–237

Imslard P, Larsen JG, Prestvk T, Sigmond E (1977) The geology and petrology of Bouvetcya, South Atlantic Ocean. Lithos 10: 213-234

Irifune T, Ringwood AE (1993) Phase transformations in subducted oceanic crust and buoyancy relationships at depths of 600-800 km in the mantle. Earth Planet Sci Letters 117:101-110

Irvine TN (1970a) Heat transfer during solidification of layered intrusions. I. Sheets and sills. Can J Earth Sci 7:1031-1061

Irvine TN (1970b) Crystallization sequences in the Muskox intrusion and other layered intrusions. 1. Olivine-pyroxene-plagioclase relations. Geol Soc S Africa Spec Publ 1:441-476

Irvine TN (1977) Origin of chromitite layers in the Muskox intrusion and other stratiform intrusions: a new interpretation. Geol 5:273-277

Irvine TN (1980) Magmatic infiltration metasomatism, double-diffusive fractiona crystallization, and adcumulus growth in the Muskox intrusion and other layered intrusions. In: Hargraves EB (ed) Physics of magma processes. Princeton Univ. Press, pp 326-383

Irvine TN, Baragar WRA (1971) A guide to the chemical classification of the common volcanic rocks. Can J Earth Sci 8: 523-548

Irvine TN, Smith CH (1967) The ultramafic rocks of the Muskox intrusion, Northwest Territories, Canada. In: Wyllie PJ (ed) Ultramafic and related rocks. Wiley, New York, pp 38-49

Irvine TN, Keith DW, Todd SG (1983) Platinum-palladium reef of the Stillwater complex, Montana: II. Origin by double diffusive convective magma mixing and implications for the Bushveld complex. Econ Geol 78:1287-1334

Irving AJ (1974) Megacrysts from the New Basalts and other basaltic rocks of southeastern Australia. Geol Soc Amer Bull 85: 1503-1514

Irving AJ (1978) A review of experimental studies of crystal/liquid trace-element partitioning. Geochim Cosmochim Acta 42: 743-770

Irving AJ (1980) Petrology and geochemistry of composite ultramafic xenoliths in alkalic basalts and implications for magmatic processes within the mantle. Amer J Sci 280A:389-426

Irving AJ, Frey FA (1984) Trace element abundances in megacrysts and their host basalts: constraints on partition coefficients and megacryst genesis. Geochim Cosmochim Acta 48: 1201-1221

Irving AJ, Green DH (1976) Geochemistry and petrogenesis of the Newer basalts of Victoria and South Australia. J Geol Soc Aust 23:45-66

Irving AJ, Price RC (1981) Geochemistry and evolution of lherzolite-bearing phonolitic lavas from Nigeria, Australia, East Germany, and New Zealand. Geochim Cosmochim Acta 45:1309-1320

Ishizaka K, Carlson RW (1983) Nd-Sr systematics of the Setouchi volcanic rocks, southwest Japan: a clue to the origin of orogenic andesite. Earth Planet Sci Letters 64:327-340

Ishizaka K, Yamaguchi M (1969) U-Th-Pb ages of sphene and zircon from the Hida metamorphic terrain, Japan. Earth Planet Sci Letters 6(3):179-185

Ishizaka K, Yanagi T (1977) K, Rb and Sr abundances and Sr isotopic composition of the Tanzawa granitic and associated gabbroic rocks, Japan: Low-potash island arc plutonic complex. Earth Planet Sci Letters 33:345-352

Ishizaka K, Yanagi T, Hayatsu K (1977) A strontium isotopic study of the volcanic rocks of the Myoko Volcano Group, central Japan. Contrib Mineral Petrol 63:295-307

Ito E, Stern RJ (1981) Oxygen and strontium isotopic investigation on the origin of volcanism in the Izu-Volcano-Mariana arc Carnegie Inst Wash Yearb 82:449-455

Ito E, Stern RJ (1986) Oxygen- and strontium isotopic investigations of subduction zone volcanism: the case of the Volcano arc and the Marianas Island arc. Earth Planet Sci Letters 76: 312-320

Ito E, White WM, Göpel C (1987) The O, Sr, Nd, and Pb isotope geochemistry of MORB. Chem Geol 62(3/4):157-176

Iwamori H (1993) Dynamic disequilibrium melting model with porous flow and diffusion-controlled chemical equilibration. Earth Planet Sci Letters 114:301-313

Jackson ED (1961) Primary textures and mineral associations in the Ultramafic Zone of the Stillwater complex, Montana. US Geol Survey Prof Paper 358

Jackson ED (1967) Ultramafic cumulates in the Stillwater, Great Dyke, and Bushveld intrusions. In: Wyllie PJ (ed) Ultramafic and related rocks. Wiley, New York, pp 20-38

Jackson ED (1968) The chromite deposits of the Stillwater complex, Montana. In: Ore Deposits of the United States (Graton-Sales Volume). Amer. Inst. Mining. Metallurgy and Petroleum Engineers, New York, pp 1495-1510

Jackson ED, Silver EA, Dalrymple GB (1972) Hawaiian-Emperor chain and its relation to Cenozoic circumpacific tectonics. Geol Soc Amer Bull 83:601-618

Jackson MC (1993) Crystal accumulation and magma mixing in the petrogenesis of tholeiitic andesite from Fukujin seamount, northern Mariana island arc. J Petrol 34:259-289

Jacobsen RE, Macleod WN, Black R (1958) Ring-complexes in the Younger Granite province of northern Nigeria. Geol Soc London Mem 1

Jacobsen SB, Pimentel-Klose MR (1988) A Nd isotopic study of the Hamersley and Michipicoten banded iron formations: The source of REE and Fe in Archean oceans. Earth Planet Sci Letters 87:29-44

Jacobsen SB, Wasserburg GJ (1979) Nd and Sr isotopic study of the Bay of Islands ophiolite complex and the evolution of the source of midocean ridge basalt. J Geophys Res 84(B13): 7429-7445

Jacquemin H, Sheppard SMF, Vidal P (1982) Isotope geochemistry (O, Sr, Pb) of the Golda Zuelva and Mboutou anorogenic complexes, North Cameroon: Mantle origin with evidence for crustal contamination. Earth Planet Sc. Letters 61:97-111

Jahn BM, Murthy VR (1975) Rb-Sr age Archean rocks from the Vermilion district, northeastern Minnesota. Geochim Cosmochim Acta 39:1679-1689

Jahn BM, Shih CY, Murthy VR (1974) Trace element geochemistry of Archean volcanic rocks. Geochim Cosmochim Acta 38: 611-627

James DE (1981) The combined use of oxygen and radiogenic isotopes as indicators of crustal contamination. Ann Rev Earth Planet Sci 9:311-344

James DE (1982) A combined O, Sr, Nd, and Pb isotopic and trace element study of crustal contamination in central Andean lavas, 1. Local geochemical variations. Earth Planet Sci Letters 57:47-62

James DE, Brooks C, Cuyubamba A (1976) Andean Cenozoic volcanism: Magma genesis in the light of strontium isotopic composition and trace-element geochemistry. Geol Soc Amer Bull 87:592-600

James DE, Murcia LA (1984) Crustal contamination in northern Andean volcanics. J Geol Soc London 141:823-830

Jaques AL (1976) An Archean tholeiitic layered sill from Mt. Kilkenny, Western Australia. Geol Soc Aust J 23:157-168

Jarrar G, Baumann A, Wachendorf H (1985) Age determinations in the Precambrian basement of the Wadi Araba area, southwest Jordan. Earth Planet Sci Letters 63:292-304

Javoy M, Weis D, Hodgson N, Agrinier P (1985) Stable and radiogenic isotope systematics in Cape Verde carbonatites and basaltic lavas (Abstract). Eos 66:1137

Javoy M, Stillmann CJ, Pineau F (1986) Oxygen and hydrogen isotope studies on the basal complexes of the Canary Islands: Implications on the conditions of their genesis. Contrib Mineral Petrol 92:225-235

Jochum KP, Arndt NT, Hoffmann AW (1991) Nb-Th-La in komatiites and basalts: Constraints on komatiite petrogenesis and mantle evolution. Earth Planet Sci Letters 107: 272-289

Johnson RC, Hills FA (1976) Precambrian geochronology and geology of the Boxelder Canyon area, northern Laramie Range, Wyoming. Geol Soc Amer Bull 87:809-317

Johnson RW (1976) Potassium variation across the New Britain volcanic arc. Earth Planet Sci Letters 31:184-191

Jolly WT, Smith RE (1972) Degradation and metamorphic differentiation of the Keweenawan tholeiitic lavas of northern Michigan, U.S.A. J Petrol 13:273-309

Jones AP, Smith JV, Hansen EC, Dawson JB (1983) Metamorphism, partial melting, and K-metasomatism of garnet-scapolite-kyanite granulite xenoliths from Lashaine, Tanzania. J Geol 91:143–165

Jones LM, Kesler SE (1980) Strontium isotopic geochemistry of intrusive rocks, Puerto Rico, Greater Antilles. Earth Planet Sci Letters 50:219–224

Jones LM, Mossman DJ (1988) The isotopic composition of strontium and the source of the North Mountain basalts, Nova Scotia. Canad J Sci 25:942–944

Jones LM, Walker RL (1972) Geochemistry of the McMurdo volcanics, Victoria land. Part 1. Strontium isotope composition. Antarctic J US 7(5):142–144

Jones LM, Walker RL, Hall BA, Borns HW Jr (1973) Origin of the Jurassic dolerites and basalts of southern Victoria Land. Antarctic J US 8:268–270

Jones LM, Walker RL, Allard GO (1974) The rubidium-strontium whole-rock age of major units of the Chibougamau greenstone belt, Quebec. Can J Earth Sci 11(11):1550–1561

Jones WB, Lippard SJ (1979) New age determination and the geology of the Kenya Rift-Kavirondo Rift junction, W. Kenya. J Geol Soc London 136:693–704

Juang WS, Chen JC (1989) Geochronology and geochemistry of volcanic rocks in northern Taiwan. Bull Central Geol Surv 5: 31–68

Juckes LM (1968) The geology of Mannefallknausane and part of Vestfjella, Dronning Maud Land. British Antarct Surv Bull 18: 65–78

Kagami H, Tainosho Y, Iizumi S, Hayama Y (1985) High initial Sr-isotope ratios of gabbro and metadiabase in the Ryoke Belt, southwest Japan. Geochem J 19:237–244

Kagami H, Honma H, Shirahase T, Nureki T (1988) Rb-Sr whole-rock isochron ages of granites from northern Shikoku and Okayama, southwest Japan: Implications for the migration of the Late Cretaceous to Paleogene igneous activity in space and time. Geochem J 22(2):69–80

Kalamarides RI (1984) Kiglapait geochemistry VI: Oxygen isotopes. Geochim Cosmochim Acta 48:1827–1836

Kalamarides RI, Berg JH, Hank RA (1987) Lateral isotopic discontinuity in the lower crust: an example from Antarctica. Science 237:1192–1194

Kalsbeek F, Hansen M (1989) Statistical analysis of Rb-Sr isotope data by the "bootstrap" method. Chem Geol (Isotope Geoscience Section) 73(4):289–298

Kalsbeek F, Pidgeon RT (1980) The geological significance of Rb-Sr whole-rock isochrons of polymetamorphic Archean gneisses, Fiskenaesset area, southern West Greenland. Earth Planet Sci Letters 50:225–237

Kalsbeek F, Pidgeon RT, Taylor PN (1987) Nagssugtoqidian mobile belt of West Greenland: A cryptic 1850 Ma suture between two Archaean continents – chemical and isotopic evidence. Earth Planet Sci Letters 85:365–385

Kalt A, Hegner E, Satir M (1997) Nd, Sr and Pb isotopic evidence for diverse lithospheric mantle sources of East African rift carbonatites. Tectonophysics 278:31–45

Kaneoka I (1974) Investigation of excess argon in ultramafic rocks from the Kola Peninsula by the ^{40}Ar/^{39}Ar method. Earth Planet Sci Letters 22:145–156

Kaneoka I (1990) Radiometric age and Sr isotope characteristics of the volcanic rocks from the Japan Sea floor. Geochem J 24(1):7–20

Kaneoka I, Matsuda J-I, Zashu S, Takahashi E, Aoki K (1978) Ar and Sr isotopes of mantle derived rocks from the Japanese Islands. Bull Volcanol 41:424–433

Kaneoka I, Notsu K, Takigami Y, Fujioka K, Sakai H (1990) Constraints on the evolution of the Japanese Sea based on ^{40}Ar-^{39}Ar ages and Sr isotopic ratios for volcanic rocks of Yamato seamount chain in the Japanese Sea. Earth Planet Sci Letters 97:211–225

Kapustin YL, Polyakov AI (1985) Carbonatite volcanoes of East Africa and the genesis of carbonatites. Internat Geol Rev 27: 434–449

Karig DE (1972) Remnant arcs. Geol Soc Amer Bull 83:1057–1068

Karlstrom KE, Dallmeyer RD, Grambling JA (1997) ^{40}Ar/^{39}Ar evidence for 1.4Ga regional metamorphism in New Mexico: Implications for thermal evolution of lithosphere in the southwestern USA. J Geol 105(2):205–224

Katerinopoulos A, Kyriakopoulos K, del Moro A, Kokkinakis A, Giannotti U (1998) Petrology, geochemistry, and Rb/Sr age determination of Hercynian granitic rocks from Thessaly, central Greece. Chemie der Erde 58:64–79

Katili JA (1975) Volcanism and plate tectonics in the Indonesian island arc. Tectonophysics 26:165–188

Kay RW, Kay SM (1986) Petrology and geochemistry of the lower continental crust: An overview. Geol Soc London Spec Publ 24:147–159

Kay RW, Kay SM (1988a) Crustal recycling and the Aleutian arc. Geochim Cosmochim Acta 52(6):1351–1360

Kay RW, Kay SM (1988b) Aleutian magmas in space and time. In: Plafker G et al. (eds) Geology of Alaska: The geology of North America. Geol Soc Amer Bull, Boulder, Colorado

Kay RW, Sun SS, Lee-Hu CN (1978) Pb and Sr isotopes in volcanic rocks from the Aleutian Islands and Pribilof Islands, Alaska. Geochim Cosmochim Acta 42:263–273

Kays MA, McBirney AR, Goles GG (1981) Xenoliths of gneisses and the conformable, clot-like granophyres in the marginal border group, Skaergaard intrusion, East Greenland. Contrib Mineral Petrol 76:265–284

Kays MA, Goles GG, Grover TW (1989) Precambrian sequence bordering the Skaergaard Intrusion. J Petrol 30(2):321–361

Kear D, Wood BL (1959) The geology and hydrogeology of Western Samoa. New Zealand Geol Surv Bull 63

Keken P van (1997) Evolution of starting mantle plumes: A comparison between numerical and laboratory models. Earth Planet Sci Letters 148:1–11

Keller J (1981) Carbonatitic volcanism in the Kaiserstuhl alkaline complex: Evidence for highly fluid carbonatitic melts at the Earth's surface. J Volcanol Geotherm Res 9:423–431

Kempe DRC, Deer WA, Wager LR (1970) Geological investigations in East Greenland. Part VIII: The petrology of the Kangerdlugssuaq alkaline intrusion, East Greenland. Medd Groenland 190(2)

Kennedy AK, Hart SR, Frey FA (1990) Composition and isotopic constraints on the petrogenesis of alkaline arc lavas: Lihir Island, Papua New Guinea. J Geophys Res 95(B5):6929–6942

Kennedy AK, Kwon S-T, Frey FA, West HB (1991) The isotopic composition of postshield lavas from Mauna Loa volcano, Hawaii. Earth Planet Sci Letters 103(1/4):339–353

Kent JT, Watson GS, Onstott TC (1990) Fitting straight lines and planes with an application to radiometric dating. Earth Planet Sci Letters 97(1/2):1–17

Kent RW, Storey M, Saunders AD (1992a) Large igneous provinces: sites of plume impact or plume incubation? Geol 20:891–894

Kent RW, Storey M, Saunders AD, Ghose NC, Kempton PD (1992b) Comment and reply on "Origin of the Rajmahal Traps and the 85° E Ridge: Preliminary reconstructions of the trace of the Crozet hotspot." Geol 20:957–959

Kenyon PM, Turcotte DL (1987) Along-strike magma mixing beneath mid-ocean ridges: Effects on isotopic ratios. Earth Planet Sci Letters 84:393–405

Kepezhinskas P, Defant MJ, Drummond MS (1996) Progressive enrichment of island arc mantle by melt-peridotite interaction inferred from Kamchatka xenoliths. Geochim Cosmochim Acta 60:1217–1229

Kerans C (1983) Timing of emplacement of the Muskox intrusion: Constraints from Coppermine homocline cover strata. Can J Earth Sci 20:673–683

Kerr AC, Kempton PD, Thompson RN (1995) Crustal assimilation during turbulent magmas ascent (ATA); new isotopic evidence from the Mull Tertiary lava succession, N.W. Scotland. Contrib Mineral Petrol 119:142–152

Kerr AC, Tarney J, Marriner GF, Klaver GT, Saunders AD, Thirlwall MF (1996a) The geochemistry and petrogenesis of the Late Cretaceous picrites and basalts of Curacao, Netherlands Antilles: Remnant of an oceanic plateau. Contrib Mineral Petrol 124:29–43

Kerr AC, Marriner GF, Arndt NT, Tarney J, Nivia A, Saunders AD, Storey M, Duncar RA (1996b) The petrogenesis of Gorgona komatiites and geochemical constraints. Lithos 37:245–260

Kerr AC, Marriner GF, Tarney J, Nivia A, Saunders AD, Thirlwall MF, Sinton CW (1997) Cretaceous basaltic terranes in western Colombia: elemental, chronological and Sr-Nd isotopic constraints on petrogenesis. J Petrol 38:677–702

Kerrich R, Kishida A, Willmore LM (1984) Timing of Abitibi belt lode gold deposits: Evidence from $^{39}Ar/^{40}Ar$ and $^{87}Rb/^{86}Sr$. Geol Assoc Canada 9:9–75

Kerrich R, Fryer EJ, King RW, Willmore LM, Hees E van (1987) Crustal outgassing and LILE enrichment in major lithosphere structures. Archean Abitibi greenstone belt: evidence on the source reservoir from strontium and carbon isotope tracers. Contrib Mineral Petrol 97:156–168

Kerridge JF, Haymon RM, Kastner M (1983) Sulfur isotope systematics at the 21° N site, East Pacific Rise. Earth Planet Sci Letters 66:91–100

Kesson SE (1973) The primary geochemistry of the Monaro alkaline volcanics, south-eastern Australia-evidence for upper mantle heterogeneity. Contrib Mineral Petrol 42:93

Khomyakov AP (1994) Ultra-agpaites: a new type of pegmatoid in agpaitic nepheline syenite intrusions. Geochem Internat 31(3):107–122

Kim KH, McMurtry GM (1991) Radial growth rates and ^{210}Pb ages of hydrothermal massive sulfides from the Juan de Fuca Ridge. Earth Planet Sci Letters 104(2/4):299–314

Kincaid C, Schilling J-G, Gable C (1996) The dynamics of off-axis plume-ridge interaction in the uppermost mantle. Earth Planet Sci Letters 137:29–43

King SD, Anderson DL (1995) An alternative mechanism of flood basalt formation. Earth Planet Sci. Letters, 136:269–279

Kinny PD (1986) 3820 Ma zircons from a tonalitic Amitsoq gneiss in the Godthåb district of southern West Greenland. Earth Planet Sci Letters 79:337–347

Kirsten T, Müller C (1967) Argon and potassium in mineral fractions of three ultramafic rocks from the Baltic shield. In: Anonymous (ed) Radioactive Dating and Methods of Low Level Counting. Internat. Atomic Energy Agency, Vienna, Austria

Kistler RW, Peterman ZE (1973) Variations in Sr, Rb, K, Na and initial $^{87}Sr/^{86}Sr$ in Mesozoic granitic rocks and intruded wall rocks in central California. Geol Soc Amer Bull 84:3489–3512

Kistler RW, Peterman ZE (1973) Reconstruction of crustal blocks of California on the basis of initial strontium isotopic compositions of Mesozoic granitic rocks. US Geol Survey Prof Paper 1071

Kistler RW, Obradovic JD, Jackson ED (1969) Isotopic ages of rocks and minerals from the Stillwater complex, Montana. J Geophys Res 74:3226–3237

Kitchen D (1984) Pyrometamorphism and the contamination of basaltic magma at Tieverogh, County Antrim. J Geol Soc London 141:773

Klein EM, Langmuir CH, Zindler A, Staudigel H, Hamelin B (1988) Isotope evidence of a mantle convection boundary at the Australian-Antarctic discordance. Nature 333:623–628

Klein EM, Langmuir CH, Staudigel H (1991) Geochemistry of basalts from the Southeast Indian Ridge, 115° E–138° E. J Geophys Res 96:2089–2107

Klemenic PM (1985) New geochronological data on volcanic rocks from northeast Sudan and their implication for crustal evolution. Precamb Res 30:263–276

Klemenic PM (1987) Variable intra-plate igneous activity in central and northeast Sudan. J African Earth Sci 6(4):465–474

Klerkx J, Deutsch S (1977) Resultats preliminaires obtenus par la methode Rb/Sr sur l'age des formations Precambrian D'Uweinat (Libye) Mus. Roy. Afr. Centr., Tervuren (Belg.) Dept. Geol. Min., Rapp. Ann., pp 84–93

Klerkx J, Deutsch S, DePaepe P (1974a) Rubidium, strontium content and strontium isotopic composition of strongly alkalic basaltic rocks from the Cape Verde Islands. Contrib Mineral Petrol 45:107–118

Klerkx J, Deutsch S, Hertogen J, DeWinter J, Gijbels R (1974b) Comments on "Evolution of Eolian Arc volcanism (southern Tyrrhenian Sea)" by F. Barberi, G. Ferrara, F. Innocenti, J. Keller, and L. Villari. Earth Planet Sci Letters 23:297–303

Klerkx J, Deutsch S, Pichler H, Zeil W (1977) Strontium isotopic composition and element data bearing on the origin of Cenozoic volcanic rocks of the Central and Southern Andes. J Volcanol Geotherm Res 2:49–71

Klewin KW, Berg JH (1990) Petrology of the Potato River layered intrusion, northern Wisconsin, USA. J Petrol 31:1115–1139

Klewin KW, Shirey SB (1992) The igneous petrology and magmatic evolution of the Midcontinent Rift System. Tectonophysics 213:33–40

Knittel U, Defant MJ (1988) Sr isotopic and trace element variations in Oligocene to Recent igneous rocks from the Philippine island arc: evidence for recent enrichment in the sub-Philippine mantle. Earth Planet Sci Letters 87:87–99

Knorring K von, Dubois CGB (1961) Carbonatite lava from the Fort Portal area of western Uganda. Nature 192:1064

Kogarko LN (1987) Alkaline rocks of the eastern part of the Baltic Shield (Kola Peninsula). In: Fitton JG, Upton BGJ (eds) Alkaline igenous rocks. Geol Soc London Special Pub 30:531–544

Kogarko LN (1993) Geochemical characteristics of oceanic carbonatites from the Cape Verde Islands. S African J Geol 96:119–125

Kogarko LN, Kramm U, Blaxland A, Grauert B, Petrovna EN (1981) Age and origin of alkaline rocks of the Khibina massif (isotopes of rubidium and strontium). Dokl Akad Nauk SSSR 260(4):1001–1004 (in Russian)

Kogarko LN, Kramm U, Grauert B (1983) New data on age and genesis of alkaline rocks of Lovozero Massif; Rb and Sr isotopes. Dokl Akad Nauk SSSR 268(4):970–972 (in Russian)

Kogiso T, Tatsumi Y, Shimoda Y, Barsczus HG (1997a) HIMU ocean island basalts in southern Polynesia: new evidence for whole-mantle scale recycling of subducted oceanic crust. J Geophys Res 102:8085–8103

Kogiso T, Tatsumi Y, Nakano S (1997b) Trace element transport during dehydration processes in the subducted oceanic crust: 1. Experiments and implications for the origin of ocean island basalts. Earth Planet Sci Letters 148:193–204

Kolker A, Lindsley DH, Hanson GN (1990) Geochemical evolution of the Maloin Ranch pluton, Laramie anorthosite complex, Wyoming: trace elements and petrogenetic models. Amer Mineral 75:572–588

Kooten GK Van (1980) Mineralogy, petrology, and geochemistry of an ultrapotassic basaltic suite, central Sierra Nevada, California, USA. J Petrol 21:651–684

Kooten GK Van (1981) Pb and Sr systematics of ultrapotassic and basaltic rocks from central Sierra Nevada, California. Contrib Mineral Petrol 76:378–385

Korsch MJ, Gulson BL (1986) Nd and Pb isotopic studies of an Archaean layered mafic-ultramafic complex, Western Australia, and implications for mantle heterogeneity. Geochim Cosmochim Acta 50:1–10

Kostitsyn YA (1989) Dealing with isochrons in the presence of geochemical dispersion. Geochem Internat 26(12):19–26

Koyaguchi T (1989) Chemical gradient at diffusive interfaces in magma chambers. Contrib Mineral Petrol 103:143–152

Krafft M, Keller J (1989) Temperature measurements in carbonatite lava lakes and flows from Oldoinyo Lengai, Tanzania. Science 245:168–169

Kramm U (1993) Mantle components of carbonatites from the Kola alkaline province, Russia and Finland: a Nd-Sr study. Eur J Mineral 5:985–989

Kramm U, Kogarko LN (1994) Nd and Sr isotope signatures of the Khibina and Lovozero agpaitic centres, Kola alkaline province, Russia. Lithos 32:225–242

Kramm U, Kogarko LN, Kononova VA, Vartiainen H (1993) The Kola alkaline province, CIS/Finland: Precise Rb-Sr ages define 380–360 Ma range of all magmatism. Lithos 30:33–44

Kramm U, Maravic HV, Morteani G (1997) Neodymium and Sr isotopic constraints on the petrogenetic relationships between carbonatite and cancrinite syenites from the Lueshe alkaline complex, East Zaire. J African Earth Sci 25:55–76

Krishnaswami S, Turekian KK, Bennett JT (1984) The behavior of ^{232}Th and ^{238}U decay chain nuclides during magma formation and volcanism. Geochim Cosmochim Acta 48:505–511

Kristoffersen Y, Haughland K (1986) Evidence for the East Antarctic plate boundary in the Weddell Sea. Nature 322:538–541

Kröner A (ed) (1981) Precambrian plate tectonics. Elsevier, Amsterdam

Kröner A, Roobol MJ, Ramsay CR, Jackson NJ (1979) Pan African ages of some gneissic rocks in the Saudi Arabian Shield. J Geol Soc 136:455–462

Kröner A, Stern RJ, Dawoud AS, Compston W, Reischmann T (1987) The Pan-African continental margin in northeastern Africa: Evidence from a geochronological study of granulites at Sabaloka, Sudan. Earth Planet Sci Letters 85:91–104

Kröner A, Linnebacher P, Stern RJ, Reischmann T, Manton W, Hussein I (1991) Evolution of Pan-African island arc assemblages in the southern Red Sea Hills, and in southwestern Arabia as exemplified by geochemistry and geochronology. Precamb Res 53:99–118

Krogh TE (1973) A low-contamination method for hydrothermal decomposition of zircon and extraction of U and Pb for isotopic age determinations. Geochim Cosmochim Acta 37:485–494

Krogh TE (1982a) Improved accuracy of U-Pb zircon dating by selection of more concordant fractions using a high-gradient magnetic separation technique. Geochim Cosmochim Acta 46:631–635

Krogh TE (1982b) Improved accuracy of U-Pb zircon dating by selection of more concordant systems using an air abrasion technique. Geochim Cosmochim Acta 46:637–649

Krogh TE (1993) High precision U-Pb ages for granulite metamorphism and deformation in the Archean Kapuskasing structural zone, Ontario: implications for structure and development of the lower crust. Earth Planet Sci Letters 119:1–18

Krogh TE, Davis GL (1971) Zircon U-Pb ages of Archean metavolcanic rocks in the Canadian Shield. Carnegie Inst. Washington, Yearbook 70:241–242

Krogh TE, Davis GL (1973) The effect of regional metamorphism on U-Pb systems in zircons and a comparison with Rb-Sr systems in the same whole rocks. Ann. Rept. Director, Geophys. Lab., Carnegie Inst. Washington, Yearbook 72:601–610

Krogh TE, Turek A (1982) Precise U-Pb zircon ages from the Gamitagama greenstone belt, southern Superior Province. Can J Earth Sci 19:859–867

Krogh TE, Corfu F, Davis DW, Dunning GR, Heaman LM, Kamo SL, Machado N, Greenough JD, Nakamura E (1987) Precise U-Pb isotopic ages of diabase dykes and mafic to ultramafic rocks using trace amounts of baddeleyite and zircon. In: Halls HC, Fahrig WF (eds) Diabase Dyke Swarms. Geol Assoc Canada, Spec Paper 34:147–152

Krogh TE, Kamo SL, Sharpton VL, Marin LE, Hildebrand AR (1993a) U-Pb ages of single shocked zircons linking distal K/T ejecta to the Chicxulub crater. Nature 366:731–734

Krogh TE, Kamo SL, Bohor BF (1993b) Fingerprinting the K/T impact site and determining the time of impact by U-Pb dating of single shocked zircons from distal ejecta. Earth Planet Sci Letters 119:425–429

Kruger FJ, Marsh JS (1982) Significance of ^{87}Sr/^{86}Sr ratios in the Merensky cyclic unit of the Bushveld complex. Nature 298:53–55

Kruger FJ, Cawthorn RG, Wash KL (1987) Strontium isotopic evidence against magma addition in the Upper Zone of the Bushveld complex. Earth Planet Sci Letters 84:51–58

Kudo AM, Brookins DG, Laughlin AW (1972) Sr isotopic disequilibrium in lherzolites from the Puerco Necks, New Mexico. Earth Planet Sci Letters 15:291–295

Kuehner SM, Edgar AD, Arima M (1981) Petrogenesis of the ultrapotassic rocks from the Leucite Hills, Wyoming. Amer Mineral 66:663–677

Kuehner SM, Laughlin JR, Grossman L, Johnson ML, Burnett DS (1989) Determination of trace element mineral/liquid partition coefficients in melilite and diopside by ion and electron microprobe techniques. Geochim Cosmochim Acta 53:3115–3130

Kullerud L (1991) On the calculation of isochrons. Chem Geol (Isotope Geoscience Section) 87:115–124

Kuno H (1966) Lateral variation of basalt magma types across continental margins and island arcs. Bull Volcanol 29:195–222

Kurasawa H (1967) The isotopic composition of lead and concentration of U, Th, and Pb in volcanic rocks from Dogo, Oki Island. Bull Volc Soc Japan 12:99–100

Kurasawa H (1984a) Strontium isotopic ratios of the volcanic rocks in the western part of the Sanin region and surrounding area, southwest Japan. Bull Volc Soc Japan (Ser II) 29:215–234

Kurasawa H (1984b) Strontium isotopic ratios of the volcanic rocks from Dogo of the Oki Islands, Japan. J Japanese Assoc Mineral Petrol Econ Geol 79:484–497

Kurasawa H, Fujinawa H, Leeman WP (1986) Calc-alkaline and tholeiitic rock series magmas coexisting within volcanoes in Japanese Island arcs–Strontium isotope study. J Geol Soc Japan 92:255–268 (in Japanese with English abstract)

Kurtz J (1983) Geochemistry of Early Mesozoic basalts from Tunisia. J African Earth Sci 1:113–125

Kurz MD, Kammer DP (1991) Isotopic evolution of Mauna Loa volcano. Earth Planet Sci Letters 103(1/4):257–269

Kurz MD, Meyer PS, Sigurdsson H (1985) Helium isotopic systematics within the neovolcanic zones of Iceland. Earth Planet Sci Letters 74:291–305

Kusakabe M, Mayeda S, Nakamura E (1990) S, O and Sr isotope systematics of active vent materials from the Mariana backarc basin spreading axis at 18° N. Earth Planet Sci Letters 100(1/3):275–282

Kwon S-T, Tilton GR, Grünenfelder MH (1989) Lead isotope relationships in carbonatites and alkalic complexes: an overview. In: Bell K (ed) Carbonatites. Unwin Hyman, London, pp 360–387

Kyle PR (1980) Development of heterogeneities in the sub-continental mantle: Evidence from the Ferrar Group, Antarctica. Contrib Mineral Petrol 73:89–104

Kyle PR (1981) Mineralogy and geochemistry of a basanite to phonolite sequence at Hut Point peninsula, Antarctica, based on core from Dry Valley Drilling Project drill holes 1, 2, and 3. J Petrol 22(4):451–500

Kyle PR, Cole JW (1974) Structural control of volcanism in the McMurdo Volcanic Group, Antarctica. Bull Volcanol 38(1):16–25

Kyle PR, Adams J, Rankin PC (1979a) Geology and petrology of the McMurdo Volcanic Group at Rainbow Ridge, Brown Peninsula, Antarctica. Geol Soc Amer Bull 90:676–688

Kyle PR, Sutter JF, Treves SB (1979b) K/Ar age determinations on drill core from DVDP holes 1 and 2. Mem Nat Inst Polar Res, Special Issue 13:214–219, Tokyo, Japan

Kyle PR, Elliot DH, Sutter JF (1981) Jurrasic Ferrar Supergroup tholeiites from the Transarctic Mountains, Antarctica, and their relationship to the initial fragmentation of Gondwana. In: Cresswell MM, Vella P (eds) Gondwana Five. Balkema, Rotterdam, pp 283–287

Kyle PR, Dibble RR, Giggenbach WF, Keys J (1982) Volcanic activity associated with the anorthoclase phonolite lava lake, Mount Erebus, Antarctica. In: Craddock C (ed) Antarctic GeoScience. Un. of Wisconsin Press, pp 735–745

Kyle PR, Pankhurst RJ, Bowman JR (1983) Isotopic and chemical variations in Kirkpatrick Basalt Group rocks from southern Victoria Land. In: Antarctic Earth Science. Australian Acad Sci, pp 234–237

Kyle PR, Pankhurst RJ, Moorbath S, Bowman JR (1987) Nature and development of an enriched mantle source for Jurassic Ferrar tholeiites, Transantarctic Mountains, Antarctica. Terra Cognita 7:614

Lalou C, Labeyrie L, Brichet E, Perez-Leclaire H (1984) The East Pacific Rise hydrothermal deposits: radiochronology of the sulfides and isotopic geochemistry of silica deposits. Bull Soc Geol France 26:9–14

Lambert DD, Morgan JW, Walker RJ, Shirey SB, Carlson RW, Zientek ML, Koski MS (1989) Rhenium-osmium and samarium-neodymium isotopic systematics of the Stillwater complex. Science 244:1169–1174

Lambert RSJ (1969) Discussion on: Tertiary granites and associated rocks of the Marsco area, Isle of Skye. Quart J Geol Soc London 124:380–383

Lang B, Steinitz G (1989) K-Ar dating of Mesozoic magmatic rocks in Israel: A review. Israel J Earth Sci 38:89–103

Langmuir CH, Bender JF, Bence AE, Hanson GN, Taylor SR (1977) Petrogenesis of basalts from the Famous area: mid-Atlantic ridge. Earth Planet Sci Letters 36:133–156

Langmuir CH, Vocke RD Jr, Hanson GN, Hart SR (1978) A general mixing equation with applications to Icelandic basalts. Earth Planet Sci Letters 37:380–392

Lanphere MA (1983) $^{87}Sr/^{86}Sr$ ratios for basalt from the Loihi seamount, Hawaii. Earth Planet Sci Letters 66:380–387

Lanphere MA (1984) $^{87}Sr/^{86}Sr$ ratios for basalt from Loihi Seamount, Hawaii Marine Geol 54:380–387

Lanphere MA, Dalrymple GB (1980) Age and strontium isotopic composition of the Honolulu volcanic Series, Oahu, Hawaii. Amer J Sci 280-A:736–751

Lanphere MA, Frey FA (1987) Geochemical evolution of Kohala Volcano, Hawaii. Contrib Mineral Petrol 95:100–113

Lanphere MA, Dalrymple GB, Smith RL (1975) K-Ar ages of Pleistocene rhyolitic volcanism in the Coso Range, California. Geol 3:339–341

Lanphere MA, Dalrymple GB, Clague DA (1980a) Rb-Sr systematics of basalt from the Hawaiian-Emperor volcanic chain. In: Jackson ED, Koisumi I et al. (eds) Initial Reports of the Deep Sea Drilling Project 55:695–706. U.S. Government Printing Office, Washington, D.C.

Lanphere MA, Cameron KL, Cameron M (1980b) Sr isotopic geochemistry of voluminous rhyolitic ignimbrites and related rocks, Batopilas area, western Mexico. Nature 286:594–596

Lanyon R, Varne R, Crawford AJ (1993) Tasmanian Tertiary basalts, the Balleny plume, and opening of the Tasman Sea (southwest Pacific Ocean). Geol 21:555–558

Lanyon R, Crawford AJ, Eggins SM (1995) Westward migration of Pacific Ocean upper mantle into the Southern Ocean region between Australia and Antarctica. Geol 23:511–514

Lapierre H, Dupu s V, Mercier de Lépinay B, Bosch D, Monié P, Tardy M, Maury RC, Hernandez J, Polvé M, Yeghicheyan D, Cotten J (1999) Late Jurassic oceanic crust and Upper Cretaceous Caribbean plateau picrite basalts exposed in the Duarte igneous complex, Hispaniola. J Geol 107:193–207

Larsen ES (1948) Batholith of southern California. Geol Soc Amer Mem 29

Larsen ES, Hurlbut CS Jr, Burgess CH, Buie BF (1941) Igneous rocks of the Highwood Mountains: Part VII. Petrology. Geol Soc Amer Bull 52:1857–1858

Larsen O (1982) The age of the Kap Washington Group of volcanics, North Greenland. Geol Soc Denmark Bull 31:49–55

Larson RL (1991a) Latest pulse of Earth: Evidence for a mid-Cretaceous superplume. Geo 19:547–550

Larson RL (1991b) Geological consequences of superplumes. Geol 19:963–966

Laskowski TE, Fluegeman RH, Grant NK (1980) Rb-Sr glauconite systematics and the uplift of the Cincinnati arch. Geol 8:368–370

Laughlin AW, Brookins DG, Kudo AM, Causey JD (1971) Chemical and strontium isotopic investigations of multramafic inclusions and basalt, Bandera Crater, New Mexico. Geochim Cosmochim Acta 35:107–113

Lawson AC (1885) Report on the geology of the Lake of the Woods region, with special reference to the Keewatin (Huronian?) Belt of the Archean rocks Geol Surv Canada Ann Rept 1:1–151

Lawson AC (1888) Report on the geology of the Rainy Lake region. Geol Surv Canada Ann Rept 3:1–182

Lawson AC (1913) The Archean geology of Rainy Lake re-studied. Geol Surv Canada Mem 40:1–115

Leat PT, Thompson RN Dickin AP, Morrison MA, Hendry GL (1989) Quaternary volcanism in northwestern Colorado: Implications for the roles of asthenosphere and lithosphere in the genesis of continental basalts. J Volcanol Geotherm Res 37:291–310

LeBas MJ, Rex DC, Stillmann CJ (1986) The early magmatic chronology of Fuerteventura Canary Islands. Geol Mag 123(3):287–298

LeBel L, Cocherie A, Baubron J-C, Fouillac AM, Hawkesworth CJ (1985) A high-K, mantle derived plutonic suite from Linga, near Arequipa (Peru). J Petrol 26(1):124–148

LeCheminant AN, Heaman LM (1989) Mackenzie igneous events, Canada: Middle Proterozoic hotspot magmatism associated with ocean opening. Earth Planet Sci Letters 96:38–48

Lee DE, Marvin RF, Stern TW, Peterman ZE (1970) Modification of potassium-argon ages by Tertiary thrusting in the Snake Range, White Pine County, Nevada. US Geol Survey Prof Paper 700-D:92–102

Lee J, Stern RJ, Bloomer SH (1995) Forty million years of magmatic evolution in the Mariana arc: The tephra glass record. J Geophys Res 100:17671–17687

Lee W-J, Wyllie PJ (1997) Liquid immiscibility between nephelinites and carbonatite from 1.0 to 2.5 Gpa compared with mantle melt compositions. Contrib Mineral Petrol 127:1–16

Leeman WP (1970) The isotopic composition of strontium in late-Cenozoic basalts from the Basin and Range province, western United States. Geochim Cosmochim Acta 34:857–872

Leeman WP (1974) Late Cenozoic alkali-rich basalt from the western Grand Canyon area, Utah and Arizona: isotopic composition of strontium. Geol Soc Amer Bull 85:1691–1696

Leeman WP (1979) Primitive lead in deep crustal xenoliths from the Snake River Plain, Idaho. Nature 281:365–366

Leeman WP (1982) Tectonic and magmatic significance of strontium isotopic variations in Cenozoic volcanic rocks from the western United States. Geol Soc Amer Bull 93:487–503

Leeman WP, Dasch EJ (1978) Strontium, lead, and oxygen isotopic investigation of the Skaergaard intrusion, East Greenland. Earth Planet Sci Letters 41:47–59

Leeman WP, Fitton JG (1989) Magmatism associated with lithosphere extension: Introduction. J Geophys Res 94(B6): 7682–7684

Leeman WP, Harry DL (1993) A binary source model for extension-related magmatism in the Great Basin, western North America. Science 262:1550–1554

Leeman WP, Manton WI (1971) Strontium isotopic composition of basaltic lavas from the Snake River Plain, southern Idaho. Earth Planet Sci Letters 11:420–434

Leeman WP, Phelps DW (1981) Partitioning of rare earths and other trace elements between sanidine and coexisting volcanic glass. J Geophys Res 86:10,193–10,199

Leeman WP, Vitaliano CJ, Prinz M (1976) Evolved lavas from the Snake River Plain: Craters of the Moor National Monument, Idaho. Contrib Mineral Petrol 56:35–60

Leeman WP, Doe BR, Whelar J (1977) Radiogenic and stable isotope studies of hot-spring deposits in Yellowstone National Park and their genetic implications. Geochem J 11:65–74

Leeman WP, Budahn JG, Gerlach DC, Smith DR, Powell BN (1980) Origin of Hawaiian tholeiites: trace element constraints. Amer J Sci 280:794–819

Leeman WP, Menzies MA, Matty DJ, Embree GF (1985) Strontium, neodymium, and lead isotopic compositions of deep crustal xenoliths from the Snake River Plain: Evidence for Archean basement. Earth Planet Sci Letters 75:354–368

Leggo PJ (1974) A geochronological study of the basement complex of Uganda. J Geol Soc London 130:263–277

Leggo PJ, Hutchison R (1968) A Rb-Sr isotope study of ultrabasic xenoliths and their basaltic host rocks from the Massif Central, France. Earth Planet Sci Letters 5:71–75

LeMaitre RW (1962) Petrology of volcanic rocks, Gough Island, South Atlantic. Geol Soc Amer Bull 73:1309–1340

LeMarchand F, Villemant B, Calas G (1987) Trace element distribution coefficients in alkaline series. Geochim Cosmochim Acta 51:1071–1081

LeMasurier WE (1990a) Late Cenozoic volcanism on the Antarctic plate: an overview In: LeMasurier WE, Thomson JW (eds) Volcanoes of the Antarctic plate and Southern Oceans. Amer. Geophys. Union, Washington, D.C. (Ant Res Ser 48, pp 1–17)

LeMasurier WE (1990b) Miocene-Pliocene centers, Hobbs Coast. In: LeMasurier WE, Thomson JW (eds) Volcanoes of the Antarctic plate and Southern Oceans. Amer. Geophys. Union, Washington, D.C. (Ant Res Ser 48, pp 1–17)

LeMasurier WE, Thomson JW (1990) Volcanoes of the Antarctic plate and Southern Oceans. Amer. Geophys. Union, Washington, D.C. (Ant Res Ser 48)

LeMasurier WE, Wade FA (1977) Volcanic history in Marie Byrd Land: implications with regard to southern hemisphere tectonic reconstructions. In: Gonzalez-Ferran O (ed) Proceedings of the International Symposium on Andean and Antarctic Volcanology Problems. Santiago, Chile

Leo GW, Hedge CE, Marvin RF (1980) Geochemistry, strontium isotope data, and potassium-argon ages of the andesite-rhyolite association in the Padang area, West Sumatra. J Volcanol Geotherm Res 7:139–156

LeRoex AP (1985) Geochemistry, mineralogy, and magmatic evolution of the basaltic and trachytic lavas from Gough Island, South Atlantic. J Petrol 26:149–186

LeRoex AP, Erlank AJ (1982) Quantitative evaluation of fractional crystallization in Bouvet Island lavas. J Volcanol Geotherm Res 13:309–338

LeRoex AP, Reid DL (1978) Geochemistry of Karroo Dolerite sills in the Calvinia District, western Cape Province, South Africa. Contrib Mineral Petrol 66:351–360

LeRoex AP, Dick HJB, Gulen L, Reid AM, Erlank AJ (1987) Local and regional heterogeneity in MORB from Mid-Atlantic Ridge between 54.5° S and 51° S: Evidence for geochemical enrichment. Geochim Cosmochim Acta 51:541–556

LeRoex AP, Cliff RA, Adair BJ (1990) Tristan da Cunha, South Atlantic-geochemistry and petrogenesis of a basanite-phonolite lava series. J Petrol 31:779–812

Lesher CE (1986) Effects of silicate liquid composition on mineral-liquid element partitioning from Soret diffusion studies. J Geophys Res 91:6123–6141

Lesher CE (1990) Decoupling of chemical and isotope exchange during magma mixing. Nature 344:235–237

Lessing P, Catanzaro EJ (1964) ^{87}Sr/^{86}Sr ratios of Hawaiian lavas. J Geophys Res 69(8):1599–1601

Levi B, Nyström JO, Thiele R, Åberg G (1988) Geochemical trends in Mesozoic-Tertiary volcanic rocks from the Andes in central Chile, and tectonic implications. J South Amer Earth Sci 1(1):63–74

Lightfoot PC, Hawkesworth CJ (1988) Origin of Deccan Trap lavas: evidence from combined trace element and Sr-, Nd- and Pb-isotope studies. Earth Planet Sci Letters 91:89–104

Lightfoot PC, Naldrett AJ (eds) (1994) Proceedings, Sudbury-Noril'sk Symposium. Ontario Geol Survey Special Pub 5

Lightfoot PC, Hawkesworth CJ, Sethna SF (1987) Petrogenesis of rhyolites and trachytes from the Deccan Trap: Sr, Nd and Pb isotope and trace element evidence. Contrib Mineral Petrol 95:44–54

Lightfoot PC, Naldrett AJ, Gorbachev NS, Doherty W, Fedorenko VA (1990) Geochemistry of the Siberian Trap of the Noril'sk area, USSR, with implications for the relative contributions of crust and mantle flood basalt magmatism. Contrib Mineral Petrol 104:631–644

Lightfoot PC, Hawkesworth CJ, Hergt J, Naldrett AJ, Gorbachev NS, Federenko VA, Doherty W (1993) Remobilisation of the continental lithosphere by a mantle plume: major-, trace-element, and Sr-, Nd-, and Pb-isotope evidence from picritic and tholeiitic lavas of the Noril'sk District, Siberian Trap, Russia. Contrib Mineral Petrol 114:171–188

Lightfoot PC, Hawkesworth CJ, Hergt J, Naldrett AJ, Gorbachev NS, Federenko VA, Doherty W (1994) Chemostratigraphy of Siberian Trap of the Noril'sk District, Russia: Implications for the evolution of flood basalt magmas. In: Lightfoot PC et al. (eds) Proceedings, Sudbury-Noril'sk Symposium. Ontario Geol Survey Special Pub 5

Lightfoot PC, Keys RR, Morrison GG, Bite A, Farrell KP (1997) Geologic and geochemical relations between the contact sublayer, inclusions, and the main mass of the Sudbury igneous complex: A case study of the Whistle Mine embayment. Econ Geol 92:647–674

Lin P-N (1992) Trace element and isotopic characteristics of western Pacific pelagic sediments: Implications for the petrogenesis of Mariana Arc magmas. Geochim Cosmochim Acta 56(4):1641–1654

Lin P-N, Stern RJ, Bloomer SH (1989) Shoshonitic volcanism in the northern Mariana arc. 2. Large-ion lithophile and rare earth element abundances: Evidence for the source of incompatible element enrichments in intraoceanic arcs. J Geophys Res 94(B4):4497–4514

Lindstrom DJ (1983) Kinetic effects on trace element partitioning. Geochim Cosmochim Acta 47:617–622

Liotard JM, Barsczus HG, Dupuy C, Dostal J (1986) Geochemistry and origin of basaltic lavas from the Marquesas archipelago, French Polynesia. Contrib Mineral Petrol 92:260–268

Lipman PW (1966) Water pressure during differentiation and crystallization of some ashflow magmas from southern Nevada. Amer J Sci 264:810–826

Lipman PW, Glazner AF (1991) Introduction to middle Tertiary Cordilleran volcanism: Magma sources and relation to regional tectonics. J Geophys Res 96:13,193–13,199

Lipman PW, Steven TA, Mehnert HH (1970) Volcanic history of the San Juan Mountains, Colorado, as indicated by potassium-argon dating. Geol Soc Amer Bull 81:2329–2352

Lipman PW, Doe BR, Hedge CE, Steven TA (1978) Petrologic evolution of the San Juan volcanic field, southwestern Colorado: Pb and Sr isotope evidence. Geol Soc Amer Bull 89:59–82

Lipman PW, Dungan M, Bachmann O (1997) Comagmatic granophyric granite in the Fish Canyon Tuff, Colorado: Implications for magma-chamber processes during a large ash-flow eruption. Geol 25(10):915–918

Lippolt HJ (1983) Distribution of volcanic activity in space and time. In: Fuchs K, Gehlen K von, Mälzer H, Murawski H, Semmel A (eds) Plateau uplift. Springer Verlag, Berlin, pp 112–120

Lippolt HJ, Wasserburg GJ (1973) Rb-Sr isochron ages of Monche Tundra rocks, Kola Peninsula, bearing excess argon. Fortschr Mineral 50:102–104

Liu C, Matsuda A, Xie G (1994) Major- and trace-element compositions of Cenozoic basalts in eastern China: Petrogenesis and mantle source. Chem Geol 114:19–42

Lloyd FE, Arima M, Edgar AD (1985) Partial melting of a phlogopite-clinopyroxenite nodule from south-west Uganda: An experimental study bearing on the origin of highly potassic continental rift volcanics. Contrib Mineral Petrol 91:330–339

Long LE, Sial AN, Nekvasil H, Borba GS (1986) Origin of granite at Cabo de Santo Agostinho, northeast Brazil. Contrib Mineral Petrol 92:341–350

Long PE (1978) Experimental determination of partition coefficients for Rb, Sr, and Ba between alkali feldspar and silicate liquid. Geochim Cosmochim Acta 42A:833–846

Lonsdale P, Hawkins JW (1985) Silicic volcanism at an off-axis geothermal field in the Mariana Trough back-arc basin. Geol Soc Amer Bull 96:940–951

Lopez-Escobar L, Moreno RH, Tagiri M, Notsu K, Onuma N (1985) Geochemistry of lavas from San Jose volcano, southern Andes (33°45' S). Geochem J (19:209–222

Loubet M, Sassi R, DiDonato G (1988) Mantle heterogeneity: a combined isotope and trace element approach and evidence for recycled continental crustal materials in some OIB sources. Earth Planet Sci Letters 89:299–315

Lowdon JA (1960) Age determinations by the Geological Survey of Canada. Report 1. Isotopic Ages. Geol Surv Canada, Paper 60-17

Lowman PD (1992) The Sudbury structure as a terrestrial mare basin. Rev Geophys 30:227–243

Lu F, Anderson AT, Davis AM (1992) Melt inclusions and crystal-liquid separation in rhyolitic magma of the Bishop Tuff. Contrib Mineral Petrol 110:113–120

Ludden JN (1978) Magmatic evolution of the basaltic shield volcanoes of Réunion Island. J Volcanol Geotherm Res 4:171–198

Ludden JN, Hubert C, Gáriepy C (1986) The tectonic evolution of the Abitibi greenstone belt of Canada. Geol Mag 123:153–166

Ludwig KR (1992) ISOPLOT: A plotting and regression program for radiogenic-isotope data: Version 2.57. US Geol Surv Open File Rept OF91-445

Ludwig KR (1997) Optimization of multicollector isotope-ratio measurement of strontium and neodymium. Chem Geol 135:325–334

Luhr JF, Aranda-Gomez JJ, Pier JG (1989) Spinel-lherzolite-bearing Quaternary volcanic centers in San Luis Potosí, Mexico. 1. Geology, mineralogy, and petrology. J Geophys Res 94(B6): 7916–7940

Lum CC, Leeman WP, Foland KA, Kargel JA, Fitton JG (1989) Isotopic variation in continental basaltic lavas as indicators of mantle heterogeneity: Examples from the western U.S. Cordillera. J Geophys Res 94(B6):7871–7884

Lussiaa-Berdou-Polve M, Vidal P (1973) Initial strontium composition of volcanic rocks from Jan Mayen and Spitzbergen. Earth Planet Sci Letters 18(2):333–338

Luttinen AV, Rämö OT, Huhma H (1998) Neodymium and strontium isotopic and trace element composition of a Mesozoic CFB suite from Dronning Maud Land, Antarctica: Implications for lithosphere and asthenosphere contributions to Karoo magmatism. Geochim Cosmochim Acta 62:2701–2714

Lutz TM, Srogi L (1986) Biased isochron ages resulting from subsolidus isotope exchange: A theoretical model and results. Chem Geol 56:63–71

Lutz TM, Foland KA, Faul H, Srogi LA (1988) The strontium and oxygen record of hydrothermal alteration of syenites, Abu Kruq complex, Egypt. Contrib Mineral Petrol 98:212–223

Maaloe S (1985) Principles of igneous petrology. Springer Verlag, Berlin

Maaloe S, Sørensen IB, Hertogen J (1986) The trachybasaltic suite of Jan Mayen. J Petrol 27:439–466

Maaloe S, James D, Smedley E Petersen S, Garman LB (1992) The Koloa volcanic suite of Kauai, Hawaii. J Petrol 33:761–784

Maas R, McCulloch MT, Campbell IH, Coad PR (1986) Sm-Nd and Rb-Sr dating of an Archean massive sulphide deposits, Kidd Creek, Ontario. Geol 14:585–588

Maboko MAH, Boelrijk NA M, Priem HNA, Verdurmen AET (1985) Zircon U-Pb and biotite Rb-Sr dating of the Wami River granulites, Eastern Granulites, Tanzania: Evidence for approximately 715 Ma old granulite-facies metamorphism and final Pan-African cooling approximately 475 Ma ago. Precamb Res 30:361–378

MacDonald GA, Katsura T (1964) Chemical composition of Hawaiian lavas. J Petrol 5:82–133

Macdougall JD (ed) (1988) Continental flood basalts. Kluwer Academic Publishers, Dordrecht

Macdougall JD, Lugmair GW (1985) Extreme isotopic homogeneity among basalts from the southern East Pacific Rise: mantle or mixing effects? Nature 313:209–211

Macdougall JD, Lugmair GW (1986) Sr and Nd isotopes in basalts from the East Pacific Rise: Significance for mantle heterogeneity. Earth Planet Sci Letters 77:273–284

Macfarlane AW, Marcet P, LeHuray AP, Petersen U (1990) Lead isotope provinces of the central Andes inferred from ores and crustal rocks. Econ Geol 85(8):1857–1880

Machado DK, Parsons WH, Richards AF, Mulford JW (1962) Capelinhos eruption of Fayal Volcano, Azores, 1957–1958. J Geophys Res 67(9):3519–3529

Machado N, Ludden JN, Brooks C (1982) Fine-scale isotopic heterogeneity in the sub-Atlantic mantle. Nature 295:226–228

Machado N, Brooks C, Hart SR (1986) Determination of initial $^{87}Sr/^{86}Sr$ and $^{143}Nd/^{144}Nd$ in primary minerals from mafic and ultramafic rocks: Experimental procedures and implications for the isotopic characteristics of the Archean mantle under the Abitibi greenstone belt, Canada. Geochim Cosmochim Acta 50:2334–2348

Mackenzie DE, Chappell BW (1972) Shoshonitic and calc-alkaline lavas from the highlands of Papua New Guinea. Contrib Mineral Petrol 35:50–62

Magaritz M, Whitford DJ, James DE (1978) Oxygen isotopes and the origin of high $^{87}Sr/^{86}Sr$ andesites. Earth Planet Sci Letters 40:220–230

Mahoney JJ, Spencer K (1991) Isotopic evidence for the origin of the Manihiki and Ontong Java oceanic plateaus. Earth Planet Sci Letters 104(3/4) 196–210

Mahoney JJ, Macdougall JD, Lugmair GW, Murali AV, Sankar Das M, Gopalan K (1982) Origin of the Deccan Trap flows at Mahabaleshwar inferred from Nd and Sr isotopic and chemical evidence. Earth Planet Sci Letters 60:47–60

Mahoney JJ, Macdougall JD, Lugmair GW, Gopalan K (1983) Kerguelen hotspot source for the Rajmahal Traps and Ninetyeast Ridge? Nature 303:385–389

Mahoney JJ, Macdougall JD, Lugmair GW, Gopalan K, Krishnamurthy P (1985) Origin of contemporaneous tholeiitic and K-rich alkali lavas: A case study form the northern Deccan Plateau, India. Earth Planet Sci Letters 72:39–53

Mahoney JJ, Natland JH, White WM, Poreda R, Bloomer SH, Fisher RL, Baxter AN (1989) Isotopic and geochemical provinces of the western Indian Ocean spreading centers. J Geophys Res 94(B4):4033–4052

Mahoney JJ, LeRoex AP, Peng Z, Fisher RL, Natland JH (1992) Western limits of the Indian Ocean MORB mantle and the origin of low $^{206}Pb/^{204}Pb$ MORB: Isotope systematics of the central southwest Indian Ridge (17°–50° E). J Geophys Res 97: 19771–19790

Mahoney JJ, Storey M, Duncan RA, Spencer KJ, Pringle M (1993) Geochemistry and geochronology of the Ontong Java Plateau. In: Pringle M, Sager W, Sliter W (eds) The Mesozoic Pacific: Geology, tectonics, and volanism. Amer. Geophys. Union, Washington, D.C. (Geophys Monograph 77, pp 233–261)

Mahoney JJ, Jones WB, Frey FA, Salters VJM, Pyle DG, Davies HL (1995) Geochemical characteristics of lava flows from Broken Ridge, the Naturaliste Plateau, and southernmost Kerguelen Plateau: Cretaceous plateau volcanism in the southeast Indian Ocean. Chem Geol 120:315–345

Mahoney JJ, Frei R, Tejada MLG, Mo XX, Leat PL, Nägler TF (1998) Tracing the Indian Ocean mantle domain through time: isotopic results from old West Indian, east Tethyan and South Pacific seafloor. J Petrol 39:1285–1306

Mahood GA (1990) Second reply to comment of R.S.J. Sparks, H.E. Huppert, and C.J.N. Wilson on "Evidence for long residence times of rhyolitic magma in the Long Valley magmatic system, and isotopic record in the precaldera lavas of Glass Mountain". Earth Planet Sci Letters 99:395–399

Mahood GA, Halliday AN (1988) Generation of high-silica rhyolite: A Nd, Sr, and O isotopic study of Sierra La Primavera, Mexican Neovolcanic belt. Contrib Mineral Petrol 100:183–191

Mahood GA, Hildreth WE (1983) Large partition coefficients for trace elements in high-silica rhyolites. Geochim Cosmochim Acta 47:11–30

Mahood GA, Stimac JA (1990) Trace-element partitioning in pantellerites and trachytes. Geochim Cosmochim Acta 54: 2257–2276

Manetti P, Capaldi C, Chiesa S, Civetta L, Conticelli S, Gasparon M, LaVolpe L, Orsi G (1991) Magmatism of the eastern Red Sea margin in the northern part of Yemen from Oligocene to present. Tectonophysics 198:181–202

Manspeizer W (ed) (1988) Triassic-Jurassic rifting; continental breakup and the origin of the Atlantic Ocean and passive margins. Elsevier, Amsterdam

Manspeizer W, Puffer JH, Cousminer HL (1978) Separation of Morocco and eastern North America: a Triassic-Liassig stratigraphic record. Geol Soc Amer Bull 89:901–920

Manton WI (1968) The origin of associated basic and acid rocks in the Lebombo-Nuanetsi igneous province, southern Africa, as implied by strontium isotopes. J Petrol 9:23–39

Mantovani MSM, Cordani UG, Roisenberg A (1985a) Geochimica isotopica em vulcanicas acidas da Bacia do Paraná, e implicacoes geneticas associadas. Revista Brasileira de Geosciencias 15(1):61–65

Mantovani MSM, Marques LS, Sousa MA, Civetta L, Atalla L, Innocenti F (1985b) Trace element and strontium isotope constraints on the origin and evolution of Paraná continental flood basalts of Santa Catarina State (Southern Brazil). J Petrol 26(1):187–209

Marcantonio F, Dickin AP, McNutt RH, Heaman LM (1988) A 1800-million-year-old Proterozoic gneiss terrane in Islay with implication for crustal structure and evolution of Britain. Nature 335:62–64

Markis J, Mohr P, Rihm R (1991) Red Sea: Birth and early history of a new oceanic basin. Tectonophysics 198(2/4):129–466

Marsh BD, Carmichael ISE (1974) Benioff zone magmatism. J Geophys Res 79:1196–1206

Marsh JS (1989) Geochemical constraints on coupled assimilation and fractional crystallization involving upper crustal compositions and continental tholeiite magma. Earth Planet Sci Letters 92(1):70-80

Martin H (1987) Petrogenesis of Archean trondhjemites, tonalites, and granodiorites from eastern Finland: Major and trace element geochemistry. J Petrol 28:921-953

Martin H, Mathias M, Simpson ESW (1960) The Damaraland subvolcanic ring complexes in southwest Africa. In: Sorgenfrei T (ed) Report of the International Geological Congress XXI, Session 13. Copenhagen, Norden, pp 156-174

Marvin RF, Byers FM Jr, Mehnert HH, Orkild PP, Stern TW (1970) Radiometric ages and stratigraphic sequence of volcanic and plutonic rocks, southern Nye and western Lincoln Counties, Nevada. Geol Soc Amer Bull 81:2657-2676

Marzouki F, Jackson NJ, Ramsay CR, Darbyshire DPF (1982) Composition, age, and origin of two Proterozoic diorite-tonalite complexes in the Arabian Shield. Precamb Res 19:31-50

Massey NWD (1983) Magma genesis in a late Proterozoic proto-oceanic rift: REE and other trace-element data from the Keweenawan Mamainse Point Formation, Ontario, Canada. Precamb Res 21:81-100

Mathez EA, Dietrich VJ, Irving AJ (1984) The geocheistry of carbon in mantle peridotites. Geochim Cosmochim Acta 48: 1849-1859

Matsuda J-I, Notsu K, Okano J, Yaskawa K, Chungue L (1984) Geochemical implications from Sr isotopes and K-Ar age determinations for the Cook-Austral Islands chain. Tectonophysics 104:145-154

Matsuhisa Y (1977) Isotope geochemistry of an island arc traverse. Geochem J 11:107

Matsuhisa Y, Kurasawa H (1983) Oxygen and strontium isotopic characteristics of calc-alkaline volcanic rocks from the central and western Japan arcs: Evaluation of contribution of crustal components to the magma. J Volcanol Geotherm Res 18:483-510

Mauche R, Faure G, Jones LM, Hoefs J (1989) Anomalous isotopic composition of Sr, Ar, and O in the Mesozoic diabase dikes of Liberia, West Africa. Contrib Mineral Petrol 101:12-18

Maury RC, Bizouard H (1974) Melting of acid xenoliths into a basanite: An approach to the possible mechanisms of crustal contamination. Contrib Mineral Petrol 48:275-286

Maury RC, Westercamp D (1985) Variations chronologiques et spatials des basalts neogenes des Petites Antilles; implications sur l'evolution d'Arc. In: Mascle A (ed) Geodynamiques des Caribes. Technip, Paris, pp 77-89

May PR (1971) Pattern of Triassic-Jurassic diabase dikes around the North Atlantic in context of predrift position of the continents. Geol Soc Amer Bull 82:1285-1291

Mayes CL, Lawver LA, Sandwell DT (1990) Tectonic history and new isochron chart of the South Pacific. J Geophys Res 95: 8543-8567

Mazzone P, Grant NG (1988) Mineralogical and isotopic evidence for phenocryst-matrix disequilibrium in the Garner Mountain andesite. Contrib Mineral Petrol 99(2):267-271

McBirney AR (1975) Differentiation of the Skaergaard intrusion. Nature 253:691-694

McBirney AR (1978) Volcanic evolution of the Cascade Range. Ann Rev Earth Planet Sci 6:437-456

McBirney AR (1979) Effects of assimilation. In: Yoder HS Jr (ed) The evolution of igneous rocks. Fiftieth Anniversary Perspectives. Princeton Un. Press, pp 307-338

McBirney AR (1984) Igneous petrology. Freeman, San Francisco

McBirney AR (1989a) (ed) (1989) Piton de la Fournaise Volcano, Réunion Island. J Volcanol Geotherm Res 36:1-232

McBirney AR (1989b) The Skaergaard Layered Series: I. Structure and average compositions. J Petrol 30:363-397

McBirney AR (1995) Mechanisms of differentiation of layered intrusions: evidence from the Skaergaard intrusion. J Geol Soc London 152:421-435

McBirney AR (1996) The Skaergaard intrusion. In: Cawthorn RG (ed) Layered intrusions. Elsevier, Amsterdam, pp 147-180

McBirney AR (1998) The Skaergaard Layered Series. Part V. Included trace elements. J Petrol 39(2):255-276

McBirney AR, Nicolas A (1997) The Skaergaard Layered Series. Part II. Magmatic flow and dynamic layering. J Petrol 38: 569-580

McBirney AR, Noyes RM (1979) Crystallization and layering of the Skaergaard intrusion. J Petrol 20:487-554

McBirney AR, Sutter JF, Naslund HR, Sutton KG, White CM (1974) Episodic volcanism in the central Oregon Cascade Range. Geol 2(12):585-589

McCall GJH (1971) Some ultrabasic and basic igneous rock occurrences in the Archaean of Western Australia. Geol Soc Australia Spec Paper 3:429-442

McCourt WJ, Aspden JA, Brook M (1984) New geological and geochronological data from the Colombian Andes: continental growth by multiple accretion. J Geol Soc London 141: 831-845

McCulloch MT, Gamble JA (1991) Geochemical and geodynamical constraints on subduction zone magmatism. Earth Planet Sci Letters 102:(3/4):358-374

McCulloch MT, Perfit MR (1981) $^{143}Nd/^{144}Nd$, $^{87}Sr/^{86}Sr$, and trace element constraints on the petrogenesis of Aleutian island arc magmas. Earth Planet Sci Letters 56:167-179

McCulloch MT, Jaques AL, Nelson DR, Lewis JD (1983a) Nd and Sr isotopes in kimberlites and lamproites from Western Australia: an enriched mantle origin. Nature 302:400-403

McDermott F, Hawkesworth CJ (1991) Th, Pb, and Sr isotope variations in young island arc volcanics and oceanic sediment. Earth Planet Sci Letters 104(1):1-15

McDonough WF, Nelson DO (1984) Geochemical constraints on magma processes in a peralkaline system: the Paisano volcano, west Texas. Geochim Cosmochim Acta 48:2443-2455

McDonough WF, McCulloch MT, Sun SS (1985) Isotopic and geochemical systematics in Tertiary-Recent basalts from southeastern Australia and implications for the evolution of the sub-continental lithosphere. Geochim Cosmochim Acta 49:2051-2067

McDougall I (1962) Differentiation of the Tasmanian dolerites: Red Hill dolerite-granophyre association. Geol Soc Amer Bull 73:279-316

McDougall I (1963a) Potassium-argon ages of some rocks from Viti Levu, Fiji. Nature 198:677

McDougall I. (1963b) Potassium-argon age measurements on do lerites from Antarctica and South Africa. J Geophys Res 68(5): 1535-1545

McDougall I (1971) The geochronology and evolution of the young volcanic island of Réunion, Indian Ocean. Geochim Cosmochim Acta 35:261-288

McDougall I (1976) Geochemistry and origin of basalt of the Columbia River Group, Oregon and Washington. Geol Soc Amer Bull 87:777-792

McDougall I, Chamalaun FH (1969) Isotopic dating and geomagnetic polarity studies on volcanic rocks from Mauritius, Indian Ocean. Geol Soc Amer Bull 80:1419-1442

McDougall I, Compston W (1965) Strontium isotope composition and potassium-rubidium ratios in some rocks from Réunion and Rodriguez, Indian Ocean. Nature 207:252-253

McDougall I, Coombs DS (1973) Potassium-argon ages for the Dunedin volcano and the outlying volcanics. New Zealand J Geol Geophys 16(2):179-188

McDougall I, Duncan RA (1988) Age progressive volcanism in the Tasmantid Seamounts. Earth Planet Sci Letters 89:207-220

McDougall I, McElhinny MW (1970) The Rajmahal traps of India-K-Ar ages and paleomagnetism. Earth Planet Sci Letters 9: 371-378

McDougall I, Ollier CD (1982) Potassium-argon ages from Tristan da Cunha, South Atlantic. Geol Mag 119:87-93

McDougall I, Schmincke H-U (1976) Geochronology of Gran Canaria, Canary Islands: Age of shield building volcanism and other magmatic phases. Bull Volcanol 40(1):1-21

McDougall I, Upton BGJ, Wadswroth WJ (1965) A geological reconnaissance of Rodriguez Island, Indian Ocean. Nature 206: 26-27

McDougall I, Morton WH, Williams MAJ (1975) Age and rates of denudation of Trap Series basalts at Blue Nile gorge, Ethiopia. Nature 254:207-209

McDougall I, Watkins ND, Kristjansson L (1976a) Geochronology and paleomagnetism of a Miocene-Pliocene lava sequence at Bessastadaá, eastern Iceland. Amer J Sci 276:1078–1095

McDougall I, Watkins ND, Walker GPL, Kristjansson L (1976b) Potassium-argon and paleomagnetic analysis of Icelandic lava flows: Limits on the age of anomaly 5. J Geophys Res 81:1505

McDougall I, Saemundsson K, Johanneson H, Watkins ND, Kristjansson L (1977) Extension of the geomagnetic polarity time scale to 6 5 m.y.: K-Ar dating, geological, and paleomagnetic study of a 3500 m lava succession in western Iceland. Geol Soc Amer Bull 88:1–15

McElhinny MW (1966) Rb-Sr and K-Ar age measurements on the Modipe gabbro of Bechuanaland and South Africa. Earth Planet Sci Letters 1:439–442

McGregor VR (1973) The early Precambrian gneisses of the Godthaab district, West Greenland. Phil Trans Roy Soc 273A:343–358

McGregor VR, Mason B (1977) Petrogenesis and geochemistry of metabasaltic and metasedimentary enclaves in the Amitsoq gneisses, West Greenland. Amer Mineralogist 62:887–902

McHone JG (1978) Distribution, orientations, and ages of mafic dikes in central New England. Geol Soc Amer Bull 89:1645–1655

McHone JG, Butler JR (1984) Mesozoic igneous provinces of New England and the opening of the North Atlantic Ocean. Geol Soc Amer Bull 95:757–765

McIntosh WC, Kyle PR, Sutter JF (1986) Paleomagnetic results from the Kirkpatrick Basalt Group, Mesa Range, North Victoria Land, Antarctica. In: Stump E (ed) Geological Investigations in Northern Victoria Land. Amer. Geophys. Union (Antarctic Research Ser 46, pp 289–303)

McIntyre GA, Brooks C, Compston W, Turek A (1966) The statistical assessment of Rb-Sr isochrons. J Geophys Res 71(22): 5459–5468

McIver JR, Gevers TW (1970) Volcanic vents below the Royal Society Range, central Victoria Land, Antarctica. Trans Geol Soc South Africa 73:65–90

McKelvey BC, Webb PN (1959) Geology of upper Taylor Glacier region, Pt. 2 of Geological investigations in southern Victoria Land, Antarctica. New Zealand J Geol Geophys 2(4):718–728

McKelvey BC, Webb PN (1962) Geology of Wright Valley, Pt. 3 of Geological investigations in southern Victoria Land, Antarctica. New Zealand J Geol Geophys 5(1):143–162

McKenzie DP, Bickle MJ (1988) The volume and composition of melt generated by extension of the lithosphere. J Petrol 29: 625–679

McNutt MK, Fischer KM (1987) The South Pacific superswell, seamounts, islands, and atolls. Geophys Monogr 43:25–34

McNutt RH, Crocket JH, Zentilli M (1975) Initial ^{87}Sr/^{86}Sr ratios of plutonic and volcanic rocks of the Central Andes between latitudes 26° and 29° South. Earth Planet Sci Letters 27:305–313

McNutt RH, Clark AH, Zentilli M (1979) Lead isotopic compositions of Andean igneous rocks, latitudes 26° and 29° S: Petrologic and metallogenic implications. Econ Geol 74(4): 827–837

Meighan IG (1979) The acid igneous rocks of the British Tertiary Province. Bull Geol Survey Great Britain 70:10–22

Meighan IG, Gibson D, Hood DN (1984) Some aspects of Tertiary acid magmatism in NE Ireland. Mineral Mag 48:351–363

Meighan IG, McCormick AC, Gibson D, Gamble JA, Graham IJ (1988) Rb-Sr isotopic age determinations and the timing of Tertiary central complex magmatism in NE Ireland. In: Morton AC, Parson LM (eds) Early Tertiary Volcanism and the opening of the NE Atlantic. Geol Soc London Special Pub 39:349–360

Meijer A (1976) Lead and strontium isotopic data bearing on the origin of volcanic rocks from the Mariana island arc system. Geol Soc Amer Bul 87:1358–1369

Meinschein WG, Barghoorn ES, Schopf JW (1964) Biological remnants in a Precambrian sediment. Science 145:262–263

Melluso L, Morra V, Brotzu P, Razafiniparany A, Ratrimo V, Razafimathatratra D (1997) Geochemistry and Sr-isotopic composition of the Late Cretaceous flood basalt sequence of northern Madagascar: petrogenetic and geodynamic implications. J African Earth Sci 24:371–390

Mengel K, Kramm U, Wedepohl KH, Gohn E (1984) Sr isotopes in peridotite xenoliths and their basaltic host rocks from the northern Hessian Depression (NW Germany). Contrib Mineral Petrol 87:369–375

Mensing TM (1987) Geology and petrogenesis of the Kirkpatrick Basalt, Pain Mesa and Solo Nunatak, northern Victoria Land, Antarctica. PhD dissertation, The Ohio State University, Columbus, Ohio

Mensing TM (1991) High-titanium basalt and dolerite clasts from the Elephant and Reckling moraines. Antarct J US 26(5): 26–27

Mensing TM (1994) Isotopic and geochemical evidence for alteration of Jurassic basalt in northern Victoria Land, Antarctica, during the Cretaceous Period. Program with Abstracts, Geol. Assoc. of Canada, A75

Mensing TM, Faure G (1996) Cretaceous alteration of Jurassic volcanic rocks, Pain Mesa, northern Victoria Land, Antarctica. Chem Geol 129:153–161

Mensing TM, Faure G, Jones LM, Bowman JR, Hoefs J (1984) Petrogenesis of the Kirkpatrick Basalt, Solo Nunatak, northern Victoria Land, Antarctica, based on isotopic compositions of strontium, oxygen, and sulfur. Contrib Mineral Petrol 87: 101–108

Mensing TM, Faure G, Jones LM, Hoefs J (1991) Stratigraphic correlation and magma evolution of the Kirkpatrick Basalt in the Mesa Range, northern Victoria Land, Antarctica. In: Ulbrich H, Rocha Campos AC (eds) Gondwana Seven Proceedings. University de São Paulo, Brazil, pp 653–667

Menzies MA (1989) Cratonic, circumcratonic, and oceanic mantle domains beneath the western United States. J Geophys Res 94(B6):7899–7915

Menzies MA, Cox KG (eds) (1988) Oceanic and continental lithosphere: Similarities and differences. J Petrol Special Volume, 445p

Menzies MA, Murthy VR (1980a) Mantle metasomatism as a precursor to the genesis of alkaline magmas-isotopic evidence. Amer J Sci 280A:622–638

Menzies MA, Murthy VR (1980b) Nd and Sr isotope geochemistry of hydrous mantle nodules and their host alkali basalts: Implications for local heterogeneities in metasomatically veined mantle. Earth Planet Sci Letters 46:323–334

Menzies MA, Wass SY (1983) CO_2-rich mantle below south-eastern Australia: rare earth element, Sr, and Nd isotope study of Cenozoic alkaline magmas and apatite-rich xenoliths, Southern Highlands Province, New South Wales, Australia. Earth Planet Sci Letters 65:287–302

Menzies MA, Leeman WP, Hawkesworth CJ (1983) Isotope geochemistry of Cenozoic volcanic rocks reveals heterogeneity below western USA. Nature 303:205–209

Menzies MA, Bosence D, El-Nakhal HA, Al-Khirbash S, Al-Kadasi MA, Al-Subbary A (1990) Lithospheric extension and opening of the Red Sea: Sediment-basalt relationships in Yemen. Terra Nova 2:340–350

Mertz DF, Devey CW, Todt W, Stoffers P, Hofmann AW (1991) Sr-Nd-Pb isotope evidence against plume-asthenosphere mixing north of Iceland. Earth Planet Sci Letters 107:243–255

Mertzman SA (1979) Strontium isotope geochemistry of a low potassium olivine tholeiite and two basalt-pyroxene andesite magma series from the Medicine Lake Highland, California. Contrib Mineral Petrol 70:81–88

Metz JM, Mahood GA (1985) Precursors to the Bishop Tuff eruption: Glass Mountain, Long Valley, California. J Geophys Res 90:11,121–11,126

Mezger K, Altherr R, Okrusch M, Henjes-Kunst F, Kreuzer H (1985) Genesis of acid/basic rock associations: A case study: The Kallithea intrusive complex, Samos, Greece. Contrib Mineral Petrol 90:353–366

Michard-Vitrac A, Lancelot J, Allègre CJ, Moorbath S (1977) U-Pb ages on zircons from early Precambrian rocks of West Greenland and the Minnesota river valley. Earth Planet Sci Letters 35:449–453

Michard A, Montigny R, Schlich R (1986) Geochemistry of the mantle beneath the Rodriguez triple junction and the South-East Indian Ridge. Earth Planet Sci Letters 78:104–114

Middlemost EAK (1975) The basalt clan. Earth Sci Rev 11:337–364

Middlemost EAK (1980) A contribution to the nomenclature and classification of volcanic rocks. Geol Mag 117:51–57

Miklius A, Flower MFJ, Huijsmans JPP, Mukasa SB, Castillo P (1991) Geochemistry of lavas from Taal Volcano, southwestern Luzon, Philippines: Evidence for multiple magma supply systems and mantle source heterogeneity. J Petrol 32(3):593–627

Milanovskiy Y (1976) Rift zones of the geologic past and their associated formations. Report 2. Internat Geol Rev 18:619–639

Miller H (1984) Orogenic development of the Argentinian/Chilean Andes during the Palaeozoic. J Geol Soc London 141:885–892

Miller JA (1964) Age determinations made on samples of basalt from the Tristan da Cunha Group and other parts of the mid-Atlantic ridge. Appendix II. In: The Volcanological Report of the Royal Society Expedition to Tristan da Cunha (1962). Phil Trans Roy Soc A256:565–569

Miller JA, Mudie JD (1961) Potassium-argon age determinations in granite from the island of Mahé in the Seychelles archipelago. Nature 192:1174–1175

Millward D, Marriner GF, Saunders AD (1984) Cretaceous tholeiitic volcanic rocks from the Western Cordillera of Colombia. J Geol Soc London 141:847–860

Milner SC, LeRoex AP (1996) Isotope characteristics of the Okenyenya igneous complex, northwestern Namibia: constraints on the composition of the early Tristan plume and the origin of the EM 1 mantle component. Earth Planet Sci Letters 141:277–291

Milner SC, Duncan AR, Ewart A (1992) Quartz latite rheoignimbrite flows of the Etendeka Formation, north-western Namibia. Bull Vulcanol 54:200–219

Milner SC, LeRoex AP, Watkins RT (1993) Rb-Sr age determinations of rocks from the Okenyenya igneous complex, northwestern Namibia. Geol Mag 130:335–343

Milner SC, Duncan AR, Ewart A, Marsh JS (1995a) Promotion of the Etendeka Formation to group status: a new integrated stratigraphy. Commun. Geol Survey Namibia 9:5–11

Milner SC, Duncan AR, Whittingham AM, Ewart A (1995b) Trans-Atlantic correlation of eruptive sequences and individual silicic volcanic units within the Paraná-Etendeka igneous province. J Volcanol Geotherm Res 69:137–157

Milner SC, LeRoex AP, O'Connor JM (1995c) Age of the Mesozoic igneous rocks in northwestern Namibia and their relationship to continental breakup. J Geol Soc London 152:97–104

Milnes AR, Cooper BJ, Cooper JA (1982) The Jurassic Nisanger Basalt of Kangaroo Island, South Australia. Trans Roy Soc South Australia 106(Pt. 1):1–13

Milton C (1968) The natrocarbonatite of Oldoinyo Lengai, Tanzania. Geol Soc Amer Abstracts for 1968, Spec Paper 121

Minster CF, Allègre CJ (1978) Systematic use of trace elements in igneous processes. Part III: Inverse problem of batch partial melting in volcanic suites. Contrib Mineral Petrol 68:37–52

Mitchell CH, Widdowson M (1991) A geological map of the southern Deccan Traps, India, and its structural implications. J Geol Soc London 148:495–505

Mitchell JG, Mohr P (1986) K-Ar systematics in Tertiary dolerites from West Connacht, Ireland. Scott J Geol 22:225–240

Mitchell JG, LeBas MJ, Zielonka J, Furnes H (1983) On dating the magmatism of Maio, Cape Verde Islands. Earth Planet Sci Letters 64:61–76

Mitchell NC, Livermore RA (1998) The present configuration of the Bouvet triple junction. Geol 26(3):267–270

Mitchell RH (1980) Pyroxenes of the Fen alkaline complex, Norway. Amer Mineral 65:45–54

Mitchell RH (1995) Kimberlites, orangeites, and related rocks. Plenum Press, New York

Mitchell RH, Brunfelt AO (1975) Rare earth element geochemistry of the Fen alkaline complex, Norway. Contrib Mineral Petrol 52:247–259

Mitchell RH, Crocket JH (1972) Isotopic composition of strontium of the Fen alkaline complex. J Petrol 13:83–97

Mitchell RH, Smith CB, Vladykin NV (1994) Isotopic composition of strontium and neodymium in potassic rocks of the Little Murun complex, Aldan Shield, Siberia. Lithos 32:243–248

Mittlefehldt DW, Miller DF (1983) Geochemistry of the Sweetwater Wash pluton, California: Implications for "anomalous" trace element behavior during differentiation of felsic magmas. Geochim Cosmochim Acta 47:109–124

Miyashiro A (1978) Nature of alkalic volcanic rock series. Contrib Mineral Petrol 66:91–104

Mohr PA (1968) The Cainozoic volcanic succession in Ethiopia. Bull Volcanol 32(1):5–14

Mohr PA (1970) Catalog of chemical analyses of rocks from the intersection of the African, Gulf of Aden, and Red Sea rift Systems. Smithsonian Contributions to the Earth Sciences, No. 2

Mohr PA (1971) Ethiopian Rift and plateaus: some volcanic petrochemical differences. J Geophys Res 76(8):1967–1984

Mohr PA (1991) Structure of Yemeni Miocene dike swarms, and emplacement of coevalgranite plutons. Tectonophysics 198:203–221

Mohr PA (1992) Nature of the crust beneath magmatically active continental rifts. Tectonophysics 213:269–284

Molzahn M, Reisberg L, Wörner G, Mezger K (1994) Isotopic studies (Sr, Nd, Pb, Os, O) of the Jurassic Ferrar flood-basalt province of northern Victoria Land, Antarctica:evidence for three stages of petrogenetic evolution? Mineral Mag 58A:623–624

Molzahn M, Reisberg L, Wörner G (1996) Os, Sr, Nd, Pb, O isotope and trace element data from the Ferrar flood basalts, Antarctica: Evidence for an enriched subcontinental lithospheric source. Earth Planet Sci Letters 144:529–546

Monaghan M, Measures C, Klein J, Middleton R (1987) ^{10}Be and ^{9}Be in a phenocryst and groundmass separates from Aleutian Arc volcanic rocks: Implications for a magmatic origin of ^{10}Be in island-arc lavas. Eos 68:1525

Montigny R, Javoy M, Allègre CJ (1969) Le problème des andésites. Étude du volcanisme Quaternaire du Costa Rica (Amérique Centrale) a l'aide des traceurs couples $^{87}Sr/^{86}Sr$ et $^{18}O/^{16}O$. Bull Geol France (7 ser) 11:794–799

Moorbath S (1975a) Evolution of Precambrian crust from strontium isotopic evidence. Nature 254:395–398

Moorbath S (1975b) Geological interpretation of whole-rock isochron dates from high grade gneiss terrains. Nature 255:391

Moorbath S (1976) Age and isotope constraints for the evolution of Archaean crust. In: Windley BF (ed) The early history of the Earth. Wiley and Sons, London, pp 351–360

Moorbath S (1978) Ages, isotopes, and evolution of Precambrian continental crust. Chem Geol 20:151–187

Moorbath S, Bell JD (1965a) Strontium isotope abundance studies and rubidium-strontium age determinations on Tertiary igneous rocks from the Isle of Skye, north-west Scotland. J Petrol 6(1):37–66

Moorbath S, Bell JD (1965b) The application of strontium isotope studies to the origin of some acid rocks in the North Atlantic Tertiary Province. Proc Geol Soc London 1621:57–59

Moorbath S, Pankhurst RJ (1976) Further rubidium-strontium age and isotopic evidence for the nature of the Late Archaean plutonic event in West Greenland. Nature 262:124

Moorbath S, Park RG (1971) The Lewisian chronology of the southern region of the Scottish mainland. Scott. J Geol 8:51–74

Moorbath S, Thompson RN (1980) Strontium isotope geochemistry and petrogenesis of the Early Tertiary lava pile of the Isle of Skye, Scotland, and other rocks of the British Tertiary Province: An example of magma-crust interaction. J Petrol 21(2):295–321

Moorbath S, Walker GPL (1965) Strontium isotope investigations of igneous rocks from Iceland. Nature 207:837–840

Moorbath S, Welke H (1969a) Lead isotope studies on igneous rocks from the Isle of Skye, Northwest Scotland. Earth Planet Sci Letters 5:217–230

Moorbath S, Welke H (1969b) Isotopic evidence for the continental affinity of the Rockall Bank, North Atlantic. Earth Planet Sci Letters 5:211–216

Moorbath S, Sigurdsson H, Goodwin R (1968) K-Ar ages of the oldest exposed rocks in Iceland. Earth Planet Sci Letters 4:197–205

Moorbath S, Welke H, Gale NH (1969) The significance of lead isotope studies in ancient, high-grade metamorphic complexes, as exemplified by the Lewisian rocks of north-west Scotland. Earth Planet Sci Letters 6:245–256

Moorbath S, O'Nions RK, Pankhurst RJ, Gale NH, McGregor VR (1972) Further rubidium-strontium age determinations on the very early Precambrian rocks of the Godthaab district, West Greenland. Nature 240:78–82

Moorbath S, O'Nions RK, Pankhurst RJ (1973) Early Archaean age for the Isua iron Formation, West Greenland. Nature 245:138–139

Moorbath S, Powell JL, Taylor PN (1975a) Isotopic evidence for the age and origin of the "grey gneiss" complex of the southern Outer Hebrides, Scotland. J Geol Soc London 131:213–222

Moorbath S, O'Nions RK, Pankhurst RJ (1975b) The evolution of early Precambrian crustal rocks at Isua, West Greenland – geochemical and isotopic evidence. Earth Planet Sci Letters 27:229–239

Moorbath S, Allaart JH, Bridgwater D, McGregor VR (1977) Rb-Sr ages of Early Archaean supracrustal rocks and Amitsoq gneisses at Isua. Nature 270:43–44

Moorbath S, Thorpe RS, Gibson IL (1978) Strontium isotope evidence for petrogenesis of Mexican andesites. Nature 271:437–439

Moorbath S, Taylor PN Goodwin R (1981) Origin of granitic magma by crustal remobilization: Rb-Sr and Pb-Pb geochronology and isotope geochemistry of the Late Archean Qôrqut Granite complex of southern West Greenland. Geochim Cosmochim Acta 45:1051–1060

Moore JG, Nakamura K, Alcaraz A (1966) The 1965 eruption of Taal Volcano. Science 151:955–960

Moore WS, Stakes D (1990) Ages of barite-sulfide chimneys from the Mariana Trough. Earth Planet Sci Letters 100:265–274

Morgan WJ (1971) Convection plumes in the lower mantle. Nature 230:42–43

Morgan WJ (1972a) Deep mantle convection plumes and plate motions. Amer Assoc Petroleum Geologists Bull 56:203–213

Morgan WJ (1972b) Plate motions and deep mantle convection. Geol Soc Amer Mem 132:203–213

Morgan WJ (1978) Rodriguez, Darwin, Amsterdam, …, a second type of hotspot island. J Geophys Res 83:5355–5360

Morgan WJ (1981) Hotspot tracks and the opening of the Atlantic and Indian Oceans. In: Emiliani C (ed) The Sea. Wiley, New York (vol VII, pp 443–487

Morgan WJ (1983) Hotspot tracks and the early rifting of the Atlantic. Tectonophysics 94:123–139

Morioka M, Kigoshi K (1975) Lead isotopes and age of Hawaiian lherzolite nodules. Earth Planet Sci Letters 25:116–120

Mørk ME (1984) Magma mixing in the post-glacial Veidivötn fissure eruption, southeast Iceland: A microprobe study of mineral ages and glass variations. Lithos 17:55–75

Morris JD, Hart SR (1983) Isotopic and incompatible element constraints on the genesis of island arc volcanics from Cold Bay and Amak Islands, Aleutians, and implications for mantle structure. Geochim Cosmochim Acta 47:2015–2030

Morris JD, Hart SR (1986) Isotopic and incompatible element constraints on the genesis of island arc volcanics from Cold Bay and Amak Island, Aleutians, and implications for mantle structure: Reply to a critical comment by M.R. Perfit and R.W. Kay. Geochim Cosmochim Acta 50:483–487

Morrison AD, Reay A (1995) Geochemistry of Ferrar Dolerite sills and dykes at Terra Cotta Mountain, south Victoria Land, Antarctica. Antarctic Science 7:73–85

Morse SA (1969) Geology of the Kiglapait layered intrusion, Labrador. Geol Soc Amer Mem 112

Morse SA (1981) Kiglapait geochemistry III: Potassium and rubidium. Geochim Cosmochim Acta 45:163–180

Morse SA (1982) Kiglapait geochemistry V: Strontium. Geochim Cosmochim Acta 45:223–234

Morse SA (1983) Strontium isotope fractionation in the Kiglapait Intrusion. Science 220:193–195

Mortimer N, Parkinson D, Raine JI, Adams CJ, Graham IJ, Oliver PJ, Palmer K (1995) Ferrar magmatic province rocks discovered in New Zealand: implications for Mesozoic Gondwana geology. Geol 23(2):185–188

Morton AC, Parson LM (eds) (1988) Early Tertiary volcanism and the opening of the NE Atlantic. Geol Soc London Special Pub 39

Morton AC, Taylor PN (1987) Lead isotope evidence for the structure of the Rockall dipping-reflector passive margin. Nature 326:381–383

Morton AC, Hitchen K, Ritchie JD, Hine NM, Whitehouse M, Carter SG (1995) Late Cretaceous basalts from Rosemary Bank, northern Rockall Trough. Geol Soc London 152:947–952

Morton JP, Long LE (1980) Rb-Sr dating of Paleozoic glauconite from the Llano region, central Texas. Geochim Cosmochim Acta 44:663–672

Moyer TC, Esperanza S (1989) Geochemical and isotopic variations in a bimodal magma system: the Kaiser Spring volcanic field, Arizona. J Geophys Res 94(B6):7841–7859

Moyer TC, Nealey LD (1989) Regional compositional variations of Late Tertiary bimodal rhyolite lavas across the Basin and Range/Colorado Plateau boundary in western Arizona. J Geophys Res 94(B6):7799–7816

Muehlenbachs K, Byerly G (1982) O^{18}-enrichment of silicic magmas caused by crystal fractionation at the Galapagos spreading center. Contrib Mineral Petrol 79:76–79

Muehlenbachs K, Anderson AT Jr, Sigvaldason GE (1974) Low-O^{18} basalts from Iceland. Geochim Cosmochim Acta 38:577–583

Mueller PA, Wooden JL (1976) Rb-Sr whole-rock age of the contact aureole of the Stillwater igneous complex, Montana. Earth Planet Sci Letters 29:384–388

Mueller PA, Peterman ZE, Granath J (1981) An Archean bimodal volcanic series, Owl Creek Mountains, Wyoming. Geol Soc Amer Abstracts with Programs 13:515

Mueller PA, Wooden JL, Schulz K, Bowes DR (1983) Incompatible-element-rich andesitic amphibolites from the Archean of Montana and Wyoming: evidence for mantle metasomatism. Geol 11:203–206

Muenow DW, Perfit MR, Aggrey KE (1991) Abundances of volatiles and genetic relationships among submarine basalts from the Woodlark Basin, southwest Pacific. Geochim Cosmochim Acta 55:2231–2239

Muir RJ, Fitches WR, Maltman AJ (1992) Rhinns complex: A missing link in the Proterozoic basement of the North Atlantic region. Geol 20:1043–1046

Mukasa SB (1986a) Common Pb isotopic compositions of the Lima, Arequipa, and Toquepala segments in the coastal batholith, Peru: Implications for magmagenesis. Geochim Cosmochim Acta 50:771–782

Mukasa SB (1986b) Zircon U-Pb ages of super-units in the coastal batholith, Peru: Implications for magmatic and tectonic processes. Geol Soc Amer Bull 97:241–254

Mukasa SB, Tilton GR (1984) Lead isotope systematics in batholitic rocks of the western and coastal Cordillera. In: Harmon RS, Barreiro BA (eds) Andean magmatism, chemical and isotopic constraints. Shiva, Cheshire, England, pp 180–250

Mukasa SB, Tilton GR (1985) Pb isotope systematics as a guide to crustal involvement in the generation of the Coastal Batholith, Peru. In: Pitcher WS et al. (eds) Magmatism at a plate edge: The Peruvian Andes. Blackie, Glasgow, Scotland, pp 235–238

Mukasa SB, Wilshire HG (1997) Isotopic and trace element compositions of upper mantle and lower crustal xenoliths, Cima volcanic Field, California: implications for the evolution of the subcontinental lithospheric mantle. J Geophys Res 102(B9): 20133–20148

Mukasa SB, McCabe R, Gill JB (1987) Pb-isotope compositions of volcanic rocks in the west and east Philippine island arcs: Presence of the Dupal isotopic anomaly Earth Planet Sci Letters 84:153–164

Mukasa SB, Henry DJ (1990) The San Nicolas batholith of coastal Peru: Early Paleozoic continental arc or continental rift magmatism? J Geol Soc London 147:27–39

Mukasa SB, Flower MFJ, Miklius A (1994) The Nd-, Sr-, and Pb-isotopic character of lavas from Taal, Laguna de Bay, and Arayat volcanoes, southwestern Luzon, Philippines: Implications for arc magma petrogenesis. Tectonophysics 235:205–221

Mukasa SB, Wilson AH, Carlson RW (1998) A multielement geochronologic study of the Great Dyke, Zimbabwe: significance of the robust and reset ages. Earth Planet Sci Letters 164: 353–370

Mungall JE, Martin RF (1994) Severe leaching of trachytic glass without devitrification, Terceira, Azores. Geochim Cosmochim Acta 58:75–83

Munizaga F, Aguirie L, Herve F (1973) Rb/Sr ages of rocks from the Chilean metamorphic basement. Earth Planet Sci Letters 18:87–92

Munksgaard NC (1984) High $\delta^{18}O$ and possible pre-eruptional Rb-Sr isochrons in cordierite-bearing Neogene volcanics from SE Spain. Contrib Mineral Petrol 87:351–358

Murata KJ, Formoso MLL, Roisenberg A (1987) Distribution of zeolites in lavas of southeastern Parana Basin, State of Rio Grande do Sul, Brazil. J Geol 95:455–468

Musselwhite DS, DePaolo DJ, McCurry M (1989) The evolution of a silicic magma system: Isotopic and chemical evidence from the Woods Mountains volcanic center, eastern California. Contrib Mineral Petrol 101:19–29

Mussett AE (1984) Time and duration of Tertiary igneous activity of Rhum and adjacent areas. Scottish J Geol 20:273–280

Mussett AE (1988) Timing and duration of activity in the British Tertiary Igneous Province. In: Morton AC, Parson LM (eds) Early Tertiary volcanism and the opening of the NE Atlantic. Geol Soc London Special Pub 39:337–348

Mussett AE, Ross JG, Gibson IL (1980) $^{40}Ar/^{39}Ar$ dates of eastern Iceland lavas. Geophys J of the R.A.S., 60:37–52

Myers JD, Marsh BD (1987) Aleutian lead isotopic data: Additional evidence for the evolution of lithospheric plumbing systems. Geochim Cosmochim Acta 51:1833–1842

Myers JD, Marsh BD, Sinha AK (1985) Strontium isotopic and selected trace element variations between two Aleutian volcanic centers (Adak and Atka): Implications for the development of arc volcanic plumbing systems. Contrib Mineral Petrol 91: 221–234

Myers JD, Marsh BD, Sinha AK (1986) Geochemical and strontium isotopic characteristics of parental Aleutian arc magmas: Evidence from the basaltic lavas of Atka. Contrib Mineral Petrol 94:1–11

Mysen BO (1982) The role of mantle anatexis. In: Thorpe S (ed) Andesites; orogenic andesites and related rocks. Wiley, Chichester, U.K., pp 489–522

Naar DF, Hey RN (1989) Recent Pacific-Easter-Nazca plate motions. In: Sinton JM (ed) Evolution of Mid-Oceanic Ridges. Amer. Geophys. Union, Washington, D.C. (Geophysical Monograph 57, pp 9–30)

Nagasawa H, Schreiber HD, Morris RV (1980) Experimental mineral/liquid partition coefficients of the rare earth elements (REE), Sc, and Sr for perovskite, spinel, and melilite. Earth Planet Sci Letters 46:431–437

Nakada S, Kamata H (1991) Temporal change in chemistry of magma source under central Kyushu, southwest Japan: progressive contamination of mantle wedge. Bull Volcanol 53(3): 182–194

Nakamura Y, Tatsumoto M (1988) Pb, Nd, and Sr isotopic evidence for a multicomponent source for rocks of Cook-Austral islands and heterogeneities of mantle plumes. Geochim Cosmochim Acta 52:2909–2924

Nakamura E, Campbell IH, McCulloch MT, Sun SS (1989) Chemical geodynamics in a back-arc region around the Sea of Japan: Implications for the genesis of alkaline basalts in Japan, Korea, and China. J Geophys Res 94(B4):4634–4654

Naldrett AJ, Lightfoot PC, Fedorenko V, Doherty W, Gorbachev NS (1992) Geology and geochemistry of intrusions and flood basalts of the Noril'sk region, USSR, with implications for the origin of the Ni-Cu ores. Econ Geol 87:975–1004

Nash WP (1972) Apatite-calcite equilibria in carbonatites: Chemistry of apatite from Iron Hill, Colorado. Geochim Cosmochim Acta 36:1313–1319

Nash WP, Crecraft HR (1985) Partition coefficients for trace elements in silicic magmas. Geochim Cosmochim Acta 49(11): 2309–2322

Naslund HR (1984) Petrology of the Upper Border Series of the Skaergaard Intrusion. J Petrol 25:185–212

Nathan S, Fruchter JS (1974) Geochemical and paleomagnetic stratigraphy of the Picture Gorge and Yakima basalts (Columbia River Group) in central Oregon. Geol Soc Amer Bull 85: 63–76

National Geographic Society (1989) Supplement to Asia-Pacific. National Geographic 176(5):558A

National Geographic Society (1990) Atlas of the world, 6th edn. Nat. Geogr. Soc., Washington D.C.

Natland JH, Turner DL (1985) Age progression and petrological development of Samoan shield volcanoes: evidence from K-Ar ages, lava compositions, and mineral studies. In: Brocher TM (ed) Geological Investigations of the Northern Melanesian Borderland. Circum-Pacific Council for Energy Resources, Houston, Texas, pp 139–171

Naylor RS, Steiger RH, Wasserburg GJ (1970) U-Th-Pb and Rb-Sr systematics in 2700×10^6-year-old plutons from the southern Wind River Range, Wyoming. Geochim Cosmochim Acta 34: 1133–1159

Neal CR, Davidson JP (1989) An unmetasomatized source of the Malaitan alnöite (Solomon Islands): Petrogenesis involving zone refining, megacryst fractionation, and assimilation of oceanic lithosphere. Geochim Cosmochim Acta 53:1975–1990

Nelson DO (1983) Implications of oxygen-isotope data and trace element modeling for a large-scale mixing model for the Columbia River Basalt. Geol 11:248–251

Nelson DO, Dasch EJ (1976) Disequilibrium of strontium isotopes between mineral phases of parental rocks during magma genesis. J Volcanol Geotherm Res 1:183–191

Nelson DR (1998) Granite-greenstone crust formation on the Archaean Earth: A consequence of two superimposed processes. Earth Planet Sci Letters 158:109–120

Nelson DR, McCulloch MT, Sun SS (1986) The origins of ultrapotassic rocks inferred from Sr, Nd, and Pb isotopes. Geochim Cosmochim Acta 50:231–246

Nelson ST, Davidson JP (1993) Interactions between mantle-derived magmas and mafic crust, Henry Mountains, Utah. J Geophys Res 98(B2):1837–1852

Nesbitt RW, Sun SS (1976) Geochemistry of Archaean spinifex textured peridotites and magnesian and low magnesian tholeiites. Earth Planet Sci Letters 31:433–453

Nesbitt RW, Giles ADT, Moore AC, Hopwood TP (1970) The Giles complex, central Australia: A stratified sequence of mafic and ultramafic intrusions. Geol Soc London Special Pub 1:547–564

Neumann E-R (1978) Petrogenesis of the Larvik ring complex in the Permian Oslo rift, Norway. In: Neumann E-R, Ramberg IB (eds) Petrology and Geochemistry of Continental Rifts. Reidel Pub. Co., Dordrecht (Proc. Nato Adv. Stud. Inst., vol I, pp 231–236)

Neumann E-R (1980) Petrogenesis of the Oslo region larvikites and associated rocks. J Petrol 21:498–531

Neumann E-R (1991) Ultramafic and mafic xenoliths from Hierro, Canary Islands: Evidence for melt infiltration in the upper mantle. Contrib Mineral Petrol 106:236–252

Neumann E-R, Wulff-Pedersen E (1997) The origin of highly silicic glass in mantle xenoliths from the Canary Island. J Petrol 38:1513–1540

Neumann E-R, Larsen BT, Sundvoll B (1985) Compositional variations among gabbroic intrusions in the Oslo rift. Lithos 18:35–59

Neumann E-R, Pallesen S, Andresen P (1986) Mass estimates of cumulates and residues after anatexis in the Oslo grapen. J Geophys Res 91:11,629–11,640

Neumann E-R, Titton GR, Tuen E (1988) Sr, Nd, and Pb isotope geochemistry of the Oslo Rift igneous province, southeast Norway. Geochim Cosmochim Acta 52:1997–2007

Neumann E-R, Sundvoll B, Overli PE (1990) A mildly depleted upper mantle beneath southeast Norway: Nd-Sr isotopic and trace-element evidence from basalts in the Permo-Carboniferous Oslo Rift. Tectonophysics 175

Neumann E-R, Olsen KH, Baldrige WS, Sundvoll B (1992) The Oslo Rift: A review. Tectonophysics 208:1–8

Newman S, Finkel RC, Macdougall JD (1984) Comparison of ^{230}Th-^{238}U disequilibrium systematics in lavas from three hotspot regions: Hawaii, Prince Edward, and Samoa. Geochim Cosmochim Acta 48:315–324

Nicholls IA, Whitford DJ (1978) Geochemical zonation in the Sunda volcanic arc, and the origin of K-rich lavas. Bull Aust Soc Explor Geophys 9:93–98

Nicholls IA, Whitford DJ, Harris KL, Taylor SR (1980) Variation in the geochemistry of mantle sources for the tholeiitic and calc-alkaline mafic magmas, western Sunda volcanic arc, Indonesia. Chem Geol 30:177–199

Nicholson H, Condomines M, Fitton JG, Fallick AE, Grönvold K, Rogers G (1991) Geochemical and isotopic evidence for crustal assimilation beneath Krafla, Iceland. J Petrol 32(5):1005-1020

Nicholson SW, Shirey SB (1990) Evidence for a Precambrian mantle plume: A Sr Nd, and Pb isotopic study of the Midcontinent Rift System in the Lake Superior region. J Geophys Res 95: 10851-10868

Nicholson SW, Cannon WF, Schulz KJ (1992) Metallogeny of the Midcontinent Rift System of North America. Precamb Res 58: 355-386

Nicolaysen LO, de Villiers JWL, Burger AJ, Strelow FWE (1958) New measurements relating to the absolute age of the Transvaal System and of the Bushveld igneous complex. Trans Geol Soc S Africa 61:137-164

Nicolussi GK, Davis AM, Pellin MJ, Lewis RS, Clayton RN, Amari S (1997) S-process zirconium in presolar silicon carbide grains. Science 277:1281-1283

Nielson DR, Stoiber RE (1973) Relationship of potassium content in andesitic lavas and depth to the seismic zone. J Geophys Res 78:6887-6892

Nier AO (1938) Isotopic constitution of Sr, Ba, Bi, Tl, and Hg. Phys Rev 54:275-278

Nisbet EG, Cheadle MJ, Arndt NT, Bickle MJ (1993) Constraining the potential temperature of the Archaean mantle: A review of the evidence from komatiites. Lithos 30:291-307

Nixon PH, Boyd FR (1979) Garnet-bearing lherzolites and discrete nodule suites from the Malaita alnöite, Solomon Islands, SW Pacific, and their bearing on oceanic mantle composition and geotherm. In: Boyd FR, Myer HOA (eds) The Mantle Sample: Inclusions in Kimberlites and other volcanics. Amer. Geophys. Union, Washington D.C., pp 400-423

Nixon PH, Mitchell RH, Rogers NW (1980) Petrogenesis of alnöitic rocks from Malaita, Solomon Islands, Melanesia. Mineral Mag 43:587-596

Nobel FA, Andriessen PAM, Hebeda EH, Priem HNA, Rondell HE (1981) Isotopic dating of the post-alpine Neogene volcanism in the Betic Cordilleras, southern Spain. Geol Mijnbouw 60: 209-214

Noble DC, Hedge CE (1969) ^{87}Sr/^{86}Sr variations within individual ash-flow sheets. US Geol Survey Prof Paper 650-C:133-139

Noble DC, Sargent KA, Mehnert HH, Ekren EB, Byers FM Jr (1968) Silent Canyon volcanic center, Nye Country, Nevada, 65-76) In: Eckel EB (ed) Nevada Test Site. Geol Soc Amer Mem 110

Noble DC, Bowman HR, Hebert AJ, Silberman ML, Herapoulos CE, Fabbi BP, Hedge CE (1975) Chemical and isotopic constraint on the origin of low-silica latite and andesite from the Andes of central Peru. Geol 3:501-504

Noe-Nygaard A, Rasmussen J (1968) Petrology of a 3000 metre sequence of basaltic lava flows in the Faeroe Islands. Lithos 1:286

Nohda S (1973) Rb-Sr dating of the Yatsushiro granite and gneiss, Kyushu, Japan. Earth Planet Sci Letters 20:140-144

Nohda S, Wasserburg GJ (1981) Nd and Sr isotopic study of volcanic rocks from Japan Earth Planet Sci Letters 52:264-276

Nohda S, Wasserburg GJ (1986) Trends of Sr and Nd isotopes through time near the Japan Sea in northeastern Japan. Earth Planet Sci Letters 78:157-167

Nohda S, Tatsumi Y, Otofuji Y, Matsuda T, Ishizaka K (1988) Asthenospheric injection and back-arc opening: isotopic evidence from northeast Japan. Chem Geol 68:317-327

Nordhoff C, Hall JN (1936) The Bounty trilogy. Little, Brown & Co., Boston (3 volumes)

Norton D, Taylor HP Jr (1979) Quantitative simulation of the hydrothermal systems of crystallizing magmas on the basis of transport theory and oxygen isotope data: an analysis of the Skaergaard intrusion. J Petrol 20(3):421-486

Notsu K (1983) Strontium isotope composition in volcanic rocks from the northeast Japan arc. J Volcanol Geotherm Res 18: 531-548

Notsu K, Lajo JA (1984) Regional variation of ^{87}Sr/^{86}Sr ratio in Late Cenozoic volcanic rocks from southern Peru. Geochem J 18:241-250

Notsu K, Ono K, Soya T (1987a) Strontium isotopic relations of bimodal volcanic rocks at Kikai volcano in the Ryukyu arc, Japan. Geol 15:345-348

Notsu K, Escobar LL, Onuma N (1987b) Along-arc variation of Sr-isotope composition in volcanic rocks from the southern Andes. Geochem J 21:307-314

Notsu K, Arakawa Y, Kobayashi T (1990) Strontium isotopic characteristics of arc volcanic rocks at the initial stage of subduction in western Japan. J Volcanol Geotherm Res 40(3):181-196

Nougier J, Cantral JM, Karche JP (1986) The Comores archipelago in the western Indian Ocean: Volcanology, geochronology, and geodynamic setting. J African Earth Sci 5:135-145

Nozawa T (1983) Felsic plutonism in Japan. Geol Soc Amer Mem 159:105-122

Nutman AP, Bridgwater D (1983) Deposition of Malene supracrustal rocks on an Amîtsoq basement in outer Ameralik, southern West Greenland Grønlands Geol Undersøgelse Rapp 112:43-51

Nye CJ, Reid MR (1986) Geochemistry of primary and least fractionated lavas from Okmok volcano, central Aleutians: Implications for arc magmagenesis. J Geophys Res 91: 10271-10289

O'Brien HE, Irving AJ, McCallum IS (1991) Eocene potassic magmatism in the Highwood Mountains, Montana: petrology, geochemistry, and tectonic implications. J Geophys Res 96: 13237-13260

O'Brien HE, Irving AJ, McCallum IS, Thirlwall MF (1995) Sr, Nd, and Pb isotopic evidence for interaction of post-subduction asthenospheric potassic mafic magmas of the Highwood Mountains, Montana, with ancient Wyoming craton lithospheric mantle. Geochim Cosmochim Acta 59:4539-4556

O'Connor JM, Duncan RA (1990) Evolution of the Walvis Ridge-Rio Grande Rise hotspot system: Implications for African and South American plate motions over plumes. J Geophys Res 95:17475-17502

O'Connor JM, LeRoex AP (1992) South Atlantic hotspot-plume systems. 1. Distribution in time and space. Earth Planet Sci Letters 113:343-364

Officer CB, Drake CL (1985) Terminal Cretaceous environmental events. Science 154:1551-1553

O'Hara MJ (1973) Non-primary magmas and dubious mantle plume beneath Iceland. Nature 243:507

O'Hara MJ (1977) Geochemical evolution during fractional crystallization of a periodically refilled magma chamber. Nature 266:503-507

O'Hara MJ (1980) Nonlinear nature of the unavoidable long-lived isotopic trace and major element contamination of a developing magma chamber. Phil Trans Roy Soc London A297: 215-227

O'Hara MJ, Matthews RE (1981) Geochemical evolution in an advancing, periodically replenished, periodically tapped, continuously fractionated magma chamber. J Geol Soc London 138:237-277

Ohtani E (1990) Majorite fractionation and genesis of komatiites in the deep mantle. Precamb Res 48:195-202

Old RA (1975) The age and field relations of the Tardree Tertiary rhyolite complex, County Antrim, N. Ireland. Bull Geol Surv GB 51:21-40

Old RA, Rex DC (1971) Rubidium and strontium age determinations of some Precambrian grantic rocks, S.E. Uganda. Geol Mag 108(5):353-360

Omar GI, Steckler MS (1995) Fission track evidence on the initial rifting of the Red Sea: Two pulses, no propagation. Science 270:1341-1344

O'Neil JR, Hedge CE, Jackson ED (1970) Isotopic investigation of xenoliths and host basalts from the Honolulu Volcanic Series. Earth Planet Sci Letters 8:253-257

O'Nions RK, Clarke DB (1972) Comparative trace element geochemistry of Tertiary basalts from Baffin Bay. Earth Planet Sci Letters 15:436-446

O'Nions RK, Grönvold K (1973) Petrogenetic relationships of acid and basic rocks in Iceland Sr isotopes and rare-earth elements in late and postglacial volcanics. Earth Planet Sci Letters 19: 397-409

O'Nions RK, Pankhurst RJ (1973) Secular variations in the Sr-isotope composition of Icelandic volcanic rocks. Earth Planet Sci Letters 21:13-21

O'Nions RK, Pankhurst RJ (1974) Petrogenetic significance of isotope and trace element variations in volcanic rocks from the Mid-Atlantic. J Petrol 15(3):603–634

O'Nions RK, Fridleifsson IB, Jakabsson SP (1973) Strontium isotopes and rare-earth elements in basalts from Heimaey and Surtsey volcanic eruptions. Nature 243:213–214

O'Nions RK, Hamilton PJ, Evensen NM (1977) Variations in $^{143}Nd/^{144}Nd$ and $^{87}Sr/^{86}Sr$ ratios in oceanic basalts. Earth Planet Sci Letters 34:13–22

Onstott TC, Dorbor J (1987) $^{40}Ar/^{39}Ar$ and paleomagnetic results from Liberia and the Precambrian APW data base for the West African Shield. J African Earth Sci 6:537–552

O'Reilly SY, Griffin WL (1984)Sr isotopic heterogeneity in primitive basaltic rocks, southeastern Australia: Correlation with mantle metasomatism. Contrib Mineral Petrol 87:220–230

Orlova MP, Zhidkov AY, Orlov DM, Zotova IF (1994) The internal structure and formation of the Synnyr alkali intrusion, north Baykal region. Geochem Internat 31(3):86–106

Othman BD, Tilton GR, Menzies MA (1990) Pb, Nd, and Sr isotopic investigations of kaersutite and clinopyroxene from ultramafic nodules and their host basalts: The nature of the sub-continental mantle. Geochim Cosmochim Acta 54: 3449–3460

Oversby VM (1971) Lead in oceanic islands. Faial, Azores, and Trinidade. Earth Planet Sci Letters 11:401–406

Oversby VM (1972) Genetic relations among the volcanic rocks of Réunion: Chemical and lead isotopic evidence. Geochim Cosmochim Acta 36:1167–1179

Oversby VM, Ewart A (1973) Lead isotopic compositions of Tonga-Kermadec volcanics and their petrologic significance. Contrib Mineral Petrol 37:181–210

Oversby VM, Lancelot J, Gast PW (1971) Isotopic composition of lead in volcanic rocks from Tenerife, Canary Islands. J Geophys Res 76(14):3402–3413

Owen LB, Faure G (1977) Rb-Sr dating of the Lackner Lake complex, northern Ontario, Canada. Precamb Res 5:299–303

Paces JB, Bell K (1989) Non-depleted sub-continental mantle beneath the Superior Province of the Canadian Shield: Nd-Sr isotopic and trace element evidence from Midcontinent Rift basalts. Geochim Cosmochim Acta 53:2023–2035

Page NJ (1971) Sulfide minerals in the G and H chromitite zones of the Stillwater complex, Montana. US Geol Survey Prof Paper 694:1–20

Page NJ, Rowe JJ, Hafty J (1976) Platinum metals in the Stillwater complex, Montana. Econ Geol 71:1352–1363

Page RW, Johnson RW (1974) Strontium isotope ratios of Quaternary volcanic rocks from Papua New Guinea. Lithos 7:91–100

Palacz ZA (1984) Isotopic and geochemical evidence for the evolution of a cyclic unit in the Rhum intrusion, north-west Scotland. Nature 307:618–620

Palacz ZA, Saunders AD (1986) Coupled trace element and isotope enrichment in the Cook-Austral-Samoa islands, southwest Pacific. Earth Planet Sci Letters 79:270–280

Palacz ZA, Tait SR (1985) Isotopic and geochemical investigation of unit 10 from the Eastern Layered Series of the Rhum intrusion, northwest Scotland. Geol Mag 122(5):485–490

Palacz ZA, Wolff JA (1989) Strontium, neodymium, and lead isotope characteristics of the Granadilla Pumice, Tenerife: a study of the causes of strontium isotope disequilibrium in felsic pyroclastic depsoits. In: Saunders AD, Norry MJ (eds) Magmatism in the ocean basins. Geol Soc London Spec Pub 42:147–160

Pallister JS (1987) Magmatic history of Red Sea rifting: perspective from the central Saudi Arabian coastal plain. Geol Soc Amer Bull 98:400–417

Pallister JS, Stacey JS, Fischer LB, Premo WR (1987) Arabian Shield opholites and Late Proterozoic microplate accretion. Geol 15: 320–323

Palmason G (ed) (1982) Continental and oceanic rifts. Amer. Geophys. Union, Washington, D.C.

Palmer AR (1983) The decade of North American geology: 1983 geologic time scale. Geol 11:503–504

Pankhurst RJ (1969) Strontium isotope studies related to petrogenesis in the Caledonian basic igneous province of NE Scotland. J Petrol 10(1):115–143

Pankhurst RJ (1977) Strontium isotope evidence for mantle events in the continental lithosphere. J Geol Soc London 134:255

Pankhurst RJ (1982) Rb-Sr geochronology of Graham Land, Antarctica. J Geol Soc London 139:701–712

Pankhurst RJ, Smellie JL (1983) K-Ar geochronology of the South Shetland Islands, Antarctica: Apparent lateral migration of Jurassic to Quaternary island arc volcanism. Earth Planet Sci Letters 66:214–222

Pankhurst RJ, Moorbath S, McGregor VR (1973a) Late event in the geologic evolution of the Godthaab district, West Greenland. Nature 243:24–26

Pankhurst RJ, Moorbath S, Rex DC, Turner G (1973b) Mineral age patterns in ca. 3700 MY old rocks from West Greenland. Earth Planet Sci Letters 20:157–170

Pankhurst RJ, Beckinsale RD, Brooks CK (1976) Strontium and oxygen isotope evidence relating to the petrogenesis of the Kangerdlugssuaq alkaline intrusion, East Greenland. Contrib Mineral Petrol 54:17–42

Pankhurst RJ, Walsh JN, Beckinsale RD, Skelhorn RR (1978) Isotopic and other geochemical evidence for the origin of the Loch Uisg granophyre, Isle of Mull, Scotland. Earth Planet Sci Letters 38:355–363

Pankhurst RJ, Weaver SD, Brook M, Saunder AD (1980) K-Ar chronology of Byers Peninsula, Livingston Island, South Shetland Islands. Brit Ant Surv Bull 49:277–282

Pankhurst RJ, Millar IL, Grunow AM, Storey BC (1993) The pre-Cenozoic magmatic history of the Thurston Island crustal block. J Geophys Res 98(B7):11835–11849

Papanastassiou DA, Wasserburg GJ (1969) Initial strontium isotopic abundances and the resolution of small time differences in the formation of planetary objects. Earth Planet Sci Letters 5:361–376

Papike JJ (1987) Proceedings of the International Mineralogical Association Symposium: "Mineralogy and Geochemistry of Granites and Pegmatites." Geochim Cosmochim Acta 51(3):387

Paslick C, Halliday AN, James D, Dawson JB (1995) Enrichment of the continental lithosphere by OIB melts: Isotopic evidence from the volcanic province of northern Tanzania. Earth Planet Sci Letters 130:109–126

Paslick C, Halliday AN, Lange RA, James D, Dawson JB (1996) Indirect crustal contamination: Evidence from isotopic and chemical disequilibria in minerals from alkali basalts and nephelinites from northern Tanzania. Contrib Mineral Petrol 125(4):277–292

Pasteels P, Villeneuve M, de Paepe P, Klerkx J (1989) Timing of the volcanism of the southern Kivu province: implications for the evolution of the western branch of the East African Rift system. Earth Planet Sci Letters 94(3/4):353–363

Paster TP, Schauwecker DS, Haskin LA (1974) The behavior of some trace elements during solidification of the Skaergaard Layered Series. Geochim Cosmochim Acta 38:1549–1577

Patchett PJ (1978) Paleomagnetism and the Grenville orogeny: New Rb-Sr ages from dolerites in Canada and Greenland. Earth Planet Sci Letters 40:349–364

Patchett PJ, Bylund G (1977) Age of Grenville belt magmatism: Rb-Sr and paleomagnetic evidence from Swedish dolerites. Earth Planet Sci Letters 35:92–104

Patchett PJ, Ruiz J (1987) Nd isotopic ages of crust formation and metamorphism in the Precambrian of eastern and southern Mexico. Contrib Mineral Petrol 96:523–528

Patchett PJ, Breeman O van, Martin RF (1979) Sr isotopes and the structural state of feldspars as indicators of post-magmatic hydrothermal activity in continental dolerites. Contrib Mineral Petrol 69:65–73

Patterson CC (1956) Age of meteorites and the earth. Geochim Cosmochim Acta 10:230–237

Paul DK (1971) Strontium isotope studies on ultramafic inclusions form Dreiser Weiher, Eifel, Germany. Contrib Mineral Petrol 34:22–28

Paul DK, Potts PJ, Rex DC, Beckinsale RD (1977) Geochemical and petrogenetic study of the Girnar igneous complex, Deccan volcanic province, India. Geol Soc Amer Bull 88:227–234

Paul DK, Kresten P, Barman TR, McNutt RH, Brunfelt AO (1984) Geochemical and petrolgoical relations in some Deccan basalts, western Maharashtra, India. J Volcanol Geotherm Res 21:165–176

Pauling L (1960) Nature of the chemical bond. 3rd edition, Cornell University Press, Ithaka, New York

Pe GG (1975) Strontium isotope ratios in volcanic rocks from the northwestern part of the Hellenic arc. Chem Geol 15:53–60

Pe GG, Gledhill A (1975) Strontium isotope ratios in volcanic rocks from the south-eastern part of the Hellenic arc. Lithos 8: 209–214

Pe-Piper G, Piper DJW (1994) Miocene magnesian andesites and dacites, Evia, Greece: Adakites associated with subducting slab detachment and extension. Lithos 31:125–140

Peate DW, Hawkesworth CJ (1996) Lithospheric to asthenospheric transition in low-Ti flood basalts from southern Paraná, Brazil. Chem Geol 127:1–24

Peate DW, Hawkesworth CJ Mantovani MSM, Shukowsky W (1990) Mantle plume and flood-basalt stratigraphy in the Paraná, South America. Geol 18:1223–1226

Peate DW, Regelous M, Rogers NW (1992a) Paraná magmatism and the opening of the South Atlantic. In: Storey BS, Alabaster T, Pankhurst RJ (eds) Magmatism and the causes of continental break-up. Geol Soc London Special Pub 68:221–240

Peate DW, Hawkesworth CJ, Mantovani MSM (1992b) Chemical stratigraphy of the Paraná lavas (South America): Classification of magma types and their spatial distribution. Bull Volcanol 55:119–139

Peate DW, Pearce JA, Hawkesworth CJ, Colley H, Edwards CMH, Hirose K (1997) Geochemical variations in Vanuatu arc lavas: The role of subducted material and a variable mantle-wedge composition. J Petrol 38(12):1331–1358

Peate DW, Hawkesworth CJ, Mantovani MSM, Rogers NW, Turner SP (1999) Petrogenesis and stratigraphy of the high-Ti/Y Urubici magma type in the Paraná flood basalt province and implications for the nature of Dupal-type mantle in the South Atlantic region. J Petrol 40:451–474

Peccerillo, A (1990) On the origin of the Italian potassic magmas - comment. Chem Geol 8:183–191

Pedersen AK (1981) Armalcolite-bearing Fe-Ti oxide assemblages in graphite-equilibrated salic volcanic rocks with native iron from Disko, central West Greenland. Contrib Mineral Petrol 77:307–324

Pedersen AK, Pedersen S (1987) Sr isotope chemistry of contaminated Tertiary volcanic rocks from Disko, central West Greenland. Bull Geol Soc Denmark 36(3/4):315–336

Pedersen S, Larsen O, Hald N, Campsie J, Bailey JC (1976) Strontium isotope and lithophile element values from the submarine Jan Mayen province. Bull Geol Soc Denmark 25:15–20

Pegram WJ (1990) Development of continental lithospheric mantle as reflected in the chemistry of the Mesozoic Appalachian tholeiites, U.S.A. Earth Planet Sci Letters 97:316–331

Pegram WJ, Register JK, Fullagar PD, Ghuma MA, Rogers JJW (1976) Pan-African ages from a Tibesti massif batholith, southern Libya Earth Planet Sci Letters 30:123–128

Peng ZG, Zartman RE Futa K, Chen D (1986) Pb-, Sr-, and Nd-isotopic systematics and chemical characteristics of Cenozoic basalts, eastern China. Chem Geol 59:3–33

Peng ZX, Mahoney J, Hooper P, Harris C, Beane J (1994) A role for lower continental crust in flood basalt genesis? Isotopic and incompatible element study of the lower six formations of the Western Deccan Traps. Geochim Cosmochim Acta 58:267–288

Peng ZX, Mahoney JJ, Hooper PR, Mcdougall JD, Krishnamurthy P (1998) Basalts of the northeastern Deccan Traps, India: Isotopic and elemental geochemistry and relation to southwestern Deccan stratigraphy. J Geophys Res 103(B12):29843–29866

Percival JA, Card KD (1983) Archean crust as revealed in the Kapuskasing uplift, Superior province, Canada. Geol 11:323–326

Perfit MR, Kay RW (1936) Comment on "Isotopic and incompatible-element constraints on the genesis of island arc volcanics from Cold Bay and Amak Island, Aleutians, and implications for mantle structure." Geochim Cosmochim Acta 50(3):477–482; Reply 483–488

Perfit MR, Brueckner H, Lawrence JR, Kay RW (1980a) Trace element and isotopic variations in a zoned pluton and associated volcanic rocks, Unalaska Island, Alaska: A model for fractionation in the Aleutian calc-alkaline suite. Contrib Mineral Petrol 73:69–87

Perfit MR, Gust DA, Bence AE, Arculus RT, Taylor SR (1980b) Chemical characteristics of island-arc basalts: Implications for mantle sources. Chem Geol 30:227–256

Perrin R (1954) Granitization, metamorphism, and volcanism. Amer J Sci 252:449–465

Peterman ZE, Heming RF (1974) $^{87}Sr/^{86}Sr$ ratios of calc-alkalic lavas from the Rabaul Caldera, Papua New Guinea. Geol Soc Amer Bull 85:1265–1268

Peterman ZE, Hildreth RA (1978) Reconnaissance geology and geochronology of the Precambrian of the Granite Mountains, Wyoming. US Geol Survey Prof Paper 1055:1–22

Peterman ZE, Hedge CE, Coleman RG, Snavely PD Jr (1967) $^{87}Sr/^{86}Sr$ ratios in some eugeosynclincal sedimentary rocks and their bearing on the origin of granitic magma in orogenic belts. Earth Planet Sci Letters 2:433–435

Peterman ZE, Lowder GG, Carmichael ISE (1970a) $^{87}Sr/^{86}Sr$ ratios of the Talasea Series, New Britain, Territory of New Guinea. Geol Soc Amer Bull 81:39–40

Peterman ZE, Carmichael ISE, Smith AL (1970b) $^{87}Sr/^{86}Sr$ ratios of Quaternary lavas of the Cascade Range, northern California. Geol Soc Amer Bull 81:311–318

Peterman ZE, Doe BR, Prostka HJ (1970c). Lead and strontium isotopes in rocks of the Absaroka volcanic field, Wyoming. Contrib Mineral Petrol 27:121–130

Peterman ZE, Carmichael ISE, Smith AL (1970d). Strontium isotopes in Quaternary basalts of southeastern California. Earth Planet Sci Letters 7:381–384

Peterman ZE, Goldich SS, Hedge CE, Yardley DH (1972) Geochronology of the Rainy Lake region, Minnesota-Ontario. Geol Soc Amer Mem 135:193–215

Peters M, Haverkamp B, Emmermann R, Kohnen H, Weber K (1991) Palaeomagnetism, K-Ar dating and geodynamic setting of igneous rocks in western and central Neuschwabenland, Antarctica. In: Thomson MRA et al. (eds) Geological evolution of Antarctica. Cambridge Un. Press, pp 549–555

Petersen TD (1990) Petrology and genesis of natrocarbonatite. Contrib Mineral Petrol 105:143–155

Petrini R, Civetta L, Piccirillo EM, Bellieni G, Comin-Chiaramonti P, Marques LS, Melfi AJ (1987) Mantle heterogeneity and crustal contamination in the genesis of low-Ti continental flood basalts from the Paraná Plateau (Brazil):Sr-Nd isotope and geochemical evidence. J Petrol 28:701–726

Petrini R, Joron JL, Ottonello G, Bonatii E, Seyler M (1988) Basaltic dykes from Zabargard Island, Red Sea: Petrology and geochemistry. Tectonophysics 150:229–248

Philippi E (1912) Geologische Beschreibung des Gaussbergs. Deutsche Südpolar-Expedition 1901–1903. Geographie und Geologie 11(1):47–71

Philpotts AR, Reichenbach I (1985) Differentiation of Mesozoic basalts of the Hartford basin, Connecticut. Geol Soc Amer Bull 96:1131–1139

Philpotts JA, Schnetzler CC (1970) Phenocryst-matrix partition coefficients for K, Rb, Sr and Ba, with applications to anorthosite and basalt genesis. Geochim Cosmochim Acta 34:307–322

Philpotts JA (1978) The law of constant rejection. Geochim Cosmochim Acta 42:909–920

Piccirillo EM, Melfi AJ (eds) (1988) The Mesozoic flood volcanism of the Paraná Basin: Petrogenetic and geophysical aspects. Inst. Astron. Geofis., Univ. of São Paulo, Brazil

Piccirillo EM, Raposo MIB, Melfi AJ, Comin-Chiaramonti P, Bellieni G, Cordani UG, Kawashita K (1987) Bimodal fissural volcanic suites from the Paraná Basin (Brazil): K-Ar age, Sr-isotopes, and geochemistry. Geochimica Brasiliensis 1(1):53–69

Piccirillo EM, Civetta L, Petrini R, Longinelli A, Bellieni G, Comin-Chiaramonti P, Marques LS, Melfi AJ (1989) Regional variations with the Paraná flood basalts (southern Brazil): Evidence for subcontinental mantle heterogeneity and crustal contamination. Chem Geol 75(1/2):103–122

Piccirillo EM, Bellieni G, Gavazzini G, Comin-Chiaramonti P, Petrini R, Melfi AJ, Pinese JPP, Zantadeschi P, De Min A (1990) Lower Cretaceous tholeiite dyke swarms from the Ponta Grossa Arch (southeast Brazil): Sr-Nd isotopes and genetic relationships with the Paraná flood volcanics. Chem Geol 89(1/2):19–48

Pickett DA, Wasserburg GJ (1989) Neodymium and strontium isotope characteristics of New Zealand granitoids and related rocks. Contrib Mineral Petrol 103:131–142

Pidgeon RT, Compston W (1965) The age and origin of the Cooma granite and its associated metamorphic zones, New South Wales. J Petrol 6:193–222

Pidgeon RT, Hopgood AM (1975) Geochronology of Archean gneisses and tonalities from north of the Frederikshåbs isblink, S.W. Greenland. Geochim Cosmochim Acta 39: 1333–1346

Pidgeon RT, Wilde SA (1990) The distribution of 3.0 Ga and 2.7 Ga volcanic episodes in the Yilgarn craton of Western Australia. Precamb Res 48:309–325

Pier JG, Podosek FA, Luhr JF, Brannon JC, Aranda-Gomez JJ (1989) Spinel-lherzolite-bearing Quaternary volcanic centers in San Luis Potosí, Mexico, 2. Sr and Nd isotopic systematics. J Geophys Res 94(B):7941–7951

Pik R, Deniel C, Coulon C, Yirgu G, Marty B (1999) Isotopic and trace-element signatures of Ethiopian flood basalts: Evidence for plume-lithosphere interactions. Geochim Cosmochim Acta 63:2263–2279

Pilger A, Rösler A (eds) (1975) Afar Depression of Ethiopia. Schweizerbart, Stuttgart, pp 296–299

Pin C, Briot D, Bassin C, Poitrasson F (1994) Concomitant separation of strontium and samarium-neodymium for isotopic analysis in silicate samples, based on specific extraction chromatography. Anal Chim Acta 298:209–217

Pitcher WS (1978) The anatomy of a batholith. J Geol Soc London 135:157–182

Pitcher WS, Bussell MA (1977) Structural control of batholith emplacement in Peru: A review. J Geol Soc London 133:249–256

Pitcher WS, Atherton MP, Cobbing EJ, Beckinsale RD (eds) (1985) Magmatism at a plate edge: The Peruvian Andes. Blackie Halsted Press, Glasgow, Scotland

Platt RG, Mitchell RH (1982) Rb-Sr geochronology of the Coldwell Complex, northwestern Ontario, Canada. Can J Earth Sci 19: 1796–1801

Pokrovskiy BG, Vinogradov VI (1991) Isotope investigations on alkalic rocks of central and western Siberia. Internat Geol Rev 33(2):122–134

Pokrovskiy BG, Zhuravlev DS (1991) New data on isotope geochemistry of Kuril island-arc effusives. Geochem Internat 28(10):101–105

Poli S, Chiesa S, Gillot P-Y, Guichard F, Vezzoli L (1989) Time dimension in the geochemical approach and hazard estimates of a volcanic area: The Isle of Ischia case (Italy). J Volcanol Geotherm Res 36:327–335

Powell JL (1965a) Low abundance of ^{87}Sr in Ontario carbonatites. Amer Mineral 50:1075–1079

Powell JL (1965b) Isotopic composition of strontium in four carbonate vein-dikes. Amer Mineral, 50:1921–1928

Powell JL (1966) Isotopic composition of strontium in carbonatites and kimberlites. Mineral. Soc. India, Internat. Mineral. Assoc. volume 58–66

Powell JL (1998) Night comes to the Cretaceous: Dinosaur extinction and the transformation of modern geology. W.H. Freeman, New York

Powell JL, Bell K (1970) Strontium isotopic studies of alkalic rocks: Localities from Australia, Spain, and the Western United States. Contrib Mineral Petrol 27:1–10

Powell JL, Bell K (1974) Isotopic composition of strontium in alkalic rocks. In: Sørensen H (ed) The alkaline rocks. Wiley, London, pp 412–421

Powell JL, DeLong SE (1966) Isotopic composition of strontium in volcanic rocks from Oahu. Science 153:1239–1242

Powell JL, Hurley PM, Fairbairn HW (1962) Isotopic composition of strontium in carbonatites. Nature 196:1085–1086

Powell JL, Faure G, Hurley PM (1965a) Strontium-87 abundance in a suite of Hawaiian rocks of varying silica content. J Geophys Res 70(6):1509–1513

Powell JL, Hurley PM, Fairbairn HW (1965b) The strontium isotopic composition and origin of carbonatites. In: Tuttle OF, Gittins J (eds) Carbonatites. Wiley InterScience, New York, pp 365–378

Powell JL, Skinner WR, Walker D (1969) Whole-rock Rb-Sr age of metasedimentary rocks below the Stillwater complex, Montana. Geol Soc Amer Bull 80:1605–1612

Prestvik T, Duncan RA (1991) The geology and age of Peter I Öy, Antarctica. Polar Res 9:89–98

Prestvik T, Barnes CG, Sundvoll B, Duncan RA (1990) Petrology of Peter I Öy (Peter I Island), West Antarctica. J Volcanol Geotherm Res 44:315–338

Prestvik T, Torske T, Sundvoll B, Karlsson H (1999) Petrology of Early Tertiary nephelinites off mid-Norway: additional evidence for an enriched endmember of the ancestral Iceland plume. Lithos 46(2):317–330

Price RC, Compston W (1973) The geochemistry of the Dunedin Volcano: Strontium isotope chemistry. Contrib Mineral Petrol 42:55–61

Price RC, Johnson RW, Gray CM, Frey FA (1985) Geochemistry of phonolites and trachytes from the summit region of Mt. Kenya. Contrib Mineral Petrol 89:394–409

Price RC, Kennedy AK, Riggs-Sneeringer M, Frey FA (1986) Geochemistry of basalts from the Indian Ocean triple junction: Implications for the generation and evolution of Indian Ocean ridge basalts. Earth Planet Sci Letters 78:379–396

Price RC, MucCulloch MT, Smith IEM, Stewart RB (1992) Pb-Nd-Sr isotopic compositions and trace element characteristics of young volcanic rocks from Egmont Volcano and comparisons with basalts and andesites from the Taupo volcanic zones, New Zealand. Geochim Cosmochim Acta 56:941–953

Pride DE, Hagner AF (1972) Geochemistry and origin of the Precambrian iron formation near Atlantic City, Fremont County, Wyoming. Econ Geol 67:329–338

Priem HNA, Boelrijk NAIM, Hebeda EH, Verdurmen EAT, Verschure RH, Oen IS, Westra L (1979) Isotope age determinations on granitic and gneissic rocks from the Ubendian-Usagaran System in southern Tanzania. Precamb Res 135:217–224

Prince LA, Hanson GN (1972) Rb-Sr isochron ages for the Giants Range granite, northeastern Minnesota. Geol Soc Amer Mem 135:217–224

Pringle M, Sager W, Sliter W (eds) (1993) The Mesozoic Pacific: Geology, tectonics and volcanism. Amer. Geophys. Union, Washington, D.C. (Geophys. Monograph 77)

Prior GT (1907) Report on the rock-specimens collected during the "Discovery" Antarctic Expedition 1901–1904. London National Antarctic Exped., Natural History, Geol 1:101–140

Provost A (1990) An improved diagram for isochron data. Chem Geol (Isotope Geoscience Section) 80:85–99

Puffer JH, Ragland PC (eds) (1992) Estern North American Mesozoic magmatism. Geol Soc Amer Spec Paper 268

Purdy JW, York D (1966) A geochronometric study of the Superior Province near Red Lake northwestern Ontario. Can J Earth Sci 3:277–286

Purdy JW, York D (1968) Rb-Sr whole-rock and K-Ar mineral ages of rocks from the Superior Province near Kirkland Lake, northeastern Ontario, Canada. Can J Earth Sci 5:699–705

Pushkar P (1968) Strontium isotopes in volcanic rocks of three island arc areas. J Geophys Res 73:2701–2714

Pushkar P, Condie KC (1973) Origin of the Quaternary basalts from the Black Rock Desert region, Utah: Strontium isotopic evidence. Geol Soc Amer Bull 84:1053–1058

Pushkar P, Stoeser DB (1975) ^{87}Sr/^{86}Sr ratios in some volcanic rocks and some semifused inclusions of the San Francisco volcanic field. Geol 6:669–671

Pushkar P, McBirney AR, Kudo AM (1972) The isotopic composition of strontium in Central American ignimbrites. Bull Volcanol 35(2):265–294

Pushkar P, Stueber AM, Tomblin JF, Julian GM (1973) Strontium isotopic ratios in volcanic rocks from St. Vincent and St. Lucia, Lesser Antilles. J Geophys Res 78(8):1279–1287

Pyatenko IK, Saprykina LG (1976) Carbonatite lavas and pyroclastics in the Paleozoic sedimentary volcanic sequence of the Kontozero District, Kola Peninsula. Dokl Akad Nauk SSSR, pp 185–187

Pyle DG, Christie DM, Mahoney JJ, Duncan RA (1995) Geochemistry and geochronology of ancient southeast Indian and southwest Pacific seafloor. J Geophys Res 100:22261–22282

Pyle DM, Pinkerton GE, Norton GE, Dawson JB (1995) The dynamics of degassing of Oldoinyo Lengai. In: Bell K, Keller J (eds) Carbonatite Volcanism; Oldoinyo Lengai and the Petrogenesis of Natrocarbonatites. IAVCEI Proceedings in Volcanology, Springer-Verlag, Berlin (vol 4, pp 37–46)

Quilty PG, Shafik S, McMinn A, Brady H, Clarke I (1983) Microfossil evidence for the age and environment of deposition of sediment of Heard and McDonald islands. In: Oliver RL, James PR, Jago,JB (eds) Antarctic Earth Science. Australian Academy of Science Canberra ACT, pp 636–639

Radain AA, Ali S, Nassef AD, Abdel-Monem AA (1987) Rb-Sr geochronology and geochemistry of plutonic rocks from Wadi Shuqub quadrangle, west-central Arabian Shield. J Afr Earth Sci 6:553–568

Rahaman MA, Bennet JB, Breemen O van, Bowden P (1984) Detailed age study cf the migration of younger granite ring complexes in rorthern Nigeria. J Geol 92:173–184

Rajamani V, Shivkumar K, Hanson GN, Shirey SB (1985) Geochemistry and petrogene is of amphibolites, Kolar schist belt, south India: Evidence for komatiitic magma derived by low percentages of melting o the mantle. J Petrol 26:99–123

Rampino MR, Stothers RB (1988) Flood basalt volcanism during the past 250 million years. Science 241:663–668

Ray GL, Shimizu N, Hart SR (1983) An ion microprobe study of the partitioning of trace elements between clinopyroxene and liquid in the system diopside-albite-anorthite. Geochim Cosmochim Acta 47:2131–2140

Ray JS (1998) Trace element and isotope evolution during concurrent assimilation, fractional crystallization, and liquid immiscibility of a carbonated silicate magma. Geochim Cosmochim Acta 62:3301–3306

Read HH (1949) A contemplation of time in plutonism. Quart J Geol Soc London 105:101–156

Reddy VV, Subbarao KV, Reddy GR, Matsuda J, Hekinian R (1978) Geochemistry of volcanics from the Ninetyeast Ridge and its vicinity in the Indian Ocean. Marine Geol 26:99–117

Reed JC Jr, Zartman RE (1973) Geochronology of Precambrian rocks of the Teton Range, Wyoming. Geol Soc Amer Bull 84: 561–582

Reid AM, Donaldson C, Brown RW, Ridley WI, Dawson JB (1975) Mineral chemistry of peridotite inclusions from Lashaine volcano, Tanzania. Phys Chem Earth 9:525

Reid AM, LeRoex AP (1988) Kaersutite-bearing xenoliths and megacrysts in volcanic rocks from the Funk Seamount in the southwest Indian Ocean. Mineral Mag 52:359–370

Reinitz IM, Turekian EK, Har BJ (1991) The behavior of uranium decay chain nuclides and thorium during the flank eruptions of Kilauea (Hawaii) between 1983 and 1985. Geochim Cosmochim Acta 55:3735–3740

Renne PR, Basu AR (1991) Rapid eruption of the Siberian traps flood basalts at the Permo-Triassic boundary. Science 253: 176–179

Renne PR, Glen JM, Milner SC, Duncan AR (1996) Age of the Etendeka flood volcanism and associated intrusions in southwestern Africa. Geol 24:659–662

Rex DC (1967) Age of a dolerite from Dronning Maud Land. British Antarctic Surv Bull 11.101

Rex DC (1972) Potassium-argon age determinations on volcanic and associated rocks from the Antarctic Peninsula and Dronning Maud Land. In: Adie RJ (ed) Antarctic Geology and Geophysics. Universitetsforlaget Oslo

Rex DC, Gledhill AR, Bridgwater D, Myers JS (1979) A Rb-Sr whole rock age of 55±7 m.y. from the Nualik plutonic centre, East Greenland. Rapp. Grønlands Geol Unders 95:102–105

Rheinisch R (1912) Petrographische Beschreibung der Gaussberg-Gesteine. Deutsche Südpolar Expedition 1901–1903. Geographie and Geologie 11(1):73–87

Rhodes JM, Dawson JB (1975) Major and trace element chemistry of peridotite inclusions from the Lashaine volcano, Tanzania. Phys Chem Earth 9:545–557

Rhodes JM, Wenz KP, Neal CA, Sparks JW, Lockwood JP (1989) Geochemical evidence for invasion of Kilauea's plumbing system by Mauna Loa magma. Nature 337:257–260

Rhodes RC (1975) New evidence for impact origin of the Bushveld Complex, South Africa. Geol 3:549–554

Richards JP, Spooner ETC (1989) Evidence for Cu-(Ag) mineralization by magmatic-meteoric fluid mixing in Keweenawan fissure veins, Mamainse Point, Ontario. Econ Geol 84:360–385

Richards JP, McCulloch MT, Chappell BW, Kerrich R (1991) Sources of metals in the Porgera gold deposit, Papua New Guinea: Evidence from alteration, isotope and noble metal geochemistry. Geochim Cosmochim Acta 55(2):565–580

Richards MA, Duncan RA, Courtillot VE (1989) Flood basalts and hot-spot tracks: Plume heads and tails. Science 246:103–106

Richardson SH (1979) Chemical variation induced by flow differentiation in an extensive Karroo dolerite sheet, southern Namibia. Geochim Cosmochim Acta 43:1433–1441

Richardson SH, Erlank AJ, Duncan AR, Reid DL (1982) Correlated Nd, Sr and Pb isotope variation in the Walvis Ridge basalts and implications for the evolution of their mantle source. Earth Planet Sci Letters 59:327–342

Richardson SH, Erlank AJ, Reid DL, Duncan AR (1984) Major and trace elements and Sr and Nd isotope geochemistry of basalts from the Deep Sea Drilling Project Leg 74 Walvis Ridge transect. Init Repts DSDP 74:739–754

Richey JE (1932) Tertiary ring structures in Great Britain. Trans Geol Soc Glasgow 19:42–140

Ridley WI (1970) The petrology of the Las Canadas volcanoes, Tenerife, Canary Islands. Contrib Mineral and Petrol 26: 124–160

Ridley WI, Dawson JB (1975) Lithophile and trace element data bearing on the origin of peridotite xenoliths, ankaramite and carbonatite from Lashaine volcano, North Tanzania. Phys Chem Earth 9:559

Ries AC, Shackleton RM, Dawoud AS (1985) Geochronology, geochemistry, and tectonics of NE Bayuda Desert, N. Sudan: Implications for the western margin of the Late Proterozoic fold belt of NE Africa. Precamb Res 30 43–62

Ringwood AE (1955) The principles governing trace element distribution during magmatic crystallization. Geochim Cosmochim Acta 7:189–202

Ripley EM, Lambert DD, Frick LR (1998) Re-Os, Sm-Nd, and Pb isotopic constraints on mantle and crustal contributions to magmatic sulfide mineralization in the Duluth Complex. Geochim Cosmochim Acta 62:3349–3366

Roberts SJ, Ruiz J (1989) Geochemistry of exposed granulite facies terrains and lower crustal xenoliths in Mexico. J Geophys Res 94(B6):7961–7974

Robertson IDM, Breemen O van (1970) The southern satellite dykes of the Great Dyke, Rhodesia. Geol Soc S Africa Spec Pub 1:621–644

Robinson GR, Froelich AJ (eds) (1985) Proceedings of the second U.S. Geological Survey workshop on the early Mesozoic basins of the eastern United States. US Geol Survey Circular 946: 1–147

Rocaboy A, Cantagrel JM, Vidal V (1987) Sr, Nd, Pb isotopic and trace element composition of the Gambier Islands (French Polynesia). Eos 68:438

Rocholl A, Stein M, Molzahn M, Hart SR, Wörner G (1995) Geochemical evolution of rift magmas by progressive tapping of a stratified mantle source beneath the Ross Sea Rift, northern Victoria Land, Antarctica. Earth Planet Sci Letters 131:207–224

Rock NMS (1976) The comparative strontium isotopic composition of alkaline rocks: New data from southern Portugal and East Africa. Contrib Mineral Petrol 56:205–228

Rock NMS (1986) The nature and origin of ultramafic lamprophyres: Alnöites and allied rocks. J Petrol 27:155–196

Rock NMS (1987) The nature and origin of lamprophyres: An overview. In: Fitton JG, Upton BGJ (eds) Alkaline igneous rocks. Geol Soc London Spec Pub 30:191–216

Rock NMS (1991) Lamprophyres. Van Nostrand Reinhold, New York

Rock NMS, Duffy TR (1986) REGRES: A Fortran-77 program to calculate nonparametric and "structural" parametric solutions to bivariate regression equations. Computers Geosci 12: 807–818

Rock NMS, Webb JA, McNaughton NJ, Bell GD (1987) Nonparametric estimation of averages and errors for small data-sets in geoscience: a proposal. Chem Geo (Isotope Geoscience Section) 66:163–177

Rodda P (1967) Outline of the geology of Viti Levu. New Zealand J Geol Geophys 10:1260–1273

Roddick JC (1978) The application of isochron diagrams in ^{40}Ar-^{39}Ar dating: A discussion. Earth Planet Sci Letters 41: 233–244

Roddick JC (1987) Generalized numerical error analysis with applications to geochronology and thermodynamics. Geochim Cosmochim Acta 51:2129–2135

Roddick JC, Compston W (1977) Strontium isotopic equilibration: A solution to a paradox. Earth Planet Sci Letters 34:238–246

Roden MF (1981) Origin of coexisting minette and ultramafic breccia, Navajo volcanic field. Contrib Mineral Petrol 77: 195–206

Roden MF, Hart SR, Frey FA, Melson WG (1984a) Sr, Nd, and Pb isotopic and REE geochemistry of St. Paul's Rocks; the metamorphic and metasomatic development of an alkali basalt mantle source. Contrib Mineral Petrol 85:376–390

Roden MF, Frey FA, Clague DA (1984b) Geochemistry of tholeiitic and alkalic lavas from the Koolau Range, Oahu: implications for Hawaiian volcanism. Earth Planet Sci Letters 69:141–158

Roden MF, Trull T, Hart SR, Frey FA (1994) New He, Nd, Pb and Sr isotopic constraints on the constitution of the Hawaiian plume: results from the Koolau volcano, Oahu, Hawaii, USA. Geochim Cosmochim Acta 58:1431–1440

Rogers G, Hawkesworth CJ (1989) A geochemical traverse across the North Chilean Andes: evidence for crustal generation from the mantle wedge. Earth Planet Sci Letters 91(3/4):271–285

Rogers NW, Setterfield TN (1994) Potassium and incompatible-element enrichment in shoshonitic lavas from the Tavua volcano, Fiji. Chem Geol 118:43–62

Rogers NW, Bachinski SW, Henderson P, Parry SJ (1982) Origin of potash-rich basic lamprophyres: trace element data from Arizona minettes. Earth Planet Sci Letters 57:305–312

Rogers NW, Parkes RJ, Hawkesworth CJ, Marsh JS (1985) The geochemistry of potassic lavas from Vulsini, Central Italy, and implications for mantle enrichment processes beneath the Roman region. Contrib Mineral Petrol 90:244–257

Ronov AB, Yaroshevsky AA (1969) Chemical composition of the Earth crust. In: Hart PJ (ed) The Earth's Crust and Upper Mantle. Amer. Geophys. Union, Washington D.C. (Monograph 13, pp 37–57)

Ronov AB, Yaroshevsky AA (1976) A new model for the chemical structure of the Earth's crust. Geochem Internat 13(6):89–121

Roobol MJ, Ramsay CR, Jackson NJ, Darbyshire DPF (1983) Late Proterozoic lavas of the central Arabian shield-evolution of an ancient volcanic arc system. J Geol Soc London 140: 185–202

Rose WI Jr, Grant NK, Hahn GA, Lange IM, Powell JL, Easter J, Degraff JM (1977) The evolution of Santa Maria volcano, Guatemala. J Geol 85:63–87

Rose WI Jr, Grant NK, Easter J (1979) Geochemistry of Los Chocoyos ash, Quezaltenango Valley, Guatemala. Geol Soc Amer Spec Paper 180:87–99

Rosing MT, Lesher CE, Bird DK (1989) Chemical modification of East Greenland Tertiary magmas by two-liquid interdiffusion. Geol 17:626–629

Rossman GR, Weis D, Wasserburg GJ (1987) Rb, Sr, Nd and Sm concentrations in quartz. Geochim Cosmochim Acta 51(9): 2325–2330

Rudnick RL, Fountain DM (1995) Nature and composition of the continental crust: A lower crustal perspective. Rev Geophys 33(3):267–309

Ruiz J, Jones LM, Kelly WC (1984) Rubidium-strontium dating of ore deposits hosted by Rb-rich rocks, using calcite and other common Sr-bearing minerals. Geol 12:259–262

Ruiz J, Patchett PJ, Ortega-Guitierrez F (1988) Proterozoic and Phanerozoic basement terranes of Mexico from Nd isotopic studies. Geol Soc Amer Bull 100:274–281

Rutherford E, Soddy F (1902) The radioactivity of thorium compounds. II. The cause and nature of radioactivity. J Chem Soc London 81:837–860

Ryabchikov ID, Boettcher AL (1980) Experimental evidence at high pressures for potassic metasomatism in the mantle of Earth. Amer Mineral 65:915–919

Ryabchikov ID, Hamilton DL (1994) Near-solidus melts in carbonatized mantle peridotites in the presence of apatite and uraninite. Geochem Internat 31:77–85

Sabine PA (1960) The geology of Rockall, North Atlantic. Bull Geol Survey of Great Britain 16:1–156

Saemundsson K, Noll H (1974) K/Ar ages of rocks from Husafell, western Iceland, and the development of the Husafell central volcano. Jökull 24:40–59

Saemundsson K, Kristjansson L, McDougall I, Watkins ND (1980) K-Ar dating, geological and paleomagnetic study of a 5 km lava succession in northern Iceland. J Geophys Res 85:3628–3646

Saether E (1957) The alkaline rock province of the Fen area in southern Norway. Det. Kgl. Norske Vidensk. Selsk. Skr. No. 1, Trondheim

Salters VJM (1996) The generation of mid-ocean ridge basalts from the Hf and Nd isotope perspective. Earth Planet Sci Letters 141:109–123

Salters VJM, Hart SR, Pantó G (1988) Origin of late Cenozoic volcanic rocks of the Carpathian arc, Hungary. In: Royden L, Horvath F (eds) The Pannonian Basin: A Study of Basin Evolution. Amer. Assoc. Petrol. Geologists, Tulsa, OK (Memoir 45, pp 279–292)

Salters VJM, Storey M, Sevigny JH, Whitechurch H (1992) Trace element and isotopic characteristics of Kerguelen-Heard Plateau basalts. Proc Ocean Drilling Prog Sci Results 120:55–62

Saunders AD, Norry MJ (eds) (1989) Magmatism in the ocean basins. Blackwell Scient. Pub., Oxford, England (Geological Soc Spec Pub 42)

Saunders AD, Tarney JT (1979) The geochemistry of basalts from a back-arc spreading center in the East Scotia Sea. Geochim Cosmochim Acta 43:555–572

Saunders AD, Storey M, Gibson IL, Leat P, Hergt J, Thompson RN (1991) Chemical and isotopic constraints on the origin of basalts from the Ninetyeast Ridge, Indian Ocean: Results from DSDP Legs 22 and 26 and ODP Leg 121. Proc Ocean Drilling Prog Sci Results 121:559–590

Schaaf P, Heinrich W, Besch T (1994) Composition and Sm-Nd isotopic data of the lower crust beneath San Luis Potosí, central Mexico: Evidence from a granulite-facies xenolith suite. Chem Geol 118:63–84

Schiano P, Clocchiatti R, Joron JL (1992) Melt and fluid inclusions in basalts and xenoliths from Tahaa Island, Society archipelago: Evidence for metasomatized upper mantle. Earth Planet Sci Letters 111:69–82

Schiano P, Clocchiatti R, Shimizu N, Maury R, Jochum KP, Hofmann AW (1995) Hydrous silica-rich melts in the subarc mantle and their relationship with erupted lavas. Nature 377: 595–600

Schilling J-G (1973a) Iceland mantle plumes: Geochemical study of the Reykjanes Ridge. Nature 242:565

Schilling J-G (1973b) Afar mantle plume: rare earth evidence. Nature (Phys Sci) 242:2–5

Schilling J-G (1991) Fluxes and excess temperatures of mantle plumes inferred from their interaction with migrating mid-ocean ridges. Nature 352:397–403

Schilling J-G, Noe-Nygaard A (1974) Faeroe-Iceland plume: rare-earth evidence. Earth Planet Sci Letters 24:1–14

Schilling J-G, Meyer PS, Kingsley RH (1982) Evolution of the Iceland hotspot. Nature 296:313–320

Schilling J-G, Zajac M, Evans R, Johnston T, White W, Devine JD, Kingsley R (1983) Petrologic and geochemical variations along the Mid-Atlantic Ridge from 27° N to 73° N. Amer J Sci 283: 510–586

Schilling J-G, Thompson G, Kingsley R, Humphris S (1985) Hotspot-migrating ridge interaction in the South Atlantic. Nature 313:187–191

Schilling J-G, Kingsley RH, Hanan BB, McCully BL (1992) Nd-Sr-Pb isotopic variations along the Gulf of Aden: evidence for mantle plume-continental lithosphere interaction. J Geophys Res 97:10927–10966

Schilling J-G, Hanan BB, McCully B, Kingsley RH, Fontignie D (1994) Influence of the Sierra Leone mantle plume on the equatorial Mid-Atlantic Ridge: A Nd-Sr-Pb isotopic study. J Geophys Res 99:12005–12028

Schiøtte L, Compston W, Bridgwater D (1988) Late Archaean ages for the deposition of clastic sediments belonging to the Malene supracrustals, southern West Greenland: Evidence from an ion probe U-Pb zircon study. Earth Planet Sci Letters 87:45-58

Schleicher H, Lippolt HJ, Raczek I (1983) Rb-Sr systematics of Permian volcanites in the Schwarzwald (SW Germany). Part II. Age of eruption and the mechanism of Rb-Sr whole rock age distortions. Contrib Mineral Petrol 84:281-291

Schleicher H, Keller J, Grauer B, Baumann A, Kramm U (1984) Sr Isotopenuntersuchungen an Alkalivulkaniten des Kaiserstuhls. Fortschritte der Mineral 62(1):207-209

Schmidt DL, Rowley PR (1986) Continental rifting and transform faulting along the Jurassic Transantarctic Rift, Antarctica. Tectonics 5(2):279-291

Schmincke H-U (1967) Stratigraphy and petrography of four Yakima basalt flows in south-central Washington. Geol Soc Amer Bull 78:1385-1422

Schmincke H-U (1973) Magmatic evolution and tectonic regime in the Canary, Madeira, and Azores island groups. Geol Soc Amer Bull 84:633-648

Schmincke H-U (1982) Volcanic and chemical evolution of the Canary Islands. In: Rad U von, Hinz K, Sarnthein M, Seibold E (eds) Geology of the Northwest African Continental Margin Springer Verlag, Berlin, pp 273-306

Schmincke H-U (1987) Field guide to Gran Canaria. Pluto Press, Wissen

Schmus WR Van (1965) The geochronology of the Blind River-Bruce Mines area, Ontario, Canada. J Geol 73:755-580

Schmus WR Van, Wetheril GW, Bickford ME (1963) Rb-Sr age determinations of the Nipissing Diabase, north shore of Lake Huron, Ontario, Canada. J Geophys Res 69(9): 5589-5593

Schmus WR Van (1971) Rb-Sr age of Middle Keweenawan rocks, Mamainse Point and vicinity, Ontario, Canada. Geol Soc Amer Bul 82:3221-3226

Schmus WR Van, Hinze WJ (1985) The midcontinent rift system. Ann Rev Earth Planet Sci 13:345-384

Schmus WR Van, Bickford ME, Zietz I (1987) Early and Middle Proterozoic provinces in the central United States. In: Kröner A (ed), Proterozoic Lithospheric Evolution. Amer Geophys Union, Washington D.C. (Geodyn Ser 1:43-68)

Schnetzler CC, Philpotts JA (1970) Partition coefficients of REE between igneous matrix material and rock-forming phenocrysts-II. Geochim Cosmochim Acta 34:331-340

Schreiner GDL (1958) Comparison of the ^{87}Rb → ^{87}Sr ages of the red granite of the Bushveld complex from measurements on the total rock and separated mineral fractions. Proc Royal Soc 245A: 112-117

Schubert G, Sandwell D (1989) Crustal volumes of the continents and of oceanic and cont nental submarine plateaus. Earth Planet Sci Letters 92:234-246

Schultz JL, Boles JR, Tilton GR (1989) Tracking calcium in the San Joaquin basin, California: A strontium isotopic study of carbonate cements at North Coles Levee. Geochim Cosmochim Acta 53(8):1991-2003

Schwartz GM, Sandberg AE (1940) Rock series in diabase sill at Duluth, Minnesota Geol Soc Amer Bull 51:1135-1172

Schwartzman DW, Giletti BJ (1968) "Excess" argon in minerals from the Stillwater complex, Montana. (Abstract). Amer Geophys Union Trans 49:359

Scott RB, Nesbitt RW, Dasch EJ, Armstrong RL (1971) A strontium isotope evolution model for Cenozoic magma genesis, eastern Great Basin, USA. Bull Volcanol 35(1):1-26

Scott-Smith BH, Skinner EMW (1984) Diamondiferous lamproites. J Geol 92:433-438

Scrutton RA (1973) The age relationship of igneous activity and continental break-up. Geol Mag 110(3):227-234

Segalstad TV (1975) Petrology of the Skien basaltic rocks, southwestern Oslo region, Norway. Lithos 12:221-239

Seki T (1978) Rb-Sr geochronology and petrogenesis of the late Mesozoic igneous rocks in the Inner Zone of the southwestern part of Japan. Memoirs Faculty of Science Kyoto University Ser Geol and Minera 45(1):71-110

Seki T (1981) Rb-Sr isochron age of the Mikuni-San rhyolites, Okayama, and geochronology of the Cretaceous volcanic activity in southwest Japan. J Geol Soc Japan 87(8):535-542

Seidemann DE, Masterson WD, Dowling MP, Turekian KK (1984) K-Ar dates and ^{40}Ar/^{39}Ar age spectra for Mesozoic basalt flows of the Hartford Basin, Connecticut and the Newark Basin, New Jersey. Geol Soc Amer Bull 95:594-598

Seidemann DE (1988) The hydrothermal addition of excess ^{40}Ar to the lava flows from the Early Jurassic in the Hartford Basin (northeastern USA): Implications for the time scale. Chem Geol (Isotope Geoscience Section) 8:37-46

Semenov YI (1994) Minerals and ores of the Khibiny-Lovozero alkali massif. Geochem Internat 31(3):160-165

Sempéré J-C, Palmer J, Christie DM, Morgan JP, Shor AN (1991) Australian-Antarctic discordance. Geol (19(5):429-432

Sen G (1983) Deccan Trap intrusion: magma mixing in the Chakhla-Delakhari sill, Chindwara District, Madhya Pradesh. J Geol Soc India 24:381

Sen G, Frey FA, Shimizu N, Leeman WP (1993) Evolution of the lithosphere beneath Oahu, Hawaii: Rare earth element abundances in mantle xenoliths. Earth Planet Sci Letters 119:53-69

Serencsits CMcC, Faul H, Foland KA, Hussein AA, Lutz TM (1981) Alkaline ring complexes in Egypt: their ages and relationship in time. J Geophys Res 86(B4):3009-3013

Sethna SF, Czygan W, Sethna BS (1987) Iron-titanium oxide geothermometry for some Deccan Trap tholeiitic basalts, India. J Geol Soc India 29:483-438

Shackleton RM, Ries AC, Coward MP, Cobbold PR (1979) Structure, metamorphism and geochronology of the Arequipa Massif and coastal Peru. J Geol Soc London 136:195-214

Shafiqullah M, Tupper WM, Cole TSJ (1970) K-Ar age of the carbonatite complex, Oka, Quebec. Canadian Mineralogist 10:541-552

Sharma M, Basu AR, Nesterenko GV (1991) Nd-Sr isotopes, petrochemistry, and origin of the Siberian flood basalt, USSR. Geochim Cosmochim Acta 55(4):1183-1192

Sharma M, Basu AR, Nesterenko GV (1992) Temporal Sr-, Nd-, and Pb-isotopic variations in the Siberian flood basalts: Implications for the plume-source characteristics. Earth Planet Sci Letters 113:365-381

Sharpe MR (1985) Strontium isotope evidence for preserved density stratification in the Main Zone of the Bushveld Complex, South Africa. Nature 316:119-126

Sharpton VL, Dalrymple GB, Marin LE, Ryder G, Schuraytz BC, Urrutia-Fucugauchi J (1992) New links between the Chicxulub impact structure and the Cretaceous/Tertiary boundary. Nature 359:819-821

Shaw DM (1953) The camouflage principle and trace element distribution in magmatic minerals. J Geol 61:142-151

Shaw DM (1970) Trace element fractionation during anatexis. Geochim Cosmochim Acta 34:237-242

Shaw DM (1976) Development of the early continental crust. Part 2. Prearchaean, Protoarchaean and later eras. In: Windley BF (ed) The early history of the Earth. Wiley, London, pp 33-58

Shaw DM, Reilly GA, Muysson JR, Pattenden GE, Campbell FE (1967) The chemical composition of the Canadian Precambrian shield. Can J Earth Sci 4:828-854

Shaw DM, Cramer JJ, Higgines MD, Truscott MG (1986) Composition of the Canadian Precambrian shield and the continental crust of the Earth. Geol Soc London Spec Publ 24:275-282

Sheraton JW, Cundari A (1980) Leucitites from Gaussberg, Antarctica. Contrib Mineral Petrol 71:417-427

Sheridan MF, Stuckless JS, Fodor RV (1970) A Tertiary silicic cauldron complex at the northern margin of the Basin and Range province, central Arizona. USA. Bull Volcanol 34:1-14

Shibata K (1974) Rb-Sr geochronology of the Hikami granite, Kitakami mountains, Japan. Geochem J 8(4):193

Shibata K, Adachi M (1974) Rb-Sr whole-rock ages of Precambrian metamorphic rocks in the Kamiaso conglomerate from central Japan. Earth Planet Sci Letters 21:277-287

Shibata K, Ishihara S (1979) Initial ^{87}Sr/^{86}Sr ratios of plutonic rocks from Japan. Contrib Mineral Petrol 70:381-390

Shibata K, Nozawa T (1986) Late Precambrian ages for granitic rocks intruding the Hida metamorphic rocks. Bull Geol Surv Japan 37:43-51

Shibata K, Nozawa T, Wanless RK (1970) Rb-Sr geochronology of the Hida metamorphic belt, Japan. Can J Earth Sci 7:1383-1401

Shibata K, Otsubo T, Maruyama T (1988) A Rb-Sr whole-rock age of the Utsubo granitic complex, Hida Mountains. Bull Geol Surv Japan 39:135-138

Shimizu H, Sangen K, Masuda A (1982) Experimental study on rare-earth element partitioning in olivine and clinopyroxene formed at 10 and 20 kbar for basaltic systems. Geochem J 16:107-117

Shimizu N (1974) An experimental study of the partitioning of K, Rb, Cs, Sr and Ba between clinopyroxene and liquid at high pressures. Geochim Cosmochim Acta 38:1789-1798

Shimizu N (1975) Geochemistry of ultramafic inclusions from Salt Lake Crater, Hawaii, and from the southern African kimberlites. Phys Chem Earth 9:655-670

Shimron AE, Brookins DG (1974) Rb-Sr radiometric age of late Precambrian fossil-bearing and associated rocks from Sinai. Earth Planet Sci Letters 24:136-140

Shinjo R (1999) Geochemistry of high Mg andesites and the tectonic evolution of the Okinawa. Trough-Ryukyu arc system. Chem Geol 157:69-88

Shirey SB (1986) Mantle heterogeneity and crustal recycling in Archean granite-greenstone belts: evidence from Nd isotopes and trace elements in the Rainy Lake area, Superior Province, Ontario, Canada. Geochim Cosmochim Acta 50:2631-2651

Shirey SB, Hanson GN (1984) Mantle-derived Archean monzodiorites and trachyandesites. Nature 310:222-224

Shirey SB, Bender JF, Langmuir CH (1987) Three-component isotope heterogeneity near the Oceanographer transform. Mid-Atlantic Ridge. Nature 325:217-223

Shirey SB, Klewin KW, Berg J, Carlson R (1994) Temporal changes in the sources of flood basalts: isotopic and trace element evidence from the 1100 Ma old Keweenawan Mamainse Point Formation, Ontario. Geochim Cosmochim Acta 58(20):4475-4490

Shuto K (1974) The strontium isotopic study of the Tertiary acid volcanic rocks from the southern part of northeast Japan. Science Report, Tokyo Kyoiku Diagaku, Section C, 12(116):75-140

Sial AN (1976) The post-Paleozoic volcanism of northeast Brazil and its tectonic significance. An Acad Bras Cienc 48:299-311

Siders MA, Elliot DH (1985) Major and trace element geochemistry of the Kirkpatrick Basalt, Mesa Range, Antarctica. Earth Planet Sci Letters 72:54-64

Sieh K, Bursik M (1986) Most recent eruptions of the Mono Craters, eastern central California. J Geophys Res 91:12539-12571

Sigmarsson O, Condomines M, Fourcade S (1992) Mantle and crustal contribution in the genesis of recent basalts from off-rift zones in Iceland: Constraints from Th, Sr, and O isotopes. Earth Planet Sci Letters 110:149-162

Sigurdsson H (1968) Petrology of acid xenoliths from Surtsey. Geol Mag 105:440-453

Sigurdsson H (1977) Generation of Icelandic rhyolites by melting of plagiogranites in the oceanic layer. Nature 269:25-28

Sigurdsson H, Tomblin JF, Brown GM, Holland JG, Arculus RJ (1973) Strongly undersaturated magmas in the Lesser Antilles Island arc. Earth Planet Sci Letters 18:285-295

Silva LC, LeBas MJ, Robertson AHF (1981) An oceanic carbonatite volcano on Santiago, Cape Verde Islands. Nature 294:644-645

Simmons EC, Lambert DD (1981) Preliminary Sr-isotope data for the Kiglapait intrusion. In: Morse SA (ed) Nain Anorthosite Project, Labrador: Field Report 1980. Univ. Mass. Geol. Geog. Dept., Contrib. 38:41-46

Simonetti A, Bell K (1994) Nd, Pb, and Sr isotopic data from the Napak carbonatite-nephelinite centre, eastern Uganda: An example of open-system fractional crystallization. Contrib Mineral Petrol 115:356-366

Simonetti A, Bell K (1995) Nd, Pb, and Sr isotopic data from the Mount Elgon volcano, eastern Uganda-western Kenya: Implications for the origin and evolution of nephelinite lavas. Lithos 36:141-153

Sims KWW, DePaolo DJ, Murrell MT, Baldridge WS, Goldstein SJ, Clague DA (1995) Mechanisms of magma generation beneath Hawaii and mid-ocean ridges: Uranium/thorium and samarium/neodymium isotopic evidence. Science 267:508-512

Sims PK, Morey GB (1972) Geology of Minnesota: A Centennial volume. Minnesota Geol. Surv., Minneapolis, Minnesota

Sinton CW, Duncan RA, Denyer P (1997) The Nicoya Peninsula, Costa Rica: A single suite of Caribbean oceanic plateau magmas. J Geophys Res 102:15507-15520

Sinton CW, Duncan RA, Storey M, Lewis J, Estrada JJ (1998) An oceanic flood basalt province within the Caribbean plate. Earth Planet Sci Letters 155:221-236

Skelhorn RR, Elwell RWD (1966) The structure and form of the granophyric quartz-dolerite intrusion, centre II. Ardnamurchan, Argyllshire. Trans Royal Soc Edinburgh 66(12):285-306

Skogseid J, Eldholm O (1987) Early Cenozoic crust at the Norwegian continental margin and the conjugate Jan Mayen Ridge. J Geophys Res 92:11471-11491

Sleep NH (1984) Tapping of magmas from ubiqitous mantle heterogeneities, an alternative to mantle plumes. J Geophys Res 89:10029-10041

Sleep NH (1992) Hotspot volcanism and mantle plumes. Ann Rev Earth Planet Sci 20:19-43

Smellie JL (1987) Geochemistry and tectonic setting of alkaline volcanic rocks in the Antarctic Peninsula: A review. J Volcanol Geotherm Res 32:269-285

Smellie JL, Pankhurst RJ, Hole MJ, Thomson JW (1988) Age distribution and eruptive conditions of late Cenozoic alkaline volcanism in the Antarctic Peninsula and eastern Ellsworth Land. Review. British Antarct Surv Bull 80:21-49

Smith AG, Hallam A (1970) Fit of the southern continents. Nature 225:139

Smith HJ, Leeman WP, Davidson J, Spivack AJ (1997) The B isotope composition of arc lavas from Martinique, Lesser Antilles. Earth Planet Sci Letters 146:303-314

Smith IE, Compston W (1982) Strontium isotopes in Cenozoic rocks from southeastern Papua New Guinea. Lithos 15:199-206

Smith WC (1924) The plutonic and hypabyssal rocks of South Victoria Land: British Antarctic ("Terra Nova") Expedition 1910-1913. Nat. History Report, Geol 1(6):167-227

Sneeringer M, Hart SR, Shimizu N (1984) Strontium and samarium diffusion in diopside. Geochim Cosmochim Acta 48(8):1589-1608

Sørensen H (ed) (1974) The alkaline rocks. Wiley, New York

Somayajulu B, Tatsumoto M, Rosholt JN (1966) Disequilibrium of ^{238}U series in basalt. Earth Planet Sci Letters 1:387-391

Song Y, Frey FA (1989) Geochemistry of periodotite xenoliths in basalt from Hannuoba, eastern China: Implications for subcontinental mantle heterogeneity. Geochim Cosmochim Acta 53:97-113

Song Y, Frey FA, Zhi X (1990) Isotopic characteristics of Hannuoba basalts, eastern China: Implications for their petrogenesis and the composition of subcontinental mantle. Chem Geol 85:35-52

Sonnenthal EL, McBirney AR (1998) The Skaergaard Layered Series; Part IV: Reaction-transport simulations of foundered blocks. J Petrol 39:633-661

Spadea P, D'Antonio M, Thirlwall MF (1996) Source characteristics of the basement rocks from the Sulu and Celebes Basins (Western Pacific): Chemical and isotopic evidence. Contrib Mineral Petrol 12:159-176

Späth A, LeRoex AP, Duncan RA (1996) The geochemistry of lavas from the Comores Archipelago, western Indian Ocean: Petrogenesis and mantle source region characteristics. J Petrol 37:961-991

Sparks RSJ (1992) Magma generation in the Earth. In: Brown G, Hawkesworth C, Wilson C (eds) Understanding the Earth. Cambridge University Press, Cambridge, pp 91-114

Sparks RSJ, Marshall LA (1986) Thermal and mechanical constraints on mixing between mafic and silicic magmas. J Volcanol Geotherm Res 29:99-124

Sparks RSJ, Huppert HE, Wilson CJN (1990) Comment on "Evidence for long residence times of rhyolitic magma in the Long Valley magmatic system: The isotopic record in precaldera lavas at Glass Mountain" by A.N. Halliday, G.A. Mahood, P. Holden, J.M. Metz, T.J. Dempster, and J.P. Davidson. Earth Planet Sci Letters 99:387-389

Spera F, DeVivo B, Ayuso RA, Belkin HE (eds) (1998) Vesuvius. J Volcanol Geotherm Res 82:1-239

Spiegelman M, Elliott T (1993) Consequences of melt transport for uranium series disequ librium in young lavas. Earth Planet Sci Letters 118:1-2c

Spooner CM, Hepworth JV, Fairbairn HW (1970) Whole-rock Rb-Sr isotopic investigation of some East Africa granulites. Geol Mag 107:511-522

Spooner ETC, Barrie CT (1993) A special issue devoted to Abitibi ore deposits in a modern context; preface. Econ Geol 88(6)

Stacey JS, Hedge CE (1984) Geochronologic and isotopic evidence for Early Proterozoic crust in the eastern Arabian Shield. Geol 12:310-313

Staudigel H, Zindler A, Hart SR, Leslie T, Chen C-Y, Clague D (1984) The isotope systematics of a juvenile intraplate volcano: Pb, Nd and Sr isotopic ratios of basalts from Loihi seamount, Hawaii. Earth Planet Sci Letters 69:13-29

Staudigel H, Park K-H, Pringle M, Rubenstone JL, Smith WHF, Zindler A (1991) The longevity of the South Pacific isotopic and thermal anomaly. Earth Planet Sci Letters 102:24-44

Stecher O, Carlson RW, Shirey SB, Bridgwater D, Nielson T (1986) Nd-isotope evidence for the evolution of metavolcanic rocks from the Archaean block of Greenland and Labrador (Abstract). Terra Cognita 6:236

Steiger RH, Jäger E (1977) Subcommission on Geochronology: Convention on the use of decay constants in geo- and cosmochronology. Earth Planet Sci Letters 36:359-362

Stein M, Goldstein SL (1996) From plume head to continental lithosphere in the Arabian-Nubian Shield. Nature 382:773-778

Stein M, Hofmann AW (1992) Fossil plume head beneath the Arabian lithosphere? Earth Planet Sci Letters 114:193-209

Stern CR, Kilian R (1996) Role of the subducted slab, mantle wedge, and continental crust in the generation of adakites from the Andean Austral Volcanic Zone. Contrib Mineral Petrol 12:263-281

Stern CR, Stroup JB (1982) Petrochemistry of the Patagonian batholith, Ultima Esperanza, Chile. In: Craddock C (ed) Antarctic GeoScience. Univ. Wisconsin Press, Madison, pp 135-147

Stern CR, Futa K, Muehlenbachs K (1984) Isotope and trace element data for orogenic andesites from the austral Andes. In: Harmon RS, Barreiro BA (eds) Andean magmatism; chemical and isotopic constraints. Shiva Pub., UK, pp 31-46

Stern RJ (1978) Agrigan: An introduction to the geology of an active volcano in the northern Mariana Island Arc. Bull Volcanol 41:1-13

Stern RJ (1979) On the origin of andesite in the northern Mariana Island Arc: Implications from Agrigan. Contrib Mineral Petrol 68:207-219

Stern RJ (1982) Strontium isotopes from circum-Pacific intra-oceanic island arcs and marginal basins: Regional variations and implications for magma-genesis. Geol Soc Amer 93:477-486

Stern RJ, Bibee LD (1984) Esmeralda Bank: Geochemistry of an active submarine volcano in the Mariana Island arc. Contrib Mineral Petrol 86:159-169

Stern RJ, Gottfried D (1986) Petrogenesis of a late Precambrian (575-600 Ma) bimodal suite in northeast Africa. Contrib Mineral Petrol 92:492-501

Stern RJ, Hedge CE (1985) Geochronology and isotopic constraints on the late Precambrian crustal evolution in the Eastern Desert of Egypt. Amer J Sci 285:97-127

Stern RJ, Manton WI (1987) Age of Feiran basement rocks, Sinai: Implications for late Precambrian crustal evolution in the northern Arabian-Nubian Shield. J Geol Soc 144:569-576

Stern RJ, Gottfried D, Hedge CE (1984) Late Precambrian rifting and crustal evolution in the Northeastern Desert of Egypt. Geol 12:168-172

Stern RJ, Bloomer SH, Lin P-N, Ito E, Morris JD (1988) Shoshonitic magmas in nascent arcs: New evidence from submarine volcanoes in the northern Marianas. Geol 16:426-430

Stern RJ, Lin P-N, Morris JD, Jackson MC, Fryer P, Bloomer SH, Ito E (1990) Enriched back-arc basin basalts from the northern Mariana Trough: Implications for magmatic evolution of back-arc basins. Earth Planet Sci Letters 100:210-225

Stern RJ, Morris JD, Bloomer SH, Hawkins JW Jr (1991) The source of the subduction component in convergent margin magmas: Trace element and radiogenic isotope evidence from Eocene boninites, Mariana forearc. Geochim Cosmochim Acta 55:1467-1481

Stern RJ, Jackson MC, Fryer P, Ito E (1993) O, Sr, Nd, and Pb isotopic composition of Kasuga Cross-Chain in the Mariana arc: A new perspective on the K-h relationship. Earth Planet Sci Letters 119:459-475

Stettler A (1977) ^{87}Rb-^{87}Sr systematics of a geothermal water-rock association in the Massif Central, France. Earth Planet Sci Letters 34:432-438

Stettler A, Allègre CJ (1979) ^{87}Rb/^{87}Sr constraints on the genesis and evolution of the continental volcanic system Cantal (France). Earth Planet Sci Letters 44:269-278

Stewart BW, DePaolo DJ (1991) Isotopic studies of processes in mafic magma chambers: 2. The Skaergaard Intrusion, East Greenland. Contrib Mineral Petrol 104:125-141

Stewart BW, DePaolo DJ (1992) Diffusive isotopic contamination of mafic magma by coexisting silicic liquid in the Muskox Intrusion. Science 225:708-710

Stewart JH, Carlson JE (1978) Generalized maps showing distribution, lithology, and age of Cenozoic igneous rocks in the western United States. In: Smith RB, Eaton GP (eds) Cenozoic tectonics and regional tectonics and regional geophysics of the Western Cordillera. Geol. Soc. Amer. (Memoir 152, pp 263-264)

Stewart JW, Evernden JF, Snelling NJ (1974) Age determinations from Andean Peru: A reconnaissance survey. Geol Soc Amer 85(7):1107-1116

Stewart K, Rogers NW (1996) Mantle plume and lithosphere contributions to basalts from southern Ethiopia. Earth Planet Sci Letters 139:195-211

Stice GD, McCoy FW Jr (1968) The geology of the Manu'a Islands, Samoa. Pacific Science 22(4):427-457

Stille P, Unruh DM, Tatsumcto M (1983) Pb, Sr, Nd, and Hf isotopic evidence of multiple sources for Oahu, Hawaii basalts. Nature 304:25-29

Stille P, Unruh DM, Tatsumcto M (1986) Pb, Sr, Nd, and Hf isotopic constraints on the origin of Hawaiian basalts and evidence for a unique mantle source. Geochim Cosmochim Acta 50:2303-2319

Stille P, Oberhänsli R, Wenger-Schenk K (1989) Hf-Nd isotopic and trace element constraints on the genesis of alkaline and calc-alkaline lamprophyres. Earth Planet Sci Letters 96:209-219

Stipp JJ, McDougall I (1968) Geochronology of the Banks Peninsula volcanoes, new Zealand. New Zealand J Geol Geophys 11(5):1239-1260

Stix J, Gorton MP (1990) Variations in trace element partition coefficients in sanidine in the Cerro Toledo rhyolite, Jemez Mountains, New Mexico: Effects of composition, temperature, and volatiles. Geochim Cosmochim Acta 54:2697-2708

Stockwell CH (1982) Proposals for time classification and correlation of Precambrian rocks and events in Canada and adjacent areas of the Canadian Shield. Part 1. Geol Surv Canada Paper 80-19

Stoeser DB, Camp VE (1985) Pan-African microplate accretion of the Arabian Shield. Geol Soc Amer Bull 96:817-826

Stoffers P, Hekinian R, Ackermand D, Binard N, Botz R, Devey CW, Hansen D, Hodkinscn R, Jeschke G, Lange J, Perre EVD, Scholten J, Schmitt M, Sedwick P, Woodhead JD (1990) Active Pitcairn hotspot found. Mar Geol 95:51-55

Stolper E, Walker D (1980) Melt density and the average composition of basalt. Contrib Mineral Petrol 74:7-12

Stone WE, Crocket JH, Dickin AP, Fleet ME (1995) Origin of Archean ferropicrites: Geochemical constraints from the Boston Creek flow, Abitibi greenstone belt, Ontario, Canada. Chem Geol 121:51-71

Storey BC, Alabaster T (1991) Tectonomagmatic controls on Gondwana break-up models: Evidence from the Proto-Pacific margin of Antarctica. Tectonics 10(6):1274-1288

Storey BC, Garretts SW (1985) Crustal growth of the Antarctic Peninsula by accretion, magmatism and extension. Geol Mag 122:5-14

Storey BC, Alabaster T, Pankhurst RJ (eds) (1993) Magmatism and the causes of continental break-up. Geol Soc London Spec Pub 68

Storey M, Saunders AD, Tarney J, Leat P, Thirlwall MF, Thompson RN, Menzies MA, Marriner GF (1988) Geochemical evidence for plume-mantle interactions beneath Kerguelen and Heard Islands, Indian Ocean. Nature 335:371–374

Storey M, Wolff JA, Norry MJ, Marriner GF (1989a) Origin of hybrid lavas from Agua de Pau volcano, São Miguel, Azores. In: Saunders AJ, Norry MJ (eds) Magmatism in the ocean basins. Geol Soc London Spec Pub 42:161–180

Storey M, Saunders AD, Tarney J, Gibson IL, Norry MJ, Thirlwall MF, Leat P, Thompson RN, Menzies MA (1989b) Contamination of Indian Ocean asthenosphere by the Kerguelen-Heard mantle plume. Nature 338:574–576

Storey M, Kent RW, Saunders AD, Salters VJM, Hergt J, Whitechurch H, Sevigny HH, Thirlwall MF, Leat P, Ghose NC, Gifford M (1992) Lower Cretaceous volcanic rocks on continental margins and their relationship to the Kerguelen Plateau. Proc ODP Sci Results 120:33–54

Stosch H-G, Lugmair GW (1986) Trace element and Sr and Nd isotope geochemistry of peridotite xenoliths from the Eifel (West Germany) and their bearing on the evolution of the subcontinental lithosphere. Earth Planet Sci Letters 80: 281–298

Stosch H-G, Seck HA (1980) Geochemistry and mineralogy of two spinel peridotite suites from Dreiser Weiher, Germany. Geochim Cosmochim Acta 44:457–470

Stosch H-G, Carlson RW, Lugmair GW (1980) Episodic mantle differentiation: Nd and Sr isotope evidence. Earth Planet Sci Letters 47:263–271

Streckeisen AL (1967) Classification and nomenclature of igneous rocks. Neues Jahrbuch Mineral Abh 107:144–240

Streckeisen AL (1976) To each plutonic rock its proper name. Earth Sci Rev 12:1–33

Strong DF, Flower MFJ (1969) The significance of sandstone inclusions in lavas of the Comores archipelago. Earth Planet Sci Letters 7:47–50

Strong DF (1972a) The petrology of lavas of Grande Comore. J Petrol 13:181–217

Strong DF (1972b) Petrology of the island of Moheli, western Indian Ocean. Geol Soc Amer Bull 83:389–406

Stuart FM, Ellam RM, Duckworth RC (1999) Metal sources in the Middle Valley massive sulphide deposit, northern Juan de Fuca Ridge: Pb isotope constraints. Chem Geol 153:213–225

Stuart R, Natland J, Glassley W (1974) Petrology of volcanic rocks recovered on DSDP Leg 19 from the North Pacific Ocean and the Bering Sea. Init Rept Deep Sea Drilling Proj 19:615–627

Stuckless JS, Ericksen RL (1976) Strontium isotopic geochemistry of the volcanic rocks and associated megacrysts and inclusions from Ross Island and vicinity, Antarctica. Contrib Mineral Petrol 58:111–127

Stuckless JS, Irving AJ (1976) Strontium isotope geochemistry of megacrysts and host basalts from southeast Australia. Geochim Cosmochim Acta 40:209–213

Stuckless JS, O'Neil JR (1973) Petrogenesis of the Superstition-Superior volcanic area as inferred from strontium- and oxygen-isotope studies. Geol Soc Amer Bull 84:1987–1998

Stuckless JS, Hedge CE, Worl RG, Simmons KR, Nkomo LT, Wenner DB (1985) Isotopic studies of the Late Archean plutonic rocks of the Wind River Range, Wyoming. Geol Soc Amer Bull 96: 850–860

Stueber AM (1969) Abundances of K, Rb, Sr and Sr isotopes in ultramafic rocks and minerals from western North Carolina. Geochim Cosmochim Acta 33:543–553

Stueber AM, Ikramuddin M (1974) Rubidium, strontium and the isotopic composition of strontium in ultramafic nodule minerals and host basalts. Geochim Cosmochim Acta 38:207–216

Stueber AM, Murthy VR (1966) Strontium isotope and alkali element abundances in ultramafic rocks. Geochim Cosmochim Acta 30:1243–1259

Stueber AM, Pushkar P, Hetherington EH (1984) A strontium isotopic study of Smackover brines and associated solids, southern Arkansas. Geochim Cosmochim Acta 48:1637–1649

Stump E, Sheridan MF, Borg SG, Sutter JF (1980) Early Miocene subglacial basalts, the East Antarctic ice sheet, and uplift of the Transantarctic Mountains. Science 207:757–759

Suárez M, Pettigrew TH (1976) An Upper Mesozoic island-arc back-arc basin in the southern Andes and South Georgia. Geol Mag 113:305–328

Subbarao KV (1972) The strontium isotopic composition of basalts from the East Pacific and Chile rises and abyssal hills in the East Pacific Ocean. Contrib Mineral Petrol 37:111–120

Subbarao KV (ed) (1988) Deccan Flood Basalts. Geol. Soc. India, Bangalore (Memoir 10)

Subbarao KV, Hedge CE (1973) K, Rb, Sr and ^{87}Sr/^{86}Sr in rocks from the mid-Indian oceanic ridge. Earth Planet Sci Letters 18:223–228

Subbarao KV, Hekinian R (1978) Alkali-enriched rocks from the central eastern Pacific Ocean. Marine Geol 26:249–268

Subbarao KV, Reddy VV (1981) Geochemical studies on oceanic basalts from the Indian Ocean. Tectonophysics 75:69–89

Subbarao KV, Clark GS, Forbes RB (1973) Strontium isotopes in some seamount basalts from the northeastern Pacific Ocean. Can J Earth Sci 10:1479–1484

Subbarao KV, Chandrasekharam D, Navaneethakrishnan P, Hooper PR (1994) Stratigraphy and structure of parts of central Deccan basalt province: Eruptive models. In: Subbarao KV (ed) Volcanism. Wiley Eastern, New Delhi, pp 321–332

Sugimura A (1967) Spatial relations of basaltic magmas in island arcs. In: Hess HH, Poldervaart A (eds) Basalts. Interscience Publishers, New York (vol II, p 537)

Sun C-O, Williams RJ, Sun SS (1974) Distribution coefficients of Eu and Sr for plagioclase-liquid and clinopyroxene-liquid equilibria in oceanic ridge basalt: an experimental study. Geochim Cosmochim Acta 38:1415–1433

Sun S-S, Hanson GN (1975a) Evolution of the mantle: geochemical evidence from alkali basalt. Geol 3:297–302

Sun S-S, Hanson GN (1975b) Origin of Ross Island basanitoids and limitations upon the heterogeneity of mantle sources for alkali basalts and nephelinites. Contrib Mineral Petrol 52: 77–106

Sun SS (1980) Lead isotopic study of young volcanic rocks from mid-ocean ridges, ocean islands, and island arcs. Trans Roy Soc London Ser A297:409–445

Sun SS, Jahn BM (1975) Lead and strontium isotopes in post-glacial basalts from Iceland. Nature 255:727–531

Sun SS, Tatsumoto M, Schilling J-G (1975) Mantle plume mixing along the Reykjanes Ridge axis: Lead isotopic evidence. Science 190:143–147

Sun SS, Nesbitt RW, Sharaskin AY (1979) Geochemical characteristics of mid-ocean ridge basalts. Earth Planet Sci Letters 44: 119–138

Sundvoll B, Larsen BT (1990) Rb-Sr isotope systematics in the magmatic rocks of the Oslo Rift. Norges Geol Undersök Bull 418:27–46

Sutherland DS (ed) (1981) Igneous rocks of the British Isles. Wiley

Sutter JF, Smith TE (1979) ^{40}Ar/^{39}Ar ages of diabase intrusions from Newark Trend basins in Connecticut and Maryland: Initiation of central Atlantic rifting. Amer J Sci 279:808–831

Swanson DA, Wright TL, Hooper PR, Bentley RD (1979) Revisions in stratigraphic nomenclature of the Columbia River Basalt Group. US Geol Surv Bull 1457G:1–59

Swisher CC, Grajales-Nishimura JM, Montanari A, Margolis SV, Claeys P, Alvarez W, Rénne P, Cedillo-Pardo E, Maurrasse FJ-MR, Curtis GH, Smit J, Williams MO (1992) Coeval ^{40}Ar/^{39}Ar ages of 65.0 million years ago from Chicxulub Crater melt rock and Cretaceous-Tertiary boundary tektites. Science 275:954–958

Tanner PWG, Pankhurst RJ, Hyden G (1982) Radiometric evidence for the age of the subduction complex in the South Orkney and South Shetland Islands, West Antarctica. J Geol Soc London 139:683–690

Taras BD, Hart SR (1987) Geochemical evolution of the New England seamount chain: Isotopic and trace-element constraints. Chem Geol 64:35–54

Tatsumi Y (1989) Migration of fluid phases and generation of basalt magmas in subduction zones. J Geophys Res 94(B4): 4697–4704

Tatsumi Y, Kogiso T (1997) Trace element transport during dehydration processes in the subducted oceanic crust: 2. Origin of chemical and physical characteristics in arc magmatism. Earth Planet Sci Letters 148:207–221

Tatsumi Y, Hamilton DL, Nesbitt RW (1986) Chemical characteristics of fluid phase released from a subducted lithosphere and origin of arc magmas: Evidence from high pressure experiments and natural rocks. J Volcanol Geotherm Res 29:293–309

Tatsumi Y, Shinjoe H, Ishizuka H, Sager WW, Klaus A (1998) Geochemical evidence for a mid-Cretaceous superplume. Geol 26:151–154

Tatsumoto M (1966) Isotopic composition of lead in volcanic rocks from Hawaii, Iwo Jima, and Japan. J Geophys Res 71(6): 1721–1733

Tatsumoto M (1969) Lead isotope study of volcanic rocks and possible ocean-floor thrusting beneath island arcs. Earth Planet Sci Letters 6:369–376

Tatsumoto M, Basu AR, Huang W, Wang J, Xie G (1992) Sr, Nd, and Pb isotopes of ultramafic xenoliths in volcanic rocks of eastern China: enriched components EM1 and EM2 in subcontinental lithosphere. Earth Planet Sci Letters 113:107–128

Taylor HP Jr (1974) The application of oxygen and hydrogen isotope studies to problems of hydrothermal alteration and ore deposition. Econ Geol 69:843–883

Taylor HP Jr (1980) The effects of assimilation of country rocks by magmas on $^{18}O/^{16}O$ and $^{87}Sr/^{86}Sr$ systematics in igneous rocks. Earth Planet Sci Letters 47:243–254

Taylor HP Jr, Epstein S (1963) $^{18}O/^{16}O$ ratios in rocks and coexisting minerals of the Skaergaard Intrusion, East Greenland. J Petrol 4(1):51–74

Taylor HP Jr, Forester RW (1971) Low O^{18} igneous rocks from the intrusive complexes of Skye, Mull, Ardnamurchan, western Scotland. J Petrol 12:465–489

Taylor HP Jr, Forester RW (1979) An oxygen and hydrogen isotope study of the Skaergaard Intrusion and its country rocks: A description of a 55 M.Y. old fossil hydrothermal system. J Petrol 29(3):355–419

Taylor HP Jr, Sheppard SMF (1986) Igneous rocks: I. Processes of isotopic fractionation and isotope systematics. In: Valley JW, Taylor HP Jr, O'Neil JR (eds) Stable Isotopes in High Temperature Geological processes. Mineral. Soc. Amer., Blacksburg, VA (Rev. in Mineralogy 16, pp 227–271)

Taylor HP Jr, Gianetti B, Turi B (1979) Oxygen isotope geochemistry of the potassic igneous rocks from the Rocamonfina Volcano, Roman comagmatic region, Italy. Earth Planet Sci Letters 46:81–106

Taylor PN, Moorbath S, Goodwin R, Petrykowski AC (1980) Crustal contamination as an indicator of the extent of Early Archaean continental crust: Pb isotopic evidence from the Late Archaean gneisses of West Greenland. Geochim Cosmochim Acta 44:1437–1453

Taylor PN, Jones NW, Moorbath S (1984) Isotopic assessment of relative contributions from crust and mantle sources to magma genesis of Precambrian granitoid rocks. Phil Trans Roy Soc London 310A:605–625

Taylor RB (1964) Geology of the Duluth gabbro complex near Duluth, Minnesota. Minn Geol Surv Bull 44

Taylor RN, Thirlwall MF, Murton BJ, Hilton DR, Gee MAM (1997) Isotopic constraints on the influence of the Icelandic plume. Earth Planet Sci Letters 148:E1–E8

Taylor SR (1964) Abundance of chemical elements in the continental crust: A new table. Geochim Cosmochim Acta 28:1273–1285

Taylor SR, McLennan SM (1985) The continental crust: Its composition and evolution. Blackwell Scient. Pub., Oxford

Taylor SR, Capp AC, Graham AL, Blake DH (1969) Trace element abundances in andesites. II: Saipan, Bougainville, and Fiji. Contrib Mineral Petrol 23:1–26

Tejada MLG, Mahoney JJ, Duncan RA, Hawkins MP (1996) Age and geochemistry of basement and alkalic rocks of Malaita and Santa Isabel, Solomon Islands, southern margin of the Ontong Plateau. J Petrol 37:361–394

Tera F, Brown L, Morris JD, Sacks S, Klein J, Middleton R (1986) Sediment incorporation in island-arc magmas: Inferences from ^{10}Be. Geochim Cosmochim Acta 50:535–550

Terakado Y, Nakamura N (1984) Nd and Sr isotopic variations in acidic rocks from Japan: Significance of upper mantle heterogeneity. Contrib Mineral Petrol 87:407–417

Terakado Y, Shimizu H, Masuda A (1988) Nd and Sr isotopic variations in acidic rocks formed under a peculiar tectonic environment in Miocene southwest Japan. Contrib Mineral Petrol 99(1):1–10

Thirlwall MF, Graham AM (1984) Evolution of high Ca, high Sr, C-series basalt from Grenada, Lesser Antilles: The effects of intracrustal contamination. J Geol Soc London 141:427–445

Thirlwall MF, Upton BGJ, Jenkins C (1994) Interaction between continental lithosphere and the Icelandic plume – Sr-Nd-Pb isotope geochemistry of Tertiary basalts, NE Greenland. J Petrol 35:839–879

Thirlwall MF, Jenkins C, Vroon PZ, Mattey DP (1997) Crustal interaction during construction of oceanic islands. Pb-Sr-Nd-O isotope geochemistry of the shield basalts of Gran Canaria, Canary Island. Chem Geol 135:233–262

Thompson RN (1982) Magmatism of the British Tertiary Province. Scott J Geol 18:49–107

Thompson RN (1985) Asthenospheric source of Ugandan ultrapotassic magma? J Geol 93:630–608

Thompson RN, Esson J, Dunham AC (1972) Major element chemical variation in the Eocene of the Isle of Skye, Scotland. J Petrol 13:219–253

Thompson RN, Morrison MA, Mattey DP, Dickin AP, Moorbath S (1980a) An assessment of the Th-Hf-Ta diagram as a discriminant for tectono-magmatic classifications and in the detection of crustal contamination of magmas. Earth Planet Sci Letters 50:1–10

Thompson RN, Gibson IL, Marriner GF, Mattey DP, Morrison MA (1980b) Trace-element evidence of multistage mantle fusion and polybaric fractional crystallization in the Palaeocene lavas of Skye, NW Scotland. J Petrol 21:265–293

Thompson RN, Dickin AP, Gibson IL, Morrison MA (1982) Elemental fingerprints of isotopic contamination of Hebridean Palaeocene mantle-derived magmas by Archaean sial. Contrib Mineral Petrol 79:159–168

Thompson RN, Morrison MA, Dickin AP, Hendry GL (1983) Continental flood basalts ... arachnids rule OK? In: Hawkesworth CJ, Norry MJ (eds) Continental basalts and mantle xenoliths. Shiva, Nantwich, pp 158–185

Thompson RN, Morrison MA, Dickin AP, Gibson IL, Harmon RS (1986) Two contrasting styles of interaction between basic magma and continental crust in the British Tertiary Province. J Geophys Res 91(6):5985–5997

Thomson JE (1953) Geology of the Mamainse Point copper area. Ontario Dept Mines Ann Rept 62(4):1–25

Thomson MRA, Crame JA, Thomson JW (1991) Geological evolution of Antarctica. Cambridge University Press

Thorpe RI (1986) U-Pb geochronology of the Coldwell Complex, northwestern Ontario, Canada: Discussion. Can J Earth Sci 23:125–126

Thorpe RS (ed) (1982) Andesites. Wiley, Chichester, U.K.

Thorpe RS, Potts PJ, Francis PW (1976) Rare earth data and petrogenesis of andesite from the north Chilean Andes. Contrib Mineral Petrol 54:65–78

Thorpe RS, Francis PW, Moorbath S (1979a) Rare earth and strontium isotope evidence concerning the petrogenesis of North Chilean ignimbrites. Earth Planet Sci Letters 42:359–367

Thorpe RS, Francis PW, Moorbath S (1979b) Strontium isotope evidence for petrogenesis of Central American andesites. Nature 227:44–145

Thorpe RS, Francis PW, O'Callaghan L (1984) Relative roles of source composition, fractional crystallization in the petrogenesis of Andean volcanic rocks. Phil Trans Roy Soc London (Ser A) 310:675–692

Thorweihe U, Schandelmeier H (eds) (1993) Geoscientific Research in Northeast Africa. Balkema, Rotterdam

Tilling RI, Rhodes JM, Sparks JW, Lockwood JP, Lipman PW (1987) Disruption of the Mauna Loa magma system by the 1868 Hawaiian earthquake; geochemical evidence. Science 235:196–198

Tilton GR (1960) Volume diffusion as a mechanism for discordant lead ages. J Geophys Res 65:2935–2949

Tilton GR, Barreiro BA (1980) Origin of lead in Andean calc-alkaline lavas, southern Peru. Science 210:1245-1247

Ting W, Rankin AH, Woolley AR (1994) Petrogenetic significance of solid carbonate inclusions in apatite of the Sukulu carbonatite, Uganda. Lithos 31:177-187

Titterington DM, Halliday AN (1979) On the fitting of parallel isochrons and the method of maximum likelihood. Chem Geol 26:183-195

Todd SG, Keith DW, LeRoy LW, Schissel DJ, Mann EL, Irvine TN (1982) The J-M platinum-palladium reef of the Stillwater complex, Montana: I. Stratigraphy and petrology. Econ Geol 77:1454-1480

Togashi S, Shirahase T, Tamanyu S (1985) Sr isotope geochemistry of voluminous rhyolite of the Tamagawa welded tuffs, related rocks and andesite of the young volcanics in Hachimantai area, northeast Japan. J Volcanol Geotherm Res 24(1/2):339-352

Tolstikhin IN, Dokuchaeva VS, Kamensky IL, Amelin YV (1992) Juvenile helium in ancient rocks: II. U-He, K-Ar, Sm-Nd, and Rb-Sr systemtics in the Monche Pluton. ^{3}He/^{4}He ratios frozen in uranium-free ultramafic rocks. Geochim Cosmochim Acta 56:987-999

Torres-Roldan RN (1979) The tectonic subdivision of the Betic zone (Betic Cordillera, southern Spain): Its significance and one possible geotectonic scenario for the westernmost alpine belt. Amer J Sci 279:20-51

Torssander P (1988) Sulfur isotope ratios of Icelandic lava incrustations and volcanic gases. J Volcanol Geotherm Res 35(3):227-236

Torssander P (1989) Sulfur isotope ratios of Icelandic rocks. Contrib Mineral Petrol 102(1):18-23

Treiman AH (1989) Carbonatite magma: properties and processes. In: Bell K (ed) Carbonatites. Unwin Hyman, London

Treves SB (1962) The geology of Cape Evans and Cape Royds, Ross Island, Antarctica. In: Wexler H, Rubin NJ, Caskey JE Jr (eds) Antarctic Research. Amer. Geophys. Union, Washington, D.C. (Monograph 7, pp 40-46)

Treves SB (1967) Volcanic rocks from the Ross Island, Marguerite Bay, and Mt. Weaver areas, Antarctica. Japanese Ant. Research Exped. Reports, Special Issue 1:136-149

Trua T, Deniel C, Mazzuoli R (1999) Crustal control in the genesis of Plio-Quaternary bimodal magmatism of the Main Ethiopian Rift (MER): Geochemical and isotopic (Sr, Nd, Pb) evidence. Chem Geol 155:201-222

Trull TW, Perfit MR, Kurz MD (1990) He and Sr isotope constraints on subduction contributions to Woodlark Basin volcanism, Solomon Islands. Geochim Cosmochim Acta 54(2):441-454

Tu K, Flower MFJ, Carlson RW, Zhang M, Xie G (1991) Sr, Nd, and Pb isotopic compositions of Hainan basalts (south China): Implications for a subcontinental lithospheric Dupal source. Geol 19:567-569

Tu K, Flower MFJ, Carlson RW, Xie G, Chen C-Y, Zhang M (1992) Magmatism in the South China Basin. 1. Isotopic and trace-element evidence for an endogenous Dupal mantle component. Chem Geol 97:47-63

Turek A, Smith TE, Huang CH (1981) Rb-Sr whole-rock geochronology of the Gamitagama area, north central Ontario. Can J Earth Sci 18:323-329

Turek A, Smith PE, Schmus WR Van (1982) Rb-Sr and U-Pb ages of volcanism and granite emplacement in the Michipicoten belt-Wawa, Ontario. Can J Earth Sci 19:1608-1626

Turek A, Smith PE, Schmus WR Van (1984) U-Pb zircon ages and the evolution of the Michipicoten plutonic-volcanic terrane of the Superior Province, Ontario. Can J Earth Sci 21:457-464

Turek A, Smith PE, Symons DTA (1985) U-Pb geochronology of the Coldwell Complex, northwestern Ontario, Canada. Can J Earth Sci 22:621-625

Turekian KK, Wedepohl KH (1961) Distribution of the elements in some major units of the Earth's crust. Geol Soc Amer Bull 72:175-192

Turekian KK (1964) The marine geochemistry of strontium. Geochim Cosmochim Acta 28:1479-1496

Turner DC, Bowden P (1979) The Ningi-Burra complex, Nigeria: Dissected calderas and migrating magmatic centre. J Geol Soc London 136:105-119

Turner DL, Jarrard RD (1982) K-Ar dating of the Cook-Austral island chain: A test of the hot-spot hypothesis. J Volcanol Geotherm Res 12:187-220

Turner FJ, Verhoogen J (1960) Igneous and metamorphic petrology, 2nd edn. McGraw-Hill, New York

Turner S, Hawkesworth CJ, Rogers NW, King P (1997) U-Th isotope disequilibria and ocean island basalt generation in the Azores. Chem Geol 239:145-164

Turpin L, Velde D, Pinte G (1988) Geochemical comparison between minettes and kersantites from the western European Hercyniam orogen: Trace element and Pb-Sr-Nd isotope constraints on their origin. Earth Planet Sci Letters 87:73-86

Twist D, French BM (1983) Voluminous acid volcanism in the Bushveld complex: A review of the Rooiberg Felsite. Bull Volcanol 46:225-242

Twist D (1985) Geochemical evolution of the Rooiberg silicic lavas in the Loskop Dam area, southeastern Brushveld. Econ Geol 80:1153-1165

Ueno N, Kaneoka I, Ozima M (1974) Isotopic ages and strontium isotopic ratios of submarine rocks in the Japan Sea. Geochem J 8:(4):157-164

Uhlig C (1907) Der sogenannte grosse ostafrikanische Graben zwischen Magad (Natron See) und Lawa ya Mueri (Manyara See). Geographische Zeit 15:478-505

Uken R, Watkeys MK (1997) Diapirism initiated by the Bushveld complex, South Africa. Geol 25(8):723-726

Ulff-Möller F (1990) Formation of native iron in sediment - contaminated magma: I. A case study of the Hanekammen complex on Disko Island, West Greenland. Geochim Cosmochim Acta 54:57-70

Umeji AC (1983) Geochemistry and mineralogy of the Freetown basic igneous complex of Sierra Leone. Chem Geol 39:17-38

Umeji AC (1985) On the beerbachites from Freetown, Sierra Leone. Geol Mag 122:667-667

Upton BGJ, Wadsworth WJ (1965) Geology of Réunion Island, Indian Ocean. Nature 207:151-154

Upton BGJ, Wadsworth WJ (1966) Basalts of Réunion Island, Indian Ocean. Bull Volcanol 29:7-24

Upton BGJ, Wadsworth WJ (1972) Aspects of magmatic evolution of Réunion Island. Phil Trans Roy Soc London A271:105-130

Upton BGJ, Wadsworth WJ, Newmann TC (1967) The petrology of Rodriguez Island, Indian Ocean. Geol Soc Amer Bull 78:1495-1506

Upton BGJ, Aspen P, Chapman NA (1983) The upper mantle and deep crust beneath the British Isles: evidence from inclusions in volcanic rocks. J Geol Soc London 140:105-122

Upton BGJ, Emeleus CH, Beckinsale RD (1984) Petrology of northern East Greenland Tertiary flood basalts.: Evidence from Hold with Hope and Wollaston Forland. J Petrol 25:151-184

Vail JR (1985) Alkaline ring complexes of Sudan. J African Earth Sci 3:51-59

Valley JW, Taylor HP Jr, O'Neil JR (eds) (1986) Stable isotopes in high temperature geological processes. Mineralogical Society of America, Washington, D.C. (Reviews in Mineralogy, vol XVI)

Vance D, Stone JOH, O'Nions RK (1989) He, Sr and Nd isotopes in xenoliths from Hawaii and other islands. Earth Planet Sci Letters 96:147-160

Vartiainen H, Woolley AR (1974) The age of the Sokli carbonatite, Finland, and some relationships of the North Atlantic alkaline igneous province. Bull Geol Soc Finland 46:81-91

Veevers JJ, Cole DI, Cowan EJ (1994) Southern Africa: Karoo Basin and Cape Fold Belt. Geol Soc Amer Mem 184:223-279

Venkatesan TR, Pande K, Gopalan K (1993) Did Deccan volcanism pre-date the Cretaceous/Tertiary transition? Earth and Planet Sci Letters 119:181-189

Venturelli G, Capedri S, Di Battistini G, Crawford A, Kogarko LN, Celestini S (1984) The ultrapotassic rocks of southeastern Spain. Lithos 7:37-54

Verma SP (1983) Magma genesis and chamber processes at Los Humeros caldera, Mexico-Nd and Sr isotopic data. Nature 301:52-55

Verma SP (1984) Sr and Nd isotopic evidence for petrogenesis of mid-Tertiary felsic volcanism in the mineral district of Zacatecas (Sierra Madre Occidental), Mexico. Isot GeoScience 2:37-53

Verma SP, Schilling J-G (1982) Galapagos hot spot-spreading center system $^{87}Sr/^{86}Sr$ and large ion lithophile element variations (85° W–101° W). J Geophys Res 87(B3):10838–10856

Verma SP, Verma MP (1986) A compilation of Sr and Nd isotope data on Mexico. J Geol Soc India 27:130–143

Vernon-Chamberlain VE, Snelling NJ (1972) Age and isotope studies on the Arena Granites of S.W. Uganda. Musée Royal de l'Afrique Annales, Ser. 1N-8°, Science Geologiques 73:4–43

Verpaelst P, Brooks C, Franconi A (1980) The 2.5 Ga Duxbury massif, Quebec: a remobilized piece of pre-3.0 Ga sialic basement (?). Can J Earth Sc 17:1–13

Verschure RH, Maijer C, Andriessen PAM, Boelrijk NAIM, Hebeda EH, Priem HNA, Verdurmen EAT (1983) Dating explosive volcanism perforating the Precambrian basement in southern Norway. Norsk Geol Unders 380:35–49

Vidal P (1992) Mantle More HIMU in the future? Geochim Cosmochim Acta 56:4295–4299

Vidal P, Clauer N (1981) Pb and Sr isotopic systematics of some basalts and sulfides from the East Pacific Rise at 21° N (project RITA). Earth Planet Sci Letters 55:237–246

Vidal P, Dosso L, Bowden P, Lameyre J (1979) Strontium isotope geochemistry in syenite-alkaline granite complexes. In: Ahrens LH (ed) Origin and Distribution of the Elements. Pergamon Press, Oxford, U.K., p 223

Vidal P, Chauvel C, Brousse R (1984) Large mantle heterogeneity beneath French Polynesia. Nature 307:536–538

Vidal P, Deniel C, Vellutini PJ Piguet P, Coulon C, Vincent J, Audin J (1991) Changes of mantle sources in the course of a rift evolution: The Afar case. Geophys Res Letters 18:1913–1916

Viljoen MJ, Viljoen RP (1969) The geology and geochemistry of the lower ultramafic unit of the Onverwacht Group and a proposed new class of igneous rocks. Geol Soc S Africa Spec Pub 2:55–85

Villemant B, Jaffrezic H, Joron JL, Treuil M (1981) Distribution coefficients of major and trace elements; fractional crystallization in the alkali basalt series of Chaine des Puys (Massif Central, France). Geochim Cosmochim Acta 45:1997–2016

Vincent EA, Crocket JH (1960) Studies in the geochemistry of gold. I. The distribution of gold in rocks and minerals of the Skaergaard intrusion, East Greenland. Geochim Cosmochim Acta 18:130–142

Vincent PM (1970) The evolution of the Tibesti volcanic province, eastern Sahara. In: Clifford TN, Gass IG (eds) African Magmatism and Tectonics. Oliver and Boyd, Edinburgh

Vinogradov AP (1959) The geochemistry of rare and dispersed elements in soils, 2nd edn. Consultants Bureau, New York

Vinogradov AP (1962) Average contents of chemical elements in the principal types of igneous rocks of the Earth's crust. Geochemistry (1962(7):641–664

Vinogradov VI, Volynets ON, Grigor'yev VS, Koloskov AV (1988) Strontium-isotope distribution in some silicic igneous rock bodies of southern Kamchatka. Doklady, Earth Science Section, 289:176–179

Visser DJL, Gruenewaldt G von (1970) Symposium on the Bushveld igneous complex and other layered intrusions. Geol Soc S Africa Spec Pub 1

Viswanathan S (1974) Basaltic komatiite occurrences in the Kolar gold field of India. Geol Mag 111:353–354

Vogel TA, Younker LW, Willband JT, Kampmueller E (1984) Magma mixing: the Marsco suite, Isle of Skye, Scotland. Contrib Mineral Petrol 87:231–241

Vogel TA, Eichelberger JC, Younker LW, Schuraytz BC, Horkowitz JP, Stockman HW, Westrich HR (1989) Petrology and emplacement dynamics of intrusive and extrusive rhyolites of Obsidian Dome, Inyo Craters volcanic chain, eastern California. J Geophys Res 94(B2):17937–17956

Volker F, McCulloch MT, Altherr R (1993) Submarine basalts from the Red Sea: new Pb, Sr, and Nd isotopic data. Geophys Res Letters 20:927–930

Vollmer R (1976) Rb-Sr and U-Th-Pb systematics of alkaline rocks: The alkaline rocks from Italy. Geochim Cosmochim Acta 40:283–296

Vollmer R (1977) Isotopic evidence for genetic relations between acid and alkaline rocks in Italy. Mineral Petrol 60:109–118

Vollmer R (1989) On the origin of the Italian potassic magmas. 1. A discussion contribution. Chem Geol 74:229–239

Vollmer R (1990) On the origin of the Italian potassic magmas – reply. Chem Geol 85:191–196

Vollmer R, Norry MJ (1983a) Possible origin of K-rich volcanic rocks from Virunga, East Africa, by metasomatism of continental crustal material: Pb, Nd, and Sr isotopic evidence. Earth and Planet Sci Letters 64 374–386

Vollmer R, Norry MJ (1983b) Unusual isotopic variations in Nyiragongo nephelinites. Nature 301:141–143

Vollmer R, Ogden P, Schilling J-G, Kingsley RH, Waggoner DG (1984) Nd and Sr isotopes in ultrapotassic volcanic rocks from the Leucite Hills, Wyoming. Contrib Mineral Petrol 87:359–368

Volpe AM, Macdougall JD, Hawkins JW (1987) Mariana trough basalts (MTS): Trace element and Sr-Nd isotopic evidence for mixing between MORB-like and arc-like melts. Earth Planet Sci Letters 82:241–254

Volpe AM, Macdougall JD, Hawkins JW (1988) Lau Basin basalt (LBB): Trace element and Sr-Nd isotopic evidence for heterogeneity in back-arc basin mantle. Earth Planet Sci Letters 90: 174–186

Volpe AM, Macdougall JD, Lugmair GW, Hawkins JW, Lonsdale P (1990) Fine-scale isotopic variation in Mariana Trough basalts: evidence for heterogeneity and a recycled component in back arc basin mantle. Earth Planet Sci Letters 100:251–264

Vroon PZ, Bergen MJ van, White WM, Varekamp JC (1993) Sr-Nd-Pb isotope systematics of the Banda Arc, Indonesia: combined subduction and assimilation of continental material. J Geophys Res 98(B12):22349–22366

Vugrinovich RG (1981) A distribution-free alternative to least-squares regression and its application to Rb/Sr isochron calculations. Math Geol 13:443–454

Vukadinovic D, Nicholls IA (1989) The petrogenesis of island arc basalts from Gunung Slamet volcano, Indonesia: Trace element and $^{87}Sr/^{86}Sr$ constraints. Geochim Cosmochim Acta 53(9):2349–2364

Wadge G, Wooden JL (1982) Late Cenozoic alkaline volcanism in the northwestern Caribbean: Tectonic setting and Sr isotopic characteristics. Earth Planet Sci Letters 57:35–46

Wager LR (1953) Layered intrusions. Dansk Geol Foren Meddelelser 12:335–349

Wager LR (1960) The major element variation of the layered series of the Skaergaard intrusion and a re-estimation of the average composition of the Hidden Layered Series and of successive residual magmas. J Petrol 1:364–398

Wager LR (1965) The form and internal structure of the alkaline Kangerdlugssuaq intrusion, East Greenland. Mineral Mag 23: 387–497

Wager LR, Brown GM (1968) Layered igneous rocks. Oliver and Boyd, Edinburgh

Wager LR, Deer WA (1939) The petrology of the Skaergaard intrusion, Kangerdlugssuaq, East Greenland. Meddelelser om Grönland 105(4)

Wager LR, Mitchell RL (1951) The distribution of trace elements during strong fractionation of basic magma – a further study of the Skaergaard Intrusion, East Greenland. Geochim Cosmochim Acta 1:129–208

Wager LR, Weedon DS, Vincent EA (1953) A granophyre from Coire Uaigneich, Isle of Skye, containing quartz paramorphs after tridymite. Mineral Mag 30:263–276

Wager LR, Vincent EA, Brown GM, Bell JD (1965) Marscoite and related rocks of the Western Redhills Complex, Isle of Skye. Phil Trans Roy Soc London 257A:273–307

Walker JA, Carr MJ, Feigenson MD, Kalamarides RI (1990) The petrogenetic significance of interstratified high and low-Ti basalts in central Nicaragua. J Petrol 31(5):1141–1164

Walker PT (1958) Study of some rocks and minerals from the Dufek Massif, Antarctica. In: Goldthwait RP (ed) USNC-IGY Antarctic glaciological data: Field work 1957 and 1958. Ohio State Univ. (Rept. Research Found 825-4, Part I, pp 209–232)

Walker RJ, Shirey SB, Stecher O (1988) Comparative Re-Os, Sm-Nd, and Rb-Sr isotope and trace element systematics for Archean komatiite flows from Munro Township, Abitibi belt, Ontario. Earth Planet Sci Letters 87:1–12

Walker RJ, Echeverría LM, Shirey SB, Horan MF (1991) Re-Os isotopic constraints on the origin of volcanic rocks, Gorgona Island, Columbia: Os isotopic evidence for ancient heterogeneities in the mantle. Contrib Mineral Petrol 107:150–162

Wallace JM, Ellam RM, Meighan IG, Lyle P, Rogers NW (1994) Sr isotope data for the Tertiary lavas of Northern Ireland: Evidence for open-system petrogenesis. J Geol Soc London 151: 869–877

Wallace P, Carmichael ISE (1989) Minette lavas and associated leucitites from the western front of the Mexican Volcanic Belt: Petrology, chemistry, and origin. Contrib Mineral Petrol 103: 470–492

Walraven F (1985) Genetic aspects of the granophyric rocks of the Bushveld complex. Econ Geol 80:1166–1180

Walraven F, Armstrong EA, Kruger FJ (1990) A chronostratigraphic framework for the north-central Kaapvaal craton, the Bushveld complex, and the Vredefort structure. Tectonophyics 171:23–48

Wanless RK, Loveridge WD (1972) Rubidium-strontium isochron age studies. Report 1. Geol Surv Canada Paper 72-23

Wanless RK, Loveridge WD (1978) Rubidium-strontium isochron age studies. Report 2 (Canadian shield). Geol Surv Canada Paper 77-14

Wanless RK, Stevens RD, Loveridge WD (1970) Anomalous parent-daughter isotopic relationships in rocks adjacent to the Grenville Front near Chibougamau, Quebec. Eclogae Geol Helv 63:345–364

Wardle RJ, Ryan B, Nunn GAG, Mengel FC (1990) Labrador segment of the Trans-Hudson orogen: crustal development through oblique convergence and collision. In: Lewry JF, Stauffer MR (eds) The Early Proterozoic Trans-Hudson Orogen of North America. Geol Assoc Canada Spec Paper 37: 353–369

Warren P (1983) Al-Sm-Eu-Sr systematics of eucrites and Moon rocks: Implications for planetary bulk compositions. Geochim Cosmochim Acta 47:1559–1571

Washington HS (1918) Italian leucite lavas as a source of potash. Metall Chem Eng 18(2):3–21

Wass SY (1980) Geochemistry and origin of xenolith-bearing and related alkali basaltic rocks from the Southern Highlands, New South Wales, Australia. Amer J Sci 280A:639–666

Wass SY, Rogers NW (1980) Mantle metasomatism-precursor to continental alkaline volcanism. Geochim Cosmochim Acta 44: 1811–1823

Wasserburg GJ, Craig H, Menard HW, Engel AEJ, Engel CJ (1963) Age and composition of a Bounty granite and age of a Seychelles Islands granite. J Geol 71:785–789

Wasserburg GJ, Albee AL, Lanphere MA (1964) Migration of radiogenic strontium during metamorphism. J Geophys Res 69(20):4395–4401

Wasserburg GJ, Jacobsen SB, DePaolo DJ, McCulloch MT, Wen T (1981) Precise determination of Sm/Nd ratios, Sm and Nd isotopic abundances in standard solutions. Geochim Cosmochim Acta 45:2311–2323

Waters AC (1961) Stratigraphic and lithologic variations in the Columbia River basalt. Amer J Sci 259:583–611

Watkins ND, Gunn BN, Nougier J, Baksi AK (1974) Kerguelen: Continental fragment or oceanic island? Geol Soc Amer Bull 85:201–212

Watkins ND, McDougall I, Nougier J (1975) Paleomagnetism and potassium-argon age of St. Paul Island, Southeastern Indian Ocean. Contrasts in geomagnetic secular variation during the Brunhes Epoch. Earth Planet Sci Letters 24:377–384

Watkins RT, LeRoex AP (1994) A reinvestigation of the Okenyenya igneous complex: A new geological map, structural interpretation and model of emplacement. Commun Geol Surv Namibia 9:13–21

Watkins RT, McDougall I, LeRoex AP (1994) K-Ar ages of the Brandberg and Okenyenya igneous complexes, north-western Namibia. Geologische Rundschau 83:348–356

Watson EB (1982) Basalt contamination by continental crust: Some experiments and models. Contrib Mineral Petrol 80: 73–87

Watson S, MacKenzie D (1991) Melt generation by plumes: A study of Hawaiian volcanism. J Petrol 32:501–537

Weaver BL (1991) The origin of ocean island basalt end-member compositions: Trace element and isotopic constraints. Earth Planet Sci Letters 104:381–197

Weaver BL, Wood DA, Tarney J, Joron JL (1986) Role of subducted sediment in the genesis of oceanic-island basalts: Geochemical evidence from South Atlantic Ocean islands. Geol 14: 275–278

Weaver BL, Wood DA, Tarney J, Joron JL (1987) Geochemistry of ocean island basalts from the South Atlantic: Ascension, Bouvet, St. Helena, Gough, and Tristan da Cunha. In: Fitton JG, Upton BGJ (eds) Alkaline igneous rocks. Geol Soc London Spec Pub 30:253–268

Weaver SD, Johnson RW (eds) (1987) Tectonic controls on magma chemistry. Elsevier, Amsterdam

Weaver SD, Pankhurst RJ (1991) A precise Rb-Sr age for the Mandamus igneous complex, North Canterbury, and regional tectonic implications. New Zealand J Geol Geophys 34:341–345

Weaver SD, Sceal JSC, Gibson IL (1972) Trace element data relevant to the origin of trachytic and pantelleritic lavas in the East African rift system. Contrib Mineral Petrol 36:181–194

Weaver SD, Saunders AD, Pankhurst RJ, Tarney J (1979) A geochemical study of magmatism associated with the initial stages of back-arc spreading. Contrib Mineral Petrol 68: 151–169

Weaver SD, Storey BC, Pankhurst RJ, Mukasa SB, DiVenere VJ, Bradshaw JD (1994) Antarctica-New Zealand rifting and Marie Byrd Land lithospheric magmatism linked to ridge subduction and mantle plume activity. Geol 22:811–814

Webb PN, McKelvey BC (1959) The geology of Victoria Dry Valley, Pt. 1 of Geological investigations in South Victoria Land, Antarctica. New Zealand J Geol Geophys 2(1):120–136

Wedepohl KH (ed) (1969 to 1978) Handbook of Geochemistry. Springer Verlag, Berlin

Wedepohl, KH (1985) Origin of the Tertiary basaltic volcanism in the northern Hessian Depression. Contrib Mineral Petrol 89: 122–143

Wegener A (1924) The origins of continents and oceans. Methuen, London

Weigand PW, Ragland PC (1970) Geochemistry of Mesozoic dolerite dikes from eastern North America. Contrib Mineral Petrol 29:195–214

Weis D, Demaiffe D (1985) A depleted mantle source for kimberlites from Zaire: Nd, Sr, and Pb isotopic evidence. Earth Planet Sci Letters 73:269–277

Weis D, Deutsch S (1984) Nd and Pb isotope evidence from the Seychelles granites and their xenoliths: Mantle origin with slight upper-crust interaction for alkaline anorogenic complexes. Isotope GeoScience 2:13–35

Weis D, Frey FA (1991) Isotope geochemistry of Ninetyeast Ridge basement basalts: Sr, Nd, and Pb evidence for involvement of the Kerguelen hot spot. Proc Ocean Drilling Program Sci Results 121:591–610

Weis D, Liegeois LP, Javoy M (1986) The Timedjelalen alkaline ring-complex and related N-S dyke swarms (Adrar des Iforas, Mali) – a Pb-Sr-O isotopic study. Chem Geol 57:201–216

Weis D, Demaiffe D, Cauet S, Javoy M (1987a) Sr, Nd, O, and H isotopic ratios in Ascension lavas and plutonic inclusions: Cogenetic origin. Earth Planet Sci Letters 82:255–268

Weis D, Liegeois JP, Black R (1987b) Tadhak alkaline ring-complex (Mali): Existence of U-Pb isochrons and "Dupal" signature 270 Ma ago. Earth Planet Sci Letters 82:316–322

Weis D, Bassias Y, Gautier I, Mennessier J-P (1989) Dupal anomaly in existence 115 Ma ago: Evidence from isotopic study of the Kerguelen Plateau (South Indian Ocean). Geochim Cosmochim Acta 53:2125–2131

Weis D, White WM, Frey FA, Duncan RA, Fisk MR (1992) The influence of mantle plumes in generation of Indian oceanic crust. In: Duncan RA, Rea DK, Kidd RB, Rad U von, Weissel JK (eds) Synthesis of Results from Scientific Drilling in the Indian Ocean. Amer. Geophys. Union, Washington, D.C. (Geophys Monogr 70, pp 57–89)

Weis D, Frey FA, Leyrit H, Gautier I (1993) Kerguelen archipelago revisited: Geochemical and isotopic study of the southeast province lavas. Earth Planet Sci Letters 118:101–119

Welke H, Moorbath S, Cumming GL, Sigurdsson H (1968) Lead isotope studies on igneous rocks from Iceland. Earth Planet Sci Letters 4:221–231

Wellman R, McDougall I (1974a) Potassium-argon ages of the Cainozoic volcanic rocks of New South Wales. J Geol Soc Australia 21:247–272

Wellman R, McDougall I (1974b) Cainozoic igneous activity in eastern Australia. Tectonophysics 23:49–65

Wells MK (1962) Structure and petrology of the Freetown layered basic complex of Sierra Leone. Overseas Geology and Mineral Resources Bulletin Supplement 4:1–115

Wen J, Bell K, Blenkinsop J (1987) Nd and Sr isotope systematics of the Oka complex, Quebec, and their bearing on the evolution of the sub-continental upper mantle. Contrib Mineral Petrol 97:433–437

Wendt I (1969) Derivaton of the formula for a regression line based on a least-square analysis. Bundesanstalt für Bodenforschung, Hannover (Internal Report)

Wendt I (1993) Isochron or mixing line. Chem Geol (Isotope Geoscience Section) 104:301–305

Wendt I, Carl C (1991) The statistical distribution of the mean squared weighted deviations. Chem Geol (Isotope Geoscience Section) 86:275–285

Wendt I, Besang C, Harre W, Kreuzer H, Lenz H, Müller P (1977) Age determinations of granitic intrusions and metamorphic events in the early Precambrian of Tanzania. 25th Inter. Geol. Congress, Montreal, Section 1, pp 195–314

West HB, Leeman WF (1987) Isotopic evolution of lavas from Haleaka crater, Hawaii. Earth Planet Sci Letters 84:211–225

West HB, Gerlach DC, Leeman WP, Garcia MO (1987) Isotopic constraints on the origin of Hawaiian lavas from the Maui volcanic complex, Hawaii. Nature 330:216–220

Wetherill GW (1956) Discordant uranium-lead ages. Trans Amer Geophys Union 37:320–325

Wetherill GW (1963) Discordant uranium-lead ages - Pt. 2. Discordant ages resulting from diffusion of lead and uranium. J Geophys Res 68:2957–2965

White CM, McBirney AR (1978) Some quantitative aspects of orogenic volcanism in the Oregon Cascades. In: Smith RB, Eaton GP (eds) Cenozoic tectonism and regional geophysics of the western Corcillera. Geol Soc Amer Memoir 152:369–288

White CM, Geist DJ, Frost CD, Verwoerd WJ (1989) Petrology of the Vandfaldsdalen macrodike, Skaergaard region, East Greenland. J Petrol 30(2) 271–298

White RS, McKenzie DP (1989) Magmatism at rift zones: The generation of volcanic continental margins and flood basalt. J Geophys Res 94(B6):7698–7729

White RS, Morton AC (1995) The Iceland plume and its influence on the evolution of the NE Atlantic. J Geol Soc London 152:935

White RS, Spence GD, Fowler SR, McKenzie DP, Westbrook GK, Bowen AN (1987) Magmatism at rifted continental margins. Nature 330:439–444

White RW (1972) Stratigraphy and structure of basins on the coast of Liberia. Liberian Geol Surv Spec Paper 1:1–18

White WM, Bryan WB (1977) Strontium isotope, K, Rb, Cs, Sr, Ba, and rare earth geochemistry of basalts from the FAMOUS area. Geol Soc Amer Bull 38:571–576

White WM, Dupré B (1986) Sediment subduction and magma genesis in the Lesser Antilles: Isotopic and trace element constraints. J Geophys Res 91 5927–5941

White WM, Hofmann AW (1978) Geochemistry of the Galápagos Islands: Implications for mantle dynamics and evolution. Carnegie Institution of Washington (Ann Rept Dept Terrest Magnetism, Yearbook 77, pp 596–606)

White WM, Hofmann AW (1982) Sr and Nd isotope geochemistry of oceanic basalts and mantle evolution. Nature 296:821–824

White WM, Patchett P (1984) Hf-Nd-Sr isotopes and incompatible-element abundances in island arcs and implication for magma origins and crust-mantle evolution. Earth Planet Sci Letters 67:167–185

White WM, Schilling J-G (1978) The nature and origin of geochemical variation in Mid-Atlantic Ridge basalts from the Central North Atlantic. Geochim Cosmochim Acta 42:1501–1516

White WM, Schilling J-G, Hart SR (1976) Evidence for the Azores mantle plume from strontium isotope geochemistry of the central North Atlantic. Nature 263:659–663

White WM, Tapia MDM, Schilling J-G (1979) The petrology and geochemistry of the Azores Islands. Contrib Mineral Petrol 69:201–213

White WM, Dupré B, Vidal P (1985) Isotope and trace elements geochemistry of sediments from the Barbados Ridge-Demerara Plain region, Atlantic Ocean. Geochim Cosmochim Acta 49:1875–1886

White WM, Hofmann AW, Puchelt H (1987) Isotope geochemistry of Pacific Mid-Ocean Ridge basalt. J Geophys Res 92(B6): 4881–4894

White WM, Cheatham MM, Duncan RA (1990) Isotope geo-chemistry of Leg 115 basalts and inferences on the history of the Réunion mantle plume. Proc Ocean Drilling Program Sci Results 115:53–61

White WM, McBirney AR, Duncan RA (1993) Petrology and geochemistry of the Galapagos Islands: Portrait of a pathological mantle plume. J Geophys Res 98:19533–19563

White WS (1966) Geologic evidence for crustal structure in the western Lake Superior basin. In: Steinhart JS, Jefferson Smith T (eds) The Earth beneath the continents. Amer. Geophys. Union, Washington, D.C. (Geophysical Monograph 10, pp 28–41)

White WS (1968) The native copper deposits of northern Michigan. In: Ridge JD (ed) Ore deposits of the United States 1933-1967. Amer. Inst. Mining Metal. Petrol. Eng., New York, pp 303–325

White WS (1971) A paleo-hydrologic model for the mineralization of the White Pine copper deposit, northern Michigan. Econ Geol 66:1–13

White WS, Wright JC (1954) The White Pine copper deposit, Ontonagan County, Michigan. Econ Geol 49:675–716

White WS, Wright JC (1966) Sulfide-mineral zoning in the basal Nonesuch Shale, northern Michigan. Econ Geol 61:1171–1190

Whitehouse MJ (1990) Isotopic evolution of the Southern Outer Hebridean Lewisian gneiss complex: Constraints on Late Archean source regions and the generation of transposed Pb-Pb palaeoisochrons. Chem Geol (Isotope Geoscience Section) 86:1–20

Whitehouse MJ, Neumann E-R (1995) Sr-Nd-Pb isotope data for ultramafic xenoliths from Hierro, Canary Islands-melt infiltration processes in the upper mantle. Contrib Mineral Petrol 119:239–246

Whitford DJ (1975) Strontium isotopic studies of the volcanic rocks of the Sunda arc, Indonesia, and their petrogenic implications. Geochim Cosmochim Acta 39:1287–1302

Whitford DJ, Duncan RA (1978) Origin of the Ninetyeast Ridge: Trace element and Sr isotope evidence Carnegie Institution of Washing (Yearbook 77, pp 606–613)

Whitford DJ, Jezek PA (1979) Origin of Late Cenozoic lavas from the Banda arc, Indonesia: Trace element and Sr isotope evidence. Contrib Mineral Petrol 68:141–150

Whitford DJ, Jezek PA (1982) Istopic constraints on the role of subducted sialic material in Indonesian island-arc magmatism. Geol Soc Amer Bull 93:504–513

Whitford DJ, Kinny PD (eds) (1991) Isotopic evolution of the mantle and crust. Australian J Earth Sci 38:523–611

Whitford DJ, White WM (1981) Neodymium isotopic composition of Quaternary island arc lavas from Indonesia. Geochim Cosmochim Acta 56:989–995

Whitford DJ, Compston W, Nicholls IA, Abbott MJ (1977) Geochemistry of late Cenozoic lavas from eastern Indonesia: Role of subducted sediments in petrogenesis. Geol 5:571–575

Whitford DJ, Nicholls IA, Taylor SR (1979) Spatial variations in the geochemistry of Quaternary lavas across the Sunda arc in Java and Bali. Contrib Mineral Petrol 70:241–456

Whitford DJ, White WM, Jezek PA (1981) Neodymium isotopic composition of Quarternary island arc lavas from Indonesia. Geochim Cosmochim Acta 45:989–995

Whitford DJ, White WM, Jezek PA (1982) Neodymium isotopic composition of Quaternary island arc lavas from Indonesia. Geochim Cosmochim Acta, 45:989–996

Whitney JA, Stormer JC Jr (1985) Mineralogy, petrology, and magmatic conditions from the Fish Canyon Tuff, central San Juan volcanic field, Colorado. J Petrol 26:726–762

Whittacker EJW, Muntus R (1970) Ionic radii for use in geo-chemistry. Geochim Cosmochim Acta, 34:945–956

Widom E, Carlson RW, Gill JB, Schmincke H-U (1997) Th-Sr-Nd-Pb isotope and trace element evidence for the origin of the São Miguel, Azores, enriched mantle source. Chem Geol 140: 49–68

Wilkinson JFG (1982) The genesis of mid-ocean ridge basalt. Earth Sci Rev 18:1–57

Wilkinson JFG (1986) Classification and average chemical compositions of common basalts and andesite. J Petrol 27:31–62

Wilkinson P, Mitchell JG, Cattermole PJ, Downie C (1986) Volcanic chronology of the Meru-Kilimanjaro region, northern Tanzania. Geol Soc London 143:601–605

Williams DAC (1972) Archean ultramafic, mafic, and associated rocks, Mt. Monger, Western Australia. J Geol Soc Australia 19: 163–188

Williams DAC, Hallberg JA (1973) Archaean layered intrusions of the Eastern Goldfields region, Western Australia. Contrib Mineral Petrol 38:45–70

Williams H (1936) Pliocene volcanoes of the Navajo-Hopi country. Geol Soc Amer Bull 47:111–172

Williams H, Turner FJ, Gilbert CM (1955) Petrography. An introduction to the study of rocks in thin section. W.H. Freeman, San Francisco

Williams PR, Harahap BH (1987) Preliminary geochemical and age data from postsubduction intrusive rocks, northwest Bornea. Australian J Earth Sci 34:405–416

Williams Q, Garnero EJ (1996) Seismic evidence for partial melt at the base of the Earth's mantle. Science 173:1528–1530

Williams RW, Gill JB (1989) Effects of partial melting on the uranium decay series. Geochim Cosmochim Acta 53:1607–1619

Williams RW, Gill JB, Bruland KW (1986) Ra-Th disequilibria systematics; timescale of carbonatite magma formation at Oldoinyo Lengai volcano, Tanzania. Geochim Cosmochim Acta 50:1249–1259

Williamson JH (1968) Least-squares fitting of a straight line. Can J Phys 46:1845–1847

Willis B, Washington HS (1924) San Felix and San Ambrosio: Their geology and petrology. Geol Soc Amer Bull 35:365–384

Willis KM, Stern RJ, Clauer N (1988) Age and geochemistry and Late Precambrian sediments of the Hammamat Series from the Northeastern Desert of Egypt. Precamb Res 42:173–188

Wilshire HG, Binns RA (1961) Basic and ultrabasic xenoliths from volcanic rocks of New South Wales. J Petrol 2:185–208

Wilshire HG, Nielson Pike JE, Meyer CE, Schwarzman EC (1980) Amphibole-rich veins in lherzolite xenoliths, Dish Hill, and Deadman Lake, California. Amer J Sci 280A:576–593

Wilshire HG, McGuire AV, Noller JS, Turrin BD (1991) Petrology of lower crustal and upper mantle xenoliths from the Cima volcanic field, California. J Petrol 32:169–200

Wilson JT (1963a) A possible origin of the Hawaiian Islands. Can J Phys 41:863–870

Wilson JT (1963b) Evidence from islands on the spreading of the ocean floor. Nature 197:536–538

Wilson M (1989) Igneous petrogenesis. Unwin Hyman, London, U.K.

Wilson M, Downes H (1991) Tertiary-Quaternary extension-related alkaline magmatism in western and central Europe. J Petrol 32:811–849

Windley BF (ed) (1976) The early history of the Earth. Wiley, New York

Windley BF (1977) The evolving continents. Wiley, London

Winn RD (1978) Upper Mesozoic flysch of Tierra del Fuego and South Georgia Island: A sedimentologic approach to lithosphere plate restoration. Geol Soc Amer Bull 89:533–547

Wise WS (1969) Origin of basaltic magmas in the Mohave Desert area, California. Contrib Mineral Petrol 23:53–64

Witt-Eickschen G, Kaminsky W, Kramm U, Harte B (1998) The nature of young vein metasomatism in the lithosphere of the West Eifel (Germany): Geochemical and istopic constraints from composite mantle xenoliths from the Meerfelder Maar. J Petrol 39(1):155–186

Wörner G (1992) Kirkpartrick lavas, Exposure Hill Formation and Ferrar sills in the Prince Albert Mountains, Victoria Land, Antarctica. Polarforschung 60(2):87–90

Wörner G, Orsi G (1992) Volcanic observations on Scott Island in the Antarctic Ocean. Polarforschung 60:82–83

Wörner G, Schmincke H-U (1984a) Mineralogical and chemical zonation of the Laacher See tephra sequence, Germany. J Petrol 25:805–835

Wörner G, Schmincke H-U (1984b) Petrogenesis of the zoned Laacher See tephra sequence. J Petrol 25:836–851

Wörner G, Wright TL (1984) Evidence for magma mixing within the Laacher See magma chamber, East Eifel, Germany. J Volcanol Geotherm Res 22:301–327

Wörner G, Schmincke H-U, Schreyer W (1982) Crustal xenoliths from the Quaternary Wehr volcano (East Eifel). Neues Jahrb Mineral Abh 144:29–35

Wörner G, Beusen JM, Duchateau N, Gijbels R, Schmincke H-U (1983) Trace element abundances and mineral/melt distribution coefficients in phonolites from the Laacher See volcano, Germany. Contrib Mineral Petrol 84:153–173

Wörner G, Staudigel H, Zindler A (1985) Isotopic constraints on open system evolution of the Laacher See magma chamber, Eifel, Germany. Earth Planet Sci Letters 75:37–49

Wörner G, Zindler A, Staudigel H, Schmincke H-U (1986) Sr, Nd, and Pb isotope geochemistry of Tertiary and Quaternary alkaline volcanic from West Germany. Earth Planet Sci Letters 79:107–119

Wörner G, Viereck L, Hertogen J, Niephaus H (1989) The Mt. Melbourne volcanic field (Victoria Land, Antarctica) II. Geochemistry and magma genesis. Geologisches Jahrbuch E38: 395–433

Woldegabriel E, Aranson JL, Walter RC (1990) Geology, geochronology and rift basin development in the central sector of the Main Ethiopian Rift. Geol Soc Amer Bull 102(4):439–457

Wood BJ, Fraser DG (1976) Elementary thermodynamics for geologists. Oxford University Press

Wood DA (1981) Partial melting models for the petrogenesis of Reykjanes Peninsula basalts, Iceland: implications for the use of trace elements, and strontium and neodymium isotope ratios to record inhomogeneities in the upper mantle. Earth Planet Sci Letters 52:183–190

Wood DA, Joron JL, Treuil M, Norry M, Tarney J (1979) Elemental and Sr isotope variations in basic lavas from Iceland and the surrounding ocean floor. The nature of the mantle source inhomogeneities. Contrib Mineral Petrol 70:319–340

Wood DA, Marsh NG, Tarney J, Joron JL, Fryer P, Treuil M (1981) Geochemistry of igneous rocks recovered from a transect across the Mariana Trough, arc, and trench, Sites 453–361, Deep Sea Drilling Project, Leg 60. Inititial Report Deep Sea Drilling Proj 60:611–646

Wooden JL, Mueller PA (1988) Pb, Sr, and Nd isotopic compositions of a suite of Late Archean igneous rocks, eastern Beartooth Mountains: implications for crust-mantle evolution. Earth Planet Sci Letters 87:59–72

Wooden JL, Mueller PA, Hunt DK, Bowes DR (1982) Geochemistry and Rb-Sr geochronology of Archean rocks from the interior of the southeastern Beartooth Mountains, Montana and Wyoming. In: Mueller PA, Wooden JL (eds) Precambrian Geology of the Beartooth Mountains. Montana Bur Mines Geol Spec Pub 84:45–56

Wooden JL, Czamanske GK, Fedorenko VA, Arndt NT, Chauvel C, Bouse RM, King B-SW, Knight RJ, Siems DF (1993) Isotopic and trace-element constraints on mantle and crustal contributions to Siberian continental flood basalts, Noril'sk area, Siberia. Geochim Cosmochim Acta 57(15):3677–3704

Woodhead JD (1989) Geochemistry of the Mariana arc (western Pacific): Source composition and processes. Chem Geol 76: 1–24

Woodhead JD (1992) Temporal geochemical evolution in oceanic intra-plate volcanics: A case study from the Marquesas (French Polynesia) and comparison with other hotspots. Contrib Mineral Petrol 1 1:458–467

Woodhead JD (1996) Extreme HIMU in an oceanic setting: the geochemistry of Mangaia Island (Polynesia), and temporal evolution of the Cook-Austral hotspot. J Volcanol Geotherm Res 72:1–19

Woodhead JD, Devey CW (993) Geochemistry of the Pitcairn seamount: I. Source character and temporal trends. Earth Planet Sci Letters :16:81–99

Woodhead JD, Fraser DG (1985) Pb, Sr, and ^{10}Be isotopic studies of volcanic rocks from the northern Mariana Islands: Implications for magma genesis and crustal recycling in the Western Pacific. Geochim Cosmochim Acta 49:1925–1930

Woodhead JD, McCulloch MT (1989) Ancient seafloor signals in Pitcairn Island lavas and evidence for large amplitude, small length-scale mantle heterogeneities. Earth Planet Sci Letters 94:257–273

Woodhead JD, Harmon RS, Fraser DG (1987) O, S, Sr, and Pb isotope variation in volcanic rocks from the northern Mariana Islands: Implications for crustal recycling in intra-oceanic arcs. Earth Planet Sci Letters 83:39–52

Woodhead JD, Greenwood P Harmon RS, Stoffers P (1993) Oxygen isotope evidence for recycled crust in the source of EM-type oceanic island basalts. Nature 362:809–813

Woolley AR (1989) The spatial and temporal distribution of carbonatites. In: Bell K (ed) Carbonatites: Genesis and evolution. Unwin Hyman, London, U.K., pp 15–37

Woolley AR, Kempe DRC (1989) Carbonatites: Nomenclature, average chemical compositions, and element distribution. In: Bell K (ed) Carbonatites: Genesis and evolution. Unwin Hyman, London, U.K., pp 1–14

Worden JM, Compston W (1973) A Rb-Sr isotopic study of weathering in the Mertondale granite, Western Australia. Geochim Cosmochim Acta 37:2567–2576

Worst BG (1958) The differentiation and structure of the Great Dyke of Southern Rhodesia. Trans Geol Soc South Africa 61:283–354

Worst BG (1960) The Great Dyke of Southern Rhodesia. Bull Geol Surv Southern Rhodesia 47:1–234

Wright AC, Kyle PR (1990) Volcanoes of the Antarctic plate and southern oceans. In: LeMasurier WE, Thomson JW (eds) Antarctic Research Series. Amer. Geophys. Union, Washington, D.C. (Monograph 48, pp 449–453)

Wright E, White WH (1987) The origin of Samoa: new evidence from Sr, Nd, and Pb isotopes. Earth Planet Sci Letters 81:151–162

Wright TL (1984) Origin of Hawaiian tholeiite: A metasomatic model. J Geophys Res 89 3233–3252

Wright TL, Grolier MJ, Swanson DA (1973) Chemical variation related to stratigraphy of the Columbia River Basalt. Geol Soc Amer Bull 84:371–386

Wright TL, Swanson DA, Duffield WA (1975) Chemical compositions of Kilauea east-rift lavas, 1968–1971. J Petrol 16:110–133

Wulff-Pedersen E, Neumann E-R, Jensen BB (1996) The upper mantle under La Palma, Canary Islands: Formation of Si-K-Na rich melt and its importance as a metasomatic agent. Contrib Mineral Petrol 125:113–139

Wyers GP, Barton M (1987) Geochemistry of a transitional ne-trachybasalt-Q-trachyte lava series from Patmos (Dodecanesos) Greece: Further evidence for fractionation, mixing, and assimilation. Contrib Mineral Petrol 97:279–291

Wyllie PJ (1961) Fusion of Torridonian sandstone by a picritic sill in Soay (Hebrides). J Petrol 2:1–37

Wyllie PJ (ed) (1967) Ultramafic and related rocks. Wiley, New York

Wyllie PJ (1977) Crustal anatexis: An experimental review. Tectonophysics 43:41–72

Wyllie PJ (1981) Plate tectonics and magma genesis. Geol Rundschau 70 128–153

Wyllie PJ (1982) Subduction products according to experimental prediction. Geol Soc Amer Bull 93:468–476

Wyllie PJ (1984) Sources of granitoid magmas at convergent plate boundaries. Phys Earth Planet Int 35:12–18

Wyllie PJ (1988) Solidus curves, mantle plumes, and magma generation beneath Hawaii. J Geophys Res 93(B5):4171–4181

Wyllie PJ (1989) Origin of carbonatites: Evidence from phase equilibrium studies. In: Bell K (ed) Carbonatites. Unwin Hyman, London, pp 15–37

Wyllie PJ, Tuttle OF (1960) The system CaO-CO$_2$-H$_2$O and the origin of carbonatites. J Petrol 1:1–46

Wyllie PJ, Tuttle OF (1962) The carbonatitic lavas. Nature 194:1269

Xie Q, Kerrich R, Fan J (1993) HFSE/REE fractionations recorded in three komatiite-basalt sequences, Archean Abitibi greenstone belt: Implications for multiple plume sources and depths. Geochim Cosmochim Acta 57:4111–4118

Xu YJ, Mercier CC, Menzies MM, Ross JV, Harte B, Lin C, Shi L (1996) K-rich glass-bearing wehrlite xenoliths from Yitong, northeastern China: Petrologic and chemical evidence for mantle metasomatism. Contrib Mineral Petrol 125:406–420

Yanagi,T (1975) Rubidium-strontium model of formation of the continental crust and the granite at the island arc. Memo Fac Sci Kyushu Un Ser D 22:37–98

Yanagai T, Ishizaka K (1978) Batch fractionation model for the evolution of volcanic rocks in an island arc: An example from central Japan. Earth Planet Sci Letters 40:252–262

Yanagai T, Arikawa H, Hamamoto R, Hirano I (1988) Petrological implications of strontium isotope compositions of the Kinpo volcanic rocks in southwest Japan: Ascent of the magma chamber by assimilating the lower crust. Geochem J 22:237–248

Yañez P, Ruiz J, Patchett PJ, Ortego-Gutierrez F, Gehrels GE (1991) Isotopic studies of the Acatlan complex, southern Mexico: Implications for Paleozoic North American tectonics. Geol Soc Amer Bull 103(6):817–828

Yen TP (1977) Rb and Sr in the Neogene basalts of Taiwan. Mem Geol Soc China 2:213–222

Yoder HS (1973) Contemporaneous basaltic and rhyolitic magmas. Amer Mineral 58:153–171

Yoder HS (1988) The great basaltic "floods". S African J Geol 91(2):139–156

York D (1966) Least-squares fitting of a straight line. Can J Phys 44:1079

York D (1967) The best isochron. Earth Planet Sci Letters 2:479–482

York D (1969) Least squares fitting of a straight line with correlated errors. Earth Planet Science Letters 5:320–324

You C-F, Castillo PR, Gieskes JM, Chan LH, Spivack AJ (1996) Trace element behavior in hydrothermal experiments: implications for fluid processes at shallow depths in subduction zones. Earth Planet Sci Letters 140:41–52

Young GM, Church WR (1966) The Huronian system in the Sudbury district and adjoining area of Ontario: A review. Proc Geol Assoc Canada 17:65–82

Yu D, Fontignie D, Schilling J-G (1997) Mantle plume-ridge interactions in the central North Atlantic: a Nd isotope study of Mid-Atlantic Ridge basalts from 30° N to 50° N. Earth Planet Sci Letters 146:259–272

Zak SL, Kamenev EA, Minakov PV, Armand AL, Mikheichev AS, Petesilie JA (1972) The Khibina Alkaline Massif. Nedra, St. Petersburg (in Russian)

Zangana NA, Downes H, Thirlwall MF, Marriner GF, Bea F (1999) Geochemical variations in peridotite xenoliths and their constituent clinopyroxenes from Ray Pic (French Massif Central): Implications for the composition of the shallow lithospheric mantle. Chem Geol 153:11–36

Zartman RE (1974) Lead isotopic provinces in the Cordillera of the western United States and their geologic significance. Econ Geol 69:792–805

Zartman RE (1988) Three decades of geochronologic studies in the New England Appalachians. Geol Soc Amer Bull 100:1168–1180

Zartman RE, Tera F (1973) Lead concentration and isotopic composition in five peridotite inclusions of probable mantle origin. Earth Planet Sci Letters 10:54–66

Zhang ZM, Liou JG, Coleman RG (1984) An outline of the plate tectonics of China. Geol Soc Amer Bull 95:295–312

Zhi X, Song Y, Frey FA, Feng J, Zhai M (1990) Geochemistry of Hannuoba basalts, eastern China: constraints on the origin of continental alkali and tholeiite basalt. Chem Geol 88:1–33

Zhou P, Mukasa SB (1997) Nd-Sr-Pb isotopic, and major- and trace-element geochemistry of Cenozoic lavas from the Khorat Plateau, Thailand: Sources and petrogenesis. Chem Geol 137:175–193

Zhou X, Armstrong RL (1982) Cenozoic volcanic rocks of eastern China-secular and geographic trends in chemistry and strontium isotopic composition. Earth Planet Sci Letters 58: 301–329

Zhuravlev DZ, Zhuravlev AZ, Chernyshev IV (1985) Isotopic zoning in the Kurile island arc indicated by ^{143}Nd/^{144}Nd and ^{87}Sr/^{86}Sr ratios. Internat Geol Rev 27:781–785

Zhuravlev DZ, Tsvetkov AA, Zhuravlev AZ, Gladkov NG, Chernyshev IV (1987) ^{143}Nd/^{144}Nd and ^{87}Sr/^{86}Sr ratios in Recent magmatic rocks of the Kurile island arc. Chem Geol (Isotope Geoscience Section) 66:227–243

Zielinski RA (1975) Trace element evaluation of a suite of rocks from Réunion Island, Indian Ocean. Geochim Cosmochim Acta 39:713–734

Zielinski RA, Frey FA (1970) Gough Island: Evaluation of a fractional crystallization model. Contrib Mineral Petrol 29:242–254

Zindler A, Hart SR (1986) Chemical geodynamics. Ann Rev Earth Planet Sci 14:493–571

Zindler A, Hart SR, Frey FA (1979) Nd and Sr isotope ratios and rare earth element abundances in Reykjanes Peninsula basalts: Evidence for mantle heterogeneity beneath Iceland. Earth Planet Sci Letters 45:249–262

Zindler A, Jagoutz E, Goldstein S (1982) Nd, Sr, and Pb scotopic systematics in a three-component mantle: A new perspective. Nature 298:519–523

Zindler A, Staudigel H, Batiza R (1984) Isotope and trace element geochemistry of young Pacific seamounts: Implications for the scale of upper mantle heterogeneity. Earth Planet Sci Letters 70:175–195

Zolutukhin VV, Al'Mukhamedov AI (1988) Traps of the Siberian platform. In: Macdougall JD (ed) Continental Flood Basalts. Kluwer Academic Publishers, Dordrecht, pp 273–310

Zubkov VS, Smirnov VN, Plyusnin GS, Al'murkhamedov YA, Nikolayev VM, Paradina LF, Kuznetsova SV (1991) First K-Ar dates and Sr isotopic compositions of basanite from the explosion pipes of the Chulym-Yenisey Basin. Doklady, Earth Science Section 307(1/6):217–220

Author Index

Explorer Ridge (Juan de Fuca Ridge) 66, 67
Explorer Seamount (Pratt-Welker Seamounts) 66

F

Faeroe Islands (North Atlantic Ocean) 196, 197,
    200–202
Faial (Azore Islands) 47, 48
Falls River (Yellowstone Caldera, Wyoming) 196
FAMOUS (Central Atlantic Ocean) 35
Fang Glacier (Mt. Erebus, Ross Island, Antarctica) 345
Fangataufa (Duke of Gloucester Islands) 69
Fantale Volcano (Ethiopia) 265
Farallon de Pajaros (*see also* Uracas, Mariana Islands)
    106
Farm Overysel (Bushveld Complex, South Africa) 369
Fataga Formation (Gran Canaria Island, Canary
    Islands) 45, 46
Fatu Hiva (Marquesas Islands) 70
Fen Complex (Norway) 319–321
Feni Island (Tabar-Feni Islands, Western Pacific
    Ocean) 141
Ferguson Island (Solomon Sea) 87, 141
Fernandina (Galapagos Islands) 63
Fernando de Noronha (South Atlantic Ocean) 38, 47,
    51, 52, 77, 224, 230, 231
Ferrar Dolerite (Transantarctic Mountains) 244, 246,
    248–254, 342, 344
Fiji Islands (Pacific Ocean) 127
Finch Mine (South Africa) 342
Fish Canyon Tuff (San Juan Mountains, Colorado) 183
Fiskenaesset (West Greenland) 396
Fitzroy Trough (Western Autsralia) 334, 340, 341, 349
Floreana (Galapagos Island) 63
Flores (Azore Islands) 48
Flores (Sunda Islands, Indonesia) 131, 134
Flores Sea (Indonesia) 131
Florida (USA) 234
Fly Highlands (Papua New Guinea) 140, 141
Fogo (Cape Verde Islands) 50
Formigas (Azore Islands) 47
Forrestal Range (Antarctica) 252, 253
Fort Portal (Uganda) 286
Fortaleza (Northeast Brazil) 228
Fortuna (Spain) 305
Fosdick Mountains (Marie Byrd Land, Antarctica) 346
Fossa Magna (Honshu, Japan) 121–124
Four Corners Area (southwestern USA) 328, 329
France 95, 307–309, 316, 317
Franklin Island (Ross Sea, Antarctica) 345
Freetown Complex (Sierra Leone) 223, 228, 231–233
Fuerteventura Island (Canary Islands) 46, 47, 50, 74
Fukujin Submarine Volcano (Mariana Islands) 108,
    109
Funk Seamount (Prince Edward Islands) 97
Futana (Vanuatu Islands) 144

G

Gabon (West Africa) 294
Gair Mesa (Mesa Range, Northern Victoria Land,
    Antarctica) 244
Galapagos Islands (Pacific Ocean) 60, 62, 63, 89
Galapagos Spreading Center 61–63
Galeras Volcano (Central Cordillera, Colombia) 159
Galunggung Volcano (Java, Indonesia) 132
Gambier (Duke of Gloucester Islands) 68, 69, 71, 74,
    337, 339
Gamitagama Greenstone Belt (Ontario) 383–385
Gardner Pinnacles (Emperor Seamounts) 84
Gaussberg (East Antarctica) 342, 348, 349
Gembudo Volcano (Kyushu, Japan) 123, 124
Genovesa (Galapacos Islands) 63
Georgia (USA) 234
Geraldine (Australia) 326
Germany 190, 286, 291, 307, 309–312, 314, 329
Geronimo Head Formation (Superstition Mountains,
    Arizona) 178
Gerro Galan Volcano (Argentina) 161, 163
Giacomini (Pratt-Welker Seamount Chain) 66
Giant's Causeway (Northern Ireland) 203
Giants Range Granite (Minnesota) 382
Gilbert Islands (Pacific Ocean) 40
Giles Complex (Musgrave Block, Australia) 375–377
Girnar Complex (Deccan Plateau, India) 213, 214
Glacier Peak (Cascade Range, Washington) 184
Glass Mountain (Long Valley, California) 172, 173
Glen Cannel Granophyre (Mull, Scotland) 201
Globe (Arizona) 331
Gnotuk Maar (Victoria, Australia) 337
Goboboseb Mountains (Namibia) 221, 222
Godthåb (West Greenland) 390, 395, 397, 398
Godthåbsfjord (West Greenland) 395, 397
Goias State (Brazil) 229
Golan Heights (Lebanon) 274
Golda Zuelva Complex (Cameroon) 227, 228
Gondar (Ethiopia) 264
Gondwana 90, 92, 189, 218, 219, 237, 242, 243, 249, 254,
    255, 346
Goodenough Island (Solomon Sea) 141
Gorda Ridge (East Pacific Rise) 60, 61
Gorgona Island (Colombia, East Pacific Ocean) 151,
    158, 385
Gosse Pile Intrusion (Australia) 375
Gough Island (South Atlantic Ocean) 36, 47, 55, 56, 77, 224
Gowganda Tillite (Huronian System, Ontario) 360
Grabben Gullen Volcanic Province (New South Wales,
    Australia) 335
Graciosa (Azore Islands) 48
Gramado Magma Type (Paraná Basalt) 219–221
Gran Canaria (Canary Islands) 44–46, 125
Grand Canyon (Arizona) 180, 330

O

Geologic Time Scale

Era	Period	Epoch	Age of Boundary (Ma)
Cenozoic	Quaternary	Holocene	0.01
		Pleistocene	1.6
	Neogene (Late Tertiary)	Pliocene	5.3
		Miocene	23.7
	Paleocene (Early Tertiary)	Oligocene	36.6
		Eocene	57.8
		Paleocene	66.4
Mesozoic	Cretaceous	Late	97.5
		Early	144
	Jurassic	Late	163
		Middle	187
		Early	208
	Triassic	Late	230
		Middle	240
		Early	245
Paleozoic	Permian	Late	258
		Early	286
	Pennsylvanian (Late Carboniferous)		320
	Mississippian (Early Carboniferous)		360
	Devonian	Late	374
		Middle	387
		Early	408
	Silurian	Late	421
		Early	438
	Ordovician	Late	458
		Middle	478
		Early	505
	Cambrian	Late	523
		Middle	540
		Early	570

Table A.1b.
The geologic time scale:
Precambrian (after Harrison
and Peterman 1982)

Eon	Era	Age of Boundary (Ma)
Proterozoic	Late	900
	Middle	1 600
	Early	2 500
Archean	Late	3 000
	Middle	3 400
	Early	3 800
Hadean		4 600

If you have any concerns about our products,
you can contact us on
ProductSafety@springernature.com

In case Publisher is established outside the EU,
the EU authorized representative is:
Springer Nature Customer Service Center GmbH
Europaplatz 3, 69115 Heidelberg, Germany

Printed by Libri Plureos GmbH
in Hamburg, Germany